Biogeography
and
Evolution
in
New Zealand

CRC Biogeography Series

Series Editor
Malte C. Ebach
University of New South Wales
School of Biological
Earth and Environmental Sciences
New South Wales
Australia

Biogeography
and
Evolution
in
New Zealand

MICHAEL HEADS

CRC Press
Taylor & Francis Group
Boca Raton London New York

CRC Press is an imprint of the
Taylor & Francis Group, an **informa** business

Cover photo caption: Mount Tasman and Aoraki (Mount Cook) from Okarito Trig. © Ian Singleton.

CRC Press
Taylor & Francis Group
6000 Broken Sound Parkway NW, Suite 300
Boca Raton, FL 33487-2742

First issued in paperback 2020

© 2017 by Taylor & Francis Group, LLC
CRC Press is an imprint of Taylor & Francis Group, an Informa business

No claim to original U.S. Government works

ISBN-13: 978-1-4987-5187-2 (hbk)
ISBN-13: 978-0-367-65838-0 (pbk)

Library of Congress Cataloging-in-Publication Data

Names: Heads, Michael J.
Title: Biogeography and evolution in New Zealand / Michael Heads.
Description: Boca Raton : Taylor & Francis, 2017. | Series: Biogeography series ; 1 | "A CRC title." | Includes bibliographical references and index.
Identifiers: LCCN 2016010188 | ISBN 9781498751872 (alk. paper)
Subjects: LCSH: Biogeography--New Zealand. | Evolution (Biology)--New Zealand. | Biotic communities--New Zealand.
Classification: LCC QH197.5 .H43 2017 | DDC 577.2/20993--dc23
LC record available at http://lccn.loc.gov/2016010188

**Visit the Taylor & Francis Web site at
http://www.taylorandfrancis.com**

**and the CRC Press Web site at
http://www.crcpress.com**

For Trace

Contents

Preface

The theme of this book is the history of life in New Zealand through geological time. The origins of the plants and animals have been controversial for over a century, but a vast amount of new information on the topic is now available thanks to recent work in molecular biology. The effects of the molecular revolution on biology resemble the tidal wave of new knowledge that followed the invention of the microscope; all fields of biology have changed in fundamental ways. The classifications, or phylogenies, of biological groups are now often based on molecular data, the DNA sequences. Wide-ranging, geographic surveys of DNA sequences in different species have revealed an unsuspected, new world of organic structure that exists at all scales, from submicroscopic to intercontinental, and have highlighted the significance of life in New Zealand.

New Zealand is home to bizarre creatures such as the kiwi and the tuatara, and the country's flora and fauna have always been recognized as distinctive and unusual. Jared Diamond (1997: 3) thought that "New Zealand is biologically the closest thing we can come to exploring another planet," and the new molecular work has confirmed this. It has revealed that many large groups of plants and animals are composed of two subgroups, with one made up of the "normal" members that are widespread and diverse around the world, and the other restricted to New Zealand. For example, the parrots comprise two main groups: the New Zealand group Strigopoidea (*Nestor* and *Strigops*) and all other parrots. Likewise, the passerines—the group of small, perching birds that make up more than half of all birds—comprise two groups: the New Zealand wrens (Acanthisittidae) and all the others. Many other examples of "globally basal endemics" are now known from New Zealand, and there are probably more than in any other region, except perhaps Madagascar. This suggests that the ancestors of many groups, such as parrots and passerines, had a widespread, global distribution and then differentiated in and around the region that became New Zealand. In this way, the new molecular phylogenies show that New Zealand can be recognized as a biological "parallel universe," one that can shed light on global evolution.

As well as examining the geographic affinities of New Zealand groups outside the country, this book focuses on the groups' distributions within the area. Apart from having so many globally basal endemics, the New Zealand biota is of special interest because many of the marine, freshwater, and terrestrial groups have such localized distributions. Groups of related species and populations also display distinctive ranges, and often these do not correlate with ecological factors. In many cases the distribution patterns can be related, not with the present environment, but with ancient landscape features. As with the globally basal endemics, the patterns appear to have been caused by large-scale, geological changes and evolutionary divergence.

The New Zealand biota has a special importance in the history of ideas about biogeography. In 1863, Darwin wrote: "I really think there is hardly a point in the world so interesting with respect to geographical distribution as New Zealand" (Burckhardt et al., 1999: 542), and since then, New Zealand has been a test case for theories of biogeography and evolution. One key question is whether most differentiation has been caused by chance, transoceanic dispersal in individual organisms, or by geological and climatic changes that have fragmented the ranges of many widespread species at the same time, a process termed "vicariance." The American ichthyologist Gareth Nelson developed the concept of vicariance in the 1970s, and it has now gained wide acceptance. In a well-known passage, Nelson (1975: 494) wrote:

> With regard to general problems of biogeography, the biota of New Zealand has been, perhaps, the most important of any in the world… all notable authorities have felt obliged to explain its history: explain New Zealand and the world falls into place around it.

One of the most important general conclusions of molecular studies is the idea that geographic distribution has an even closer relationship with evolution and phylogeny than was thought. When molecular classifications of groups began to be published, no one knew which morphological features, if any, would agree with the new groupings of species. Would the molecular classifications be compatible with variation in features used in traditional classifications, such as flowers in angiosperms, genitalia in insects, or teeth in mammals? Or would the key features instead be characters that had been neglected by earlier workers, such as microscopic structure, biochemistry, or behavior?

Some molecular groupings of plants and animals have turned out to correspond well with traditional taxonomic arrangements, but many molecular groups are new. These new groups often do not coincide with variation in traditional taxonomic characters or even with more cryptic features. Instead, another feature has proved to be of unexpected relevance: geographic distribution. Almost all of the molecular groups have distinctive geographic ranges and well-defined spatial structures; the precise, dovetailing distributions of related genetic groups are indicated in the molecular work with even more clarity than they were in morphological studies. Sequencing studies show that precise geographic structure occurs in most organisms, even in marine groups and microbes.

The general phenomenon of deep geographic structure in molecular phylogenies only started to become apparent around the year 2000, and as yet it has seldom been discussed. Nevertheless, it is seen across all the molecular work. No one expected geography to be "the character," but the fact that it is constitutes one of the most striking results of the whole molecular research program, and it raises many new questions. Why does the molecular variation exhibit such deep spatial structure, even in organisms such as microbes and marine groups? What does this tell us about phylogeny, speciation, and evolution in general? How can this new observation be utilized in practice, for example, in estimating the ages of groups or in conserving biodiversity?

If distribution patterns were the result of chance dispersal, there would be no reason to expect such clear-cut geographic structure in so many groups. As it is, the observed patterns suggest that evolution proceeds by phases of geographic differentiation (vicariance) alternating with phases of range expansion and geographic overlap (dispersal). In this model, both processes are determined, not by chance, but by geological and climatic changes that have affected whole communities. This would also explain another interesting fact: the repetition of the same distributional breaks and centers of endemism in many groups that have different ecologies and means of dispersal. Distribution is related to and caused by evolution, and so it can be used to tell us about evolution.

The interpretation of the New Zealand biota has been controversial ever since the first modern studies. While the new molecular data have answered many questions, several old problems remain unresolved, and there are many new ones. This book attempts to provide answers to these by integrating the strongest results of the molecular genetics—the new classifications—with the latest results from tectonic geology. This integration produces conclusions about the factors of space and time in the evolution of New Zealand groups, and these are then combined with recent ideas on mutation to produce a new synthesis of evolution in the region.

This book concludes that for most plants and animals, New Zealand is neither a "sink" for dispersal from elsewhere nor a "center of origin" of groups that have dispersed to other regions. The biota instead reflects differentiation that has developed in widespread or even global groups, by vicariance between New Zealand and other regions. The modern New Zealand biota is much older than was thought—Mesozoic, rather than Cenozoic—and has developed in tandem with geological change.

The first two chapters of this book outline general aspects of analyzing evolution in time and space. Chapter 3 introduces some main aspects of New Zealand geology; more detailed references to regional aspects of geology are made in later chapters. Chapter 4 introduces the New Zealand biota and its global significance. Chapters 5 through 9 provide a regional treatment of New Zealand biogeography and highlight localities of particular significance. Chapters 10 through 12 analyze case studies of particular groups of plants and animals, with reference to the localities discussed earlier. Chapters 13 and 14 discuss general morphological trends, and Chapters 15 and 16 consider the morphological evolution of New Zealand birds and mammals in their biogeographic context. Chapter 17 presents general conclusions and indicates some applications to conservation practice.

Acknowledgments

I'm very grateful for the help and encouragement I've received from friends and colleagues, especially Lynne Parenti (Washington, DC); John Grehan (Buffalo); Isolda Luna-Vega and Juan Morrone (Mexico City); Amparo Echeverry (Medellín); Jürg de Marmels (Maracay); Mauro Cavalcanti (Rio de Janeiro); Guilherme Ribeiro (São Paulo); Jorge Crisci (Buenos Aires); Andres Moreira-Muñoz (Santiago); Alan Myers (Cork); Robin Bruce (London); Pierre Jolivet (Paris); Pauline Ladiges and Gareth Nelson (Melbourne); Malte Ebach, Tony Gill, and David Mabberley (Sydney); Robin Craw (Whangarei); Rhys Gardner and Milen Marinov (Auckland); Karin Mahlfeld, Frank Climo, and George Gibbs (Wellington); Brian Patrick (Christchurch); and Peter Johnson, Patricio Saldivia, John Steel, and the late Bastow Wilson (Dunedin).

Abbreviations

s.lat. = sensu lato, in a broad sense
s.str. = sensu stricto, in a narrow sense
1 Ma = 1 million years ago (a date)
1 Myr = 1 million years (a period of time)

1

The Spatial Basis of Biogeography

To do science is to search for repeated patterns, not simply to accumulate facts, and to do the science of geographical ecology is to search for patterns of plant and animal life that can be put on a map.

MacArthur, 1972: 1

Until recently, phylogeography has not taken its "geography" base seriously enough… particularly in light of rampant, blind application of molecular clock methodologies to complex geographic and historical questions…. The result is a field that seriously needs to broaden its vision…. A fundamental concern is that phylogeographic approaches have focused largely on the "phylo" part and less on the "geography."

Peterson and Lieberman, 2012: 1

This book discusses many distribution patterns seen in and around New Zealand, and explanations are suggested for most of these. The Earth and its living layer exist together and interact in complex ways that are studied by ecologists. In addition, the Earth and its life *evolve* together. This concept was controversial when it was first proposed (Croizat, 1964) and is still disputed (McDowall, 2010a,b), but it has gained wide acceptance and is now often quoted as an anonymous maxim (e.g., Carroll, 2010: 126). Many aspects of distribution—elevational and latitudinal ranges, for example—display obvious correlations with aspects of the environment. Yet many distributions are *not* explained by current ecological variables. These probably result from changes over evolutionary time, and they have been accounted for either by Earth's history, as in Croizat's (1964) model, or by chance dispersal, a process that does not depend on geological change.

The Spatial Component of Evolution: Basic Concepts and Methods of Analysis

Biological groups, especially close relatives, are often allopatric; that is, they occupy different areas. Allopatric groups may be separated by a gap (they are "disjunct"), but their distributions often meet without any obvious physical barrier between them. In other groups, the distributions overlap—they show varying degrees of sympatry. Biogeography needs to explain both allopatry and sympatry, but allopatry is a distinctive, fundamental phenomenon seen in most groups, and many authors have accepted its primary significance in evolution. Among the New Zealand gentians, for example, "allopatry is much more frequent than sympatry" (Glenny, 2004).

Species Concepts

The concept of species is controversial, and the particular concept that is used in a biological analysis is often regarded as critical. This book uses the species concept outlined by Darwin (1859: 32): "I look at the term species, as one arbitrarily given for the sake of convenience to a set of individuals closely resembling each other, and that it does not essentially differ from the term variety." This is very different from the neodarwinian concept, in which species have a fundamental importance. In Darwin's model, species are not regarded as special compared with groups at other ranks, above or below the species level (Heads, 2014). As Dennett (1996: 95) wrote, discussing the definition of species: "Where should we draw the line? Darwin shows that we don't need to draw the line in an essentialist way to get on with our science… I am inclined to interpret the persisting debates [about species definitions] as more a matter of Aristotelan tidiness than a useful disciplinary trait." Dobzhansky (1937: 312) argued in a similar vein: "Species is a stage in a process, not a static unit." (In contrast, the neodarwinian view stressed that "a species is not a stage of a process, but the result of a process" [Mayr, 1942: 119].)

In the Darwinian view, accepted here, the key problem in evolution and biogeography is not speciation, but the process of differentiation in general, whether at, above, or below the level of species. In cases of clinal variation, there are no distinct groups at any rank, although clines are just as significant for biogeography as distinct phylogenetic breaks and often occur at the same locality.

Dispersal and Vicariance

In the traditional model, all groups attain their distribution by dispersal, radiating out from a localized center of origin (Matthew, 1915). In a vicariance model, two or more groups instead attain their individual distributions, not by expanding outward from a center of origin but by the division (vicariance) of a widespread ancestor. (Of course, all groups originate in a particular area or center, even in a vicariance event. But the term "center of origin," as used in the literature and in this book, refers to the restricted area inferred in the first model.)

For example, consider a genus comprising one species in the east of a continent and one in the west, with the two meeting in the middle. In dispersal theory, each species originated at an independent point and has spread out by dispersal from there, eventually meeting the other species. In a vicariance interpretation, an ancestor that was widespread in both the east and west of the continent has split down the middle, and the mutual boundary of the species represents this fracture. Each species has evolved over a broad front and neither has attained its distribution by spread from a "center of origin."

On an even larger scale, a global group may divide into, say, northern hemisphere and southern hemisphere descendants, neither of which has migrated to its current position. The two main phylogenetic groups, or clades, of the cypress family, Cupressaceae, provide a good example. The subfamily Cupressoideae is regarded as Laurasian, while the subfamily Callitroideae is Gondwanan, and the break between the two is attributed to vicariance (Mao et al., 2012; Yang et al., 2012). Vicariance has been defined as "allopatric (geographical) *speciation* caused by the origination of a barrier within the range of the ancestral species" (Crisp et al., 2011; italics added); it can also lead to differentiation above (or below) species level.

Darwin (1859: 352) established modern dispersal theory in the first edition of the *Origin of Species*. The idea proposes that if a group is present in more than one area, it must have *moved* to all these areas (except for one—the center of origin). In the fifth edition of his book, Darwin (1869: 467) emphasized this view, arguing that "not only all the individuals of the same species, but that allied species, although now inhabiting the most distant points, have proceeded from a single area—the birthplace of their early progenitor." This view denies the possibility of vicariance, in other words, that the group has been present in more than one area ever since its origin, and that it evolved by diverging from another group that occupied another area. In both the Darwinian and the neodarwinian syntheses, allopatric differentiation is caused by chance dispersal from a center of origin, while the overlap of clades is also caused by dispersal. In vicariance biogeography, allopatric differentiation is caused by vicariance, while overlap is caused by dispersal.

The Modern Synthesis (Huxley, 1942; Mayr, 1942) has provided the evolutionary paradigm for most studies over the last 70 years, and its approach to evolution in space and time is based on Darwin's ideas as presented in the work of Matthew (1915). Allopatry is explained by dispersal, and vicariance is regarded as very rare. Mayr (1965) concluded: "Quite obviously, except for a few extreme [i.e., local] endemics, every species is a colonizer because it would not have the range it has, if it had not spread there by range expansion, by 'colonization', from some original place of origin." New clades did not inherit their range from their ancestor.

Most biologists and also leading philosophers have followed Mayr. In a discussion on the New Zealand flora, Winkworth et al. (1999: 1324) wrote that the "extraordinary evolutionary importance of recent long-distance dispersal" is "clearly evident from molecular data"; it is "obvious." McGlone et al. (2001: 209) argued that "long-distance dispersal has played a key role in the development of the New Zealand flora, and... is important in explaining patterns of geographic distribution and endemism within the archipelago."

Dennett (1996: 99) wrote that in many cases, "perhaps almost all," speciation occurs by a process "in which a small group—maybe a single mating pair—wander off and start a lineage that becomes reproductively isolated." In dispersal theory, the groups of a region are there because they have dispersed there, not because they evolved there.

Distribution patterns often display striking repetition even among groups with different means of dispersal and ecology. These could all be the result of chance dispersal and adaptation, as theory suggests. Yet the principle of multiple working hypotheses (Chamberlin, 1965; cf. Mill, 1869) indicates that it would be desirable to have an alternative explanation—any explanation—that accounts for the evidence and does not break the rules of logic. In practice, a vicariance model provides an alternative. This was rejected by Mayr (1982), but views have changed since the topic began to be discussed in the 1970s. The importance of vicariance is now acknowledged in an increasing number of studies (more than 2000 in 2014; Google Scholar).

Until the 1970s, biogeographic studies assumed that a center of origin and chance dispersal (along with extinction) explained the distribution of most groups. In contrast, most models now use both chance dispersal and vicariance. The approach favored here relies less on particular events in single groups and instead proposes alternating, community-wide phases of vicariance and range expansion using normal means of dispersal.

"Dispersal biogeography" is a research program that explains the geographic distribution of organisms by chance dispersal, while "vicariance biogeography" explains distribution with reference to geological events (Gillespie et al., 2012). Ree and Smith (2008) observed that "historical biogeography is grounded in the notion that Earth history has profoundly influenced the geographic ranges of species," but this in fact only applies to vicariance biogeography, where both vicariance and overlap (range expansion) are attributed to geological events. In dispersal biogeography, the same pattern in different groups is attributed to individual dispersal events at different times, not to single causes such as geological events.

Authors sometimes suggest that "vicariance... can occur only after the range of a species has already expanded via dispersal [range expansion]" (Crisp et al., 2011). This will be true in some cases, but in many cases, the direct ancestor will itself have attained its wide distribution not by range expansion, but by vicariance, splitting off from a group with an even wider range.

An Example of Vicariance and Subsequent Dispersal: Psilotaceae

The fern family Psilotaceae comprises two genera:

> *Tmesipteris*: Throughout mainland New Zealand, the Auckland Islands, New Caledonia, and the eastern seaboard of Australia, north to New Guinea and the Philippines, and east to Fiji, Samoa, the Society Islands, and the Marquesas Islands.
>
> *Psilotum*: Pantropical, south to the central North Island.

Tmesipteris occurs on its own in the southern parts of its range (Tasmania, southern North Island, South Island, Stewart Island, Auckland Islands), while *Psilotum* occurs on its own through most of the tropics. The two genera overlap in an area extending from the Philippines south to the central North Island and east to the central Pacific, but despite this, the two genera display widespread allopatry.

One dispersal account of the group suggested that *Tmesipteris* had a restricted center of origin, from which it "probably spread through much of Australia and what is now New Zealand and New Caledonia prior to the breakup of Gondwanaland" (Brownsey and Lovis, 1987: 445). Still, the authors admitted that its absence from South America is then "not readily explained." A dispersalist study of *Psilotum* based on phylogeny would also find a localized center of origin for that genus.

In contrast with the center-of-origin approach, a vicariance model accepts that the absence of *Tmesipteris* from South America, for example, is only relative; the genus as such is absent, but it is represented there by its closest ally, or sister group, *Psilotum*. The two genera show a high level of allopatry, and this is consistent with their origin by vicariance. *Psilotum* is pantropical, suggesting that the initial break with *Tmesipteris* occurred somewhere in their overlap zone, Philippines–New Zealand–central Pacific, a region that has undergone extreme tectonic disturbance through most of the Phanerozoic. Following their allopatric differentiation, subsequent overlap of the two genera can be explained by range expansion (dispersal) in one or both genera, but there is no need to postulate any range expansion of either genus outside this region. How the *Psilotum* + *Tmesipteris* common ancestor attained its distribution is another question, which, as usual, would need to consider the group's sister.

Two Different Concepts of "Dispersal"

The term "dispersal" has many different meanings, including daily migration, annual migration, colonization of areas within the group's range, range expansion, chance dispersal involving speciation, and other processes. Because of this ambiguity, it is not surprising that discussions and debates about "dispersal" can be confusing. Two of the processes termed dispersal are of special relevance here.

"Normal" Dispersal: An Observed Ecological Process

"Normal" dispersal—the physical movement of organisms—is observed every day. Normal dispersal can take place over short distances, as in small, flightless invertebrates, or very long distances, as in seabirds, but this dispersal in itself does not lead to speciation. In a typical example, weeds may invade a garden, but they do not speciate there. The details of "normal dispersal" tend to be discussed in the ecology literature. Ecologists use the term "long-distance dispersal" to mean the normal, long-distance movement of individuals, without implying that there has been any speciation.

Chance Dispersal: An Inferred Mode of Speciation

In contrast with normal, observed dispersal, the inferred process of "chance dispersal" often refers to a mode of speciation, and the term is used in this way in texts on biogeography, systematics, and evolution. Crisp et al. (2011) defined "long-distance dispersal and establishment" as "allopatric (geographical) speciation caused by an exceptional dispersal event…." In this process, an individual disperses by chance over a feature that otherwise acts as a barrier for the group. The concept is not the same as the normal long-distance dispersal observed in, say, albatrosses, which occurs every day and does not lead to speciation or to any taxonomic differentiation.

Long-distance dispersal leading to speciation has not been observed. Authors often *infer* long-distance dispersal, calculated using phylogenies and center-of-origin programs, as "demonstrated" or "revealed" (e.g., Voelker et al., 2014; Müller et al., 2015), but this is true only in a theoretical framework, unlike observations on normal, ecological dispersal. A phylogeny indicates relationships and spatial patterns, not the mode of differentiation or speciation.

In many cases, "long-distance dispersal" is not an accurate description of speciation by chance dispersal, as the distance thought to have been crossed is short. Long-distance dispersal is often proposed to account for trans-oceanic affinities, but it is also proposed for differentiation across rivers (e.g., in the primates of the Amazon basin) and even smaller barriers. The key concept in speciation by long-distance dispersal is not the long distance, but the exceptional, unpredictable, or *chance* nature of the process. Only one or a few founders disperse across the barrier, the rest of the group does not. Thus, the process of long-distance dispersal is sometimes described as "chance dispersal" or "jump dispersal."

Whatever term is used, the process of chance dispersal has often been proposed for New Zealand plants and animals. Landis et al. (2008) (reiterating the conclusions of Campbell and Landis, 2001, and Waters and Craw, 2006) considered that "the entire New Zealand terrestrial biota actually became established by accidental colonists since the Oligocene…." An "accident" is an irregularity in a structure, or an event that is unexpected or without apparent cause. Accidental or chance dispersal implies events that are not related to other factors, have no obvious cause, and are unpredictable. Again, this process stands in contrast with the ordinary dispersal seen every day, which obeys normal laws of probability. For example, the probability of seed dispersal decreases in a predictable way with distance from the parent, depending on the type of seed and other measurable factors.

Vicariance biogeography uses three processes to explain distributions: vicariance (to explain differentiation and allopatry), range expansion by normal dispersal (to explain overlap), and extinction. Many biogeographic studies use these processes but also infer speciation by chance dispersal, whereas this is rejected here as unnecessary. It is also rejected by many geneticists who have studied speciation (see the section "Founder Speciation" below).

Waters and Craw (2006) wrote that "...the difficulty some researchers [no references cited] have accepting dispersal seems to be rooted more in philosophy than biological reality: they simply consider dispersal events to be unimportant, ad hoc, and untestable phenomena." Although no references were cited, our research group is implied. The reason no references were supplied is that our group has never suggested this. It is a straw man argument—no one has ever believed that dispersal is unimportant. Dispersal events are of fundamental importance for organisms' survival, and they can be observed every day. All individual organisms have dispersed to where they are now. In addition, many groups have expanded (or contracted) their range; this is an important process and, despite Waters and Craw's (2006) view, it is open to investigation and testing. However, *chance* processes of dispersal cannot explain repeated distribution patterns in unrelated clades with different ecology and means of dispersal. (Chance dispersal would tend to give rise to unique distribution patterns, and these are not dealt with here. Instead, the focus is on some of the most obvious patterns that are repeated in many different groups.)

In addition to its different concept of dispersal, vicariance theory has a different, nontraditional *interpretation* of the fossil record. Again, our critics' response is not to critique the proposal, but to claim that we ignore the fossil record; this is another straw man and an insult rather than an objection. (For replies to critics, see Heads, 2015a,b.)

Authors such as Wen et al. (2013: 913) have also suggested that in vicariance analyses (Craw et al., 1999; Parenti and Ebach, 2009; Heads, 2012a), dispersal has been "condemned as untestable and unimportant," but, again, this is not correct. The studies cited by Wen et al. (2013) all *invoked* dispersal to account for overlap, while rejecting chance dispersal as a mode of speciation. It is obvious that dispersal is important—individual organisms all disperse, and if vicariance were the only biogeographic process, there would only be one species of organism in any area. Vicariance theory infers general phases of vicariance alternating with phases of dispersal and range expansion, with the phases caused, not by chance, but by tectonic and climatic changes. The concept of chance dispersal as a particular process can then be rejected, as it is not needed. It was never a real explanation, and almost 1500 years ago, Procopius described: "... the thing which we too often refer to as chance, simply because we do not know what makes events follow the course we see them follow" (Procopius, AD 550/1966: 61; sect. 4.45).

Some More on Normal, Observed Dispersal and Inferred Chance Dispersal

Although the term "dispersal" is ambiguous, in most cases, the context suggests which process is being referred to. For example, Gillespie et al. (2012) wrote: "Actual observations of long-distance dispersal remain limited, although recent technological advances are making long-distance tagging more feasible (e.g., tracking bird migratory routes)." It is clear that this refers to normal "ecological" dispersal. On the other hand, elsewhere in their paper, Gillespie et al. (2012) referred to long-distance dispersal as an explanation for endemism on islands, and this refers to inferred, chance dispersal (a mode of speciation). In the same way, many other studies conflate the two processes: normal, observed dispersal, and chance dispersal as inferred from phylogenies. In dispersal biogeography, speciation by chance dispersal is interpreted as a particular case of normal dispersal by physical movement—the two processes are "parts of the same continuum" (Trewick and Morgan-Richards, 2009). There is room for confusion in this approach though. For example, a dispersal kernel is the function that describes the probability of normal dispersal of seeds or juveniles over different distances: the probability of dispersal decreases with increasing distance from the parent. Crisp et al. (2011) referred to dispersal kernels as "probability distributions of long-distance dispersal and establishment" (i.e., for speciation, see the earlier discussion), but dispersal kernels describe the probability of normal, observed dispersal, not of chance dispersal with speciation.

In dispersal biogeography, speciation is caused by the physical movement of plants and animals in one-off events. The key feature of the process—the fact that the events are singular—is attributed to chance. In contrast, vicariance biogeography uses physical movement by normal dispersal to explain range overlap. Differentiation is instead attributed to a cessation (or at least a decreasing rate) of movement, and this change is attributed not to chance but to geological change. Phases of range expansion by normal dispersal are also explained by geological change.

In "chance dispersal," a single dispersal event occurs on its own, not in conjunction with dispersal events over similar areas in other groups. (Similar areas are thought to have developed at different times.) The process of chance dispersal is thought to explain how organisms that undergo normal dispersal over centimeters can undergo intercontinental leaps of 10,000 km and more, often just once in their entire phylogenetic history. In contrast with chance dispersal, normal dispersal events, including range expansion, are correlated with many other events. For example, if a new island is formed, it is colonized by terrestrial groups from other land nearby using their normal means of dispersal. This process will not comprise a single dispersal event by a single taxon, but many events by many of the pioneer, that is, weedy species, from the surrounding source areas. In the same way, if two blocks of crust (terranes) or volcanic island arcs converge and fuse, the biotas will also converge and mix by normal ecological dispersal. If two volcanic arcs or arc segments are rifted apart, the two communities will diverge as they become less connected by normal dispersal.

Vicariance biogeography explains change in distributions by change in the environment. Two large-scale tectonic processes that affect distribution are plate divergence (which can often lead to phylogenetic divergence) and plate convergence (which can lead to interdigitation or overlap of biotas). The mixing of biotas because of tectonic convergence is not the result of "chance dispersal," which can operate over any distance, at any time. Instead, the mixing occurs by normal dispersal; in this process, if the distance between two land masses decreases, dispersal of the terrestrial populations between the land masses becomes more likely.

In dispersal biogeography, distributions are thought to develop *after* the phylogeny, with groups spreading out from their center of origin. This model is supported by ecologists (Levin, 2000), paleontologists (Eldredge et al., 2005), and systematists (Swenson et al., 2012). In contrast, in a vicariance model, the distribution develops at the same time as the phylogeny, by *in-situ* differentiation. In this way, a group can have a wide distribution at the time of its origin. Subsequent range expansion or extinction may or may not occur. In this model, distribution has much closer links with phylogeny than it does in dispersal theory, and this is compatible with the geographic structure that is being discovered in so many phylogenies. In a vicariance model, many aspects of distribution represent inherited information (in a center-of-origin model, this is only true for point endemics), and so biogeographic information has potential applications in tectonics. One example is in the debate about whether or not a "West Caledonian fault" exists in New Caledonia (Heads, 2008a).

In one well-documented pattern, groups have disjunct distributions that span the Tasman Sea. Examples include the lichen *Siphula elixii*, found in the northwestern South Island of New Zealand (Denniston Plateau) and in southwestern Tasmania (Galloway, 2007). Dispersal theory assumes that one or other of the two localities is the center of origin, and the main aim of the research program is to find out which one. The program also aims to find out the *means* by which the group dispersed from the center of origin to the other locality. An alternative, vicariance approach does not assume a localized center of origin within the group's range and instead looks outside the group. The main question in this research program is: "What is the sister group of the Tasmania + Denniston clade?" If it occurs in, say, mainland Australia, then the Tasmania + Denniston group and its sister group are both likely to have originated by simple vicariance, with the disjunct group having its "center of origin" in Tasmania + Denniston.

Many modern studies date clades using the fossil record of the clades, the age of strata and islands where the clades occur, and single tectonic events, such as the uplift of the Panama Isthmus, that may have divided clades (in the Pacific and Atlantic). In this approach, there is no need to map clades in order to study their origin, as the distribution is a secondary feature. Yet more and more molecular analyses are now mapping clades anyway, as the geographic patterns are often among the most interesting results. As deep, intricate geographic structure is reported in more and more molecular studies, chance dispersal becomes less and less likely.

Dispersal and Rare or Unique Events

Dawkins (2009) discussed modes of speciation and wrote that sometimes a barrier arises, dividing populations (vicariance), but "More usually, the barrier was there all along, and it is the animals themselves that cross it, in a rare freak event" (p. 257). In this model, "freak dispersal events" (p. 258) comprise the main mode of speciation. For example, Galapagos finches were "presumably blown across [from the mainland], perhaps by a freak storm" (p. 258). It is evident that dispersal theory does not explain common patterns observed in biogeography by referring to normal processes observed in ecology. Instead, the explanations propose rare, or even unique, events that are unrelated to any other factor or event. A typical example might infer a trans-Pacific dispersal by rafting or wind transport that only happened once in, say, 50 million years. Even freak storms that only happened once every 1000 years would be much too frequent to explain a pair of sister groups with one member on each side of the Pacific and a divergence date at, say, 50 Ma. Since that time, there will have been 50,000 freak storms, but there is evidence for just one dispersal event. This is a problem for dispersal theory. In order to deal with it, dispersalists have concluded that a group's *normal means of dispersal* (the observed "morphological dispersal syndrome") are not relevant to the inferred chance dispersal that led to its origin. Instead, *nonstandard means* are invoked (Higgins et al., 2003).

In many animals, occasional, vagrant individuals are observed outside their usual range, a phenomenon that is often thought to constitute evidence supporting chance dispersal. Yet for landbirds, Lees and Gilroy (2014) showed that "variation in the propensity for vagrancy to oceanic islands world-wide is a *surprisingly poor predictor* of colonization success across species on a global scale… Contrary to expectations, we find that the capacity for trans-oceanic dispersal *may not* be a key determinant of oceanic island colonization amongst landbirds" [italics added]. This observation is surprising in the dispersal paradigm, but it is consistent with vicariance theory.

Vicariance biogeography accepts that individuals move within their group's range, that groups can expand their range, and that both processes take place by normal means of dispersal. It also accepts that allopatry is caused by vicariance. This means that chance dispersal *as a mode of speciation* is unnecessary; it is possible to explain the biogeography of the Australasian region using allopatric differentiation and normal dispersal only, without invoking any chance dispersal (Heads, 2014). The vicariance model proposed for the region concluded that the last great modernization of the biota took place in the pre-breakup (mid-Cretaceous) phase of rifting, uplift, regression of inland seas, and magmatism.

As indicated, vicariance theory accepts that allopatry (caused by vicariance) and overlap (caused by range expansion) are both caused by geological or climatic events, rather than by chance. Anthropogenic change can often transform the environment, and in this way act as a "geological" factor. For example, the cattle egret, *Bubulcus ibis*, has a close relationship with cattle (*Bubulcus* means "cowherd"), eating the ticks and flies on the cattle, along with insects that the cattle disturb. *B. ibis* is native to parts of Africa, Europe, and Asia, but since the nineteenth century it has expanded its range into southern Africa, America, Australia, and New Zealand, following the introduction of cattle in these areas. The range expansion of *Bubulcus* (and cattle) was not the result of chance dispersal, but was caused by anthropogenic changes to the ecosystem.

Founder Speciation

In founder speciation, a founder becomes isolated from its parent population by dispersing over a "barrier," and it then differentiates to become a new species. Mayr (1954) attributed the speciation to a "genetic revolution" that he theorized would be produced by the founder effect. The founder effect itself is well established, but most geneticists now reject the idea that it can lead to a "genetic revolution" or to speciation (Coyne and Orr, 2004: 401; Orr, 2005; Crow, 2008; Whitlock, 2009; discussions in Heads, 2012a: 42; 2014: 26). Modern dispersal theorists suggest instead that founder speciation is the result of adaptation to the new environmental conditions causing *ecological* speciation, but this has its own problems (see the later section "Ecological Speciation and Sympatric Speciation: A Critique"). In practice, the biogeographic patterns that were explained by founder speciation can be interpreted in other ways, as the next example illustrates.

In founder speciation, widespread species give rise to local species by chance dispersal and "budding off" (whether the genetic change is the result of founder effect and genetic revolution, or ecological speciation). This has been termed "peripatric speciation" (Mayr, 1954) and "speciation by colonization" (Hennig, 1966). As a typical example, Mayr cited birds of paradise; he proposed that the *Paradisaea apoda* complex, widespread through mainland New Guinea, gave rise to *P. rubra*, restricted to small islands off the western tip of New Guinea. Nevertheless, in molecular phylogenies, *P. rubra* is not nested in the widespread *P. apoda* group, as the peripatric theory would predict. Instead it is sister to *P. decora*, endemic to small islands off the eastern tip of New Guinea, 2000 km away (Irestedt et al., 2009). This is incompatible with peripatric speciation, but it is consistent with vicariance in a widespread ancestor that has undergone phylogenetic breaks and displacement around the major strike-slip faults in the region, in particular the Sorong fault (Heads, 2014: 336).

It is sometimes suggested that "budding speciation contains the unique signature that early in the speciation process, sisters should have overlapping or adjacent ranges with very different sizes" (Anacker and Strauss, 2014). Yet this is also consistent with vicariance, as there is no reason why, during a vicariance event, an ancestor should be divided into two equal halves. The sister groups in many examples of vicariance have very different range sizes. For example, the clade *Paradisaea rubra* + *P. decora* on the offshore New Guinea islands has a small range, but it is sister to (not nested in) the *P. apoda* complex, which has a much greater range through mainland New Guinea.

Origin of Distributional Overlap

In dispersal theory, both the allopatry and the overlap of clades' distributions are attributed to dispersal. A single pattern of overlap repeated in many groups is thought to have developed by dispersal in the different groups at different times. In this model, each individual case of distributional overlap is thought to be idiosyncratic and not related to other events, as with the concept of speciation by chance dispersal.

Instead, in the approach used here, allopatry is explained by vicariance, while overlap is attributed to dispersal (range expansion). Where a pattern of overlap occurs, it is manifested in many members of a biota, not just individual cases. This suggests that overlap develops in general, community-wide phases that are caused by specific geological and climatic changes, not by chance. For example, some of the last great phases of overlap can be attributed to the last great increases in sea level, in the Cretaceous (Heads, 2012a, 2014).

The focus here is on allopatry, whether this is complete or incomplete, as the allopatry reflects evolutionary (vicariance) events in a group's history. With events further back in time and phylogeny, there is more blurring of boundaries and increased overlap among clades, and the analysis becomes more difficult.

Means of Dispersal, Observed Dispersal, and Inferred Dispersal: Some Case Studies

Any biogeographic synthesis needs to integrate historical and ecological information. Ecological factors, such as dispersal probability kernels, are often included in tests of alternative dispersal hypotheses (Crisp et al., 2011). But the relationship between normal, observed dispersal (and its probability kernels) and the chance dispersal inferred from phylogenies is problematic, and the two are often not related in any obvious way. This is illustrated in the following examples.

The Hawaiian Islands Biota and Metapopulations

Chance dispersal theory is often supported by referring to young volcanic islands, such as the Hawaiian Islands and others in the central Pacific. These islands are important as classic cases for biogeographic theory in general, and they also have a special relevance for this book as many groups are restricted to New Zealand and young, volcanic islands in the central Pacific.

The Hawaiian Islands have never been joined to a continent. Because of this, traditional theory proposed that the islands were populated by organisms dispersing over thousands of kilometers from mainland Asia and America (Wilson, 2001). Propagules arriving in Hawaii are thought to have given rise to new species, either by the founder effect or by ecological speciation.

In an alternative model, many groups that are now Hawaiian endemics evolved on former high islands that occurred in the region before the modern Hawaiian Islands existed (Heads, 2012a). For example, what were once high islands are now represented south of Hawaii by atolls, such as the Line Islands, and north of Hawaii by seamounts, such as the flat-topped Musicians group. The endemic biotas of these original high islands could have developed by vicariance, mediated by processes such as the rifting of volcanic island archipelagos and subsidence caused by the weight of the volcanics. Within active volcanic archipelagos such the Hawaiian chain, volcanic loading leads to increased channel width and is an important, but neglected mechanism of vicariance (Heads, 2012a). The taxa, including endemics, that were already in the region on prior islands before the Hawaiian islands appeared would have colonized these new islands, when they erupted, by normal dispersal—without differentiation and as part of a whole community. With the complete subsidence of the high islands in the Line Islands and the Musicians seamounts, the taxa would have been left as endemics on Hawaii.

It has been suggested that there was a period from 34 to 30 Ma when none of the Hawaiian Islands were emergent, but the methods used to predict the past heights of the islands underestimated the heights of the present islands by over 1000 m. This means that they probably also underestimated the heights of the former islands (Heads, 2012a: 318).

In the "hop-skotch" process suggested here, clades consist of dynamic metapopulations or populations of populations, with groups persisting in a region by hopping from one island to another as new islands appear. The process differs from long-distance, chance dispersal as there is no mainland source. An island's biota is sometimes thought to have been derived from that of another extant island, but, again, this assumption is unwarranted. Instead, the species could have come from former, now-submerged islands.

Outside the Hawaiian Islands, metapopulation dynamics has been used to explain the distribution of groups in the southwest Pacific. In oribatid mites, *Austronothrus* is known from New Zealand, Norfolk Island, and Borneo, while the *Crotonia unguifera* species group has a similar, arcuate distribution in New Zealand, Norfolk Island, Vanuatu, and the Philippines (Luzon). Colloff and Cameron (2014) wrote that the pattern "is indicative of a curved, linear track; consistent with the accretion of island arcs and volcanic terranes around the plate margins of the Pacific Ocean, with older taxa persisting on younger islands though localised dispersal within island arc metapopulations…" The patterns are consistent with "differentiation of old metapopulations by vicariance as plates drifted apart, older volcanic islands subsided and new ones emerged…."

Many features of the Pacific seafloor now lie thousands of meters below the sea surface, and it is hard to believe that they were once emergent. O'Grady et al. (2012) suggested that mapping and interpreting the 2000, 4000, and 5000 m isobaths in a discussion of central Pacific biogeography (Heads, 2012a) was "disingenuous," as sea level has not dropped by more than 100 m or so. But the authors overlooked the thousands of meters of subsidence that the Pacific seafloor itself has undergone. As the seafloor has drifted away from the spreading ridge that produced it—the East Pacific Rise—it has cooled (increasing its density) over tens of millions of years and subsided (Smith and Sandwell, 1997). This has led to the submergence of most of the islands perched on it. Evidence for the subsidence is seen in the numerous atolls (in which the coral reefs have grown with subsidence) and flat-topped seamounts (guyots) in the region. These are former high islands that were planed flat to sea level by erosion and later sank far below sea level during tectonic subsidence.

Advocates of dispersal theory for the central Pacific Islands accept very long-distance dispersal, for example, from Asia and America to the Hawaiian Islands. Yet at the same time, they deny the possibility of local dispersal among islands in the region that would allow metapopulations to persist in active archipelagos over long periods of time, more or less *in situ*. This position is not tenable, and it is rejected here.

For the terrestrial fauna of Hawaii, Gillespie et al. (2012) found that more of the flighted, wind-dispersed taxa have affinities to the east (America) than with the west (Asia), and this is compatible with dispersal to Hawaii by the prevailing easterly storms. In contrast, all the flightless groups of animals that were sampled (these are inferred to disperse by rafting) have western affinities, in line with the main ocean currents. These results would appear to be good evidence for dispersal to Hawaii from the continents. Yet for plants, the pattern is reversed—there are more wind-dispersed groups with western affinities than with eastern affinities (Gillespie et al., 2012). This sort of incongruence is good evidence against dispersal theory as a general explanation for the biota, but was overlooked by the authors. Thus chance dispersal is more or less immunized from falsification, as its advocates propose that the process can operate just as well against prevailing winds and currents as with them.

Trans-Tasman Sea Affinities

Typical cases of normal dispersal include the many plants and animals that are blown across the Tasman Sea, from Australia to New Zealand, with the prevailing winds. This is expected, given the organisms' physiology and ecology. Following the extensive disturbance by European agriculture and the creation of new types of habitat, some of these groups, such as the welcome swallow (*Hirundo tahitica*, Hirundinidae), have established stable populations in New Zealand. It is likely that the same groups also established populations there in the past, following large-scale disturbance by volcanism, for example.

Other examples of normal dispersal and establishment mediated by ecological change include the spur-winged plover *Vanellus (Lobibyx) miles* (Charadriidae) of New Guinea and Australia, which was self-introduced to New Zealand in the 1860s. The species is "highly dispersive... responding to new food sources in recently constructed and temporary wetland, leaving as soon as they dry out" (Piersma, 1996b: 418). Likewise, *Charadrius (Elseyornis) melanops* (Charadriidae) of Australia and Tasmania was self-introduced to New Zealand in 1954 and was breeding by 1961. Recent range expansion of this bird in Australia and New Zealand was facilitated by the construction of artificial wetlands, such as farm dams and sewage ponds (Piersma, 1996b: 440), and *C. melanops* is now the most widespread wader in Australia.

In other groups, such as whales, sharks, and seabirds, regular journeys of individuals or populations across the Tasman Sea are even more frequent (they are often annual), and again, this is a part of each group's normal ecology. Given this sort of dispersal, the question with respect to New Zealand *endemism* is: "What could cause a group to *stop* migrating across the Tasman and thus permit differentiation?" The mechanism proposed in dispersal theory is chance, while the mechanism adopted here is geological or climatic change.

Prevailing winds across the Tasman Sea are from Australia to New Zealand, and this has been thought to account for the biotic similarities between the two. Normal ecological dispersal across the Tasman would be viable as a general model explaining the New Zealand biota if this were a subset of the Australian biota. In fact, though, the two biotas are very different, and the differences cannot be explained by normal dispersal.

Dispersal from Australia to New Zealand is often observed in terrestrial groups, such as birds and Lepidoptera, in the same direction as the prevailing winds. In contrast, observed dispersal in these groups from New Zealand to Australia is very unusual, except in the annual migrations of some seabirds. For example, in butterflies, *all* trans-Tasman dispersal is from Australia to New Zealand; there have been no observed cases of dispersal in the opposite direction (Gibbs, 1980: 34).

Despite these observations on actual dispersal, *chance* dispersal from New Zealand to Australia is *inferred* for many groups, based on their phylogeny. For a sample of animal phylogenies, Sanmartín and Ronquist (2004: Fig. 9) calculated that dispersal from New Zealand to Australia (against the wind) is even more frequent than from Australia to New Zealand (7.8 events versus 5.8). An updated sample of plant phylogenies showed the opposite trend, 6.7 versus 21 (Sanmartín et al., 2007), but the authors accepted chance dispersal from New Zealand to Australia in many plants. This has also been accepted in many other studies (e.g., Wardle, 1978; Wagstaff and Garnock-Jones, 2000; von Hagen and Kadereit, 2001; Lockhart et al., 2001; Wagstaff and Wege, 2002; Winkworth et al., 2002a,b, 2005; Meudt and Simpson, 2006; Bush et al., 2009; Pirie et al., 2009; Himmelreich et al., 2012).

In the liverwort family Schistochilaceae, Sun et al. (2014) proposed a center-of-origin/dispersal model and wrote (p. 197): "The [inferred] trans-Tasman dispersal was all westward, a pattern opposite to the predominant eastward dispersal pattern seen in various plant groups... the inference of westward trans-Tasman dispersal in *Schistochila* suggests that prevailing wind direction was not a governing factor in dispersal direction."

Sun et al. (2014) also admitted that "*Schistochila* appears to be unpromising for migration over long distances..." because of its dioecy and lack of asexual reproduction. Thus, the inferred dispersal is not related to the weather patterns or to the observed means of dispersal. The authors argued that their results "demonstrate successful long distance dispersal despite these features," but this was only because of the programs they used, and these are discussed in later sections, below.

Trans-Pacific Affinities

Studies on a sample of animal phylogenies inferred that dispersal from New Zealand to South America (with the prevailing winds) is more common than from South America to New Zealand (7.7 events versus 3.5; Sanmartín and Ronquist, 2004). This would make sense. Yet for plants, the figures were 3.2 versus 4.8 (Sanmartín et al., 2007)—dispersal against the prevailing winds is more common.

To summarize for trans-Tasman animal groups and for trans-Pacific plant groups, inferred dispersal against the wind is more common than dispersal with the wind. "Antiwind" dispersal is also inferred for many trans-Tasman plants and trans-Pacific animals, in contrast with the number of *observed* antiwind dispersal events in butterflies—zero.

Sanmartín et al. (2007) tested the prediction that for plants, there is an eastward bias in chance long-distance dispersal among Australia, New Zealand, and South America, in line with the prevailing winds and currents. This prediction was falsified, at least for inferred dispersal across the South Pacific. So the question then is: "How does the proposed antiwind dispersal operate?"

Sanmartín et al. (2007) suggested that transport by seabirds could explain long-distance dispersal of seeds that lack obvious adaptations to wind dispersal. Yet they gave no details of which seabirds might have transported which plants. In addition, the same patterns seen in the trans-Pacific seed plants are repeated in a large number of invertebrates and seedless plants that are not transported by birds.

Sanmartín et al. (2007) also suggested another explanation for the apparent antiwind dispersal that proposes even longer, eastward dispersal routes with the winds and currents. In this model, dispersal from New Zealand to Australia, for example, would not have been direct, but would have proceeded from New Zealand to South America and eastward past Africa, then on to Australasia. Still, many groups are unlikely to survive this circumglobal journey. It also fails to explain the precise break seen in many groups between New Zealand/Lord Howe Rise, and the east coast of Australia. It is difficult to see how a New Zealand group could migrate eastward and almost circle the globe, over South America and Africa, before settling in Australia, almost at the same place it started from.

So the question remains: "What is the explanation for the high levels of inferred antiwind dispersal?" Sanmartín et al. (2007: 15) concluded: "Our current understanding of dispersal processes is still fragmentary and we need to understand the underlying mechanisms in much more detail... we will need larger data sets and additional methodological advances in order to test fully these dispersal patterns and infer processes from phylogenetic data." The problem is unlikely to be the result of insufficient data, as there is such a large amount already available. This indicates that the problem lies with the methods used, and in particular, certain assumptions that were inherited from the Modern Synthesis. The vicariance approach adopts a different set of assumptions, namely, that evolution and distributions can be explained by alternating phases of vicariance and range expansion alone; that a group's normal, observed means of dispersal are the ones used in everyday movement, annual migrations, and range expansion; and that speciation by chance dispersal is unnecessary.

Observed Dispersal and Inferred Dispersal: A Case Study in a Group of Alpine Plants, the *Hebe* Complex

Normal, observed dispersal and inferred, chance dispersal have both been examined in detail in one New Zealand group, the shrubs and subshrubs of the *Hebe* complex (*Veronica* s.lat., Plantaginaceae) (Pufal and Garnock-Jones, 2010). In at least 10, mainly alpine species, seeds disperse from a hygrochastic capsule that opens when moistened by rain. Raindrops that then fall into the open capsules splash droplets out, taking the seeds with them. Observations on five species with hygrochastic capsules indicated that the average dispersal distance was 13 cm, and the greatest distance covered by a single dispersal event was 1.1 m. Pufal and Garnock-Jones (2010) suggested that secondary dispersal in many types of alpine vegetation is unlikely, because of the microtopography of small cracks and crevices.

Pufal and Garnock-Jones (2010) proposed that hygrochastic capsules have evolved in the *Hebe* complex as part of an "antitelechoric" strategy; the capsules "enforce short-distance dispersal." Nevertheless, based on a molecular phylogeny, the authors also accepted that two species in the complex (*V. ciliolata* and *V. densifolia*) have dispersed ~1700 km from New Zealand to Australia, against the prevailing winds.

Each of these species occurs in both countries, but each belongs to a larger clade that is otherwise endemic to New Zealand. Because of this, the authors inferred a center of origin there.

Dispersal of these alpine plants from New Zealand to Australia across the Tasman Basin would represent a leap of a magnitude 1.5 million times greater than the maximum observed dispersal distance. Meudt and Bayly (2008) concluded that the mechanisms of the process in the two species are "unknown" and "hard to envisage." This contradicts the idea that normal, ecological dispersal and inferred, long-distance dispersal are variants of the same process. Pufal and Garnock-Jones (2010) agreed, and they argued that the trans-Tasman migration is not the result of the normal, observed dispersal. Instead, it "is probably due to chance secondary dispersal, as rare long-distance dispersal is more often *not related* to the primary [mode of] dispersal" [italics added].

In contrast with this conclusion, McGlone et al. (2001: 209) argued that while "long-distance dispersal clearly has an element of chance… long-distance dispersal is also a strongly nonrandom process in that it selects for certain types of propagules and plant types." Nevertheless, recent work instead supports Pufal and Garnock-Jones' (2010) conclusion that inferred events of long-distance dispersal are *not* related to propagule type. For example, Christenhusz and Chase (2013) discussed *Nicotiana* sect. *Suaveolentes*, the disjunct between South Africa and Australia, and wrote: "The seeds of *Nicotiana* are small, but not wind-dispersed, and the age of section *Suaveolentes* definitely rules out other explanations; it seems clear that long distances across oceans are being covered by taxa *without any clear adaptations* for travelling such distances" [italics added]. But do the clock ages "definitely" rule out other possibilities? Is it really "clear" that dispersal has occurred? Does biogeographic analysis need to rely on a process of dispersal for which there is no known mechanism and which is so difficult to investigate?

Another example concerns island-endemic species. These are often expected to have lower dispersal ability than their nonendemic congeners on the same island. Yet a study of measured dispersal values in plants from the Canary Islands found that "in many cases, endemic species had… the same or better dispersal ability than their non-endemic congeners" (Vazačová and Münzbergová, 2014).

Higgins et al. (2003) also concluded that "the relationship between morphologically defined dispersal syndrome and long-distance dispersal is poor… the MDS [morphological dispersal syndrome] is not informative in the context of LDD [long-distance dispersal]." All these authors acknowledge that chance dispersal, with its obscure mechanisms, is not related to normal, observed dispersal and its means (such as the hygrochastic fruit wall), whether these are morphological, physiological, or behavioral. Higgins et al. (2003) concluded that some modes of dispersal "seem *inherently untractable* (e.g., the occasional seed dispersed on an oceanic raft). The multitude of mechanisms of LDD means that it is difficult to exclude the possibility that a species has access to an undetected LDD mechanism" [italics added]. Mechanisms that operate just once, at a random time in the history of a lineage (perhaps millions of years), and are not correlated with any other parameter are impossible to exclude. Yet, as Higgins et al. (2003) concluded, "the fact that the world's flora is not cosmopolitan reminds us that, although exceptional dispersal events do occur, not every species ends up realizing its potential for the exceptional." In fact, most plant groups have distributions structured around the same biogeographic and phylogenetic breaks, or nodes, that recur in a large number of other groups, and this suggests that exceptional events of chance dispersal and speciation occur only in very few clades, if at all (Heads, 2004). In contrast, range expansion by normal means of dispersal can take place in many groups during phases of general mobilism, but here, the dispersal is triggered by geological and climatic events, not by chance.

To summarize, chance dispersal is an accepted part of dispersal theory. Nevertheless, inferred chance dispersal shows no correlation with a group's normal means of dispersal. It is not related to physical or biological factors and can take place in any direction at any time; it is often proposed to have occurred just once in the lifetime of a lineage, which may be millions of years.

Danthonioid Grasses and Dispersal by Serendipity

The grass subfamily Danthonioideae has a more or less global distribution, and in New Zealand, component genera such as *Chionochloa* dominate extensive areas of grassland. Linder et al. (2013) inferred chance, long-distance dispersal for the group, including many trans-oceanic dispersal events. The authors concluded that the dispersal direction is not related to wind direction or any other observable factor, and so they attributed it to "serendipity." Yet they also wrote: "There is an increasing appreciation that much long-distance dispersal is concordant, and that there are patterns in both the direction and the frequency of dispersal." How does a chance process driven by serendipity produce concordant, repeated patterns? For example, Danthonioideae are absent in several areas, including Japan, Hawaii, southern coastal Greenland, and Iceland, where the climate is suitable for the group. These areas of absence are all in the northern hemisphere, while the basal endemics in the group are mainly southern, and it is not necessary to attribute this simple, standard geographic structure to chance.

Linder et al. (2013) concluded that for Danthonioideae, the "most parsimonious" interpretation of the biogeographic history requires 27 long-distance dispersal events across ocean basins in different directions, determined by serendipity. In contrast, a vicariance explanation does not rely on serendipity as a cause of main patterns and instead proposes geological change as the mechanism.

New Analytical Methods (DIVA, DEC, and Others) Based on Old Concepts of Local Centers of Origin and Chance Dispersal

New data from well-sampled studies are valuable, but the production of more and more new data documenting the repetition of striking patterns that are themselves not explained is unsatisfactory. For example, a pattern such as the New Zealand–New Guinea connection is already well documented in descriptive studies of many groups, including the gray warbler, *Gerygone*; the New Zealand tui, *Prosthemadera*; and the bellbird, *Anthornis* (Heads, 2014: Figs. 6.5 and 6.6). These and other examples are discussed in later sections. But although the New Zealand–New Guinea pattern has by now been well documented over many years and in a wide range of groups, its recognition as a standard pattern and its interpretation have both been neglected. There has been no discussion of it in the mainstream literature other than brief references to "chance dispersal."

As different authors have observed, the major advances in biology over recent decades have been technological rather than conceptual (Crisci, 2006; Karl et al., 2012). The revolution in molecular biology has produced a vast amount of new, high-quality data on the phylogeny and distribution of clades. In contrast, contemporary advances in biogeographic methods that analyze these data are represented by software that simply formalizes and incorporates key concepts of the Modern Synthesis.

One of the main questions in traditional theory was: "How can a group's center of origin be located?" Darwin (1859) argued that the more advanced forms of a group would occupy its center of origin, while primitive forms would be forced to migrate away. In contrast, Mayr (1942) proposed that primitive forms occupy the center of origin, while advanced forms move out and invade new areas and niches. Mayr's view was accepted by cladists (Hennig, 1966) and phylogeneticists (Avise, 2000); Katinas et al. (2013) wrote that Hennig's *progression rule* "establishes that the primitive members of a taxon are found closer to its center of origin than more derived ones...."

Was Darwin or Mayr correct? Do advanced or primitive groups occupy the center of origin? The question is seldom discussed, although it represents a fundamental problem in dispersal theory. The problem does not arise in vicariance theory, as the concept of center of origin is not accepted. In modern dispersal theory, centers of origin are indicated, not by primitive groups but by "basal" groups, but this concept is problematic and is discussed next.

Do Basal Clades Occupy Centers of Origin?

The basal node, or primary division, in a phylogeny connects two sister groups that are equally basal. A group that is less diverse than its sister group is often referred to as the "basal clade," but the term is confusing and the relationship often misinterpreted. Authors often assume that a "basal clade" is older than its sister, that it is more divergent, that it has more primitive morphology than its sister, and that it occupies a center of origin from which its sister dispersed. But there is no reason to assume that any of these are true. (According to Grandcolas et al. [2014: 658], I suggested [in Heads, 2009b] that the basal species in a group occupies the "center of origin" of the group, but this is not what I wrote.)

A typical example from the current literature concerns a group of Bignoniaceae found in Madagascar and other Indian Ocean islands (Comoros, Mauritius, Seychelles). Callmander et al. (2016) concluded: "The phylogenetic framework shows *Phylloctenium*, which is endemic to Madagascar and restricted to dry ecosystems, as basal and sister to the rest of the tribe, suggesting Madagascar to be the centre of origin of this clade."

In another example, Goremykin et al. (2013) suggested that *Trithuria*, Nymphaeales, and *Amborella* form "the most basal lineage of flowering plants." (For a critique of the phylogeny and confirmation that *Amborella* on its own is basal, see Drew et al., 2014 and Simmons and Gatesy, 2015.) Goremykin et al. wrote that their phylogeny "indicates that aquatic and herbaceous species [*Trithuria* and Nymphaeales] dominate the *earliest extant lineage* of flowering plants" [italics added]. But the group proposed as basal—*Trithuria*, Nymphaeales, and *Amborella*—is not the earliest lineage, as it has the same age as its sister—all the other angiosperms. On the basis of the phylogeny alone, there is no reason to think that either one is more or less primitive than the other, or that the aquatic–herbaceous habit is primitive.

In another example, Gray and Atkinson (2003) gave a phylogeny for Indo-European languages in which Hittite is sister to all the others, and within the others, Tocharian is sister to the rest. This can be expressed as Hittite (Tocharian (rest)), or in a more graphic form:

Hittite.
　　Tocharian.
　　The rest.

Gray and Atkinson (2003) referred to Hittite and Tocharian as "the languages at the base of the tree," but this overlooks the fact that the sister group of Tocharian (the rest of the Indo-European languages—Hindi, Greek, English, etc.) is just as basal as Tocharian. Gray and Atkinson (2003) proposed that Hittite, an extinct language of Anatolia, occupied the center of origin of the Indo-European group, but there is no reason to infer this from the phylogeny, as its sister group (Tocharian and "the rest") is just as ancient.

Do Paraphyletic, Basal Grades Occupy Centers of Origin?

In modern treatments, a "basal" *grade* in a phylogeny—a sequence of basal clades—is often regarded as primitive in its morphology and ecology, and it is thought to occupy the center of origin. Consider a clade made up of five groups, each found in area A or B, and with a taxon-area phylogeny: A (A (A (A, B))). The four groups in area A do not comprise a monophyletic group (as B is not included), but form a paraphyletic complex, a basal grade. For this pattern, standard theory will always infer a center of origin in A, where the basal grade is located, and will propose a single, chance dispersal event to B. This view is incorporated in programs such as DIVA (Ronquist, 1997) and LAGRANGE (Ree and Smith, 2008)—these will always calculate a center of origin in A followed by dispersal to B. But instead, the pattern could have been caused by repeated vicariance in an ancestor already present in both A and B, at a break in A, followed by local overlap in A (Heads, 2009b). There is no dispersal between A and B.

This simple process is illustrated here with a hypothetical example of the A (A (A (A, B))) pattern, with A represented by Australia and B by New Zealand. The process begins with a widespread Australia–New Zealand ancestor (Figure 1.1a). The ancestor then differentiates around a break in central Australia caused by mid-Cretaceous marine transgressions dividing the continent (Figure 1.1b). Late Cretaceous rifting in the Tasman Basin then created a disjunction across the basin (Figure 1.1c). In a final phase, range expansion around the primary node in central Australia led to secondary overlap there (Figure 1.1d). (A node is the junction or break zone of two sister clades; it is displayed in a phylogeny and is also evident in geography, on the ground [Heads, 2004].)

The vicariance model shown here indicates that, despite the presence of the basal grade in Australia, there is no need to invoke a center of origin in either Australia or New Zealand, or to infer any dispersal between the two. Australia is not necessarily a center of origin, and is just as likely to have been a center of differentiation in a widespread ancestor. The widespread ancestor has undergone repeated differentiation events at breaks in Australia, and this has produced the paraphyletic basal grade there. The process is probably quite common, as tectonic events in a region are often episodic.

Standard dispersal theory interprets the A (A (A (A, B))) area phylogeny as representing dispersal from A to B. Nevertheless, chance dispersal can happen anywhere, in any direction, and so the same area phylogeny could instead indicate a center of origin in B, with four dispersal events from B to A (Cook and Crisp, 2005; Sanmartín et al., 2007). The studies suggested that this would be more likely if the prevailing winds were from B to A, for example. Sanmartín et al. (2007) wrote that the problem indicates "the limitations of our current methods and their assumptions," but the problem only arises in dispersal theory, not in a vicariance model, and so it is not considered further here.

In the general case of an area/clade phylogeny: A (B (C (D, E))), the pattern is often thought to represent a *series of dispersal events*, from A to B to C to D to E (as in the "progression rule" of Hennig, 1966). Yet the pattern can also be explained by a *series of differentiation events* in a widespread ancestor: first, between A and the rest, then between B and CDE, then between C and DE, and finally between D and E. Figure 1.1 is put first because it is the most important in the book.

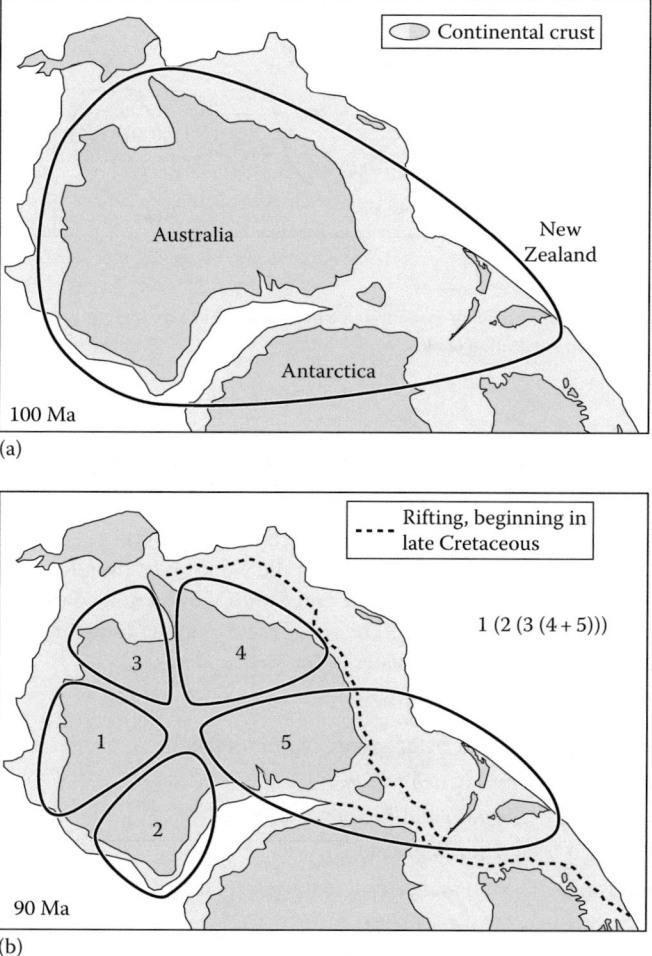

(a)

(b)

FIGURE 1.1 Reconstruction of evolution in a hypothetical group. (a) A widespread ancestor. (b) Differentiation around a break in central Australia prior to opening of the Tasman Basin. *(Continued)*

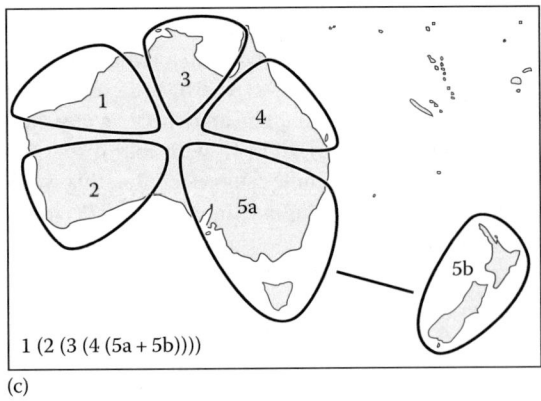

(c)

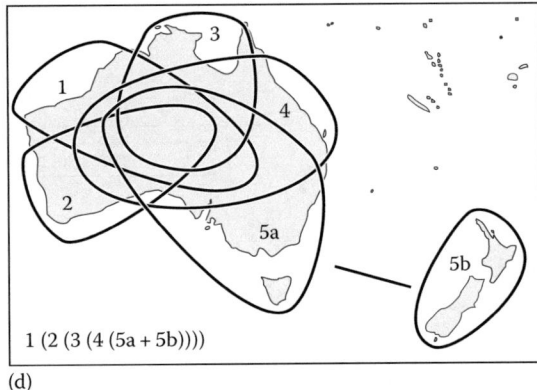

(d)

FIGURE 1.1 (*Continued*) Reconstruction of evolution in a hypothetical group. (c) Rupture of clade 5 with opening of the Tasman Basin. (d) Overlap of the Australian clades.

Interpretation of Basal Grades Restricted to Particular Regions: Examples from Plants

The grammitid ferns form a clade in Polypodiaceae comprising ~900 species. A recent study of the group found a basal, paraphyletic complex centered in the Neotropics (Sundue et al., 2014). The authors carried out a DEC analysis in LAGRANGE (Ree and Smith, 2008), and so they inferred long-distance dispersal from a Neotropical center of origin. The study recovered the following phylogeny (note that the first clade is sister to the rest, the second is sister to the rest, and so on):

Neotropics, Nearctic (outgroup and sister group, *Niphidium* etc.).
  Neotropics, Hawaii and Africa (*Cochlidium* etc.).
    Neotropics, Africa (*Alansmia* and *Leucotrichum*).
      Neotropics, Africa (*Ceradenia* etc.).
        Neotropics, Hawaii and Africa (*Mycopteris* etc.).
          Neotropics (*Moranopteris*).
            E Asia (India–NE Australia–Japan) (*Chrysogrammitis*).
              E Asia (India–NE Australia–Japan) (*Calymmodon* etc.).

> E Asia plus Pacific Islands (incl. Hawaii) (*Themelium* etc.).
>
> **Circumaustral**: Chile, South Africa, SE Australia, and New Zealand (also Falkland, Tristan da Cunha, Gough, South Georgia, Marion, Crozet, Kerguelen, and Amsterdam Islands) (*Notogrammitis*).
>
> **Lord Howe Island** ("*Grammitis*" *diminuta*) and **New Caledonia** ("*G.*" *deplanchei* and "*G.*" *pseudaustralis*).

Sundue et al. (2014) inferred a center of origin in the Neotropics, followed by long-distance dispersal to Asia, then from there to Australasia, and from there to the rest of the circumaustral region. Yet the phylogeny is also compatible with a global ancestor that underwent vicariance. In this model, the first differentiation events occurred around the Neotropics and Africa, and these were followed by overlap within these regions. In this model, the circumaustral clade originated *in situ*, by simple vicariance with the Lord Howe/New Caledonia clade, and the ancestor of these two originated by vicariance with the East Asia plus Pacific Islands clade. This indicates that a main phylogenetic and geographic node separating these three clades is located in or around the Tasman Sea.

Another example of a paraphyletic, basal grade endemic to a single region occurs in the orchid *Spiranthes*. This genus is widespread in the Americas, Eurasia, and Australasia, including New Zealand. Based on a phylogenetic study, Dueck et al. (2014) concluded: "the cladograms we have presented indicate little doubt that America represents the center of origin for the genus and that Old World species of *Spiranthes* are derived from New World taxa." Nevertheless, this is only because most of the clades, including a basal paraphyletic grade, are American, and so an alternative vicariance model is also possible.

In the heath family, Ericaceae, the tribe Styphelieae has eight species in New Zealand. Puente-Lelièvre (2012) found that each of these is a "distinct lineage that is nested within an Australian clade." Yet these clades all include New Zealand members, and so they are not "Australian clades"—they are Australia–New Zealand clades—and for the reasons just discussed, there is no need to assume a center of origin in either Australia or New Zealand.

A clade in the tree *Vitex* (Lamiaceae) has the following phylogeny (Pratt, 2013):

> New Caledonia (*V. collina* p.p.).
> > New Caledonia (*V. collina* p.p. + *V. unifolia*).
> > New Zealand + Australia (*V. lucens*, puriri, + *V. lignum-vitae*).

Again, the basal grade in New Caledonia does not necessarily mean that the region was a center of origin for the New Zealand and Australian species. Instead, New Caledonia can be interpreted as a center of differentiation in an ancestral complex that was already widespread around the Tasman region.

In the Plantaginaceae, Albach et al. (2004) discussed a long dispute about *Veronica* plus allies in the northern hemisphere, and the closely related *Hebe* complex in the southern hemisphere. Which area is the center of origin? The authors presented a molecular phylogeny in which the northern hemisphere genera form a basal paraphyletic complex. The southern group is nested in this. Albach and Chase (2004: 183) concluded: "The debate about an origin in the Southern... or Northern hemisphere... has been ended in favour of the latter." The debate has been "settled"; the ancestor spread "from Eurasia to Australia and onto New Zealand" (Albach et al., 2004: 432).

In their proposed solution, Albach and colleagues assumed that the group has a classic, restricted center of origin, but this is not necessary, and the authors did not address this question (they overlooked the many vicariance analyses of the group). They also assumed that the center of origin occurs in the locality of a paraphyletic, basal grade, in this case, with *Veronica* and allies mainly in the north. Again, this cannot be assumed.

Albach et al. (2004) argued that the phylogeny, with a basal, northern grade and the nested, largely allopatric, southern group, indicated the "derivation of the latter from within the former" (p. 429). Instead of this ancestor–descendant relationship between the main, extant groups, a vicariance model proposes a widespread common ancestor for all these, followed by their allopatric differentiation.

In the cosmopolitan plant family Asteraceae (Compositae), the five basal clades are all South American and constitute a basal grade. This means that a South American center of origin for the family is calculated, whether the "ancestral areas" approach (Bremer, 1992) or "parsimony optimization" (Maddison and Maddison, 1992) is used (Katinas et al., 2013). Nevertheless, it is just as possible that the pattern was caused by five vicariance events at breaks in or around South America, in an ancestor that was already worldwide.

Within the Asteraceae, the diverse, global tribe Astereae is well represented in New Zealand. One study of the tribe adopted traditional theory and concluded:

> The presence at the base of the tree of two South African genera, *Printzia* and *Denekia*, would indicate that the tribe indeed originated on that continent... The second lineage to diverge is the Chinese genus *Nannoglottis*, which suggests long-distance dispersal or rafting of members of Astereae to eastern Asia early in the evolution of the tribe.... (Brouillet et al., 2009: 618)

Instead, while the tribe must have attained part of its worldwide range by dispersal (creating the overlap with its sister group), this probably occurred *before* the differentiation of the modern clades in the tribe, as they show similar patterns. Good examples are seen in and around the Pacific (Nesom, 1994; Heads, 1998b).

Interpretation of Basal Grades Restricted to Particular Regions: Examples from Animals

In gastropods, the whelk genus *Cominella* has 13 species in New Zealand, where it is widespread and occurs from intertidal sites down to 600 m (Powell, 1979). It also occurs in Australia (six species) and Norfolk Island (one species), and the following phylogeny has been presented (Donald et al., 2015). NZ = New Zealand.

NZ (NZ (NZ (NZ (Australia and Norfolk I.)))).

Donald et al. (2015) calculated a center of origin in New Zealand, followed by dispersal to Australia. Yet this was based on the assumption that a paraphyletic basal grade occupies the center of origin, and the phylogeny is just as compatible with vicariance in a widespread, trans-Tasman ancestor. Following early breaks in or around New Zealand, there has been extensive overlap there.

In other gastropods, the superfamily Amphiboloidea includes snails that are often dominant on mudflats and salt marsh and in mangrove. The group has a typical Tethyan range from Australasia through southern Asia to the Persian Gulf, and it has the following phylogeny (Golding, 2012):

Northern Australia (*Maningrida*).
New Zealand (*Amphibola*).
Southern and eastern Australia and Tasmania (*Phyllomedusa*).
Australia to Persian Gulf (the remaining four genera).

Golding et al. (2012) regarded the phylogeny as "compelling evidence" that the group had a center of origin in Australasia, and this was corroborated by a center-of-origin program (RASP; Yu et al., 2010, 2011). Nevertheless, the program and the conclusion assume that a paraphyletic basal grade is a center of origin. Instead, Australasia may be a center of differentiation in a widespread ancestor. The high level of overall allopatry among the four amphiboloid clades is consistent with an origin by simple vicariance, with subsequent overlap restricted to Australia. (As discussed below, lumping the first three clades together as "Australasian" obscures the critical allopatry.)

A clade of six octopuses (Octopodidae) has the following phylogeny (Strugnell et al., 2008):

Southern Ocean (*Adelieledone*).
Southern Ocean (*Pareledone*).
Southern Ocean (*Megaleledone*).
Subcosmopolitan (*Graneledone*, *Velodona*, and *Thaumeledone*).

The "Southern Ocean" clades occur on the continental shelf around Antarctica and form a basal grade. The subcosmopolitan lineage inhabits the deep sea and extends from Antarctica north to New Zealand (the most diverse area), the north Pacific, the north Atlantic, and the southwest Indian Ocean. For the complex as a whole, the authors inferred a center of origin in Antarctica and in shelf habitat, followed by a radiation northward and into the deep sea. In an alternative model, both the Antarctic shelf and the areas that are now deep sea were already occupied by the group before its differentiation into the modern clades. In this vicariance process, the differentiation would have taken place at three breaks, all located in or around Antarctica. The modern ocean basins (Atlantic, Indian, and Pacific) are each occupied by distinct clades of the deep-sea group, consistent with allopatry caused by vicariance.

In spiders, most members of Linyphiidae subfamily Mynogleninae occur in tropical Africa (mountains of Cameroon and East Africa) and New Zealand (main islands, Antipodes and Chatham Islands). There are also a few species on Réunion, Tasmania, and Fiji, but these were not sampled in the study by Frick and Scharff (2013). The group as a whole has a typical Indian Ocean distribution, straddling the margin between the African and Indo-Australian plates. Frick and Scharff (2013) found the following phylogeny (NZ = New Zealand):

NZ.
  NZ.
    NZ.
      NZ.
        NZ + Africa.
        NZ + Africa.

The authors concluded that this "…suggests a single origin of mynoglenines in New Zealand with two dispersal events to Africa, and does not support Gondwana origin." Yet the topology is also consistent with repeated vicariance of an ancestor that was already widespread in Africa + New Zealand, at a break in or near New Zealand, followed by a final break between two Africa + New Zealand clades (e.g., between one in the north and one in the south).

Apart from the basal paraphyletic grade in New Zealand, the Mynogleninae are also of special interest through their trans-Indian Ocean distribution. Frick and Scharff (2013) considered that the total absence of Mynogleninae from South America, southern Africa, and most of Australia is real, and not a sampling artifact, and that this "does not fit the traditional Gondwana distribution patterns." Here, the authors assumed that the prebreakup Gondwanan biota was homogeneous, and this is a standard idea. For example, Cowen (2013: 241) wrote that "New Zealand was part of Gondwana until the Cretaceous, and it had a normal fauna at that time." In this view, the absence of southern groups from one or more of the southern continents is problematic for a vicariance model (Friedman et al., 2013). Yet there are very few clades endemic to, and found throughout, any of the modern continents, and there is no reason to think that the Gondwanan biota was homogeneous. As Buckley et al. (2015) wrote: "Undoubtedly, biotic regionalisation existed across Gondwana, and this may have influenced the absence of some lineages from modern New Zealand."

In the beetle genus *Phanodesta* (Trogossitidae), the species from New Zealand, New Caledonia, and Lord Howe Island form a basal grade (formerly treated as *Lepirina*). The sole remaining species, *Phanodesta variegata* of Juan Fernández Islands, has a derived, deeply nested position in the phylogeny, and Leschen and Lackner (2013) interpreted this as "evidence for long-distance dispersal from Australasia." Nevertheless, the observed phylogeny would also result if an Australasia–Juan Fernández ancestor underwent its first differentiation events in Australasia, and these were followed by a break between the Juan Fernández species and its sister, *P. carinata* of northern New Zealand (Three Kings Islands). An Australasia–Juan Fernández ancestor in *Phanodesta* is consistent with a standard pattern; there are many plants and animals restricted to Australasia–Juan Fernández (Heads, 2014: 226–227). Examples not listed in that reference include nine lichen species (Galloway, 2007) and the beaked whale *Mesoplodon traversii*.

One last example can be cited. Mascarene Islands skinks are nested phylogenetically within an otherwise Australasian clade (Austin and Arnold, 2006), and because of the basal grade in Australasia, Waters and Craw (2006) assumed a center of origin there. They wrote: "It therefore appears that long-distance rafting from Australasia may best explain their colonization of the western Indian Ocean. In light of this finding, trans-Tasman dispersal of skinks may also be biogeographically plausible." But invoking a center of origin for the Mascarene skinks in Australia is unnecessary, and so there is no justification for trans-Indian Ocean dispersal or for trans-Tasman dispersal.

Coding the Areas Used in Ancestral Areas Programs

Many groups comprise subgroups with simple, allopatric distributions, as in Figure 1.1b, where the five hypothetical clades have the distributions and phylogeny: NW Australia (SW Australia (north-central Australia (NE Australia (SE Australia + New Zealand)))). For allopatric groups such as these, some programs will recover a widespread ancestor followed by vicariance. However, the coding of the distributions is important. For example, if the geographic areas "Australia" and "New Zealand" are accepted, *a priori*, as the biogeographic areas, the phylogeny would be coded as Australia (Australia (Australia (Australia (Australia + New Zealand)))), and the perfect allopatry would be hidden. With this coding, all programs would calculate a center of origin (in Australia). Many studies have employed geographic areas such as "New Zealand," "Australia," "New Guinea," "New Caledonia," and "Fiji" as the unit areas in biogeographic analysis, but all these areas are both biological and tectonic composites, and this has led to flawed results.

A study of the tree *Podocarpus* (Podocarpaceae) found the following clade, with the phylogeny and areas coded as indicated in nonbold type (Quiroga et al., 2016). Other details on distribution are added here in bold, and these indicate significant allopatry in "Australia" and "South America."

> *P. salignus*, South America **(Chile, 35°–42° S)**, + *P. smithii* Australia **(NE Queensland)**.
>> *P. nubigenus*, South America **(Chile, 38°–53° S)**.
>>> *P. lawrencei*, Australia **(Tasmania and New South Wales)**.
>>>> *P. cunninghamii*, NZ.
>>>>> *P. acutifolius*, NZ, + *P. totara*, NZ.
>>>>> *P. gnidioides*, New Caledonia, + *P. nivalis*, NZ.

An analysis using LAGRANGE (Ree and Smith, 2008) inferred a center of origin in South America.

A Case Study from Liverworts

The liverwort family Schistochilaceae has a wide distribution through the southern hemisphere, in southern South America, southern Africa, Australasia, Fiji, and Samoa; it also extends north to China and Japan. The highest species diversity occurs in New Zealand. Sun et al. (2014) sequenced 53 of the 80 species and found a complex area phylogeny, with many widespread clades. These included as many as six New Zealand–southern South America groups.

Sun et al. (2014) analyzed their data using three ancestral area programs, DIVA, DEC, and MCMC: all programs indicated a center of origin for Schistochilaceae in southern South America and New Zealand, probably including West Antarctica. The authors concluded that New Zealand is an "early divergence and dispersal center from which subsequent dispersals occurred to South America, Tasmania, Asia, the Pacific, and South East and North East Australia." Sun et al. (2014) admitted that because of its dioecy and lack of asexual reproduction, "*Schistochila* appears to be unpromising for migration over long distances... However, our results demonstrate successful long-distance dispersal despite these features." Nevertheless, this is only because the programs that the study used will automatically find a center of origin at the location of a paraphyletic basal grade.

Summary on Ancestral Areas Programs

Recent biogeographic programs differ in whether, and to what extent, they allow ancestors to occupy multiple areas, and this is a necessary prerequisite for vicariance (Sanmartín et al., 2007; Crisp et al., 2011). For example, the popular DEC model (Ree and Smith, 2008) stipulates that if an ancestor is widespread across two or more areas, lineage divergence can only occur between a single area and the rest of the range, or between a single area and the entire range. It rules out standard vicariance *a priori*, as it does not allow the subdivision, in a single step, of a widespread ancestral range (comprising more than three unit areas) into two allopatric daughter ranges, each comprising more than one unit area (Lamm and Redelings, 2009; Matzke, 2013). The model requires secondary dispersal and extinction events to explain this common case, although normal vicariance is a much simpler explanation.

As Lamm and Redelings (2009) also observed, many analyses assume that the ancestral range of a group should be similar in spatial extent to the ranges of the living species. Again, this rules out simple vicariance, *a priori*, even for common cases such as widespread, global groups made up of many species, each with local or regional ranges (see Heads, 2012a, Chapter 6).

The new programs are based on the key concepts of traditional neodarwinism: centers of origin (located where primitive species occur) and speciation by chance dispersal (Mayr, 1942; Hennig, 1966). This reversion in the new, molecular work to the earlier, interpretive concepts of the Modern Synthesis is a distinctive feature. In one example, a study of evolution in the New Zealand alpine plant *Pachycladon* (Brassicaceae) adopted a center-of-origin approach, and as the authors commented, this is "directly aligned with Simpson's [1953a] view of adaptive radiations" (Joly et al., 2014). The idea of evolutionary "radiation" is based on the core concept of a point center of origin in morphology, ecology, and biogeography.

Kodandaramaiah (2011) criticized biogeographic programs and suggested that "almost all analytical methods in historical biogeography are strongly biased toward inferring vicariance." In fact, though, all the current methods that are popular in phylogenetic studies (DIVA, DEC, and their derivatives) will always find centers of origin where there are paraphyletic basal groups, and infer dispersal from there. These results are treated as "evidence," and Kodandaramaiah (2011) concluded: "Although it was previously believed that vicariance was the predominant mode of speciation, mounting evidence now indicates that speciation by dispersal is common." Yet this evidence is all derived from two sources: spatial studies using DIVA or DEC, and fossil-calibrated clock studies that are also biased toward center-of-origin/dispersal models, as they treat estimates of minimum clade age as estimates of maximum age. In contrast, empirical studies of molecular clade distributions in marine biology and in microbiology are indicating the true extent of allopatry and are causing paradigm shifts away from chance dispersal and toward vicariance (see the section "High Levels of Geographic Structure in Observed Clades" later in this chapter).

Ecological Speciation and Sympatric Speciation: A Critique

Traditional neodarwinian theory explained many cases of allopatry by invoking a process of long-distance dispersal across a barrier, followed by founder effect speciation. A genetic founder effect is well known, but geneticists now argue that it will not, in itself, lead to speciation (see the section "Founder Speciation"). In response to this problem, modern dispersal theory still invokes founder *events*, but does not rely on the founder effect to explain speciation. Instead, the model proposes long-distance dispersal followed by ecological speciation. In this process, species that now occupy different localities diverged because of environmental differences between the localities and the effects of natural selection. The habitats may be allopatric, but this is not necessary for ecological speciation to take place. In this process, the habitats occupied by the new species at the time of their divergence can even show geographic overlap. In contrast, the allopatric speciation model interprets divergence between populations as the result of geographic isolation, not differential selection (Mayr, 1942).

Ecological speciation was advocated by Julian Huxley in his book *Evolution: The Modern Synthesis* (Huxley, 1942, Chapter 6), but it was rejected by Mayr (1942). Likewise, a landmark monograph of New Zealand harvestmen concluded: "The present data provide no evidence to support the occurrence of ecological speciation but strongly suggest that geographical speciation would account for the development of most if not all of the abundant fauna recorded from New Zealand" (Forster, 1954: 30). Despite this, the concept has had a resurgence, and Seehausen (2013) referred to "the birth of ecological speciation" in the late 1980s. Many cases of ecological speciation have now been proposed (Ingram et al., 2012), although most examples concern infraspecific forms—putative incipient species—rather than new species.

The recent support for ecological speciation is unexpected. This is because Mayr (1942, 1954) introduced founder effect speciation for the very reason that he regarded ecological speciation among the New Guinea islands he was studying as so unlikely. Mayr had a wide field experience in the islands, and he observed that the ecological differences within each island were greater than any differences between them. This conclusion and Mayr's consequent rejection of simple ecological speciation are both accepted here, although the mechanism that he did propose—founder effect speciation—is not. Instead, complex tectonic activity in and around New Guinea (not well understood when Mayr was writing) can explain the patterns on its own. In the test case provided by this classic locality for biogeographic theory, there is no need for either founder effect speciation or ecological speciation.

Of course, the plants and animals in a region sort themselves out into the different habitats that are available, but this does not necessarily mean they have originated by ecological speciation (Heads, 2012a, Chapter 1). Even sister species often have ecological differences, but on its own, this is not good evidence for ecological speciation; the process would be expected to produce sister species with different ecology but the same geographic distribution, and this is a very unusual pattern.

Sympatric Speciation and the Rarity of Sympatric Sister Species

If an ancestor with a distribution area A undergoes geographic differentiation, one descendant evolves in one part of area A and the other descendant evolves in the other part—they are both allopatric. Allopatric groups that have evolved in this way may display ecological differences, but often they do not (this "niche conservatism" is discussed in more detail in the next section).

In an alternative to geographic differentiation, the ancestor in area A undergoes ecological differentiation; the descendant clades develop different ecology but they have the *same geographic distribution*, the area A—that is, they are sympatric in a strict sense. For example, consider a biogeographic region comprising an area of endemism in hill forest 50 km across. Sympatric, ecological speciation might lead to an ancestor that was endemic and widespread through the area breaking down into one endemic species on ridges and a sister species with the same geographic distribution, but inhabiting valley floors. Yet while there are countless examples of groups that are sympatric in the strict sense (this is one of the main bases of comparative biogeography), sister species that have the same distribution, and are not just overlapping, are very rare (Dawson, 2012). This suggests that sympatric speciation is uncommon (Fitzpatrick et al., 2009; Heads, 2012a, Chapter 1; Heads, 2014, Chapter 2). Most putative examples of sympatric speciation are sister species that overlap only in part (Steele et al., 2009) or show concentric distributions (Bird et al., 2011). In these cases, the clades have significant differences in their distributions.

Clades are sometimes described as sympatric if they display *any* overlap in their geographic range (e.g., Anacker and Strauss, 2014), but a distinction is drawn here between true sympatry and geographic overlap. In Californian plants, Anacker and Strauss (2014) found significant overlap between sister species, but, as they acknowledged, this can be the result of secondary range expansion following initial allopatry. There is no need for sympatric or ecological speciation.

Suggested Cases of Sympatric Sister Species

The best evidence for ecological, nongeographic speciation would be sister species that differ only in their ecology and not in their geographic distribution. As already noted, these are very rare. To demonstrate pure, ecological speciation, geographic factors should be ruled out, but most papers supporting ecological speciation (e.g., Shafer and Wolf, 2013) do not provide maps or good information on distribution, and so the evidence is difficult to assess. The same is true for many suggested cases of sympatric speciation. For example, two reef fishes, *Hexagrammos agrammus* and *H. otakii*, in the Sea of Japan have been interpreted as the result of sympatric speciation (Crow et al., 2010). The former species occurs in seaweed beds, while the latter is found at deeper, rocky sites. Nevertheless, a detailed distribution map was not presented, and this would be needed to assess the level of sympatry.

Studies of particular clades that do include the required biogeographic information demonstrate that many supposed cases of sympatric and ecological speciation have little support (Heads, 2012a). For example, Grossenbacher et al. (2014) regarded species that display *allopatric* nesting within the range of sister species as "sympatric" with the sisters. In the New Zealand intertidal limpet *Notoacmea*, Krug (2011) suggested: "One recently derived pair of sister species was sympatric even at the microhabitat level in New Zealand." Yet distribution maps indicate that the two members of all sister pairs in *Notoacmea* are allopatric: *N. badia* and *N. potae*, *N. daedala* and *N. elongata* (with overlap at only 2 out of 18 sites), *N. rapida* and *N. scapha*, *N. pileopsis* and *N. sturnus*, and *N. parviconoidea* and *N. subantarctica* (Nakano et al., 2009). This high level of allopatry between close relatives is common in plants and animals.

The marine gobies *Clevelandia* and *Eucyclogobius* are both monotypic and form a clade. They have been described as sympatric (Dawson et al., 2002), but the former extends from the Canada/U.S. border to southern Baja California, while the latter is restricted to California state (from the northern to the southern border) (GBIF, 2016). The great difference between the distributions indicates that they were not just the result of sympatric, ecological differentiation over the same area, that is, the area occupied by the common ancestor.

Recent papers have presented information on the degree of overlap between closest relatives. A study of the reef fish *Pomacanthus* proposed that 80% of the sister species demonstrate complete or substantial (>85%) range overlap (Hodge et al., 2013). Yet measuring the areal extent of a clade's distribution is not straightforward. Does a clade found more or less throughout a country at hundreds of sites have the same area as a clade known from only 10 sites scattered throughout the country? Quantifying the extent of overlap is also problematic. For example, consider a hypothetical pair of sister clades that display complete overlap in New Zealand; one has an additional record in Mauritius, comprising just 1% of its distribution area, and the other has a record on Juan Fernández Islands, again comprising just 1% of its total area. In terms of area, the two are 98% sympatric, but a global perspective suggests that the two clades are more or less allopatric; they have Indian Ocean and Pacific Ocean distributions, respectively, and the overlap in New Zealand is a secondary feature.

Hodge et al. (2013: Fig. 2) mapped two species of *Pomacanthus* found mainly along the east coast of tropical America and calculated that the species overlap for 99.1% of their area. However, *P. paru* has additional records on islands in the Atlantic (Fernando de Noronha and Ascension), where *P. arcuatus* is absent. In terms of area, these records are negligible, but in the biogeographic and evolutionary context, they are significant and probably represent a trace of the original allopatry.

The other maps presented by Hodge et al. (2013) also indicate that this "island effect" is important. *Pomacanthus imperator* ranges all the way from Africa to French Polynesia and was said to show 95.2% overlap with its sister species, *P. annularis*. Yet this is only found between the Maldives and the Solomon Islands (Hodge et al., 2013: Fig. 3c)—a very different distribution. Likewise, *P. chrysurus* is restricted to east Africa and Madagascar and displays 98.1% overlap with its sister *P. semicirculatus*, but this species extends from eastern and southern Africa all the way to Tonga (Figure 3f).

In most cases, sister species have significant differences in their distribution, even when there is considerable overlap. The simplest explanation is that the differences reflect the original allopatry of the phylogenetic break. For example, in New Zealand, the three species of *Raukaua* (Araliaceae) occur in the three main islands and display extensive overlap. Nevertheless, *R. simplex* is not in Northland, while *R. edgerleyi* is absent from large parts of the eastern South Island (between Dunedin and Kaikoura) (NZPCN, 2016*). Both these areas are important as centers of endemism in other groups, suggesting that one or both areas represent the original area of the third species, *R. anomalus*, prior to the overlap.

Sympatric speciation is sometimes inferred for sister species both endemic to a single island or a habitat island, such as a lake. Yet biotas in these habitats have often been derived from prior islands or lakes that existed in the vicinity. One supposed case of sympatric speciation concerns two sister species of palms endemic to Lord Howe Island (Savolainen et al., 2006). The inference of sympatric speciation depended on the speciation date. Yet this was calibrated by assuming that clades endemic to the Mascarene Islands could be no older than their island, and this has been contradicted by the minimum ages produced in clock studies (Heads, 2011). The geological and biological contexts indicate that the Lord Howe palms probably diverged on prior volcanic islands in the vicinity, not on the present island (Stuessy, 2006; Heads, 2014: 50).

---

* The distribution maps at New Zealand Plant Conservation Network are problematic and cannot be taken at face value; records from some collectors (Rance, Thorsen, and Barkla) are reliable, but data from others (Druce, Simpson, Jane, Donaghy, National Vegetation Survey) seem to have been corrupted and are not used here.

The same conclusion also applies to other plants on Lord Howe Island (Papadopulos et al., 2011; Heads, 2014: 50). These include a monophyletic clade of three species of *Asplenium* ferns endemic to Lord Howe Island and attributed to sympatric speciation by Ohlsen et al. (2015).

The three-spined sticklebacks (*Gasterosteus aculeatus*) are another well-known example in which sympatric, ecological speciation has been proposed. Distinct limnetic (surface water) and benthic (lake floor) forms of this fish coexist in coastal lakes of British Columbia. Nevertheless, in molecular phylogenies limnetic and benthic forms from the same lake are not sister groups and are instead closer, respectively, to limnetic and benthic forms of other lakes (Taylor and McPhail, 2000). This indicates that the limnetic and benthic forms diverged prior to the formation of the modern lakes.

Many sister orders or families are cosmopolitan and sympatric, and sister genera often have distributions that show broad overlap, but sister species seldom display true sympatry, and sister clades below species level are hardly ever sympatric. A general pattern of increasing geographic overlap with clade age has been described in terrestrial groups (Barraclough and Vogler, 2000; Kamilar et al., 2009) and also marine groups (Frey, 2010; Quenouille et al., 2011). This indicates that the predominant mode of speciation is allopatric, and that any overlap between sister groups develops after their origin.

Niche Conservatism: Lack of Niche Divergence between Sister Clades

The rarity of sister species that are sympatric in the strict sense suggests that most speciation is allopatric and is caused by geography rather than ecology. In addition, many studies have documented sister species that are allopatric and do not show ecological niche divergence. Again, these support the primacy of allopatry in speciation.

Sister species with the same niche are said to display phylogenetic niche conservatism, as they probably maintain the same niche as their common ancestor. One of the first references to this is in Darwin's manuscript titled "Natural Selection." Here he described cases "in which each district has a representative [allopatric] species, filling as far as we can perceive the same place in the economy of nature" (Stauffer, 1975: 202).

Species that display niche conservatism have not invaded new niches, but have instead *inherited* their niche from their ancestor (Odling-Smee, 2010). This process of "ecological inheritance" can be compared with the inheritance of *distributions* that occurs in vicariance, but not in center-of-origin theory. In the latter, a new species attains its distribution by colonization, not by inheritance.

The traditional model suggests that new species evolve by invading new areas and new niches; this has been termed the CODA model, as it is based on center of origin, dispersal, and adaptation (Lomolino and Brown, 2009). Instead, Wiens (2004) suggested that lineages tend to *maintain* their ancestral ecological niche and it is their *failure* to adapt to a new environment—a barrier—that often isolates incipient species and begins the process of speciation. For example, in North American salamanders:

> ...niche conservatism, rather than niche divergence, plays the primary role in promoting allopatric speciation... vicariance may often occur when environmental change in the geographic space between two sets of populations (i.e., development of a geographic barrier) occurs more rapidly than adaptation to these new ecological conditions, resulting in the fragmentation of a species' ancestral geographic range [central populations go extinct]... Thus, in contrast to ecological models of nonallopatric speciation, incipient species may become geographically isolated *because of the ecological similarity of their populations, rather than their ecological differences....* (Kozak and Wiens, 2006: 2604; italics added)

In another example, studies of the Mexican Jay (*Aphelocoma ultramarina*) gave results that conflict with a model of ecological speciation, as "in most cases, the allopatric environments they [sister lineages in *A. ultramarina*] occupy are not significantly more divergent than expected under a null model" (McCormack et al., 2009). In the alpine Primulaceae of Europe, "a large majority of sister-species pairs... are strictly allopatric and show little differences in substrate and climatic preferences... Allopatric speciation with little niche divergence appears to have been by far the most common mode of speciation..."

(Boucher et al., 2016). These are the sorts of observations that led field biologists such as Mayr (1942, 1954) to reject simple ecological speciation, even though they were well aware of ecological sorting.

Peterson (2011) reviewed the topic of ecological niche differentiation during invasion or speciation and found little or no evidence for it. He pointed out that niche conservatism is widespread, although many methodological complications have obscured this. "In particular," he wrote: "niche models are frequently over-interpreted: too often, they are based on limited occurrence data in high-dimensional environmental spaces, and cannot be interpreted robustly to indicate niche differentiation." Peterson also suggested that niche conservatism tends to break down over time; while no ecological "signal" associated with speciation is discernible, differentiation between the niches of sister groups can develop later.

This whole line of thinking is an old one. As Zink (2013) stressed, "vertebrate biologists have observed for decades that sister species are nearly always allopatric." He discussed the factors that have been proposed to keep sister species allopatric: not enough time for overlap, adaptation to habitat (suggesting original, ecological divergence), or the fact that the species are too similar to coexist. Zink (2013) constructed climatic niche models for conspecific avian sister clades on either side of two biogeographic barriers in the southwestern United States and Mexico. Niche divergences between the groups were insignificant and no more divergent than would be expected by the environments available to each. This indicates that the ancestral niches have been conserved in the descendant groups, and that climatic niche divergence does not explain the groups' allopatric ranges. These results agree with those from an earlier study that constructed niche models for sister species of birds, mammals, and butterflies located north and south of the Isthmus of Tehuantepec in Mexico (Peterson et al., 1999). The niche models for each sister species did not differ, suggesting that niches do not diverge until well after speciation.

Sister clades might be allopatric for the simple reason that they have not had sufficient time to expand into each other's ranges, but Zink (2013) calculated that the bird clades he studied are too old for this to be the reason. This means that the sister groups are either adapted to different, allopatric environments, or they are too similar in their ecology and morphology to coexist, and competition enforces allopatry. Although the clades occur in regions with differing climates, they use areas of climate space in their respective ranges that are more similar than expected by chance. Again, this is support for the importance of niche conservatism; the sister groups inhabit ancestral portions of climatic niche space that they have inherited, not invaded.

A final example of niche evolution and distribution comes from a study of Californian plants. Here Anacker and Strauss (2014) "found pervasive evidence for niche conservatism among sisters" and found no evidence of character displacement in overlapping sister groups.

Niche Differences between Species That Have Not Resulted from Ecological Speciation

Sister species often display ecological differences, but these are not necessarily caused by ecological speciation. As Boucher et al. (2014) observed: "Given that climate always exhibits a strong spatial structure, we suggest that speciation will almost always produce sister species with different mean climatic niches. However, this initial divergence implies no evolutionary change in physiology or ecology, it could simply be a by-product of geographic separation."

Hendry (2009) wrote that "support for ecological speciation is so often asserted in the literature that one can get the impression of ubiquity." Despite this, Hendry failed to find simple and strong signatures of ecological speciation in two famous cases, guppies (*Poecilia reticulata*) and sticklebacks (*Gasterosteus aculeatus*). He concluded:

> Many studies seem to be taking "some data consistent with ecological speciation" to mean "our study provides evidence of ecological speciation."... Essentially all published studies of ecological speciation purport to be confirmatory, [but] many cases of divergent selection and adaptive divergence are associated with only weak to modest levels of reproductive isolation... Ecological speciation is sometimes inferred simply when different populations or species show adaptive divergence... Also, it is often inferred simply because different species reside in different ecological environments. A major limitation here is that the *ecological divergence may have occurred after speciation*.... (Hendry, 2009: 1385; italics added)

For example, a population of animals separated by geological rifting into eastern and western areas, each with similar habitat, can undergo phylogenetic divergence and speciation. Later, if there is uplift of the rift shoulders, the west may become wet and the east dry, for example. This would probably lead to local adaptation, but the original speciation was caused by geographic isolation, not by the different conditions or adaptations.

Conclusions on Ecological Differentiation

It is now known that allopatric populations can fix different mutations even under uniform selection (e.g., Sobel et al., 2010). Recent work in molecular genetics and genomics suggests that mutation is not random, and that the role of selection in evolution is much less than has been thought. Selection results in secondary pruning of lineages, but nonrandom mutation is the primary factor determining the course of evolution (Nei, 2013). This idea supports Darwin's (1872, 1874) later conclusions on "laws of growth" and subsequent ideas on phylogenetic constraint, but contradicts the basic assumptions of the Modern Synthesis (Heads, 2009a).

Nevertheless, most modern ecologists still attribute great significance to ecology, and many favor sympatric, ecological speciation. In the Modern Synthesis, a group's habitat determines its morphology, physiology, behavior, and distribution. Instead, in the model used here, the distribution of a group, along with its morphology, physiology, and behavior, determines its habitat.

In addition to environmental factors, the prior genetic–geographic structure of the ancestor is also important in speciation. Population genetic models illustrate how new species can emerge when a species range becomes very large compared with the dispersal distances of its individuals (Hoelzer et al., 2008), a genetic effect termed isolation by distance. In real-life examples, Wang, Glor, and Losos (2013a) found that geographic isolation (isolation by distance) in *Anolis* lizards explained much more genetic divergence than ecological isolation did. In the anoles, "despite the proposed ubiquity of ecological divergence, nonecological factors play the dominant role in the evolution of spatial genetic divergence." Other cases of nonadaptive speciation without selection include speciation by hybridization, polyploidy, and different kinds of mutations; for example, nonadaptive speciation can take place in snails by left–right reversal (Hoso, 2012).

With respect to distribution, allopatric patterns could be explained by allopatric speciation, and overlap could be explained by sympatric, ecological speciation. But the studies just discussed suggest that the latter process is problematic, and geographic overlap is instead attributed here to range expansion. The New Zealand case studies presented in this book examine whether actual patterns of allopatry and overlap can be explained with reference to vicariance and range expansion, with both processes mediated by specific geological events.

The Relationship between Biogeography and Ecology[*]

As discussed, at least one architect of the Modern Synthesis supported the idea of ecological *speciation* (Huxley, 1942). Nevertheless, Mayr (1942) and recent authors have pointed out fundamental problems with this model. In a parallel development, the traditional idea that ecological factors determine the *geographic distribution* of groups has been contradicted by recent work on species distribution models, and this is discussed here.

What, then, is the exact significance of ecology in determining distribution? Some of the most obvious spatial patterns in biology can be explained by more or less obvious ecological factors. For example, bands of particular species occur at different heights along the seashore or on a mountain range. But do ecological factors also determine a group's *geographic* distribution?

Ecology and biogeography both investigate biological differentiation in space and time, and the only real difference is the scale at which this is studied. For example, consider the biogeographic region cited

---

[*] This section incorporates material first published in the *Biological Journal of the Linnean Society* (Heads, 2015c) and reproduced here with permission of John Wiley & Sons.

earlier, an area of hill forest comprising a center of endemism 50 km across. Distribution at this large scale involves endemic species and phylogenetic factors; outside the region, equivalent habitats include different species. *Within* the area of endemism, species will occur in suitable habitats, for example, in lowland or montane sites, by streamsides or on ridges. Distribution at this medium scale is determined by ecology. At the smallest scales, stochastic factors can determine whether a tree is found at a certain spot or, say, a few centimeters to the left.

Does Ecology Determine Geographic Distribution?

All naturalists know that unusual habitats often harbor unusual species, and that many species are restricted, within their distribution region, to certain habitats. But these observations do not mean that the habitat of a species determines its geographic distribution, as most species do not occur in some regions even though suitable habitat is present there. For example, most species of southern African deserts do not occur in American deserts, or even in north African deserts. Likewise, a New Zealand species found in alpine bogs of Otago may not occur in similar alpine bogs in South America or even in Canterbury; the species only occurs in alpine bogs, but why is it restricted to those particular alpine bogs? These sorts of patterns indicate that ecology is not the primary factor determining geographic distribution.

In some cases, distribution within New Zealand is likely to be determined by ecology and the distribution of habitats. For example, many groups of cooler habitats range north to a line: Mount Taranaki (Egmont)–Lake Taupo–East Cape, which crosses the North Island and marks the northern limit of higher mountains and alpine habitat. (These groups often extend a few kilometers north of Taupo on the Hauhungaroa Range, e.g., *Chionochloa rubra* subsp. *rubra*; NZPCN, 2016.) Examples include

The lichen *Labyrintha* (Galloway, 2007).

The fern *Grammitis armstrongii* (= *Notogrammitis crassior*) (Parris and Given, 1976).

The angiosperm clades *Ranunculus* sect. *Pseudadonis* (Hörandl and Emadzade, 2012; NZPCN, 2016), *Bulbinella* (Moore, 1964), *Zotovia* (Edgar and Connor, 1998), *Anisotome* (Dawson, 1961), and *Abrotanella* (Heads, 1998), as well as *Poa novae-zelandiae* (Edgar, 1986), *Hebe odora, H. tetragona, H. venustula* (Bayly and Kellow, 2006), *Gentianella grisebachii* (Glenny, 2004), and *Acaena profundeincisa* (Macmillan, 1983).

The milllipede *Eumastigonus hemmingseni* (Iulomorphidae) (Johns, 2015).

The cicada group *Kikihia* clade 3 (Marshall et al., 2008, 2009).

A diverse clade in the skink genus *Oligosoma* (*O. grande, O. nigriplantare,* and relatives) (Chapple et al., 2009).

The northern boundary in these groups at the Egmont–East Cape line suggests a simple, ecological explanation, as it coincides with the northern limit of high mountains. (On the other hand, similar limits at a Cape Egmont–East Cape line also occur in coastal fishes; examples include the northern limits of *Callorhinchus milii, Neophrynichthys latus,* and *Bovichtus variegatus* and the southern limit in *Favonigobius exquisitus*; Francis, 2012.)

In contrast with distributions that seem to show ecological boundaries, many New Zealand groups are endemic to areas that do not appear to have distinctive environmental conditions. In addition, many of the most distinctive distribution patterns, such as disjunctions at the Alpine fault, are shared by species that have very different ecology (alpine, lowland, etc.), and so the pattern cannot be explained by obvious ecological factors. This sort of nonecological, biogeographic break can be seen even within a single mountain range, and examples are discussed later in this book. On a single mountain though, species distribution appears to be determined by ecological factors alone.

Ecological factors can eliminate a group from a site, but they do not determine which groups are in the region to begin with. At the regional scale, New Zealand is significant not just for its endemics but also for the groups that are absent, despite the presence of suitable habitat there. For example, it is one of the few areas in the world where the aquatic plant *Ceratophyllum* (Ceratophyllaceae) is not indigenous, despite the fact that it has been introduced to New Zealand and is now a weed there. Why was

Ceratophyllum (along with many others) absent to start with, before anthropogenic change took place? The pattern cannot be explained by ecology. The same is true for New Zealand endemics that become weeds elsewhere, such as two of the three stick insect species in Britain (Lee, 1999), and the freshwater snail *Potamopyrgus antipodarum*, which now reaches very high densities in many parts of the United States. Ecological and physiological factors explain why introduced *Ceratophyllum* survives in New Zealand and why the former New Zealand endemics now thrive overseas, but they do not explain the original distributions of the groups.

History of Ideas on Ecology and Distribution

Early Ideas on Ecology and Geographic Distribution

The idea that current climate underpins organic distribution was favored by the geographers of ancient Greece and Rome, although they were already aware of other potential causes. For example, Strabo (64 BC–AD 24) knew about tectonic changes in the landscape (1.3.4–5). In his discussion of fossil oyster shells and others, he inferred former inland seas and concluded that "the beds of the sea themselves sometimes rise, and, on the other hand, sometimes sink." He also realized that topographic changes can result from earthquakes (1.3.17–20; cf. 9.5.2) (translations from Strabo, 1917–1933). Strabo also acknowledged the importance of endemism, and he referred to the distinctive biotas of India (15.1.13–21), Ethiopia (17.2.2), Egypt (17.2.4), and Libya (17.3.4), for example. But he did not use tectonic change to explain evolution or distribution. Instead, he relied on aspects of the current environment and above all on climate; the different temperatures of the climate zones "have a strong bearing on the organisations of animals and plants" (2.3.1). Strabo cited the work by Poseidonius (now lost) criticizing "those who divided the inhabited world into continents in the way they did, instead of by certain circles parallel to the equator (through means of which they could have indicated variations in animals, plants and climates, because some of these belong peculiarly to the frigid zone and others to the torrid zone)" (2.3.7; cf. 2.2.3).

Pliny the Elder (AD 23–79) also accepted this idea of climatic determinism. For example, he wrote that one particular simian (probably the gelada baboon) "is said to be unable to live in any other climate but that of its native country, Ethiopia" (Pliny the Elder, 1938–1963, 8: 80).

The ancient idea that climatic factors determine geographic distribution has persisted over the millennia, and it is still accepted in most ecological work. It is based on a traditional focus on the physiognomy, or gross structure, of organisms and vegetation types, rather than on systematic groupings. Elucidation of the latter required a more detailed knowledge of morphology than the Greeks and Romans had, and this only developed in the Renaissance.

Taxonomic Biogeography Since Its Origin in the Sixteenth Century

In contrast with ecological regions, which are based on environment and vegetation type, the biogeographic regions that have been proposed by taxonomists are defined, not by community physiognomy but by "natural groups"—clades or taxa. The regions that these delimit show marked differences from ecological regions or biomes, such as desert and rain forest.

Modern taxonomy began in the Renaissance, and the first floras and faunas appeared in northern Italy in the sixteenth century. These works listed all the known species, not just the useful ones as in the earlier herbals and bestiaries. The new approach represents an important psychological break and the birth of modern systematics. Cesalpino wrote the first flora in 1583 and was regarded by Linnaeus as the first, true systematist (quotation in Whewell, 1847/1967). The new research program relied on detailed examinations of animal and plant structure, and the revolution in biology is obvious in the illustrations of individual organisms. Over a short space of time, these changed from the medieval representations—like something a child might draw—to depictions by artists such as Leonardo da Vinci, Albrecht Dürer, and Jacopo Ligozzi that would be accepted for publication today in any leading systematics journal. Research in systematics took place both in the field and with collections, living and preserved. Courses were taught at the universities, with a focus on the groups, their features, their ecology, and their distributions.

The new science soon advanced far beyond the superficial, medieval knowledge of plants, animals, and their environments.

The ancient, pre-Renaissance approach to biology examined global and local variations only in terms of the gross structure of the organism and of the biomes. In contrast, modern (Renaissance) systematics aimed to study all aspects of an organism, with detailed observations and dissections, and this led to the recognition of more or less natural clades. A clade's physiognomy indicates something about its native habitat, but not its geographic distribution. It is often possible to see from its general appearance that a plant specimen is from rainforest, for example, but unless the particular group is known, it is impossible to say whether it is from America, Africa, or Asia.

The new, systematic approach in biology, in conjunction with the new phase of global exploration, demonstrated that the distributions of clades (unlike the biomes) do *not* show a simple relationship with climate. In practice, most clades have distributions with intricate configurations and precise boundaries that are repeated in many groups, but do not coincide with climatic zones. Wallace's line is the best-known example, but similar cases occur in all parts of the world. After three centuries of collecting and studying by the new "systematists," Darwin (1859: 346) was able to conclude the following:

> In considering the distribution of organic beings over the face of the globe, the first great fact which strikes us is, that neither the similarity nor the dissimilarity of the inhabitants of various regions can be accounted for by their climatal and other physical conditions.

This is a revolutionary idea: the distribution of groups is *not* explained by environmental factors, and some other "historical" factor is implicated. Darwin used geographic distribution as key evidence for evolution, and this is the approach adopted here.

Modern Ecological Biogeography and the Niche Theory of Distribution

One of the most influential modern ecologists was Joseph Grinnell, who was based in California. He proposed that the niche of a species, especially the climate where it could live, determined its *geographic distribution* (Grinnell, 1914, 1917, 1924). Likewise, William Diller Matthew, who worked in New York, was probably the most influential modern biogeographer, and he regarded climate as the main determinant of *evolution* (Matthew, 1915). (This was in opposition to the "internalist" views that prevailed in paleontology at the time; Griesemer, 1990.) As Griesemer (1990) emphasized, "the parallels between Matthew the paleontologist and Grinnell the ecologist working on problems of distribution and environment are striking." It was these two authors' views on evolution, ecology and biogeography (and not the work of the contemporary geneticists) that formed the basis of the Modern Synthesis. Climate, or environmental conditions in general, was seen as determining evolution and distribution at local and global scales.

Grinnell (1914) proposed that species ranges are delimited by environmental barriers created by current conditions. The barriers are formed by land and water boundaries or by changes in temperature, humidity, food supply, and available breeding sites. The barriers act by "checking the spread of species of birds and mammals," and "the ranges of all birds and mammals may be accounted for by one or more of the factors indicated…" (Grinnell, 1914: 253). Applying the idea on a broader scale, Grinnell (1924) wrote that different physical environments "circulate about over the surface of the earth, and the species occupying them are thrust or pushed about, *herded* as it were, hither and thither…" [italics added].

Most modern ecologists agree with Grinnell; for example, Higgins et al. (2012) wrote that "the distributions of species are determined by the distributions of the environmental conditions where they can persist." Yet there are problems with this idea. For example, there are very few pantropical or pantemperate species other than introduced weeds. In answer to this, ecologists often admit that climate determines distribution *only within a single biogeographic region*. But the delimitation of regions is controversial. Wallace's regions are often cited, but in Africa, for example, Wallace's "Ethiopian Region," the biota in the drier areas south of the rainforest is very different from that in the drier areas north of the rainforest, despite their similar environments and location within what is supposed to be a single region.

The Chasm between Ecological and Systematic Biogeography

Since the rise of modern ecology at the start of the twentieth century, an intellectual chasm has developed between the biogeography produced by ecologists and that produced by systematists and phylogeneticists. The two forms of the subject have been developed by different authors in different journals, using different concepts and terminologies (Wiens and Donoghue, 2004). Peterson et al. (2011) wrote that "the fields have long been paradoxically disparate and distant from one another" (cf. Wiens, 2011a), and Peterson et al. discussed the differences between "the two biogeographies." The two forms of the subject often overlook each other's results, and many ecological papers on species diversity and distribution ignore evolutionary aspects of the patterns altogether (Wiens, 2011b).

In contrast with the systematists, ecologists have tended to ignore geological development and the historical aspect of biogeographic problems, such as why the species and genera of similar habitats in America, Africa, and Asia are so different. At a smaller scale, the ecological theory of island biogeography also overlooks evolution and the geological context—the structure and history of the volcanic centers that have formed the world's islands and archipelagos. While ecologists have focused on the present environment, systematists have continued to investigate taxonomy, biogeography, and history at all scales. This has led to the discovery that significant boundaries underlie such apparent ecological entities as "the tropical American rainforest," "the African savanna," and "the New Zealand alpine zone."

In general, modern ecology has failed to follow Darwin in engaging with the immense body of knowledge on clade distributions built up by systematics since the Renaissance. Instead, it has concentrated on large-scale biomes and small-scale plots, and has assumed that the same factors governing the distribution of biomes also determine those of clades. For example, Araújo and Peterson (2012) wrote that the distributional limits of species match particular combinations of climate variables, and that climate determines species' distributions. They wrote that criticisms of this "are bold, as they fly in the face of the long-established idea that climate governs species' ranges at broad extents (e.g., von Humboldt and Bonpland, 1805…)." Nevertheless, this idea has been contradicted by systematic biogeographers and evolutionists, such as Darwin (1859), who understood that climate does *not* account for many aspects of clade distribution (see the quotation given earlier). The idea is also undermined by the fact that modern, climate-based species distribution models often do *not* make accurate predictions about distributions, and this is discussed next.

Climate and Distribution

Species distribution models (SDMs) or "niche models" aim to reconstruct the ecological requirements of species and so predict their geographic distributions. If the known localities of a species experience rainfall of x, temperature of y, and so on, it is assumed that the total range of the species will be made up of all the areas with those conditions, and no other areas. Until now, most SDMs have been based on climate and have incorporated Grinnell's (1917) ecological theory of distribution; they assume that species distributions are determined by the current environment, rather than historical factors. SDMs based on current conditions are often described as predicting the actual distributions of a species (Elith and Leathwick, 2009), but in fact they only predict areas of *potential* distribution, areas with environments comparable to those of its known localities. (This corresponds to the difference between a group's realized niche, seen in its actual distribution, and its fundamental niche, reflected in its potential distribution.) A number of studies, discussed below, have shown that there is often a great difference between the actual distribution of a species and its potential distribution.

Climate-Based SDMs Often Do Not Work

One of the most interesting insights of the climate-based SDMs is that their predictions of many species distributions are so poor. Examples are discussed elsewhere (Heads, 2015c), and some New Zealand cases are cited in this section. These results are significant as they undermine the idea of ecological determinism. The use of SDMs is widespread, but Soberón and Nakamura (2009) concluded that "important conceptual issues in this field remain confused" (cf. Araújo and Guisan, 2006). In particular, the relationship between climate and species distributions has become controversial in ecology, and there is a growing appreciation of the insight gained by the early systematists—that distribution is not caused simply by climate.

Many niche models overpredict species distributions (Heads, 2015c). In a New Zealand example, a niche model for the tree *Pittosporum cornifolium* (Pittosporaceae) predicted that it would occur in the northeastern South Island (north of Banks Peninsula) and the east coast of the North Island, but it is in fact absent from both areas (Clarkson et al., 2012). This region of anomalous absence coincides with a center of endemism for many other groups, such as *Hebe parviflora* (Plantaginaceae) (Bayly and Kellow, 2006), and it also coincides with an important feature of the structural geology, the East Coast Basin (now filled with sediment). Most areas of endemism are also areas of anomalous absence for other groups (Heads, 2004), and this, along with the overprediction of the niche model, suggests that the absence of *Pittosporum cornifolium* from the northeastern region is not caused by current environmental conditions, but by phylogeny and past environments. The species as such is absent from the region, but it is represented there by related species.

Ecological niche models for the New Zealand species of *Pachycladon* (Brassicaceae) were prepared based on features of climate and soil, and the predicted ranges of all the species overestimated the actual distributions (Joly et al., 2014: Fig. 3). For example, *P. cheesemanii* of the eastern South Island was predicted to extend north of the Alpine fault, in the Richmond Mountains, *P. crenatus* of Fiordland was predicted to be widespread in the Southern Alps, *P. latisiliquum* of northwest Nelson was predicted to occur on the central North Island volcanoes, *P. exile* of North Otago was predicted to occur in Central Otago, and *P. wallii* of Central Otago was predicted to occur in the Kaikoura ranges. Not one of these predictions is correct, but, as usual, all the areas of absence are important centers of endemism in other groups. This is good evidence that phylogenetic, historical factors, rather than present ecology, are responsible for the regional differentiation.

Many other distribution models for both terrestrial and marine groups have made similar overpredictions, and these undermine the idea of ecological determinism. The failure of so many models suggests that species are not "herded" around the globe, or even their own continent, by environmental conditions. The extensive SDM literature constitutes a large-scale test of Darwin's (1859) and Grinnell's (1917, 1924) conflicting ideas on the significance of climate for distribution. The studies contradict Grinnell's idea that current ecology determines distribution and instead support Darwin's (1859) conclusion: ecology ("climatal and other physical conditions") does not determine geographic distribution. Instead, evolution and phylogeny are the key factors.

Niche Variation in Space and Time

Niche models assume that unless a species evolves, its niche is constant over space and time. Nevertheless, there is good evidence that many species occupy different niches in different areas and at different times (review by Heads, 2015c). The habitat or niche of a species may show regional variation because of genetic differentiation within the species, and the literature of applied biology includes many examples of intraspecific geographic variation in niches (Holt, 2009). This is often the case in species with strong phylogeographic structure, with different genotypes in different areas. Yet niche variation can also occur in populations of a species *without* any evident genetic structure, and the niche can vary across a geographic range because of long-term changes in the landscape. For example, if a mountain range is uplifted beneath one area of a lowland community, the species in this part (minus some extinctions) will become montane, and will occupy a niche that differs from the ancestral one.

Geographic variation in a group's niche is common in higher-level clades. For example, most Salicorniaceae grow in coastal salt marsh or low-altitude desert, but in the Andes, which have undergone rapid uplift, they occur in the alpine zone. Caddisflies, members of the insect order Trichoptera, are terrestrial except for a single marine family, Chathamiidae. This is restricted to tidal pools of New Zealand and New South Wales, at the margins of an ocean basin that has been rifted open.

Geographic variation in a group's niche is also observed in introduced species. Many species become more abundant in regions where they are introduced than in their native habitat. For example, as mentioned earlier, two of the three stick insect species present in Britain are New Zealand endemics that have become weedy in their new locality (Lee, 1999), and so there is no real reason to assume that species must be best adapted to their own region.

Species niches can also change through time. For example, phylogeographic evidence indicates that three forest trees of southeastern Australia (*Atherosperma moschatum*, *Eucalyptus regnans*, and *Nothofagus cunninghamii*) survived the last glacial maximum (LGM) in multiple refugia. Worth et al. (2014) modeled the current distribution of each species and projected those models onto LGM climates. The LGM models failed to predict survival of the study species in the refugia identified from genetic evidence, apart from those in perhumid western Tasmania, and so the authors concluded that the niches of the species may have changed since the LGM.

Biotic Factors and Distribution

Given the acknowledged failure of SDMs based on abiotic factors, ecologists are now attempting to integrate biotic processes (competition, predation, parasitism, dispersal ability, etc.) into the models. As yet there have been few process-based SDMs, and there are many problems with the approach. In any case, Warren et al. (2014) wrote that while the geographic distributions of species have often been attributed to biological and ecological processes, this is often unwarranted. While both abiotic and biotic factors can determine structural aspects of the vegetation in an area, they probably do not account for the plant and animal *clades* that are present or their boundaries. With clades, historical factors come into play.

Historical Factors and Distribution

If a group's current niche does not determine its geographic distribution, what is the relationship between the two? Drake (2013) wrote that this question is "one of the most fraught in ecology," and he suggested that "the root of this controversy is the lack of truly clear concepts...." Critical examination of the basic concepts in the field has long been neglected; Peterson et al. (2011) wrote that in previous work on niches and distributions, "conceptual and methodological rigour took back seat to rapid development of software and data resources...."

One conceptual blind spot in ecological niche models has been the neglect of history. Nevertheless, some ecologists now recognize that the models need to move beyond present "conditions," both abiotic and biotic, and to incorporate history, over geological time. In practice though, most ecologists regard historical aspects as even more difficult to incorporate in SDMs than biotic processes, and so they are overlooked.

The New Caledonian plant *Amborella* (Amborellaceae) is a good example illustrating the potential significance of historical factors in governing distribution. It has a special interest as it is probably the sister group of all other extant angiosperms. Poncet et al. (2013) provided a niche model for the distribution of the single species on its island, Grande Terre, and concluded (p. 11):

> The predictive ability of the model was very good throughout the Central East mainland zone, but *Amborella* was predicted in the northern part of the island where this plant has not been reported. Furthermore, no significant barrier was detected based on habitat suitability that could explain the genetic differentiation across the area....
>
> Interestingly, although *Amborella* has never been encountered in the Panié Range or at Roches Ouaïème [north-eastern Grande Terre], we found this area to be suitable, without any clear geological [i.e., lithological] barrier. This absence of populations in this region suggests that constraints other than the current bioclimate may have limited the distribution over time.

Other studies of *Amborella* have shown that it is restricted to the basement terranes of New Caledonia (Heads, 2008b, 2012a), located in central Grande Terre, and this suggests that the plant's absence in northern parts of the island reflects the tectonic history, rather than aspects of climate or lithology. The northeastern parts of the island represent a separate island arc terrane that has been accreted to the basement terranes. The fact that *Amborella* is absent in the northeast, despite suitable conditions there, indicates that it was not present on the arc before it accreted.

In a New Zealand example, the carabid beetle *Mecodema tenaki* is endemic to the North Cape region of the North Island (Unuwhao to Rangiora Bay). A detailed study demonstrated that neither the community composition of the forest nor soil properties were good predictors of the beetle's presence. Nevertheless, all the beetles observed were located at sites underlain by rocks of the Parengarenga Group, mainly the Kaurahoupo Conglomerate (Ball et al., 2013). This is a Miocene volcanic formation, and the pattern suggests an historical explanation, with the beetle's distribution related to the volcanism or associated changes rather than to present environmental conditions.

Formal Acknowledgment of Historical Factors in Distribution

Peterson (2009) described a general model for distribution based on a "BAM diagram." This predicts that species are distributed where three areas, B, A, and M, intersect. B comprises areas where there are suitable biotic factors, A defines areas where there are suitable abiotic factors, and M ("movement") is made up of the areas that are *accessible* to the species.

Schurr et al. (2012) also stressed the importance of the accessible area for distribution: "limited dispersal and migration can cause a species to be absent from geographical regions and from parts of the niche space in which it could in principle show positive population growth." But while limited dispersal can sometimes explain why a species remains absent in an area, it does not explain why it was absent there and present somewhere else to begin with, at the origin of the species. Peterson (2009) wrote that the "accessible area" is determined by some combination of "present day dispersal ability and historical range shifts," but to begin with, it is determined by the range that a group held at its origin. This original distribution may or may not undergo secondary modification by further phylogeny, extinction, or expansion, but in many cases, the distribution is the result of phylogeny and vicariance alone, without any physical movement.

Some Applications of Vicariance-Based Ecology

The Neutral Theory of Biodiversity

The evidence from SDMs cited earlier suggests that a group's niche does not necessarily determine its geographic distribution. In a similar way, neutral theory suggests that differences among groups' niches do not determine patterns of local abundance, at the scale of ecological plots (Hubbell, 2001). Neutral theory was proposed as an alternative to niche theory, as an explanation of species coexistence. The approach is valid and interesting in so far as it rejects niche theory. Nevertheless, neutral theory as presented by Hubbell (2001) is based on traditional island biogeography (equilibrium theory, with dispersal from a mainland), which neglects biological history over geological time. In addition, while the original form of neutral theory accepted the idea of "point-mutation speciation" (i.e., speciation at a center of origin), this has been identified as a key weakness (Kopp, 2010). Kopp replaced point centers of origin with a "random-fission" model of speciation, or in other words vicariance. This modification is supported by a large amount of evidence supporting differentiation by fission, rather than at point centers, but the evidence also shows that the site of fission is not random. The location of a break probably depends instead on spatial variation in the genome of the ancestor and on the location of geomorphological changes.

Ecology and Biogeography in New Zealand

The average temperature in the far north of mainland New Zealand, around Kaitaia, is about 16°C, while in the south, around Invercargill and Stewart Island, it is about 10°C. Early workers observed this and, following the ancient approach of Poseidonius (as recorded in Strabo, 2.3.7), suggested biogeographic regions that comprise latitudinal belts. Colenso (1868) proposed six latitudinal belts for the North Island, based on plant distributions and the idea that temperature is of fundamental importance.

In contrast, Cheeseman (1906) followed the "systematics" tradition in his New Zealand flora and described species distributions with reference to actual collection localities. (This is more useful for biogeography than citations of limits as latitudes, as many standard distribution patterns are not latitudinal.)

The current New Zealand flora (Allan, 1961) reverted to the earlier, "ecological" scheme and described species distributions in terms of their latitudinal range. Allan, following the ecologist Cockayne (1928), never accepted the Renaissance approach to systematics and distribution, as epitomized in Darwin's (1859) and Cheeseman's (1905) rejection of ecological factors as fundamental. Instead, Cockayne and Allan followed the ancient, ecological biogeographers and supported the ecological approach of "bio-systematics"; Cockayne described their aim: "to drive a nail in the coffin" of the earlier herbarium taxonomy (from a 1926 letter; Thomson, 1990).

A critique of ecology as the sole determinant of distribution does not mean that ecological differences do not exist. For example, species and other clades occupy particular types of environment within the New Zealand region. Leathwick (1995) examined data from some 10,000 forest plots, each 0.4 ha in area, in the North and South Islands. Physical conditions at the sites were compared with the presence/absence of 33 widespread New Zealand tree species, allowing a comparison of the species' climatic relationships. Most angiosperm trees other than *Nothofagus* species reach their greatest abundance in warm, moist environments with high levels of solar radiation. *Nothofagus* species tend to occur in cooler and/or lower insolation environments. Coniferous species have a wide variation in their climatic optima, but tend to occur in sites characterized by geomorphic disturbance or harsh edaphic conditions.

These examples show that local levels of abundance can reflect the current environment, but the geographic distributions of species are not as easy to explain. For example, *Nothofagus truncata* has a major disjunction between northern Westland and the Jackson River, along the Alpine fault. Another plant, the subshrub *Kelleria dieffenbachii* (Thymelaeaceae), occurs in New Guinea, Australia, and New Zealand, including the Auckland Islands. Nevertheless, it is absent from Stewart Island and Bluff, on the adjacent South Island coast, and these are the only known localities of a close relative, *K. lyallii*. This suggests that the allopatry is the result of an old, phylogenetic break, and it is difficult to explain by current ecological factors.

Species distribution models often work well within a biogeographic region, or area of endemism, but in New Zealand there are a large number of these, and they are often very local. Even a localized center of endemism can include several distinct environments. For example, a small area, 12 km across, around Cromwell and the adjacent Dunstan Mountains in Central Otago includes lowland endemics as well as alpine endemics. The former include the lowland scarab beetle *Prodontria lewisi* (Barratt, 2007) and the lichen *Acarospora otagoensis* (Galloway, 2007), both at Cromwell; the latter include the alpine herb *Myosotis albosericea* in the southern Dunstan Mountains.

The Concept of "Center of Origin" in Ecological Theory

The vicariance model of allopatric speciation was developed through the late 1960s and 1970s. Yet many ecological analyses and standard texts have overlooked this and have maintained, following the Modern Synthesis, that a species attains its distribution by range expansion from a localized center of origin (e.g., Levin, 2000; Gaston, 2003: 81). Kamino et al. (2012) agreed: "recently originated species are expected to have restricted ranges and not be in equilibrium with the environment, if they lacked the time to disperse to their whole potential distribution." Cox and Moore (2010: 204) wrote: "Let us imagine that a species has recently evolved. It is likely, to begin with, to expand its area of distribution or range until it meets barriers of one kind or another...." But range expansion does not necessarily happen following vicariance—it may or may not occur, and in many cases, a new species will already have a wide range at the time of its origin. If an ancestral complex with a global distribution evolves by breaking down into, say, two allopatric groups, one in the northern hemisphere and one in the southern hemisphere, neither one may expand its range, although both have very wide distributions. In a vicariance model, the distribution of a new clade is just as likely to contract as to expand.

In ecological work, the assumption of a center of origin is often unspoken, but overlooking the original distribution of clades leads to the idea that their current distributions are always caused by range expansion meeting ecological limits. For example, Gaston (2009a) considered "the fundamental question of why, at some point, species no longer evolve the ability to overcome the factors constraining their distributions and thus fail *to continue* to spread" [italics added]. Excoffier et al. (2009) suggested that "range expansions have occurred recurrently in the history of most if not all species," but there is no evidence for this; many species that are allopatric with their sister groups have probably never spread since their origin. Ecologists have proposed that random but spatially constrained dispersal would, in the absence of directional constraints on dispersal, lead to *circular* species ranges (as discussed by Baselga et al., 2012). Again, this assumes a point center of origin.

The idea of a restricted center of origin is also seen in another suggestion by ecologists: that the degree to which a species fills its potential range is related to the age of the species. Paul et al. (2009) supported this idea (although Schurr et al., 2012, contradicted it), and near the end of their paper, they admitted the possibility that species could begin their existence with large ranges. This was because "many models" (a reference to vicariance) propose that speciation splits *prior distributions*. Thus, as stressed, a clade may have a large distributional area at the time of its origin.

The Ecological Center of Origin in Practice

Evolutionary differentiation can proceed without a center of origin, as in vicariance, and so a center of origin cannot be assumed. Following their origin, many groups do expand their geographic range, and this (rather than sympatric speciation) is regarded here as the main cause of overlap. The physiology and morphology of a group often change during evolution, and so the ecology of a group can also change. This process is often analyzed in terms of a center of origin in "ecological space."

For example, consider a genus that comprises four species, three with ecology *a* (by streamsides, say) and one with ecology *b* (on ridges, say). The species have an ecological phylogeny: *a* (*a* (*a* (*a*, *b*))) (i.e., the first species in habitat *a* is sister to the rest). "Optimization" models assume that the original ecology was *a* (because it is the ecology of a basal, paraphyletic grade), that at some stage the genus has dispersed into ecology *b*, and that it has then adapted to the conditions there. In this way, new groups are *wedged* (Darwin, 1859) into their new environment. Yet in a vicariance model, the ancestor already had a wide ecological range *a* + *b* (i.e., by streamsides *and* on ridges), and there does not need to be an ecological center of origin in *a* or in *b*. In this case, the descendant inherits its ecological niche and habitat, rather than invading it.

Groups can also expand their ecological range, just as they can expand their geographic distribution. For example, changes in the environment or in the morphology can cause a group's ecological range to increase or decrease (or go extinct). This model differs from the neodarwinian point of view; in that model, a group invades a niche or habitat first, and *then* evolves suitable morphology and physiology (Dawkins, 2005).

General Conclusions on Biogeography and Ecology

Two of the most obvious attributes of a species are its present habitat and its biology, including its means of dispersal. Making observations on these features is often straightforward, and it is not surprising that they have been assumed to cause distribution. Nevertheless, in many cases, models based on these factors do not work in practice; they fail to explain what is known about concrete examples of distribution. As a review by one ecologist concluded: "It has often proven frustratingly difficult to explain what determines the limits of a particular species at a given place and time" (Gaston, 2009b). The idea that historical factors can be as significant as current conditions in determining distribution, or even override them, should allow a new synthesis of ecology and biogeography. This was prevented by "hard" niche theory—the old idea that the distribution of clades is determined by environmental conditions.

Different clades occupy different habitats, and ecology is often an excellent guide to a group's precise location within its geographic area. Nevertheless, many groups include a small number of anomalous populations found in atypical environments, such as coastal populations in a group that is generally montane, or vice versa. These can indicate phylogenetic or historical effects and ecological lag, with many populations having a relictual ecology and surviving in suboptimal habitat. Populations with anomalous ecology are often located in particular geographic areas, such as Cape Campbell and Shag Point in the South Island, and their distributions often coincide with tectonic features. This suggests that ecological patterns can reflect events in geological time, and that ecological lag in clades and even communities can persist for millions of years.

To summarize, ecology prevents groups from establishing in some areas, but does not explain why groups occur where they do to begin with. A species is present at a site because it is in the species pool of the region, and the different habitats in a region draw their species from this pool. A group is often eliminated from a site by local conditions, but these do not determine the composition of the regional pool. Within a geographic region, a species may occur only on mountains, for example, but its montane ecology does not explain why it occurs on those particular mountains and not others. The same is true for the distribution of "rainforest species," "volcanic island species," or species of any other habitat. A group's biogeography—its geographic distribution—determines the particular range of habitats and niches that are available to it, and if it is viable in at least one of these, it will survive.

In modern niche theory, as presented by Peterson et al. (2011) for example, dispersal is fundamental while vicariance is insignificant, and this has always been the standard view of ecological biogeography. Classical authors such as Strabo had a good understanding of tectonic changes in the Earth, and also knew that different areas have different species. In addition, they were not averse to the idea of evolution. They were vague about its possible effects though, and instead of following this line of thought, they stressed the relationship between climate and distribution. This ancient teaching is expressed in the ideas of Grinnell (1914, 1917) and Matthew (1915), in the Modern Synthesis, and in contemporary research. The theory suggests that a new species invades its habitat from a *center of origin*, by *dispersal* and *adaptation*. Instead, in vicariance theory, a new species *inherits* its geographic area, along with habitat. In a vicariance process, one of the main factors that determines the distribution of a group is the distribution of its ancestor. If a global clade splits into one group in the north and one in the south, each of these has a very large range at the time of its origin. If a group in ancestral New Zealand splits into western and eastern descendants, these begin their existence with smaller ranges. In a vicariance event, a group's distribution and its original ecology, along with those of its sister group, can be inherited from its ancestor. Later, a group can undergo secondary expansion or contraction. Nevertheless, fossil-calibrated molecular clock studies suggest that many groups' original distributions, along with their allopatry with relatives, have persisted for millions of years, and so current distribution can represent inherited information.

Through the twentieth century, reductionist, population thinking became popular in evolution and systematics. Yet as well as focusing on the population level, biologists interested in gaining a broader perspective on species can also look "up" to phenomena at the genus and family levels. As Wiens et al. (2010) wrote, "many of the traits and patterns [studied by ecologists and conservation biologists] may have ancient roots that go far deeper than the species and ecological conditions seen today." In other words, ancient features survive in modern groups and habitats.

In the process of hybridism and also in the inheritance of ancestral polymorphism (which existed before the differentiation of the modern groups began), evolution proceeds by the recombination of ancestral characters rather than the evolution of new characters. In this way, inherited characters and their distributions can be much older than the modern forms that carry them. In a similar way, an ancestor that had a widespread geographic range, and also occupied a wide range of habitats, leaves descendants that each inherit only a part of the ancestor's geographic and ecological ranges. For example, a group of trees with one species in well-drained forest and one in swamp forest may be derived from an ancestor that lived in both types of habitat, and possibly others as well, with each of the two modern species having inherited its habitat type rather than invading it.

Many aspects of distribution, especially at smaller scales within a biogeographic region, are governed by ecology. In contrast, important patterns discussed in this book, such as the break at the New Zealand Alpine fault, are shared by groups that have different ecologies, including plants and animals, lowland and alpine clades, and reef organisms and terrestrial groups. The distributions and distributional breaks that are presented do not have obvious explanations in terms of current environmental factors. The field of macroecology deals with large-scale patterns caused by current ecological factors. This book deals instead with the historical development of genetic, geographic patterns. These occur from a global scale down to the microbiogeographic scale of a few kilometers and have no apparent ecological cause. In most cases, these patterns coincide instead with important tectonic features.

The High Levels of Geographic Structure Observed in Clades

Any distribution can be explained either by chance dispersal from a center of origin, or by vicariance of a widespread ancestor and range expansion. Dispersal and vicariance models make different predictions though. Chance dispersal can occur in any group, at any time and in any direction, and so, general, repeated patterns are not expected. Vicariance predicts that episodes of environmental change will affect entire communities at the same time. Thus, a vicariance model predicts that there will be high levels of geographic structure, and that similar distributions and biogeographic–phylogenetic nodes will be repeated in different groups with different ecologies and means of dispersal.

The High Levels of Geographic Structure and Repeated Distribution Patterns in Clades

More and more molecular sequencing studies are documenting simple and repetitive biogeographic distributions. Instead of being unique or even rare, the same phylogenetic–geographic nodes and disjunctions recur in many groups and in members of different communities. This indicates that the patterns reflect a general cause, one which is not related to the ecology of the individual groups or to chance. One possibility is suggested by the fact that phylogenetic–geographic nodes are often located at tectonic features, fundamental structures in the Earth's crust. Important biogeographic nodes are characterized by distribution limits, endemism, anomalous absences of groups, and disjunct distribution records. In many cases, the same node functions as both a center and a margin of distribution, as a locality of paradoxical and "astonishing" forms, and as a site of phylogenetic incongruence and "unusual hybrids" that replace the normal species (Heads, 2004).

Not all biologists accept that the neatly dovetailing geographic structure of clades indicates any particular process. An anonymous reviewer of one of my manuscripts agreed that high levels of geographic structure are being revealed in the molecular clades, but argued that this just proves the great powers of chance. Others have argued that chance dispersal is not completely random, and so it can lead to patterns. Yet an event that happens only once in millions of years and is not related to any other biological or physical factor, or event, is, for all intents and purposes, random, and it is not clear how such a mechanism could produce repeated patterns.

Chance events of dispersal and extinction occur everywhere, all the time, but they do not explain the main trends and patterns in biogeography. Of course, events with a probability of one in a million do occur. But only the most common distribution patterns that are repeated in many groups, not unusual or unique distributions, are investigated in this book. The events that explain the common patterns are community-wide processes that have affected large numbers of plants and animals, not freak events in single clades. No two distributions are exactly the same, and even in repetitions of the same basic pattern, the details show almost infinite variation. Nevertheless, the main aspects, such as the nodes, limits, and disjunctions, are few in number and repetitive. Since these features recur in large numbers of groups, they are amenable to analysis. Some of the most common patterns in New Zealand are discussed in later chapters, and are illustrated with one or two of the clearest and best-documented examples. Some of the standard patterns, such as disjunction along the New Zealand Alpine fault, are already documented in hundreds of clades.

Two examples of biogeographic structures are discussed next, with examples. Neither pattern is predicted in a dispersal model. The two patterns are: single, monophyletic clades of terrestrial groups that are good dispersers but are restricted to particular islands, and island groups that are sister to (not nested in) widespread, diverse mainland groups. In order to account for the patterns, dispersal theory needs to propose ad hoc hypotheses in addition to simple dispersal.

Single, Monophyletic Groups of Good Dispersers in Island Regions: Long-Distance Dispersal Events That Occur Just Once in Geological Time

Long-distance dispersal theory "posits that, driven by rare events (such as storms or tsunamis), organisms have been carried across gaps, such as oceans, that are not normally traversed" (Crisp et al., 2011). This process will occur, of course, but does it explain the observed patterns? As discussed earlier, events that are rare in ecological time can be common in geological time. For example, even if eastward trans-Tasman dispersal was successful in a group, say, once every 1000 years—a very rare event in ecological time—in just one million years, it would occur 1000 times, and 1000 New Zealand forms would each be nested separately in a parent clade of Australia. This is quite different from the phylogeny that is actually observed in the groups that have been sequenced; in many cases, the New Zealand members of a group form a single, monophyletic clade.

Storms or rafting events that are rare even in geological time, occurring, say, once in 100,000 years, are still too frequent to explain the many monophyletic groups in New Zealand that are dated to earlier than 10 Ma, for example. Events that are unique—not just rare—in the entire history of a clade are required to explain these patterns. Although these dispersal events only occur once in the clade, they must also occur—just once—in each of the large numbers of clades that show similar patterns, such as trans-Tasman disjunction.

Chance dispersal has often been justified on the basis that over the immense span of geological time, even the most unlikely events will occur. Pole (1994) saw oceans "not as barriers to plant dispersal but as hurdles which, given enough time, are overcome." In practice, though, the phylogenies imply that many groups "disperse" into an area, such as New Zealand, just once in many millions of years. What is the reason for this, when all groups have effective means of dispersal? There is no obvious answer, and this suggests that the differentiation of New Zealand groups was caused not by dispersal, but by the *lack* of dispersal.

New Zealand Lizards

Two groups of lizards are represented in New Zealand, geckos and skinks, and both provide typical examples of monophyletic island clades. Geckos (infraorder Gekkota) date back to at least the mid-Cretaceous (a fossil in amber is dated at 97–110 Ma; Arnold and Poinar, 2008), and all New Zealand geckos form a monophyletic clade in the family Gekkonidae (Nielsen et al., 2011). Based on an inferred age of this clade (40 Ma), the authors proposed trans-oceanic dispersal from Australia, but they did not explain why this process has occurred just once in the ~100 million years of gecko history.

The skink family, Scincidae, has been dated as mid-Jurassic (~170 Ma; Hedges and Kumar, 2009), and the New Zealand members form a monophyletic clade (Hare et al., 2008; Chapple et al., 2009: Fig. 4). Based on an inferred age of the clade (22 Ma), Chapple et al. accepted long-distance trans-oceanic dispersal to New Zealand but gave no explanation as to why this has only occurred once in ~170 million years of skink history.

In both of the studies just cited, the inference of chance dispersal was based on the young ages calculated for the New Zealand lizards. The ages in turn were based on the fossil record, and also by assuming that the supposed emergence of New Caledonia from complete submergence at 37 Ma gives the maximum age of endemic lizards there. Yet the fossils will only give minimum ages, and the idea that New Caledonia underwent complete submergence relies on dubious geological assumptions and extrapolations (Heads, 2014).

New studies have found that fossils dated as Miocene in age (16–19 Ma) are "very similar to (if not indistinguishable from)" extant New Zealand skinks and geckos (Lee et al., 2009). The authors inferred "long-term conservatism of the New Zealand reptile fauna." If the New Zealand skinks and geckos each originated by vicariance with relatives elsewhere, this would provide a simple explanation for the monophyly of each clade.

Many geckos and skinks tolerate high levels of disturbance and have a weedy or pioneer ecology. Both families are widespread throughout the Pacific Islands and have excellent means of trans-oceanic dispersal. Yet this ecology does not explain the fact that the New Zealand contingents of each family are monophyletic and instead makes it more enigmatic, at least in a dispersal framework. Competitive exclusion might be suggested as a mechanism, with clades that arrived first excluding others, but multiple clades of geckos and skinks coexist elsewhere.

Tribe Sicyoeae (Cucurbitaceae) in Australasia

This is a tribe in the gourd and cucumber family and has most of its diversity in warmer parts of the Americas. There is also a single Old World clade comprising three species of *Sicyos* in Australia and

New Zealand. The members of the tribe are pioneers in disturbed areas and the fruit bear spines. A molecular clock study dated the origin of the tribe at 32 Ma (Sebastian et al., 2012), and one obvious question is why this group, despite its ecology and effective means of dispersal, has only dispersed to the vast area of the Old World once, in such a long period of time. In the dispersal model, this is attributed to chance, while in vicariance, it is attributed to phylogeny, mediated by geological events.

The weedy ecology of New Zealand *Sicyos*, geckos, skinks, and many others has enabled them to survive high levels of disturbance in the region caused by large-scale rifting, mountain building, marine flooding, and volcanism. On the other hand, many individual clades of pioneer groups in New Zealand are local or regional endemics, as in many lizards. Other examples include the "fire-weed" species of New Zealand *Celmisia* (Asteraceae); these can become dominant in native vegetation after it is burnt for farming (Wall, 1926; Allan, 1961: 656). In Canterbury, *C. spectabilis* has "taken full advantage of the destruction of the forests" by burning, and it is abundant in areas such as the Ashburton and Rakaia Valleys (Given, 1968: 35). Likewise, *C. viscosa* acts as a fire-weed in burnt alpine vegetation in Central Otago. To summarize, groups with pioneer, weedy ecology can also be ancient relics that maintain ancient distributions.

Reciprocal Monophyly of Island Clades and Diverse, Widespread Mainland Clades

The New Zealand Biota is Not a Subset of the Australian Biota

In dispersal theory, local groups that have widespread, allopatric relatives are thought to have evolved by dispersing and "budding off" from the widespread sister group, with the widespread sister group remaining the same. In the classic examples, island taxa with mainland relatives were attributed to chance dispersal from the mainland to the island (Mayr, 1942). Instead of the group in the local region budding off from the group with the widespread range, a vicariance model proposes an ancestor that was already present in both areas before differentiation. In a widespread ancestor that was undergoing active evolution, divergence occurred at the break between the two descendants, and both descendants, not just the local one, differ from the ancestor.

Molecular studies have now been carried out on many groups on mainlands and islands, and these have found interesting results. Not only are many island clades diverse but monophyletic, as discussed, they are often not nested in localized, mainland species. Instead they are *sister* to widespread, diverse mainland clades. This is well documented in, for example, the many New Zealand groups that are sister to global clades (Heads, 2009c).

Dispersal theory suggests that the current New Zealand flora is "a specialized cool temperate subset of the Australian flora" (McGlone et al., 2001: 208). Yet molecular phylogenetic data now show that a large number of New Zealand endemic plants—including ferns (*Loxoma*), lowland herbs (*Myosotidium*), forest trees (*Alseuosmia, Ixerba, Rhabdothamnus*), alpine herbs (*Hectorella*), and many others—are not nested in Australian clades. Another typical example is the forest orchid *Winika*, endemic to the New Zealand mainland (North, South, and Stewart Islands). *Winika* is sister to a diverse clade (including at least 24 species, mostly in *Dendrobium*), that is widespread from southeastern Australia to Lord Howe Island, New Caledonia, Fiji, New Guinea, Vietnam, and Taiwan (Papadopulos et al., 2011, Appendix). Why is the New Zealand clade, *Winika*, sister to (not nested in) a group that is so diverse and widespread? A model based on dispersal from Australia to New Zealand would require additional ad hoc hypotheses to explain the pattern, but vicariance in a widespread ancestor is a simple explanation. If this were accepted, the focus of enquiry would shift from locating a center of origin to understanding the reason for the break between mainland New Zealand and the sister group's area. This would require a consideration of regional tectonics.

Another example similar to that of *Winika* is *Alectryon* (Sapindaceae), a genus of trees found in Malesia, Australasia, and the Pacific Islands. There is a single species in New Zealand, *A. excelsus*, and this might be expected to be the result of dispersal from, say, southeastern Australia. Yet *A. excelsus* is not nested in a clade from there or anywhere else in Australia. Instead, it is sister to a diverse clade that is widespread through Australia (the Pilbara, central areas, north and east coasts), also New Guinea, and probably (based on morphology), New Caledonia and Hawaii (Edwards and Gadek, 2001).

The New Zealand *Alectryon excelsus* could still be the result of dispersal from Australia, for example, if an ancestor in, say, New South Wales has died out. Yet there is no evidence for this and so the theory would be ad hoc, while the fact that the pattern is a common one, repeated in many groups, means that a general explanation is desirable. Another possibility is that a chance dispersal event occurred before any of the other differentiation in the widespread, diverse sister. But again, why did the dispersal event only occur once, and why did it occur right at the base of the phylogeny? Dispersal theory does not explain the cooccurrence of these two features, but it is the standard signature of vicariance. The presence of both phenomena would be interesting enough in a single New Zealand endemic, but the same pattern is repeated in many groups. Again this indicates that the differentiation was not the result of single, chance events in many different clades, but of a single vicariance event that affected a whole biota.

As with plants, dispersal theorists have suggested that New Zealand's avifauna is a subset of Australia's (Trewick and Gibb, 2010: 242). Yet in molecular phylogenies of ratite birds, the New Zealand kiwi, *Apteryx*, is sister to *Aepyornis* of Madagascar (Mitchell et al., 2014a), while the New Zealand gray warblers (species of *Gerygone*) and honeyeaters (*Anthornis* and *Prosthemadera*) each have their sisters in New Guinea (Heads, 2014). Several other groups of New Zealand birds, including clades of seabirds, Gruiformes, parrots, and passerines, are also not nested in Australian groups or even sister to them, but are basal in global complexes (this will be discussed in Chapter 4).

In a debate in geology, Shervais (2001) argued that formation of the interesting rock sequences known as ophiolites is not a stochastic event, but rather a natural consequence of the tectonic setting (above a subduction zone). Likewise, in vicariance biogeography, cessation of ordinary dispersal (leading to allopatry) and increase in ordinary dispersal (leading to overlap) need not be attributed to chance, as both are natural consequences of the tectonic and climatic settings. For example, phases of overlap often occur during marine incursions, when a coastline and its biota move, together, into the inland parts of a continent. There is considerable evidence that this process has taken place in all the continents (Heads, 2012a, 2014), and it probably also occurred in New Zealand.

Biogeographic Parallels between the Southwest Pacific and the Caribbean: More Islands with "Basal" Groups

The Caribbean islands are a geological and biological analogue of the islands in the southwest Pacific. Molecular studies have indicated that both regions include many endemics with widespread, diverse sisters, in contrast with the predictions of dispersal theory. For example, the shrub *Brunfelsia* (Solanaceae) comprises two large clades. The first, with 30 species, is widespread through tropical South America, while the second, with 23 species, is endemic to the Greater and Lesser Antilles (Filipowicz and Renner, 2012). The two clades show precise allopatry. In dispersal theory, the island clade should instead be nested in a mainland group at a local center of origin, probably somewhere along the Caribbean coast of South America. As it is, the clades in the two regions are said to display reciprocal monophyly, and the main division is a simple South America–Antilles break that corresponds with a major tectonic feature, the Caribbean plate margin.

In an example from birds, a diverse clade of passerine families (Calyptophilidae, Phaenicophilidae, Nesospingidae, and Spindalidae) is endemic to the Greater Antilles region, including the Bahamas, Caymans, and Cozumel Island. This island-endemic group is sister to a large complex found through the Americas including the Greater Antilles (Icteridae—27 genera, Parulidae—17 genera, also Teretistridae, Icteriidae, and Zeledoniidae) (Barker et al., 2013). The phylogeny indicates that the island group is not derived from a local center of origin on the mainland, such as Florida, Mexico, or Venezuela. Instead, the pattern is compatible with an early origin by vicariance of a pan-American group, followed by secondary overlap in the Greater Antilles. In one clade of the American complex, Teretistridae of Cuba are sister to Icteridae, widespread through the Americas. This implies that two basal breaks took place in a pan-American ancestor around the Greater Antilles, with the second following a phase of overlap.

The anoles are a group of Neotropical iguanian lizards. There are about 400 species, including many Caribbean endemics, and they have been the subject of many ecological and biogeographic studies. A revision of the group concluded:

> We accept the argument that vicariance (and accretion) is the appropriate null hypothesis, and that dispersal is the *ad hoc* explanation invoked for exceptional cases… We continue to be concerned that overexposure of data that document recent dispersal has diverted attention from extensive data that are consistent with vicariant (and accretion) events in anole evolution. (Nicholson et al., 2012: 69)

These observations apply just as well to the southwest Pacific biota, where the common patterns are consistent with vicariance and accretion events. The parallels between the southwest Pacific and the Caribbean suggest a general, global mechanism of evolution that integrates geological and biological processes, and does not depend on chance dispersal to explain patterns that are repeated in many different groups.

Geographic Structure in Microorganisms, Spore Plants, Estuarine Groups, and Marine Groups: Toward a Trans-Realm Biogeography

Molecular studies have revealed unexpected, high levels of geographic structure, not only in terrestrial plants and animals, but also in microorganisms and marine groups that have microscopic, pelagic larvae. These recent discoveries are causing profound changes in the fields of microbiology, aquatic biology, and marine biology (Heads, 2012a, Chapter 1; reviews in Fontaneto, 2011).

Vicariance in Microorganisms

In microbial biogeography, the traditional paradigm was that "everything is everywhere and the environment selects"; in other words, distributions reflect ecological factors only. Many empirical results conflict with this assumption though, and microbiologists "might be on the cusp of a complete reversal of opinion on this point" (O'Malley, 2007). Williams (2011) argued, against the traditional idea, that in microbes "everything is endemic." For example, studies on diatoms have described deep geographic structure, with marked endemism and allopatry (Vyverman et al., 2007). Work on the ciliate genus *Semispathidium* concluded that "the vicariance speciation model is applicable to protists" (Foissner et al., 2010), and a global analysis of regional endemism in another ciliate, *Lagenophrys*, indicated standard continental and intercontinental distributions (Mayén-Estrada and Aguilar-Aguilar, 2012).

Molecular surveys of soil rotifers found "surprising" and "unexpected" structure (Robeson et al., 2011), and the authors wrote:

> Local communities show *strong spatial structuring*, whereas more distant communities are very different from one another, even in similar environments… This world view directly contradicts the idea of EisE [everything is everywhere] and suggests that the diversity of microbial eukaryotes such as rotifers may be vast beyond our imagining, especially given that endemic microbes may have *species ranges of about 100 m*… Bdelloids are known to produce small, resistant resting stages… that should disperse easily by wind. The fact that our results show that bdelloids are not widely distributed implies that other microbial eukaryotes with less resistant stages… should have even more geographically restricted distributions… the ability to form resistant stages probably evolved not for dispersal but to survive periods of unfavorable environmental conditions (i.e., dry and cold conditions)… thereby maintaining unique local communities. (Robeson et al., 2011: 4407–4408; italics added)

Allopatry even exists in bacteria, and in a study of the global genus *Chroococcidiopsis*, Bahl et al. (2011) concluded:

> Temporally scaled phylogenetic analyses showed no evidence of recent inter-regional gene flow, indicating populations have not shared common ancestry since before the formation of modern continents... (p. 1). Although a desert cyanobacterium such as *Chroococcidiopsis* may notionally possess the characteristics for ubiquitous dispersal due to aeolian transport of desert particulates and its desiccation/radiation tolerance..., this is not reflected in contemporary colonization [i.e. distribution] patterns. The existence of regionally isolated gene pools of hot and cold *Chroococcidiopsis* variants strongly supports the concept that widespread contemporary dispersal is not common, and that relationships reflect ancient historical legacies. Although strong selection for hot and cold variants has occurred, this pre-dates contemporary climatic selective pressures.... (p. 5)

In contrast with the work in microbiology, many recent papers on higher plants and animals have accepted ubiquitous chance dispersal and ecological speciation (see earlier sections). In this way, the field is arriving at the same point that the microbiologists, having absorbed the lesson of the molecular biogeographic structure, are now leaving behind.

Vicariance in Spore Plants

Vicariance in Fungi

Berbee and Taylor (2010) have reviewed studies on molecular biogeography and evolution in fungi. Earlier workers had broad conceptions of fungal species and their geographic ranges. Because fungi disperse by spores that are carried over long distances, it was argued that the barriers that stopped gene flow in plants and animals would not restrict the travel of fungi. But in fact, molecular studies have demonstrated that many fungal species, perhaps most, have restricted distributions. In one example that surprised mycologists, *Fusarium graminearum*, which causes the important disease wheat scab, was shown to consist of seven phylogenetic species. Only one of the seven had a wide distribution across the northern hemisphere. Each of the remaining six was restricted to a single continent. In other case studies, molecular data have helped define species of *Neurospora*, *Lentinula*, *Saccharomyces*, and *Schizophyllum* and have indicated that range restriction, not global distribution, is the rule.

Vicariance in Ferns

Ferns, like fungi, disperse by spores, and so it has always been assumed that their distribution patterns are the result of dispersal. Nevertheless, a review of global, molecular biogeography in scaly tree ferns (Cyatheaceae) found evidence for Gondwanan vicariance and limited trans-oceanic dispersal (Korall and Pryor, 2014). These authors wrote that "our understanding of fern biogeography is undergoing a paradigm shift where vicariance versus long-distance dispersal scenarios both need to be carefully evaluated..." Several examples of fern distributions in and around New Zealand are cited in this book, and simple vicariance explanations are suggested.

Vicariance in Estuarine Groups

As with the microbes and spore plants, small estuarine invertebrates were thought to have great powers of dispersal, and so it was assumed that a vicariance model would not apply to them. Nevertheless, sequencing studies have again found results that support vicariance. For example, Knox et al. (2011) examined two species of estuarine amphipods, *Paracorophium excavatum* and *P. lucasi*, both endemic to the North and South Islands of New Zealand. Sequence divergences of 12.8% were detected between the species, but divergences of up to 11.7% were also observed among well-supported clades *within* the species, suggesting the possibility of cryptic species. The authors concluded that genetic structure in *Paracorophium* "appears to represent prolonged isolation and allopatric evolutionary processes..."

For example, the four main clades in *P. lucasi* are allopatric, and occur, respectively, in northern Northland, southern Northland–Bay of Plenty, eastern North Island, and western North Island–northern South Island. Knox et al. (2011) concluded that "the overall genetic structure in *Paracorophium* species appears to have resulted from periods of vicariance."

Vicariance in Marine Groups

Unexpected Levels of Geographic Structure in Marine Groups

Many marine groups have large population sizes and high dispersal potential, as in microbes, and so theory predicted that they would have weak population differentiation. Marine species often have planktonic larvae and were always thought to lack genetic discontinuities, even over large geographic scales. Despite this, increasing numbers of molecular studies have found genetic discontinuities and high levels of geographic structure, including much endemism and allopatry (review in Heads, 2005a; Wörheide et al., 2008; Gerstein and Moore, 2011).

A study of New Zealand sea lions (*Phocarctos*) concluded that "the apparent disconnect between dispersal ability per se and realized distributions may help to explain the paradox of spatial structuring of intrinsically dispersive biotas" (Collins et al., 2014). This lack of relationship between a clade's means of dispersal and its actual distribution has been a main theme of panbiogeography. It is becoming even clearer now, with molecular surveys documenting the fine details of geographic differentiation.

As with the work in microbiology, the molecular surveys of marine taxa demonstrate that high dispersal potential does not always lead to high rates of realized dispersal. Long-term natal philopatry, in which organisms return to their birthplace to reproduce, is known in sea birds such as albatrosses, and it is now reported in sharks (Feldheim et al., 2014). The new findings have led to a reassessment of marine biogeography and have implications for the management and conservation of marine fish stocks.

Some of the most significant biological disturbances on Indo-Pacific coral reefs are outbreaks of the fecund, corallivorous crown-of-thorns starfish, *Acanthaster planci*. Based on the high dispersal potential of *A. planci*, successive outbreaks within and across regions were assumed to spread via the planktonic larvae released from a primary outbreak. However, molecular studies now show that *A. planci* populations have surprising levels of geographic structure (Timmers et al., 2012). Population structure occurs among regions in the central Pacific, among archipelagos within regions, among some islands within archipelagos, and even among some sites around the same island. The absence of shared haplotypes between regions indicates that there is almost no long-distance exchange among archipelagos. Thus, the phenomenon of outbreaks occurring at similar times in disjunct areas is best explained by similar climatic, ecological, or anthropogenic conditions, not by planktonic dispersal of larvae.

Many similar examples of genetic structure in marine groups are now documented. In the widespread pearl oyster genus, *Pinctada*, the absence of overlapping distributions between sister lineages and the observed isolation by distance suggest that allopatry is the prevailing speciation mode. This is unexpected, as the species have a long larval phase lasting between 16 and 30 days (Cunha et al., 2011). Likewise, despite its potential for long-range dispersal, the seagrass *Zostera marina* in California maintains significant population structure at all spatial scales, ranging from meters to tens of kilometers (Kamel et al., 2012). These results demonstrate that in marine groups, as in terrestrial groups, effective dispersal and range expansion can occur at different times, but cannot be assumed—based on the means of dispersal—to be occurring at all times.

In marine biogeography, "The hypothesis that pelagic larval duration (PLD) influences range size in marine species with a benthic adult stage and a pelagic larval period is intuitively attractive" (Mora et al., 2012). Yet many studies have tried to find a relationship and failed. In a broad variety of seafloor species, Lester et al. (2007) found that "dispersal [PLD] is not a general determinant of range size." The meta-analysis of Weersing and Toonen (2009) also refuted the idea that PLD is a good predictor of gene flow and population structure in marine systems. Mora et al. (2012) collected data on PLD, range sizes, and evolutionary ages for seven tropical fish families, but even after controlling for evolutionary age, they found that PLD has an insignificant or very small effect on the range size. Mora et al. concluded: "Identifying the factors limiting the geographical extent of reef fish species remains a challenge to reef fish biogeography; yet our study suggests that the focus should be shifted to attributes other than dispersal over ecological (i.e., PLD) or evolutionary (i.e., species age) time." This is consistent with a vicariance history for marine groups.

Vicariance Models for Marine Groups

Some studies have explained the unexpected new patterns of structure and allopatry in marine clades by chance dispersal. But many others have taken the opportunity to reassess the fundamental assumptions of marine biogeography, and marine biologists have begun to explore vicariance models (Heads, 2005a). It might seem unlikely that vicariance would occur in marine organisms, yet populations of open ocean groups can be divided, for example, by new marine currents or uplifted land. Shallow-water clades, such as reef fishes and other benthic (seafloor) organisms, require substrate near sea level, and so rifting or convergence of landmasses will affect these groups. For example, hydrothermal vent communities on spreading ridges are well known for their distinctive faunas, and their distributions are consistent with a vicariance history mediated by normal plate tectonics processes (Moalic et al., 2012). In the flatfish *Symphurus*, an undescribed species has a distribution range that extends 1500 km along the Tonga–Kermadec arc (part of the Pacific plate margin) and occurs at hydrothermal vents near seamount summits (species A; Tunnicliffe et al., 2010). The species is sister to *S. thermophilus*, found in similar habitat along another part of the Pacific plate margin, the Mariana arc. The Pacific plate has grown from its beginnings at a triple junction, and its margin has also grown, with parts of the active margin often rifting from other parts. This would provide a simple mechanism for vicariance in *Symphurus* species and other plate margin endemics.

Many authors have cited the closure of Tethys and the Panama Isthmus as processes that would have caused vicariance in global marine groups (e.g., Stelbrink et al., 2010). Lowered Pleistocene sea levels have also created new land and are often used to explain more local patterns. Many other tectonic changes have also affected the Earth's marine biota, and New Zealand examples are discussed in the following chapters. Relevant tectonic changes in the New Zealand region include the rifting and opening of several basins (Tasman Basin, Great South Basin, Emerald Basin, the basin between New Zealand and Antarctica, the Bounty Trough, and the South Fiji Basin). These rifting events can account for many examples of marine allopatry in nearshore groups. For example, southern South America and the Antarctic Peninsula rifted apart in the Oligocene, and a study of the echinoderms in the region concluded that the tectonic history has been an important factor regulating the faunal assemblage (Barboza et al., 2011). The authors noted that "most patterns of distribution of marine benthic invertebrates are commonly explained by vicariant speciation caused by plate tectonics." In addition to divergent processes such as seafloor spreading, tectonic convergence events, such as the collision of the Hikurangi Plateau with New Zealand, will lead to biogeographic juxtaposition in marine communities.

In marine Crustacea, molecular studies concluded that the widespread lobster *Metanephrops* originated in the Cretaceous (Chan et al., 2009). Its diversification on different continental shelves was attributed to vicariance and range expansion associated with the breakup of Gondwana. The authors interpreted *M. challengeri*, endemic to New Zealand, as an ancient lineage that became isolated following continental drift.

The gastropod *Echinolittorina* includes many allopatric clades. For a sample of Indo-Pacific species, Reid et al. (2006) wrote: "Neither intraspecific phylogeographical structure nor boundaries between species show obvious correspondence with prevailing ocean currents.... This suggests that phylogeographical patterns are not determined simply by present-day dispersal, but that historical and ecological explanations must also be considered." The authors proposed vicariant separation during times of low sea level.

The fish *Chrysiptera rex* (Pomacentridae) inhabits coral reefs through the Indo-Pacific, and it possesses three distinct color variants that are congruent with genetic clades (Drew et al., 2010). In the west Pacific, each of these color variants is associated with a specific geographic region (the South China Sea, the Philippines, and Indonesia), despite the species having a long pelagic larval duration of 15–20 days. As Drew et al. (2010) noted, the geological history of the region has included large-scale plate movements as well as episodic changes in eustatic sea level, and this history has provided ample opportunities for both vicariance and dispersal.

The typical form of *Sargassum hemiphyllum*, a brown alga, is common in Japan and Korea, while another variety, var. *chinense*, occurs on the coast of southern China. The divergence between the two has been attributed to vicariance caused by Miocene tectonics in the Sea of Japan and South China Sea (Cheang et al., 2010). In *Tridacna maxima*, the giant clam of the Indo-west Pacific, low levels of gene flow and deep evolutionary lineages are "most probably due to vicariance" (Nuryanto and Kochzius, 2009). The evolution and distribution of polychaete worm clades in the north Atlantic have been explained by shared history and vicariant events (Jolly et al., 2006).

Chaenogobius annularis is a Japanese goby of rocky intertidal sites, and a local study indicated genetic subdivision even within an area 20 km across (Hirase et al., 2012b). Another study showed that genetic differentiation also occurs at larger scales, and that the species consists of two main geographic clades, one found on the Pacific coast of Japan and the other on the Sea of Japan coast (Hirase et al., 2012a). The authors concluded that this reflects low dispersal ability and vicariance caused by environmental changes around the Japanese archipelago.

In a final marine example, the hydroid *Plumularia setacea* is found in very shallow water and has a near-cosmopolitan distribution. In a sequencing study, Schuchert (2014) found that:

> Almost all sampled regions had only private haplotypes and the resulting trees split into a multitude of geographically delimited lineages… (p. 1) The present results suggest that the geographic scale of genetic structuring in marine hydrozoans can be much finer than was expected. In *P. setacea*, the presence of only private lineages in almost all examined localities is reminiscent of the diversification of terrestrial biota of oceanic islands. (p. 8)

New Zealand samples formed a trans-Pacific clade with samples from Chile, Argentina, and the San Juan Islands, Washington. This clade is sister to an Indian Ocean clade (Mozambique Channel and Réunion Island).

Local Endemism in Marine Animals: New Zealand Mollusks

The molecular results from marine organisms confirm and clarify what has long been suspected from traditional studies: many marine groups display high levels of local endemism and precise allopatry, despite their habitat. The marine mollusks of New Zealand provide a good case study. For example, the gastropod *Subonoba* (Rissoidae) has local endemics on each of Macquarie Island, the Auckland Islands, Campbell Island, the Bounty Islands, the Snares (three endemics), Stewart Island, the Chatham Islands (two endemics), Awanui Heads, Tom Bowling Bay, and Three Kings Islands (Powell, 1979).

The numbers of marine mollusk species and subspecies restricted to a localized area of New Zealand are listed next. The data were compiled from the latest treatment of the fauna (Powell, 1979). Many subsequent studies on different molluscan groups have also recorded local endemics in New Zealand (e.g., Marshall, 1995a,b, 1998, 1999, 2006; Geiger, 2012), but these are not included in the brief, indicative analysis given here. Only those localities with 5 or more endemic species or subspecies are listed; areas with more than 20 endemics are in boldface. (All records are from varying distances off the land locality cited.) Note that "Fiordland" and "eastern Otago" are much larger than the other areas listed.

Macquarie Island 18.

Auckland Islands 22.

Campbell Island 7.

Bounty Islands 14.

The Snares 15.

Stewart Island 18.

Fiordland 38.

Dunedin/Otago Peninsula 14.

Otago Heads 6.

Eastern Otago 24.

Oamaru 6.

Mernoo Bank/Chatham Rise (not including Chatham Islands) 17.

Chatham Islands 59.

Cook Strait 22.

Bay of Plenty (Tauranga, Mayor Island) 8.

Auckland/Hauraki Gulf (not Barrier Islands) 11.

Barrier Islands 18 (Little Barrier 3; Great Barrier 15).

The Poor Knights 14.

Whangaroa Harbour/Whangaroa Bay 6.

Ahipara 6.

Mangonui Harbour/Doubtless Bay/Rangaunu Bay 16.

North Cape/Tom Bowling Bay 30.

Spirits Bay/Cape Maria van Diemen 9.

Three Kings Islands/Cape Maria van Diemen/Spirits Bay/Tom Bowling Bay 9.

Three Kings Islands 122.

Kermadec Islands 68 (Brook, 1998).

High levels of local endemism are indicated in the subantarctic islands, the southern South Island, Cook Strait, Chatham Islands, the northern North Island, Three Kings Islands (with the maximum level recorded), and the Kermadec Islands. Levels are much lower in central New Zealand (Westland, Canterbury, Nelson, Marlborough, most of the North Island), except at Cook Strait. Very localized endemism occurs in the North Cape/Cape Maria van Diemen block.

Toward a Trans-Realm Biogeography

Habitats in the marine and terrestrial realms differ in obvious ways, but many of the same biological principles apply in both. For example, as in terrestrial groups, the level of geographic overlap between marine sister groups increases with node age, suggesting that differentiation is mainly allopatric (Frey, 2010; Quenouille et al., 2011). Frey wrote that if this is correct, the majority of sister species in her study group, the gastropod *Nerita*, have maintained allopatry over millions of years.

Dawson (2009) suggested focusing on the fundamental similarities between marine and terrestrial biogeographies, rather than the differences, and advocated the development of a "trans-realm" comparative biogeography. He observed that many marine and terrestrial animals agree in their patterns of variation and distribution, and he argued that comparative study will provide reciprocal illumination. By now there are many well-documented parallels between the realms at a wide range of scales. In Japan, for example, a phylogeographic break in a marine red alga between Honshu and Hokkaido coincides with a break in terrestrial beetles and is "most likely due to a vicariant event" (Yang et al., 2008).

Dawson and Hamner (2008) pointed out that oceanic environments and their resident populations have been seen as few, large, and homogeneous, in contrast with the high levels of structure in terrestrial ecosystems. Yet molecular work is now showing similar levels of geographic structure in marine groups. The historical dichotomy between studies of evolution in terrestrial and marine systems is epitomized by the contrasting approach to islands. While many endemic marine species are known from oceanic islands, there has been a long tradition of excluding island biogeography theory from marine biogeography, and vice versa. This separation is not maintained here though; New Zealand has a longer coastline than the conterminous United States, and its marine groups display some of its most interesting distributions.

2

*Analyzing the Timeline of Evolution**

The topologies or branching sequences of many molecular phylogenies are of special interest. The clades indicate standard, repeated distribution patterns and precise, interlocking biogeographic structure. In contrast, molecular clock dates are often unreliable, because of problems with calibration, assignment of priors in Bayesian analyses, models of rate changes, and other constraints. Molecular clock estimates of clade ages are sometimes regarded as "trustworthy" (Ree and Sanmartín, 2009), but the dating is probably the weakest part of the molecular enterprise. A detailed treatment of angiosperms cited molecular clock dates for many clades, but warned that these "should all be treated with extreme caution" (Stevens, 2014). Later, Stevens (2016) referred again to the clade dates and wrote: "most must be seriously inaccurate."

The timeline of phylogeny is a key topic in evolutionary studies. The chronologies of evolution that are supported in the Modern Synthesis and in current molecular biology are similar; both are calibrated with the fossil record. Although this chronology has long been the "official" one (e.g., it is depicted in murals in lecture halls around the world), many phylogeneticists now acknowledge serious problems in relying on the fossil record to calibrate evolution. Dawkins (2005: 462) wrote that "the fossil record, even at the best of times, can be a fickle witness…. [but] we could calibrate the evolutionary clock on parts of evolution where the fossil record is good…." Yet it is not clear how we would know which parts are good, as "better than other parts" does not mean "good" in an absolute sense.

Whether they are based on morphological or molecular data, phylogenies that are fossil calibrated have implied Cenozoic dates for many families and most genera. These clade ages have been used to rule out evolution by orderly, community-wide processes, such as Mesozoic vicariance associated with continental rifting. Instead, the young ages have been used to support chance, trans-oceanic dispersal in the Cenozoic.

Knapp (2013: 1183) summarized the situation:

> Because dated molecular phylogenies revealed ages of taxon splits much younger than could be accounted for by simple vicariance, long-distance dispersal of organisms became a stock explanation for unexpected patterns. However, many outside the field of molecular phylogenetics fail to appreciate that these dates, calculated with a "molecular clock" and calibrated with fossils, are minimum dates and come with big error bars. Vicariance is fighting back, too, with the advent of molecular panbiogeography….

Interpreting the Fossil Record

Most groups are not represented in the fossil record at all, and even large, family-level groups can be absent in the fossil record over large areas. Of the 108 angiosperm families indigenous in New Zealand, 33 (31%) are either not recorded in the country's fossil record at all (19 families) or have Pleistocene fossils only (14 families) (Lee et al., 2001). The fossil record in most of the other families is probably also very deficient.

Recent work has developed the idea that the age of a clade does not equal the age of its oldest fossil; in other words, the fossil record cannot be read literally. In an extinct clade, the youngest fossils *predate*

---

* This chapter incorporates material first published in the *Journal of Biogeography* (Heads, 2012b) and reproduced here with permission of John Wiley & Sons.

the extinction (Signor and Lipps, 1982). Likewise, the oldest fossils of a clade *postdate* its origin and only give a minimum age for the clade. This has been termed the Sppil–Rongis effect (the converse of the Signor–Lipps effect). Dornburg et al. (2011) noted: "the taphonomic bias in the fossil record (Sppil-Rongis effect) increases the probability of fossil preservation toward the present, with large gaps often artificially truncating the distribution of lineages at deeper time scales."

Despite this limitation, the authors of the Modern Synthesis accepted that the fossil record provides the best source of information on evolutionary chronology, or even the only source (Simpson, 1944). Recent authors have agreed: "molecular clocks require fossil calibration… Directly or indirectly, all molecular clock analyses rely on palaeontological data for calibration" (Donoghue and Benton, 2007). "For any molecular dating study, a basic requirement is at least one fossil constraint to anchor the estimated divergence times onto the timeline of the Earth" (Hug and Roger, 2007). "Fossil calibrations are essential when dating evolutionary events" (Wilkinson et al., 2011).

For algae, Martin and Zuccarello (2012) suggested that "while all calibrations have to be tentative… due to the lack of good fossil evidence in many algal groups, even rough estimates can be informative into processes leading to divergence in genera." But fossil-calibrated dates are not even rough estimates of clade age—they are only estimates of minimum clade ages.

Fossil calibrations work well in certain hard-bodied marine groups, but there is excellent fossil evidence, supplemented by molecular studies, for very large gaps in the fossil record of terrestrial groups. For example, there are groups known only from modern species and Cretaceous fossils (S. Heads, 2008). New fossils often push back the known age of groups by tens of millions of years; for example, the oldest known fossil snakes were ~100 Myr old, until Caldwell et al. (2015) described snake fossils dated as 167 Ma.

Despite the claims made by the defenders of the fossil record, there are alternative methods of dating, and many recent authors studying evolutionary events have avoided relying on fossils to give maximum clade age. Instead, they have dated phylogenetic breaks (and thus clades) by dating the tectonic events that coincide spatially with them (a detailed discussion is given later in this chapter). This tectonic-vicariance approach does not reject the fossil record, but uses it to provide minimum, not maximum, clade ages. (In fact, changes in evolutionary rate would mean that a fossil calibration for a clade does not even provide accurate minimum ages for other clades in a phylogeny.)

Placing Fossils in a Phylogeny

Apart from sampling error in the fossil record and the Sppil–Rongis effect, identifying fossils often presents serious difficulties, especially in older material. Molecular work has rejected many long-established groups that were based on traditional morphology. For example, in plants, the dicotyledons are now thought to include monocotyledons, ratite birds now include tinamous, and the mammalian order Artiodactyla now includes Cetacea. Countless other cases are known at lower taxonomic levels. Yet fossil-calibrated molecular biogeographic studies depend on the correct identification of fossils and their assignment to a position in a phylogeny, and this is done using traditional morphological homologies. Morphological analyses of living taxa have been wrong so often that it seems strange to base the molecular clock calibration entirely on morphological analysis of fossil material, a much more difficult task. In many groups, identification and phylogenetic placement become more difficult in earlier members. For example, the Mesozoic members of a group can look very different from the extant members.

Biologists have often assumed that fossil taxa must be primitive and basal with respect to living relatives, and so morphological characters that support this may be selected for phylogeny reconstruction. Pennington et al. (2004) noted a tendency in many studies to assign fossils to the stem of the clade that they belong to, rather than the crown group (the youngest clade including all the extant members), and, as they emphasized, this will lead to underestimates of divergence times. Smith et al. (2010: 5897) also criticized "the default practice of assigning fossils to the stem of the most inclusive crown clade to which they probably belong, thereby possibly biasing estimated ages (possibly throughout the tree) to be younger." A typical example concerns the geckos (Gekkota) and their oldest fossil, the mid-Cretaceous *Cretaceogekko*. Studies of geckos in Australia (Pepper et al., 2011) and New Zealand (Nielsen et al., 2011) have used this fossil to calibrate the base of the gecko tree. Nevertheless, while *Cretaceogekko* is the oldest known gecko fossil, the only analysis of the genus gave no indication that it is basal in the Gekkota phylogeny (Arnold and Poinar, 2008).

Wilf and Escapa (2015) discussed southern hemisphere biogeography and stressed that "molecular dates are extremely sensitive to placements of calibrating fossils at stem vs crown nodes... and to choices of methods and calibration scenarios." As they observed, "the convention of placing calibrations at stem nodes, unless they are explicitly resolved into a crown group, seems to cause significant directional bias. This procedure... forces crown nodes to be younger than the calibration fossil, whose real evolutionary position was either in the crown or along its subtending branch, not at a stem node...." Wilf and Escapa (2015: 288) concluded:

> Our results strongly suggest that the recent emphasis in the literature on post-Gondwanan dispersal... is partly based on megabiased clocks.... we urge significantly greater caution when using molecular dating to interpret the biological impacts of geological events... If our findings apply to a broader spectrum of organisms, there would be profound consequences for the general understanding of evolutionary rates... Gondwanan history remains fundamental to understanding Southern Hemisphere plant radiations....

Informal Transmogrification of Minimum (Fossil-Calibrated) Clade Ages into Maximum Clade Ages

In many recent studies, the age of the oldest fossil of a clade—an estimate of *minimum* clade age—is converted into an estimate of *maximum* clade age. The conversion is often not discussed or even mentioned, and so it can be referred to as a transmogrification (Heads, 2012c). In earlier studies, the transmogrification was informal; authors gave a literal reading of the fossil record and equated the age of a clade with the age of its earliest known fossil, "perhaps adding a safety margin of a few million years" (Soligo et al., 2007: 30). In this approach, groups that have their oldest known fossil in the Eocene, for example, such as bats or modern primates, could, at a stretch, have evolved in the Paleocene, but a Cretaceous origin would be ruled out. One literalist study estimated that the probability of primates existing at 80 Ma (Late Cretaceous) was one in 200 million (Gingerich and Uhen, 1998), although fossil-calibrated molecular clock analyses have since calculated that primates did exist then (Janečka et al., 2007; Fabre et al., 2009).

Two Case Studies of Informal Transmogrification

Grammitid Ferns (Polypodiaceae)

As discussed earlier (Chapter 1), the grammitid ferns include many trans-oceanic affinities (Sundue et al., 2014). For example, *Notogrammitis* has a typical circumaustral distribution in Chile, South Africa, southeastern Australia, and New Zealand (11 species), as well as many subantarctic islands.

Sundue et al. (2014) inferred a center of origin in the Neotropics followed by dispersal to Asia and Australasia, but this was based on the flawed assumption that a paraphyletic basal grade indicates a center of origin. To date the clades, Sundue et al. (2014) used fossil calibrations from Schuettpelz and Pryer (2009). The latter authors treated their calibrations (with one exception) as *minimum* clade ages, and this is correct. However, based on the calibrations, Sundue et al. (2010) calculated that the family originated in (not in or before) the Eocene. In other words, they treated the age as a *maximum*, even though it was based on calibrations that were minimum ages, and there is no logical justification for this. Likewise, the circumaustral *Notogrammitis* was proposed to have had an origin in (not in or before) the Miocene, long after Gondwana breakup. Based on this young age, a tectonic origin of the trans-oceanic disjunctions was ruled out.

Despite the young dates proposed for the genus by Sundue et al. (2014), *Notogrammitis* displays significant disjunctions along major faults in New Zealand (*N. gunnii*) and in New Guinea ("*Grammitis*" *ceratocarpa*; Heads, 2003: 422). These and other patterns in Malesian members of the genus (Heads,

2003: 373, 395, 425) can be explained by the direct effect of tectonic change, but only if the group is older than suggested by Sundue et al. (2014). The sister group of the circumaustral *Notogrammitis* is endemic to New Caledonia and Lord Howe Island, where *Notogrammitis* itself is absent (Sundue et al., 2014). This allopatry is consistent with the origin of both clades by simple vicariance of a panaustral ancestor at a break in the Tasman Sea region.

Nothofagaceae

This family comprises a single genus, *Nothofagus*, with four subgenera. (The subgenera have been treated at generic rank by Heenan and Smissen, 2013, but there is no absolute difference between subgenera and genera, and so the only effect of this change is to introduce an extensive new synonymy.) Knapp et al. (2005) provided what they described as "unequivocal molecular clock evidence" that trans-Tasman Sea distributions in *Nothofagus* "can only be explained by long-distance dispersal." This was based on stipulating a 75 Ma age for the four subgenera in the family, and this in turn was based on the age of the oldest fossil in each of the four (75 Ma). A second constraint estimated the age of the split between *N. cunninghamii* and *N. moorei* as 20 Ma, based on the age of the oldest fossils of these species (20 Ma). The authors treated these fossil ages—minimum clade ages—as maximum clade ages, and only in this way were they able to conclude that vicariance "can be rejected on the basis of the divergence dates."

Transmogrification of Clade Ages in a Bayesian Framework: The Problem of the Priors

Following the earlier, informal transmogrifications of clade age, formal transmogrification is now often carried out in Bayesian analyses. Many papers annotate fossil-calibrated phylogenies with minimum and maximum estimates of clade ages, given in the form of 95% credibility intervals. In other words, fossil-calibrated ages have been converted from minimum estimates into maximum estimates that have statistical support. How exactly is this achieved? In these studies, the transmogrification is carried out in a Bayesian framework, using programs such as BEAST (Drummond and Rambaut, 2007). The key point is that specific prior probability distributions (priors) are assigned to the node ages used as calibration points before any analysis is carried out. In other words, a clade's oldest known fossil is often used to calibrate the phylogeny of a group, and the actual age of the clade, or rather the probabilities of different possible ages, is specified as prior. For a particular clade with an oldest *fossil* age of, say, 10 Ma, possible *clade* ages and their probabilities could be specified as, say, 10 Ma (90%), 20 Ma (50%), and 30 Ma (10%). As an alternative, a steeper probability/age curve could be set, prior to any analysis, specifying clade ages of 10 Ma (99%), 11 Ma (10%), and 12 Ma (0.5%).

The first question that needs to be asked of any Bayesian analysis is: "How are the priors selected?" For a given fossil age, normal, lognormal, gamma, or exponential curves are often specified as priors for probability/age curves, and these give rapidly decreasing probabilities for older clade ages. An exponential prior will provide particularly young clade ages. It assigns the highest probability to a clade age that is the *same* as that of its oldest known fossil age, with the decreasing probabilities for older ages following a steep, exponential curve. Ho and Phillips (2009: 372) warned that exponential priors should be used only "when there is strong expectation that the oldest fossil lies very close to the divergence event." Nevertheless, the traditional approach—a literal reading of the fossil record—*always* expects clade age to reflect the oldest known fossil age (although this has been refuted in many individual cases), and so many authors now adopt exponential priors (review in Heads, 2012c).

The priors, or prior probability curves, for clade ages in Bayesian analyses represent "sources of knowledge such as expert interpretation of the fossil record" (Drummond and Rambaut, 2007). Yet they can also introduce error, by incorporating literal interpretations of the fossil record; in this approach, the oldest fossil age provides the most likely clade age. By selecting appropriate priors, young clade ages with narrow credibility intervals can be generated and traditional theory supported.

Several authors have now recognized the problem with the priors. Ho (2014) admitted that "choosing the parameters of these distributions can be a difficult exercise." He noted that there are formalized methods that allow priors for the actual clade age to be based on the known fossil data. However, these methods simply use the density of the fossil record to extrapolate back beyond the record to give the likely clade ages, and so are unreliable.

Ho and Duchêne (2014) admitted that while the choice of priors "can have a substantial impact on the estimates of node ages…," it "is often subjective and based on an overall interpretation of factors that are difficult to quantify individually." Hipsley and Müller (2014) also admitted that selecting node age priors in fossil-based calibrations "is often arbitrary or idiosyncratic at best…"; despite this, they supported the method.

Berbee and Taylor (2010) also stressed that "different priors for node ages had a striking effect on our results….," but "whether or not it is possible to select reasonable priors is a serious concern in Bayesian analysis (Felsenstein, 2004)." Parham et al. (2012: 352) observed: "Most studies use a Bayesian framework for estimating divergence dates with probability curves between minimum and maximum bounds… but there is presently no practical way to estimate curve parameters [i.e., priors]." Warnock et al. (2015) wrote that while "Bayesian posterior estimates of divergence times are extremely sensitive to the time priors" (p. 1), "there is frequently no material basis for selecting among the parameters" (p. 6), and the priors are "invariably established without justification" (p. 2). Lee and Skinner (2011: 540) agreed: "current practice often consists of little more than educated guesswork." Pirie and Doyle (2012: 108–109) concluded that sporadic preservation in the fossil record "may be effectively impossible to model" and that priors "should… be regarded as highly arbitrary"; their value "is something we simply cannot know."

As Parham et al. (2012) concluded: "the fact that a widely applied methodology is subjected to such ambiguous assumptions that have a major impact on results… is a major limitation of molecular divergence dating studies" (p. 352). Parham et al. suggested that authors should adopt maximum bounds that are "soft and liberal…" (p. 352), but many studies fail to do so and instead use narrow, exponential priors. In the softest, most liberal approach, the fossil dates are treated as minimum ages, and this is the procedure adopted here.

Wilkinson et al. (2011: 28) also criticized the current approach, as adopted in studies on primates (Chatterjee et al., 2009). They wrote that the young age estimated for the main divergence in primates (strepsirrhines versus haplorhines):

> may be due to the use of two exponential distributions [priors]… These distributions… implicitly assume that the true age is close to the minima and unlikely to be much older than those minima. This assumption, we feel, is unlikely to be warranted, as it does not take account of the sizable gaps that exist in the primate fossil record.

The gaps are not infinite though and possible maximum ages can be suggested, based on tectonics. For example, in primates, the oldest known fossils are from the Paleogene, and molecular clocks suggest a Cretaceous origin, but tectonic-vicariance calibration (which assumes that the Madagascar and American endemic clades have evolved because of rifting) suggests a Jurassic age (Heads, 2012a).

Wheat and Wahlberg (2013) discussed "the dangers of over-confident date estimates" and showed how the literature presents "temporal estimates likely harboring substantial errors… Deciding upon which priors to use is very challenging,…" The authors wrote that "one of the most influential comparative genomic studies of the past decade" (Clark et al., 2008) estimated the age of the *Drosophila* clade (63 Myr old) and divergences among the species. Wheat and Wahlberg (2013) wrote that Clark et al. did this "(i) using only pairwise sequence comparisons for a small set of genes, (ii) assuming a strict molecular clock, and (iii) using a rate of molecular evolution based upon simple assumptions of colonization and island formation in the Hawaiian Islands… Each part of this approach is now known to be fraught with error…." Wheat and Wahlberg (2013) instead calculated that *Drosophila* was 70 Myr older than Clark et al. (2008) had proposed, with a first divergence within the genus at ~112 Ma.

Campanulid angiosperms are a large group of eudicots including Aquifoliales, Asterales, Apiales, and Dipsacales. A Bayesian study found that the group likely originated at ~105 Ma, and not before 115 Ma (95% credibility interval or highest posterior density [HPD] interval) (Beaulieu et al., 2013). Yet this was based on the prior assumption that the age of the member groups used for calibration could be no more than 5 Myr older than their oldest fossil. Such a small difference between the oldest fossil date and the actual clade age is unlikely, and the calculated dates based on this prior are not accepted here. By stipulating the alternative assumption that the groups were no more than 15 Myr older than their oldest fossil—still a rather small difference—the authors calculated an oldest date of 130 Ma for campanulids. They suggested that this date was too *old*, as it was older than the oldest eudicot fossil. This reasoning is not logical though, as campanulids could be much older than both the oldest campanulid fossil and the oldest eudicot fossil.

Bayesian analyses that stipulate appropriate priors will "validate" young ages for clades, a key component of Modern Synthesis biogeography. These clade ages can then be used to "rule out" earlier vicariance. This whole process is then said to provide "evidence" supporting a center of origin/dispersal model. As in traditional transmogrification, the age of the oldest known fossil in a clade is converted from a *minimum* clade age into an estimate of *maximum* clade age, and the Bayesian framework adds a gloss of respectability to the process. The HPD intervals provide a false illusion of statistical support, and the calibrations, together with the maximum clade ages based on them, are likely to be gross underestimates of clade age.

Examples of Bayesian Dating Studies

The following studies all deal with groups represented in New Zealand; they provide typical examples of Bayesian transmogrification and the problem of the priors (others are given in Heads, 2012c). In all cases, clade ages estimated using Cenozoic fossils and steep priors are used to rule out Mesozoic clade ages and vicariance associated with Gondwana breakup.

Monimiaceae

Based on the oldest known fossils of Monimiaceae, dated at 87–83 Ma and 83–71 Ma, Renner et al. (2010) stipulated an age of 83 Ma with a normally distributed prior and a standard deviation of just 1.5 Myr on the crown group node. This gave an age for the family (95% HPD interval) of 80.5–85.5 Ma. Based on fossils dated at 34–28 Ma, they also placed a normally distributed prior of 30 Ma on the divergence of *Xymalos*, with a standard deviation of 1.5 Myr. This gave an age (95% HPD interval) of 33–28 Ma (Oligocene).

Using these dates to calibrate the phylogeny, Renner et al. (2010) calculated that the New Zealand–New Caledonia clade *Hedycarya arborea* + *Kibaropsis* separated from its sister group (the remaining *Hedycarya* species + *Levieria* in New Guinea, eastern Australia, and New Caledonia) at 24 Ma (95% HPD, 35–14 Ma). This young date rules out the possibility of a vicariance origin for the New Zealand–New Caledonia group (either with continental drift or predrift rifting) and supports trans-oceanic dispersal as the favored explanation. Yet it is based on the priors that were stipulated, not on evidence.

Winteraceae

By setting narrow priors, Thomas et al. (2014) calculated that Winteraceae, with an oldest fossil dated at 125 Ma, had an actual age of 127.91 Ma or, at most (95% HPD), 133 Ma. Based on the priors, the authors also calculated very young ages for the clades. For example, the New Zealand *Pseudowintera* was dated at just 45 Ma (95% HPD, ~33–57 Ma), younger than the rifting of the Tasman Basin. Because of the young clade ages, the authors were able to conclude that "… the Winteraceae of Zealandia have descended from at least four separate long-distance dispersers from Australia: the ancestors of *Bubbia comptonii* [New Caledonia]; the main New Caledonian clade; *Pseudowintera* in New Zealand; and *Bubbia howeana* on Lord Howe Island."

The study by Thomas et al. (2014) did not mention, let alone explain, the most obvious feature of the distribution of Winteraceae: the very high degree of allopatry among the three main clades and the sister-group, Canellaceae (Heads, 2012a: Fig. 9-5; this is based on phylogenies with the same topology as the one in Thomas et al., 2014). The simplest explanation for the allopatry is that the differentiation between Canellaceae and Winteraceae, and also within Winteraceae, was by simple vicariance (with minor local overlap developing later), and that the priors stipulated by Thomas et al. were misleading.

Orchidaceae

In Orchidaceae, the oldest fossils of *Dendrobium* (dated at 23 Ma) and *Earina* (dated at 25 Ma) occur in New Zealand, and both genera are extant there. Instead of treating the fossil ages as minimum clade ages, Gustafsson et al. (2010) stipulated priors for these fossils and another orchid fossil that gave a date for the break between *Dendrobium* and *Earina* at 25–40 Ma (95% HPD). The authors calculated dates for other orchid genera found in New Zealand and elsewhere: *Pterostylis* split from its sister at 40 Ma, *Microtis* at 31 Ma, and *Spiranthes* at 17 Ma (sampling of further genera will probably make these younger). These young dates falsified a vicariance explanation for New Zealand members of the genera. The authors assigned monocots a maximum age of 120 Ma, corresponding to the age of the oldest known monocot fossils. They acknowledged that this constraint "may be questionable since fossils generally provide minimal ages, but in the absence of further evidence such upper bounds are technically advantageous." In fact, the procedure is essential if absolute ages are to be inferred from fossils, but, as the authors admitted, the method is questionable.

Unambiguous orchid fossils are very rare, and only three are known (Gustafsson et al., 2010). Yet there is a vast amount of other evidence that can be used to calibrate the timeline of evolution in the family. Every clade of orchids has a distribution that is defined by phylogenetic and spatial breaks, and these can be correlated with events in tectonic or climatic history, giving tens of thousands of potential calibrations.

Alstroemeriaceae

The monocot family Alstroemeriaceae is distributed in South and Central America, New Zealand, and Australia. Dispersal of the component genus *Luzuriaga* across the Southern Ocean from Patagonia to New Zealand has been suggested (Conran et al., 2014a), but this was based on the young date of origin estimated for the genus (48–54 Ma; Eocene). This was calculated by setting the age of the related genus *Smilax* at 46 Ma (95% confidence interval 37.2–55.8 Ma), as the oldest *Smilax*-like fossils are known from strata dated at 37.2–48.6 Ma and 48.6–55.8 Ma (all in the Eocene). This procedure stipulates that *Smilax* can be no older than the Eocene, the epoch in which its oldest fossil occurs. This is not logical (the age of the base of the Eocene is irrelevant), but it is the only basis for the *Luzuriaga* date. Another study of the family used a similar approach (Chacón et al., 2012).

Haloragaceae

The genus *Haloragis* has a standard distribution in Australia, New Zealand, New Caledonia, and Juan Fernández Islands. A clock study stipulated very narrow priors for the two fossil calibrations used (with standard deviations of just 1.0 and 0.9 Myr) (Chen et al., 2014a). This resulted in very young ages for the Juan Fernández species, and so the trans-Pacific disjunction was attributed to long-distance dispersal.

Malvaceae

Wagstaff and Tate (2011) calibrated evolution in Australasian Malvaceae by assuming that eastern Australian clades diverged from western Australian clades in the Miocene, and that endemic forms on the Chatham Islands of New Zealand could be no older than proposed ages for the modern islands, 1–3 Ma. Both of these assumptions can be questioned. The first is rejected elsewhere (Heads, 2014), and the second is contradicted by the growing list of Chatham Islands endemics that have been dated in clock studies as older than the islands (Trewick, 2000; Liggins et al., 2008a; Heenan et al., 2010; Buckley and Leschen, 2013). For their third calibration point, Wagstaff and Tate (2011) used the oldest fossils of tribe Malvaceae in Australasia and South America, dated as Eocene. They transmogrified this date into an estimate of most likely maximum clade age, specifying an exponential prior. Based on these three doubtful calibrations, the authors deduced a Miocene age for the New Zealand Malvaceae, ruled out a vicariance origin for the group, and supported the traditional center of origin/chance dispersal model.

Campanulaceae

Prebble et al. (2011) calibrated a phylogeny for *Wahlenbergia* (Campanulaceae), with an emphasis on the New Zealand species. Three calibration points were used. The oldest fossil of *Wahlenbergia* in New Zealand is pollen from the Waipipian, a stage in the mid-Pliocene dated as 3.0–3.6 Ma. The prior for the Waipipian fossil's clade, a group of New Zealand–Australia–South Africa clade, was assigned a value of 3.3 Ma, a standard deviation of 0.2 Myr, and a normal distribution. These settings gave a 95% HPD interval of 3.0–3.6 Ma for the clade age. Here, the fossil age or minimum clade age was transmogrified into an estimate of maximum clade age that coincides with that of the geological stage from which the fossil is known. There is no logical reason for adopting this approach, although it was also used in the Alstroemeriaceae study cited earlier. (A global study of Apiales adopted a similar, arbitrary approach, stipulating in the priors that clades could not be older than the subepoch, e.g., Late Paleocene, in which their oldest fossils were found; Nicolas and Plunkett, 2014.)

Prebble et al. (2011) treated a second fossil calibration for *Wahlenbergia* in the same way. A Campanulaceae seed from the Miocene Upper Karpatian stage of Europe is dated as 16.5–17.5 Ma. Priors for its clade were assigned a mean value of 17 Ma, a normal distribution, and a standard deviation of just 0.4 Myr. This was done in order to give a 95% HPD interval of 16.34–17.66 Ma for the clade age, corresponding to the age range of the Upper Karpatian.

A third node in the Prebble et al. study, the split between the Campanulaceae and the Lobeliaceae, was dated in a previous fossil-calibrated analysis at 52 Ma (Wikström et al., 2001). As priors for this node, Prebble et al. assigned a normal distribution, a mean age of 52 Ma, and a standard deviation of 3 Myr. This gave an actual age for the Campanulaceae of 47.1–56.93 Ma (95% HPD interval).

The ages that Prebble et al. (2011) calculated were young enough to rule out a vicariance origin for southern hemisphere affinities in *Wahlenbergia*. Thus, the authors treated the clade ages as evidence for a single Pliocene dispersal event from Africa to Australasia, over ~8000 km of open ocean.

Calceolariaceae

The genus *Jovellana* (Calceolariaceae) is distributed in New Zealand and Chile. A fossil-calibrated study calculated a Neogene age for the genus and ruled out vicariance (Nylinder et al., 2012). Yet the authors reached this conclusion only because they stipulated in the priors that clade age could be no more than 5 Myr older than oldest fossil age.

Goodeniaceae

Goodeniaceae (Asterales) are most diverse in Australia and are represented in New Zealand by *Selliera* on the mainland and *Scaevola* on the Kermadec Islands. In their study of the family, Jabaily et al. (2014) admitted that "…early fossil evidence from the Asterales is limited." Nevertheless, they relied on a fossil to calibrate their phylogeny. The fossil is *Raiguenrayun*, dated at 47.5 Ma and identified as a member of Asteraceae, a close relative of Goodeniaceae (Barreda et al., 2012). Asteraceae has two basal clades, Barnadesioideae and all the rest, and *Raiguenrayun* is placed in "the rest." For the basal node in Asteraceae, between the two basal clades, Jabaily et al. (2014) stipulated as priors an exponential age distribution with a minimum of 47.5 Ma and standard deviation of just 6 Myr. Selecting these priors allowed the authors to conclude that "dispersal [in Goodeniaceae] is inferred for the extra-Australian lineages due to their recent origins." Using ancestral area methods (DEC, S-DIVA) similar to those discussed in Chapter 1, the authors calculated a center of origin for the family in Australia.

Asteraceae

Swenson et al. (2012) calibrated a phylogeny for Asteraceae using the fossil *Raiguenrayun*, dated at 47.5 Ma (see last section). The authors stipulated several different priors for the clade age of the fossil (without giving the reason for their selection), and calculated median ages for the basal node of

Asteraceae between 31 and 55 Ma (95% HPD). In this way, they were able to rule out any vicariance in the family associated with the breakup of Gondwana.

Another study calibrated a phylogeny for Asteraceae with two calibration points (Sancho et al., 2014). The first was the fossil *Raiguenrayun* dated at 47.5 Ma. The authors assigned an exponential prior distribution with a minimum age of 47.5 Myr and a standard deviation of 6 Myr, which gives a maximum possible age for the main Asteraceae clade (95% HPD) of 65 Myr. The second calibration point stipulated that endemic species on the Chatham Islands could be no older than the islands themselves, assumed to be 1–3 Ma (this is dubious; see the section "Malvaceae"). Based on the calibrations, the authors calculated that the trans-South Pacific genus *Lagenophora* originated at just ~16 Ma (early Miocene), and so they attributed the trans-oceanic disjunction to long-distance dispersal. However, the calibrations were undermined by the recent discovery that Late Cretaceous fossils (76–66 Ma) belong to an extant genus of Asteraceae, *Dasyphyllum* (Barreda et al., 2015).

Other Widespread Plant Groups

In other studies on global or pantropical groups, Bayesian transmogrifications of the kind just described have produced fossil-calibrated, maximum ages for the groups (with 95% HPD support) that are younger than the Atlantic, Indian, and Pacific basins. Based on these dates, the intercontinental disjunctions have been attributed to chance, trans-oceanic dispersal. Widespread clades present in New Zealand that have been analyzed in this way include the pantropical subfamily Chrysophylloideae (Sapotaceae) (Bartish et al., 2011), the global tribe Ranunculeae (Emadzade and Hörandl, 2011), and the genus *Leptinella* (Asteraceae) (Himmelreich et al., 2012).

Specifying Bayesian Priors and Their Parameters

Anyone reviewing the Bayesian dating studies would conclude that few, if any, geographic disjunctions date back to the breakup of Gondwana, at least for genera and tribes. Nevertheless, while this view is the current consensus, it is not based on any new data or analysis and instead reflects the formal imposition of a prior belief—that fossil age is more or less equal to clade age. In these recent studies, fossil-calibrated minimum clade ages are converted into maximum clade ages that have good statistical support, but only by decree, and the potential gaps in the fossil record (the Sppil–Rongis effect) are swept under the carpet. The alternative method advocated here instead integrates data from tectonics, biogeography, and the fossil record, with fossil data used to provide minimum ages.

The comments given earlier are not meant as a rejection of Bayesian analysis per se, but of the particular priors selected in current biogeographic work. Treating fossil ages only as minimum clade ages, as suggested here, is equivalent to using flat priors. The imposition of steep, nonflat priors for fossil-based clade ages is unjustified and unnecessary, and leads to erroneous conclusions about the formation of biogeographic patterns. The impact of these conclusions on ecological and evolutionary interpretations has been serious. Authors have felt obliged to reject simple tectonic explanations for general distribution patterns, and instead favor chance processes and unknown ecological factors as causative.

Integrating Living and Fossil Records

In dispersal theory, "the fossil record is emerging again as being crucially important in biogeography" (Crisp et al., 2011). No biogeographer has ever suggested ignoring the fossil record, but a cornerstone of the Modern Synthesis was the idea that fossils are always more important than living groups in reconstructing evolution. For example, Darlington (1957/1966: 320) wrote that the fossil record "allows an almost magical view into the past," and Briggs (1974: 249) used the same words. Instead, the fossil record of a group is treated here as no more and no less important than the living one. Although the extant record is the focus of this book, in the discussions of clades, fossils from outside the extant range are cited wherever they are known. In most cases, the age of a clade's oldest known fossil is not mentioned as it only provides a minimum age, but these are often valuable and some are cited. Fossils often have more significance for understanding phylogeny and morphological trends than for dating, although living specimens show many aspects of morphology that are absent in fossils.

In contrast with the identification of fossils, which is often controversial, molecular work is leading to stable phylogenies of living groups. A common feature of these phylogenies is the high level of morphological parallelism that is displayed. These imply that many characters, even complex ones that were assumed to define monophyletic groups, have evolved many times. This means that early, fragmentary fossils—wood, pollen, or bones from the Cretaceous, for example—are less likely to belong to the extant group that they most resemble.

Protecting Dispersal Theory from Falsification by Early New Zealand Fossils of Modern Groups

Although dispersal theory treats the fossil record as "crucially important" and even "magical" (see last section), it disregards fossils that contradict the theory. For example, old New Zealand fossils of a clade that is still extant in the region are of special interest. Yet in dispersal studies, these fossils are dismissed as irrelevant if they are older than the molecular clock date for the New Zealand clade. The fossils are attributed to an early invasion, but the clade is proposed to have then died out in New Zealand, and the modern New Zealand members are attributed to another, recent invasion. This approach protects the theory of recent dispersal from falsification by early fossils of modern groups. New Zealand examples treated in this way include the following:

- In clades of dominant trees, *Nothofagus* subgenus *Lophozonia* (Nothofagaceae) (Cook and Crisp, 2005) and *Agathis* (Araucariaceae) (Biffin et al., 2010) were both attributed to post-Cretaceous dispersal, despite having Cretaceous fossils in New Zealand (the species-level affinities of the fossils are ambiguous).

- New Zealand Richeeae (Ericaceae) include macrofossil material (*Richeaphyllum*) dated as 20–25 Ma (its precise affinities within the group are ambiguous) and *Dracophyllum*-type pollen dated at 47–40 Ma (Jordan et al., 2010). Nevertheless, a molecular clock age of ~7 Ma was accepted for extant New Zealand Richeeae, and the fossils were attributed to extinct lineages.

- New Zealand Styphelieae (Ericaceae) have New Zealand fossils dated at 20–23 Ma but a molecular clock age of 7 Ma, and so extinction and recolonization have been proposed (Puente-Lelièvre et al., 2013).

- The monocot herb *Luzuriaga* (Alstroemeriaceae) has fossil material in New Zealand dated at 23 Ma, while the extant New Zealand species of *Luzuriaga* was dated in a clock study as 2.9 Ma (Chacón et al., 2012). Chacón et al. concluded: "Like so many other New Zealand clades... [*Luzuriaga*] went extinct, perhaps during times of submergence, only to reach New Zealand again by long-distance dispersal from southern Chile" (cf. Conran et al., 2014a).

- In the cod genus *Micromesistius* (Gadidae) of southern and northern oceans, New Zealand fossils are older than the molecular clock dates for the southern clade, and so Halvorsen et al. (2012) inferred extinction of the first lineage in the south with two subsequent dispersals to the southern hemisphere.

In all these cases, the clock dates are calibrated with fossils and represent minimum ages. Yet instead of acknowledging this (and so admitting the possibility of vicariance), the authors accept the dates as maximum ages, rule out vicariance, and propose the ad hoc hypothesis of extinction and recolonization.

Crisp et al. (2011) suggested that the "biotic turnover" inferred for New Zealand, with extinction and then recolonization, has been overlooked by earlier workers because fossils of extinct lineages have been misassigned to younger, related lineages. In practice, though, the documented bias in identifying older fossils works in the other direction—older fossils are assumed to be ancient, basal, groups or stem groups (see the earlier section "Placing Fossils in a Phylogeny").

Protecting Dispersal Theory from Old New Zealand Endemics with No Extralimital Fossils

Old fossils threaten a dispersal history for the southwest Pacific biota, as do old endemics in New Zealand (such as the basal passerines, Acanthisittidae) and in New Caledonia (such as the basal angiosperms,

Amborellaceae). These might be thought to have evolved in their region, rather than dispersing there. However, dispersal theory is protected from falsification by these groups through the proposal that populations once existed on mainland Australia or Asia (the center of origin) but have since gone extinct there, even though there are no mainland fossils. But there is no evidence for this, and the suggestion is ad hoc.

Both vicariance biogeography and dispersal theory agree that "lack of fossil evidence of a former presence of a taxon in a given area, should not be accepted *prima facie* as evidence that it was always absent" (Crisp et al., 2011). Yet, as discussed earlier, dispersal theory contradicts this cautious approach by treating the age of the oldest known fossil of a clade as the maximum, or near-maximum, age of the clade. This will provide support for young ages and trans-oceanic chance dispersal, and in this case, dispersal theory accepts that the absence of fossils is real.

On the other hand, if the absence of fossils contradicts dispersal theory, the theory rejects the absence as ambiguous. For example, *Amborella* from New Caledonia is often accepted as the sister group of all other angiosperms (Drew et al., 2014; Simmons and Gatesy, 2015), and it has no known fossils (including pollen) anywhere else in the world. This complete absence of such an ancient group over such a wide area is interesting, but although Crisp et al. (2011) regarded the fossil record in general as "crucially important," they concluded that the distribution of *Amborella*, along with many similar examples in the Tasman region, is "shrouded in mystery." This is because the existence of such old groups, together with their absence, both living and fossil, outside the region, conflicts with the standard dispersal theory: "New Zealand and New Caledonia resemble 'oceanic' islands with young, immigrant biota" (Crisp et al., 2011). Thus, the authors argued that lineages such as *Amborella* "have probably been subject to considerable extinction and… are essentially uninformative about biogeographical history." Swenson et al. (2013a) agreed that *"Amborella's* presence in New Caledonia remains enigmatic and conveys no biogeographical information until new [fossil] discoveries are available."

Nevertheless, while all older groups have been subject to the ravages of extinction, this does not mean they should be ignored. The juxtaposition of the basal angiosperm, *Amborella*, in New Caledonia and the basal passerine clade, Acanthisittidae, in neighboring New Zealand is "enigmatic" and "uninformative" in dispersal theory; otherwise, it is a valuable clue. In fact, there is a great concentration of endemic groups in the Tasman region that have global sisters (Heads, 2009c), and as a general pattern, this needs to be explained. It suggests that the region was the site of the primary phylogenetic/biogeographic break in angiosperms, passerines, and other groups, and this is accepted by authors who use the clades and tectonics of the region to calibrate phylogenies (e.g., Moyle et al., 2012; Ericson et al., 2014). In this evolutionary model, the absence of any *Amborella* or acanthisittid fossils outside their known range reflects actual absence. Extralimital records—either living or fossil—could yet be found, but that does not mean that current knowledge of the living and fossil distributions must be treated as uninformative.

Biogeography and Extinction: Dealing with the Possible Extinction of a Clade in Areas Where It Has No Fossils

The allopatry that develops by vicariance is often modified by subsequent range expansion and contraction, or even extinction. Local, regional, or global range expansion is a normal occurrence, as is local or more widespread extirpation of populations and clades. Extinction cannot be ignored in biogeographic studies of modern groups, and many groups have fossils outside their extant range. Many of these are well documented and many are cited throughout this book. But how can biogeographic analysis deal with the possible extinction of groups from areas where there is no fossil evidence? Were all groups once widespread throughout the world, with their modern distributions produced by extinction alone? If this were the case, any distribution could be explained by chance extinction of the group everywhere outside its known range, even without fossil evidence (cf. Crisp et al., 2011, on *Amborella*). For example, the pattern in a worldwide group comprising 100 allopatric species would be accounted for by the former worldwide distribution of all 100 species, followed by the extinction, in each species' current area of endemism, of the other 99 species. This is very unlikely though; extinction does not provide a good explanation for the precise spatial breaks between allopatric sister groups, and these are a main focus of the present study.

Crisp et al. (2011) wrote that "extinction has long been acknowledged as a key determinant of observable biogeographical patterns, but is often considered intractable and ignored." But the authors did not suggest how to deal with the possible extinction of a group in areas where it has no fossil record. Instead, they argued that "only the fossil record (if available) can provide evidence of former occurrences of taxa in areas where they are now extinct." There are many other sorts of evidence that can be considered though, and interpretation of extinction even without a fossil record is not impossible.

Records from *Nothofagus* (Nothofagaceae) are useful here. Members of the genus are trees that are dominant over large areas and produce great amounts of pollen (they are wind pollinated). The genus has an excellent fossil record, with all extant subgenera known back to the Cretaceous. Comparison of the fossil and living distributions, together with the centers of diversity in the subgenera, indicates that extinction has tended to wipe out groups in areas where the groups were less diverse to start with (Heads, 2006a). One example is *Nothofagus* sect. *Brassospora*. The main center of diversity for *Brassospora* (for living species and for fossil species) is in New Guinea and New Caledonia, and this is allopatric with the centers of diversity in the other subgenera in New Zealand and South America. The fossil record indicates that there were never more than a few species of *Brassospora* in New Zealand. If the subgenera were allopatric in origin—which is likely given their allopatric centers of fossil and living diversity—*Brassospora* was probably present in New Zealand by secondary range expansion. In any case, all populations of this minor outlier of diversity in New Zealand are now extinct. In a similar way, a group with 1000 species in Australia and one in New Zealand will tend to go extinct in New Zealand before it does in Australia. The case of *Nothofagus* suggests that extinction does not occur in random areas, but in areas of lower diversity.

Is Evolution Clocklike?

Nineteenth-century evolutionists were well aware that groups sharing the *same* biogeographic pattern have *different* degrees of differentiation. Groups with the Australasia–Patagonia disjunction, for example, include species, subgenera, and genera. Likewise, modern clock studies of 75 plant pairs with this disjunction showed a continuum of degree of differentiation, with calculated ages ranging from 1 to 40 Ma (Winkworth et al., 2015). This and similar examples suggest that all main distribution patterns display the same feature (Heads, 2012a: 62).

Avise (1992) summarized the situation and interpreted it as a paradox: "...concordant phylogeographic patterns among independently evolving species provide evidence of similar vicariant histories... However, the heterogeneity of observed genetic distances and inferred speciation times are difficult to accommodate under a uniform molecular clock" (p. 63).

The usual response has been to ignore the evidence for vicariance, accept the molecular clock dates, and explain the patterns using chance dispersal.

Degree of differentiation is sometimes attributed to time since divergence, as in the molecular clock. Yet the fact that many groups have the *same* distribution but *different* degrees of difference suggests an alternative. Based on this phenomenon and on a good knowledge of geology, Hutton (1872) was able to conclude that "[degree of] differentiation of form, even in closely allied species, is evidently a very fallacious guide in judging of lapse of time."

The nineteenth-century authors reasoned that because the component groups in a single biogeographic pattern have differing degrees of difference (branch lengths), evolutionary rates could differ in different groups. They rejected an evolutionary clock model and instead suggested that a single geological event would cause many biological breaks. Thus, Hutton (1896) agreed with T.H. Huxley (1873) that Pacific biogeography has been mediated by tectonic change. In contrast, Matthew (1915) and the authors of the Modern Synthesis assumed a dispersal model and an evolutionary clock, with degree of differentiation related in a simple way to time elapsed since a clade's origin. Different branch lengths in a single biogeographic pattern were thus attributed to chance dispersal events in the different groups taking place at different times.

For example, Mayr (1931: 9) discussed the unusual avifauna of Rennell Island, in the Solomon Islands, and wrote: "The different degree of speciation suggests that the time of immigration has not been the same for all the species." Applying the same idea to the birds of New Caledonia, Hawaii, and other islands, he wrote that "strikingly different degrees of differentiation indicate colonization at different ages" (Mayr, 1944: 186).

This evolutionary clock idea is often applied to New Zealand. For example, in meliphagid birds, the degree of difference between mainland subspecies and Chatham Islands subspecies is greater in bell-birds (*Anthornis melanura*) than in tuis (*Prosthemadera novaeseelandiae*), and so Bartle and Sagar (1987) suggested that the tui was the more recent arrival on the Chathams.

Developing a critical approach to the molecular clock means questioning what the degree of differentiation of a clade (its branch length) represents. Is branch length related to time since divergence, or does it instead reflect prior features of genome architecture that determine the inherent evolvability of a clade?

Goldberg and Trewick (2011) described a wide range of divergence between Chatham Island insects and their sister groups on mainland New Zealand, from 0% in some groups up to 14% in the stag beetle *Geodorcus*. As the authors wrote, "perhaps this wide range in divergence estimates reflects differences in biology [e.g., evolvability] rather than history, or a combination of the two." It is now accepted that evolutionary rates can display extreme variation, and so there could be large drops in rates throughout large groups after a general phase of differentiation. Degree of differentiation would then be a function of the *time* that passed during a differentiation event, not the *age* of the event.

Molecular studies also suggest that the different degrees of divergence seen at any biogeographic node are not related to means of dispersal. Teske et al. (2014: 392) wrote: "Based on mitochondrial DNA data, many organisms with ranges spanning multiple biogeographical regions exhibit genetic structure across the transition zones between these regions, while others appear to be genetically homogeneous. No clear link has been found between the presence or absence of such spatial genetic discontinuities and species' dispersal potential...."

Calibrating Phylogenies Using Tectonics and Vicariance

Most groups have no fossil record, and in many groups fossils are sparse. In addition, assigning the fossils that do exist to a phylogeny can be very difficult. Stipulating just how much older than its oldest fossil a group can be (the "problem of the priors") is also problematic. Another approach to dating clades does not use fossils to give maximum ages, but instead relies on spatial coincidence between clade distributions and well-dated tectonic features. Most clades exhibit distinct distributional breaks with close relatives and among their own subclades, as well as other, well-defined geographic structure. As discussed earlier, one of the main discoveries in molecular systematics is the ubiquity of this structure, even in groups such as bacteria, fungi, and marine organisms. Tectonic calibration exploits this important finding.

Berbee and Taylor (2010: 2) wrote:

> Fossils must be used to tie a molecular phylogeny to a geologic time scale,... Molecular clocks calibrated by fossils are the only available tools to estimate timing of evolutionary events in fossil-poor groups, such as fungi. Alas, fossil evidence remains scanty and substitution rates change chaotically from lineage to lineage,... developing better analytical methods may not be easy....

Instead of relying on the poor fossil record, though, the beautiful patterns of allopatry discovered in the new molecular work on fungi and other groups can be utilized to date clades.

In many cases, the phylogenetic/biogeographic break zones show a close spatial association with major tectonic structures that are well dated, and this relationship can be used to calibrate phylogenies. In this method, breaks between sister groups are attributed to vicariance mediated by tectonic or climatic changes in the same location.

The problem of the priors does not arise in tectonic calibration, as estimates of the ages of tectonic events are not minimum ages, and a survey of 613 papers published over the period 2007–2013 showed that 15% used geological calibrations (Hipsley and Müller, 2014). Fossil calibrations should be used to give minimum ages only, and, despite the temptation, these should not be converted into absolute ages by means of dubious priors.

The problems associated with converting fossil age into maximum clade age are complex and perhaps insurmountable; the potential error in the dates is difficult to estimate and is often high. In contrast, tectonic and magmatic events have been assigned absolute dates using radiometric techniques, and the error in these dates can be very small (often <0.05%; Wilf and Escapa, 2014). Calibrating the age of clades (their phylogenetic break zones) with tectonic events that occurred along the same belt avoids the problems of estimating clade age using fossils alone. The phylogenetic topology provides relative dates for the different nodes, while geology provides the sequence and absolute dates (with errors) for associated tectonic events. Any available fossils are used to give minimum, not maximum clade ages.

Tectonic Calibration and the Passerine Birds

Passerines include about half of all bird species, and their oldest known fossils are dated as Eocene (Australia). There is more reliable material from the Oligocene (Europe). Although the earliest European Passeriformes "already closely resemble their modern counterparts" (Mayr, 2009: 190), paleontologists have concluded that passerines diverged only in the Cenozoic (Mayr, 2013). Passerines display a morphological, or at least osteological, uniformity, and Mayr (2013) argued that a Cretaceous divergence "would imply an unprecedented evolutionary stasis for more than 80 million years in one of the most species-rich group of endothermic vertebrates."

Instead of depending on the oldest fossil ages and accepting that the group itself could not be much older than these, molecular studies have exploited the strong biogeographic structure of passerines and utilized tectonic-vicariance calibrations (Ericson et al., 2014). Although theory suggests that passerines have dispersed, obscuring their biogeographic patterns, in practice, birds "surprisingly often exhibit strong biogeographic patterns that are closely linked to their evolutionary history. This is true regardless of whether we study phylogeographic patterns within a species or geographic distributions of families and other higher-level taxa" (Ericson et al., 2014). Many other ornithologists have also commented on the paradoxical, sedentary nature of bird species (see quotations in Heads, 2012a: 278).

In contrast with the paleontological analyses, Ericson et al. (2014) calculated that passerines had a Cretaceous origin and are much older than the oldest fossil. The authors concluded:

> Another observation consistent with a Cretaceous origin of the passerines is that their phylogenetic relationships reveal a biogeographic pattern that has a clear Gondwanan signature. The basal oscines, the New World suboscines and the Old World suboscines, are all confined to continents that were once part of Gondwana.... The current distributions of the basal oscines, New World suboscines and Old World suboscines may be important in revealing the earliest history of the passerines, and this may also be true for the current distribution of the New Zealand wrens... A Cretaceous origin for the passerine radiation has far-reaching implications, as it suggests the fossil record is severely incomplete.... (Ericson et al., 2014: 11–12).

Authors discussing the evolution of New Zealand birds have often assumed that the fossil record is more or less reliable, and so they have read it literally, accepting that passerines and other orders could not be Cretaceous. For example, Fleming (1982: 317) discussed the cuckoo genus *Chrysococcyx* (*Chalcites*), present in Australia, New Zealand, and Melanesia (Heads, 2014, Figure 6.15). He concluded: "A hypothesis that the disjunct distribution of *Chrysococcyx* dates from Mesozoic Gondwanaland (Marchant, 1972) has little to recommend it, because evolution of the genus was closely related to that of its passerine hosts, which did not exist in the Mesozoic." Marchant's hypothesis does have something to recommend it, though, as it explains a standard pattern without using chance.

Criticism of Tectonic Calibration

Tectonic-vicariance calibration is a new field, and some applications have been more convincing than others. In one of the best-known cases, Atlantic/Pacific breaks in phylogenies have often been attributed to the final rise of the Panama isthmus, although this is likely to be based on an oversimplification (Heads, 2005a,b). In New Zealand, phylogenetic boundaries at the Nokomai–Nevis–Cardrona fault system of Central Otago have been attributed to the Neogene reverse faulting there (Waters et al., 2001), but this is probably reactivation of an older structure (Beanland and Barrow-Hurlbert, 1988), and at least some boundaries at this feature may date back to its earlier, Cretaceous origin. The geological history of many regions, including Panama and Central Otago, is complex, and the same geographic zone has been subjected to geological activities at different times. Nevertheless, the best way to assess the overall value of any new methodology is to try it out in a range of situations.

Ho et al. (2015) suggested that "a potential disadvantage of relying on biogeographic calibrations is that they are rarely available for multiple nodes in a phylogenetic tree of interest." This is incorrect though, and one of the main reasons for using tectonic calibration is that so many phylogenies include a wealth of potential calibrations at different levels in the hierarchy. A good example is the phylogeny of primates (Heads, 2012a). Many other examples of phylogenies that include multiple nodes associated with different tectonic features are discussed in this book (see also Heads, 2014).

In chance dispersal theory, a repeated pattern of allopatric differentiation at a single break zone is attributed to chance events having occurred there at different times in different groups, not by a single vicariance event. In this model, any apparent spatial correlation between the biogeography and tectonics is illusory. Thus, tectonic calibration has been criticized by supporters of chance dispersal theory, because it assumes vicariance, rather than testing it (Kodandaramaiah, 2011). Yet vicariance has already been tested many times. In any case, the results from studies that use tectonic-vicariance dating for a wide range of groups in a broad region (e.g., Heads, 2012a, 2014) could either be coherent and explicable (with more basal nodes related to older tectonic events), or chaotic and inexplicable, and so these studies provide a higher-level test of a general vicariance model.

Stelbrink et al. (2012) criticized tectonic calibration because it accepts a vicariance model of speciation rather than chance dispersal, but this is the whole point of the approach. There is considerable evidence, including the precision of the geographic patterns and their repetition in different groups, that speciation does not occur by chance dispersal. The increasing level of allopatry in lower-rank groups also indicates their original differentiation by vicariance. Thus, it makes sense to at least try out a method of analysis that does not assume chance dispersal and does not use chance events to explain repeated patterns.

Kodandaramaiah (2011) argued that a "major drawback of vicariant calibrations is the problem of circular reasoning when the resulting estimates are used to infer ages of biogeographic events… I argue that fossil-based dating is a superior alternative." This criticism is unfounded though, as the use of vicariance events to provide dates is based on a simple deduction: if the phylogenetic break was caused by a spatially coincident tectonic feature, and this is dated at x million years old, then the phylogenetic break is also x million years old. There is no circular reasoning here, whether or not particular cases of tectonic-vicariance calibration are incorrect. If a model of vicariance caused by tectonics is adopted, deductions about the age of a break can be tested against each other and other events. This follows a normal process of investigation based on the proposal of hypotheses, the testing of predictions deduced from the hypotheses, and the framing of new hypotheses.

Some members of the dispersal school have argued that using tectonic calibrations "would make biogeography less of a science and more of a religion" (Stelbrink et al., 2012), but despite the rhetoric, an increasing number of molecular studies are using tectonic-vicariance events to calibrate phylogenies. Examples include work on carabid beetles (Sota et al., 2005), tenebrionid beetles (Papadopoulou et al., 2010), Onychophora (Allwood et al., 2010), cichlid fishes (Genner et al., 2007; Azuma et al., 2008), galaxiid fishes in New Zealand (Burridge et al., 2008, 2012; Craw et al., 2008; Waters et al., 2010), rhinophrynid-pipid frogs (Bewick et al., 2012), passerine birds (Cracraft and Barker, 2009; Ericsson et al., 2014), timaliid birds (Moyle et al., 2012), birds of paradise (Irestedt et al., 2009), and crows (Jønsson et al., 2012). A recent review concluded that "biogeographic [tectonic and climatic] calibrations have the potential to make a valuable and effective contribution…" as they "provide a rich source of calibrating information" (Ho et al., 2015: 5).

In their review of studies on the New Zealand avifauna, Trewick and Gibb (2010) discussed methods for dating clades. They wrote: "Paradoxically, a growing recognition of the considerable importance of dispersal in founding isolated populations… and thus biotas… has occurred at the same time as some phylogeneticists have turned to vicariance events (for want of alternatives) to date speciation and diversi-fication of a variety of different taxa and clades" (Trewick and Gibb, 2010: 243). Yet phylogeneticists' use of tectonic calibrations would only be paradoxical if they had accepted chance dispersal as a valid mode of speciation to begin with. More interesting paradoxes in molecular biology include the discovery that many marine organisms show high, unsuspected levels of allopatric differentiation, despite having pelagic larvae (Heads, 2005a). The growing use of tectonic calibrations that Trewick and Gibb (2010) observed is not because of any lack of alternatives, as there is always a fossil record in a related taxon (even if this is remote in phylogeny or geography). Likewise, there is always a related taxon endemic to an island that can be dated. Despite these options being available, authors are now starting to use tectonic calibration, as they judge it to be a better method. Trewick and Gibb (2010) criticized several studies for using tectonic calibrations, although these studies could have erred, if at all, by not using *enough* tectonic calibrations. Trewick and Gibb themselves did not offer any suggestion as to how phylogenies should be calibrated.

Studies on New Zealand groups that have used tectonic calibration include work on freshwater fishes. Burridge et al. (2008) used geological dates of changes in river drainages and isolation of fish populations to calibrate rates of molecular evolution. As they wrote: "This method utilizes precise spatiotemporal disrup-tions of linear freshwater systems and hence avoids many of the limitations associated with typical DNA calibration methods involving fossil data or island formation." There are many problems in synthesizing tectonics and biogeography in this way, but the general approach is more realistic than the earlier methods.

Even paleontologists have agreed that tectonic-vicariance dating can be useful for calibrating phylog-enies. For example, a paper by Marshall (1990) included a section titled "The value of biogeography." In this, he proposed: "The dating of events that led to geographic isolation provides a way of estimating lower bounds on divergence times, which circumvents the need to deal directly with the fossil record and the difficulties associated with the taxonomic assignment of underived and fragmentary fossils." "However," he concluded, "for many groups there are no relevant biogeographic events that bear on times of origin...." This problem does not apply in New Zealand though, as most groups have well-defined boundaries that coincide with general biogeographic breaks and also with major geological fea-tures. These features have been studied in some detail by geologists, and dating the active phases of the features has been a main focus of enquiry. Thus, the increasing quantity of well-supported biogeographic data from the molecular work, together with the information on well-dated geological events, can be utilized to provide new syntheses.

Geology and geological calibration is not always straightforward. For example, some studies have calibrated phylogenies by assuming that island endemics were no older than the geological formation of their island. However, studies using independent calibrations show that many island endemics—whether on real islands or "habitat islands"—are much older than the island where they are now found (Heads, 2011). This is probably because former islands existed in the vicinity. For example, a biogeographic analysis of the Three Kings Islands in northern New Zealand accepted that the dynamic geological processes in the region mean that older lineages could be surviving there on younger islands, and island age may not correlate well with the age of endemics (Buckley and Leschen, 2013).

The best geological calibrations do not depend on stratigraphy and the age of strata or volcanic islands, but on underlying tectonic structures, such as the plate margins that have generated many islands. Continuous volcanism has occurred along the southwest Pacific plate margin ever since it began to migrate away from Australia and into the Pacific in the Cretaceous. This means it is likely that groups on the cur-rent, Cenozoic islands (Vanuatu, Fiji, Tonga, and the Kermadec Islands) have persisted as metapopulations along the arc ever since migration of the subduction zone and its arc began. New volcanic islands are always appearing along active subduction zones, and these are soon colonized from neighboring islands in the chain. In this way, clades can persist on the islands along an active arc, even if it is rifted apart.

How Old Are Modern Species?

The neodarwinian synthesis followed Matthew (1915) in proposing that most extant species originated in the Pleistocene, since 2.5 Ma (Mayr, 1942, 1963; Mayr and Diamond, 2001). This has been ques-tioned in recent years, but it is still sometimes accepted for New Zealand groups. In the genus *Myosotis*

(Boraginaceae), for example, one clade (sect. *Exarrhena* "austral clade") comprises 47 species from New Zealand, Australia, New Guinea, and South America, 42 of which are endemic to New Zealand. Meudt et al. (2015) suggested that the clade originated and diversified in the Pleistocene (but see Heads, 2014: 102). Likewise, for *Hebe* (Plantaginaceae), *Coprosma* (Rubiaceae), and *Celmisia* (Asteraceae), Burrows (2011b) wrote: "The underlying impetus for evolutionary radiation... was the provision of many, diverse habitats, through the dynamic climatic and geomorphic events [uplift, volcanism etc.] of the Plio-Pleistocene."

In contrast with the traditional idea of Pleistocene speciation, molecular clock studies have instead concluded that many species originated in the Miocene, dating back to 23 Ma. This is a step in the right direction, but the fossil-calibrated clock dates are still only minimum possible ages. Critical clock studies carried out in conjunction with biogeographic analyses suggest that species are even older.

Relevant New Zealand examples are seen in the moth family Micropterigidae. This group is of special interest as it is sister to all other Lepidoptera. The "basal" clade within Micropterigidae (the "Australian group") is endemic to Australia, mainland New Zealand, and New Caledonia (Gibbs, 2010, 2014). Based on the age of the family's oldest fossils (136 Ma) and the distribution patterns of the members, Gibbs (2014) concluded that "trans-oceanic dispersal has not contributed to, nor disrupted, [the moths'] current distribution pattern," and that trans-Tasman relationships in the family "represent vicariant processes that were probably established around the time of the opening of the Tasman Sea."

Micropterigidae are represented in New Zealand by the monotypic *Zealandopterix* and 18 species of *Sabatinca*. In *Sabatinca*, Gibbs (2014) found seven pairs of sister species, and, apart from the case of *S. aemula/S. chrysargyra*, clock studies showed that members of each pair diverged before the Miocene (Gibbs and Lees, 2014). Thus, Gibbs (2014) concluded that the species do not conform "to the prevailing mountain-building or glaciation hypotheses offered by many recent phylogeographic studies of New Zealand biota, e.g., the review by Wallis and Trewick (2009)."

The Modern Synthesis argued that most species were Pleistocene, while recent molecular work and paleontology suggests they are often Miocene, but biogeographic analyses presented in this volume suggest that some are even older—Paleogene or even Cretaceous. Many trans-Tasman and trans-Pacific affinities within single species would be explicable if the species are this old, but is there any other evidence?

The monocot genus *Ripogonum* (Ripogonaceae) is a rainforest liane, in which early Eocene fossil leaves from Patagonia (*R. americanum*) are "closely comparable" (Carpenter et al., 2014: 1) or "virtually identical" (p. 4) to the leaves of the extant species *R. album* of Australia and New Guinea and *R. scandens* of New Zealand. "The likeness between fossil and extant specimens extends to details of fine venation, including the presence of ultimate vein loops. Indeed, we contend that a modern leaf [of *R. americanum*] with the same details as preserved in the fossils would readily be accepted as belonging to either of these two extant species" (Carpenter et al., 2014: 3). A common fossil taxon in Miocene deposits of Otago "appears superficially to be close to the extant New Zealand *Ripogonum scandens*" (Conran et al., 2015). Fossil material of *Ripogonum* is also known from the early Eocene of New Zealand, and this supports the idea that Ripogonaceae display ancient Gondwanan connections, as in many other Liliales (Carpenter et al., 2014).

Some hard-bodied, marine groups have a good fossil record, and among these are individual species with a long record. For example, the New Zealand marine gastropod *Scissurella condita* (Scissurellidae) is known as fossils from Early Jurassic strata through to Miocene strata (Geiger, 2012), indicating that the species existed for at least 150 Myr.

Molecular clock studies have also indicated that species can have Paleogene or even Cretaceous ages. The expected phylogeny for the New Zealand frog genus *Leiopelma* is: (*L. hochstetteri* (*L. archeyi* (*L. hamiltoni, L. pakeka*))). Carr et al. (2015) calculated the divergence date between *L. hochstetteri* and *L. archeyi*—that is, the age of the *L. hochstetteri* lineage and of its sister group, as 67 Ma, in the late Cretaceous. This was based on fossil calibrations and so represents a minimum possible age for the species *L. hochstetteri*. This is much older than suggested by one recent view: "species longevity for higher organisms, including vertebrates, varies between about four and ten million years" (Campbell, 2013: 46).

Assessing Analytical Methods in Biogeography: Do They Explain Classic Problems or Lead to New Ideas?

> I quite agree with what you say on Lieut. Hutton's review (who he is, I know not): it struck me as very original: he is one of the very few who see that the change of species cannot be directly proved & that the doctrine must sink or swim according as it groups & explains phenomena. It is really curious how few judge it in this way, which is clearly the right way.

> Darwin in a letter to J.D. Hooker, April 23, 1861; see Burckhardt et al., 1994. F.W. Hutton was Professor of Natural Science at what became the University of Otago, New Zealand.

Neither the chance dispersal model nor the vicariance/range expansion model can be proved or disproved in a mathematical sense, but they can be assessed by the efficiency with which they group and explain phenomena, and provide solutions to concrete biogeographic problems. Does chance dispersal provide coherent, testable explanations for common, repeated patterns? Or does it just explain them away? Does the theory lead to any novel, bold predictions that go beyond what is already accepted as true? Crisp et al. (2011) observed that "developing synergies between biogeography, ecology, molecular dating and palaeontology are providing novel data and hypothesis-testing opportunities." There is no doubt that the synergies are providing impressive technical advances and a mountain of new data, but are they leading to any new ideas?

Crisp et al. (2011) suggested one possibility: in the new studies "understanding of how lineages became distributed as they are has changed dramatically... it has been learned that the geographical evolution of biota has been driven by a greater diversity of processes with a more complex history than under a simple vicariance (or dispersal) paradigm." But this is not new; all workers have accepted both vicariance and dispersal, at least in some sense. Dispersal theorists have accepted that vicariance has occurred in some cases of local endemism (Mayr, 1965, as quoted earlier), and no one has ever suggested that evolution works by pure vicariance, as that would lead to every single area having a single, endemic clade.

Crisp et al. (2011) also wrote that new methods have led to "renewed recognition that ecological factors (e.g., climatic tolerance and dispersal limitation) underlie deep historical events (i.e., speciation, extinction and distributional change)...." This is not a new concept; the same views are presented in a founding document of the Modern Synthesis (Huxley, 1942, in the sections on "ecological speciation").

As another possible novelty, Crisp et al. (2011) proposed that "new approaches are challenging the classic 'Gondwana paradigm'... Surprisingly, most trans-oceanic plant disjunctions... and many of those in animal taxa... have been determined to be asynchronous or too young to be fully explained by the break up of Gondwana." Again, this repeats the Modern Synthesis view: if the southern continents were ever joined, "they apparently separated too long ago to leave any recognizable traces of the union in the distributions of existing plants and animals..." (Darlington, 1964: 1085). The "new" views are in fact the same as the Modern Synthesis position. This is not surprising, as they are both based on the same conversion of fossil-calibrated clade ages (minimum dates) into maximum dates, and on the same assumptions about centers of origin and speciation by chance dispersal.

To summarize, molecular work over the last decade or two has made stunning technical advances and produced a fabulous wealth of new data, but in most studies, the concepts used to interpret these data are inherited from the Modern Synthesis. This may be changing though. Since the 1970s, the process of vicariance has been accepted by many authors, and a growing number of contemporary authors are using vicariance in work on molecular dating.

Analysis of Space and Time in Evolution

Many contemporary studies can be described as "plug-and-play" biogeography (L. Parenti, pers. comm. 2012). They calculate (1) a phylogenetic tree; (2) a model of the spatial evolution, using programs such as DIVA or LAGRANGE; and (3) a timeline for the phylogeny, using a program such as BEAST. DIVA and

LAGRANGE will provide a center of origin (e.g., at the location of a paraphyletic basal grade), and, by stipulating appropriate priors, BEAST can be made to support young ages for clades. The results support the Modern Synthesis: the fossil record is the best (or only) guide to the timeline of evolution, modern intercontinental groups developed after the Mesozoic, and distribution is the result of chance dispersal from localized centers of origin.

If inferences made in molecular phylogenetic studies about both time (using fossil-based Bayesian dates) and space (using ancestral area programs) are so flawed, why does the present book utilize these same studies? The reason is that the data they provide on clade *phylogeny* and *distribution* are invaluable. The molecular revolution means that biologists are no longer forced to depend on fossils to establish the timeline of evolution, as they can harvest the new data on clade distributions. These can be integrated with the results from an equivalent revolution in tectonics, brought about by new radiometric dating techniques. In this way, the most reliable findings of molecular biology can be combined with geology to move beyond the Modern Synthesis and its dependence on chance to explain repeated patterns.

What is fact and what is hypothesis in biogeography? The "facts" presented here for discussion are distributions of molecular clades or, if these are not available, distributions of morphological clades. Of course, all these represent hypotheses, and there are many aspects that are still unclear. Sampling is often sparse, and statistical support of proposed groups is often imperfect. Nevertheless, these groups are accepted here as facts, at least for the purposes of discussion.

Beyond the level of phylogenies of groups are hypotheses about biogeography. How can these be tested? Science is supposed to "test" hypotheses, but what counts as a test or a valid test? Medieval scholars tested hypotheses by referring to scripture. Many modern programs that are used to analyze biogeographic patterns (DIVA, DEC, etc.) do not provide good tests of hypotheses about processes (vicariance, dispersal extinction), because they are based on flawed assumptions.

In this book, hypotheses about general distribution patterns and their origins are tested by making comparisons among large numbers of plants and animals, and with the major features of structural geology. The main question is: "Can a coherent synthesis of evolution be constructed in which the clades are dated as no younger than their youngest fossil-calibrated age, and the standard phylogenetic–geographic breaks are explained by major tectonic events in the same localities?" As a first step to answering this question, the next chapter summarizes the tectonic history of New Zealand.

3

New Zealand Geology

In this chapter we take a break from biology and instead consider the main aspects of New Zealand geology, from the Mesozoic through to the present. This history forms the basis for the discussions of plant and animal distributions that are given, along with additional geological data, in subsequent chapters.

The New Zealand Plateau or Zealandia

New Zealand is the subaerial remnant (~10%) of an otherwise submerged block of continental crust, the New Zealand Plateau (Farquhar, 1906) or Zealandia (Mortimer and Campbell, 2014). The continent is well defined by the 2000 and 2500 m isobaths and extends from the subantarctic islands to New Caledonia (Figure 3.1). This large block straddles the active margin between the Indo-Australian and Pacific plates. (In the figure, the barbs along the subduction zones are on the overriding plate and the points on the barbs indicate the direction of subduction beneath it.) The Indo-Australian plate is moving northward, and the Pacific plate is moving eastward. The New Zealand part of the plate margin comprises two subduction zones of opposite polarity, the Hikurangi margin in the north and Puysegur–Fiordland margin in the south. These are connected by a transform fault system made up of the Alpine fault zone and the Marlborough fault system. Subduction of the Pacific plate beneath the North Island terminates at the southern end of the Hikurangi Trough, where the thick continental crust of the Chatham Rise intersects the margin. (See the "Glossary of geological terms" for definitions of "subduction," "transform fault," and others.)

The migration of subduction zones is one of the most important processes in southwest Pacific tectonics. Figure 3.1 shows the series of basins that have opened behind the Australia/Pacific plate margin as this has migrated eastward into the Pacific. Reading from the oldest to the youngest, these are the Tasman, South Loyalty, Coral Sea, New Caledonia, North Loyalty, South Fiji, North Fiji, and Lau basins.

In addition to the current, active subduction zones, Figure 3.1 also indicates extinct subduction zones (as lines with open barbs) cutting through the North Island and along the northern margin of the Chatham Rise. The earlier activity at these zones has brought about profound changes in geology and biology, and their history is discussed in more detail in the following sections.

Another important feature in the region is the Louisville Ridge (Figure 3.1), a combination of short seamount chains, isolated peaks, and flat guyots (Pushcharovsky, 2011). The seamounts at the northern end of the Louisville Ridge are the oldest, and these are now being subducted beneath the Indo-Australian plate. The guyot morphology suggests that the Louisville Ridge seamounts were emergent in the past, and the presence of vesicular basalts there also indicates subaerial eruption (Frost, 2010). These former islands, together with former islands on the Hikurangi Plateau, and islands such as Rapa and Tahiti in southeastern Polynesia, would have provided habitat for the many groups endemic to the New Zealand–southeastern Polynesia region.

Most of Zealandia is made up of continental crust that was rifted from mainland Gondwana. It has undergone dramatic deformation through the Cenozoic, and its modern shape is very different from the one it had when it separated from Gondwana. In addition, the original Zealandia did not include ophiolite sequences that accreted in the Cenozoic; these make up large parts of the North Island and New Caledonia. As for the large Hikurangi Plateau, it is not clear whether or not Zealandia, when it broke away from Gondwana, included this structure. Different authors have suggested that the Hikurangi

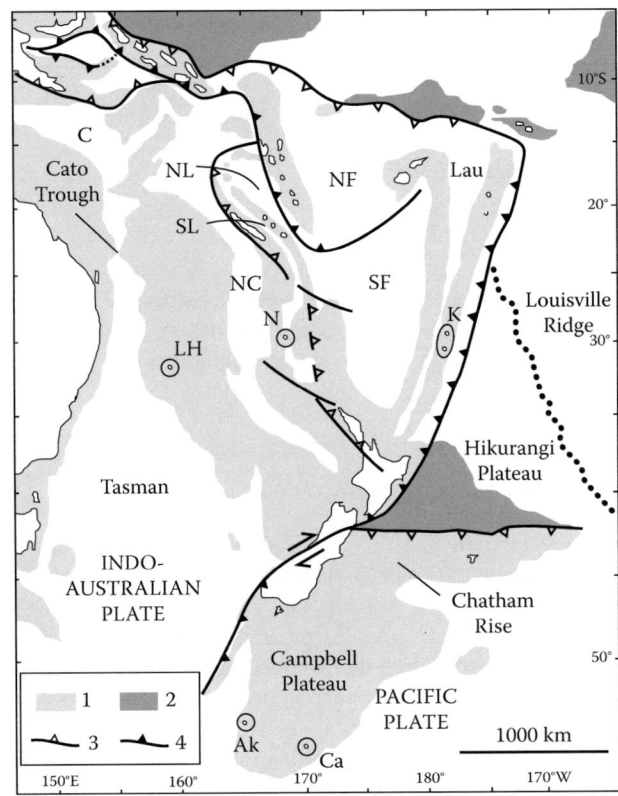

FIGURE 3.1 The southwest Pacific tectonic setting (Schellart et al., 2006; Frost, 2010; Cato Trough area from Mortimer and Campbell, 2014): 1, continental/island arc crust; 2, oceanic plateau; 3, inactive subduction zone; 4, active subduction zone. Ak, Auckland Islands; Ca, Campbell Island; K, Kermadec Islands; LH, Lord Howe Island; N, Norfolk Island. Basins between the Tasman and Lau basins as follows: C, Coral Sea; NC, New Caledonia; NF, North Fiji; NL, North Loyalty; SF, South Fiji; SL, South Loyalty. (Reprinted from *Earth-Sci. Rev.*, 76, Schellart, W.P., Lister, G.S., and Toy, V.G., A Late Cretaceous and Cenozoic reconstruction of the Southwest Pacific region: Tectonics controlled by subduction and slab roll-back processes, 191–233, Copyright 2006, with permission from Elsevier.)

Plateau arrived at 86 Ma (Late Cretaceous, around the same time as breakup) (Worthington et al., 2006), 15–20 Myr before this (Reyners et al., 2011), or 10 Myr after (Cooper and Ireland, 2014).

It is often stated that New Zealand is a center of endemism (e.g., Myers et al., 2000), but this is problematic. The political entity, which includes the Kermadec Islands but not Norfolk Island, for example, has no biogeographic significance. Likewise, the geological entity Zealandia—the New Zealand Plateau—does not correspond with an important area of endemism. Most groups that are distributed right across it, from the Campbell Plateau in southern New Zealand to New Caledonia, are also known from Fiji (at least as fossils, e.g., the bird *Coenocorypha*) or eastern Australia (e.g., the plant *Dracophyllum*). Nevertheless, rifting at the margins of Zealandia, both before and during breakup (in the west at the Tasman Sea and in the south between Campbell Plateau and Marie Byrd Land, Antarctica), probably did cause many disjunctions.

The part of Gondwana that later rifted off as Zealandia was built up by the accretion of terranes—fault-bounded blocks of crust with independent histories—to the margin of the supercontinent. The terrane accretion occurred by addition of incoming terranes during phases of plate convergence. The allochthonous, accreted terranes of the New Zealand mainland are mapped in Figure 3.2 (from GNS Science, 2016); the autochthonous sedimentary and volcanic rocks that covered parts of the basement through the Cenozoic are not shown. (Allochthonous rocks have formed elsewhere and been moved to their current position, autochthonous rocks have developed *in situ*.)

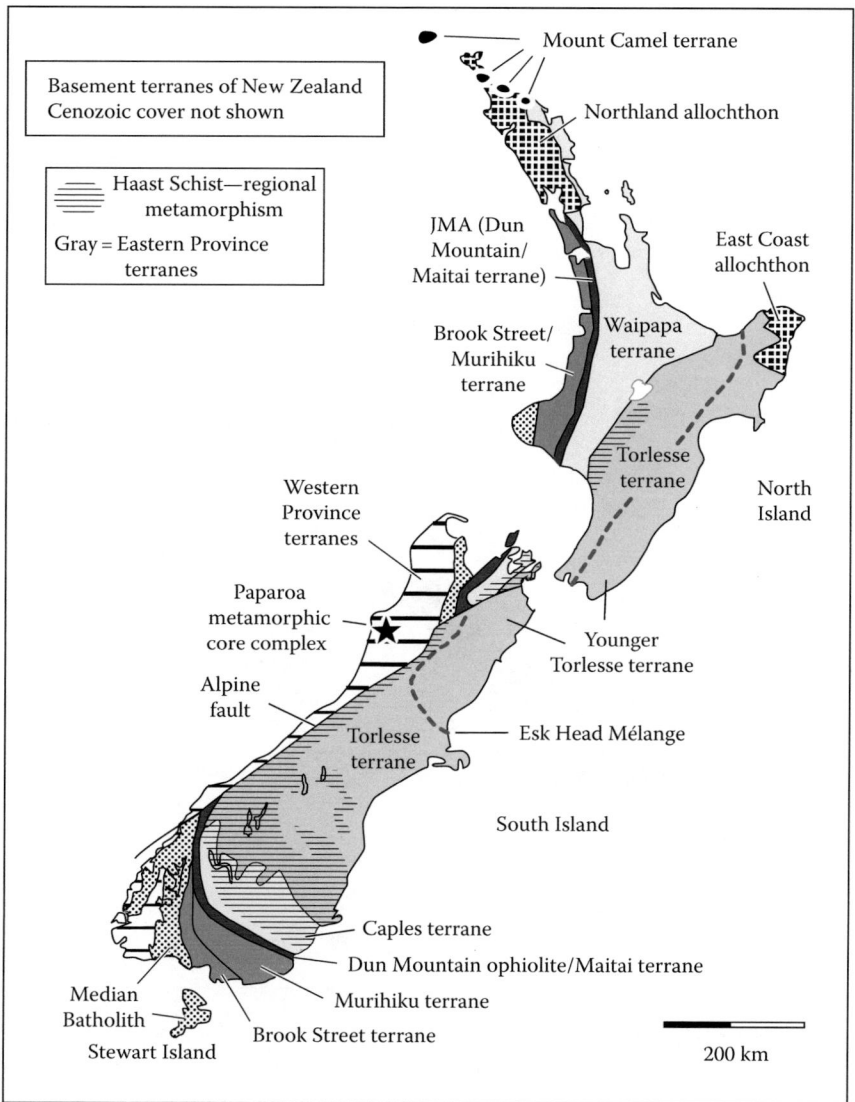

FIGURE 3.2 Basement terranes of New Zealand (Late Cretaceous and Cenozoic cover not shown). JMA, Junction magnetic anomaly. All the terranes are allochthonous, while the Median Batholith and the Haast Schist are autochthonous features. The Haast Schist group is a belt of regional metamorphism that comprises the Otago, Alpine, Marlborough, and Kaimanawa Schists. The accretion of the allochthonous terranes, together with the emplacement of the Median Batholith and the Haast Schist, has developed during convergence. The Paparoa metamorphic core complex represents the beginning of a new tectonic regime of extension, beginning at ~100 Ma. (Simplified from: www.gns.cri.nz/Home/Our-Science/Earth-Science/Regional-Geology/Geological-Origins/Basement-terranes-of-New-Zealand.)

The main topographic features of the New Zealand mainland are the three large islands, the two straits that separate them, the volcanoes of the North Island (some active), and the Southern Alps of the South Island. This range reaches 3754 m (12,316′) at Aoraki/Mount Cook, the highest mountain in New Zealand.

The South Island is of special interest for biogeography because of its biological and geological diversity. The main geological structures include a plate margin (the Alpine fault), a long-lived belt of Mesozoic magmatism (Median Batholith), and a disrupted belt of ultramafic rocks (the Dun Mountain ophiolite in the Maitai terrane). In the North Island, the Median Batholith and the Dun Mountain ophiolite are buried under younger rock, while the plate margin lies off the east coast (Figure 3.2).

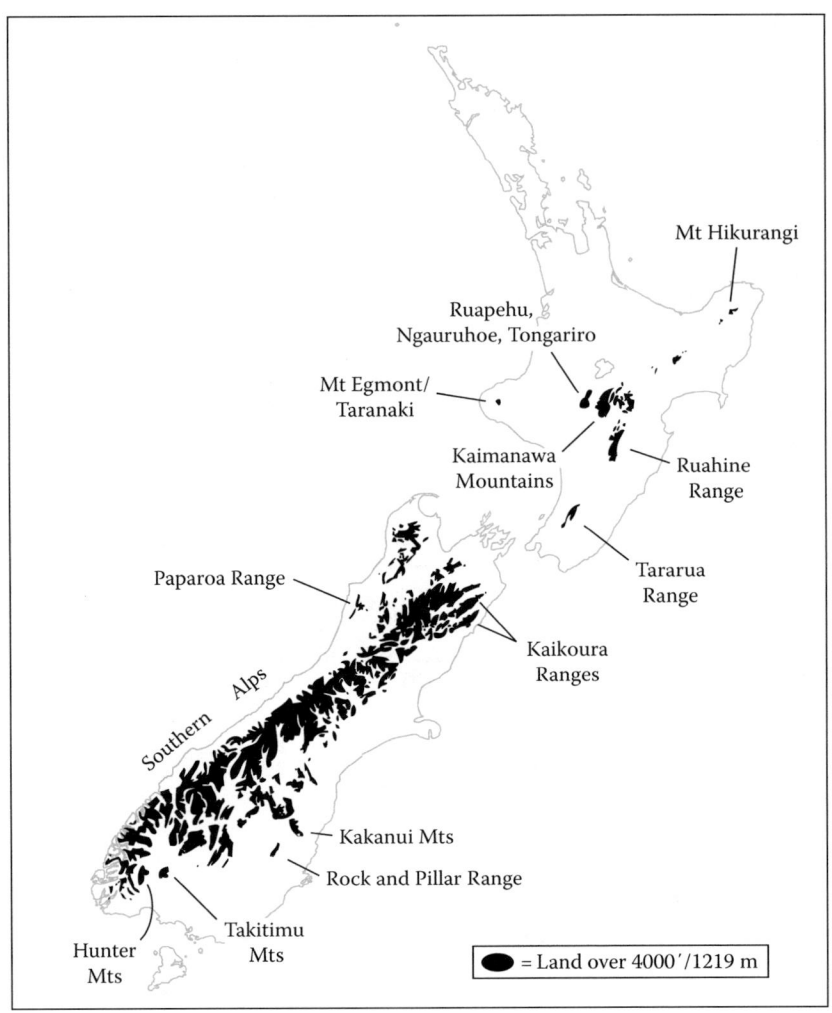

FIGURE 3.3 New Zealand elevation.

The province names are often used in the following discussions, and the informal provinces are shown in Figure 3.4.

The Paleozoic history of New Zealand is not considered here, as the main biogeographic patterns are considered to have developed after that time. The Mesozoic–Cenozoic history includes three main episodes, consecutive phases of convergence, extension, and convergence (Mortimer, 2006):

1. 230–100 Ma (Triassic–mid-Cretaceous): **Convergence** and subduction beneath the "New Zealand" part of the Gondwana margin. This caused terrane accretion, Median Batholith magmatism, and the Rangitata orogeny.
2. 100–25 Ma (mid-Cretaceous–Oligocene): **Extension** and rifting. This developed into seafloor spreading to the south, west, and north of New Zealand, from 83 Ma (mid-Cretaceous) to 25 Ma (latest Oligocene). From 83 to 55 Ma, Zealandia broke away from the rest of Gondwanaland with the opening of the Tasman Sea and Southern Ocean basins.
3. 25–0 Ma (latest Oligocene–present): **Convergence** and subsequent mountain building.

FIGURE 3.4   Districts of New Zealand (there are no formal provinces, but the regions shown are widely used).

A critical turning point in New Zealand's tectonic history occurred at the Early Cretaceous–Late Cretaceous boundary (~100 Ma) (Figure 3.5). Particular features associated with this break also have special significance for biogeography, as discussed in later chapters, and so it is stressed here. Prior to ~100 Ma, convergence and terrane accretion in the Rangitata orogeny uplifted ancestral New Zealand and created a vast, Andean-style mountain range. This tectonic regime came to an end at ~100 Ma, when there was a rapid switch to extension, rifting, and continental drift. Reverse faulting was replaced with normal faulting (Figure 3.6). (Reverse faulting tends to cause shortening and thickening of the crustal layers and can lead to mountain uplift, while normal faulting causes crustal extension and thinning, with valley formation.)

The mid-Cretaceous change in tectonic regime from convergence to extension is marked by a major angular unconformity that separates the older, strongly deformed "basement" rocks (the terranes mapped in Figure 3.1) from the younger, less deformed "cover" strata (Laird and Bradshaw, 2004). (Unconformities are physical surfaces dividing superposed strata of different ages. They are caused by events such as uplift that either prevent deposition or remove deposited sediment. Although unconformities represent strati-graphic absences, they are of special tectonic and biogeographic interest.)

Mortimer and Campbell (2014: 124) wrote that the basement rocks in New Zealand are "low, often hidden and occur beneath other more visible rocks. They are generally old and hard, and include plutonic and metamorphic rocks. Cover rocks drape over and cover older basement rocks. They are usually young and soft, and are mainly layered, sedimentary rocks."

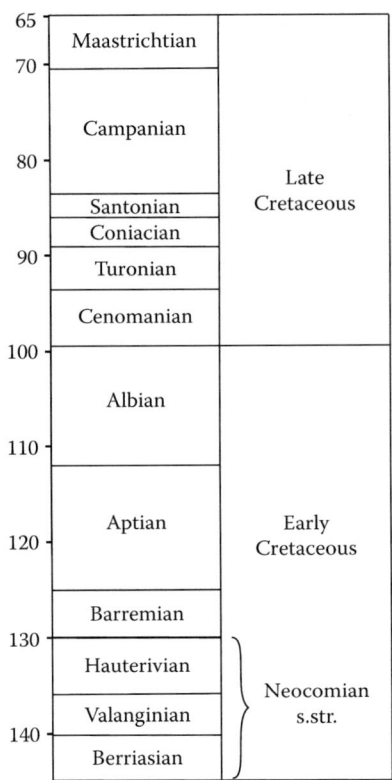

FIGURE 3.5   The ages of the Cretaceous period.

Fission track dating indicates three major phases of uplift in New Zealand since the Paleozoic (Kamp et al., 1989). At the Jurassic–Cretaceous transition, the Otago Schist belt was uplifted in association with the Rangitata orogeny. Western parts of the Otago Schist were uplifted again in the Late Cretaceous. Overall, the amount of Mesozoic uplift ranged from minimal amounts north of Arthur's Pass, to ~3 km near Mount Cook, to 10 km in the south at Lake Wanaka (Tippet and Kamp, 1993). Finally, in the Neogene, there has been widespread uplift in the Kaikoura orogeny (from ~10 Ma onward).

Cretaceous events are often seen as the key to understanding New Zealand geology (Campbell and Hutching, 2007: 112). It is suggested here that they are just as significant for understanding biogeography and evolution in the region. The Cretaceous period is known worldwide for its warm climate, shallow epicontinental seas, continental rifting and breakup, and giant oil fields. In the mid-Cretaceous, the upper ocean was 8°C–20°C warmer than in modern times (Bice et al., 2003), and extensive marine incursions are recorded worldwide. In addition to its geological interest, many New Zealand features that were emplaced or activated in the mid-Cretaceous (including plutons, metamorphic core complexes, normal faults, synclines, basins, and hydrocarbon reservoirs) coincide with well-known biogeographic boundaries.

Terrane Accretion and Orogeny in Mesozoic New Zealand

As a result of plate convergence and subduction at the Pacific margin of Gondwana, the New Zealand Plateau built up by terrane accretion through the Paleozoic and Mesozoic (Wandres and Bradshaw, 2005). The terranes formed in the ocean off Gondwana and were made up of sediments deposited on the ocean floor, other ocean floor materials, seamounts, and volcanic islands. Following the formation of the

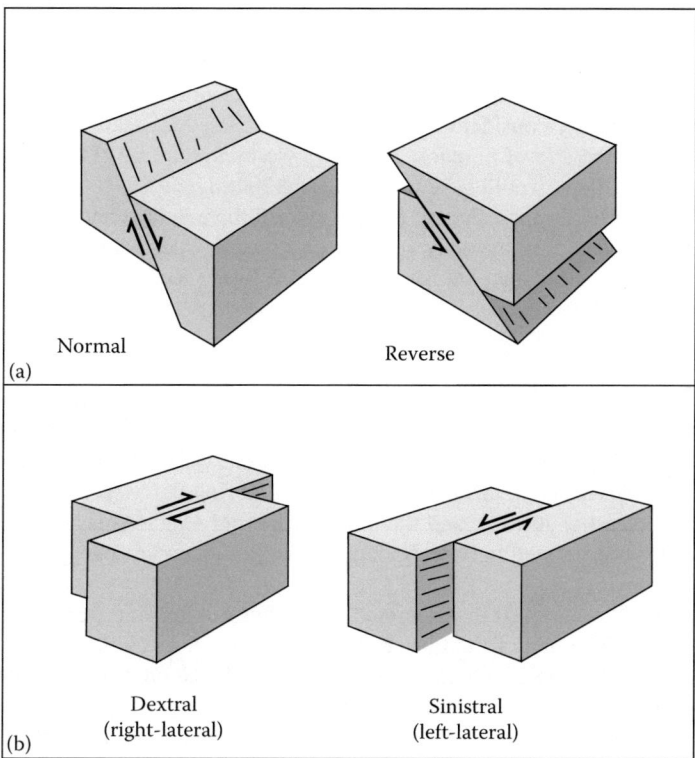

FIGURE 3.6 Types of faults: (a) dip-slip faults and (b) strike-slip faults (see also the "Glossary of Geological Terms"). (After van der Pluijm, B.A. and Marshak, M., *Earth Structure: An Introduction to Structural Geology and Tectonics*, 2nd ed., Norton, New York, 2004.)

terranes, they were accreted to the Gondwana mainland. Accretion occurred at a long-lived subduction zone that dipped beneath the Gondwana margin (Australia–New Zealand–western Antarctica–western South America), and this belt formed the Australides orogen. A well-known section of the margin in the South Island preserves a series of juxtaposed terranes (Figure 3.1). Westward subduction beneath most of the "New Zealand" region ceased at ~100–105 Ma in the mid-Cretaceous, and did not begin again until 80 Myr later.

The basement terranes of New Zealand can be grouped into three provinces (Figure 3.2): the allochthonous Western Province (older rocks), the allochthonous Eastern Province (younger rocks), and an autochthonous magmatic belt, the Median Batholith that developed along the margin of the two allochthonous provinces.

Accretion of the Western Province Terranes

The Western Province terranes are exposed in the western South Island. They formed through the Paleozoic (from the Cambrian to Carboniferous), and were accreted along the Gondwana margin during Paleozoic orogeny.

Emplacement of the Median Batholith

The Median Batholith of western New Zealand is a belt of intrusive rocks that represents a long-lived, autochthonous magmatic arc. This was active along the Gondwana convergent margin from the Triassic (230 Ma) to the mid-Cretaceous (105 Ma). It is mainly composed of Cretaceous granite (Campbell and Hutching, 2007), and the largest pulses of batholith growth occurred toward the end of its activity, from 145 to 105 Ma (Mortimer and Campbell, 2014: 135). The last component to be emplaced was the Separation Point Suite of Fiordland and Nelson (130–105 Ma).

By ~120 Ma, both magmatism and shortening of the crust had thickened the Zealandia part of the Gondwana margin to at least 45 km (Klepeis and King, 2009). Continental crust has a lower density than oceanic crust, and so buoyancy considerations mean that the overthickened crust along the continental margin would have been the site of a major, Andean-style mountain range (Beggs et al., 2008). This range, the Zealandia Cordillera, would have been at least 800 km long.

The partial melting of Gondwana's Pacific margin caused the emplacement of the Median Batholith and, at the same time, the felsic Whitsunday volcanic province in Queensland. The two make up a large igneous province, a linear belt of magmatism that extends from Queensland through New Zealand to West Antarctica (Thurston Island).

Accretion of the Eastern Province Terranes

The New Zealand Eastern Province terranes were formed over a long period from the Permian to Cretaceous and were accreted to the mainland in the Mesozoic. Around the Pacific Gondwana margin as a whole (the Australides orogen), terranes were accreted during the Mesozoic in two main episodes: one in the later Triassic/earlier Jurassic and the other in the mid-Cretaceous (Vaughan and Livermore, 2005). The first event is represented in New Zealand by the Rangitata I Orogeny and in West Antarctica by the Peninsula Orogeny.

New Zealand Eastern Province terranes accreted to the Western Province terranes and lie outboard of them, on the Pacific side. The first of the Eastern Province terranes to collide was the Brook Street terrane (Figure 3.2), representing an island arc, that docked with the Gondwana margin in the Triassic; the junction is now covered by the Median Batholith (Davey, 2005).

A second phase of accretion of Eastern Province terranes occurred through the later Jurassic and Early Cretaceous. This coincides with the Rangitata II Orogeny in New Zealand and the Palmer Land event in the Antarctic Peninsula (Vaughan and Livermore, 2005). In the first event of this orogeny, rocks of second arc-associated terrane, the Murihiku, were thrust across the eastern margin of the Brook Street island arc terrane during the Middle to Late Jurassic (~160–140 Ma) (Landis et al., 1999). The Murihiku terrane had been formed by volcaniclastic sedimentation on the flanks of an arc over a period of about 120 million years (Late Permian to Early Cretaceous, ~260–140 Ma), but the source of the Murihiku terrane was probably *not* the Brook Street terrane.

The main feature of the Murihiku terrane is a great fold, the Southland syncline, that is one of the most prominent geomorphic features in the southern South Island. The fold is asymmetric, with a subvertical northeastern limb and a gently dipping to subhorizontal southwestern limb that includes smaller, "parasitic" folds. The syncline formed in the mid-Cretaceous. This date and the southwest vergence of the fold suggest that it developed during accretion of the Caples and Torlesse terranes. The Kawhia syncline is an equivalent fold in the North Island component of the Murihiku terrane.

Outboard of the Murihiku is a third arc-derived terrane, the Maitai, that includes the Dun Mountain ophiolite belt. The juxtaposition of the Murihiku and Maitai–Dun Mountain terranes probably took place in the Cretaceous (Turnbull and Allibone, 2003). In the South Island, the ophiolite belt is associated with the Junction Magnetic Anomaly. In the North Island, the ophiolite is not exposed, but the magnetic anomaly can be detected along the west of the North Island and out to sea west of Kaitaia. The ophiolite is assumed to be present here, buried beneath other rocks (the southern part is shown in Figure 3.2).

The Maitai–Dun Mountain terrane was formed in the Permian–Triassic, and geochemical signatures of the ophiolite indicate both spreading ridge and subduction zone affinities (Jugum et al., 2006). The Dun Mountain belt has been tectonically emplaced with the Maitai terrane during the Rangitata orogeny. In the Neogene, the ophiolite, along with all the other terranes, has been bent and broken by substantial deformation.

Outboard of the Maitai–Dun Mountain terrane are the Caples terrane and the Torlesse terrane complex (Figure 3.2), both composed of Permian to Early Cretaceous graywackes and argillites. The more inboard Caples was island arc derived, while the Torlesse is made up of deep-sea sediments (these are quartzo-feldspathic, indicating a continental source) and seamount material. The latter component is negligible in terms of volume, but has important implications for the history of terrestrial and reef communities. Collision between the Torlesse and Caples terranes led to regional metamorphism (the Haast Schist belt) overprinting the terrane boundary. In Otago, the schist represents the deepest parts of a late Paleozoic–Mesozoic accretionary wedge that have been exhumed (Mortimer, 2003). Juxtaposition of the Torlesse and Caples terranes with the already accreted Brook Street, Murihiku, and Maitai terranes had occurred by the end of the Early Cretaceous.

The culmination of the Rangitata II Orogeny is evident in widespread Aptian and Albian angular unconformities in sedimentary rocks of eastern New Zealand, and these all indicate erosion and uplift. Magmatism and contractional deformation culminated at 116–105 Ma, with a last deformation at ~101–102 Ma (Albian). Coeval mid-Cretaceous magmatism and metamorphism is also recorded in the east Pacific, in Patagonia and in Mexico (Vaughan and Livermore, 2005).

Geologists have emphasized that the Median Batholith and the Eastern Province terranes do not represent a single, Mesozoic subduction–arc complex (Mortimer et al., 2002). The terranes have originated in different areas and accreted to the Gondwana margin at different times. The accretion tectonics were complex and included strike-slip (horizontal) movement and even extension, as well as convergence with uplift. All three processes can have important effects on the landscape and its living populations.

To summarize, the Rangitata orogeny was driven by subduction and terrane accretion, and led to uplift and other changes along the eastern edge of Gondwana. The process culminated through the Early Cretaceous. The orogeny created a major Andean-style mountain range in a belt of land that later formed the core of modern New Zealand. Rifting followed in the Late Cretaceous, and the two processes—orogeny and rifting—turned the topography of the New Zealand region inside out, producing land where there had been sea, and sea where there had been land. It is suggested in this book that the events produced a similar, profound revolution in the biota. Yet the Rangitata orogeny was not the "origin" of New Zealand in an absolute sense, just the last major modernization of a region and its biota. Through the orogeny, terranes and their biotas were juxtaposed and amalgamated; they were also deformed, metamorphosed, and uplifted. Following the orogeny, the proto-Zealandia region underwent a phase of extension and rifting that continued for as long as 20 Myr before the crust broke completely and seafloor spreading began in and around the Tasman Basin. This calved-off and isolated sinuous continental ribbons that are now aligned more or less parallel to the eastern Australian margin, and are each separated from the others by new oceanic basins (Figure 3.1) (Whattam et al., 2008).

Subduction Zones and Island Arcs

The boundaries of accreted terranes represent old, convergent plate margins—former subduction zones where the terranes have docked with other terranes and orogeny has developed. Subduction zones (Figure 3.7) are "interior Earth systems of unparalleled scale and complexity" (Stern, 2002). They form at convergent plate margins, where they generate volcanic arcs. Depending on the context, this can take place either in continental crust (e.g., in the Median Batholith or along the Andes) or in oceanic crust (producing island arcs). Subduction zones are also sites of collision between blocks of continental crust (with low relative density), volcanic arcs, and volcanic plateaus. Sometimes these are large and buoyant enough to choke the subduction. Older, denser oceanic crust and most of its seamounts will be subducted, although subducting seamounts can cause large earthquakes.

Subduction zones are the most important sites of crust removal as well as crust generation, with the latter caused by the volcanism that occurs above the subduction. Among the most obvious products of the southwest Pacific subduction zone are the island arc systems. These are enormous tracts of thickened crust that are hundreds to thousands of kilometers long, several hundred kilometers wide, and up to 35 km thick (Stern, 2010). Continental crust, as in Zealandia, is a mosaic of orogenic belts composed of island arcs and seafloor rocks that have been welded together.

Convergent margins are also important for biological evolution and coincide with important areas of endemism. "Piling up" of biodiversity at the southwest Pacific subduction zone has been suggested for marine mollusks of New Caledonia (Marshall, 2001), and this has probably also occurred in the New Zealand region. Island arcs can accrete at convergent margins and orogens, but they can also undergo

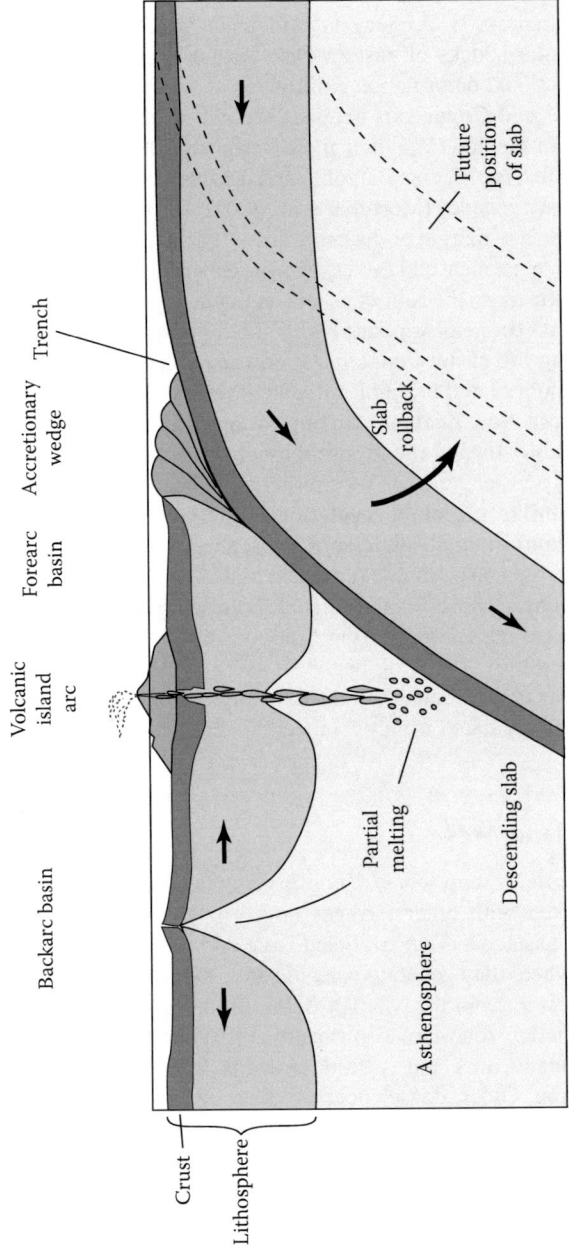

FIGURE 3.7 Cross section through a subduction zone, with a descending slab (plate) of lithosphere and a backarc basin with its spreading center (ridge). The lithosphere forms the tectonic plates, and comprises the upper mantle and the crust. Small arrows indicate plate motion the large arrow indicates slab rollback. The future position of the slab following rollback is indicated in broken lines. (Based on van der Pluijm, B.A. and Marshak, M., *Earth Structure: An Introduction to Structural Geology and Tectonics*, 2nd ed., Norton, New York, 2004.)

extension, displaying evidence such as normal faults, intra-arc rifts, and backarc basins, and these features often coincide with breaks in marine and terrestrial clades.

Several subduction zones, some extinct, are known in the southwest Pacific (Figure 3.1), and these are thought to have been derived from a single initial belt. The different zones have been active at different times through the Cenozoic, and their history is complex (Schellart et al., 2009). Controversial topics include their location, polarity, and lateral migration, the possible subduction of ocean basins, the mechanism by which ophiolite sequences have been emplaced, and the causes of widespread Cenozoic magmatism located far from the plate margin. All these processes have implications for biological evolution in the region.

The Culmination of Convergence and the Switch to Extension at ~100 Ma

The crust in the New Zealand region shows evidence of great thickening and shortening—effects of accretion and orogeny—but there have also been phases of large-scale extension and continental rifting. The most obvious sign of the latter is the development of half-grabens, but in addition much of the crust is thinner than in unextended orogens, and the inferred Mesozoic arc–trench gap is greater (Mortimer et al., 2002).

Most convergent margins undergo cycles, with phases of shortening and thickening of the crust followed by phases of crustal extension and thinning. One such cycle can take several tens of millions of years to complete (Klepeis and King, 2009). The peak of thermotectonic activity on the Zealandia margin occurred from 130 to 100 Ma; this represented the end of convergence and, at the same time, the beginning of breakup in this part of Gondwana (Mortimer and Cooper, 2004). Following the switch from convergence to extension at 100 Ma, Zealandia underwent continental rifting for 20 Myr until this developed into complete fracturing of the crust by seafloor spreading (84 Ma). During the long-lasting, 20 Myr phase of predrift extension, the Andean-style Zealandia Cordillera that had formed during convergence was destroyed by rifting and subsidence. The formation of the cordillera and its subsequent destruction would have had a great impact on the regional biota, before the modified biota was then rifted apart by seafloor spreading.

To summarize, convergence along the Gondwanan margin operated until ~100 Ma, when subduction ceased and extension began. The reason for this switch—the key event in New Zealand's history—is still unresolved, but the following mechanisms have been suggested.

Arrival of a Spreading Ridge at the Gondwana-Pacific Subduction Zone

A spreading center (divergent plate margin) arrived at the Zealandia region in the mid-Cretaceous, and the possible effects of this have often been discussed. The Mesozoic–Cenozoic history of the Pacific region is dominated by the growth of the Pacific plate. The plate formed in the Jurassic at a triple junction of three spreading ridges in the Cook Islands region (Figure 3.8). To begin with, the plate was surrounded by the Izanagi, Farallon, and Phoenix plates (Müller et al., 2008). With continued activity on the three spreading ridges that bounded it, the Pacific plate underwent tremendous growth, which was accommodated by subduction of the Phoenix plate beneath Gondwana. At around 100 Ma, the Pacific–Phoenix spreading ridge converged on the trench at the Zealandia margin. The approach, collision, and subduction of the spreading ridge at the trench could have ended subduction and led to the development of extension in the region (Bradshaw, 1989). (In a similar way, Mortimer et al., 2006, speculated that the West Wishbone Ridge, east of Chatham Rise, was rifted following the arrival of an intraoceanic spreading center after 115 Ma.) Impingement of this active spreading center against the Gondwana margin and its southwestward propagation led to widespread magmatism and extension, and ultimately to the continental split between Zealandia and West Antarctica across basement trends.

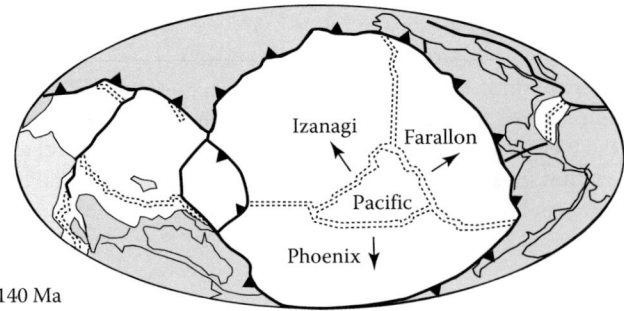

FIGURE 3.8   Reconstruction of the ocean floor at 140 Ma, The Izanagi, Farallon, Phoenix, and Pacific plates are shown soon after the origin of the Pacific plate at a triple junction of three spreading ridges. Arrows indicate the outward migration of the spreading ridges bounding the Pacific plate. Oceanic crust white (Pacific and Tethys), continental crust gray, active mid-ocean ridges as dotted lines, subduction zones as solid lines with barbs. (From Müller, R.D., et al., *Science*, 319, 1357, 2008.)

Slab Capture of Subducting Phoenix Plate Slabs by the Pacific Plate

Another proposed mechanism for the end of subduction and the beginning of extension invokes the process of "slab capture" (Luyendyk, 1995). Young basalt (<10 Myr old) that has just been produced at a spreading ridge is warm, buoyant, and unsubductable, and so if a spreading ridge and its young basalt margins approach a trench, subduction and spreading can both cease. In this case, the spreading ridge solidifies, and the two plates that were diverging away from it are fused into one. The slab that was subducting can now take on the motion of the plate that captured it by the fusion. In this way, a subduction zone can evolve into a region of extension or a transform system with strike-slip deformation.

As the Phoenix plate subducted beneath Zealandia and Marie Byrd Land in the mid-Cretaceous, the Pacific–Phoenix ridge converged on the east Gondwana margin. Spreading and subduction slowed, and stopped at 110–105 Ma. The end of spreading meant that the subducting slab (the Phoenix plate) became welded to the Pacific plate, with the captured part of the Phoenix plate fracturing into microplates (Luyendyk, 1995). The result is shown in Figure 3.9 (from McCoy-West et al., 2010). The motion of the Pacific plate changed direction at ~99 Ma (Veevers, 2000), and as it proceeded to move northward, it pulled the captured, partly subducted microplates with it. This exerted a basal traction on the overlying Gondwana margin and resulted in the extension of Zealandia and Marie Byrd Land. The northward motion of the Pacific plate accelerated in the Late Cretaceous and reached velocities of >100 km/Myr just prior to the start of seafloor spreading between New Zealand and Marie Byrd Land at ~84 Ma (McCoy-West et al., 2010).

Within Zealandia, notable extension events associated with magmatism and core complex exhumation (at 101, 97, and 82 Ma) are all oriented at ~30° oblique to the paleotrench (Tulloch et al., 2009). This is consistent with the idea that intracontinental rifting in Zealandia was controlled by its capture by the Pacific plate, seafloor spreading between Zealandia and West Antarctica, or both of these.

Slab Window Formation

Slab windows form when a spreading ridge is subducted and torn apart, creating a gap, or window, between its two subducting slabs, and this can lead to rifting. Tulloch et al. (2009) regarded slab window formation, along with mantle plumes, as less likely explanations for the prebreakup rifting. However, Timm et al. (2010) proposed that slab detachment along the New Zealand part of the Gondwana margin opened a slab window by 97 Ma, allowing hot, deeper mantle to upwell beneath the margin. This upwelling led to intracontinental thinning and rifting, prior to seafloor spreading between Zealandia and West Antarctica at 84 Ma.

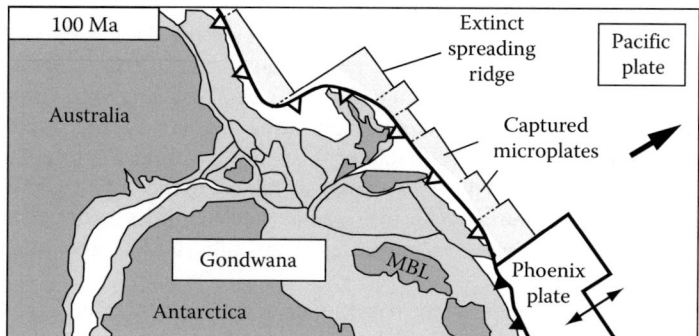

FIGURE 3.9 Reconstruction showing the East Gondwana margin at 100 Ma, MBL, Marie Byrd Land. Thick line with double arrows indicates active spreading ridge, line with black triangles indicates active subduction zone, line with white triangles indicates extinct subduction zone. Large arrow shows Pacific plate motion before its change to northward. When the spreading and the subduction ceased, the captured microplates and New Zealand became fused to the Pacific plate. (Based on McCoy-West, A.J., et al., *Journal of Petrology*, 51, 2003, 2010).

Arrival of the Hikurangi Plateau at the Chatham Rise

The Hikurangi Plateau is a large, thick sequence of basalt that contacts the Chatham Rise, an extinct part of the Gondwana-Pacific subduction zone, and is subducting beneath the current plate boundary at the Hikurangi trench (Figure 3.10, from Hoernle et al., 2010). The plateau is thought to be a large fragment of the giant, mid-Pacific plateau that was erupted in the Early Cretaceous (~120 Ma) and later broke apart. About 20–30 Myr after the Hikurangi Plateau formed, it arrived at the New Zealand part of the Gondwana margin. It began to subduct, but because of its size and buoyancy it choked the system, and subduction ceased (Mortimer, 2004; Schellart et al., 2006; Davy et al., 2008; Cluzel et al., 2010a). The arrival of the plateau at Zealandia has been dated at either 72 Ma (Cooper and Ireland, 2014, coeval with metamorphism of the Pounamu terrane, an exotic terrane in the Alpine Schist), at ~86 Ma (Worthington et al., 2006), or at ~100–105 Ma (Henig and Luyendyk, 2007; Castillo et al., 2009; Reyners et al., 2011).

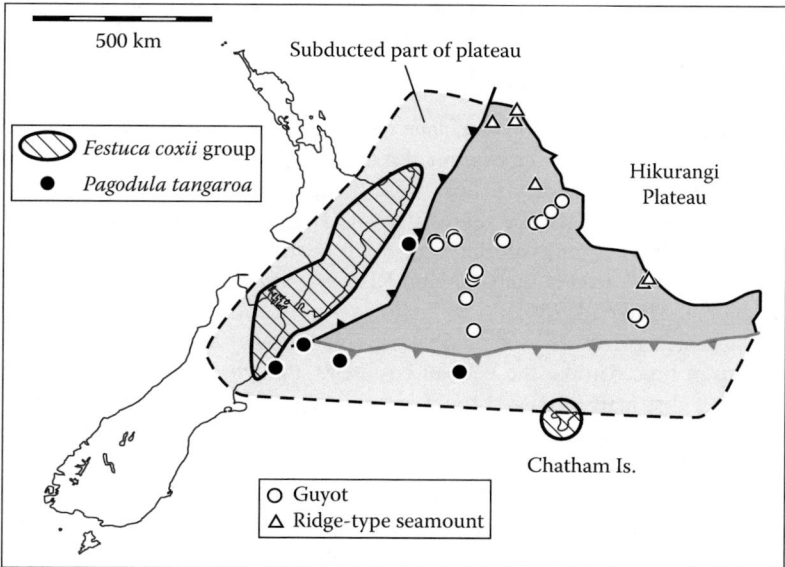

FIGURE 3.10 Hikurangi Plateau. Black line with barbs indicates active subduction zone; gray line with barbs indicates extinct subduction zone. Distribution of the *Festuca coxii* group (Poaceae) and *Pagodula tangaroa* also shown. (Data from Hoernle, K., et al., *Geochim. Cosmochim. Acta*, 74, 7196, 2010; Lloyd, K.M., et al., *Aliso*, 23, 406, 2007; Marshall, B.A., and Houart, R., *N. Z. J. Geol. Geophys.*, 54, 89, 2011.)

The collision of such a large structure as the Hikurangi Plateau with the Gondwana margin would have had profound effects throughout proto-Zealandia. By choking the Chatham Rise section of the subduction zone, it is likely that the arrival of the plateau caused the development of the Emerald fracture zone, the precursor of the Alpine fault (Figure 3.2). The Zealandia–Hikurangi Plateau collision could have also led to spreading between the Chatham Rise and the Campbell Plateau in the Bounty Trough (90–83 Ma) and caused the metamorphism of the Alpine Schist at 86 Ma. More distant effects are also possible; the arrival of the plateau may be related to the fan-shaped opening of the Late Cretaceous marginal basins, beginning with the Tasman and South Loyalty basins (Schellart et al., 2006). In this model, the subduction zone moved eastward into the Pacific while rotating around a pivot at the Hikurangi Plateau.

The mid-Cretaceous eruption of the giant basalt plateau in the central Pacific was Earth's largest known magmatic event, and it covered ~1% of the planet's surface with volcanism (Fitton et al., 2004; Timm et al., 2011). After its eruption, the plateau was broken up by seafloor spreading and rifted apart, creating the Hikurangi Plateau, the Ontong Java Plateau (now located next to the Solomon Islands), and the Manihiki Plateau (Cook Islands) (visible at the top of Figure 3.1). The history of these plateaus is of special interest for biogeography.

The Ontong Java Plateau is constructed from submarine volcanics and also a thick succession of volcaniclastic rocks that resulted from subaerial eruptions (Thordarson, 2004). Fossilized or carbonized wood fragments have been found near the bottom of four of the eruptive members (Fitton et al., 2004). Likewise, on Manihiki Plateau, the basement and later volcaniclastic layers were formed by subaerial and shallow-water eruptions (Ai et al., 2008).

The Hikurangi Plateau has been described by Hoernle et al. (2010). In its geochemistry, the basement shows similar characteristics to the Ontong Java Plateau lavas. The most prominent features on the Hikurangi Plateau are the ~30 large, guyot and ridge-type seamounts (Figure 3.10). The guyots in the interior of the plateau are characterized by flat erosional tops, some of which are >50 km in diameter along their longest axis and ~1000 m higher than the plateau basement. If sediment thickness is subtracted, the guyots have heights of ≥2000 m above the igneous basement. Rock samples from the guyots indicate shallow-water to subaerial eruptions. The steep-sided, circular or oval, flat-topped morphology is also consistent with the idea that the guyots are the bases of former island volcanoes, with the flat tops formed by erosion of the island to sea level. The seamounts have later subsided to 2000 m below sea level. Eight sites have been sampled by drilling on the Ontong Java Plateau and one on the Manihiki Plateau. As yet, the volcanics on the Hikurangi Plateau that were associated with plateau formation have only been sampled by dredging, so it is likely that additional areas of subaerial volcanism will be identified (Hoernle et al., 2010).

Figure 3.10 also shows the distribution of a clade of grasses, the *Festuca coxii* group (Poaceae) (Lloyd et al., 2007). The restricted range can be explained if the group colonized the Gondwana mainland from the Hikurangi Plateau and its islands, as these were being subducted. The distribution of the marine gastropod *Pagodula tangaroa* at the Chatham Rise and Hikurangi trench (Figure 3.10; Marshall and Houart, 2011) can be explained in a similar way. With continued subduction and plate convergence, groups have been "piled up" around the Chatham Rise and the North Island, just as marine mollusks have in New Caledonia (Marshall, 2001).

Dates for the basement of the Hikurangi Plateau (118–96 Ma) and for lavas from the seamounts (99–67 Ma) overlap in time. Unlike the plateau basement, though, the seamounts are derived from a mantle source with a distinctive (HIMU-type) isotopic signature. Rocks with similar compositions occur in alkaline igneous complexes in the northeastern South Island (90–100 Ma), Chatham Islands (85–82 Ma), and Marie Byrd Land. This HIMU province supports the idea that the Hikurangi Plateau had arrived at the Gondwana margin by ~100 Ma (Hoernle et al., 2010).

The Hikurangi Plateau/Chatham Rise collision and perhaps also plume activity beneath Marie Byrd Land could have contributed to the Zealandia–West Antarctica breakup (Hoernle et al., 2010). The Hikurangi collision shut down subduction along the Zealandia part of the Gondwana margin and caused detachment of the descending slab. This window in turn allowed hotter mantle to upwell, first beneath the Chatham Rise and Bounty Trough, and later beneath the Campbell Plateau (cf. Mechanism 3, above). Mortimer and Campbell (2014: 145) wrote: "the old Pacific slab, pinned by the Hikurangi Plateau, dropped off into the mantle and hot, runny, asthenosphere mantle flowed up into that space like a fountain, heated and weakened the crust, [and] exerted tensional forces...."

Orogenic Collapse

As another possible mechanism for the change to extension, Mortimer and Campbell (2014: 145) cited orogenic collapse, "in which the crust of the Cordillera Zealandia was so thick and high that it spread out sideways under its own weight...." The temporal relationship between the gravitational collapse of the cordillera and extension in the Tasman Sea, Ross Sea, and West Antarctic rift system suggests that these are all linked (Rey and Müller, 2010), although the collapse is probably an effect rather than a cause.

Changing Motions of the Australia and Pacific Plates at ~100 Ma

In the southwest Pacific, westward subduction of oceanic lithosphere beneath the Gondwana margin continued through most of the Paleozoic and Mesozoic until ~100 Ma and was more or less head-on ("Chilean type"; Veevers, 2000). At ~100 Ma there was a change in relative plate motion, and from 100 to 43 Ma subduction (north of New Zealand) was sinistral oblique ("Mariana type"). Also at 100 Ma Australia broke with Antarctica by predrift rifting, and there is an associated mid-Cretaceous unconformity. These mid-Cretaceous events produced the continental margins of modern Australia, as well as the extensive chain of highlands along the eastern Australian seaboard. (In New Zealand, the equivalent Early Cretaceous mountains of the Rangitata II orogeny have been destroyed and replaced by the Late Miocene–Recent mountains of the Kaikoura orogeny.)

In one reconstruction, East Gondwana moved *eastward* toward the Pacific–Phoenix plates from 135 to 100 Ma, while the Zealandia Cordillera developed along the convergent margin (Figure 3.11, from Rey and Müller, 2010). Plate motion slowed from 115 to 100 Ma, and from 100 to 90 Ma Gondwana remained stationary, but at ~90 Ma Australia began to move *northwestward*, away from Antarctica. Rey and Müller (2010) proposed that the eastward deceleration of Gondwana led to slowed subduction rates, lowering the dynamic support of the Zealandia Cordillera and causing its gravitational collapse. When Australia's motion then switched from eastward (nearly normal to the trench) to northwestward (nearly parallel to the trench), subduction ceased, and rifting in the Tasman Sea, Ross Sea, and West Antarctic rift system was initiated. Subsequent seafloor spreading led to the fragmentation of the plate margin and dispersion of microcontinents.

This biphasic model (Figure 3.11) fits with the two stages of Late Cretaceous rifting in proto-New Zealand (Kula et al., 2007). In the first phase, from ~100 to 88 Ma, there was northward propagation of continental extension and the Tasman spreading center, as recorded in the Paparoa metamorphic core complex of Westland, New Zealand. In the second phase, from ~89 to 82 Ma, extension occurred between the Campbell Plateau of southern New Zealand and Marie Byrd Land in Antarctica. The rifting developed into the Pacific–Antarctic spreading center and the first oceanic crust adjacent to the Campbell Plateau formed at ~84 Ma.

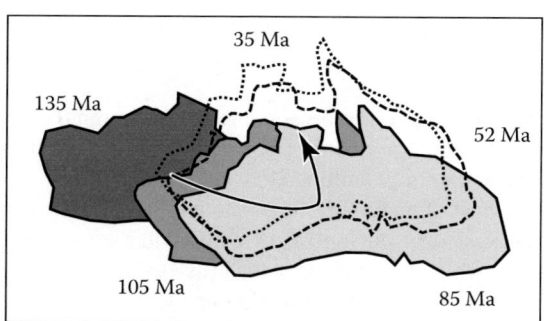

FIGURE 3.11   Australia's trajectory from 135 to 35 Ma. (From Rey, P.F. and Müller, R.D., *Nature Geosci.*, 3, 257, 2010.)

100–84 Ma: Predrift Rifting, Magmatism, and Metamorphism, Including Formation of Core Complexes

Predrift Rifting

In order to simplify the calculations, early models of extensional margin development considered rifting as if it were instantaneous. In fact, rifts in the interior of continents that evolve to form large ocean basins typically last for 30–80 Myr before complete rupture of the continent and the onset of seafloor spreading (Umhoefer, 2011). In extreme cases, the predrift, rifting stage of continental breakup can last for up to 200 Myr and include several, discrete rifting episodes (Miall, 2008). Although biogeographers often treat Gondwana breakup as a simple fracture, it is important to remember that New Zealand underwent ~20 Myr of continental rifting between the first extensional events and the beginning of seafloor spreading and continental drift.

Biogeographers often assume that large-scale vicariance around Zealandia has only developed among the modern geographic units—Australia, New Zealand, New Caledonia, and so on. For example, the basal angiosperm, *Amborella* in New Caledonia, has been thought to "fail the test of a vicariance explanation" (Crisp et al., 2011), as it is not the same age as New Caledonia's isolation from Australia and New Zealand, but is older. This approach is unrealistic as it overlooks the continental rifting that took place along the Gondwana margin for tens of millions of years *before* final breakup. These events must have had an enormous impact on the biota and would have facilitated differentiation in large numbers of groups, including angiosperms and passerine birds.

While it is clear that subduction beneath proto-New Zealand ceased at some point in the mid-Cretaceous, the exact time is uncertain, although there is a consensus age of ~100 Ma (Tulloch et al., 2009). This divides the youngest Median Batholith plutons and the youngest Eastern Province sediments deposited in a trench environment from the oldest basaltic magmatism with intraplate, rather than plate margin, chemistry. Nevertheless, ages as old as 128 Ma (the age of the last subduction-related, I-type magmatism in the Median Batholith) or as young as 82 Ma (the age of the last thrusting in the Eastern Province accretionary prism) could also indicate the end of subduction (Tulloch et al., 2009). Rifting in Zealandia is thought to have commenced by ~110 Ma (Sprung et al., 2007). The rifting culminated in seafloor spreading and the formation of the Tasman Sea between ~85 Ma and ~60–53 Ma. Zealandia's separation from present-day Antarctica occurred first at the Chatham Rise margin (at ~118 to ~83 Ma) and then at the Campbell Plateau margin (~83 to ~79 Ma) (Eagles et al., 2004).

As mentioned earlier, the mid-Cretaceous change from compression to extension is marked in stratigraphy by a major angular unconformity that separates the "basement" from the "cover" strata. The youngest basement strata contain Albian fossils and the youngest associated zircons have been dated by radiometric techniques at ~100 Ma (Laird and Bradshaw, 2004). In general, the oldest strata overlying the unconformity contain fossils of similar Albian age, and the oldest radiometric dates also give similar dates of ~100 Ma, indicating a very rapid transition between the two tectonic regimes.

Matthews et al. (2011) found good evidence for a plate reorganization at a global scale that occurred over a period of 3–8 Myr between 110 and 90 Ma, and was more dramatic than the better-known reorganization event at 50 Ma. Developments in the Pacific-Gondwana Australides orogen in general were similar to those seen in New Zealand; mid-Cretaceous compression was followed (from ~100 Ma on) by extension or transtension (transcurrent deformation with extension), often on a spectacular scale (Vaughan and Livermore, 2005).

In the New Zealand part of the Australides orogen, the dramatic, rapid switch to extensional tectonics at ~100 Ma led to widespread continental rifting. The extension occurred throughout the eastern margin of the Australian plate and caused normal faulting associated with rift basins (grabens and half-grabens), magmatism, and the emplacement of metamorphic core complexes (Deckert et al., 2002).

Metamorphic core complexes form during major continental extension as compressional welts—thickened ridges or domes—in which lower strata in the crust undergo rapid uplift and are then subjected to extension and denudation (Figure 3.12). Granite plutons intrude the lower part of core complexes at different stages of the deformation.

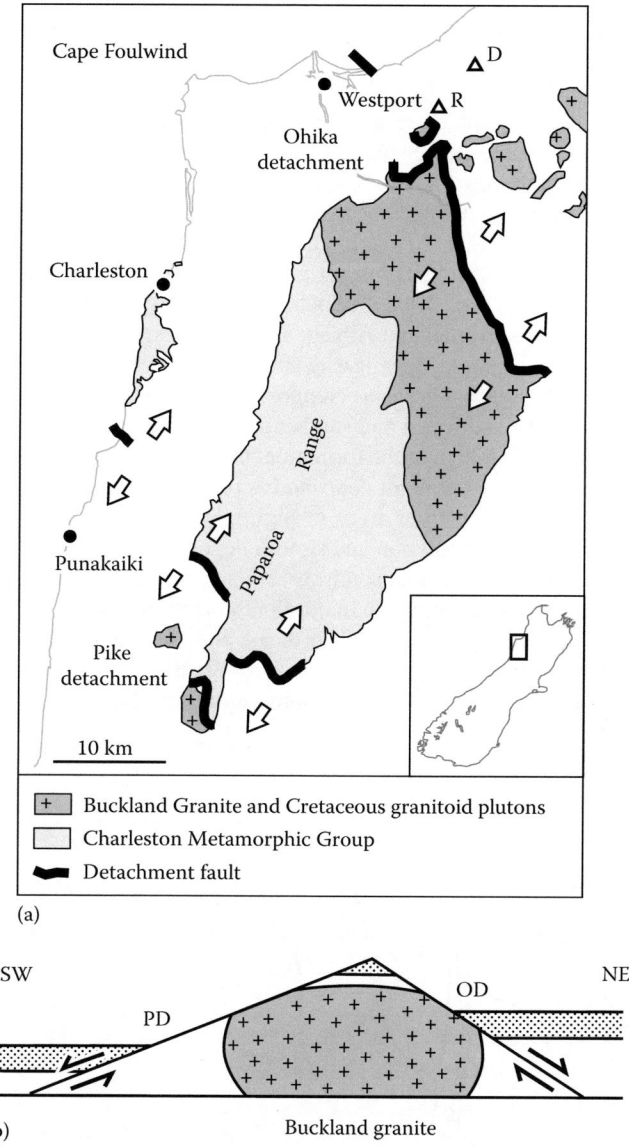

FIGURE 3.12 (a) Generalized map of the Paparoa metamorphic core complex. D, Denniston Plateau; R, Mount Rochfort. (b) Section of the core complex, with partly eroded upper plate. The detachment faults show opposite sense of shear. PD, Pike detachment; OD, Ohika detachment. (From Schulte, D., *Kinematics of the Paparoa Metamorphic Core Complex, West Coast, South Island, New Zealand*, MSc thesis, University of Canterbury, New Zealand, 2011.)

The topographic and biological effects of the new extensional regime, which included magmatism, uplift, and rifting, would have been profound. In flat terrain, vertical fault displacements of less than a meter can cause major ecological change over large areas. Even in terrain with more relief, as in hills and mountains, fault displacements of tens of meters can cause a significant environmental change. If a population is already genetically "primed" to break up and diverge into different forms in a region, this differentiation may be triggered at faults, belts of faults (fault zones), and other faulted structures such as core complexes. As the extension of Zealandia continued, whole mountain ranges were destroyed, and this must have caused revolutionary changes in the regional biology.

The distributions of many molecular clades in Zealandia show a close geographic relationship with the extensional structures, suggesting that the switch from compression to extension caused phylogenetic differentiation. For example, an important biogeographic break occurs at the Waihemo fault zone in Otago, corresponding with large-scale fault displacement there in the mid-Cretaceous. In Westland, a node representing both a biogeographic break and a center of endemism occurs at the northern Paparoa Range, coinciding with a zone of extension and core complex emplacement.

In the first stages of the post-100 Ma extensional regime, metamorphic core complexes developed in the Western Province, while reactivated shear zones, faulting, and regional uplift are evident in the Eastern Province (Cox and Sutherland, 2007). In Otago, for example, coarse, locally derived sediments were deposited from 112 Ma in large, rift-related grabens that opened in the basement (Figure 3.13, from Litchfield, 2001; Mortimer et al., 2002; Tulloch et al., 2009). The grabens formed on the margins of the schist belt during the latest Aptian and Albian, and similar structures are found throughout proto-New Zealand at this time. One example, the mid-late Cretaceous Kyeburn Formation of Otago, is 4000 m thick, and includes conglomerates and breccias composed of coarse, terrestrial detritus. The sediments arrived from upland areas to the southwest and southeast and were deposited in a fault-angle depression (Bishop et al., 1976). In the upper 500 m of the formation, the conglomerates become much more angular and consist mainly of Otago Schist fragments deposited as talus. The largest clasts are 2.5 m in diameter, indicating that the talus was derived from areas of high relief while the schist basement (Figure 3.2) was being exhumed. The Kyeburn Formation and similar deposits of conglomerate and talus breccias in east Otago (Henley Breccia), Westland, and southwest Nelson represent piedmont sediments shed from developing fault scarps as movement began on major South Island faults.

Some of this mid-Cretaceous faulting strikes northeast and some northwest. Mortimer et al. (2002: 361) wrote that the tectonic complexity of the southeastern South Island has developed by both dip-slip and strike-slip faulting, and many of the terrane-bounding faults (Figure 3.2) are likely to have been reactivated over time.

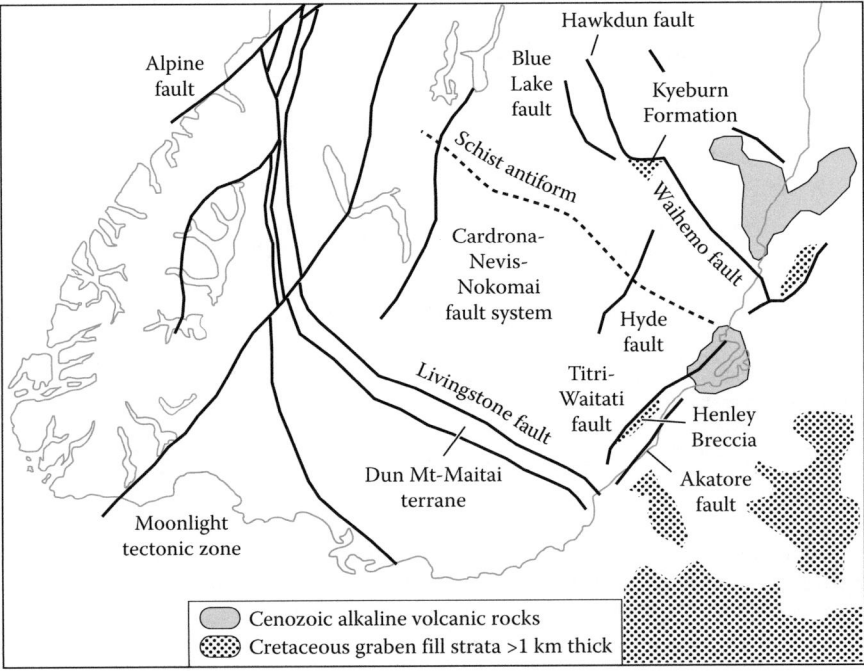

FIGURE 3.13　Geology of Otago, showing the mid-Cretaceous breccias and some of the major faults. (From Mortimer, N. et al., *N.Z.J. Geol. Geophys.*, 45, 349, 2002; Litchfield, N.J., *N.Z.J. Geol. Geophys.*, 44, 517, 2001; Tulloch, A. et al., *Geol. Soc., London, Special Publ.*, 321, 89, 2009.)

Plutonism, Metamorphism, and Core Complex Emplacement

The mid-Cretaceous events in New Zealand included widespread magmatism, and major plutonic complexes were emplaced at 120–105 Ma: the Paparoa Batholith (the central part of a metamorphic core complex), the Hohonu Batholith, the Big Deep Granite of the Karamea Batholith in the upper Buller Gorge, and the final components of the Median Batholith (the Separation Point suite of Fiordland and Nelson) (Wandres and Bradshaw, 2005). Mid-Cretaceous metamorphic events have been dated in Fiordland (116–105 Ma), the Southern Alps (100–71 Ma), and in Wellington (100–85 Ma) (Mortimer and Cooper, 2004; Vaughan and Livermore, 2005).

In the mid-Cretaceous, the tectonic regime at the margin of western New Zealand changed from one of crustal shortening, thickening and arc magmatism, to one dominated by extension and core complex emplacement (Klepeis et al., 2007). By ~120 Ma, the crust was >45 km thick, but soon after this, extensional structures including the Paparoa core complex began to form in the orogen. This occurred at the same time as the emplacement of the last arc plutons and the end of subduction beneath New Zealand.

The Paparoa metamorphic core complex (Figure 3.12) formed between 110 and 90 Ma by rifting associated with extension, orogenic collapse, metamorphism, and granite plutonism. Metamorphic core complexes are localized zones (<100 km wide) of extreme extension and are the precursors of continental breakup (Klepeis et al., 2007). During core complex formation, plates of lower crust are dragged rapidly to the surface from beneath the fracturing, extending upper crust (Lister and Davis, 1989). The lower plates are uplifted and unroofed below normal, extensional detachment faults (décollements) or ductile shear zones. The two detachment systems are bivergent, that is, they are both normal, but have opposite senses of shear. Given the combination of localized extension and rapid uplift, it is not surprising that many metamorphic core complexes, including the Taranaki basin, the Paparoa Range, Fiordland, Dunstan Mountains, and southeast coast of Stewart Island (Tulloch et al., 2014), are associated with biogeographic boundaries, disjunctions, and centers of endemism.

In the lower plate of the Paparoa core complex, the emplacement and metamorphism of the Charleston Orthogneiss occurred at 118 and 107 Ma, respectively (Sagar and Palin, 2011). Most of the extension in the Paparoa core complex was accommodated on the southwest-dipping Pike Detachment and was active as early as 116 ± 6 Ma (Ring et al., 2006). Mylonitic rocks were uplifted to the surface and eroded into evolving half-grabens by 105–100 Ma. Overall, the active phase of normal faulting was from ~115–95 Ma (Schulte, 2011). After exhumation, the core complex was again buried to depths of at least ~5 km and reheated. Subsequent exhumation from ~90 to ~80 Ma was probably related to Early Campanian extension induced by seafloor spreading in the Tasman Basin (Schulte, 2011).

Studies on the Mount Irene shear zone in western Fiordland indicate that the switch from a convergent to an extensional tectonic regime took place between 111 and 108 Ma (Scott and Cooper, 2006). Fiordland has also been interpreted as a Cretaceous metamorphic core complex, with two primary components: an elongate core of high-pressure, granulite facies rocks (Western Fiordland Orthogneiss) and, overlying this, a mid-Paleozoic plutonic-metasedimentary complex (Gibson et al., 1988; Gibson and Ireland, 1995; Klepeis et al., 2007). However, other studies have interpreted the extensional shear zones separating the orthogneiss from adjacent rocks as discontinuous, later features, rather than detachments between the upper and lower plates of a core complex (Allibone et al., 2009a,b). In any case, the shear zones are major features, and deformation along them would have been significant for the Fiordland biota, which includes many endemics.

At 115–95 Ma, about the same time as the core complexes were emplaced in proto-New Zealand, other core complexes formed in Marie Byrd Land, along with mafic dikes and A-type (intraplate) granitoids that flank the West Antarctic rift system (Rey and Müller, 2010).

The Haast Schist

The Haast Schist (Figure 3.2) includes the Otago, Alpine and Marlborough Schists in the South Island, the Kaimanawa Schist in the central North Island, and, perhaps, the Chatham Islands Schist. The rocks show distinctive folding, host much endemism, and have a complex, controversial history. The schist

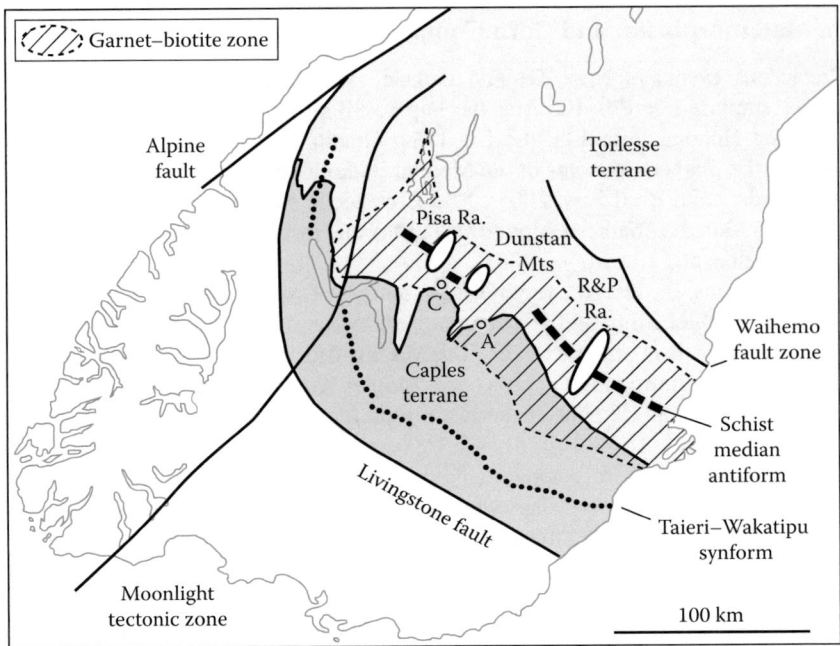

FIGURE 3.14 The Torlesse and Caples terranes. The boundary of the two terranes is overprinted by the metamorphism of the Otago Schist (only its antiformal core zone is shown here). Cretaceous metamorphism in the schist ranges from prehnite–pumpellyite facies on the flanks to greenschist facies (garnet–biotite zone) in the core. To the west, the rocks have been affected by Alpine metamorphism; these Alpine schists are not shown here (see Figure 3.2). (From Forster, M.A., and Lister, G.S., *Aust. J. Earth Sci.*, 50, 181, 2003; Mortimer, N., *Amer. J. Sci.* 303, 603, 2003).

makes up a metamorphic belt ~150 km wide, with a central core represented by a garnet–biotite zone (Figure 3.14). It overprints the boundary of the Torlesse and Caples terranes and lies parallel to the old Pacific subduction zone. The Torlesse and Caples terranes were scraped off the Pacific plate as it subducted, and they formed an accretionary wedge above the trench, with the Caples terrane thrust over the Torlesse. Structural thickening of the wedge during the Jurassic–Cretaceous formed the schist belt, which has been deformed and then exhumed (unroofed). The mode of origin of the metamorphic fabric during the subduction history is the topic of current investigation (Forster and Lister, 2003; Rahl et al., 2011).

One of the structural features in the southern part of the Otago Schist, formed from metamorphosed Caples terrane, is the Taieri–Wakatipu synform (Figure 3.14). This downfold is equivalent to the Goulter synform of the Marlborough Schist. Another feature in the central core of the schist is a median anti-form, an upfold. Both these folds are large, Cretaceous structures that define the broad architecture of the schist belt. The median antiform is a structural high that extends through the Rock and Pillar Range and the Pisa Range, and includes the most intensely metamorphosed and deformed rocks that are exposed in the Otago Schist (Mortimer, 2003; Rahl et al., 2011). The metamorphic grade in the schist belt increases from low-grade prehnite–pumpellyite facies on the outer flanks to higher-grade greenschist facies (with chlorite to biotite-garnet zones) in the antiformal core. The origin of the central belt of uplift in the Otago Schist—the antiform—is controversial. Some authors suggest it was caused by extension, in a metamorphic core complex, while others suggest it was caused by contraction.

The Otago Schist as a Metamorphic Core Complex

In its earlier history, the Otago Schist displays evidence of shortening and deformation associated with terrane collision and the Late Jurassic–Early Cretaceous Rangitata orogeny. Forster and Lister (2003) accepted this model for the genesis of the schist, but they argued that it did not explain the median antiform, the "metamorphic welt." The authors recognized ductile shear zones along both margins of the antiform, both with normal orientation, and so they concluded that the earlier shortening was followed by a phase of extension. They proposed that the elongate, southwest to northeast trending, dome structures of the ranges along the antiformal axis (Mount Cardrona, Pisa Range, southern Dunstan Mountains, Rock and Pillar Range) are parts of a metamorphic core complex, formed and exhumed during mid-Cretaceous extension. This structure, along with the coeval Paparoa core complex and the Bounty Trough, was the main focus of the regional extension that was occurring at this time throughout Zealandia.

Exhumation of the Otago Schist antiform at the shear zones began at 122 Ma, with final exhumation of the antiformal core at 110 Ma (Forster and Lister, 2003). A period of shortening took place after the shear zones had formed, and similar switches in tectonic mode (compression–extension–compression) are seen in the Paparoa core complex and in northeastern New Caledonia (Rawling and Lister, 2002). This sort of dynamism and oscillating orogenesis would have had profound effects on the ecology and evolution of the regional biota.

The continental extension across the New Zealand region may have more or less ceased when the Tasman Sea began to open. The shear zones in the schist belt display both cross-belt and belt-parallel stretching lineations. The former were caused by rollback of the subducting Pacific slab, while the latter developed during prebreakup extension associated with rifting between Australia and Antarctica.

The antiform (the "Otago forearc high" of Rahl et al., 2011) is of potential biogeographic significance, as it coincides with many centers and boundaries of endemism. The underlying mechanism for the antiform's uplift is of interest for biogeography, as the different mechanisms that have been proposed make different predictions about timing and rates of evolution. Final exposure of the schist is indicated by schist clasts in Albian breccias (Figure 3.13), but dating the earlier stages of the uplift is more difficult. Deckert et al. (2002) concluded that exhumation of the Otago Schist either occurred very slowly (~200 m/Myr) and continuously from 190 to 110 Ma, or was punctuated by a phase of more rapid exhumation (at rates of 600–1000 m/Myr) after ~135 Ma when the shear zones formed.

Gray and Foster (2004) dated main metamorphism and deformation in the Otago Schist from ~160 to 140 Ma (in the low-grade flanks) through to 120 Ma (shear zone deformation). The metamorphism was followed either by very gradual cooling or no cooling, implying little uplift, until about 110–100 Ma when there was rapid, extensional exhumation of the metamorphic core.

Mortensen et al. (2010) suggested a broad time continuum for orogenic gold mineralization associated with the evolution of the Otago Schist belt, but also two dominant, short-lived events at 142–135 Ma (in the low-grade flanks) and 106–101 Ma (in the antiformal core). The two pulses probably reflect thermal events during which most of the gold was extracted, transported, and deposited.

Metamorphism Developing with Progressive Accretion and Underplating of the Torlesse Terrane

Some models of the schist belt evolution accept discrete orogenic episodes. In these models, metamorphism and deformation is attributed to an inferred collision "event" between the Caples and Torlesse terrane, and this formed the accretionary wedge in the Early, Middle, or Late Jurassic. In contrast, Rahl et al. (2011) regarded metamorphism and deformation as developing in a steady fashion during convergence on the Gondwanan margin, and continuing through the full history of the wedge (290–105 Ma). They noted: "The nearly continuous record of arc magmatism and progressive younging of sediments towards the toe of the Otago wedge both suggest a gradual evolution and progressive accretion of the Torlesse, rather than discrete orogenic episodes."

The region between a subduction trench and its volcanic arc is often 2–300 km across and constitutes the forearc basin (Figure 3.7). Many forearc basins develop a regional upwarp, a structural high that is induced during subduction. The ultimate cause is probably accretion and underplating (accumulation on the base of the crust). Rahl et al. (2011) used this process, rather than extension and core complex formation, to explain the uplift of the Otago Schist antiform. Whatever the cause, the uplift of the Otago antiform is associated with high levels of diversity, endemism, and boundaries of distribution.

Studies of strain through the Caples–Torlesse subduction wedge showed that regional-scale deformation *lacked extension*, indicating that underplating, uplift, and erosion caused the exhumation of the high-grade rocks (Rahl et al., 2011). In this model, accretion in the schist occurred by more or less *continuous underplating*, at depths of about 25–30 km. This caused the surface uplift that pushed the forearc—the median antiform—high above sea level, leading to subaerial erosion and terrestrial sedimentation.

Cooling ages around 115–110 Ma (Gray and Foster, 2002; Forster and Lister, 2003) indicate that the shear zones bounding the Central Otago antiform were initiated with the onset of the rifting that developed throughout New Zealand at this time. In contrast, Rahl et al. (2011) stressed that slow erosion of an emergent forearc high was a persistent feature through the history of the Otago accretionary wedge.

With continued accretion, the Livingstone fault (Figure 3.14) between the Caples and Maitai terranes was overturned (it is now north-dipping) and the large, southwest-vergent fold of the Southland Syncline was formed.

The Alpine Schist

The Alpine Schist is a 20 km wide strip located along the central Southern Alps, north of the Otago Schist (Figure 3.2). Metamorphic grade increases westward, from prehnite–pumpellyite facies east of the Main Divide, to upper greenschist and amphibolite facies within 1–2 km of the Alpine fault (this is higher than the Otago greenschist facies) (Mortimer, 2003; Murphy, 2010). The more intense metamorphism reflects Neogene exhumation of these rocks from greater crustal depths with increasing proximity to the Alpine fault.

Metamorphism of the Alpine Schist began at ~100 Ma, with a major episode of high-grade mineral growth at ~86 Ma (Vry et al., 2004). Vry et al. concluded that plate convergence along the New Zealand portion of the Gondwana margin continued after ~105 Ma and culminated in the oblique collision of the Hikurangi Plateau. The metamorphism of the Alpine Schist is evidence of that hit.

Post-100 Ma Felsic (Silicic) Magmatism

Magmatic activity related to extension and rifting, rather than subduction, began in New Zealand and on the Lord Howe Rise as early as 112 Ma and continued until 82 Ma (Tulloch et al., 2009; Cluzel et al., 2010b). Rift-related magmatism that *preceded* breakup migrated eastward from eastern Australia (130–95 Ma) to New Zealand (112–82 Ma) and New Caledonia (89–83 Ma), generating large volumes of silicic magma (Cluzel et al., 2010a). In contrast, when the stretched continental crust finally broke and seafloor spreading began, the different oceanic basins west and south of Zealandia opened at about the same time, ~83 Ma. This separated the New Zealand region of continental crust (Zealandia, including New Caledonia) from Australia and Antarctica. The large numbers of clades disjunct between these regions indicate that much of the biota had already differentiated by the start of seafloor spreading and so could have taken place during the Rangitata orogeny or the predrift rifting.

The events in Cretaceous New Zealand have equivalents in Antarctica. In Marie Byrd Land, magmatism related to subduction and orogeny ceased in the Early Cretaceous (110 Ma), and rift-related magmatism had begun by 101 Ma (McCoy-West et al., 2010). In New Zealand, over a similar period (112–82 Ma), there was a similar transition in the felsic magmatic rocks from I-type geochemistry, indicating subduction and orogeny, to A-type chemistry, indicating anorogenic conditions at some distance from a plate boundary (Tulloch et al., 2009). The change was attributed to the progressive thinning of the Zealandia continental crust; this meant that over time there was less opportunity for crustal contamination of the magma. The I-type to A-type transition and the emplacement of mafic dikes chart the transition from subduction to an extensional regime.

The New Zealand felsic magmatic rocks (rhyolites, tuffs, and granites) formed during the period 112–82 Ma were emplaced in distinct episodes (Tulloch et al., 2009). The earlier ones, dated at 112 Ma, are known only in association with the Waihemo fault zone in Otago (Figure 3.14). These overlap in age with plutons in the Median Batholith, and although this is too far away to share a common magma source, both features coincide with many biogeographic breaks. Younger felsic magmatics, formed at 101–97 Ma, occur across Zealandia from near the paleotrench to the continental interior, and these indicate widespread, more or less simultaneous extension.

Post-100 Ma Mafic (Mainly Alkaline) Volcanism

During the change from subduction to extension at ~100 Ma, subduction-related volcanism was replaced by intraplate, alkaline basaltic volcanism. Igneous activity linked to continental rifting lasted for ~20 million years before seafloor spreading began, with an apparent peak in activity between 100 and 90 Ma (Timm et al., 2010).

Intraplate volcanism continued through the Cenozoic in many parts of New Zealand—North and South Islands, Chatham Rise, Campbell Plateau—and in some areas it persisted over long periods of time. In the Chatham Islands, for example, it is recorded in the Late Cretaceous, Eocene, and Pliocene. This intraplate volcanism has not been caused by a localized hotspot, the usual explanation for volcanism not located at plate margin. Instead, it represents part of a long-lived (~100 Myr), diffuse, alkaline magmatic province that includes eastern New Guinea, eastern Australia and Tasmania, New Zealand, and West Antarctica (Finn et al., 2005). Soon after 100 Ma, a great change in the isotopic composition of intraplate basalts took place through the region (Crawford et al., 2003); all basaltic magmatism in the region since 100 Ma has had a distinctive HIMU isotopic and trace element signature. A general tectonic model for the region should account for three factors: the post-100 Ma extension, the alkaline magmatic province and its HIMU signature, and the seafloor spreading at 84 Ma.

The origin of intraplate volcanism in general is a subject of current debate (Timm et al., 2010). For continental areas, the classic models invoke either mantle plumes or extensional thinning, often preceding continental rifting. Yet neither of these models explains the great longevity or the irregular distribution of intraplate volcanism in Zealandia.

In New Zealand, mid-Cretaceous to Late Cretaceous alkaline igneous rocks are exposed in several places in the South Island and in offshore islands, including the Chatham Islands. Magmatism increased at ~30 Ma, but most centers are younger than 15 Ma and Recent activity is reported in Northland (Finn et al., 2005). Repeated phases of volcanism have produced large, composite shield volcanoes at Dunedin and Banks Peninsula, regions associated with local vicariance and endemism, while volcanism on the Campbell Plateau and Chatham Rise has been important in providing localized, island habitat for terrestrial groups.

Possible Causes of the Intraplate Volcanism

Mantle Plumes

The seafloor spreading between Zealandia and mainland Gondwana is explained in some models by the formation of a mantle plume beneath Marie Byrd Land during the Late Cretaceous (see discussion in Timm et al., 2010). Studies proposing mantle plumes have cited the lack of evidence for significant regional extension in New Zealand during the Cenozoic (this would have been a possible cause of the volcanism), as well as the enriched isotopic signatures of the alkaline rocks. Local hot spots and also hot lines with origins in the mantle have been suggested (Finn et al., 2005).

Yet despite these proposals, there is no geophysical evidence (e.g., from seismic tomography) that plume-like structures exist beneath New Zealand (Hoernle et al., 2006). Also, the intraplate volcanism does not show age progressions that agree with the direction and rate of plate motion, again contradicting a plume origin (Hoernle et al., 2006; Timm et al., 2010). For example, at Banks Peninsula, the volcanic region is ~50 km across, and volcanism persisted there for ~6 Myr. The volcanic region centered on Dunedin, Oamaru-Kakanui, and Waipiata (by Ranfurly) is ~100 km across, and volcanism here lasted ~30 Myr. In these two time intervals, the plate would have moved over the mantle by ≥300 km and ≥1800 km, respectively, contradicting a mantle plume origin for the volcanism. In addition, while regional uplift is well documented in central Zealandia, there is no evidence for the large igneous province that indicates the arrival of a mantle plume head.

The HIMU signatures in the volcanics, cited earlier, have been related to mantle plumes, but the origin of HIMU signatures in alkaline magmas is controversial (McCoy-West et al., 2010). It has been attributed to chemically modified oceanic lithosphere that was subducted and stored in the deep mantle, aged (long periods of isolation are required to generate the isotopic signatures), and then transported back to the surface via mantle plumes (Panter et al., 2006). However, HIMU sources may also exist within subcontinental lithospheric mantle, generated by the infiltration of metasomatic fluids and melts. The origin of HIMU signatures remains uncertain, and they are not sufficiently diagnostic to define mantle plumes beneath a region (Finn et al., 2005).

Extension and Thinning

Tectonic thinning of the crust has been suggested as a cause of the intraplate volcanism, but there is no close correlation between the volcanism and local extensional events (Timm et al., 2010). Major continental rifting in Zealandia ended in the mid-Cretaceous, and there is little evidence for large-scale extension and thinning in the Cenozoic. Localized extension took place during some episodes of volcanic activity, but it is unlikely that Cenozoic extension caused enough lithospheric thinning to generate the amount of melting required (Hoernle et al., 2006). Timm et al. (2010) found dates suggesting that the intensity of intraplate volcanism on Zealandia has increased during the latest period of compressional tectonics (i.e., since ~22 Ma), again ruling out extensional tectonics as the controlling factor.

Multiple, Small Lithospheric Detachments

Diffuse volcanism is known from many continental areas around the world and has been explained by mantle plumes, groups of hundreds of hot spots, tectonic thinning, rifting, and strike-slip faulting. Yet none of these provide an adequate explanation for the Zealandia volcanism. Authors have instead suggested that sudden detachment and sinking of subducted slabs in the Late Cretaceous triggered lateral and vertical flow of mantle material (Hoernle et al., 2006; Timm et al., 2010). As a portion of the lithosphere is stripped off and sinks, asthenosphere can well up into the resulting cavity, melting by decompression. Regions of Zealandia's lithospheric base are thought to contain large amounts of material accreted by underplating during subduction. Later, multiple small detachments of this material from the lithospheric keel could explain how asthenospheric upwelling can be localized at specific regions of the plate over long periods of time. In other words, the volcanism is a natural, but indirect consequence of subduction.

Tectonic History in the Southwest Pacific Since 90 Ma

Soon after the introduction of plate tectonics theory, geologists realized that oceanic trenches of active subduction zones, and the hinge of the descending slab at the top of the trench, do not always maintain the same position and can move significant distances (Stegman et al., 2006). A significant number of trenches migrate in a direction *opposite* to the motion of the subducting plate, in a retrograde direction, as in a retreating wave on a shore (Figure 3.7). The subduction zone at the Gondwana-Pacific margin is one example; it appears to have retreated into the Pacific through the Late Cretaceous and Cenozoic. This migration process has been termed rollback or retreat of the trench, slab, or hinge. The negative buoyancy of the descending slab means that it is likely to sink vertically (rather than subducting forward and downward) and to migrate away from the upper plate (Heuret and Lallemand, 2005). The motion of the upper (overriding) plate at the trench is also important in determining this slab rollback. Suggested rates of rollback range from 2 to 15 cm/year, and the retreat of the Indian/Pacific plate margin into the Pacific has been a continuous process since rifting began behind the margin at ~100 Ma.

The evolution of the southwest Pacific region since extension and rifting began has been reconstructed using plate kinematics, and the model is shown in Figure 3.15 (from Schellart et al., 2006; Schellart, 2007). Australia has a fixed position in the diagrams. (For a detailed review of the breakup of Gondwana as a whole and the rifting around Antarctica, see Veevers, 2012.)

The main feature of the model illustrated here is the continued, eastward migration (rollback) of the active margin between the Indo-Australian plate and the Pacific plate. The margin is defined by the Pacific subduction zone with its trench and its arc.

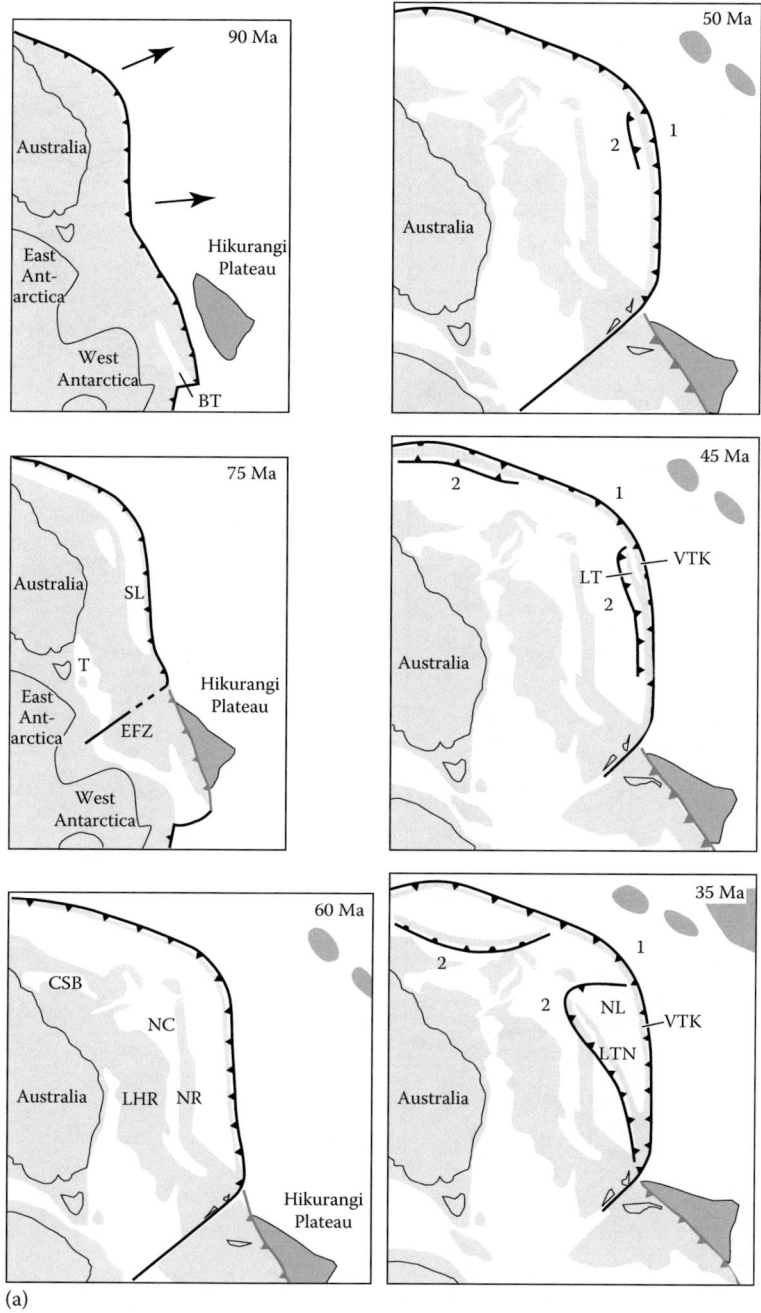

FIGURE 3.15 Reconstruction of the Southwest Pacific region for the Late Cretaceous period onward. The reference frame is Australia fixed. Light gray indicates continental/arc crust. Dark gray indicates oceanic plateaus. Geographic outlines are shown to help identify the location of the present-day coastline but have no paleogeographic significance. Barbs along the subduction zones are shown on the overriding plate. Volcanic arcs occur along the subduction zones on the overriding plate, ~200 km back from the trench. (a) Arrows (90 Ma) indicate migration of the subduction zone by slab rollback. 1, original Pacific subduction zone, 2, New Caledonia subduction zone, BT, Bounty Trough; CSB, Coral Sea basin; EFZ, Emerald fracture zone; LHR, Lord Howe Rise; LT, Loyalty–Three Kings ridge; LTN, Loyalty–Three Kings–Northland ridge; NC, New Caledonia basin; NL, North Loyalty basin; NR, Norfolk Ridge; SL, South Loyalty basin; T, Tasman basin; VTK, Vitiaz–Tonga–Kermadec ridge; NL, North Loyalty basin. (Simplified from Schellart et al., *Earth-Sci. Rev. 76*, 191, 2006.) *(Continued)*

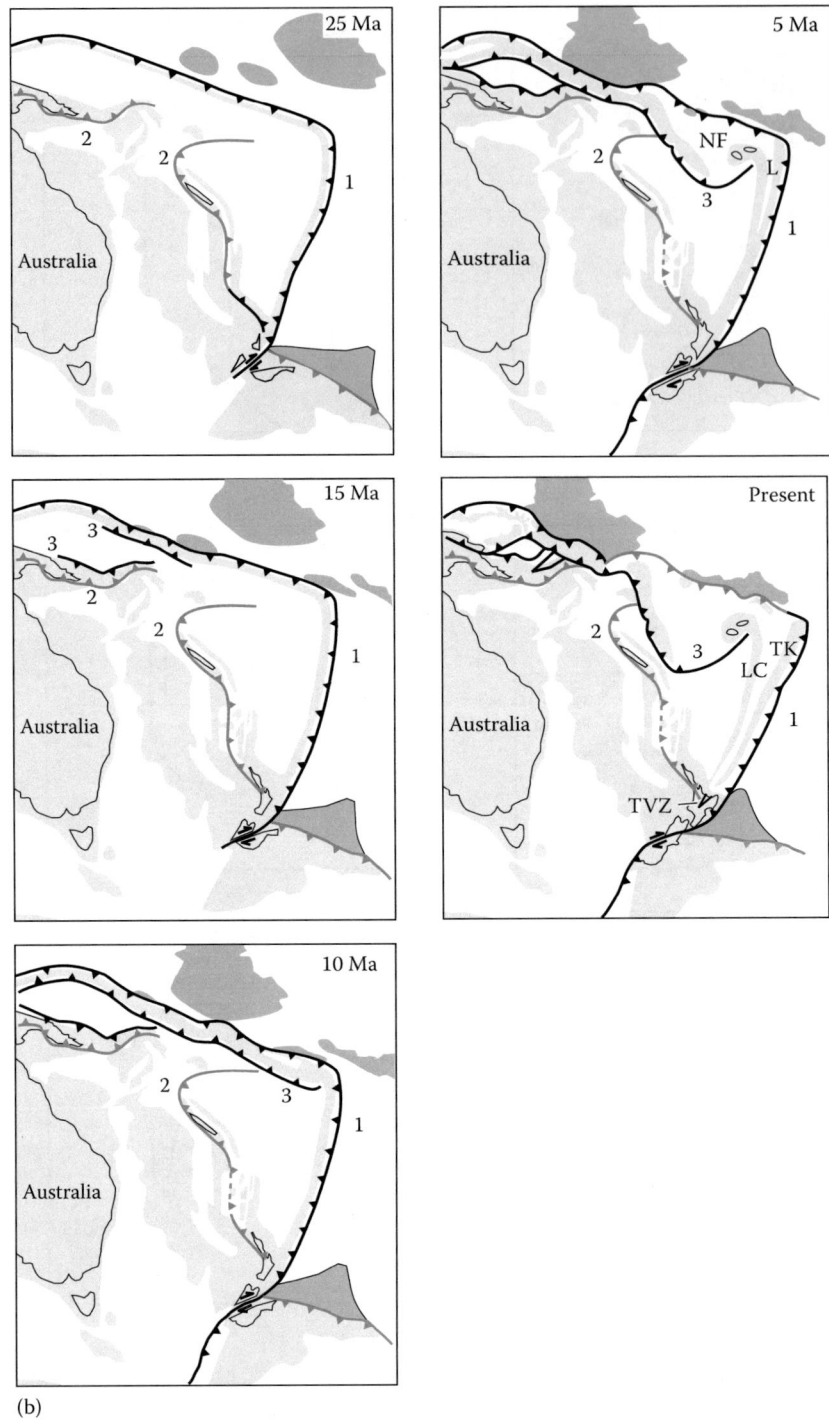

(b)

FIGURE 3.15 (*Continued*)   Reconstruction of the Southwest Pacific region for the Late Cretaceous period onward. The reference frame is Australia-fixed. Light gray indicates continental/arc crust. Dark gray indicates oceanic plateaus. Geographic outlines are shown to help identify the location of the present-day coastline but have no paleogeographic significance. Barbs along the subduction zones are shown on the overriding plate. Volcanic arcs occur along the subduction zones on the overriding plate, ~200 km back from the trench. (b) 1, 2, 3, indicate first, second and third generation subduction zones; L, Lau basin; LC, Lau -Colville ridge; NF, North Fiji basin; TK, Tonga -Kermadec ridge; TVZ, Taupo volcanic zone. (Simplified from Schellart et al., *Earth-Sci. Rev. 76*, 191, 2006.)

The eastward rollback of the southwest Pacific subduction zone began following the emplacement of a silicic large igneous province (Whitsunday volcanic province and Median Batholith) beneath the margin, beginning at ~132 Ma (Whattam, 2009). Since then, periodic opening and closing of ocean basins has been a feature of the region. What were once two major ocean basins, the South Loyalty Basin (south of the Loyalty Islands) and later the North Loyalty basin (north of the islands), were created by seafloor spreading, but were later largely destroyed (closed) by subduction.

90 Ma (Late Cretaceous)

The incipient rollback of the subduction zone (indicated by the arrows in Figure 3.15) caused extension in the backarc (see 75 Ma). (The backarc is the region behind a volcanic arc, on the side away from the trench; Figure 3.7.) By 90 Ma, the Bounty Trough between the Chatham Rise and the Campbell Plateau had already opened, and extensional rifts caused by normal faulting were developing throughout the Gondwana margin (not shown). The Hikurangi plateau is depicted here approaching the Chatham Rise, although some authors suggest that they had already collided by this time.

75 Ma (Late Cretaceous)

The Hikurangi Plateau–Gondwana collision led to the formation of the Emerald fracture zone, a precursor of the Alpine fault.

At ~85 Ma, the Tasman-Coral Sea basin (initially two separate basins) and the South Loyalty Basin began to open behind the subduction zone, as it migrated eastward into the Pacific. Both basins eventually extended from the Papuan Peninsula to northern New Zealand (cf. Crawford et al., 2003). (The basins indicated in Figure 3.15 opened at linear spreading ridges; these are not shown in the figures.)

60 Ma (Paleocene)

By this time, the New Caledonia basin (NC) had opened between Lord Howe Rise (LHR) and Norfolk Ridge (MR).

50 Ma (Early Eocene)

In addition to the initial southwest Pacific subduction zone that dips westward and has retreated eastward, the model illustrated here proposes a second subduction zone that developed inboard of the first, in the South Loyalty Basin. This New Caledonia subduction zone (labeled "2" in the figure) is now extinct, but when active it extended from Vanuatu and New Caledonia to northern New Zealand. In contrast with the first subduction zone, the second was northeast dipping. This second subduction zone seems more feasible than one alternative—rapid polarity reversal along the long Gondwana-Pacific subduction zone. Tomography studies have identified a high-velocity anomaly, interpreted as a fossil subducted slab, in the lower mantle below the Tasman Basin. This is situated at the location and depth predicted if there was a single, northeast-dipping New Caledonia subduction zone in the middle Cenozoic (Schellart et al., 2009).

45 Ma (Middle Eocene)

As the New Caledonia subduction zone developed, it rolled back to the west, toward Australia, and seafloor spreading occurred in its backarc zone. This caused the ridge that the first Pacific arc was built on to be rifted lengthwise, separating the landward Loyalty–Three Kings Ridge–Northland Plateau arc (LTN) from the seaward Vitiaz–Tonga–Kermadec arc (VTK). To begin with, the two arcs would have shared a similar biota inherited from the ancestral arc, but continued extension and the opening of the North Loyalty–South Fiji Basin between the arcs would have led to differentiation and disjunction. For example, many groups, including clades of plants, snakes, and other animals, link the Loyalty Islands of New Caledonia more closely with the Vitiaz arc islands (Bismarck Archipelago, Solomon Islands, Vanuatu, Fiji, and Tonga) than with the New Caledonian mainland (Heads, 2014: 293).

At the same time as the "second-generation" subduction zone was developing at New Caledonia, another new subduction zone was forming behind the Pacific arc, north of New Guinea.

35 Ma (Late Eocene)

The new subduction zone induced arc volcanism along the d'Entrecasteaux–Loyalty–Three Kings Ridge–Northland Plateau seamount chain (LTN). The trench and the arc retreated southwestward, causing the opening of the North Loyalty–South Fiji backarc basin (NL) and the closure of the South Loyalty Basin, which was destroyed by subduction. Eventually the southwestward rollback led to the collision of the Loyalty arc with the continental crust of New Caledonia/Norfolk Ridge. This caused the southwest-directed ramping-up (obduction) of arc-related ophiolites onto New Caledonia at 38–34 Ma.

25 Ma (Oligocene)

Continued westward migration of the Loyalty–Three Kings ridge led to collision and ophiolite obduction in the northern North Island at 25–22 Ma. High-pressure metamorphic rocks have been documented west of the Loyalty Ridge (eclogites and blueschists in northeastern New Caledonia) and west of the Three Kings Ridge (amphibolites and schists) (Schellart, 2007). Further south, central New Zealand began to undergo transpressional (transcurrent and compressional) deformation.

15 Ma (Middle Miocene)

A third generation of subduction zones (3) developed northeast of New Guinea, the northern one developing alongside the primary Pacific arc. The South Fiji Basin continued to open, and there was also extension and subsidence within the Norfolk Ridge.

10 Ma (Late Miocene)

One of the New Guinea subduction zones propagated southeastward along the Melanesian arc: New Britain–Solomon Islands–Vanuatu to Fiji. Subduction began south of New Zealand.

5 Ma (Pliocene)

Southwestward rollback of the new Melanesian subduction zone (3) began to open the North Fiji basin (NF) and consumed much of the North Loyalty basin. Continued eastward rollback of the primary Pacific subduction zone led to further seafloor spreading in its backarc, opening the Lau basin (L).

Present

Backarc seafloor spreading in the Lau basin is separating the Lau–Colville ridge (LC) from the Tonga–Kermadec ridge (TK). This rifting extends into the central North Island as the Taupo volcanic zone.

Slab Rollback and Extensional Opening of Backarc Basins

The opening of the many rifted basins in the Southwest Pacific—from the oldest, in the west (the Tasman Basin), to the youngest, in the east (the Lau basin)—is the probable cause of much vicariance. In a similar way, the closure of the basins would have led to overlap, or at least juxtaposition of biotas. The basins have formed in the backarc regions, and this backarc extension has equal significance for geology and biology. Backarc basins are built behind migrating island arcs, which lengthen and increase their curvature through time (Hamilton, 1988). The critical feature for backarc basin formation is the migration of the arc.

Hamilton (2007) argued that the conventional model of plate tectonics driven by bottom-heated, whole-mantle convection does not account for observed plate interactions. Instead, he suggested, hinge rollback is the critical factor in viable mechanisms. The Pacific basin plates spread rapidly yet shrink by rollback, whereas the subduction-free Atlantic widens by slow mid-ocean spreading. Hamilton (2007) argued that these and other first-order features of global tectonics cannot be explained by conventional models.

Volcanic arcs are produced in the overriding plate of a convergent margin when subduction is rapid enough (more than ~2 cm/year) for long enough (Stern, 2010). Intraoceanic island arc systems tend to form where old (Cretaceous or older), dense oceanic crust is subducted. Such lithosphere is prone to sinking vertically, causing rollback of the trench and its arc. With continued rollback, the arc will eventually collide with another arc, a plateau, or a continent. Subduction of *older* oceanic lithosphere favors rollback and development of an overall extensional strain regime above the subducting slab; this leads to the opening of backarc basins. In contrast, subduction of *younger* oceanic lithosphere creates a compressional convergent margin with the arc system migrating forward, away from the ocean basin.

In *retreating* orogens, as in the southwest Pacific, the velocity of the overriding plate (here the Indo-Australian) toward the hinge is less than the hinge retreat. Thus, retreating orogens undergo long-term extension in response to the slab rollback, and they are characterized by the opening of backarc basins. In contrast, in *advancing* orogens, such as the Andes, the overriding (South American) plate is advancing toward the downgoing plate faster than the hinge is retreating (Cawood et al., 2009). Advancing orogens are characterized by foreland basins in the backarc region, on the continent (termed the retro-arc). In contrast with backarc extension, which is horizontal, foreland basin tectonics involve vertical, flexural displacements that create belts of low hills, for example, in western Amazonia. Foreland basin dynamics, as in Amazonian Peru and Ecuador, and the extensional tectonics of backarc basins, as in the southwest Pacific, can both generate very high levels of biodiversity (Heads, 2012a).

Backarc Rifting as a Mode of Vicariance in Arc Biotas

If an archipelago is rifted apart, there will be little phylogenetic impact on groups that are already local endemics on individual islands. In contrast, any metapopulations that are widespread through the archipelago will be sundered, and so vicariance is possible. Backarc basins form by seafloor spreading and this can occur at a rate of 1–15 cm/year (some of the highest rates on Earth are recorded in the Lau basin). In just one million years, this will result in a basin 10–150 km wide.

Backarc tectonics in the Zealandia region since 100 Ma have been influenced by rollback of the different subduction zones, as well as underlying changes in plate motion and the arrival of the Hikurangi Plateau. Rapid retreat of the southwest Pacific subduction zone into the Pacific began in the Late Cretaceous, and by 85 Ma the original Pacific arc was separated from the Gondwana mainland by backarc spreading (Figure 3.15). In its early stages, the arc extended around the Gondwana coast from proto-New Guinea to proto-New Zealand, but with eastward retreat through the Cenozoic it has moved ~2000 km into the Pacific, pivoting at the Hikurangi plateau, and its convexity has increased. Volcanism along the arc would have been more or less continuous, with new islands constantly being produced, and both terrestrial and shallow-water marine organisms would have inhabited the arc as it moved east. This history accounts for high levels of endemism and diversity on the islands, for example, the plant family Degeneriaceae, endemic to Fiji, and *Scaevola gracilis* (Goodeniaceae), endemic to the Kermadec and Tonga Islands.

Slab rollback can cause basin formation in the backarc region, and this can lead to intra-arc extension splitting an arc lengthwise. In this case, both the descendant arcs can inherit at least part of the original biota. In Figure 3.15, the second-generation subduction zones with their arcs and then the third all developed alongside prior subduction zones and were then separated from them by backarc spreading. For example, continued southwestward retreat of the southern Solomon Islands–Vanuatu subduction zone (from 10 Ma) led to spreading in the backarc region (accommodated by transcurrent rifting along the Hunter fracture zone) and this formed the North Fiji Basin. The process has rifted apart the Vanuatu and Fiji sections of the arc that were once juxtaposed, and the two archipelagos together make up an important center of endemism. Many species of trees, for example, are only known from these two archipelagos (Smith, 1979–1996), consistent with the formation of the disjunctions by the backarc extension.

In the most recent example of lengthwise rifting, at about 7 Ma, the Tonga–Kermadec arc was rifted from the Lau–Colville arc, and the two have been drifting apart as slab rollback into the Pacific continues. The Tongan islands have become separated from the Lau group in eastern Fiji, and the original Tonga–Lau biota has been rifted apart. This explains why the plants and animals of the Lau group include so many Tongan and central Pacific elements that are absent in the main islands of Fiji, west of Lau (Smith, 1979–1996). The same backarc basin is propagating southward as a rift through the central North Island, the Taupo volcanic zone (Figure 3.15: Present).

The Waipounamu Erosion Surface

The extension that began in the mid-Cretaceous caused New Zealand to subside, and this continued through the Paleogene. Widespread erosion of the Otago Schist and the Albian graben breccias resulted in the peneplanation of much of eastern Otago. From the Silver Peaks near Dunedin, west to the Garvie Mountains and the Pisa Range, the Waipounamu erosion surface forms the conspicuous, flat tops of the block-faulted ranges. The surface has been preserved as a more or less horizontal feature, despite the Neogene uplift of the ranges (Cox and Sutherland, 2007; Rahl et al., 2011: Fig. 3). The erosion surface is probably composite and diachronous, with different parts eroded at different times (Bishop, 1994; LeMasurier and Landis, 1996), and by subaerial processes as well as marine transgression. A prominent unconformity surface seen in northwest Nelson and also on the seafloor off the South Island is thought to be a lateral equivalent of the Waipounamu erosion surface, and has been used to date its initial features at 105–85 Ma (Deckert et al., 2002). The overall age, or ages, of the erosion are more difficult to establish; for example, the oldest sedimentary rocks resting on the Waipounamu erosion surface in inland Otago (near the Dunstan Mountains) are Miocene.

At the Titri fault system, 10–15 km inland of the southeastern Otago coast (Figure 3.13), a major erosion surface separates the schist basement and the Henley Breccia from the overlying sequence (Litchfield, 2001). The latter comprises Late Cretaceous fluvial sandstones, mudstones, and coal measures, and Paleocene to middle Miocene marine sandstones, limestones, mudstones, and greensands. Here the surface was formed by fluvial erosion before the marine transgression.

Landis et al. (2008) recognized both a Cretaceous peneplain of terrestrial origin, and the Waipounamu erosion surface, of marine origin. The two unconformities are subparallel and in many localities the Waipounamu surface truncates the Cretaceous peneplain. LeMasurier and Landis (1996) admitted: "There is ample room for debate about how much levelling was fluvial and how much marine." The surface, or surfaces, developed over a long time period, from 85 to 22 Ma (Landis et al., 2008), and, to summarize, was probably caused by different processes at different times and in different areas.

Whether fluvial or marine, "it is clear that the Waipounamu surface formed at, or very near, sea-level" (LeMasurier and Landis, 1996). In many areas the covering sediments have been eroded away and it is not known how old they were, or whether they were fluvial or marine. In other areas, both fluvial and marine sediments are present; these indicate low-lying land with deltas and swamp forest and also narrow arms of the sea. The Gore Lignite Measures, for example, represent a marine delta with interdigitating marine and terrestrial conditions. There was continuous deltaic sedimentation in the area, although there may not have been continuous sedimentation at any one point (Pole, 2010). The marine incursions and the complex, dynamic landscapes with shifting coastlines would have caused differentiation in some groups and range expansion, overlap, and hybridism in others, while some, less weedy groups were probably extirpated.

Oligocene Flooding

The subsidence across Zealandia led to marine incursion in inland New Zealand that reached its maximum extent in the Late Oligocene. Some stratigraphers have even suggested that "there *may* not have been any land at all" (Campbell and Landis, 2001: 6). There is considerable biological, paleontological, and geological evidence for persistent land through the Oligocene though (this is reviewed in Chapter 8). Some relevant aspects of the limestones that were deposited during the marine incursions are discussed next.

Amuri Limestone and the Marshall Paraconformity

The Amuri Limestone forms the most widespread and distinctive Cenozoic strata along eastern New Zealand. It occurs intermittently for over 700 km from the southeastern North Island to the Waitaki River (Lewis, 1992), and it illustrates several key themes of mid-Cenozoic stratigraphy, tectonics, and biogeography.

Most of the Amuri Limestone is white biomicrite composed of coccolithophorid and foraminiferal fossils in a matrix of calcium carbonate mud. Both the base and top of the unit are time transgressive and are younger toward the south. In the southern North Island and in Marlborough, sedimentation began in Late Cretaceous or Paleocene time; in northern Canterbury, it began in the late Eocene; further south, in southern Canterbury and north Otago, most of the sequence is lower to mid-Oligocene. In the Clarence Valley, the Amuri Limestone reaches thicknesses of 400 m and includes spectacular sections (Rattenbury et al., 2006).

The top of the Amuri Limestone is capped by a mid-Oligocene angular unconformity, the "Marshall Paraconformity," that separates the limestone from the overlying strata. This break is usually interpreted as a current-induced erosion surface. It was first described as a paraconformity, that is, a subtle unconformity in which the strata are parallel and there is little erosion, but most authors now recognize it as an obvious unconformity (Thompson et al., 2014).

Tinto et al. (2011) wrote: "There is evidence for basin-scale tectonic activity before and after the Marshall Paraconformity,... any regional event to which the Marshall Paraconformity can be attributed must have occurred between 30–27.3 Ma... and have largely ceased by ~25 Ma when there was widespread limestone deposition across the basin...." The timing of resumption of deposition, at the end of the Oligocene, coincides with the change from extension to transpression in New Zealand plate tectonics.

Lewis (1992) concluded that the Marshall unconformity surface showed differential relief and that this developed by uplift, folding, and differential erosion. There was local emergence above sea level at the crests of the anticlines. Areas where the Amuri Limestone has been totally removed were inferred to reflect highs that were emergent for long enough (3–4 Myr at ~30 Ma) to allow total dissolution. The scale of the fold deformation was not substantial (the folds have amplitudes of tens of meters and wavelengths of kilometers), but in the low-lying landscape this would have been significant for the local ecology. The regional rise from at least outer shelf depths, then resubmergence after unconformity development, was an important event in the history of the offshore Canterbury Basin. Evidence of a correlative compressive tectonic regime is also present in North Otago, Canterbury, Marlborough, and the North Island.

In contrast with this analysis by Lewis (1992), Fulthorpe et al. (1996) suggested that the mid-Oligocene unconformity was caused by eustatic sea-level fall, not tectonics, and that the hiatus was nondepositional rather than erosional. Nevertheless, Lever (2007) replied that there are multiple unconformities from Eocene to Miocene, both on land and offshore, and that their ages often do not match the timing of falls in global sea level. This means they are more likely to have been caused by local tectonic movements. In addition, if global sea-level falls have caused the unconformities, or if they were formed by sediment starvation at maximum transgression, then the ages of the unconformities should match across the region. In fact, the unconformities do not appear to be synchronous. Again, this suggests that local tectonic or volcanic events have been responsible.

The great differences in thickness of mid-Cenozoic sediment that accumulated in adjacent subbasins across the South Island indicate that some kind of differential subsidence was occurring. Volcanism was occurring in numerous basins during this time, and in the Oamaru region of north Otago, volcanic uplift and subsidence can explain many of the unconformity surfaces (Lever, 2007).

Although the Oligocene in New Zealand is often considered to have been a period of passive subsidence with little or no tectonic activity, Lever (2007) concluded that there was significant deformation across the South Island throughout this epoch. There were marked differences in subsidence rates and evidence of areas where there are clear episodes of folding, faulting, and uplift, along with areas of volcanism. These localized tectonic events would explain the unconformities. Lever suggested dispensing with the term Marshall Paraconformity in the South Island, as it is an oversimplification of a complex history in which multiple unconformities developed. Lever's (2007) paleogeographic reconstruction indicated that areas that were "probably land" occurred throughout the Oligocene.

The Otekaike Limestone, lying above the Amuri limestone, occurs between the Waitaki River and Kakanui Range in north Otago and is dated as Late Oligocene–Early Miocene. Karstic features developed in the limestone following subaerial emergence and dissolution, and the erosion surface represents another unconformity above the Marshall Paraconformity (Lewis and Belliss, 1984). The widespread emergence in the Waitaki region during the Late Oligocene was a significant regional event, and it indicates a rapid sequence of emergence and resubmergence. Again, this contradicts the idea of a single, New Zealand-wide transgression followed by a great regression.

Cenozoic Deformation

Through the Cenozoic, New Zealand underwent marine incursions and also tectonic deformation. The scale and rate of deformation accelerated in the Neogene; this led to the bending of the orogen, horizontal and vertical displacement on the Alpine fault, and the development of distributed shear throughout large parts of the country. These topics are discussed later in the book, in the section on the Alpine fault (Chapter 9).

Kaikoura Orogeny

The Southern Alps of the central South Island were formed during the Kaikoura orogeny that effectively began in the late Miocene, although its earliest signs are evident earlier in the Miocene. At this time, the long period of extension across Zealandia that began at ~100 came to an end and was replaced with renewed compression. The change of tectonic regime was the result of changes in the orientation of continental collision between the Pacific and Australian plates. As a result, displacement on the Alpine fault changed from transcurrent to transpressive at ~5 Ma. Before this time, displacement along the plate boundary had been largely transcurrent for more than 10 Myr (Davey et al., 2007). Growth of the Southern Alps has been caused by the consequent reverse slip (Figure 3.6) on the Alpine fault (~5 mm/year from GPS data), and shortening of the crust east of the fault (Wallace et al., 2012). Stratigraphic data show that initial uplift in Central Otago took place at ~10 Ma (Craw et al., 2013). Fission track analyses indicate the earliest signs of uplift at 7–8 Ma at the southern end of the Alps (Haast Pass), at 5 Ma at the northern end (Arthur's Pass), and at 3 Ma along the southeastern margin, with greatest rates of uplift occurring over the last 1 Myr (Tippett and Kamp, 1993). Rapid sedimentation on the West Coast, indicating uplift, began at ~5 Ma (Sutherland, 1996).

Taupo Volcanic Zone

Subduction-related arcs stabilized in and around the North Island along the Three Kings Ridge–Northland Plateau trend from 23 to 18 Ma, along the Lau–Colville–Taranaki trend from 17 to 6 Ma, and along the Tonga–Kermadec–Taupo arc from 2 Ma to the present (Mortimer et al., 2010). Over the last 2 Myr, the backarc basin opening between the Lau–Colville Ridge and the Tonga–Kermadec arc has propagated southward and created a large, V-shaped rift in the central North Island, the Taupo Volcanic Zone (Figure 3.15: Present). This rift is a region of intense silicic volcanism and rapid extension of continental crust (Wilson et al., 1995). It contains New Zealand's largest lake, Lake Taupo (a caldera), and its only active volcanoes (Figure 3.3).

4

Introduction to the New Zealand Biota and Its Geography

This chapter discusses the two main models that have been used to interpret the New Zealand biota. It also introduces some of the most distinctive components of the biota and presents case studies illustrating the intercontinental affinities of the region.

The History of Ideas on New Zealand Biogeography

Many studies have addressed the question as to whether New Zealand groups have originated by chance, overwater dispersal, or by vicariance resulting from geological change. The debate goes back to the nineteenth century, when Wallace (1881) supported chance dispersal and Hutton (1884–1885, 1896) argued for vicariance.

Dispersal Models for the New Zealand Biota

Matthew's (1915) account supporting dispersal theory was very influential, and the ideas it proposed about global biogeography and evolutionary chronology later formed the basis of the Modern Synthesis. Ever since Matthew's paper was published, dispersal has been the standard explanation for the New Zealand biota (e.g., Oliver, 1925). Workers in this tradition have presented a biogeographic narrative in which the vast majority of New Zealand groups, or even all of them, have originated by dispersal from some other region (Fleming, 1979; Pole, 1994).

Matthew's (1915) ideas have persisted into the molecular era. For example, Winkworth et al. (2002b: 518) wrote that sequencing studies have provided "compelling evidence" for the importance of dispersal in establishing plant distributions in the southwestern Pacific, with "basal" species indicating the centers of origin (p. 516). Knapp et al. (2007) cited "overwhelming evidence" for the importance of late Cenozoic dispersal to New Zealand. Yet all the modern evidence for dispersal has been generated by analyses that will always find a center of origin (by assuming that this occurs at the site of basal grades), and that will always calculate young ages for clades (by stipulating in the priors that groups cannot be much older than their oldest fossil) (see Chapters 1 and 2). Reliance on these flawed methods has produced the modern support for chance dispersal, and it has also led to the avoidance of any in-depth engagement with the actual details of distribution and geology.

Modern supporters of the dispersal model sometimes assert, rather than demonstrate, that problematic aspects are "clear." For example, McGlone (2005) wrote that "the Australasian fossil record is clearly in conflict with a pure vicariance interpretation for many taxa," but no examples were cited. For the New Zealand flora, the "extraordinary evolutionary importance of recent long-distance dispersal" is "clearly evident from molecular data"; it is "obvious" (Winkworth et al., 1999: 1324). Waters and Craw (2006) wrote that "molecular methods now provide a clear means of distinguishing between recent dispersal versus ancient vicariance... even under very rough molecular calibrations"; in fact, the dates are problematic, and the problem of the priors means the potential error in them is unknown (see Chapter 2). McDowall (2008) argued that most of New Zealand's biota "clearly" arrived by dispersal following the rifting of Zealandia from Gondwana, and he concluded that recent (postrift) dispersal is "demonstrated" by "molecular information." Nevertheless, molecular information on its own does not provide the age of a group; sequence data need to be interpreted and the phylogenies need to be calibrated. McDowall

(2008) did not discuss the evolutionary models or the calibrations adopted in the molecular studies, and so his assessment of the work was unconvincing.

Kelly and Sullivan (2010: 3) suggested that "recent advances in paleontology and molecular phylogenetics have revealed in unprecedented detail how species have come and gone throughout New Zealand's past." Yet, in practice, the fossil-based model has failed to provide answers to many concrete questions in biogeography, and Kelly and Sullivan (2010) ended up depending on "extraordinary dispersal abilities" to explain the biota. In this approach, standard patterns of distribution are explained, not with reference to averages and normal processes repeated in large numbers of groups, but by rare or even unique events that are not correlated with any other factor, either physical or biological.

Campbell (2013) provided support for Matthew's (1915) and Fleming's (1979) dispersal theory, and argued that most groups in New Zealand have dispersed there since the marine flooding in the Oligocene. He suggested that "birds are probably responsible for transport and introduction of many small invertebrates (mollusks and crustaceans), and maybe vertebrates too (frogs)" (p. 51). He wrote that "more than 95% of all New Zealand native biota is descended from Australian ancestors," but no source was cited, and a review of molecular studies on New Zealand clades found that only 42 of the 101 groups studied had been assigned a center of origin in Australia (Wallis and Trewick, 2009). (Most of these 42 groups are discussed in this book, and the center-of-origin models are questioned.)

Following Campbell (2013), Mortimer and Campbell (2014: 229) wrote: "research has shown that more than 95% of the native biota of New Zealand is derived from ancestors in Australia and, furthermore, that these organisms dispersed to New Zealand within the past 20 million years or so." Again, the source of the 95% figure was not specified; with respect to the clade ages, the treatment of fossil dates as maximum ages was discussed in Chapter 2.

In their review of molecular studies, Wallis and Trewick (2009) listed the close affinities that have been confirmed between New Zealand groups and their sister groups in Australia, New Caledonia, other Pacific islands, South America, and other areas around the Pacific margin. (Southern Africa, Madagascar, central Asia, and the Mediterranean could be added to this list.) "Consequently," Wallis and Trewick wrote, "a consensus is emerging that dispersal has been a major process leading to the formation of the flora and fauna" (p. 3552). Yet the far-flung connections of New Zealand groups are just as compatible with ordinary, allopatric differentiation of global groups as with one-off, long-distance colonizing flights. The consensus among dispersalists relies entirely on the evidence generated by the center-of-origin programs and the dating programs.

In a review of molecular clock analyses, Wallis and Trewick (2009) proposed that, at most, only 10% of the New Zealand groups studied date back to the splitting of New Zealand from Gondwana. They concluded that "When genetic distance information alone is used, we need to be certain that differentiation is sufficiently low to implicate dispersal rather than Gondwanan origins..., which it clearly is in many cases" (p. 3557). Nevertheless, the calibrations and priors used in the different studies are critical for assessing this, and although Wallis and Trewick (2009) cited the "lack of good calibrations in many cases," they did not discuss them.

The "Paleozealandic Element" in Dispersal Theory

Dispersal theory has interpreted the New Zealand biota as a series of superimposed "elements" or "strata" that invaded the region at different times (Fleming, 1979). The oldest was termed the Paleozealandic element (Cockayne, 1921). Taxa in this group were thought to be the oldest extant New Zealand groups, and were assumed to have migrated into the New Zealand area before the breakup of Gondwana. Fleming (1979) treated the tuatara, the short-tailed bat *Mystacina*, and the ratite birds as Paleozealandic because of their phylogenetic isolation. He interpreted all the other extant New Zealand vertebrates as later, Cenozoic immigrants derived by long-distance dispersal, mainly from Australia. This was based on a literal reading of the fossil record—the time of a group's first appearance in the New Zealand fossil record was thought to indicate when the group migrated to New Zealand (Fleming, 1979: 107, 110).

Many molecular studies still depend on a similar method of interpreting the fossil record in order to calibrate phylogenies and establish which groups can, or cannot, be Paleozealandic (Wallis and Trewick,

2009). For example, Wagstaff and Breitwieser (2002) reviewed molecular studies on New Zealand Asteraceae. They concluded: "New Zealand separated from Gondwana some 80 million years ago and was isolated by the southern oceans for about 40 million years before the Asteraceae first appeared in the fossil record. Hence each of the major clades of New Zealand Asteraceae may relate to dispersal events across an oceanic barrier." (This can be compared with the Bayesian study of the family by Swenson et al., 2012, cited in Chapter 2.)

In the Asteraceae, lack of recognized fossils from the Mesozoic meant that the breakup of Gondwana was assumed to have no direct relevance. Yet there are many reasons why a group that is diverse and abundant now might not be preserved or recognized in the Mesozoic fossil record. Basing assumptions about age on this absence is dangerous and unnecessary. In any case, recent studies indicate that Late Cretaceous fossils (76–66 Ma) belong to an extant genus of Asteraceae (Barreda et al., 2015), confirming the predictions made by biogeographers.

The traditional dispersal theory for New Zealand accepted that a small Paleozealandic element was the result of vicariance. One group of dispersal theorists has argued that there is no convincing evidence for *any* New Zealand group having had a vicariant, Gondwanan origin (Pole, 1994, 2001; McGlone, 2005; Waters and Craw, 2006; Trewick et al., 2007; Goldberg et al., 2008; Landis et al., 2008), but this conclusion has been questioned by many other authors, including dispersalists. McDowall (2008), for example, followed Fleming (1979) in accepting that a small number of New Zealand taxa, the Paleozealandic element, originated prior to Gondwana breakup. He wrote (McDowall, 2010b: 1–2):

> I am not as ready as some to say "goodbye" to Gondwana and to attribute the entire biota to dispersal derivations (McGlone, 2005, and see Wallis and Trewick, 2009). This might seem surprising for a life-long dispersalist... It is possible that there was an ancient role for Gondwana in the freshwater fish fauna... a few elements in the freshwater fish fauna may reflect ancient Gondwanan origins, e.g. perhaps the species of the non-diadromous "pencil galaxias" complex (a group of small, subalpine species...).

McDowall (2010b) also accepted other New Zealand clades as Paleozealandic, including acanthisittid wrens, geckos, *Leiopelma* frogs, and several invertebrate groups: Onychophora, freshwater bivalves (Hyriidae), the freshwater crayfish *Paranephrops* (Parastacidae), its temnocephalid commensals (Platyhelminthes), freshwater insects such as *Nannochorista* (Mecoptera; Nannochoristidae) occur in Australia, Tasmania, New Zealand, Chile and Argentina), and perhaps aquatic crustaceans such as the phreatoicid isopods. In plants, McDowall (2010b) accepted *Agathis* (Araucariaceae) as Paleozealandic, but suggested that most modern plants, including *Nothofagus* (Nothofagaceae: Fagales), have reached New Zealand by dispersal (despite the existence of Cretaceous fossils of *Nothofagus* in New Zealand).

Tennyson (2010) referred harvestmen (Pettalidae; Boyer and Giribet, 2009), "basal" moths (Micropterigidae), and wetas (Orthoptera: Anostostomatidae) to the Paleozealandic element. Other authors have added certain aquatic insects (Gibbs, 2006), earthworms (Buckley et al., 2011), and an archaic, nonvolant mammal known from Early Miocene fossils (Worthy et al., 2006). Terrestrial turtles are another group only recorded in New Zealand as Early Miocene fossils, and Worthy et al. (2011b) attributed the group's presence to vicariance. The authors wrote that many of the Early Miocene clades, including a diverse herpetofauna, "represent lineages endemic to New Zealand and had poor dispersal capabilities, supporting the long held view that a part of the Zealandian fauna was vicariant in origin."

The Paleozealandic element and its composition remain controversial, but in dispersal theory, the group comprises, at most, a minute fraction of the biota with little relevance to the main biogeographic processes. McDowall (2008: 207) dismissed the groups as "rather idiosyncratic" taxa that "have little to tell us about our history." The patterns discussed in this book suggest the alternative idea that the "Paleozealandic element" is an artifact. The biogeographic patterns seen in the supposed "Paleozealandic" groups are no different from those of countless other indigenous New Zealand taxa, and the concept of a small, relictual "Paleozealandic element" whose members migrated to the New Zealand region before Gondwana breakup is not accepted here.

With the rise of plate tectonics theory in the 1960s and 1970s, continental drift became accepted and provided a simple mechanism for vicariance. Yet instead of rejecting the centers of origin and migration routes of the old biogeographic synthesis, dispersalists simply transferred these to Pangaea and Gondwana. Even Paleozealandic elements had ancestors that were assumed to have dispersed into the New Zealand region, rather than evolving there. For example, instead of seeing widespread southern (Gondwanan) groups as the result of vicariance with northern (Laurasian) sister groups, the southern groups were explained by dispersal from northern centers of origin (cf. Winkworth et al., 2002a, on *Myosotis*). Depending on the group, dispersal was inferred either before the rifting of Gondwana and Laurasia (with dispersal over land) or after rifting (with dispersal over water). In this way, the core concept of center of origin was preserved intact, despite the revolution in the Earth sciences.

Problems with the Dispersal Model: Paradox and Perversity in the New Zealand Biota

Many New Zealand endemics have global sister groups (see following sections) and provide an important indication of how distinctive the biota is. They are not predicted in a dispersal model; instead, they appear to be paradoxical and need to be explained by ad hoc hypotheses. Another paradox in dispersal theory is the unexpected absence of many groups in New Zealand. For example, the widespread coffee family (Rubiaceae) is represented in New Zealand by *Galium* and *Coprosma*. The latter is one of the most diverse plant genera in the country and probably the most ubiquitous. On the other hand, the tea family, Theaceae, is notable by its complete absence in New Zealand, both living and fossil, although it is diverse and widespread in montane New Guinea, Indonesia, China, Japan, and the Americas. These presences and absences in New Zealand are not correlated with any obvious features of ecology or means of dispersal, and they are attributed here to historical factors.

The absence of many Australian groups in the New Zealand biota is of special interest. In dispersal theory, the New Zealand biota is a subset of Australia's biota (Chapter 1), and oceans such as the Tasman are not barriers to dispersal, but rather "hurdles which, given enough time, are overcome" (Pole, 1994). Yet many authors have noted surprising absences from New Zealand of groups that are diverse and widespread in Australia. Leiopelmatid frogs occur in New Zealand and North America, and are sister to all other extant frogs. Dispersal theorists admit that the presence of leiopelmatids in New Zealand, and the *complete absence* of "all other frogs" there—despite this clade's high diversity in Australia—"might seem odd" (Waters and Craw, 2006). Yet it is only odd if dispersal theory is accepted; if it is not, the pattern can be treated as a typical case of allopatry caused by simple vicariance.

In addition to the large clade of frogs, many other groups are diverse in Australia but absent in New Zealand, and some of these are excellent dispersers. For example, in birds, Hutton (1872) thought it remarkable that New Zealand has no cockatoos (Cacatuidae) or grass-parakeets (*Neophema*, *Polytelis*, etc.), as these are so common in Australia and Tasmania. Orchids produce large amounts of dust-like seed, but Rupp and Hatch (1945) emphasized that genera such as *Diuris* or *Cymbidium*, which are abundant in Australia, are absent in New Zealand. It is now known that the mycorrhizal fungus of both genera (*Tulasnella calospora*; Warcup, 1971; Nontachaiyapoom et al., 2010) is present in New Zealand (Watkins, 2012), and so the absence of the orchids remains a paradox for dispersal theory.

In another example, the peregrine falcon (*Falco peregrinus*) has great powers of flight and is widespread through the New World and the Old World, including Australia and New Caledonia. Yet, "perversely," it does not occur in New Zealand (Trewick and Gibb, 2010: 235). This pattern of absence is significant as it is repeated so often; many groups that are otherwise found worldwide are absent in the indigenous New Zealand biota. A plant example is the family Ceratophyllaceae. Many of these absences reflect simple allopatry. For example, *Falco peregrinus* as such is absent from New Zealand, but it is represented there by the endemic *F. novaeseelandiae*, suggesting that the absence of the former is the result of phylogeny, not the vagaries of chance dispersal. The absence of *Falco peregrinus* from New Zealand can always be explained by citing extinction in New Zealand or competition with

F. novaeseelandiae and lack of niche space, but both these suggestions are ad hoc, and there is no actual evidence for either. As many as six species of *Falco* overlap in southeastern Australia, indicating that coexistence is not a problem there.

Trewick and Gibb (2010: 243–244) found another apparent paradox and wrote:

> A putative [paleo-] Zealandian element might include currently unplaced lineages (e.g., *Aptornis* [a fossil bird]), and taxa on long branches with nodes basal to their respective group (e.g., large parrots, New Zealand wrens, New Zealand pigeon). But a paradox exists, as at least some of these lineages include populations on young offshore islands [Norfolk Island, Chatham Islands] that indicate a retained capacity for dispersal and colonization over substantial areas of ocean.

Yet it is well known that many young islands host old endemics, often because of former islands in their vicinity (Heads, 2011). The assumption that the young, offshore islands have been colonized only from the mainland overlooks the many former islands on outlying parts of the New Zealand plateau that have been submerged by rifting, subsidence, and erosion.

To summarize, chance dispersal theory regards some of the most distinctive aspects of the New Zealand biota as paradoxical, odd, or perverse and is unable to explain them. This indicates the weakness of the theory. Rather than being paradoxes, distributions in groups such as the frogs, orchids, and birds that have just been mentioned all shed valuable light on evolution in and around New Zealand.

A Vicariance Model of New Zealand Biogeography

Although the dispersal paradigm of New Zealand biogeography has remained dominant over the last century, there have always been dissenting opinions. For example, in botany, the fossil record "suggests that the composition and diversity of the New Zealand flora during the Cenozoic matched that of Australia, and there may be little requirement to invoke trans-Tasman dispersal in order to source the modern angiosperm flora of New Zealand" (Lee et al., 2001: 353). In zoological work, Edgecombe and Giribet (2008: 11) wrote: "It is currently popular to view long distance dispersal as the main driving force shaping diversity in New Zealand," but in their own study of trans-Tasman centipedes, these authors inferred ancient vicariance.

There are still many problems with vicariance models. The mechanisms for vicariance in the Tasman region proposed in many studies have often been limited to the breakup of Gondwana. These studies have all assumed the biogeographic unity of Zealandia, and most have overlooked prebreakup events such as intracontinental rifting. Nevertheless, ancestral biotas were in the region long before Gondwana breakup, and they experienced a complex history of tectonic disturbance through the Paleozoic, Mesozoic, and Cenozoic. For example, the Tasmania–New Zealand center of endemism itself cannot be the result of the breakup of Gondwana, but it could have been caused by a prior break between Tasmania plus New Zealand on one hand, and mainland Australia on the other. Groups such as the basal parrots (*Nestor* and *Strigops* of New Zealand), the basal passerines (Acanthisittidae of New Zealand), and the basal angiosperms (*Amborella* of New Caledonia) probably all differentiated from their sisters long before seafloor spreading began in the Tasman.

The examples discussed next illustrate the model of spatial evolution described in Chapter 1. Three processes are accepted here: the development of allopatry (including speciation) by vicariance, movement of individuals and range expansion by normal means of dispersal, and extinction. Sympatric speciation appears to be much rarer than has been thought, and it is unlikely to explain the shared, sympatric patterns of many groups. Chance dispersal is rejected here as a mechanism of speciation.

In the model used here, allopatry is caused by vicariance (not chance dispersal), and overlap is the result of range expansion by normal means of dispersal (not chance dispersal). Both processes are determined by tectonic and climatic change, together with evolutionary development. In this model, chance events are not used to explain patterns shared by groups with different ecologies and means of dispersal.

The New Zealand Biota and Its Global Significance

The special biogeographic significance of New Zealand was recognized by T.H. Huxley (1868, 1873). In his analysis of extant terrestrial vertebrates, he proposed four main regions:

New Zealand.

Australia, New Guinea, and the Philippines.

South and Central America north to Mexico.

The rest of the world.

A modern study of species distributions in amphibians, birds, and mammals found four main regions, with the following arrangement (Holt et al., 2013: Fig. S1):

Australasia: Australia + New Zealand.

Latin America: South America, Central America, and the Caribbean.

Old World: Africa (including the Arabian Peninsula), Madagascar, India, SE Asia, and Oceania.

North temperate: Nearctic, Palearctic, and Sino-Japanese.

As in Huxley's (1868, 1873) analysis, the results highlight the key significance of Australasia. The "basal" region has two components, Australia and New Zealand (the study did not include reptiles, in which the New Zealand tuatara, *Sphenodon*, is sister to all other snakes and lizards).

New Zealand endemic taxa are of special interest, and many have a high taxonomic rank. For example, the mite *Chirophagoides* represents an endemic New Zealand subfamily, while the fly *Mystacinobia* represents an endemic family. Thus, the fly might be regarded as older than the mite and perhaps more worthy of conservation. But neither of these rankings add any empirical information to the fact that both groups are endemic clades; the rankings only reflect the traditional conventions in mite taxonomy and fly taxonomy, and a family of flies cannot be regarded as having any more, or less, significance than a subfamily of mites. Instead of focusing on the *rank* of the endemic clades, the clades can be examined in relation to their sister groups, and this is the approach taken in the following section.

Globally Basal Endemics in New Zealand

Many New Zealand endemics belong to diverse groups that also have endemics in many other countries. For example, many New Zealand forms have sister groups in Australia or New Caledonia, and so have no particular phylogenetic significance at a global scale. In contrast, some of the most interesting endemics in the New Zealand region have sister groups that are diverse and distributed more or less worldwide.

Many groups are endemic in the lands and islands surrounding the Tasman and Coral Sea, including New Zealand. Of these endemics, as many as 58 have sister groups that are circumglobal or at least pantropical (Heads, 2014, and see in the following text). For example, the dolphin *Tursiops australis* of Victoria, Australia, is sister to the rest of the genus, and these other species occur worldwide outside polar seas (Moura et al., 2013).

Of the "globally basal" Tasman endemics, 38 are in New Zealand (listed in the following text), and 22 are endemic there. These groups display high levels of biodiversity, as they combine a localized distribution with a diverse, more or less worldwide sister group (Heads, 2014). In this way, they have special significance for biogeography, phylogeny, ecology, and conservation, and represent the region's biological "crown jewels." They are not just local endemics, of whatever taxonomic rank, but are alternative versions of common, worldwide groups. In this way, they justify Diamond's (1997) view that New Zealand's flora and fauna are the nearest thing on Earth to an alien biota.

The New Zealand endemics in the following list are indicated with asterisks.

Flora

*1. The red alga *Apophlaea* is endemic to the New Zealand mainland, the subantarctic islands, and the Chatham Islands. Its sister is a clade of *Hildenbrandia* that is widespread throughout the north and south temperate zones (Sherwood and Sheath, 1999, 2003).

*2. The red alga *Dione* of the South Island is sister to the rest of the global order Bangiales (Sutherland et al., 2011).

*3. The red alga *Minerva* of New Zealand is sister to the rest of the Bangiales except *Dione* (Sutherland et al., 2011).

*4. The red alga *Lysithea* of the New Zealand subantarctic islands is sister to a global clade of Bangiales (Sutherland et al., 2011). It is recorded from Macquarie, Auckland, and Antipodes Islands (W. Nelson, pers. comm., November 23, 2011).

5. The red alga *Lembergia* Saenger (*Lenormandia allanii* Lindauer) is known from the Three Kings Islands and Aupouri Peninsula in the far north of New Zealand. It is sister to *Sonderella* of Victoria, and the two genera form a separate tribe, the Sonderelleae, in Rhodomelaceae (Phillips, 2001). Choi et al. (2002) found that *Sonderella* is basal in Rhodomelaceae, the largest family of red algae and distributed worldwide (*Lembergia* was not sampled).

6. In coralline red algae (subclass Corallinophycidae, four orders), *Corallinapetra novae-zelandiae* from northern New Zealand (Stephenson Island in Whangaroa Bay) and an undescribed form from southeastern South Australia (Beachport) is sister to the cosmopolitan clade of Hapalidiales plus Corallinales (Nelson et al., 2015).

7. The liverwort *Goebeliella* occurs in New Zealand (main islands, Auckland Islands, Chatham Islands) and New Caledonia. Its sister is a diverse, cosmopolitan group comprising Radulaceae, Frullaniaceae, Jubulaceae, and Lejeuneaceae (He-Nygrén et al., 2006; D. Glenny, pers. comm., September 12, 2007).

8. The liverwort *Dinckleria* ("*Proskauera*") of New Zealand and southeastern Australia is sister to the cosmopolitan *Plagiochila*, the largest liverwort genus (Heinrichs et al., 2006; Engel and Heinrichs, 2008). In New Zealand, *Dinckleria* occurs in the main islands, the subantarctic islands, and Chatham Islands. In Australia, the genus is in Tasmania and at the McPherson–Macleay Overlap (the McPherson Range–Macleay River region of central eastern Australia).

9. *Xeronema* (Asparagales) comprises *X. callistemon*, found on the Poor Knights Islands and Hen and Chicken Islands of northern New Zealand, and *X. moorei* of New Caledonia. It is sister to a large, cosmopolitan clade of Asparagales (Amaryllidaceae, Agavaceae) (Fay et al., 2000; Janssen and Bremer, 2004). (Conran et al., 2015, described dispersed cuticle fragments from Miocene sediments of Southland as "*Xeronema*-like," but this needs further study.)

10. In grasses (Poaceae), the trans-Tasman *Chionochloa* (New Zealand, Lord Howe Island, and Mount Kosciuszko in southeastern Australia) is sister to a global group comprising most of the subfamily Danthonioideae (Linder et al., 2010; Heads, 2014: 14, 216).

11. *Ranunculus* sect. *Pseudadonis* (Ranunculaceae) is diverse in the New Zealand mountains; it also has one species on Mount Kosciuszko, southeastern Australia, and one in Tasmania. Its sister is the worldwide clade *Hecatonia* + *Batrachium* (Hörandl and Emadzade, 2012).

*12. In morphological studies, *Myosurus minimus* subsp. *novae-zelandiae* of North and South Islands is sister to subsp. *minimus* of South Africa, Eurasia, Australia, and North America (Garnock-Jones, 1986).

13. *Drosera arcturi* (Droseraceae) of New Zealand, southeastern Australia, and Tasmania is sister to all other *Drosera* species (except *D. regia*), a global clade (Rivadavia et al., 2003).

14. The tree *Pennantia* (Pennantiaceae) is in New Zealand (South Island, North Island, Three Kings Islands), Norfolk Island, and along the eastern seaboard of Australia (from Sydney to Cairns) (Gardner and de Lange, 2002). The genus is sister to a cosmopolitan clade including Torricelliaceae, Griseliniaceae, Pittosporaceae, Apiaceae, and Araliaceae (Stevens, 2016).

15. The shrub *Teucridium* (Lamiaceae) of central New Zealand (North and South Islands) is sister to *Oncinocalyx* of the McPherson–Macleay Overlap (Cantino et al., 1999). This pair is sister to *Teucrium* (~300 species), which is almost cosmopolitan but absent in New Zealand.

16. A diverse clade of Asteraceae (*Pleurophyllum, Damnamenia, Celmisia, Pachystegia,* and *Olearia p.p.*) occurs in New Zealand (including the subantarctic and Chatham Islands), southeastern Australia, and New Guinea. Its sister is a global group (comprising the rest of *Olearia,* also *Aster, Brachycome, Vittadinia, Lagenophora,* etc.) (Brouillet et al., 2009). (The map in Heads, 2014: Fig. 6.9, was based on an earlier arrangement with the Australasian group including the South American *Chiliotrichum.*)

Fauna

1. In mollusks, Athoracophoridae is a family of terrestrial slugs and occurs in New Zealand (including the subantarctic and Chatham Islands), the Melanesian islands, and eastern Australia. The group is sister to the cosmopolitan Succineidae (Wade et al., 2006).

2. The centipede *Craterostigmus* has one species in Tasmania and the other in New Zealand (North, South, and Stewart Islands; Edgecombe and Giribet, 2008). Some parts of the genome show *Craterostigmus* as sister to all other extant centipedes, while others have it near the base but still sister to a cosmopolitan group (Giribet and Edgecombe, 2006).

*3. The mite *Chirophagoides* (Sarcoptidae) of New Zealand lives in association with the bat *Mystacina.* It is sister to a clade comprising *Chirnyssoides* of tropical America, and *Notoedres,* a diverse group found worldwide but known in New Zealand only from introduced mammals (Klompen, 1992).

*4. In Opiliones, the genus *Synthetonychia* (Synthetonychiidae, 14 species) is endemic to New Zealand (Stewart Island, western South Island, Marlborough Sounds, and Wellington to Taihape) (Forster, 1954). The family is the sister group of all other Laniatores, a worldwide group (Giribet and Sharma, 2015).

*5. In aphids (Hemiptera: Aphididae), *Aphis coprosmae* of the South Island has a worldwide sister group (Kim et al., 2011).

*6. Another clade of New Zealand aphids, comprising *Paradoxaphis, Aphis healeyi,* and *A. cottieri* (the "Southern Hemisphere clade"), also has a worldwide sister group (Kim et al., 2011).

7. The dipteran family Mystacinobiidae comprises a single, wingless genus, *Mystacinobia,* associated with the New Zealand bat *Mystacina* (cf. no. 3). The sister group of *Mystacinobia* is an undescribed species ("McAlpine's fly") from southeastern Australia and the two form a Tasman Basin clade that is sister to a worldwide complex, the family Sarcophagidae (Kutty et al., 2010; distributions from T. Pape, pers. comm., July 30, 2013).

*8. In the worldwide dipteran family Fanniidae, one of the New Zealand members, *Zealandofannia,* is monotypic and endemic there and only known from *Mystacina* guano (cf. numbers 3 and 6). It is sister to a worldwide clade (all the remaining Fanniidae except *Australofannia, Piezura,* and *Euryomma*) (Domínguez and Pont, 2014).

*9. In the littoral beetle *Cafius* (Staphylinidae), a clade of four mainland New Zealand species ("clade B") has a cosmopolitan sister group, at least in two of the three phylogenies presented (Jeon et al., 2012; K.-J. Ahn, pers. comm., December 26, 2012).

*10. In Lepidoptera, the family Mnesarchaeidae is endemic to New Zealand (North and South Islands) and is sister to the more or less global group Hepialoidea (ghost moths; not in the Caribbean, West Africa, or Madagascar) (Gibbs, 1990a, 2006; pers. comm., August 7, 2007; J.R. Grehan, pers. comm., October 5, 2007).

*11. In Hymenoptera, the family Maamingidae is endemic to New Zealand (North and South Islands; Early et al., 2001) and is sister to the global family Diapriidae, including Mymarommatidae (Munro et al., 2011) (or to Diapriidae including Monomachidae; Castro and Dowton, 2006).

*12. The fish *Cheimarrichthys* is endemic to New Zealand (North and South Islands). It is diadromous, spending part of its life cycle in freshwater and part in the sea, near the coast. It is sister to a diverse, global clade of marine fishes, Leptoscopidae plus Pinguipedidae (Smith and Craig, 2007).

13. In reptiles, the New Zealand tuatara *Sphenodon* (along with its widespread, extinct relatives, the other members of order Rhynchocephalia) is sister to all extant lizards and snakes (order Squamata).

*14. In ducks (Anatidae), *Hymenolaimus* (blue duck) of the North and South Islands is sister to *Sarkidiornis* of tropical America, Africa, Madagascar, India, and southern China (Robertson and Goldstien, 2012).

*15. In Gruiformes s.str., *Aptornis*, the adzebill, was a flightless, turkey-sized bird of the North and South Islands that went extinct in historical times. Its combination of morphological features is "truly remarkable" (Livezey, 2011), and it makes up a clade with Rallidae (cosmopolitan) and Heliornithidae (pantropical). The three groups form a tritomy in analyses that have included a small amount of molecular data from *Aptornis* (Houde, 2009). Whether *Aptornis* is sister to Rallidae, to Heliornithidae, or to both, it is sister to a globally widespread group.

16. Aegothelidae (Aegotheliformes), the owlet-nightjars, are endemic to the Tasman–Coral Sea region. They are known from New Guinea (including the D'Entrecasteaux Archipelago, and Maluku, but not the Bismarck Archipelago), New Caledonia, eastern Australia, Tasmania, and New Zealand (extinct in the last locality by ~AD 1200). The family (the only one in its order) is sister to a cosmopolitan clade, the Apodiformes (swifts and hummingbirds) (Pacheco et al., 2011).

*17. In the seabird *Puffinus* (Procellariidae), a clade of North and South Islands (*P. huttoni, P. gavia*, and, based on morphology, the fossil *P. spelaeus*) is sister to a group of 23 species and subspecies with a global distribution (Austin et al., 2004).

*18. In the wader family Charadriidae (Charadriiformes), *Charadrius bicinctus* (banded dotterel; North, South, Chatham, and Auckland Islands), *C. obscurus* (New Zealand dotterel; North, South, and Stewart Islands), and *C. frontalis* ("*Anarhynchus*," the wrybill plover of Canterbury riverbeds) form a New Zealand clade that is sister to a worldwide complex of 14 *Charadrius* species (dos Remedios et al., 2015).

*19. In parrots (order Psittaciformes), the New Zealand superfamily Strigopoidea is sister to the rest (Rheindt et al., 2014). The superfamily comprises two genera, *Strigops* (kakapo; South Island and Stewart Island) and *Nestor* (kaka and kea; South Island, Stewart Island, and North Island; present on Norfolk and Chatham Islands in historical times but now extinct there).

*20. In passerine birds, the New Zealand wren family Acanthisittidae (North, South, and Stewart Islands), with two extant species, is sister to the rest (Cracraft and Barker, 2009; Aggerbeck et al., 2014). Two acanthisittid species went extinct in historical times and fossil representatives are also known, but only from New Zealand (Worthy et al., 2010).

*21. The two, diverse, worldwide clades of passerines are the sister groups Passerida and the Corvida. In the Passerida, the New Zealand clade Callaeidae (Callaeatidae) plus *Notiomystis* is sister to the rest (Ewen et al., 2006; Driskell et al., 2007; Aggerbeck et al., 2014). The clade was widespread in North, South, and Stewart Islands until historical times, but is now rare.

*22. In the Corvida ("core Corvoidea"), the Mohouidae of New Zealand (comprising *Mohoua* s.lat.: North, South, and Stewart Islands) are sister to all the rest (Aggerbeck et al., 2014; Jønsson et al., 2016).

In other Tasman groups, traditional morphology suggests a global sister, but further work is needed to confirm this. For example, the glow-worm *Arachnocampa* (Diptera: Keroplatidae) has two main clades: one in New Zealand (North and South Islands) and one along the eastern seaboard of Australia from Tasmania to northern Queensland (Baker et al., 2008). The genus makes up the Arachnocampinae, one of four subfamilies in Keroplatidae. Of the other subfamilies, Keroplatinae and Macrocerinae are each diverse and worldwide, while Sciarokeroplatinae comprises a single species from China. It is quite possible that the sister group of *Arachnocampa* will prove to be a worldwide clade.

Likewise, in wasps, some unusual characters suggest that *Kiwigaster* (Braconidae) of New Zealand might be the sister group of the rest of its subfamily, the worldwide Microgastrinae, although comprehensive phylogenetic analyses have not yet been carried out (Fernández-Triana et al., 2011).

Interpretation of the Globally Basal Groups

The Tasman region and New Zealand endemics with global sisters are of obvious interest to conservationists, but how should they be interpreted in evolutionary terms? Consider the two clades of globally basal aphids cited as numbers 5 and 6 in the "Fauna" list. Their significance was recognized in the genus name *Paradoxaphis*, and by Teulon et al. (2003: 10), who wrote:

> Accepted dogma is that the southern Aphidinae (i.e., *Aphis, Paradoxaphis, Euschizaphis, Casimira*) are descendants of recent chance trans-tropical immigrants from the Northern Hemisphere… However, recent molecular work has found that a group of four New Zealand endemic aphids belonging to the genera *Aphis* and *Paradoxaphis* form a highly supported lineage (possibly basal in the tribe Aphidini)… These results place this New Zealand group as central to the evolution of the species-rich Aphidinae, which contains many agricultural pests….

Note that New Zealand's status as "central to the evolution" of the Aphidinae need not mean that New Zealand is a "center of origin" for the group.

Among other New Zealand aphids, *Neophyllaphis* (feeding on *Podocarpus*) and *Sensoriaphis* (feeding on *Nothofagus*) have been cited as primitive genera with Gondwanan distributions (Teulon et al., 2003). *Neophyllaphis* is a close morphological match of one of the oldest fossils in Aphididae, dated as Late Cretaceous.

All the Tasman region clades listed, including 22 New Zealand endemics, have global sister groups and so cannot be explained by dispersal to the region without also invoking other ad hoc hypotheses. The simplest explanation is that they have been derived when global groups developed primary phylogenetic breaks in the Tasman or New Zealand regions. This has been followed in most cases by local overlap, but in *Teucridum*, the sister group is absent in New Zealand. The tuatara (*Sphenodon*), one of the groups, has widespread fossil relatives (perhaps the result of secondary range expansion, overlapping with squamates), but none of the others have a known fossil record outside their present range.

In many cases, a group of plants or animals has a smaller range than its sister, but this does not necessarily imply any extinction or chance dispersal. The pattern will be observed whenever a phylogenetic break does not occur in the exact center of the ancestor's range. For example, the New Zealand wrens are the sister of all other passerines. Trewick and Gibb (2010) suggested that the wrens could have evolved in a wider area of Gondwana before the separation of Zealandia, and later went extinct everywhere except New Zealand. Yet this is ad hoc and unnecessary if the split between the New Zealand wrens and the rest was a case of simple vicariance. If this occurred in a more or less global ancestor, between the area that later became New Zealand and the rest of the world, there is no need to invoke any extinction, only range expansion of the "other passerines" clade into pre- or proto-New Zealand. Patterns of differentiation in the "other passerines" suggest that this expansion developed, at the latest, during the mid-Cretaceous, a time of major predrift rifting, core complex emplacement, volcanism, and marine incursion. The overall pattern is repeated in many New Zealand groups with a global sister, and so the simplest possible explanation is desirable.

Many New Zealand groups have sisters that, while not global, are still very widespread, and so again, it is simpler to explain the evolution in these clades by vicariance than by dispersal. For example, the New Zealand brachiopod *Novocrania huttoni* (Craniidae) is sister to a group that is widespread through the Pacific (Japan to Antarctica, the central Pacific, South America) and in South Africa (Cohen et al., 2014). The group is one of the six main clades in the family Craniidae. The authors concluded: "Overall, it seems unlikely that dispersal from a centre or centres of origin has been the most important long-term control on craniid distribution. Intrinsic reproductive factors like those already discussed and tectonic vicariance (continental drift) are more likely to have been important."

To summarize, New Zealand, along with eastern Australia, New Caledonia, and New Guinea, is part of the Tasman–Coral Sea region, an area characterized by a high number of endemics with global sister groups. This does not mean that the region has been a center of origin for the global groups, but it can be explained if the region represents the basal node in the groups. The region thus forms a key break zone for global biogeography and also a major tectonic boundary (between the Indo-Australian and Pacific plates). The simplest explanation is that the two phenomena are related.

Intercontinental Affinities of the New Zealand Biota

At some level in the phylogenetic hierarchy, all members of the New Zealand biota, apart from the globally basal groups, display phylogenetic affinities with particular localities outside the region. Some of these are introduced next. A more detailed study of New Zealand's biogeographic affinities with other regions (Heads, 2014) was based on molecular work. The present volume considers New Zealand in greater detail, and focuses on distribution patterns within the country. These all have close relationships with broader affinities outside the country, and in many cases different parts of New Zealand are more closely related to regions outside the country than with other areas within it.

The distributions discussed in this and the following chapters are based on the molecular evidence where this is available, but morphological studies have been used for information on areas and groups where molecular surveys have not been carried out.

Simple Allopatry: The Mite Harvestmen (Opiliones, Suborder Cyphophthalmi)

In Opiliones, or harvestmen, the basal node separates the mite harvestmen (suborder Cyphophthalmi) from the diverse, cosmopolitan, typical harvestmen. The mite harvestmen are litter-dwellers just a few millimeters long and over recent years have been the topic of numerous, detailed studies by Giribet and colleagues. Local New Zealand clades are discussed in subsequent chapters, and the global differentiation of the group is introduced here.

The mite harvestmen as a whole have a simple distribution; they are a global group in which the main branches display perfect allopatry, with no overlap. The three main clades are endemic to southern, northern, and tropical regions, respectively, and display simple allopatry (Figure 4.1; Giribet et al., 2012).

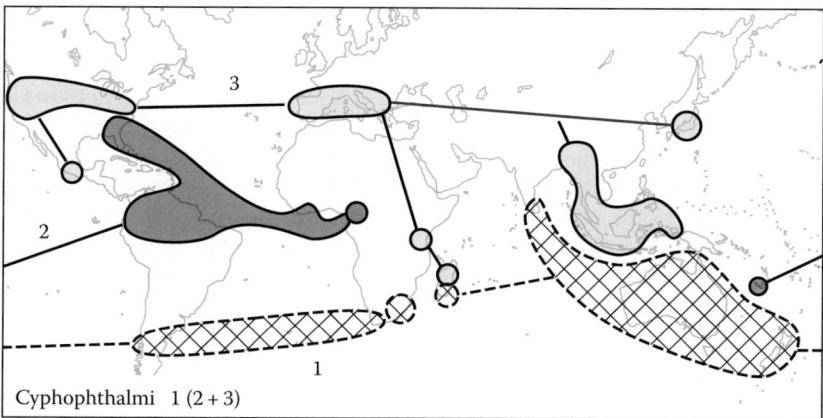

FIGURE 4.1   Distribution of Opiliones suborder Cyphophthalmi. 1, (cross hatching) Infraorder Scopulophthalmi: Pettalidae; 2, (dark gray) Infraorder Sternophthalmi: Troglosironidae (New Caledonia) + Neogoveidae (America and Africa) + Ogoveidae (Cameroon); 3, (light gray) Infraorder Boreophthalmi: Sironidae (America, Mediterranean, Madagascar, Japan) + Stylocellidae (Southeast Asia). The genera in Kenya and Madagascar have not yet been sequenced and their position in the phylogeny is based on morphological analyses. (Data from Giribet, G. et al., *Biol. J. Linn. Soc.*, 105, 92, 2012.)

Nevertheless, the "southern" clade is not simply southern, as there is a significant locality in Sri Lanka, and the "northern" clade extends south to Madagascar. The "tropical" clade is restricted to warmer areas, but it is not in East Africa or Asia. All three clades include trans-Atlantic affinities, and there are major breaks among the three in the central Africa/Madagascar region, and along an arc: New Caledonia–New Guinea–Indonesia. The mutual allopatry of all three clades, and also the six families within them (see caption to Figure 4.1), is consistent with an origin by simple vicariance of a cosmopolitan ancestor. Any significant range expansion of any of the clades following this vicariance would have led to overlap and disturbed the allopatry. The simplest explanation is that differentiation among the three main clades took place before continental breakup in the Mesozoic rifted apart the individual clades.

Cyphophthalmi are distributed worldwide, but have a patchy, "skeletal" distribution. Their sister group, comprising all the other harvestmen (three suborders), has a genuine cosmopolitan distribution. Critical questions here are as follows: How did the two groups differentiate to begin with? When did the great overlap between the two groups develop? The overlap has been extensive, yet although Cyphophthalmi are widespread, there are no records from much of central Africa and South America, as well as most of New Guinea and the Caribbean. One possibility is that this broad area of absence represents the original distribution of typical harvestmen.

The fossil record of Cyphophthalmi is scarce, although a specimen in Early Cretaceous amber from Burma probably belongs to one of the modern families, Stylocellidae. In contrast, fossils of typical harvestmen are known from the Paleozoic, and a clock study estimated that the families of Cyphophthalmi diverged in the Mesozoic (Giribet et al., 2012). Giribet et al. concluded that "the biogeographical data show a strong correlation between relatedness and formerly adjacent landmasses, and oceanic dispersal does not need to be postulated to explain disjunct distributions, especially when considering the time of divergence." The authors also found that the faunas of the current landmasses are not monophyletic units, and that cladogenesis occurred prior to the breakup of Gondwana. These important conclusions probably apply to many other groups in New Zealand.

The allopatry of the three main Cyphophthalmi clades and the six families is consistent with an origin by simple vicariance, more or less *in situ*, from a global ancestor. Despite this, Giribet et al. (2012) assumed a much more complex history, in which each individual family evolved at a restricted center of origin. These were located using a standard center-of-origin program (DEC implemented in LAGRANGE; Ree and Smith, 2008; see Chapter 1). As a result, the authors proposed separate dispersal scenarios to explain the simple, allopatric pattern. For example, the authors proposed an origin for the southeast Asian family, Stylocellidae, in the Malay Peninsula, followed by dispersal north to China and east to New Guinea. They suggested that the family "includes the only reported cases of possible trans-oceanic dispersal in Cyphophthalmi." This was invoked to explain the family's presence in areas such as the Bird's Head Peninsula of northern New Guinea. Yet the global distribution of the Cyphophthalmi clades suggests that the boundary between Stylocellidae + Troglosironidae in New Guinea, and so on, and their sister group Pettalidae in Queensland, represents the phylogenetic break at the origin of the groups, rather than a subsequent juxtaposition.

For the "southern" group, Pettalidae, Giribet et al. (2012) calculated a center of origin either in South Africa, in South Africa + Sri Lanka, or in South Africa + New Zealand. Yet a center of origin restricted to any one of these would not account for the allopatry of the three trans-Atlantic clades in the suborder, while a vicariance model would. Boyer and Giribet (2007) inferred a vicariance model for the diversification *within* Pettalidae, and the evidence suggests that this also applies to the origin of the group itself.

The New Caledonian Troglosironidae is related to forms in Colombia, giving a trans-tropical Pacific connection between Zealandia and *tropical* South America. This affinity differs from the well-known connection between Australasia and *southern* South America. Nevertheless, it is a standard one, and examples are given elsewhere (Heads, 2014: 271–274). Another example overlooked in that list occurs in *Oxalis* (Oxalidaceae), where the following species pair has been proposed in morphological work (Lourteig, 1979):

> *O. exilis*: Australia, New Zealand (Kermadec Is. to Stewart I.), Norfolk Island, and New Caledonia.

> *O. filiformis*: Ecuador, Colombia, and Costa Rica (GBIF, 2016).

In a similar case, the moss *Bucklandiella angustissima* (Grimmiaceae) is known from New Zealand (South, Auckland, Campbell, and Macquarie Islands) and the Ecuadorian Andes (Bednarek-Ochyra and

Ochyra, 2011). The trans-tropical Pacific connection has been related to central Pacific tectonics and magmatism in the Jurassic-Cretaceous, followed by terrane accretion at the Pacific margins (Heads, 2012a, 2014).

Giribet et al. (2012) studied habitat suitability models for the four main clades in Cyphophthalmi (clades 1 and 2 in Figure 4.1, plus the two families in clade 3). In all four, they found that the actual clade distribution is much more restricted than the potential distribution, as defined by the modeled suitable habitat. Thus, the ecology of the groups does not determine their biogeography and the result "constitutes further evidence for the old cladogenesis and low dispersal abilities of Cyphophthalmi… This pattern also corroborates the hypothesis that tectonic movements and vicariance events have defined distributions and driven diversification in this group of soil arthropods" (Giribet et al., 2012).

Despite this support for tectonics and vicariance, Giribet et al. (2012) did not adopt an overall vicariance model for the allopatric families. In their dispersal model for the families, they found that "mysteries remain because certain temperate clades have migrated to warmer climates (e.g., *Pettalus* in Sri Lanka or *Austropurcellia* in Queensland, Australia)…." A migration model is not adopted here, and so these two genera are not regarded as mysteries, but as normal allopatric vicariants. Their distribution can be explained most easily, not by climatic change or migrations, but by phylogenetic breaks related to tectonics. The clades are not simply "temperate" or "tropical," as each of the three main clades occurs in both temperate and tropical regions. Clades have persisted in their original areas where there has been any viable habitat, through periods of great tectonic and climatic change.

To summarize, a simple vicariance model for the precise allopatry of the mite harvestmen predicts that the groups have survived within their respective regions wherever suitable ecology has been present, but that the regions themselves were determined by geology and phylogeny, not by ecological factors or migration. The New Zealand clade is part of a group that spans Indian, Pacific, and Atlantic Ocean basins, but with a distribution that cannot be described as simply "southern" or "temperate."

Global Allopatry and Local Overlap: New Zealand Shags (Phalacrocoracidae)

In many global groups, the component clades display a high level of allopatry, but some interdigitation or overlap at the break zones blurs the margins, at least locally. This sort of pattern is a little more complex than the simple allopatry seen in Cyphophthalmi, and it is illustrated here with the shags (cormorants).

Of the 359 known species of seabirds, 84 (23%) breed in the New Zealand archipelago and 35 (9.7%) are endemic there. This is the highest regional diversity of seabirds in the world (Taylor, 2000). (Species numbers will be higher now than in 2000 as the result of taxonomic inflation, but the proportions will be similar.) For the main seabird groups in New Zealand, the numbers of species breeding there, compared with the number of species worldwide, are as follows: Diomedeidae (albatrosses), 13/24 = 54%; Procellariidae (petrels and shearwaters), 28/70 = 40%; Spheniscidae (penguins), 6/17 = 35%; Phalacrocoracidae (shags), 12/39 = 31%; and Sternidae (terns and noddies), 10/43 = 23%.

Phalacrocoracidae have four main clades with the following phylogeny (Kennedy and Spencer, 2014) and distributions (Orta, 1992; Holdaway et al., 2001; Nelson, 2005):

> Old World (Africa and southeastern Europe to southern Asia and Australasia, including mainland. New Zealand) (*Microcarbo*, 5 species).
>> Chile and Peru (*Poikilocarbo*, 1 species).
>>> Old World to New Zealand and North America (*Phalacrocorax* s.lat., 15 species).
>>> Southern Ocean plus Americas (*Leucocarbo* s.lat., 19 species).

The phylogeny is well supported except for the position of the Mediterranean *Gulosus aristotelis*, which is still unclear; the bird could belong to *Phalacrocorax* (where it is mapped here) or to *Leucocarbo*, or be sister to both.

In New Zealand, *Microcarbo* is represented by a single species (*M. melanoleucos*). The two species-rich clades of shags, *Phalacrocorax* and *Leucocarbo*, both occur in New Zealand and are

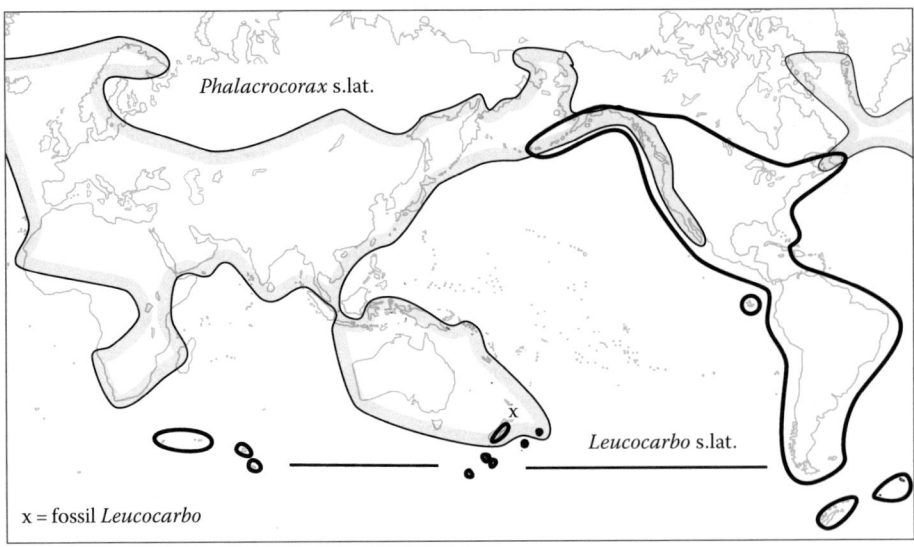

FIGURE 4.2   Distribution of the shag genera *Phalacrocorax* and *Leucocarbo* (Phalacrocoracidae), breeding records only. (Phylogeny from Kennedy et al., 2000, 2009; distributions from Orta, J., Phalacrocoracidae, in *Handbook of the Birds of the World*, Vol. 1, eds. J. del Hoyo, A. Elliott, and J. Sargatal, Lynx Edicions, Barcelona, Spain, 1992, 326–353; Holdaway, R.N. et al., *N. Z. J. Zool.*, 28, 119, 2001; Nelson, J.B., *Pelicans, Cormorants, and Their Relatives: The Pelecaniformes*, Oxford University Press, Oxford, UK, 2005.)

mapped in Figure 4.2 (breeding records only). They are both widespread globally but are mainly allopatric, with overlap restricted to New Zealand and the margins of North America. The break between them is not related to any obvious climatic factors, and any ecological explanation for their different distributions is obscure. *Phalacrocorax* cannot be described simply as "warmth loving," as it occurs in Alaska and Finland; *Leucocarbo* is not simply "subantarctic" as it is widespread in the American tropics. The overall allopatry between the two is probably not the result of competitive exclusion, as the groups overlap in New Zealand and North America.

A dispersal model prepared using one of the programs written for the purpose might calculate a center of origin for *Phalacrocorax* at a point in the northern hemisphere, for example, with dispersal south to southern Africa and Australasia. A center of origin for *Leucocarbo* might be proposed in the southern hemisphere, with dispersal northward through America. Where the two genera met, some overlap has developed. To begin with, the two point centers of origin would have been connected by a long-distance dispersal event, with one of the two sister genera being derived from the other. The dispersal event would need to have occurred soon after the origin of the first group, before it had undergone any differentiation, otherwise the second group would be nested in it, rather than the two being sisters. (A dispersal model assumes that the dispersal event did not occur *before* either of the two genera existed, as that would indicate they originated by vicariance.)

A vicariance model instead proposes that the *Phalacrocorax* and *Leucocarbo* originated by allopatric differentiation of a more or less global ancestor at breaks in the southwest and northeast Pacific. Subsequent range expansion has led to some secondary geographic overlap around the break zones.

Phalacrocorax is represented in New Zealand by five species. The genus occurs on the mainland (three species) and the Chatham Islands (two species, one widespread through the Old World), but is absent from the subantarctic islands. In contrast, *Leucocarbo* has seven New Zealand species, with only two on the mainland (extant in South and Stewart Islands, fossil in the North Island) and five on offshore islands, including endemics on Macquarie, Auckland, Campbell, Bounty, and Chatham Islands. Thus within New Zealand, the diversity in *Phalacrocorax* is concentrated on the mainland, while the diversity in *Leucocarbo* is concentrated on the islands south and east of New Zealand. These offshore islands are located on the Campbell Plateau and Chatham Rise, large areas of sunken continental crust.

The regional differentiation in shags between the mainland and the Campbell Plateau conforms to a common pattern, one that can be explained by the opening of the Great South Basin.

Distribution Straddling the Ocean Basins and Their Spreading Centers

Instead of examining the distributions of clades with a particular rank—species or genera, for example—the affinities of New Zealand groups can be traced back in the phylogeny until intercontinental relationships are indicated. This sort of analysis shows that New Zealand clades belong to groups that straddle one or more of the main ocean basins. These include the Indian, Pacific, and Atlantic basins, as well as the former basins of the Tethys seas that are now uplifted or subducted (Heads, 2014). The Tethyan affinities extend from the southwest Pacific through central Asia to the Mediterranean.

In many groups, the main clades of a group are allopatric and based around different ocean basins. For example, the main clade of ratites, with six families, comprises a trans-Indian Ocean group (including the kiwis, Apterygidae) and its sister, a trans-Pacific group (including the moas, Dinornithidae) (Heads, 2014: Fig. 3.21; Mitchell et al., 2014a). Overlap between the two clades occurs only in New Zealand, and so the pattern can be explained by vicariance of an Indo-Pacific ancestor, with subsequent, local dispersal within the New Zealand region. A similar pattern occurs in podocarp trees, with one clade comprising a trans-Indian Ocean group (*Afrocarpus* + *Nageia*) and its trans-Pacific sister (*Retrophyllum*, present in New Zealand as fossils) (Heads, 2014: Fig. 3.20).

The ground-dwelling leafhoppers, family Myerslopiidae (Hemiptera), are another typical example. There are three main clades (morphological studies by Hamilton, 1999):

Indian Ocean: Madagascar, Australia, and New Caledonia (Sagmatiini).

Pacific Ocean: Chile and New Zealand (Myerslopiini).

Juan Fernández Islands (Evansiolini).

Other patterns involve distribution across different combinations of basins. In the grass *Festuca*, New Zealand members belong to two clades that are not sister groups (Lloyd et al., 2007; Inda et al., 2008). One of the New Zealand clades (*F. multinodis*, *F. luciarum*, *F. coxii*, and *F. ultramafica*) is endemic to New Zealand and belongs to a group otherwise known from the western Mediterranean and Macaronesia. This gives a typical trans-Tethyan connection. The second group in New Zealand comprises five New Zealand endemics (*F. actae*, *F. deflexa*, *F. madida*, *F. mathewsii*, and *F. novae-zelandiae*) and the circumantarctic *F. contracta* (Patagonia and Falkland, Kerguelen, and Macquarie Islands). This group is sister to a clade of western North America (*F. californica* of the U.S. West Coast, from Washington to southern California, and *F. hintoniana* of Mexico). This gives a typical Southern Ocean–northeast Pacific affinity, as seen in other groups such as the seaweed *Macrocystis* (Heads, 2014: 118).

The next sections give examples of distribution around the respective ocean basins, starting with the Indian Ocean and working east.

Indian Ocean Groups: Examples from Ratites, Ducks, Beetles, and Plants

Many New Zealand groups span the Indian Ocean basin and the spreading center that has created it (Heads, 2014). Cases cited so far include the podocarps and leafhoppers referred to in the last section, and also the spiders of Linyphiidae, subfamily Mynogleninae (New Zealand, Fiji, Tasmania, Réunion, and tropical Africa) (Chapter 1). Other examples include the land snail family Rhytididae, known from New Zealand (with the giant land snail *Paryphanta*), Australasia, and South Africa (Moussali and Herbert, 2016). In carabid beetles, members of the subtribe Anillina with grooved elytra occur in Australia, New Zealand, and Madagascar (Sokolov, 2015). New Zealand lichens include 15 trans-Indian Ocean species (Galloway, 2007). In angiosperms, *Sarcocornia blackiana* and *S. quinqueflora* (Amaranthaceae) are each distributed in western and eastern Australia, New Zealand, and New Caledonia. Together with *S. globosa* of Western Australia, they form a clade that is sister to *S. xerophila* of South Africa (Steffen et al., 2015).

One particular trans-Indian Ocean basin pattern is exemplified by the kiwis (Apterygidae). They are endemic to New Zealand, while their sister group, Aepyornithidae (the extinct elephant birds), is restricted to Madagascar (Mitchell et al., 2014a). Likewise, in ducks, a widespread New Zealand clade (*Anas chlorotis, A. aucklandica, A. nesiotis*, and the extinct *A. chathamica*) is sister to *Anas bernieri* of Madagascar (Mitchell et al., 2014c).

The study of the *Anas* ducks was calibrated with a Miocene fossil, *A. soporata*. Mitchell et al. (2014c) wrote: "based on its plesiomorphic morphology, we judged *A. soporata* to be an outgroup to extant *Anas*," but no other details were given. As the authors admitted, "we consider *Anas soporata* to be a stem member of the genus; however, if this interpretation is incorrect and *A. soporata* is in fact crown-*Anas* then this would lead us to underestimate divergence dates." This illustrates the difficulty of placing a fossil at a precise point on a phylogeny, and the implications that this has for dating. Another problem is the question: just how much older than the fossil can the clade be (see Chapter 2)? Mitchell et al. (2014c) stipulated, as a prior, that the group the Miocene fossil belongs to could be no older than the exact base of the Oligocene (33.9 Ma). But why pick that age? Why not 34.0 Ma? Why not the middle of the Eocene, or the base of the Eocene, or the Cretaceous? There is no reason; the approach is arbitrary and illustrates the "problem of the priors" in dating clades.

Another distribution matching the ratite and duck examples is seen in the beetle family Chaetosomatidae, restricted to Madagascar and New Zealand (Leschen et al., 2003). Examples in plants include *Korthalsella salicornioides* (Viscaceae), present in Madagascar, New Zealand, and New Caledonia (Aubréville et al., 1967; Molvray, 1997). In the large, global genus *Euphorbia* (Euphorbiaceae), an earlier morphological study (Radcliffe-Smith, 1983) concluded that the sole New Zealand species, *E. glauca* of coastal sands and rocks, belonged to sect. *Esula*, not to sect. *Balsamis* (where it had been placed for more than a century). Radcliffe-Smith regarded *E. glauca* as closest to *E. borbonensis* of Réunion, indicating a typical trans-Indian Ocean pattern. A molecular study confirmed that *E. glauca* belongs to sect. *Esula*, but did not sample *E. borbonensis*; *E. glauca* was placed as sister to *E. emirnensis* of Madagascar, maintaining the trans-Indian Ocean affinity (Riina et al., 2013).

Indian Ocean patterns often include endemism around the Tasman and Coral Seas. A good example illustrating the relationship of the three areas is the New Zealand mangrove *Avicennia marina* (Acanthaceae), which comprises three allopatric forms (Duke, 1991):

Tasman Sea (New Zealand, New Caledonia, and SE Australia north to McPherson–Macleay Overlap) (var. *australasica*).

Coral Sea, Arafura Sea, and the Philippines (Mindanao) (var. *eucalyptifolia*).

Indian Ocean (South Africa to India, Borneo, and *Western* Australia) (var. *marina*).

In contrast, the close relative *A. germinans* is disjunct across the Atlantic Ocean (West Africa, Caribbean, Pacific coast of Central America, and Colombia).

Tethyan Groups

An example of Tethyan distribution was cited earlier: the gastropod superfamily Amphiboloidea that ranges from Australasia to the Persian Gulf (Heads, 2014). In a plant example, the tribe Aciphylleae (Apiaceae) of New Zealand and southeastern Australia forms a clade with the *Acrotrema* group of China and the Himalayas, and the tribe Smyrnieae of Iran to Europe and the Mediterranean (Spalik et al., 2010).

A group in *Veronica* s.lat. (Plantaginaceae; Albach and Meudt, 2010) comprises three groups with distributions based around the Tethyan and Pacific basins:

Tethyan: Afghanistan, Iran, and Europe (subg. *Pocilla* + subg. *Pentasepala*).

Tethyan: Turkey—Spain (subg. *Chamaepithoides*).

South Pacific: Australasia + Patagonia (subg. *Pseudoveronica*, incl. *Hebe* and others).

The third is allopatric with the first two, and the three clades can be derived by vicariance of an ancestor that was widespread around the Tethyan and Pacific basins (but absent in Africa and most of America), with subsequent overlap restricted to the Mediterranean region.

Distribution around the Tasman Sea

As with all major biogeographic nodes, the Tasman–Coral Sea region represents a center of endemism, an area of absence, a distributional break zone, and an area of overlap. Local endemics in the region include the globally basal groups discussed earlier. Groups that are endemic to the region and also widespread through it include the leaf-veined slug family Athoracophoridae, sister to the worldwide Succineidae. Another example is the moss *Cyathophorum bulbosum*, found in *eastern* Australia, *eastern* New Guinea, Lord Howe Island, and New Zealand (including Auckland and Chatham Islands) (Kruijer, 2002). Other Tasman region endemics are more restricted. The standard trans-Tasman center of endemism comprising New Zealand and southeast Australia has been referred to already. Some authors have argued that while Zealandia began rifting from Australia (Gondwana) at 84 Ma, complete rupture between the two did not take place until 55 Ma, and so phylogenetic differentiation between Zealandia and Australian groups should be dated to 55 Ma (Selvatti et al., 2015). But at the Cato Trough, off Queensland, the continental blocks of Zealandia and Australia are separated by just 25 km (Figure 3.1). The "complete separation" is a minor aspect of the geography, and is less important for the geological and biological history of the region than the large-scale rifting in the Tasman Basins. In a dispersal model, plants and animals were dispersing all around Gondwana before the complete separation of Zealandia and Australia (the biota was homogeneous) and were crossing the "bridge" at the Cato Trough region up until that time. In contrast, a vicariance model predicts that there was relative immobilism and regional heterogeneity in the predrift Gondwana biota, and that the rifting that began at 84 Ma led to differentiation between southeastern Australia and Zealandia.

Another connection, one that is seldom discussed, links New Zealand and New Guinea and straddles both the Tasman Sea and the Coral Sea basins. This occurs in clades of the following groups: the gray warbler *Gerygone* (Nyári and Joseph, 2012; concatenated phylogeny), meliphagid birds (including the New Zealand tui, *Prosthemadera*, and bellbird, *Anthornis*) (Joseph et al., 2014), the grass *Poa* (Birch et al., 2014), the *Hebe* complex (Plantaginaceae), *Coprosma* (Rubiaceae), *Abrotanella* (Asteraceae), *Myosotis* (Boraginaceae), *Carpodetus* (Rousseaceae), *Cyathea* (Cyatheaceae), and others (Heads, 2014: 219).

Many cases are also documented in morphological studies. One example involves *Kelleria multiflora* (Thymelaeaceae) of the northern and western South Island, and the sole New Zealand member of the genus with a pedicel. Elsewhere, pedicels are known only in *K. patula* of New Guinea and *K. ericoides* of New Guinea and Borneo (Heads, 1990b). In lichens, *Bryoria indonesica* is in New Zealand, New Guinea, and Borneo, while *Pseudocyphellaria carpoloma* is in New Zealand, Norfolk Island, and Papua New Guinea (Galloway, 2007).

The moss *Cyrtopus* is recorded in New Zealand and on Normanby Island, eastern Papua New Guinea, with an early record from Hawaii. Its sister genus, *Bescherellia*, is neatly allopatric and occurs in the Philippines (Mindanao), mainland New Guinea, eastern Australian highlands, New Caledonia, and Fiji (Sastre-de Jesús, 1987; Fife, 2015).

The pseudoscorpion *Smeringochernes* is recorded in New Zealand, Solomon Islands, New Guinea, Caroline Islands, and the Mariana Islands (Beier, 1976). In a possible example that needs confirmation, the laughing owl of New Zealand (*Sceloglaux*: Strigidae) is "perhaps most closely related to *Uroglaux*" of New Guinea (Marks et al., 1999: 239). Finally, the cuckoo *Chalcites lucidus lucidus* (Cuculidae) undertakes annual migrations between New Zealand and the islands off eastern New Guinea (Heads, 2014, Figure 6.15). One possible explanation for the direct, trans-oceanic route is that the birds are following an old migration route along former coastlines (Hutton, 1872).

Phylogenetic/Biogeographic Breaks in the Tasman Basin: An Example from Cuckoos

Phylogenetic breaks or disjunctions across the Tasman Sea are seen in many groups, including those endemic to the margins of the basin, for example, in Tasmania and New Zealand. The trans-Tasman disjunctions in these localized groups are often attributed to dispersal from one side to the other.

Yet similar breaks around the spreading center also occur in groups that are widespread in the South Pacific as a whole.

One example is a clade of cuckoos (Cuculidae) that comprises two monotypic genera (phylogeny from Sorenson and Payne, 2005; distribution from IUCN, 2016):

> *Scythrops novaehollandiae*: Northern and eastern Australia (breeding), winter migrant to Sulawesi, New Guinea, and the Bismarck Archipelago.
>
> *Urodynamis taitensis* (formerly placed in *Eudynamys*): New Zealand (breeding), winter migrant to the Bismarck Archipelago, Solomon Islands, Vanuatu, Fiji, and east through the Pacific to the Marquesas Islands.

The break between the two components is located at the plate boundary (spreading center) in the Tasman Basin, and it extends north to New Britain (Bismarck Archipelago). The result is a simple pattern of vicariance: Australia–New Guinea versus New Zealand–Pacific Islands. Most of the land present in the second of these two regions is formed by volcanic islands that have never been attached to a continent. Nevertheless, the Pacific has been a major center of volcanism ever since the Pacific plate formed in the Jurassic. The dynamic island systems would have always been inhabited by plants and animals, and the very large volcanic plateaus that are now submerged in the region include fossil wood (Chapter 3).

A Break around the Tasman Region Plus Secondary Overlap There: Stylidiaceae

In most cases, overlap of close relatives in the Tasman region is consistent with secondary range expansion in the region following a phylogenetic break there. An example of this is the plant family Stylidiaceae. It belongs to a large clade in Asterales with the following phylogeny (Tank and Donoghue, 2010):

> A **SW Pacific** complex with a center of diversity in New Caledonia (Alseuosmiaceae, Phellinaceae, and Argophyllacae) (Heads, 2014: Fig. 3–12).
>
> An **Indo-Pacific** group, most diverse in the Tasman region but absent in New Caledonia (Stylidiaceae) (Figure 4.3).
>
> A **cosmopolitan** clade (Asteraceae, Goodeniaceae and relatives; not mapped).

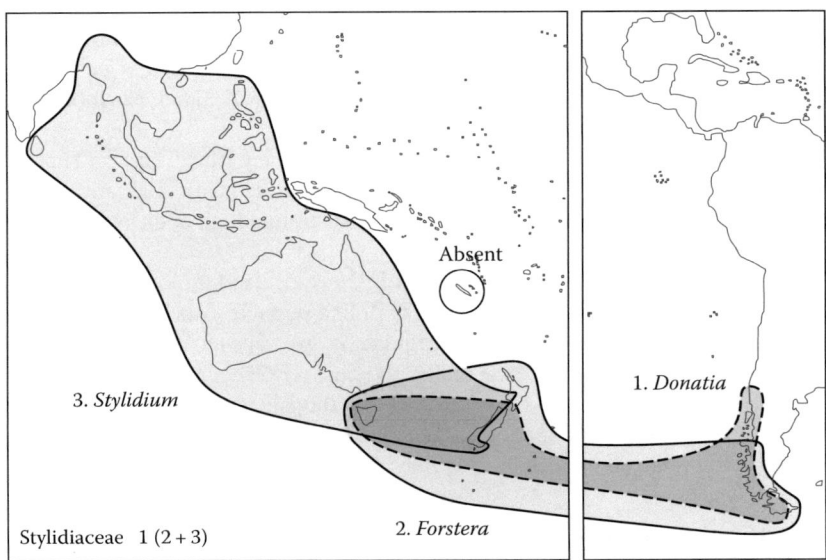

FIGURE 4.3 Distribution of Stylidiaceae: *Stylidium* (including *Oreostylidium* and *Levenhookia*) and *Forstera* (including *Phyllachne*). (Data from Wagstaff, S.J. and Wege, J., *Am. J. Bot.*, 89, 865, 2002; Heads, M., *J. Biogeogr.*, 37, 1179, 2010.)

The phylogeny indicates basal breaks around the Tasman region. The first two clades show significant allopatry. For example, only the first clade is present in New Caledonia (its center of diversity) and central New Guinea, whereas only the second group occurs in India, the New Zealand subantarctic islands, and South America. Both clades overlap in eastern Australia and mainland New Zealand.

Stylidiaceae (Figure 4.3) includes two genera found around the margins of the South Pacific basin—*Forstera* (including *Phyllachne*) and *Donatia*. The third genus in the family, *Stylidium* (including *Oreostylidium*), is distributed around the eastern Indian Ocean (Wagstaff and Wege, 2002). *Stylidium* extends across the old Indian plate–Australian plate margin (now inactive), but does not occur west of the main Indian ridge. All three genera overlap around the Tasman Basin: in Tasmania, South Island, and Stewart Island. Rather than seeing this overlap area as a center of origin for the family, the overall allopatry of the Indian Ocean *Stylidium* with its sister group, the trans-Pacific *Forstera*, is stressed here. The area of overlap is restricted relative to the whole distribution, and before the opening of the Tasman Basin it would have been even narrower. The overlap of the Indian and Pacific genera is interpreted here as a secondary feature that developed after their vicariance.

Within the area of overlap between *Stylidium* s.lat. and *Forstera* s.lat., allopatry is evident at local scales, reflecting the underlying, intercontinental allopatry of the genera. The Indo-Australian *Stylidium* has a single, distinctive species in New Zealand, and occurs mainly in the west (in the east, it extends north only to Mihiwaka near Dunedin; personal observations). The South Pacific *Forstera* is present in western New Zealand together with *Stylidium*, but *Forstera* also occurs in the New Zealand subantarctic islands and along a strip through central New Zealand (North Otago, Canterbury, Marlborough, and eastern North Island) (Glenny, 2009), where *Stylidium* is absent. Even within the potential area of overlap, *Stylidium* occupies several significant localities where *Forstera* is absent (Inch Clutha, Awarua Plain, and the Breaksea Islands off eastern Stewart Island). Around Westport (Townson, 1906; Glenny, 2009) and on Stewart Island (Wilson, 1987), *Forstera* replaces *Stylidium* at higher altitudes (in the Westport district, the mutual boundary occurs at an elevation of about 600 m). In Tasmania, *Stylidium* is widespread, but *Forstera* is restricted to the west. To summarize, overall, *Stylidium* is an Indian Ocean group of warmer habitats, *Forstera* is a cold-tolerant Pacific group. In the overlap area, there is microvicariance in ecology and distribution.

Overlap around the Tasman in Liliales: Colchicaceae and Alstroemeriaceae

In the lilies, the family Colchicaceae is in North America and widespread in the Old World, extending to New Zealand (Figure 4.4). The family's sister group is Alstroemeriaceae s.lat., a trans-Pacific group present in southeastern Australia, New Zealand, and South and Central America (Mennes et al., 2015; Stevens, 2016). The two families together cover most of the world, and over the vast majority of their distribution they are allopatric. The simplest explanation for this is that both families originated by vicariance of a global ancestor. Local overlap occurs only around the Tasman Basin (Figure 4.4), and the simplest explanation for this is secondary dispersal.

In the region of overlap with Colchicaceae, the Alstroemeriaceae have one genus in New Zealand (*Luzuriaga*, shared with South America) and another, its sister, endemic to Australia (*Drymophila*). This means that if the overlap of the two families was the result of dispersal in Alstroemeriaceae, it probably developed by normal means (not by chance dispersal) before the rifting of the Tasman Sea caused the break between *Luzuriaga* and *Drymophila*.

The local distribution of the Colchicaceae and Alstroemeriaceae in New Zealand (not shown in Figure 4.4) is also of interest. In the South Island, the Pacific group Alstroemeriaceae occurs (with *Luzuriaga*) only in the west and south, and grows in forest. In contrast, the Indian-Atlantic group Colchicaceae is in the South Island (with *Wurmbea*) but is restricted to the east (in Canterbury and North Otago), and grows in open habitat: tussock grassland and the edges of swamps and lakes (Moore and Edgar, 1970, under "*Iphigenia*"). Thus, the two families are allopatric on an intercontinental scale, and within their restricted area of overlap differ in their local distributions and ecology.

The sister of Colchicaceae + Alstroemeriaceae is another Tasman group, *Petermannia* of the McPherson–Macleay Overlap in eastern Australia (Figure 4.4). The distribution of Alstroemeriaceae suggests that the breakup of these three groups dates to *before* the opening of the Tasman Basin. Prebreakup

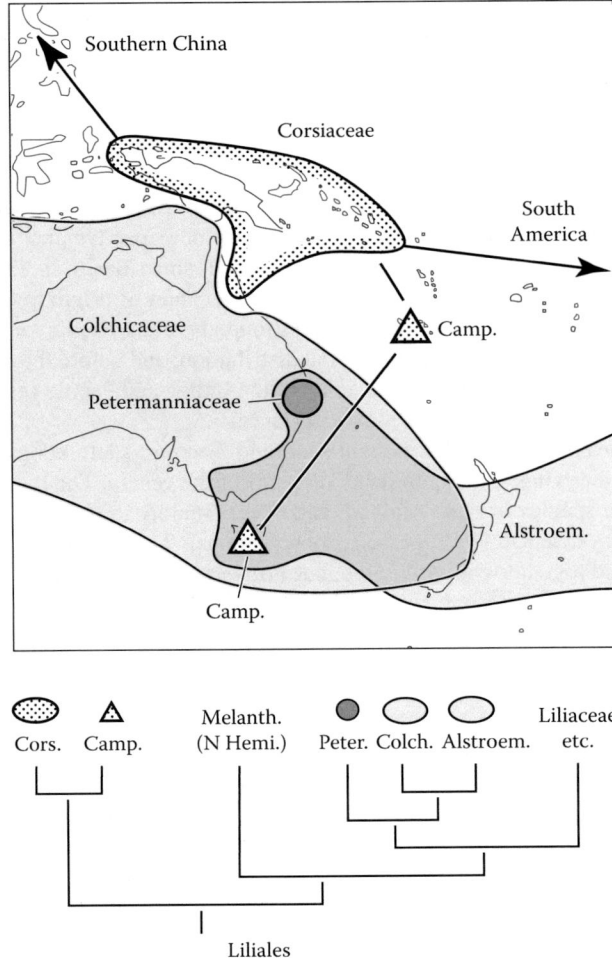

FIGURE 4.4 Distribution of five of the main clades in Liliales. Cors., Corsiaceae; Camp., Campynemataceae; Melanth., Melanthiaceae; Peter., Petermanniaceae; Colch., Colchicaceae; Alstroem., Alstroemeriaceae. The clade labeled "Liliaceae, etc." comprises Liliaceae, Smilacaceae, Ripogonaceae, and Philesiaceae. (Phylogeny from Mennes, C.B. et al., *J. Biogeogr.*, 42, 1123, 2015; distribution from Stevens, P.F., Angiosperm phylogeny website, www.mobot.org/MOBOT/Research/APweb/, accessed May, 2016.)

vicariance around the Tasman was also important in the most basal differentiation of the order Liliales, as one of its three clades, Campynemataceae, is endemic to New Caledonia and Tasmania. This means that the notable *absence* of Colchicaceae and Alstroemeriaceae from New Caledonia is likely to be phylogenetic rather than ecological (or the result of chance extinction).

The distribution of Campynemataceae and Petermanniaceae and the region of overlap between Colchicaceae and Alstroemeriaceae are widespread around the Tasman Sea. Before rifting in the Tasman Basin, though, they would have all defined a narrow, linear region.

Colchicaceae are represented in New Zealand by a single genus, *Wurmbea*, which has the following phylogeny (Case et al., 2008):

South Africa.
    A paraphyletic grade in Western Australia.
        South Australia.
        New South Wales and New Zealand (South Island).

The sister group of *Wurmbea* is from South Africa, and so, because of the paraphyletic basal grade there, Case et al. (2008) concluded that *Wurmbea* had a center of origin in South Africa and dispersed, once, across the Indian Ocean to Australia. Nevertheless, they admitted that "paradoxically, such long-distance dispersal is inconsistent with several features of colchicoid life-history" (p. 149), for example, seed dispersal is probably very local. "Hence, the mechanism(s) by which long-distance dispersal was achieved in *Wurmbea*, and in Colchicaceae as a whole, remains an enigmatic problem" (p. 149). Likewise, in another trans-Indian Ocean group, the tribe Arctotidinae (Asteraceae), "the mode of dispersal is difficult to envisage" (McKenzie and Barker, 2008). These examples illustrate the difficulty dispersal theory often has in providing cogent solutions to basic biogeographic problems.

As discussed already, dispersal theory often concludes that distributions are paradoxical and enigmatic, and can only be explained by extraordinary events. Instead of considering alternative explanations for the general pattern, such as vicariance, Case et al. (2008) explained the trans-Indian Ocean enigma in *Wurmbea* by proposing additional chance dispersal. They wrote that long-distance dispersal from South Africa to Western Australia "was clearly not a unique event in *Wurmbea* biogeographic history, as [the New Zealand member] dispersed from within the Australian *Wurmbea* clade to the South Island of New Zealand" (p. 149) (cf. Waters and Craw, 2006, on trans-Indian Ocean and trans-Tasman skinks; see Chapter 1).

Neither trans-Indian Ocean nor trans-Tasman Sea dispersal is necessary. Instead, the phylogeny of *Wurmbea* and the others is just as likely to reflect an eastward sequence of differentiation events in already widespread ancestors. In *Wurmbea*, differentiation has proceeded from the Indian Ocean (between South Africa and Western Australia), to central Australia (between southwestern and southeastern Australia), and finally to the Tasman, separating the New Zealand taxon from relatives in New South Wales.

The sister group of Colchicaceae is the South Pacific family Alstroemeriaceae, which comprises two main clades (Chacón et al., 2012; distributions from Moreira-Muñoz, 2007):

Mexico to Tierra del Fuego, many species in the Andes, also in eastern Brazil (*Alstroemeria* and *Bomarea*).
Chile from latitude 35°–55° S and New Zealand (*Luzuriaga*); Australia (*Drymophila*).

The break between the two clades occurs in or around Chile, as in *Ourisia* (Plantaginaceae), marsupials, and others (Heads, 2014; Figures 3.6 and 3.8), and there has been subsequent overlap there. Following this primary break, the trans-South Pacific affinity has been rifted apart during the breakup of Gondwana.

A similar pattern occurs in the liverwort genus *Nothoceros* (Villarreal and Renner, 2014). There are two main groups: New Zealand + Patagonia and southern United States to southern Brazil. The main break is located between Patagonia and southern Brazil, and this occurred before the break in the trans-Pacific group.

In Colchicaceae, Conran et al. (2014a) proposed trans-oceanic dispersal of *Luzuriaga* from a center of origin in Patagonia to New Zealand, but only because of the young age that was estimated for the genus. This was calculated by stipulating that the related genus *Smilax* could be no older than the geological formation in which its oldest fossils occur, an arbitrary procedure that is rejected here (see Chapter 2).

Overlap around the Tasman in a Clade of Thymelaeaceae

The following widespread clade of Thymelaeaceae displays a high level of allopatry between the two sister groups (Motsi et al., 2010).

1. **South Pacific and Atlantic basins**: The clade includes *Kelleria* of eastern Australia, New Zealand (including the subantarctic Auckland Islands), New Guinea, and Borneo. Its sister, *Drapetes*, is in Patagonia and the Falkland Islands, and the two form a typical South Pacific affinity. This clade then connects across the Atlantic with African groups of the polyphyletic "*Gnidia*."

2. **Indian Ocean basin**: This clade includes *Pimelea* in Australasia, from New Zealand to New Guinea and Western Australia (but not the subantarctic islands or Borneo). The sister of *Pimelea* is a clade of "*Gnidia*," across the Indian Ocean in India and Africa.

The two clades overlap only around the Tasman Basin (in southeastern Australia, Tasmania, and mainland New Zealand), as in *Stylidium* + *Forstera*, and Alstroemeriaceae + Colchicaceae. The region again represents the main phylogenetic and geographic break zone in a widespread southern group.

Trans-South Pacific Disjunctions: Examples from New Zealand Hymenoptera

Distributions that span the southern South Pacific are well known and are often illustrated with the tree *Nothofagus*; so far, they have been cited here in flowering plants (Stylidiaceae, Figure 4.2; Alstroemeriaceae, Figure 4.3), harvestmen (Figure 4.1), and shags (Figure 4.2). An example from Myrtaceae is the following clade (Thornhill et al., 2015):

> *Pimenta* p.p.: Tropical South America and the West Indies, + *Campomanesia,* Uruguay to Colombia.
>
> *Ugni*: Central Chile to Mexico.
>
>> *Myrteola*: Western South America from the Tierra del Fuego and Falkland Islands along the Andes to Venezuela, also the mountains of southern Venezuela.
>>
>> *Lophomyrtus* (incl. *Neomyrtus*): New Zealand.

Theory suggests that the paraphyletic, basal grade in South America occupies the center of origin, with the New Zealand contingent being the result of secondary dispersal. Instead, South America could represent a center of basal differentiation in a widespread, common ancestor, and this is consistent with the very different distributions of the three South American clades.

In less well-known groups such as dandelions, for example, *Taraxacum* subsect. *Antarctica* (Asteraceae) occurs in southwestern and southeastern Australia, New Zealand, and South America, from the Falkland Islands and Fuegia along the Andes to southern Peru (Doll, 1982). In New Zealand lichens, there are many trans-Pacific species (Galloway, 2007). In the crane flies (Tipulomorpha) alone, more than 700 species belong to Australasia–South America clades (Ribeiro and Eterovic, 2011). Other, less well-known examples occur in the insect order Hymenoptera, and several of these are described next.

The order Hymenoptera is made up of the basal clade Symphyta (sawflies), a large complex of parasitic and parasitoid groups ("Parasitica") that has turned out to be paraphyletic, and the clade Aculeata (bees, ants, and true wasps). The hymenopteran fauna of New Zealand is anomalous, as Symphyta and Aculeata, which are diverse elsewhere, are depauperate, while the parasitic groups show high diversity. Despite these differences, all three traditional groups of Hymenoptera include standard South Pacific disjunctions.

In Symphyta, morphological studies of *Guiglia* (Orussidae) indicate a basal break between the Chilean *G. chiliensis* and the remaining species in New Zealand, Fiji, and Australia (Vilhelmsen, 2004, 2007). This South Pacific group has an allopatric sister to the north, a trans-tropical Pacific clade present in Japan, Laos, Malaysia, Indonesia, the Philippines, Papua New Guinea (*Stirocorsia*), and warm America, from Argentina to the United States (five genera, four of these in Costa Rica/Panama) (Vilhelmsen, 2007).

Guiglia and its sister group define an area of endemism: South Pacific + tropical Pacific. In contrast, another orussid, *Orussus loriae* of New Guinea, is sister to a group of African species, and this gives a typical *Indian Ocean* disjunction. Vilhelmsen (2004) concluded that the common ancestor of the family Orussidae:

> ...was probably widespread, the initial splitting events taking place prior to or coinciding with the separation of Laurasia from Gondwana. Later putative vicariance events can be correlated with the gradual breakup of Gondwana... The minimum age of the common ancestor of the Orussidae is >180 Myr when estimated from the biogeographical pattern, >95 Myr when estimated from the phylogenetic position of the fossils; the earlier date is considered to be the most likely.

The parasitoid groups of New Zealand Hymenoptera show high diversity (in Diapriidae and Mymaridae) and also high-level endemism, with Maamingidae being a New Zealand endemic family (Early et al., 2001; Fernández-Triana et al., 2011). With their high diversity, the parasitoid groups provide valuable information on regional differentiation. For example, in New Zealand, Diapriidae subfam. Ambositrinae have disjunct centers of diversity in Nelson and west Otago (Heads, 1997). The parasitoid groups also include standard South Pacific disjunctions, introduced next.

The three parasitoid families just cited, Diapriidae, Mymaridae, and Maamingidae, all belong to a diverse, monophyletic complex with more than 25,000 species. The basic geographic structure of the complex is simple though; all three of the main clades have Australasian or Australasian–South/Central American clades at or near their base. The phylogeny is as follows (Munro et al., 2011):

1. Monomachidae: South Pacific (Australia, New Guinea, southern South America to Mexico; 29 species).
2. Diaprioidea: Global (~2,400 species described), comprising:

   Maamingidae: New Zealand (North and South Islands).

   Diapriidae (including Mymarommatidae): Global.
3. Chalcidoidea: Global and diverse (~23,000 species described), comprising:

   Mymaridae: Cosmopolitan.

   Rotoitidae: New Zealand and Chile (Chiloé Island and vicinity).

   All remaining Chalcidoidea: Cosmopolitan.

All the clades listed here, apart from the New Zealand endemic Maamingidae, include South Pacific disjunctions. For example, in the Diapriidae of southern Chile, there are "relatively few faunal connections to the rest of the Neotropics but endemism is high and there are distinct Gondwanic ties to New Zealand, Australia, and (rarely) South Africa" (Masner and Garcia, 2002).

In other parasitoid Hymenoptera, the clade Ichneumonidae + Braconidae form the sister group of Aculeata, indicating that the parasitoids are a paraphyletic complex. The ichneumonids are one of the most diverse and well-known groups of Hymenoptera. Most of the 37 subfamilies have most of their genera in the Palearctic region, but the subfamilies Pedunculinae, Eucerotinae, and Labeninae each have their main distribution and their basal clades in the southern hemisphere (Gauld and Wahl, 2000, 2002). The phylogenetic position of these southern groups is also significant, as one of them, Labeninae, is often recovered at the base of the ichneumonid phylogeny, or basal with the Xoridinae (Ward, 2011). In Labeninae, most genera and species are confined to Australasia and South America, and often inhabit temperate forests dominated by *Nothofagus, Araucaria,* and *Podocarpus* (Ward, 2011). *Certonotus*, for example, is known from Tasmania, eastern Australia, New Guinea, New Zealand, and southern South America, with most records from *Nothofagus* forest (Porter, 1981). The single New Zealand species, *C. fractinervis*, is one of the most distinctive insects in the country because of its large size. Morphological studies indicate that its sister is *C. vestigator* of New Guinea (Gauld and Wahl, 2000), the same relationship that is seen in gray warblers, tuis and bellbirds, and many others (see the earlier section "Distribution around the Tasman Sea").

In Braconidae (sister of Ichneumonidae), the subfamily Doryctinae include a large South American clade (32 genera) that is sister to one in Australia (four genera). "The divergence time estimates suggest that diversification in the subfamily could have in part occurred as a result of continental break-up events that took place in the southern hemisphere" (Zaldivar-Riverón et al., 2008).

In Aculeata, the bee (Apoidea) fauna of Australia, New Guinea, and New Zealand is the most distinctive in the world, as it is dominated by members of one family, Colletidae. More than half of all the bee species in Australia and 90% of those in New Zealand are colletids (Michener, 2007; Chenoweth and Schwarz, 2011; Groom and Schwarz, 2011). Colletidae, with ~100 genera and 2500 species, are distributed worldwide, but the centers of subfamily diversity are all southern. They occur in the Australian region, southern Africa (especially the Cape region), and southern South America. There are only two globally widespread genera, *Colletes* and *Hylaeus*, and there are no endemic genera in the Nearctic, Palearctic, or Oriental regions. The oldest bee fossil is Early Cretaceous and this gives a useful minimum age for the group. At least one colletid clade straddles the South Pacific basin, as a group from New Zealand and Australia ("clade 3"—*Leioproctus* and others—in subfamily Neopasiphaeinae) is sister to a group in South America (Almeida et al., 2012).

Some Gondwanan Pseudoscorpions

In a study of pseudoscorpions, Harvey (1996a) listed what he interpreted as Gondwanan groups, and these include several genera represented in New Zealand. Together these illustrate the standard connections with South Africa, Australia, New Guinea, New Caledonia, and South America:

> *Austrochthonius*: South Africa, Australia, New Zealand, and South America.
>
> *Sathrochthonius*: Australia, New Zealand, New Caledonia, and South America, north to Venezuela.
>
> *Synsphyronus*: Australia, New Zealand, and New Caledonia.
>
> *Protochelifer*: Australia and New Zealand.
>
> *Philomaoria*: New Zealand and New Caledonia.
>
> *Reischekia*: New Zealand (two endemic species) and western New Guinea (one species).

Harvey (1996b) also discussed several pseudoscorpion groups that occur around the rim of the Indian Ocean (western Australia, India, southern Africa, and Madagascar), but which are generally *lacking* in eastern Australia, New Zealand, and South America. As usual, a center of endemism also acts, in other groups, as a center of absence.

Panaustral + Central Pacific Distribution

Many groups have distributions that include trans-South Pacific or panaustral components, but also occur in the central Pacific, a region that was never part of Gondwana. The main tectonic events in the region include the growth of the Pacific plate, with its margins migrating outward (Figure 3.8), and the eruption of the large igneous plateaus, including the Hikurangi Plateau.

The Schefflera Group (Araliaceae)

This group of araliad trees has three components: one in Africa and Madagascar (*Cussonia*), one in Australia and Chile (*Motherwellia, Cephalaralia,* and *"Raukaua"*), and one in New Zealand and the Pacific islands (*Schefflera* s.str., *Raukaua* s.str., and *Cheirodendron*) (Figure 4.5; Frodin et al., 2010; Mitchell et al., 2012). (The connection between Australia and Chile shown in Figure 4.5 approximates the great circle connection between the two areas that runs south of New Zealand.)

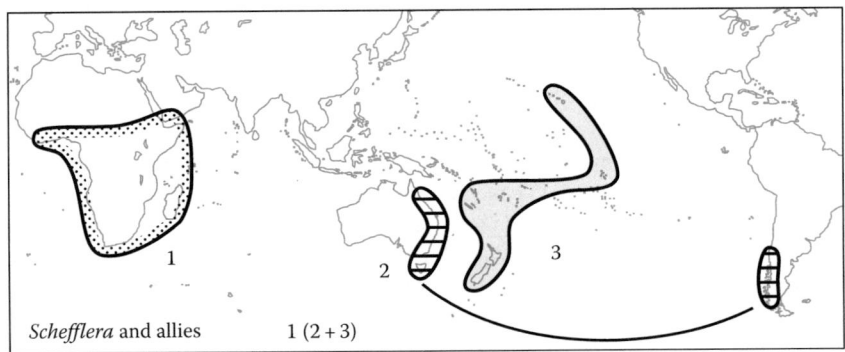

FIGURE 4.5 Distribution of *Schefflera* and allies (Araliaceae), with phylogeny indicated by nested numbers. 1, *Cussonia*; 2, *Motherwellia, Cephalaralia,* and *"Raukaua"* p.p.; 3, *Raukaua* s.str., *Schefflera* s.str., and *Cheirodendron*. (Data from Frodin, D.G. et al., *Plant Diversity Evol.* 128, 561, 2010; Mitchell, A.D. et al., *Aust. Syst. Bot.*, 25, 432, 2012.)

The three clades in the *Schefflera* group display simple allopatry, consistent with a vicariance history. An analysis of the *Schefflera* group using center-of-origin programs inferred an Australasian center of origin (Nicolas and Plunkett, 2014), but vicariance is a much simpler explanation that does not require any chance dispersal.

The members of the New Zealand–central Pacific clade have probably developed as dynamic meta-populations, persisting *in situ* by constantly invading new islands as these appeared within the region. Other examples of the distribution include the following:

A clade of *Hebe* (Plantaginaceae): New Zealand and Rapa Island, southeastern Polynesia (Heads, 2014: 105).

A clade of *Fuchsia* (Onagraceae): New Zealand and Tahiti (Heads, 2014: 224).

Melicope section *Melicope* s.str. (Rutaceae): New Zealand, the Kermadec Islands, Rapa Island, and Tahiti (Appelhans et al., 2014a,b).

A well-defined group in *Freycinetia* (Pandanaceae): New Zealand, Norfolk Island, New Caledonia, Rarotonga, Rapa, and Hawaii (Stone, 1973).

The New Zealand–central Pacific clade of Araliaceae shown in Figure 4.5 (*Schefflera* s.str., *Raukaua* s.str., and *Cheirodendron*) is allopatric with its relatives and also displays internal allopatry. For example, *Cheirodendron* is the only member of the group in the Hawaiian and Marquesas Islands, and it is endemic there. The two archipelagos form an important center of endemism for many groups (Heads, 2012a). (For another clade of Araliaceae that is widespread, diverse, and endemic in the central Pacific, see Heads, 2012a: Fig. 6–13).

Metrosidereae (Myrtaceae)

The tribe Metrosidereae includes trees and lianes that are abundant in many New Zealand forests (Figure 4.6; Wright et al., 2003; Wilson et al., 2005; Wilson, 2011). The tribe has a standard Indo-Pacific distribution, with a notable absence in mainland Australia, although there are fossils in Tasmania (Tarran et al., 2016). The trans-Pacific disjunction is repeated in the family Myrtaceae, as it is also displayed in *Lophomyrtus* and *Myrteola*.

The sister group of Metrosidereae consists of Backhousieae, in eastern and northern Australia, and Syzygieae, widespread through the tropical and southern parts of the Old World (in New Zealand south to Rarangi near Blenheim) (Thornhill et al., 2015).

The Metrosidereae as such are absent in mainland Australia, and this might be attributed to extinction there following Neogene aridification. Nevertheless, the tribe's sister group has a main center of diversity

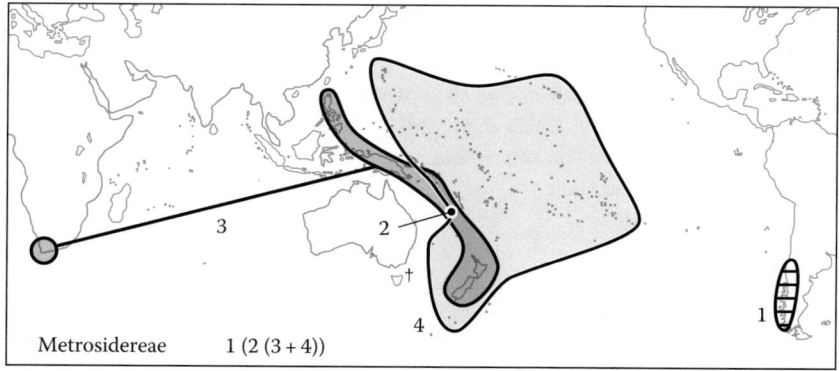

FIGURE 4.6 Distribution of tribe Metrosidereae (Myrtaceae). The phylogeny is indicated by nested numbers: 1, *Tepualia*; 2, *Carpolepis*; 3, *Mearnsia*; 4, *Metrosideros* s.str.; †, fossils in Tasmania. (Data from Wright, S. et al., *Proc. Nat. Acad. Sci. USA*, 97, 4118, 2000; Wright, S. et al., *Evolution*, 57, 2893, 2003; Wilson, P.G. et al., *Plant Syst. Evol.*, 251, 3, 2005; Wilson, P.G., Myrtaceae, in *The Families and Genera of Vascular Plants*, Vol. X: *Flowering Plants Eudicots: Sapindales, Cucurbitales, Myrtaceae*, ed. K. Kubitzki, Springer, Heidelberg, Germany, 2011, pp. 212–271.)

in mainland Australia (one of the two main clades, Backhousieae, is endemic there), and so it is possible that Metrosidereae have never occurred there.

The absence of Metrosidereae in Australia, where its sister group is present, is matched by the complete absence of its sister group in southern Zealandia (south of the Marlborough Sounds) and the New World, where Metrosidereae are diverse and abundant. Thus, despite the extensive overlap between the two clades, there is a significant allopatry between them.

Allopatry is also conspicuous *within* the tribe Metrosidereae. The first node in the phylogeny separates Chile from the rest. The second break separates a New Caledonia endemic from a widespread Indo-west Pacific group. Within this last group, a clade that is disjunct across the Indian Ocean (*Mearnsia*) is sister to a widespread central and west Pacific group (*Metrosideros*). The two genera display a high level of allopatry, and their overlap is restricted to parts of the southwest Pacific (mainland New Zealand, New Caledonia, and the Solomon Islands). The simplest assumption is that this zone lies around the site of the initial break between the genera. Most *Mearnsia* species are lianes, while *Metrosideros* species are trees and shrubs; the two conform to a common pattern in which clades comprise trees with orthotropic (erect) trunks in the central Pacific, and lianes and shrubs with plagiotropic (horizontal) shoot axes around the western margin of the Pacific.

Within the Pacific genus *Metrosideros*, the clades also display a high level of allopatry. For example, all the species of the Hawaiian and Marquesas Islands form a clade that is endemic there (Heads, 2012a: Figs. 7–8), as in the *Schefflera* group just discussed. In this way, Metrosidereae display simple allopatry at tribal, generic, and subgeneric levels, consistent with the orderly, *in-situ* breakdown of a widespread ancestor, with only a minor amount of local range expansion. This contrasts with a dispersal model for the group and a center of origin in the New Zealand region (Simpson, 2005).

Trewick et al. (2007: 4) suggested that New Zealand–Pacific island groups such as *Metrosideros* have "obvious recent dispersal histories." Nevertheless, this assumes that groups endemic to young islands, such as the Kermadec Islands at the plate margin north of New Zealand, are no older than the islands themselves. Nevertheless, new islands are constantly being formed around active plate margins, and clades endemic there are likely to be older than the present islands that they occupy.

Gardner et al. (2004: 410) concluded that "*Metrosideros* is clearly very efficiently dispersed by wind, as affirmed by its presence on the Auckland Is. and Kermadec Is. and in Remote Oceania." All individual plants of *Metrosideros*, whether on continental crust in New Zealand or on a remote volcanic island, have dispersed to their particular spot. Yet this process—physical movement by normal means of dispersal—does not explain the distribution of the main clades in the genus, and their precise allopatry in different sectors of the Indo-Pacific region. All the islands occupied by *Metrosideros* occur in areas where there have been many phases of island formation and destruction, and so it is not necessary to propose long-distance dispersal of the genus from distant continental crust in New Zealand, for example.

Panaustral + Central Pacific Distribution, and the New Zealand "Star" Pattern: The Case of Asteliaceae

The monocot family Asteliaceae illustrates several of New Zealand's main biogeographic connections (Figure 4.7; Moore and Edgar, 1970; Birch et al., 2012). One clade comprises the localized Australian genera *Neoastelia* and *Milligania*. The other clade, the genus *Astelia*, is made up of one trans-Indian Ocean clade and one trans-Pacific clade.

Origin of Asteliaceae

The origin of a group cannot be established by looking at it in isolation; the group's relatives also need to be considered. The sister group of Asteliaceae is the family Hypoxidaceae, which is widespread through warmer America, Africa, tropical Asia, and Australia, but has most records in areas where Asteliaceae are absent. Conversely, there are no indigenous Hypoxidaceae in the South Pacific islands east of New Caledonia or in Patagonia, where Asteliaceae are diverse (Smithsonian Institution, 2016; Stevens, 2016). In New Zealand, there is only a single species of Hypoxidaceae, while in contrast, Asteliaceae have their main diversity there. This suggests that the two families have diverged in allopatry, with some subsequent overlap.

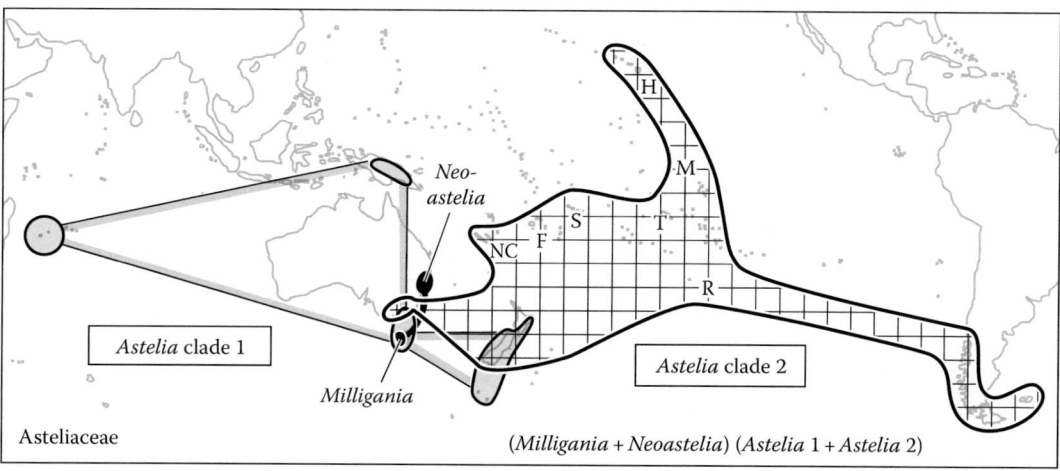

FIGURE 4.7   Distribution of Asteliaceae. F, Fiji; H, Hawaiian Islands; M, Marquesas Islands; NC, New Caledonia; R, Rapa Island; S, Samoa; T, Tahiti. (Data from Birch, J.L. et al., *Mol. Phylogen. Evol.*, 65, 102, 2012.)

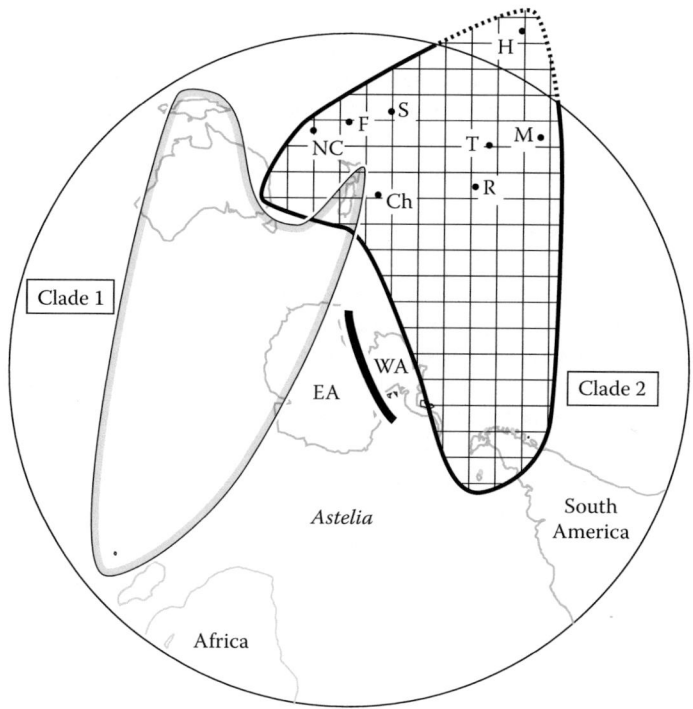

FIGURE 4.8   Distribution of Asteliaceae shown on a polar projection, with reconstruction of preglaciation distribution in Antarctica. Bold line = boundary between cratonic east Antarctica and the accreted terranes of West Antarctica. Ch, Chatham Islands; EA, East Antarctica; F, Fiji; H, Hawaiian Islands; M, Marquesas Islands; NC, New Caledonia; R, Rapa Island; S, Samoa; T, Tahiti; WA, West Antarctica. (Based on Skottsberg, C., *Trans. R. Soc. N. Z.*, 67, 218, 1922, Figure 1; Birch, J.L. et al., *Mol. Phylogen. Evol.*, 65, 102, 2012.)

Differentiation in Asteliaceae

Following the vicariant origin of Asteliaceae as a widespread Indo-Pacific group, the family itself underwent a break at what is now the western margin of the Tasman Basin (in southeastern Australia) that separated *Neoastelia + Milligania*, from *Astelia* (Birch et al., 2012). This was followed by a second break around the Tasman Basin, and this caused the divergence of *Astelia* into two clades, one distributed across the Indian Ocean and the other spanning the Pacific. A vicariance model predicts that this primary phylogenetic break in the Tasman region extended in some way through Antarctica, possibly between East Antarctica and West Antarctica (Figure 4.8).

Subsequent overlap among the clades of Asteliaceae has been quite local, and it is restricted to a small part of the overall range: central New Zealand, Victoria, and Tasmania. The overall pattern in the family can be compared with the Indian Ocean *Mearnsia* and the Pacific Ocean *Metrosideros* overlapping only in New Zealand, New Caledonia, and the Solomon Islands (Figure 4.6).

The Asteliaceae have all four main clades present in southeastern Australia, and so center-of-origin analyses (using S-DIVA and DEC) located a center of origin there (Birch and Keeley, 2013). This is only one possibility though, and a vicariance model for Asteliaceae accounts for numerous aspects of the group's distribution that the center-of-origin model does not. These include, for example, the extensive allopatry between Asteliaceae and Hypoxidaceae, the complete allopatry between *Neoastelia* and the remaining Asteliaceae, and the high level of allopatry between the Indian Ocean and Pacific Ocean clades of *Astelia*. Distribution *within* the Pacific clade of *Astelia* is examined in the following sections, but first, a distinctive pattern that occurs in the genus is introduced in general terms.

The New Zealand "Star" Pattern

One distinctive and common distribution pattern involves groups comprising several clades, each present in New Zealand and one other area. Examples include the *Hebe* complex (Plantaginaceae; Heads, 2014: Fig. 6.8). (In the diverse "*Olearia* clade 2" (Asteraceae), the New Zealand members display separate connections with Australia, New Guinea, and possibly Rapa Island, but perhaps not with South America as was once thought; Brouillet et al., 2009; cf. Heads, 2014: Fig. 6.9.) The New Zealand "star" pattern is also well documented in morphological analyses of groups that have not yet been sequenced, such as the regional members of *Euphrasia* (Orobanchaceae) (Heads, 1994a). A generalized, hypothetical example of this common "star" pattern is illustrated in Figure 4.9 (cf. Figure 1.1).

The ancestral area programs in current use interpret the star pattern as the result of dispersal from a center of origin in New Zealand. Nevertheless, a vicariance model for the star pattern is also possible, beginning with an ancestor that was more widespread than the "star." This ancestor gave rise to the group with the "star" group, and also its sister, which, in many cases, is more or less allopatric with the "star" group. (In Figure 4.10a, present-day coastlines are shown for reference only, in order to indicate the location of the groups. At the time of the break between the "star" group and its sister, the geography would have been very different.) The direct ancestor of the "star" group then undergoes vicariance around a main node, or break, located in the ancestral New Zealand region (Figure 4.10b), and this gives

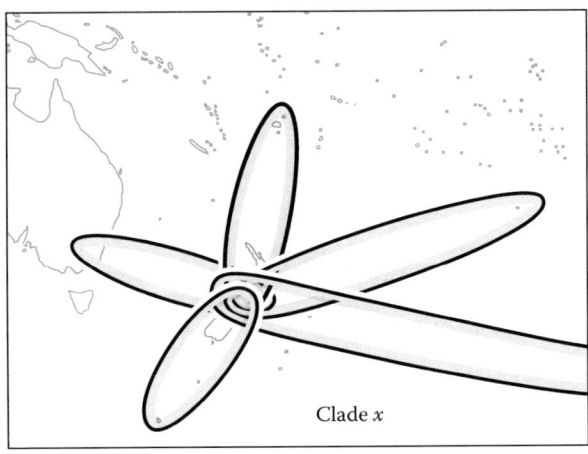

Clade *x*

FIGURE 4.9   A generalized, hypothetical example of a common pattern, the New Zealand "star pattern."

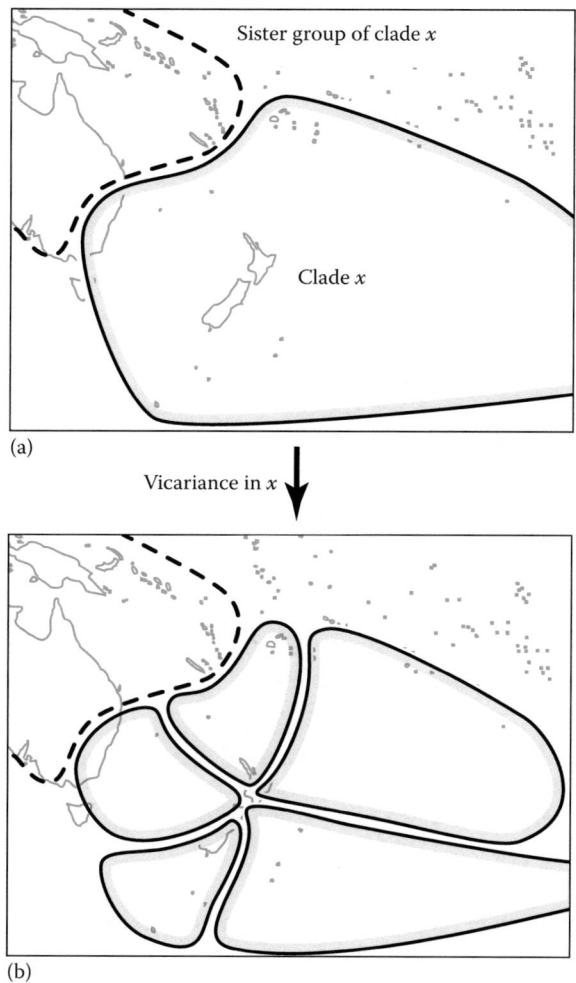

FIGURE 4.10   A vicariance model for the New Zealand star pattern. (a) A widespread clade, *x*, originates by vicariance with its sister group to the northwest (present-day coastlines shown for reference only). (b) Clade *x* undergoes vicariance at a main node in the ancestral New Zealand region. Subsequent rifting around Zealandia, local extinction in the region, and overlap of the clades within New Zealand result in the star pattern shown in Figure 4.9.

rise to the several clades forming the star. This differentiation is followed by seafloor spreading and other geological changes around New Zealand. In the final phase, local extinction around Zealandia (caused by subsidence and other processes) and overlap of the clades within New Zealand will produce the star pattern (Figure 4.9). Note that the group as a whole has itself originated by vicariance with its sister group in the northwest.

Within *Astelia*, the two main clades, around the Indian and Pacific basins, are almost allopatric. There is also extensive allopatry among the subclades (subgenera and sections) of the Pacific group, and these form a version of the "star" pattern with New Zealand at the center (Figure 4.11, based on the phylogeny in Birch et al., 2012). There is also allopatry *within* these groups, but this is not mapped here. For example, all the eastern Polynesian species in Hawaii, the Marquesas Islands, and Rapa Island form a single clade, resembling Hawaii–Marquesas clades in Araliaceae (*Cheirodendron*) and Myrtaceae (*Metrosideros* clade). Figure 4.12 shows a reconstruction of the *Astelia* clade distributions at the time of the clades' origin and before their overlap. (Present-day coastlines are shown for reference only.)

Each of the clades in the *Astelia* star pattern has a standard range: New Zealand–subantarctic islands, New Zealand–SE Australia, New Zealand–New Caledonia, New Zealand–Fiji, New Zealand–Tahiti,

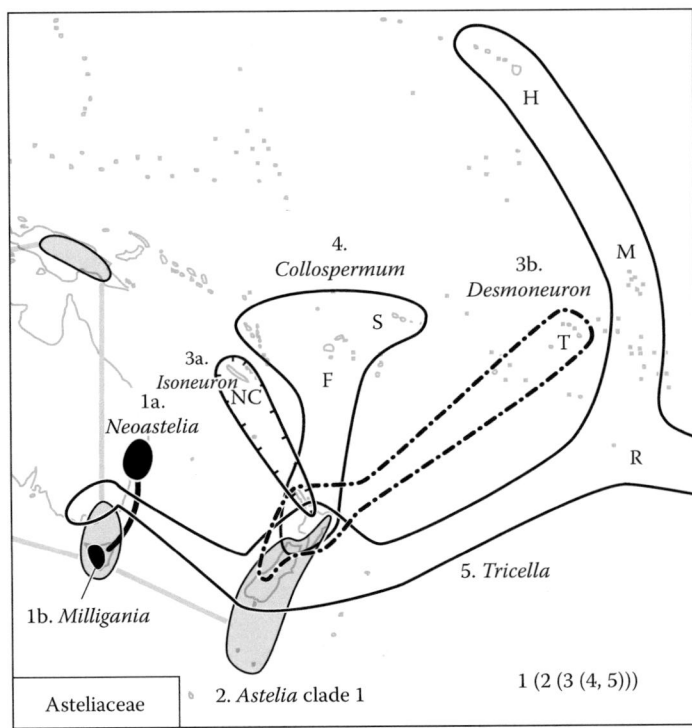

FIGURE 4.11 Distribution of Asteliaceae in the southwest Pacific and New Zealand. F, Fiji; H, Hawaiian Islands; M, Marquesas Islands; NC, New Caledonia; R, Rapa Island; S, Samoa; T, Tahiti. (Data from Moore, L.B. and Edgar, E., *Flora of New Zealand*, Vol. 2: *Indigenous Tracheophyta, Monocotyledones Except Gramineae*, Government Printer, Wellington, New Zealand, 1970; Birch, J.L. et al., *Mol. Phylogen. Evol.*, 65, 102, 2012.)

and New Zealand–Rapa. Each of these individual patterns is repeated in many other plants and animals. For example, in freshwater gastropods (Tateidae), "the species from Fiji were closer related to New Zealand than to Australian or New Caledonian taxa, which is rather exceptional" (Zielske and Haase, 2014). The break between the Society Islands (Tahiti) on one hand and the Marquesas and Hawaiian Islands on the other also occurs in *Metrosideros* (Myrtaceae) (Heads, 2012a: Fig. 7.8) and in *Melicope* (Rutaceae) (Heads, 2012a: 423). The break in *Astelia* between New Caledonia and Fiji is a standard one and is useful for calibrating the timeline of the evolution. Disjunctions at the New Zealand Alpine fault are also useful for calibration, and these occur in *A. subulata* (subg. *Astelia*), *A. nivicola* var. *moriceae* (subg. *Tricella*), and *A. linearis* var. *linearis* (subg. *Tricella*) (Heads, 1998a).

Center-of-origin analyses of *Astelia* inferred dispersal from New Zealand to each of the outlying areas (Birch and Keeley, 2013), but this does not explain why the subgroups display such standard patterns or why they are allopatric outside New Zealand. Why is each subgroup in the star pattern restricted to its own, small area outside New Zealand? Dispersal theory explains this by suggesting that long-distance dispersal happened just once in each of the different subgroups; this is expected because long-distance dispersal is a very unlikely process. In contrast, vicariance theory stresses that the pattern can be explained by rifting or uplift between the subgroups and a degree of genetic flexibility in the regional ancestor. A vicariance model accounts for allopatry outside New Zealand, and it only requires dispersal of Asteliaceae by normal, observed means of dispersal *within* New Zealand, to explain the overlap there. A vicariance model also accounts for the similar star patterns in the *Hebe* complex, *Olearia*, and others.

In New Zealand, as many as four named clades of *Astelia* overlap in central parts of the country (Nelson, Marlborough, and the eastern North Island). In more outlying areas of the New Zealand plateau, only one of the main clades is in the Auckland and Campbell Islands (a member of the Indian Ocean group),

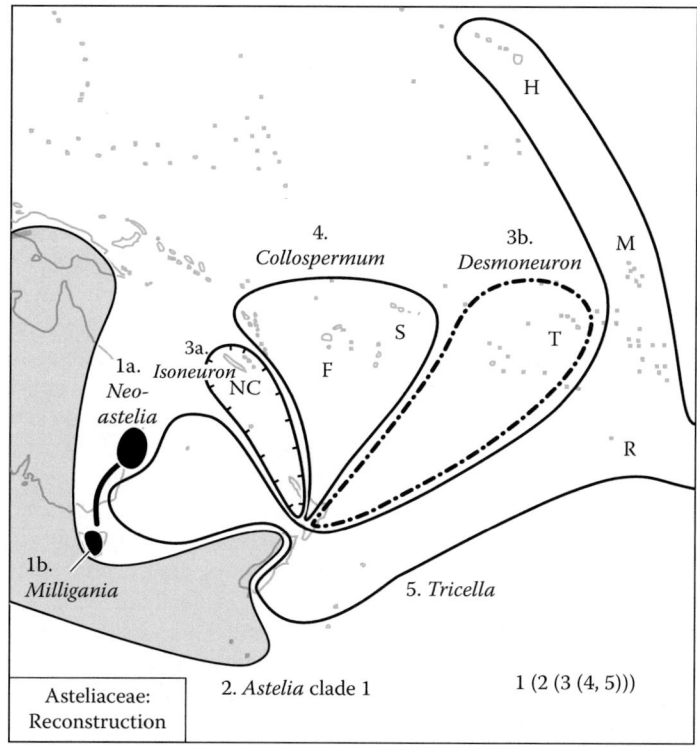

FIGURE 4.12 Reconstruction of Asteliaceae distribution in the southwest Pacific before secondary overlap of clades in New Zealand and southeast Australia (present-day coastlines shown for reference only). F, Fiji; H, Hawaiian Islands; M, Marquesas Islands; NC, New Caledonia; R, Rapa Island; S, Samoa; T, Tahiti.

one in the Chatham Islands (related to species in New Zealand, Rapa Island, etc.), and one in the Three Kings Islands (related to species in New Zealand, Fiji, etc.).

The phylogenetic position of the South American sect. *Micrastelia* is still unresolved. Maximum parsimony and Bayesian analyses indicate that it is sister to sects. *Desmoneuron* and *Isoneuron*, but in maximum likelihood analyses it appears as sister to the clade containing *Collospermum* and subg. *Tricella*. Birch et al. (2012) regarded the position of the group as uncertain. The maximum likelihood arrangement is illustrated here (Figure 4.11), as it results in a simple, allopatric pattern.

Another Pacific Group: *Coprosma* (Rubiaceae)

The herbs, shrubs, and trees in *Coprosma* s.str. are distributed from Australia and Indonesia, east through the Pacific islands, including Hawaii and Rapa, to Juan Fernández Islands, with a concentration of species in New Zealand. The distribution is similar to that of the trans-Pacific clades in the *Schefflera* group (clades 2 and 3; Figure 4.5) and in *Astelia*. Cantley et al. (2014) discussed *Coprosma* and wrote:

> Most archipelagos of the Pacific Ocean, including the Hawaiian Islands, are volcanic in origin and were never connected to continental land masses. The derivation of the Hawaiian flora is entirely the result of long-distance dispersal and *in-situ* speciation from various source areas, including the Americas, Asia and islands of Oceania.

Yet the model these authors presented overlooked the past geography of the region, in particular, the countless former high islands and archipelagos in the Pacific. These indicate that the biogeography of the groups present in the region was established, not on the geography of today, but on a past geography (Heads, 2012a). For example, just because *C. ernodeoides* of Hawaii has its closest relatives in New Zealand, Australia, and New Guinea does not mean it has dispersed from there. Cantley et al. (2014) concluded:

> It is… clear that dispersal from New Zealand to other locations in the Pacific was fundamental to the biogeography of Pacific *Coprosma*…. *Coprosma* is… capable of dispersal over thousands of kilometres….

Instead, the shared patterns indicate that *Coprosma*, like *Astelia* and the *Schefflera* complex, has occupied the South Pacific region ever since the origin of the group. *Coprosma* and its allies in and around the Pacific region have their closest relatives in Australia and in Africa, with the allopatric groups forming the southern tribe Anthospermeae (Heads, 1996; Bremer and Eriksson, 2009). The simplest explanation is that the different groups, including *Coprosma*, originated by vicariance of an Anthospermeae ancestor that was already widespread in the southern hemisphere. (The next question—explaining how the Anthospermeae ancestor attained its widespread distribution—would require a consideration of its own sister group.) Thus, the distribution of *Coprosma* can be accounted for in the same way as the distributions of *Schefflera*, *Astelia*, *Metrosideros*, and many others, and there is no need to use chance processes to explain repeated patterns. Some of these patterns are well known and constitute classic paradoxes in dispersal theory, for example, the disjunct affinity between the Marquesas and Hawaiian Islands seen in *Coprosma*, *Astelia*, *Metrosideros*, and *Melicope* (Rutaceae) (Heads, 2012a; Cantley et al., 2014).

In *Coprosma*, Cantley et al. (2014) provided support for the widespread, central Pacific group, subgenus. *Lucidae*, that is distributed on and within a vast triangle: New Zealand (all the large leaved species plus *C. propinqua*)–Juan Fernández Islands–Hawaii (Heads, 1996). This was recovered as one of the four clades in *Coprosma* s.str., and it is the only one present in the many island groups *between* New Zealand and Hawaii. (Other groups occur within these two areas, in Australia, and in Malesia.) This makes *Coprosma* subgenus *Lucidae* equivalent to *Metrosideros* s.str.; it is another example of a diverse group centered on Polynesia.

Muehlenbeckia (Polygonaceae): Another "Star Pattern"

A clade in Polygonaceae comprises two components with the following distributions (Schuster et al., 2011a,b):

> Trans-South Pacific: Papua New Guinea, Solomon Islands, Australia, Tasmania, New Zealand (five species), and South and Central America (*Muehlenbeckia*).
>
> Northern hemisphere: North Africa, Europe, Asia, and North America (*Fallopia*).

The simple allopatry is consistent with an origin of the two groups by vicariance of a widespread ancestor. The phylogeny proposed for *Muehlenbeckia* (Schuster et al., 2011a), still with limited statistical support, is as follows (NZ = New Zealand):

> NZ and Lord Howe Island (*M. complexa*), sister to a species of New Guinea (accreted terrane) and Solomon Islands (*M. platyclada*).
>> NZ (*M. australis*), sister to a clade in South America.
>>> NZ (*M. ephedroides*), sister to a clade in NZ (*M. axillaris*) + SE Australia (*M. tuggeranong*).
>>> NZ (*M. astonii*), sister to a clade found in all states of Australia and in southern New Guinea (Merauke).

All four clades occur in New Zealand and in one other area, giving a "star" pattern, as in *Astelia*. Just as *Muehlenbeckia* is allopatric with its sister group, the clades *within Muehlenbeckia* show extensive allopatry, at least outside New Zealand. This is consistent with an origin of the clades by vicariance, followed by secondary overlap within New Zealand. The connection in the first clade between the New Guinea region and Zealandia is not evident in *Astelia*, but is reported in molecular and morphological studies of many other clades (listed earlier, in the section "Distribution around the Tasman Sea").

The "Home Advantage" or "Founder-Takes-All" Model of Biogeography

Many groups, including the *Hebe* complex (Heads, 2014: Figs. 6.8 and 6.9), *Astelia* (Figure 4.11), and *Muehlenbeckia* (last section), display a "star" pattern, with a focus on New Zealand. Explanations of the pattern must account for two main aspects: the extensive allopatry of the clades evident through most of the groups' vast distributions, and the overlap of the clades in a single, restricted area.

With respect to the first aspect, simple allopatry with only one clade of a group in each area is an automatic result of vicariance. It is not so easy to explain by chance dispersal on its own, and other, additional factors need to be proposed. Recent dispersal theory has attributed allopatry to the home-advantage that a founder obtains (Waters et al., 2013). In this "founder-takes-all" model, one lineage colonizes a new region, and other lineages are then prevented from establishing there by competition. Priority effects are often observed during invasion by a weedy population, but this does not mean that the same process of physical movement has occurred in the phylogenetic deployment of clades, as their distributions can develop by evolution alone (vicariance), without physical movement. In addition, the dispersal/founder-takes-all model is contradicted by groups such as *Astelia*, *Olearia*, the *Hebe* complex, and *Muehlenbeckia*, as multiple clades overlap in New Zealand. Likewise, one study illustrated the "founder-takes-all" idea with a cover photo illustrating king penguins (*Aptenodytes patagonicus*) on Marion Island (Waters et al., 2013), yet gentoo penguins (*Pygoscelis papua*), rockhopper penguins (*Eudyptes chrysocome*), and Macaroni penguins (*E. chrysolophus*) also breed there.

Evolutionary priority effects on community structure have sometimes been suggested. For example, in the Fiordland mountains, plant clades dated as older showed greater dominance and occupied a greater proportion of niche space than younger clades (W.G. Lee et al., 2012b; Tanentzap et al., 2015). This was attributed to priority effects, with groups that dispersed earlier to New Zealand establishing dominance. It is possible that the result is an artifact though, as dominant groups are more likely to produce fossils and so will appear to be older if the fossil record is read literally. A large, dominant, wind-pollinated plant such as *Chionochloa* (Poaceae) will produce much greater amounts of pollen (and hence fossils) than a small, much less common, insect-pollinated plant such as *Euphrasia* (Orobanchaceae). It is not surprising that a fossil-based date for the former genus in New Zealand is 20.0 Ma, for the latter 5.7 Ma (Tanentzap et al., 2015).

In any case, the home-advantage/founder-takes-all model does not explain how the clades in groups with "star" patterns diverged. What is the reason for this striking pattern? The extensive overlap of the clades in New Zealand indicates that ecological factors such as competition are not the primary cause of the overall allopatry. One simple explanation is that the clades evolved by vicariance, before the breakup of Gondwana and the central Pacific plateaus. At a later stage, the clades underwent local range expansion in New Zealand (and, in the case of *Astelia*, a small area in Victoria), leading to local overlap.

The case of *Astelia* illustrates the general principle that biogeographic analysis needs to explain series of patterns together, and that focusing on the distributions of single clades in isolation can be misleading. For example, a classic question in zoology is: "Why are strepsirrhines (lemurs) present on Madagascar, while platyrrhines (monkeys) are absent there?" Ali and Huber (2010) gave an ingenious explanation for the presence of lemurs that invoked changes in the currents between mainland Africa and Madagascar, but this does not explain the absence of monkeys, and so it does not provide a general explanation for the pattern as a whole. In the same way, a "founder-takes-all" model for lemurs on Madagascar, which would propose that the presence of strepsirrhines (lemurs) prevented haplorhines (monkeys) establishing, does not explain the overall pattern as the two groups display extensive overlap in the Old World. The distributions are:

Haplorhines (monkeys, apes): Present in America and the Old World (not in Madagascar).

Strepsirrhines (lemurs, etc.): (Not in America) present in the Old World including Madagascar.

The striking symmetry can be explained by simple vicariance of a widespread ancestor, giving a single clade in South America and in Madagascar, followed by overlap within mainland Africa and Asia. In the same way, an efficient model of biogeography for many New Zealand groups needs to explain why an endemic, A, is present *and also* why its sister group, B, is absent.

New Zealand Groups with Pacific, Not Gondwanan, Affinities: The Basal Frogs

Ancient groups represented in New Zealand, such as *Schefflera*, *Metrosideros*, and *Astelia*, also occur in the central Pacific, a region that never formed part of Gondwana. Other New Zealand clades display affinities with groups located around parts of the Pacific margin, such as Japan and western North America, and these areas were also never part of Gondwana. An example is seen in the plant *Montia* (Montiaceae), with a clade of Australian and New Zealand species that has its sister group (*M. howellii*) in California and Oregon (O'Quinn and Hufford, 2005). The basal frogs, Leiopelmatidae, have a similar trans-Pacific range, as the family comprises *Leiopelma* in New Zealand and its sister *Ascaphus* in the western United States and southwestern Canada, mainly on the accreted terranes. Waters and Craw (2006: 353) wrote that divergence date of *Leiopelma* from *Ascaphus* "easily precedes the separation of New Zealand from Gondwana and is thus biogeographically uninformative for southern relationships." Nevertheless, Gondwana breakup was not the only geological event in the southern hemisphere, and many biogeographic patterns were already established before breakup, as shown by fossil groups.

The sister group of *Leiopelma* in New Zealand and *Ascaphus* in western North America comprises all other extant frogs. (Fossil genera from the Jurassic of Argentina—*Vieraella* and *Notobatrachus*—have sometimes been placed in the *Leiopelma–Ascaphus* clade, but this was rejected by Gao and Wang, 2001, and Marjanović and Laurin, 2013.) Thus, extant frogs contain two primary clades: Leiopelmatidae, in New Zealand and western North America (with no extralimital fossils), and "other frogs," found in most parts of the world but *not* in New Zealand. What could be the reason for this impressive distributional symmetry? Worthy et al. (2013) emphasized that the New Zealand fossil record includes *Leiopelma* (Early Miocene), but no evidence of any other frogs, although as many as five frog families are extant in Australia, and five species have been introduced and naturalized in New Zealand. Simple competitive exclusion is an unlikely explanation for the allopatry, as both leiopelmatids and "other frogs" coexist in western North America.

Waters and Craw (2006: 353) admitted that the presence of the leiopelmatids in New Zealand and the absence of other frogs there "might seem odd," but it would only be odd in a chance dispersal paradigm—in a vicariance process, this sort of allopatry is the usual outcome. Countless groups are present in New Zealand, while their sister group is absent there and present somewhere else. Although Waters and Craw thought the pattern in frogs might seem odd, they also argued that it "does not *prove* a New Zealand vicariant origin for *Leiopelma*" [italics in original]. They did not offer any other explanation though. Of course, vicariance cannot be proved in a mathematical sense, but as Darwin admitted, even evolution itself cannot be directly proved, and "the doctrine must sink or swim according as it groups & explains phenomena" (as quoted in Chapter 2). The same applies to biogeography, and if dispersal theory cannot explain the New Zealand frogs, then the theory "sinks." The simplest explanation for the two clades of extant frogs, Leiopelmatidae and the others, is that they originated by vicariance (before the breakup of Gondwana and the accretion of the North American terranes), and that the overlap in western North America is the result of secondary overlap.

It is possible that Leiopelmatidae occur somewhere outside their known range, or were present there but have gone extinct, or that other frogs could be found, living or fossil, in some unexplored part of New Zealand. But these are theoretical possibilities, not actual phenomena, and it is the latter that require explanation. The known distribution of Leiopelmatidae conforms to a standard, trans-Pacific pattern observed in many groups, and there is no need to invoke undiscovered populations or extinction elsewhere in order to explain it. As with dispersal, extinction happens all the time and, at some stage, in all groups. Yet proposing unknown extinction events in particular groups to explain vicariance patterns on an ad hoc basis, without evidence, is untestable and unnecessary.

Summary of the Intercontinental Affinities of New Zealand Biota

While biogeographers often describe the tectonic development of the Tasman region as "Zealandia drifted away from Gondwana," the history is much more complex than this. Panbiogeographic analyses of the New Zealand biota (e.g., Heads, 1990a; Craw et al., 1999) have emphasized vicariance and discussed the role of Gondwana, but they have not explained "the entire New Zealand biota in terms of

Gondwanan break-up and rafting of the fragments to their present day positions," as Winkworth et al. (2002b: 514) suggested they did. This would mean overlooking the many non-Gondwanan groups in New Zealand, such as trans-tropical Pacific and Tethyan groups. These would have had no direct connections with Gondwana before the biotas were fused, with the terranes, to form modern New Zealand. New Zealand is a biological and geological composite of different elements—Gondwanan, Tethyan, and the Pacific—and so "New Zealand" does not exist as a biogeographic or geological entity. The same is also true for "New Caledonia," "New Guinea," "Australia," the Neotropical region, and many others; these are all composites that have been built up through terrane accretion.

Although the Gondwanan origin of much of New Zealand is undisputed, large parts have not been derived from Gondwanan, to the west, but from the central Pacific, in the northeast. These include great regions of anomalous, thickened crust—the large igneous plateaus—that were erupted, at least in part, subaerially, and include fossil wood. Geologists have stressed their importance for tectonics (they represent the largest eruptive event known), but the significance of the central Pacific biota for global biogeography has been neglected.

The Hikurangi Plateau (Figure 3.10) formed in the central Pacific and collided with Zealandia at about the same time as, or soon before, Zealandia was rifted from Gondwana. Such a dramatic event as the accretion of the Hikurangi Plateau (with its seamounts) must have had significant effects on the biogeography of the region. The Northland-East Coast allochthon and the associated Loyalty–Three Kings Ridge is another Pacific oceanic terrane that collided with New Zealand only after Gondwana breakup. It now forms a large part of the North Island, and its biogeography is discussed in the next chapter.

5

Biogeography of Northern New Zealand: The Offshore Islands, the Northland–East Coast Allochthon, and the Taupo Volcanic Zone

The New Zealand mainland comprises North, South, and Stewart Islands, and the biogeography of this central region is treated later, in Chapters 7 through 11. Before approaching this complex topic, the biogeography of the outlying parts of the New Zealand plateau is examined, beginning in the north (this chapter) and then moving south (the next chapter). New Caledonia is of special interest, and the offshore islands on the northern part of the New Zealand plateau, while small, host a diverse flora and fauna with many endemics as well as interesting connections with areas outside New Zealand.

New Zealand—New Caledonia Connections

The New Caledonian biota is well known for its many distinctive features. For example, no other region with such a small area possesses such a rich and distinctive conifer flora, which includes 13 genera, two endemic (Jaffré, 1995). Affinities between New Zealand and New Caledonia were mapped earlier, in plants such as *Mearnsia* (Figure 4.6) and *Astelia* (Figure 4.11) (see also Heads, 2014). The New Zealand–New Caledonia pattern was discussed in more detail elsewhere (Heads, 2010 and Suppl.), and many examples were cited. A study of *Syzygium* (Myrtaceae) (Thornhill et al., 2015) found a clade with a similar distribution and a phylogeny:

> *S. maire*: New Zealand.
> > *S. fullagarii*: Lord Howe I.
> > > *S. multipetalum*: New Caledonia.
> > > > *S. arboreum*: New Caledonia.
> > > > *S. kuebinense*: New Caledonia.

Examples from Weevils

In insects, New Zealand–New Caledonia sister lineages are known in cicadas, stick insects, dung beetles, some of the Anostostomatidae wetas (Buckley et al., 2015), and in Lepidoptera (the *Sabatinca incongruella* group of Micropterigidae; Gibbs and Lees, 2014).

The connection is also evident in many other groups, and the fungus weevils, Anthribidae, provide an impressive, well-documented example. They are one of the basal families in the weevils (superfamily Curculionoidea), which have an overall phylogeny (Marvaldi et al., 2009):

> Anthribidae + Nemonychidae.
> > Belidae.
> > > Attelabidae.
> > > > Caridae.
> > > > > Brentidae.
> > > > > Curculionidae.

Of the 22 anthribid genera in New Zealand, 11 are endemic and not closely related to any others else-where; Holloway (1982) regarded these as part of the "archaic (endemic) element of New Zealand biota." Nine of the genera (including an "archaic" Three Kings Islands endemic, *Tribasileus*) are either shared with the region extending northward from New Caledonia to Malaysia or have affinities in that area, as follows (Holloway, 1982; updates in Kuschel, 2003):

1. *Androporus*: New Zealand (one species) and New Caledonia (three species).
2. *Dasyanthribus*: New Zealand (Three Kings Islands and Northland south to Mayor Island; one species) and New Caledonia (one species).
3. *Hoherius* of New Zealand appears to be closest to *Proscoporhinus* from New Caledonia.
4. *Lawsonia*: New Zealand (one species) and New Caledonia (probably one species).
5. *Lophus* of New Zealand shares many of its characters with *Perroudius* from New Caledonia.
6. *Liromus*: New Zealand (one or more species) and New Caledonia (one or more species).
7. *Micranthribus*: New Zealand (one species) and New Caledonia (one species).
8. *Tribasileus* is a Three Kings endemic (one of the "archaic endemics") in which some characters of the female genitalia are shared with the New Caledonian *Anthribisomus*, although the two genera are otherwise very dissimilar.
9. *Hoplorhaphus* of New Zealand has affinities with *Eczesaris*, known from New Guinea and Malaysia. (A similar New Zealand–New Guinea connection is recorded in many groups, e.g., passerine birds; Heads, 2014.)

In two other anthribid genera, New Zealand members have affinities with both New Caledonian and Australian groups:

Helmoreus: New Zealand (one species), New Caledonia (seven species), and Australia (one species). It is related to *Plintheria* of New Guinea.

Cacephatus: New Zealand subregion (six species), Lord Howe Island (one species), Norfolk Island (one species), New Caledonia (three species), and southeastern Australia (one species).

The final member of New Zealand Anthribidae, *Lichenobius*, differs from the others as it has no direct affinities with New Caledonia or New Guinea groups. It comprises a flightless species on each of the Bounty, Snares/Stewart, and Chatham Islands and may have a distant affinity with *Xynotropis* of Tasmania.

To summarize, in its anthribid fauna New Zealand shows a primary link with New Caledonia and also has affinities with southeastern Australia/Tasmania. In contrast, it does not have direct relationships with Chile (Kuschel, 2003).

The anthribids are a diverse, worldwide group. The sister group is the much less diverse family Nemonychidae, found mainly in the north and south temperate zones, but with some species in the tropics (Kuschel and Leschen, 2010). Nemonychids feed on gymnosperms. They are most diverse in Australasia, South America, and Central America, with about 70% of the species present there. All four New Zealand species are associated with Podocarpaceae (including Phyllocladaceae), and they are closer to the Chilean species (found on Podocarpaceae) than to the Australian ones (found on Araucariaceae).

The anthribids (feeding on fungi) and nemonychids (feeding on gymnosperms) form a clade that is sister to all other weevils. There are four other curculionoid groups that are not true weevils (Curculionidae). One is the family Belidae; in this group the New Zealand members of one subfamily (Belinae) display affinities with Australia, while the single New Zealand species of *Aralius* (subfamily Aglycyderinae) belongs to an otherwise New Caledonian genus (Kuschel, 2003). Another group of "basal" weevils, the family Brentidae, includes a typical New Zealand–New Caledonia connection: *Neocyba* is distributed in northern New Zealand (from the Three Kings Islands to Punakaiki) and is a close relative of the New Caledonian *Rhadinocyba*.

Kuschel (2003) summarized the basal, paraphyletic grade of weevils (Orthoceri): "Out of 32 genera of native orthocerous weevils in New Zealand, 24 (75%) are endemic. The closest biogeographic relationship of this fauna is first with New Caledonia, then with Australia, and then with the area north-west of New Caledonia to Sulawesi and eastward across the southern Pacific to Chile."

In true weevils (Curculionidae), New Zealand–New Caledonia connections are seen in the following genera (Lyal, 1993b):

Strongylopterus: New Zealand, New Caledonia (including the Art Islands), New Guinea, Fiji, Juan Fernández Islands.

Rhynchodes: New Zealand, New Caledonia (including Lifu).

Scelodolichus: New Zealand, New Caledonia.

Outside the weevil complex, many other beetles show similar connections. For example, *Kuschelengis* (Erotylidae) is endemic to New Zealand (North Island, Nelson) and New Caledonia (Skelley and Leschen, 2007).

Differences between New Caledonia and New Zealand

Although one study (Sanmartín and Ronquist, 2004) failed to recover the New Zealand–New Caledonia grouping, this was an artifact of the small sample size, and most other authors have accepted a close biological affinity between the two countries (Heads, 2014). Yet there are also fundamental differences between New Zealand and New Caledonia. This is obvious in diverse, abundant groups such as the Rubiaceae, for example. In this family, trees, shrubs, and herbs of the tribe Anthospermeae, especially *Coprosma*, are abundant in most New Zealand vegetation. Balancing this, the many other tribes of Rubiaceae are represented only by the subcosmopolitan *Galium*. In contrast, New Caledonia has only a single species of Anthospermeae, *Normandia neocaledonica*, but has 26 genera (3 endemic) in other tribes of Rubiaceae. The great difference is not simply because New Caledonia is warmer than New Zealand; although true *Coprosma* is absent in New Caledonia, it surrounds the country, as it inhabits New Guinea and Australia, New Zealand, and Vanuatu. (The same distinctive pattern is seen in the passerine bird *Petroica*; Heads, 2014.) These distributions are consistent with the development, between New Caledonia and northern New Zealand, of early breaks in widespread ancestors.

Lord Howe Island and Norfolk Island

Lord Howe and Norfolk Islands are young, volcanic islands located on strips of the New Zealand continental plateau that are otherwise submerged (Figure 3.1). The biota of Norfolk Island and nearby Phillip Island includes such distinctive groups as the parrot *Nestor productus* (extinct by 1851). This is a member of the basal parrot group (*Nestor* and *Strigops*) of New Zealand, and is related to *N. meridionalis*, but the produced upper mandible of the beak is distinctive (Oliver, 1955). Norfolk Island formed at ~3 Ma and is located on the Norfolk Ridge, the ribbon of Mesozoic crust that extends north from New Zealand to New Caledonia. Lord Howe Island formed at 7 Ma on the Lord Howe Rise, another strip of continental crust extending northwestward from New Zealand. As with Norfolk, Lord Howe Island is well known for its endemics and interesting biogeographic connections (Heads, 2014: Figs. 3–15 and 6–12). During the breakup of the Gondwana margin, Lord Howe Rise and the Norfolk Ridge were first rifted from Australia (with the rest of Zealandia) and then from each other.

Lord Howe and Norfolk Islands have high levels of endemism. Lord Howe endemics include five plant genera: *Lepidorrhachis*, *Hedyscepe* and *Howea* (Arecaceae), *Negria* (Gesneriaceae), and *Lordhowea* (Asteraceae). The plants of Norfolk and nearby Phillip Island include two endemic genera (*Streblorrhiza*, Fabaceae, and *Ungeria*, Malvaceae, both monotypic), along with 44 endemic species and subspecies (de Lange et al., 2005). It is likely that many of the endemics are older than the present islands, for example, the Norfolk endemic whelk *Cominella norfolkensis* has been dated as 10.2–30.1 Ma (Donald et al., 2015). How can this high level of endemism and early divergence be squared with the young age and small size of the islands?

Colloff (2011a) described a new genus of oribatid mite endemic to Norfolk Island, and he concluded (p. 17) that groups there: "… may have existed as a series of metapopulations and thus may be very considerably older than the age of any individual island… The age of an endemic species on a volcanic island should not be considered congruent with island age, nor long-distance dispersal the only plausible explanation for its presence." The large, former island located on Norfolk Ridge near Norfolk Island (Figure 5.2) is one potential source of the endemic Norfolk Island biota. The island was destroyed by extension during the opening of the Norfolk Basin between 25 and 15 Ma (Figure 3.15). (Figure 5.2 shows the modern limits of the remains of the former island; before the extension, the island itself would have been smaller than this.) Meffre et al. (2007) illustrated a well-preserved leaf fossil from seafloor rocks in the area occupied by the former island.

In a similar way, the groups that *survive* today on Lord Howe Island as an endemic there probably *evolved* on other, former islands in the vicinity (there are many seamounts), or even on the emergent continental crust of the Lord Howe Rise before it was submerged (Heads, 2014: 50).

In many trans-Tasman groups, populations from the eastern and western margins of the Tasman Basin are linked via records on Norfolk Island, Lord Howe Island, or both. For example, the tree *Planchonella costata* (Sapotaceae) of northern New Zealand and Norfolk Island is sister to *P. eerwah*, recorded in northeastern Australia (McPherson–Macleay Overlap to Cape York) (Swenson et al., 2013b). Before disruption by the opening of the Tasman Basin, the distribution would have been more or less linear.

In another example, the freshwater crab *Amarinus lacustris* (Hymenosomatidae) is known from northern North Island, Norfolk and Lord Howe Islands, and southeastern Australia (Victoria, South Australia, Tasmania, King Island) (Lucas, 1980). There are several morphological differences between populations from New Zealand + Norfolk Island, and Australia + Tasmania. As in the *Planchonella* group, before disruption by the opening of the Tasman Basin, the distribution would have been more or less linear.

The Skinks of Lord Howe Island, Norfolk Island, and New Zealand

The New Zealand skinks belong to a single clade with the following phylogeny (Chapple et al., 2009):

New Caledonia (*Marmorosphax* + *Lioscincus* + *Caledoniscincus* + *Nannoscincus*).
　　Norfolk Island + Lord Howe Island (*Cyclodina*).
　　　　New Zealand: northern North Island and offshore islands (*Oligosoma smithii* and *O. microlepis*).
　　　　New Zealand: widespread (all other *Oligosoma* species).

The phylogeny shows a north-to-south geographic sequence, and, as usual, this could reflect either a series of founder events or a sequence of north-to-south differentiation events in a widespread, common ancestor.

The basement terranes of New Caledonia are composed of Mesozoic continental crust, but some authors have suggested that the entire island was covered by the sea for 20 Myr in the Paleocene and Eocene, and only emerged at 37 Ma (Grandcolas et al., 2008). Chapple et al. (2009) accepted this as the age of the island, and so they assumed that the New Caledonian skink clade has a maximum age of 37 Ma (late Eocene). This date was then used to calibrate a molecular clock for the skinks in the region. Based on the calibration, the Lord Howe Island/Norfolk Island representative of the clade, *Cyclodina*, was dated at 25 Ma and so is much older than the individual islands it is endemic to (7 and 3 Ma). Chapple et al. (2009) made the reasonable suggestion that earlier in its history *Cyclodina* had survived on former volcanic islands, now represented by submarine seamounts, present along the Lord Howe and Norfolk ridges. Yet the calibration for the study as a whole was based on the assumption that the New Caledonia members could *not* have survived on now-sunken islands around New Caledonia. If the Lord Howe–Norfolk species survived on former islands in their vicinity, New Caledonian members would have probably done the same, as New Caledonia is a much larger island surrounded by shallow seamounts, ridges, and plateaus.

The New Caledonian biota includes many old endemics, such as *Amborella*, the sister group of all other angiosperms, and these must have survived somewhere through the Paleogene (cf. Sharma and

Giribet, 2009). In such a dynamic region, tectonic activity, including uplift and subsidence, is often local-ized, and without a continuous marine stratum of a single age it is impossible for geologists to eliminate the possibility of any former land. Only very small islets are required for the survival of groups such as plants, invertebrates, and skinks. For example, Matapia Island in northern New Zealand has an area of just 2 ha, but as many as three *Oligosoma* species coexist there (Parrish and Anderson, 1999).

Biologists who support the total submersion theory for New Caledonia have admitted that this means the presence of numerous, old groups ("relicts") there is "puzzling" (Murienne, 2009a: 1434; Grandcolas et al., 2008: 3312). Other biologists have accepted that some populations must have persisted through the inundation, either in refugia on the mainland, or on former, offshore islands (Lowry, 1998; Ladiges and Cantrill, 2007; Jolivet, 2008; Jolivet and Verma, 2008a,b; cf. Murienne, 2009b: 644).

Chapple et al. (2009) concluded that *Oligosoma* of mainland New Zealand diverged from its sister group at ~23 Ma following long-distance dispersal from Lord Howe or Norfolk Island, as the chronology ruled out earlier vicariance. But, as indicated, the chronology was based on the assumption that New Caledonia taxa could not have survived on former land nearby, although the authors accepted that the Lord Howe–Norfolk species did just this. This means that the sole argument for long-distance dispersal to New Zealand is not convincing, and so vicariance at a node between Lord Howe–Norfolk and main-land New Zealand is a viable mode of origin for the mainland skinks.

The Islands off Northland

Many plants and animals are endemic on the small islands found off the northeastern coast of New Zealand; these include the Three Kings, Poor Knights, Hen and Chickens, and Barrier Islands (Figure 5.1). Clades here often have close relationships, not with groups on the New Zealand mainland, but with clades in other regions, including Australia, Norfolk Island, New Caledonia, and the Kermadec Islands. Since the mid-Cretaceous, the tectonic history of northeastern New Zealand has been dominated by the arrival of the Hikurangi Plateau at the eastern North Island, the emplacement of the Northland–East Coast alloch-thon, and activity along an important fault zone, the Vening Meinesz transform margin.

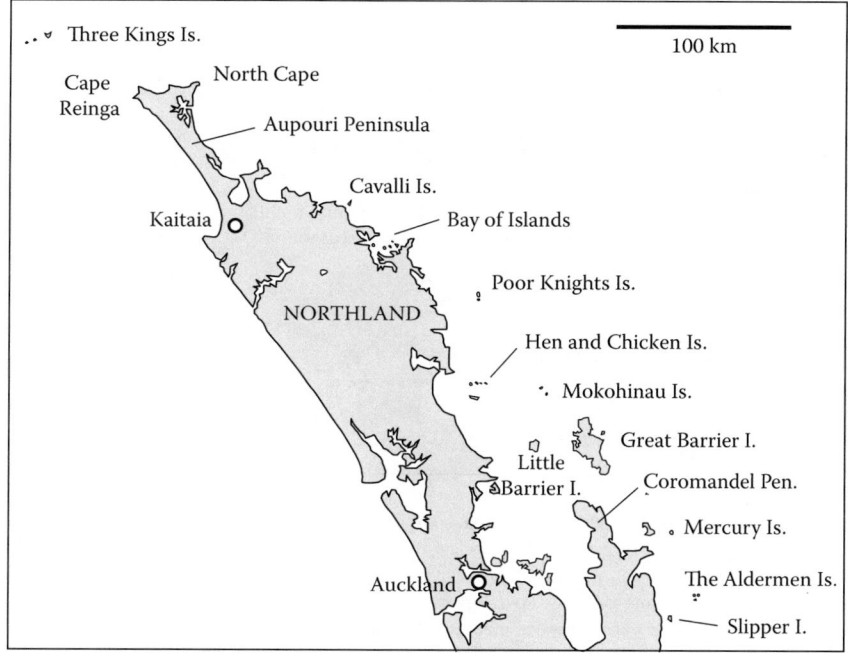

FIGURE 5.1   Northern New Zealand, showing localities mentioned in the text.

The islets off the northeastern North Island make up a center of endemism that lies between continental crust (Northland and Norfolk Ridge) to the west and an extinct, submerged island arc to the east. The island arc is represented by a chain of volcanic seamounts, the Loyalty–Three Kings Ridge–Poor Knights seamounts (= Northland Plateau seamounts) (Figure 5.2). (Note that the Three Kings Ridge lies ~200 km east of the Three Kings Islands, and the Poor Knights seamounts are a similar distance east of the Poor Knights Islands.) The Loyalty–Three Kings arc has not been erupted at the main, west-dipping Pacific subduction zone, but behind a later, *east*-dipping subduction zone (shown in Figure 5.2 with open barbs). As illustrated in Figure 3.15, a northeast-dipping subduction to the east of New Caledonia was initiated as early at 55 Ma (Cluzel et al., 2006). This is consistent with the migration of the Tonga–Kermadec subduction system into the Pacific, rotating in a windscreen-wiper fashion with a pivot at the Hikurangi Plateau as seafloor spreading propagated southward in the South Fiji Basin (DiCaprio et al., 2009).

The continental ribbon of the Norfolk Ridge and the Loyalty–Three Kings seamount chain both extend, in parallel, from northern New Zealand to New Caledonia and Vanuatu. At least some peaks of the seamount chain have flat tops, signifying wave erosion of former islands to sea level followed by subsidence (Herzer et al., 2009). In addition, at least one large island existed on Norfolk Ridge (near Norfolk Island), as mentioned earlier.

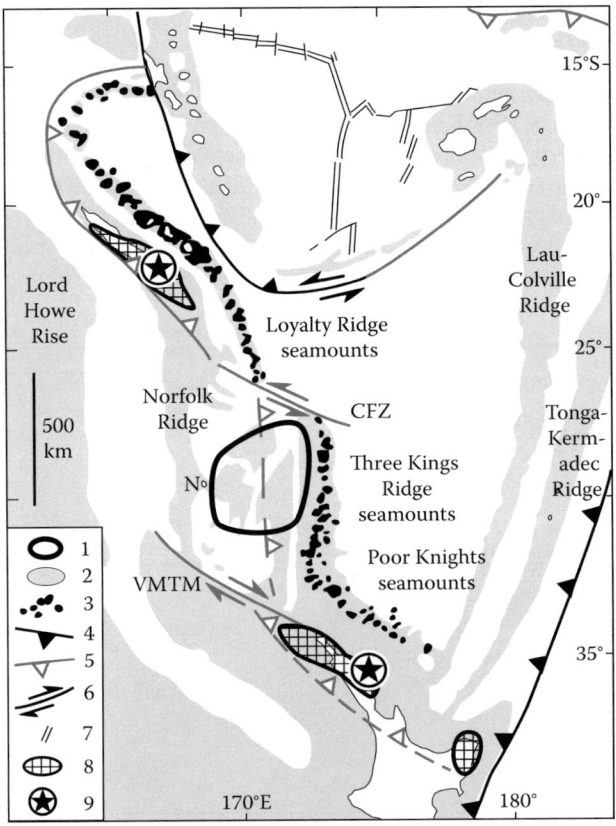

FIGURE 5.2 New Caledonia and northern New Zealand: CFZ, Cook Fracture Zone; N, Norfolk Island; VMTM, Vening Meines transform margin. 1, area emergent 38–21 Ma (destroyed by subsequent extension) (Meffre et al., 2007); 2, continental/arc crust; 3, seamounts (subduction-induced arc volcanics); 4, subduction zone; 5, New Caledonia fossil subduction zone; 6, strike-slip fault; 7, spreading ridge; 8, Poya terrane in New Caledonia, Northland and East Cape allochthons in New Zealand; 9, the monocot *Xeronema* (Xeronemataceae). (Reprinted from *Earth Planet. Sci. Lett.*, 278, Schellart, W.P., Kennett, B.L.N., Spakman, W., and Amaru, M., Plate reconstructions and tomography reveal a fossil lower mantle slab below the Tasman Sea, 143–151, Copyright 2009, with permission from Elsevier.)

Endemism on the Northeastern Islands

It is not surprising that the small islets off the northern North Island host endemics, but it is interesting that many of the offshore island clades are endemic to more than one of the island groups. For example, *Xeronema callistemon* (Xeronemataceae) is a distinctive monocot herb endemic to the Poor Knights Islands and also the Hen and Chickens (Chapter 4). *Hoheria equitum* (Malvaceae) is a tree with the same distribution (Heads, 2000). In a simple dispersal model, an offshore island species is assumed to be derived from the mainland and nested in a group from northern, mainland New Zealand. Instead, the Poor Knights-Hen and Chickens species of *Xeronema* is sister to a clade in New Caledonia, and *Hoheria equitum* is sister to a clade of five *Hoheria* species distributed through most of the North and South Islands (Wagstaff and Tate, 2011). Both patterns are compatible with vicariance early in the history of the groups, rather than dispersal from a mainland center of origin.

A similar case is seen in the fern *Asplenium pauperequitum*, which is endemic to the Poor Knights, the Mokohinau Islands (60 km southeast of the Poor Knights), and the Chatham Islands (Cameron et al., 2006). It is sister to—not nested in—*A. flabellifolium*, which is widespread in mainland Australia and New Zealand (Ohlsen et al., 2015). Although *A. flabellifolium* is present on some offshore islands, including Great Barrier, it is conspicuous by its absence on the Poor Knights, Mokohinau, and Chatham Islands, where *A. pauperequitum* is endemic (Wright, 1980; Brownsey and Smith-Dodsworth, 1989; Perrie and Brownsey, 2005; Heads, 2014: Fig. 6.24). The simplest explanation for these absences is that *A. flabellifolium* and *A. pauperequitum* differentiated by vicariance.

"Horstian" Distribution Off Eastern Northland and Auckland

The Poor Knights and the Hen and Chickens belong to a fringing chain of islands that runs from the Three Kings Islands to the Bay of Plenty (only the main, northern islands are shown in Figure 5.1). The chain has many endemics, and although the islands have different geological origins, they are united by their biogeography. In many cases, but not all, endemics on the islands have obvious relatives on the mainland.

As indicated earlier, many endemics on the northeastern islands occur on more than one of the island groups. Many of these plants and animals have a common type of distribution that has been termed "horstian." In this pattern, a clade is found along a belt of islands fringing a mainland, but does not occur on the mainland, even though the islands are closer to the mainland than they are to each other. For example, *Senecio marotiri* (Asteraceae) occurs on Whakairipiha Island in the Bay of Islands, the Poor Knights, the Hen and Chickens, and the Coromandel islands (Webb, 1988; de Lange and Cameron, 1999). Another example is the Three Kings–Poor Knights pattern discussed in more detail later. Elsewhere, classic examples of horstian distribution are documented for island clades off northern New Guinea and off northern Venezuela. This sort of pattern can develop in simple horst/graben systems, with vertical movement of fault blocks, but in practice the tectonics are often more complex and have included horizontal as well as vertical displacement. Because of this complexity, the term "horstian" distribution is simplistic, but it is used here as the type of pattern is both common and distinctive, and no other term has been suggested.

A related pattern in northern New Zealand is seen in taxa that have most records on the islands but are also present on a few headlands and peninsulas of the adjacent mainland. One example is the large shrub or small tree *Nestegis apetala* (Oleaceae), on small islands from Norfolk Island to the Mercury Islands, but also on the mainland at Whangarei (Figure 5.3; Eagle, 2006). Other *Nestegis* species are widespread in New Zealand from North Cape to the northern South Island, but none are on Norfolk or the Three Kings Islands.

In another example, the shrub *Hebe bollonsii* (Plantaginaceae) occurs on the Hen and Chickens, the Poor Knights, and also along a short stretch of the mainland coast opposite the Poor Knights (Mimiwhangata to Tutukaka) (Bayly and Kellow, 2006). *H. pubescens* is recorded from Great Barrier, Little Barrier, and Mokohinau Islands, and also Coromandel Peninsula (Bayly and Kellow, 2006).

The spider *Aparua* (Dipluridae) has a similar type of distribution. North of Auckland the genus has records at Three Kings Islands (*A. regia*), Spirits Bay on the mainland (*A. hollowayi*), Houhora on the

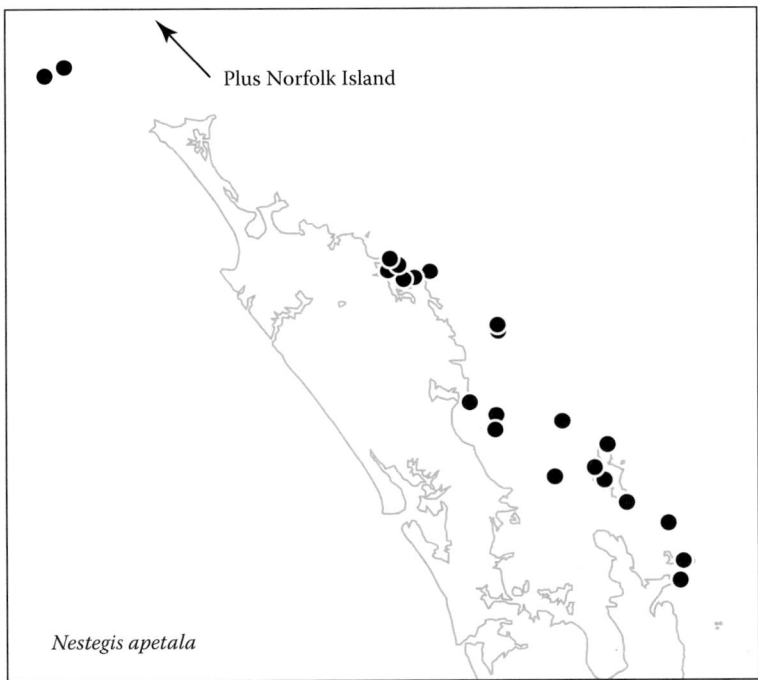

Plus Norfolk Island

Nestegis apetala

FIGURE 5.3   Distribution of *Nestegis apetala* (Oleaceae). (Data from Eagle, A., *Eagle's Complete Trees and Shrubs of New Zealand*, Te Papa Press, Wellington, New Zealand, 2006; NZPCN, New Zealand Plant Conservation Network, 2016, www.nzpcn.org.nz.)

mainland and the Poor Knights (*A. houhora*), the Hen and Chickens (*A. taranga*), Little Barrier Island (*A. hapua*), and Cuvier Island (*A. puna*) (Forster and Wilton, 1968; Court, 1982). There are also mainland records further south.

Many marine clades have distributions along the east coast of Northland that are the same, or similar, to the horstian distributions seen in terrestrial groups. Some of these marine groups found along the fringing islands migrate from the tropics in warm summers, for example, the fish species *Synodus similis* (Synodontidae) and *Aulacocephalus temminckii* (Serranidae) (Francis, 2012). The marine clades also include groups that are endemic to the region, including the following coastal fishes (Francis, 2012):

Odax cyanoallix (Labridae): Three Kings Is. to the Poor Knights.

Pempheris adspersa (Pempheridae): Three Kings Is. to East Cape.

Girella fimbriata (Girellidae): Kermadec Is., North Cape to Bay of Plenty.

Chrysiptera rapanui (Pomacentridae): Kermadec Is., North Cape to the Poor Knights. This yellow form is endemic; the other, less colorful form occurs at Easter Island.

Parma kermadecensis (Pomacentridae): Kermadec Is., North Cape to Whangarei.

The distributions have not been documented in detail, but it is likely that some of these forms are restricted to the islands.

The hagfishes are the jawless sister group of all other vertebrates and have an interesting biogeography (Cavalcanti and Gallo, 2007). The largest hagfish known is *Eptatretus goliath*, a giant, fluorescent pink species 1.2 m long, caught east of the Poor Knights Islands at the head of Hauraki Canyon (Mincarone and Stewart, 2006). This is at the edge of the continental slope, near the sunken Northland Plateau.

Connections of the "Horstian" Distributions with the Norfolk Ridge and Other Regions

The coastal fish *Chironemus microlepis* (Chironemidae) is recorded near the New Zealand mainland only from the Poor Knights and is also known from Lord Howe, Norfolk, and Kermadec Islands (Meléndez and Dyer, 2010). The disjunctions among the last three island groups can be explained by the seafloor spreading that has opened the South Fiji and New Caledonia basins (Figure 3.15). The disjunction between the Poor Knights and the continental Norfolk Ridge is seen in other groups, such as the monocot *Xeronema* (stars in Figure 5.2). As mentioned earlier, this has one species on the Poor Knights and the Hen and Chickens, and one in New Caledonia, on the ophiolite there. The two species form the Xeronemataceae, and this northern Zealandia group is sister to a diverse, global complex of families, the Asparagales. Figure 5.2 shows the New Caledonia ophiolite and also the equivalent terranes in New Zealand, the Northland, and East Coast allochthons.

Moore (1957) cited the similar open, rocky habitats of *Xeronema* in the Poor Knights and Hen and Chickens (at coastal elevations) and in New Caledonia (at 1500 m). She also wrote that "no morphological feature suggests an explanation for the peculiar distribution of the genus"; in other words, the means of dispersal seemed inadequate for long-distance dispersal. Instead of developing theories on dispersal, she compared the distribution with those of other "horstian" taxa, such as the tree *Meryta sinclairii* (Araliaceae) on the Hen and Chickens and Three Kings Islands. The other species of *Meryta* are in New Caledonia (as with *Xeronema*) and other southwest Pacific islands (Heads, 2012a: Fig. 6–13).

The connection between the northeastern horsts and the Norfolk Ridge (New Caledonia) seen in *Xeronema* is repeated in *Nestegis apetala* (Oleaceae), extending to Norfolk Island (Figure 5.3). The same connection occurs in the genus *Hymenanthera* (Violaceae). *H. novae-zelandiae* is endemic to northern New Zealand islands (Three Kings, Poor Knights, and others) and some headlands on the mainland, and its sister is *H. latifolius* of Norfolk Island (Mitchell et al., 2009). Likewise *Senecio australis* (Asteraceae) is recorded from Mokohinau and Great Barrier Islands, and also Norfolk Island (de Lange et al., 2014).

Other groups have horstian distributions off Northland and connections with regions outside New Zealand. These include the tree *Meryta sinclairii* (Araliaceae) on the Three Kings and Hen and Chickens, with a sister in Fiji, and the following lichens (Galloway, 2007):

Roccellina exspectata (the only member of the genus in New Zealand): Three Kings Is., by Cape Karikari (Black Rocks off Moturoa Is.), Poor Knights, Hen and Chickens Is. Also Curtis I. in Bass Strait, between Tasmania and mainland Australia.

Pertusaria thiospoda: Near Cape Reinga (Tapotopotu Bay), Whangarei, Rangitoto I., Great Mercury I., and Slipper I. (southern Coromandel Peninsula), and on the mangrove *Avicennia marina*. Also Norfolk I., Lord Howe I. and Australia.

Pertusaria xanthoplaca: Hen and Chicken Is., Little Barrier I., Great Barrier I., and Great Mercury I. (on coastal rocks). Also Norfolk I, Lord Howe I., eastern Australia, and Papua New Guinea.

Three Kings Islands and the Mount Camel Terrane

The Three Kings Islands are the northernmost of the fringing islands and lie 55 km off the mainland. They have some of the most distinctive endemics in the region, although they are only 4.9 km² in extent. The islands are the emergent part of the much larger Three Kings Plateau, which is 10,000 km² in area and less than 500 m deep. The age of the Karetu Trough, between the plateau and the mainland, is not clear, but it has been regarded as Pliocene (literature review in Buckley and Leschen, 2013). The geology and biology of the islands and the Mount Camel terrane that they are part of are discussed next.

Three Kings Islands and the Mount Camel Terrane: Geology

Two basement terranes are recognized in Northland, the Mount Camel terrane and the much more extensive Waipapa terrane (Figure 5.4, from Whattam et al., 2008). The Mount Camel terrane is exposed on the Three Kings Islands, The Bluff on 90 Mile Beach, Mount Camel (at Houhora Harbour), Karikari Peninsula, and Whangaroa Harbour. Local endemics occur at all of these, and there are many on the Three Kings Islands.

The Mount Camel terrane comprises Cretaceous to early Cenozoic terrigenous mudstones and sandstones, with intercalated basaltic and felsic volcanics (Nicholson and Black, 2004). Correlated rocks occur in the Nouméa Basin of New Caledonia. One question still to be resolved is whether the Mount Camel terrane is autochthonous and has been affected by the Northland allochthon passing over it (Figure 5.4), or whether it is in fact part of the allochthon, with the real basement hidden beneath it (Toy and Spörli, 2008). In any case, the terrane has a specific, linear distribution and has had a distinctive tectonic and magmatic history. It also hosts a distinctive biota.

The Mount Camel terrane volcanics have formed in an arc environment, but they also show strong continental signatures in their chemistry (Nicholson and Black, 2004; Toy and Spörli, 2008). The Three Kings Islands lavas fall into two groups (Nicholson et al., 2008). Group 1 comprises Late Cretaceous continental arc lavas. Group 2 lavas overlie these and are themselves cut by Miocene dikes, and so they could have been generated by Eocene–Oligocene subduction along the Loyalty–Three Kings Ridge. The Group 2 lavas are the first on-land evidence for the southern extension of the Loyalty–Three Kings arc system, active from the later Cretaceous to the early Cenozoic.

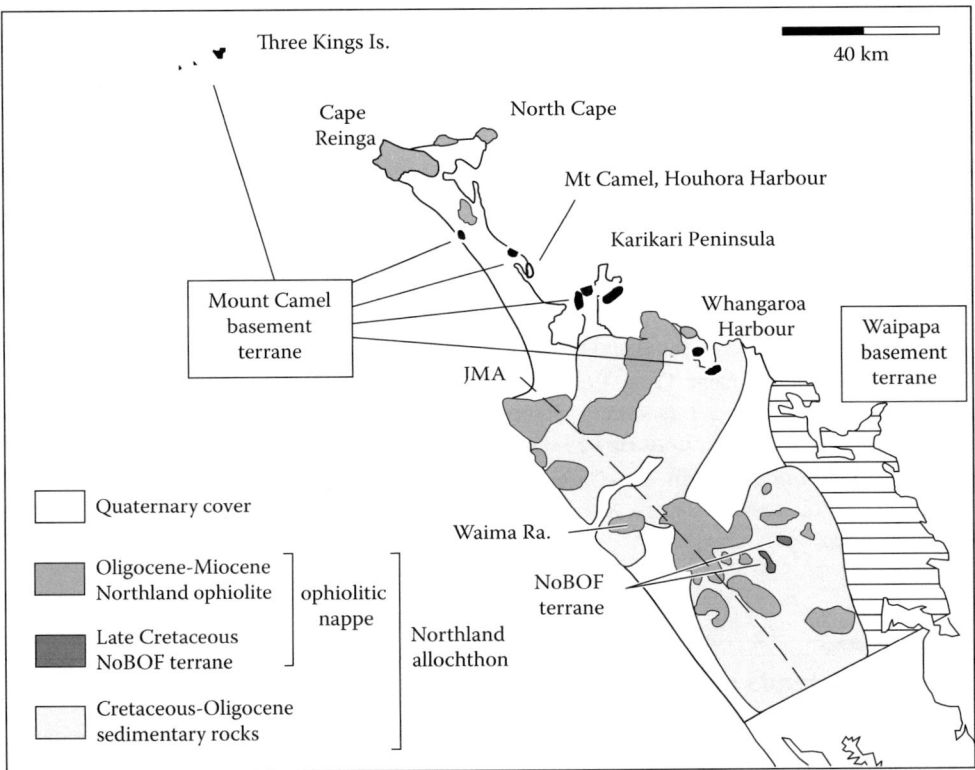

FIGURE 5.4 Simplified geological sketch map of Northland. (Miocene volcanics not shown.) JMA, Junction magnetic anomaly, representing the buried Dun Mountain ophiolite belt; NoBOF terrane, Northland basaltic ocean-floor terrane. (From Whattam, S.A. et al., *Geochem. Geophys. Geosyst.*, 9, Q03021, 2008.)

The earlier Mount Camel lavas and the Nouméa Basin lavas of New Caledonia have similar geochemistry, and their ages overlap (Nicholson et al., 2011). The data indicate active, continental arc volcanism along the New Zealand–New Caledonia margin of Gondwana at ~101–103 Ma. This coincides with the critical event in New Zealand tectonic history (Chapter 3), the switch from compressional tectonics to extension. The younger Mount Camel terrane lavas also have equivalents in the Nouméa Basin, and both were probably generated during westward rollback of the east-dipping subduction zone at 35–15 Ma (Figure 3.15).

Three Kings Islands and the Mount Camel Terrane: Biota

Despite the small size of the Three Kings Islands, the biota on this part of the Mount Camel terrane features several endemic genera. These include the tree *Elingamita* (Myrsinaceae); the stick insect *Pseudoclitarchus* (Buckley and Leschen, 2013); the heteropterans *Paratruncala* and *Basileobius* (Larivière and Larochelle, 2004); and eight beetle genera: *Gourlayia*, *Kiwiharpalus*, and *Maoriharpalus* (Carabidae) (Larochelle and Larivière, 2007); *Heterodoxa* and *Pseudopisalia* (Staphylinidae); *Partystona* and *Zomedes* (Tenebrionidae); and *Tribasileus* (Anthribidae) (Leschen et al., 2003).

One critical question is whether the Three Kings endemics are primary endemics, that were always restricted to the islands, or secondary endemics, restricted to the Three Kings because of extinction elsewhere by introduced mammalian predators. Marris and Johnson (2010) discussed this with reference to the Three Kings endemic click beetle, *Amychus manawatawhi* (Elateridae). They interpreted this as a primary endemic species. If it were a secondary endemic, other populations would be expected on other predator-free, northern offshore islands. Introduced mammalian predators—mice and rats—occur on the Chatham Islands, where *Amychus candezei* (keyed with *A. manawatawhi*) is endemic. Again, this suggests the beetles have not been extirpated from some regions by predation.

In addition to its high levels of endemism, the Three Kings biota is characterized by notable absences, even in marine groups. For example, in coastal fishes, spotties (*Notolabrus celidotus*: Labridae) are widespread and abundant all around the New Zealand mainland and the Chatham Islands, but "surprisingly" (Francis, 2012: 9) are not found at the Three Kings. (The related *N. fucicola* [banded wrasse] is present at the Three Kings and is also widespread through New Zealand.) Other "peculiarities" (Francis, 2012: 9) at the Three Kings include the absence of butterfish (*Odax pullus*: Labridae). This is replaced on the islands by the blue-finned butterfish, *O. cyanoallix*, which is endemic to the coastal region from the Three Kings to the Poor Knights.

The endemism on the Three Kings is often attributed to their isolation from the mainland by the sea strait, but some of the most distinctive endemics are marine; for example, a red alga endemic to the Three Kings, *Skeletonella* (Ceramiaceae), represents a distinct tribe (Millar and de Clerck, 2007).

Relationships of the Three Kings Biota Outside New Zealand

Many plants and animals appear to contradict a model of Neogene vicariance between the Three Kings and Northland. These include Three Kings endemics whose closest relatives are not restricted to Northland, the North Island, or even New Zealand, but to complexes that are more widespread. For example, the Three Kings endemic beetle *Phanodesta manawatawhi* (Trogossitidae) is sister to a trans-Pacific clade present through New Zealand (including the Three Kings), the Chatham Islands, and Juan Fernández Islands (Leschen and Lackner, 2013). This suggests an old break in a widespread Pacific group.

The Three Kings biota also has direct biogeographic connections beyond New Zealand. First, there is a connection with Australia. In one example, the beetle *Scabritiopsis* (Staphylinidae) is endemic to the Three Kings and is morphologically close to the Australian genus *Scabritia* (Théry and Leschen, 2013). The same connection is displayed in the following lichens (Galloway, 2007):

Pertusaria petrophyes: Three Kings Is., Lord Howe I., Victoria, New South Wales, Queensland, Papua New Guinea.

Pertusaria subisidiosa: Three Kings Is., New South Wales and Queensland.

Xanthoparmelia norcapnodes: Three Kings Is. and southeastern Australia.

The angiosperm *Tecomanthe* (Bignoniaceae) has one endemic species on the Three Kings Islands, one at the McPherson–Macleay Overlap (including Fraser Island) and the Cairns region, and three in New Guinea and the Solomon Islands (van Steenis, 1977). (Reichgelt et al., 2013, cited Miocene macrofossils of *Tecomanthe* [p. 40] or "Bignoniaceae aff. *Tecomanthe*" [p. 42] from Otago; Conran et al., 2014b, cited it as "?*Tecomanthe* ?*Deplanchea*." Reichgelt et al., 2013, also cited *Meryta* macrofossils, although the review by Conran et al., 2014b, did not refer to this. The determinations would mean that *Tecomanthe* and *Meryta* are secondary relics on Three Kings and Three Kings–Poor Knights, respectively, but a full description of the material is needed.) Likewise, the Three Kings earthworm *Megascolides tasmani* is sister to *Digaster lingi* from Queensland (Buckley et al., 2011).

The beetle family Cucujidae has a more or less worldwide distribution (including the Arctic), but in New Zealand it is restricted to the Three Kings Islands, where it is represented by the genus *Platisus*. Elsewhere, *Platisus* occurs only on the east coast of mainland Australia and in Tasmania (Watt et al., 2001; Buckley and Leschen, 2013).

Biotic connections between the Three Kings and western Zealandia are exemplified by *Basilioterpa* (Hemiptera: Cercopidae), which comprises one endemic species on the Three Kings Islands and two on Lord Howe Island (Larivière et al., 2010). Likewise, the plant *Macropiper melchior* (Piperaceae) of the Three Kings was keyed with *M. hooglandii* of Lord Howe Island, not with the mainland New Zealand species (Gardner, 1997a). A similar distribution is seen in the hemipteran genus *Lissaptera* (Aradidae), known from the Three Kings Islands, Northland, and Lord Howe Island (Larivière and Larochelle, 2004).

Other Three Kings groups have their nearest relatives along the Norfolk Ridge, on Norfolk Island or New Caledonia. The lichen *Pannaria subcrustacea* is known from the Three Kings, New Caledonia, and northern Queensland (Galloway, 2007). The pseudoscorpion *Apatochernes turbotti* occurs on the Three Kings Islands and North Cape to the Bay of Islands (Beier, 1976). The species is related to *A. posticus* of Norfolk Island. (The pair is replaced further south in Northland by local endemics, such as *A. vastus* of Omahuta, and other species that are more widespread in the region.)

The tree *Pennantia* (Pennantiaceae) comprises four species that have been arranged as follows in morphological studies by Gardner and de Lange (2002):

Eastern Australia: Sydney to Cairns (*P. cunninghamii*).

North, South, and Stewart Islands (*P. corymbosa*).

Three Kings Islands (*P. baylisiana*).

Norfolk Island (*P. endlicheri*).

The Three Kings *Pennantia baylisiana* is sister to the species on Norfolk Island, not to the much closer one on mainland New Zealand. *P. corymbosa* is widespread through the mainland in many climates and vegetation types, but is recorded north only to Kaitaia, at the southern end of Aupouri Peninsula. This suggests a phylogenetic–biogeographic node between Kaitaia and the Three Kings.

In other Three Kings groups, the connections are with areas further east. *Carex elingamita* (Cyperaceae) of the Three Kings is closest to *C. kermadecensis* of the Kermadec Islands (morphological studies; Hamlin, 1958; Moore and Edgar, 1970). Likewise, the tree *Myrsine oliveri* (Myrsinaceae) of the Three Kings is sister to *M. kermadecensis* of the Kermadec Islands (Heenan et al., 2010; Papadopulos et al., 2011). The marine gastropod *Larocheopsis* (Larocheidae) comprises two species: *L. amplexa* of the Three Kings, and *L. macrostoma*, from a seamount on the Kermadec Ridge, 130 km south of Esperance Rock (Geiger, 2012, pers. comm., April 19, 2015). This Three Kings–Kermadecs disjunction can be explained by the opening of the South Fiji Basin (Figure 3.15).

The trees in *Streblus* sect. *Paratrophis* (Moraceae) (now often treated as the genus *Paratrophis*) include a group that appeared in Corner's (1962) key as follows:

3a: *S. heterophyllus* (now treated as *Paratrophis microphylla* and *P. banksii*). New Zealand: North and South Islands, but not on the Three Kings Is.

3b: A group of four species, with one each on: the Three Kings Is. (*S. smithii*); the Solomon Islands; Fiji, Samoa, Niue, and the Cook Is.; and Tahiti.

As in the *Carex* and *Myrsine* species just cited, the Three Kings form is allied with others further east, rather than with species on the New Zealand mainland.

The traditional explanation for endemism on the Three Kings Islands and Aupouri Peninsula proposes that developments in Neogene paleogeography, such as former arms of the sea that created islands, have caused all the patterns. Nevertheless, the distributions just discussed suggest that the Three Kings biota has also had significant connections with areas outside New Zealand. These include the Loyalty–Three Kings arc and the New Caledonian parts of the continental region represented by the Mount Camel terrane.

Relationships between the Three Kings Islands and Mainland New Zealand

The precise nature of biogeographic relationships between the Three Kings Islands and the mainland is of special interest as it could shed light on possible origins of the islands' biota. The nearest part of the mainland to the Three Kings is the North Cape–Unuwhao–Cape Reinga area (Figure 5.1). The many endemics in this block include the tree *Metrosideros bartlettii* at Radar Bush, Kohuronaki Bush, and Unuwhao; the bug *Modicarventus* (Heteroptera) at Unuwhao and Whareana (Larivière and Larochelle, 2004); and the stick insect *Tepakiphasma*, restricted to Te Paki (Buckley and Bradler, 2010).

Buckley and Leschen (2013) studied several Three Kings insects and their relatives and found two patterns. In the first, Three Kings groups had sisters in the North Cape region. Examples include clades in the beetles *Brachynopus latus* (Staphylinidae), *Epistranus lawsoni* (Zopheridae), and *Syrphetodes* (Ulodidae). (A subsequent, more detailed study instead found the Three Kings *Syrphetodes* species, *S. insularis*, to be sister to a widespread Northland clade; Leschen and Buckley, 2015.)

The widespread affinities of some Three Kings–North Cape groups indicate that the break responsible for this area of endemism had more than just local significance. For example, in geckos, an undescribed species of *Dactylocnemis* endemic to the Three Kings is sister to an undescribed species from the North Cape region, and the pair is sister to a clade that is widespread through the western North Island (Nielsen et al., 2011). Likewise, in *Epistranus lawsoni*, the Three Kings/North Cape clade is sister to one from the South Island (Buckley and Leschen, 2013).

The idea of an early break is also indicated in the second pattern described by Buckley and Leschen (2013), in which Three Kings lineages have sisters that are widespread through all or much of New Zealand. Examples include:

The plant *Pittosporum fairchildii* (Pittosporaceae) of the Three Kings is sister to a widespread clade of 13 species in the North, South, Stewart, and Kermadec Islands (Carrodus, 2009: Fig. 2.7).

The stick insect *Pseudoclitarchus* of the Three Kings is sister to *Clitarchus* of North, South, and Stewart Islands.

A form of the beetle *Brachynopus scutellaris* (Staphylinidae) endemic to the Three Kings has its sister group in North, South, and Chatham Islands.

Morphology-based phylogenies have found similar patterns:

The plant *Brachyglottis arborescens* (Asteraceae) of the Three Kings is probably closest to *B. repanda*, widespread in the North Island and northern South Island, but not on the Three Kings (NZPCN, 2016).

Mecodema regulus (Carabidae) of the Three Kings is sister to a clade distributed from Northland (Ahipara) to Canterbury (Banks Peninsula) (Seldon and Leschen, 2011; Seldon et al., 2012).

Coptomma marrisi (Cerambycidae) of the Three Kings is sister to *C. sticticum* of the North Island and *C. lineatum*, widespread in the North Island and Nelson (north and west of the Wairau fault) (Song and Wang, 2003).

Neocyba, in the "basal weevil" family Brentidae, comprises one species endemic to the Three Kings and one that ranges from Northland (Coopers Beach) to Punakaiki in Westland (Kuschel, 2003).

Molecular studies have indicated other Three Kings in which the sister groups are widespread in the North Island:

Tarphiomimus sp. nov. (Coleoptera: Zopheridae) of the Three Kings is sister to a group found throughout the North Island (Buckley and Leschen, 2013).

The skink *Oligosoma fallai* of the Three Kings is sister to a clade that is widespread through the North Island (*O. ornata, O. alani, O. macgregori, O. oliveri, O. whittakeri, O. townsi*) (Chapple et al., 2009).

The land snail *Rhytidarex* (Rhytididae) of the Three Kings is sister to a clade (*Amborhytida, Schizoglossa,* and *Paryphanta* s.str.) that extends through much of the North Island (Spencer et al., 2006).

In plants, the *Hebe* complex (Plantaginaceae) is represented on the Three Kings by a single, endemic species, *H. insularis,* that has complex affinities and has not been sequenced (Bayly and Kellow, 2006). No other member of the diverse *Hebe* complex is present on the Three Kings, even though the group includes many weedy species. To summarize, the great biological difference between the Three Kings Islands and the rest of New Zealand suggests that the Three Kings have preserved parts of an old biota of the Loyalty–Three Kings arc or the Mount Camel terrane.

As indicated, the North Cape area is also an important center of endemism. It is now covered by the Northland allochthon (Figure 5.4), but it may preserve elements of an earlier biota of the basement terranes, or the allochthon itself could have been colonized by Loyalty–Three Kings groups before accretion. The tectonic relationship between the Loyalty–Three Kings arc and the allochthon plus its ophiolite is discussed later.

Other Mount Camel Terrane Endemism

The Three Kings Islands are part of the Mount Camel terrane, also known from Mount Camel, Karikari Peninsula, and Whangaroa Harbour (Figure 5.4). Groups on the terrane include local endemics in the land snail genus *Cytora* (Pupinidae); these occur on the Three Kings (*C. hirsutissima, C. filicosta*), Mount Camel (*C. houhora*), and Karikari Peninsula (*C. parrishi*) (Marshall and Barker, 2007). Endemic *Cytora* species also occur in the North Cape–Cape Reinga area (*C. gardneri, C. hispida, C. kerrana, C. lignaria*).

In the land snail genus *Allodiscus* (Charopidae), the type species and its nearest allies occur in the following areas (Marshall and Barker, 2008) (Mount Camel terrane areas and their endemics are in bold):

Three Kings Islands (*A. cassandra*).

North Cape–Cape Reinga area (*A. pumilus* and *A. spiritus*).

Mt Camel (*A. camelinus*).

Karikari Peninsula (*A. fallax*).

Poor Knights Islands (*A. cooperi*).

Maunganui Bluff, Northland west coast (*A. venulatus*).

Widespread in parts of the North Island, but not on Aupouri Peninsula (*A. conopeus* and *A. dimorphus*).

Other Mount Camel terrane endemism occurs in the *Coprosma rhamnoides* complex (Rubiaceae). This is widespread and very common throughout mainland New Zealand, but the sole glabrous member is an undescribed species endemic to Whangaroa Harbour (illustrated by Eagle, 2006, as *C. neglecta* "ii"). It grows in lowland forest, right to the edge of the mangrove (personal observations).

The Northland shrub *Hebe diosmifolia* has two races, one diploid and one tetraploid. Murray et al. (1989) wrote that the two "do not show a very clear pattern of [north-south] geographic separation," but their records of tetraploids at Te Paki, Rarawa (Houhora Harbour), and Karikari Peninsula, with diploids to the *west*, indicate a boundary near the Mount Camel terrane.

In New Zealand, the fern *Davallia* comprises a single endemic species with two subspecies: *D. tasmanii* subsp. *tasmanii* on the Three Kings and *D. tasmanii* subsp. *cristata* in Puketi forest, 12 km south of Whangaroa Harbour and the exposures of Mount Camel terrane (von Konrat et al., 1999).

Possible Mount Camel terrane groups in the marine biota include the red alga *Lembergia* Saenger (= *Lenormandia allanii* Lindauer). This is endemic to the Three Kings Islands and the mainland from Cape Maria van Diemen to Reef Point (i.e., Tauroa, near Ahipara) and Houhora Harbour (Mount Camel terrane). As mentioned earlier (Chapter 4), *Lembergia* plus *Sonderella* of Victoria form a clade that is basal in Rhodomelaceae, the largest family of red algae and one that is distributed worldwide.

The Vening Meinesz Transform Margin and the Northland Plateau

Regional Geology

Distribution patterns along the Poor Knights–North Cape–Three Kings belt coincide with one of the main tectonic features in the region, the Vening Meinesz transform margin (Figure 5.5; Herzer et al., 2009a). The opening of the South Loyalty basin and, later, the Norfolk Basin included major strike-slip displacement at the basins' southern margin, along the Vening Meinesz transform (Figure 5.2).

The Poor Knights Islands and the nearby Hen and Chickens are young volcanic islands formed in the Early and Late Miocene. Their endemism and connections with New Caledonia rather than the New Zealand mainland (e.g., *Xeronema*, Chapter 4) suggest that they have inherited some of their biota from the sunken Northland Plateau or another, nonmainland source. Lowered sea levels during the Pleistocene mean that the Poor Knights were connected to the mainland at this time (Brook and McArdle, 1999).

Croizat (1964: 159) discussed the distribution patterns along the northeastern islands, and he wrote that the region provides "some of the finest examples of horstian distribution on record." Likewise, the region's geology includes "one of the finest examples of a continent-backarc transform" (Herzer and Mascle, 1996), the Vening Meinesz transform. This is a belt up to 50 km wide that includes linear fault

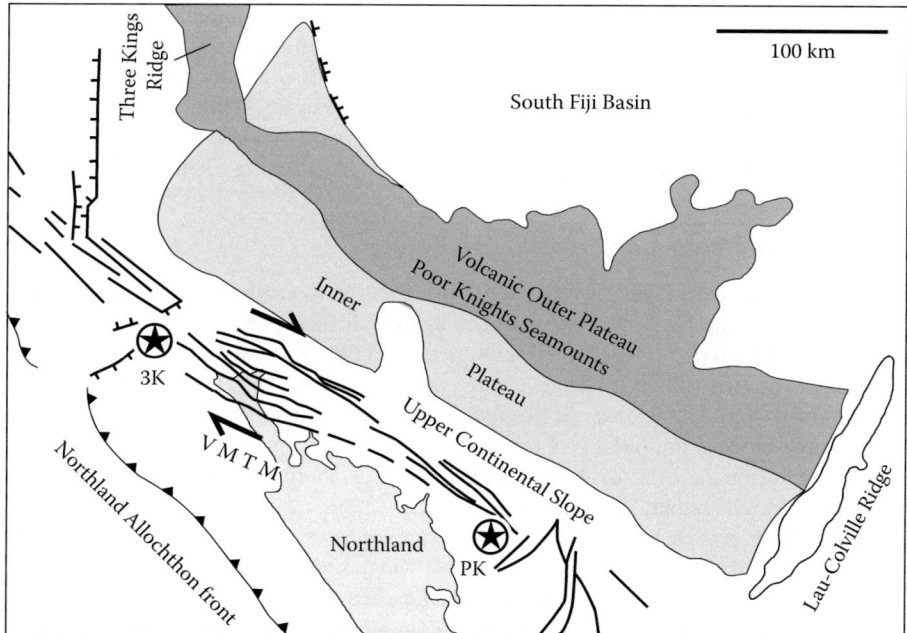

FIGURE 5.5 Northland Plateau (gray) and the faults of the Vening Meinesz transform margin (VMTM). Normal faults hatched. 3K, Three Kings Islands; PK, Poor Knights Islands. (Based on Herzer, R.H. et al., *Marine Geophys. Res.*, 30, 21, 2009a; Herzer, R. et al., *Geol. Soc. N. Z. Miscell. Pub.*, 126, 22, 2009b.)

scarps up to 2000 m high. The transform was initiated as a dextral strike-slip fault zone (cf. "Glossary of Geological terms"), and deformation along it and the now-submerged Northland Plateau lasted into the Middle Miocene. Earlier authors regarded the belt as a fracture zone, but it is now interpreted as a complex, volcanic, transform—an old plate margin (see the offset in the New Caledonia subduction zone along the Vening Meinesz transform in Figure 5.2; Herzer et al., 2009a). With its history of strike-slip, volcanism, obduction, and subsidence, it is not surprising that this margin is associated with such high levels of endemism, both terrestrial and marine, as well as disjunction.

The outer Northland Plateau (Figure 5.5), seaward of the Vening Meinesz transform, is undeformed and bears many seamounts, including the Poor Knights seamount chain that strikes northwest–southeast. The inner Northland Plateau has undergone strong deformation caused by Early Miocene dextral shearing along the Vening Meinesz transform. The inner plateau has volcanic seamounts comparable to those found on the outer plateau and also includes the northeastern margin of the Northland allochthon (Herzer et al., 2009a; Schellart, 2012).

The Norfolk backarc basin opened in the Miocene by means of transform faults at its northern and southern margins, the Cook fracture zone and Vening Meinesz transform (Figure 5.2). Before the basin opened, the Three Kings Ridge lay alongside the Norfolk Ridge, but as the Norfolk basin rifted open in the Miocene (from 23 to 15 Ma), the Three Kings Ridge was carried 290 km southeastward (Figure 3.15). This is equivalent to the distance between the Three Kings and Poor Knights Islands, 275 km. Herzer et al. (2009b) suggested that this movement of the Three Kings Ridge would have transported terrestrial biota southeastward with the plate, and the dextral strike-slip movement could also have caused disjunction.

The outer part of the Northland Plateau is volcanic, as with the inner part, and includes a distinctive seamount chain, the Northland Plateau or Poor Knights seamounts. This is probably an extension of the Three Kings Ridge (Figure 5.5, cf. Figure 5.2). Dredged rocks on the Northland Plateau chain are dated as 32–18 Ma (Mortimer et al., 2007). Herzer et al. (2009a) cited evidence of shallow water and subsidence on Three Kings Ridge. In the outer part of the Northland Plateau, a few large seamounts have flattish tops and were probably eroded to sea level. On the inner plateau, tilted flat tops on acoustic basement fault blocks suggest that many parts of the plateau were above sea level before they were planed off in the Early Miocene (Herzer et al., 2009a). Later these were lowered to 500–1500 m below sea level. Faulting, differential uplift, and subsidence were common in the Early Miocene, and an emergent or shallow inner plateau accounts for the presence of otherwise unexplained detrital conglomerates and pebbles.

Herzer et al. (2009a) considered two models for the Northland/South Fiji Basin region. In one, the Loyalty–Three Kings–Northland Plateau–Colville–Lau arcs formed a single arc that collided with northern New Zealand in the Oligocene. In the second model (Figure 3.15), the Loyalty–Three Kings–Northland Plateau ridge collided with Northland, and the South Fiji Basin then formed by the Colville–Lau arc "unzipping" from the first arc, with backarc spreading accommodated along the Vening Meinesz transform.

The Three Kings and Poor Knights Islands: Biology

Distributions on the northern islands are sometimes attributed to early Maori translocations of plants and animals, but Brook and McArdle (1999) argued against this as an explanation for the endemic Poor Knights land snail *Placostylus hongii*. Many groups are too old to be explained by Maori translocation and are even older than the Miocene islands of the Poor Knights and the Hen and Chickens themselves. The monocot family Xeronemataceae, as discussed already, is endemic to these two groups and New Caledonia. Xeronemataceae are basal in a diverse, global clade of monocots (13 families) and must be much older than Miocene; a fossil-calibrated molecular clock study estimated that the family diverged at ~100 Ma (Janssen and Bremer, 2004).

As with Three Kings groups, many Poor Knights endemics have widespread sisters. In one clade of the diverse species *Brachynopus latus* (Staphylinidae), a Poor Knights group has its sister on the Three Kings and the northern two-thirds of the North Island (Buckley and Leschen, 2013). The Poor Knights beetle *Mecodema ponaiti* (Carabidae) is sister to the rest of the *M. curvidens* group, widespread from the Three Kings Islands to Banks Peninsula (morphology-based analyses by Seldon and Leschen, 2011, and Seldon et al., 2012).

Endemism on the Poor Knights and the other offshore islands is sometimes attributed to extinction of mainland populations. It is obvious that there has been massive, anthropogenic extinction, and fossils show that several large weevils have been extirpated from parts of Canterbury (Kuschel and Worthy, 1996). Yet extirpation by mammals on the mainland does not explain the high levels of endemism seen in the smaller invertebrates on the islands (Watt, 1982) or in marine groups.

In addition, extirpation on the mainland does not account for the vicariance often observed between mainland and island sister groups. For example, the giant weta *Deinacrida heteracantha* (Orthoptera: Anostostomatidae) is New Zealand's largest insect. In the nineteenth century, it was distributed through mainland Northland and on the Barrier Islands (map in Gibbs, 2001). But since then it has been wiped out by introduced pests everywhere except the rat-free Little Barrier Island. This once widespread species is sister to *D. fallai*, which is endemic to the Poor Knights Islands. Unlike *D. heteracantha*, it has no mainland records, and the allopatry between the two species suggests that it was never there.

Three Kings–Poor Knights Disjunction

Many of the Northland horstian distributions involve the Poor Knights. A typical example is the land snail *Cytora motu*, endemic to the Poor Knights Islands and an islet, Battleship Rock, in the outer Bay of Islands, 50 km to the northeast (Marshall and Barker, 2007). The species is allopatric with its closest relative, *C. torquillinum*, which is widespread on the mainland from Hokianga to Hawkes Bay, and on many offshore islands from the Hen and Chickens to Mayor Island.

One distinctive horstian distribution is seen in groups disjunct between the Three Kings Islands and, 275 km away, the Poor Knights, and Hen and Chickens. The land snail faunas of the Three Kings, Poor Knights, and Mokohinau Islands share certain features and comprised a group in a classification (Barker, 2005). In plants, the tree *Meryta sinclairii* (Araliaceae) is known in the wild only from the Three Kings Islands and the Hen and Chickens, and has a sister in Fiji (Heads, 2012a). Its occurrence on the Hen and Chickens is sometimes attributed to Maori translocation of a Three Kings endemic, but similar patterns are documented in other plants, including the following:

> *Macropiper excelsum* subsp. *peltatum* (Piperaceae): Three Kings, Poor Knights, Mokohinau, and Great Barrier Is. It is replaced on the mainland by *M. excelsum* subsp. *excelsum* (Gardner, 1997a).
>
> *Coprosma macrocarpa* subsp. *macrocarpa* (Rubiaceae): Three Kings and Poor Knights Is. This is replaced on the mainland by *C. macrocarpa* subsp. *minor* (Gardner and Heads, 2004).
>
> In New Zealand flax, *Phormium tenax* (Xanthorrhoeaceae), plants from the Three Kings and Poor Knights Is. are distinguished by their orange or salmon-pink flowers, and leaves with a golden-yellow margin (Smissen and Heenan, 2008).

Other groups have similar distributions, but also include records at North Cape/Cape Reinga:

> *Cordyline obtecta* (Laxmanniaceae): Norfolk I., Three Kings, North Cape and nearby Murimotu Island, and Poor Knights (de Lange et al., 2005; NZPCN, 2016).
>
> The lichen *Haematomma fenzlianum*: Three Kings, Pandora Beach (10 km east of Cape Reinga), Poor Knights, and Hen and Chickens (Galloway, 2007).
>
> The marine gastropod *Thoristella davegibbsi* (Trochidae) occurs off the Three Kings, Cape Reinga, and the Poor Knights. The species is allopatric with its close relative *T. carmesina*, which is widespread from North Cape to East Cape, but not recorded from the Three Kings or Poor Knights (Marshall, 1998).

In spiders, *Hypodrassodes* sp. (Gnaphosidae) from the Poor Knights seems most closely related to *H. insulana* of the Three Kings (Court, 1982). In other arthropods, Watt (1982) cited the following groups with Poor Knights–Three Kings disjunctions:

> *Anisolabis* sp. (Dermaptera) from Poor Knights is "either a geographical race of *A. kaspar*, previously only known from the Three Kings Islands, or a closely related Poor Knights endemic species."
>
> *Myerslopia* sp. (Hemiptera: Myerlopiidae) "may be a Poor Knights endemic. It is closely related to *M. triregia* of the Three Kings Islands. Members of this genus are flightless...."

Arthracanthus sp. 2. (Coleoptera: Melyridae) only known from the Poor Knights and Three Kings Islands.

Holoparamecus sp. 4 (Coleoptera: Merophysiidae) is "apparently a Poor Knights endemic closest to *Holoparamecus* sp. 2, an undescribed species from the Three Kings Islands."

Lissotes mangonuiensis (Coleoptera: Lucanidae) from Poor Knights to Kaitaia is related to *L. oconnori* of Kaitaia to North Cape and *L. triregius* of the Three Kings Islands.

At least five true weevils (Curculionidae) show similar patterns:

Hadracalles fuliginosus: Three Kings, Poor Knights, Hen and Chickens, and Aldermen Is. off the Coromandel Peninsula (Early, 1995).

Clypeolus veratrus: Three Kings, Poor Knights, Mokohinau Is., Hauraki Gulf Is., and Auckland (Kuschel, 1982; Lyal, 1993a,b).

Praolepra sp.: Three Kings, Poor Knights, Hen and Chickens (Kuschel, 1982).

Psepholax sp. (related to *P. sulcatus*): Three Kings, "close to" an unnamed form on the Poor Knights (Lyal, 1993a) ("the two island forms differ from the mainland form in similar ways").

Anagotus turbotti: Three Kings, Poor Knights, Hen and Chickens. Although this is a large species, there are no known fossils on the mainland.

Similar disjunctions occur in many marine groups, such as the chiton *Chiton aorangi*, endemic to the Three Kings, the Poor Knights, and the Aldermen Islands (Creese and O'Neill, 1987).

Three Kings–Poor Knights connections are also seen in character geography that underlies the phylogeny. The carabid beetle *Mecodema ponaiti* is endemic to the two Poor Knights Islands, where it has two variants (Seldon and Leschen, 2011). The variant from Aorangi Island has similar external morphological features to those of *M. manaia* (Whangarei Heads), whereas the variant from Tawhiti Rahi Island is similar in its external morphology to *M. regulus*, endemic to the Three Kings Islands.

Another Mount Camel Terrane–Poor Knights Distribution

In New Zealand *Myrsine* (Myrsinaceae), the largest species is the tree *M. aquilonia*. The species is not on the Three Kings Islands, but as with the Three Kings–Poor Knights groups, it links the Mount Camel terrane biota with the Poor Knights. It is restricted to four localities, all in Northland: Te Arai by The Bluff on 90 Mile Beach (Mount Camel terrane), Rangaunu Harbour by Karikari Peninsula (near Mount Camel terrane), the Poor Knights Islands, and the mainland opposite the Poor Knights (Heenan and de Lange, 2004).

On the Poor Knights, *Myrsine aquilonia* grows in coastal forest, around cliffs, and on rock stacks, but in Rangaunu Harbour it occurs in a tidal, estuarine habitat—on a shell bank among mangroves and the saltmarsh plant *Sarcocornia* (Amaranthaceae). This recalls the habitat of the *Coprosma* at Whangaroa Harbour (mentioned earlier in this chapter), and it indicates that the Mount Camel terrane biota could have survived in earlier, disturbed maritime habitat. Molecular studies have shown *M. aquilonia* embedded in the widespread New Zealand complex, *M. divaricata* (Heenan et al., 2010), but whatever the taxonomic status of *M. aquilonia* should be, the biogeography, morphology, and ecology of this group of populations are all distinctive and need to be explained.

The Northland and East Coast Allochthons

The Northland and East Coast regions of the North Island have been covered with allochthonous terranes (Figure 5.4). These formed elsewhere and have been moved to their current position in the Cenozoic, following the accretion of the New Zealand basement terranes in the Paleozoic and Mesozoic, and the Hikurangi Plateau in the latest Cretaceous. The Northland and East Coast allochthons have been emplaced from the northeast and cover much of the northern North Island. These regions are introduced here with a consideration of their biology.

Biology of Northland–East Coast

Northland endemics are abundant, while East Coast endemics are fewer. In plants, *Libertia cranwelliae* (Iridaceae) is endemic to East Cape (Awatere and Kopuapounamu valleys) (Blanchon et al., 2002), and *Hebe tairawhiti* (Plantaginaceae) is endemic to the East Coast allochthon (from East Cape to Gisborne; Bayly and Kellow, 2006), as is *Plantago picta* (Plantaginaceae) (Meudt, 2012). The last species is not derived from within a North Island clade, but is instead sister to *P. raoulii*, widespread in the Three Kings, North, South, Stewart, and Chatham Islands (Tay et al., 2010).

Animals endemic to the East Cape area include the beetles *Duvaliomimus obscurus*, *D. orientalis*, *D. crypticus* (Carabidae) (Townsend, 2010), and *Mitophyllus solox* (Lucanidae) (Holloway, 2007). The *Sagola pulchra* species group (Staphylinidae) comprises four species: *S. pulchra*, which extends from Auckland to the Bay of Plenty, and three local endemics in the northern East Coast (Park and Carlton, 2014):

Sagola ramsayi: Lake Waikaremoana.

S. poortmani: Mount Hikurangi.

S. keejeongi: East Cape (Lottin Point, Hovells Watching Dog).

Clades in Northland often have affinities with groups in East Cape, with disjunctions across the Bay of Plenty. In harvestmen, "The origin of the East Cape fauna is closely linked with that of the North Auckland fauna with which it is closely related" (Forster, 1954: 308). Other examples include the following:

The shrub *Pimelea sporadica* (Thymelaeaceae): Northland and, disjunct, from the East Cape area to Gisborne. The species grows on ultramafic and sandstone rocks, on dune sand, and in shrubland (Burrows, 2009a).

The tree *Beilschmiedia tarairi* (Lauraceae): Northland, Auckland, and Coromandel Peninsula, and disjunct in the East Cape region. The similar *B. tawaroa* is restricted to coastal parts of the same regions and has the same disjunction across the Bay of Plenty (Wright, 1984). Both are lowland forest trees.

The marine gastropod *Calliostoma penniketi* (Calliostomatidae): Three Kings and disjunct off East Cape (Marshall, 1995a).

The marine gastropod *Sigapatella terraenovae* (Calyptraeidae): Three Kings and North Cape, and disjunct off East Cape (Marshall, 2003).

The land snail *Phenacharopa novoseelandica*: Eastern Northland, along the east coast from Gisborne to Wellington, and also on Motunau Island, Canterbury (Climo, 1969; F.M. Climo, pers. comm., August 27, 2015).

The beetle *Brachynopus apicellus* (Staphylinidae): Kaitaia to Auckland, and disjunct at East Cape (Löbl and Leschen, 2003).

These Northland–East Cape distributions run parallel with the Vening Meinesz transform margin, with the extinct subduction zone (which has encroached from the northeast), and with the "horstian" distributions on the offshore islands. Other groups show distributions on the mainland that are also oriented with the extinct subduction zone. In the land snail genus *Chaureopa*, species distributions form a series of parallel arcs from Northland to East Cape (Climo, 1985: Fig. 9).

Some of the Northland–East Cape groups are related, not to other New Zealand clades, but to clades in areas such as New Caledonia. This is documented in the true wetas of the orthopteran family Anostostomatidae (king crickets, etc.). The weta genera *Motuweta* and *Anisoura* are New Zealand endemics that occur in Northland, the East Cape region, and, disjunct, on the Mercury Islands (Figure 5.6). They are the New Zealand representatives of a group termed the "tusked wetas," and they form a clade with *Aistus* and *Carcinopsis* of New Caledonia (Pratt et al., 2008).

The weta family Anostostomatidae as a whole has a typical Gondwanan distribution in South Africa, Madagascar, India, Australasia, and Central and South America (Pratt et al., 2008). Within New Zealand it has been regarded as a "Paleozealandic element," and Pratt et al. (2008) accepted that some of its New Zealand clades may have formed before Gondwana breakup.

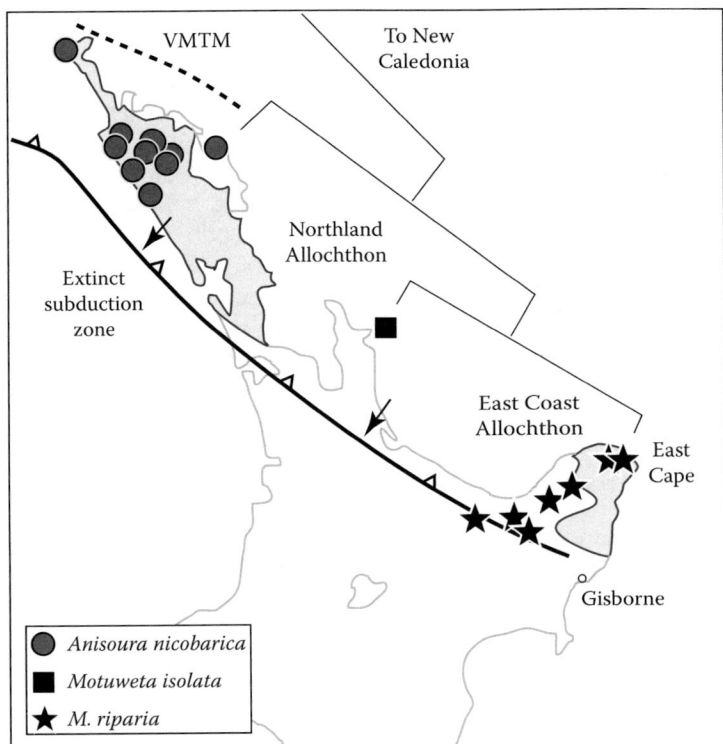

FIGURE 5.6   The New Zealand tusked wetas, *Motuweta* and *Anisoura* (Orthoptera: Anostostomatidae) (distributions from Gibbs, 2001; phylogeny from Pratt et al., 2008) and related allochthonous terranes (Cluzel et al., 2010b). VMTM, Vening Meinesz transform margin. Arrows indicate the former southeastward rollback of the subduction zone, now extinct. (Reprinted from Heads, M., *Biogeography of Australasia: A Molecular Analysis*, Cambridge University Press, Cambridge, UK, 2014. With permission from Cambridge University Press.)

Liebherr et al. (2011) accepted that the weta assemblage is probably of Jurassic age, based on an Upper Jurassic fossil wing fragment from Port Waikato, New Zealand, described as *Notohagla mauii* (Grant-Mackie et al., 1996). *Notohagla* belongs to the Prophalangopsidae, a member of a broader group with the phylogeny: Anostostomatidae (Prophalangopsidae (Haglidae + Gryllidae)). This implies that the Anostostomatidae are no younger than the Jurassic Prophalangopsidae. Trewick and Morgan-Richards (2005) calculated much younger ages for the weta lineage, but Liebherr et al. (2011) did not accept these, as they were based on a questionable evolutionary rate (2.3% divergence/Myr, from Brower, 1994). The derivation of this rate has been criticized elsewhere (see Papadopoulou et al., 2010; Heads, 2012a). Undescribed fossils that are "putatively weta" (Pratt et al., 2008) are also known from Australia.

With a minimum Jurassic age, the direct ancestors of the northern New Zealand–New Caledonia tusked wetas (Figure 5.6) are likely to have evolved before the origin and emplacement of the alloch-thon. The arc associated with the allochthon—the Three Kings–Poor Knights seamount chain—and the Northland Plateau itself are now submerged, but the allochthon would have inherited some of their biota as they were sinking and the allochthon was being obducted onto the continental crust.

Another group with significant evolution at East Cape is the stick insect *Clitarchus hookeri* (Buckley et al., 2010). It comprises a paraphyletic basal grade in western North Island (17 clades, mainly allopatric), then "Clade Q," distributed as follows:

East Cape.
 NE Coromandel Peninsula.
 Widespread in eastern North and South Islands (from Hawkes Bay to Dunedin).

Differentiation at the first two nodes has developed around the East Coast–Northland allochthons, implying early breaks and long-term persistence *in situ*.

In the brown seaweed *Carpophyllum maschalocarpum*, populations from the Bay of Islands (Cape Wiwiki) and Poor Knights were found to group with populations from East Cape and Gisborne, not with the other populations that occur along the coast between these regions (Buchanan and Zuccarello, 2012). The pattern was attributed to "leap-frog" dispersal but could instead reflect the tectonic emplacement of the Northland and East Coast allochthons.

Northland and East Coast Allochthons: Geology

Northland and East Cape are linked in their tectonic structure by the large, related allochthons that have been emplaced there. The Northland and East Coast allochthons both include ophiolite suites, or "ophiolites," distinctive rock sequences that are found in collisional orogens. The sequence in complete ophiolites comprises, from the bottom to top: ultramafic rocks (peridotite) representing upper mantle, intrusive gabbros, sheeted (vertical, parallel) dikes, extruded basalts (pillow lavas), and deep sea sediments. Ophiolites are small parts of ocean basins that have been ramped up onto continental crust, or obducted, at subduction zones. They often occur in mountain belts that have formed during collisions associated with subduction. (The mode of emplacement of ophiolites on land is tectonic, in contrast with marine sedimentary rocks on continental crust, which have formed by sedimentation during marine transgressions.) Reconstructions of global tectonics before the current ocean basins formed depend on interpretations of the ophiolite record (Metcalf and Shervais, 2008).

Ophiolites and other allochthons are well known in terrestrial biogeography as centers of endemism. Examples include the Dun Mountain ophiolite of the South Island and the New Caledonian ophiolite (Figure 5.2). Many endemics of ophiolite suites occur on the ultramafic rocks of the sequence, and these are often cited as "ultramafic endemics." In broader terms they can be regarded as ophiolite endemics, related to the tectonic history of arcs and their backarc basins rather than to their soil chemistry (Heads, 2014: Chapter 8).

Although ophiolites are composed of ocean floor, and so might not be expected to have had any association with land, they have often formed in close association with subduction zones and island arcs. Ophiolites may develop behind the arc with respect to the trench—in the backarc region—or in front, in the forearc (Figure 3.7). Most are thought to form in the forearc of nascent island arcs (Metcalf and Shervais, 2008). Geologists are interested in ophiolites as they help locate prior subduction zones and ocean basins that have otherwise been subducted. Terrestrial biologists are interested in the ophiolites as they preserve biotic traces of the island arcs that have formed with them.

A belt of ophiolites runs parallel with the Loyalty–Three Kings arc and is probably related to it (Figure 5.2). The belt has been obducted onto the continental crust of New Caledonia, Northland, and East Cape and hosts many endemics. These were probably inherited from islands on the mainland basement (as the ophiolite was obducted over it); from the islands of the Loyalty–Three Kings–Northland Plateau arc before these were submerged as seamounts; from the sunken, landward part of the Northland Plateau; and from other islands that were subducted beneath the ophiolite. The current Loyalty Islands have anomalous, high levels of diversity and the same arc (the Loyalty–Three Kings–Northland Plateau chain) might also be expected to have had high diversity further south.

The Northland ophiolite massifs occur at the North Cape–Unuwhao–Cape Reinga area and also further south, along with associated sedimentary components of the allochthon (Figure 5.4). The massifs are rootless and allochthonous. Ultramafic rocks, representing the base of the ophiolite suite, are present only at North Cape itself and 25 km south of there at Tangoake (Whattam, 2009). Many endemics occur at North Cape and elsewhere on the allochthon.

Along the East Coast, autochthonous Early Cretaceous–Oligocene graywackes and mudstones (including Torlesse terrane) are overthrust by the Early Cretaceous–Oligocene sedimentary and igneous rocks (Matakaoa Volcanics) of the East Coast allochthon (Brathwaite et al., 2008). The Matakaoa Complex and also the Tangihua Volcanics of the Northland allochthon have probably been generated in a single backarc basin (Cluzel et al., 2010b).

The Northland and East Coast allochthons are thought to have formed in the Cretaceous–Oligocene and to have been obducted onto the continental crust in the Oligocene. They have been described as "series of sedimentary and ophiolitic nappes, which contain the dismembered Cretaceo-Paleogene sedimentary continental margin as well as thrust sheets of backarc-basin mafic complexes" (Herzer et al., 2009a). The history of the ophiolites is critical in any tectonic model for the southwest Pacific but is the topic of current debate. The age of formation is controversial, and estimates as to how far the ophiolites have traveled vary from hundreds of kilometers to almost no distance at all (Herzer et al., 2009a). Rait (2000) estimated that the rocks in the East Coast allochthon have been moved ~300 km from their original site of deposition, and Cassidy (1993) presented paleomagnetic evidence suggesting that the Northland ophiolite had been translated southward hundreds of kilometers, from 15° to 20° lat. Yet later paleomagnetic studies found that the Northland rocks formed much closer to their present location (Whattam et al., 2005).

What *is* agreed is that the emplacement of the allochthons occurred in the Late Oligocene–Early Miocene and was from northeast to southwest. The emplacement of the allochthons, along with the associated Miocene volcanism on the Northland arc (not shown here) and the beginnings of the Kaikoura orogeny, signaled the resumption of convergent tectonics in on-land New Zealand. This was after a long period (~84 to 52 Ma) in which only extensional structures developed (Toy and Spörli, 2008).

Emplacement of the Ophiolite at the East-Dipping New Caledonia Subduction Zone

The origin of ophiolites is enigmatic. The sheeted dike complexes of typical ophiolites indicate that they originated by seafloor spreading, at oceanic ridges, yet ophiolites have geochemical properties that are only found in subduction environments, especially in forearcs.

The mode of emplacement of ophiolites on land is also controversial. The tectonic model for the southwest Pacific illustrated earlier (Figure 3.15) features an east-dipping subduction zone that developed in the Cenozoic, inboard of the original Pacific subduction zone. Aitchison et al. (1995) proposed this subduction zone for New Caledonia, and subsequent work has suggested that the same zone extended to New Zealand (Crawford et al., 2003; Schellart et al., 2006) and perhaps New Guinea (Whattam et al., 2008). The subduction zone has consumed a large ocean basin, with its seamounts and islands (but perhaps not all of its biota), before continent–island arc collision ended its activity. Northland allochthon emplacement and Three Kings Ridge–Northland Plateau volcanism were both related to the shutting down of this northeast-dipping subduction system at the end of the Oligocene.

In New Caledonia, the buoyant continental crust of Norfolk Ridge was carried by the down-going plate into the New Caledonia subduction zone and jammed it, causing subduction to cease. After the continental crust that had started to subduct stalled, it rebounded upward and the ophiolite in the forearc above was lifted up onto the continental crust. In one model the ophiolite had formed in the forearc at the time of subduction initiation; the New Guinea and New Zealand ophiolites formed and were emplaced in the same way. In this model, the New Caledonia–New Zealand ophiolite belt represents the forearc of the Loyalty–Three Kings Ridge arc.

Ages of the Ophiolite

The basalts and intercalated sediments that comprise the Northland and East Coast ophiolites have been dated by fossils as Late Cretaceous to Early Eocene, and by radiometry as mid-Cretaceous to Oligocene (~100–27 Ma). In the past, the ages have been interpreted as representing a single suite, with the older ages representing formation, and the younger radiometric ages reflecting alteration events (Nicholson et al., 2007). In contrast, Whattam et al. (2005, 2006, 2008) and Whattam (2009) found mainly Oligocene radiometric dates for the Northland ophiolite, and they questioned the usual idea that it formed in the Late Cretaceous–Eocene. They proposed instead that it formed in the Oligocene and was obducted soon after. They interpreted the earlier ages as representing a separate, underlying terrane, the Northland basaltic ocean floor (NoBOF) terrane, that was dated as Late Cretaceous–Paleocene (Figure 5.4). In Whattam and colleagues' model, this NoBOF terrane (Albian; 112–100 Ma) is equivalent to the Poya Terrane of New Caledonia (Campanian; 84–71 Ma) (dates from Cluzel et al., 2001). The Northland ophiolite in the strict sense is a later structure and an equivalent of the New Caledonia and New Guinea ophiolites.

In a new model, Whattam (2009) concluded that the ophiolites formed in the forearc in the first stages of a collision between an arc (New Guinea–Loyalty–Three Kings Ridge) and a continent (Australia–Zealandia) and were emplaced soon after. The collision has been diachronous, beginning in the north at New Guinea, proceeding through New Caledonia and ending in northern New Zealand (Schellart, 2007; Whattam et al., 2008; Herzer et al., 2009a).

Did the Northland–East Coast Ophiolite Form in a Backarc or a Forearc?

The Northland ophiolitic nappe as a whole includes basalts of Cretaceous and Late Oligocene ages, with small quantities of Paleocene alkaline basalts that represent intraplate, oceanic islands or seamounts (Herzer et al., 2009a). The basalts have a suprasubduction zone chemistry and a transitional affinity from mid-ocean ridge basalt to volcanic arc basalt. They were thought to have formed in a backarc basin (Whattam et al., 2004, 2005; Nicholson et al., 2007), but other studies have suggested a forearc origin (Whattam et al., 2006, 2008). (It is difficult to distinguish forearc basalts from backarc basalts on the basis of geochemistry alone; Whattam, 2009.) Whattam et al. (2008) and Cluzel et al. (2010b) concluded that the mafic allochthons of New Caledonia, Northland, and the East Coast mainly represent the obducted forearc of the Loyalty–Northland Plateau arc, although some older parts of the allochthons formed earlier as backarc lithosphere.

The occurrence of Late Cretaceous–Oligocene oceanic crust with suprasubduction zone affinities to the east of the Norfolk Ridge poses the problem of the missing Late Cretaceous–Oligocene arc (Cluzel et al., 2010b). The model shown earlier (Figure 3.15, from Schellart et al., 2006) implies that this has been buried under the developing Loyalty–Three Kings Ridge, although the oldest rocks so far drilled there (at the northern end of the ridge, on the d'Entrecasteaux Ridge) are Eocene.

Have Old Endemics on the Ophiolite "Floated" onto It from the Basement during Obduction?

One of the most obvious biogeographic patterns in Northland is the west–east differentiation across the province seen in many groups. For example, in the subshrub *Pimelea prostrata* (Thymelaeaceae), subsp. *seismica* occurs in *western* Northland (along the coast), in northwest Nelson and in Wellington, while subsp. *thermalis* is in *central* Northland and south to the central North Island (Burrows, 2009a). The same east–west division in Northland occurs in the two forms of *P. villosa*, with subsp. *arenaria* in the west and subsp. *villosa* in the east (Burrows, 2009b). The east–west break could be the result of differentiation during the southwestward rollback of the Loyalty–Three Kings subduction zone through the Northland region (see the reconstruction for 35–15 Ma in Figure 3.15). But at least some of the east–west differentiation may have an earlier origin.

One of the best-known centers of endemism in inland Northland, the Waima Range (Figure 5.4), is relevant to the east–west division of the province. The Waima Range is formed from Northland ophiolite, and its wet, montane forest includes three locally endemic trees: *Coprosma waima* (Rubiaceae), *Olearia crebra* (Asteraceae), and *Ackama nubicola* (Cunoniaceae). Such a high level of local endemism in mainland New Zealand trees is unique, and it has been described as an enigma (de Lange et al., 2002). It is true that the Waima Range (781 m) is the highest point in Northland, but it is only 11 m higher than Tutamoe, and its high rainfall is not unusual; the massifs from Tutamoe to Mataraua, also Puketi and Maungataniwha, have annual rainfall >2000 mm, which is wetter than the Waima Range (NIWA, 2016). Likewise the lithology of the Waima area is undistinguished from large parts of Northland.

Hebe perbella (Plantaginaceae) is a shrub known from rocky, open sites on the Waima Range and also on other ophiolite massifs further north along the west coast (Ahipara, Warawara), giving a linear distribution (de Lange, 1998; de Lange and Rolfe, 2008). This belt coincides with the underlying Murihiku basement terrane. The Murihiku terrane lies west of the Junction Magnetic Anomaly (representing the Dun Mountain terrane), which runs Auckland–Ahipara (Figure 5.4). If the Waima endemics or their direct ancestors originally occupied the Murihiku terrane, they must have "floated" onto other, higher strata as the Northland allochthon was thrust over the Murihiku. Geological events such as terrane accretion and phases of volcanism are not instantaneous. If they are "messy" enough, some populations will be able to survive the events more or less *in situ*, for example, by constantly colonizing new lava flows from older ones, or, as in Northland, by colonizing habitat islands on the new thrust sheets (nappes) from habitat islands on the nearby basement as the nappes are obducted.

Hebe perbella is unusual as some inflorescences include peloric flowers, or peloria (de Lange, 1998; de Lange and Rolfe, 2008). Peloria represent a distinctive mutation in which several flowers are fused into a single structure with a new symmetry. In New Zealand the mutation is recorded in just two cases: in *Hebe perbella* above the Murihiku terrane of Northland, and in *Mazus arenarius* (Mazaceae) on the Murihiku terrane of Southland (Heads, 1994a). This parallelism on the Murihiku terrane has perhaps evolved with the mid-Cretaceous folds that formed in the terrane during the collision of the Caples–Torlesse terranes with the already accreted Brook Street, Murihiku, and Maitai terranes, and has been inherited by modern forms.

Northland Miocene Volcanism

Following emplacement of the Northland ophiolite, volcanic massifs were erupted through Northland in the Early Miocene (not mapped here). These are often spectacular, as at Bream Head, or extensive, as in the upper Waipoua River and Tutamoe Plateau. It has been suggested that the massifs formed in a "volcanic belt" or "Northland volcanic arc," although this is controversial.

Many of the Northland endemics on the Miocene volcanics would have been "inherited" by the youngest volcanic strata, after having persisted as metapopulations more or less *in situ* through the individual phases of the volcanism. The Miocene volcanism in Northland has been interpreted as forming two parallel belts, one along the northeastern margin of Northland and the other along the southwestern margin (Herzer et al., 2009a), but in Schellart's (2007) model the true arc of the now extinct, east-dipping subduction zone is the Three Kings Ridge–Northland Plateau seamount chain. In this model, the Northland volcanic belt is not a true arc, but the result of a subducting slab being detached from the rest of its plate at the east-dipping subduction zone (Schellart et al., 2006; Schellart, 2007). (If a subducting slab breaks, tears, or sinks vertically, the disturbance can lead to volcanism in the crust above.)

Alternative Geological Models for Northland

Whattam et al. (2004, 2006, 2008) and Whattam (2009) revised models for the history of the Northland ophiolite (often referred to as the Tangihua complex or Tangihua ophiolite of the Northland allochthon). They suggested that the Northland ophiolite formed in the Oligocene–Miocene (32–15 Ma), near the site of its late Oligocene–Early Miocene obduction. This is in contrast to previous studies that suggested that the ophiolite is older (Cretaceous–Paleocene) and far traveled. Whattam et al. (2006) accepted that the later stages of ophiolite formation and earliest stages of the Northland arc (25–11 Ma) overlap in time as well as space.

In their model, Whattam et al. (2008) proposed that the ophiolite nappes in New Guinea, New Caledonia, and New Zealand each comprise an older part, made of basaltic ocean floor (BOF) (with mid-ocean ridge basalt affinity), and a younger, suprasubduction zone (SSZ) ophiolite (Figure 5.4). BOF and SSZ terranes are represented in New Guinea by the Emo metamorphic terrane and Papuan Ultramafic Belt, in New Caledonia by the Poya terrane and the New Caledonia ophiolite, and in Northland by the Northland BOF terrane and the Northland ophiolite. Whattam et al. (2008) proposed that a single, more or less continuous marginal basin extended from the Papuan Peninsula via New Caledonia to Northland. This massive backarc basin formed a "giant" South Loyalty Basin that was later subducted. It is this backarc basin in which the underlying basaltic ocean floor terranes (the Emo, Poya, and Northland BOF terranes) are thought to have formed, before the New Caledonia subduction zone developed in the basin and generated its SSZ ophiolites. The ophiolite emplacement began in the Emo backarc basin at 65 Ma, in the South Loyalty Basin at 55 Ma, and in the South Fiji Basin/Northland basin at 37 Ma.

Whattam et al. (2008) accepted that the new, east-dipping subduction zone initiated in what had been a backarc, the South Loyalty Basin, between older (Cretaceous–Paleocene) lithosphere in the west (Poya terrane/Northland BOF terrane) and younger rocks in the east. The ophiolites formed in the eastern part, in the new forearc, and were obducted soon (<20 Myr) after, above the same east-dipping subduction zone where they formed. The processes of ophiolite formation and emplacement were essentially a single event. The ophiolite obduction in northern New Zealand occurred around the same time as the "soft" collision between the Hikurangi Plateau and the northern New Zealand margin (~25 Ma; Figure 3.15) (Whattam et al., 2005).

Whattam et al. (2008) used a "subduction infancy" model (Shervais, 2001; Stern, 2004) to explain the origin of ophiolites at the New Caledonia subduction zone. During subduction zone initiation, old, dense, oceanic lithosphere begins to sink into the asthenosphere and the disturbance leads to ophiolite formation above the subduction. Ophiolites can form in the backarc or the forearc, but they probably develop most often in the latter.

The Northland Basaltic Ocean Floor Terrane: A Poya Terrane Analogue?

Whattam et al. (2008) accepted a separate "Northland basaltic ocean floor (NoBOF) Terrane" for the older rocks in the Northland allochthon (Cretaceous–Paleocene; 80–55 Ma). The evidence from these earlier rocks—whether they are the unaltered ophiolite (Nicholson et al., 2007, 2011) or a separate terrane (Whattam et al., 2008)—is of special interest for both biogeography and tectonics.

Whattam et al. (2004) wrote that accommodating Cretaceous–Eocene oceanic crust adjacent to the Northland Peninsula in the late Oligocene has been a major obstacle to developing a model for the emplacement of the Northland Ophiolite. In the models of Schellart et al. (2006) and Whattam et al. (2008), the Northland basaltic ocean floor terrane and its equivalents in New Caledonia (Poya terrane) and New Guinea (Emo terrane) were generated in the 80–55 Ma backarc basin of the *Pacific* subduction zone (Figure 3.15). The terranes formed by fragmentation of a Cretaceous arc constructed on thinned continental crust at the leading edge of the Australian Plate. The BOF terranes were eventually obducted together with the ophiolites that formed in the forearc of the *New Caledonia* subduction zone.

Subduction between New Caledonia and New Guinea

The area between New Caledonia and the accreted terrane belt of New Guinea is one of the most critical areas for southwest Pacific biogeography and tectonics, but it is complex and not well understood. The models for this area presented by Crawford et al. (2003), Schellart et al. (2006), and Whattam et al. (2008) have broad similarities but the reconstructions differ. The region holds special importance for New Zealand biogeography as there are many direct New Zealand–New Guinea connections that do not include New Caledonia (e.g., in honeyeaters and acanthizid warblers), as well as New Zealand–New Caledonia–New Guinea affinities.

Schellart et al. (2006) accepted that northeast-dipping subduction began in the South Loyalty Basin at 55–50 Ma and that the closure of the basin resulted in the diachronous emplacement of Cretaceous–Paleocene ophiolitic nappes onto New Caledonia (40–34 Ma) and then onto Northland (24–21 Ma) (Figures 3.7 and 5.2). Whattam et al. (2008) suggested that this does not account for the northeast-dipping subduction that established off Papua New Guinea by at least 65–60 Ma. This resulted in the emplacement of the extensive Papuan Ultramafic Belt (59–58 Ma). In the model of Whattam et al. (2008), the Papuan, New Caledonian, and Northland ophiolites all formed in a collision zone soon after northeastward or eastward subduction was initiated and were emplaced soon after their formation. Activity at this zone migrated southward along the entire eastern margin of the Australian plate. The ophiolites represent the forearc of an extensive intraoceanic arc system that was continuous from mainland Papua New Guinea to New Caledonia and Northland.

In the model of Schellart et al. (2006), the New Caledonia subduction zone was never connected with the New Guinea subduction zone, but in the model of Whattam et al. (2008) they are part of the same structure. This last model is consistent with biogeography, as there are many direct connections between

New Caledonia and the Papuan Peninsula. Nevertheless, these could also be explained by backarc basin formation behind the primary Pacific subduction zone, as this also extended from New Guinea to New Zealand in the Schellart et al. (2006) model.

Responses to the Whattam et al. Model

In contrast with Whattam et al. (2006, 2008), Nicholson et al. (2007) concluded that radiometric and paleontological ages for the Tangihua Complex (Northland ophiolite) date three separate events in one terrane (rather than two terranes): formation in the late Cretaceous (c. 100 Ma), thermal alteration at ~50 Ma, and thermal alteration at ~30 Ma. The third event agrees with geological evidence for the age of emplacement of the allochthon (Late Oligocene to earliest Miocene), and it suggests that the emplacement caused the thermal alteration.

Cluzel et al. (2010b) found the Whattam et al. (2008) model problematic as no large Oligocene basin is known to have existed north of New Zealand, and because the obducted slab did not include young, hot lithosphere—there is no evidence of a metamorphosed sole in the Tangihua Complex (Northland ophiolite) or in the Matakaoa Volcanics of the East Coast. In addition, no fossils younger than Early Eocene have been found within the mafic terranes.

The Taupo Volcanic Zone

The rift of the Taupo volcanic zone (Figure 3.15: Present) includes high volcanoes, a large caldera (Lake Taupo), and widespread geothermal activity. The volcanism began at ~1.5–1.9 Ma (Briggs et al., 2005). The zone represents the opening of a backarc basin between the Lau–Colville Ridge and the Tonga–Kermadec ridge at ~8 mm/year, and its propagation from the Bay of Plenty through the central North Island. The rift cuts across the northwest–southeast trends seen in the region in the southern end of the New Caledonia subduction zone, the Vening Meinesz fracture zone, and the Northland–East Coast allochthons.

The New Zealand pohutukawa, *Metrosideros excelsa* (Myrtaceae), is a well-known coastal tree that grows near the beach (cf. Figure 4.6). But as Dieffenbach (1843) observed, the species also occurs around the inland Lake Taupo and Lake Rotorua. These anomalous inland sites are much colder than the species' other habitats, on the coasts and offshore islands of the North Island (north of New Plymouth on the west coast and Gisborne on the east coast). Dieffenbach (1843) attributed the tree's presence at the inland sites not to chance dispersal or current ecology, but to historical, geological factors. He saw the inland populations as evidence that the sea had once extended into the central part of the North Island. In fact the line Rotorua–Taupo more or less defines the major volcanic rift in the central North Island, an incipient backarc basin. After surviving, and probably benefiting from, the marine incursions of the Oligocene, the pohutukawa trees have also survived the Neogene volcanism, with populations propagating quickly enough for the metapopulation to survive on the volcanic deposits as these have accumulated.

The ecology of the beach tree pohutukawa raises the general question: "What is significance of the shore, its biota, and its ecology over an evolutionary time scale?" The shore is the Earth's great ecological margin, and it is characterized by periodic inundation and the high levels of disturbance that this causes. The biota of the beach, mangrove, and saltmarsh is more or less weedy and includes many widespread clades, but there are also regional endemics, such as *Metrosideros excelsa* in New Zealand. Mangrove forest is sometimes regarded as depauperate, but the landward margin is often diverse (Heads, 2003, 2006b). Examples of mangrove endemism include the lichen *Lecidea aucklandica* (Galloway, 2007) and the mite *Acalitus avicenniae* (Manson, 1984), both restricted to Northland and Auckland. As Dieffenbach (1843) recognized, with changes in sea level, both the shore and its biota will undergo changes in distribution.

The rift of the Taupo volcanic zone includes many endemics. For example, at the northern end of the zone, *Kunzea salterae* (Myrtaceae) is endemic to Whale Island, where it sometimes grows in active geothermal sites (de Lange, 2014). *K. tenuicaulis* is confined to active geothermal habitats from Lake Rotoiti south to the southern edge of Lake Taupo, and de Lange (2014: 74) discussed the differentiation of this species. He wrote: "In view of the geologically recent (estimated to be a maximum of 2 million years old…) habitats this species occupies, this molecular divergence from all other members of the *K. ericoides* complex is considered remarkable." Nevertheless, the volcanic arc that this "volcano weed" occupies (not the modern volcanogenic habitats themselves) dates back to long before 2 Ma, and, as a general principle, groups can be much older than the particular strata where they are now endemic. Although the current backarc rift dates back only to the beginning of the Pleistocene, volcanic and geothermal habitat has been a feature of the subduction zone in the region since the Cretaceous (Figure 3.15). This means that the volcano weeds found along the rift were probably associated with prior volcanism in the region before the development of the rift and are older than the habitat they currently occupy.

Groups associated with the volcanism include large trees, such as rimu, *Dacrydium cupressinum* (Podocarpaceae). This is abundant through New Zealand, especially on the West Coast and on the volcanic plateau of the central North Island. The seedlings cannot tolerate the shade of a closed canopy, but where forest has been destroyed, such as by fire or volcanic eruptions, the species can regenerate strongly (Dawson and Lucas, 2011: 62).

Another large tree that thrives around areas of volcanism and other disturbance is *Libocedrus bidwillii* (Cupressaceae). On Mount Hauhungatahi, by Ruapehu and ~70 km from the center of the large, 1718 Taupo eruption, *Libocedrus bidwillii* (Cupressaceae) is the dominant tree at elevations over 1000 m. Pollen studies showed that "Following the eruption *Libocedrus bidwillii* expanded in all sites" (Horrocks and Ogden, 1998). The eruption appears to have facilitated an expansion by providing abundant, open sites for establishment.

The shrubs in the plant family Ericaceae tolerate acid soils and are often associated with volcanic landscapes (Heads, 2003). One New Zealand example, *Gaultheria oppositifolia*, occurs in the North Island in rocky and open places, from the standard boundary: Mount Taranaki–National Park (Waimarino)–East Cape, north to Tarawera, Matamata and Mayor Island, with a stronghold in the Rotorua–Taupo region (Franklin, 1962). This distribution is aligned with the central volcanic plateau, part of a backarc rift that is dominated by volcanic activity. This is consistent with the species and its ancestors' having occupied young volcanics in the region for many millions of years.

Animal groups endemic to the volcanic rift include three staphylinid beetles (Park and Carlton, 2014):

Sagola monticola: Mount Ngauruhoe, Desert Road below Ngauruhoe (Mangatawai Stream), and Rotorua.

Sagola gilae: 10 miles north of Waiouru, 3 km south of Turangi.

Sagola boudreauxae: Rurima Island off Whakatane, Bay of Plenty.

Endemism on the Volcanoes

The active volcanoes of the Taupo volcanic zone include the highest mountain in the North Island, Ruapehu (2797 m), which began erupting at ~250,000 BP. The volcanoes are much higher than the neighboring Kaimanawa Mountains (1726 m), where Haast Schist basement is exposed. As the backarc Taupo rift opened and the volcanic edifices built up, they would have inherited montane and alpine biota already in the region on the older ranges nearby. For example, the herb *Ourisia vulcanica* (Plantaginaceae) is restricted to the Kaimanawa Mountains and the volcanoes, where it grows at and above the alpine tree line in volcanic ash, scoria, and pumice. Other alpine species, such as the subshrubs *Parahebe spathulata* and *P. hookeriana* (Plantaginaceae) occur on the volcanoes, the Kaimanawa Mountains, and adjacent ranges, north to Maungapohatu (*P. spathulata*) and Hikurangi (*P. hookeriana*) (Garnock-Jones and Lloyd, 1994). The subshrub *Pimelea microphylla* (Thymelaeaceae) occurs on high ridges of the Kaimanawa and Kaweka Ranges, up to 1650 m on the volcanoes (Tongariro, Ngauruhoe, and Ruapehu), and also in bare areas of the adjacent Rangipo (Onetapu) Desert (Burrows, 2009b). It grows in bare or sparsely vegetated scoria, on lapilli and coarser volcanic debris, in crevices on solid volcanic rock, and, in the neighboring ranges, on old tephra.

Endemism on the volcanoes is sometimes cited, but its significance has been neglected. McDowall (1996) wrote: "… there is some, though very slight, evidence for invertebrate endemicity associated with the volcanic region but little to suggest a fauna of any antiquity that might have survived the AD 186 Taupo eruption." This downplays the endemism in the region though.

The lichen *Stereocaulon wadei*: Ruapehu (near Whakapapa; Galloway, 2007).

The land snail *Allodiscus tongariro*: Whakapapaiti, on the southwestern slopes of Ruapehu (Marshall and Barker, 2008).

The stiletto fly *Anabarhynchus monticola* (Therevidae): Tongariro and Ruapehu (Lyneborg, 1992).

The caddis fly (Trichoptera) *Paroxyethira zoae*: Ruapehu and 10 km to its northwest on Hauhungatahi. The latter is an eroded volcano dated at ~300,000 BP (Ward and Henderson, 2004).

The other conspicuous volcano in the North Island, Mount Taranaki/Egmont, formed around 70,000 BP. It also hosts endemics, such as the lichen *Psoroma cyanosorediatum* (Galloway, 2007) and the herb *Wahlenbergia pygmaea* subsp. *drucei* (Campanulaceae) (Petterson, 1997). The shrub *Melicytus drucei* (Violaceae) is endemic to two nearby volcanoes that are also part of the Taranaki Volcanic Succession: Mount Pouakai (250,000 BP, on the northern flanks of Mount Taranaki) and the Kaitake Range (575,000 BP) (Molloy and Clarkson, 1996). The endemic invertebrates on Mount Taranaki include two spiders, a tachinid fly, a scarabaeid beetle, three moths (a geometrid, a noctuid, and a tortricid), and probably an undescribed stonefly (Fox, 1982). Fox discussed the strong biogeographic affinities between the northwest Nelson, Taranaki, and Ruapehu areas, and he cited two noctuids, a geometrid and a carabid that are each restricted to Egmont and Ruapehu. He also cited species that occur only on Egmont and in the South Island. As he concluded: "the Egmont National Park holds a very important place in the ecology and biogeography of New Zealand, but it is not a simple story." Other young volcanoes, such as Kilimanjaro in East Africa, have endemics that have been dated as older than the rocks (Heads, 2012a). This indicates that clades endemic to volcanoes predate the mountains and have remained more or less *in situ*, as dynamic metapopulations, while the volcanic edifices were built up beneath them by repeated eruptions. Regional endemics present on Mount Taranaki that are likely to have been inherited by the mountain include *Sagola egmontensis* (Coleoptera: Staphylinidae), known from Mount Egmont and, 70 km to the northeast, on the Miocene mudstones of Mount Messenger (Park and Carlton, 2014).

Kermadec Islands

The Kermadec Islands (Figure 3.1) are young, volcanic islands that formed in the Pleistocene, but they have many endemics. These include 68 of the 358 marine mollusks present (Brook, 1998), 5 of the 13 echinoderms (Morton, 2004), the plant species *Coprosma petiolata* (Rubiaceae), *Metrosideros kermadecensis* (Myrtaceae) and *Hebe breviracemosa* (Plantaginaceae), the coastal fish *Enneapterygius kermadecensis* (Tripterygiidae; Francis, 2012), the seabird *Pelagodroma albiclunis* (Hydrobatidae), and the parakeet *Cyanoramphus novaezelandiae cyanurus*.

In the usual model, the direct ancestors of the endemics have migrated to the islands from the New Zealand mainland or from other, extant Pacific islands. Bayly and Kellow (2006: 25) wrote that "Hebes show clear evidence of a capacity for long-distance dispersal—for example, through their occurrence on the Pleistocene age Kermadec Islands."

An alternative possibility is that the Kermadecs groups have inhabited the active Tonga–Kermadec arc ever since its formation near Australia, surviving as a metapopulation on the individually ephemeral islands that have been produced there. This is consistent with the historical geology, and with Hooker's (1857) observations on the flora. He noted that the affinities are "most strong" with New Zealand and that "It is extremely difficult to account for this similarity of vegetation by transport; added to which, the prevailing winds blow from the north-west... If their presence [New Zealand ferns in the Kermadecs] is to be accounted for wholly by trans-oceanic transport... why has there been no addition of some of the many Fiji or New Caledonian island ferns...?" (p. 126).

In the reconstruction illustrated earlier (Figure 3.15), the Tonga–Kermadec arc is part of the original Pacific arc that has migrated away from the Gondwana (Australia) margin into the Pacific. The South Fiji Basin was formed in the later Cenozoic by the rifting apart of the Loyalty–Three Kings Ridge–Northland Plateau arc and the Tonga–Kermadec arc, as the latter rotated into the Pacific.

This migration and rifting of arcs provide a simple explanation for the disjunct distribution of groups that surround the basin. Examples already cited in this chapter include the fish *Chironemus microlepis* at the Poor Knights, Lord Howe, Norfolk, and Kermadec Islands, and affinities in *Carex* and in *Myrsine* that link the Three Kings and Kermadec Islands. In other examples, the palm *Rhopalostylis baueri* is endemic to Norfolk and Kermadec Islands (de Lange et al., 2005). The halophytic herb *Samolus repens* var. *strictus* (Primulaceae) is on Norfolk Island, Three Kings Islands, North Cape, Karikari Peninsula, Poor Knights, and Mokohinau Islands, and also the Kermadec Islands (de Lange and Cameron, 1999).

Scaevola gracilis (Goodeniaceae) is endemic to the Tonga–Kermadec arc, where it inhabits the Kermadec Islands of Raoul, North Meyer, and Macauley Islands, and the Tongan islands of Kao and Tofua (Sykes, 1998). Although activity along the Tonga–Kermadec subduction zone dates back to the Cretaceous (Figure 3.15), the current Kermadec Islands are all young volcanoes (formed since 1.6 Ma), and Raoul is active, as is Tofua in the Tongan group. *Scaevola gracilis* is most abundant on Raoul Island, where the plants form dense stands on open pumice slopes in the crater and on the rim (Sykes, 1998). The species is also common along the crater rim on Tofua. Given its biogeography and ecology, *S. gracilis* can be interpreted as a typical subduction zone weed. It belongs to an unresolved clade otherwise known from southwestern and southeastern Australia (phylogeny from Howarth et al., 2003; distributions from GBIF, 2016). The plant's biogeography and ecology suggest that its origin dates back to the initial roll-back of the Pacific subduction zone and its arc away from the Gondwana margin in the Cretaceous (Figure 3.15). (For the broader biogeography of *Scaevola*, see Heads, 2010.)

A Trans-Tropical Pacific Group in Northern New Zealand: *Sicyos* (Cucurbitaceae)

The tribe Sicyoeae (Cucurbitaceae) has a standard trans-tropical Pacific distribution (Heads, 2012a: 302). The tribe has most of its diversity in warmer parts of the Americas, but it also has a small, Old World clade in Australasia. This last group has three species: *Sicyos undara* in northeastern Queensland, *S. australis* in eastern Australia and New Zealand, and *S. mawhai* on islands off northeastern New Zealand (Figure 5.7; Telford et al., 2012). The Australasian clade has its sister in northern Argentina/

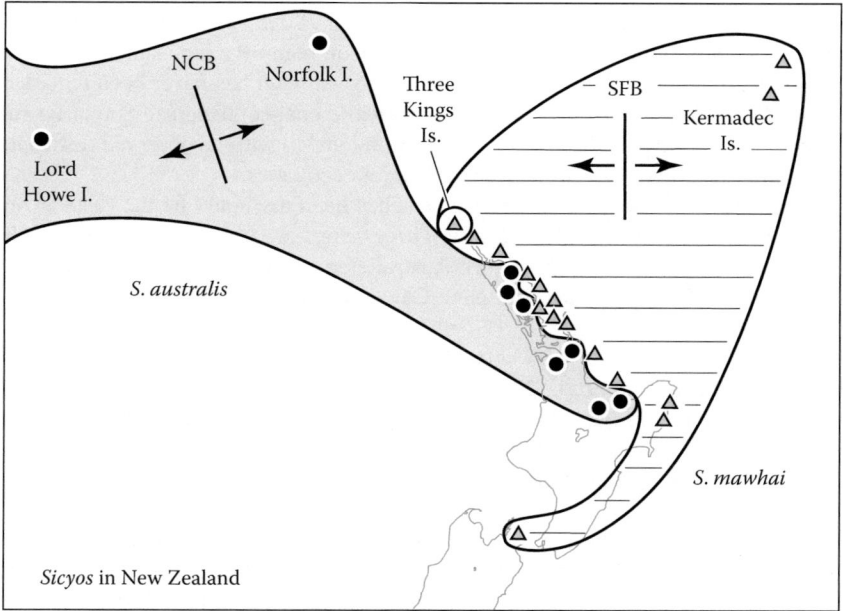

FIGURE 5.7 *Sicyos* (Cucurbitaceae) in New Zealand. NCB, New Caledonia basin; SFB, South Fiji Basin. (Data from Telford, I.R.H. et al., *Aust. Syst. Bot.*, 25, 188, 2012.)

Bolivia (*S. malvifolius*) and New Mexico/Mexico (*S. ampelophyllus* and *S. laciniatus*); this gives a distribution in three disjunct regions (phylogeny from Sebastian et al., 2012; distributions from GBIF, 2016).

Members of the Australasian group are herbaceous climbers that inhabit disturbed areas, and their weedy, pioneer ecology is shared with many other Cucurbitaceae. The Australasian plants have fruit with barbed spines, and Sebastian et al. (2012) suggested that fruit have dispersed across the Pacific by adhering to migrating seabirds. The authors cited the storm-petrel *Pelagodroma marina maoriana* (Hydrobatidae) as a possible vector, as it migrates between Australasia (where it breeds) and the eastern tropical Pacific.

The Australasian clade of *Sicyos* is monophyletic, and there is no fossil evidence of any other *Sicyos* clades in the region. In dispersal theory, this means that dispersal from America to Australasia has succeeded just once in the history of the tribe, which is thought to have evolved at 32 Ma (Sebastian et al., 2012). The mechanism for long-distance dispersal that the authors suggested appears to be feasible, but it does not explain why there is evidence for only one dispersal to the entire Old World in 32 Myr. It also does not explain why so many similar, trans-tropical Pacific distributions occur in groups that are not associated with seabirds (Heads, 2014). These facts suggest that the distributions of the seabird and the *Sicyos* clade (along with many others) have a common cause, and that the latter is not the result of the former.

Trans-tropical Pacific biogeographic connections between Australasia and tropical America are documented in many groups (Heads, 2012a, 2014). One possible explanation is chance dispersal. Another possibility is that metapopulations of weedy groups have survived in the central Pacific on islands located around different kinds of volcanic centers. These have been active since the Mesozoic, producing many archipelagos and also the large Cretaceous plateaus that are now embedded in Australasia and tropical America. The trans-tropical Pacific distribution of the tribe Sicyoeae, together with its pioneer, maritime ecology, suggests that these plants would have survived on and around the Pacific plateaus and arcs, both before and after their accretion.

In molecular phylogenies the trans-Tasman *Sicyos australis* is sister to *S. mawhai*, found on many islands off Northland and a few mainland beaches, plus the Kermadec Islands (Figure 5.7; Sebastian et al., 2012). Both species have a weedy ecology and tolerate high levels of disturbance. Despite this, they maintain a precise boundary within New Zealand and also outside the country.

Sicyos mawhai is recorded in the nesting grounds of seabirds in the genera *Pterodroma* and *Puffinus* s.lat. One of these, *Puffinus (Ardenna) bulleri*, breeds only on the Poor Knights Islands, where *S. mawhai* occurs. There are estimated to be 2.5 million individuals of *P. bulleri* (Marchant and Higgins, 1990), and the birds migrate around the Pacific to Australia, Japan, many Pacific islands, and the west coasts of North and South America (GBIF, 2016). Despite this, *S. mawhai* has never been recorded outside its coherent, well-defined New Zealand range. As usual, while chance dispersal cannot be ruled out, the group's inferred means of dispersal show no obvious relationship with its observed distribution, and the distribution is shared with many groups that have different ecologies.

In a vicariance model, the range of *Sicyos mawhai* has been disrupted by the eastward migration of the Tonga–Kermadec arc away from the Loyalty–Three Kings arc to its current position (Figure 3.15). Earlier, the range of *Sicyos australis* in eastern Australia and New Zealand has been disrupted by the opening of the Tasman Basin and also the New Caledonia basin (between Norfolk Ridge and Lord Howe Rise).

The basal node in *S. mawhai* is a break between the Three Kings Islands and the rest of the localities (morphological study by Telford et al., 2012). This break and the boundary between *S. australis* and *S. mawhai* coincide with the Vening Meinesz transform margin (Figure 5.5). In a broader, regional context, the orientation of the boundary can be compared with Northland and East Coast allochthons and the buried, extinct subduction zone proposed to explain them (Figure 5.6).

The division between *S. australis* and *S. mawhai* occurred before the Late Cretaceous opening of the Tasman and New Caledonia basins rifted *S. australis* apart. This coincides with the volcanism in the Mount Camel terrane (aerial to subaerial eruptions at 108–119 Ma; Nicholson et al., 2008) and the Nouméa Basin, which in turn coincides with the putative basal position of the Three Kings population in *S. mawhai*. It is also just a few million years after the eruption of the mid-Pacific plateaus. The original break took place between *S. australis* on the Norfolk Ridge (Gondwana) and *S. mawhai* on the Pacific arc (and the Loyalty–Three Kings arc which separated from the Pacific arc). The original break thus corresponds with the rifting between the Pacific arc and the mainland, which opened the South Loyalty Basin (Figure 3.15). This occurred around the same time as the opening of the Tasman Basin (84 Ma). Later the South Loyalty basin closed, bringing the two species together again.

S. mawhai has an anomalous southern locality at Cook Strait and this is consistent with the Cenozoic deformation in central New Zealand that pulled northeastern areas into the Cook Strait region (see Chapter 9). The old coastline that ran Northland–East Cape–Cook Strait region–Chatham Islands originally followed a straight line, but through the Neogene it has been bent into its present Z-shape.

6

Biogeography of New Zealand's Subantarctic Islands and the Chatham Islands

The New Zealand Plateau, or Zealandia, makes up the Earth's largest submerged continent. The northern part includes Lord Howe Rise, Norfolk Ridge, and New Caledonia. On the drowned, southern part—the Campbell Plateau—the small, scattered subantarctic islands are the only emergent land remaining (Figure 6.1). This chapter discusses the southern region, along with the Chatham Rise and its only emergent part, the Chatham Islands.

Campbell Plateau and New Zealand's Subantarctic Islands

Endemism and Absence in the Subantarctic Region

Of the Tasman region endemics that have global sisters (Chapter 4), at least three occur on the Campbell Plateau islands. Athoracophorid land slugs are represented by endemic species on Macquarie, Auckland, Campbell, The Snares, and Chatham Islands (Burton, 1963, 1980). *Chionochloa* (Poaceae) is represented by *C. antarctica*, endemic on Auckland and Campbell Islands. *Ranunculus* sect. *Pseudadonis* (Ranunculaceae) includes *R. pinguis*, also endemic on Auckland and Campbell Islands.

Many Campbell Plateau groups link the islands in specific patterns of endemism, with Auckland Islands–Campbell Island being one of the most common. Another endemic there is the distinctive shrub *Hebe benthamii* (Plantaginaceae). It is not sister to, or nested in, a southern mainland clade, as dispersal theory would predict. Instead, it is sister to a diverse complex (all of *Hebe* except *H. macrantha*) that is distributed throughout the New Zealand region, on Rapa Island (southeastern Polynesia), and in Patagonia (Wagstaff et al., 2002). In a similar way, *Plantago aucklandica* of the Auckland Islands is not nested in a local, southern New Zealand clade but is sister to *P. obconica* (South Island) and *P. lanigera* (North, South, and Stewart Islands) (Tay et al., 2010).

The biota of the Campbell Plateau is well known for its high level of endemism. In the Auckland Islands beetle fauna, the presence of four endemic genera and the diversity evident in several other genera has been described as "odd, and... not consistent with the present land area of the Auckland Islands" (Michaux and Leschen, 2005: 101). In Diptera, seven genera are endemic on the Auckland Islands, one on Campbell Island, three on both Auckland and Campbell Islands, and one on the Bounty Islands.

In beetles, the Carabidae alone have four genera endemic to the subantarctic islands region. Three are monotypes: *Calathosoma* (Auckland Islands), *Synteratus* (The Snares), and *Bountya* (Bounty Islands). The last two do not have close relatives on the New Zealand mainland. *Synteratus* may be more closely related to the Tasmanian *Sloanella* than to the New Zealand *Oopterus* or *Zolus* (Johns, 1974).

The fourth carabid, *Loxomerus*, is endemic to the Auckland and Antipodes Islands with four species (Larochelle and Larivière, 2013), and it is keyed with *Taenarthrus* of South and Stewart Islands (Johns,

167

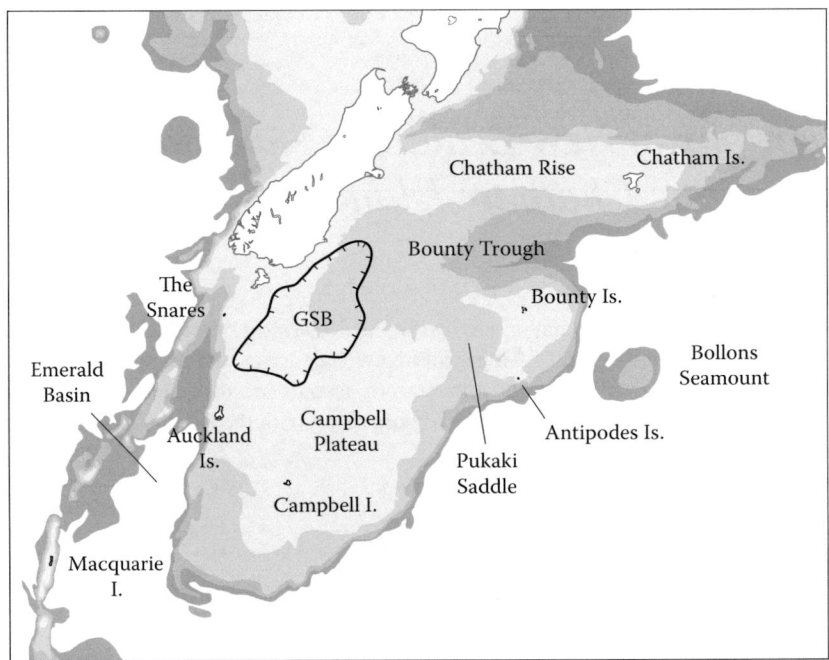

FIGURE 6.1 The southern part of the New Zealand Plateau, including the Campbell Plateau (with Auckland and Campbell Islands) and the Chatham Rise (with the Chatham Islands). The 1000, 2000, 3000, and 4000 m isobaths are indicated; the area of continental crust is delimited approximately by the 2500 m contour. GSB, Great South Basin.

2010). The subantarctic *Loxomerus* and *Calathosoma*, plus the mainland *Taenarthrus*, form the New Zealand contingent of tribe Migadopini, well known to biogeographers for its trans-Pacific distribution in Australia, New Zealand, and southern South America.

Hymenoptera, Lepidoptera, and spiders also have endemic genera on the subantarctic islands and, in addition, all show a distinctive absence there of groups common in New Zealand (Michaux and Leschen, 2005). In Plecoptera (stoneflies), *Aucklandobius* (four species) is endemic to Auckland Islands and *Rungaperla* (two species) to Campbell Island (McLellan, 2006). In other invertebrates, *Argaplana* is an endemic genus of terrestrial flatworms (Geoplanidae) on Campbell Island that constitutes its own tribe, Argaplanini. Another geoplanid genus, *Marionfyfea*, is endemic to Auckland and Campbell Islands (Winsor, 2011).

Apart from the high level of endemism, the biota of the Campbell Plateau is notable for the many groups that are not only endemic there but are also widespread across the region. For example, the small herb *Stellaria decipiens* (Caryophyllaceae) is endemic to Macquarie, Auckland, Campbell, The Snares, and Antipodes Islands. The dipteran *Apetaenus* subg. *Macrocanace* (Canacidae) is found on The Snares, Auckland, Campbell, Bounty, and Antipodes Islands (Harrison, 1976; McAlpine, 2007). Despite its wide distribution, maritime ecology, and evident means of dispersal, this group of beach flies is absent from Stewart and South Islands. Likewise, the marine isopod genus *Sporonana* consists of one endemic species on Macquarie Island (at 5–20 m depth), one on Campbell Island (4 m), and one on The Snares (intertidal) (Just and Wilson, 2004).

Apart from its fascinating endemics, the Campbell Plateau is also a center of unexpected absences. *Celmisia* (Asteraceae) is one of the most widespread and diverse plant genera in the New Zealand alpine zone, but "surprisingly" (Given, 1975) it is not on the subantarctic islands, where the cool climate should suit it. This ecological anomaly is probably the result of phylogenetic history, rather than ecological factors. While *Celmisia* and its sister group (*Olearia* p.p. and *Pachystegia*) are both absent from the subantarctic islands, their mutual sister group (*Pleurophyllum* s.lat.) has most of its diversity there. The whole complex was cited earlier because of its global sister group (Chapter 4). It has the following phylogeny (Wagstaff et al., 2011):

Pleurophyllum s.lat. (including *Damnamenia* and "macrocephalous *Olearia*"): There are 11 species: **most diverse on the subantarctic islands** and the Chatham Islands. There are also three species in Fiordland, one of which (*O. colensoi*) extends northward through the alpine zone to the central North Island.

Celmisia: New Zealand mainland (61 species, mostly alpine) and SE Australia/Tasmania (9 species, all alpine); **not on the subantarctic islands**.

Olearia p.p. and *Pachystegia*: New Zealand mainland (all the *Olearia* species there not included in the macrocephalous clade), Chatham Islands, New Guinea (Cross et al., 2002), and SE Australia (altogether 38 species); **not on the subantarctic islands**.

The absence of the diverse and widespread *Celmisia* + *Olearia* p.p. + *Pachystegia* clade in the subantarctic region is not absolute—the group *as such* is absent, but it is represented there by a close ally, *Pleurophyllum* s.lat.

Geology of the Campbell Plateau: A Submerged Block of Continental Crust

Most of the islands on the Campbell Plateau are mainly composed of Cenozoic volcanics. Volcanism on the Auckland Islands, for example, has been dated at 12–37 Ma (Hoernle et al., 2006). The islands themselves and their biota are sometimes regarded as having the same age (e.g., McDowall, 2010a: 197). Yet continental basement terrane is also exposed on many of the islands, and Campbell Island has the oldest known rock in New Zealand (Precambrian schist; Campbell and Hutching, 2007). Mesozoic granite of the Median Batholith is exposed on The Snares, Auckland, and Bounty Islands, and Mesozoic schist basement is exposed on the Chatham Islands. The buoyancy of the continental crust that the plateau is formed from means it is likely that other, earlier islands also existed.

Granite has a density of 2.7 g/cm^3, but the same amount of iron (density 7.9 g/cm^3) weighs nearly three times as much. Mafic rocks such as basalt contain magnesium and iron (hence the term "mafic") and are more dense (>3 g/cm^3) than felsic rocks such as granite. Felsic rocks form "continental crust"; this is buoyant in the lithosphere and is usually emergent above the ocean, while mafic rocks form "oceanic crust" that is denser and usually lies below sea level. If continental crust is extended and thinned, it becomes less buoyant, in the same way that a small iceberg does not project as far above sea level as a large one. The continental crust of the Campbell Plateau is much thinner than normal continental crust, and so the extension that has caused this has caused the submergence of the plateau.

Terrestrial groups would have occurred across the whole plateau before its submergence and on any subsequent volcanic islands. There is evidence for past volcanic islands as well as the current ones. For example, on the Pukaki Rise in the central Campbell Plateau (between the Auckland Islands and Pukaki Saddle; Figure 6.1), there are several volcanic structures with bases ~1 km in diameter and flat tops just ~60 m below sea level (Timm et al., 2010).

The nature and timing of the extension of the plateau are debated; the thinning could be the result of Early Cretaceous extension or the Late Cretaceous breakup of New Zealand from Antarctica (Grobys et al., 2009). Whether general thinning of the Campbell Plateau crust took place at the same time as the mid-Cretaceous extension of the Bounty Trough and Great South Basin (Figure 6.2) is also debated. Grobys et al. (2009) suggested that an early phase of extension across the Campbell Plateau (Early Cretaceous at ~135–110 Ma, or even in Jurassic time) predated the opening of the Great South Basin (this structural feature is now filled with sediment).

Extension of the New Zealand Plateau is relevant to biogeography not only as a mechanism for subsidence, flooding, and extinction but also as a means of vicariance. On the Campbell Plateau, populations

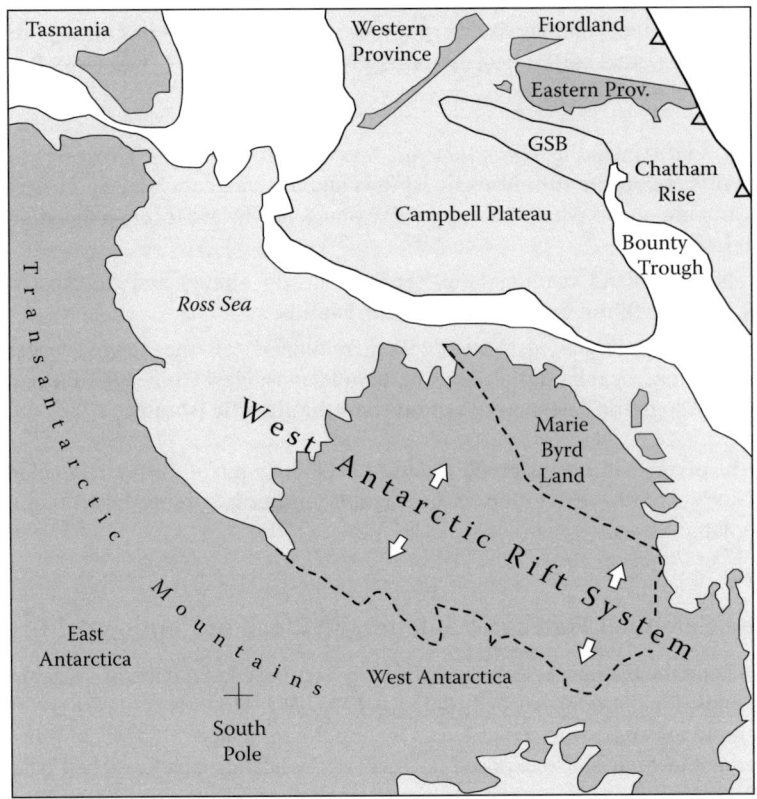

FIGURE 6.2   Reconstruction of the part of the east Gondwana margin at ~90 Ma, soon before complete seafloor spreading developed between the Campbell Plateau and Antarctica (Siddoway, 2008). The current 1500 m isobath, corresponding to the rifted margin, is shown. GSB, Great South Basin. Broken line indicates the Cretaceous WARS, extension indicated by arrows. (Reprinted from Heads, M., *Biogeography of Australasia: A Molecular Analysis*, Cambridge University Press, Cambridge, UK, 2014. With permission from Cambridge University Press.)

of terrestrial and reef organisms have been separated by extension in the Great South Basin, the Bounty Trough, and at the Pukaki Saddle (between the Bounty Platform and the western Campbell Plateau). These are shown in Figure 6.1, along with Bollons Seamount, a large block that has been rifted from the main continental mass (Rowden et al., 2005).

The Campbell Plateau and Antarctica

Prior to Gondwana breakup, the Campbell Plateau and Marie Byrd Land (West Antarctica) formed a continuous part of the East Gondwana convergent margin (Figure 6.2). Episodes of arc magmatism are recorded in Marie Byrd Land and in New Zealand, as oceanic crust from the pre-Pacific basin encroached on Gondwana and subducted beneath it (Siddoway, 2008). At about 100 Ma, transtension–extension developed in the backarc region as the result of changing plate motions. This stretching of the crust produced, or contributed to, the intracontinental West Antarctic Rift System (WARS) (Figure 6.2), the Great South Basin–Campbell Plateau extensional province (Figure 6.2), and extensional features on mainland New Zealand (see Chapter 3).

The WARS is the result of late Mesozoic and Cenozoic extension between East and West Antarctica, and it is one of the largest active continental rift systems on Earth (Cande et al., 2000). Part of the rift shoulder along the margin of the WARS is formed by the Transantarctic Mountains, a major range in which uplift continued into mid-Cenozoic time. A direct relationship between the Transantarctic

Mountains and the extensional structures is equivocal (Davey, 2010). Bialas et al. (2007) suggested that the whole WARS–Transantarctic Mountains region was already a high-elevation plateau with thicker than normal crust before the onset of continental extension. An alternative model proposed that uplift of the Transantarctic Mountains occurred in an overall extensional environment during formation of the WARS (Van Wijk et al., 2008).

The WARS is the product of multiple phases of deformation from the Jurassic to the present, but the most dramatic extension occurred in Cretaceous time (Siddoway, 2008). The main rifting phase between 105 and 90 Ma resulted in >100% extension across the Ross Sea and central West Antarctica. West Antarctica–New Zealand breakup is distinguished as a separate event developing from 83–70 Ma and at the same time as the opening of the Tasman Basin.

Rifting in the Great South Basin and the Bounty Trough, and Seafloor Spreading between the Chatham Rise and Antarctica

The Great South Basin (Figures 6.1 and 6.2) extends north into basins off Canterbury and northeast into the Bounty Trough. Killops et al. (1997) concluded that the coal-bearing Cretaceous sediments in the Great South Basin and the Canterbury Basin have similar oil and gas potential to their counterparts in the Taranaki Basin, which is the main, current source of petroleum in New Zealand. Katz (1968) suggested that environmental conditions for the formation of bituminous sequences are often created at the "turning points in paleogeographic history," and these same turning points have also been critical for biogeography.

The Campbell Plateau, Lord Howe Rise, the Challenger Plateau, and the Ross Sea—continental plateaus that are now submerged—all underwent extension in the Early Cretaceous at ~120 Ma. Forster and Lister (2003) proposed a first extension in the South Island at ~110 Ma. Grobys et al. (2009) attributed the opening of the Great South Basin to extension between the Campbell Plateau and the South Island at ~120–110 Ma (Cretaceous, Aptian–Albian). This followed earlier extension across the Campbell Plateau (beginning at 135 Ma), and preceded further rifting that separated New Zealand from Australia and Antarctica at ~85 Ma. Following the extensional opening of the Great South Basin by normal faulting, sediments dated at 86.5–84.0 Ma show the beginning of a marine influence and indicate that the basin began to subside at this time.

A major fault system bounds the northwestern margin of the Great South Basin and continues south as the Sisters shear zone (Tulloch et al., 2006; Kula et al., 2007). This extensional detachment fault system is exposed on land in southeastern Stewart Island. It represents the boundary where mainland New Zealand (including most of Stewart Island) was rifted from the Campbell Plateau, and the break is now interpreted as part of a metamorphic core complex (Mortimer and Campbell, 2014: 145; Tulloch et al., 2014). Apatite fission track ages in the Sisters shear zone cluster around 85–75 Ma. This reflects the final stages of continental breakup, just before and perhaps during the initiation of seafloor spreading between New Zealand and Antarctica (Ring et al., 2015).

The Bounty Trough is ~1000 km long and 350 km wide and separates the Campbell Plateau from the Chatham Rise to the north (Figures 6.1 and 6.2). It is a large, aborted, continental rift system that is comparable with the East African rift (Grobys et al., 2007). The trough opened in the Cretaceous, after extension in the WARS, as an early stage of the continental breakup of New Zealand from Antarctica.

After 20 Myr of prebreakup rifting in Zealandia, extension became focused in areas such as the Great South Basin and the Bounty Trough (Mortimer et al., 2002). This was the earliest rifting along the Pacific–Antarctic plate boundary and had begun by 90 Ma (Eagles et al., 2004). The Chatham Rise began to be rifted from West Antarctica at about the same time (Larter et al., 2002).

Rifting in the Bounty Trough and Great South Basin ceased at 83 Ma, when complete seafloor spreading began southeast of the Chatham Islands, separating New Zealand and Antarctica at the Pacific–Antarctic ridge. The Campbell Plateau started to separate from West Antarctica soon after this (83.0–79.1 Ma), and the opening of the Tasman Basin also began at this time.

To summarize, the rifting occurred in the sequence: (1) WARS (rifting aborted), (2) Great South Basin and Bounty Trough (rifting aborted), (3) Southern Ocean between New Zealand and Antarctica, and Tasman Basin between New Zealand and Australia (rifting developed into seafloor spreading). The rifting, whether aborted or developing into seafloor spreading, has been critical for biological evolution in the region. For example, it is a likely mechanism for vicariance in vertebrates that display natal philopatry, returning to the same location to breed every year. This will even apply to marine groups, such as seabirds and marine mammals, that have excellent means of dispersal but breed on land.

Penguins

Penguins are conspicuous in the New Zealand subantarctic islands. They illustrate an important break in the region between Macquarie Island in the southwest and the other islands on the main, continental part of the New Zealand Plateau. The six extant genera of penguins have a phylogeny and breeding localities as follows (Baker et al., 2006):

> *Aptenodytes*: **Antarctica**, also Falkland, Kerguelen, and Macquarie Islands (there is a Pliocene fossil from Canterbury, New Zealand, but there is no evidence that this is from a breeding site).
>
> *Pygoscelis*: **Antarctica**, also Falkland, Kerguelen, and Macquarie Islands.
>
> *Spheniscus*, *Eudyptula*, *Megadyptes*, and *Eudyptes*: Southern Australasia (south to Auckland and Campbell Islands), South Africa, and South America (north to the Galapagos Islands), also southern Indian and Atlantic Ocean islands (Kerguelen, Falkland Islands, etc.). **Not in Antarctica**.

There is a paraphyletic, basal grade restricted to Antarctica (*Aptenodytes* and *Pygoscelis*), and so Baker et al. (2006) proposed that the modern penguins had a center of origin there, followed by northward dispersal. On the other hand, Ksepka et al. (2006) incorporated fossils in their analysis of the group and showed that the "basal" groups are New Zealand fossil taxa; because of this they inferred a center of origin in Australia/New Zealand. In fact, as pointed out in Chapter 1, there is no reason to assume that basal clades or grades occupy a center of origin, and so there is no need for a center of origin of penguins either in Antarctica or in Australasia. The main aspect of the molecular phylogeny is the allopatry between the first two clades and the third, and this is compatible with a widespread southern hemisphere ancestor differentiating into *Aptenodytes* and *Pygoscelis* in the south (later overlapping there), and the others in the north.

The break in penguins between the two southern genera and the four northern genera coincides with the line: southern margin of Zealandia–Kerguelen Islands–Falkland Islands, and this suggests that the phylogenetic break was caused by the new, circumantarctic spreading ridges in the Southern Ocean. Activity on these would have separated breeding sites on the rifted continental margins of Campbell Plateau, and the Falkland Islands in the north, from those of Kerguelen and Antarctica in the south. There are the only two areas of overlap: the southern genera overlap with the northern clade (*Eudyptes*) on the Kerguelen Islands, indicating local range expansion southward of *Eudyptes*; the two southern genera also overlap with the northern clade on the Falkland Islands, indicating northward range expansion of the two southern genera.

Macquarie Island lies north of the Pacific–Antarctic plate margin on a separate plate margin, and the presence of the Antarctic clade there probably reflects the complex history of the plate margins in the Macquarie Ridge area. The original spreading center between the Antarctic and Pacific plates was continuous with the Tasman spreading center, and it only joined with the Antarctic–Australia margin in the Eocene (Schellart et al., 2006; cf. Figure 6.2).

Fordyce and Ksepka (2012) inferred a center of origin for penguins in Zealandia. Based on the fact that the oldest known penguin fossils, the two species of *Waimanu* (61.6–58 Ma), are from New Zealand, they referred to "the end of the Cretaceous [as the time that] penguins presumably got their start..." (p. 44) and wrote that "having gotten their sea-legs in Zealandia, penguins soon expanded their domain dramatically, dispersing across thousands of miles and into new climate zones..." (p. 45).

The specimens of *Waimanu* "show clearly that the penguin lineage was established not long after the KT boundary event…" (Fordyce, 2010: 80).

Ksepka (2010) was more cautious, though, and wrote:

> *Waimanu manneringi* [61.6 Ma] is in fact the oldest penguin we know of. But it is highly unlikely it was actually the first penguin. The rock record is incomplete, and there is a roughly 10 *million* year gap between *Waimanu tuatahi* [58–60 Ma] and the next oldest penguin fossil, showing we are missing big pieces of penguin history—probably on both sides of the 60 million year mark.

Other groups with similar patterns to that of the penguins include the icefish (Perciformes suborder Notothenioidei). There are 132 species, and most are found on the continental shelf around Antarctica where they dominate the fish fauna. However, the suborder also extends north to temperate latitudes, with Bovichtidae endemic in Australia, New Zealand, southern South America, and islands in the southern Atlantic and Indian Oceans. Pseudaphritidae are restricted to Australia, and Eleginopidae to Patagonia. Matschiner et al. (2011) concluded that the early divergences in the Notothenioidei are "congruent with vicariant speciation and the breakup of Gondwana."

Eared Seals: Otariidae

The biogeographic break seen in penguins between Antarctica and Macquarie Island on one hand and the Campbell Plateau on the other also occurs in the eared seals, Otariidae. In the clade formed by this family and its sister group, Odobenidae (walrus), three northern groups make up a paraphyletic basal grade, with the last northern group being sister to a southern hemisphere clade, as follows (Higdon et al., 2007; Yonezawa et al., 2009; distributions from Brownell et al., 1974; Wynen et al., 2001):

Circum-Arctic (*Odobenus*, walrus).

   North Pacific (*Callorhinus*).

      North Pacific. North Pacific (*Eumetopias*); + Japan, California, Baja California and the offshore Guadalupe Island, Galapagos Islands, and Isla de la Plata off Ecuador (*Zalophus*).

      Southern hemisphere. Antarctica, South America (north to Peru), South Africa, Australia, and New Zealand, with outliers in the Galapagos Islands, Guadalupe Island, and Baja California (*Otaria, Phocarctos, Neophoca, Arctocephalus*).

As with the penguins, there is no need to assume that the paraphyletic basal grade—in this case in the northern hemisphere—represents a center of origin (cf. Figure 1.1). The break between the southern hemisphere clade and its northern sister group occurs at their only area of overlap: Baja California–Guadalupe Island (off Baja California)–Galapagos Islands. This area coincides with the *eastern* margin of the Pacific plate, while the main break in the penguins coincides with the *southern* margin of the Pacific plate.

The southern hemisphere clade of Otariidae is structured as follows (Yonezawa et al., 2009):

1. *Otaria flavescens*: South America (north to Ecuador and Uruguay).
2. *Arctocephalus tropicalis*: Tristan da Cunha east across the Indian Ocean to Macquarie Island; + *A. pusillus*: SW Africa and SE Australia.
3. *Phocarctos hookeri*: Auckland Island, Campbell Island, Otago Peninsula, and Chatham Islands (extinct), formerly more widespread around the New Zealand mainland; + *Neophoca cinera*: south Australia and southwestern Australia.
   4a. *Arctocephalus philippii*: Juan Fernández Islands and Desventuradas Islands off Chile; + *A. townsendi*: Guadalupe Island, off Baja California.
   b. *A. gazella*: Antarctic Peninsula east via South Georgia and Kerguelen to Macquarie Island.
   c. *A. forsteri*: SW Australia to Tasmania and New Zealand (South Island, Auckland, Campbell, and Chatham Islands); + *A. australis*: South America (north to Peru and southern Brazil); + *A. galapagoensis*: Galapagos Islands.

The four main clades show extensive overlap, and the distribution is not discussed further here. The break between the last three species (*A. forsteri* group) and *A. gazella* occurs between the Campbell Plateau, New Zealand, Australia, and South America/Galapagos on one hand and Antarctica–Macquarie Island (*A. gazella*) on the other. The break corresponds with the Antarctic plate margin, as in the main penguin clades (see note on Macquarie Island and the Antarctic plate margin earlier in this chapter, in the section "Penguins").

The *Pleurophyllum* Group (Asteraceae)

Differentiation between Campbell Plateau groups and mainland New Zealand sisters is a common pattern. It is illustrated here with the *Pleurophyllum* group of Asteraceae (*Pleurophyllum* s.lat.), mentioned earlier (Chapter 4). The group has three main clades (Figure 6.3; Wagstaff et al., 2011):

> *Damnamenia*. Woody-based herbs of Auckland and Campbell Islands.
> *Pleurophyllum* s.str. Megaherbs of Macquarie, Auckland, Campbell, and Antipodes Islands.
> *Olearia* "macrocephalous clade." Trees of North, South, Stewart, Chatham, and NE Auckland Islands.

The group as a whole is most diverse on the Campbell Plateau. On the mainland, outside Fiordland, there is just one species—*O. colensoi*, found through the South and North Islands (Wagstaff et al., 2011). The phylogeny indicates that the group, mainly in the subantarctic, cannot be simply derived from a localized mainland clade as dispersal theory would predict, without invoking other ad hoc hypotheses. In contrast, a vicariance origin *in situ* is straightforward, as the group's sister (*Olearia* p.p., *Celmisia* and *Pachystegia*) is widespread in mainland New Zealand, the Chatham Islands, Australia, and New Guinea, but absent on the subantarctic islands (see phylogeny given earlier, in the section "Endemism and Absence in the Subantarctic Region"). (*Pacifigeron* of Rapa Island is possibly related to the widespread mainland group but has not been sequenced.)

Within the *Pleurophyllum* group, *Damnamenia* of Auckland and Campbell Islands overlaps completely with its sister (*Olearia* "macrocephalous clade" and *Pleurophyllum* s.str.), implying local overlap on the Auckland–Campbell Island sector. The biogeography is not examined further here, as the precise

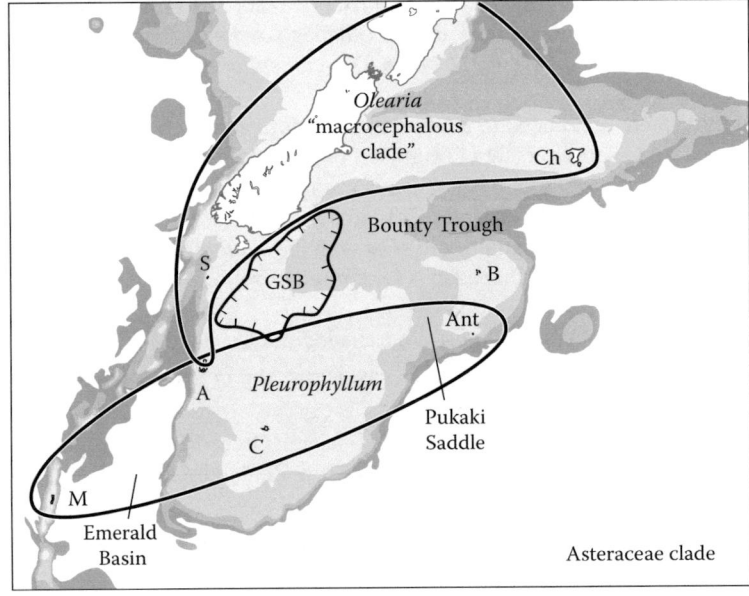

FIGURE 6.3 Distribution of *Pleurophyllum* and its sister group *Olearia* "macrocephalous clade" (Asteraceae). A, Auckland Islands; Ant, Antipodes Islands; B, Bounty Islands; C, Campbell Island; Ch, Chatham Islands; GSB, Great South Basin; M, Macquarie Island; S, The Snares. (Data from Wagstaff, S.J. et al., *Am. J. Bot.*, 98, 62, 2011.)

locality records that would be required for this are not available. In contrast, *Pleurophyllum* and "macrocephalous *Olearia*" themselves display a simple pattern, with a high level of allopatry between the former group in the south and the latter in the north. The two groups occur together only on Auckland Island, and even within this island the distributions differ, as *Pleurophyllum* is widespread, while "macrocephalous *Olearia*" is restricted to the northeastern corner. ("Macrocephalous *Olearia*" is represented here by *O. lyallii*, also found in The Snares.)

Auckland Island is one of the main centers of endemism in the subantarctic islands, but it is also an important phylogenetic break, as in *Pleurophyllum*/"macrocephalous *Olearia*." In their study of the whole *Pleurophyllum* group, Wagstaff et al. (2011) assumed that the basal clade, *Damnamenia*, occupied the center of origin. Based on this, they proposed dispersal from the subantarctic islands to the New Zealand mainland, then from the mainland to the Chatham Islands, and finally back to the subantarctic islands. Yet they did not refer to, or explain, the most striking aspect of the distribution: the mutual allopatry of "macrocephalous *Olearia*" and *Pleurophyllum*. From the Auckland Islands the break between these two clades extends northeast to the area between the Chatham Islands and the Bounty/Antipodes Islands. Thus, the break coincides with the Great South Basin and the Bounty Trough, and it represents a large-scale phylogenetic division on the Campbell Plateau. The simplest explanation is that the break was caused by the rifting open of the Great South Basin and the Bounty Trough in the Late Cretaceous (90–84 Ma).

A similar break to the one between *Pleurophyllum* and "macrocephalous *Olearia*" occurs in *Juncus* (Juncaceae). *J. scheuchzerioides* is recorded from Macquarie, Auckland, Campbell, and Antipodes Islands, as well as in Patagonia, South Georgia, and the Kerguelen Islands. Its closest relative appears to be *J. pusillus*, found on the Auckland Islands and through the New Zealand mainland (North, South, and Stewart Islands) (Moore and Edgar, 1970). Here, again, the split between the New Zealand mainland species and the one on the Campbell Plateau and South America is located at Auckland Islands–Great South Basin.

Within *Pleurophyllum* itself, the two clades break up the subantarctic range into two sectors, one in the north and one in the south, with overlap on Auckland and Campbell Islands (Wagstaff et al., 2011):

Auckland, Campbell, and Macquarie Islands (*P. hookeri*).
Auckland, Campbell, and Antipodes Islands (*P. criniferum, P. speciosum*).

Thus, the zone Auckland Islands–Campbell Island, which is an important center of endemism (e.g., *Damnamenia, Chionochloa antarctica, Ranunculus pinguis, Hebe benthamii*, the geoplanid *Marionfyfea*), also acts as a biogeographic margin. The distributions suggest that *Pleurophyllum* has survived the submergence of the Campbell Plateau, and Wagstaff et al. (2011) accepted it as a paleoendemic clade that persisted on the islands through the last glacial maximum. It is now accepted that many groups in the subantarctic region and even Antarctica survived *in situ* through the Pleistocene ice ages. Fraser et al. (2012) concluded:

> …numerous recent studies provide evidence of deeply divergent lineages unique to Antarctica, indicating glacial survival in fragmented habitats followed by postglacial expansion, and pointing to long-term persistence of terrestrial taxa, such as arthropods, on the Antarctic continent. Although a few highly dispersive marine species have been able to recolonise postglacially, most surviving high-latitude taxa appear to have persisted throughout glacial maxima in local refugia.

Sites of local survival would have included small ice-free areas, for example, nunataks in alpine regions such as the Transantarctic Mountains, where the endemics include the montane oribatid mite family Maudheimiidae.

Subantarctic and Mainland Clades: Examples from New Zealand *Poa* (Poaceae)

In the grass genus *Poa*, Edgar (1986; Edgar and Connor, 1999) proposed an informal group of New Zealand species made up of two subgroups, and these are mapped in Figure 6.4. "Clade A1" (species 1–3) is endemic to the New Zealand subantarctic islands. (*P. cookii* of Macquarie Island was thought to

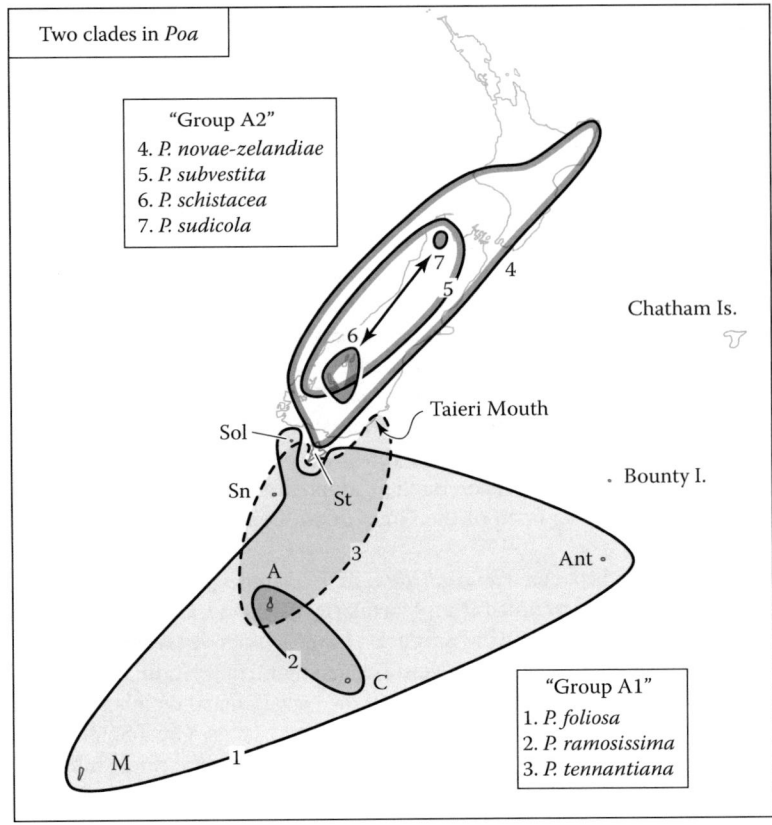

FIGURE 6.4 Two clades in *Poa* (Poaceae). Group A1, gray fill. Group A2, no fill. Arrows indicate morphological affinity between *P. sudicola* of northwest Nelson and *P. schistacea* of Otago. A, Auckland Islands; Ant, Antipodes Islands; C, Campbell Island; M, Macquarie Island; Sn, The Snares; Sol, Solander Island; St, Stewart Island. (Data from Edgar, E., *N. Z. J. Bot.*, 24, 425, 1986; Edgar, E. and Connor, H.E., *N. Z. J. Bot.*, 37, 63, 1999; Edgar, E. and Connor, H.E., *Flora of New Zealand*, Vol. 5: *Gramineae*, Government Printer, Wellington, New Zealand, 2000; Gillespie, L.J. et al., *Aust. Syst. Bot.*, 22, 413, 2009; Wilson, H.D., *N. Z. J. Bot.*, 1, 81, 1987.)

belong here but is now placed with another group.) "Clade A2" (species 4–7), in contrast, is a mainland New Zealand group. Molecular work (Gillespie et al., 2009; Birch et al., 2014) indicated that the groupings are not entirely satisfactory, but *Poa* is diverse in New Zealand (38 species), and so far the sampling is not sufficient to indicate a new arrangement. In any case, the distributions of the individual species are of interest.

"Clade A1" (with *P. tennantiana* and *P. foliosa*) and "clade A2" (*P. novaezelandiae*) show dove-tailing allopatry at a mutual boundary in Stewart Island. Clade A1 is represented on islets west and northeast of Stewart Island, but on Stewart Island itself it is only in the south and is replaced in the north, on Mount Anglem, by "Clade A2."

In the South Island, *Poa* "Clade A2" includes a pair of disjunct, local endemics that are putative sister species (Edgar and Connor, 1999): *P. schistacea* of Central Otago schist and *P. sudicola* of northwest Nelson. The two species are separated by a large disjunction that corresponds with displacement along the Alpine fault. *P. sudicola* is a subalpine species restricted to limestone scree and steep mudstone slopes in the Matiri Range and on Pike Peak (Allen Range). The ecology suggests that it is a coastal species stranded inland, and *P. schistacea* probably had a similar origin; with the erosion of superficial limestone, the species has been redeposited back on the basement schist.

A similar distribution to that of the subantarctic *Poa tennantiana* and *P. foliosa* is seen in *Stilbocarpa* (Apiaceae). This genus occurs on Macquarie, Auckland, Campbell, The Snares, and Antipodes Islands, islets adjacent to Stewart Island, Stewart Island itself, and Coal Island in Preservation Inlet, Fiordland (Mitchell et al., 1999). Note that the subantarctic *Poa* clade, *Stilbocarpa*, and *Pleurophyllum* are all widespread on the Campbell Plateau but are absent from the Chatham Islands. This pattern is also common in marine groups, for example, the clingfish *Gastrocymba quadriradiata* (Gobiesocidae), which is restricted to The Snares, Auckland, Campbell, Bounty, and Antipodes Islands (Francis, 1995).

Several plants of the subantarctic region are referred to as "megaherbs." These include the endemic genera *Stilbocarpa* (Apiaceae) and *Pleurophyllum* (Asteraceae), and also species of *Bulbinella* (Asphodelaceae) and *Anisotome* (Apiaceae). In the Chatham Islands, the most obvious example is the endemic genus *Myosotidium* (Boraginaceae). None of these megaherbs are large compared with tropical herbs, such as banana plants. Nevertheless, they are an order of magnitude larger than related groups on mainland New Zealand, and it is this difference that needs to be accounted for. The different explanations offered for this enigma depend on the biogeographic and evolutionary models that are accepted for the Campbell Plateau and Chatham Rise region in general.

Albatrosses

Albatrosses (*Diomedea* s.str.) comprise two main clades. The first (*D. sanfordi* and *D. epomophora*) is distributed in the subantarctic islands, near Dunedin, and on the Chatham Islands. This touches the mainland at a single point on Otago Peninsula, just east of the Titri-Waitati fault (Figure 3.13). The sister group includes all the other *Diomedea* species and has a widespread southern hemisphere distribution (Kennedy and Page, 2002). The albatross phylogeny indicates a basal break around the Campbell Plateau and is consistent with the *D. sanfordi* + *D. epomophora* clade evolving in its subantarctic range, rather than migrating there.

Diomedea sanfordi breeds at Auckland Islands (Enderby Island only), on the South Island mainland at Otago Peninsula, and at the Chatham Islands, while its southern sister *D. epomophora* breeds on Auckland Islands (Auckland, Adams, and Enderby Islands) and Campbell Island. The break between the two species, as in *Pleurophyllum* versus "macrocephalous *Olearia*," coincides with the Great South Basin (Figure 6.3). The overlap of the two species is restricted to Enderby Island, at the same northeastern Auckland Islands node seen in the two plant groups. The localized distribution of this albatross clade on the South Island, on the margins of the Great South Basin, resembles that of the subantarctic *Poa* clade at Taieri Mouth (Figure 6.4).

The Snares

Auckland and Campbell Islands have a diverse terrestrial biota, but even the smaller islands on the Campbell Plateau host significant endemism. The Snares have a land area of just 3.5 km², but the biota includes endemic birds such as *Eudyptes robustus* (Spheniscidae), *Coenocorypha huegeli* (Scolopacidae), and *Petroica dannefaerdi* (Petroicidae) (the latter being completely black, as in *P. traversi* of the Chatham Islands). Other endemics on The Snares include the beetle genus *Synteratus* (Carabidae) (Larochelle and Larivière, 2001) and the lichen *Solenopsora sordida*, the sole representative of its genus in New Zealand (Galloway, 2007).

There are many connections between The Snares and the southwestern South Island. In weevils, *Lyperobius nesidiotes* of The Snares forms a clade with *L. coxalis* and *L. australis* of Fiordland (Craw, 1999). New Zealand terrestrial leeches include *Ornithobdella*, found on penguins and known only from The Snares and Solander Island to the north. The related *Hirudobdella* comprises *H. benhami* in forest at Caswell Sound, Fiordland, and *H. antipodum* on the Open Bay Islands, southern Westland (Miller, 1999). *Ornithobdella* and *Hirudobdella* are placed with *Aethobdella* of Victoria to Queensland, and the three form a trans-Tasman clade, the Ornithobdellidae (Richardson, 1979).

New Zealand has 758 indigenous seaweed species, and 265 (35%) of these are endemic (Hurd et al., 2004). (Large red algae are the most diverse and endemic.) Local, allopatric endemism is seen in many of these. For example, in the brown alga *Lessonia*, the southwest Pacific populations make up a monophyletic clade with the following species (Martin and Zuccarello, 2012):

L. tholiformis: Chatham Islands.

L. variegata: Mainland New Zealand—a paraphyletic complex with four allopatric clades.

L. corrugata: Tasmania.

L. adamsiae: The Snares.

L. brevifolia: Auckland, Campbell, Antipodes, and Bounty Islands (forming a clade with the last two species).

Bounty and Antipodes Islands

The Bounty Islands (1.4 km^2) and Antipodes Islands (22 km^2) are the emergent parts of the Bounty Platform and host many endemics. The Bounty Islands themselves are mostly composed of Early Jurassic granodiorite that forms part of the Median Batholith. The exposed rocks on the Antipodes are all young volcanics erupted from 5 to 1 Ma. Nevertheless, a review of the Antipodes Islands arthropods found "surprising" levels of endemism (Marris, 2000). The Bounty Islands also harbor intriguing endemics such as the shag *Leucocarbo ranfurlyi* (Phalacrocoracidae). Rather than being nested in a mainland group, it is sister to a diverse complex found in North and South Islands (*L. carunculatus*), Stewart Island to Otago Peninsula (*L. chalconotus*), Chatham Islands (*L. onslowi*), Auckland Islands (*L. colensoi*), and Campbell Island (*L. campbelli*) (Figure 4.2; Kennedy and Spencer, 2014).

Pseudhelops *(Coleoptera, Tenebrionidae)*

The beetle *Pseudhelops* is distributed mainly on the subantarctic islands, while its closest relative is *Cerodolus* of mainland New Zealand (Watt, 1971). A morphological study found three clades in *Pseudhelops* (Figure 6.5; Leschen et al., 2011):

The Snares + Chatham Islands (two species).

Stewart Island (one species).

Five species on, respectively, Auckland Islands, Auckland Islands, Campbell Island, Antipodes Islands, and Bounty + Antipodes Islands.

The three clades form an allopatric set of affinities striking east–northeast. A similar connection is evident in the plant *Myrsine chathamica* (Myrsinaceae), known from Stewart Island and the Chatham Islands. The Snares–Chathams clade in *Pseudhelops* is separated from the diverse southern clade by the Great South Basin and the Bounty Trough.

Leschen et al. (2011) wrote that the young age of the volcanic rocks on Antipodes Islands seems to conflict with the presence of two *Pseudhelops* species there. They suggested that the distribution could be accounted for by submerged land on the eastern Campbell Plateau. As further support for this, they cited the beetle genus *Bountya* (Carabidae) and other arthropods endemic to the Bounty Islands. *Bountya* is sister to a diverse clade present in Australia and Patagonia, but not New Zealand (Liebherr et al., 2011), probably indicating a subantarctic connection between Australia and Patagonia.

Groups on The Snares with relatives further east on the Bounty, Antipodes, and Chatham Islands, as in *Pseudhelops*, are separated by the Great South Basin (Figure 6.5), suggesting that the vicariance here was caused by the basin's formation. Other groups that could have been affected in this way include the carabid beetles *Diglymma castigatum* on The Snares and Bounty Islands, and *Oopterus clivinoides* on The Snares, Auckland, Campbell, and Antipodes Islands (Larochelle and Larivière, 2001).

Cyanoramphus *Parrots (Psittacidae)*

Cyanoramphus is the southernmost parrot genus, with a distribution extending from Macquarie Island (extinct) and through New Zealand to New Caledonia and the Society Islands (extinct). Its distribution, along with that of its closest relatives in New Caledonia, Fiji and Tonga, is mapped elsewhere (Heads, 2012a:

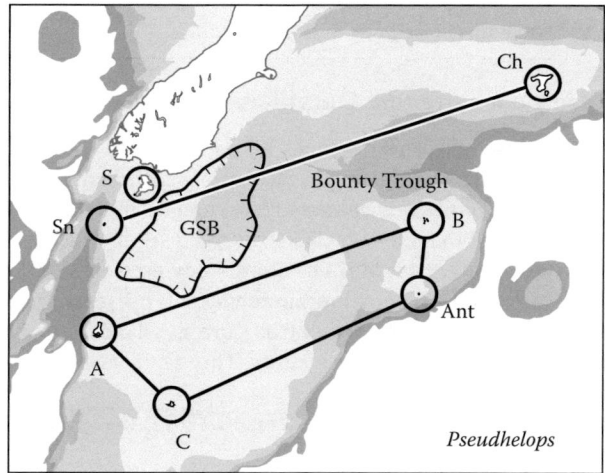

FIGURE 6.5 Distribution of *Pseudhelops* (Tenebrionidae). A, Auckland Islands; Ant, Antipodes Islands; B, Bounty Islands; C, Campbell Island; Ch, Chatham Islands; GSB, Great South Basin; M, Macquarie Island; S, Stewart Island; Sn, The Snares. (Data from Leschen, R.A.B. et al., *N. Z. Entomol.*, 34, 12, 2011.)

Fig. 6–11). It is likely to have evolved *in situ*, by vicariance with its relatives to the north and west around a break zone: New Caledonia–Society Islands (Heads, 2012a: Fig. 6–11).

The phylogeny of the genus is given here (Boon et al., 2001; Chambers and Boon, 2005). Neither the Society Islands species nor the Macquarie Island species (*C. erythrotis*) have been sequenced, and the latter is placed here on the basis of morphology. There is also fossil material of *Cyanoramphus* from Campbell Island, but its specific affinities are unknown (Holdaway et al., 2010). Mainland distributions are shown in bold:

New Caledonia (*C. saisseti*).

    Chatham Islands (*C. forbesi*).

        Norfolk Island (*C. cookii*).

           Antipodes Island (*C. unicolor*).

               Macquarie, Auckland, Antipodes, Chatham, **Stewart, South, North** (extinct), Kermadec, and Lord Howe (extinct) Islands (*C. erythrotis, C. novaezelandiae, C. malherbi, C. hochstetteri*).

               Auckland, **Stewart, South, and North Islands** (*C. auriceps*).

The basal differentiation has occurred at nodes around an outer arc: New Caledonia, Chatham, Norfolk, and Antipodes Islands. The species at the last locality has separated from a widespread clade distributed in the subantarctic islands, through the mainland, and north to Lord Howe and the Kermadec Islands. If the outlying Chathams, Norfolk, and Antipodes clades had dispersed to these outer islands from the mainland, they would be nested in the main, widespread clade; in fact, they are basal to it. In a vicariance model, differentiation (rather than physical migration) has begun around the outer margins of the New Zealand Plateau and has then worked its way into the central part, mainland New Zealand.

Invasion of the mainland from the subantarctic islands, or vice versa, by theoretical colonizing flights is not a satisfactory explanation for the clades of *Cyanoramphus*, for the differentiation between the subantarctic beetle *Pseudhelops* and its mainland sister *Cerodolus*, or for the subantarctic megaherbs and their mainland relatives. Nevertheless, metapopulation processes in many species do include normal dispersal among widespread parts of the New Zealand archipelago. If a population is extirpated on one island, that island can be reinvaded from another. For example, breeding populations of sea lions (*Phocarctos hookeri*) were wiped out on the mainland by human hunting, but since 1994 a few individuals, presumably from the subantarctic islands, have been breeding on the South Island at Otago Peninsula (Collins et al., 2014). This (more or less) observed dispersal within a species, and within its known range, should be distinguished from the inferred dispersal that the traditional model uses to explain all biotic differentiation between the subantarctic islands and the mainland.

A Basal Kelp in the Antipodes Islands: Durvillaea "Species A"

In the large kelp *Durvillaea* (Phaeophyceae), "species A" is endemic to Antipodes Islands (Fraser et al., 2010). It is not nested in a mainland group, but is instead sister to a clade distributed across the rest of the New Zealand Plateau (cf. *Cyanoramphus* discussed earlier) and around the Southern Ocean. In a clock study, Fraser et al. (2010) calculated that the two clades diverged at ~10 Ma, although they noted that the Antipodes Islands themselves formed at <5 Ma. The clock study used fossil-calibrated rates, and so the dates it produced are at best (assuming a clock) minimum ages. This means that the Antipodes Islands *Durvillaea* could indeed be older than the rocks of its current islands, and suggests that it survived around other, former islands and coastlines in the vicinity. This would also explain the Antipodes Islands clades of beetles and parrots cited earlier.

In *Cellana* limpets (Gastropoda, Nacellidae), Reisser et al. (2011) found a clade with two components:

Auckland, Campbell, The Snares, Stewart, and South and North Islands (subspecies of *Cellana strigilis* s.str., plus *C. denticulata*).

Bounty, Antipodes, and Chatham Islands (subspecies of *Cellana* "*strigilis*").

The two clades are allopatric, but show an east–west pattern, unlike the north–south patterns discussed earlier. The break coincides with the Pukaki Saddle, west of the Bounty and Antipodes Islands.

Connections between the Subantarctic Islands and the Chatham Islands

Many New Zealand groups occur on the Campbell Plateau islands and the Chatham Islands to the north, but not on the mainland. For example, the plant *Colobanthus muscoides* (Caryophyllaceae) can be found on Macquarie, Auckland, Campbell, The Snares, Antipodes, and Chatham Islands, but not on the mainland (Allan, 1961). *Senecio radiolatus* (Asteraceae) comprises two subspecies: subsp. *radiolatus* from the Chatham Islands and subsp. *antipodus* on the Antipodes Islands (C.J. Webb in Connor and Edgar, 1987). The beetle *Lichenobius* (Anthribidae) consists of one species on The Snares and Big South Cape Island (off Stewart Island), one on the Bounty Islands, and one on the Chatham Islands (Holloway, 1982). The flea *Parapsyllus nestoris antichthones* is recorded on the Bounty, Antipodes, and Chatham Islands (Smit, 1979).

Sedges (Cyperaceae)

In the sedges, *Carex ternaria* (Cyperaceae) is endemic to Auckland, Antipodes, and Chatham Islands and is related to mainland species. Moore and Edgar (1970) placed it in a keyed group (formerly *C. ternaria* s.lat.) with three other species—*C. lessoniana*, *C. geminata*, and *C. coriacea*—each found in North and South Islands, with *C. geminata* and *C. coriacea* also in the Stewart Island. *C. ternaria* is larger than the three mainland species (cf. the *Poa* clades in Figure 6.4) and has larger spikes and more conspicuous awns. In a similar pattern, *Carex sectoides* is endemic to the Antipodes and Chatham Islands and was keyed with *C. secta*, found throughout the North, South, Stewart, and Chatham Islands (Moore and Edgar, 1970).

Puccinellia (Poaceae)

The saltmarsh grass *Puccinellia walkeri* comprises three allopatric subspecies with distributions that can be interpreted as parallel bands (Edgar, 1996):

Stewart Island, coastal South Island to Banks Peninsula, and Cook Strait (subsp. *walkeri*).

Auckland, Campbell, and Chatham Islands (subsp. *chathamica*).

Antipodes Islands (subsp. *antipoda*).

The Auckland and Campbell Islands plants connect with populations on the Chatham Islands, not with those on the much closer Antipodes Islands. This affinity, along with the endemism on the Antipodes in other groups, and at least one endemic group (*Durvillaea* sp. A) being dated as older than the islands (see section "Bounty and Antipodes Islands"), suggests that these groups all had wider distributions across the southeastern parts of the Campbell Plateau.

Mollymawks (Thalassarche, *Diomedeidae*)

The last terrestrial life surviving on offshore, rocky stacks before these are completely eroded away includes groups such as lichens, Diptera, and seabirds, and so these provide valuable bio-geographic data on past New Zealand landscapes. Mollymawks (*Thalassarche*) are members of the albatross family and are one of the many groups that exhibit fine-scale allopatry among the Campbell Plateau islands.

Molecular studies on *Thalassarche* showed the following phylogeny for the main clade (breeding records only; Robertson and Nunn, 1998; Kennedy and Page, 2002):

Macquarie, Campbell, and Antipodes Islands (*T. chrysostoma*, *T. melanophrys*, and *T. impavida*, the first two of which also breed in the southern Indian and Atlantic Oceans).

The Snares and Solander Islands (by Stewart Island), Chatham Islands, and Three Kings Islands (*T. bulleri*).

Auckland and Antipodes Islands (*T. steadi*) + Tasmania (*T. cauta*).

Crozet (southern Indian Ocean), The Snares, Bounty, and Chatham Islands (*T. salvini* and *T. eremita*).

In this group, the representatives on The Snares, Auckland, and Campbell Islands each belong to separate tracks. An outer arc occupied by the last clade, in The Snares–Bounty–Chathams Islands, overlaps on the Chathams with an inner arc skirting the mainland in The Snares–Chathams–Three Kings Islands (*T. bulleri*). Again, the patterns are consistent with the idea that the modern clades differentiated following extension and the rifting apart of nesting sites on the Campbell Plateau (with breaks at the Great South Basin, Bounty Trough, Emerald Basin, and Pukaki Saddle, Figure 6.1).

Another "outer arc" track around New Zealand mainland is illustrated by the gastropod *Larochea miranda* (Larocheidae), present on the Auckland Islands, Chatham Islands, and Three Kings Islands to East Cape (Geiger, 2012). The moss *Fissidens strictus* (Fissidentaceae) is in Victoria, Tasmania, northern North Island (north of lat. 37° S), and the Auckland Islands. Beever (2014) wrote that "the lack of any records from the mid latitudes of N.Z. is puzzling"; the distribution suggests possible extinction on Chatham Rise/Chatham Islands. Part of the disjunction can be explained by Alpine fault movement.

Subantarctic Pipits (Anthus, *Motacillidae*)

Anthus is a passerine, a landbird, but its different forms link the subantarctic and Chatham Islands in the same way that the mollymawks do. *Anthus novaeseelandiae* s.str. occurs in Australia, New Zealand, and New Guinea (Christidis and Boles, 1994). In New Zealand, authors have recognized one or two subspecies on the New Zealand mainland, one on the Auckland and Campbell Islands, one on the Antipodes Islands, and one on the Chatham Islands (see review in Foggo et al., 1997). In contrast, allozyme and morphometric data indicated that there are two New Zealand species, one on the mainland and one on all the offshore islands (Foggo et al., 1997; cf. Oliver, 1955).

Taxonomists have tended to recognize any populations on New Zealand offshore islands as separate entities because they have assumed a model of dispersal radiating from the mainland. Yet this process would produce endemics on individual islands, not a monophyletic Campbell Plateau/Chatham Rise clade found on all the offshore islands, as in *Anthus*. Foggo et al. (1997: 371) wrote that their discovery:

> …is surprising because most of the islands are at least as distant from each other as they are from the New Zealand mainland. If colonization across water is assumed, then the mainland must be a far more likely source because the number of New Zealand Pipits there is orders of magnitude greater than those of the islands.

As indicated earlier, Campbell Plateau clades are likely to be relics, dating to before the submergence of the plateau, and so colonization across water from the mainland is not necessary. The new arrangement in *Anthus*, with a break between the mainland on one hand and the Campbell Plateau/Chatham Rise on the other, shows a surprising incongruence with the dispersal model, and this was emphasized by Holdaway et al. (2001). They commented that "such a distribution of species is anomalous in comparison to that of other taxa, and begs the question as to how such a situation could have arisen…." Heather and Robertson (2005: 379) suggested a "double invasion" of the New Zealand region, but this is unnecessary and does not explain the simple allopatry between the two clades.

Affinities between Campbell Plateau and Areas Outside New Zealand

Tasmania, Patagonia, and the North Pacific

In Orthoptera, there are endemic genera of cave crickets (Rhaphidophoridae) on the Auckland Islands (*Dendroplectron*), Campbell Island (*Notoplectron*), Bounty Islands (*Ischyroplectron*), and The Snares/Antipodes Islands (*Insulanoplectron*) (Michaux and Leschen, 2005). In the same family, *Tasmanoplectron* of Tasman Island, off southeastern Tasmania, "has no close affinities with any known Australian Rhaphidophoridae, but it shares certain characters with 9 New Zealand and subantarctic islands genera." These include a unique character shared only with *Insulanoplectron* of The Snares and Antipodes Islands (Richards, 1971: 591). Given the paleogeography, this sort of link between the Campbell Plateau and Tasmania might be expected.

In beetles, nine genera cited by Michaux and Leschen (2005) link the subantarctic islands with the southern South Island, while seven link the subantarctic region with Patagonia. In other connections, the beetle *Antarcticodomus* (Salpingidae, Aegialitinae) comprises five species on the Auckland, Campbell, The Snares, Stewart, Bounty, Antipodes, and Chatham Islands. The remaining members of Aegialitinae occur around the margin of the North Pacific, from the Kuril Islands in northeastern Japan to California. Other beetles that are endemic to the Campbell Plateau and belong to otherwise Holarctic groups include the staphylinid genera *Baeostethus* and *Stylogymnusa* (Leschen et al., 2002).

Abalone (Haliotis, Gastropoda): Campbell Plateau Differentiation in an Indian Ocean Group

Haliotis or abalone is a global genus of large, intertidal gastropods made up of an Indian Ocean group and a northern group (Degnan et al., 2006). The Indian Ocean group comprises two clades:

South Africa to India and New Zealand: Many species.
New Zealand: Two species, *H. iris* (paua) and *H. virginea*.

The two clades overlap in New Zealand, but they show a significant difference in their distribution there. The second clade, *H. iris* + *H. virginea*, is represented on the main islands and also on the Campbell Plateau, where *H. virginea huttoni* is endemic to Auckland and Campbell Islands. In contrast, its widespread Indian Ocean sister group occurs in New Zealand south only to The Snares Islands (Powell, 1979). This is consistent with an original break between a Campbell Plateau group and an Indian Ocean clade, with the break located somewhere near the Great South Basin. Following the break, secondary overlap has developed around the main islands.

Tectonic Extension on the Campbell Plateau and the Origins of the Endemism There

To summarize this chapter so far, the plants and animals on the islands of the Campbell Plateau and the Chatham Rise are the relics of an ancestral biota that was almost completely destroyed by subsidence and marine flooding. The submergence of the plateau occurred because the New Zealand crust was stretched and thinned. Eventually it broke, and Zealandia drifted away from Gondwana, but by that time the groups of the Campbell Plateau had already diverged from mainland relatives, at boundaries such as the Great South Basin. There has been no mainland center of origin or dispersal from the mainland to the islands, but instead a break, located between what is now an island region (Campbell Plateau) and what is now the New Zealand mainland.

The plants and animals present on the subantarctic islands of the Atlantic, Indian, and Pacific Oceans have often been attributed to postglacial dispersal from a mainland source, but not always. One study of endemism in subantarctic plants concluded instead that the different regional floras had survived in local refugia during the last glacial maximum (van der Putten et al., 2009). A molecular clock study of Asteraceae in the New Zealand subantarctic islands reached similar conclusions, and it proposed that many plant groups inhabited the circumantarctic region until the late Cenozoic (Wagstaff et al., 2007). The authors concluded that "the distinctive flora of the subantarctic islands may harbour some of the last remnants of this once diverse flora" (p. 8).

Many biogeographic studies of the New Zealand subantarctic islands have concentrated on how particular groups have dispersed to the islands from the mainland. Yet the groups that are both endemic to the subantarctic region and widespread there suggest that there is a general problem, a Campbell Plateau biota. The question is then: "How did this biota differentiate from that on the mainland (and in Antarctica)?" Part of the answer probably lies in the extension and rifting that occurred in the buildup to seafloor spreading.

The Great South Basin and "Sliver" Distributions on the Mainland

The largest sedimentary basin on the Campbell Plateau is the Great South Basin off the southeastern South Island (Killops et al., 1997). The basin itself has filled with sedimentary strata and is not obvious in the seafloor bathymetry, but it represents a large-scale, intracontinental rift located between the South Island and the main area of the Campbell Plateau. The basin has a small outlier onshore in southeastern Otago, at the Cretaceous Titri fault (Figure 3.13; Mortimer et al., 2002). This geological boundary recalls mainland "slivers" of subantarctic groups found in this region, as seen in the grass *Poa tennantiana* at Taieri Mouth (Figure 6.4) and the albatross *Diomedea sanfordi* on Otago Peninsula. Other examples include the supralittoral amphipod *Orchestia aucklandiae* (Talitridae) on Auckland, Campbell, and Stewart Islands, and Otago Peninsula (Papanui Beach) (Hurley, 1957). Two carabid beetles also illustrate the pattern:

> *Mecodema alternans*: The Snares, Stewart Island, Bluff, Dunedin, and Chatham Islands (Larochelle and Larivière, 2001).
>
> *Kenodactylus*: Auckland Islands, Campbell Island, The Snares, Antipodes Islands, Stewart Islands, and Dunedin, as well as Patagonia, Falkland Islands, and South Georgia (Townsend, 2010).

Chatham Islands

The Chatham Islands are the only emergent parts of the Chatham Rise, a ribbon of continental crust that extends 1000 km eastward from the New Zealand mainland. The islands are well known for their many endemics, such as the plant *Myosotidium* (Boraginaceae) and the insect *Chathamaka* (Hemiptera) (Larivière et al., 2010). (*Embergeria* in the Asteraceae was regarded as a second angiosperm genus endemic to the Chatham Islands, but it is now thought to be sister to the mainland *Kirkianella*, with the pair deeply nested in *Sonchus*; Mejías and Kim, 2012.) In birds, "the degree of radiation and endemism of both marine and terrestrial species within the [Chathams] group sets it apart from all other islands in the New Zealand region" (Holdaway et al., 2001: 151).

Geographic Affinities of Chatham Islands Endemics

While some Chatham Islands clades are closest to those of the New Zealand mainland, others are related to groups on the subantarctic islands (see earlier). Others have relatives on the islands of northeastern New Zealand and the central Pacific. For example, the Chatham Islands endemics *Hebe chathamica* and *H. dieffenbachii* (Plantaginaceae) are most closely allied with *H. rapensis* of Rapa Island (southeastern Polynesia) (Bayly and Kellow, 2006), while the beetle *Rhantus schauinslandi* (Dytiscidae) is more closely related to species such as *R. vitiensis* of Fiji than to species on the New Zealand mainland (Ordish, 1989).

In rails, the extinct *Gallirallus* ("*Nesolimnas*") *dieffenbachii* of the Chatham Islands is sister to (not nested in) a diverse, west Pacific clade that ranges from the Philippines to Lord Howe, New Zealand, and the Cook Islands with the following species (Garcia-R et al., 2014):

> *G. philippensis*: Philippines and New Guinea (possibly its original distribution), extending to Samoa, Australia, and New Zealand.
>
> *G. pendiculentus*: Mariana Islands.
>
> *G. sylvestris*: Lord Howe Island.
>
> *G. owstoni*: Guam.
>
> *G. wakensis* (extinct): Wake Island (an emergent part of the mid-Pacific seamounts in the north Pacific and southwest of the Hawaiian Islands).
>
> *G. ripleyi*: Cook Islands.
>
> *G. insignis*: New Britain.
>
> *G. woodfordi*: Solomon Islands.
>
> *G. rovianae*: Solomon Islands.

The phylogeny means that *G. dieffenbachii* cannot be easily derived from any of these species.

Garcia-R et al. (2016) placed *G. dieffenbachii* in a tritomy with the aforementioned clade and also *G. okinawae* plus *G. torquatus* of Asia. The sister to this whole clade is *G.* ("*Cabalus*") *modestus* (extinct) of the Chatham Islands. The two Chatham Islands species form a basal paraphyletic grade that cannot be easily derived from a center of origin in Asia, Oceania, or Australia.

One of the most distinctive Chatham Islands plants is the endemic genus *Myosotidium* (Boraginaceae). The single species has been decimated by sheep and pigs, but it was "originally an abundant coastal plant, usually growing in sandy soil just above highwater mark… In several localities it once formed an unbroken line for miles together on the seashore" (Cheeseman, 1914: Pl. 146, p. 2).

Heenan et al. (2010) found that this potential mangrove associate was closest to *Omphalodes nitida* of Spain and Portugal. This would give a standard Tethyan disjunction, as in *Scleranthus*, *Ceratocephala*, and so on (Heads, 2014). Another study instead found *Myosotidium* to be the sister of a Chilean species and retrieved the following clade in *Omphalodes* (Weigend et al., 2013):

> NE Mexico: *Omphalodes aliena.*
>
> Chatham Islands: *Myosotidium.*
>
> Chile: *Cynoglossum paniculatum.*

This gives a standard New Zealand–Chile–Mexico ("Pacific triangle") distribution (Heads, 2014). The study by Heenan et al. (2010) did not include *O. aliena* or *C. paniculatum*, while the study by Weigend et al. (2013) did not include *O. nitida*, and so the sister group of *Myosotidium* is still uncertain. Affinities with the Mediterranean and Chile are both possible.

Marine groups endemic on the Chatham Islands include dominant reef organisms such as the brown algae *Durvillaea chathamensis*, *Lessonia tholiformis*, and *Landsburgia myricaefolia*. In the red alga *Gigartina*, a Chatham Islands clade is basal in an Indian Ocean group (Nelson and Broom, 2008):

> *Gigartina grandifida*: Chatham Islands.
> *G. radula*: South Africa.
> *G. atropurpurea*: North and South Islands.

These far-flung affinities in *Gigartina*, *Hebe*, and *Myosotidium* could be interpreted as the result of chance dispersal to the Chatham Islands from different parts of the Indian, Pacific, and Tethys basins. Landis et al. (2008) suggested that "there is certainty that the entire Chatham Islands biota (prior to the arrival of people) is derived from long-distance dispersal, all within the last two million years." Yet this overlooks the seamounts in the vicinity, and in an alternative model, the affinities of the Chathams clades instead reflect the affinities of a prior, much larger biota in the broader Chatham Islands region. Most of this biota has been destroyed during the submergence of Zealandia.

Chatham Islands Endemics with Sister Groups That Are Diverse and Widespread on the New Zealand Mainland

In many groups, a clade on the Chatham Islands is not *nested in* a clade that is widespread on mainland New Zealand, as predicted in dispersal theory, but is *sister* to such a clade. For example, in *Sonchus* s.lat. (Asteraceae), "*Embergeria*" of the Chathams is sister to "*Kirkianella*" of the North, South, and Three Kings Islands (Mejías and Kim, 2012; both are nested in *Sonchus*).

In the beetle *Brachynopus scutellaris* (Staphylinidae), the Chatham Islands clade is sister to a diverse clade in the North and South Islands (Buckley and Leschen, 2013).

In insects, the cosmopolitan butterfly genus *Vanessa* (red admirals, Nymphalidae) has just two species in New Zealand: *V. gonerilla* on the mainland and *V. ida* on the Chatham Islands. Patrick and Patrick (2012: 128) wrote: "It is surprising that a distinct red admiral has evolved on the Chatham Island group considering the possibility that a strong-flying species like the New Zealand red admiral would continue to re-invade." They suggested that the disjunction between the two species was caused by major geographic changes along the Chatham Rise, rather than an overseas dispersal event.

Curtis (2011) surveyed six species of dune invertebrates that are widespread in the North, South, and Chatham Islands. In all but one of the species, the Chatham Islands contingents were monophyletic. Two of the species, *Hydrellia enderbii* (Diptera) and *Phycosecis limbata* (Coleoptera), each had the phylogeny: (Chatham Islands (North Island and South Island)), again contradicting a model of simple dispersal from the mainland.

In *Cyanoramphus* parakeets (see earlier section "Bounty and Antipodes Islands"), the Chatham Islands species *C. forbesi* is sister to the six remaining New Zealand species, which range from Macquarie Island north through the mainland and Chatham Islands to Lord Howe and the Kermadec Islands (Boon et al., 2001). In passerines, the black tomtit *P. traversi* of the Chathams is sister to *P. macrocephala*, sampled across its range on the mainland (North Island to The Snares) and the Chathams (Miller and Lambert, 2006).

A similar example occurs in ducks (Anatidae). *Anas chathamica* (formerly *Pachyanas*) was endemic to the Chatham Islands until it went extinct between the thirteenth and fifteenth centuries. Molecular studies showed that it was sister to a diverse clade (brown teal and relatives) that is widespread through New Zealand from the Auckland and Campbell Islands to the northern North Island (Mitchell et al., 2014c). (There is also evidence of an extinct teal endemic to Macquarie Island; Holdaway et al., 2001.) The New Zealand clade as a whole belongs to a standard Indian Ocean affinity, with the following phylogeny (Mitchell et al., 2014c; cf. Robertson and Goldstein, 2012):

> Australia (*A. castanea*) + Australia–New Zealand (*A. gracilis*).
> Madagascar, in mangrove forest (*A. bernieri*).
> Chatham Islands (*A. chathamica*; extinct).

NZ mainland, widespread and extinct in the Chatham Islands, in mangrove, etc. (*A. chlorotis,* brown teal).

Auckland Islands (*A. aucklandica*).

Campbell Islands (*A. nesiotis*).

In passerines, the family Locustellidae is represented in New Zealand by *Megalurus* (*Bowdleria*) *rufescens* of the Chatham Islands, and its sister group *M. punctatus* of the New Zealand mainland and The Snares (five allopatric subspecies). *M. rufescens* is sister to the whole of *M. punctatus* (Lanfear and Bromham, 2011), not to any particular clade within it, as would be expected in a dispersal model.

Likewise, in the pigeon genus *Hemiphaga*, the Chatham Islands representative, *H. chathamensis*, is sister to the polymorphic *H. novaeseelandiae* found through mainland New Zealand, Norfolk Island, and the Kermadec Islands (Goldberg et al., 2011).

Morphological studies suggest a similar pattern may occur in the bellbird, *Anthornis melanura* (Meliphagidae). The Chatham Islands subspecies has yellow eyes, whereas the other three subspecies, ranging from the Auckland Islands to the Three Kings Islands, all have red eyes (Bartle and Sagar, 1987). Molecular work is needed to resolve the phylogeny.

In plants, *Aciphylla traversii* and *A. dieffenbachii* (Apiaceae) form a Chatham Islands clade, and their sister group is a diverse complex (10 species) found throughout the South Island and north to Mount Hikurangi (Radford et al., 2001). Likewise, in spiders, *Dolomedes schauinslandi* of the Chatham Islands is sister to the rest of the genus, a group of three species on the mainland (Vink and Dupérré, 2010).

Absences from the Chatham Islands

In addition to its unusual endemics, the Chatham Islands biota is distinguished by conspicuous absences of widespread New Zealand groups. Examples include the two *Poa* clades shown in Figure 6.4. In Foraminifera, Hayward et al. (1997) noted curious absences on the Chathams of some common New Zealand brackish water species. At least 14 intertidal fish species that are common and widespread on the mainland appear to be absent from the Chatham Islands, "despite extensive areas of suitable habitat" (Paulin and Roberts, 1992: Plate 4E). In fact, many shore species that are common and ubiquitous in New Zealand are absent in the Chathams, despite prolonged pelagic development and favorable currents. Knox (1963) referred to this as "puzzling," although the enigma only arises if the Chathams biota is assumed to be a derived subset of the mainland biota.

Molecular Clock Dates of Chatham Islands Groups: Old Taxa Endemic to the Young Chatham Islands

Young marine strata occur at what is currently the highest point of the Chatham Islands, and geologists have concluded that the entire group was submerged from 6 Ma until 3 Ma (Campbell, 2008; Campbell et al., 2008, 2009; Landis et al., 2008). Nevertheless, several plants endemic to the Chatham Islands have been dated in clock studies as older than 3 Ma (Heenan et al., 2010):

Hymenanthera (Melicytus) chathamica (Violaceae): 3.6–4.7 Ma.

Embergeria grandifolia (Asteraceae): 3.5–7.8 Ma.

Sporadanthus traversii (Restionaceae): 5.2–5.9 Ma.

Myosotidium hortensia (Boraginaceae): 7.0–14 Ma.

In animals, an endemic clade of Chatham Island beetles (*Geodorcus capito* + *G. sororum*, Lucanidae) was dated at 6 Ma (Trewick, 2000). The widespread New Zealand beetle *Brachynopus scutellaris* (Staphylinidae) includes an endemic Chatham Islands clade with an estimated age of 10.17 Ma (0.95 credible intervals, 4.67–16.27) (Buckley and Leschen, 2013). The endemic Chatham Islands skink *Oligosoma nigriplantare nigriplantare* was dated at 5.9–7.3 Ma (Liggins et al., 2008a).

The phylogenies of all these plants and animals (except *Brachynopus* and *Oligosoma*) were calibrated in the cited studies using either fossils, which give minimum clade ages, or using the ages of islands in the regions Norfolk (3 Ma), Lord Howe (7 Ma), and Kermadec Islands (2 Ma). These calibrations were used to date endemics on these islands that are related to Chatham Islands taxa. This latter procedure assumes that island-endemic taxa can be no older than their islands, although the ages that Heenan et al. (2010) calculated for the four Chatham Islands species listed earlier contradict this principle. Also, groups endemic to Norfolk and Lord Howe Islands, such as the skink *Cyclodina*, have been shown, using external calibrations, to be much older (minimum age 25 Ma) than the islands themselves (Chapple et al., 2009). This means the island-endemic plants there that have been used for calibration could also have survived in the area on prior islands, long before their current islands existed.

Heenan et al. (2010) were correct to emphasize that the molecular divergence ages they gave "should be considered as minimum ages." This means that many other Chatham Island endemics with younger dates, in addition to those listed earlier, are likely to be older than the current islands. Heenan et al. (2010) and Buckley et al. (2015) suggested that the Chatham Islands clades dated as older than 3 Ma (listed earlier) mean that there could have been emergent land in the Chatham Islands before 3 Ma, formed by some of the eruptions, or on the Chatham Rise. (They did not mention the guyots on the Chatham Rise and the Hikurangi Plateau.)

These ideas contradict the conclusions of some geologists on complete submergence, but they are well supported. The principle also applies to earlier times in the Cenozoic, when it is even harder to deduce paleogeography from geological evidence alone. The Chatham Islands region has a long history of volcanism dating back to the Mesozoic at least, and so it is possible that plants and animals have survived in the region on and around volcanoes, new and old, since before Gondwana breakup.

The new molecular clock dates undermine calibrations that assume Chatham Islands clades have a maximum possible age of 3 Ma because this is the supposed date of emergence of the islands. Goldberg et al. (2014) assumed this in dating a phylogeny for the diverse New Zealand carabid beetle *Mecodema*. Likewise, in plants, the timeline of evolution in New Zealand Malveae has been calibrated by assuming (along with two other calibration points) that the Chatham Islands endemic clade is no older than 1–3 Ma (Wagstaff and Tate, 2011; cf. Heads, 2014).

The basis for this calibration is the proposal by Landis et al. (2008): "there is certainty that the entire Chatham Islands biota (prior to the arrival of people) is derived from long-distance dispersal, all within the last two million years." De Queiroz (2014: 240) discussed the question whether any Mesozoic, Gondwanan lineages have persisted on the Chatham Islands to the present and wrote that "the rocks tell us the answer is 'not likely.'" In fact, a brief review of the regional geology indicates several features that are consistent with the modern biota's derivation from a Mesozoic Gondwanan margin community.

Geology of the Chatham Islands, Chatham Rise, and Hikurangi Plateau

Based on stratigraphic evidence from the Chatham Islands—in particular, the young age and marine origin of the highest elevation rocks—geologists have inferred complete submersion of the present Chatham Islands until 2 or 3 Ma (Campbell, 2008; Campbell et al., 2008, 2009). Nevertheless, several features suggest the existence of other prior islands in the vicinity of the Chathams: the buoyancy of the Chatham Rise continental crust, the repeated phases of volcanism, the many flat-topped seamounts in the area, significant differences in the tectonics of different parts of the islands, molecular clock dates for several groups, and the endemism and phylogenetic relationships of the living taxa (including several that are basal to widespread New Zealand Plateau groups).

Precise paleogeographic data are seldom available for large areas that are now submerged, but in the Chathams region there are already many indications as to where past land existed. The Chatham Islands are an emergent, volcanic part of the Chatham Rise, itself formed from Jurassic continental basement (Chatham Schist), and this is exposed in the northern part of the main island.

Volcanism

Following the deposition of terrestrial sandstones at 100 Ma, since the Late Cretaceous there have been repeated phases of volcanism in the Chatham Islands (Campbell et al., 2008). This is typical for Zealandia, where Cenozoic intraplate volcanism "was nearly continuous and widely dispersed" (Hoernle et al., 2006).

In the Chathams, widespread eruptions at 80–70 Ma formed the main southern part of Chatham Island, now much eroded. Sporadic but widespread eruptions at 63–55 Ma (Red Bluff tuff) "may have formed an island" (Campbell, 2008: 38). Eruptions at 42–34 Ma produced Mount Chudleigh (which "may have formed a small island"), and further eruptions occurred at 6 Ma and 5–3 Ma.

The eruptive phases cited here are only those whose products are exposed on the present islands. Other signs of former eruptions have either been eroded away, are buried, or lie out to sea. Rowden et al. (2005) mapped 812 seamounts on the submerged part of the New Zealand Plateau, and these included about 40 located around the rim of the Chatham Rise, with a strong concentration southeast of the Chatham Islands.

Multibeam bathymetric surveys east of the South Island revealed numerous submarine volcanoes on the Chatham Rise and evidence of submarine erosion on its southern margin (Collins et al., 2011). The largest volcanic cones are ~2000 m in diameter, and some stand as high as 400 m above the surrounding seafloor. The tops of most of the volcanic cones are flat, indicating erosion to sea level. Other submerged features on the Chatham Rise that are probably former islands include the flat-topped Mernoo Bank, 51 m deep at its highest point. Holdaway et al. (2001: 151) wrote that "the former presence of islands between the South Island and the Chathams, where the Veryan and Mernoo banks now stand, show that not all [Chatham Islands] species would necessarily have had to cross the present distance from the mainland." Yet rather than being stepping stones for dispersal from the South Island, the former islands (Veryan Bank is near Mernoo Bank) would have maintained their own biota. These would have colonized new land appearing at the Chatham Islands, and when Veryan and Mernoo Banks were submerged, any endemic clades would be left as endemics on the Chathams. Given the active volcanism and the normal means of dispersal of plants and animals, it is likely that the biotas of many other former islands (now flat-topped seamounts) around the Chathams underwent the same process.

Tectonism in and around the Chatham Islands

Apart from the volcanism and subsidence around the Chatham Islands, several cryptic faults have been active within the Chatham Islands. These mean that stratigraphic evidence for submersion from one area, such as the young marine sediments at the highest point, does not necessarily apply to the region as a whole for any one period of time. For example, one major fault must be responsible for the uplift of the basement schist in northern Chatham Island, although the fault is not exposed (Campbell, 2008). Displacement on other large-scale faults in the region would have also led to differential uplift and subsidence. Holt (2008) recorded considerable variation in uplift rates across Chatham Island and over very short distances (including 10-fold differences in rates over just 400 m), and these variations "cannot yet be fully explained." Again, these could reflect activity on the basement fault between the Chatham Schist to the north and younger rocks to the south. Uplift rates are higher in the north and south and less in the central parts.

Holt (2008) described the "poor understanding of the characteristics and history of tectonics and uplift of the Chatham Islands area… Chatham Island tectonic history is not resolved… the tectonic history may be quite complicated. The northern, central, and southern regions of Chatham Island behave differently in terms of deformation" (p. 139). Again the tectonism means that the evidence for submergence of part of the Chatham Islands does not constitute evidence for submergence of the whole archipelago.

Chatham Islands and the Hikurangi Plateau

Through most of the Mesozoic, a subduction zone was active along the northern margin of the Chatham Rise. The Hikurangi Plateau formed at ~123 Ma in the central Pacific as a large igneous province (much larger than the New Zealand mainland), and it has encroached on the Zealand part of the Gondwana margin from there. Following its arrival and collision with the Chatham Rise at some time in the late Cretaceous it began to subduct, but after subducting 50–100 km, it choked the Chatham subduction zone (Figure 3.1). Plate movement was transferred to the Emerald fracture zone, southwest of New Zealand, and later, with renewed convergence, the Alpine fault (Figure 3.15; Schellart et al., 2006).

Although most of the Hikurangi Plateau was erupted under shallow water, there are numerous, large seamounts on the plateau and these can be up to 50 km across. Most are flat-topped guyots that were previously emergent islands (Figure 3.10). The Hikurangi Plateau probably formed as part of the greater Ontong Java Plateau, which later rifted apart. The Ontong Java Plateau itself (now adjacent to the Solomon Islands; Figure 3.15: Present) includes material from subaerial eruptions and fossil wood in intercalated sedimentary strata. The Manihiki Plateau of the Cook islands was also derived from the original single plateau, and at least its early stages are known to have been subaerial (Chapter 3).

Many taxa of the Chatham Islands, the North Island East Coast region, and Northland have central Pacific affinities, and this is consistent with their having been "scraped off" the descending Hikurangi Plateau at the subduction zone. The clades have piled up around the Chatham Rise and the North Island as the biogeographic equivalent of an accretionary wedge, as suggested for marine mollusks of New Caledonia (Marshall, 2001). The *Festuca coxii* group (Poaceae) ranges from the North Canterbury coast (Waipara River) northeast to East Cape and is also on the Chatham Islands (Figure 3.10) (Lloyd et al., 2007). This is the sort of distribution that would be expected if the Hikurangi Plateau and its islands contributed groups to New Zealand as they subducted. The process would also account for the Chatham Islands groups with sister groups in the central Pacific (affinities with Rapa Island and with Fiji were cited earlier).

Summary on Vicariance and the Chatham Islands

Based on stratigraphic evidence, in particular the young age of the highest elevation rocks, geologists have inferred complete submersion of the present Chatham Islands 4 Ma, with emergence at 2 or 3 Ma. Many studies, for example, de Queiroz (2014), have accepted this as the oldest age of any islands in the region. Nevertheless, de Queiroz did not mention the clock ages of Chatham Islands endemics that predate the current islands; the many Chatham Islands groups that are sister to, not nested in, diverse mainland complexes; the continued phases of volcanism in the Chatham Islands; the many guyots surrounding the islands; the evidence for major local tectonism that would have raised or lowered different parts of the islands; the Hikurangi Plateau (with its many guyots) that crashed into the Chatham Rise after arriving from the central Pacific; or the affinities between many Chatham Islands groups and relatives in the central Pacific.

In this way, de Queiroz (2014) and other dispersal theorists have overlooked all the relevant evidence except for fossil-calibrated clock dates, and these are treated, illogically, as *maximum* clade ages. It was only by the use of this reasoning that de Queiroz was able to conclude: "there is no evidence whatsoever of Gondwanan relict species on the Chathams" (p. 242). In fact, there is a wide range of evidence suggesting that the direct ancestors of the current species lived in the area before the modern species or the islands that currently host them existed, and before Gondwana breakup.

Affinities of the Chatham Islands Biota with Northern and Southern Parts of Mainland New Zealand

One distinctive feature of the Chatham Islands biota is its strange combination of affinities with the far north and the far south of mainland New Zealand. Southern connections in beetles (*Pseudhelops*; Figure 6.5), mollymawks (*Thalassarche*), and albatrosses (*Diomedea*) have already been cited. Likewise, in shags, *Leucarbo chalconotus* of Stewart Island to Otago Peninsula is sister to *L. onslowi* of the Chatham Islands (Kennedy and Spencer, 2014). In plants, *Urtica australis* (Urticaceae) is known from Auckland, Campbell, and Antipodes Islands, southwestern Fiordland (Chalky Island), islands in Foveaux Strait, and the Chatham Islands. *Dracophyllum paludosum* (Ericaceae) of the Chatham Islands is related to *D. scoparium* of Campbell Island (the two were synonymized by Venter, 2009), and *Brachyglottis huntii* (Asteraceae) of the Chatham Islands is related to *B. stewartiae* of The Snares, Stewart Island, and Solander Island (Allan, 1961; Heenan et al., 2010). The harvestman genus *Neonuncia* is known from the Chatham Islands, Auckland Islands, Campbell Island, and Fiordland (Forster, 1954).

Chatham Islands taxa with affinities instead in Northland include several mosses:

Archidium elatum (Archidiaceae): Northland (Ahipara, Moturoa Island) and the Chatham Islands (Fife, 2014b). This New Zealand endemic is the only member of Archidiaceae in the country.

Calymperes tenerum (Calymperaceae): Pantropical; in New Zealand at the Kermadec Islands, Northland (Te Paki), and the Chatham Islands (Fife, 2014c).

Fissidens integerrimus (Fissidentaceae): Victoria, Tasmania, Northland, Auckland (Waitakere Ranges), and the Chatham Islands (Beever, 2014).

Fissidens oblongifolius (Fissidentaceae): Tasmania, mainland Australia, New Caledonia, Vanuatu, Northland (including offshore islands), and the Chatham Islands (Beever, 2014).

Macromitrium brevicaule (Orthotrichaceae): Australia, Kermadec Islands, Northland to the Coromandel Peninsula, and the Chatham Islands (NZPCN, 2016).

In ferns, *Asplenium pauperequitum* is restricted to the Poor Knights and Hen and Chickens in Northland and the Chatham Islands (Cameron et al., 2006) and is allopatric with its western sister group (Chapter 5). In angiosperms, *Sporadanthus* (Restionaceae) is in Australia, northern North Island, and the Chatham Islands (de Lange et al., 1999; Heenan et al., 2010). *Myrsine oliveri* (= *Rapanea dentata*) (Myrsinaceae) of the Three Kings Islands was regarded by Oliver (1948) as closest to *M. chathamica* of the Chatham Islands.

In animals, the pseudoscorpion *Maorichthonius* (monotypic) is recorded from the northern North Island (Leigh, Hauraki Gulf) and the Chatham Islands (Beier, 1976). The carabid beetle *Bembidion albescens* is in the Chatham Islands, Coromandel Peninsula, and Northland (Larochelle and Larivière, 2001).

The mollymawk *Thalassarche bulleri* comprises two subspecies: *T. b. bulleri*, which breeds on The Snares and Solander Islands, and *T. b. platei*, which breeds on the Chathams and the Three Kings Islands (Robertson et al., 2007). A similar outer arc connecting the Chathams with the far north and the subantarctic region is seen in the fly *Fannia anthracinalis* (Fanniidae), recorded from the Three Kings, Chatham, Auckland, and Campbell Islands. Its closest relatives are not the mainland species, but *F. mangarensis* of the Chathams, *F. laqueorum* of The Snares, and members of the *F. anthracina* group in southern South America (Domínguez and Pont, 2014). This recalls the possible affinities of *Myosotidium* with *Cynoglossum paniculatum* of Chile (see earlier section "Geographic Affinities of Chatham Islands Endemics").

In the bellbird, *Anthornis melanura* (Meliphagidae), the plumage on the head is blue in the Poor Knights and Chathams subspecies, but violet in all other populations (Bartle and Sagar, 1987). (The eye is yellow on the Chatham Islands, red elsewhere.)

The southern affinity of Chathams clades with groups on the Campbell Plateau has been discussed already. Before the Neogene deformation of central New Zealand, the northern groups would have occurred along the coastline that followed a straight line from Northland to Cook Strait and the Chatham Islands. When the Cook Strait region was pulled into central New Zealand the straight line was bent into a Z-shape with the first bend at East Cape and the second at Marlborough. Distributions along the arc were disrupted, leading to disjunctions among islands along the Chatham Rise and off Northland.

An Example of Character Geography: The Conspicuous Development of Tree Architecture and the Paucity of Divaricate Shrubs in the Chatham Islands Flora

In addition to its many endemics, the Chatham Islands flora is distinctive through the architecture of the plants. These include the tallest tree in the diverse, South Pacific *Hebe* complex, *H. barkeri* (Plantaginaceae); the tallest tree in the South Pacific genus *Olearia*, *O. traversii* (Asteraceae) (Dawson and Lucas, 2011); the largest tree in the Pacific genus *Coprosma*, *C. chathamica* (Rubiaceae) (up to 20 m tall and 1 m diameter; Dawson and Lucas, 2011); and the only tree in the South Pacific genus *Corokia*, *C. macrocarpa* (Argophyllaceae) (Dawson and Lucas, 2011). In a similar way, the Chatham Islands endemic *Leptinella featherstonii* (Asteraceae) differs from all other species of *Leptinella* and the allied *Cotula*, which are prostrate herbs, through its suberect habit and subwoody stems (Lloyd, 1972).

These examples lie east of the New Zealand mainland, toward the central Pacific, and recall the many other groups that are trees on Pacific islands such as Hawaii and the Juan Fernández Islands, but are shrubs or herbs elsewhere (Heads, 2012a: 403). (For the development of central Pacific trees in the Myrtaceae tribe Metrosidereae, see Chapter 4.) The concentration of "tree genes" in the central Pacific region may be much older than the particular clades of trees in which they are now maintained.

The large trees that distinguish the Chatham Islands flora are matched by another feature of the area, the scarcity of "divaricate shrubs." Plants with divaricate architecture are characterized by the differentiation of long shoots that abort and short shoots with suppressed internodes. They are diverse and abundant on the New Zealand mainland, but on the Chatham Islands there are only two species: *Coprosma propinqua* and *C. acerosa* (Rubiaceae). This differentiation is even seen within a single species, *Plagianthus regius* (Malvaceae), as mainland populations have a divaricate "juvenile" phase, while Chatham Island plants do not (Wagstaff and Tate, 2011).

Aspects of the Chatham Islands Fauna

The invertebrate fauna of the Chatham Islands is diverse and distinctive. For example, the land snails show many differences from their mainland relatives, and Mahlfeld (2008) suggested that islands in the region have existed for much longer than the 2–3 Myr proposed by some geologists.

The Chatham Islands insect fauna includes ~800 species (making up 8% of the described New Zealand fauna), and about 20%–25% of these are endemic (Dugdale and Emberson, 2008). Complementing this, many insect groups that are abundant on the mainland are absent from the Chatham Island, despite the presence of their host plants there. Dugdale and Emberson (2008: 117) wrote that "there are conflicting interpretations of this enigma and no satisfactory explanation," and the origin of the Chathams insect fauna itself is "enigmatic" (Dugdale and Emberson, 1996).

Nevertheless, the geographic affinities of the insects appear to be the same as those in the plants. This is evident in Emberson's (1998) analysis of the Chatham Islands beetle fauna, which is summarized here. The beetles include taxa with central Pacific affinities, such as *Rhantus schauinslandi* (Dytiscidae); this is more closely related to species such as *R. vitiensis* of Fiji than to New Zealand mainland species (Ordish, 1989). Southern affinities are exemplified by *Pseudhelops* (Tenebrionidae) (Figure 6.5) and *Antarcticodomus* (Salpingidae), both present in the Chathams and subantarctic islands. An unnamed species of *Microcryptorhynchus* (Curculionidae) from the Chatham Islands is near *M. latitarsis* of the Auckland Islands, Stewart Island, and Fiordland. *Exeiratus* (Curculionidae) comprises one species on the Chathams and others from the southern South Island (Dunedin, Southland, Fiordland), Stewart Island, The Snares, Auckland Islands, and Tasmania. *Patellitergum* (Curculionidae), one of four beetle genera endemic to the Chathams, has no obvious relatives.

The birds of the Chatham Islands are also diverse. Until historical times the fauna included two endemic forms of rails (*Diaphorapteryx* and *Cabalus,* both now placed in *Gallirallus*) as well as the duck "*Pachyanas*" (now placed in *Anas*) (Holdaway et al., 2001). All three are now extinct. Among the interesting fossils on the islands are bones from a giant bird, possibly >1.5 m tall, that are dated from the Cretaceous/Paleogene boundary (64–65 Ma; Stilwell and Consoli, 2012). The same strata in which the bird bones were found also include fossils of theropod dinosaurs (Stilwell et al., 2006). The authors wrote that the creatures inhabited a coastal, temperate environment in a tectonically dynamic, volcanic landscape with eroding hills (horsts) adjacent to flood plains and deltas. The forest was dominated by conifers and lycopods.

Larger Size of Chatham Islands Birds

The Chatham Islands flora includes a conspicuous development of tree architecture, and megaherbs such as *Myosotidium* appear to be more common on the subantarctic and Chatham Islands than on the New Zealand mainland. A similar phenomenon is seen in birds. The following endemic Chatham Islands clades are larger than their mainland relatives ([†] indicates the bird is extinct): *Anas chathamica*[†], *Haematopus chathamensis, Gallirallus dieffenbachii*[†], *Fulica chathamensis*[†], *Hemiphaga chathamensis*[†], *Cyanoramphus novaezelandiae chathamensis, Gerygone albofrontata, Anthornis melanocephala*[†],

Prosthemadera novaeseelandiae, *Petroica macrocephala*, and *Corvus moriorum*[†] (Oliver, 1955; Worthy and Holdaway, 2002; Scofield and Stephenson, 2013; Mitchell et al., 2014c). The extinct Chatham Islands parrot (*Nestor chathamensis*) was distinguished from its sister taxon, the New Zealand kaka (*N. meridionalis*), by a longer beak, broader pelvis, and larger femora (Wood et al., 2014b).

Cyanoramphus forbesi of the Chatham Islands is larger than the mainland *C. auriceps*, which has often been regarded as its closest relative (e.g., Collar, 1997). However, in the only molecular phylogeny available (Boon et al., 2001), *C. forbesi* is sister to a group that includes larger species (*C. unicolor* and *C. novaezelandiae*) as well as smaller ones (*C. auriceps* and *C. malherbi*).

In *Megalurus* (*Bowdleria*), there is no overall size difference between mainland and Chatham Islands forms, but in *M. rufescens* of the Chathams, "the pelvis and hindlimb elements are larger and very much more robust… [and the] wing elements and pectoral girdle reduced," compared with the mainland *M. punctatus* (Olson, 1990b).

In *Rhipidura*, there is no apparent difference between mainland and Chatham Islands forms, while in *Anthus*, the form from the Chathams and subantarctic islands has longer bills but shorter wings (Foggo et al., 1997). A positive exception to the trend is seen in *Coenocorypha*, as the extant Chatham Islands form, *C. pusilla* is the *smallest* of the New Zealand forms. Nevertheless, the extinct *C. chathamica* of the Chatham Islands appears to have been larger than the other New Zealand species (Oliver, 1955). Another exception is a Chatham Islands penguin; this is known from Holocene fossils and nested phylogenetically in *Megadyptes antipodes* of the mainland and subantarctic islands. The fossil form is 40% smaller than extant individuals (Wood et al., 2014a).

It is interesting to note that the mainland representatives of all these groups are widespread from the north to the south. If they were present only in the north or the south, it would be easier to invoke an ecological explanation for the increased size of the Chathams forms. As it is, there is no simple environmental contrast between the diverse habitats of the mainland birds and those in the Chatham Islands. For the large *Anas chathamica*, Mitchell et al. (2014c) suggested that "part of this size increase may be related to territoriality," but they did not explain why it would occur on the Chatham Islands in particular.

Williams et al. (2014) regarded the larger size of the Chatham Islands birds as conforming to an "island rule" for vertebrates. This is "a graded trend from gigantism in the smaller species to dwarfism in the larger species" (Lomolino, 2005). Yet, while Lomolino found that this was a global pattern, the survey by Meiri et al. (2008) found that it was not, and the island rule does seem to be an oversimplification. For example, the extinct duck *Anas chathamica* of the Chatham Islands is larger than its mainland relative, *A. chlorotis*, but on the other hand, *Anas aucklandica* and *A. nesiotis* of Auckland and Campbell Islands are much *smaller* than the mainland species. Meiri et al. (2008) concluded: "Instead of a rule, size evolution on islands is likely to be governed by the biotic and abiotic characteristics of different islands, the biology of the species in question and contingency." In addition, the biogeographic and evolutionary history of the particular islands should be considered.

Differentiation within the Chatham Islands Region

Biogeographic differentiation is evident within the Chatham Islands region, but it has not been the subject of any detailed study. Some birds, such as shore plover, red-crowned parakeet, tomtit, black "robin," and tui, were formerly widespread on all the islands in the archipelago, but were eliminated on the main island by introduced predators and are now restricted to cat-free Mangere Island, south of the main island. Nevertheless, other birds on the southern islands (Pitt, Mangere, Little Mangere, Southeast Island) have no early records on Chatham Island. Likewise, the plants *Lepidium oblitum* and *L. panniforme* (Brassicacerae) are endemic to Mangere Island, where they replace the two other Chatham Islands species (*L. rekohuense* and *L. oligodontum*) (de Lange et al., 2013).

The petrel *Pterodroma axillaris* breeds only on Southeast Island off Pitt Island, and there are fossils on Mangere. (The species has recently been introduced to Pitt Island and Chatham Island; IUCN, 2016.) In contrast, the allopatric *Pterodroma magentae* breeds only on Chatham Island. In New Zealand, *Pterodroma nigripennis* breeds only in the Chathams, on Southeast and Mangere Islands, and possibly Star Keys (Imber, 1994). The parakeet *Cyanoramphus forbesi* has only been recorded breeding on Pitt, Mangere, and Little Mangere (Oliver, 1955; Robertson et al., 2007). The species is sister to a clade that is widespread from Macquarie Island to the Kermadec Islands (Boon et al., 2001).

"Outer arc" distributions in the eastern Chatham Islands occur in plants such as *Leptinella featherstonii* (Asteraceae), with an overall eastern distribution in the archipelago and distinctive populations on the Forty Fours and on The Pyramid, off Pitt Island (Lloyd, 1982). *Lepidium rekohuense* (Brassicaceae; de Lange et al., 2013) is restricted to Rabbit Island off Pitt Island, the Forty Fours, and the north coast of the main island. Although land birds such as the pipit *Anthus novaeseelandiae* (Motacillidae) breed on the main islands and also the Forty Fours, the gulls *Larus dominicanus* and *Chroicocephalus bulleri* visit the Forty Fours but only breed on the main islands. The five mollymawks (*Thalassarche* species) that are present in the Chatham Islands all have breeding records restricted to the outer arc. Royal albatrosses (*Diomedea sanfordi*) breed in the Chathams only on islets north of the main island and on the Forty Fours, while the Antipodean albatross *D. antipodensis* breeds only on an islet southwest of the main island (Robertson et al., 2007).

Is Mainland New Zealand a Center of Origin for the Offshore Island Biota?

In the earlier dispersal-based accounts and also in modern phylogeography, most groups on the islands off mainland New Zealand are attributed to dispersal from a center of origin in New Zealand. This approach is evident in studies of pigeons, for example.

Hemiphaga (Columbidae)

The pigeon genus *Hemiphaga* comprises two species in traditional taxonomy, and this is supported in molecular work (Goldberg et al., 2011):

> *H. novaeseelandiae*: One subspecies widespread on the New Zealand mainland (North, South, and Stewart Islands) and the other on Norfolk Island. There are also early sight records and fossil material from the Kermadec Islands (Holdaway et al., 2001).
>
> *H. chathamensis*: Chatham Islands.

Fleming (1976) presented a theory of biogeographic radiation for *Hemiphaga* and other bird groups. In this model, each island clade represented a separate dispersal event from a center of origin on the New Zealand mainland; Fleming's figure, showing *Hemiphaga*, *Prosthemadera* (a honeyeater), and *Nestor* (a parrot), is reproduced here (Figure 6.6).

A detailed molecular study of *Hemiphaga* sampled populations from Norfolk Island, the Chatham Islands, and nine mainland localities from Northland to Stewart Island (Goldberg et al., 2011). The authors concluded that "tentative rooting of the *Hemiphaga* clade with cyt *b* data indicates exchange between mainland New Zealand and the Chatham Islands prior to colonization of Norfolk Island." This is not what the evidence suggests though, as the cytochrome *b* phylogeny is: Chathams (Norfolk + New Zealand), in line with the traditional taxonomy. This phylogeny indicates one of the following:

- Vicariance between (Chathams) and (NZ + Norfolk).
- Dispersal from the Chathams to NZ + Norfolk.
- Dispersal from NZ + Norfolk to the Chathams.

Dispersal from New Zealand to the Chathams and then dispersal from New Zealand to Norfolk, as Goldberg et al. (2011) suggested, would instead give a phylogeny NZ[1] (NZ[2] + Chathams (NZ[3] + Norfolk)) or a more complex variant of this, where NZ[1], NZ[2], and NZ[3] are different clades from mainland New Zealand. But the wide sampling that Goldberg et al. (2011) carried out on the mainland is valuable

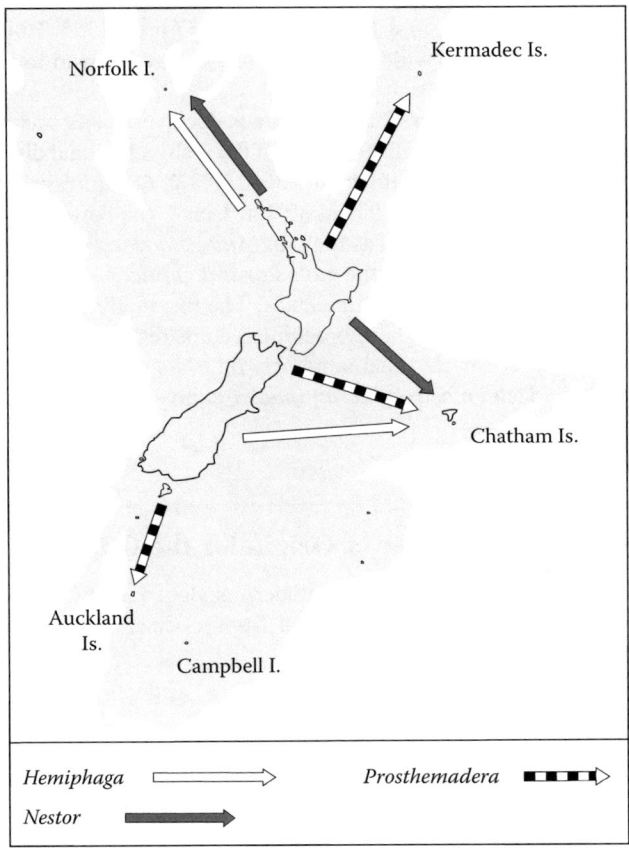

FIGURE 6.6 A dispersal model of bird distribution on the New Zealand Plateau (gray area = 3000 m isobath). (Based on Fleming, C.A., *Tuatara*, 22, 30, 1976.)

precisely because it helps to exclude this possibility—the study showed that the mainland form is a monophyletic clade, not a paraphyletic grade. The finding of mainland monophyly is excellent evidence *against* the dispersal model, which cannot be saved without invoking additional, ad hoc theories of nonrandom extinction or inadequate sampling.

One interpretation of the *Hemiphaga* distribution is Fleming's (1976) model, as supported by Goldberg (2011). Another model, based on the molecular phylogeny, would involve vicariance between mainland New Zealand plus Norfolk Island on one hand versus the Chatham Islands on the other. The widespread mainland New Zealand subspecies is allopatric with its subspecies in Norfolk Island; this species as a whole is allopatric with its sister species in the Chathams, and this entire group, *Hemiphaga*, is allopatric with its sister group, located to its west and north. The phylogeny (from Gibb and Penny, 2010) is as follows:

> *Hemiphaga*: New Zealand.
>> *Lopholaimus*: Eastern Australia coast from Cape York Peninsula to Tasmania.
>> *Gymnophaps*: Maluku (Moluccas), New Guinea, Bismarck Archipelago, and Solomon Islands.

The three genera form a typical Tasman Sea clade, and the complete absence from New Caledonia is a normal variant. As with the species of *Hemiphaga*, the genera display precise allopatry, indicating that no range expansion has taken place after their origin. The sister group of the three genera is *Ptilinopus* s.lat., an Indo-Pacific genus with interesting diversity in New Caledonia (Heads, 2012a). *Ptilinopus* overlaps with the *Hemiphaga–Lopholaimus–Gymnophaps* clade in eastern Australia and parts of Melanesia, indicating range expansion in one or both groups, probably following an original break around New Caledonia.

To summarize, *Hemiphaga* shows allopatry at subspecies level, species level, and generic level. Beyond this, there is some overlap between *Hemiphaga–Lopholaimus–Gymnophaps* and *Ptilinopus* (but not in New Zealand), indicating the breakdown of allopatry between these older clades. The simplest explanation for the overall pattern is vicariance of a widespread ancestor already present in Australia, Zealandia, and Melanesia, with some secondary range expansion between *Hemiphaga–Lopholaimus–Gymnophaps* and *Ptilinopus*. The only reason for interpreting the New Zealand offshore island groups by radiation from a mainland center of origin would be if the offshore island groups were no older than the present islands they occupy, but there is no reason to assume this.

New Zealand Cicadas (Hemiptera, Auchenorrhyncha, Cicadidae)

Most members of suborder Auchenorrhyncha (cicadas and spittlebugs) that have been recovered from Dominican amber (Oligocene–Miocene) and even Baltic amber (Eocene) are "virtually indistinguishable" from modern forms (Dietrich, 2003). In Cicadidae (cicadas), the New Zealand members belong to the tribe Cicadettini, which is present in all faunal regions except the Neotropics. Arensburger et al. (2004a: 564) suggested that the group's presence in South America "would be expected from an ancient Gondwana distribution," but there is no reason to assume that the Gondwana biota was homogeneous before breakup. Buckley et al. (2015) wrote: "Undoubtedly, biotic regionalisation existed across Gondwana, and this may have influenced the absence of some lineages from modern New Zealand." In the same way, before breakup, some Gondwanan lineages are likely to have been present in New Zealand but absent from the region that became South America.

The New Zealand cicadas are present throughout the mainland and also occur on Norfolk, Kermadec, and Chatham Islands. They belong to two separate clades in tribe Cicadettini (Marshall et al., 2015):

1. New Zealand mainland (*Notopsalta* and *Amphipsalta*). Sister to a diverse, widespread clade of Australia, Asia, and North America.
2. New Zealand mainland, Norfolk, Kermadec, and Chatham Islands (*Maoricicada*, *Kikihia*, and *Rhodopsalta*) and New Caledonia (*Pauropsalta*, *Myersalna*, and *Rouxina*). Sister to a widespread group of Africa, Europe, Asia, and Australia.

The simplest explanation for both is that two widespread intercontinental groups each had their basal break in or around the region that became Zealandia. Given the phylogeny, it is difficult to interpret either of the New Zealand groups as the result of dispersal to New Zealand.

Nevertheless, a long-distance dispersal model for the cicadas, with two invasions of New Zealand, has been proposed (Arensburger et al., 2004a). This was based solely on the young age of the clades that was estimated (11.6 Ma), but the calibration was not specified (the authors cited two "independent geological calibrations" given as a personal communication from C. Simon). Arensburger et al. (2004a: 558) described the inferred invasion of the New Zealand cicadas as "a remarkable feat for an insect not generally known for its dispersal abilities."

One of the New Zealand cicadas, *Kikihia*, occurs throughout the mainland, as well as Norfolk, Kermadec, and Chatham Islands, and a phylogeographic analysis "demonstrates that they have shown incredible dispersal abilities" by colonizing these outer islands (Arensburger et al., 2004a: 565). This assumption is unjustified though, as the islands could have been colonized from former nearby islands by normal, short-distance dispersal. In contrast with a dispersal model, a vicariance history for these cicadas does not require incredible processes, just the continuous arc volcanism observed along the Tonga–Kermadec subduction zone, and the intraplate volcanism that has occurred around the Chatham Rise and the Norfolk Ridge.

In another paper on the cicadas, Arensburger et al. (2004b: 1769) reiterated that "the invasion of Norfolk, the Kermadecs and Chatham Islands had to have occurred through long-distance dispersal." As in the study cited earlier (Arensburger et al., 2004a), this inference was based entirely on the young ages estimated for the clades. Arensburger et al. (2004b) calibrated the phylogeny using the endemic

Norfolk Island species, the age of which "can be reasonably estimated from the geological age of the island" (3–2.3 Ma) (p. 1773). The authors argued that the island's "volcanic origin and the absence of nearby islands requires that its biota be the descendants of long-distance dispersing ancestors" (p. 1775).

As with any new substrate, fresh lava is colonized from somewhere else, but often from nearby localities rather than from further away. Dispersal onto the new Norfolk Island was probably from prior islands in the vicinity, and there is excellent evidence for a large, former island nearby (Figure 5.2; Meffre et al., 2007). Arensburger et al. (2004b) accepted that the present Chatham Islands were completely submerged in the Oligocene and so must have been colonized from mainland. Yet given the many seamounts around the Chathams, the repeated phases of volcanism there since the Cretaceous, the tectonism within the islands, the clock dates for endemic taxa that are older than the islands, and the biogeographic connections with localities such as southeastern Polynesia, it is probable that there was always some land present in the region.

In a recent study of the New Zealand cicadas, Marshall et al. (2015) used COI rates estimated in other studies, and this generated "unexpectedly old" age estimates for the group. The authors were more cautious than in their previous papers about their dating estimates, and they echoed the call by Sauquet et al. (2012) for "increased background research… at all stages of the calibration process." In addition to problems with calibration, "… even a strong rate prior (i.e., a well-tested clock with a narrow confidence interval) is no guarantee of sensible results because so much uncertainty can exist in the estimate of the molecular substitution process. Decisions regarding the taxon sample, substitution model, partitioning scheme, and Bayesian priors can substantially influence the outcome."

Outer Arcs on the New Zealand Plateau: Albatrosses and Insects

Biogeographic outer arcs on the New Zealand Plateau were cited earlier in groups such as the mollymawk *Thalassarche bulleri*, breeding on Stewart, Chatham, and Three Kings Islands, and the fly *Fannia anthracinalis* (Fanniidae), present on the Auckland and Campbell, Chatham, and Three Kings Islands (see earlier section "Affinities of the Chatham Islands Biota with Northern and Southern Parts of Mainland New Zealand"). In a similar way, the weevil *Microcryptorhynchus suillus* is distributed in a great arc around the mainland, on Auckland and Campbell Islands, Codfish Island off Stewart Island, the Chatham Islands, and Northland (Emberson, 1998).

Similar outer arcs may also occur in megascolecid earthworms, although more sampling is needed; Buckley et al. (2011) found sister affinities between *Rhododrilus cockaynei* of the Auckland Islands and *Rhododrilus* "nov. sp. 2" from the Three Kings Islands. The authors suggested that "the close sister group relationships between earthworm species on widely spaced islands, which are often oceanic and positioned on different geological terranes… are consistent with dispersal playing a role in colonization." Yet the distributions are also consistent with those of other groups, such as the mollymawks, flies, and beetles cited earlier. This general pattern combines two patterns already discussed, the first running between the subantarctics and the Chatham Rise, the second connecting the Chatham Rise and Hikurangi Plateau with East Cape and Northland. Although the outer island groups have relatives or even sister groups on the mainland, there is no need to explain their distribution by dispersal from there.

A "Star Pattern" in Zealandia: A Group of Filmy Ferns

The fern *Hymenophyllum* (Hymenophyllaceae) includes a clade that is endemic to the New Zealand Plateau and widespread there. The group has the following phylogeny (Papadopulos et al., 2011: Appendix) (N = North Island, S = South Island, St = Stewart Island):

New Caledonia (*Hymenophyllum paniense*).
   N, S, St, Chatham Islands (*H. scabrum*).
      N, S, St, Auckland and Campbell Islands (*H. villosum*).
      N, S, St, Three Kings Islands (*H. sanguinolentum*).

The first break is a simple, allopatric split between New Caledonia and New Zealand. In contrast, the last three species show extensive overlap on mainland New Zealand, but also display perfect allopatry among the offshore islands. This is a more localized version of the "star pattern" of distribution (Chapter 4), and the allopatry can be interpreted as a trace of the original vicariance. In this model, an eastern entity (on Chatham Islands and eastern New Zealand) has diverged from a western entity, which itself then diverged into a group in northern New Zealand–Three Kings Islands, and a group in southern New Zealand–subantarctic islands. A subsequent phase of overlap has occurred throughout what became mainland New Zealand, and only there. This suggests that the overlap took place following the isolation of the offshore island regions from the mainland by extension-induced subsidence and flooding.

7

Biogeography of Stewart Island and the Southern South Island: Mesozoic–Paleogene Geology

Having dealt with the northern and southern offshore islands in the last two chapters, we are now in a position to consider distribution in mainland New Zealand. This region includes many endemics, such as the acanthisittid wrens, that have global sisters. Other high-level groups that are endemic to the mainland and widespread there include the beetle family Cyclaxyridae (North, South, and Stewart Islands). Its sister group is either Tasmosalpingidae in Tasmania and Victoria, or Lamingtoniidae at the McPherson–Macleay Overlap (New South Wales–Queensland border) (Gimmel et al., 2010).

This chapter and the next examine some of the main aspects of biogeography in a transect across Stewart Island and South Island, from the southwest to the northeast. The focus is on patterns that coincide with tectonic developments of the later Mesozoic and the Paleogene (Paleocene to Oligocene). Some of the most obvious tectonic features in New Zealand, such as the Alpine fault and the mountains of the Kaikoura orogeny, only developed in the Neogene (Miocene to the present) and these, along with their associated biogeography, are discussed in Chapter 9.

New Zealand Basement Terranes and Their Origin

The discussion of New Zealand geology in Chapter 3 focused on the main phases of deformation in New Zealand: (1) convergence (with terrane accretion) until 100 Ma, (2) extension and rifting from 100 to 25 Ma, and (3) renewed convergence from 25 Ma (latest Oligocene). One important topic that was not discussed concerns the origin and location of the basement terranes, before they were translated and accreted to Gondwana in the Rangitata orogeny.

The basement terranes of New Zealand (Figure 3.2) comprise the allochthonous Western Province (older rocks), the allochthonous Eastern Province (younger rocks), and an autochthonous magmatic belt, the Median Batholith, which developed along the margin of the first two provinces. In the Western Province, the rocks and fossils show affinities with Antarctica, Tasmania, and the rest of the Lachlan orogeny of southeastern Australia (Campbell and Hutching, 2007; Adams et al., 2009a).

New Zealand Eastern Province rocks have possible source areas further north in eastern Australia, in the New England orogen (Adams et al., 2007), although Wandres and Bradshaw (2005) suggested an Antarctica source for the Rakaia terrane (the older part of the Torlesse, Figure 3.2). (Combined detrital zircon age data for all the Torlesse group terranes "reveal an essential unity"; Adams et al., 2009b.) In the model of Adams et al. (2007), the North Island Waipapa terrane formed off northeastern New South Wales, while other Eastern Province terranes formed off the northeastern Queensland/New Guinea margin. The Eastern Province terranes have then been translated southward, moving parallel to the margin, to their present position, where they have been sutured to the Western Province rocks. In this model, most of the terranes that make up modern New Zealand (but not the northeastern allochthons and the Hikurangi Plateau) were initially dispersed in a belt 6000 km long (or even longer in the model of Wandres and Bradshaw, 2005). This belt was located off Gondwana and stretched from New Guinea to south of Tasmania. Following margin-parallel translation, the terranes were amalgamated in the Early Cretaceous to form proto-Zealandia. Tethyan–Eastern Province connections in both Mesozoic fossils and extant taxa are likely to have developed during this complex process. This model also has possible implications for the biogeographic arc: New Zealand–New Caledonia–New Guinea, although in many groups this arc may have developed later, during slab rollback (Figure 3.15).

Before the fusion of Western and Eastern Province terranes, other terranes then present in what became the New Zealand region have drifted to South America (Adams, 2010). In this model, terranes that formed as depocenters off Queensland were translated in a dextral sense along the Gondwana coast to form New Zealand, while terranes that formed as depocenters within what is now New Zealand have been translated along the coast in a similar way to form large parts of Patagonia.

In the Eastern Province, the Torlesse and Caples terranes both exhibit dual affinities. Both include a wedge of clastic material with fossil faunas having "Gondwana margin" ("Austrozean") affinities. Both terranes also include oceanic sequences (complete with guyots) that have been introduced tectonically by seafloor spreading (Adams et al., 1998), and these sequences have faunas with "Tethyan" affinities. These affinities are best seen in the North Island Waipapa terrane (equivalent to the Caples terrane) (Adams et al., 1998) and in the Te Akatarawa miniterrane (Cawood et al., 2002) of the Canterbury Torlesse. For example, fusulinid foraminifera in the Waipapa terrane show standard Tethyan distribution ranges, with records from New Zealand, Southeast Asia, and the Mediterranean. Other taxa have similar distributions with additional Pacific localities in Japan and northwestern North America (e.g., Oregon and British Columbia; Leven and Grant-Mackie, 1997). These suggest a Tethyan/Pacific derivation of the Waipapa terrane, possibly from the Phoenix plate (Spörli, 2006). This plate extended across a large part of the Pacific region before it was destroyed by subduction (Figure 3.8). Likewise, rare Late Jurassic limestones in the Pahau terrane (younger Torlesse) have rich fossil faunas that relate more to faunas of the Panthalassa Ocean than to Gondwana margin faunas. (Panthalassa was the precursor of the modern Pacific Ocean that existed in the region before the Pacific plate and basin formed.) The limestones are interpreted as remnants of guyots that have been scraped off and juxtaposed during accretion (Campbell and Hutching, 2007: 111).

The dual Gondwanan and Tethyan affinities of the Mesozoic fossils and extant groups, along with the globally basal groups of the Tasman–Coral Sea region, all reflect the history of the New Zealand terranes and their relationship with eastern Tethys.

The patterns cited in this chapter and the next two are presented here with minimum documentation, in order to give a brief introduction to the main tectonic and biogeographic features. The ways in which these features occur together in single groups are illustrated in subsequent chapters, devoted to case studies of particular plants and animals.

Western Province Terranes

The first applications of terrane theory to biogeography were introduced in work carried out in New Zealand (Craw, 1988; Heads, 1990a). Research on biogeography and terrane tectonics has continued, and some of the most interesting recent studies have been on the Mediterranean region (e.g., Magri et al., 2007; Bidegaray-Batista and Arnedo, 2011). The biotas of the New Zealand terranes show great differences. For example, the Western Province terranes account for much less land area than the Eastern Province terranes (Figure 3.2), but they include many unusual endemics with Gondwanan affinities, contrasting with the Tethyan affinities of many Eastern Province groups.

In the Western Province, northwest Nelson is well known for its large number of endemics. These include lichens such as *Hertella neozelandica* (Mount Arthur; the only record of the genus in New Zealand), *Pertusaria flavovelata* (endemic to Mount Arthur), and *Pertusaria hadrospora* (endemic to Kaihoka Lakes, Tasman Mountains, Lake Cobb, and the Arthur Range) (Galloway, 2007). Examples of endemic angiosperms include *Montia drucei* (Portulacaceae; Mount Arthur) (NZPCN, 2016), *Bulbinella talbotii* (Asphodelaceae; Gouland Downs), *Dracophyllum ophioliticum* (Ericaceae; Cobb and Takaka Valleys, only on ultramafics; Venter, 2002), *D. marmoricola* (around Mount Arthur, only on marble; Venter, 2002), and *D. trimorphum* (Whanganui Inlet to Puponga, sea level to 150 m, never more than 1 km from the sea, and often exposed to salt spray; Venter, 2009). As the last species indicates, the region includes coastal endemics as well as alpine ones. Endemic invertebrates include *Sagola kahurangiensis* of Mount Arthur (Coleoptera, Staphylinidae) (Park and Carlton, 2014) and the genus *Pholeodytes* (Coleoptera, Carabidae), which has five species, all restricted to caves in northwest Nelson (Larochelle and Larivière, 2005).

The other main center of endemism in the Western Province is Fiordland. The endemics are far too many to list, but examples include the lichens *Placopsis murrayi* (east and west of the main divide) and *Porina psilocarpa* (Breaksea Sound) (Galloway, 2007). *Celmisia holosericea* (Asteraceae) more or less defines Fiordland, being present in all the mountains north to the Darrans, and also at lower altitude at West Cape (NZPCN, 2016). Other endemics include the caddis fly genus *Erichorema*, only known from Secretary Island (Ward et al., 2004). In harvestmen (order Opiliones), Fiordland has the most distinctive and diverse fauna in New Zealand (Forster, 1975). In contrast, the spider family Dipluridae is diverse through most of New Zealand but is not recorded from Fiordland (Forster and Wilton, 1968: 558). The lichen *Ramalina* is similar, as it is diverse throughout New Zealand but absent in Fiordland; it grows right up to the eastern boundary at Lakes Monowai, Manapouri, and Te Anau (Bannister et al., 2004).

Forster and Forster (1974) wrote that the reasons for the distinctive Fiordland biota "are not completely clear but are thought to relate to marine transgressions which at times during the geological history have isolated the forests of this area" (p. 2801). In this way, the authors suggested that the Fiordland biota is the result, not of ecology but of paleogeographic changes. Modern geology suggests that these were caused not just by marine transgressions but also by a tectonic history in which Fiordland developed independently from the rest of the South Island. This would help to explain the endemism in Fiordland and also the many conspicuous absences there.

Fiordland endemics include terrestrial groups and also marine taxa such as the fish *Fiordichthys slartibartfasti* (Bythitidae). This is known only from Milford Sound, Preservation Inlet, and Bradshaw Sound, giving one of the most restricted distributions of any coastal fish in mainland New Zealand (Roberts et al., 2005). Little is known of the biology of *F. slartibartfasti*, and it could be an "emergent species," living at shallow depths in the fiords but present in much deeper water outside the fiords (Roberts et al., 2005). This phenomenon occurs in several groups, including the fish *Lepidoperca tasmanica* (Serranidae); in Fiordland this occurs within diving depth, while at its other New Zealand localities (The Snares, Otago, Chatham Islands), it occurs deeper than 150 m (Francis, 2012). The pattern requires a general explanation, such as uplift in Fiordland.

Forster and Forster (1985) cited the close biogeographic relationship between Fiordland and the subantarctic islands, and there is also a close relationship between Fiordland and Stewart Island. This is seen in the distributions of endemics, and also in absences, as illustrated by the following coastal fishes (Francis, 2012):

Zeus faber (John Dory, Zeidae): Widespread around the Three Kings, North, South, and Chatham Islands, but absent from Fiordland and Stewart Island.

Forsterygion gymnotum (Tripterygiidae): Widespread around North and South Islands, but absent from Fiordland and Stewart Island.

Allomycterus pilatus (Diodontidae): Widespread around Three Kings, North, and South Islands, but absent from Fiordland and Stewart Island.

Pagrus auratus (snapper; Sparidae): Widespread around Three Kings, North, South, and Chatham Islands, but absent from Fiordland and Stewart Island.

At its northern margins, Fiordland meets Mount Aspiring National Park between the head of Lake Te Anau and the head of Lake Wakatipu. Many groups show biogeographic breaks in this region (cf. Figure 3.13). For example, *Celmisia coriacea* (Asteraceae) is widespread in Fiordland (north to Mackinnon Pass) and the nearby Longwood Range. It is allopatric with its close relative *C. armstrongii*, which extends north from the Humboldt Mountains (Mount Bonpland, just south of Mount Aspiring National Park), along the Southern Alps and into Nelson (Given, 1980). The break between the two species coincides with the major faults bounding the Dun Mountain/Maitai terrane (Livingstone fault and others; Figure 3.13).

Mark (1977) mapped nine *Celmisia* species in Mount Aspiring National Park, and he noted (p. 47) that the distribution limits of *C. armstrongii* and four other large species "may be quite sharp for no apparent reason." Likewise, Holloway (1936) described rapid biogeographic turnover in subalpine shrublands around the Joe River. This complex area would repay further studies and detailed field mapping.

Western Province—Australia Affinities

Many groups are restricted to Western Province terranes and Australia. The Fiordland endemic marine fish *Fiordichthys slartibartfasti*, cited earlier, is a typical example—it belongs to a genus that is represented elsewhere only by an endemic species in Victoria.

A similar case occurs in the beetle tribe Camiarini (family Leiodidae). This has a typical trans-South Pacific distribution with records in Victoria (one genus, Victoria), New Zealand (five genera), and southern South America (one genus) (Seago et al., 2015). One of the New Zealand genera, *Camiarodes*, is a Fiordland endemic (Percy Saddle to Wilmot Pass), and the authors suggested that it may be the sister of the Victorian genus, *Camisolus*.

Western Province–Australia patterns can be broken down into three main types, as illustrated by the following lichens (Galloway, 2007):

Nelson—Australia/Tasmania:

> *Pannaria centrifuga*: Nelson (Long Lake, Tasman Mountains), Lord Howe Island, New South Wales, and Tasmania.
>
> *Porina constrictospora*: Nelson (Pakawau, Golden Bay) and Tasmania.
>
> *Pycnothelia caliginosa*: Nelson (Cobb Valley, Tasman Mountains, Denniston Plateau) and Tasmania.
>
> *Ramalina* stuartii: Nelson (Mount Arthur) and Australia.
>
> *Thysanothecium hookeri*: Northland, Nelson (Puponga), and Australia (WA, Vic, NSW, Tas).

Denniston Plateau—Australia/Tasmania:

> *Siphula elixii*: Denniston Plateau and SW Tasmania.
>
> *Xanthoparmelia isidiotegeta*: Denniston Plateau, Tasmania, Victoria, New South Wales, and Queensland.
>
> *Xanthoparmelia philippsiana*: Denniston Plateau, Macquarie Island, and Tasmania.
>
> *Mycoblastus kalioruber*: Denniston Plateau and Tasmania (Ludwig and Kantvilas, 2015).
>
> *Pertusaria flavoexpansa*: Stewart Island, Denniston Plateau, and Tasmania (Ludwig and Kantvilas, 2015).

Fiordland—Australia/Tasmania:

> *Collema quadriloculare*: Fiordland (Homer Tunnel) and Tasmania.
>
> *Ephebe fruticosa*: Fiordland (Lake Thompson) and Tasmania.
>
> *Porina aptrootii*: Fiordland (Doubtful Sound, Deep Cove) and Tasmania.
>
> *Porina kantvilasii*: Fiordland (Breaksea Sound), Campbell Island, Tasmania, and SW Australia.

A Distinctive Nelson Beetle, *Orthoglymma* (Carabidae)

Studies on carabid beetles concluded that New Zealand members of the tribe Broscini and many other taxa already in existence when the Tasman Sea opened, "display high fidelity to Gondwanan terranes comprising the older portions of New Zealand" (Liebherr et al., 2011). An example from Broscini is the genus *Orthoglymma*. This is endemic to Western Province terranes of northwest Nelson (Little Wanganui River) and is sister to a diverse complex of genera found in Australasia and Patagonia (including the trans-Tasman *Percosoma* group).

Bountya, endemic to the subantarctic Bounty Islands (Median Batholith), is another member of Broscini. Its sister group is a diverse Australian–Patagonian group that is absent from mainland New Zealand.

Liebherr et al. (2011: 395) concluded:

> The hypothesized Gondwanan-aged taxa demonstrate inordinate fidelity to the Gondwanan-aged geological terranes that constitute the western portions of New Zealand, especially in the South Island. Persistence of these relicts through a hypothesized "Oligocene drowning" event is the most parsimonious explanation for the concentration of Gondwanan relicts in the Nelson, Buller and Fiordland districts of the South Island [all in the Western Province].

Northwest Nelson, as exemplified by *Orthoglymma*, is one of New Zealand's main centers of endemism. It is delimited by the mid-Cretaceous Paparoa metamorphic core complex to the south, and a mid-Cretaceous metamorphic core complex in the Taranaki basin to the north (Tulloch et al., 2014).

Western Distribution in Disjunct "Outer Arcs"

Clades distributed in outer arcs around the east of Zealandia (e.g., subantarctic islands–Chatham Islands–Three Kings Islands) were discussed earlier. Many other groups are distributed in disjunct, outer arcs around the *western* margin of Zealandia and have most of their records on Western Province terranes. The affinities can be disjunct between the subantarctic islands and Fiordland, between Fiordland and Nelson (along the Alpine fault), and between Nelson and Tasmania (Heads, 1990a). The two examples shown in Figure 7.1 are the liverwort *Acromastigum mooreanum*, also present in Tasmania (Engel and Glenny, 2008), and the monocot *Astelia subulata* (Asteliaceae) (NZPCN, 2016).

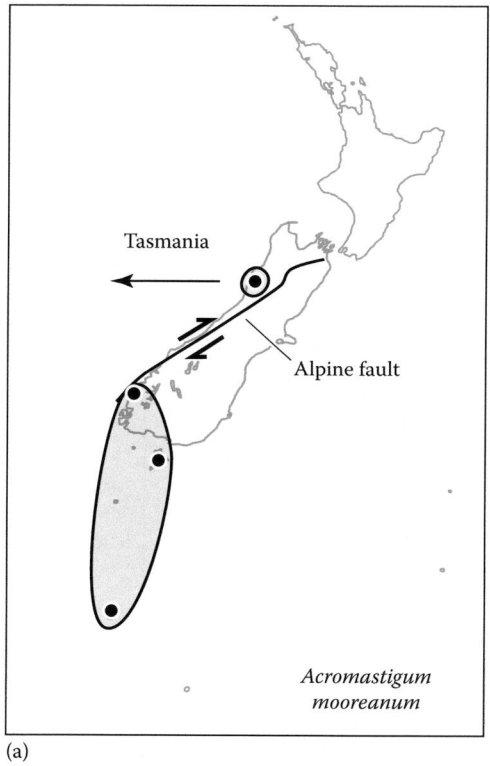
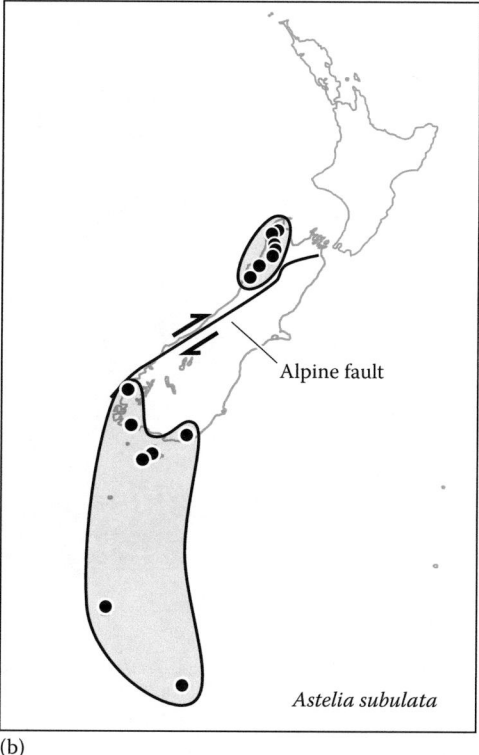

(a)   (b)

FIGURE 7.1   (a) Distribution of the liverwort *Acromastigum mooreanum*. (Data from Engel, J.J. and Glenny, D., *A Flora of the Liverworts and Hornworts of New Zealand*, Vol. 1, Missouri Botanical Garden Press, St. Louis, MO, 2008.) (b) Distribution of the monocot *Astelia subulata* (Asteliaceae). (From Heads, M., *N. Z. J. Zool.*, 16, 549, 1990a; NZPCN, New Zealand plant conservation network, 2016, www.nzpcn.org.nz.)

The Boundary between Western and Eastern Provinces
and Mid-Cretaceous Activity There

The Western Province comprises two terranes—Buller and Takaka—of early Palaeozoic age (Figure 7.2). Each of these has equivalents in Australia and Antarctica (Scott, 2013). In New Zealand, the boundary between the two is formed by the Anatoki fault in Nelson and the Old Quarry fault in Fiordland. The Anatoki fault originated as a mid-Palaeozoic structure, but it has had a polyphasic history that includes reactivation in the Cretaceous and Cenozoic (Scott, 2013). This reactivation and the associated magmatism are associated with important biogeographic boundaries; the main centers of endemism in the Western Province include southwestern Fiordland (Buller terrane) and the Mount Arthur—Mount Owen sector of northwest Nelson (Takaka terrane).

Eastern Province terranes differ from those of the Western Province, as they are younger (Late Palaeozoic–Mesozoic) and they were accreted to Gondwana later. The boundary between Western and Eastern Provinces had a Mesozoic origin, but, as with the Buller–Takaka terrane boundary, this tectonic feature has had a polyphasic history (Scott, 2013). The boundary has been obscured by large-scale plutonism of the Median Batholith, but occurs at the contact between the Takaka terrane (Western Province)

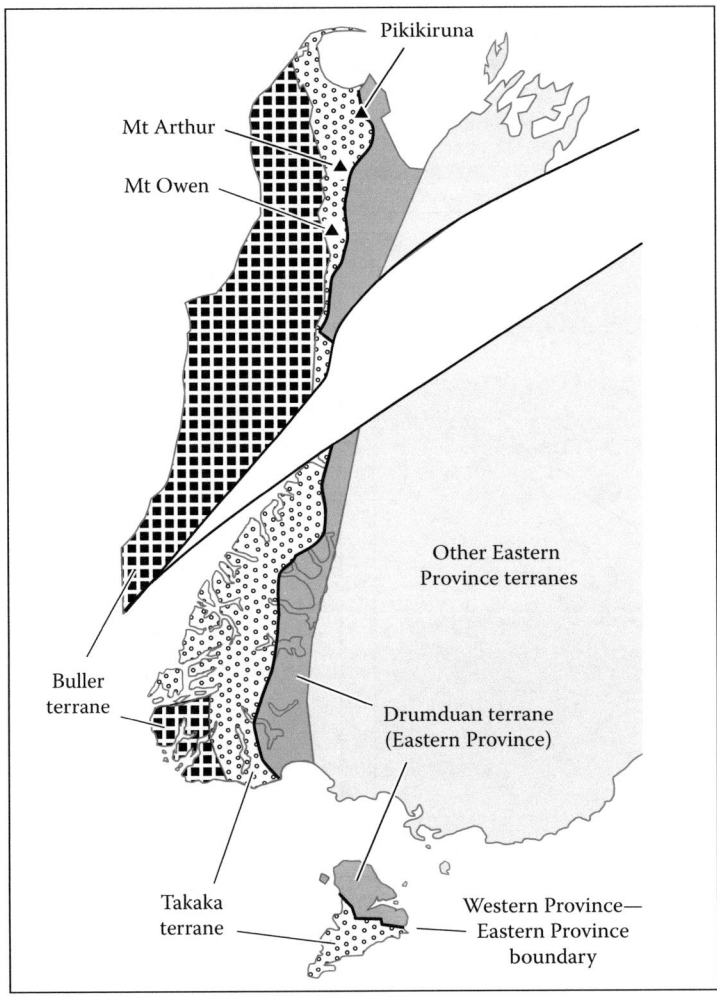

FIGURE 7.2 New Zealand terranes and the boundary between Western and Eastern Provinces. Basement terranes only are shown, without the intrusives of the Median Batholith or later sedimentary strata. (Based on Scott, J.M., *N. Z. J. Geol. Geophys.*, 56, 276, 2013; Allibone, A.H. et al., *N. Z. J. Geol. Geophys.*, 50, 283, 2007.)

and the Drumduan terrane (Eastern Province) (Figure 7.2, based on Scott, 2013). The Drumduan ter-rane forms central Nelson, *eastern* Fiordland, and *northern* Stewart Island. The same differentiation in Fiordland and Stewart Island is seen in many biological distributions. The Drumduan terrane is intruded by the Darran Suite and related rocks, and these units show the same eastern Fiordland–Stewart Island distribution (mapped by Mortimer, 2014).

The Western Province–Eastern Province boundary is indicated by a series of early–mid-Cretaceous ductile shear zones. These include the Gutter shear zone/Escarpment fault in Stewart Island, the Grebe mylonite zone in eastern Fiordland, and the Wainui shear zone in Nelson (Allibone and Tulloch, 2008; Scott, 2013). (Schulte et al., 2014, mapped the Grebe belt, the Paparoa metamorphic core complex, and the Sisters shear zone in Stewart Island as "major extensional detachments.")

The shear zone at the Western Province–Eastern Province boundary mark a lithological and metamorphic discontinuity created by convergent deformation at 128–115 Ma (Scott, 2013). The event was probably caused by the collision of a volcanic island arc (Eastern Province terrane(s)) with the continent (Western Province) (Scott et al., 2009), and this first phase of deformation led to the emplacement of large volumes of magma through the continental crust. A second phase of convergent deformation took place at 115–100 Ma (with the last Separation Point Suite intrusions at 105 Ma), and this reworked the now-juxtaposed Western and Eastern Province rocks. At this time in Stewart Island, all rocks south of the Escarpment fault were exhumed simultaneously at ~110–100 Ma (Scott, 2013).

Coprosma talbrockiei and the Western Province Terranes of Northwest Nelson

Most members of the plant family Rubiaceae are found in the tropics, but the wind-pollinated tribe Anthospermeae is southern, with component groups centered in Africa, in Australia, and around the South Pacific. The tribe is not linked to a localized center of origin; instead, its sister group comprises five tribes: Argostemmateae, Paederieae, Putorieae, Theligoneae, and Rubieae (Bremer and Eriksson, 2009). This large sister group is widespread globally, but is most diverse in tropical and northern hemisphere regions, and is less diverse in the southern hemisphere. The overall pattern is consistent with an *in-situ* origin of Anthospermeae, by vicariance, prior to the breakup of Gondwana.

The phylogeny within Anthospermeae is as follows (Anderson et al., 2001):

Carpacoce: South Africa.

Anthospermum: Africa, Madagascar, and Yemen; + *Galopina*, South Africa; + *Phyllis*, Macaronesia.

Pomax: Australia (widespread through the mainland).

Opercularia: Australia (widespread in the southwest and the east).

Leptostigma: Trans-South Pacific (SE Australia, New Zealand, and disjunct in Chile, Peru, and Colombia).

Coprosma complex: Trans-tropical and South Pacific (from southern China and Sumatra to central and South America, also Tristan da Cunha).

Coprosma is ubiquitous in New Zealand and has its center of diversity there (Chapter 4). Groupings within the Anthospermeae are not yet resolved (Bremer and Eriksson, 2009; Cantley et al., 2014), but the *Coprosma* complex appears to be structured as follows (Figure 7.3a; Markey et al., 2004; distributions from Heads, 1996 and Thompson, 2010):

Trans-Tasman Sea. "*Coprosma*" *talbrockiei* of NW Nelson, + *Durringtonia* of the New South Wales–Queensland border (McPherson–Macleay Overlap [MMO]). This pair is sister to *Normandia* of New Caledonia.

Trans-Pacific Ocean. *Coprosma* and *Nertera*. China and Sumatra, across the Pacific via New Zealand, Hawaii, etc., to Mexico, the Antilles, the Andes, and Tristan da Cunha. This clade is *not* in the MMO or New Caledonia.

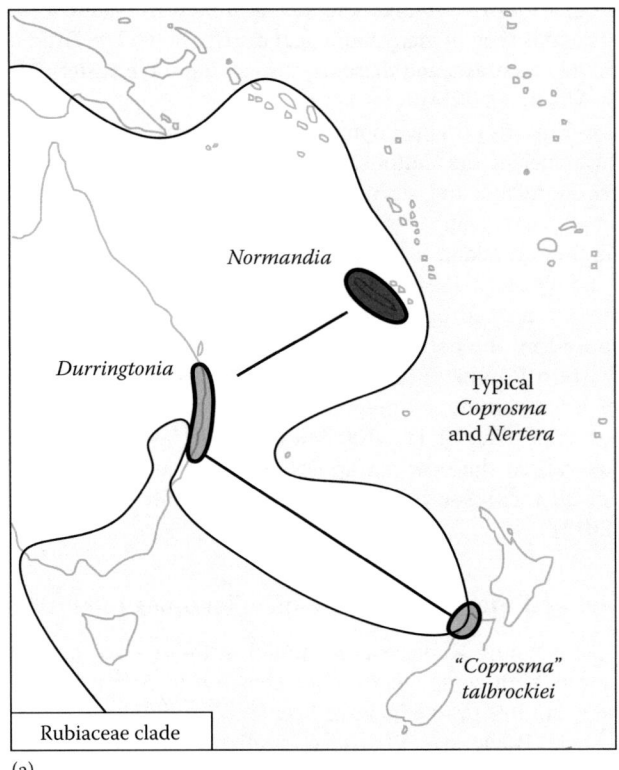

(a)

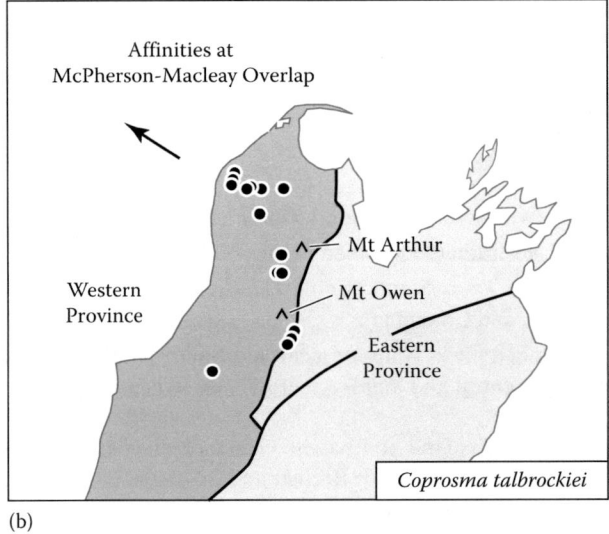

(b)

FIGURE 7.3 (a) Distribution of a clade in Rubiaceae. (Data from Markey, A. et al., *Coprosma talbrockiei*: An oddball sheds light on the *Coprosminae*, Abstract of paper presented at *SYSTANZ Meeting*, Whakapapa Village, New Zealand, 2004, www.math.canterbury.ac.nz/bio/pages/SYSTANZ/meeting/SYSTANZ_2004_programme.pdf; Thompson, I.R., *Muelleria*, 28, 29, 2010. (b) Distribution of "*Coprosma*" *talbrockiei*. (Data from New Zealand plant conservation network, 2016, www.nzpcn.org.nz.)

An earlier prediction, based on biogeographic patterns, suggested: "It would not be surprising if *Durringtonia* turns up in the swamps and heaths of New Zealand's west coast" (Heads, 1996: 398), and so it is interesting to find that its closest relative is endemic there.

The two clades in the *Coprosma* complex are mainly allopatric. Overlap is restricted to northwest Nelson (Western Province), where "*Coprosma*" *talbrockiei* is endemic (Figure 7.3b) and *Coprosma* s.str. is widespread. The absence of *Coprosma* from New Caledonia and the MMO is difficult to explain in terms of current ecology. Yet the distribution as a whole is compatible with an origin of the *Normandia* and *Coprosma* groups by simple vicariance at the Western Province–Eastern Province margin and at the MMO. The distribution of the *Normandia* group in New Caledonia, the MMO, and New Zealand, now forms a dogleg, but before seafloor spreading opened the Tasman Sea basin, the three localities lay in a gently curved arc along what is now the coast of Australia. The Whitsunday/Median Batholith belt of magmatism developed along the same arc, and it provides a mechanism for the differentiation of the groups there. This would apply whether *Durringtonia*, *Normandia*, and *Coprosma talbrockiei* form a clade or a basal, paraphyletic grade. After the origin of the modern genera, local range expansion of *Coprosma* in northwest Nelson has resulted in secondary overlap there.

Western Province and Eastern Province Differentiation in *Lyperobius* Weevils

The boundary between Western and Eastern Provinces also corresponds to an important boundary in many animals. For example, in Nelson, a localized belt of endemism for several land snail species occurs along the western margin of the Eastern Province, near Takaka. It follows the line: Pikikiruna–Gorge Creek–Clifton–Tata Islands (Climo, 1971; Mahlfeld, 2005).

Many Western Province clades are separated from Eastern Province relatives by a phylogenetic and biogeographic break at the province boundary. As mentioned, this was active in the mid-Cretaceous. A good example of a break here is seen in the giant weevils or speargrass weevils, *Lyperobius* (Curculionidae). One clade is restricted to the Western Province and includes endemics on The Snares, on the Buller terrane in southwestern Fiordland, and on the Takaka terrane in central Fiordland (Figure 7.4; Craw, 1999). Morphological analyses indicated that the group's sister comprises three species located in the Eastern Province.

One member of the Eastern Province clade, *Lyperobius patricki*, has its southern limit at the Waihemo fault zone. This is another important mid-Cretaceous boundary (Figure 3.13) and is discussed below but can be introduced here. In Figure 7.4 and subsequent maps, the Waihemo fault zone is shown continuing inland as the Blue Lake fault zone, along the southwestern margin of the St Bathans Range (Figure 3.13). The fault zone formed in the Cretaceous as part of a major crustal boundary between Otago Schist and Torlesse graywacke (Henne et al., 2011).

In a related example from plants, *Hebe pauciflora* (Plantaginaceae) is restricted to the Western Province part of Fiordland and extends right up to the boundary with the Eastern Province (Bayly and Kellow, 2006). The species is distinctive and its affinities are obscure; Moore (in Allan, 1961) placed it with "Buxifoliatae" (*H. pauciramosa* and others), while Bayly and Kellow (2006) have it in a group of its own. In any case, it is largely allopatric with *H. pauciramosa*, which occurs in northern Stewart Island, *eastern and southwestern* Fiordland, and north to Nelson.

Stewart Island

The Stewart Island biota is distinguished by high levels of endemism, including 28 endemic species of vascular plants (Wilson, 1987: 84). There are also striking absences; in plant life there is no *Nothofagus*, for example. This particular absence has been regarded as "anomalous" (Wilson, 1987) and "surprising" (Dawson and Lucas, 2011: 34). It has sometimes been attributed to glaciation, but Wilson (1987: 83) wrote: "Available information does not favour the suggestion of complete deforestation during the Pleistocene... most of the species now on Stewart Island probably persisted through the glacial phases. Indeed, conditions during glacial times may have been marginally more favourable for *Nothofagus* than they are now." The absence of *Nothofagus* is part of a shared pattern, as other common New Zealand

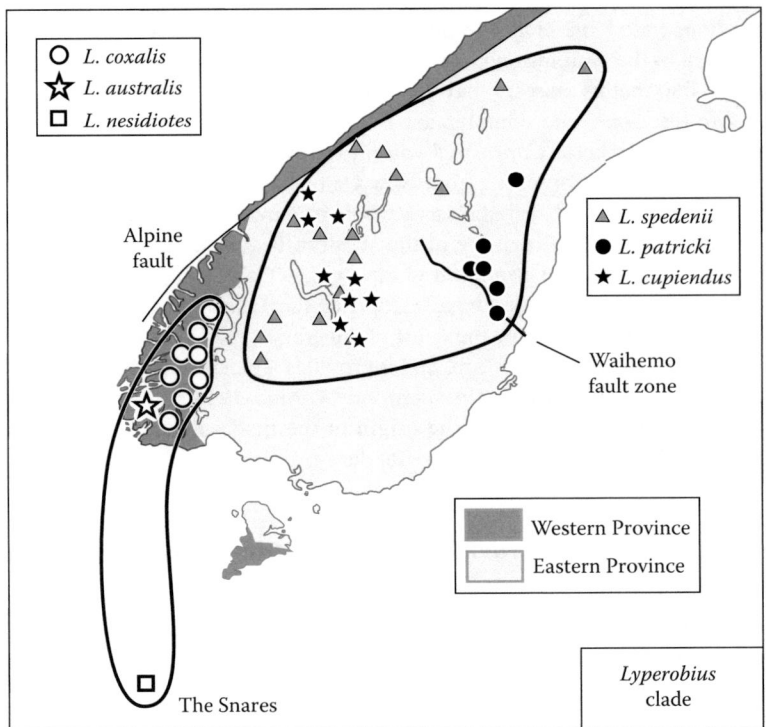

FIGURE 7.4 Distribution of two sister clades in *Lyperobius* (Curculionidae). (Data from Craw, R.C., *Fauna N. Z.*, 39, 1, 1999.)

trees are absent from Stewart Island. These include *Phyllocladus* (Podocarpaceae), *Sophora* (Fabaceae), *Libocedrus* (Cupressaceae), *Kunzea* (Myrtaceae), and *Melicytus ramiflorus* (Violaceae). In bird life, the fantail, *Rhipidura*, is widespread through the North and South Islands, but there are no breeding records on Stewart Island (Robertson et al., 2007).

Wilson (1987: 84) concluded that Stewart Island includes two distinct floristic districts: the Mount Anglem region in the north and the Tin Range area in the south. These are separated by the main tectonic division in New Zealand, the boundary between Eastern Province and Western Province (Scott, 2013). In Stewart Island, this is represented by the Gutter shear zone and the associated Escarpment fault (Figure 7.5; based on Scott, 2013). The belt in Stewart Island is up to 10 km wide, wider than it is in Fiordland. As Wilson (1987: 84) wrote, "the existence of both local endemics and vicarious species pairs suggests that the distinctness of the two regions is of long standing… and predates the Pleistocene glaciation."

One southern Stewart Island endemic, *Aciphylla stannensis*, has its closest relative (*A. trifoliolata*) not in northern Stewart Island, but in northwest Nelson (Figure 7.5; Dawson, 1980). Both species occur on Western Province rocks near the margin with the Eastern Province. In contrast, many Eastern Province groups, such as *Carmichaelia petriei*, extend south and west only to *northern* Stewart Island and *eastern* Fiordland (Figure 7.5; Heenan, 1996a). Groups such as these, along with the geology, show Stewart Island as a compound, a fragment of the margin rather than a homogeneous entity in itself.

In vascular plants, 80 species (many of which are widespread in the North and South Islands), many genera, and 7 families occur in Stewart Island only north of the Escarpment fault/Gutter shear zone (data mostly from Wilson, 1987; some records from NZPCN, 2016; data for individual groups from Connor, 1991; Edgar, 1996; Petterson, 1997; Blanchon et al., 2002; Bayly and Kellow, 2006; Glenny, 2009; Meudt, 2012). The families are Adiantaceae, Iridaceae, Euphorbiaceae, Malvaceae, Fabaceae, Celastraceae (*Stackhousia*), and Pennantiaceae. The northern species include alpines (e.g., *Ourisia*

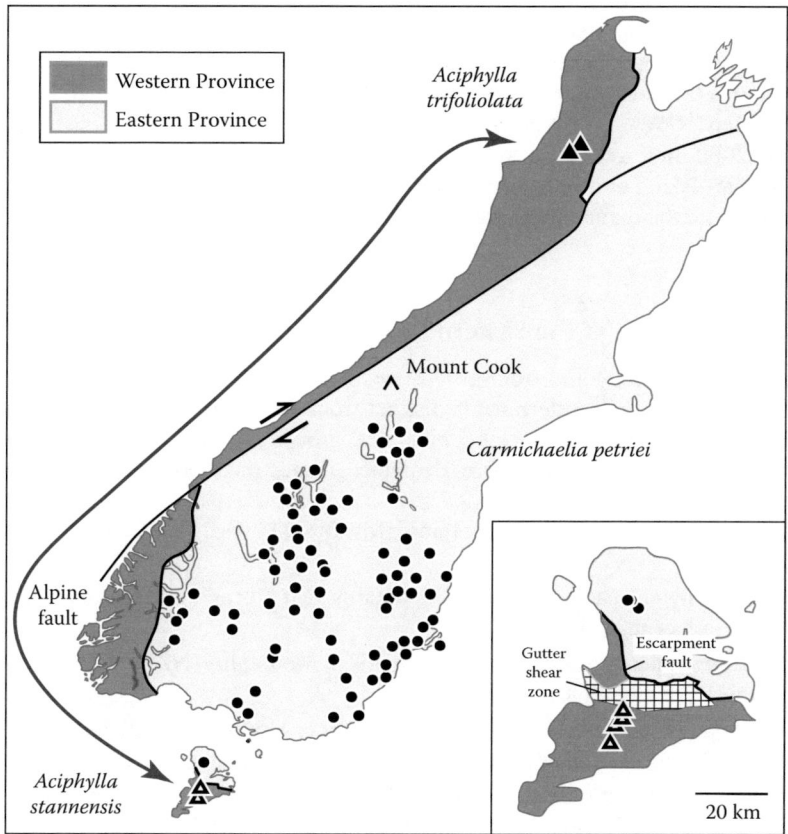

FIGURE 7.5 New Zealand's tectonic Western Province, with the distribution of a sister pair in *Aciphylla* (Apiaceae), and Eastern Province, with the distribution of *Carmichaelia petriei* Fabaceae. (Data from Dawson, J.W., *N. Z. J. Bot.*, 18, 115, 1980; Heenan, P.B., *N. Z. J. Bot.*, 34, 299, 1996b.)

caespitosa, Abrotanella pusilla, Hebe hectorii*) as well as lowland groups (e.g., *Euphorbia glauca* and *Hebe salicifolia*). Several plants range from the North Island south to eastern Fiordland and northern Stewart Island: *Wahlenbergia violacea* (formerly *W. gracilis*), *Argyrotegium mackayi* (formerly *Gnaphalium mackayi*), and *Raoulia hookeri*.

The flora south of the Escarpment fault, on the Western Province part of Stewart Island, is not simply an impoverished version of the northern flora; it also includes species that are not in northern Stewart Island, but do occur in the western South Island (e.g., *Anisotome haastii*, Apiaceae). Other southern Stewart Island plants are found elsewhere only in the subantarctic islands. For example, *Lepidium limenophylax* (Brassicaceae) is on the Auckland Islands, the Snares, and the Muttonbird (Titi) Islands off southwestern Stewart Island (de Lange et al., 2013). Southern Stewart Island also has its own endemics, including *Ranunculus stylosus*, *R. viridis* (Ranunculaceae), and *Aciphylla stannensis* (Apiaceae).

In terms of its climate, Stewart Island does not have distinct, warmer sectors in the north and cooler sectors in the south (although the north tends to be sunnier than the south), but instead a western sector

---

* Hector was a founder of New Zealand science, and many taxa, from mosses to whales, are named after him. Earlier I argued against changing the spelling of the common New Zealand species epithet "hectori" (now spelled "hectorii" in plants) to the classical Latin form "hectoris," in order to avoid unnecessary nomenclatural novelty and confusion (Heads, 1998b). The current code of botanical nomenclature advocates the use of hectoris, but only as a recommendation, not as a binding "article" (see Recomm. 60C, which deals with the spelling of names). However, Heenan et al. (2008a) have followed Recomm. 60C (in changing the spelling of *Olearia traversii* to *O. traversiorum*) and, based on this approach, "hectoris" should be used for plants originally named as "hectori" or later "hectorii." Instead, the standardized spelling "hectorii" is used in this book.

("zone E" of Tomlinson, 1976, shared with Fiordland and the west coast) and an eastern sector ("zone G2," shared with southern Southland). This is because the east is sheltered from the southwesterly winds. Rainfall in the north and south is similar; the main rainfall difference is between the drier lowlands and the wetter mountains (Mounts Anglem and Rakeahua in the north and Mount Allen in the southern Tin Range). Another map (NIWA, 2016) shows that north and south Stewart Islands both have a mean annual rainfall of 1500–2000 mm, as in eastern Fiordland and the Catlins. The upper Freshwater Valley and southwestern Stewart Island are wetter, with a rainfall of 2000–2500 mm. Again, these climatic variations do not account for the main, north–south biogeographic differentiation, which are better explained by the tectonic history.

Eastern Fiordland as Part of the Eastern Province

Information both from geology and from groups such as *Carmichaelia petriei* (Figure 7.5) indicates that Eastern Fiordland (Drumduan terrane) is distinct from the rest of Fiordland and is instead part of the Eastern Province (Figure 7.3). Many other biological groups are present in Fiordland, but only in the eastern part, and extend northeast from there. Examples include the following:

Solorina crocea (lichen): Hunter Mountains (Mt Burns), north to Canterbury and Nelson (Galloway, 2007).

Kelleria villosa (Thymelaeaceae): Hunter Mountains (Mt Burns) and Takahe Valley, north to Nelson and Marlborough (Heads, 1990).

Kelleria paludosa: Upper Grebe Valley and Murchison Mountains (NZPCN, 2016), north to central Canterbury (Heads, 1990).

Pimelea oreophila (Thymelaeaceae): Hump Ridge, Hunter Mountains, and Murchison Mountains, north to the volcanic plateau of the central North Island (Burrows, 2011a).

Coprosma rubra (Rubiaceae): Back Valley (north of Mt Titiroa), north through the eastern South Island and North Island to the Raukumara Range (NZPCN, 2016).

Olearia bullata (Asteraceae): Hunter Mountains (Green Lake), Kepler Mountains (Fowler Pass), and Murchison Mountains, widespread in the eastern South Island (Heads, 1998b).

Celmisia viscosa (Asteraceae): Hunter Mountains (Mt Burns), Kepler Mountains, and Murchison Mountains, north to Nelson (NZPCN, 2016).

Haastia (Asteraceae, three species): Hunter Mountains (Mt Burns), Mt Titiroa, and the Murchison Mountains, north to Nelson and Marlborough (NZPCN, 2016).

Hebe buchananii (Plantaginaceae): Hunter Mountains (Green Lake) and Homer Saddle, north to Mount Cook (Bayly and Kellow, 2006).

Parahebe canescens (Plantaginaceae): Eastern shores of Lake Manapouri to North Canterbury (Garnock-Jones and Lloyd, 2004).

Hierodoris extensilis (Lepidoptera, Oecophoridae): Hunter Mountains (Mt Burns) and Mt Titiroa, on granite sand plains. It is related to three species: *H. polita* of central and eastern Otago; *H. frigida* of central and eastern Otago, disjunct in Nelson; and *H. gerontion* of the Eyre, Garvie, Hector, and Old Man Ranges (Hoare, 2005, 2012). None of these three is in Fiordland.

Coprosma rigida (Rubiaceae) appears to be a classic "mainland New Zealand" endemic, as it is widespread through North, South, and Stewart Islands and is absent from the Chathams. It is often abundant and occurs in both forest and shrubland, including rough farmland and from lowland to montane elevations. Nevertheless, the records point to interesting absences in parts of the mainland (NZPCN, 2016). For 140 km through most of Fiordland (between Preservation Inlet and Catseye Bay by George Sound), *C. rigida* is present only along the eastern strip. In a similar way, it is absent in and around the Paparoa Range, in northwest Nelson west of a line, Matiri Range–Cobb Reservoir, and on Mount Egmont/Taranaki. The species is otherwise widespread in the North Island, north to Aupouri Peninsula, but it is absent from the standard biogeographic arc: Coromandel Peninsula, Great and Little Barrier Islands, Poor Knights, Hen and Chickens, North Cape, and Three Kings Islands. This connects with the absence in the Chatham Islands.

As well as being a boundary for clades, eastern Fiordland also acts as a mutual boundary for close relatives. Examples include the clades of *Lyperobius* (Figure 7.4) and also the following plants:

> *Bulbinella gibbsii* (Xanthorrhoeaceae): Fiordland and Stewart Island. It overlaps with *B. angustifolia* only in the Hunter Mountains, and the latter species extends from there to Dunedin and Marlborough (NZPCN, 2016).
>
> *Celmisia semicordata* subsp. *stricta* (Asteraceae): Eastern Fiordland (Hunter Mountains), the Eyre Mountains, and Central Otago (Given, 1980). The other subspecies of *C. semicordata* lie to the east (in Central and eastern Otago; subsp. *aurigans*) and to the west (western Fiordland and further north; subsp. *semicordata*).

The tussock grass *Chionochloa* dominates many New Zealand landscapes. There are four main clades in New Zealand, and the most species rich of these illustrates the significance of eastern Fiordland. Its molecular phylogeny and distribution are structured as follows (Pirie et al., 2010):

> *C. pallens cadens*: Fiordland to Mount Cook and Franz Josef.
>
> *C. vireta*: Fiordland to "South Canterbury," + *C. frigida*, SE Australia (Kosciuszko area).
>
> *C. nivifera*: Eastern Fiordland (Mount Titiroa and Cleughearn Peak to the Kaherekoau Mountains, with an outlying record from the Turret Range, 20 km further west; Connor and Lloyd, 2004).
>
> A clade of 10 species and subspecies (*C. conspicua*, *C. flavescens*, etc.), widespread from Stewart Island to Northland.

The node between eastern Fiordland (*C. nivifera*) and its sister area, all mainland New Zealand, represents a main break in the phylogeny and biogeography. The break here is preceded only by breaks involving Fiordland, the Mount Cook/Glaciers region, and southeastern Australia.

Paparoa Range and Papahaua Range

One of the key phylogenetic and biogeographic nodes in the Western Province occurs near Westport and Cape Foulwind. This boundary and center of endemism coincides with an important tectonic feature, the Paparoa metamorphic core complex (Figure 3.12). The biogeographic patterns here involve the Paparoa Range and, across the Buller River, the Papahaua Range (Mount Rochfort and Denniston Plateau to Mount Stockton). The core complex and its environs mark a center of extension and uplift associated with prebreakup rifting in the mid-Cretaceous. The main hinge of extension in the core complex (lying between the two bivergent detachment faults to the north and south) runs between the Buckland Peaks in the northern Paparoa Range and a point on the coast, ~8 km south of Cape Foulwind (Schulte, 2011; Schulte et al., 2014).

The Paparoa Range and neighboring lowland regions form a distributional limit in many groups, for example:

> *Schizaea bifida* (Schizaeaceae): Three Kings Islands south to Charleston, by the Paparoa Range (Brownsey and Perrie, 2014).
>
> *Astelia skottsbergii* (Asteliaceae): Mountains of northwest Nelson (e.g., Mt Arthur) to the Paparoa Range (Moore and Edgar, 1970).
>
> *Kunzea ericoides* s.str. (Myrtaceae): Northern Marlborough and Nelson to the Paparoa Range (de Lange, 2014).
>
> *Pimelea longifolia* (Thymelaeaceae): North Island to the Paparoa Range (Burrows, 2008).

Epacris pauciflora (Ericaceae): Northern North Island and northwest Nelson to the Paparoa Range (Allan, 1961; NZPCN, 2016).

Dracophyllum pubescens (Ericaceae): Northwest Nelson to the Paparoa Range (Venter, 2009).

Hebe townsonii (Plantaginaceae): Northwest Nelson (Mount Burnett, Arthur Range) to Punakaiki, by the Paparoa Range (Bayly and Kellow, 2006).

Celmisia dubia (Asteraceae): Northwest Nelson to the Paparoa Range (Allan, 1961).

Celmisia dallii (Asteraceae): Northwest Nelson to the Paparoa Range (Given, 1968).

The land snail *Allodiscus austrodimorphus*: Fiordland, north to Punakaiki (Marshall and Barker, 2008).

The land snail *Allodiscus chion*: North Island south to Punakaiki (Marshall and Barker, 2008).

Exsul (Diptera, Muscidae), a spectacular alpine fly: Northern Fiordland (Milford Track, Milford Sound) to Fox and Franz Josef Glaciers, Arthur's Pass, and the Paparoa Range (Patrick, 1996).

Kiwitrechus (Carabidae): NW Nelson to the Paparoa Range (Mount Dewar) (Larochelle and Larivière, 2007).

Neocyba (Coleoptera, Brentidae): Three Kings Islands, south to Punakaiki (its closest relative is *Rhadinocyba* of New Caledonia) (Kuschel, 2003).

Nunnea (Coleoptera, Staphylinidae): NW Nelson and Marlborough Sounds, south to Punakaiki (Park and Carlton, 2015a).

The region around the core complex is also an area of endemism for both lowland and alpine groups, including the following:

The lichen *Porina diffluens*: Punakaiki, by the coast (Galloway, 2007).

Cephaloziella aenigmata (Marchantiophyta): Paparoa Range (Engel and Glenny, 2008).

Chionochloa juncea (Poaceae): Denniston Plateau (Connor, 1991).

Celmisia morganii (Asteraceae): Lower Buller Valley and Ngakawau Gorge (Given, 1980).

The land snails *Allodiscus punakaiki* and *A. laganus* (Charopidae): Punakaiki (Marshall and Barker, 2008).

The stiletto fly *Anabarhynchus diversicolor* (Therevidae): Punakaiki (Lyneborg, 1992).

The weta *Deinacrida talpa* (Orthoptera, Anostostomatidae): Paparoa Range (Gibbs, 2001).

The beetle *Priasilpha embersoni* (Coleoptera, Priasilphidae): Paparoa Range (Leschen and Michaux, 2005).

Baeocera benolivia (Coleoptera, Staphylinidae): Capleston and Inangahua (Löbl and Leschen, 2003).

Neoferonia prasignis (Coleoptera, Carabidae): Westport and Stockton, near Granity (Johns, 2015).

The Paparoa area is also a center of disjunction, as in the following:

Paparoa disjunct to NW Nelson:

Hebe ochracea (Plantaginaceae): Northwest Nelson (Mount Arthur to Mount Owen), disjunct at the northern Paparoa Range (Bayly and Kellow, 2006).

Dracophyllum elegantissimum (Ericaceae): Northwest Nelson and Heaphy track, disjunct at Mt Rochfort/northern Paparoa Range (Venter, 2004).

Paparoa disjunct to the southern South Island (*along the Alpine fault*):

Acromastigum cunninghamii (Marchantiophyta): North Island to Stockton Plateau and Mount Glasgow, disjunct at Haast Pass and Fiordland (Engel and Glenny, 2008).

Gentianella scopulorum (Gentianaceae): Charleston (by the Paparoa Range), disjunct with its sister species, *G. saxosa* of the South Island south coast (Glenny, 2004).

Astelia subulata (Asteliaceae): Northwest Nelson to the Paparoa Range, disjunct in Fiordland, the Catlins, Stewart Island, and the subantarctic islands (Figure 7.1; NZPCN, 2016).

Vesicaperla substirpes (Plecoptera): Paparoa Range, disjunct in Central Otago (McLellan, 1993).

Mitophyllus insignis (Coleoptera, Lucanidae): Northwest Nelson and the Paparoa Range, disjunct at Lake McKenzie in the Humboldt Mountains (northwest Otago) (Holloway, 2007).

Apterachalcus (Diptera, Dolichopodidae): Paparoa Range, disjunct on Stewart, Auckland, and Campbell Islands (Bickel, 1991). This is the only member of its global family that has lost its wings and halteres.

Paparoa disjunct to Australia/Tasmania:

The fern *Sticherus urceolatus* (Gleicheniaceae): Takaka and Stockton Plateau, also in Tasmania, Victoria, and New South Wales (Brownsey et al., 2013).

Neogrollea (Marchantiophyta): Lowlands from Charleston to the Stockton Plateau (in pakihi swamp with *Empodisma* and *Gleichenia*), also in Tasmania (Engel and Glenny, 2008).

Epilobium gunnianum (Onagraceae): Northwest Nelson to near Westport, as well as Tasmania, eastern Victoria, and eastern New South Wales.

Other groups lying north of the Paparoa region, in northwest Nelson, also show connections across the Tasman Basin. For example, *Poranthera alpina* (Phyllanthaceae) of northwest Nelson (Mount Arthur, Mount Owen, etc.) is closely allied to *P. petalifera* of Tasmania (Orchard and Davies, 1985).

Alpine fault disjunction + intercontinental disjunction:

Archeophylla schusteri (Marchantiophyta): Campbell Island, Stewart Island, Fiordland (Manapouri), disjunct at the Paparoa Range, Nelson (Arthur Range), and Mount Ruapehu. The two other species in *Archaeophylla* are from southern South America (Engel and Glenny, 2008).

Lepidozia glaucophylla (Marchantiophyta): Fiordland (Deep Cove, Borland Burn, Mackinnon Pass, Lake Marion), disjunct at the Paparoa Range, also southeastern Australia and Tasmania (Engel and Glenny, 2008).

Telaranea inaequalis (Marchantiophyta): Stewart Island, Fiordland (Bowen Falls), disjunct at Stockton Plateau, also present in Tasmania (Engel and Glenny, 2008).

Acromastigum cavifolium (Marchantiophyta): Stewart Island, Fiordland (Henry Pass, Secretary Island), and Jackson Bay, disjunct at the Paparoa Range (also Westport, Stockton, and Denniston Plateaus), Mount Moehau (Coromandel), and Tasmania (Engel and Glenny, 2008).

Acromastigum mooreanum (Marchantiophyta): Auckland Islands, Stewart Island, and Fiordland (Secretary Island), disjunct at the Paparoa Range and Tasmania (Figure 7.1; Engel and Glenny, 2008).

Sphaerothorax (Coleoptera, Clambidae): Otago and Southland, disjunct at the Paparoa Range and northward in New Zealand; also in Australia and South America (Endrödy-Younga, 1995).

Dendroblax (Coleoptera, Lucanidae): Otago (Ben Lomond by Queenstown), disjunct at the Paparoa Range and northward in New Zealand. Its closest relatives are *Lamprima* of Australia, Lord Howe Island, Norfolk Island, and New Guinea, and *Streptocerus* of Chile (Holloway, 2007).

The plant family Loganiaceae includes an interesting case involving endemism around the Paparoa and Papahaua Ranges, and also trans-Tasman disjunction (Gibbons et al., 2014). In this example, *Schizacme helmsii* (Loganiaceae) is endemic to *the northern and southern* Paparoa Range, Denniston

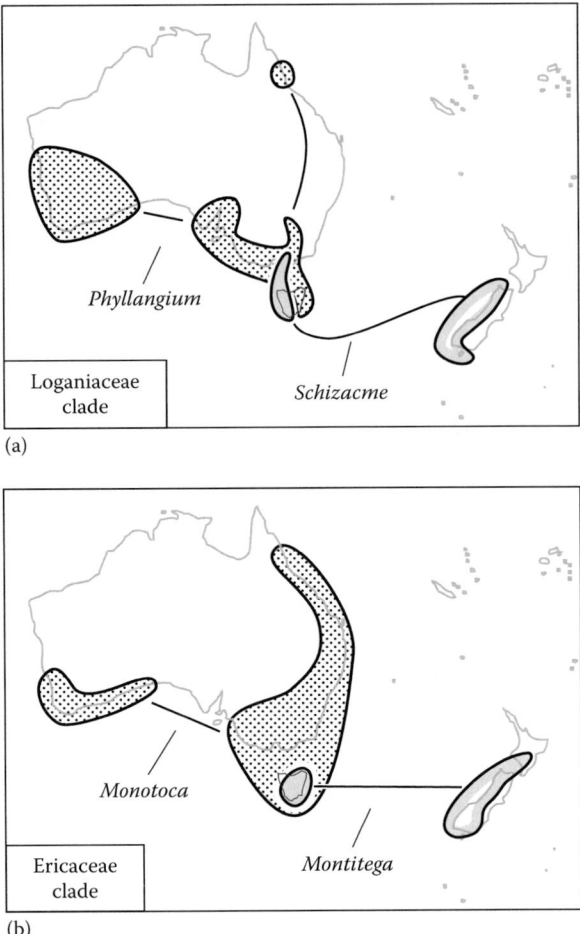

FIGURE 7.6   (a) Distribution of a clade in Loganiaceae: *Phyllangium* and *Schizacme*. (From Gibbons, K.L. et al., *Telopea*, 17, 363, 2014; GBIF, Global biodiversity information facility, http://data.gbif.org/, 2016.) (b) Distribution of a clade in Ericaceae: *Monotoca* and *Montitega*. (Reprinted from Heads, M., *Biogeography of Australasia: A Molecular Analysis*, Cambridge University Press, Cambridge, UK, 2014. With permission.)

Plateau, and Stockton Plateau. It is sister to the two Australian species of *Schizacme* (found in Tasmania and Victoria) rather than the two other New Zealand species. *Schizacme* as a whole is sister to *Phyllangium*, and the two genera have the following distributions (Figure 7.6a, from Gibbons et al., 2014; GBIF, 2016):

> *Schizacme*: New Zealand, W Tasmania, and Victoria.
>
> *Phyllangium*: SW Australia, SE Australia and E Tasmania, and NE Queensland.

The two genera show a high level of allopatry, and any geographic overlap is restricted to local areas of Tasmania and Victoria. Gibbons et al. (2014) concluded: "The phylogenies inferred in this study suggest *Schizacme* might have originated in New Zealand as a result of long distance dispersal of a common ancestor of *Schizacme* and *Phyllangium* from Australia to New Zealand [followed by dispersal of *Schizacme* back to Australia]." Nevertheless, the phylogenies are also consistent with an origin of the two genera by vicariance. This would have taken place in a widespread, prebreakup, Australia–New Zealand ancestor somewhere near the Great Dividing Range of Australia, followed by local overlap *within* Victoria and Tasmania. A vicariance model for *Phyllangium* and *Schizacme* predicts that the pattern should be repeated in other groups, and this is seen, for example, in *Monotoca* and its sister *Montitega* (Ericaceae) (Figure 7.6b; Heads, 2014).

The primary phylogenetic break in *Schizacme* is not at the Tasman Sea, but between the clade of Victoria, Tasmania, Paparoa Range, Denniston Plateau, and Stockton Plateau on one hand and the clade of western South Island and Stewart Island on the other. There is minor overlap only in the northern Paparoa Range (Gibbons et al., 2014). Thus, the main break in *Schizacme*, somewhere around the Paparoa Range, has taken place before the break in the Tasman Basin rifted the Australia–Paparoa Range clade apart. The latter event occurred at 85 Ma, with the opening of the Tasman, suggesting that the former event occurred with the extension and uplift of the Paparoa metamorphic core complex at ~100 Ma.

Murihiku Terrane and the Southland Syncline

The rest of this chapter focuses on biogeography in the Eastern Province. The area is dominated by the terranes of the Torlesse group, but several other terranes are exposed in Southland. The first of these, moving east from the Western Province rocks, is the Drumduan terrane, discussed earlier. Next is the Brook Street island arc terrane, in which strata with decreasing age show an increasing influence of continental crust (Scott, 2013). This reflects the collision of the arc with the continent. The terrane includes important biogeographic localities such as Bluff, the Longwood Range, and the Takitimu Mountains.

Next in the sequence is the Murihiku terrane, which is distinguished by the Southland syncline, a great fold that developed in Cretaceous time. The axis of the syncline coincides with northern or southern biogeographic limits in many groups (Heads and Patrick, 2003). For example, in *Mazus* (Mazaceae), the syncline axis marks the break between *M. arenarius* and *M. radicans* (Figure 7.7; Heenan et al., 1996). *Mazus arenarius* and *Hebe perbella* in Northland (Chapter 5) are the only New Zealand plants with peloric flowers (Heads, 1994a); the effects of this mutation resemble the telescoping and fusion of parts that produced the angiosperm flower, and so it has general significance for angiosperm evolution.

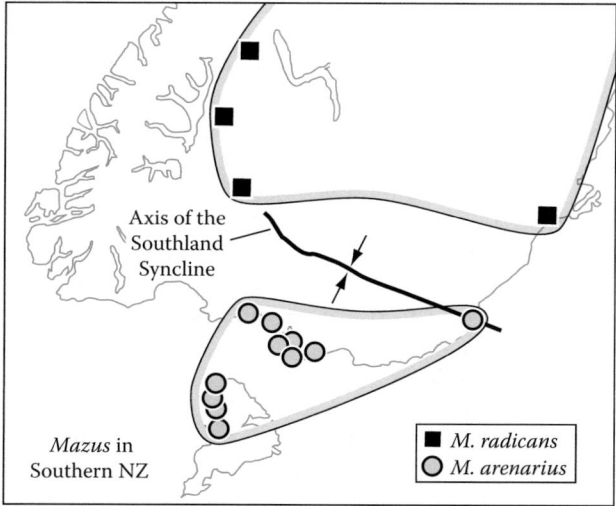

FIGURE 7.7 Distribution of *Mazus* (Mazaceae) in southern New Zealand. (Data from Heenan, P.B. et al., *N. Z. J. Bot.*, 34, 33, 1996.)

The Otago Schist

The Otago Schist (Figure 3.14) forms a northwest trending belt 150 km wide, in which metamorphism overprints the Torlesse terrane/Caples terrane boundary. (Only the central core of the belt, the garnet–biotite zone, is shown in Figure 3.14.) Two major structures in the schist were cited earlier (see Chapter 3): the Taieri–Wakatipu synform and the median antiform. The first runs between the northern and southern parts of the Blue Mountains (Turnbull and Allibone, 2003) and coincides with a biogeographic break there in insects and plants (Patrick et al., 1985). A similar break occurs at the median antiform, on the Rock and Pillar Range, as seen in the alpine weta *Hemideina maori* (Anostostomatidae). The species has two molecular clades and these correspond with color morphs (King et al., 2003):

Southern Rock and Pillar Range (black morph).
Northern Rock and Pillar Range, Otago, Canterbury, and Marlborough (yellow morph).

In addition, dark individuals have been reported from the Lammermoor and Umbrella Ranges, at the southern edge of the species range, although these have not been sequenced. The pattern indicates a north–south break at the axial antiform of the schist, along the line: Rock and Pillar Range–Pisa Range. The break has developed with the extension and uplift at the Cretaceous antiform, long before the uplift of the modern, block-faulted mountain ranges (Rock and Pillar Range, Pisa Range, etc.) in the Kaikoura orogeny. This history has meant that the populations of the region that became the Rock and Pillar Range do not form a monophyletic clade. This is also true for other block-faulted ranges of Otago. For example, the Blue Mountains biota includes the stonefly *Taraperla johnsi*, known only from the southern end of the range (Rankleburn headwaters) and Maungatua (McLellan, 2003), while the northern end of the range has its own endemics, such as the moth *Aoraia oreobolae* (Hepialidae; Dugdale, 1994).

Moonlight Tectonic Zone

The Moonlight tectonic zone (MTZ) is an aulacogen, or failed rift (Kamp, 1986), that cuts across the basement terranes of Southland and northwestern Otago (Figure 3.14). An arm of the sea extended inland along the rift, and deposits of Oligocene marine sediments are known from Bob's Cove, where the rift crosses Lake Wakatipu. The MTZ also coincides with an important regional boundary in biogeography (Heads, 1990a). Taxa with breaks here include species of *Sabatinca* in the "basal" family of Lepidoptera, Micropterigidae (Figure 7.8; Gibbs, 2014).

The MTZ is a zone of weakness that has accommodated part of the plate boundary strain, both before and after the development of the Alpine fault (White, 2002). North of Lake Wanaka, the MTZ meets the Siberia fault zone and the faults that define the upper Haast and Landsborough valleys (White, 2002). The MTZ formed part of the Challenger Rift System, a continental feature 1200 km long and 100–200 km wide that was active through western New Zealand during Eocene and Oligocene time (Kamp, 1986). Extension led to the opening of the rift, thinning of the crust, and the subsidence of the New Zealand Plateau. Later disruption of the sedimentary beds in the rift indicates that transcurrent displacement on the Alpine fault began in the Early Miocene (23 Ma). Deformation along the MTZ part of the rift was discussed by Turnbull et al. (1975), and Turnbull (1980) suggested that the MTZ has probably traversed the area from pre-Cenozoic time onward.

Groups with mutual boundaries at the MTZ include the harvestmen *Neopurcellia* (east to the MTZ) and *Rakaia* (west to the MTZ) (Boyer and Giribet, 2009). In the alpine plant *Celmisia sessiliflora* (Asteraceae), Given (1968) recognized a Fiordland form, a northern South Island form, and a Central Otago form, with the latter ranging west to Coronet Peak, 10 km east of the MTZ.

Western limits at the MTZ occur in several beetles, including *Oregus* (Carabidae) (Figure 7.9; Pawson and Emberson, 2001; Pawson et al., 2003), *Mimopeus impressifrons* (Tenebrionidae) (Figure 7.10; Johns, 2015), and *Sagola anisarthra* (Staphylinidae) (Figure 7.10; Park and Carlton, 2014). Plants with western limits at the MTZ include the genus *Leonohebe* (Plantaginaceae; Bayly and Kellow, 2006), *Hebe crawii*,

H. pimeleoides and the *H. annulata* group (Plantaginaceae; Heads, 1992, 1994c), *Poa senex* (Poaceae) (Edgar and Connor, 2000), *Coprosma obconica* (Rubiaceae) (de Lange and Gardner, 2002), *Coprosma pedicellata* (Molloy et al., 1999a), *Coprosma virescens* (NZPCN, 2016), *Celmisia prorepens* (NZPCN, 2016), *Olearia hectorii* (Heads, 1998b), *Gentianella corymbifera* subsp. *corymbifera* (Glenny, 2004), and the distinctive shrub *Carmichaelia crassicaulis* (Fabaceae; Heenan and Barkla, 2007). Taxa with an eastern limit at the MTZ include the plant *Hebe subalpina* (Plantaginaceae; Bayly and Kellow, 2006) and the beetle genus *Chalcodrya* (Chalcodryidae; Watt, 1974).

The alpine scree weta *Deinacrida connectens* (Orthoptera, Anostostomatidae) and its immediate allies extend west to the Moonlight tectonic zone (Figure 7.11; Gibbs, 2001; Trewick and Wallis, 2001). Molecular studies have indicated that the Stewart Island and Nelson endemics form one of seven clades in *D. connectens* s.lat. (the other six having allopatric ranges from Southland to Nelson) (Trewick and Wallis, 2001).

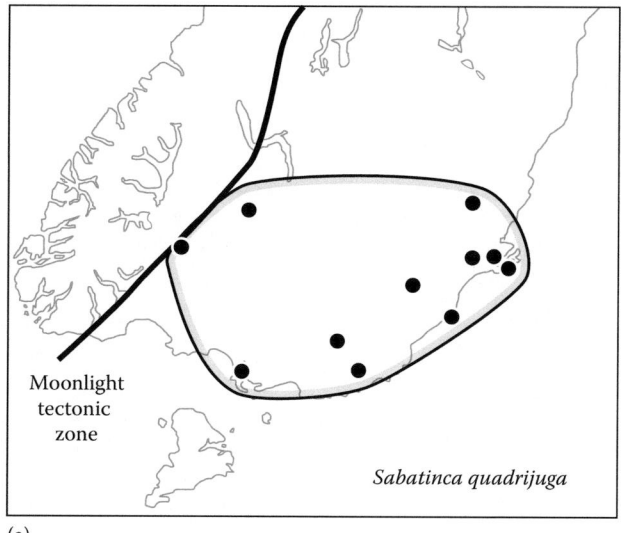

(a)

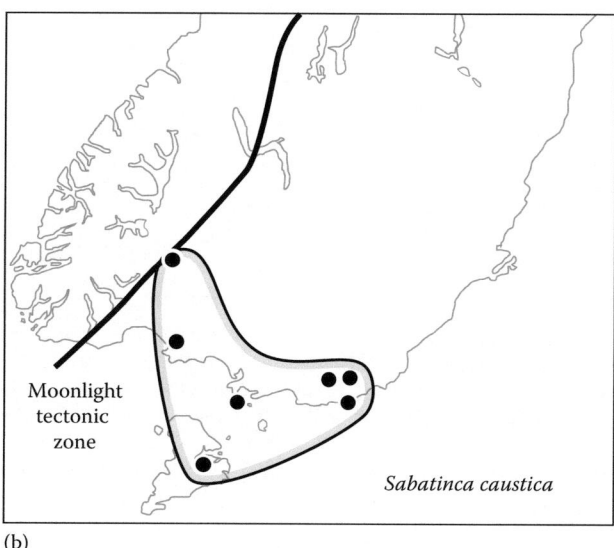

(b)

FIGURE 7.8 Distribution of three *Sabatinca* species (Lepidoptera, Micropterigidae). The dots in Gibbs (2014) were mapped on a 10′ × 10′ grid; here the dots near the Moonlight tectonic zone have been relocated slightly to the actual localities, as listed by Gibbs. (a) *S. quadrijuga* and (b) *S. caustica*. (Data from Gibbs, G.W., *Fauna N.Z.*, 72, 1, 2014.)   *(Continued)*

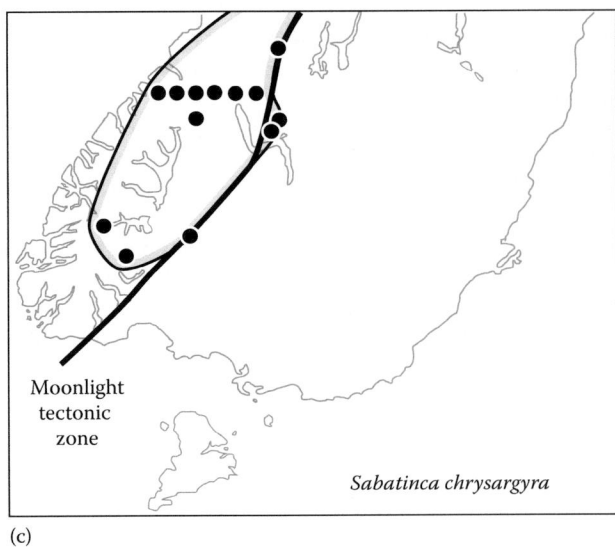

(c)

FIGURE 7.8 (*Continued*) Distribution of three *Sabatinca* species (Lepidoptera, Micropterigidae). The dots in Gibbs (2014) were mapped on a 10′ × 10′ grid; here the dots near the Moonlight tectonic zone have been relocated slightly to the actual localities, as listed by Gibbs. (c) *S. chrysargyra* (only the southern part of the range is shown). (Data from Gibbs, G.W., *Fauna N.Z.*, 72, 1, 2014.)

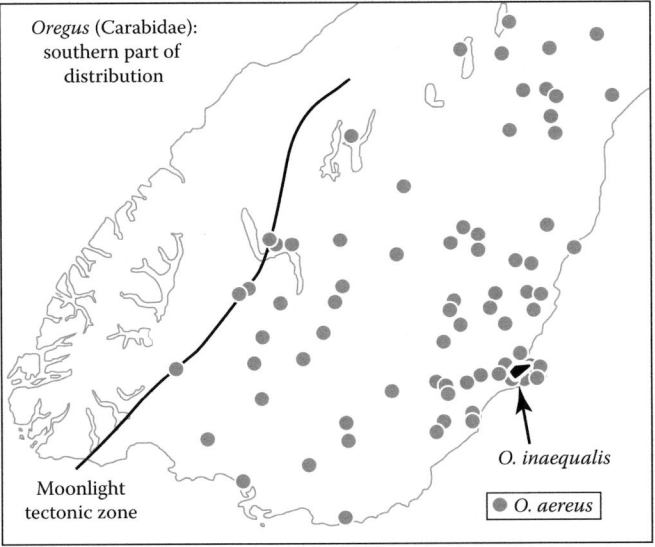

FIGURE 7.9 Distribution of *Oregus* (Carabidae) (only the southern part of the range is shown), with *O. inaequalis* endemic at Dunedin, and the widespread *O. aereus*. (Data from Pawson, S.M. and Emberson, R.M., *Oregus inaequalis Castelnau, its distribution and abundance at Swampy Summit Otago*, DOC Science Internal Series 6, New Zealand Department of Conservation, Wellington, New Zealand, 2001; Pawson, S.M. et al., *Invert. Syst.* 17, 625, 2003.)

Along with the break at the MTZ, this group also shows transcurrent disjunction along the Alpine fault and a distributional limit at the Waihemo fault zone; these are discussed in more detail in the following sections.

Pontodrilus (Annelida, Megascolecidae)

Species of the earthworm *Pontodrilus* occur in different parts of the world and most inhabit seashores in the intertidal zone. *P. lacustris* is unusual as it is known only from an inland population, near the western

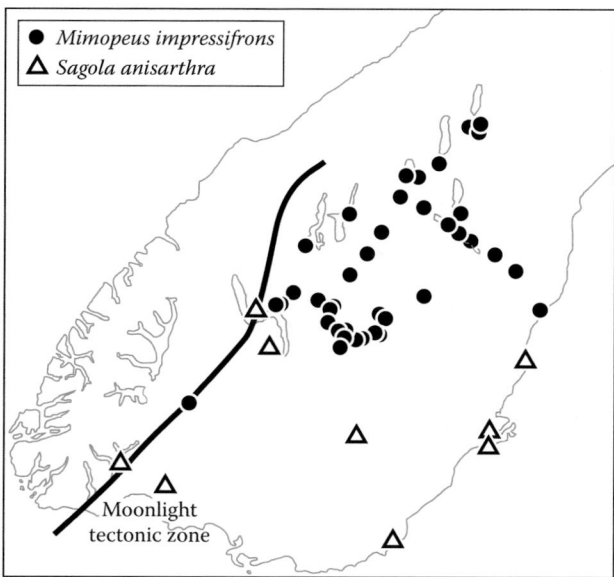

FIGURE 7.10. Distribution of *Mimopeus impressifrons* (Coleoptera, Tenebrionidae) and *Sagola anisarthra* (Coleoptera, Staphylinidae). (Data from Johns, P., New Zealand biodiversity. http://nzbiodiversity.com, accessed December, 2015; Park, J.-S. and Carlton, C.E., *The Coleopterists' Bulletin*, 68, monograph 4, 1–156, 2014.)

shore of Lake Wakatipu below Mount Nicholas (Lee, 1959). This site is located on the MTZ. Lee found its presence there "difficult to explain," as it is so far from any other *Pontodrilus* (the nearest is on the Chatham Islands), but the record is a typical case of a marine form stranded inland following the retreat of epicontinental seas. As mentioned earlier, Oligocene marine sediments are preserved along the MTZ at Lake Wakatipu.

Maoricicada campbelli (Hemiptera, Cicadidae)

The cicada genus *Maoricicada* belongs to a diverse New Zealand–New Caledonia complex of genera (Chapter 6). Buckley et al. (2001) studied the widespread New Zealand *Maoricicada campbelli* and aimed to evaluate biogeographic hypotheses for the alpine biota, in particular, disjunction at the Alpine fault. The fault is only one of many relevant tectonic structures though. For example, in both Fiordland and Nelson, *M. campbelli* (Fleming, 1971) has its western limit at the Median Batholith (Figure 7.12 shows the Fiordland boundary; collection localities from Hill et al., 2009, Appendix). Another main break within *M. campbelli* separates the "Central Otago clade" and "northern clade 1," and this coincides with the Moonlight tectonic zone, especially at Lake Wakatipu and southwest of there.

Maoricicada campbelli occurs in North and South Islands and its basal phylogenetic split separates the "Central Otago clade" from its sister, the "northern clade" (Figure 7.12; the northern clade has four allopatric subgroups; only the southern two are shown in the figure). Buckley et al. (2001) referred to the deep genetic divergence at the basal node in the species, which is "surprisingly large" and exceeds that between other *Maoricicada* species. The authors also stressed the narrow geographic separation between the clades at Lake Wakatipu and discussed the break: "We do not believe that natural selection is capable of maintaining such high levels of mtDNA variation over such a small geographic range (i.e., in the Lake Wakatipu region)" (p. 1403). Instead, they proposed that the most likely explanation for the break is an "ancient cladogenetic event," which is reasonable. They did not propose any event in particular, but the location of the break suggests it was related to deformation along the MTZ. In eastern Otago, the southern limit of the "northern clade 2" is near Ranfurly, at the Waihemo fault zone.

Buckley et al. (2001) suggested that the "northern clade" migrated down from the north of the South Island and surrounded the Central Otago clade. In a vicariance model, the main clades in *M. campbelli*

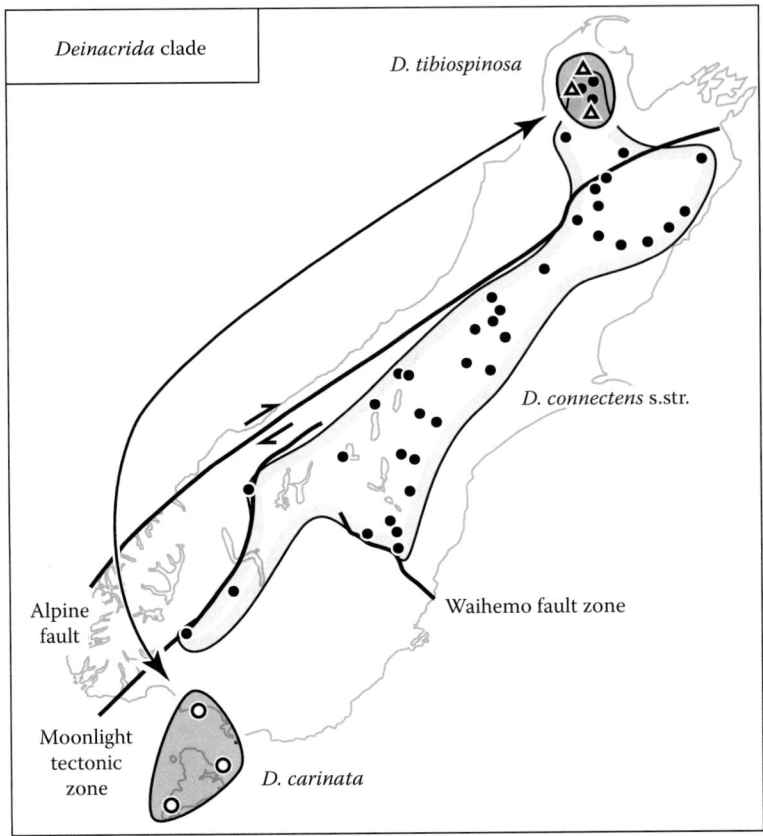

FIGURE 7.11   Distribution of a clade in *Deinacrida*: *D. connectens*, *D. tibiospinosa*, and *D. carinata* (Orthoptera, Anostostomatidae). (Data from Gibbs, G.W., Habitats and biogeography of New Zealand's deinacridine and tusked weta species, in *The Biology of Wetas, King Crickets and Their Allies*, ed. L.H. Field, CABI Publishing, Oxford, UK, 2001, pp. 35–56; Trewick, S.A. and Wallis, G.P., *Evolution*, 55, 2170, 2001.)

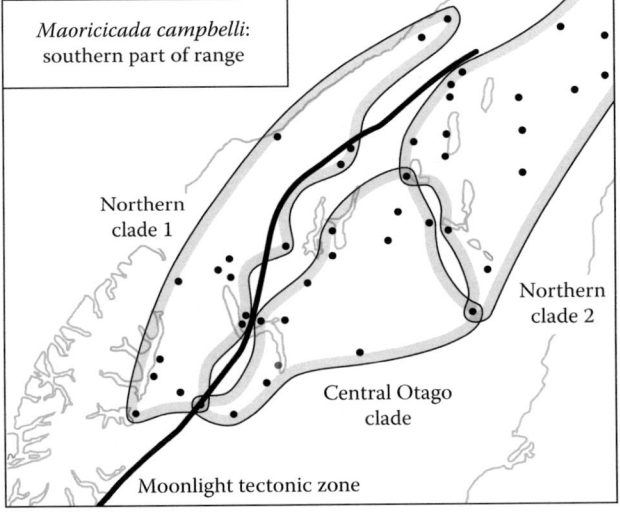

FIGURE 7.12   Distribution of the cicada *Maoricicada campbelli* in the southern part of its range. (Data from Buckley, T.R. et al., *Evolution*, 55, 1395, 2001; Buckley, T.R. and Simon, S., *N. Z. Entomol.*, 33, 118, 2007.)

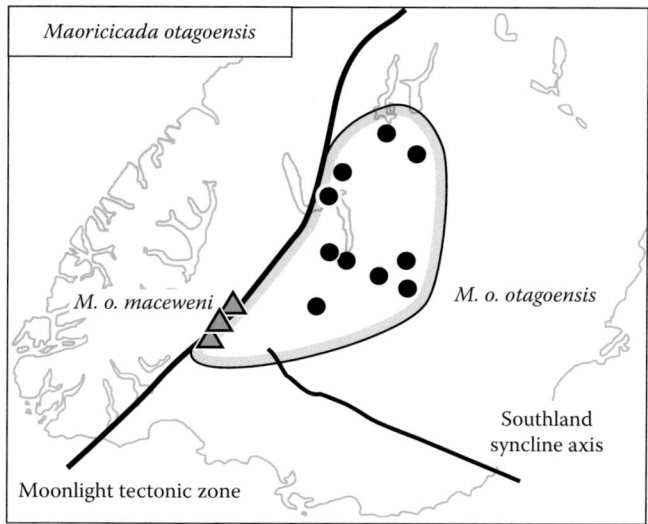

FIGURE 7.13 Distribution of the cicada *Maoricicada otagoensis*. (Data from Dugdale, J.S. and Fleming, C.A., *N. Z. J. Zool.*, 5, 295, 1978.)

have evolved by simple allopatric differentiation at the MTZ and the Waihemo fault zone. Other species of *Maoricicada* that have a boundary at the MTZ include *M. otagoensis* (Figure 7.13), *M. clamitans*, and *M. nigra* (Dugdale and Fleming, 1978). *M. otagoensis otagoensis* is another Central Otago endemic, comparable with the "Central Otago clade" of the *M. campbelli* group.

Calibrating Clocks in Maoricicada Using the Age of Habitats: "Mountain Age" and "Island Age" Methods

Buckley et al. (2001) calibrated the timeline for phylogeny in *Maoricicada campbelli* by assuming that *Maoricicada* was no older than *the current alpine habitat* (~5 Ma), as the genus is mainly alpine. This makes the assumption that taxa currently located in the alpine zone require alpine habitat, and this is not necessarily correct. In any case, *M. lindsayi* occurs at 0–500 m elevation, and *M. campbelli* itself is recorded from 10 to 1580 m (Buckley and Simon, 2007). In fact, most of the "alpine" species of *Maoricicada* are known from some localities below the alpine zone; 10 out of 14 have populations at less than 1000 m (around the level of the tree line) and several have populations in the lowlands (Dugdale and Fleming, 1978). These records suggest that *Maoricicada* and at least some of its diversity existed *before* the present alpine environment developed, surviving in lowland areas of open vegetation. The diverse ancestral complex would have been inherited by the mountains of the Kaikoura orogeny, as they rose. If this was the history, it would undermine a habitat calibration based on the age of the mountains.

Based on the "mountain age" *habitat* calibration, Buckley et al. (2001) calculated that the earliest divergence within *M. campbelli* occurred at 2.3 Ma, and they concluded: "We cannot reconcile the earliest date of divergence within *M. campbelli* with rifting along the Alpine Fault, a process that began to accelerate 10 million years ago." Yet the young age they calculated for divergence in *M. campbelli* is based on the doubtful assumption they made about the age of the genus.

Buckley et al. (2001) assumed that alpine taxa in *Maoricicada* could be no older than their alpine *habitats*. In a subsequent study of the genus, Buckley and Simon (2007) abandoned this method. Instead, though, they calibrated a molecular clock for *Maoricicada* by assuming that related cicadas, endemic on Norfolk Island and Kermadec Islands, could be no older than their current *islands*. But islands are just another habitat type within a region, as are mountains, leaf surfaces, or puddles. The existence of *prior* islands in the region means that calibration using island age, will, as with calibration using habitat age, give dates for clades that are too young by an indeterminate amount. Buckley and Simon (2007) found that *Maoricicada* diverged only in the mid-Miocene. This date is too young (by how much is not known), but at least it confirms that *Maoricicada* existed before the origin of high montane habitats.

Hill et al. (2009) carried out further studies on *Maoricicada campbelli* and confirmed that the main break in *M. campbelli* is between a Central Otago clade and the rest. They wrote "we can only speculate as to what process initiated both the divergence of *M. campbelli* from the rest of *Maoricicada* at <2.6 Ma, and the main *M. campbelli* split at 1.5–2.1 Ma" (p. 13). As discussed, the "island age" calibration (from Buckley and Simon, 2007) that these dates depend on is flawed.

The authors based their conclusions on these dates, but they did not refer to major geological features at the boundaries, such as the Moonlight tectonic zone, and instead suggested that increased sedimentation of river gravel caused by rising mountains produced habitat "into which *M. campbelli* evolved" (p. 14). The new sedimentation and new habitat might explain local range expansion or contraction in the group during the Kaikoura orogeny, but they do not account for the *location* of the main break in the species.

Other Northeast-Trending Features in Otago

Other large-scale fault systems in central and east Otago strike northeast, along with the Moonlight tectonic zone (Figure 3.13), and at least three of these coincide with important biogeographic breaks. These tectonic features are introduced next.

Nevis–Cardrona Fault System

The Nevis–Cardrona Fault System, including the Nokomai fault (Figure 3.13; Beanland and Barrow-Hurlbert, 1988; Kerr et al., 2000), marks a significant biogeographic boundary in freshwater fishes. It coincides with the split between the species *Galaxias* "southern" (to the west) and *Galaxias* "D" (to the east), and in *G. gollumoides* one of the two main clades is endemic to the Nevis, while the other is widespread through Southland (Waters et al., 2001, 2010).

There is also a major east/west break at the Nevis–Cardrona fault in the skink *Oligosoma maccannii* (between clade 1 + 2 and clade 3 + 4; O'Neill et al., 2008). The same region is also significant for plants; for example, *Carex pterocarpa* (Cyperaceae) extends from the Old Man Range of South Canterbury, south and west through Central Otago to the Nokomai area (Gow cirque) (Dickinson et al., 1998).

A critical question for the biogeography of the fault system is the time of its origin. Geological studies concluded that the faults in the Nevis–Cardrona Fault System, mainly located in the Otago Schist, are older structures that have been reactivated (Beanland and Barrow-Hurlbert, 1988). Turnbull (1981) discussed post-metamorphic folding and faulting in the schist. He found that considerable fault movement took place in the Cromwell area prior to the Miocene deposition of Manuherikia Group sediments over a peneplain schist surface. This faulting was accompanied by mesoscopic kink folding and warping and possibly by macroscopic warping. The faulting, folding, and warping "probably dates from the Cretaceous Rangitata Orogeny" (Turnbull, 1981). A separate phase of reverse faulting took place much later in the Piocene–Pleistocene, during the Kaikoura orogeny. In summary, the Nevis–Cardrona faults themselves (not their Neogene reverse displacement) formed with the mid-Cretaceous folding in the schist and the Rangitata orogeny.

Titri–Leith–Waitati Fault Zone and Dunedin Endemism

On the east coast of Otago, endemism at Dunedin is recorded in groups such the beetle *Oregus inaequalis* (Figure 7.9, from Pawson et al., 2003). Its sister group is *O. aereus* (Goldberg et al., 2014), and this is widespread in Otago. The land snail *Alsolemia* (Charopidae) shows a similar pattern. It comprises two species: *A. cresswelli* of north Dunedin (Black's Bush, the type locality, and Bethunes Gully; D. Roscoe pers. comm., 2013) and *A. monoplax* of the southern South Island and Codfish Island (Figure 7.14; Climo, 1989, and F. Climo pers. comm., 2013).

Oregus inaequalis, Alsolemia cresswelli, and other local endemics are restricted to Dunedin, on and east of the Titri–Leith–Waitati fault zone (Figures 3.13 and 7.14). This feature strikes northeast along the

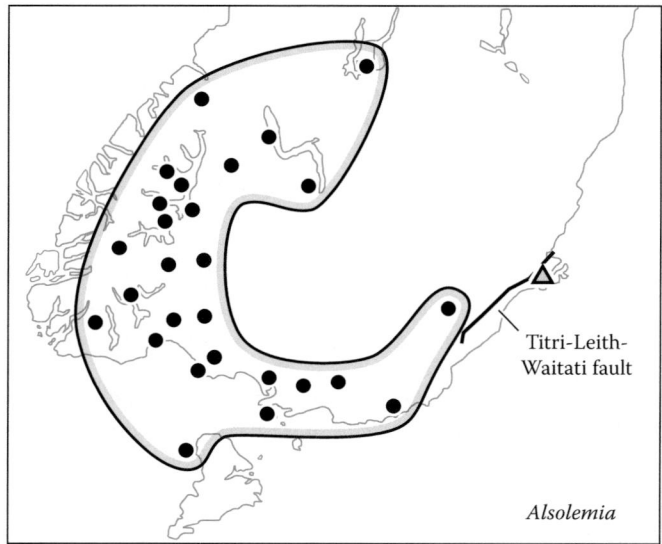

FIGURE 7.14   Distribution of the land snail genus *Alsolemia* (Charopidae). Dots, *A. monoplax*; triangle, *A. cresswelli.* (Data from Climo, F.M., *N. Z. J. Zool.*, 16, 587, 1989.)

eastern side of the Taieri Plain, across the eastern slopes of Flagstaff and Swampy Summit above Leith Valley, and north to Waitati (Price and Coombs, 1975; Bishop and Turnbull, 1996). Mid-Cretaceous Henley Breccia is preserved in the fault angle.

In former times, before submergence of the continental plateau, Dunedin groups, including the endemics, probably ranged further east. The Saunders Ridges, located 10 km off the Dunedin coast, could represent part of an ancient barrier island complex (Gorman et al., 2013). Likewise, the remains of small volcanoes with wave-planed tops, dated at 35–30 Ma, occur on the submerged continental shelf along the east Otago coast around Oamaru (Cas et al., 1989; Hicks, 2014).

Other Dunedin endemics, all found east of the Titri–Leith–Waitati fault zone, include the following lichens (Galloway, 2007):

Anisomeridium laevigatum: Otago Peninsula, Bethunes Gully.

Encephalographa otagensis: St Clair, Black Head, Green Island Bluff, on coastal rocks. (This is the sole New Zealand representative of the genus.)

Endococcus ramalinarius: Green Island Bush. (There is similar material from Concepción, Chile.)

Phoma dubia: Pelichet Bay (now Logan Park.)

Porina otagensis: Bethune's Gully, by the Mt Cargill track.

Strigula johnsonii: Morrison's Creek and Mt Cargill.

Dunedin endemics in other groups include the plant *Helichrysum selago* var. *tumidum* (Asteraceae) on Otago Peninsula (Allan, 1961); the harvestman *Prasma sorenseni regalia* of Dunedin city (Royal Terrace) and Warrington (Forster, 1954); the harvestman *Forsteropsalis marplesi* of Leith Valley and Leith Saddle (closest to *F. chiltoni* of Stewart Island) (Taylor, 2011); the dipteran *Anabarhynchus fuscofemoratus* (Therevidae) on Otago Peninsula (Harris, 2006); another undescribed fly *Zelandotipula* sp. d (Tipulidae), known only from Leith Valley (Collier, 1992); the plecopteran *Nesoperla patricki* of Swampy Summit (McLellan, 2003); the trichopteran *Olinga fumosa* of Swampy Summit; the trichopteran *Pseudoeconesus paludis* of Swampy Summit and Otago Peninsula (Peat and Patrick, 2014); and an undescribed butterfly, a *Lycaena* species (Lycaenidae) from Pine Hill and Port Chalmers (R.C. Craw, pers. comm., 2014; Patrick and Patrick, 2012). South of Dunedin, but still east of the Titri fault, another undescribed *Lycaena* species is endemic at Chrystalls Beach (R.C. Craw, pers. comm., 2014; Patrick and Patrick, 2012). This is situated on the narrow sliver of land east of the Akatore fault (Figure 3.13).

Complementing the Dunedin endemics are notable absences in the biota. For example, "the absence of *Celatoblatta* cockroaches and *Megadromus* beetles is one of the strange anomalies of the Otago Peninsula" (Johns, 2015, under *Celatoblatta subcorticaria*).

Some of the eastern endemics have apparent disjunctions. For example, the beetle *Mitophyllus fusculus* (Lucanidae) is known from the Catlins (Hunters Hills), near Taieri Mouth, and Moeraki, from near sea level to 550 m (Holloway, 2007). *Acaena pallida* (Rosaceae) is in the southeast at Stewart Island, the Invercargill area, the Catlins, and Dunedin east of the Titri–Leith–Waitati fault, but also has disjunct populations at the Chatham Islands, Cook Strait, and southeastern Australia (Macmillan, 1991; NZPCN, 2016).

In contrast with all these groups, which are found east of the Titri–Leith–Waitati fault, typical Central Otago elements instead range west of the fault, for example, *Celmisia prorepens* (Asteraceae) extends from Maungatua (near Dunedin west of the fault) and Rock and Pillar Range, west to the Eyre Mountains and the Moonlight tectonic zone.

Groups with northern limits at Dunedin include the albatross *Diomedea sanfordi*, the beetle *Kenodactylus*, the amphipod *Orchestia aucklandiae* (all cited in Chapter 6), and the coastal fish *Notothenia microlepidota* (Nototheniidae; Francis, 2012). Other limits at Dunedin include the following.

The Little Blue Penguin Eudyptula

Several groups otherwise restricted to the subantarctic are represented by "sliver" distributions on the mainland around the Titri fault. This is seen in the plant *Poa* (Figure 6.4), in the seabird *Diomedea*, and in the amphipod *Orchestia aucklandiae* (Chapter 6). Breaks in biogeography coinciding with northeast-striking faults are also evident in the little blue penguin, *Eudyptula*. This is a polymorphic group in which several subspecies and either one or two species have been recognized. The two main clades are distributed as follows (Peucker et al., 2009; Grosser et al., 2015):

Australia, Tasmania, and New Zealand: SE South Island coast only (Catlins, Otago Peninsula, Oamaru, and Motunau I) ("ASENZ" clade = *E. novaehollandiae*).

All mainland New Zealand (including Stewart Island, Oamaru, and Banks Peninsula) ("NZO" clade = *E. minor*).

Peucker et al. (2009) wrote: "There is no evidence of a geographic barrier that accounts for the localized presence of the ASENZ clade within New Zealand, and it may be slowly expanding its range." Nevertheless, the genetic break between the ASENZ and NZO clades coincides spatially with the Titri–Waitati fault, with the former to the east and the latter to the west.

With respect to timing, the break between the two clades probably developed prior to the Tasman seafloor spreading at 84 Ma that ruptured the ASENZ clade. The Titri–Waitati fault again satisfies this requirement as it was active in the mid-Cretaceous (as indicated by the Aptian–Albian Henley Breccia in the fault angle). In the mid-Cretaceous, the fault underwent normal (extensional) displacement, and this would have pulled apart crustal blocks as well as penguin breeding sites.

Such a vicariance model for *Eudyptula* suggests natal philopatry in the early forms, with the Africa–Australia–southeastern South Island populations splitting off as a *southern* vicariant of the main New Zealand clade. On a pre-Miocene geography, before transcurrent displacement along the Alpine fault (Figure 6.2), the boundary is more or less a straight line. Suitable marine habitat would have already been available on the Campbell Plateau because of the opening of the Bounty Trough.

The Penguins Megadyptes *and* Eudyptes

The break in *Eudyptula* that coincides with the Titri–Waitati fault can be compared with the east/west break between breeding localities in two other penguins, *Megadyptes* and *Eudyptes*. *Megadyptes* breeds on the New Zealand subantarctic islands (Auckland, Campbell, Snares), Stewart Island, and the east coast of the South Island (breeding from Stewart Island to Banks Peninsula). Vagrants are known north to East Cape and at the Chatham Islands, and there are Holocene fossils on the Chathams (Wood et al., 2014a).

Ekman (1953) regarded penguins as "true marine animals," and he found the restricted distribution of *Megadyptes* "remarkable."

The sister group of the eastern *Megadyptes* is the western *Eudyptes* (Baker et al., 2006). This breeds on subantarctic islands in the Indian and Atlantic Oceans, and in the New Zealand region from the subantarctic islands north to Stewart Island, Solander Island, Fiordland, and Westland. The breeding records overlap with those of *Megadyptes* on the subantarctic islands and western Stewart Island/Codfish Island, but north of there the two genera are allopatric, with *Eudyptes* in the west and *Megadyptes* in the east (Robertson et al., 2007). The *Megadyptes/Eudyptes* split indicates a possible break at the Great South Basin and the Titri–Waitati fault, mid-Cretaceous extensional features.

Gore–Rock and Pillar Range–Kyeburn

Many Otago groups have an eastern or western boundary at a line striking northeast—Fortrose (Toetoes Bay)–Gore–Rock and Rillar Range–Kyeburn—and many are endemic there. The boundary lies along the Hyde fault, at the eastern side of the Rock and Pillar Range (Figure 3.13), but continues further southwest and northeast. Examples include the following Curculionidae (Barratt and Kuschel, 1996):

Irenimus similis: Rock and Pillar and Lammermoor Range.

I. patricki: Rock and Pillar and Lammermoor Range.

I. stolidus: Maungatua, Rock and Pillar, Taieri Plain, Strath Taieri (Hyde), Macrae's Flat, Palmerston, and also Lammermoor Range.

I. aemulator: Gore, Rock and Pillar Range, Kyeburn, and Macrae's Flat.

I. vastator: Fortrose.

In caddisflies (Trichoptera), *Neurochorema pilosum* has a similar distribution, at Clydevale, Middlemarch, and Macrae's Flat (Heads and Patrick, 2003: Fig. 13).

Waihemo Fault Zone

The Waihemo fault zone, by the Shag River in the southeastern South Island, is a normal fault system that developed in the mid-Cretaceous (Tulloch et al., 2009). It also forms a biogeographic boundary, and examples have already been referred to in weevils (Figure 7.4) and wetas (Figure 7.11). The Waihemo fault zone meets the Hawkdun and Blue Lake faults at its northwestern end, runs offshore at Shag Point, and is aligned with the southern edge of the Bounty Trough (Curran et al., 2010).

The Waihemo fault zone separates many Otago groups to the south from Canterbury/Marlborough groups to the north. The Hawkdun Range, Mount Ida Range, St Mary Range, Danseys Pass, and the Kakanui Mountains form an uplifted belt along the northern margin of the Waihemo fault zone–Hawkdun fault, and this is the southern limit for many groups. Plant genera ranging south to here include the following:

The herb *Notothlaspi* (Brassicaceae): Nelson and Marlborough south to the Waihemo fault zone (Mt Ida; Petrie, 1896).

The shrub *Exocarpos* (Santalaceae): Nelson and Marlborough south to the Waihemo fault zone (head of the Eweburn; Petrie, 1896; personal observations, 1984). The genus is absent from the North Island, where the only other New Zealand member of Santalaceae, *Mida*, is endemic (North Cape to Wellington).

Other groups ranging south to the Waihemo fault zone include the following:

The lichen *Xanthoparmelia pictada*: Inland Kaikoura Range, Torlesse Range, Kakanui Mountains, Mt Ida, and Mt St Mary (Galloway, 2007).

The shrub *Hebe pinguifolia* (Plantaginaceae): Richmond Range south to Mount Arnould, St Bathans Range, Hawkdun Range, and Mount Pisgah (Heads, 1993; Bayly and Kellow, 2006).

The herb *Leptinella atrata* s.str. (Asteraceae): Torlesse and Craigieburn Ranges, south to Mount Kyeburn and Dansey's Pass (Lloyd, 1972; Himmelreich et al., 2012).

The butterfly *Percnodaimon pluto* (Nymphalidae): Northeastern South Island, south to Mount Kyeburn and Mount Prospect (east of Lake Hawea) (Patrick and Patrick, 2012).

The butterfly *Percnodaimon* "sp. 4": Southern Canterbury, south to the St Bathans, Hawkdun, Ida, Kyeburn, and Kakanui Ranges (Patrick and Patrick, 2012).

These ranges are also centers of endemism; for example, *Prodontria patricki* (Scarabaeidae) is endemic to montane sites in the Hawkdun Range and at Dansey's Pass (Heads and Patrick, 2003).

Finally, the Waihemo fault zone is also a northeastern limit for Central Otago groups, such as the butterfly *Lycaena* sp. (Lycaenidae) ("Waihemo boulder"). It is recorded from Strath Taieri, Waihemo, Pig Root, and Shag Valley (Patrick and Patrick, 2012).

Around the Waihemo fault zone, field biologists have concentrated on the montane groups, but the biota of lower elevation sites is also informative. For example, Awahokomo Creek on the northern slopes of Mt St Mary/Kohurau represents both a southern limit (*Gentianella calcis*, Gentianaceae; *Pimelea declivis*, Thymelaeaceae) and a center of endemism (*G. calcis* subsp. *calcis*, *Carmichaelia hollowayi*, Fabaceae, and *Poa spania*, Poaceae; Molloy et al., 1999b). *Pachycladon exile* (Brassicaceae) is only known from there, Cape Wanbrow on the coast (Molloy et al., 1999a), and the "Otepopo" (Waianakarua) River at the eastern end of the Kakanui Mountains (Petrie, 1896). *P. exile* grows on various types of parent material, including limestone, tuff, breccia, and alluvium (Heenan and Mitchell, 2003), indicating that the endemism and boundaries in the Awahokomo/Otepopo region are not simply determined by the limestone soils.

The Waihemo fault zone, like the Moonlight tectonic zone, is a mid-Cretaceous extensional feature that has broad regional significance for biogeography. It is thought to have developed at ~112–113 Ma, based on dates of the associated Kyeburn Formation and Shag Point ignimbrite (Tulloch et al., 2009; Henne et al., 2011). The fault development followed the collision and suturing of the Torlesse and Caples terranes; this caused regional metamorphism at the suture zone (the Haast Schist) and the last phase of the Rangitata orogeny.

The schist was later exhumed during Aptian–Albian rifting either by crustal thickening (underplating) or by extension associated with rifting between Australia and Antarctica. The first sedimentary record of the exposed schist is in the Kyeburn Formation, preserved at the Waihemo fault zone. The Aptian–Albian age for the exhumation of the schist and the inception of the Waihemo fault zone provide a date for the significant biological differentiation there.

The greatest fault movements on the Waihemo fault zone took place during the mid-Cretaceous, when the high-rank schist in the southwest formed the upthrow side of what was then a normal (extensional) fault system. Further north, the Waitaki fault system along the Waitaki valley is also made up of normal faults that formed at this time. The Waihemo fault zone was later reactivated after Late Miocene time, in the Kaikoura orogeny. In this phase, movement took place in the opposite direction, with the northeastern side being elevated by reverse (compressional) faulting to raise the modern Kakanui Range (Mutch, 1963).

The Aoraki Node

The *Lobelia macrodon* group (Campanulaceae) is widespread through much of the South Island, although it is absent in the Paparoa Range and in Stewart Island, despite suitable habitat there (Figure 7.15; Knox et al., 2008). (An example of *endemism* along this sector of absence is the dolichopodid fly *Apterachalcus*, present at the Paparoa Range, Stewart Island, Auckland, and Campbell Islands; see the earlier section in this chapter "Paparoa Range and Papahaua Range"; Bickel, 1991.)

Within the *Lobelia macrodon* group, the three clades meet at a node near Aoraki, or Mount Cook. In the traditional biogeographic model, this sort of pattern would be explained either by dispersal *to* Aoraki from refugia in the far south and the far north of the South Island (Fleming, 1979) or by dispersal *from* a

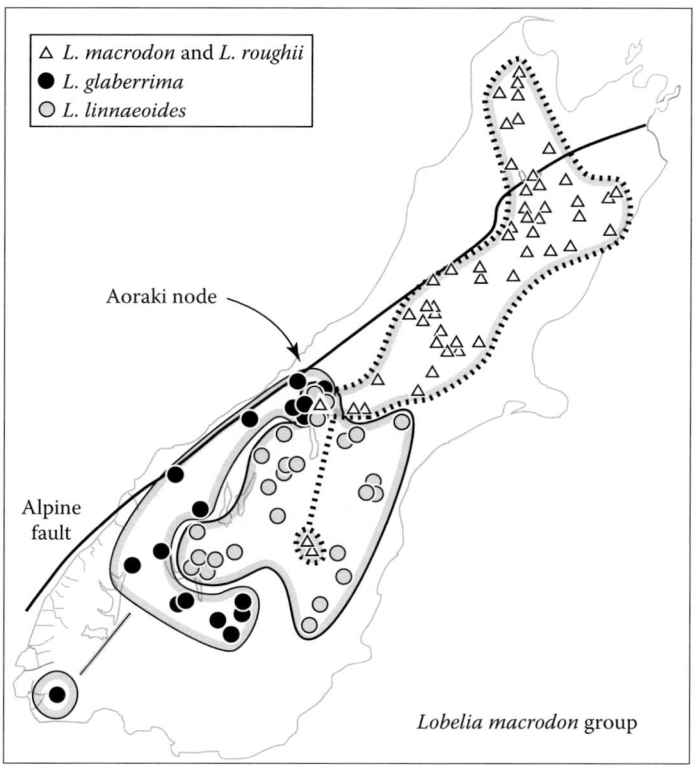

FIGURE 7.15 Distribution of the *Lobelia macrodon* clade (Campanulaceae). (Data from Knox, E.B. et al., *N. Z. J. Bot.*, 46, 77, 2008.)

center of origin at Aoraki. Instead, it could have resulted from differentiation of an already widespread South Island ancestor around a phylogenetic and biogeographic break at, or near, Aoraki. This vicariance model is consistent with the many groups that show geographic limits or endemism at Aoraki. The model implies that the clades have survived in multiple microrefugia throughout the South Island, not just in the far north and south of the South Island.

The Aoraki node is also documented in the rock wren *Xenicus gilviventris*, one of two extant members of Acanthisittidae. This species is restricted to the South Island and is an alpine species only known from above 900 m. It comprises two main clades, one ranging from Fiordland to Mount Cook and the other from Mount Cook to northwest Nelson (Weston and Robertson, 2015).

The Mount Cook region (Figure 7.16) appears as a northern limit in *Carmichaelia petriei* (Figure 7.5), *Lobelia glaberrima* (Figure 7.15), and other well-collected plants such as the following:

Hebe buchananii (Plantaginaceae): Longwood Range and eastern Fiordland to Mount Cook (Malte Brun) (Bayly and Kellow, 2006).

Epilobium purpuratum (Onagraceae): Fiordland–Rock and Pillar Range–Mount Cook (Raven and Raven, 1976).

Dracophyllum prostratum (Ericaceae): Fiordland–eastern Otago, north to Mount Cook (Venter, 2009).

Gingidia amphistoma (Apiaceae): Forbes Mountains and Mount Aspiring north to the Mount Cook Range and the nearby Malte Brun Range (Heenan et al., 2013b).

The caddis fly *Paroxyethira pounamu* ranges from Fiordland north to the Mount Cook node (Irishman Creek by Lake Pukaki) (Ward and Henderson, 2004). In other records from insects (all from Johns, 2015),

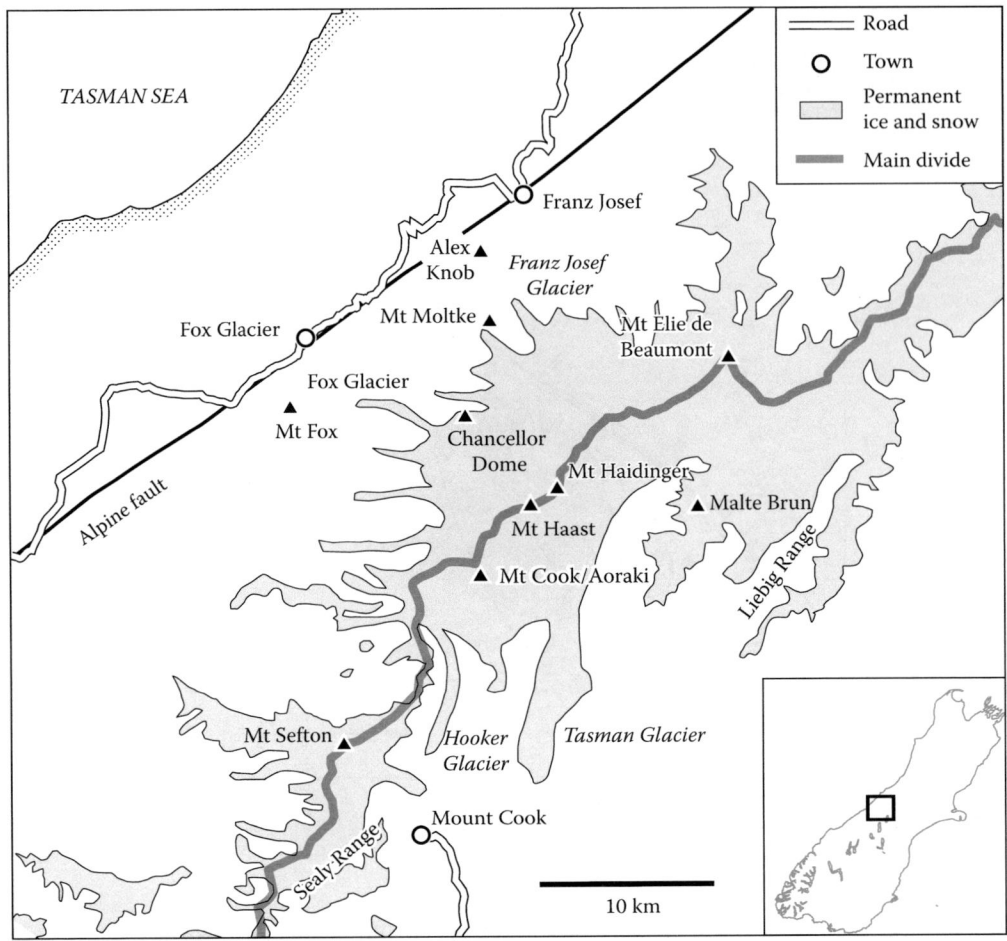

FIGURE 7.16 The Mount Cook region of the South Island, showing the area of permanent ice and snow (small, ice-free areas within this are not shown) and some of the localities mentioned in the text.

Megadromus bullatus (Carabidae) is a spectacular, well-collected beetle known from Stewart Island, eastern Fiordland, and Central Otago, north to Mount Cook (5 km south of Mount Cook village). Another carabid, *Holcaspis ohauensis*, is recorded at Mount Cook (Mount Cook Station–McLeod Creek) and around Lake Ohau. *Mimopeus impressifrons* (Tenebrionidae) ranges west to the Moonlight tectonic zone (at the Takitimu Range and Frankton) and north to the southern ends of Lakes Pukaki and Tekapo. The cockroach *Celatoblatta anisoptera* (Blattidae) has limits at Alexandra (Clutha Gorge), junction of east and west Matukituki valleys (Moonlight tectonic zone), Mount Cook (Hooker Valley, Malte Brun), and on the east coast at Oamaru.

Other groups link Aoraki with the Mount Torlesse region to the northeast. For example, the lichen *Xanthoparmelia malcolmii* is recorded from the Kea Point trail near Mount Cook village and from the Torlesse Range (Galloway, 2007).

Although many authors have focused on the high levels of endemism in the northern and southern thirds of the South Island, there is also significant endemism in the central provinces of Westland and Canterbury, and these contradict a simple "glacial elimination" model for the central South Island. Endemism in the Aoraki/Mount Cook region in particular has been neglected but is now recorded in several groups:

The lichen *Lecidea lapicida* var. *maungahukae*: Mounts Cook, Haast, and Haidinger, 2925–3555 m (Galloway, 2007).

The lichen *Xanthoparmelia bulfiniana*: Tasman Valley, on moraine debris (Galloway, 2007).

The liverwort *Andrewsianthus hodgsoniae*: Sealy Range (Engel and Glenny, 2008).

The angiosperm *Aciphylla inermis* (Apiaceae): Sealy Range (Allan, 1961; possibly a hybrid).

The land snail *Allodiscus cryptobidens*: Hooker Valley (Marshall and Barker, 2008).

The land snail *Allodiscus godeti*: Hooker Valley (Marshall and Barker, 2008).

The dipteran *Anabarhynchus harrisi* (Therevidae): Tasman Valley (Harris, 2006).

The hymenopteran *Shireplitis meriadoci* (Braconidae): Tokea Point (Fernández-Triana et al., 2013).

An undescribed species of the moth *Dichromodes* (Geometridae): Known only from 3000 m on Mount Cook (B. Patrick, pers. comm., April 10, 2015).

Other endemics in the Mount Cook region extend from east to west of the *main divide*, but do not occur west of the *Alpine fault*. An example is the moth *Hierodoris eremita* (Oecophoridae), with records from the Liebig Range, Ball Glacier, Hooker Valley, Sealy Range (all east of the main divide), Fox Glacier, and Alex Knob, above Franz Josef Glacier (west of the main divide) (Hoare, 2005).

Likewise, the hydraenid beetle *Podaena mariae* is restricted to Black Birch Stream near Mount Cook village (east of the divide) and near Fox Glacier (west of the divide) (Delgado and Palma, 2010).

Mount Cook endemics can have regional significance. In the tree *Pseudopanax crassifolius* (Araliaceae), a population from Mount Cook is sister to all the other sampled populations, distributed throughout the rest of the mainland and also (as *P. "chathamicus"*) on Chatham Islands (Perrie and Shepherd, 2009).

Westland Endemism around the Fox and Franz Josef Glaciers

Several forms are restricted to a narrow sliver west of the main divide but east of the Alpine fault, located around Fox and Franz Josef glaciers, and ~10 km north of Mount Cook:

The harvestman *Synthetonychia glacialis*: Franz Josef (Forster, 1954).

Vesicaperla townsendi (Plecoptera): Glacier Road, Franz Josef, at the Alex Knob turnoff (McLellan, 1977).

Spaniocerca hamishi (Plecoptera): Franz Josef Glacier ("base of waterfall"), Alex Knob above Franz Josef Glacier, Fox Glacier ("streams near terminal"), and intermediate localities (McLellan, 2000).

Lyperobius glacialis (Coleoptera, Curculionidae): mountains surrounding Franz Josef and Fox Glaciers—Mt Moltke, Castle Rocks, and Chancellor Shelf (Craw, 1999).

Odontria convexa (Coleoptera, Scarabaeidae): Franz Josef (Lake Wombat and an early collection from "Waiho") and near Fox Glacier village (Johns, 2015).

Taenarthrus gelidimontanus (Coleoptera, Carabidae): above Franz Josef Glacier (Alex Knob) and Fox Glacier (Chancellor Hut) (Johns, 2010).

Rossjoycea (Coleoptera, Carabidae), a monotypic genus: rocky areas of shrubland and grassland above Franz Josef Glacier (Larochelle and Larivière, 2013).

Asaphodes glaciata (Lepidoptera, Geometridae): Mount Moltke above Franz Josef Glacier and, 7 km away, above Fox Glacier (Chancellor Dome, 1250–1550 m; B. Patrick, pers. comm., April 13, 2015).

Other glacier endemics in the Westland "gap" include the monotypic dipteran genus *Zelandochlus*, also known as the glacier midge or ice worm. It is endemic to Franz Josef and Fox glaciers, where it lives in ice caves and meltwater pools (Boothroyd and Cranston, 1999). (The glacier midge, *Z. latipalpis*, was later shown to be an apterous species of the widespread austral genus *Parochlus*; Cranston et al., 2010.) A fossil-calibrated time tree indicated that the species diverged from its sister, in Chile, in the Oligocene (minimum age), long before glaciations (Cranston et al., 2010). This provides good evidence that what are now glacier endemics were *pre*adapted to glacial conditions, and that they occupied their geographic location long before the current conditions developed, rather than invading the habitat. During glacial advances, taxa such as the glacier endemics probably expanded their range, rather than undergoing range contraction (or even extirpation) as in the usual distribution model for the region.

This center of alpine endemism around Franz Josef and Fox also marks the northern limit of a Westland subalpine plant, *Hebe mooreae* var. *telmata* (Wardle, 1975; Heads, 1992, 1994c). The precise northern limit lies between Fox and Franz Josef glaciers, in the head of the Waikukupa Valley. From here, the plant's distribution extends just 50 km southwest along the mountains via Mount Fox to the Douglas Range (in the Paringa–Mahitahi area of south Westland).

8

Biogeography of the Northern South Island, and Its Mesozoic–Paleogene Geology

This chapter continues the biogeographic transect across the South Island, examining some of the main distribution patterns from central Canterbury to Cook Strait, and focusing on Mesozoic and Paleogene geology.

The Torlesse Node

Castle Hill basin is a tectonic depression in central Canterbury enclosed by the Torlesse and Craigieburn Ranges and drained by the headwaters of Broken River. The region, termed here the Torlesse node, is significant for biogeography, but is not a major center of endemism comparable to those associated with the great tectonic features of Nelson and Otago. It includes several endemics and disjuncts and also acts as a distributional boundary in many groups. For example, it forms the mutual boundary of the subshrubs *Pimelea dura* and *P. mesoa* (Thymelaeaceae) (Figure 8.1; Burrows, 2011a). (These two plants belong to a group that also includes *P. notia*, with a northern limit at the Waihemo fault zone.) In an example from the fauna, the millipede *Dityloura unicostata* extends from southern Canterbury north to the Torlesse node, and *D. edaphica* extends from Marlborough south to the node; the two show local overlap on the eastern flanks of the Torlesse Range (Johns, 1979). Groups in the Castle Hill/Torlesse Range area display connections and disjunctions with Marlborough, northwest Nelson, and Central Otago.

In Castle Hill basin, Late Cretaceous coal measures including sandstone, mudstone, and muddy coal lie with angular unconformity on Torlesse terrane basement. The coal measures are covered by glauconitic sandstone (greensand) deposited during a Late Cretaceous–Eocene marine transgression (Gage, 1970). The coal measures and the greensand are diachronous and mark a slow, westward migration of the coastline over a land surface of low relief. Eocene uplift interrupted deposition in the east and south, but elsewhere deposition proceeded until the mid-Oligocene, when limestone was deposited and submarine volcanoes were active. Sedimentation then ceased in the center of the area, while there was renewed elevation and tilting in the south and east. Marine sediments of the Enys Formation accumulated early in the Miocene, but the sea finally withdrew later in the Miocene and Early Pliocene.

Cockayne (1899, 1906) cited several common coastal plants found inland at the lower Waimakariri Gorge, and he attributed their presence there to the "former extension of the coast-line inland." Cockayne (1906) also discussed other localities, such as Castle Hill and Weka Pass, where coastal plants occur on inland limestone, and he concluded: "If such a distribution is correlated with the marine origin of the rocks, then it is evident that species can exist under special conditions for enormous periods of time." Cockayne (1906) recognized that the observed pattern was not caused by the ecological or physiological effects of the limestone (some of the species are not restricted to limestone), but by its history and tectonic context. The same process of inland stranding could also mean that the shore biota of the Late Cretaceous–Eocene inland seas, along with members of the forest represented by the coal, has direct descendants in the present landscape.

As the inland seas receded, limestones were exposed subaerially and would have been immediately colonized by pioneer plants and animals. The limestones continued to be uplifted, with their biological cargo of coastal weeds. During this process, erosion has continued, and in many areas the limestone strata have been completely removed, revealing the basement. As the limestones were eroded, the metapopulations

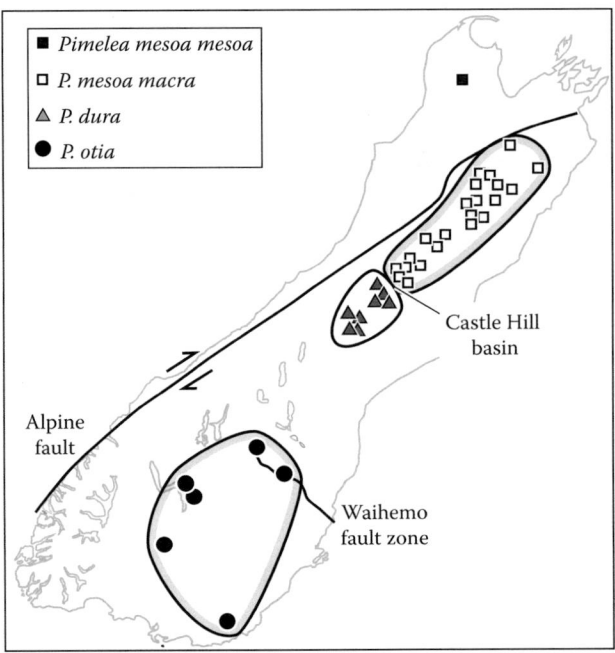

FIGURE 8.1 Distribution of four *Pimelea* species (Thymelaeaceae) with abaxially hairy leaves. (Data from Burrows, C.J., *N. Z. J. Bot.* 49, 41, 2011a.)

persisted *in situ*, and were eventually "deposited" down onto the basement. This is the reverse of the process in which old metapopulations "float" more or less *in situ* as younger strata are deposited.

Endemics at the Torlesse node, as well as connections with the south, northeast, and northwest, include the following examples.

Torlesse Node Endemics:

Rhizolecia (lichen): A monospecific genus endemic to Castle Hill and nearby Cave Stream and Flock Hill (Galloway, 2007).

Menegazzia pulchra (lichen): Hawdon River, Klondyke Corner (upper Waimakariri), Craigieburn Range (Galloway, 2007).

Xanthoparmelia olivetoricella (lichen): Craigieburn Range. Related to *X. adpicta* to the north (northwest Nelson, Kaweka Range) and to *X. plana* to the south (Ahuriri Valley) (Galloway, 2007).

Myosotis traversii var. *cinerascens* (Boraginaceae): Castle Hill Basin on limestone shingle-slips (Allan, 1961).

The millipede *Icosidesmus saxatilis* (Dalodesmidae): Torlesse and Craigieburn Ranges (Johns, 1964).

Anabarhynchus embersoni (Diptera: Therevidae): Craigieburn Range (Lyneborg, 1992).

Taenarthrus pluriciliatus (Carabidae): Craigieburn Range (Johns, 2010).

The geographic limits of the endemism are not always clear-cut. The alpine subshrub *Leonohebe tetrasticha* (Plantaginaceae) has most of its records in the Torlesse–Craigieburn region, with outlying populations around this area at Mount Somers, Arthur's Pass, and the Puketeraki Range (Bayly and Kellow, 2006). However, "features approaching" *L. tetrasticha* occur in its widespread relative *L. cheesemanii* as far afield as the Kirkliston, Liebig, and Amuri Ranges (Bayly and Kellow, 2006). *L. cheesemanii* extends from south Canterbury to Marlborough, but is replaced in the center of this range, the Torlesse region, by *L. tetrasticha*. The centripetal concentration of *L. tetrasticha* characters at a node in the Torlesse area is consistent with evolution around shrinking inland seas.

Torlesse Node to Canterbury and Otago
Examples of this track include the following:

Gentianella corymbifera subsp. *gracilis* (Gentianaceae): Castle Hill Basin to the Eyre Mountains (Figure 8.2; Glenny, 2004).

Ourisia caespitosa var. *gracilis* (Plantaginaceae): Torlesse Range to the Eyre Mountains (Figure 8.3). This was mapped in a morphological study (Heads, 1994b), but was not accepted as distinct from the type variety by Meudt (2006). A subsequent, well-sampled molecular study found good support for the clade and showed it as sister to the rest of the species (Meudt et al., 2009; Appendices).

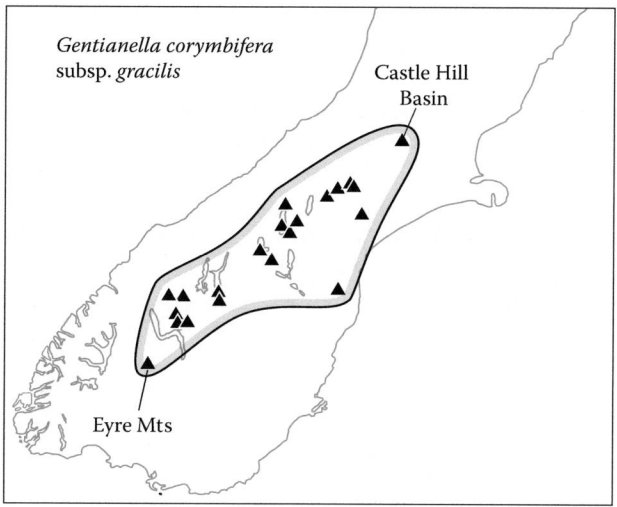

FIGURE 8.2 Distribution of *Gentianella corymbifera* var. *gracilis* (Gentianaceae). (Data from Glenny, D., *N. Z. J. Bot.*, 42, 361, 2004.)

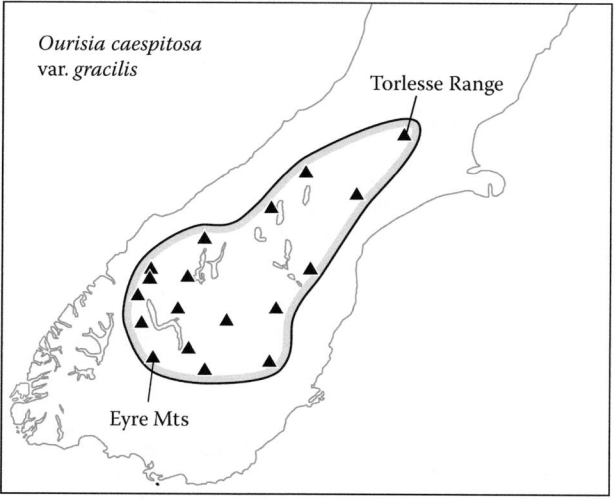

FIGURE 8.3 Distribution of *Ourisia caespitosa* var. *gracilis* (Plantaginaceae). (Data from Heads, M., *Candollea* 49, 23, 1994b.)

Olearia fimbriata (Asteraceae): Broken River south to Central Otago and Otago Peninsula (Heads, 1998b).

Austrosimulium alveolatum (Diptera: Simuliidae): Porters Pass (Torlesse Range)–Lake Heron (Craig et al., 2012).

Izatha psychra (Lepidoptera: Oecophoridae): Porters Pass and Lake Pukaki (Hoare, 2010).

Torlesse Node to Marlborough:

Epilobium brevipes (Onagraceae): Porters Pass to Awatere Valley (Raven and Raven, 1976; NZPCN, 2016).

Myosotis colensoi (Boraginaceae): Castle Hill basin and disjunct in Marlborough at the Ben More and Chalk Ranges, near Cape Campbell (Figure 8.4; Druce and Williams, 1989). The last taxonomic work on the species (Moore in Allan, 1961) placed *M. colensoi* with *M. cheesemanii* of the Pisa Range–northern Dunstan Mountains, east of the Moonlight tectonic zone.

Hebe glaucophylla (Plantaginaceae): Castle Hill to Kaikoura Ranges (Figure 8.5; Bayly and Kellow, 2006).

Hebe brachysiphon (Plantaginaceae): North from the Torlesse region (and Mt Hutt) to the Wairau/Alpine fault. This fault forms the boundary with its putative sister group, *H. divaricata*, which replaces it in northern Marlborough and Nelson (Bayly and Kellow, 2006).

Parahebe (*Heliohebe*) *raoulii* (Plantaginaceae): Broken River/Rakaia Gorge to Clarence Valley, including a subspecies around Mount Cass and Weka Pass (Garnock-Jones, 1993).

Australopyrum calcis (Gramineae): Mount Torlesse and Castle Hill to North Canterbury (subsp. *calcis*) and Leatham River, Marlborough (subsp. *optatum*) (Molloy, 1994).

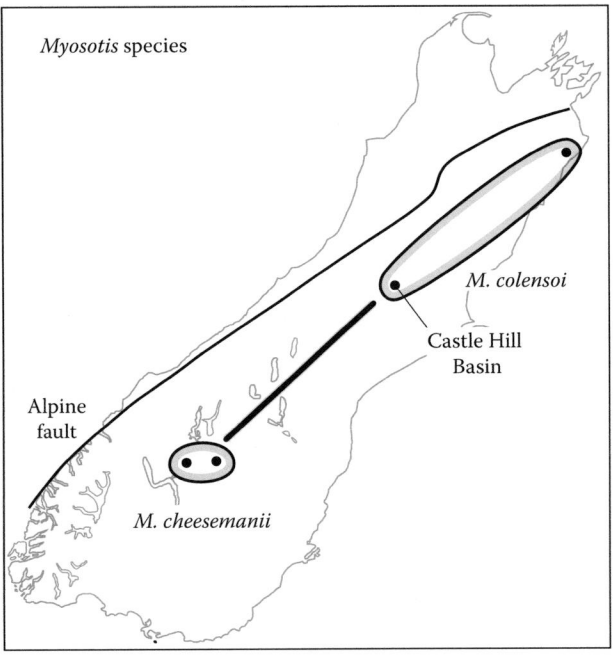

FIGURE 8.4 Distribution of *Myosotis cheesemanii* and *M. colensoi* (Boraginaceae). (Data from Allan, H.H., *Flora of New Zealand*, Vol. 1, Government Printer, Wellington, New Zealand, 1961; Druce, A.P. and Williams, P.A., *N. Z. J. Bot.*, 27, 167, 1989; Heads, M. and Patrick, B., The biogeography of southern New Zealand, in *The Natural History of Southern New Zealand*, eds. J. Darby, R.E. Fordyce, A. Mark, K. Probert, and C. Townsend, University of Otago Press, Dunedin, New Zealand, 2003, pp. 89–100.)

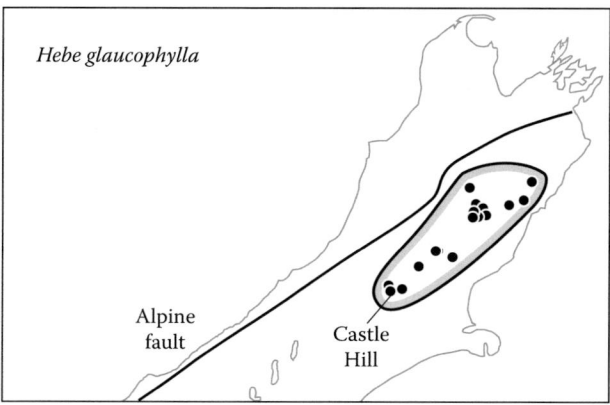

FIGURE 8.5 Distribution of *Hebe glaucophylla* (Plantaginaceae). (Data from Bayly, M. and Kellow, A., *An Illustrated Guide to New Zealand Hebes*, Te Papa Press, Wellington, New Zealand, 2006.)

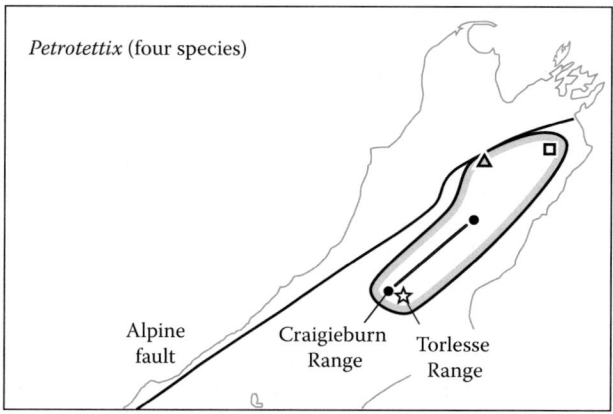

FIGURE 8.6 Distribution of *Petrotettix* (Rhaphidophoridae). (Data from Richards, A.M., *J. R. Soc. N. Z.*, 2, 151, 1972.)

The alpine cave weta *Petrotettix* (Orthoptera: Rhaphidophoridae) is endemic to New Zealand and is related to other New Zealand genera (Figure 8.6; Richards, 1972). It comprises four species, with one each at Craigieburn Range; Torlesse Range north to Mt St Patrick (near Hanmer, north Canterbury); Black Birch Range, lower Awatere Valley; and Travers Range (by the Nelson lakes, just south of the Wairau fault).

> *Icosidesmus cismontanus* (Diplopoda: Dalodesmidae): Craigieburn Range and Arthur's Pass to Seaward and Inland Kaikoura Ranges (Tapuaenuku) (Johns, 2015).
>
> *Arthurdendyus latissimus* (Turbellaria: Geoplanidae): Porters Pass, and junction of the Waiau and Hope rivers (Johns, 2015).
>
> *Austrosimulium albovelatum* (Diptera: Simuliidae): Torlesse center (Porters Pass to Rangitata River, 250–616 m) and disjunct at Kaikoura at sea level (Craig et al., 2012).

Torlesse Node to Northwest Nelson:

> *Ischalis dugdalei* (Geometridae): Mount Torlesse to NW Nelson (Weintraub and Scoble, 2004).
>
> *Lyperobius fallax* (Curculionidae): Broken River basin to NW Nelson (Craw, 1999).
>
> *Taenarthrus pluriciliatus* (Carabidae) of the Craigieburn Range "in many characters, shows a close relationship with *T. aenigmaticus* of Mount Arthur, NW Nelson" (Johns, 2010; Figure 8.7).

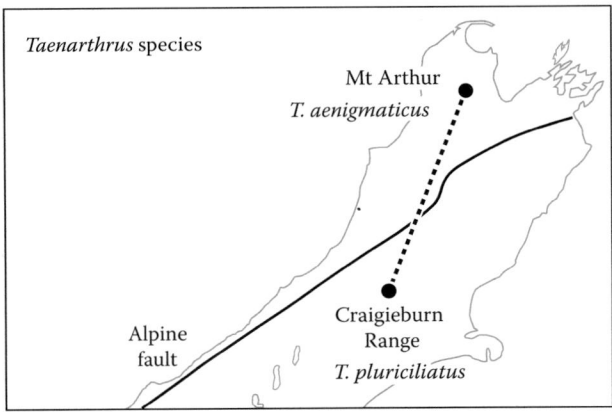

FIGURE 8.7 Distribution of *Taenarthrus* (Carabidae) species. (Data from Johns, P.M., *Rec. Canterbury Museum*, 24, 39, 2010.)

Torlesse Node to Northwest Nelson and Southern Marlborough:

> *Gentianella tenuifolia* (Gentianaceae): Craigieburn Range to NW Nelson and Marlborough (Glenny, 2004).
>
> *Odontria halli* (Scarabaeidae): Lake Coleridge and Castle Hill basin to NW Nelson and Marlborough (Johns, 2015).

East Coast and Inland to Torlesse Node
(For an interpretation of this pattern, see the discussion of Cockayne's work earlier in this section):

> *Pleioplectron simplex* (Orthoptera: Rhaphidophoridae): Invercargill (Otatara) and Dunedin to north Canterbury (Hanmer, at the Hope fault), and inland to Castle Hill Basin (Johns, 2015).
>
> *Holcaspis intermittens* (Carabidae): Banks Peninsula inland to Castle Hill Basin (Johns, 2015).
>
> *Hadramphus tuberculatus* (Curculionidae): Inland to Mt Oakden near Lake Coleridge (Craw, 1999).

The affinities of many Torlesse groups with areas outside New Zealand can also shed light on their local history, and a good example is the grass *Australopyrum*. This is endemic to southeastern Australia, eastern New Guinea (Papuan Peninsula), and New Zealand, where it extends from the Torlesse region to Marlborough (Connor et al., 1993). As with many groups on the Torlesse terrane, *Australopyrum* has Tethyan affinities; it has been proposed as sister to *Taeniatherum* of central Asia, the Mediterranean, and the United States (Petersen et al., 2006), or as a member of a clade with *Dasypyrum* of the Mediterranean and *Secale* of the northern hemisphere (Mason-Gamer, 2005).

In New Zealand, *Australopyrum* is represented by the calciphile *A. calcis* (Torlesse–Marlborough). In this species, subsp. *calcis* grows on Mesozoic limestone exposed for at least 2 Myr, while subsp. *optatum* is on Cenozoic limestone that was exposed only in the Pleistocene (Molloy, 1994). These particular outcrops are young, and in earlier times the plants would have probably occupied other base-rich habitats. As different limestone strata become exposed within the grass's range by fault movement, uplift, and erosion, they are likely to be colonized, until they in turn are eroded away. In this way, metapopulations survive more or less *in situ*, in the region, by a perpetual hopscotch.

One main biogeographic question for the Torlesse region concerns the three separate connections of the biota, with Central Otago, Kaikoura, and northwest Nelson, and whether these developed together or at different times. The history of marine incursions into the region from the Late Cretaceous has been complex, and dissecting out their effects will require detailed mapping of populations.

Many groups have distribution limits around the Arthur's Pass area, which acts as a higher altitude component of the Torlesse node. For example, *Traversia* (Asteraceae) is distributed from Nelson and Marlborough south to Arthur's Pass, while *Hectorella* (Montiaceae) and *Olearia moschata* (Asteraceae) extend from Fiordland north to Arthur's Pass (http://nzflora.landcareresearch.co.nz/; NZPCN, 2016).

The Kaikoura Node, Southern Marlborough

The Wairau fault in Marlborough is the northern continuation of the Alpine fault. About 135 km of strike-slip displacement on the Wairau fault has separated the Jurassic Esk Head terrane into a Marlborough component (Esk Head Mélange) and a southern North Island component (Rimutaka Mélange) (Figure 3.2). The Wairau fault also acts as a boundary between a center of endemism in northern Marlborough (Marlborough Sounds and the Richmond Range) and another in southern Marlborough that includes the Inland and Seaward Kaikoura Ranges. Cockayne (1928) proposed a "North-eastern botanical district" that extends from the Wairau River to the Hurunui River in northern Canterbury (Figure 8.8). The biogeography of this region is of special interest, as the modern distributions appear to be older than the young mountains that dominate the landscape.

One of the best-known endemics in southern Marlborough is *Pachystegia* (Asteraceae: Astereae) (Figure 8.9), distributed from the Wairau fault south to the Lowry Peaks Range, 20 km south of the Hope fault (Molloy, 2001). The genus comprises five species of distinctive, long-lived shrubs that have large, leathery leaves and are often cultivated. The plants are conspicuous on exposed bluffs and cliffs above the coast road and may be subjected to persistent salt spray. Most plants are confined to steep, rocky sites, but one population of *P. minor* grows in stable riverbed alluvium (Jordan Stream, Puhi Puhi River), and *P. insignis* is recorded on stable coastal sands (Valhalla Stream by Kekerengu). The genus is recorded up to 1200 m elevation (Molloy and Simpson, 1980). The four species surround Mount Tapuaenuku (Inland Kaikoura Range), and the general significance of this important locality is discussed later.

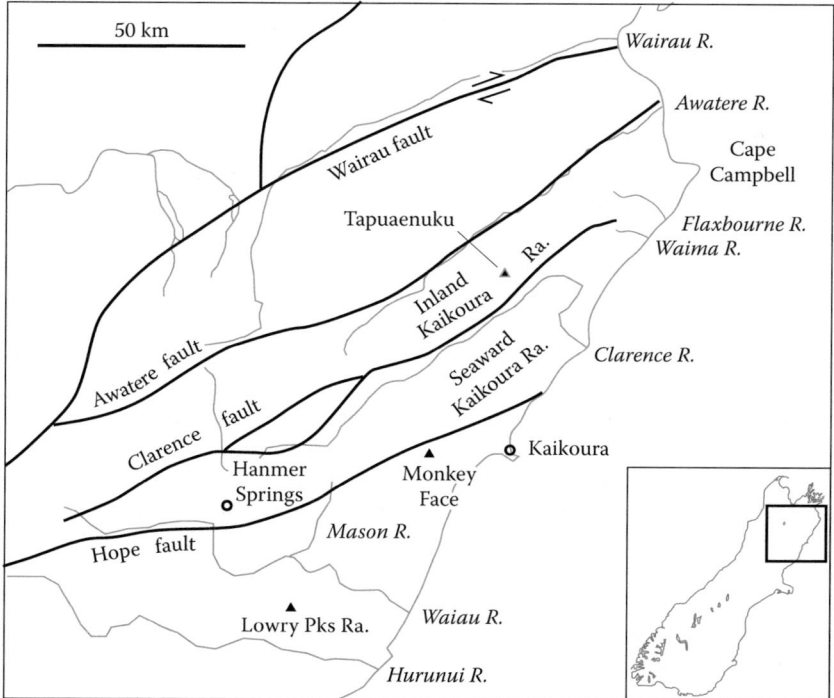

FIGURE 8.8 Locality map of the Kaikoura region.

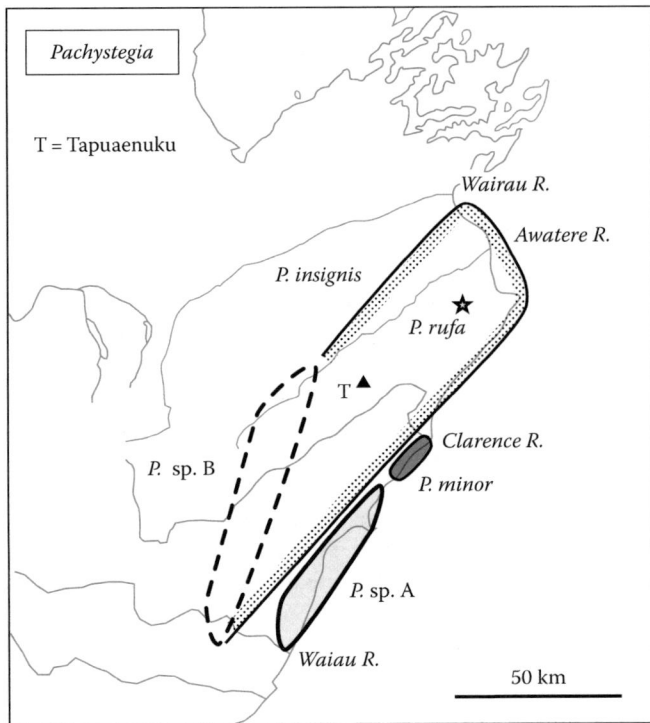

FIGURE 8.9   Distribution of *Pachystegia* (Asteraceae). (Data from Molloy, B.P.J., *Pachystegia rufa* and allied rock daisies, Conservation Advisory Science Notes 336, Department of Conservation, Wellington, New Zealand, 2001; Molloy, B.P.J. and Simpson, M.J.A., *N. Z. J. Ecol.*, 3, 1, 1980.)

Pachystegia is not just a local group nested within a South Island or even a New Zealand clade. Instead, it is sister to a diverse complex (*Olearia* sect. *Divaricaster* and other *Olearia* species) that is widespread through mainland New Zealand, the Chatham Islands, and southeastern Australia (Cross et al., 2002; Brouillet et al., 2009; Wagstaff et al., 2011). This is compatible with a divergence of *Pachystegia* from its sister before the opening of the Tasman Basin, which rifted the sister group apart. An origin caused instead by the uplift of the Kaikoura Ranges in the Pliocene would require dispersal to Australia against the prevailing winds, and a very young origin for diverse clades in the sister group, such as *O.* sect. *Divaricaster*.

Apart from *Pachystegia*, other distinctive shrubs endemic to southern Marlborough include *Parahebe hulkeana* (Plantaginaceae) (Figure 8.10) and *Brachyglottis monroi* (Asteraceae) (Awatere catchment to Mason River). These, along with *Pachystegia insignis*, are the most important canopy species in 3 m tall "bluff scrub" and are seldom found outside this association (Wardle, 1971). As in *Pachystegia*, *Parahebe hulkeana* and its relatives (the *Heliohebe* group) form a set of affinities surrounding Tapuaenuku. In a similar way, *Pimelea sericeovillosa* is separated from its close relative *P. aridula* by a boundary at Tapuaenuku (Burrows, 2011b).

In the New Zealand brooms (Fabaceae), there is a concentration of local endemics in the southern Marlborough region. These include a tree broom with distinctive, weeping architecture (*Chordospartium stevensonii*) and four allies of this plant (*Chordospartium muritai, Carmichaelia glabrescens, C. carmichaeliae*, and *C. astonii*). Other plants that are endemic in the region include:

Wahlenbergia albomarginata subsp. *flexilis* (Campanulaceae): Beaches from south of Cape Campbell to just south of Kaikoura Peninsula, and from the coast to the Clarence Fault (Petterson, 1997).

Epilobium forbesii (Onagraceae): Awatere Valley to Hope fault.

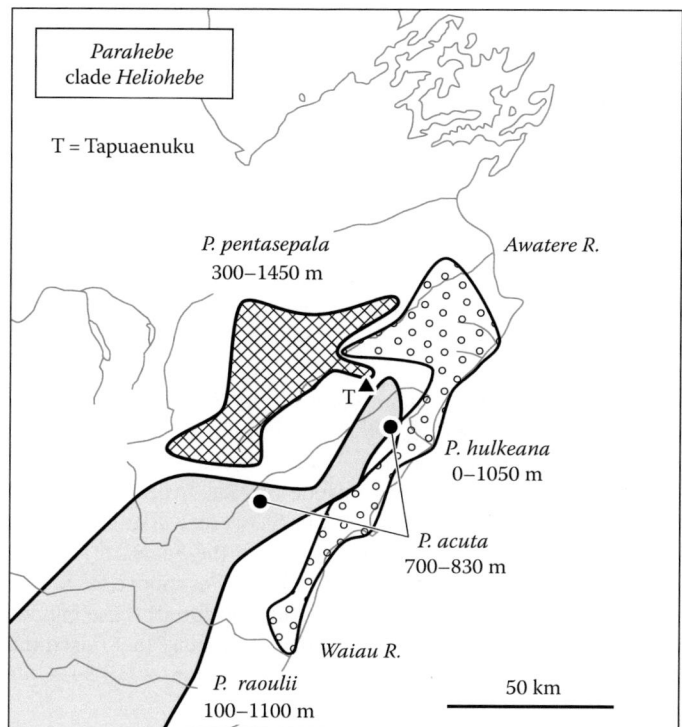

FIGURE 8.10 Distribution of *Parahebe* clade *Heliohebe* (Plantaginaceae). (Data from Garnock-Jones, P.J., *N. Z. J. Botany*, 31, 323, 1993.)

Pimelea concinna (Thymelaeaceae): Marlborough west of Tapuaenuku (Burrows, 2011b).

Gentianella magnifica (Gentianaceae): Rachel Range (head of Awatere Valley) and Hanmer Range (Glenny, 2004).

Olearia coriacea (Asteraceae): Awatere Valley to Jack's Pass.

Celmisia cockayneana (Asteraceae): Wairau Valley to Hope fault.

C. insignis: Wairau Valley to Awatere Valley (Given, 1980).

C. monroi: Wairau fault to Hope fault (Given, 1980; NZPCN, 2016).

Leucogenes neglecta (Asteraceae): Mountains between the Wairau and Awatere Rivers (Molloy, 1995).

Helichrysum coralloides (Asteraceae): Wairau fault to Hope fault.

Rachelia (a monotypic genus in the *Raoulia* group) (Asteraceae): Awatere Valley to Hope fault.

The *Hebe* complex (Plantaginaceae) alone includes eight examples:

Hebe salicornioides: Clarence River (upper Wairau but not the main Wairau mountains) to Hope fault.

Hebe decumbens: Wairau fault to Hope fault.

Hebe ramosissima: Awatere Valley to Hope fault.

Hebe rupicola: Awatere Valley to Mason River.

Parahebe martinii: Wairau Valley (Waihopai River) south to Mason River.

P. (Heliohebe) pentasepala: Wairau Valley (Leatham), Awatere, and Clarence Valleys.

P. (Heliohebe) acuta: Seaward Kaikoura Range.

P. (Heliohebe) hulkeana: Awatere Valley to Waiau River mouth (including coastal cliffs).

The last three species (mapped in Figure 8.10, from Garnock-Jones, 1993) are members of the *Heliohebe* clade (treated here as a group in *Parahebe*; cf. Heads, 1994c, 2014; Albach and Meudt, 2010: Fig. 3). *Heliohebe* extends south to the Torlesse node (Rakaia River) and Banks Peninsula, but has its main diversity in the Kaikoura region.

The notes given here describe Marlborough south of the Wairau fault as a center of endemism. However, it is also a significant center of absence in groups such as the plant genus *Forstera* (Glenny, 2009) and the harvestman family Pettalidae (Boyer and Giribet, 2009).

The genus *Haastia* (Asteraceae: Senecioneae) is made up of subshrubs and cushion plants that are endemic to the South Island. All three species are restricted to the alpine zone. The genus is allied to *Brachyglottis* and others in a South Africa–Australasia–Chile group (Pelser et al., 2010). There is some evidence that the sister group of *Haastia* might be *Urostemon*, which is endemic to and widespread through the North Island (Pelser et al., 2010). The two genera look very different, but this pairing would give a standard break at Cook Strait. *Haastia* is not endemic to the Kaikoura Ranges, but all three of its species occur between there and Arthur's Pass: *H. pulvinaris* (vegetable sheep) is restricted to this region; *H. recurva* also extends south to Mount Hutt; and *H. sinclairii* extends south to *eastern* Fiordland. *H. sinclairii* is also in northwest Nelson (Todd, 1996; NZPCN, 2016).

Animal clades endemic to the Kaikoura area include the seabird *Puffinus huttoni* (Procellariidae), which breeds only in the mountains of the Inland and Seaward Kaikoura Ranges. Another example is the moth *Pseudocoremia hudsoni* (Geometridae), found in the Seaward Kaikoura Range, at Mount Robert by Lake Rotoiti (right by the Wairau fault), and at Jack's Pass above Hanmer Springs (right by the Hope fault) (Stephens et al., 2007). The traditional explanation for all these endemics in the Kaikoura mountains is that they evolved during the Neogene Kaikoura orogeny that raised the mountains. But, as indicated earlier, at least one of the endemics, *Pachystegia*, is sister to a diverse, widespread Australia/ New Zealand group, and a Neogene origin is unlikely.

Distribution in an Inland Strip: Central Otago–Southern Marlborough

Many South Island groups have distributions following a central strip. This extends along the axis of the island from central Southland and Central Otago, via the Torlesse node, to southern Marlborough (south of the Wairau fault). The cause of this common pattern is not as obvious as it might seem.

Is Distribution in the South Island Central Strip and around the Moonlight Tectonic Zone Caused by Current Climate?

The usual explanation for the "central strip" pattern is that it reflects climate, as the belt includes the driest areas in New Zealand. In addition to the north–south variation in temperature in New Zealand, there is also a rapid west–east change in rainfall. The west coast of the South Island and the mountains that run along the coast, the Southern Alps, are the wettest areas in the country because of the orographic effect. East of the alps, the inland parts of the central South Island lie in a rain shadow and include the driest areas in the country, Central Otago and south Canterbury. These have many endemics, but this is true for most areas in New Zealand, whether or not they have a distinctive climate.

Areas with similar rainfall run parallel with the Southern Alps, as do many clade distributions. But it is not always clear whether the clades follow the rainfall patterns, or whether both the clades and the rainfall have a common cause, such as the orogeny. For example, the Moonlight tectonic zone runs parallel with the Southern Alps, and for part of its length also lies near the 1600 mm isohyet. Different authors have attributed the important biogeographic boundary here to the change in rainfall (Dugdale and Fleming, 1978, on *Maoricicada otagoensis*; see Figure 7.13). Nevertheless, the distributions show a closer spatial correlation with the tectonic boundary. In addition, breaks at this boundary have been dated as older than the orogeny and orographic patters in rainfall. In the harvestmen, the break between *Neopurcellia* plus relatives (west of the MTZ), and *Rakaia* plus relatives (east of the MTZ) has been dated as Cretaceous (Giribet et al., 2012). This is long before the Miocene orogeny began to create the modern mountains and the orographic climate, but it is the same age as the MTZ.

In most South Island distribution patterns, such as disjunction along the Alpine fault, there is no obvious correlation with climate. Maps of rainfall and soil-moisture deficit indicate that areas southwest of a line: Dunedin–Queenstown are much wetter than areas to the northeast (NIWA, 2016), yet many species on the "central strip" range both north and south of this boundary.

Groups such as the shrub *Hebe subalpina* (Plantaginaceae) are distributed in an "alpine strip" along the wet mountains that lie west of the dry "central strip" (Bayly and Kellow, 2006). Yet if the distribution simply reflected rainfall, the species would continue north into the wet mountains of northwest Nelson (and Taranaki), where it is in fact absent, and it would drop out further east, in the drier Spenser and St Arnaud Ranges, where it is in fact present. The actual distributional limit coincides, not with rainfall, but with the Alpine–Wairau fault. Although the overall range coincides with tectonic and phylogenetic boundaries, *within* this area the distribution of *H. subalpina* is determined by ecological factors—it is restricted to subalpine sites.

Some authors have rejected climate as an explanation for plant distributions in the South Island and have instead proposed that variation in the lithology and geochemistry of the substrate is critical (Mitchell and Heenan, 2002; Heenan and Mitchell, 2003). This is appealing because the different substrates have abrupt boundaries, as in the species distributions and unlike the smoother climatic gradients. The issue is complex though, as the lithological breaks occur at terrane boundaries, and these are also important tectonic features (old plate margins).

Central Strip Distribution Hitting the Coast in Marlborough

Many South Island, central strip groups that otherwise occur at subalpine elevations have low-elevation records in Marlborough, and these are concentrated around sites such as the Flaxbourne and Waima River mouths (Figure 8.8). Examples include the following plants:

Epilobium rostratum (Onagraceae): Central Otago (Alexandra) to Tapuaenuku and the lower Clarence Valley (Mead Stream) (Raven and Raven, 1976; Druce and Williams, 1989).

Pimelea declivis (Thymelaeaceae): Inland north Otago (Mt St Mary/Kohurau) to Waima River mouth (Burrows, 2011a).

Pimelea traversii: Central Southland (Livingstone Mountains) and Central Otago to the mouths of the Flaxbourne and Waima Rivers (Burrows, 2008).

Pimelea aridula: Central Otago to Flaxbourne River mouth, with a large gap in the range filled by the related *P. sericeovillosa* (Burrows, 2011b).

These species range from inland localities in the southern South Island along the central strip to Marlborough, where their distributions drop in elevation and intersect the coast. This common pattern was described by Martin (1938: 418) in his study on the Marlborough flora. Under the heading "Subalpine plants at low levels," he wrote:

An association of plants occurs in several places in eastern Marlborough less than 100 feet above sea-level (e.g. near the mouth of the Flaxbourne River, in the gorge of Woodside Creek at Wharanui, or on coastal cliffs near Lake Grassmere), which is noteworthy for the high percentage of plants most commonly subalpine. It invariably occurs on limestone rock or rubble, and very commonly in association with *Pachystegia insignis*. The following species are elsewhere almost invariably subalpine or upper montane:
Anisotome filifolia, Pimelea traversii, P. sericeo-villosa [s.lat.], *Poa acicularifolia, Ranunculus lobulatus* [now *R. insignis*], *Carmichaelia monroi, Hymenanthera alpina* (only one specimen observed), and *Senecio* [now *Brachyglottis*] *lagopus*.
Near Flaxbourne mouth, *Carmichaelia monroi* [this population now segregated as *C. astonii*] forms practically a pure association.

One of these "subalpine plants" sometimes found at sea level, *Poa acicularifolia*, comprises two subspecies (Edgar, 1986): *P. a. acicularifolia* of Marlborough and Canterbury, growing on limestone and mat-forming, and *P. a. ophitalis* of Marlborough and Nelson, growing on ultramafic rocks and tuft-forming. (For the relationships between the "basiphilous" floras of limestone and of ultramafic substrate, see Heads, 2014: 201, 308; for plagiotropic plant architecture on limestone, see Heads, 2006b.)

The same Flaxbourne River region that includes the anomalous lowland populations also includes endemics. A typical example, *Dichelachne lautumia* (Poaceae), is restricted to a limestone quarry near the mouth of the Flaxbourne River, at 100 m above sea level (Edgar and Connor, 1999). In *Wahlenbergia matthewsii* (Campanulaceae), blue-tinted flowers occur only along the coast from Flaxbourne River to the mouth of the Clarence River (Petterson, 1997).

In New Zealand *Celmisia*, there are about 62 species (NPCN, 2016). Most are alpine, but there are seven lowland and coastal species (Given, 1968). These occur on Stewart Island and the Catlins coast of south-east Otago, which might be expected in terms of climate, but also in northeast Otago, Banks Peninsula, Kaikoura, Ngakawau (north of Westport), Coromandel Peninsula, and Auckland. *Celmisia monroi* is on the coast at Flaxbourne River mouth and extends inland to the alpine zone of Marlborough. Given (1968: 37) wrote: "A problem is posed by the occurrence of celmisia plants at low levels between Kaikoura and Blenheim. Much of the area in which this occurs is limestone and several of the plants found here show calciphilic tendencies, e.g., *Gentiana astonii*, *Wahlenbergia matthewsii* and *Myosotis arnoldi*. Probably open habitats have been maintained for a very long time on the limestone, allowing a scrub and herb vegetation to develop." Nevertheless, there are many different kinds of open habitat in the lowlands and coastal areas, not just on limestone. The otherwise alpine groups occur at sea level along the Kaikoura coast because this is where an otherwise alpine track intersects the current coastline. The plants' ancestors were probably already in the area since before the Amuri Limestone was deposited. Ancestral clades have probably expanded their range with the marine transgressions, with modern groups differentiating during the recession of the seas.

To summarize, the central strip distributions appear to have been formed before the present climatic regime or topography existed. Some time after their origin they have been deformed by regional tectonics, including uplift in the south, and subsidence around Marlborough and Cook Strait. Central strip distribution around the Torlesse region was discussed in the last section, and the biological patterns of affinity, endemism, and so on were related to prior marine incursions, as suggested by Cockayne (1906).

Geology of the Kaikoura Region, Southern Marlborough

The geology of southern Marlborough has been reviewed by Rattenbury et al. (2006). The basement throughout the region is formed from the Pahau (younger Torlesse) terrane. This terrane amalgamated with the Rakaia (older Torlesse) terrane to the south during convergent tectonism of the mid-Jurassic to mid-Cretaceous. The junction between the two is represented by the Esk Head Mélange (Figures 3.2 and 8.11).

Following terrane amalgamation and uplift, subduction ceased in the mid-Cretaceous and was followed by a prolonged period of crustal extension (Chapter 3). This resulted in the opening of the Tasman Sea and, in southern Marlborough, formation of new basins and sediment deposition in these. Further south, in Canterbury strata, only the very youngest stages of the Cretaceous are present, but in Marlborough the Cretaceous sequence is much thicker and contains many older strata.

The development of the mid-Cretaceous basins in the Kaikoura region was accompanied by significant igneous activity (Figure 8.11), but this was followed by relative quiescence from the end of the Cretaceous. From this time, regional subsidence and marine transgression are signaled by the slow accumulation of widespread, fine-grained clastic and carbonate rocks. Later, in the Miocene, the first signs of the Kaikoura orogeny appeared.

Events Following the End of Subduction in the Albian

Subduction at the Hikurangi margin ceased at the end of the Early Cretaceous (Albian), and around the beginning of the Late Cretaceous (Cenomanian) the compressional regime was replaced by one of regional extension. In addition to the opening of the Bounty Trough and the Tasman Sea, the extension is indicated in the New Zealand mainland by (1) stretching of the basement, with widespread normal faulting and half graben formation (in the Kaikoura region this is seen in the Clarence Valley); (2) emplacement

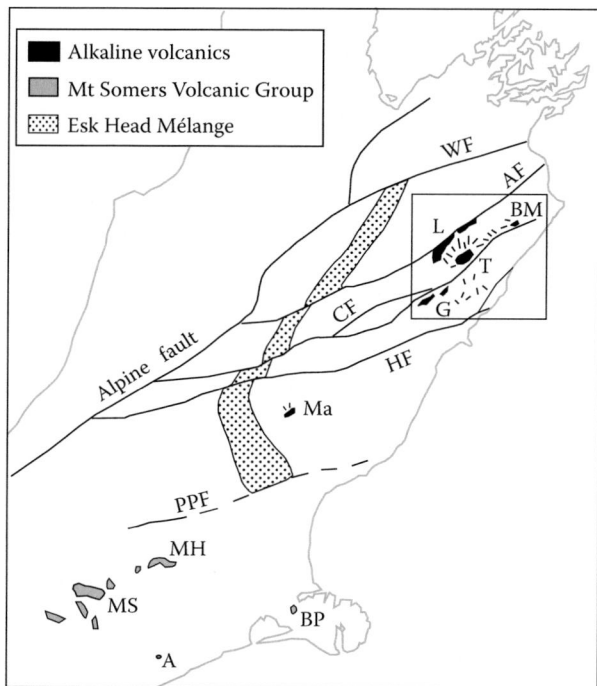

FIGURE 8.11   Distribution of Cretaceous igneous rocks in the South Island, east of the Alpine fault. Cenomanian alkaline volcanics (black): BM, Blue Mountain Complex; G, Gridiron volcanics; L, Lookout volcanics; Ma, Mandamus Igneous Complex; T, Tapuaenuku Igneous Complex. Dike swarms and minor intrusions are also shown. Mt Somers Volcanics Group (gray): A, Ashburton drillhole; BP, Banks Peninsula; MH, Malvern Hills; MS, Mt Somers. Major faults: AF, Awatere fault; CF, Clarence fault; HF, Hope fault; PPF, Porters Pass fault; WF, Wairau fault. (From Weaver, S.D. and Pankhurst, R.J., *N. Z. J. Geol. Geophys.*, 34, 341, 1991.)

of metamorphic core complexes and similar structures in the Paparoa Range, Fiordland, and Central Otago; (3) intrusion of silicic plutons in western regions and alkaline igneous complexes in eastern areas; and (4) extensive alkaline volcanism, including the Lookout volcanics in Marlborough.

The faulting, basin formation, igneous activity, and associated uplift were the immediate precursors to the opening of the Tasman Basin. Local and regional distributions suggest that all four processes have been significant for biogeography.

Normal Faulting and Basin Formation

In Marlborough, the change from convergent margin tectonics to regional extension in the Albian led to normal faulting and the development of new basins. For example, in the lower Clarence Valley, the Ouse fault is thought to have been a major normal fault in the mid-Cretaceous (Crampton and Laird, 1997). Much later, in the Neogene Kaikoura orogeny, it was reactivated as a reverse fault; the Otago faults have a similar history (Chapter 3).

Sedimentary sequences dated from mid-Cretaceous to Paleogene are preserved along the southern sides of the Awatere, Clarence, and Hope faults and represent the southernmost deposition in the newly developed East Coast Basin. This basin extends from East Cape to Kaikoura and lies in the frontal arc of the Hikurangi subduction zone (Figure 3.10); its inland margin is formed by the North Island shear belt.

The formation of the East Coast Basin was complex; even when extension was developing in the southern half, compression persisted in the northern half, around East Cape (Crampton and Laird, 1997). (Possible explanations for the continued compression include the collapse and extension of overthickened or elevated crust—a likely product of prolonged subduction.)

Volcanism

The Wallow Group of southeastern Marlborough (earliest Late Cretaceous) unconformably overlies the Pahau basement terrane. The group includes fluvial, terrestrial, and shallow marine deposits that interfinger with basalt flows (Rattenbury et al., 2006). The Wallow Group extends from the Awatere Valley to Monkey Face, just south of the Hope fault (Figure 8.8). This is the boundary between the East Coast and Canterbury Basins.

Within the Wallow Group, the Gridiron Formation (Figure 8.12) includes subaerial basalt flows, pillow basalt and volcanogenic conglomerate, breccias, dikes, and sills. It is locally interbedded with nonmarine sandstone, siltstone, coal seams, and lake silt deposits (Warder Formation). The presence of fossilized plant roots and ripple cross-lamination in the coal measures indicates deposition in a terrestrial environment subject to periodic inundation, probably a fluvial, coastal plain (Crampton, 1988). The associated Bluff Sandstone was deposited in a shallow marine fan-delta in a subsiding basin (Rattenbury et al., 2006).

Beu et al. (2014) documented the freshwater mollusks of the Warder Formation. They wrote: "The presence of a hyriid freshwater mussel in the Clarence Valley fauna and the continuous Jurassic-Recent presence of Hyriidae in New Zealand suggest that most if not all New Zealand freshwater Mollusca evolved on Zealandia from Gondwanan stock." They criticized the idea of dispersal from Australia, and

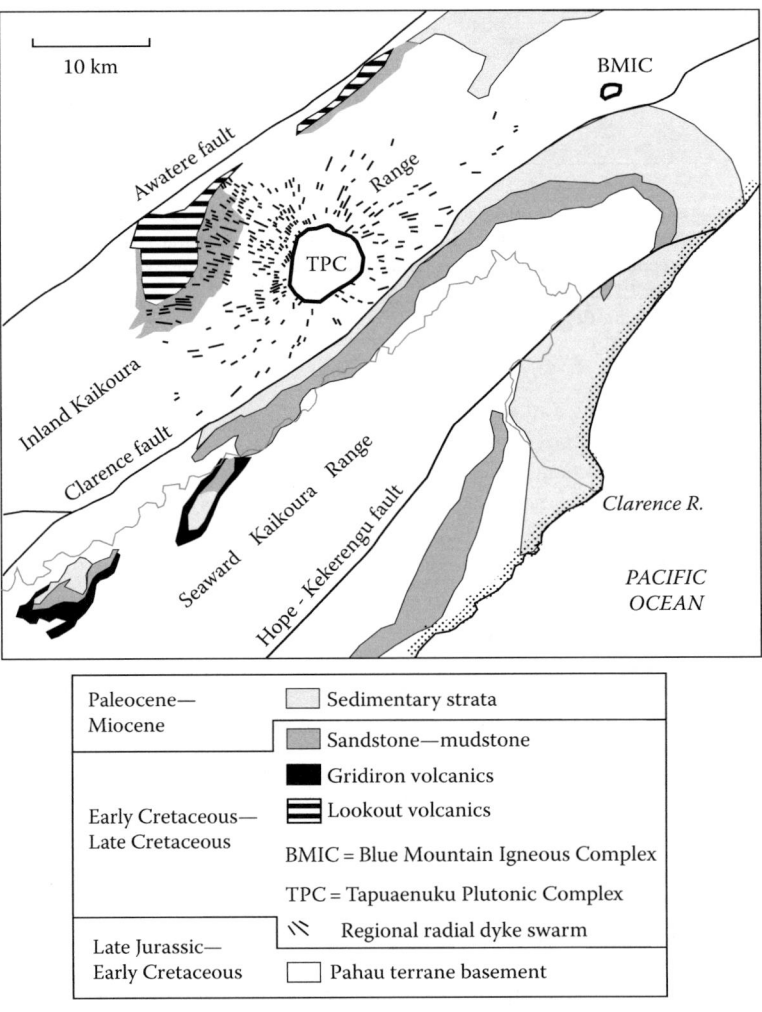

FIGURE 8.12 Simplified geological map of the inland Kaikoura region. (From McCoy-West, A.J. et al., *J. Petrol.*, 51, 2003, 2010.)

also the idea of complete Oligocene submergence. They concluded: "…much of the freshwater molluscan fauna of the southern landmasses resulted from evolution and extinction after fragmentation of an originally Gondwanan fauna…."

The Lookout Formation comprises basal pebble conglomerate and coal measures overlain by extensive volcanic flows. The Lookout volcanics (Figure 8.12) are mafic-ultramafic alkaline volcanic rocks erupted during a period of widespread intraplate volcanism and extension (~100–82 Ma; McCoy-West et al., 2010). These are the oldest samples of the HIMU magmatic mega-province of the southwest Pacific (Chapter 3), and they show petrological similarities to the Tapuaenuku Intrusive Complex, discussed next. The Lookout and Gridiron volcanics are inferred to have been fed by dikes emanating from the Tapuaenuku intrusion during the earliest Late Cretaceous.

Intrusion

Mount Tapuaenuku is the highest mountain in the Kaikoura ranges. It represents an important node for biogeography, and it also has a particular geological significance. Along the axis of the Inland Kaikoura Range, the upper 600 m, including Mount Tapuaenuku, consists of a layered, multiphase intrusion, the Tapuaenuku Igneous Complex (Figure 8.12). This is a mafic-ultramafic alkaline structure intruded at a shallow level into unconsolidated Torlesse sediments (Baker, 1994). The main, central body is a sub-circular pluton, 7 km in diameter. It is surrounded by a regional dike swarm—the only one in New Zealand—that comprises more than 500 dikes, each 46–60 cm thick, radiating for more than 10 km out from the central pluton (Challis, 1961; Rattenbury et al., 2006). The original volcano itself has been eroded away, leaving only the roots, although the lava flows of the Lookout and Gridiron volcanics probably represent fissure eruptions from the dikes. Crystals of apatite (mainly calcium phosphate) up to 1 cm long make up 20% of the rocks in the igneous complex (Williams, 1989). The central pluton was intruded at 96 Ma, the dikes between 100 and 60 Ma. The Tapuaenuku Igneous Complex is the earliest product of the intraplate alkaline igneous activity that has peppered southern New Zealand from the mid-Cretaceous through to the present (Baker et al., 1994).

The Tapuaenuku Igneous Complex is just one component of a more widespread igneous province. The Mandamus Igneous Complex is located 110 km southwest of Tapuaenuku, at the southern end of the Tekoa Range (just north of Hurunui River) (Figure 8.11; Weaver and Pankhurst, 1991). It is dated at 97 Ma, in close agreement with ages of the Tapuaenuku Igneous Complex, and it includes similar strata. It is the southernmost representative of a group of intrusive complexes that cut Torlesse rocks and are characterized by alkaline gabbros and related rocks. Apart from the Tapuaenuku and Mandamus complexes, these also include the Blue Mountain Complex in the Inland Kaikoura Range, and numerous small intrusions in the Seaward Kaikoura Range. Dikes at Cape Palliser, southeastern North Island, have a similar petrography and may be the same age. Even excepting these Cape Palliser rocks, the belt of igneous complexes and associated volcanics extends for about 150 km through Marlborough to North Canterbury along a northeast–southwest trend. As with the earlier Esk Head Mélange (Figure 3.2), this belt—the Central Marlborough igneous province—probably developed as a more or less linear feature parallel to the mid-Cretaceous Zealandia margin and oriented northwest–southeast.

Similar alkaline igneous rocks are known from Malvern Hills (by Whitecliffs), Mount Somers (60 km southwest of the Torlesse Range), a drillhole at Ashburton, Banks Peninsula (Figure 8.11), and also the Chatham Islands (Baker et al., 1994). This distribution recalls groups such as the bluff weta, *Deinacrida elegans*, known from the Kaikoura Ranges and Mount Somers, and the *Heliohebe* clade in *Parahebe* (shown in part in Figure 8.10), distributed from Marlborough to Banks Peninsula.

Uplift

Throughout New Zealand, the basement and the covering strata are separated by a widespread, mid-Cretaceous unconformity, indicating uplift and a break in sedimentation. In Marlborough, a break is apparent in many locations, but it is not always easy to define. In these cases, "The diachronous nature of the unconformity surface… probably reflects ongoing deposition in *localised, syntectonic basins* adjacent to *areas of uplift and erosion*" (Rattenbury et al., 2006: 29; italics added). A dynamic system of localized subsidence and uplift such as this will have had direct ecological and evolutionary effects on living communities, and the impact will be even greater where the tectonic activity is episodic.

Fission track dating indicates two major cooling events (uplifts) in Marlborough, one in the mid-Cretaceous (~100 Ma) and the other beginning in the Early Miocene (~20 Ma) (Kao, 2001). Regional updoming and uplift that accompanied the mid-Cretaceous extension and magmatism are evident in Clarence Valley, where the marine Split Rock Formation was raised above sea level and covered with a terrestrial sequence (Baker and Seward, 1996).

Latest Cretaceous and Paleogene Subsidence

The terrestrial sequence covering the Split Rock Formation was in turn draped with shallow marine deposits, and these represent the onset of a long period of transgressive marine sedimentation from Late Cretaceous to Oligocene. The marine incursion resulted from slow, thermal subsidence following the mid-Cretaceous magmatism and rifting. West of the Marlborough platform and the Hurunui high (a western extension of the Chatham Rise), a long-lasting embayment, 120 km wide, possibly linked the East Coast and Canterbury basins in latest Cretaceous time (Santonian–Maastrichtian) (Crampton et al., 2003). Shallow marine strata in southern Marlborough and northern Canterbury suggest that this furnished at least periodic marine connections across the emergent, proto-New Zealand. The northern basins continued to extend southward through the Paleogene and deposited a blanket of carbonate rocks, the Amuri Limestone (Chapter 3).

In the Early Oligocene, a period of erosion and/or nondeposition (the Marshall Paraconformity) was accompanied by mild deformation and initiation of the modern plate boundary (Rattenbury et al., 2006). In the Neogene, the Kaikoura region experienced considerable shortening (50% in some areas), sizeable fault displacements (up to 20–30 km), and large tectonic rotations (e.g., 100°) about vertical axes, and these have caused great changes in the topography (Crampton et al., 2003). The rotational deformation through the Neogene affected much of the North Island and northeastern South Island, but not the stable cores of Fiordland and Nelson.

The Cook Strait Node

Cook Strait separates North and South Islands, and is 26 km wide. The biotas of the two main islands show great differences, and many authors have emphasized phylogenetic breaks at Cook Strait (Engler, 1882; Wallis and Trewick, 2009; Buckley et al., 2011). Nevertheless, as Marshall et al. (2009: 2006) observed, the exact significance of the strait for allopatric divergence is "not clear." This is because the strait itself has only existed since the mid- to late Pleistocene (450,000 years b.p.; Lewis et al., 1994), and so it is probably too young to have caused most of the North Island/South Island differences.

A typical example of a group with a southern limit at Cook Strait is the moss *Hylocomium splendens* (Hylocomiaceae). This is a bipolar species found in New Zealand along the eastern North Island in "high-altitude and wind-swept locations" (Mount Hikurangi, Kaweka Range, Mount Ruapehu, Kaimanawa, Ruahine, and Tararua Ranges) (Fife, 2014d). Fife described its apparent absence from the South Island as "remarkable."

A typical example of a group with a northern boundary at Cook Strait is the sandfly *Austrosimulium ungulatum* (Simuliidae). This is widespread throughout Stewart Island and the South Island, north to northwest Nelson and the Marlborough Sounds. In many areas it is abundant, but it is completely absent from the North Island (Craig et al., 2012).

In an intercontinental example, Southern Ocean bryozoans comprise 1681 species (Barnes and Griffiths, 2007). In the order Cheilostomata, the South Island and South Africa are the main hotspots of endemism, and New Caledonia has three times as many species as the Galapagos. In contrast, the order Cyclostomatida has its two main hotspots in the North Island and southeastern Australia, and the Galapagos Islands have twice as many species as New Caledonia. Here, the South Island appears as part of an Indian Ocean center of diversity, and the North Island as part of a Pacific Ocean center.

At a local scale, there are many species breaks at Cook Strait, for example, in the cicada genus *Maoricicada*. *M. campbelli* was discussed earlier with reference to the Moonlight and Waihemo tectonic zones. *M myersi* is endemic north of Cook Strait, while its sister species *M. lindsayi* occurs south of the strait (Marshall et al., 2009).

Of the eight main clades in the cicada *Kikihia*, three are restricted to north of the strait (*K. scutellaris*, *K. cauta*, and clade 4), and three (*K. horologium*, clade 2, and clade 5) are only found south of the strait (Marshall et al., 2008). Clade 1 and clade 3 (the *K. subalpina* complex) both occur in North and South Islands, but each includes a genetic break at the strait.

Phylogenetic breaks at Cook Strait in skinks (Greaves et al., 2008) and in galaxiid fishes (Waters et al., 2006) have been dated as Pliocene (minimum age), and so these predate the strait in its modern form. In contrast, Marshall et al. (2009) suggested that the Cook Strait breaks in *Kikihia subalpina* clades date only to the mid- to late Pleistocene, around the time that the strait appeared. Their study relied on a "standard" evolutionary rate used for insects (2.3% divergence/Myr), but the calibration that this is based on has been criticized (Heads, 2012a).

In an earlier study, Marshall et al. (2008) dated the first split in *Kikihia* at 6.6 Ma, with most species originating in the Pleistocene. They calibrated the phylogeny by assuming the endemic *K. convicta* on Norfolk Island was no older than the island, and that *K. subalpina* in the North Island was no older than the subalpine habitats there (these appeared with uplift and volcanism at 1.2 Ma). Both these dates are likely to be underestimates of clade age (see Chapter 5).

The New Zealand freshwater amphipod *Paracalliope fluviatilis* displays deep genetic divergence (up to 26%) between lineages from different locations (Hogg et al., 2006). The group represents a complex of at least four distinct species, although none of these correspond to any discernible morphological variation. Hogg et al. (2006) reported that dispersal in the group among isolated habitats appears minimal, and they found allopatric clades with the following distributions and phylogeny:

Southern North I. (Wellington).
South I.
    SW North I.
    NW North I.
    NE North I.

The first break is at Cook Strait, between Wellington (by Cook Strait) and the rest. The next break is also at Cook Strait, between the South Island clade and the rest, in the North Island. Instead of proposing chance dispersal across the strait, Hogg et al. (2006: 243) suggested that "Such clear genetic breaks may reflect the turbulent geological history of this region."

It is natural to assume that a Cook Strait boundary in a terrestrial group is the result of the sea barrier. Yet many southern groups reach their northern limit at the *northern* side of the strait (e.g., the cicada *Amphipsalta strepitans*; Marshall et al., 2012). In a similar way, many northern groups have their southern limits at the southern side of the strait. For example, *Hebe parviflora* (Plantaginaceae) is widespread in the eastern North Island and has its southern limit along the northeastern shores of the South Island (Marlborough Sounds, Cape Campbell) (Bayly and Kellow, 2006). These distributions indicate that it is the Cook Strait region—not the strait itself—that marks the phylogenetic break.

In a similar pattern to that of these last, terrestrial groups, Ross et al. (2012) reported a Cook Strait break in the estuarine bivalve, *Austrovenus stutchburyi*, but noted that the break does not coincide exactly with the modern strait. Instead, northwest Nelson specimens belong to the North Island clade, and Wellington specimens are in the South Island clade. Again, the pattern suggests that the modern topography of the Cook Strait region is not relevant to the biogeographic break.

Goldstien et al. (2006) studied three, unrelated limpet species from the intertidal zone of New Zealand, and in each one found a genetic disjunction in central New Zealand. The authors concluded:

> It is clear that there is a phylogeographic break between the North and South Islands but whether this is due to the "narrows" of Cook Strait, the complex "Greater Cook Strait" or the east coast environment remains unclear, particularly because two South Island populations within and adjacent to Cook Strait Narrows (French Pass and Cape Campbell) display genetic connection to the North Island (Goldstien et al., 2006: 3264)

All the indications are that this complex, greater Cook Strait node is older than the current geography.

As with most main nodes, the greater Cook Strait area is a distributional margin and also a center of endemism (Heads, 1990a). The endemics include lichens such as *Pertusaria parvula*: Chetwode Islands, and *Rinodina subtubulata*: Kapiti Island and Wellington on coastal rocks (Galloway, 2007). Endemic plants include *Rytidosperma petrosum* (Poaceae) on Kapiti Island, Stephen's Island, and D'Urville Island (Edgar and Connor, 2000). The Cook Strait endemic shrub *Hymenanthera* aff. *novae-zelandiae* (Violaceae) is sister to two other Cook Strait endemics, *H. obovatus* and *H.* aff. *obovatus* "Coast" (Mitchell et al., 2009). The Cook Strait endemic herb *Lepidium obtusatum* (Brassicaceae) occurs at the entrance to Wellington Harbour on coastal rocks and cliffs, and beaches at high tide mark; it is distinct in its group through its rhizomatous habit (de Lange et al., 2013).

The trees and shrubs of *Sophora* (Fabaceae) include a clade comprising the following three species (Heenan et al., 2001):

Sophora molloyi: Cook Strait region (outer Marlborough Sounds, Kapiti Island, and headlands of the south Wellington coast). A shrub with decumbent or prostrate stems, but no rhizomes.

S. longicarinata: Northwest Nelson and northern Marlborough. A tree with rhizomes.

S. microphylla: North and South Islands, widespread. A tree with no rhizomes. There is a "juvenile" (basal) growth stage with divaricate branching that is absent in the last two species.

Animals that are endemic around Cook Strait include the following (Johns, 2015):

Mimopeus buchanani (Coleoptera: Tenebrionidae): Within a triangle: Stephens I.–Maud I.–Titahi Bay, including Trio and Chetwode Is.

Megadromus bucolicus (Coleoptera: Carabidae): Islets in the outer Marlborough Sounds (Stephens, Trio, and Chetwode Is., Kokomahua I. near the mouth of Queen Charlotte Sound, and Maud I.). Closest to *M. capito*, which is widespread in the North Island.

Hemiandrus bilobatus (Orthoptera: Anostostomatidae): In a triangle: Brothers Is.–Waikanae–Baring Head, including records from Wellington, Porirua, and Upper Hutt.

Phases of active phylogeny at the Cook Strait node also explain the many Wellington endemics, which include, for example, the shrub *Pimelea cryptica* (Thymelaeaceae): Wellington, near sea level (Burrows, 2011a), and the beetle *Holloceratognathus passaliformis* (Lucanidae): Orongorongo, Wainuiomata, and Eastbourne (Holloway, 2007).

The phenomenon of South Island montane distributions intersecting the coastline in Marlborough also occurs in other parts of the Cook Strait region. In butterflies, *Lycaena salustius* (Lycaenidae) is ubiquitous through the New Zealand mainland to 2000 m elevation. Gibbs (1980: 142) cited four "forms" distributed in, respectively, the North Island; coasts around Wellington and islands of Cook Strait; South Island lowlands; and South Island above tree line. The South Island alpine form is closest to the Cook Strait form, not to the lowland South Island form.

Members of the Cook Strait biota also display many disjunctions. For example:

Acaena pallida (Rosaceae): Stewart Island to Dunedin, disjunct at Cook Strait, also in Tasmania and New South Wales (Macmillan, 1991).

Atriplex buchananii (Amaranthaceae): Stewart Island to Dunedin (also Central Otago), between the mouths of the Rangitata and Rakaia Rivers, and disjunct at Cook Strait (Marlborough Sounds, Wellington) (Troughton and Card, 1974; NZPCN, 2016).

Other disjunctions link the northeast horsts and Cook Strait, as in *Sicyos mawhai* (Figure 5.7). For the moss *Ischyrodon lepturus* (Fabroniaceae), Fife (2014a) wrote: "Curiously, on the main islands of N.Z. there are two distinct centers of distribution: in the Hauraki Gulf and Cook Strait/Nelson regions." (Hauraki Gulf records are from the Hen and Chickens, Little Barrier, and Great Barrier Islands; Cook Strait/Nelson records are from Titahi Bay, Island Bay, Kapiti I., Stephens I., D'Urville

I. Chetwode Is, Motuara I.; Puponga.) The species also occurs in the Chatham Islands, Australia, and probably South Africa.

There are at least two distinct connections between the South and North Islands across Cook Strait. One is the line: northwest Nelson–Marlborough north of the Wairau fault–western North Island, and the other is the line: Kaikoura–eastern/central North Island. These were illustrated in studies of earthworms that indicated two distinct sets of trans-Cook Strait affinities: one in the west, D'Urville Island–Waikanae, and the other in the east, Canterbury–eastern Wairarapa (Lee, 1959).

An example of the first connection is a group of *Myosotis* (Boraginaceae) species restricted to limestone and ultramafic substrates (Moore in Allan, 1961):

M. concinna: NW Nelson (Mt. Arthur, Mt. Owen). On limestone.

M. monroi: E Nelson (Red Hills and Dun Mt.). On ultramafics.

M. eximia: Ruahine Range, Kaimanawa Mountains. On limestone bluffs and talus.

M. saxosa: Hawkes Bay (Titiokura), and *M. amabilis*: Mt. Hikurangi (sometimes included in *M. saxosa*). Usually on limestone or similar calcareous sediments (NZPCN, 2016).

The second connection is seen in species such as *Carmichaelia nana* (Heenan, 1995) and *Convolvulus waitaha*: Central Otago, eastern Canterbury, eastern Marlborough south of Wairau, Wellington, and Hawkes Bay (Heenan et al., 2003). Both tracks straddle Cook Strait, but they are more or less allopatric.

Breaks at Cook Strait and the Alpine Fault, Incomplete Lineage Sorting, and Microrefugia: The Case of *Pseudopanax ferox* (Araliaceae)

Systematists sample many genes because they are searching for the species tree, not to obtain a diversity of gene trees (Edwards, 2009). Yet the different gene trees that are found for a clade give a wealth of data on spatial evolution, and much of this is lost when the information from the different genes is lumped together to give a single phylogeny. For example, in the New Zealand tree *Pseudopanax ferox* (Araliaceae) two different parts of the genome display different patterns of variation (Shepherd and Perrie, 2011). Sequences from chloroplast DNA show a primary break between northeast and southwest, while nuclear sequences have a break between northwest and southeast (Figure 8.13). The patterns are "incongruent," and this is a problem for hierarchical classification. Nevertheless, the results illustrate two of the main biogeographic patterns in mainland New Zealand. In the chloroplast data, there is a primary split at Cook Strait, while the nuclear sequences indicate a break that coincides with the Alpine fault and the North Island shear belt. Both breaks are standard biogeographic patterns and together form an X-shaped structure. This resembles a trend seen in the many New Zealand faults that strike either northeast or northwest.

How can the incongruent divisions of *Pseudopanax ferox* evident in the chloroplast and nuclear DNA be explained? One possibility is that the species has inherited earlier polymorphism that *already* existed in the ancestor before the descendant groups began to differentiate (incomplete lineage sorting). This process conflicts with the traditional model, in which an ancestor is always homogeneous, not polymorphic, when differentiation of the modern clades begins, but it is now often proposed in molecular studies.

The inheritance of ancestral polymorphism has important implications for evolutionary inference. For example, cyanogenesis occurs in about 10% of vascular plant species, but these do not form a monophyletic clade. Because of this, the feature is thought to have evolved independently in various families. But if the last common ancestor of vascular plants was *already* polymorphic for cyanogenesis before the ancestor began to differentiate into the modern families, it is possible that cyanogenesis evolved only once.

Pseudopanax ferox is a lowland tree, and it might be assumed that it was extirpated in cooler parts of the country during the Pleistocene glaciations. However, Shepherd and Perrie (2011) wrote that the genetic data are consistent with the species' persistence through the Last Glacial Maxima, even at high latitudes in the southeastern South Island. Thus, the tree has preserved the regional pattern of differentiation by surviving through periods of adverse climate in multiple microrefugia. To summarize, the biogeographic

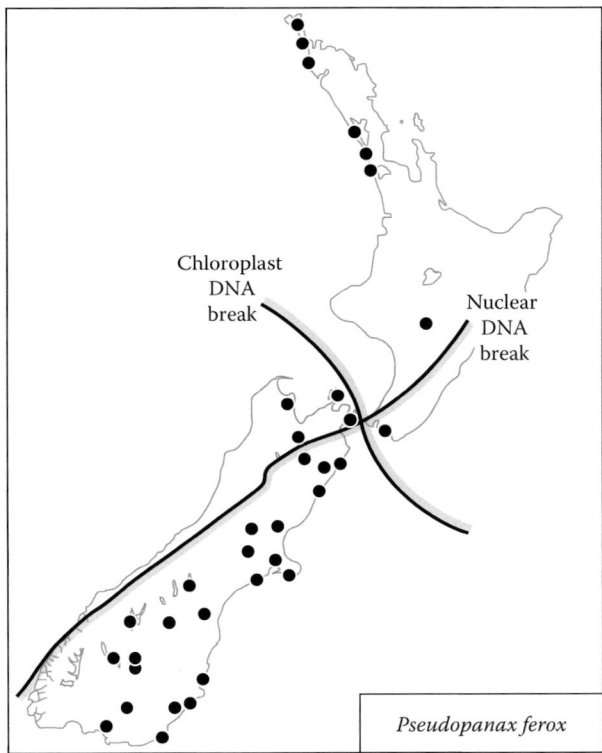

FIGURE 8.13   Distribution of *Pseudopanax ferox* (Araliaceae), showing the differentiation in nuclear DNA and chloroplast DNA. (Data from Shepherd, L.D. and Perrie, L.R., *Mol. Ecol.*, 20, 1389, 2011.)

pattern in *Pseudopanax ferox* is incompatible with either chance dispersal or Pleistocene events, and Shepherd and Perrie (2011) concluded: "Our results indicate that the disjunct distribution is a product of vicariance rather than long-distance dispersal."

Another possible case of incomplete lineage sorting is seen in the moth *Planotortrix avicenniae* (Tortricidae). In the initial description, the species was regarded as a mangrove endemic that ranges from Northland (Kerikeri) to Tauranga (Dugdale, 1990). A subsequent sequencing study by Langhoff et al. (2009) concluded instead that the species was polyphyletic. The authors wrote: "There are many possible reasons for the lack of species monophyly in the COI data set for *Ctenopseustis* and *Planotortrix*. Biological explanations include incomplete lineage sorting, introgression and horizontal gene transfer."

Incomplete lineage sorting means that the distinctive morphological characters of the mangrove form, *Planotortrix avicenniae*, can represent an ancestral polymorphism that occurred in *some* populations of a widespread ancestor, with the characters carried over into the modern "species" or group of populations. In this case, the characters of the mangrove form would be ancient and older than the modern species, rather than being the result of penetration of several modern species into the mangrove, followed by morphological convergence among them.

Oligocene Marine Transgression

Seafloor spreading around Zealandia developed in the Late Cretaceous, and this was associated with on-land extension and thinning. Continental crust is less dense than oceanic crust, and so the thinning led to subsidence of the land, just as a small iceberg is less emergent than a large one.

The subsidence resulted in major marine incursions, and characteristic sedimentary strata show that large parts of New Zealand were covered by shallow or very shallow seas from the mid-Oligocene to the Early Miocene. The flooding culminated in the Late Oligocene during the Waitakian Stage (25.2–21.7 Ma).

Earlier estimates of the land area that remained emergent at the peak of the marine transgression varied from ~20% to 50% of the present land area (review in Heads, 1998a).

Pole (1994), arguing that the entire New Zealand flora is the result of dispersal, suggested that "If New Zealand was completely submerged, then all these organisms [tuataras, moas, kiwis, *Nothofagus*, conifers] had to disperse there over a wide ocean gap." He also suggested that "arguments by geologists against total submergence are usually intuitive, based on biogeographic assumptions...." Campbell and Landis (2001) took up these suggestions and concluded that at some point Zealandia was completely submerged (cf. Campbell and Hutching, 2007; Landis et al., 2008). Campbell et al. (2008: 44) wrote that "a reasonable case can be made for complete submergence of Zealandia."

To prove that there was, for a time, no emergent land would require a single, continuous stratum of marine sediments of the same age, or a well-defined marine erosion surface, over the entire New Zealand plateau. Such a stratum or surface does not exist. This could be because it has been removed in all the areas where it is now absent by erosion, as inferred by supporters of complete submergence, or because it never existed in some areas. Likewise, to prove that there was always some land present through the Late Oligocene, a continuous sequence of terrestrial sediments would be needed for the whole time, and nothing like this exists. In other words, there is no incontrovertible, stratigraphic proof for either scenario. Nevertheless, there is a large amount of other evidence that suggests there was always at least some land present, even at the height of the marine incursions.

Tectonic Evidence for Oligocene Land

The Oligocene is often considered to have been a period of passive subsidence with little or no tectonic activity, but Lever (2007) concluded that during this time there was significant deformation across the South Island (see Chapter 3). The folds have amplitudes of only tens of meters, but in the low-lying landscape even localized, small-scale subsidence and uplift would have been significant for biology. Along with episodes of folding, faulting, and uplift through the Oligocene, there were phases of volcanism. Lever's (2007) paleogeographic reconstruction depicts areas that were "probably land" occurring throughout the Oligocene.

Lithological Evidence for Oligocene Land

The main geological reason that Campbell and Landis (2001: 7) gave for proposing total submersion is that the Oligocene limestone "is largely devoid of sediments eroded from the land." Mortimer and Campbell (2014: 156) wrote that at the time of maximum flooding: "the [terrigenous] sediment tap was almost turned completely off. Limestones... were mostly deposited with only small amounts of sand and silt." Yet they also admitted (p. 207) that recent analyses of rocks formed at or near the time of maximum marine flooding "seem to indicate the presence of land."

A special journal issue considered the maximum extent of the Oligocene inundation and included detailed studies of different regions in New Zealand (Bassett et al., 2014). Most of these presented lithological and other geological evidence for continuous land. Oligocene and Early Miocene limestones sampled from Southland to Northland contain significant amounts of siliciclastic sand, supporting the presence of emergent land during maximum transgression (Mortimer and Strong, 2014).

Near Kokonga, in eastern Central Otago, limestone correlated with the Otekaike Limestone of North Otago is interpreted as a rocky shoreline facies formed near the time of maximum marine transgression (Scott et al., 2014). In the Waitaki–Oamaru region, impure wackestones and calcareous siltstones contain terrigenous material derived from low-relief landmasses (Thompson et al., 2014).

West of the Paparoa Range, there is compelling evidence for nonmarine, Oligocene environments associated with the Paparoa core complex (Riordan et al., 2014). In the Taranaki basin, thick clastic sediments and ongoing tectonism indicate land throughout Oligocene and Early Miocene (Strogen et al., 2014).

In the central western North Island, reverse faulting ensured that land persisted through Late Oligocene and Early Miocene (Kamp et al., 2014). In the East Cape region, the Ihungia igneous conglomerate was derived from an island or islands that were emergent in the Early Miocene and formed from obducted ophiolite. In this region, similar island formation was also possible in the Oligocene (Marsaglia et al., 2014).

Mortimer and Campbell (2014: 157) defended the idea of total submersion and wrote:

> In many places there are indeed sandstone layers in the limestones. The sand composition varies from basin to basin, which arguably indicates local Oligocene sand sources from different islands. The Oligocene limestones themselves are not totally pure, but contain a substantial sand impurity... However,... it does not totally settle the matter: the sandstone and limestone sampling is scattered in both space and in Oligocene time. Because we only know the ages of the samples to the nearest 2–3 million years, this means we may have missed a short-lived total flooding event, even though the generalised evidence suggests land.

Mortimer and Campbell (2014: 205) concluded that "the geological record is frustrating because available age control is insufficiently complete, accurate or precise" to prove or disprove total submersion in a definitive way. Because of this, data from biology have a critical significance.

Paleontological Evidence for Oligocene Land

The southern South Island is one of the key areas in the "total submersion" debate. Strata formed from marine sediments show that many parts of eastern Otago were submerged, and mid-Cenozoic marine sediments are known from the east coast inland as far as Naseby. Further west, sedimentary rocks at Bob's Cove, Lake Wakatipu, indicate a former arm of the sea there, but this probably connected to the south, via the Moonlight tectonic zone. Known incursion in Southland was much less extensive (Reay, 2003).

Lee et al. (2013b) cited sedimentological and paleontological evidence for forested land, estuaries, and rocky and sandy shores in Otago and Southland during maximum marine transgression. They proposed the presence of a large island (~140 km × 100 km) that had shorelines near Kokonga, Naseby, Queenstown, on the margin of the subsiding Waiau Basin, and near Waikaia and Waikaka (both north of Gore). A meandering river system drained south into the Pomahaka embayment. In addition, the Pomahaka coastline was likely contiguous with the Catlins block, a substantial 80 × 70 km region of uplifted Murihiku terrane basement that probably extended south to Stewart Island. The western margin of the Catlins block formed a ridge and valley system that was coastline during periods of higher sea level. Other islands are documented in western Otago. At Waimumu, near Gore, fossils indicate rocky shore, sandy beach, and estuarine habitats in Late Oligocene times (Buckeridge et al., 2014; Lee et al., 2014c), and include wood, seeds, and pollen from many species of rainforest plants (Conran et al., 2014c).

Collectively, these islands covered an area of at least 20,000 km^2, about the same size as present-day New Zealand (Lee et al., 2013b). Oligocene sea level was relatively stable, fluctuating by about 30 m, and this allowed periodic connection of the basement terrane islands separated by shallow drowned valleys. These large, subtropical islands provided a wide variety of habitats able to support a diverse biota. In addition, continued volcanism during the Oligocene flooding would have provided significant sites for terrestrial metapopulations.

High Diversity and Endemism in Early Miocene Fossil Assemblages, Soon after Maximum Flooding

Paleontologists have observed that the Early Miocene fossil record of New Zealand includes such a rich diversity of endemic, terrestrial organisms that the biota is unlikely to have originated following complete flooding in the Late Oligocene—the time available for dispersal into New Zealand and development of the endemism is too short (Lee et al., 2007a,b). Instead, paleontologists have supported a "Moa's Ark" model of Zealandia, with the original biota resulting from vicariance and surviving, in part, into the Miocene and the present (Worthy et al., 2011a).

Worthy et al. (2006) reported a mouse-sized, terrestrial mammal, from Early Miocene (16–19 Ma) fossils in Central Otago. These were found in the Bannockburn Formation, near St Bathans, and the biota there has now been the subject of many studies. Worthy et al. (2006) described the mammal as "archaic," as it was neither marsupial nor placental. They concluded that the Miocene mammal, the tuatara, the New Zealand frog *Leiopelma*, and the acanthisittid wrens, all "imply that at least some land remained emergent during the Oligocene drowning of New Zealand" (p. 19422).

Conran et al. (2009: 473) described Early Miocene orchid fossils from New Zealand and found that they are closely related to extant forms. The authors concluded:

> The presence of a diverse and complex subtropical rainforest including epiphytic orchids surrounding a freshwater lake in Early Miocene Otago supports the assertion that there was land in the New Zealand region throughout the Cenozoic (Lee et al., 2007a), whereas theories that would argue for postdrowning recolonization by New Zealand endemic taxa are much less plausible. (Pole, 1994; Campbell and Hutching, 2007)

No pre-Pleistocene fossil material of *Sphenodon*, the tuatara, was known until Early Miocene fossils were described from Central Otago (Jones et al., 2009). The authors described the fossils as "consistent with the view that the ancestors of *Sphenodon* have been on the landmass since it separated from the rest of Gondwana..." (p. 1385). The fossils are compatible with either Cretaceous vicariance, or with dispersal following Oligocene drowning, but the finding reduces the time available for dispersal to a narrow window of perhaps just 3 Myr. Jones et al. (2009) argued that "The trans-oceanic capabilities of modern *Sphenodon* are questionable," and that "It currently seems more likely that some local land surface persisted during the Oligocene" (p. 1388).

Tennyson et al. (2010) described Early Miocene moa fossils from the Bannockburn Formation, and they reviewed other "Gondwanan" vertebrates from the same formation. They concluded that this evidence makes total submersion unlikely:

> Ancestors of New Zealand frogs, geckos (diplodactylines), tuatara, moa, New Zealand wrens (Acanthisittidae) and the unusual ground-dwelling mammal [Worthy et al., 2006] would all have needed to colonize since the "drowning"... and where would they have come from when there are no known Neogene source populations...? In short, almost all archetypical Gondwanan terrestrial vertebrate taxa known from New Zealand's Recent fauna are now known to have had ancestors present in Zealandia in the late Early Miocene. If Zealandia was completely submerged, as Landis et al. (2008) and others contend, then all such taxa would need to have dispersed to Zealandia in as little as three million years and no such taxa would have arrived in the last 16 million years.

A fossil parrot, *Nelepsittacus*, from the Early Miocene of St Bathans, has been allied with the extant *Nestor* (Worthy et al., 2011b). The authors noted that the St Bathans parrot fauna, with only nestorine parrots represented, has nothing in common with the Australian parrot fauna, which includes cacatuids and a diversity of psittacid genera. They concluded:

> The New Zealand terrestrial vertebrate fauna, at a time minimally 3 Ma after the maximal marine inundation of Zealandia in the late Oligocene, was highly endemic, with no close relationship to the closest faunas in Australia. This high degree of endemism strongly suggests that the Zealandian terrestrial biota persisted, at least in part, through the Oligocene highstand in sea level.

One of the mollusks recorded from the St Bathans fauna is the freshwater snail *Latia*, a New Zealand endemic that has its sister group, *Chilina*, in southern South America. Marshall (2011) argued that the two genera originated by vicariance. Together, they form a clade that is sister to all remaining air-breathing, freshwater snails, a diverse, worldwide group (Hygrophila) of four superfamilies (Jörger et al., 2010). In addition to their "basal" phylogenetic position as a pair, *Latia* and *Chilina* both lack a free larval phase, and *Chilina* is known from fossils back to the Eocene.

Beu et al. (2014) wrote that "*Latia*, Hyriidae and freshwater crayfish [*Paranephrops*] demonstrate that at least some New Zealand extant freshwater taxa have lived on Zealandia continuously since before it began to separate from the rest of Gondwana. At least some Zealandian islands must have remained above the sea throughout mid-Cretaceous to present time."

Worthy et al. (2011c) concluded that the Early Miocene St Bathans fauna refutes a total drowning of Zealandia during the Oligo-Miocene. They cited four aspects of the fauna:

1. The presence of high endemicity at all taxon levels, with no species and only one genus (the global *Palaelodus*, a large, extinct bird), shared with Australia. Notably there are no shared genera of anseriforms despite equivalent-aged lacustrine faunas dominated by waterfowl on both sides of the Tasman Sea.

2. The presence of most of the iconic old endemics of New Zealand long assumed to have a vicariant origin (e.g., sphenodontids, leiopelmatids, dinornithiforms and galaxiids; Lee et al., 2007b).

3. The presence of the endemic freshwater limpet *Latia*, a taxon highly unlikely to be able to disperse across oceans and with an ancient, molecular-based sister relationship to South American forms.

4. The importance of subsequent extinction, in combination with dispersal of new lineages, as a faunal modifier. Extinct lineages include major, high level taxa, for example, crocodilians, turtles (large, nonmarine species), neobatrachian frogs, swiftlets, palaelodids, and several mammals.

Other studies on the St Bathans fauna have described the diverse freshwater fishes (Schwarzhans et al., 2012) and documented a large, new *Mystacina* bat (Hand et al., 2015). Apart from the vertebrates and plants, the Early Miocene fossils from Otago also include many invertebrates, and these are currently being described (Kaulfuss et al., 2010, 2011, 2014).

Worthy et al. (2013a) concluded:

> The Zealandian terrestrial fauna, at a time perhaps just 4 million years after maximum inundation during the Oligocene marine transgression, not only had all the key elements of the modern New Zealand biota, but several other probably endemic family-group taxa, including a bat (Hand et al., 2007) and a terrestrial mammal (Worthy et al., 2006). In the subsequent 16 million years, probably not one of the extant iconic family-group taxa dispersed to New Zealand.

To summarize, the Early Miocene biota is already known to have been diverse, and it is likely to have been even more diverse than the modern biota because of Pleistocene extinctions. Its great diversity and endemism make it unlikely that the Miocene biota as a whole resulted from long-distance dispersal following Late Oligocene drowning.

Lack of Biotic Turnover between Pre- and Postinundation Fossil Biotas

The total submersion model predicts great differences between preflood biotas and postflood biotas. Landis et al. (2008) suggested that there was a "a dramatic change in flora from Oligocene to Miocene time with almost total turnover bar a few exceptions," but this claim was refuted by Pole (2010). Another study compared two fossil fern assemblages, one pre-Oligocene (Southland, latest Eocene) and one post-Oligocene (Central Otago, Miocene) (Homes and Lee, 2011). The two have at least four species in common, and the authors concluded: "The presence of this fern association at two localities widely separated in time suggests the presence of a land-mass in the New Zealand region during the Oligocene."

A detailed review of plant macrofossils and pollen from the Eocene, Oligocene, and Miocene strata of New Zealand concluded:

> The forest floras of these mid-Cenozoic New Zealand sites traverse the period of maximal marine transgression… but provide *no indication of total species turnover indicative of complete submergence at the end of the Oligocene.* [The study sites] show that there were considerable similarities in the floras present on either side of the land area minimum event…. Although it cannot be proven with certainty, the most parsimonious explanation for the pattern observed here is that there was land present continuously somewhere in the region…. (Lee et al., 2012a, p. 254; italics added)

Biological Evidence for Survival of New Zealand-Endemic Clades through the Oligocene

Mortimer and Campbell (2014: 205) wrote that "The biological evidence for permanent land appears to be more convincing than the geological evidence, but on close inspection, it too is wanting in detail. The fossil record is just too weak." But the biological evidence includes much more than the fossil record; the extant biota has contributed a wealth of data on the submersion debate.

The total submersion theory predicts that terrestrial life in New Zealand was destroyed during the flooding, and was then replaced, when land reemerged, by dispersal from other continents (Campbell, 2013; Mortimer and Campbell, 2014: 229). Nevertheless, Campbell also considered problems that this theory has in accounting for biogeography:

> If life were so easily moved about, though, why is it that the world is not more mixed up? Why are there not more Australian animals… in New Zealand? Conversely, why does New Zealand have such strong endemism? These are fair questions, and they are common criticisms of the drowning hypothesis. Again, as a geologist I am not the one to answer it… perhaps there is no need to answer it: perhaps *it really is just a matter of chance, of luck….* (Campbell, 2013: 52; italics added)

Rather than avoiding valid biogeographic questions in this way, other authors have considered data from biogeography together with paleontological and geological information. These studies have found that the facts are incompatible with total submergence and the subsequent arrival of the entire terrestrial biota (e.g., Craw et al., 1999; Lee et al., 2001; Gibbs, 2006; Knapp et al., 2007; Edgecombe and Giribet, 2008; Bunce et al., 2009; Cree, 2014: 507). As Wallis and Trewick (2009: 3) wrote, "Total inundation is hard for biologists to reconcile with some apparently archaic elements of the fauna."

For example, the New Zealand tuatara (*Sphenodon*) is a member of the order Rhynchocephalia, which is known elsewhere only from Mesozoic fossils. Cree (2014: 71) argued against a post-Oligocene dispersal to New Zealand, as (1) no source population is known, despite extensive vertebrate fossils in Australia; (2) further dispersal, for example, to the Chatham Islands, has been unsuccessful (*Sphenodon* does not enter seawater voluntarily, and has high water loss through the skin); (3) dispersal of many terrestrial reptiles, such as agamid lizards and snakes, from Australia to New Zealand has not been recorded; (4) fossils show that *Sphenodon* was already present in New Zealand in the Early Miocene (19–16 Ma), soon after the flooding; and (5) *Sphenodon* can survive in very large numbers on very small islands. Cree (2014: 507) concluded: "From a biological perspective the evidence is strong that [through the Cenozoic] New Zealand remained an emergent archipelago…."

Absences of Old, Widespread Groups in the New Zealand Biota

Campbell and Landis (2001) based their idea of total submersion on the lack of geological evidence for continuous land and on biogeography. In particular, they stressed certain absences in the post-Oligocene New Zealand biota:

> When Zealandia split from Gondwana, dinosaurs, turtles and flying reptiles were among the cargo, and *surely the full spectrum of other Cretaceous land animals was present as well –* mammals, lizards, snakes, amphibians… We know what put paid to the dinosaurs (effects of a major meteorite impact), but we need another mechanism to rid Zealandia… of all other animals. Total submergence of the land would do the trick. (p. 7; italics added)

But this assumes that the biota of Gondwana was homogeneous just prior to breakup. No other continent has a homogeneous biota, and there is no evidence to suggest that Zealandia carried the "full spectrum" of Gondwanan Cretaceous biota following rifting. As Buckley et al. (2015) wrote, "Undoubtedly, biotic regionalisation existed across Gondwana, and this may have influenced the absence of some lineages from modern New Zealand."

Landis et al. (2008) wrote that "In certain respects, the modern biota of New Zealand demands this perspective: the idea of total submergence and hence wholesale destruction of terrestrial life." Yet the sole piece of evidence from the modern biota that Landis et al. (2008) offered for total submergence was the absence of native terrestrial mammals, and postflood, fossil material of a terrestrial, nonvolant mammal has been described from the Miocene (Worthy et al., 2006). Landis et al. (2008) referred to this and described the animal as "rodent-like," but the only morphological analysis of the material concluded that it belonged to an archaic group that had already evolved before placentals and marsupials diverged (Worthy et al., 2006).

Mortimer and Campbell (2014: 207–208) wrote that "We… need an explanation of some of absences in the biota, such as the lack of any native terrestrial ground-dwelling mammals and the absence of any archaic freshwater fish." But austral family Galaxiidae, which was thought to a member of the order Osmeriformes, is now thought to be sister to a diverse, global complex (Protacanthopterygii s.lat.) that includes five other orders (Li et al., 2010): Stomiiformes, Argentiniformes, Osmeriformes, Salmoniformes, and Esociformes (the last two have Cretaceous fossils). This means that Galaxiidae are much older and also more significant for fish evolution than was thought.

Mortimer and Campbell's (2014: 207–208) comment about the absence of archaic freshwater fish in New Zealand is probably a reference to Osteoglossiformes, a group present in Africa, southern Asia, Australia, and South America. The authors suggested that this and other groups are absent from New Zealand because they were extirpated there during flooding, and this is one possibility. Another possibility is that absences in the biota are the result of allopatry, and that old groups of fishes such as Osteoglossiformes were never present as such in the region.

Another example involves the torrentfish *Cheimarrichthys*. This is endemic to New Zealand (North and South Islands), where it spends part of its life cycle in freshwater and part in the sea, near the coast. It is sister to a diverse, global clade of fishes, Leptoscopidae plus Pinguipedidae, that occurs in inshore, marine waters (Chapter 4). The absence of the latter, widespread clade from New Zealand rivers and the endemism there of its sister group, the torrentfish, can both be explained by simple vicariance between the widespread clade and its New Zealand sister group. There is no need to propose any extirpation of the widespread group in New Zealand.

In a similar case, the New Zealand frog *Leiopelma* plus *Ascaphus* of the western United States together form the sister group of all other extant frogs. The absence of the clade "all other frogs" from New Zealand is notable, as the group is so diverse and, outside New Zealand, has a global distribution. Yet the absence of "all other frogs" from New Zealand can be explained by simple vicariance with its sister group, and so, again, there is no need to invoke extirpation by Oligocene flooding.

Pre-Oligocene Molecular Clock Dates for New Zealand-Endemic Clades

Many New Zealand endemics are now thought to have survived in New Zealand through the Oligocene drowning because molecular clock studies have indicated that they predate the flooding. The clock dates in all these studies were calibrated with fossils or island ages, and so they are valid as minimum ages (assuming clock-like evolution). The groups thought to be older than the flooding include liverworts (five clades of Schistochilaceae; Sun et al., 2014), monocots (Asteliaceae; Birch et al., 2012), eudicots (Proteaceae; Barker et al., 2007), earthworms (Buckley et al., 2011), freshwater decapods (Parastacidae; Toon et al., 2010), freshwater isopods (Wilson, 2008), Onychophora (Giribet and Boyer, 2010), Opiliones (Pettalidae; Giribet et al., 2012), centipedes (Giribet and Boyer, 2010), Orthoptera (groups of wetas in Anostostomatidae: *Hemiandrus*, *Motuweta*, *Deinacrida*, and *Hemideina*; Trewick and Morgan-Richards, 2005), dipterans, including Chironomidae subfam. Orthocladiinae (Krosch et al., 2011; Krosch and Cranston, 2013) and Chironomidae subfam. Podonominae (Cranston et al., 2010), beetles (Liebherr et al., 2011), Lepidoptera (*Sabatinca*; Gibbs and Lees, 2014), frogs (*Leiopelma*; Roelants et al., 2007), geckos (Nielsen et al., 2011), ratite birds (moas; Bunce et al., 2009), and two lineages of acanthisittid wrens (the Stephen's Island wren *Traversia*, and the common ancestor of *Xenicus*, *Pachyplichas*, and *Acanthisitta*) (Mitchell et al., 2014b). For the wrens, Mitchell et al. wrote that the dates provide "compelling evidence against complete submergence, as the acanthisittids are ill-suited to long-range, over-water dispersal…," and this probably applies to many of the others as well (cf. Mitchell et al., 2016).

Allwood et al. (2010) referred to the "growing list of taxa" with molecular dates that predate the Oligocene drowning, and Sharma and Wheeler (2013) agreed: "Multiple published cases of ancient lineages with pre-Oligocene history refute the hypothesis of complete inundation of New Zealand." They cited the harvestmen, Diptera, frogs, and geckos.

Dispersal theory has responded to these important molecular clock results by suggesting that the New Zealand groups only appear to be old (pre-Oligocene) because, after they dispersed to New Zealand, their young sister groups on other land masses have gone extinct. Nevertheless, Allwood et al. (2010) replied: "as this list [of pre-Oligocene New Zealand endemics] grows… these ad hoc arguments become less tenable."

A Case Study: Agathis *and Its Possible Survival through the Oligocene Flooding*

Agathis (Araucariaceae) is a genus of large trees and is widespread around the southwest Pacific (Heads, 2014: Fig. 6–17). It is also known from Eocene fossils in Patagonia (Wilf et al., 2014). In a molecular phylogeny of the genus, the main node corresponds with the South Loyalty Basin, as this feature separates clades of *Agathis* endemic to the south (New Zealand; basal), west (New Caledonia and Australia to Sumatra), and east (Vanuatu and Fiji) (Knapp et al., 2007; Heads, 2014: Fig. 6.18).

Agathis is extant in New Zealand with a single species, *A. australis* (kauri), and this dominates many forests of the northern North Island. In earlier studies, New Zealand kauri forest was treated as a separate category from beech forest and mixed podocarp-angiosperm forest, but site-level floristic analysis shows that kauri forest is best seen as a form of podocarp forest (Wiser et al., 2011). This is consistent with the recent discovery that Araucariaceae (including kauri) and Podocarpaceae form a clade; the two forest types they dominate represent two different facies of "arapod forest" (Heads, 2014).

A South American fossil species, *Agathis zamunerae*, occurs in Eocene strata dated as 52.2 Ma. This plant is of special interest as it belongs to the crown group of *Agathis* (the youngest clade including all the extant members), probably in a derived position (Wilf et al., 2014). As Wilf and Escapa (2014) stressed, the fossil is much older than recent estimates of the *Agathis* crown group age (23 Ma).

A Cretaceous fossil *Agathis* from the South Island, *A. seymouricum*, is known from a foliage-bearing shoot dated at 100 Ma. Its precise affinities are controversial (Biffin et al., 2010), but it has been regarded as closer in its morphology to the New Zealand *A. australis* than to any other extant or fossil species (Stöckler et al., 2002). Waters and Craw (2006: 353) suggested that the fossil "clearly predates New Zealand's separation from Gondwana, so it seems highly unlikely this ancient '*Agathis*' lineage was restricted to New Zealand at that time." Yet this assumes that the Gondwana biota was homogeneous and that a taxon known from one area must have been found throughout the supercontinent. Again, there is no evidence for this (see Chapter 1), and there is no evidence that *A. seymouricum* or *A. australis* ever occurred outside New Zealand. Stöckler et al. (2002) also cited the fossil resin found at many Eocene, Oligocene, and Miocene localities in New Zealand, and its chemistry resembles that of *A. australis* resin.

Stöckler et al. (2002) found that *Agathis australis* is the sister group of all the other *Agathis* clades, not the sister group of an Australian or New Caledonian species as predicted in dispersal theory. They concluded that "dispersal from Australia is an unlikely explanation for the origin of New Zealand *Agathis*" (p. 831). This finding, together with the fossil data, "provides strong evidence that New Zealand was not completely submerged during the Oligocene."

A subsequent molecular study by Knapp et al. (2007) confirmed that *Agathis australis* is basal in the genus. Using fossil calibrations, they calculated that *A. australis* and the rest of the genus had already diverged by the Early Oligocene or Eocene, and they stressed that these are minimum dates. Knapp et al. (2007) inferred a vicariant origin of the trans-Tasman connection and the survival of *Agathis* in New Zealand through the Oligocene drowning. They admitted that an *A. australis* ancestor could have existed earlier in Australia and gone extinct there without leaving fossils, as suggested by Waters and Craw (2006), but they regarded this idea as ad hoc and untestable.

Does Accepting Mid-Oligocene Land Depend on a Circular Argument?

Many authors have accepted that there was always some land present in New Zealand throughout the Cenozoic, but Waters and Craw (2006) claimed that their reasoning was circular—that biologists have relied on geologists' support for land, while the only evidence that geologists have for any land is biological.

As discussed earlier, there is now good geological evidence for persistent land, and, in any case, many biologists have based their claims for persistent land, not on geological reconstructions but on biological evidence. This has been sourced from fossils (e.g., Pole, 2010; Worthy et al., 2011a,b,c), molecular clock dates (e.g., Birch et al., 2012), and biogeographic patterns that correlate with pre-Oligocene tectonics (e.g., Heads, 1990a, this book). In cases such as the total submersion debate, where geological evidence is lacking or equivocal, biologists often prefer to stress biological evidence and place less weight on tentative or controversial geological models.

Waters and Craw (2006) concluded: "the geological 'evidence' [for Oligocene land] has largely been driven by the general assumption that New Zealand has a Gondwanan biota… which results in circular logic." Waters and Craw (2006) suggested that Fleming committed this error: "Fleming (1962a, 1979) had no problem in mapping such [Cenozoic] islands because he was a biogeographer and believed that a 'Gondwana' fauna and flora provided the extra evidence needed to discount complete Oligocene drowning." Far from being circular, though, Fleming's (1962a, 1979) ideas on particular areas of Oligocene land were logical and evidence-based, and they have been supported by many recent studies in paleontology, molecular biology, and biogeography (see citations given earlier). In addition, there is lithological evidence for land; for example, quartz grains in many of the Oligocene limestones (Bassett et al., 2014).

Waters and Craw (2006) concluded: "While we are not specifically rejecting the possibility of a Gondwanan ancestry for some elements of New Zealand's flora and fauna, we caution researchers against a circular approach to this question. Analysis of geological evidence in isolation provides no evidence for continuous terrestrial landscapes during New Zealand's late Oligocene." Yet apart from the geological evidence that does indicate land, there is also no geological evidence for continuous sea at any one time, and Waters and Craw (2006) themselves admitted that "Available geological data… neither confirm nor reject Oligocene drowning of New Zealand."

Waters and Craw (2006) argued that privileging biological information over geological knowledge is circular, but in the case they were discussing—the Oligocene drowning of New Zealand—they admitted that geological knowledge cannot answer the question. For this reason, stressing biological data is a logical approach, and it has made positive contributions to the current understanding of Oligocene history.

Equivocation in Claims for Complete Submergence

Landis et al. (2006) wrote that marine inundation in New Zealand "may well have culminated in complete submergence c. 25 Ma," but supporters of total submersion have followed Waters and Craw (2006) in admitting that geological data "can neither confirm nor deny" the existence of some land throughout the Cenozoic (Campbell and Hutching, 2007: 167; Trewick et al., 2007).

For the Oligocene drowning, Landis et al. (2008) argued that "no permanent or persistent land areas can be identified" from geological evidence, and so "until proven otherwise, Zealandia was totally submerged..." (p. 191). Nevertheless, they also admitted that "it is not possible to totally exclude the existence of a few small islands" (p. 191). Furthermore, "Occasional ephemeral islands appeared with Paleogene rifting... Islands may also have formed in association with Oligocene submarine volcanism" (p. 185).

The authors who support the idea of complete submergence have conceded that biological arguments can be made for the continuous presence of islands. For example, Campbell (2013: 36) described the "bold new paradigm [which] suggested that the ancestors of all the native terrestrial biota have somehow arrived... only within the past 23 million years" (p. 37), but he also admitted that "We could not prove that there were no islands at the time of maximum flooding... There probably were some ephemeral short-lived islands... There may have even been a continuum of short-lived islands...." (p. 38). Thus, as Liebherr et al. (2011) commented, "Although total inundation has been argued for repeatedly, protagonists of complete drowning invariably equivocate with regard to the possibility that isolated islands persisted [through the Cenozoic]."

Current Views on the Total Submersion Model

Campbell and Landis (2001: 7) wrote that "The idea of a clean slate in the Oligocene appeals to many geologists and palaeontologists" (p. 7), but they did not cite references, and Mortimer and Campbell (2014: 157) admitted that most geologists now accept that there was some continuous land. Apart from paleontologists, the only biologists who have accepted the possibility of total submersion are Trewick and Paterson (in Landis et al., 2006, 2008; Trewick et al., 2007) and Waters (in Waters and Craw, 2006). Many other biologists have examined the question, and they have all rejected the idea of total submersion (see citations given earlier).

Survival of Clades through the Oligocene as Dynamic Metapopulations

Continuity of some land through the Oligocene does not mean that land has persisted at any one point; instead, emergent land probably occurred at different places at different times. In this environment, clades would survive as dynamic complexes of populations.

For example, in the dipteran family Empididae, the "*Chelipoda*-like group" is confined to Zealandia (New Zealand, Lord Howe Island, New Caledonia)—continental crust that rifted from Gondwana—and Vanuatu, part of the Pacific island arc that rifted from Gondwana in the Cretaceous. Plant (2010) proposed that the group is "a relictual Gondwanan element that has survived Oligocene drowning as metapopulations persisting *in situ* on ephemeral islands along arcs, ridges and buoyant crustal blocks...." Many groups survive as metapopulations on a local, ecological scale; examples include weedy species that invade forest gaps caused by tree falls, land slides and so on. A similar process is suggested here for regional endemics; these have survived more or less *in situ* in their region by constantly dispersing to newly emerged islands.

Campbell et al. (2008: 45) wrote that "If it [Zealandia] was completely immersed, what happened to terrestrial life? It either perished or survived on ephemeral short-lived islands for which there is little geological evidence." The evidence from geology, paleontology, and biogeography discussed earlier suggests that the latter option is the most probable. While some biologists have accepted that the entire terrestrial biota was wiped out, most have accepted that communities survived the flooding on small, ephemeral islands. Even small islands can maintain very diverse biotas, as seen in Stewart Island, the islands off Northland, and New Caledonia, for example. This indicates that many terrestrial taxa endemic to areas around active plate margins do not require land that is persistent over geological time, and small islands in the shallow New Zealand seas would have allowed the survival of taxa through the Oligocene. In addition, the constantly changing coastlines would have facilitated local differentiation.

Evolution in the Oligocene

For some New Zealand groups, the marine incursions of the Oligocene would have been a "major environmental crisis" (Stöckler et al., 2002: 831). In other groups, the flooding, along with localized tectonism mentioned earlier, would have favored diversification. In New Zealand skinks, for example, Hickson et al. (2000) proposed rapid allopatric speciation during the Oligocene, as New Zealand was fragmented into many low-lying islands. This history explains why the modern biota of New Zealand can be characterized as one of uplifted wetlands and associated better-drained habitats; the lowland, montane, and alpine swamps and bogs represent the derivatives of Oligocene mangrove, saltmarsh, and swamp forest communities (Heads, 1990a, 1998b). The survival of pre-Oligocene taxa and distribution patterns can be reconciled with widespread Oligocene seas if these were shallow, and if coastlines in the archipelago were continuously shifting.

The total submersion hypothesis recalls the early twentieth-century debate about Greenland-style ice sheets that, according to some authors, completely covered the South Island in the Pleistocene. The biologists of the time responded that this was incompatible with the high levels of South Island endemism (Thomson, 1909), and, as mentioned earlier in this chapter, many microrefugia probably persisted through the glaciation. In a similar way, the persistence of metapopulations in dynamic microrefugia would have enabled the survival of the main biogeographic patterns on the New Zealand archipelago through the Oligocene transgression.

The Last, Great Modernization of the New Zealand Biota Was Pre-Miocene

The diverse Early Miocene fossils refute the total submergence idea. They are also significant as they show that the main components of a "modern" biota already existed at 19–16 Ma, and so the biota itself must have originated (or been modernized) at some time before this. As Tennyson et al. (2009) wrote, "The St Bathans Fauna reveals that, in the Early Miocene, Zealandia had a vertebrate fauna that *in many ways was similar to the present one....* The major faunal change since then has been the extinction of several major groups...." [italics added]. Likewise, a study of a well-preserved, Miocene fossil plants from Foulden Maar, Otago, dated at 23 Ma found that reproductive features across 23 families and 30 genera of forest angiosperms and conifers have remained constant for more than 20 Myr. The insect-, wind-, and bird-pollinated flowers, and wind- and bird-dispersed diaspores in the fossil flora all indicate remarkable conservatism over the last 20 Myr, "despite widespread environmental and biotic change" (Conran et al., 2014b). The environment of the plants has undergone substantial change, but the features have not, and the last, great modernization of the biota in which the "modern" groups originated must date back to earlier times.

Conclusions on the Total Submersion Theory

Vicariance theory predicts that New Zealand was not completely submerged in the Cenozoic. The total submersion idea received a great deal of publicity when it was first introduced, as it would falsify a vicariance origin of the New Zealand biota and support chance dispersal. Nevertheless, it now has little support among geologists, paleontologists, and biologists.

New Zealand, New Caledonia, Hawaii, and Jamaica are all important centers of endemism, but chance dispersal theory has accepted that the terrestrial biotas of these islands were drowned during marine incursions in the Paleogene. However, the biogeographic and tectonic contexts of all these examples indicate the persistence of metapopulations through the marine incursions. For example, a study of beetle distributions concluded that a total submersion of New Zealand and New Caledonia in the Cenozoic "seems improbable" (Jolivet and Verma, 2010).

Of course, all the major geological events in New Zealand would have brought about the extinction of many populations and even whole clades. This would have occurred, for example, during marine incursions (Oligocene), orogeny (Cretaceous, Neogene), ice ages (Pleistocene), and volcanism (at many times). The New Zealand region has undergone intense disturbance through much of its history, and the fossil record preserves many lineages that have gone extinct in New Zealand

(Lee et al., 2012a). In the extant biota, several woody plant families that are diverse elsewhere are each represented in New Zealand by just a single species (Monimiaceae, Passifloraceae, Meliaceae, Sapotaceae, Nyctaginaceae, Bignoniaceae, Gesneriaceae) or two species (Proteaceae, Santalaceae, Sapindaceae). This provides good evidence for an important relictual component in the biota. Nevertheless, despite the abundant evidence for large-scale extinction, inferring the *complete* extinction of the terrestrial biota at any one time is unwarranted.

9

Biogeography and Neogene Geology of Mainland New Zealand: Alpine Fault Strike-Slip, Kaikoura Orogeny, and Pleistocene Glaciation

For a long period through the later Paleozoic and Mesozoic, convergence and subduction beneath the Zealandia margin of Gondwana caused terrane accretion and generated the Median Batholith. At ~100 Ma, in the mid-Cretaceous, subduction ceased and extension began. Figure 9.1 (from Tulloch et al., 2009) is a reconstruction for a time soon after Gondwana breakup. From then until ~20 Ma, the structure of the New Zealand region remained more or less the same and was based around an inactive subduction zone with its extinct arc (the Median Batholith). These two features, along with the Eastern and Western province terranes, were more or less straight, not bent into a reverse-S shape as they are today. This remained the case until the Early Miocene, when changes in relative plate motion led to renewed convergence, the development of the Alpine fault, and massive deformation of New Zealand geography.

The reconstruction in Figure 9.1 includes modern coastlines to indicate where the different parts of the crust were; it does not imply any ancient coastline. The Zealandia continent was bounded by the Pacific subduction zone in the northeast and later by the rifted margins in the south and west (a small part is visible on the figure). It was emergent, in large part, until the Oligocene. The reconstruction indicates that the mainland region has undergone great distortion during the renewed convergence, and so the original areas of the crust as indicated by the coastlines are not strictly accurate. For example, the Median Batholith and Dun Mountain ophiolite belt are depicted in Figure 9.1 with their original, unbent positions, and all the other terranes would lie parallel with these, but the compressed part of the crust shown in the ellipse can be imagined as formerly (before contraction) extending westward, as indicated by the arrow. Some of the compressional deformation has been more or less continuous through the crust or has taken place on many small faults, but much of the plate motion has been taken up by new, strike-slip displacement on the Alpine fault (cf. Figure 3.15).

A model of Neogene tectonic development in New Zealand is illustrated in Figure 9.2 (from Kamp, 1987). The marker terrane shown in black is the Dun Mountain ophiolite belt; its reverse-S shape illustrates Neogene bending, and its fracture at the Alpine fault illustrates the strike-slip displacement here. South of the great bend in the Alpine fault, the fault marks the Indo-Australia/Pacific plate margin.

Cenozoic Deformation in New Zealand

The Emerald Fracture Zone and Extensional Deformation in the Paleogene

Following the collision of the Hikurangi Plateau with the Chatham Rise at some time in the Late Cretaceous, the Emerald fracture zone propagated through Zealandia (Figure 3.15, 75 Ma). By the end of the Cretaceous, it had developed into a transform zone that followed the Australia/Pacific plate boundary (Figure 3.15; see 75 Ma reconstruction). Nevertheless, before 45 Ma both sides of Zealandia (on the Australian plate and the Pacific plate, respectively) were moving north at about the same speed, and so there was little relative movement between them.

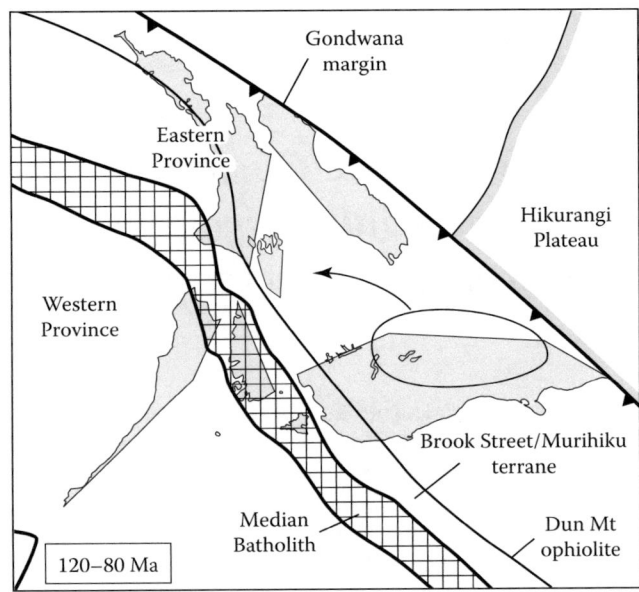

FIGURE 9.1 Reconstruction of Zealandia at 120–80 Ma. The modern coastline is included to indicate the ancient location of the blocks, not the ancient coastline. In addition, there has been continuous deformation, and in the past the region in the ellipse extended westward in the direction of the arrow. The original (unbent) location of the Dun Mountain ophiolite is shown in Otago (the slight bend in the North Island has not been straightened). (Based on Tulloch, A.J. et al., *Geol. Soci., London, Special Publ.* 321, 89, 2009.)

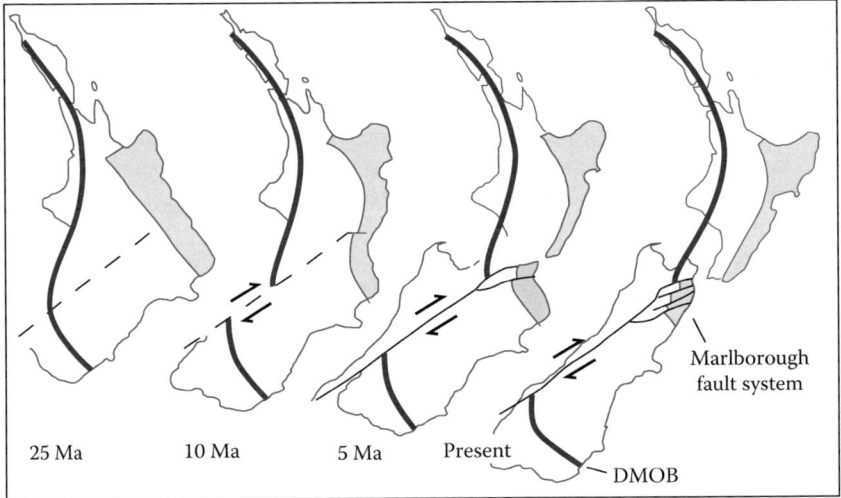

FIGURE 9.2 A model of the age and origin of recurved arcs in New Zealand in relation to movement on the Australia/Pacific plate boundary, including the Alpine fault sector. Solid black line (DMOB) indicates the Dun Mountain ophiolite belt and the associated magnetic anomaly. Reconstruction at 25 Ma shows the future position of the Alpine fault as a broken line. Gray indicates the Pahau subterrane (younger Torlesse terrane). (From Kamp, P.J.J., *J. Geol. Soc. (London)*, 144, 641, 1987.)

Activity along the plate boundary through Zealandia began to develop at ~45 Ma, when the Emerald fracture zone evolved into a divergent margin (the Resolution Ridge) west of the Campbell Plateau (Figure 3.15). The seafloor spreading there would have caused vicariance between clades on the Macquarie Ridge and their relatives on the Campbell Plateau. Much of the new sea floor generated in the Emerald Basin at this time was later subducted at the Puysegur section of the plate margin. This means that the regional biota has experienced both the opening and the subsequent closing of a large basin, as in the South Loyalty Basin, and this must have affected the regional biology.

The Emerald fracture zone proceeded to develop through central New Zealand as a divergent margin, and rifting was under way by 40 Ma (Middle Eocene). This continental rift system was defined by a belt of normal faults and was active until the mid-Oligocene (Kamp and Furlong, 2010). Although this new plate boundary was initiated in the Eocene, the pole of relative rotation between the two plates (Indo-Australian and Pacific) remained very close to the South Island, and so there was still little relative plate motion within New Zealand (Sutherland, 1999; Cox and Sutherland, 2007). Rifting continued until 30 Ma, but in the Late Oligocene (29–24 Ma), the pole of Australia/Pacific relative rotation migrated southeast. This caused the rates of plate motion within New Zealand to increase through the Miocene, and there was a rapid transition along the rift system from extension to transpression.

Transpressional Deformation in the Neogene, the Alpine Fault, and the Orocline

In the latest Miocene, increased convergence in the relative plate motion converted the northern part of the Emerald fracture zone—the Alpine fault—into an active, dextral-reverse structure, that is, it is undergoing a combination of transcurrent and reverse (compressional) displacement (Figure 3.15). The compressional component of displacement on the fault caused the uplift of the Southern Alps. Overall, the deformation along the plate boundary changed from transtension to more or less pure transcurrent (strike-slip) displacement and then to transpression.

Since motion on the new plate margin began through New Zealand, the ancient convergent margin of northeast Zealandia (which was more or less straight) has been distorted by ~850 km of dextral displacement between the Pacific and Australian plates (Molnar et al., 1999; Figure 3.15). About 460 km of this (54%) has occurred on the Alpine fault. The remaining 400 km, just under half of the total displacement, has been accommodated by bending and distributed deformation (dextral transpressional shear) in the crust. The current rate of dextral strike-slip on the Alpine fault (between Milford Sound and Hokitika) is ~27 mm/year (Norris and Cooper, 2001). This represents 70%–75% of the strike-slip component of current interplate motion (37 mm/year). The rate of strike-slip on the Alpine fault shows a substantial drop north of the junction with the Hope fault.

Since about 45 Ma, plate motion through New Zealand has caused continental shear and clockwise rotation, and this has resulted in a distinctive orocline (Turner et al., 2012; Mortimer, 2014). This is the reverse-S or Z-shaped double bend seen in the Dun Mountain ophiolite belt and the other basement terranes (Figure 9.2). The central limb of the orocline strikes northeast and appears to be rotated clockwise by ~90° relative to the exterior limbs. The Alpine fault cuts and offsets the innermost part of the central limb. Most of the bend of the orocline has developed since the Eocene, although Cretaceous extension and rifting, especially in the Bounty and New Caledonia troughs (at 85–105 Ma), also caused some of the deflection in the basement trends (Mortimer, 2014).

The clockwise rotational deformation of the orocline has affected much of northern and eastern New Zealand. The rotation has occurred around a stationary pivot in northwest Nelson (Turner et al., 2012). Although this region has been affected by reverse faulting and folding since the late Eocene, it has remained attached to the Australian plate. As a strong backstop, it has avoided the complex, distributed deformation and tectonic rotations experienced by areas to the east and north. The rigidity of northwest Nelson reflects the strength of its crystalline basement rocks and the relative weakness of the accreted metasedimentary terranes to the east.

The east coast of the North Island—the forearc of the Hikurangi subduction margin—is rotating rapidly (~3°–4°/Myr, clockwise). The transition from rapid forearc rotation in the eastern North Island to a strike-slip dominated plate boundary in the South Island occurs via a crustal-scale hinge in northeastern Marlborough (Lamb, 2011; Wallace et al., 2012).

Cutout block reconstructions of North Island deformation since 20 Ma are illustrated in Figure 9.3 (Lamb, 2011). Many of the boundaries between the blocks are represented by major faults, but these are not shown in the figure. The Esk Head Mélange, which separates the older and younger Torlesse terranes (Figures 3.2 and 8.11), is included in Figure 9.3 as a marker terrane. The Gondwana-Pacific subduction

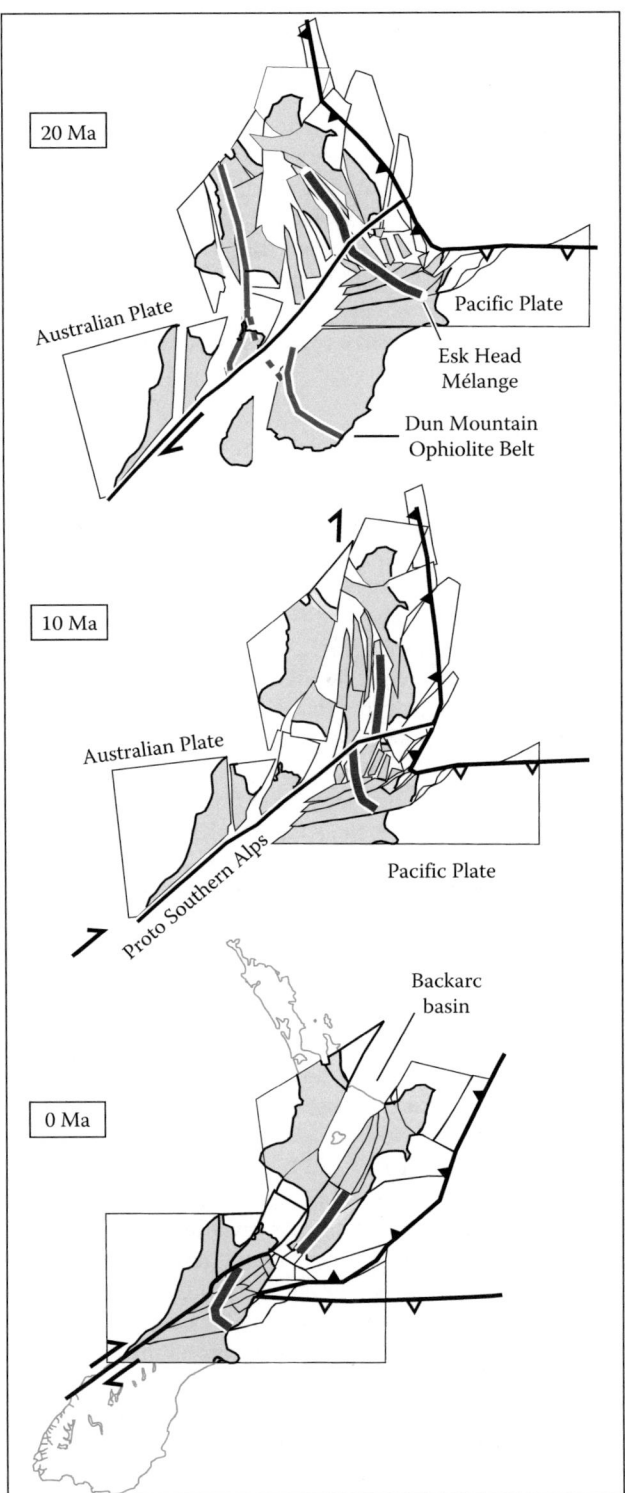

FIGURE 9.3 "Cutout" block reconstructions for the Neogene evolution of the Hikurangi margin. The reconstructions are based around the rotation of several key blocks for which there is good paleomagnetic data. The modern coastline is included to indicate the ancient location of the blocks, not the ancient coastline. (Simplified from Lamb, S., *Tectonophysics*, 509, 135, 2011.)

zone formed as a more or less straight feature, but along the sector from East Cape to Cook Strait, southern Marlborough, and the Chatham Rise, it has been bent into an acute angle, and this has been dragged into a node at Kaikoura.

Northland and East Cape to Cook Strait and Kaikoura

Many groups with strongholds in Northland, the northeastern islands, and East Cape (along the horstian tracks) also include populations further south along the deformed, rotated, eastern seaboard of the North Island. This region is made up of the exhumed, rotated, and sheared East Coast Basin. This is a sediment-filled forearc basin located between the Hikurangi trench and the volcanic arc (central North Island–Kermadec Islands–Tonga). It extends south from East Cape to southern Marlborough. *Sicyos*, with its disjunct, southern records around Cook Strait, displays a variation of this pattern, in which large disjunctions are typical (Figure 5.7).

Hebe parviflora (Plantaginaceae) is another group disjunct between Northland (Russell, Bream Head, Hen and Chickens, Great Barrier Island) and the East Coast Basin; it also occurs around Lake Taupo and south to Marlborough (Bayly and Kellow, 2006). It is one of just two tree species in the diverse *Hebe* complex; the other is *H. barkeri* of the Chatham Islands. This "outer arc" of trees in the *Hebe* complex is replaced on the rest of the mainland (most of the South Island, Stewart Island, and the western North Island) by different kinds of shrubs, subshrubs, and cushion plants.

The East Coast Basin has been active ever since the Late Cretaceous. In the Oligocene–Miocene, marine transgressions extended across exhumed parts of the basin, from the east coast west to Taihape, before the rise of the axial Ruahine Mountains. This would have led to range expansion of coastal elements inland. Following marine regression and the uplift of the mountains, some populations appear to have been left stranded inland, west of the mountains. This process explains certain land-locked plant records near Taihape, such as outliers of the eastern *Olearia gardneri* (Asteraceae), and anomalous, inland populations of the coastal shrub *O. solandri* (Heads, 1998b). The inland records of *Hebe parviflora* around Lake Taupo could either be a result of this process or of marine transgression along the Taupo volcanic zone.

Northland to Cook Strait/Kaikoura and the Chatham Islands

This pattern is similar to the last, with disjunct records from Northland to Cook Strait, but there are also populations on the Chatham Islands. The distributions resemble those of other groups found around the margin of the Hikurangi Plateau (Figure 3.10), but they also occur in Northland and so they show the Z-shaped deformation caused by the plate movement.

The click beetle *Amychus* (Elateridae) comprises three species with the following distributions (Marris and Johnson, 2010):

Three Kings Islands (*A. manawatahi*).

Chatham Islands (*A. candezei*).

Islands in Marlborough Sounds and Cook Strait (Maud Island, Sentinel Rock, Stephens Island, The Brothers, Trio Islands, Chetwode Islands); also known from subfossils at Waikari, north Canterbury (*A. granulatus*).

The Three Kings species and the Chatham Islands species are keyed together. The overall distribution of the genus (northeastern islands–Cook Strait–Chatham Islands) displays a pronounced bend into Cook Strait, with the distribution following the modern Hikurangi subduction zone and the old Chatham Rise subduction zone. Prior to the Cenozoic deformation, this track would have formed a linear or gently curved distribution along the ancient subduction zone.

The pipefish *Stigmatopora nigra* (Syngnathidae) inhabits shallow marine waters (0–9 m deep) of southern Australia and northern New Zealand, where it repeats the distribution of the click beetle *Amychus*; it is known from the northeastern offshore islands, the Cook Strait region, and the Chatham Islands (Roberts, 1991).

Biogeography at the Alpine Fault: Boundaries and Disjunctions at a Transform Fault, with Distributions Preserved in Multiple Microrefugia

Christenhusz and Chase (2013) wrote that "in explaining general biogeographical patterns, the major vicariant factor is continental drift (plate tectonics)," but there is much more to plate tectonics than continental drift—that is just one of its many results. For example, plate tectonics can also cause major *intracontinental* fault displacement and rifting (e.g., along the Amazon), orogeny, sediment production and deposition, volcanism, subsidence, and marine transgression, along with many other processes. All of these can lead to vicariance or overlap of communities within a continent.

In New Zealand, the boundary between the Indo-Australian plate and the Pacific plate changes its polarity from west-dipping subduction beneath the North Island (Hikurangi trench), to pure strike-slip (Marlborough Fault System) and oblique continental collision (Alpine fault) in the central part of the country, and to east-dipping subduction below the southwestern South Island (Puysegur trench) (Figure 3.1). In the central and southern South Island, the plate boundary is represented by the Alpine fault. Strike-slip displacement on the fault began in the Early Miocene, at ~23 Ma, but 420 km of the 470 km total displacement has occurred since 11–16 Ma (Cutten, 1979; Wallis and Trewick, 2009).

This large-scale feature also represents a major biogeographic break. The carabid beetle *Plocamostethus planiusculus* is a well-collected group that illustrates this. The species is widespread from the southern North Island south to the Paparoa Range and the Alpine fault (Maruia Springs) (Figure 9.4; Johns, 2015). Johns commented:

> In Nelson-Marlborough its southeastern boundary follows closely the Great Alpine Fault and although there may be forest on either side of the faultline this beetle rarely crosses it to occupy a seemingly identical habitat. Its southern boundary in northern Westland [Paparoa Range] is also an environmental enigma as there is no vegetational or distinct climatic change associated with that boundary.

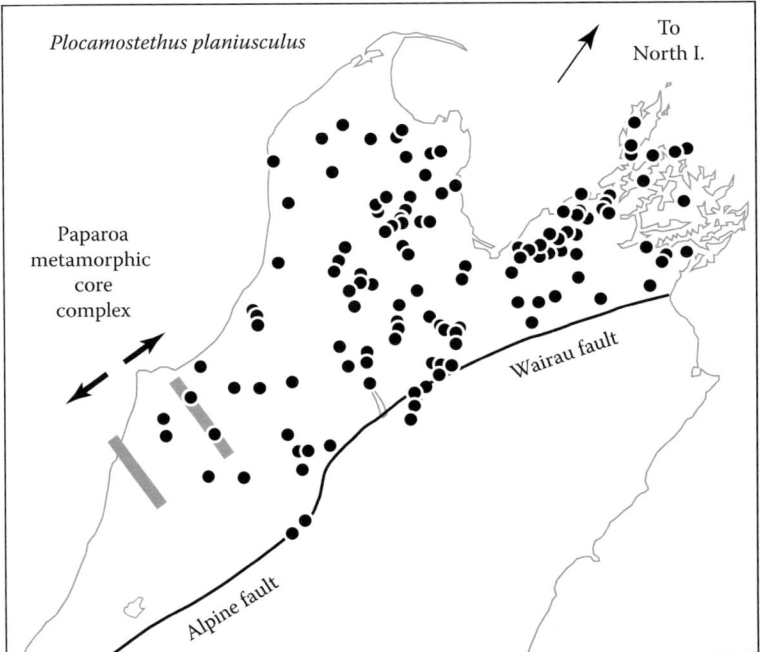

FIGURE 9.4 Distribution of *Plocamostethus planiusculus* (Carabidae) (South Island records only). Paparoa metamorphic core complex. (From Schulte, D., Kinematics of the Paparoa metamorphic core complex, West Coast, South Island, New Zealand. MSc thesis, University of Canterbury, Christchurch, New Zealand, 2011.) Solid gray lines indicate the bivergent detachment faults; arrows indicate the movement of the upper plate that unroofed the lower plate. (Data from Johns, P.M., New Zealand biodiversity. http://nzbiodiversity.com, accessed December, 2015.)

A similar plant example is *Lagarostrobos colensoi* (Podocarpaceae). In the South Island, this species is more or less restricted to areas west of the Alpine fault and ranges right up to it (NZPCN, 2016). For ~400 km, from Cascade River to Maruia Springs, the fault marks the precise eastern limit. The sister group of the species is in Tasmania.

Geological and Biological Offset along the Alpine Fault

Apart from having its southern boundary at the Paparoa Range and the Alpine fault, *Plocamostethus planiusculus* (Figure 9.4) has affinities further south that are disjunct along the fault. Johns (2015) wrote that "A close relative *P. scribae* is found on the other side of the Great Alpine Fault some 420 km to the south, on Secretary Island, Fiordland National Park perhaps indicating that the two species arose as the two sides of the fault separated." These interesting results have come from detailed morphological work, field studies, and mapping, rather than molecular surveys, but molecular work is indicating similar disjunctions.

The New Zealand basement terranes show great offsets at the Alpine fault; the Maitai, Murihiku, and Caples terranes are each offset by 440–470 km, depending on which geological marker is selected. This separation is the maximum possible strike-slip displacement on the fault. The true strike-slip component is probably less than this distance, which is exaggerated by ~90 km of shortening (Walcott, 1998). The Esk Head Mélange in the northeastern South Island (Figures 8.11 and 9.3) is separated by ~170 km from its North Island equivalent at Turakirae Head-Rimutaka Range (Cox and Sutherland, 2007).

In the South Island, highest levels of species diversity occur in the north (Nelson) and south (Otago and Southland), rather than in central areas of the island, where the highest mountains are found. Traditional theory attributed this to glaciation in the central areas causing extirpation of populations and clades there. Neiman and Lively (2004) suggested that Pleistocene glaciations will have produced a *gap* in the central areas, while movement on the Alpine fault and the rise of the Southern Alps will lead to a simple *east/west split* in taxa across the main divide. In fact, fault movement can also lead to a distribution gap, as displacement on the Alpine fault has a horizontal, strike-slip component as well as a vertical component. Biological evidence suggests that the strike-slip has produced disjunctions, or distribution gaps, in the central South Island, as in the disjunction between *Plocamostethus planiusculus* in the northern South Island and *P. scribae* in Fiordland. Other examples involving groups at the Paparoa Range area were cited earlier (Chapter 7).

Alpine Fault Disjunction in Animals

A distribution gap in the central South Island occurs in many taxa. In aphids, Teulon et al. (2003) accepted that recent, chance dispersal to New Zealand does not account for several New Zealand groups, in particular, two intriguing clades that have global sisters. These have phylogenies (Kim et al., 2011) and distributions (Teulon et al., 2003) as follows.

The first of the groups with a global sister is *Aphis coprosmae* of Nelson province: Aniseed Valley (10 km south of Nelson city) and Lake Rotoroa (on the Wairau fault extension of the Alpine fault). The second (the "southern hemisphere clade") is endemic to North and South Islands and has a phylogeny:

> *Aphis healeyi*: Widespread in the South Island, also central North Island (Taihape).
>> *Paradoxaphis plagianthi*: Christchurch area.
>>> *P. aristoteliae*: Gore (central Southland); disjunct at Lake Rotoroa (Nelson).
>>> *Aphis cottieri*: Fiordland (Lake Te Anau) to Banks Peninsula.

Teulon et al. (2003) noted that *Paradoxaphis aristoteliae* "has been found at both ends of the South Island but, despite some effort, no further populations have been located." The great gap in the distribution coincides with the Alpine fault. The northern limit of this species is Lake Rotoroa, on the Alpine fault, and *Aphis coprosmae* has a southern limit at the same locality.

In many cases, the gap in an Alpine fault disjunction is filled by a close relative. For example, the gap in *Paradoxaphis aristoteliae* is filled, in large part, by its sister group, *Aphis cottieri*.

In the beetle genus *Syrphetodes* (Ulodidae), molecular studies showed *S. occiduus* of Chancellor Shelf (above Fox Glacier) and Temple Basin (near Arthur's Pass) to be sister to *S. marrisi* of Mount Domett in northwest Nelson, 210 km to the north (Leschen and Buckley, 2015). The pair is disjunct along the Alpine fault, and the gap is filled by its sister group, comprising *S. carinatus*, Victoria Range; *S. melanopogon*, Paparoa Range (Mount Dewar); and *S. defectus*, Paparoa Range (Buckland Peaks).

Alpine Fault Disjunction in Plants

Similar Alpine fault disjunctions occur in many plants. The shrub *Pimelea suteri* (Thymelaeaceae) is disjunct between the northern and southern South Island, mainly on the ultramafic rocks of the Dun Mountain ophiolite belt (Figure 9.5; Burrows, 2011a). Burrows (2011a) found the pattern "of great biogeographic interest," but "difficult to explain." He considered the possibility that the species has died out in central areas or evolved independently in the north and the south, but he cited strong counterarguments to both these ideas. One possibility that he did not mention is passive, tectonic transport of the plants and their terranes, during horizontal, strike-slip displacement on the Alpine fault. *Pimelea pseudolyallii*, a close relative of *P. suteri*, is also mapped in Figure 9.5 (Burrows, 2011a). It fills part of the gap in the range of *P. suteri* and illustrates a typical North Otago–Marlborough distribution pattern, with a southern limit at the Waihemo fault zone.

Libertia (Iridaceae; 12 species) occurs in New Guinea, Australia, Tasmania, New Zealand, and Andean South America, "indicating a possible Gondwanic origin" (Blanchon et al., 2002). One New Zealand species, *Libertia peregrinans*, is widespread, but has a typical distribution gap in the central South Island (Figure 9.6). The gap is filled by a possible sister group, *L. ixioides* of coastal to upland habitats. Collections of *L. peregrinans* from Southland and the Chatham Islands are distinct through their longer, curled pedicels, and this affinity is indicated in Figure 9.6.

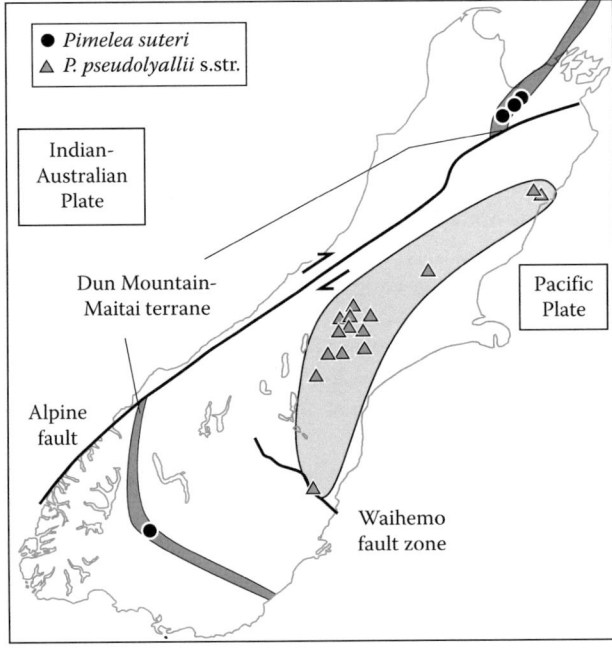

FIGURE 9.5 Distribution of *Pimelea suteri* and *P. pseudolyallii* s.str. (Thymelaeaceae). (Data from Burrows, C.J., *N. Z. J. Bot.*, 49, 41, 2011a.)

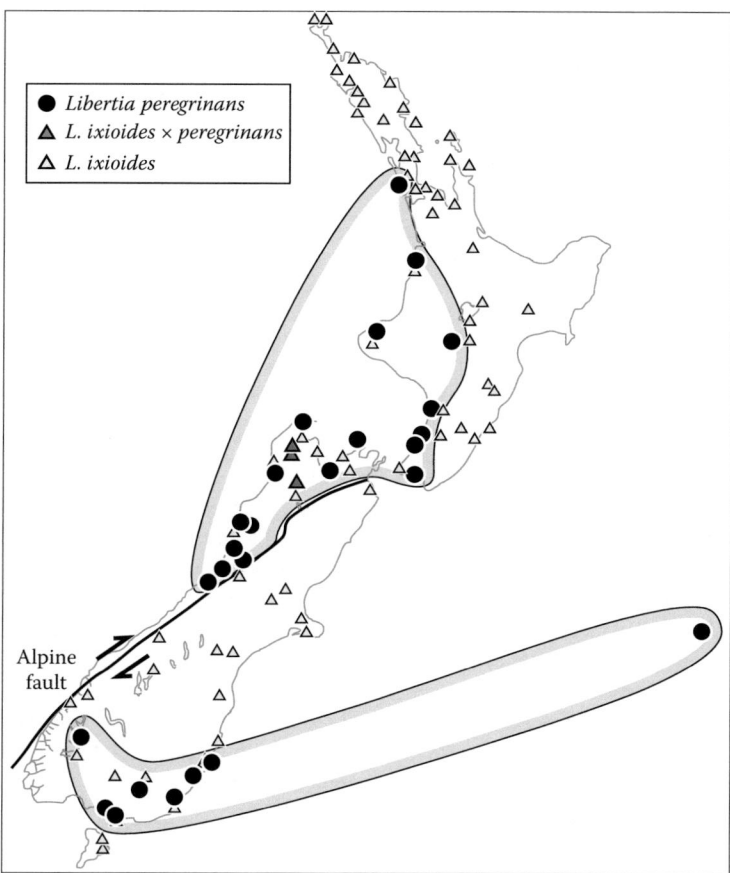

FIGURE 9.6 A pair of putative sister species in *Libertia* (Iridaceae) and their possible hybrid. (Data from Blanchon, D.J. et al., *N. Z. J. Bot.*, 40, 437, 2002.)

The disjunctions mapped here coincide with the Alpine fault, as the disjunctions have the same dextral sense as the fault movement. Many groups have similar distributions, as they are present in Nelson and Fiordland/Otago, but display anomalous absences in Westland. This last area is often referred to as the "beech gap," although this is a confusing term as southern beech (*Nothofagus*: Nothofagaceae) is rare there, rather than absent. *N. menziesii* occurs in central Westland, in Westland National Park (Karangarua River) (Wardle, 1980). There are also early records of *N. fusca* and *N. solandri* at Okarito (Hamilton, 1878), and fossil *Nothofagus* pollen from the Late Pleistocene is recorded from Westland (Newnham et al., 2007; Ryan et al., 2012).

Nevertheless, one species of beech, *Nothofagus truncata*, does have a true disjunction of 320 km between northern Westland and the Jackson River in the south, right on the Alpine fault. This is a typical example of a distributional gap along the fault in a lowland forest tree, and similar gaps exist in plants and animals of all habitats up to the alpine zone, above tree line.

Molecular studies have indicated or suggested an Alpine fault disjunction in many groups. For example, an ETS phylogeny of *Lepidium* (Brassicaceae) showed a clade distributed from Stewart Island to Banks Peninsula and Bounty Island (*L. juvencum*, *L. crassum*, *L. aegrum*, *L. seditiosum*), and also ~1500 km to the north in the Kermadec Islands (*L. castellanum*) (de Lange et al., 2013).

In *Festuca* (Poaceae), *F. deflexa* ranges from northwest Nelson to a southern limit right at the Alpine fault (Springs Junction), while its sister group (*F. matthewsii* subsp. *pisamontis* plus *F. m.* subsp. *latifundii*) is disjunct further south, in Central Otago and southern Canterbury (Lloyd et al., 2007). (This clade had low statistical support and further work is needed, but the distribution is a standard one.)

Other groups that have not been the subject of detailed phylogenetic work also include possible examples of Alpine fault disjunction. For example, four species of *Carex* (Cyperaceae) make up a closely related group that is endemic to the South Island (Moore and Edgar, 1970). The group comprises two putative sister pairs, *C. libera/C. edgarae* and *C. filamentosa/C. uncifolia*, with the following distributions:

C. libera: **NW Nelson** (including Mt Arthur); keyed with *C. edgarae*.

C. edgarae: Western **Central Otago** between Lakes Wakatipu and Wanaka (type from Nevis Valley).

C. uncifolia: Central North Island (Volcanic Plateau, Hauhangatahi Range) and eastern South Island: Richmond Range south via Canterbury (Hakatere, upper Ashburton River) to Southland (West Dome; NZPCN, 2016); keyed with *C. filamentosa*.

C. filamentosa: Southland and Stewart Island.

The gap in the first pair between Nelson and Otago, along the Alpine fault, is filled by the second pair.

Alpine Fault Disjunctions in Alpine Celmisias

The most important and characteristic plants of the New Zealand alpine zone are snowgrasses in the genus *Chionochloa* and the herbs, subshrubs, and cushions of *Celmisia* (Asteraceae) (Mark and Adams, 1973). On a single mountain, Mount Valiant in Arthur's Pass National Park, 18 *Celmisia* species occur within an altitudinal range of 600 m (Given and Gray, 1986). *Celmisia* is a southern, prostrate, alpine version of the erect shrubs and trees seen in its sister group, *Olearia* p.p. + *Pachystegia* (Chapter 6), which extends north to New Guinea. Plants of the alpine *Celmisia hectorii*, for example, can form extensive, woody, nonrooting mats over bare rock.

Celmisia hectorii is one of the most prominent plants in the alpine fell-fields and snowbanks of Otago. It extends north to Mount Cook, and both the species and its immediate relatives (series *Lignosae*) are absent beyond there. Nevertheless, the *Lignosae* reappear in Nelson. The details of the group, based on morphological studies, are as follows (Given, 1968, 1969, 1971):

Series *Lignosae*: Ericoid to erect subshrubs in subalpine to high-alpine vegetation (Mark and Adams, 1973)

Southern species (*not necessarily a clade*):

C. ramulosa: Fiordland and Central Otago, to south Canterbury (Hunters Hills).

C. philocremna: Central Otago (Eyre Mountains).

C. hectorii: Fiordland and Central Otago north to **Mount Cook**.

Northern species (*not necessarily a clade*):

C. lateralis: Central and southern **Paparoa Range and, 50 km east, Victoria Range**, north to Raglan Range and NW Nelson.

C. rupestris: NW Nelson.

C. gibbsii: Northern Paparoa Range, NW Nelson.

There is a ~250 km gap in the distribution between Mount Cook and the Paparoa/Victoria Ranges. This cannot be explained by local extinction caused by glaciation, as the plants favor habitat around glaciers. Given (1968: 24) wrote: "… since all members of the group of species clustered about *C. ramulosa* are alpine and tend to be high alpine, it is surprising that they do not occur on some of the high mountains in the centre of the island in a relict 'nunatak' pattern."

The gap includes large areas of suitable, alpine habitat, and this is occupied by the clade's putative sister group, series *Angustifoliatae* (*C. angustifolia*, *C. brevifolia*, and *C. walkeri*). These are procumbent, mat-forming subshrubs that extend from Nelson to Fiordland.

In other groups, the gap located north of Mount Cook is the site of local endemics. These include the alpine buttercup *Ranunculus godleyanus*, distributed at high elevations from Mount Cook to Arthur's Pass.

The same gap that is seen in series *Lignosae* between Mount Cook and Paparoa/Victoria Range also occurs in other alpine clades. *Hebe hectorii* s.lat. (Plantaginaceae) can dominate subalpine shrubland where it does occur, but it is absent between Mount Cook and the Paparoa Range (Bayly and Kellow, 2006). Some authors have attributed the gap in *H. hectorii* to glaciations (Wagstaff and Wardle, 1999), but these alpine plants are under snow for much of the year and thrive around glaciers, avalanche chutes, and associated habitats. McGlone et al. (2001) suggested that the gap is caused by the continuous instability of the area, with repeated, heavy glaciation during glacials and high erosion rates during interglacials. This instability is not restricted to the mountains between Mount Cook and the Paparoa Range, though—it also occurs around Mount Cook and to the south, where *H. hectorii* is present. McGlone et al. (2001) suggested that "a complex of reasons related to climate, geological events and ecological attributes are involved, and each disjunct species probably has a distribution caused by a unique combination of factors…." The details of the distributions may be unique, but the Mount Cook–Paparoa Range disjunction and similar variants are seen many groups. Another example is the group of robust shrubs and small trees: *Dracophyllum fiordense* (Fiordland)–*D. menziesii* (Fiordland–Fox Glacier)–*D. townsonii* (Paparoa Range and northwest Nelson) (Venter, 2009).

The phylogenetic affinity across the distributional gap in *Celmisia* series *Lignosae*, tabulated earlier, probably does not lie between what are the closest populations geographically, *C. lateralis* at the Paparoa and Victoria Ranges, and *C. hectorii* at Mount Cook. Instead, in its morphology, *C. lateralis* of northern Westland appears closest to *C. ramulosa* of Fiordland (and Otago) (Allan, 1961; cf. Given, 1969). These areas were juxtaposed before Alpine fault displacement (Figure 9.3). Molecular study and detailed distribution maps would be desirable.

There are at least two other possible Alpine fault disjunctions in *Celmisia* (Given, 1969):

> Section *Serratae*. This comprises one southern species, *Celmisia holosericea*, in Fiordland south of Homer Saddle, and three species from the Paparoa Range northward (*C. dallii*: Paparoa Range–western Nelson: *C. hieracifolia*: Nelson, Marlborough Sounds, North Island; *C. cockayniana*: Kaikoura Ranges). The Fiordland species is coastal to alpine, those in the north are subalpine to alpine.

> *Celmisia petriei* (sect. *Pelliculatae*). This is known from Fiordland and Otago, and also northwest Nelson, in subalpine to low alpine vegetation (Heads, 1998a: Fig. 4K).

A large part of the gap in the range of the last species, *C. petriei*, is filled by its likely sister group, *Celmisia lyallii* of eastern Fiordland (Mount Burns, upper Grebe Valley, Kepler Mountains) to north Canterbury (Given, 1968; NZPCN, 2016). Both are subalpine-low alpine species.

To summarize, the distributions in *Celmisia* are typical examples of Alpine fault disjunction, and they demonstrate that the pattern occurs in alpine as well as lowland groups. This is further evidence that the pattern requires a general explanation.

Other Aspects of Strike-Slip Displacement on the Alpine Fault

Many well-collected taxa are disjunct and have distribution boundaries that coincide closely with the Alpine fault (Heads, 1998a). This precision, and the fact that the disjunction is a general pattern, can be accounted for by the massive strike-slip displacement that has taken place on the fault (Heads, 1998a; Heads and Craw, 2004). This dislocation has had the direct effect of pulling apart both plant and animal populations.

Distributions of some taxa have a clean break at the fault, whereas others appear to have been "stretched" along the fault with lateral movement. There is no particular reason why the disjunct taxa should have invaded the other side of the fault as it moved; taxa do not necessarily invade other communities, whether or not they are being pulled apart on a fault. (The fault zone itself, a belt 1–200 m across, is often marked by swampy flats with open *Coprosma* shrubland and weedy species with widespread New Zealand distributions.)

Eighty taxa showing disjunction along the Alpine fault have been mapped (Heads, 1998a) and an additional 41 listed (Heads and Craw, 2004). All these examples and others cited next give a total of 225 taxa disjunct along the fault.

Fault disjunctions that have not been cited before include 15 genera in basidiomycete fungi (*Stromatoscypha*, *Dictyopanus*, *Dermoloma*, *Delicatala*, *Porpoloma*, *Pseudoarmillariella*, *Urosporellina*, *Macowanites*, *Zelleromyces*, *Coltriciella*, *Rogersella*, *Phaeomara*, *Radulomyces*, *Tubilicrinis*, *Austroboletus*), and three in ascomycetes (*Drepanopeziza*, *Cryptohymenium*, *Trochila*) (http://nzfungi2.landcareresearch.co.nz/), 1 in red algae (*Psaromenia*; D'Archino et al., 2010), 29 examples in lichens (Galloway, 2007; one, *Peltigera nana*, mapped in Galloway, 2000), one in mosses (Kruijer, 2002), one (*Acromastigum cunninghamii*) in liverworts (Engel and Glenny, 2008), and in angiosperms *Drosera pygmaea* (Droseraceae) (Salmon, 2001), *Olearia quinquevulnera* (Asteraceae) (Heenan, 2005), *Dracophyllum politum* (Ericaceae), and *D. fiordense* + *D. menziesii* + *D. townsonii* (Venter, 2009).

In animals, five additional examples of fault disjunction have been documented in nematodes (Wouts, 2006), three in Acari Prostigmata (Fan and Zhang, 2005), three in Acari Oribatida (moss mites) (Colloff, 2011b, 2015), one in Collembola (Steens et al., 2007), eight in Hemiptera suborder Heteroptera (Larivière and Larochelle, 2004), one in Hemiptera superfamily Coccoidea (Henderson, 2011), six in Hemiptera suborder Auchenorrhyncha (Larivière et al., 2010), one in cave wetas, Rhaphidophoridae (Cook et al., 2010), *Anabarhynchus major* (Therevidae) (Lyneborg, 1992), five in stag beetles (Lucanidae) (Holloway, 2007), two in staphylinid beetles (*Leschenea*: Park and Carlton, 2015b; *Sagola turretensis*: Park and Carlton, 2014), one in Trichoptera (Ward and Henderson, 2004), two in Lepidoptera (Hoare, 2005, 2010), and two in Hymenoptera (Berry, 1999).

In land snails, *Canallodiscus* is disjunct between Fiordland and northwest Nelson. The genus, disjunct in the New Zealand Western Province, has been compared with *Danielleilona* of northeastern Queensland (Marshall and Barker, 2008). In other land snails, Climo and Mahlfeld (2011) mapped Alpine fault disjunction in members of the genus *Kokopapa* (Punctidae) and six other groups, and gave a detailed discussion of the phenomenon.

In recent studies, the weta genus *Maotoweta* (Orthoptera: Rhaphidophoridae) was recorded from the Takitimu Mountains, Lake Te Au in Fiordland, and Mount Arthur in northwest Nelson (Johns and Cook, 2014), and at least three fault disjunctions were documented in the micropterigid moth *Sabatinca* (Gibbs, 2014). *S. quadrijuga* (Otago and Southland; Figure 7.8a) is sister to *S. aurantissima* of northern Westland and Nelson, while *S. caustica* (Southland; Figure 7.8b) is sister to *S. chalcophanes* of central Westland and Nelson (Gibbs, 2014). The breaks in both pairs were dated in a fossil-calibrated clock study to around 25 Ma (minimum clade age), and so as Gibbs (2014) wrote, neither break "conforms to the prevailing mountain-building or glaciation hypotheses offered by many recent phylogeographic studies of New Zealand biota e.g., review by Wallis and Trewick (2009)." A third Alpine fault disjunction in *Sabatinca* is seen in *S. weheka*, disjunct between Secretary Island in Fiordland, and Lake Matheson in central Westland.

Groups with Alpine fault disjunction often have very localized distributions. This might be the result of poor collecting, especially in less well-studied groups. An example is the lichen *Usnea sphacelata*, known in New Zealand only from one small area in northwest Nelson (Lake Cobb) and another in northwest Otago (Mount Aspiring, the nearby Mount Avalanche, and also Mount Head and Sir William Peak in the Forbes Mountains). Nevertheless, the areas in the gap include the glaciers region, Mount Cook and Arthur's Pass, which have all been surveyed, and Galloway (2007) concluded that *Usnea sphacelata* "is genuinely rare in New Zealand." The species is also present in the Arctic, Antarctica, North America, and South America.

There is some morphological evidence for Alpine fault disjunction in several of the Tasman region endemics that have global sisters, although this needs to be tested with sequencing. Possible examples occur in the red alga *Apophlaea*, in the athoracophorid slugs (between *Pseudaneitea gravisulca* of Picton and *P. powelli* of Coronet Peak, Eglinton Valley and Te Anau; Burton, 1963, 1980), in micropterigid and mnesarchaeid moths (mapped and discussed by Gibbs, 1990a), and in the New Zealand wren *Acanthisitta* (Heads, 1998a).

The *differentiation* between two clades now disjunct along the Alpine fault may have been caused by rifting in the precursors of the Alpine fault, either in the Paleogene or Late Cretaceous (Figure 3.15).

This would predate the great *disjunction* between the clades, caused by the Neogene strike-slip that would have also led to another phase of differentiation.

Explaining distributional disjunctions of plants and animals by strike-slip displacement along major faults, usually transforms, is not a new idea. Outside New Zealand, it has been proposed for plant and animal taxa in the Caribbean (Durham, 1985), southern California and northwestern Mexico (Axelrod, 1986; Minckley et al., 1986), Sumatra (Heads, 2003), New Guinea (Heads, 1999, 2001, 2003), New Caledonia (Heads, 2008a), and Patagonia (Heads, 1999).

A Small Number of Macrorefugia or Multiple Microrefugia?

Earlier theory attributed the central South Island gap seen in groups of *Nothofagus, Celmisia, Libertia,* and others to the extirpation of central populations by alpine uplift or glaciation (Willett, 1951). Disjunct clades were thought to have been restricted during glaciations to a few, large "refugia" in the northern and southern ends of the South Island, and following the last glaciations some clades have failed to disperse back into the center. Nevertheless, the disjunct groups include alpine clades that flourish around glaciers. Other groups include local endemics restricted to sites *within* the "beech gap," and "The presence of an extensive endemic forest-dwelling fauna in areas postulated as periglacial suggests that Willet's [1951] assumptions of the effect of glaciation on the fauna were too sweeping" (Forster, 1954: 307). These observations indicate problems with the refugium theory.

Based on distribution patterns, biogeographers have sometimes criticized standard refugium theory, with its few, large glacial refugia, and instead proposed the idea of many, small, ecological islands where groups have survived, persisting as metapopulations during glaciations (Heads, 1994d: 117). Although the idea of multiple microrefugia was controversial, it is now accepted more often. Several molecular studies mentioned in the last chapter rejected the concept of a few, large Pleistocene refugia in favor of many, much smaller microrefugia. Mee and Moore (2013: 837) wrote:

> The consensus view of Pleistocene biogeography was, until recently, that most mid- to high latitude Northern Hemisphere species survived the last glaciation in refugia south of the major ice sheets… Accumulating evidence of the importance of small, isolated refugia, however, has led to the replacement of this simple conclusion by a much more pluralistic understanding… Recently, the concept of *microrefugia* (a.k.a. cryptic refugia or climate relicts) has emerged to describe glacial refugia that are distinct from conventional refugia or macrorefugia.

One source of evidence supporting multiple microrefugia in New Zealand comes from subfossils. Beetle remains from across New Zealand imply that habitat conversion during glaciation was not as dramatic as that inferred from the pollen record, with many Pleistocene fossil beetles suggesting persistence of a mosaic of small forest patches (Buckley et al., 2015).

This idea has important implications for interpreting New Zealand distributions, as survival in the microrefugia would preserve local endemism and also more widespread biogeographic structure through the ice ages. The glaciations have caused great changes at a local scale but have not had the profound regional effect on distribution that was thought. Many groups, even within species, are much older than the glaciations and they, together with their overall distributions, date back to pre-Pleistocene times.

In a paper with the provocative title "Looking beyond glacial refugia," Marske et al. (2011) described marked differentiation in two forest beetles: *Epistranus lawsoni* and *Pristoderus bakewelli* (Zopheridae). In the latter, one of the main clades shows a conspicuous disjunction along the Alpine fault, while the other fills the gap (Figure 9.7). In both species, Marske et al. (2011) found intense population structuring over a limited spatial extent, and they observed that this is a recurrent pattern in leaf litter fauna. Divergence dates for *E. lawsoni* and *P. bakewelli* are consistent with the topographic evolution of New Zealand over the last ~10–20 Ma, whereas the last glacial maximum "does not appear to have left a marked signature."

Using the minimum divergence dates for both beetle species, Marske et al. (2011) were able to conclude that the earlier "tectonic evolution of New Zealand had a greater impact on genetic diversity

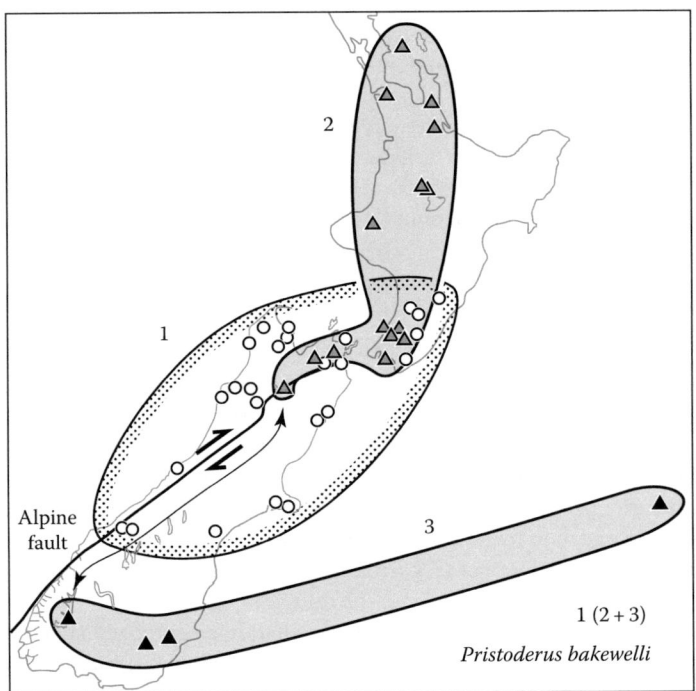

FIGURE 9.7   Distribution of *Pristoderus bakewelli* (Coleoptera: Zopheridae), with the main clades. (Data from Marske, K.A. et al., *Mol. Phylogenet. Evol.*, 59, 89, 2011.)

than more recent climatic crises, although isolation in multiple glacial refugia would have maintained tectonically-driven phylogeographic patterns." This model is accepted here for the New Zealand biota as a whole. It helps to explain common patterns such as the Otago–Chatham Islands connection in *Pristoderus* (Figure 9.7), also seen in groups such as the irises (*Libertia*; Figure 9.6) and the albatrosses (*Diomedea*; Chapter 6). In another example, the carabid beetle *Mecyclothorax* is represented in New Zealand by three endemic species: one widespread (*M. rotundicollis*), one in Central Otago (*M. otagoensis*), and one, sister to the last, on the Chatham Islands (*M. oopteroides*) (Liebherr and Marris, 2009).

The disjunct clade of *Pristoderus* has an overall distribution matched by the hemipteran *Fusilaspis* (Henderson, 2011), present from the Three Kings Islands south to the Paparoa Range, and disjunct in the south at The Snares, Bluff, Dunedin, and the Chatham Islands. Likewise, the pseudoscorpion *Nelsoninus* is a monotypic genus endemic to New Zealand and recorded from Nelson (Abel Tasman National Park), Fiordland (Hidden Falls in the lower Hollyford), and the Chatham Islands (Beier, 1976).

Groups in Which the Westland Biotic Gap Is Filled by a Sister Group

In many groups with a disjunction at the Alpine fault, the gap is not absolute but is filled by a sister group. This allopatry, at least in the gap, can be explained as a relic of the original vicariance. The pattern was indicated earlier, in the alpine plant *Celmisia* series *Lignosae* and the forest beetle *Pristoderus bakewelli*.

Another example cited above is a clade of shrubs and small trees in *Dracophyllum* (Ericaceae). It occurs from Stewart Island and Fiordland north to Fox Glacier (*D. fiordense* and *D. menziesii*) and also, disjunct across a gap of 140 km, from Greymouth and the Paparoa Range to northwest Nelson (Mount Burnett, Knuckle Hill) (*D. townsonii*) (Venter, 2009). The gap along the Alpine fault in central Westland is occupied by the related tree *D. traversii*, which is widespread from Otago to Northland.

The moss genus *Catharomnion* (Hypopterygiaceae) occurs in the North Island, the Chatham Islands, and northwestern South Island south to Punakaiki; it also has a disjunct record along the fault in

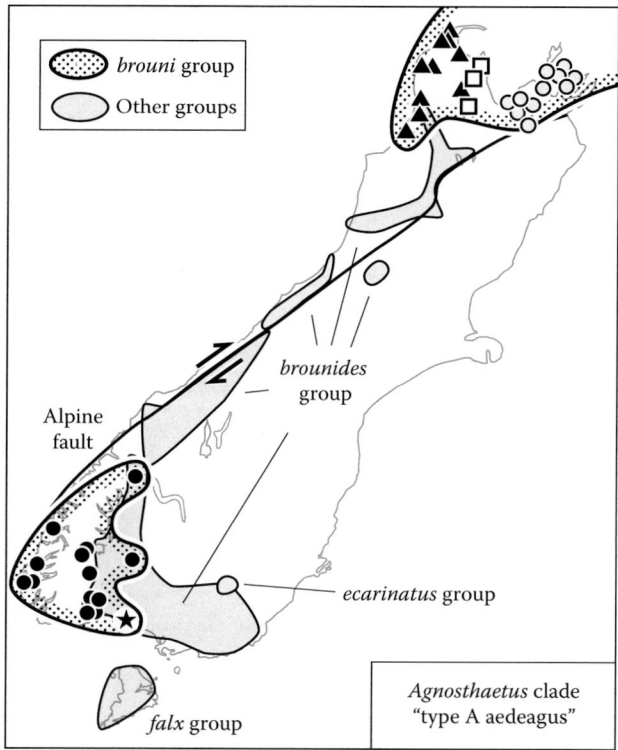

FIGURE 9.8 Distribution of *Agnosthaetus* "type A aedeagus" clade (Staphylinidae). (Catlins record of *brouni* group not included.) (Data from Clarke, D.J., *Coleopterists' Bulletin* 65 (monograph 4), 1, 2011; Clarke, D.J., pers. comm., 2013.)

northwestern Otago (Beans Burn). The gap is filled by its sister group, *Canalohypopterygium*, which has a widespread distribution through New Zealand from Northland to the Auckland Islands (Kruijer, 2002: Maps 7 and 8).

In the beetle *Agnosthaetus* (Staphylinidae), the *brouni* group is disjunct along the fault, while the other groups in its clade ("type A aedeagus") occur throughout the gap in its range (Figure 9.8; Clarke, 2011). They also show limited overlap with it, consistent with an origin by vicariance.

Other taxa in which Alpine fault gaps are filled by the sister group include the fish *Galaxias postvectis* (Figure 9.9). The species is diadromous (it has a freshwater stage and a marine stage in its life cycle), but despite this potential for dispersal it has a conspicuous distributional gap in Fiordland (McDowall, 2010b). In this case, the gap is filled by the sister group (all the other New Zealand *Galaxias* species except *G. fasciatus* and *G. argenteus*), which is widespread through New Zealand (Waters et al., 2010; Burridge et al., 2012).

A similar gap is seen in another diadromous galaxiid, *G. fasciatus* (McDowall, 2002), and in another freshwater group, the crayfish *Paranephrops* (McDowall, 2005; Apte et al., 2007). The Fiordland gap in *Paranephrops* has been attributed to the rise of the main divide, but the zone where the crayfish is absent (between Resolution Island and Jackson Bay) crosses the Alpine fault, not the main divide.

In addition to these lowland, diadromous fishes and lowland–montane crustaceans, a similar gap across Fiordland is seen in the alpine herb *Ourisia macrocarpa* subsp. *calycina* (Plantaginaceae). The gap is again filled by the sister group (*O. crosbyi* and others; Meudt, 2006; Meudt and Simpson, 2006). An even greater disjunction along the fault occurs in *Ourisia modesta* (absent between southern Fiordland and Nelson), with the gap filled by the sister group (*O. glandulosa*, *O. confertifolia*, and *O. spathulata*). In morphological studies of other plants, this replacement pattern is seen in *Poa* (Figure 6.4) and in the *Leucopogon fraseri* complex (Ericaceae; Dawson and Heenan, 2004).

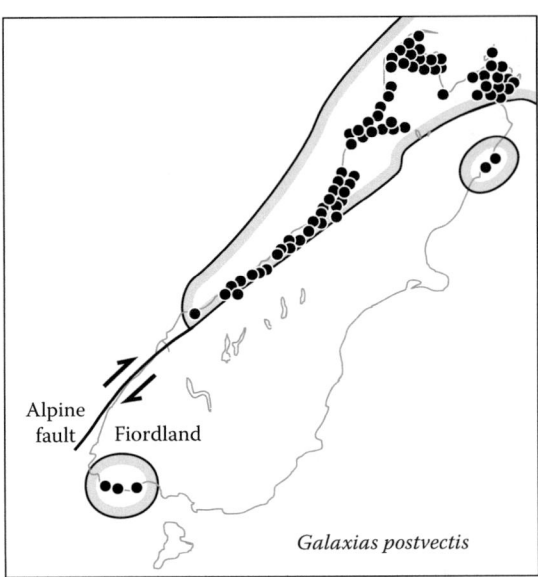

FIGURE 9.9   Distribution of *Galaxias postvectis* (Galaxiidae). (Data from McDowall, R.M., *New Zealand Freshwater Fishes: An Historical and Ecological Biogeography*, Springer, Berlin, Germany, 2010b.)

Disjunction across Fiordland and Stewart Island in Dolphins

Disjunction and differentiation across the Fiordland gap is also seen in Hector's dolphin, *Cephalorhynchus hectori*. This is the sole New Zealand representative of a southern hemisphere clade, *Cephalorhynchus* + *Lagenorhynchus*. In chance dispersal theory, "Highly mobile marine species in areas with no obvious geographic barriers are expected to show low levels of genetic differentiation" (Ansmann et al., 2012). Nevertheless, molecular work is revealing high levels of structure in marine groups (see Chapter 1), including cetaceans. For example, in the dolphins of the Hawaiian Islands, species of *Tursiops* and *Stenella* are made up of genetic clusters that each display site fidelity (Heads, 2012a). Right whales returning to the New Zealand calving ground are isolated from others in the Indo-Pacific region by maternal site fidelity maintained over an evolutionary timescale (Carroll et al., 2012). If the breeding sites of a metapopulation that shows site fidelity or natal philopatry are rifted apart, the group can differentiate by vicariance, and this process would account for patterns, such as the Fiordland gap, shared by marine mammals, diadromous fishes, and alpine herbs.

Hector's dolphin, *Cephalorhynchus hectori*, is endemic to the New Zealand mainland (absent in Fiordland and Stewart Island), and most populations occur within a few kilometers of the shore. Sequencing studies found five allopatric groups (Hamner et al., 2012):

Southern South Island (Te Waewae Bay at the mouth of the Waiau River).

Southern South Island (Toetoes Bay/Porpoise Bay at the mouth of the Mataura River).

Western South Island (northwest Nelson to Jackson Bay).

Eastern South Island (eastern Marlborough Sounds to Dunedin).

North Island ("Maui's dolphin").

Hamner et al. (2012) stressed the high degree of differentiation between the Te Waewae Bay and Toetoes Bay populations despite their separation by only ~100 km of coastline. This local differentiation around Foveaux Strait is located next to the area of the species' complete absence, in Fiordland and Stewart Island. The gap repeats the pattern of *Galaxias postvectis* (Figure 9.9) and is compatible with Alpine fault displacement.

In another subfamily of Delphinidae, the bottlenose dolphin *Tursiops aduncus* is an Indian Ocean clade. It ranges from the Red Sea and east Africa to China, north Australia and New Zealand. A study at Moreton Bay, Queensland, documented very local differentiation; the population there comprises one genetic cluster in shallow, southern areas and the other in the deeper waters of the northern bay (Ansmann et al., 2012). Around New Zealand, *T. aduncus* occurs in three disjunct coastal regions: Northland, Marlborough Sounds, and Fiordland (Tezanos-Pinto et al., 2009), and the disjunction between the last two is compatible with displacement along the Alpine fault. Phylogenetic reconstructions did not show a pattern of reciprocal monophyly or fixed nucleotide differences among populations. Nevertheless, there were strong frequency differences, and most haplotypes were found in only one of the three regions.

Alpine Fault Disjunction and Transoceanic Disjunction

Alpine fault disjunctions are well-documented in stag beetles, Lucanidae. One putative clade comprises *Geodorcus philpotti* of Fiordland, and two species disjunct along the fault in northern Westland: *G. montivagus* of the Victoria Range and *G. servandus* of Mount Tuhua (Holloway, 2007). Another clade of Lucanidae illustrates a common variant of the pattern as it combines an Alpine fault disjunction with trans-Tasman and trans-Pacific disjunctions (Figure 9.10; Holloway, 2007):

New Zealand, **disjunct along the fault** (*Dendroblax*).

Southeastern Australia, New Guinea, Lord Howe, and Norfolk Islands (*Lamprima*).

Southern Chile and adjacent Argentina (*Streptocerus*).

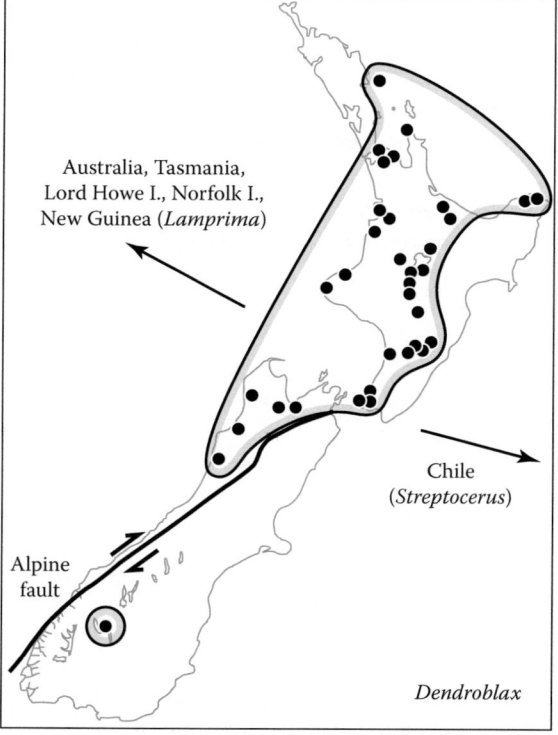

FIGURE 9.10 Distribution of *Dendroblax* (Lucanidae). (Data from Holloway, B.A., *Fauna N. Z.*, 61, 1, 2007.)

Kim and Farrell (2015) found *Lamprima* and *Streptocerus* to be sisters within subfamily Lampriminae, but they did not sample *Dendroblax*. They concluded that the main divergences in Lampriminae "were caused by vicariance events following the Gondwanan break-up."

The gap in the distribution of *Dendroblax* extends from Lake Wakatipu (Ben Lomond by Queenstown) to the Paparoa Range (Chapter 7). Another beetle genus, *Sphaerothorax* (Clambidae), has a similar range to that of *Dendroblax* and its allies. It occurs in New Zealand, where it is disjunct between Otago and the Paparoa Range, and it is also in Australia and southern South America; Endrödy-Younga (1995) concluded that it was a Gondwanan relic.

In similar patterns with both trans-Tasman and Alpine fault disjunctions, the lichen *Lecidea lygomma* var. *crassilabra* is in Otago (Remarkables, Old Man Range, Rock and Pillar Range, Maungatua), Nelson (Mount Aorere, Cobb Valley), Australia and Patagonia (Galloway, 2007). Other lichens distributed in New Zealand disjunct along the Alpine fault, and also in Australia include *Austropeltum* (a monospecific genus), *Chrooodiscus megalophthalmus*, *Cladia inflata*, *Icmadophila splachnirima*, *Menegazzia eperforata*, and *Pannaria delicata* (Galloway, 2007).

In ferns, the genus *Sticherus* (Gleicheniaceae) includes a clade structured as follows (Brownsey et al., 2013) (New Zealand localities are in bold, major disjunctions are indicated by //):

> *S. tener*: **Fiordland** // **Denniston Plateau** // Tasmania, southern Victoria.
>
> *S. urceolatus*: **Takaka, Stockton Plateau** // Tasmania, Victoria, New South Wales.
>
> *S. flabellatus*: **northern North Island** // New Caledonia // eastern Australian seaboard (Tasmania to Cape York Peninsula), New Guinea.

Brownsey et al. (2013) concluded that "It seems almost certain that [*S. tener* and *S. urceolatus*] have arrived in New Zealand as a result of long-distance dispersal from Australia," but they did not explain how they reached this conclusion. Chance dispersal does not explain the precise allopatry of the three species in New Zealand, the Alpine fault/trans-Tasman disjunction in *S. tener*, the standard Nelson–trans-Tasman disjunction in *S. urceolatus*, or the trans-Tasman-Coral Sea distribution of *S. flabellatus*. The authors suggested that "these examples add to the growing body of evidence for trans-Tasman dispersals in both directions," but the only evidence they presented were the distributions themselves. The fact that the patterns are standard, shared by a large number of organisms with different ecology and means of dispersal, is good evidence for vicariance.

Alpine fault + trans-Tasman distributions also occur in:

> The moss *Dicranoloma trichopodum*: Tasmania, southern Stewart Island (Tin Range), western Fiordland (Dusky Sound), disjunct at Blackball, Lake Hochstetter, the Paparoa Range and the North Island (Klazenga, 2003).
>
> The angiosperm clade *Kelleria bogongensis* of Victoria, + *K. laxa* of New Zealand (Heads, 1990b; Marks and Walsh, 2014).
>
> The orchid genus *Acianthus*: East Australia, New Caledonia and New Zealand, disjunct between Lake Te Anau and Poerua River (Wardle, 1975; NZPCN, 2016).
>
> The fern *Notogrammitis gunnii*: Southern South Island (Fiordland, Waitutu Forest, Central Otago–Eyre Mountains, Umbrella Mountains), northwest Nelson, and Tasmania (Perrie and Parris, 2012; NZPCN, 2016).

In similar patterns, the assassin spider *Aotearoa* (Mecysmaucheniidae) was thought to be endemic to Fiordland, with a single species, until Toft (2012) found a second, closely related species near the Stockton Plateau, Nelson. As Toft noted, "A disjunct biogeographical pattern between Fiordland and NW Nelson is not unexpected and has been recorded for a range of other species...." The gap in the range of *Aotearoa* is filled by its sister, the New Zealand endemic *Zearchaea* (Fiordland via Canterbury to Masterton; Forster, 1955). The remaining members of the family occur on the Juan Fernández Islands, southern South America, and the Falkland Islands (Wood et al., 2013; World Spider Catalog, 2016).

The beetle *Tormus* (Hydrophilidae) of New Zealand is a related case. It comprises *T. posticalis*, found east of the Alpine fault, in Fiordland, and *T. helmsi*, located mostly west of the fault, from Haast to the Coromandel Peninsula (Fikáček et al., 2013: Fig. 16). In the North Island, *T. helmsi* is restricted to the western half of the island. Overall, the boundary between the two species corresponds with the Alpine fault, although this pattern breaks down in a restricted area around Haast. The genus is probably sister to *Afrotormus* of South Africa, indicating a trans-Indian Ocean affinity.

Could the Westland Gap Have Been Caused by Glaciation?

In Matthew's (1915) approach, modern distribution has been caused by Pleistocene events, especially glaciation. In the past, the biotic gap in Westland along the Alpine fault has been attributed to extinction by glaciation and uplift in area of the gap, but several observations contradict this:

- The disjunct taxa show a wide ecological range and include intertidal algae and shorefishes, lowland taxa, and also plants and invertebrates of the alpine zone. Many of the groups thrive in habitats around glaciers, and so extinction by glaciation is unlikely.
- In many groups, the gap is filled by the sister group.
- There are many endemics in the gap.

None of these would be expected if the north/south disjunctions at the Alpine fault were caused by glaciation, but all three are explicable if the gap is the result of strike-slip displacement. The glaciation theory is also contradicted by the fact that Fiordland, in the southwestern South Island, underwent extensive glaciation but has many distinctive endemics.

Criticism of Strike-Slip Displacement in Living Communities

The idea that rocks and living communities have both undergone lateral displacement along the Alpine fault was criticized by Wallis and Trewick (2001). They concluded that distribution data can only be used to frame evolutionary and biogeographic hypotheses, not to test them. Yet many authors, including Charles Darwin, have acknowledged the significance of distribution, and have accepted it as evidence in its own right. In an 1845 letter to Hooker, Darwin referred to "…that grand subject, that almost keystone of the laws of creation, Geographical Distribution" (Burckhardt and Smith, 1987: 139), and he used distributional data both in framing and in testing hypotheses. In any case, Wallis and Trewick (2001) concluded their critique of the Alpine fault hypothesis of community disruption by accepting that it could explain the distribution of some taxa.

In another study, Trewick and Wallis (2001) examined a selection of widespread South Island invertebrates in order to investigate the possible effects of fault displacement. Rather than testing the idea with detailed mapping (this would require field work in mountainous areas with difficult access), they used a molecular clock approach. They found that most of the clades are older than the Pleistocene. This indicates that glacial extirpation has *not* been responsible for the biotic gap, and the authors accepted that taxa have survived the Pleistocene glaciations more or less *in situ*, perhaps on nonglaciated ridges above glaciated basins. They argued against the Alpine fault hypothesis, as they found that clades are not old enough, but most of the fault movement has been more recent than they assumed (Heads and Craw, 2004). In addition, Trewick and Wallis (2001) relied on the standard insect rate (2.3% divergence per million years) criticized elsewhere (Heads, 2012a), and the clades are probably older than they indicated. In a later paper, Wallis and Trewick (2009) admitted that the Alpine fault hypothesis was a "bold suggestion," but they concluded that "Despite its appeal there are numerous problems" (p. 3562). The only problem they mentioned is that the disjunct groups have been estimated to be too young, but these dates relied on flawed calibrations.

Trewick and Wallis (2001) suggested that the Alpine fault disjunction was caused by greater uplift and consequent extinction in the central areas, but transcurrent movement also occurred with uplift, and extinction does not account for the presence of sister groups or endemism in the gap. Trewick and Wallis (2001) also found that in many cases populations in Nelson and Otago are not sister groups, and they suggested that this is incompatible with the fault hypothesis. Yet there is no need for groups that were adjacent before fault movement, for example, in Nelson and Otago, to be sister groups, and so they will not be sister groups after fault movement. The fault hypothesis is compatible with a range of different distributions, and the distribution of clades after fault movement depends on their prior distribution. For example, taxa that were widespread before fault movement will be widespread after.

The Alpine fault pattern, as with all common biogeographic patterns, is seen in many taxa with different ecologies and phylogenetic branch lengths (degrees of differentiation). This suggests that the distribution pattern is not determined by ecology and that branch length in the different groups is not determined simply by time.

Case Studies of Groups at the Alpine Fault

The following groups illustrate different aspects of the Alpine fault disjunction and related patterns.

New Zealand Gentians (Gentianella: *Gentianaceae)*

The affinities of New Zealand *Gentianella* with South American and Australian plants have been discussed elsewhere (Heads, 2014: 103). For the New Zealand members, Glenny (2004) proposed a center of origin in the southern South Island, with four dispersal events from there to the North Island. Two additional long-distance dispersal events were used to explain affinities between South Island and subantarctic islands taxa, and between North Island and Chatham Islands taxa.

This dispersal interpretation depends on the idea that the genus is no older than its oldest New Zealand fossils (1.6–2.6 Ma). However, gentians are small, alpine, insect-pollinated herbs that produce minute amounts of pollen, and fossil preservation in mountain ranges is unusual. In addition, many aspects of the group's biogeography are difficult to explain if the group really is this young. If the genus in fact had a Cretaceous origin, as many aspects of the global distribution indicate, a simple vicariance history for the subantarctic and Chatham Islands representatives would be possible.

Within the South Island, several dispersal events have been proposed, for example, to explain the disjunction between *G. saxosa* of Fiordland and Stewart Island to the Catlins (False Islet) and its sister *G. scopulorum* of Charleston, by the Paparoa Range (Glenny, 2004). Dispersal is unnecessary though, as this is a typical Alpine fault disjunction, and it indicates that the group has been in New Zealand much longer than the fossil evidence indicates.

Freshwater Snails

Haase et al. (2007) sequenced disjunct clades of hydrobiid freshwater gastropods and accepted displacement along the Alpine fault as an explanation for their distribution and phylogeny.

Onychophora

The South Island onychophoran *Peripatoides* comprises three well-supported clades, one in Nelson west of the Alpine fault and two in Southland and Otago east of the fault (Trewick and Wallis, 2001). The authors concluded that "the geographic distribution of these taxa is consistent with Alpine fault vicariance. However, the fact that genetic distances are as high among the Southland-Otago taxa as between either of these and the Nelson taxon is not predicted by fault vicariance" (p. 2173). Nevertheless, an earlier break in the southern group, between the Southland and Otago clades, is compatible with another, later break being caused by Alpine fault displacement. It is not suggested that the Alpine fault explains every biogeographic pattern in New Zealand; the earlier break in the Otago-Southland clade is more likely to be related to the Southland syncline (cf. Trewick and Wallis, 2001: Fig. 3).

Pettalid Harvestmen (Opiliones): Nodes at Aoraki, Moonlight
Tectonic Zone, Waihemo Fault Zone, and Alpine Fault

The main clades of mite harvestmen, Cyphophthalmi, illustrate global allopatry, consistent with their origin by simple vicariance (Figure 4.1). The family Pettalidae (cross-hatching in Figure 4.1) occurs in all the former areas of Gondwana (except Antarctica), and the family's presence in New Zealand "is undoubtedly due to vicariance rather than dispersal" (Boyer et al., 2007: 11).

Boyer and Giribet (2007) accepted that evolution *within* the family Pettalidae has also been by vicariance. Basal differentiation occurred around South Africa, where there are two main clades. *Parapurcellia* from eastern South Africa is the sister to all other pettalid species (cf. Danthonioideae; Heads, 2014: 14), while *Purcellia* and *Speleosiro*, from southwestern South Africa, have weak support as sister to *Chileogovea*, from Chile (combined molecular data; Giribet et al., 2012). The remaining genera of Pettalidae have good support, but, as yet, the relationships among the genera do not.

In New Zealand, the Pettalidae are widespread, but they are absent from the far north, East Cape, Marlborough south of the Wairau fault, and large parts of Otago and Fiordland (Figure 9.11; Boyer and Giribet, 2009). New Zealand has three genera of Pettalidae—*Aoraki*, *Rakaia*, and *Neopurcellia*—and these occur in three different regions. Generic overlap is restricted to central New Zealand: Nelson, Marlborough Sounds, and the central North Island. The southwestern *Neopurcellia* is separated from *Aoraki* at the Aoraki node and from *Rakaia* at the Moonlight tectonic zone.

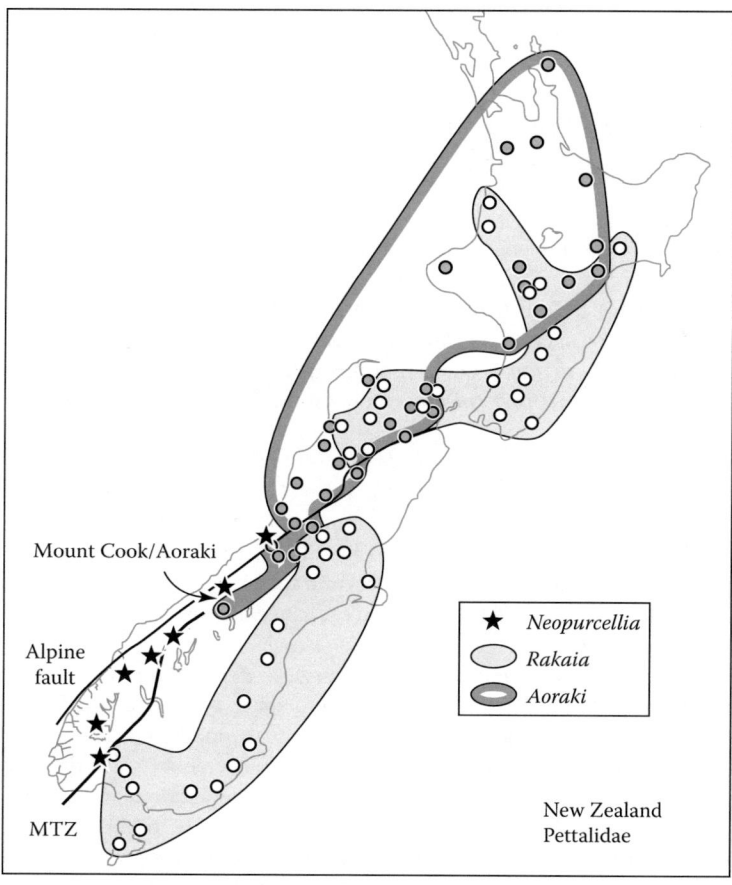

FIGURE 9.11 Distribution of the Pettalidae (Opiliones) in New Zealand. MTZ, Moonlight tectonic zone. (Data from Boyer, S.L. and Giribet, G., *J. Biogeogr.*, 36, 1084, 2009.)

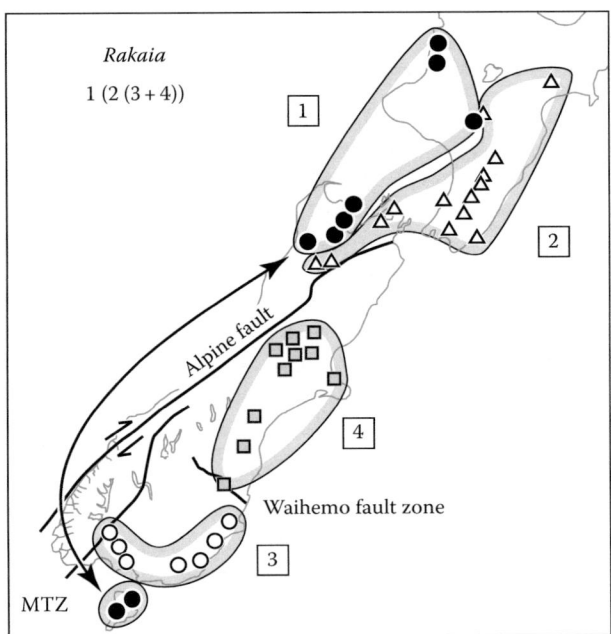

FIGURE 9.12   Distribution of *Rakaia* (Opiliones: Pettalidae). The basal clade (black dots) is disjunct along the Alpine fault. MTZ, Moonlight tectonic zone. (Data from Boyer, S.L. and Giribet, G., *J. Biogeogr.*, 36, 1084, 2009.)

In the pettalid genus *Rakaia* (Figure 9.12), the basal clade 1 (black dots in the figure) is disjunct along the Alpine fault (Boyer and Giribet, 2009). The Alpine fault also marks the break between clade 2 of *Rakaia* and clade 3 + 4. The break between clades 3 and 4 corresponds with another major tectonic structure, the Waihemo fault zone.

In another clade of pettalids, the species complex termed *Aoraki denticulata* s.lat., there are at least two different clades that show signs of disjunction on the Alpine fault (Figure 9.13; Boyer et al., 2007; Fernández and Giribet, 2014). The distributions include two parallel disjunctions, and these are explicable if differentiation within the species had already occurred before transcurrent movement on the fault. The clade in the Arthur's Pass region occurs on both sides of the main geographic divide (fine dashed line in Figure 9.13), indicating that *vertical* movement on the fault has elevated populations but has had little phylogenetic effect.

In the traditional taxonomic group *Aoraki d. denticulata* (all the populations north of Arthur's Pass, a paraphyletic complex) the degree of genetic diversity is "extraordinarily high" (Boyer et al., 2007); Fernández and Giribet (2014) found that *A. denticulata* has "the deepest genetic structure known to us for a terrestrial invertebrate [species], as measured by COI distances. The species can be divided into multiple clades and subclades that could constitute cryptic species…." Boyer et al. concluded that *A. denticulata* is "extremely old," that there is little genetic exchange among populations, and that the animals are "spectacularly poor or slow dispersers."

Boyer et al. (2007) calculated divergence dates within *A. d. denticulata* of 1.6–12.8 Ma. The calculations were based on a rate derived for beetles (based on habitat types and fossil pollen; Farrell, 2001) and the "standard" 2.3%/Myr divergence rate often used for arthropods. The authors wrote that the dates within *A. d. denticulata* "clearly" postdate fault-mediated displacement (which they cited as 25–20 Ma), but the calibrations are questionable (Heads, 2012a), the other populations in Nelson and Marlborough need to be sampled, and most of the fault movement occurred after 11–16 Ma. For these reasons, passive disruption by fault movement remains a possible explanation for the geography of the "double disjunctions" pattern (Figure 9.13), and no other mechanism has been suggested.

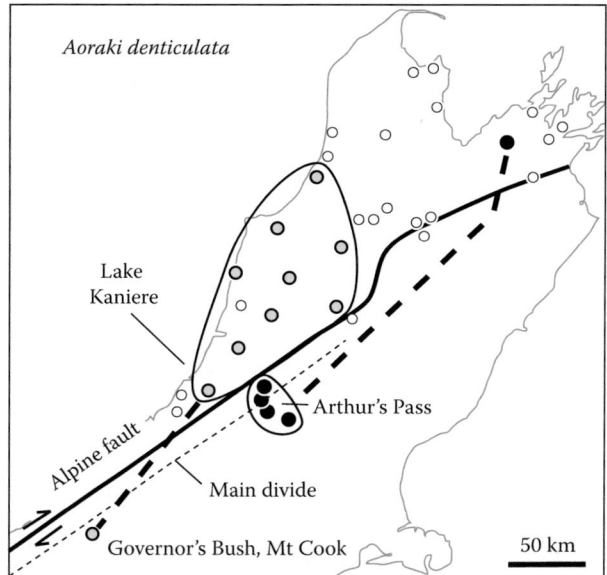

FIGURE 9.13 Distribution of *Aoraki denticulata* (Opiliones: Pettalidae) showing two separate clades (black dots, gray dots) and other populations (some not sequenced; open circles). (Data from Boyer, S.L. et al., *Mol. Ecol.*, 16, 4999, 2007; Fernández, R. and G. Giribet, *Invertebr. Syst.*, 28, 401, 2014.) The populations at Arthur's Pass were earlier treated as *A. denticulata major*, the one at Governor's Bush, Mount Cook, was treated as a separate species, *A. longitarsus*. (From Heads, M., South Pacific biogeography, tectonic calibration, and pre-drift tectonics: Cladogenesis in *Abrotanella* (Asteraceae). *Biological Journal of the Linnean Society*. 2012c. 107. 938–952. Copyright John Wiley & Sons. Reproduced with permission.)

Fernández and Giribet (2014) proposed that "The Bayesian phylogeographic [phylogenetic] analysis suggests that the ancestral populations of *A. denticulata* started to radiate from an area that now constitutes the Alps, subsequently moving toward the coast…," and they proposed an "historical scenario of stepping stone colonisation." But the phylogeny does not provide any real evidence for this, and the allopatry is consistent with simple vicariance. The authors wrote that "the absence of proper calibrations and informative markers behaving in a clock-like fashion prevents us from linking the evolution of the group to specific geological events," yet there are conspicuous boundaries and disjunctions that display a precise match with the Alpine–Wairau fault.

Plecoptera (Stoneflies)

In stoneflies, the genera *Notonemoura*, *Halticoperla*, and *Vesicaperla* are each disjunct along the Alpine fault (mapped in Heads, 1998a). McCulloch et al. (2010) studied the last two of these and also four other stonefly genera that show disjunctions, or at least a phylogenetic break, in the central South Island. Based on clock estimates of clade ages, the authors attributed the break to a single event and concluded: "If standard insect COI calibrations are applicable here (e.g., 2%–3% divergence per million years…) this putative vicariant event likely occurred in the late Pliocene" (p. 2042). Unlike Trewick and Wallis (2001), who accepted that taxa have survived the Pleistocene glaciations more or less *in situ*, McCulloch et al. (2010) attributed the differentiation in the stoneflies to glaciation. Nevertheless, although the calibrations were critical for the conclusions, they were not discussed.

The six stonefly genera that McCulloch et al. (2010) studied displayed levels of genetic divergence across the central South Island gap (between congeneric species) ranging from 0.074 (in *Halticoperla*) up to 0.091 (in *Cristaperla*). The general idea of a constant rate and the particular standard evolutionary rates that were used are not accepted here, but even if they were correct, the rates that McCulloch et al. (2010)

proposed—2% or 3% divergence per million years—give dates for the cross-gap differentiation that are older than the glaciations. New Zealand's earliest glaciation (the Ross Glaciation) occurred at 2.4–2.5 Ma (Suggate, 2004), and this means that of the 12 divergence dates for the stonefly genera (McCulloch et al., 2010), only one is compatible with glaciation—the least diverged pair (in *Halticoperla*), assuming the highest rate:

	Divergence	Divergence Date Assuming 2% Rate (Ma)	Divergence Date Assuming 3% Rate (Ma)
Halticoperla	0.074	3.7	2.46
Spaniocerca	0.076	3.8	2.53
Vesicaperla	0.078	3.9	2.60
Holcoperla	0.088	4.4	2.93
Apteryoperla	0.089	4.45	2.97
Cristaperla	0.091	4.55	3.03

Even assuming that glaciation caused divergence in *Halticoperla*, this leaves all the other divergences unexplained.

Further evidence that glaciations have not wiped out central populations or led to divergence is provided by two stonefly species that are endemic in the Westland gap, *Vesicaperla townsendi* and *Spaniocerca hamishi* (Chapter 7). These have distributions "suggesting the possibility of localized glacial refugia in this region" (McCulloch et al., 2010: 2039).

Grasshoppers and Wetas (Orthoptera)

Wallis and Trewick (2001: 607) ended their critique of the Alpine fault hypothesis by admitting that "some" disjunct pairs could owe their disjunction to strike-slip movement on the fault. This implies that the whole community these groups belonged to was also pulled apart.

Trewick and Wallis (2001) agreed that groups such as the grasshopper *Alpinacris* have distribution patterns that fit with the Alpine fault hypothesis, and this is also true for other orthopterans. In Rhaphidophoridae, the cave weta *Pachyrhamma* (including *Gymnoplectron* and *Turbottoplectron*) is disjunct between the North Island, Nelson, and Fiordland (Cook et al., 2010), and *Maotoweta* is disjunct between the Takitimu Mountains, Fiordland, and Nelson (John and Cook, 2014).

Another family of Orthoptera, the Anostostomatidae, includes the giant wetas. One of these, *Deinacrida connectens*, is the alpine scree weta, mentioned earlier for its boundary at the Moonlight tectonic zone (Chapter 7). Another species, *D. pluvialis*, extends from Fiordland to Canterbury east of the Alpine fault (although some localities are west of the main divide; Gibbs, 1999). Its sister is *D. talpa* of the Paparoa Range, west of the fault, and the pair provides "yet another example of biological disjunction along this fault" (Gibbs, 2001: 52). There is a second Alpine fault disjunction in *Deinacrida*, as the northwest Nelson *D. tibiospina* is sister to *D. carinata* of Stewart Island and coastal Southland (Figure 7.11; Gibbs, 2001, 2006; Trewick and Wallis, 2001).

These patterns in alpine grasshoppers, cave wetas, and giant wetas all suggest that phylogeny in New Zealand Orthoptera has been associated with large-scale fault activity. Molecular biogeographers have tended to overlook this sort of biogeographic-tectonic pattern, but only because it is not consistent with molecular clock dates. For example, *Deinacrida connectens* is distributed east of the Alpine fault, except in Marlborough and Nelson where it occurs on both sides of the fault (Figure 7.11). There is a phylogenetic break at the fault, and Trewick et al. (2000: 663) noted that:

> Given their close proximity, it is remarkable that there is no evidence of exchange between Marlborough (lineage A) and Nelson (lineage C). Interestingly, these two regions (Marlborough and Nelson) are bisected by the Alpine fault (with the exception of Red Hills)... [but] despite an apparent fit of the geographical patterns [with the Alpine fault hypothesis], this model is not compatible with the depth of genetic structure observed here.

Yet the dating estimate relied on the "standard" insect divergence rate of 2.3%/Myr, and this in turn is based on dubious assumptions (Heads, 2012a).

Zizina *Butterflies (Lycaenidae)*

The butterfly *Zizina oxleyi* is distributed in the *eastern* South Island and extends west of the Alpine fault only in northwest Nelson (it is also in the central North Island). Its sister group, *Z. otis*, occurs in the *western* South Island (also North Island) and west to West Africa (Yago et al., 2008; distributions from Gibbs, 1980). The two groups overlap in the northern South Island and the North Island.

Samples sequenced by Gillespie et al. (2013) showed that the western and eastern clades overlap between Puponga in northwest Nelson (the northwestern limit of the mainly eastern *Z. oxleyi*) and Springs Junction (Marble Hill), right at the Alpine fault (the southern limit of the western *Z. otis*). Given the intercontinental distribution of *Z. otis*, the break between this species and *Z. oxleyi*, at or around the fault (there is no disjunction), probably occurred before the transcurrent movement developed along it in the Neogene.

New Zealand Frogs

The only amphibians native to New Zealand, either living or fossil, are members of the endemic frog *Leiopelma*. As discussed earlier, *Leiopelma* plus *Ascaphus* of the western United States form the sister group of all other extant frogs (Chapter 4). The distribution of this clade within New Zealand is also of interest, as the available records of living and fossil *Leiopelma* (New Zealand Department of Conservation, 2016) indicate a typical Alpine fault disjunction (Figure 9.14).

Skinks

A clade of skinks, *Oligosoma infrapunctatum* and its allies, shows two affinities that are disjunct in parallel across the Alpine fault, as in *Aoraki denticulata* (Figure 9.15; Chapple and Patterson, 2007; Bell and Patterson, 2008; Chapple et al., 2009; Patterson and Bell, 2009). The group also includes other important patterns, for example, *O. otagense* is a classic Central Otago endemic (cf. Figure 7.13). It is distributed between the Moonlight tectonic zone and the Waihemo fault zone.

In another skink, the widespread *Oligosoma nigriplantare polychroma*, most of the genetic variation is partitioned across the Alpine fault. Clock estimates of genetic divergence dates within the form (minimum clade ages) predated Pleistocene glaciations (Liggins et al., 2008b: 3677). Liggins et al. concluded that the Alpine fault initiated allopatric divergence in *O. n. polychroma*, and they stressed the importance of the fault in the evolution of New Zealand's biota.

Kiwis *(Ratitae:* Apteryx*)*

The global biogeography of ratite birds is discussed elsewhere (Heads, 2014, 2015a). There are four main groups: one in mainland Africa and Eurasia (ostriches), one in South America (rheas), one disjunct across the Indian Ocean from Madagascar to Australasia (elephant birds, cassowaries, emus, and kiwis), and one, sister to the last, disjunct across the Pacific (moas and tinamous). There is also a fossil group in the north temperate zone (lithornithids).

The only areas where any of the nine families overlap are New Zealand (kiwis and moas) and southern South America (rheas and tinamous). Thus, the simplest explanation for the global distribution is simple vicariance of a worldwide group, followed by local overlap by range expansion within New Zealand and southern South America.

Mitchell et al. (2014a) suggested that the ratite phylogeny is incompatible with vicariance, but this was because they assumed that the only possible mechanism for vicariance is Gondwana breakup. The division between the trans-Indian and trans-Pacific clades in ratites and many other groups was probably caused by predrift rifting in or around Zealandia.

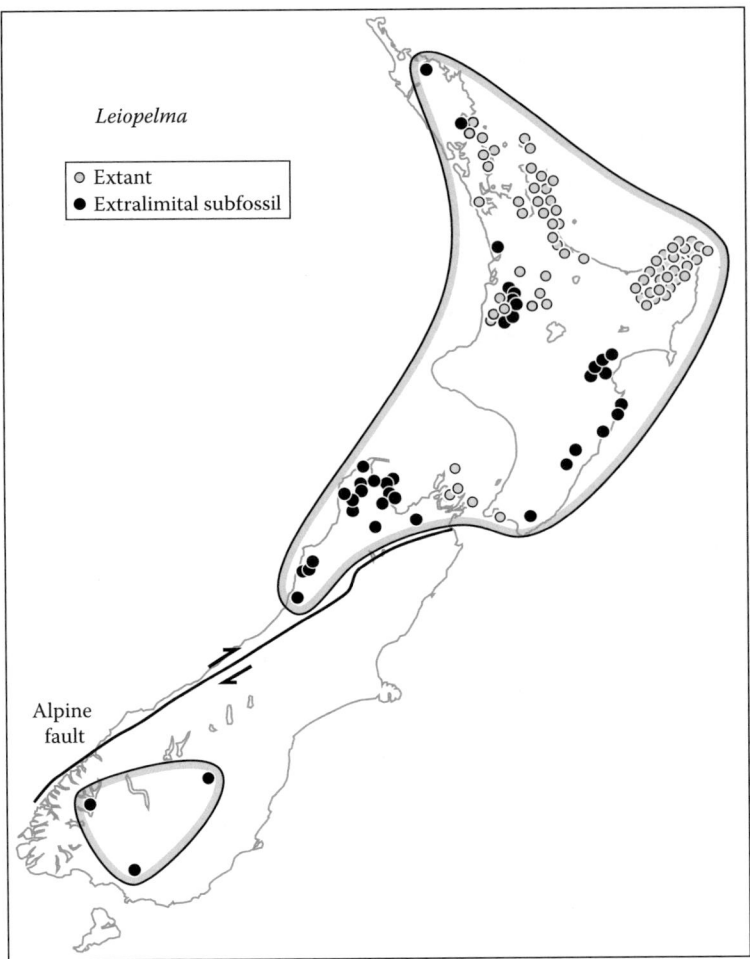

FIGURE 9.14 Distribution of *Leiopelma* (Leiopelmatidae), showing living and fossil records (seven species). (Data from New Zealand Department of Conservation, *Atlas of the amphibians and reptiles of New Zealand*, www.doc.govt.nz/our-work/reptiles-and-frogs-distribution/atlas/,2016; Worthy, T.H. et al., *J. R. Soc. N. Z.*, 43, 211, 2013a.)

The pre-Quaternary fossil record of kiwis consists of two bones (a femur and a quadrate) from the Early Miocene of Central Otago (Worthy et al., 2013b). These were assigned to a new genus, *Proapteryx*, that is more similar to *Apteryx* than any other clade, although it probably only weighed about a third as much as the smallest extant kiwi (*A. owenii*).

Worthy et al. (2013b) wrote: "*Proapteryx* reveals that ancestral kiwi were probably small in the Early Miocene," but no evidence was presented indicating that *Proapteryx* was ancestral to any extant kiwi clade. The authors also wrote: "The small size and slenderness of the femur makes it distinctly possible that *Proapteryx* was volant, supporting an overwater dispersal origin to New Zealand of kiwi that was independent of moa." But whether or not *Proapteryx* or ancestral kiwis were volant, there is no evidence that kiwis originated following overwater dispersal. Most birds can fly, but this does not mean that a particular clade must have originated by dispersal.

The kiwi, *Apteryx*, comprises two main clades, brown and spotted. Sequencing studies of extant and subfossil populations show that the brown kiwi clade is divided into two allopatric groups (Figure 9.16). Their mutual boundary lies along the Alpine fault rather than the main geographic divide (Shepherd et al., 2012).

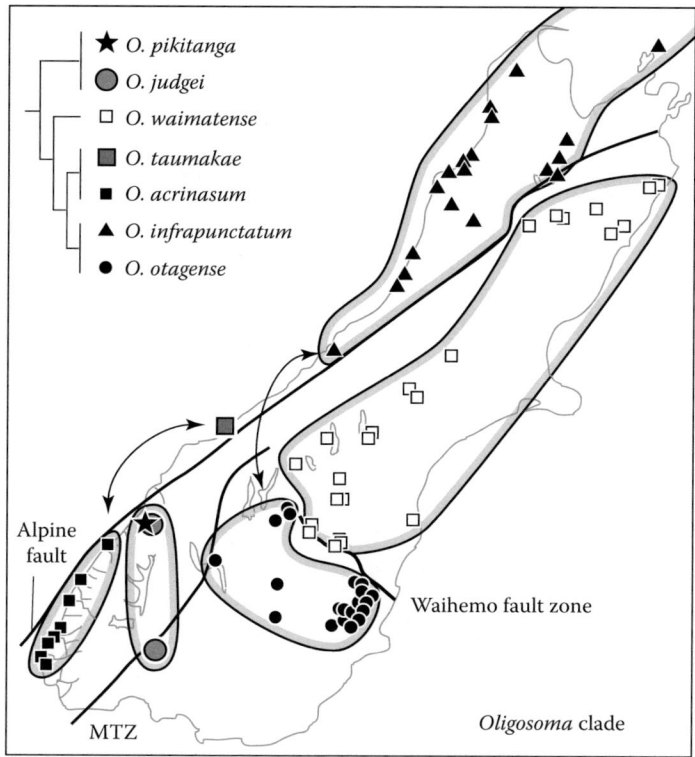

FIGURE 9.15 Distribution and phylogeny of *Oligosoma infrapunctatum* and allies (Scincidae). (Data from Chapple and Patterson, 2007; Bell and Patterson, 2008; Chapple et al., 2009; Patterson and Bell, 2009.) MTZ, Moonlight tectonic zone. (From Heads, M., South Pacific biogeography, tectonic calibration, and pre-drift tectonics: Cladogenesis in *Abrotanella* (Asteraceae). *Biological Journal of the Linnean Society.* 2012c. 107. 938–952. Copyright Wiley-VCH Verlag GmbH & Co. KGaA. Reproduced with permission.)

Baker et al. (1995) interpreted the southeastern clade (*Apteryx australis* and others) as a remnant of the "original ancestral population" that moved north from Fiordland to Haast and then Okarito, where the population diverged and later colonized the North Island. Yet the evidence is also consistent with an alternative model, in which the ancestral brown kiwi population was already widespread before the split. The main phylogenetic break occurs between populations of the *A. australis* clade at the Arawata Valley and Haast Range (near Haast village) (Holzapfel et al., 2008), and members of the northern *A. mantelli* clade at Okarito. This break corresponds with the plate boundary, the Alpine fault (not the main divide watershed), although this has not been recognized in prior studies. Most populations of the southern *Apteryx australis* group lie southeast of the main geographic divide. However, the population at Haast and many in Fiordland lie *west* of the main divide but *east* of the Alpine fault (Figure 9.16), and so the main divide does not correspond with the phylogenetic break. Wallis and Trewick (2009) discussed the basal split in the brown kiwi clade as Pliocene and "more likely to reflect uplift than glaciation" (p. 3562). Yet if the break were caused by uplift, it would be located at the main geographic divide, not at the fault. The observed pattern indicates that strike-slip fault movement is a possible explanation.

Megalapteryx *Moas (Ratitae)*

The other New Zealand ratites, the moas, are giant birds that went extinct in historical times, at some time after ~560 BP (Rawlence and Cooper, 2013). Molecular studies of subfossil moa material have supported the idea of many refugia during the Pleistocene glaciations, even in colder, montane areas. For example, the upland moa genus *Megalapteryx* shows high levels of genetic diversity relative to the

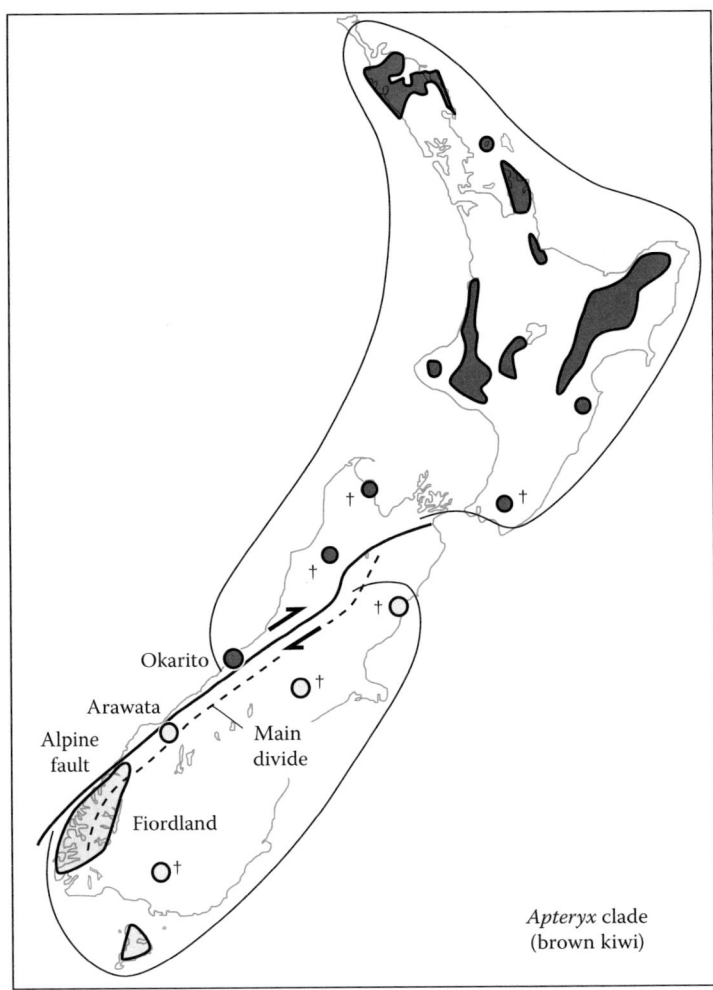

FIGURE 9.16 Distribution of the two clades of brown kiwi (*Apteryx* spp.: Apterygidae): *A. australis* species group (light gray) southeast of the Alpine fault, and *A. mantelli* species group (dark gray) northwest of the fault (Shepherd et al., 2012). Dagger symbols = sequenced, subfossil samples (population now extinct). (Reprinted from Heads, M., *Biogeography of Australasia: A Molecular Analysis*, Cambridge University Press, Cambridge, UK, 2014. With permission.)

other genera, and this "likely relates to the persistence of upland/montane habitat during both glacial and interglacial periods" (Bunce et al., 2009: 20650). With respect to the Oligocene flooding, Bunce et al. (2009: 20649) agreed that "The survival of many endemic vertebrates preserved as early Miocene fossils strongly argues against total submergence." Fossil eggshells and "tantalizing fragments" from the Miocene (19–16 Ma) of Central Otago indicate the presence of at least two large, flightless ratite taxa within a few million years of the proposed drowning event (Worthy et al., 2013b).

Megalapteryx comprises four allopatric clades, with one each in Fiordland, Otago, Nelson, and Marlborough/northern Canterbury (Figure 9.17). Bunce et al. (2009: 20650) suggested that the boundaries correspond to "barriers which effectively split the uplands of the South Island into four regions during the Pleistocene." The pattern is also compatible with Alpine fault displacement, as the Fiordland clade is sister to the Nelson one.

Bunce et al. (2009) thought the large disjunction between these two regions was caused by Pleistocene glaciation, but *Megalapteryx* favored upland, colder habitat, and during colder phases its range would probably expand rather than contract. This suggests that the Fiordland–Nelson break was instead the result of Alpine fault strike-slip. As Bunce et al. (2009) acknowledged, "The phylogeographic distribution of

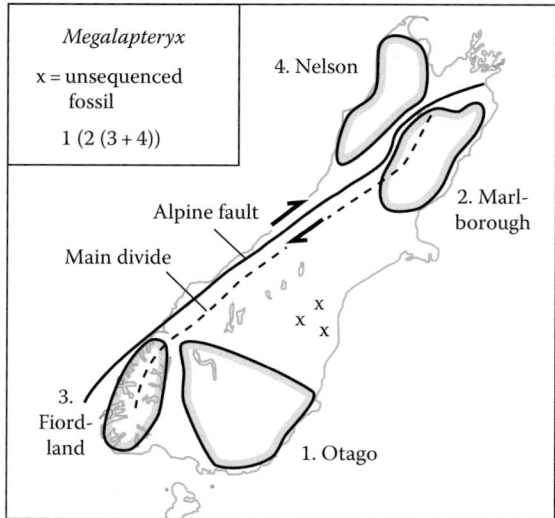

FIGURE 9.17 Distribution of the four clades in the moa genus *Megalapteryx* (Dinornithidae) (Worthy, 1997; Bunce et al., 2009). (Reprinted from Heads, M., *Aust. Syst. Bot.*, 27, 241, 2015a. With permission from Australian Systematic Botany.)

the *Megalapteryx* clades does not match their phylogenetic relationships" (Supporting Information, p. 2). This is because the Fiordland clade is sister to the Nelson clade (along the Alpine fault), not the adjacent Otago clade (Routeburn to southeastern Otago). This implies that the (Fiordland + Nelson) and Otago populations broke apart at a node somewhere between Fiordland and Otago, before the Alpine fault rifted the two components of the Fiordland + Nelson clade.

Bunce et al. (2009) concluded by comparing the *Megalapteryx* phylogeny with that of the alpine scree weta, *Deinacrida connectens* (Figure 7.11). They cited "striking similarities in the phylogeographic structure of the clades, suggesting common barriers to geneflow in these two upland species" (p. 20850). In both groups, the main phylogenetic break coincides with the Alpine–Wairau fault.

Another moa, *Pachyornis australis* is also disjunct between the northern South Island (northwest Nelson–Charleston) and the southern South Island (Albert Town–Bluff) (map in Worthy and Holdaway, 2002: 595). It is absent at intermediate fossil moa sites, and the gap along the east coast is filled by *P. elephantopus*. This is a typical disjunction along the Alpine fault, as in the Fiordland–Nelson clade of *Megalapteryx*.

The New Zealand Wrens, Acanthisittidae

The basal passerine family Acanthisittidae has two extant species. Oliver (1955) concluded that one of these, *Acanthisitta chloris*, the rifleman, has distinctive populations (subsp. *citrina*) in Fiordland/northwest Otago, and, disjunct, in Nelson. Later authors have not recognized this form, but a molecular study would be of great interest.

Alpine Fault Distributions Other than Simple Disjunction

Many groups do not exhibit a disjunction at the Alpine fault, with a complete break in the range, but have other patterns that can be explained by movement on the fault. The distribution of a group before fault movement determines the pattern observed after fault movement, as in the following examples:

- Taxa that were widespread and undifferentiated before fault movement can remain widespread after disruption and show no evidence for disruption.
- Taxa that were restricted to one or other side of the fault before fault movement will not show any disruption after fault movement.

- Taxa that had narrow ranges on both sides of the fault before fault movement can show complete disjunction after movement. These act as suitable "marker taxa" illustrating processes, just as narrow geological units, such as the Dun Mountain ophiolite and the Esk Head Mélange, are often used as marker terranes (Figure 9.3).

- Taxa with northern or southern limits that have been disrupted by the fault will also show distinctive patterns after fault movement, although there will be no complete gap in the distribution. These patterns were not cited in earlier studies (Heads, 1998a; Heads and Craw, 2004), and they are not as striking as simple disjunctions. They occur as frequently as the disjunctions though and are discussed next.

Alpine Fault Disruption of a Southern Boundary: Groups Present in Nelson and Fiordland, Absent in Westland and Stewart Island

Many plants and animals have a widespread range east of the Alpine fault in the eastern South Island and also occur in Nelson and Fiordland, but have anomalous absences west of the fault in *Westland* and also in *Stewart Island*. This pattern can be explained as the result of the Alpine fault disrupting a southern boundary. Examples from morphological studies include the widespread, subalpine shrub *Hebe mooreae* (Plantaginaceae; Figure 9.18). Alpine fault disjunction is suggested by the anomalous absences in the Paparoa Range and Stewart Island. It is also indicated by the two disjunct populations in which the plants have adaxial stomata, at Caswell Sound in Fiordland and on the Denniston Plateau in northern Westland (Heads, 1992; Bayly and Kellow, 2006). This variation is clinal, and rare individual plants elsewhere can have adaxial stomata, but the extreme populations at the two cited localities form a significant pattern.

In southern Westland, *Hebe mooreae* extends west across the fault onto the Jackson Formation at Jackson Bay (Figure 9.18). Similar exceptional records west of the fault occur near there in other groups (e.g., the beetles *Tormus posticalis*; Fikáček et al., 2013, and *Agnosthaetus* "*brounides* group," Figure 9.8; and the plants *Celmisia markii*; Lee and Given, 1984, and *Pimelea gnidioides*; Burrows, 2008). These

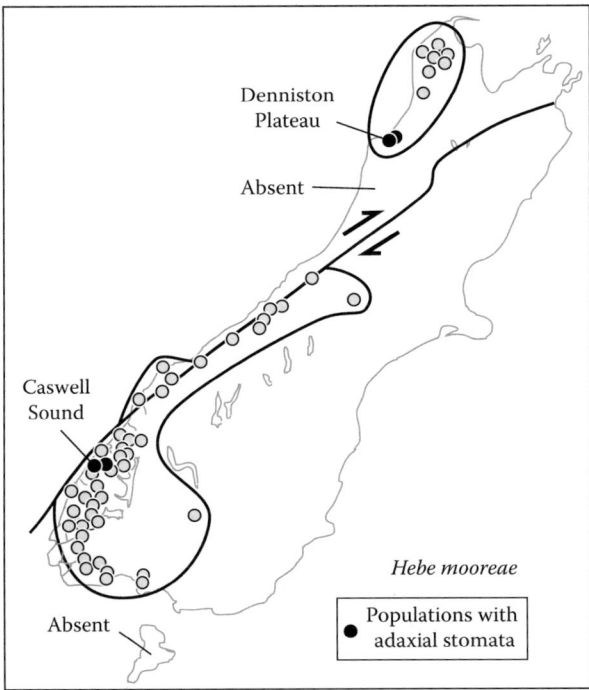

FIGURE 9.18   Distribution of *Hebe mooreae* (Plantaginaceae). (Data from Heads, M., *Candollea*, 47, 583, 1992; Bayly, M. and Kellow, A., *An Illustrated Guide to New Zealand Hebes*, Te Papa Press, Wellington, New Zealand, 2006.)

could reflect the fact that the Jackson Formation is allochthonous and has been emplaced from the south-east after the mid-Miocene (Sutherland et al., 1996).

Another possible explanation for the anomalous populations west of the Alpine fault involves a structure there that was interpreted as a terminal moraine, but is in fact a large rock avalanche triggered by Alpine fault movement at AD 660 (Barth, 2014). The avalanche originated in the mountains immediately east of the fault, and extended 4 km west of the fault, down the Cascade Valley. Plants are often carried considerable distances downslope by ordinary, snow avalanches (e.g., in the European Alps; Gaussen, 1954: 21), and this mechanism might also apply to rock avalanches.

The shrub *Coprosma fowerakeri* (Rubiaceae) is another, well-collected group that displays disruption of the southern boundary by Alpine fault movement, as in *Hebe mooreae* (Figure 9.19a; Norton and de Lange, 2003). Note the absence from the Paparoa Range and Stewart Island. Other cases are seen in the fern *Grammitis patagonica* (Parris and Given, 1976) and the angiosperms *Epilobium gracilipes* and *E. insulare* (Onagraceae; Raven and Raven, 1976), *Acaena profundeincisa* (Rosaceae; Macmillan, 1983), *Kelleria croizatii* and *K. multiflora* (Thymelaeaceae; Heads, 1990a), *Hebe macrantha* (Plantaginaceae; Bayly and Kellow, 2006), the *Lobelia macrodon* group (Campanulaceae; Knox et al., 2008), *Pachycladon* (Brassicaceae; Heenan and Mitchell, 2003), and *Chionohebe* (= *Veronica* "snow hebes group," Plantaginaceae; Meudt, 2008).

Animal species with the *Coprosma fowerakeri* pattern include:

Cryptodacne synthetica (Coleoptera: Erotylidae): In Nelson, Canterbury, and Fiordland, but absent from Westland and Stewart I (Skelley and Leschen, 2007).

Zelandobius patricki (Plecoptera): In Nelson, Canterbury, Otago, and Fiordland, but absent from Westland and Stewart I (McLellan, 1993).

Animal genera with the pattern include the nematode *Blandicephalanema* (Wouts, 2006); the coccoid hemipterans *Ctenochiton* and *Kalasiris* (Hodgson and Henderson, 2000); the heteropterans *Chinamyersia*, *Cydnochoerus*, *Polyozus*, *Brentiscerus*, and *Saldula* (Larivière and Larochelle, 2004); the hymenopteran *Aspilota* (Braconidae; Berry, 2007); and the nymphalid butterfly *Argyrophenga* (Craw, 1978).

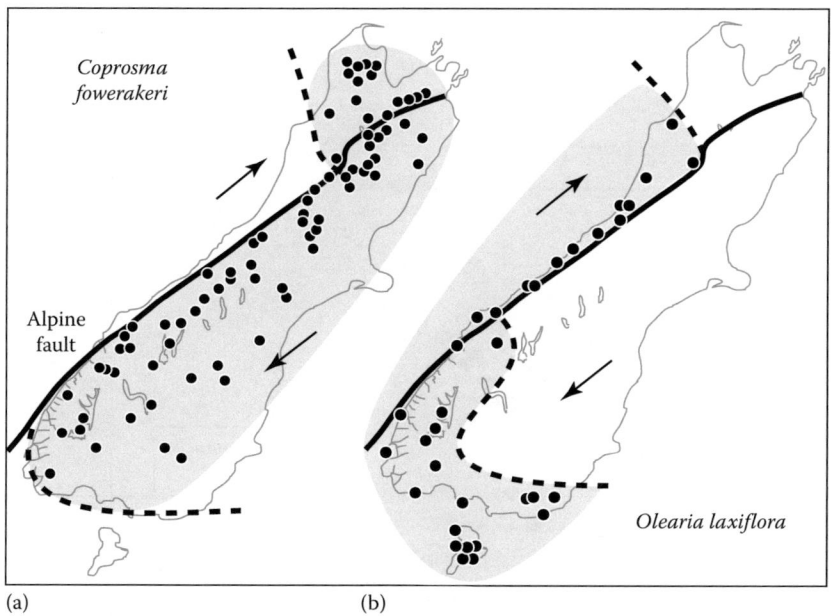

(a)   (b)

FIGURE 9.19 Two types of Alpine fault disruption. (a) Disruption of a southern limit, in *Coprosma fowerakeri* (Rubiaceae). (Data from Norton, D.A., and de Lange, P.J., *N. Z. J. Bot.*, 41, 223, 2003.) (b) Disruption of a northern limit, in *Olearia laxiflora* (Asteraceae). (Data from Heads, M., *Bot. J. Linnean Soc.*, 127, 239, 1998b.)

Alpine Fault Disruption of a Northern Boundary: Groups Present in the Southwestern and Southeastern South Island and through Westland

In this pattern, groups show anomalous absences in northern Otago, Canterbury, and Marlborough, and often northern Nelson. The distribution can be explained by fault disruption of a prior northern boundary. Examples are documented in morphological studies on the plant *Olearia laxiflora* (Asteraceae) (Figure 9.19b; Heads, 1998b) and the lucanid beetle *Geodorcus helmsi* (Holloway, 2007).

The Alpine Fault as an East/West Boundary: The *Lobelia ionantha* Group (Campanulaceae)

Some taxa that were widespread east and west of the fault before fault movement appear to have differentiated, with fault movement, into one clade west of the fault and one clade to the east. In this case, there is a phylogenetic break at the fault. No disjunction is evident, but this would be obscured by the widespread distribution of the group as a whole.

The pattern is illustrated here with a group of New Zealand herbs, the *Lobelia ionantha* group (including *Isotoma* and *Hypsela* p.p.). The basal node in a molecular phylogeny separates *L. ionantha*, found southeast of the Alpine fault, from the other three species, located northwest of the fault (Figure 9.20). Explaining the history of the group, Heenan et al. (2008b: 94) wrote:

> Dispersal from Australia of an *Isotoma fluviatilis*-like ancestral species, estimated at about seven million years ago (E.B. Knox unpubl. data), is inferred to have initially colonised the South Island… [Based on the phylogeny], the colonising lineage evolved into *L. ionantha*, occupying mid-elevation sites throughout the length of the South Island east of the Alpine Fault. This subsequently dispersed to the North Island and evolved into *L. carens*… The diversification on the South Island was evidently predicated on crossing the Alpine Fault [from east to west].

The distribution does indicate that the basal differentiation took place at the Alpine fault, but dispersal across the fault is not necessary and, on its own, does not provide a convincing explanation for the pattern.

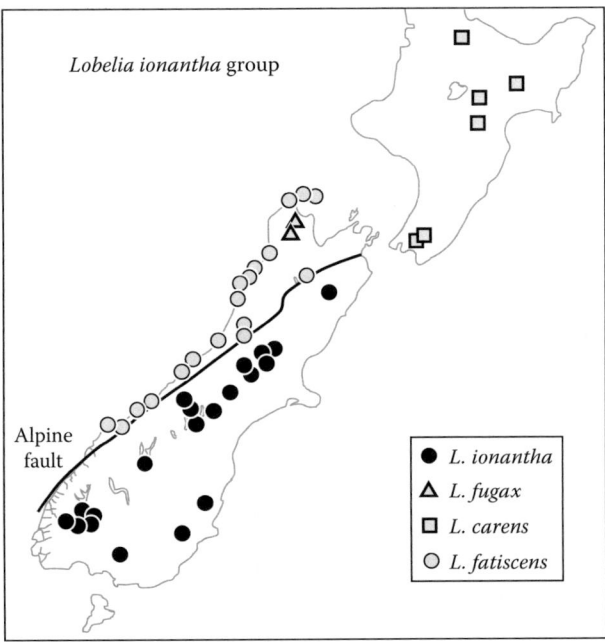

FIGURE 9.20   Distribution of the *Lobelia ionantha* clade (Campanulaceae). (Data from Heenan, P.B. et al., *N. Z. J. Bot.*, 46, 567, 2008.)

The clades on either side are monophyletic, and, without adding extra, ad hoc hypotheses, this implies a single dispersal event. Yet there is no obvious reason why dispersal would only occur once, as the fault is hundreds of kilometers long and does not form a geographic barrier. Thus, the phylogenetic break at the fault would be best explained by some sort of geological or climatic change, and fault movement rupturing populations is the most obvious mechanism. This would also explain why the boundaries of the eastern and western species coincide so closely with the fault, from one end of the island to the other. An Alpine fault origin for the basal break would imply that the break occurred earlier than 7 Ma, the date suggested for the group's dispersal to New Zealand (Heenan et al., 2008b). It would also imply that "*Isotoma*" *fluviatilis* of eastern Australia is a simple vicariant of its sister, the *Lobelia ionantha* group, not an ancestor.

The Alpine Fault and Community Ecology

Plot-scale ecological analyses indicated that two woody plant communities, "shrubland type 1" (with *Dracophyllum uniflorum*, *Schoenus pauciflorus*, etc.) and (with one exception in northern Westland) "forest type 7" (with *Hoheria glabrata*, *Coprosma pseudocuneata*, etc.), both have the western limit of their distribution at the Alpine fault (Wiser et al., 2011).

Distributions Mapped on a 27 Ma Reconstruction

Some of the groups that were mapped earlier are shown here again on a reconstruction of New Zealand at 27 Ma, prior to significant Neogene deformation and Alpine fault displacement (Kamp and Furlong, 2010; cf. Turner et al., 2012). The underlying structure at 27 Ma was similar to that at Gondwana breakup (Figure 9.1), although some convergence had already occurred, Fiordland had rotated clockwise, and the Emerald Basin had opened.

Reconstructions for *Hebe mooreae* (Figure 9.21, cf. Figure 9.18) and *Dendroblax* (Figure 9.22, cf. Figure 9.10) illustrate simple patterns. Maps of *Pimelea suteri* and *Libertia peregrinans* (Figure 9.23, cf. Figures 9.5 and 9.6) highlight the absence of these groups from Fiordland.

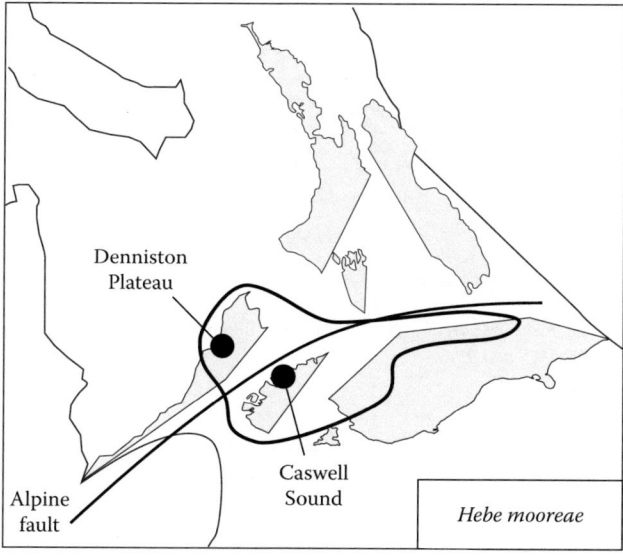

FIGURE 9.21 Distribution of *Hebe mooreae* (Plantaginaceae) on a 27 Ma reconstruction (Kamp and Furlong, 2010) (cf. Figure 9.18). (Base map from Kamp, P.J.J. and Furlong, K.P., Tectono-sedimentary framework for exploration in New Zealand basins. New Zealand Petroleum Conference 2010 Proceedings, Auckland, New Zealand. New Zealand Ministry of Economic Development.)

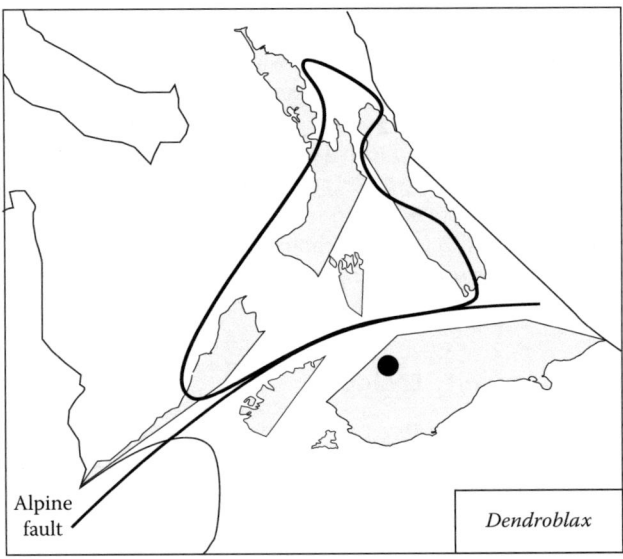

FIGURE 9.22   Distribution of *Dendroblax* (Lucanidae) on a 27 Ma reconstruction (cf. Figure 9.10).

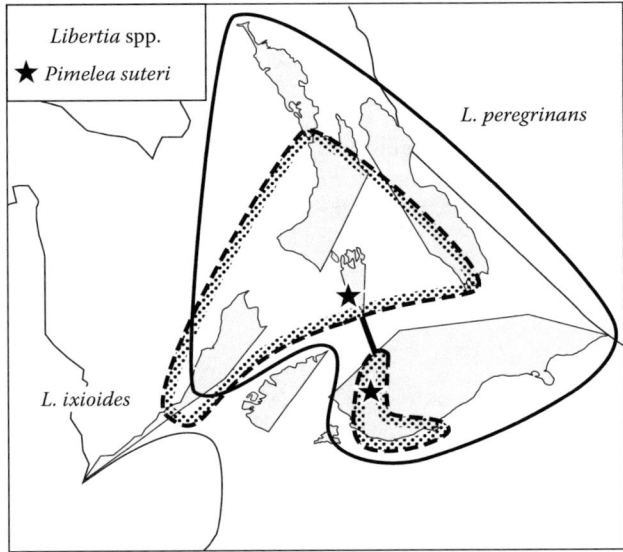

FIGURE 9.23   Distribution of *Pimelea suteri* (Thymelaeaceae), *Libertia peregrinans*, and *L. ixioides* (Iridaceae) on a 27 Ma reconstruction (cf. Figures 9.5 and 9.6).

The two *Poa* clades mapped in Figure 9.24 (cf. Figure 6.4) are not immediate sister groups but illustrate how Campbell Plateau clades can be separated from mainland relatives by the opening of the Emerald Basin and the Bounty Trough. In the mainland group, the northwest Nelson clade connects with its Central Otago sister group across a section of crust between Fiordland and the Marlborough Sounds.

In comparison with their skewed, modern form, the distributions of *Olearia laxiflora* and *Coprosma fowerakeri* plotted on the reconstruction are more or less elliptical (Figure 9.25, cf. Figure 9.19).

The *Oligosoma* clade (Figure 9.26, cf. Figure 9.15) has a complex arrangement on the modern topography, but the reconstructed distribution is a simple, more or less linear pattern.

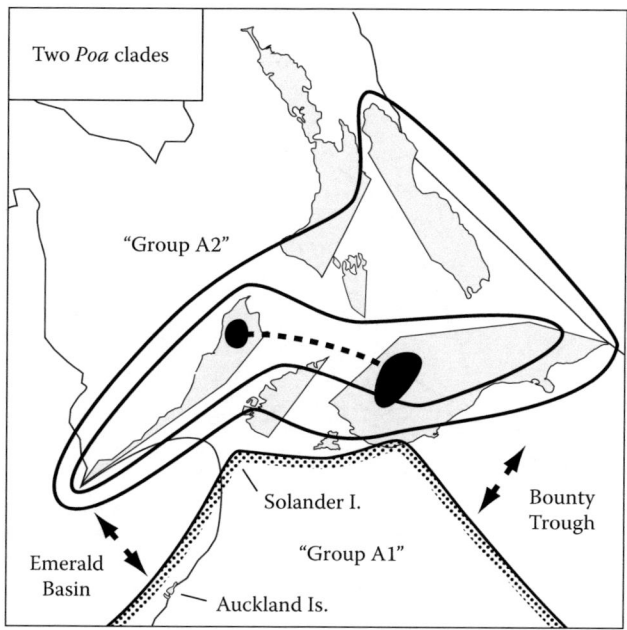

FIGURE 9.24   Distribution of two *Poa* clades, Groups A1 and A2 (Poaceae) on a 27 Ma reconstruction (cf. Figure 6.4).

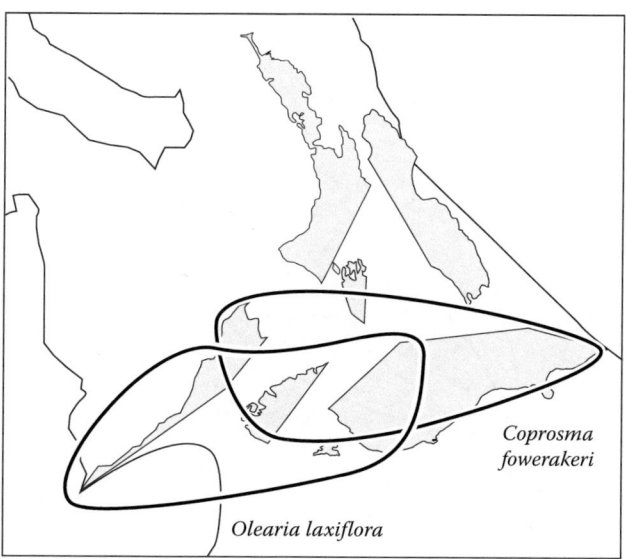

FIGURE 9.25   Distribution of *Olearia laxiflora* (Asteraceae) and *Coprosma fowerakeri* (Rubiaceae) on a 27 Ma reconstruction (cf. Figure 9.19).

In the *Agnosthaetus* "type A aedeagus clade" (Figure 9.27, cf. Figure 9.8), the fault disjunction in the *brouni* group is filled by the *brounides* group. This gives a concentric arrangement in the reconstruction, with the *brounides* group lying along a central strip. The reconstruction assumes that the *brouni* clade has been separated by the opening of the Emerald Basin.

The New Zealand mite harvestmen, all in Pettalidae (Figure 9.28, cf. Figure 9.11), are another group in which a complex distribution has a concentric arrangement in the reconstruction. One clade (*Neopurcellia*) occupies a central sector, as in the *Agnosthaetus brounides* group (Figure 9.27). In

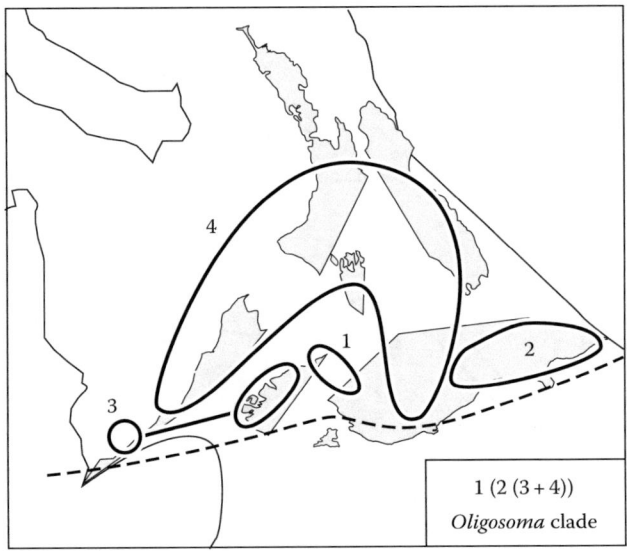

FIGURE 9.26   Distribution of a clade in *Oligosoma* (Scincidae) on a 27 Ma reconstruction (cf. Figure 9.15).

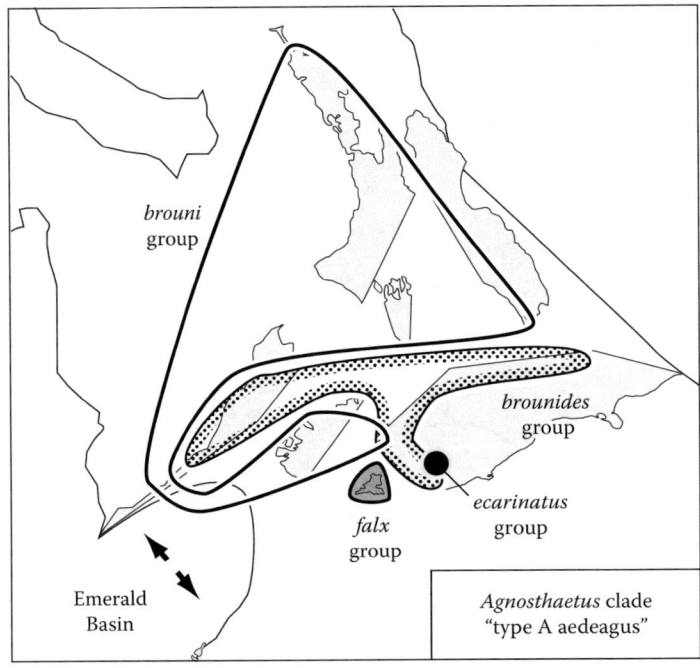

FIGURE 9.27   Distribution of a clade in *Agnosthaetus* (Staphylinidae) on a 27 Ma reconstruction (cf. Figure 9.8).

both cases, the central group has a distribution: Westland–Mount Cook, oblique to the Zealandia-Pacific margin and with a similar strike to that seen in many prebreakup extensional features (Tulloch et al., 2009: Fig. 10). The geological trend may be related to seafloor spreading between Australia and Antarctica.

The Stewart Island and southern Westland clade of harvestmen is depicted here in the same way as in *Agnosthaetus*, running south of Fiordland and not, as in *Libertia*, to the north (Figure 9.23). This has a

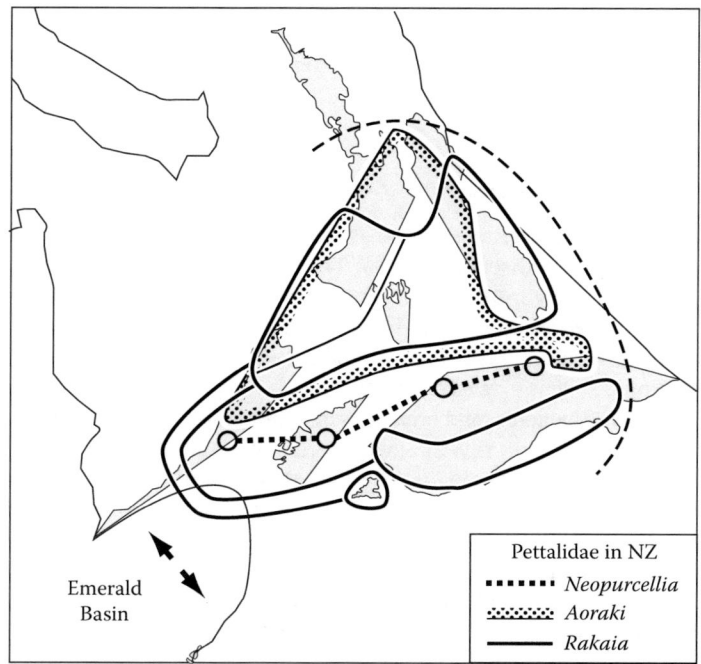

FIGURE 9.28   Distribution of New Zealand mite harvestmen (Pettalidae) on a 27 Ma reconstruction (cf. Figure 9.11).

precedent in groups such as the liverwort *Castanoclobos*, a distinctive genus known only from Stewart Island and the lower Cascade area, west of Martyr Saddle (Teer Plateau; Engel and Glenny, 2008). In the harvestmen, the biogeography of the central *Neopurcellia* is of special interest, but its phylogenetic position is still unresolved; in different analyses, it appears as the sister of the New Zealand *Aoraki*, as the sister to an Indian Ocean clade in New Zealand, Australia, and Sri Lanka, or to a trans-Atlantic clade in South Africa and Chile (Giribet et al., 2012).

Distributions That Coincide with the Alpine Fault Rather Than Rainfall Differences

Lobelia ionantha is distributed from wet Fiordland to drier Canterbury and Marlborough (Figure 9.19) but is not found across the Alpine fault in northwest Nelson. This is similar to the distribution of *Hebe subalpina*, described in the last chapter. This species is abundant in subalpine shrubland from Fiordland along the Southern Alps to the Nelson lakes, and its localities include some of the wettest areas on Earth. Nevertheless, if the distribution conformed to patterns of rainfall, *H. subalpina* would extend north into the wet Paparoa Range and northwest Nelson, where it is absent, and it would be absent further east in the drier Spenser and St Arnaud Ranges of southeast Nelson, where it is present. In other words, the species does not cross the Alpine–Wairau fault into northwest Nelson, but extends into drier country on the same side of the fault (Bayly and Kellow, 2006). The distribution coincides with the fault, not with the main divide or with rainfall.

North of the Wairau fault, *H. subalpina* is replaced by its close relative *H. urvilleana* (Bryant Range to D'Urville Island). In northwest Nelson, it is replaced by the closely related *H. calcicola*. East of the Moonlight tectonic zone, *H. subalpina* is replaced by its close relative *H. rakaiensis* (identification of Forbes Mountains material is controversial). The breaks between *H. subalpina* and its three relatives are all aligned with tectonic features rather than aspects of the current environment.

In sedges, *Uncinia purpurata* (Cyperaceae) repeats the basic pattern seen in *Lobelia ionantha* and *Hebe subalpina*. It is recorded from eastern Fiordland, east to Dunedin (the type is from Signal Hill), and north via Canterbury and Marlborough to the Tararua Range near Wellington (NZPCN, 2016). The distribution includes wet localities (in Fiordland) as well as dry sites (in Marlborough), but follows a standard pattern in being restricted to areas southeast of the Alpine–Wairau fault. This is despite the presence of apparently suitable habitat on the other side of the fault, and it suggests that historical, not just ecological, factors are responsible for the distribution.

The Kaikoura Orogeny and the New Zealand Alpine Biota

The Late Jurassic–Early Cretaceous Rangitata orogeny caused a complete revolution in the tectonics and biogeography of New Zealand, as discussed earlier. The next orogeny in New Zealand, and the last one, was the Kaikoura orogeny (Chapter 3). This began in the Early Miocene with uplift accelerating from ~5 Ma; high mountains were present by about that time. Uplift took place along the Alpine fault, and the main geographic divide now runs parallel to the fault, just east of it (Figure 3.3).

The New Zealand alpine zone is extensive, and its biota is diverse. For example, the alpine zone occupies about one-fifth of New Zealand's land area but supports about two-fifths of the country's species of moths (Dugdale, 1974). It is also very rich in other invertebrates and plants, and has intriguing, endemic species of lizards and birds. Dugdale (1974) described this high diversity and endemism, despite the young age of the mountains, as "the riddle of the alpine biota." What are the origins of this biota, and what is the answer to the riddle?

Many New Zealand groups include both lowland and alpine endemics, and several of the most interesting New Zealand endemics—those with global sister groups—occur in the alpine zone. For example, in passerines, the acanthisittids have two extant species. The rifleman, *Acanthisitta*, is in lowland and montane forest, while the rock wren, *Xenicus gilviventris*, is alpine. Of the three basal parrots, the kakapo (*Strigops*) is flightless and nocturnal, and favors forest margins, while the kaka (*Nestor meridionalis*) is in lowland and montane forest, and the kea (*Nestor notabilis*) is subalpine to alpine. *Craterostigmus*, the basal centipede, occurs in Tasmania, and in New Zealand from coastal bush at 5 m elevation to alpine habitat at 1700 m (G. Edgecombe, pers. comm., July 24, 2007).

The athoracophorid slugs are a Tasman-Coral Sea group (Heads, 2014), and in New Zealand they include both lowland and alpine populations. Their sister group is the cosmopolitan family Succineidae, land snails that occur in swampy areas, by rivers, lakes, and the seashore. The single New Zealand representative, *Austrosuccinea*, occurs with grasses on coastal foredunes (Powell, 1979). Athoracophorids are not found in this habitat but occur in forest and at higher elevations. An alpine population survives at 1500 m near Gertrude Saddle, Fiordland. Here, the animals can be found in summer on algae-covered rockfaces, under large rims of melting ice overhanging rock faces (personal observations). The athoracophorids are widespread and diverse in New Zealand, from lowland to alpine, whereas their biogeographic and phylogenetic vicariant, succineids, are represented in New Zealand only around coastal dunes. Athoracophorids are endemic to the Tasman-Coral Sea region, while succineids are worldwide. Within New Zealand, athoracophorids have 24 named species, while there is just a single succineid (Powell, 1979).

Traditional Ideas on the Alpine Biota: Young Taxa on Young Mountains

Dispersal theory proposes that the alpine biota is young and has evolved after the Kaikoura orogeny mountains were uplifted, colonizing the new habitat from elsewhere and adapting to it (Fleming, 1979). This theory has been maintained in modern molecular studies. Winkworth et al. (2005) discussed the New Zealand mountain flora and concluded that fossil pollen studies and fossil-calibrated phylogenies "provide convincing evidence that many alpine plant lineages are recent additions to the New Zealand flora" (pp. 241–242). The authors inferred that radiation of the alpine plants occurred in the Pliocene and Pleistocene. Nevertheless, these dates are all fossil-based, and small, insect-pollinated, alpine taxa with sparse populations are notorious for their low rates of fossilization; mountains in general are zones of erosion rather than deposition.

Another study of New Zealand alpine plants also used fossil-calibrated phylogenies, and, again, found young dates for genera, for example, *Ourisia* 0.8 Ma, *Pachycladon* 1.6 Ma, *Gentianella* 2.1 Ma, *Abrotanella* 4.2 Ma, and *Euphrasia* 5.7 Ma (W.G. Lee et al., 2012). Biogeographic patterns in these groups, such as trans-oceanic disjunction, Alpine fault disjunction and others (Heads, 2014) are all consistent with the clades' actual ages being at least an order of magnitude greater than this (two orders greater in *Ourisia*).

Linder (2008: 3101) suggested that "in general, the alpine biota of New Zealand is Pliocene in age (Buckley and Simon, 2007)," and Wallis and Trewick (2009) agreed; the Late Pliocene emergence of the Southern Alps and associated ranges "has clearly been accompanied by lineage radiation and speciation (Buckley and Simon, 2007)" (p. 3563). Yet the chronology is not as well established as it might seem. Buckley and Simon (2007) calibrated their cicada phylogeny using the assumption that a Norfolk Island species could be no older than the island, but there is now excellent geological evidence for former, large islands in the vicinity (Meffre et al., 2007). In biology, clock studies found that the skink clades endemic to the island were much older than the island itself (Chapter 5). Biogeographic patterns in New Zealand, including the broader distributions considered already as well as more local distributions considered later, also suggest that many of the New Zealand alpine clades are older than their alpine habitat.

Alternative Ideas on the Alpine Biota: Old Taxa on Young Mountains

Dispersal theorists have argued: "There is little doubt that mountains are a geologically recent feature of New Zealand" (Winkworth et al., 2005: 238), and so they have suggested that the alpine biota is recent too. In fact, orogenies have taken place in New Zealand in Paleozoic, Mesozoic, and Cenozoic time, and land with some relief has existed throughout the region's history.

Many authors have assumed that New Zealand groups now restricted to alpine habitat must have evolved after the formation of that alpine habitat during the latest orogeny, in the Neogene (McCulloch et al., 2010: 3; Heenan and McGlone, 2013: 111; Meudt et al., 2015). This is questionable though, as the niche of a clade can change over time (Chapter 1); if a clade is restricted to a lowland area that is uplifted, its environment and niche will change (assuming it survives), even if there is no evolution. Many alpine species thrive when cultivated at sea level, suggesting that physical conditions are not the direct cause of their present alpine habitat. High mountains are not necessary for "alpine" types of communities, as these often survive at low elevation, for example, in areas with extreme soil or climate. Finally, alpine habitats, as with oceanic islands and ecological "islands" in general, can host endemics that are much older than the habitats themselves, as shown by the minimum clade ages calculated in molecular clock studies.

The cushion plant *Hectorella* (Portulacaceae) is a good example of an old group endemic to young mountains. It is a monotypic genus and is found from Fiordland to Central Otago and Arthur's Pass. It is restricted to the alpine zone and usually grows in high-alpine sites, where it is one of the most characteristic plants of fell-field and cushion vegetation (Mark and Adams, 1973). Its sister group is *Lyallia* of the Kerguelen Islands in the southern Indian Ocean; these are situated on a large block of submerged continental crust (the second largest in the world, after Zealandia). A clock study found that the two genera diverged in the Miocene or Oligocene (18.6 ± 7.2 Myr using *trn*K/*mat*K, or 25.6 ± 4.3 Myr using *rbc*L) (Wagstaff and Hennion, 2007). These are minimum, fossil-calibrated dates, and the genera could be much older, but the dates confirm that *Hectorella* evolved long before the alpine habitats that it is now restricted to.

The large tussock grass *Chionochloa* occurs in New Zealand (most species), Lord Howe Island (one species) and Mount Kosciuszko (one species). It dominates large areas of grassland, especially in the alpine zone, but has a diverse, global sister group and is likely to be much older than the mountains. Fossil-calibrated clock calculations estimated that it diverged at 19 Ma (Pirie et al., 2009). This minimum date, along with the existence of many open-habitat endemics, suggests that New Zealand was never completely forested.

Wardle (1968: 120) wrote that "Cool climates and plants adapted to them are usually considered to be of very late Tertiary or Quaternary origin in New Zealand, but there is evidence of much greater antiquity in a number of genera and subgeneric groups." Wardle suggested that before the Pliocene uplift of high mountains, these taxa survived on wet, infertile soils. In a similar way, taxa that now inhabit alpine scree slopes, fell-fields, and other well-drained sites would have occupied lowland gravels and rocky sites. In Kerguelen, *Lyallia* occupies fell-field and moraine from the seashore up to 300 m (Hennion and Walton, 1997).

During the Kaikoura orogeny, uplift began at 7–8 Ma in the south and propagated northward and eastward, with rapid sedimentation in the West Coast region beginning at ~5 Ma (Chapter 3). Substantial surface relief formed only after the overlying soft rock was removed and erosion-resistant greywacke was exposed (Heenan and McGlone, 2013). Heenan and McGlone suggested that open, nonforested habitats such as rock bluffs, tussock grasslands, and riverbeds would have been available from about 4.0 to 3.0 Ma. Alpine habitats began to form at about 1.9 Ma and became a permanent feature of the Southern Alps from about 0.95 Ma; Heenan and McGlone (2013) inferred that "Specialist alpine plants confined to alpine habitats can have evolved only within this period." Nevertheless, this assumes that alpine plants need alpine habitat. New Zealand has many endemic groups found not only in open habitat at lower elevations but also in the alpine zone, and these groups are likely to be much older than their present habitats. Likewise, groups restricted to the alpine zone, such as the cushion plant *Hectorella*, are probably millions of years older than their present habitat.

The highest mountains in the North Island are volcanoes, and as mountains are even younger than the main New Zealand ranges (probably less than 200,000 years old; Heenan and McGlone, 2013). Despite this, they still host distinctive endemics, including land snails (see Chapter 5). These and similar "volcano weeds" thrive in areas of volcanism throughout the Pacific region and are probably much older than the individual volcanoes they currently occupy (Heads, 2003). Groups can even survive as endemics on single volcanoes, persisting *in situ* as a dynamic metapopulation while the volcanic edifice is built up by repeated eruptions.

A final observation on areas of endemism supports alternative explanations for the alpine biota: in many cases, the same geographic center of endemism, such as Charleston/Paparoa Range, or Cromwell/Dunstan Muntains, includes both *lowland* local endemics and *alpine* local endemics. Again, this indicates that the area of endemism and the endemics themselves have resulted from events that predate the Kaikoura uplift.

Adaptation and Preadaptation in Alpine Groups

In dispersal theory, adaptation to a group's new, alpine environment is a key step in its establishment and evolution. In contrast, if populations become alpine by passive uplift, *pre*adaptation may determine which groups survive. A preadaptation is a feature that did not evolve as an adaptation, but turned out to be adaptive and enhance survival in the organism's new environment.

Possible cases of preadaptation can be seen in groups such as the cicada *Maoricicada*. This well-studied genus has been mentioned already in relation to its New Zealand–New Caledonia clade, its absence from Fiordland, and its phylogenetic breaks at the Moonlight and Waihemo tectonic zones (Chapters 6 and 7).

Maoricicada occurs in the North and South Islands, and from sea level to the high-alpine zone, where it occupies boulder and herb fields adjacent to the summer snow line. This alpine habitat is a very unusual one for cicadas (Buckley and Simon, 2007), suggesting that the presence of *Maoricicada* there was caused by rapid uplift in the Kaikoura orogeny. Methods for calibrating clocks using mountain age were abandoned by Buckley and Simon (2007); as discussed earlier, these authors concluded that the insects were older than the mountains. They suggested that "the ancestral *Maoricicada*" was already "preadapted to the alpine environment because it existed before the origin of high mountainous habitats, dwelt at mid-to-low altitudes, and yet possessed the classic alpine insect adaptations of heavy pubescence and dark coloration." Preadaptation also explains most aspects of alpine plant morphology, which are also found in members of the lowland flora. These features could date back to earlier phases of mountain building, but they are not necessarily adaptations for either mountains or lowland.

The scree weta *Deinacrida connectens* (Figure 7.11) is another well-studied alpine insect. It inhabits montane to alpine scree slopes at elevations above 770 m, often around 1500 m (Gibbs, 2001). (Published records from "3600 m" are probably an error for 3600 ft; G. Gibbs, pers. comm., June 17, 2011.) One molecular study on *D. connectens* suggested that "The absence of mountains until recent geological times means that few, if any, *alpine-adapted* taxa existed in New Zealand prior to the late Pliocene/

Pleistocene" (Trewick et al., 2000: 658; italics added). It has also been suggested that "adaptation to the alpine zone [in *D. connectens*] presumably explains the high degree of phylogeographic structuring" (Trewick and Wallis, 2001).

Yet these inferences about age and adaptation overlook possible preadaptation, as seen in *Maoricicada*. The alpine wetas in the genus *Deinacrida* survive being frozen solid because of their body-fluid chemistry. This might be regarded as an adaptation, developed during or after the invasion of the new mountains as a result of the changed living conditions. Nevertheless, Gibbs (2006) stressed that the same body-fluid chemistry present in alpine species also occurs in lowland weta species. Its occurrence in the alpine groups enabled them to survive the uplift where other insects perished, but the biochemistry is not an adaptation to the new environment; instead, it is a preadaptation.

In the beetle genus *Syrphetodes* (Ulodidae), molecular clock dates for the nodes were calculated, and all species divergences were found to lie in the Miocene (i.e., were older than 5.3 Ma) (Leschen and Buckley, 2015). Thus, all six alpine species predate the development of their current alpine habitats in the South Island (1.9 Ma; Heenan and McGlone, 2013). Again, the species that later became alpine had already evolved morphological and physiological features that would enable survival in alpine conditions *before* the species and their communities were uplifted in the Kaikoura orogeny.

Tectonic Uplift of Communities and the Origin of the Alpine Biota

The standard theory for the origin of alpine biota proposes that mountains are uplifted and then they are colonized. Fleming (1982: 302) suggested that "the invasion of the subalpine zone by forest organisms" is well illustrated by the New Zealand wrens *Xenicus*, the parrot *Nestor*, and many plants. He wrote: "The newly established alpine and subalpine zones on the rising mountains gave opportunity for the development of an alpine biota to exploit its resources, through colonisation by plants and animals from the forest below, from lowland rock habitats, or from subpolar environments overseas." In this dispersal model of the New Zealand alpine biota, uplift has created "open niches" (Winkworth et al., 2005) that taxa colonize from a lowland center of origin. Another study adopting this model wrote:

> Insect species that *colonise these [alpine] environments* contend with a variety of challenges,...
> Alpine taxa in some insect clades form monophyletic groups to the exclusion of lowland species
> (i.e. the alpine *Syrphetodes* clade [Coleoptera: Ulodidae]); however, in other insect lineages that
> include multiple alpine species, these alpine taxa do not form monophyletic groups... These phy-
> logenetic patterns raise the possibility of multiple, independent *colonisations of the alpine zone*.
> (Buckley et al., 2015: 6–7; italics added)

This is a form of "vacuum biogeography" (Platnick and Nelson, 1978), in which the areas now occupied by mountains are supposed to have been bereft of organisms during uplift. In fact, as the mountains rise, the habitats are never empty, and the dispersal model overlooks the fact that plants and animals always *already* occupy an area before it is uplifted to alpine elevations. The dispersal model fails to consider the direct effects of uplift on communities, and in New Zealand uplift has been rapid; communities have been lifted from sea level to the alpine zone in less than one million years.

Tolmachev (1970) described the process of formation of highland floras, in which "the height rises with its vegetational cover, namely with its flora, which will ultimately become orophytic...." Tolmachev emphasized the significance of the "Floral and vegetational character of a certain area on the Earth's crust before its rise to the ultimate altitude." He wrote: "The importance of this moment is frequently not fully appreciated in investigations of the origin of orophytic [mountain] floras, *although their nature (especially that of the more recent formations) will largely depend on it*" [italics added].

It is now often accepted that mountain ranges such as the Andes have risen together with the populations that were already in the area before orogeny, and so the montane flora and fauna have evolved by passive uplift (Croizat, 1976; Craw et al., 1999; Ribas et al., 2007; Thomas et al., 2008). The process has also been discussed with reference to Malesia (Heads, 2003: 346), and Toussaint et al. (2014) accepted that New Guinea diving beetles (*Exocelina*: Dytiscidae) might have "diversified and undergone passive uplift to the high altitudes."

In New Zealand, distributions of plants and animals suggest that the Mesozoic Rangitata orogeny had profound effects on biogeography. In contrast, the distributions indicate that the biological effects of the Neogene Kaikoura orogeny were restricted to extinction, and the uplift and stranding of communities.

The idea that plants have been stranded inland in New Zealand following geological changes was proposed by Dieffenbach (1843) (see Chapter 5), and also by Kirk (1871), who explained the presence of maritime plants in the Waikato district, far beyond the range of the tide, in this way. Kirk wrote:

> It is readily admitted that littoral plants may occasionally be found in inland situations from accidental causes, but in the present case the number of species, and the wide area over which they collectively extend, afford forcible proof that the cause of their growth must be found in the district having been formerly a shallow estuary, probably connected with the Frith of Thames.

Kirk (1872) interpreted the littoral plants in the North Island lake district in the same way. He concluded that all these cases demonstrate "the importance of geological change as an agent in the distribution of vegetable life, a fact which has been almost lost sight of by phyto-geographical students."

As mentioned earlier (Chapter 8), Cockayne (1899, 1901, 1906) also cited New Zealand examples of coastal plants being stranded inland, and he recognized that the inland distribution of coastal groups on limestone is related to the origin of the rocks. Coastal plants have invaded inland areas during subsidence and marine transgressions, and as soon as the limestone has been exposed by being lifted out of the sea it has been colonized by shore plants.

Much of New Zealand has been inundated by the sea at one time or another, and stranding of populations inland during cycles of subsidence and uplift, and variations in global sea level, has probably occurred in many areas. For example, bottlenose dolphins (*Tursiops* sp.) inhabit the freshwater Lake McKerrow in Fiordland (Currey et al., 2007), and these are probably relictual from when the lake was a saltwater fiord. In the same way, the plant pingao (*Desmoschoenus spiralis*: Cyperaceae) is usually found in maritime sites, but it also occurs 17 km inland at head of Lake McKerrow (P. Johnson, pers. comm., 2014). These are obvious examples, but other cases are probably more subtle.

During the late Cenozoic Kaikoura orogeny, many taxa would have been extirpated during the uplift, at least locally, while others with morphological and physiological preadaptations have survived. Groups that have survived uplift include many shrubs and subshrubs, especially mat plants and cushion plants (Wagstaff and Wege, 2002). Plants with this sort of architecture, such as *Raoulia* species (Asteraceae), grow in the lowlands, on beaches, and in riverbed shingle and are also common in the gravels and rocky fell-fields of the high-alpine zone (Mark and Adams, 1973; Ward, 1993).

In a similar case from Japan, a seashore-derived land snail fauna occurs at high-elevation limestone sites in the Ogasawara (Bonin) Islands, and Wada and Chiba (2011) regarded the community as a "seashore in the mountain." The authors suggested that the species have migrated from the coastal to the inland habitat, but there is no real need for this. The fauna could have become montane by simple uplift, in the same way that the limestone and the old seashore itself "migrated" into the mountains. Earlier, the limestone would have been colonized by plants and animals as soon as it was raised above sea level.

Most "alpine" groups retain close phylogenetic connections with maritime allies. The only alpine angiosperm in New Zealand without close relatives on the coast there is *Hectorella* (Montiaceae), but as mentioned, its sister group (*Lyallia*) grows in fell-field and moraine from the seashore to 300 m on Kerguelen Island (Hennion and Walton, 1997). A similar ecophyletic series ranging from the coast to the alpine occurs in *Kunzea* (Myrtaceae); this genus grows in mangrove (*K. linearis* populations at Whangaroa Harbour), at inland sites (several species), and up to the alpine zone, with herbs and grasses (*K. serotina* at 2000 m) (de Lange, 2014).

Many groups that are typical of high-elevation sites have occasional populations in the lowlands. For example, the upland moa *Megalapteryx* is known from many subalpine sites, but also has lowland records at Punakaiki and in eastern Otago (Worthy and Holdaway, 2002; Wood, 2009). In the New Zealand wrens, *Xenicus gilviventris* is well known as an alpine species, yet it reaches the coast west of the divide, "where boulder-strewn rivers provide suitable habitat..." (Worthy and Holdaway, 2002: 425). In addition, it was once seen "on driftwood and rocks on a beach at Okarito" (Higgins et al., 2001: 82) and also "appears to have formerly lived in lowland forest" (Worthy and Holdaway, 2002: 425).

The kea, the parrot *Nestor notabilis*, is a conspicuous feature of the alpine zone in the Southern Alps (in the North Island, it is represented by Quaternary fossils). It is the only known alpine parrot. Nevertheless, it is often found below tree line, occasionally to below 100 m (Holdaway and Worthy, 1993) and sometimes on coastal flats and fiord shores (Edgar, 1972). Keas are omnivorous, even eating items such as leather boots and rubber seals of car windscreens, and this generalist, opportunistic feeding would have contributed to their ability to survive uplift in the Kaikoura orogeny.

Birds of Paradox: Seabirds Nesting in Mountains

Birds exhibit some of the most interesting ecological and evolutionary links between the sea and the mountains in New Zealand. *Puffinus huttoni* (Procellariidae) is a seabird (a shearwater, not a puffin), but it breeds only at two sites high in the Kaikoura ranges of the South Island. The species belongs to a global clade, *Puffinus* s.str., that has the following phylogeny (from Austin et al., 2004; breeding localities only are cited; in winter, many species migrate long distances to warmer areas):

East Pacific. *P. nativitatis*: Hawaiian Is., Line Is. (Christmas I.), Tuamotu Is.; *P. subalaris*: Galapagos Is.

New Zealand. Three species with allopatric breeding ranges (Holdaway and Worthy, 1994):

P. huttoni: **Seaward and Inland Kaikoura Ranges**, subalpine and alpine sites (1200–1800 m), migrates in winter to all around the Australian coast.

P. gavia: **Three Kings Is. to Marlborough Sounds**, migrates to SE Australia.

P. spelaeus: Known from fossils in **NW South I. (Takaka to Punakaiki)**, extinct in historical times.

Widespread globally in warmer areas including New Zealand, but not the east Pacific). *P. lherminieri/P. assimilis* complex (4 species and 13 subspecies; Austin et al., 2004).

The first break isolated the central/eastern Pacific group from the rest, and the second isolated the New Zealand *P. huttoni-gavia-spelaeus* group from the "global" group. This was followed by differentiation of species in all three clades.

From the Miocene onward, coastal populations at Kaikoura have been "caught" in the uplift that occurred there. The birds' inherent philopatry, together with the imperceptible rate of rise, means that the populations breeding sites would have remained *in situ* during uplift. The uplift rate, ~1 cm/year, would have raised the birds' breeding sites from the coast into the alpine zone in just 120,000 years (or somewhat longer, as this overlooks the effects of erosion).

The boundary between landbirds and seabirds is blurred by groups such as typical shorebirds, a group of families, Charadrii, in Charadriiformes. These are mainly found on seashores, but there are also inland representatives. These include the black-stilt (*Himantopus novaezelandiae*: Recurvirostridae), which breeds in inland Canterbury in shingle along the braided riverbeds (it was more widespread in historical times). The wrybill (*Charadrius frontalis*: Charadriidae) breeds in inland Canterbury and also near Glenorchy in northwest Otago (Robertson et al., 2007). Another inland shorebird is *Ibidorhyncha* (comprising the family Ibidorhynchidae), which inhabits shingle riverbanks in the Himalayas and the high plateaus of central Asia. All three cases can be explained as coastal groups that have been stranded inland by marine incursion followed by uplift and marine regression.

The best-known seabirds in New Zealand belong to another group of Charadriiformes, the gulls (Laridae). There are three species in the country. The first is the large *Larus dominicanus*, the black-backed or kelp gull. It occurs in South America (north to northern Peru and Rio), southern Africa, southern Australasia, the subantarctic islands, and the Antarctic Peninsula. It breeds on islands, coasts, pastures, lava fields, islands in rivers, and in urban areas (Burger and Gochfeld, 1996). In New Zealand, the species breeds on coastal rocks and beaches, as well as inland on shingle riverbeds. Nevertheless, it also occurs at higher altitudes. It breeds at 5000'/1500 m on mountainside near Lake Wakatipu, and there is a large colony at 3000'/900 m in the Paparoa Range and at a similar height on the Rock and Pillar Range (Oliver, 1955). Breeding populations are also recorded high up in the Dunstan Mountains and the Pisa Range of Central Otago.

These distributions can be explained by the former stranding of populations inland, following marine retreat. Later, the relictual inland populations would have been caught in Neogene uplift. A similar history would also account for *Larus modestus* breeding in interior montane deserts of northern Chile that are 100 km from the coast (Burger and Gochfeld, 1996: 576, 579).

The two other gulls in New Zealand are much smaller than *Larus dominicanus* and have gray backs. They form a clade with an Australian species (Pons et al., 2005). The phylogeny is as follows:

Chroicocephalus bulleri (black-billed gull): New Zealand, "breeds almost exclusively **inland**, mainly on rocky islands in fast-flowing braided rivers and along lake shores, rarely attempting to breed at coastal sites" (Burger and Gochfeld, 1996: 615; cf. Robertson et al., 2007).

C. scopulinus (red-billed gull): New Zealand, breeds mainly at **coastal** sites, but also inland at Rotorua (where the coastal tree *Metrosideros excelsa* also has inland records; see Chapter 5).

C. novaehollandiae: Australia and New Caledonia. Breeding at **coastal** sites in southwestern Australia, eastern Australia, and New Caledonia, and at **inland** sites in the southeastern quarter of Australia (including Tasmania) (Higgins and Davies, 1996).

The phylogeny implies that the *inland* New Zealand member split from the *coastal* New Zealand–Australia group before the latter group was itself rifted apart, and the latter event can be attributed to the opening of the Tasman Sea.

Oystercatchers (Haematopodidae) are mainly coastal, but unrelated species breed inland in four regions:

Eurasia (where *Haematopus ostralegus longipes* breeds only in inland sites).

Magellania (*H. leucopodus*).

Australia ("a few" *H. longirostris*) (Hockey, 1996).

New Zealand, where the South Island endemic *H. finschi* has most breeding records inland, along riverbeds (Oliver, 1955; Robertson et al., 2007). It also breeds at higher altitudes in the Dunstan Mountains and Pisa Range of Central Otago (Peat and Patrick, 1999).

Another Seabird Paradox: Mobile Birds with Localized, Philopatric Populations

The idea that such mobile creatures as seabirds would maintain precise philopatry in the face of great topographic changes is counterintuitive. Nevertheless, it accounts for many phenomena that are otherwise difficult to explain. For example, Wiley et al. (2012) studied Hawaiian petrels (*Pterodroma phaeopygia sandwichensis*: Procellariidae) nesting on the islands of Hawaii and Kauai. Genetic analyses indicated strong differentiation between the two colonies, while isotope studies showed they were feeding in different areas. Kauai birds provision chicks with prey derived from near, or north of, the Hawaiian Islands, while Hawaii birds obtain prey from the equatorial Pacific. Wiley et al. (2012) wrote that *P. p. sandwichensis* exemplifies the "seabird paradox," as individuals can travel more than 10,000 km on a single foraging trip, but on Hawaii and Kauai differentiated populations breed no more than 500 km apart.

The "seabird paradox" was a phrase used earlier by Milot et al. (2008), when they observed that the same birds that travel great distances to forage are reluctant to disperse even to nearby sites to breed. This apparent inconsistency is evident in the many sequencing studies on marine groups that have found unsuspected, high levels of geographic structure (Chapter 1). Within the paradigm of chance dispersal, the intricate geographic allopatry is seen as a paradox or anomaly. The "seabird paradox" is another reflection of the underlying problem—the inability of the theory to explain the facts. Some biologists do not accept this and instead see the great geographic structure of the molecular clades as proof of the great powers of chance (Linkem et al., 2013; cf. Heads, 2014: 380). Despite this, most authors have admitted there is a problem, and marine biology and microbiology are both undergoing paradigm shifts brought on by the molecular data.

To summarize evolution in the seabirds: current ecological factors and niche modeling do not account for many aspects of clades, such as the breeding site of *Puffinus huttoni* high in the Kaikoura ranges, and its basal position in a diverse, worldwide group. Instead, these simple patterns can be explained by tectonic history.

Kaikoura Orogeny Uplift and Associated Deposition of Gravels

Uplift of the Kaikoura orogeny mountains produced new alpine habitat, but it also had other ecological effects. One of the most significant was the production of great volumes of gravel that were deposited as alluvial sheets on lowland areas. These created the Canterbury plains. Excavations in a gravel pit in Christchurch uncovered stumps of 30 trees up to 5′/1.5 m in diameter, mainly of *Podocarpus totara*; these indicate the submergence of an ancient forest under 12′/3.7 m of gravels (Speight, 1917). The alluvial deposits must have had a profound influence on the prior vegetation and would have led to much extinction, but other groups would have benefited.

One plant that survived is the small shrub *Olearia adenocarpa* (Asteraceae) (mapped in Heads, 1998b, as *O.* "Harewood population"). It inhabits recent gravels and sands laid down by the Waimakariri and Rakaia rivers (Heenan and Molloy, 2004). Deposition of these gravels began with the Kaikoura uplift, but the species endemic here are likely to be much older than the uplift. The present habitat of *O. adenocarpa* indicates that it would have occupied sites around the coast, on fresh gravel and sand deposits, long before the recent strata that it now inhabits existed. Cockayne (1899, 1906) interpreted the limestone plants of inland Canterbury as old, coastal relics (see Chapter 8), and *O. adenocarpa* can be interpreted in the same way. The relatives of *O. adenocarpa* in Central Otago appear to have had a similar evolutionary and ecological history on and around the alluvial and marine sediments of that region (Heads, 1998b).

Pleistocene Refugium Theory in New Zealand: Glaciation and Centers of Origin, or Metapopulations in Multiple Microrefugia?

Pleistocene events and refugium theory have been referred to several times in the discussion so far. The term "refugium" is often used in biology, but is misleading in a subtle way, as it implies that groups present there have fled or dispersed there. In most cases, there is no reason to assume that they are anything but local populations that would have been there anyway; they did not come from anywhere else. The language and metaphors of biogeography often reflect unjustified assumptions about dispersal, and refugia are best interpreted simply as "centers of survival."

Pleistocene refugium theory is important because the Modern Synthesis attributed most aspects of distribution in the modern biota to Pleistocene cooling (Matthew, 1915; Mayr and Diamond, 2001). Dussex et al. (2014) suggested: "It is now well established that Pleistocene climate cycling had an important influence on the distribution and divergence of several taxa in New Zealand...." Nevertheless, Gibbs (2006: 86) was more sceptical and concluded: "perhaps there is a tendency to overstate the importance of the Pleistocene in New Zealand biogeography."

Advocates of the traditional model suggest that patterns of endemism and disjunction in the New Zealand flora "have arisen mainly through Pleistocene extinctions, speciation and dispersal..." (McGlone et al., 2001; cf. Fleming, 1979). Nevertheless, the minimum clade ages produced in molecular dating studies often contradict this. For example, Futuyma (2010) wrote that "Klicka and Zink (1997) called the supposed prevalence of Pleistocene speciation a 'failed paradigm' for birds, setting off a debate that seems to have settled on the conclusion that speciation rates were not elevated during the Pleistocene...."

Dramatic evidence of Pleistocene glaciation can be seen in Fiordland as the ice has scoured valleys and fiords in the hard rock, and these have retained their U-shaped cross sections after the ice receded. It is not surprising that earlier authors followed Matthew (1915) in seeing the glaciations as determining all aspects of the modern biota. The usual paradigm for explaining biogeographic patterns within New Zealand has been the Pleistocene refugium theory (Fleming, 1979).

In a nineteenth-century debate cited earlier, geologists proposed that ice sheets had covered the South Island and wiped out terrestrial life (as in the modern "total submergence" theory for the Oligocene). Nevertheless, biologists replied that complete ice cover in the region was incompatible with the numerous local endemics (Cockburn-Hood, 1877; Thomson, 1909; Cockayne, 1921). Fiordland, for example, has many distinctive endemics, although Pleistocene glaciers extended down to the sea. Geologists no longer propose a continuous ice sheet over the entire South Island, but the idea of almost continuous ice cover over much of the central South Island and Fiordland has persisted.

Numbers of endemic species are higher in the north of the South Island (Nelson) and in the south (Otago) than in central areas (Westland and Canterbury; Heads, 1997). The traditional theory proposed that organisms were extirpated from the central areas by the Pleistocene ice, while a small number of refugia existed outside the central areas. Taxa were isolated in the refugia by the glaciations and speciated there; following the last glacial maximum, they have recolonized the central South Island (Fleming, 1979).

Several molecular studies have followed earlier workers in accepting that areas of diversity represent Pleistocene refugia. In *Metrosideros* s.str. (Myrtaceae), Gardner et al. (2004) inferred that the areas of high haplotype diversity (northwest Nelson, Coromandel Peninsula + Barrier Islands, and Northland) were Pleistocene refugia. No formal dating analyses were carried out. Instead, the age was inferred because of McGlone's (1985) proposal that the areas were Pleistocene refugia, and the only reason he had for suggesting this was that they were areas of diversity. Other studies have relied on questionable calibrations. Neiman and Lively (2004) attributed differentiation in a New Zealand freshwater snail, *Potamopyrgus antipodarum*, to Pleistocene glaciations. They calibrated the phylogeny with (1) the "standard" insect rate (2.3%/Myr), (2) rates from another gastropod, *Nucella*, based on the fossil record and differentiation at Panama Isthmus, and (3) rates from endemics on Tahiti, calculated by assuming that these were no older than the current islands. These three rates are all likely to be overestimates, and so when used for clock calibration they will give clade ages that are too young (Heads, 2012a).

Survival of Groups as Metapopulations in Multiple Microrefugia

In contrast with theoretical approaches to biogeography based on Pleistocene refugium theory, comparative studies of distribution patterns in the South Island have indicated a large number of biogeographic nodes and multiple microrefugia (Heads, 1994d). This concept is now being developed in molecular studies. Although the clade ages suggested in molecular work are probably still much too young, some will be valid as minimum ages, and they already indicate that many species and distinct geographic groups are older than the Pleistocene. Because of this, authors have proposed that groups survived through the glaciations, existing as multiple, small populations.

The microrefugia would have included small areas of forest preserved in gullies in grassland, and small areas of alpine vegetation preserved on ice-free nunataks in glaciated areas, for example. Although most of the taxa that are restricted to Pleistocene microrefugia probably did not *speciate* there, molecular techniques have confirmed the location of many microrefugia where groups survived.

One example concerns ferns. Shepherd et al. (2007: 4539) wrote that:

> Fern species might be expected to exhibit little phylogeographical structure because their plentiful production of wind-dispersed spores … may result in high levels of gene flow. Nevertheless, the few previous phylogeographical studies in ferns using chloroplast DNA… have reported genetic differences between populations, as has the sole study of nuclear microsatellites…

Asplenium hookerianum is a fern of the forest understory and is widespread though New Zealand. Shepherd et al. (2007) discovered patterns of variation within it that are consistent with numerous refugia extending south to the southern South Island. Most populations, including those near areas that were glaciated, contained multiple and, in many cases, unique haplotypes. The authors concluded (p. 4544): "It seems probable that the contemporary phylogeography of *A. hookerianum* reflects long-term survival of populations more or less *in situ*, including through the last glacial. This is supported by the multiple populations with endemic haplotypes, including populations in regions not thought to have encompassed major LGM [last glacial maximum] refugia."

Molecular studies in the New Zealand *Ranunculus lyallii* (Ranunculaceae) also indicate geographic clades and localized distribution of some haplotypes, consistent with the survival of populations in multiple refugia (Lehnebach, 2008; cf. Lockhart et al., 2001; Winkworth et al., 2005). In studies on *Pachycladon* (Brassicaceae), Heenan and Mitchell (2003) supported the idea that microrefugia (nunataks) existed even in the Mount Cook area of the central South Island. This is the highest part of New Zealand and is often assumed to have been covered by an ice sheet in the Pleistocene. Other groups that are thought to have survived in microrefugia include moas (Bunce et al., 2009).

The concept of many microrefugia also solves some problems interpreting New Zealand palynological data. These were thought to indicate population expansion of trees, including forest dominants, at the end of the last glacial, but at rates that seemed too rapid to be feasible (Shepherd et al., 2007). Taken together, the biogeographic, molecular, and palynological data suggest that forest expanded from numerous microsites already within a region, rather than as a single front advancing into the wider region from elsewhere. The same explanation accounts for the postglacial "invasion" of oaks in Britain (Heads, 2014: 62).

In the New Zealand skink *Oligosoma infrapunctatum*, earlier authors thought patterns of regional differentiation were caused by Pleistocene glaciations. Instead, molecular studies concluded that the lineages had already diverged before the onset of the glaciations (Greaves et al., 2008). Overall, the species ranges from the northwestern South Island through the North Island (Figure 9.15), and there is evidence for four main clades. These are all allopatric, except in a zone of overlap along the coastal strip below the Paparoa Range, from Westport to Greymouth. In the earlier paradigm, this region might have been interpreted as a refugium or center of origin, but the new ideas on dating suggest it is an area of differentiation in an already widespread ancestor, with the break blurred by subsequent local overlap. The overall pattern through central New Zealand has been preserved by the survival of populations in multiple microrefugia.

During the glacial maximums, the southern South Island was thought to have been covered by ice and periglacial, open vegetation, but molecular studies now indicate that forest habitat persisted. Marshall et al. (2009: 1997) wrote that "there appears to be growing evidence for forest refugia in the southern South Island," and they cited genetic evidence from both New Zealand Onychophora (*Ooperipatellus* and *Peripatoides*) for forest refugia in central and southern South Island (Trewick and Wallis, 2001). Marshall et al. (2009: 1997) also wrote that "the significance of these fragments varies across taxa"; some clades that were restricted to refugia show signs of differentiation there, while others have not differentiated. This difference probably reflects differences in the groups' prior genetic architecture and evolvability (Heads, 2012a).

In the New Zealand cicada *Kikihia* (Cicadidae), Arensburger et al. (2004b) accepted that the majority of speciation events occurred long before the Pleistocene glaciations, and so these are unlikely to have influenced speciation or even distribution: "evidence that glaciers have had a major influence on modern *Kikihia* distributions is thin" (p. 1779). In *Kikihia subalpina*, a species of subalpine forest edges, Marshall et al. (2009) found a main split between North Island and South Island clades, with a break at Cook Strait. Within the South Island, there are three main clades: one in northwest Nelson, one in Marlborough north of the Wairau fault, and one widespread. Marshall et al. (2009) regarded this as evidence that the widespread clade spread from glacial refugia in Nelson and Marlborough, but it could just as easily reflect original allopatry and local range expansion of the widespread form *into* Nelson and Marlborough. If true forest taxa, including onychophorans, were able to survive the ice ages in the south (see last paragraph), subalpine species such as *K. subalpina* would have also been able to survive there. In another cicada, *Maoricicada campbelli*, Hill et al. (2009) concluded that populations had survived the glacial ages in many refugia, even within the central South Island "biotic gap."

All these ideas conflict with the traditional chronology in which the Pleistocene ice ages were crucial for global evolution and biogeography. The molecular evidence suggests that the effects of the ice ages were restricted to local extirpation and range expansion, and that any phylogenetic changes were minor.

Differentiation in the Alpine Scree Weta (*Deinacrida connectens*: Anostostomatidae): Pleistocene or Pre-Pleistocene?

Trewick and Wallis (2001) titled their paper "Bridging the 'beech-gap': New Zealand invertebrate phylogeography implicates Pleistocene glaciations and Pliocene isolation." Nevertheless, they concluded that glaciation was *not* important: "Given the local intensity of glaciations in central South Island, it is striking that so many taxa do *not* show any evidence of lasting impact" (p. 2178; italics added).

In contrast, Chinn and Gemmell (2004) supported the more orthodox idea that the alpine scree weta *Deinacrida connectens* (Figure 7.11) is "a clear example of glacial mediated genetic structure." However, their source for this was Trewick (2001), and in fact Trewick did not state this. Instead, he concluded that the species radiated in response to an earlier event—Pliocene orogeny. Trewick (2001: 291) also observed that "Two lineages [of *D. connectens*] are endemic to the central South Island, an area regarded as species poor due to glacial-extirpation of much of the biota. It appears that *D. connectens* survived across much of the South Island in a mosaic of ecological, rather than one or few, regional refugia."

The notion of survival within glaciated regions through the Pleistocene has even been advanced for Arctic groups (Trewick and Wallis, 2001). Trewick et al. (2000) proposed that "*D. connectens* speciated prior to Pleistocene climate cooling" and the high degree of phylogeographic structuring is "consistent with survival through the Pleistocene in rocky alpine environments in the vicinity of glaciers (arêtes, ridges) and above glacier basins." Likewise, in a general survey, Wallis and Trewick (2009: 18) concluded: "It is clear that most New Zealand taxa harbour phylogeographic structure that must have survived most or all of the Pleistocene glaciations. The New Zealand pattern suggests many refugia...."

Work in other parts of the world has reached similar conclusions. The peninsulas of southern Europe have long been recognized as Pleistocene refugia, but molecular work now also supports the presence of cryptic glacial refugia in *northern* Europe (Ruiz-González et al., 2013; Heads, 2014: 64). In southern Patagonia, *Hordeum* species (Poaceae) survived the Pleistocene *in situ*, without genetic or spatial restriction (Jakob et al., 2009). To summarize: for the many groups that have persisted as metapopulations in multiple microrefugia, the Pleistocene glaciations had little long-term effect on biogeography.

10

Case Studies of New Zealand Plants

This chapter presents case studies of some New Zealand plants, with a focus on clades rather than areas. The distributions are discussed and analyzed with reference to some of the principles and main nodes introduced in previous chapters.

Wahlenbergia (Campanulaceae)

Wahlenbergia is a genus of small herbs and subshrubs with ~260 species and is distributed through the Old World and South America. There is a strong concentration of species in South Africa (~81% of the species) and Australasia (~13%) (Prebble et al., 2012). There are 10 species in New Zealand, and these are widespread in open habitats, especially beaches, rocky sites, and alpine vegetation (Petterson, 1997).

As mentioned already, the fossil record of *Wahlenbergia* is very poor (Chapter 2), but there is no reason to assume that the group itself is young. Although there are no published accounts of *Wahlenbergia* pollen in Australia older than Quaternary, Prebble et al. (2011) wrote, "the absence of evidence cannot be taken as evidence of the absence of *Wahlenbergia* from Australia before this." Despite this, Prebble et al. (2011) relied on the meager fossil record of the family, using a Miocene seed from Europe to calibrate a phylogeny of the group and rule out early vicariance.

Prebble et al. (2011) focused on the Australasian members, but they also examined African and European species and proposed a phylogeny:

    Africa: Basal paraphyletic grade.
        Africa + Europe.
        Africa + Australasia.

The Australasian clade is nested in a group otherwise found in Africa and Europe (mainly in southern Africa), and so theory suggests a single long-distance dispersal event from Africa or Europe to Australasia (Prebble et al., 2011). Instead, a vicariance model infers a sequence of differentiation events in a widespread Indian Ocean group. The sequence began in southern Africa (producing the basal clades), moved east to the Indian Ocean (separating the Australasian and African clades), and, finally, moved to the Tasman Sea (separating Australian and New Zealand clades). This follows the same sequence as the breakup of Gondwana.

The Two Main Clades of *Wahlenbergia* in New Zealand

The "Australasian" clade of *Wahlenbergia* includes two clades that are present in New Zealand: a more widespread, radiate clade of New Zealand (mainly in the north), Australia, and (possibly introduced) southern Africa, and a rhizomatous clade endemic to southern New Zealand. Both groups were recognized in an earlier, morphological study (Petterson, 1997), but their sister groups are still not resolved. The two groups and their New Zealand distributions are as follows (Petterson, 1997; Prebble et al., 2011, 2012):

Mainly northern, radicate clade: North Island, South Island, and Stewart Island, islands off northeastern North Island, Chatham Islands, Kermadec Islands, and Tonga. Species from Lord Howe Island may belong here. Clade also in Australia and possibly South Africa. Plants have a single taproot and shoots resprout from the stock.

Southern, rhizomatous clade: North Island (central and southern parts), South Island, and Stewart Island. Plants produce rhizomes.

The radicate clade includes five species from the North, South, Stewart, and Chatham Islands (*Wahlenbergia vernicosa*, *W. ramosa*, *W. violacea*, *W. akaroa*, *W. rupestris*). Within New Zealand, the clade is mainly northern; in the South Island, it occurs only in the lowlands and there are no inland endemics. *W. vernicosa* occurs in the northern North Island, from Three Kings Islands south to Kawhia Harbour and Mayor Island. It is present on many offshore islands, including Poor Knights, Hen and Chickens, Mokohinau, Great Barrier, Rakitu, and Rangitoto Islands, and it is also on the Chatham Islands (a putative population from the Kermadec Islands belongs to a separate clade [Prebble et al., 2012]). In its morphology, *W. vernicosa* is closest to *W. insulae-howei* of Lord Howe Island.

The rhizomatous clade of New Zealand *Wahlenbergia* comprises five species (*W. albomarginata*, *W. cartilaginea*, *W. pygmaea*, *W. congesta*, *W. matthewsii*), with a center of diversity south of that in the radicate clade. The rhizomatous group is widespread in the South Island and Stewart Island, but extends north only to the central part of the North Island (Huiarau Range). In the South Island, it includes distinctive inland endemics, as well as an endemic, blue-flowered form of *W. matthewsii* on coastal limestone at Flaxbourne–Clarence (cf. Chapter 8). It is also well represented in the alpine zone, unlike the radicate clade. Relationships within the group are not yet resolved, and detailed mapping with further phylogenetic studies should give interesting results.

The only consistently coastal member of the rhizomatous clade is *W. congesta*. It occurs along the coast from northwest Nelson south to the Alpine fault (Poison Bay near Milford Sound; Petterson, 1997: 32), and beyond there it is absent through Fiordland. Nevertheless, it reappears at Waitutu and extends around the coast past Invercargill to Fortrose. A similar Fiordland gap is recorded in other groups, such as *Galaxias postvectis* (Figure 9.9), and can be explained by Alpine fault displacement.

Among the other rhizomatous forms is *W. albomarginata* subspecies *decora*, a high-altitude plant (not sequenced) of Nelson and Otago, with a distribution disjunct along the Alpine fault (Petterson, 1997). Petterson mapped subsp. *decora* and putative sister forms in relation to the Alpine, Awatere, and Clarence faults. Alpine fault displacement would account for the Fiordland–Nelson gap in subsp. *decora*, as well as the Fiordland gap in *W. congesta*.

Alpine fault disjunction provides a useful Miocene calibration point for many groups (Heads and Craw, 2004). Prebble et al. (2011) instead accepted a Cenozoic origin and dispersal history for *Wahlenbergia*, but this was based on the Cenozoic fossil record. For example, in order to date the *Wahlenbergia* clade found in New Zealand, Prebble et al. (2011) used the oldest known fossil pollen of the genus there, although this dates only to the Pliocene. The group itself is probably much older than this, as small, insect-pollinated plants such as *Wahlenbergia* produce minute amounts of pollen that is seldom preserved. Prebble et al. (2011) admitted that the oldest fossil record of the genus in Australia "cannot be taken as evidence of the absence of *Wahlenbergia* from Australia before this," and the same applies to New Zealand.

Although the radicate and rhizomatous clades show significant differences in their distributions, they overlap in the southern North Island and in areas of the South Island below 500 m elevation. Nevertheless, much of the overlap is caused by a single radicate species, *W. violacea*. Petterson (1997) noted that some botanists have regarded this as adventive in New Zealand because it is often found in waste places and as a pioneer after fire and roadworks. She noted that the same is true for *Wahlenbergia gracilis* in New Caledonia, *W. marginata* in Japan, and *W. quadrifida* and *W. communis* in Australia. Yet *W. violacea* was collected during Cook's first voyage to the Pacific in 1769, and Petterson regarded it as indigenous. Its presence in *southern* New Zealand, though, is likely to be the result of range expansion as it overlaps there with members of the rhizomatous clade. Many groups of plants and animals in the Pacific, including endemics, are more or less "weedy." Rather than indicating recent immigration, a degree of weediness would have been a prerequisite for surviving the rifting, volcanism, marine incursions, orogeny, and other great disturbances of the region.

Onagraceae in New Zealand

This worldwide family is represented in New Zealand by *Fuchsia* (mainly woody plants) and *Epilobium* (herbs and subshrubs).

Fuchsia (Onagraceae) and Its Global Affinities

Fuchsia excorticata is a widespread tree in New Zealand and is codominant in some forests, often by streams. The evolutionary history of this species is traced here not by looking for a center of origin *within* its range—the traditional approach—but by seeing the taxon in its phylogenetic context and comparing its distribution with those of its relatives.

There are two other New Zealand species of *Fuchsia*, and the genus also occurs in Tahiti, South America, Central America, and Mexico. Fossil material is recorded in Australia. *Fuchsia* is in the southern hemisphere and the tropics, while its sister group is a widespread, northern hemisphere genus, *Circaea*. The pair has the following position in the Onagraceae family (phylogeny from Levin et al., 2004; distributions from Wagner et al., 2007):

> Subcosmopolitan, mainly tropical (*Ludwigia*).
>> Mexico to Costa Rica (*Hauya*).
>>> **Northern hemisphere (*Circaea*) + trans-South Pacific (*Fuchsia*).**
>>>> Mexico to Panama (*Lopezia* and *Megacorax*).
>>>>> Mexico (*Gongylocarpus*).
>>>>>> World (*Epilobium* and *Chamerion*).
>>>>>>> Mexico: Baja California (*Xylonagra*).
>>>>>>> Americas: North, South, and Central America, including Pacific and Atlantic coasts, also Greater Antilles (all other species of Onagraceae: *Clarkia*, *Oenothera*, etc.).

This phylogeny can be summarized in very general terms: world; Mexico and Central America; world; Mexico and Central America; Mexico; world; Mexico; Americas. The main questions concern the origin of the widespread groups and the series of divergence events around Mexico and Central America.

The overlap among the globally widespread clades has resulted from large-scale range expansions that have obscured the original allopatry. However, the complete absence of the *Circaea* + *Fuchsia* clade from Africa and Madagascar is a probable trace of this.

Four of the groups in the Onagraceae repeat a distinctive, standard pattern in which endemics from Mexico (above all, western Mexico) and Central America have worldwide sister groups (Heads, 2009c). Dispersal theory accounts for the pattern by inferring centers of origin in Mexico and Central America. Instead, in vicariance theory, the region represents a zone of episodic *differentiation* in ancestors that were already widespread through the Americas or around the globe. Episodic tectonics were a feature of the central boundary in Mexico, which separates the western Guerrero terrane from the eastern craton. Alternating phases of compression/accretion and transtension/rifting occurred along this belt through the Jurassic and Cretaceous (Centeno-García et al., 2008: Fig. 3).

Subsequent studies have made a slight modification to the phylogeny of Onagraceae given earlier (from Levin et al., 2004); they have proposed a new monophyletic clade in which *Hauya* of Mexico and Central America is sister to *Circaea* + *Fuchsia* (Ford and Gottlieb, 2007; cf. Berry et al., 2004). If Mexico is a zone of repeated fracturing in widespread ancestors, rather than a center of origin, this would explain the break there between *Hauya* and its widespread sister group (whether this is the rest of Onagraceae except *Ludwigia* [as in Levin et al., 2004], or just *Fuchsia* + *Circaea*) (Figure 10.1).

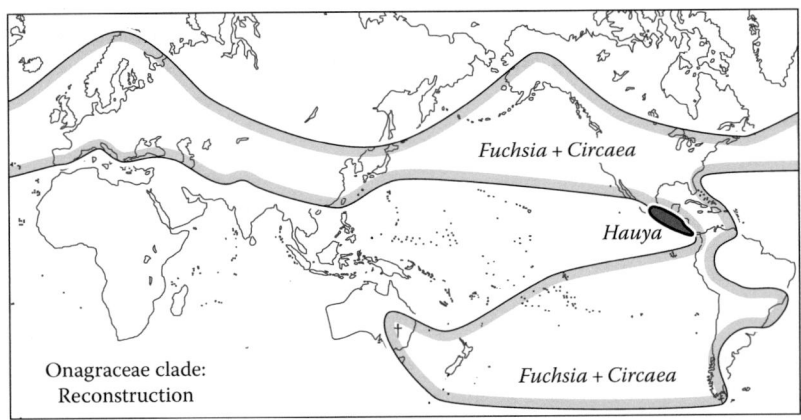

FIGURE 10.1   Reconstructed distribution of the *Fuchsia + Circaea* ancestor and its possible sister group, *Hauya* (Onagraceae).

A series of repeated fracturing in Mexico would also explain the next break there, between the southern *Fuchsia* (trees, shrubs, and lianes) and the northern *Circaea* (herbs) (Figure 10.2) (fossils of *Circaea* and *Fuchsia* are known, but only from within the extant range of their respective genus; Xie et al., 2009). The geographic boundary between *Fuchsia* and *Circaea* does not represent a simple, ecological boundary between a tropical group in southern Mexico and a temperate group in northern Mexico, as the southern component, *Fuchsia*, extends to Tierra del Fuego. Thus, it is more likely that the break had a tectonic origin, and it coincides with the Mojave-Sonora megashear. This belt of strike-slip deformation crosses northern Mexico from the Pacific to the Atlantic coasts and underwent ~800 km of left-lateral displacement in the Late Jurassic (Campos-Enriquez et al., 2011). Its possible effects on the regional biogeography have been discussed by Souza et al. (2006). The megashear is part of the Basin and Range province of southwestern North America. In this area, extension has produced distinctive topography, with northwest trending mountain ranges and valleys bounded by normal faults, and metamorphic core complexes.

Authors studying the northern *Circaea* and the southern *Fuchsia* have tended to overlook the simple allopatry of the two genera and have instead accepted a center of origin (Berry et al., 2004; Xie et al., 2009). Berry et al. (2004) wrote:

> It is unclear from our results where *Fuchsia* most likely originated, but [in light of the phylogeny] it seems just as likely that the genus arose in northern South America or even in southern North America or Mesoamerica, rather than in southern South America, as previously hypothesized.

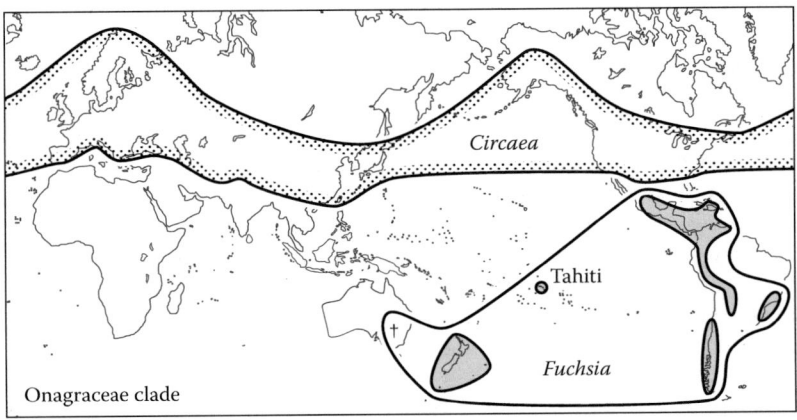

FIGURE 10.2   Distribution of *Fuchsia* (dagger symbol indicates fossil record) and *Circaea* (Onagraceae). (Data from Ford, V.S. and Gottlieb, L.D., *Syst. Bot.*, 32, 348, 2007; Wagner, W.L. et al., *Syst. Bot. Monogr.*, 83, 1, 2007.)

Wagner et al. (2007) concluded, "the strong support for a sister relationship between *Fuchsia* and *Circaea*, and the early branching of the Central American *Hauya*, suggests the possibility that *Fuchsia* might have diverged in the north rather than in the south, as previously hypothesized."

Instead of having a restricted center of origin in the north or the south, it is suggested here that *Fuchsia* differentiated *in situ*, as a central and South Pacific vicariant of the northern *Circaea*, and from an ancestor that was already widespread through many parts of the world. In *Circaea*, Xie et al. (2009) found a western North American clade to be basal, and so they inferred a western North American center of origin for this genus. Nevertheless, there is no actual evidence for range expansion in *Circaea* itself or in *Fuchsia* as they are allopatric sisters. In contrast, at an earlier stage in phylogeny, the common ancestor of *Circaea* + *Fuchsia* probably did expand its range along the Pacific seaboard of Mexico and Central America, as it overlaps there with *Hauya*.

Differentiation within *Fuchsia*

Following the origin of *Fuchsia* and *Circaea* by simple vicariance, *Fuchsia* itself has differentiated into three primary clades. The first two are both in America ("1" and "2," stippled and dark gray in Figure 10.3) while the third ("3" + "4" + "5," light gray in Figure 10.3) is a trans-Pacific group in America, Tahiti, and New Zealand. There is also fossil material of the genus in Australia, but its affinities within the group are obscure. As usual, the basal paraphyletic grade in America need not indicate a center of origin. The main break in the trans-Pacific clade lies between west and east Pacific, and coincides with the spreading center at the East Pacific Rise. This has produced the Pacific plate to the west and the Farallon Plate (with its modern derivatives) to the east.

The opening of the Pacific basin was more complex than the rifting in the Atlantic, as it included the formation and growth of a whole new plate, the Pacific plate. Ever since this formed in the Jurassic, the central Pacific has been a major zone of volcanism. The East Pacific Rise is the main spreading center in the central Pacific, and the seafloor spreading there has probably caused vicariance in groups on the oceanic plates. As with many extant Pacific groups, the earlier groups would have survived as metapopulations on systems of individually ephemeral islands, including high islands now represented by atolls and seamounts.

Godley and Berry (1995) proposed that *Fuchsia* was present in the central Pacific either as the result of recent dispersal or because "it was present on older islands that have since subsided." In a fossil-calibrated, molecular clock study, Berry et al. (2004) calculated that the endemic Tahitian species evolved at 8 Ma,

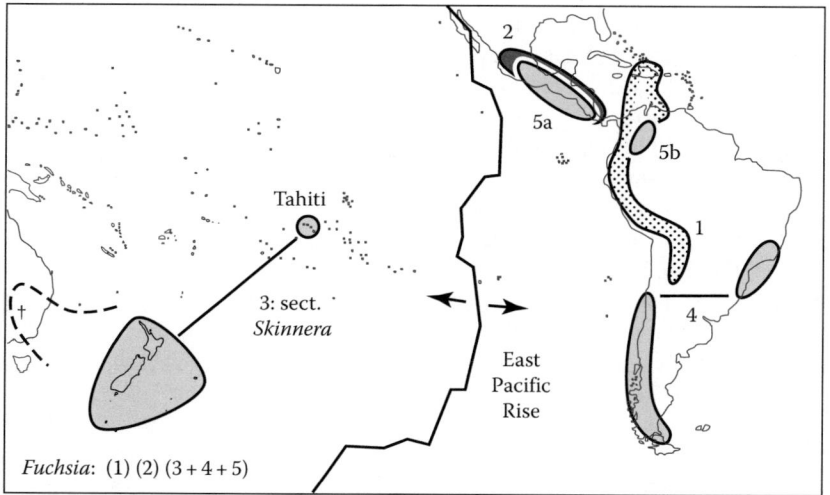

FIGURE 10.3 Distribution and phylogeny of *Fuchsia* (Onagraceae). The affinities of the fossil representative from Australia (dagger symbol) are not resolved. (Data from Godley, E.J. and Berry, P.E., *Anns. Missouri Bot. Gard.*, 82, 473, 1995; Berry, P.E., *Amer. J. Bot.*, 91, 601, 2004.)

long before the island of Tahiti itself formed (at 1.4 Ma), and the authors again discussed the significance of older islands that are now submerged.

Differentiation between east and west Pacific members of some groups probably dates to the final rifting of Australasia and South America from Antarctica, in the Late Cretaceous. But for groups in the central Pacific, such as *Fuchsia*, the evidence suggests that rifting in the Early Cretaceous was more important. The other main breaks in the trans-Pacific *Fuchsia* clade—between northern Chile/south-eastern Brazil and Colombia and between Colombia and western Panama—correspond to the rise of the Andes, beginning in the Cretaceous.

The oldest known fossils of Onagraceae are from the Late Cretaceous of North America (the fossils have been compared with *Oenothera*) and South America (compared with *Ludwigia*) (Martin, 2003). The oldest fossil material of *Fuchsia* (distinctive pollen) is from the early Oligocene of Australia (Martin, 2003). A New Zealand macrofossil from the early Miocene has floral structures that are "remarkably similar to those of modern New Zealand *Fuchsia*" (Lee et al., 2013a).

The sister group of Onagraceae, Lythraceae, also includes significant fossil records. The genera *Lythrum* and *Peplis* form distal branches (Morris, 2007), but despite this, both include Cretaceous fossils—recent discoveries extended the known fossil records of the two genera by 70 Myr (Grímsson et al., 2011). The information is valuable as it provides minimum ages for these derived clades. As Grímsson et al. (2011) concluded, "The appearance of *Lythrum* and *Peplis* in North America and *Peplis* in Asia at approximately the same interval in the mid Late Cretaceous points to an already wide geographical distribution by then."

Differentiation *in* Fuchsia *Section* Skinnera

The western clade of *Fuchsia*, section *Skinnera*, occurs in New Zealand and the Society Islands (Tahiti). This distribution duplicates that of the parrot *Cyanoramphus*, also in Zealandia and the Society Islands (Raiatea) (Heads, 2012a: Fig. 6–11), and the weevil *Andracalles* (Curculionidae), in Zealandia (North, South, Lord Howe, and Norfolk Islands), Samoa, Niue, and Tahiti (Lyal, 1993b). *Fuchsia* sect. *Skinnera* is characterized by its southwest Pacific distribution and its blue pollen. It is also notable for its diversity of shoot architecture (plants include barely woody subshrubs, lianes, and large trees) and its ecological range (plants extend from the shore to upper montane forest) (Godley and Berry, 1995).

The basal taxon in sect. *Skinnera* is a rare plant of Northland, *Fuchsia procumbens* ("1" in Figure 10.4a). Unlike most *Fuchsia* species, which are erect shrubs or small trees, *F. procumbens* is a prostrate or scandent subshrub. Its stems are scarcely woody and often rooting, and include plagiotropic (horizontal) lateral branches that develop from the cotyledonary nodes. *F. procumbens* grows in sandy, gravelly, or rocky places only a little above high tide mark, and the plants are sometimes submerged by exceptionally high tides (Allan, 1961). The species is a mangrove associate in a broad sense, and it inhabits disturbed habitat. *F. procumbens* is sister to the two New Zealand species plus *F. cyrtandroides* of Tahiti (wet montane forest, at 1150–2000 m elevation).

The ecophyletic series from mangrove to montane forest that is displayed in *Fuchsia* sect. *Skinnera* requires an explanation, along with its standard geographic distribution. Any groups that persisted as metapopulations around the central Pacific volcanism, colonizing subaerial volcanic plateaus and islands before these subsided, would have required a weedy, mangrove-associate ecology. This means that the ancestral habitat in *Skinnera* is more likely to have been the maritime strand vegetation occupied by *F. procumbens* than the tropical-montane rainforest of *F. cyrtandroides*. Other *Fuchsia* ancestors probably occupied marginal habitat around the shallow basins that were the immediate precursors of the Andes. As soon as the land that would become the Andes rose out of the sea in the Cretaceous, it would have been colonized by the weedy biota already present around the shores in the region.

Apart from the mangrove herb, *F. procumbens*, there are two other species of *Fuchsia* in New Zealand. *F. perscandens* ("3" in Figure 10.4a) is a woody liane with stems up to 6.5 cm in diameter. Young plants can trail along the ground for 1–2 m, rooting and producing lateral branches (Godley and

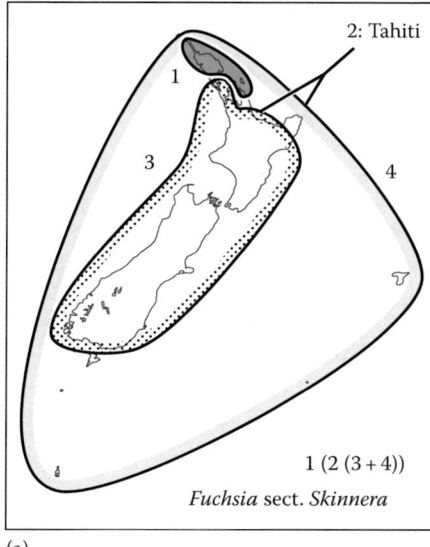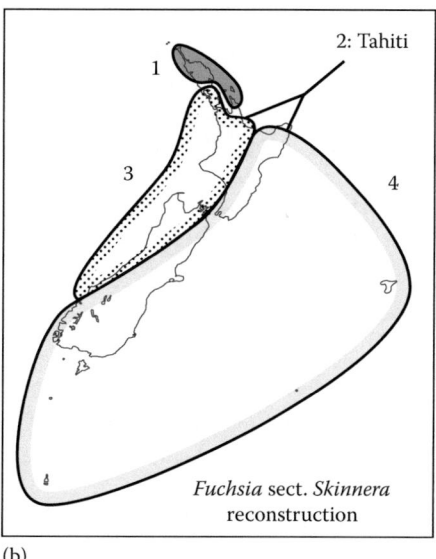

FIGURE 10.4   (a) Distribution and phylogeny of *Fuchsia* sect. *Skinnera* (Onagraceae). 1, *F. procumbens*; 2, *F. cyrtandroides*; 3, *F. perscandens*; 4, *F. excorticata*. (Data from Godley, E.J. and Berry, P.E., *Ann. Missouri Botanical Garden*, 82, 473, 1995.) (b) Reconstruction of original allopatry in *Fuchsia* sect. *Skinnera*. Species as in (a).

Berry, 1995). The species is a southern vicariant of *F. procumbens*, although its sister is *F. excorticata* (next). *F. perscandens* is absent from several parts of eastern New Zealand, including eastern Northland (one record at Wairua), the Barrier Islands, Coromandel Peninsula, East Cape, and Chatham Islands. In contrast, *F. excorticata* occurs in all these areas. *F. perscandens* is recorded in "lowland forest, especially marginal" (Allan, 1961). Habitats include coastal shrubland, limestone, river terraces, and podocarp swamp forest (Godley and Berry, 1995).

F. excorticata ("4" in Figure 10.4a) is widespread in New Zealand and is the only large tree in the genus; it can grow up to 13 m tall with a trunk up to 70 cm in diameter (Godley and Berry, 1995). Allan (1961) recorded it as "common in lowland to lower montane forest, especially marginally." The marginal ecology of *F. excorticata* and *F. perscandens*, the swamp forest and limestone habitat in the latter, and the shoreline habitat of *F. procumbens* are all consistent with a common ancestor that was a weedy mangrove associate.

F. excorticata of New Zealand and *F. cyrtandroides* of Tahiti are both trees and are found in wet forest, where they are most common along streams. In forest around Dunedin, *F. excorticata* forms twisted, horizontal trunks up to 50 cm in diameter that extend for several meters, a meter or so above the ground (personal observations). In other Myrtales, *Metrosideros umbellata* (Myrtaceae) on the Auckland Islands can have the same, "horizontal tree" architecture (personal observations). In *Fuchsia*, as in *Metrosideros* and its sister *Mearnsia*, the central Pacific forms (in French Polynesia and New Zealand) are trees while their western vicariants (in northern New Zealand, New Caledonia, etc.) are lianes. In *Fuchsia* and in *Metrosideros* s.lat., both lianes and trees occur together in New Zealand.

Fuchsia excorticata completely overlaps the range of both *F. perscandens* and *F. procumbens*, and this could be the result of the first species expanding its range westward. One possible reconstruction of the original allopatry based on this assumption is shown in Figure 10.4b.

Fuchsia has a coastal endemic in Northland and a montane one in Tahiti, and so its complete absence from the Three Kings Islands, Poor Knights Islands, and Hen and Chickens Islands, and also from New Caledonia, is unexplained by ecology. Yet this same arc of absence is a zone of endemism for taxa such as *Xeronema* (Figure 5.2), and another member of Onagraceae, *Epilobium*, also has an interesting absence from New Caledonia. This suggests that the absence could be the result of phylogeny.

Differentiation in *Epilobium*

The two genera of Onagraceae present in New Zealand are *Fuchsia* and *Epilobium*. Members of *Fuchsia* are woody, forest plants; in contrast, *Epilobium* species are herbaceous and occupy open habitat. *Epilobium* is more or less cosmopolitan and is by far the most species-rich genus in the Onagraceae. Of special interest here is the fact that there are more *Epilobium* species in New Zealand than in any other area of comparable size. In addition, the New Zealand species display a much greater ecological diversity than elsewhere, growing in all types of habitat from the shore to the high alpine (Raven and Raven, 1976). What is the reason for this concentration of evolutionary and ecological diversity in New Zealand?

Epilobium has two main clades with the following distributions (Baum et al., 1994):

> Western North America and western South America (six sections). All six sections are present in western North America, three are also present in southern Argentina (Chubut), and one of these is also widespread in Chile.
>
> Widespread globally, except in the central Pacific (sect. *Epilobium*). This clade includes 150 of the 165 species in the genus, and it occurs on all continents except Antarctica.

In the second clade, Raven and Raven (1976) inferred an interesting *lack* of affinity between Australasian and South American species. Two possible connections were suggested by earlier authors, but both were rejected by Raven and Raven. One of the cases is *Epilobium hirtigerum*, which is found in Australasia and South America. This is a weedy species and may have been introduced to South America, but the plants there differ in their morphology (the hairs are more glandular than in Australasian specimens). Raven and Raven (1976) described *E. hirtigerum* as "recently introduced" in South America, although they cited collections made there by Commerson in 1768. Raven and Raven also disputed Skottsberg's (1906) proposal of a direct affinity between creeping species of Chile and New Zealand.

If there really is no trans-South Pacific connection in *Epilobium*, the genus comprises a disjunct group in western North America and western South America, and its sister, widespread around the Atlantic and Indian Oceans, east to New Zealand (the same pattern is seen in *Lilaeopsis* of the Apiaceae; Bone et al., 2011). It is significant that there are no indigenous *Epilobium* species through most of the central Pacific east of New Guinea, Australia, and New Zealand. Despite this, introduced, weedy species of *Epilobium* now flourish in the islands, and so the absence was probably caused by phylogeny rather than ecology. This implies that the Pacific area of absence in *Epilobium* is filled by some related group, such as the *Fuchsia–Circaea* group (cited earlier) in southeastern Polynesia.

The Australasian Clade of Epilobium

The globally widespread sect. *Epilobium* is represented in Australia, New Guinea, and New Zealand by 45 indigenous species that formed a clade in morphological analyses (Raven and Raven, 1976). Raven and Raven suggested that seed morphology links this clade with plants in Eurasia (the Philippines, Sumatra, etc.), rather than with New World plants, but a molecular study of sect. *Epilobium* is needed (at higher levels in the genus, ideas on seed evolution conflict with molecular phylogenies; Baum et al., 1994).

The Australasian clade has its main center of diversity in New Zealand, with 37 species (35 endemic). As already emphasized, there are more *Epilobium* species in New Zealand than in any other area of comparable size, and within New Zealand the species are most diverse in the northern South Island (Heads, 1997). This is a common pattern, and Nelson and Marlborough are well known for their high diversity and endemism in many groups.

Despite the New Zealand diversity in *Epilobium*, the genus is absent on and east of a line: Solomon Islands–New Caledonia–Norfolk Island. This notable absence extends to the islands off northern New Zealand; despite there being 13 species in Northland, there are none on the Poor Knights or Three Kings Islands, and only one on the Hen and Chickens Islands (the widespread *Epilobium billardierianum*). The Poor Knights and the Hen and Chickens Islands are both only about 25 km offshore, while the Three Kings Islands are about 50 km offshore. It is doubtful that *Epilobium* is so scarce in these islands because of inadequate means of dispersal or other aspects of its ecology.

In contrast with the situation in the northeastern islands, in southern New Zealand *Epilobium* is well represented on the offshore islands. There are 13 indigenous species on the Chatham Islands, 4 on Auckland and Campbell Islands (1 endemic there), and two on Macquarie Island. These islands are, respectively, 700 km off Wellington, and 400, 600, and 1000 km off Stewart Island. These distributions suggest that the unusual paucity of *Epilobium* and *Fuchsia* on the northeastern New Zealand islands (and the presence of many local endemics there in other groups) has been caused by phylogenetic breaks, rather than present-day topography and ecology.

Within the Myrtales, Onagraceae and Lythraceae form a clade that is sister to Myrtaceae, Melastomataceae, and others (Stevens, 2016). Within the order, the two main groups around the Tasman region are Onagraceae (656 species worldwide; Stevens, 2016) and Myrtaceae (4620 species). In Onagraceae, New Caledonia has only two members (two *Ludwigia* species, neither endemic), whereas it has 236 species of Myrtaceae (all but two endemic) and 21 genera, 6 endemic. The great difference, along with the striking paucity of Onagraceae on the islands of northeastern New Zealand, can be explained as a trace of early vicariance with Myrtaceae.

Vicariance between Onagraceae and relatives, and Myrtaceae and relatives could also help account for the great diversity of Onagraceae and *Epilobium* (37 species) in New Zealand, and the low numbers of Myrtaceae there (living and fossil). This lack of diversity in Myrtaceae is conspicuous when compared with the floras of Australia, New Guinea, and New Caledonia. Raven and Raven (1976) attributed the diversity in New Zealand *Epilobium* to the development of open habitat in the country, but there is also plenty of open habitat in Australia (which has only nine indigenous *Epilobium* species) and in New Guinea (only four species). This indicates that the New Zealand center of diversity in *Epilobium* instead represents an early development in the genus and the expression of a genetic potential present in the region. In the same way that an ancestor can already be polymorphic before the modern groups begin to diverge, an ancestor can have a greater or lesser potential for diversification, depending on its genetic architecture. In the New Zealand region, the ancestral complex of Myrtales developed mainly into Onagraceae; in Australia and New Caledonia, it evolved mainly into Myrtaceae.

Biogeography in the Australasian Clade of Epilobium

Rather than accepting a vicariance origin for the Australasian clade of *Epilobium*, Raven and Raven (1976) proposed dispersal from Asia into Australasia. Nevertheless, they noted that the New Guinea species are not morphologically intermediate between Asian and Australasian plants, as might be expected in this model. Thus, the authors suggested that one of the New Guinea species (*E. keysseri*) is derived from a form of southern Australia (*E. billardierianum* subsp. *cinereum*), while the other three New Guinea species are related to New Zealand species. These affinities follow two separate tracks: one along in the west Tasman margin (southeastern Australia–New Guinea) and one along the eastern Tasman (New Zealand–New Guinea). Both are common patterns (Heads, 2014), suggesting that the groups displaying these differentiated by simple vicariance in an ancestral biota caused by slab rollback eastward from Australia (Figure 3.15).

Architecture in the Australasian Clade of Epilobium

In the Australasian clade of *Epilobium*, Raven and Raven (1976) postulated that the original members were more or less erect plants, and, that unlike the Eurasian species, they produced runners. (Runners in extant *Epilobium* species are either leafy and develop from the base of the shoot system or they are leafless and grow just below the soil surface.) Raven and Raven (1976) also suggested that the ancestral Australasian forms grew in moist habitat. Simple developments from these ancestral conditions—plants that have runners and grow in moist habitat—would account for the observed variation.

Basal runners are still found in about 12 Australasian species that otherwise closely resemble Eurasian species. Two alternative developments from this condition can be inferred: suppression of the horizontal (plagiotropic) runners, and suppression of the erect (orthotropic) shoot system. It is likely that each of these simple processes occurred many times.

Suppression of the runners in *Epilobium* results in a compact, clumped habit, and the development of a taproot. This allows the plants to survive in drought-prone sites such as shingle slides and dry riverbeds. Suppression of the erect shoot system leads, ultimately, to complete suppression of this system (with *opposite* leaves) and its terminating inflorescence (with *spiral* phyllotaxis). The architecture becomes increasingly dominated by the basal runners. In the extreme condition, the runners bear all the flowers, which are located singly in leaf axils, and phyllotaxis in this flowering portion of the stem remains strictly opposite. The creeping, rooting, shoot continues growing beyond the flowering portion, and so the inflorescence (in the sense of "the flowering section of the shoot") is intercalary rather than terminal as in the other species.

An intercalary inflorescence is present in about 12 widespread New Zealand *Epilobium* species (*E. brunnescens*, etc.), one Australian species, and *E. conjungens* in Tierra del Fuego (West and Raven, 1977). These species do not form a monophyletic group according to Raven and Raven (1976), but the first split in their key separates members with intercalary inflorescences from all other Australasian species as the architecture is so distinctive. The authors wrote, "There is presumably an ecological solution as to why this peculiar plant form, unknown elsewhere in the world should have evolved repeatedly in Australasia," but they did not suggest any.

The concentration of species with an intercalary inflorescence in Australasia, together with the character's disjunct occurrence in Tierra del Fuego, suggests that the present distribution of the intercalary inflorescence is the result of historical factors rather than ecological ones. Whether the group in which it occurs is polyphyletic or monophyletic, the suppression of the main, orthotropic shoot system, leaving just runners, reflects a long-term trend. The same trend is observed in many plant groups in many different habitats, and it is probably caused by the playing out of prior, genetic biases rather than the action of natural selection. Nevertheless, the trend has been viable in New Zealand *Epilobium*, and the resultant, low, creeping architecture has turned out to be preadapted for life in montane, alpine, and other open sites.

New Zealand Asteraceae (1): The Tribe Anthemideae and Its Genus *Leptinella*

The Age of Asteraceae

The family Asteraceae has 25,000 species, more than any other angiosperm family except orchids. The group is deeply nested in the angiosperm phylogeny, and it also has a poor fossil record. Based on these observations, dispersal theory proposes that the group evolved only in the Cenozoic, and that its many trans-oceanic disjunctions are the result of chance dispersal (Chapter 2). In their influential review, Raven and Axelrod (1974) wrote that "no fossil pollen of the vast family Asteraceae is known prior to the uppermost Oligocene, despite extensive search," and that this suggests "no more than a mid-Oligocene age for the family." However, Barreda et al. (2012) described a well-preserved fossil of the family from the Eocene of Patagonia, and for Sancho et al. (2014), this "suggests that the family was part of an ancient flora that inhabited southern Gondwana before the establishment of oceanic barriers to dispersal."

The idea that Asteraceae evolved before Gondwana breakup is accepted here, but it does not depend on the fossil record. Long before the Eocene fossil was discovered, specialists in the family supported the idea of early vicariance on the basis of the distribution patterns. Bentham (1873) found that these "point to a very wide dispersion of the original stock of the family at a very early period, when the physical configuration of the surface of the globe must have been very different from what it is now...." As Bentham stressed, "the absence of [fossil members] is no proof of their non-existence at various geological periods." Turner (1977) reached the same conclusion: "the family [Asteraceae] is a very old one among angiosperms generally, [its] origin stems back to at least the Cretaceous." These authors' conclusions were supported in later work on the family, again based on evidence from biogeography as well as the fossil record (Nesom, 1994; Heads, 1998b, 1999).

Barreda et al. (2015) identified Late Cretaceous fossils from the Antarctic Peninsula and New Zealand (*Tubulifloridites lilliei* type A) as belonging to an extant genus of Asteraceae, *Dasyphyllum*, previously only known from tropical South America. The Antarctic fossils are dated as 76–66 Ma, 20 m.y. older than the previously accepted oldest fossils of the family (macrofossils from the Eocene of Patagonia; Barreda et al., 2012). This finding conflicts with the usual idea of a Cenozoic origin for the family (based on interpretations of the fossil record), but it is predicted by the vicariance model presented here. In particular, it is consistent with the idea that differentiation within extant genera of Asteraceae and related families had already occurred by the Late Cretaceous.

The Center of Origin of Asteraceae

The worldwide family Asteraceae has its sister group, Calyceraceae, in southern South America. Assuming an origin by simple allopatric differentiation, the Asteraceae and Calyceraceae each evolved from a common, worldwide ancestor at a break in or around southern South America, and this was followed by local range expansion by dispersal in southern South America. There is no need for Asteraceae to have had a localized center of origin. The original area of Asteraceae was the whole world except for an unspecified part of southern South America, where the sister group, Calyceraceae, had its origin. (For the differentiation of the main clades in Asteraceae, see Heads, 2014: 9.)

Tribe Anthemideae

The following sections examine the distribution in New Zealand of members of Asteraceae that belong to three of the tribes: Anthemideae, Astereae, and Senecioneae. The first tribe, Anthemideae, is a diverse group (109 genera, ~1740 species) with a marked concentration of basal clades in the southern hemisphere. Members of the tribe are widespread through New Zealand, from maritime sites to the high-alpine zone. All the New Zealand representatives belong to the genus *Leptinella*.

The Anthemideae have a phylogeny as follows (Himmelreich et al., 2008):

1. **Global, but mainly in southern hemisphere.** South Africa (*Inezia, Hilliardia, Adenanthellum, Hippia, Lidbeckia, Thaminophyllum,* and *Osmitopsis*) and South Africa to Zambia (*Schistostephium*), plus a mainly panaustral clade in southern Africa + Australasia + South America (one or two northern hemisphere species) (*Cotula, Leptinella, Soliva,* etc.).
2. **Southern Africa** (*Inulanthera*).
3. **Southern Africa** (*Ursinia*).
4. **Southern Africa, Mediterranean, Asia, and North America** (many genera: *Artemisia, Achillea, Chrysanthemum,* etc.).

This is the arrangement based on nuclear DNA; chloroplast DNA sequences instead show clades 2 and 3 as sisters. In any case, the pattern is consistent with differentiation having developed in a global ancestor around breaks in southern Africa. At the first of these, the mainly *southern hemisphere* clade 1 has diverged from a clade (2–4) distributed instead along an arc: *southern Africa–Eurasia–North America*. (The two clades illustrate two of the world's main global distribution patterns.) Subsequent breaks in southern Africa have separated clades 2, 3, and 4, and later there has been local overlap in southern Africa.

Himmelreich et al. (2008) noted the simple biogeographic pattern in the tribe Anthemideae, with southern groups basal in the phylogeny, and this is important for orienting the *sequence of differentiation*. The authors suggested that *physical movement* of tribe members into the northern hemisphere is "clearly demonstrated" by the paraphyletic basal group in southern Africa, but this reasoning is not accepted here (cf. Figure 1.1). As Himmelreich et al. (2008) noted, a southern African center of origin has also been proposed for two other tribes in Asteraceae (Gnaphalieae and Astereae), but in these cases, as in Anthemideae, southern Africa can be interpreted as a center of differentiation rather than a center of origin.

The disjunct, mainly panaustral group of Anthemideae in clade 1 (see phylogeny) is structured as follows (Himmelreich et al., 2008):

Soliva: South America.

    Cotula: Mainly southern Africa, some species in Australia and South America, one or two in the northern hemisphere.

    Leptinella: Mainly Australasia, one species also in the southern Indian Ocean (Kerguelen, Crozet, and Marion Islands), one in Patagonia and the Falkland Islands.

Overall, *Soliva* is concentrated in South America, *Cotula* in southern Africa, and *Leptinella* in Australasia. In dispersal theory, "The position of *Cotula*, *Leptinella*, and *Soliva* in both reconstructions… indicates that there was a likely dispersal event of *Cotula* out of Southern Africa into Australia and New Zealand (*Leptinella*), and at two times (*Soliva*, *Leptinella*) into South America…" (Himmelreich et al., 2008: 143). Instead, the allopatry that exists among the three genera, at least with respect to their main massings, is consistent with an origin by vicariance. A dispersal model does not explain the great diversity of *Leptinella* in New Zealand, coupled with the absence of endemic, and perhaps even indigenous, *Cotula* there. (This is despite the fact that *Cotula* species introduced there thrive as weeds.)

To summarize: clade 1 in Anthemideae originated by vicariance with clades 2–4. Within clade 1, the widespread *Soliva–Cotula–Leptinella* clade originated by vicariance with its South African relatives. Within the *Soliva–Cotula–Leptinella* clade, the genera maintain a high level of allopatry, again indicating an origin by vicariance.

Leptinella: A Southern Hemisphere Genus That Is Diverse through New Zealand

Leptinella (Figure 10.5) is a well-marked clade formerly included in the genus *Cotula*. *Leptinella* includes 42 species and subspecies, and 29 of these are in New Zealand (28 are endemic) (Lloyd, 1972; Himmelreich et al., 2012, 2014). The *Soliva–Cotula–Leptinella* clade includes many invasive weeds, and the indigenous New Zealand members also display a tendency to occupy disturbed sites. The group has a wide ecological range, at least in open habitats, and extends from maritime cliffs and salt marsh to high-alpine sites.

Phylogeny in Leptinella

Basal differentiation in *Leptinella* has occurred at main nodes around the Tasman Basin, Central Otago–southeastern Fiordland, and Marlborough–Chatham Rise (Figure 10.5). The early phylogenetic divergence at these nodes can be explained by the large-scale extension and volcanism that took place in

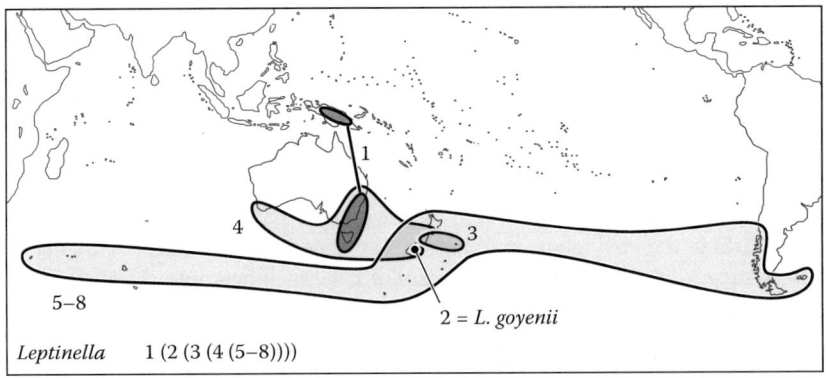

FIGURE 10.5 Distribution and phylogeny of *Leptinella* (Asteraceae). (Data from Himmelreich, S. et al., *Mol. Phylogenet. Evol.*, 65, 464, 2012.)

these regions during the Mid- and Late Cretaceous. *Leptinella* has the following phylogeny (Himmelreich et al., 2012). Localities outside New Zealand are shown in bold.

1. **SE Australia (Tasmania to McPherson–Macleay Overlap) and New Guinea**.
     2. Central Otago + SE Fiordland (*goyenii*).
          3. Outer Chatham Islands (*featherstonii*), Canterbury to Tararua Range (*pyrethrifolia* var. *pyrethrifolia*, *pyrethrifolia* var. *linearifolia*, *atrata* subsp. *luteola*).
               4. Fiordland to Canterbury, also Wellington (*maniototo*), **SE Australia, Tasmania, and SW Australia**.
                    5. South Island (*pectinata* subsp. *pectinata*, *pectinata* subsp. *villosa*, *atrata* subsp. *atrata*, *dendyi*, *conjuncta*, *albida*).
                         6. **Kerguelen**, New Zealand subantarctic islands (*plumosa*, *lanata*); Wellington to Christchurch (*nana*, *minor*, *filiformis*).
                              7. Auckland Is. and Chatham Is. (*potentillina*).
                                   8. Stewart Island to Northland Island and Chatham Island (*dioica*, *calcarea*, *squalida*, *serrulata*, *traillii*, *scariosa*, *rotundata*, *pusilla*, *intermedia*, *dispersa*, *tenella*), also **South America** (*scariosa*). (The Chatham Island population belongs to the North Island subspecies of *squalida*.)

Leptinella *Clade 1*

The basal break separates a clade from southeastern Australia and New Guinea (clade 1 in Figure 10.5) from the rest, with overlap restricted to southeastern Australia. In New Guinea, *Leptinella* occurs in the mountains and shows a typical disjunction along the craton margin (Heads, 1999: Fig. 6B).

Leptinella *Clade 2:* Leptinella goyenii

The second node in the phylogeny separates Clade 2 (*Leptinella goyenii*) of Central Otago and southeastern Fiordland from the rest of the genus (Figure 10.6). *L. goyenii* has a distinctive morphology and a restricted, high-alpine distribution. In traditional theory, the species would be interpreted as a recent derivative of one of the lowland forms, following the uplift of the current mountains. Yet the phylogeny indicates that its sister group is diverse and widespread around the southern continents, and this implies a deeper history for *L. goyenii*.

The fossil record of the Asteraceae is very deficient, and there is no known fossil material of *Leptinella* or any southern hemisphere members of its tribe, Anthemideae (Himmelreich et al., 2012). A molecular clock calculation was calibrated with fossils from other groups in the family, and this estimated that *L. goyenii* split from its sister at ~10 Ma (Himmelreich et al., 2012). As with Anthemideae, the other groups of Asteraceae are likely to be underrepresented in the fossil record, and so *L. goyenii* probably diverged long before 10 Ma. Nevertheless, even at a minimum age, the 10 Ma date confirms that the plant is much older than the high-alpine habitat it occupies, as this only formed after 1 Ma (Heenan and McGlone, 2013). This is another illustration that a clade can be much older than the particular strata or habitat where it is endemic (cf. *Hectorella*, Chapter 9).

Leptinella goyenii occupies cushion vegetation in the most exposed and frost-heaved sites on the Central Otago summits, one of the most severe environments of any New Zealand alpine plant (Mark and Adams, 1973). The species is distinct in its woody stems, sessile inflorescences, and small, sessile leaves, formed in large part from the "unterblatt" (the proximal leaf zone), rather than the "oberblatt" (the distal leaf zone) as in the rest of the genus (Lloyd, 1972: Fig. 5). These features, along with its ability to tolerate high levels of disturbance, mean that it would have been preadapted to survive the uplift of its habitat.

The trans-oceanic affinities in *Leptinella* are interpreted here to mean that the genus is Cretaceous, at least, and this would be compatible with *L. goyenii* having evolved between the mid-Cretaceous antiform in the Central Otago schist (Pisa Range, Rock and Pillar Range, etc.) and the Hunter, Kepler, and Murchison

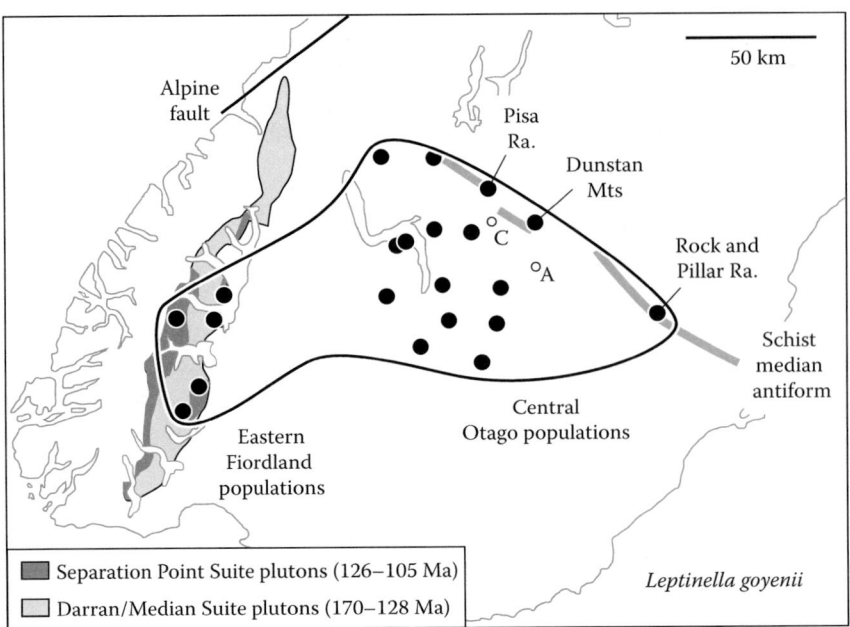

FIGURE 10.6 Distribution of *Leptinella* clade 2—*L. goyenii* (Asteraceae). C, Cromwell, A, Alexandra. The eastern Fiordland belt of plutons shown here is part of the Median Batholith. (Data from Allibone, A.H. and Tulloch, A.J., *N. Z. J. Geol. Geophys.*, 51, 115, 2008; Mark, A.F. and Bliss, L.C., *N. Z. J. Bot.* 8, 381, 1970; New Zealand plant conservation network. www.nzpcn.org.nz, 2016.)

Mountains in southeastern Fiordland (Figure 10.6). These Fiordland ranges form part of the Drumduan terrane (Figure 7.2) that was intruded by the Median Batholith (Figure 3.2). The last phase of this activity took place in the mid-Cretaceous, when the plutons of the Separation Point Suite were emplaced.

A large number of groups are Central Otago endemics, extending west only to the Thomson or Eyre Mountains and the limit of the Caples terrane, at the Livingstone fault. Examples include the plants *Kelleria childii, Anisotome brevistylus, A. cauticola, Ourisia glandulosa, Chionohebe thomsonii, Hebe propinqua, Abrotanella inconspicua,* and *Celmisia prorepens.* These are all absent from Fiordland, which has its own endemics.

In contrast, groups such as *L. goyenii* (Figure 10.6) link Central Otago and the Eyre Mountains with the eastern belt of Fiordland (Drumduan terrane and the outboard belt of the Median Batholith), including the Hunter Mountains. This interesting local pattern is well documented but seldom discussed. It is a special case of distribution with western limits in eastern Fiordland (cf. Figure 7.5). Apart from *L. goyenii,* the pattern is seen in the following plants:

> *Pimelea poppelwellii* (Thymelaeaceae): Hunter Mountains and Central Otago (Burrows, 2008).
>
> *Myosotis pulvinaris* (Boraginaceae): Hunter Mountains to Central Otago (Mark and Adams, 1973).
>
> *Gentianella amabilis* (Gentianaceae): Hunter Mountains (Mount Burns) and Central Otago (Glenny, 2004).
>
> *Hebe imbricata* (Plantaginaceae): Hunter Mountains (Green Lake), Eyre Mountains, and Central Otago (Bayly and Kellow, 2006).

Leptinella *Clades 3–8*

Clade 3 in *Leptinella* (Figure 10.5) occurs on the Chatham Islands and the central mainland, from Canterbury (Two Thumb Range) to the Tararua Range. It is more or less allopatric with clade 4 of the New Zealand mainland and Australia; overlap between the two is restricted to a small area of central Canterbury, southwest of the Torlesse Range.

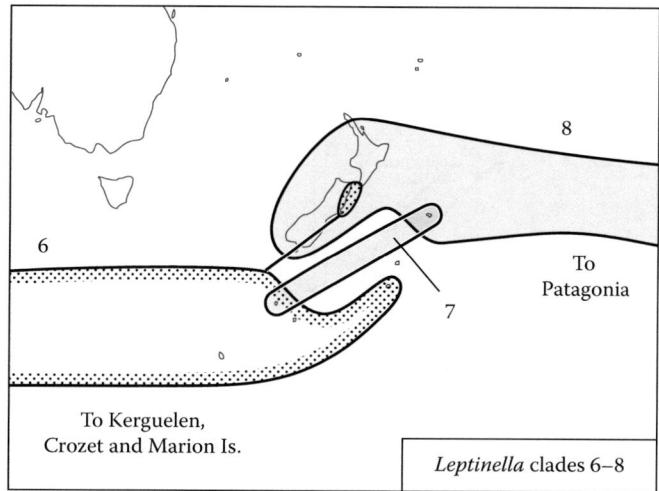

FIGURE 10.7 Distribution of *Leptinella* clades 6, 7, and 8 (clade 5, South Island, is not shown here). (Data from Himmelreich, S. et al., *Mol. Phylogenet. Evol.*, 65, 464, 2012.)

Clade 4 in *Leptinella* (Figure 10.5) has a phylogeny:

Fiordland to Canterbury (Lake Lyndon by Torlesse Ra.), also southern Wairarapa. Inland sites (*L. maniototo*).

> SE Australia and Tasmania.
>> SE Australia and Tasmania.
>> SW Western Australia.

In dispersal theory, the New Zealand representative would be expected to be nested in a group from center of origin in southeastern Australia. Instead, it is basal to a widespread Australian complex, and its origin predates the southwestern Australia–southwestern Australia split.

Leptinella clade 5 is restricted to the South Island (it is not mapped here).

Leptinella clades 6–8 (Figure 10.5) extend from Marion, Crozet, and Kerguelen Islands through Australasia to Patagonia (the connection between New Zealand and Kerguelen, two submerged blocks of continental crust, also occurs in *Hectorella–Lyallia*; Chapter 9). Clade 6 is a trans-Indian Ocean group, while clade 8 is a trans-Pacific group (Figure 10.7), a pattern also seen in *Astelia* (Figure 4.7), podocarps and ratites (Heads, 2014). In *Leptinella*, the junction between the Indian Ocean clade 6 and the Pacific Ocean clade 8 is "sutured" by clade 7 (*L. potentillina*) on the Auckland Islands and Chatham Islands (for this subantarctic islands—Chatham Islands affinity, see Chapter 6).

New Zealand Asteraceae (2): *Solenogyne* and *Lagenophora* (Astereae)

The tribe Astereae includes the *Celmisia* group, which is restricted to eastern Australia, New Guinea, and New Zealand, and is sister to a diverse, worldwide clade (Chapter 4). Members of this group are much more conspicuous in the New Zealand vegetation than the four other genera of Astereae present, but these other, smaller plants display biogeographic patterns that are just as interesting.

Two of these other genera are *Lagenophora* and *Solenogyne*, small herbs that form a clade (Sancho et al., 2014; Figure 10.8). *Lagenophora* includes a trans-South Pacific disjunction, the timing of

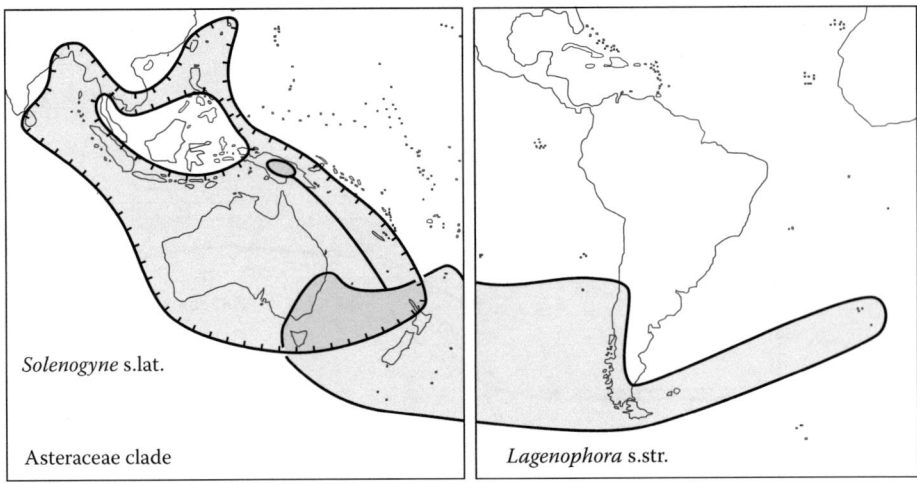

FIGURE 10.8 Distribution of the sister pair *Solenogyne* s.lat. and *Lagenophora* s.str. (Asteraceae). (Data from Moreira-Muñoz, A., Plant geography of Chile: An essay on postmodern biogeography, Unpublished PhD thesis, Friedrich-Alexander University Erlangen-Nürnberg, Bavaria, Germany, 2007; Sancho, G. et al., *Bot. J. Linnean Soc.*, 177, 78, 2014.)

which was referred to earlier; Chapter 2). The two genera have the following distributions (areas where only one clade occurs are in bold; Aus = Australia, NG = New Guinea, NC = New Caledonia, NZ = New Zealand):

Solenogyne s.lat. (including *Lagenophora lanata*, *L. gracilis*, and *L. huegelii*): **Sri Lanka, India, Japan, and SE Asia via Indonesia and the Philippines to SW** and SE Aus, NG, NC, NZ (Northland only).

Lagenophora s.str.: NG, SE Aus, mainland NZ south to **subantarctic islands, South America**.

Only the first clade occurs in Asia and Western Australia, while only the second occurs in New Zealand south of Northland and in South America. The two overlap in the Coral Sea–Tasman Sea region (New Guinea, southeastern Australia, and Northland). The pattern is consistent with an original break zone in the Tasman region having divided a widespread Indo-Pacific group into an Indian Ocean group and a Pacific Ocean group. Local range expansion (dispersal) around the break zone has led to the overlap of the two clades. The pattern is seen in many other groups, such as *Stylidium* and *Forstera* (Figure 4.3).

The different ranges of *Solenogyne* and *Lagenophora* are not just the result of ecological differences, with the former occupying warmer habitats and the latter cooler ones. For example, *Lagenophora* is in the Kermadec Islands, well to the north of *Solenogyne* in Tasmania.

Lagenophora comprises two primary clades, named the "New Zealand clade" and the "South American clade" in the study by Sancho et al. (2014). The first is most diverse in New Zealand, the second in South America.

Lagenophora "New Zealand Clade"

This is a Tasman Sea–Coral Sea clade, distributed from New Guinea to the New Zealand subantarctic islands. The three component groups are listed next. All three overlap through the New Zealand mainland, but the second and third show interesting differences outside the mainland and are mapped in Figure 10.9 (N, North Island; S, South Island; St, Stewart Island).

Group 1. New Zealand mainland (N, S, St), also Three Kings, Chatham and Kermadec Is.

Group 2. New Zealand mainland (N, S, St), SE Australia, and New Guinea.

Group 3. New Zealand mainland (N, S), subantarctic islands, Chatham Islands, Kermadec Islands.

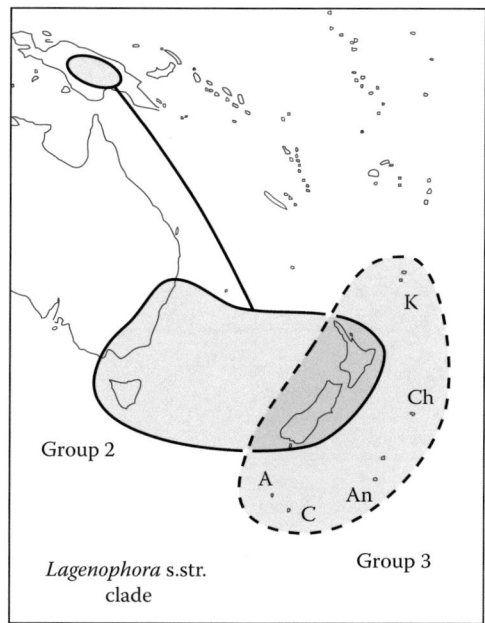

FIGURE 10.9 Distribution of a clade in *Lagenophora* s.str. (Asteraceae). A, Auckland Islands; An, Antipodes Islands; C, Campbell Island; Ch, Chatham Islands; K, Kermadec Islands. (Data from Sancho, G. et al., *Bot. J. Linn. Soc.* 177, 78, 2014.)

Groups 2 and 3 display significant allopatry and they overlap only on mainland New Zealand. This is consistent with an ancestor that was widespread around what became the Coral Sea–Tasman Sea region and then broke apart into western and eastern groups. Later, these developed overlap within New Zealand.

There is also allopatry within the groups. Group 2 consists of two primary subclades that meet at a node around the Coromandel Peninsula. (The first is formed by *Lagenophora stipitata* from southeastern Australia, Tasmania, New Guinea, Northland, and south to Coromandel Peninsula. The second comprises *L. pinnatifida* of the South Island and north to the central North Island and *L. strangulata* of the South Island and north to Coromandel Peninsula.)

Lagenophora "South American Clade"

The "New Zealand clade" of the Tasman–Coral Sea region is sister to a group found in Juan Fernández Islands, southern South America, and also the South Atlantic islands of Tristan da Cunha and Gough. The typical trans-South Pacific connection is duplicated in other Asteraceae, such as the last group treated (*Leptinella*) and also the next group, *Abrotanella*.

New Zealand Asteraceae (3): Pre- and Postdrift Tectonics and Evolution in *Abrotanella* (Senecioneae)

Abrotanella (Asteraceae) is a genus of small, alpine subshrubs. As with many groups, it is endemic to Australasia and southern South America, and has a 9000 km disjunction across the South Pacific (Figure 10.10). This is an example of the pattern usually exemplified by the tree *Nothofagus* (Nothofagaceae: Fagales). The most obvious question here is, "What is the reason for the disjunction?" In *Abrotanella*, it has been attributed to the Cretaceous breakup of Gondwana (Heads, 2011), although other studies have suggested this is impossible as it would imply that the Asteraceae are 1.5 billion years old (Swenson et al., 2012). This argument would be correct if evolutionary rate conforms to a clock or relaxed clock model, but if evolution is not clocklike, it is possible that the genus evolved in the

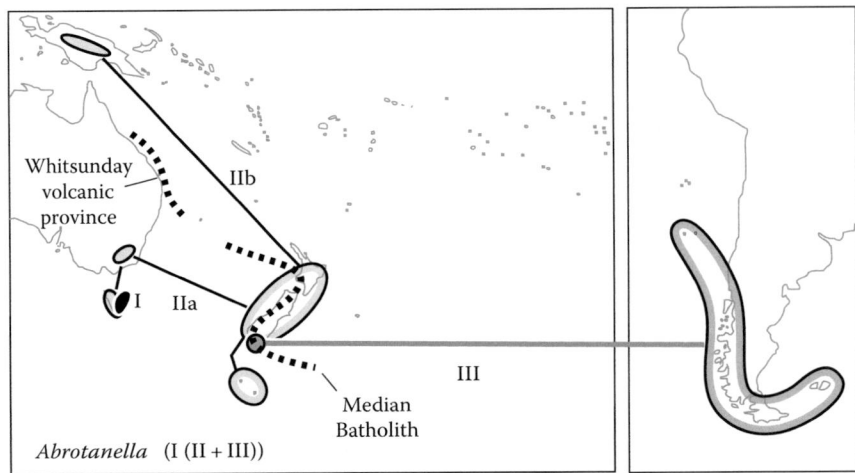

FIGURE 10.10 Distribution of *Abrotanella* (Asteraceae), showing clades I (black), II (light gray), and III (dark gray). Phylogeny indicated with nested numbers. (Distributions from Swenson, 1995; phylogeny from Swenson et al., 2012.) Whitsunday volcanic province/Median Batholith as dashed line (from Mortimer et al., 2005; Campbell and Hutching, 2007). (From Heads, M., South Pacific biogeography, tectonic calibration, and pre-drift tectonics: Cladogenesis in *Abrotanella* (Asteraceae). *Biol. J. Linnean Soc.* 2012c. 107. 938–952. Copyright John Wiley & Sons. Reproduced with permission.)

Cretaceous and the family in, for example, the Jurassic. Unlike clocklike models of evolution, nonclock, multiclock, and "very relaxed" clock models of evolution accept that significant changes in evolutionary rate can occur both within and between lineages, and that most of the differentiation in a large group can happen over a short time.

The case of *Abrotanella* is examined here in more detail. In particular, a recent molecular phylogeny of the group is used to test whether the sequence of phylogenetic breaks in the genus corresponds with the sequence of main tectonic and magmatic events in the region.

This section incorporates material first published in the *Biological Journal of the Linnean Society* (Heads, 2012c) and reproduced here with permission of John Wiley & Sons.

Timing Evolution in *Abrotanella*

Implicit Conversion of Fossil-Calibrated, Minimum Clade Ages into Maximum Clade Ages

Wagstaff et al. (2006) gave a molecular phylogeny for *Abrotanella* and accepted that "The first appearance [of a clade] in the fossil record imposes only minimum age constraints." Nevertheless, later in their paper, they overlooked this principle and transmogrified their fossil-calibrated, minimum ages for *Abrotanella* into maximum ages. They concluded that the genus "initially diverged during [not before] the Miocene" (p. 100) and that species radiations occurred "about [not 'before about'] 3.1 million years ago" (p. 100). By treating the minimum dates as maximum possible dates, they were able to rule out earlier vicariance: the disjunctions "*must* reflect long-distance dispersal" (p. 104), and "radiations" in the genus "*undoubtedly* reflect long-distance dispersal..." (p. 95; italics added).

Transmogrifying Divergence Dates for Abrotanella *in a Bayesian Framework*

In their study of *Abrotanella*, Swenson et al. (2012) converted fossil-calibrated dates (minimum ages) into maximum clade ages by using BEAST and specifying appropriate priors for the calibrations (see Chapter 1). Based on these calibrations, young clade ages were calculated. Swenson et al.

(2012) suggested that the method has the "advantage" that it will always provide maximum clade ages (with 95% credibility), and that it moves beyond the "simplistic" (but logical) use of minimum ages. Assigning only minimum ages to fossils, as is done in this book, provides no constraint on maximum ages, and so it is seen either as a problem (or "simplistic") in the traditional approach. Yet the question for Bayesian analysis is, "How are the priors selected?" As discussed in Chapter 2, several perceptive authors have already pointed out this "problem of the priors" in dating phylogenies (Lee and Skinner, 2011; Wilkinson et al., 2011; Parham et al., 2012; Pirie and Doyle, 2012; Warnock et al., 2015).

Using Island-Age Calibrations to Corroborate Bayesian Transmogrification

Wagstaff et al. (2006) and Swenson et al. (2012) compared their fossil-calibrated evolutionary rates in *Abrotanella* with rates estimated for the genera *Dendroseris* and *Robinsonia* (Asteraceae), both endemic to the Juan Fernández Islands. They found that the rates calculated for these genera corresponded quite closely to the rates calculated for *Abrotanella* and considered that the dates provided mutual support for each other.

Nevertheless, the rates calculated for the Juan Fernández Islands genera are probably overestimates, as the genera were assumed to have differentiated from their respective ancestors only after the formation of the islands. This assumption is not warranted since other taxa endemic to Juan Fernández Islands, such as the plant *Lactoris* (Aristolochiaceae), have fossils dated as Cretaceous (Gamerro and Barreda, 2008). (*Lactoris* happens to have very distinctive, abundant pollen; the pollen of other Juan Fernández Islands groups is not as distinctive and might not be recognized so easily in fossil material.) In many other cases around the world, young islands host endemics that are much older based on minimum ages from clock studies (Heads, 2011). Thus, the fossil-calibrated dates for *Abrotanella* and the island-age calibrated dates for *Dendroseris* and *Robinsonia* are both likely to be underestimates of actual age, and their mutual corroboration is unconvincing.

Spatial Evolution in *Abrotanella*: Tectonics and Biogeographic Breaks

An earlier biogeographic analysis of *Abrotanella* (Heads, 1999) was based on a morphological phylogeny (Swenson, 1995) and showed that the clade distributions were clear-cut, with simple, dovetailing allopatry. The molecular phylogeny (Swenson et al., 2012) presented patterns of allopatry that are even simpler (Figure 10.10).

In order to decide whether a model of evolution calibrated with fossils, or one calibrated with tectonics, works best in practice, both need to be tried out. Swenson et al. (2012) tested the fossil-based model in *Abrotanella*, converting the fossil-calibrated dates into maximum clade ages using selected Bayesian priors, and explained the distribution in the genus as the result of chance dispersal. They rejected a tectonic model of evolution in theory (relying on a clock model of evolution), although they only examined one tectonic break (seafloor spreading at 84 Ma causing the trans-Pacific disjunction), and they did not attempt to produce a general tectonic model for the genus. A more detailed test of tectonic/biogeographic calibration would be desirable. Is a coherent integration of evolution in *Abrotanella* with regional tectonics even possible? Such a model would have the advantages that it would not require the use of chance dispersal to explain repeated patterns, and it would not require the illogical transmogrification of fossil-calibrated clade ages from minimum dates into maximum dates.

In their study of *Abrotanella*, Swenson et al. (2012) contributed a valuable molecular phylogeny, but they did not examine the distributions of the clades in any detail and did not map them. Tectonic calibration requires analysis of the distributions, and so these are mapped here. The species distributions (Swenson, 1995; Heads, 1999) are mapped together in the groups according to the molecular phylogeny. In the approach used here, the clades are mapped, rather than being assigned to *a priori*, physiographic areas such as "New Zealand" or "Tasmania." These areas are polyphyletic and their use would confuse the analysis (Heads, 1999). Instead, the emphasis is on the geography of the clades and, in particular, the phylogenetic breaks.

Differentiation between Abrotanella *and Its Sister Group*

The South Pacific disjunctions in *Abrotanella* (Figure 10.10) are dramatic, but they are only one aspect of its distribution. In both dispersal and vicariance theory, the gaps in the range are interpreted as secondary features that developed after the genus had already evolved. In other words, seafloor spreading in the South Pacific has separated Australia, New Zealand, and South America from Antarctica and each other, and these well-studied events explain the *ruptures* in the distribution of *Abrotanella*. But they do not account for the *overall distribution* itself, with its western limit in Australasia and its eastern limit in southern South America. The rifting in the Pacific cannot explain why *Abrotanella* is absent from Africa or the northern Andes. Thus, even if the exact date of the disjunctions were known, this would only provide a minimum age for the origin of the genus as a whole.

Explaining the origin of any group requires comparison with the sister group. The South Pacific *Abrotanella* is sister to a diverse, cosmopolitan clade, the rest of the tribe Senecioneae, with about three thousand species (Pelser et al., 2010). In the same way, the South Pacific disjunct *Nothofagus* (Nothofagaceae) is sister to a diverse, cosmopolitan clade, all the other Fagales (Stevens, 2016). This pattern is a common one, repeated in the following groups:

In monocots, the South Pacific clade Ripogonaceae (Australasia and, fossil, in Chile) + Philesiaceae (Chile) is sister to the cosmopolitan Liliaceae + Smilacaceae (Stevens, 2016).

In spiders, the South Pacific clade Austrochiloidea (Australasia and southern South America) is sister to the cosmopolitan Araneoclada (~30,000 species; Selden and Penney, 2010).

In frogs, the South Pacific clade Myobatrachidae (Australia and New Guinea) + Calyptocephalellidae (central and southern Chile) is sister to the more or less global Hyloidea (Hylidae, Bufonidae, etc.) (Pyron and Wiens, 2011).

In birds, the South Pacific clade Pedionomidae (southeastern Australia) + Thinocoridae (Patagonia to the Peruvian Andes) is sister to a pantropical clade, Jacanidae + Rostratulidae (Fain and Houde, 2007).

How old could these South Pacific groups and their distributions be? Unlike the small cushion and mat plants of *Abrotanella*, which produce minute amounts of pollen, *Nothofagus* species are forest giants, dominating forest over large regions and producing immense quantities of pollen. It is not surprising that *Abrotanella* has no fossil record (as with many groups in its family), while in *Nothofagus* all four extant subgenera are known from fossils dating back to the Cretaceous. This gives a useful minimum date for *Nothofagus* that could also apply to other groups with similar distributions, such as *Abrotanella*.

The typical South Pacific distribution—not just the later disjunctions within it—is restricted to the Pacific margin of Gondwana, comprising Australasia, western Antarctica, and western South America. This region coincides with a major orogenic belt, the Australides, that does not include the old, cratonic core of Gondwana (Western Australia, India, Africa, and cratonic, eastern South America). Active terrane accretion and uplift took place along the Australides belt from Neoproterozoic to late Mesozoic time (Vaughan et al., 2005). One of the last, large-scale events was the New Zealand Rangitata orogeny (Jurassic–Early Cretaceous), and this, together with related events in other parts of the Australides, would have led to the origin of the South Pacific groups (although not their later disjunctions). The timing of the origin of *Abrotanella* is not well constrained because of repeated reactivation in the orogenic belt, but one or more major events of the Rangitata orogeny were probably the cause of the differentiation. During the orogeny, the New Zealand basement terranes were juxtaposed, fused, and uplifted, and the geography underwent revolutionary changes.

As mentioned, the origin of *Abrotanella* is related to the origin of its sister group, the rest of the diverse, cosmopolitan Senecioneae. The basal node in the tribe separates *Abrotanella* from the rest, while the next few nodes in the group shed indirect light on *Abrotanella*. A recent phylogeny of Senecioneae is as follows (Pelser et al., 2010, combined data set):

Abrotanella: Australasia and Patagonia.

Capelio: South Africa (SW Cape region).

Chersodoma: South America (NW Argentina to northern Peru).

All other Senecioneae: Cosmopolitan.

A dispersal model might infer a South Pacific center of origin, with subsequent dispersal across the Indian Ocean to South Africa, then across the Atlantic to South America, and from there to the world. A vicariance model is more straightforward, and proposes differentiation of an already global ancestor, first around the South Pacific, then around South Africa, and then around the region that began to rise as the Andes in the Late Jurassic or Early Cretaceous. The three basal nodes are all southern and allopatric, and could have been active around the same time. Subsequent overlap has developed between the cosmopolitan clade and all three basal clades, but the simple allopatry of the first three probably reflects their original vicariance.

The overlap of the fourth clade of Senecioneae with the others may have occurred soon after the events of the first three nodes, and so it would coincide with the widespread epicontinental flooding of the Cretaceous. If the overlap developed in *mid-Cretaceous* time, it would be compatible with *Late Cretaceous* rifting producing South Pacific disjunctions in both *Abrotanella* and in the fourth, cosmopolitan, group of Senecioneae.

One of just three clades in this last group is made up of *Brachyglottis* and its allies, and again this has a standard, disjunct distribution in the southern hemisphere. The *Brachyglottis* clade has originated after *Abrotanella* and *Capelio* in the phylogeny, but shows similar breaks around South Africa, New Guinea, and South America, indicating repeated episodes of differentiation at these localities. The phylogeny of the *Brachyglottis* complex is as follows (Pelser et al., 2010):

South Africa (east of the southwestern Cape and *Capelio*) (*Caputia*).

New Zealand, montane to alpine: South and Stewart Islands (*Dolichoglottis*).

New Guinea (*Papuacalia*), New Zealand (*Brachyglottis, Haastia, Urostemon, Traversia*), SE Australia and Tasmania (*Bedfordia, Centropappus*), and central Chile (*Acrisione*).

The break between the New Zealand *Dolichoglottis* and its trans-Pacific, trans-Tasman sister group occurred no later than the disjunctions in that sister group. This implies that *Dolichoglottis* is much older than the Neogene mountains it now inhabits, and this is probably also true for *Abrotanella*. If the overlap of the four main Senecioneae clades occurred in mid-Cretaceous, the basal nodes, including the origin of *Abrotanella*, could reflect Jurassic–Early Cretaceous events such as the Rangitata orogeny.

Differentiation of the Three Main Clades in Abrotanella

It is difficult to establish the precise time of origin of *Abrotanella*. Dating differentiation within the genus is more straightforward though. The three main clades (I, II, and III in Figure 10.10) are almost completely allopatric. Local overlap occurs only in central Tasmania and on one mountain in Stewart Island (Heads, 1999).

The main breaks dividing the clades do not correlate with ocean basins of today. Instead, one break (between clade I and clades II + III) is located in Tasmania, while the other (between clades II and III) is in Stewart Island. The phylogeny indicates that these breaks occurred before rifting in the Tasman disrupted clade II and rifting in the Pacific split clade III. Thus, the breaks can be attributed to the Early Cretaceous prebreakup extension and plutonism that developed in and around Tasmania and Stewart Island (Allibone and Tulloch, 2004). Leaman (2003) cited an "igneous event patchily recorded in Tasmania at 100 Ma," and Kohn et al. (2002) proposed significant cooling (uplift) there at 110–90 Ma. This igneous activity and tectonism was a precursor to final rifting at 84 Ma.

The overall distribution of *Abrotanella* (Figure 10.10) conforms to a "star pattern" seen in other groups, such as *Astelia* (Figure 4.11). Another example is the *Hebe* complex (Plantaginaceae) (Figure 10.11, phylogeny from Albach and Meudt, 2010). As in *Astelia* and the *Hebe* complex, related, disjunct clades of *Abrotanella* are allopatric outside New Zealand but overlap within the country.

The *Hebe* group (sometimes treated as part of *Veronica*) has ~100 species. It is much more diverse than *Abrotanella* and includes lowland trees as well as alpine subshrubs. It also has a more complete

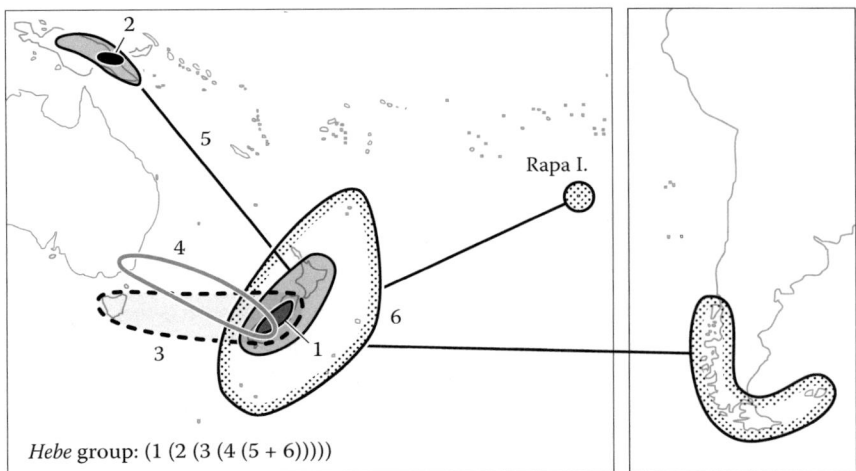

Hebe group: (1 (2 (3 (4 (5 + 6)))))

FIGURE 10.11   Distribution of *Hebe* and allies (Plantaginaceae): 1, *Leonohebe*; 2, *Detzneria*; 3, *Chionohebe* (including *Parahebe planopetiolata*); 4, *Hebejeebie* (including *Parahebe spectabilis*); 5, *Parahebe*; 6, *Hebe*. Phylogeny indicated with nested numbers. (Phylogeny from Albach and Meudt, 2010: Figs. 1–3.) (From Heads, M., South Pacific biogeography, tectonic calibration, and pre-drift tectonics: Cladogenesis in *Abrotanella* (Asteraceae). *Biol. J. Linnean Soc.* 2012c. 107. 938–952. Copyright John Wiley & Sons. Reproduced with permission.)

"star pattern" than *Abrotanella*, with clades present on the Chatham and Kermadec Islands, in the east and north of New Zealand, and on Rapa Island in the central Pacific. Nevertheless, the alpine-subantarctic *Abrotanella* is likely to be absent from these warm, low-altitude islands for simple ecological reasons, and the distributions of the two groups show many parallels. Both have an overall distribution in Australasia and southern South America; both are absent in New Caledonia and western Australia; and both comprise trans-Tasman, trans-Coral Sea, and trans-Pacific disjunct clades.

A similar setup is seen in the subshrubs and small trees in *Olearia* "clade 2" (Asteraceae) (Heads, 2014; Figure 6.9). It includes a trans-Tasman clade (*Celmisia*, *Pachystegia*, and some *Olearia* species), a New Zealand–New Guinea affinity (the *Olearia arborescens* group and *O. velutina*), and a widespread southern New Zealand group (*Pleurophyllum* and relatives). (It probably includes the *Chiliotrichum* group of South America, although this has been questioned. *Pacifigeron* from Rapa Island may also be related but has not been sequenced.)

The similar distributions (including trans-Tasman, trans-Coral Sea, trans-Pacific groups) suggest that the main clades within *Astelia*, *Abrotanella*, and the *Hebe* group and *Olearia* "clade 2" differentiated at about the same time, during a phase of prebreakup extension. This occurred before (perhaps soon before) each of the main subgeneric clades was sundered by seafloor spreading. Predrift divergence has occurred at breaks in or around New Zealand, with subsequent local overlap.

Differentiation within the Widespread Tasman Group of Abrotanella (*Clade II*)

In clade I of *Abrotanella* (Tasmania) and in clade III (Stewart Island–South America), the distribution is fairly straightforward (Figure 10.10). In clade II, a trans-Tasman group, it is more complex. Clade II comprises the *pusilla* group and the *linearis* group (Figure 10.12a and b). Both are in New Zealand, but outside New Zealand they display an intriguing pattern of allopatry: only the first is in mainland Australia, Tasmania, Central Otago, and the New Zealand subantarctic islands, while only the second is in New Guinea and central North Island. The two clades overlap in Stewart Island, western South Island, and southern North Island. Their distributions within the main region of overlap, the South Island, are shown in more detail in Figure 10.12c and d.

Outside New Zealand, the break between the *pusilla* group and the *linearis* group coincides with a large igneous province. This is a vast belt of magmatism comprising the Whitsunday volcanic province of Australia, the Median Batholith of New Zealand, and similar granites in Thurston Island, West

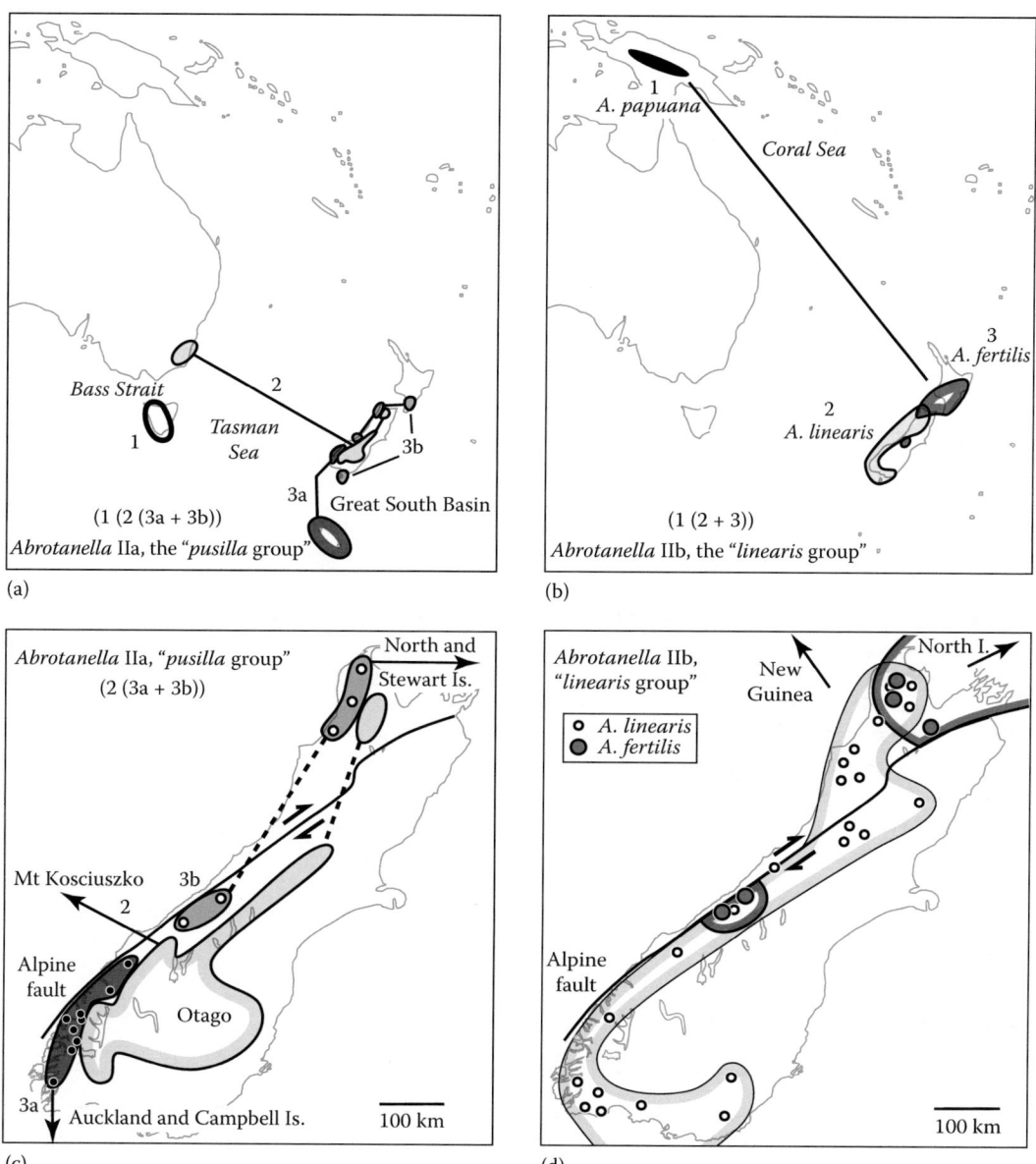

FIGURE 10.12 Distribution of *Abrotanella* "clade II" (Asteraceae). (a) The "*pusilla* group." (b) The "*linearis* group." (c) The "*pusilla* group" in the South Island. (d) The "*linearis* group" in the South Island. The disjunct southern populations of *A. fertilis* are at the Copland Range and at Lake Sweeney, 2–3 km east of the Alpine fault. (Species distributions from Swenson, 1995; Phylogeny from Swenson et al., 2012.) (From Heads, M., South Pacific biogeography, tectonic calibration, and pre-drift tectonics: Cladogenesis in *Abrotanella* (Asteraceae). *Biol. J. Linnean Soc.* 2012c. 107. 938–952. Copyright John Wiley & Sons. Reproduced with permission.)

Antarctica (broken line in Figure 10.10; Antarctica localities not shown; from Rey and Müller, 2010). In New Zealand, the belt corresponds with the boundary between Western and Eastern Provinces. The significance of this break in delimiting the distributions of weevils and alpine plants within the South Island was illustrated earlier (Figures 7.3, 7.4, and 10.6).

In Australia, the Whitsunday volcanic province is made up of mid-Cretaceous silicic volcanics with a main period of eruption from 120 to 105 Ma (Bryan and Ernst, 2008). The last phase of Median Batholith magmatism occurred at the same time, with belts of granite batholiths emplaced from 125 to 105 Ma. This extension-related, prebreakup magmatism developed near the Gondwana margin and more or less parallel to it. But the magmatic igneous belt followed a somewhat different line from the spreading center that developed soon after and led to the opening of the Tasman and Coral Seas. The prebreakup magmatic belt is aligned with the break separating the *pusilla* and *linearis* groups and also explains the node between clades II and III, at Stewart Island.

The *pusilla* and *linearis* groups (Figure 10.12) are allopatric in eastern South Island, but overlap in western South Island, Stewart Island, and southern North Island. The overlap developed after the phylogenetic break between the two groups (attributed here to mid-Cretaceous magmatism), but before Miocene displacement on the Alpine fault (discussed later). During this period of time—Late Cretaceous and Paleogene—there was crustal extension, seafloor spreading, subsidence, and widespread marine transgressions in many parts of the New Zealand plateau (Zealandia). It is possible that high levels of environmental disturbance associated with marine transgressions facilitated the overlap.

Within the *pusilla* group, there is perfect allopatry among clades 1, 2, 3a, and 3b (Figure 10.12a and c), despite the fact that the last three of these "overlap" in the South Island. The first node corresponds to the Bass Strait basins, which were most active in the Early Cretaceous. The two remaining clades, 2 and 3, form parallel arcs along western New Zealand, although there is a gap in the western clade 3 around the Haast River, and the Central Otago part of clade 2 connects with Mount Kosciuszko in southeastern Australia. Clade 3a connects Fiordland with the New Zealand subantarctic islands. Clade 2 (*Abrotanella caespitosa* group) and clade 3b (*A. pusilla*) are both disjunct along the Alpine fault, where strike-slip displacement began in the Miocene.

Within the *linearis* group (Figure 10.12b and d), the three clades are mainly allopatric, with possible overlap in northwest Nelson and Westland. *Abrotanella fertilis* is disjunct along the Alpine fault.

Three of the four New Zealand clades in the *pusilla* and *linearis* groups (clade II; Figure 10.10c and d) are disjunct along the Alpine fault. This pattern, with the offsets having the same, dextral sense as the displacement on the fault, is documented in more than 200 other groups (Chapter 9). In contrast with chance dispersal or ecological explanations, a vicariance model predicts that similar breaks will recur in different groups with different ecology and means of dispersal. Groups that are disjunct along the Alpine fault include invertebrates from lowland forest (the harvestmen), vertebrates that occupy open vegetation, from lowland to alpine elevations (the skinks), and alpine plants (*Abrotanella*).

Evolution in *Abrotanella*: A Synthesis

Reconstructions of Abrotanella *before Neogene Deformation*

Distributions of *Abrotanella* are plotted here on reconstructions for 27 Ma (Kamp and Furlong, 2010). The reconstruction for the *pusilla* group (clade IIa) is straightforward, and it indicates that Central Otago connects with Mount Kosciuszko via northwest Nelson (Figure 10.13), not directly (Figure 10.12c). The reconstruction for the *linearis* group (clade IIb) also results in a simple arrangement, with the distribution restored to form northern and southern clades (Figure 10.14).

Allopatry and Overlap in Abrotanella

One of the most obvious features in the biogeography of *Abrotanella* is the high level of allopatry seen at different levels throughout the phylogeny. There are also two important phases of regional overlap. The overall pattern can be summarized as follows:

Allopatry between the genus and the three other main clades of Senecioneae, followed by complete overlap of the basal three clades by the fourth, diverse clade of Senecioneae.

Allopatry among the three main clades of the genus (I, II, and III) at breaks in Tasmania and Stewart Island (Figure 10.10), followed by minor local overlap.

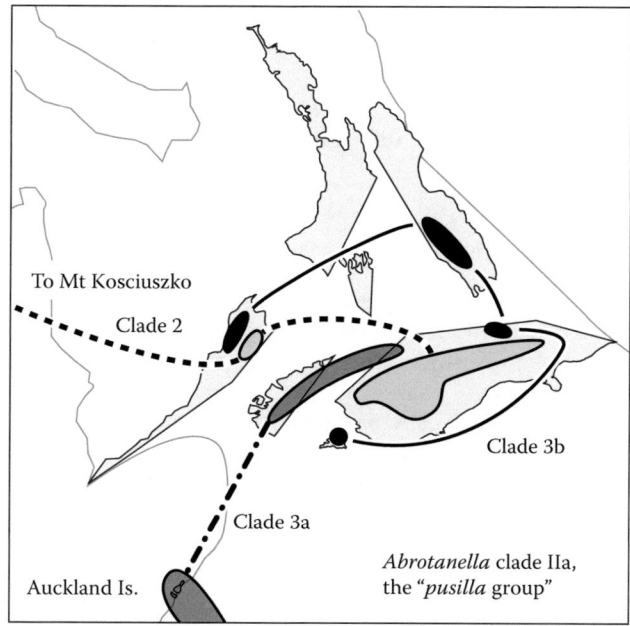

FIGURE 10.13 Distribution of *Abrotanella* clade IIa (Asteraceae) on a 27 Ma reconstruction (cf. Figure 10.12a,c). (Basemap from Kamp, P.J.J. and K.P. Furlong, Tectono-sedimentary framework for exploration in New Zealand basins, http://www.nzpam.govt.nz/cms/pdf-library/petroleum-conferences-1/2010-nzpc-technical-posterspapers/P12_Kamp-Furlong_paper.pdf, 2010.)

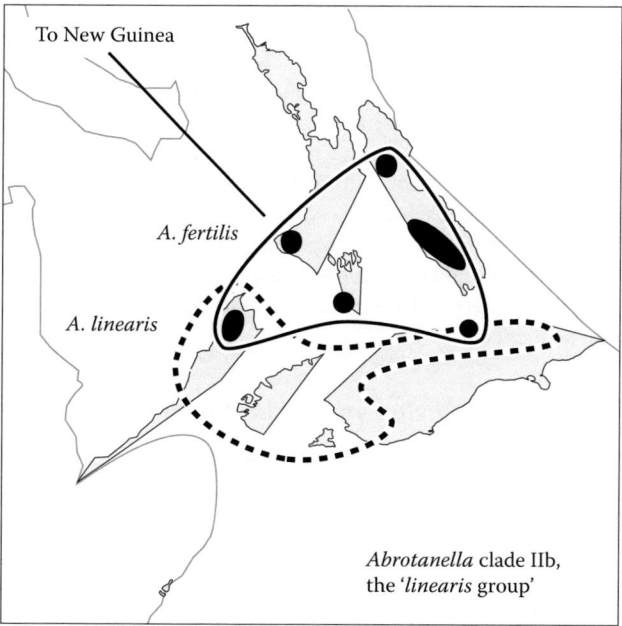

FIGURE 10.14 Distribution of *Abrotanella* clade IIb (Asteraceae) on a 27 Ma reconstruction (cf. Figure 10.12b,d). (Basemap from Kamp, P.J.J. and K.P. Furlong, Tectono-sedimentary framework for exploration in New Zealand basins, http://www.nzpam.govt.nz/cms/pdf-library/petroleum-conferences-1/2010-nzpc-technical-posterspapers/P12_Kamp-Furlong_paper.pdf, 2010.)

Allopatry in the widespread Tasman clade (clade II) between the *pusilla* group (IIa) and the *linearis* group (IIb) (Figure 10.12); subsequent overlap in western South Island, Stewart Island, and southern North Island.

Allopatry within the *pusilla* group (Figure 10.12c), perfect allopatry among all three clades.

Allopatry within *linearis* group (Figure 10.12d), clades mostly allopatric; possible local overlap in northwest Nelson and Westland.

The New Zealand Clades of Abrotanella

Four trans-oceanic clades of *Abrotanella* (Figure 10.10) straddle ocean basin spreading centers and overlap in central New Zealand. These have their respective extra-New Zealand distributions in the subantarctic islands, southeastern Australia, New Guinea, and South America, and also have different distributions within New Zealand.

Traditional theory explains the "star" pattern by proposing a center of origin in New Zealand, followed by chance dispersal to the remaining localities. The distribution pattern can also be explained by differentiation of a widespread Australasian-Patagonian ancestor into the modern clades. In this vicariance model, differentiation occurred during prebreakup extension (late Early Cretaceous) in what became the New Zealand region and elsewhere. This was followed by limited local *overlap* of the clades in proto-New Zealand, and also *rifting* of the clades between New Zealand and other localities with seafloor spreading. Seafloor spreading took place in the Late Cretaceous–Eocene and formed the basins in Bass Strait, the Tasman and Coral Seas, and the Southern Ocean. The last main phase of biogeographic evolution in *Abrotanella* took place with the rupturing of New Zealand clades by Neogene strike-slip displacement.

Differences between South Pacific Groups: Abrotanella and the Hebe Complex

The distributions of the clades in *Abrotanella* and in the *Hebe* complex show marked similarities, as discussed earlier (Figures 10.10 and 10.11). One interesting difference between the two groups concerns the phylogeny. In *Abrotanella*, the trans-Pacific clade III differentiates before the trans-Tasman (IIa) and trans-Coral Sea (IIb) clades, while in the *Hebe* group the trans-Tasman clade differentiates before the trans-Coral Sea and trans-Pacific groups. This incongruence might be attributed to chance dispersal or extinction, but other alternatives include more or less simultaneous developments during prebreakup tectonics. For example, during the last phase of Median Batholith magmatism, belts of granitic plutons were intruded. At the same time, metamorphic core complexes were exhumed across the South Island (Forster and Lister, 2003). The activity at both the granite belts and the core complexes through the mid-Cretaceous (Aptian-Albian) could have led to the origin of what became trans-Tasman, trans-Coral, and trans-South Pacific clades in different groups, such as *Abrotanella* and the *Hebe* complex, *before* final breakup affected clades in both. Episodic differentiation at several nodes over the same time period would have led to similar distributions developing in the groups but in different phylogenetic sequences.

Regional Tectonic History and Phylogeny in Abrotanella

The main tectonic developments in eastern Australasia since the mid-Mesozoic are listed next. These are described in terms of the *age* of the events, the *tectonic regime* in place, the nature of the *tectonic events*, and spatially related *phylogenetic–biogeographic breaks* in *Abrotanella*:

1. Jurassic–Early Cretaceous. Compression. Terranes that were spread along ~6000 km of the Gondwana margin (from New Guinea to Tasmania and possibly Antarctica) were translated along the margin, juxtaposed, and sutured to form proto-New Zealand (Rangitata orogeny). The timing and location of these events are consistent with their having caused the origin of *Abrotanella*. Events that occurred about the same time, or soon after, around South Africa and South America caused the next breaks in the remaining Senecioneae.

2. End of Early Cretaceous. Extension. Prebreakup magmatism and faulting in Tasmania and Stewart Island. First two nodes in *Abrotanella* (separating clades I, II, and III; Figure 10.10).

3. End of Early Cretaceous. Extension. Prebreakup magmatism in Queensland (silicic volcanism) and New Zealand (granite intrusion). Break in clade II between the *pusilla* group (IIa) and the *linearis* group (IIb).

4. Mid–Late Cretaceous. Extension. Rifting in Bass Strait, Tasman, and Great South Basins. Rifting in clade II.

5. Late Cretaceous. Extension. Seafloor spreading in the Southern Ocean separated New Zealand, Antarctica, and South America. Rifting in clade III.

6. Miocene. Transpression (oblique compression). Strike-slip displacement on the Pacific/Indo-Australian plate margin at the Alpine fault. Rifting of three New Zealand groups in clade II along the Alpine fault (Figure 10.12c,d).

7. Pliocene onwards. Transpression. Uplift of the modern New Zealand mountains (Kaikoura orogeny), with the main divide running parallel with the Alpine fault and several kilometers east of it. Little obvious effect on the biogeography of *Abrotanella* apart from the uplift of many populations.

Together, the main tectonic events in the region account for the main phylogenetic and geographic breaks in the *Abrotanella* phylogeny. The main nodes in the phylogeny form a coherent, sequential pattern that tracks the main phases of tectonic history step by step. In addition, the tectonic calibration implies that the main phase of overlap (between the *pusilla* and *linearis* groups) occurred in the mid-Cretaceous–Paleogene, coinciding with global marine transgressions. Other features of the biogeography that are not examined here also coincide with tectonic features and are standard patterns in many other groups (Heads, 1999). For example, in New Guinea, the northern boundary of the genus corresponds with the edge of the Australian craton. The group does not occur in the mountains of the accreted terranes, in northern New Guinea, despite apparently suitable habitat there. In Patagonia, *Abrotanella submarginata* shows disjunction along the strike-slip margin of the South America and Scotia plates, and, as in the South Island, the disjunction correlates with the geological displacement.

The Case of *Abrotanella*: Some Implications for Biogeographic Methods

One explanation for the precise interlocking of geographic clades in *Abrotanella*—from tribal level down to species level—is chance dispersal. Based on fossil-calibrated, post-Cretaceous dates for *Abrotanella*, Wagstaff et al. (2006) inferred a "convoluted" history of chance, trans-oceanic dispersal. Yet the actual distribution patterns (not discussed by Wagstaff et al.) indicate high levels of allopatry throughout the phylogeny and could hardly be less convoluted. The only evidence for dispersal is some local clade overlap, mainly in the western South Island. The mechanism for this range expansion would have been the plants' normal, observed means of dispersal deployed by whole populations, not the cryptic, "nonstandard" mechanisms required for one-off, trans-oceanic dispersal events over thousands of kilometers.

Swenson and Bremer (1997) concluded:

> … *Abrotanella* must have reached New Zealand by long-distance dispersal long after its separation from other continents… Possible long-distance dispersal events of *Abrotanella* are hard to envision, however,…such occasions must be extremely rare, otherwise there would be a lot of distributional noise and we would not be able to perceive any general biogeographic pattern.

This last point is crucial—the biogeographic pattern has such a simple, precise structure, that any chance events in its development must have been very rare indeed. In fact, the proposed chance dispersal events, at the base of the phylogeny, have been unique in geological time: once across the Pacific, once between New Zealand and New Guinea, once across the Tasman, and once across the Great South Basin. Why have the different clades each crossed a different basin? And why have they each crossed their respective basin only once? As stressed in Chapter 1, extremely rare events (e.g., tsunamis, megastorms, floating islands with living vegetation) that only happen, say, once in a thousand years could in theory have allowed diaspores to cross the Tasman or Pacific basins and establish on the other side. However, such rare events would have occurred ~65,000 times since the Cretaceous (5,000 times since the Miocene) and so would have occurred much too often to explain the simple allopatry.

The proposed trans-Pacific, trans-Tasman, and other transport events in *Abrotanella* have each happened just once in the entire history of the genus, and, as Swenson and Bremer (1997) concluded, they are hard to envision. They are one-off events that occur at random and are not associated with any other factor. The probability of their ever happening is vanishingly small. Yet despite this, the same, distinctive "star" pattern seen in *Abrotanella*, with a trans-Tasman affinity, a trans-Coral Sea affinity, and a trans-Pacific affinity, recurs in other, unrelated clades, including *Hebe* and *Astelia*, that all have different ecology and means of dispersal.

The allopatry of the "star pattern" clades outside New Zealand and their overlap in New Zealand mean that founder-advantage theory does not account for the clades' origin or distribution. Overall, the evidence is instead best explained by predrift vicariance followed by seafloor spreading and, later, local range expansion. The three phases were caused, not by chance, but by large-scale, geological and climatic change affecting whole communities.

Phases of range expansion, as with phases of vicariance, are not determined by chance, but by large-scale environmental changes. For example, the last, great sea level maxima are recorded in the mid-Cretaceous and were caused by a combination of tectonic and climatic factors. In models calibrated with phylogenetic/tectonic events that occurred before and after the flooding, the flooding itself coincides with the last, great phase of biogeographic overlap in the Amazon and Congo basins (Heads, 2012a). This is also the case in Pacific groups, such as *Abrotanella*, that indicate a major overlap of clades in the mid-Cretaceous.

Is Evolution Clocklike?

A typical distribution pattern such as Australasia–southern South America is displayed in many organisms with a wide range of means of dispersal and ecology, indicating that these factors are not the reason for the disjunction. Likewise, it has long been known that the disjunction also occurs in groups with very different taxonomic rank, from species groups (e.g., *Abrotanella* clade III), to genera (e.g., *Nothofagus*) to families (e.g., Gesneriaceae). The component groups of the pattern show a great range in their degree of differentiation, and this is probably true for all important distribution patterns, for example, disjunctions across the Tasman Sea. As mentioned already, early naturalists such as Hutton (1872) recognized this as good evidence that the rate of evolution has changed between and within lineages (Chapter 2).

Estimating Sampling Error in the Fossil Record

Abrotanella has no fossil record. Based on the fragmentary fossil record of its family, Asteraceae, Swenson et al. (2012) estimated that the genus originated at ~38 Ma, and this suggests a gap in the record of *Abrotanella* of 38 Myr. The tectonics-calibrated analysis given earlier suggests instead that *Abrotanella* originated at ~145–120 Ma, indicating an equivalent gap in the fossil record.

Abrotanella *and Tectonic–Vicariance Calibration*

Dispersal theory suggests that the best way to understand evolution in *Abrotanella* is to look at the fossil record and explain differentiation by chance dispersal. Swenson et al. (2012) adopted this approach, and they rejected the idea of tectonic/biogeographic calibration, but without giving it a test—they did not examine the biogeography of the group or map the clades. Swenson et al. did not consider any tectonic events apart from the rifting of the New Zealand plateau from Gondwana, yet the main phylogenetic and geographic nodes in *Abrotanella* indicate that prebreakup tectonics (extension, volcanism, and plutonism) and postbreakup tectonics (strike-slip displacement) are just as important as the seafloor spreading.

Many studies now avoid the use of fossils to give maximum clade ages and instead use tectonic calibrations (see Chapter 2). Most of these studies still use only single tectonic calibrations that are often remote from the study group, in geography and in phylogeny. For example, birds of paradise (Paradisaeidae) are found mainly in New Guinea. Irestedt et al. (2009) calibrated a phylogeny of the group by attributing the basal break in passerines to the rifting between Gondwana and New Zealand, where the basal passerine group is endemic. The results were very interesting, but they could be developed further by including the many tectonic/biogeographic calibrations that are possible in the family Paradisaeidae itself (Heads, 2002). This is the method used here for *Abrotanella*; multiple nodes are dated using tectonics, and this involves a broad engagement of molecular phylogeny and biogeography with structural geology.

There are two possible approaches in biogeographic analysis, and these are illustrated in the two models for *Abrotanella*. The first, chance dispersal, implies that evolutionary rates do not show extreme changes, and that overall the fossil record is more or less accurate. It predicts no shared patterns; because of their different branch lengths, the components of a single biogeographic pattern are inferred to have developed the pattern at different times, not because of a single, causative event. The second option—phases of vicariance and phases of overlap—implies significant changes in evolutionary rate, so that accepting both prebreakup and breakup vicariance in *Abrotanella* does not need to mean that Asteraceae are 1.5 billion years old, for example. This model is consistent with the idea that fossil record of terrestrial life has massive gaps, as indicated when extant groups with no known fossil record turn up in Cretaceous rocks (Heads, 2012a).

Predrift Rifting and Austrochiloid Spiders: Australia–New Zealand; Tasmania–South America

Many other groups have patterns that suggest predrift vicariance, as in *Abrotanella*. The craneflies (Tipulidae and allies) are a good example (Ribeiro and Eterovic, 2011; Heads, 2014: 84). In spiders, the Austrochiloidea of Australasia and southern South America were mentioned earlier because of their standard range and their global sister. Austrochiloidea comprise two families, distributed as follows (Forster et al., 1987; World Spider Catalog, 2016):

Trans-South Pacific. Tasmania and southern South America (Austrochilidae).

Trans-Tasman. Eastern mainland Australia (Queensland to Victoria) and New Zealand (South and Stewart Islands) (Gradungulidae).

The two families show precise allopatry. This can be explained by a first phase of vicariance, before a second round of vicariance led to the breaks *within* each family, in the Pacific and Tasman Basins, respectively. As in *Abrotanella*, the overall pattern can be explained by initial, predrift vicariance, including a break between Tasmania and mainland Australia that coincides with the Cretaceous basins around Bass Strait (Heads, 2014).

11

Some More Case Studies of New Zealand Plants

This chapter includes case studies from four other plant families in New Zealand: Ranunculaceae, Fabaceae, Apiaceae, and Brassicaceae. The main themes are the same as in the last chapter: intercontinental affinities of New Zealand groups can be correlated with the groups' distributions within New Zealand, and spatial evolution at both scales can be explained with reference to major events in tectonic history.

Ranunculaceae in New Zealand

New Zealand has six indigenous genera of Ranunculaceae. *Clematis* is a forest liane, while the others are herbs that grow in open, nonforest vegetation. *Clematis* is a cosmopolitan genus and the affinities of the New Zealand members require further study; the five other genera display trans-Indian, trans-Tethyan, and trans-Pacific affinities, and these illustrate the main global connections of the New Zealand biota.

Caltha

The genus *Caltha* has an interesting, disjunct distribution. It is widespread through the north temperate zone, and it also has a monophyletic clade in the Andes and Australasia. The phylogeny is as follows (Schuettpelz and Hoot, 2004; Cheng and Xie, 2014) ("North America" = United States and Canada):

North America.
　　North America and Eurasia.
　　　　North America.
　　　　　　Northern and Southern Andes.
　　　　　　　　Southern Andes.
　　　　　　　　　　New Zealand (North, South, and Stewart Islands) + SE Australia.

Dispersal theory predicts a center of origin in the northern hemisphere because of the basal, paraphyletic group there (Schuettpelz and Hoot, 2004; Cheng and Xie, 2014). Nevertheless, while the initial breaks took place in the north, the genus may have already attained its wide distribution by the time these occurred. In this case, the phylogeny would represent a series of differentiation events rather than dispersal events.

A key problem for the southern hemisphere members is the cause of the break between the southern Andean–Australasian clade and its sister group in the southern and northern Andes. This occurred before the Australasia–southern Andes break (beginning in the mid-Cretaceous) and could have taken place at central Chilean nodes, as in *Ourisia* (Plantaginaceae) and marsupials (Heads, 2014; Figures 3.8 and 3.9). In *Caltha*, the initial differentiation was followed by overlap in the southern Andes. The pan-Andean–Australasian group forms a clade with North American species, giving a disjunct triangle: Andes–Australasia–North America. This "Pacific triangle" distribution includes disjunctions across Central America and the Caribbean, as well as across the South Pacific, and it is repeated in many other groups (Heads, 2014: 115).

As for timing, Schuettpelz and Hoot (2004) proposed a vicariance model to explain the disjunction in the South Pacific clade (following dispersal of the genus from the north). They accepted that *Caltha* had originated by the mid-Cretaceous, and this is accepted here. (Cheng and Xie, 2014, estimated a much younger age, but this was based on fossil calibrations.)

Anemone, and Possible Differentiation between East and West Antarctica

In *Anemone*, most of the 200 species occur in the northern hemisphere, but there are a few in the south, including one in New Zealand (South Island and the Tararua Range, southern North Island). As shown in Figure 11.1, the genus has two branches (Schuettpelz et al., 2002; Hoot et al., 2012):

> Subg. *Anemonidium*. Northern hemisphere plus South Pacific: *Anemone tenuicaulis* of New Zealand and its sister group, *A. antucensis* of central Chile (Biobío) and adjacent Argentina (Neuquén).
>
> Subg. *Anemone*. There are two main clades: sect. *Pulsatilloides* in the southern hemisphere (plus Mexico) and its sister group, comprising the remaining sections in the northern hemisphere. The latter also includes two small, southern clades in Africa and South America (indicated on Figure 11.1 with gray lines).

The two subgenera show great overlap in the northern hemisphere, but almost complete allopatry in the south. One simple explanation is that the two subgenera diverged as Indian/Atlantic basin and Pacific basin vicariants, with subsequent range expansion in the northern hemisphere and around central Chile/ Neuquén. The timing of this overlap is obscure, but it could have occurred as recently as the recession of the Pleistocene glaciers. The original phylogenetic break between the two subgenera took place between Tasmania and New Zealand, and somewhere between central Chile and the rest of South America. The simplest explanation for the break in the Tasman is the Cretaceous opening of the Tasman Basin.

The different extant distributions of the two subgenera of *Anemone* suggest that these groups also had distinct ranges in Antarctica prior to the Neogene glaciation; one possible reconstruction is illustrated in Figure 11.2. In this model, the trans-Pacific subgenus *Anemonidium* occurred in West Antarctica, the Atlantic/Indian Ocean subgenus *Anemone* in East Antarctica. These two parts of Antarctica—the non-cratonic West Antarctica and cratonic East Antarctica—are separated by the West Antarctic Rift System (cf. Figure 6.2).

The overall pattern in the genus *Anemone* might be attributed to dispersal of the two southern clades from the north, but this would not account for the southern clades' allopatry. Schuettpelz et al. (2002: 149) concluded that "achene morphology and the relatively restricted ranges of many of the anemones in question make long-distance dispersal events unlikely. Therefore a more parsimonious explanation invokes a vicariance model... *Anemone* could have already undergone considerable radiation by

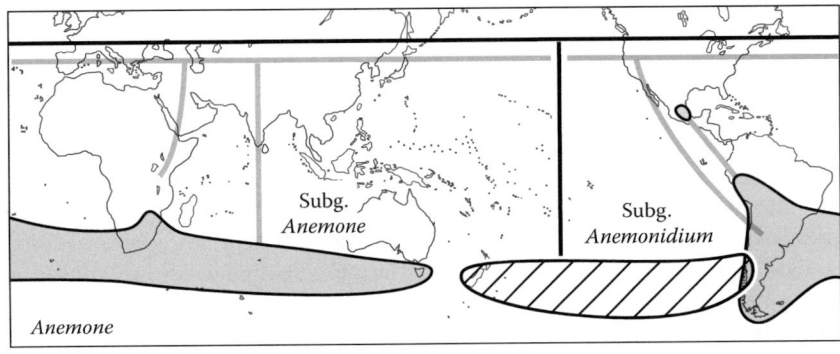

FIGURE 11.1 Distribution of *Anemone* (Ranunculaceae), showing the two main clades. Northern hemisphere distributions indicated by horizontal lines. (Data from Hoot, S.B. et al., *Syst. Bot.*, 37, 139, 2012.)

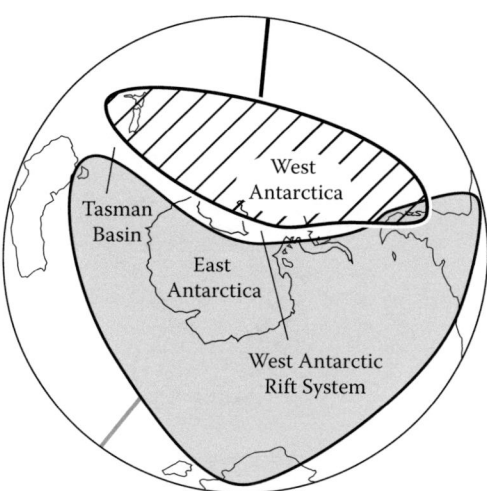

FIGURE 11.2   A reconstruction of the distribution of *Anemone* before glaciation in Antarctica (cf. Figure 11.1).

~100 mya." Schuettpelz et al. (2002) drew the interesting comparison between the pattern in *Anemone* and that of the family Proteaceae, which also includes both Pacific groups and Indian Ocean groups (Heads, 2014; Figure 4.4).

The overall distribution of *Anemone* comprises widespread groups in the northern hemisphere, and restricted outliers in the southern hemisphere. All members, except for a small number of derived members, occur together in a single region—the north—and in traditional theory, this is assumed to be the center of origin. A vicariance model, as proposed here, is an alternative model, and similar patterns occur in other groups. For example, *Euphorbia* subg. *Esula* (Euphorbiaceae) is widespread globally with 480 species, but the only species native to South America is *E. philippiana* of central Chile, widely disjunct from the rest of the group. The only other southern hemisphere species of *Esula* form an Indian Ocean clade, comprising *E. glauca* of New Zealand, its sister, *E. emirnensis* of Madagascar, and others in eastern Africa and Iran.

At some point, *Euphorbia philippiana* and its northern relatives have broken apart from the Indian Ocean clade and *its* northern relatives. As in *Anemone*, the break has been followed at some stage by overlap in the northern hemisphere, but the allopatry has been preserved in the south.

In a similar case, the liverwort *Scapania* has been described as a northern temperate group (Heinrichs et al., 2011), although it has several southern hemisphere species, including two in New Zealand. A molecular study of the genus found that the southern lineages are nested in what are otherwise northern hemisphere clades (Heinrichs et al., 2011). The authors supported the traditional theory, a northern center of origin, as they assumed that the location of a basal, paraphyletic grade represents a center of origin.

Clematis

Clematis is a worldwide genus with about 250 species, and it is closely related to *Anemone*. In New Zealand, *Clematis* is present on the mainland and the Three Kings Islands with about 10 species, but their relationships within the genus and with each other are still unresolved.

Two of the New Zealand species are mapped here (from Heenan and Cartman, 2000). *Clematis petriei* has a typical Castle Hill basin–Inland Kaikoura Range distribution (Figure 11.3, cf. Figures 7.17 and 7.20 through 7.22). *Clematis marmoraria* is restricted to northwest Nelson (Arthur Range), another well-known center of endemism (cf. Figures 7.3, 7.5, 7.11, 8.1, 8.7, 9.8, 9.20). *C. marmoraria* differs from the other New Zealand species in its alpine habitat and its nonlianoid, subshrub habit. It is the smallest known species of *Clematis*, and "at first glance it could be mistaken for a species of *Anemone* or *Ranunculus*…" (Grey-Wilson, 1987). A molecular study of the group would be worthwhile.

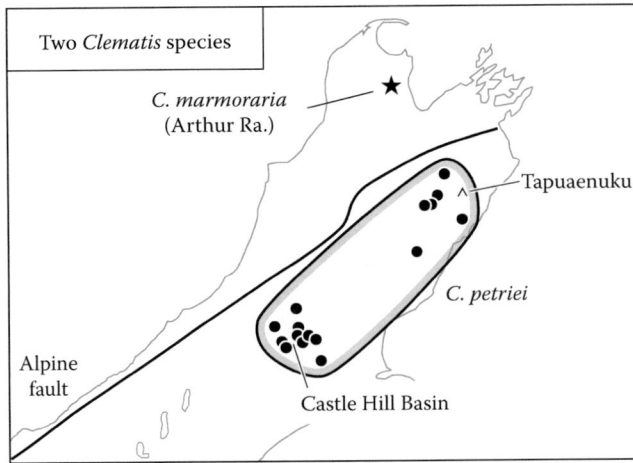

FIGURE 11.3   Distribution of *Clematis* species (Ranunculaceae). (Data from Heenan, P.B. and Cartman, J., *N. Z. J. Bot.*, 38, 575, 2000.)

Another New Zealand group, the *Clematis forsteri* complex, displays breaks, clines, and is endemic around the Cook Strait node. Plants with simple leaflets are restricted to the North Island and the northern South Island, while plants with pinnate leaflets are in the South Island and also the Cook Strait region of the North Island (Heenan and Cartman, 2000). The two main leaflet types are almost allopatric in their distribution, but overlap around Cook Strait and also around Mount Arthur in northwest Nelson. One form in the complex (*C. hookeriana*) is endemic to the greater Cook Strait area (Chapter 8): around Wellington; on Stephens, Trio, and Kapiti Islands in Cook Strait; and in the lower Awatere Valley in the South Island.

Ceratocephala

Ceratocephala occurs in New Zealand and from Kashmir to the Mediterranean (Smissen et al., 2003). It is a typical example of a Tethyan group, along with other New Zealand–Mediterranean taxa such as *Scleranthus* (Caryophyllaceae) and the tribe Aciphylleae (Apiaceae) (Heads, 2014). New Zealand has a single species of *Ceratocephala*, the endemic *C. pungens*. It occurs in lowland Central Otago and southwestern Canterbury (mapped by Garnock-Jones, 1984; there are also records from Middlemarch—the type locality, and from Kurow; Healy, 1948).

Myosurus and Central Otago halophytes

Myosurus is sister to *Ceratocephala* and is widespread through the north and south temperate zones. (For maps of these two genera, *Ranunculus*, and the remaining genera in tribe Ranunculeae, see Emadzade et al., 2011.) The only form of *Myosurus* in New Zealand, the endemic *M. minimus* subsp. *novae-zelandiae*, is closest to subsp. *minimus* of South Africa, Eurasia, North America, and Australia (Garnock-Jones, 1986). In *M. m.* subsp. *minimus* and in most species of *Myosurus* the flower has an elongate receptacle and as a result resembles the inflorescence of *Houttuynia* (Saururaceae) (Stevens, 2016). In contrast, the receptacle of *M. minimus* subsp. *novae-zelandiae* is very short. The telescoping and reduction evident here is the same trend that produced the typical angiosperm flower, with a receptacle that is more or less flat.

In New Zealand, *Myosurus* ranges from maritime and littoral sites along the east coast (Dunedin, mouth of the Waitaki River, Christchurch, Wellington), inland to the shores of Lake Manapouri, the Mackenzie basin in South Canterbury, and inland Marlborough. It does not occur west of the Alpine fault and is absent in lowland areas of Nelson, such as the Moutere depression, that have suitable climate and substrate. It also occurs in the North Island, but only around Wellington.

In a similar pattern, the indigenous New Zealand species of *Convolvulus* (Convolvulaceae) range from Central Otago (Kawarau Gorge, near the schist antiform) to the Wairau fault, and also the eastern North Island (Wellington, Cape Palliser, Hawkes Bay) (Heenan et al., 2003). These distributions lie parallel with the mountains because they are aligned with the fault, but the distribution and the fault probably both predate the mountains. This is indicated by the absence of *Myosurus* and *Convolvulus* west of the fault in Marlborough and Nelson. The distributional track intersects the coast around Cape Campbell and reappears in the North Island around Wellington.

Myosurus often grows on saline soils and is one of 15 native halophytes—mainly herbs—growing in the inland basins of Central Otago (Rogers et al., 2000). *Myosurus* occurs in shallow depressions that are seasonally wet, on "mildly sodic knolls and mildly saline plains and pans." Given the geological history, it is probable that the saline biota itself is derived from earlier marine incursions. Following the retreat of the coastline, its biota has persisted *in situ* as metapopulations on saline and sodic sites.

Reconstructing the woody vegetation of lowland Central Otago prior to anthropogenic disturbance is difficult. Nevertheless, shrubs such as *Pimelea*, *Hebe*, *Melicytus*, *Discaria*, *Carmichaelia*, and New Zealand "*Helichrysum*" occur in relict shrublands on lower valley slopes, and this suggests they formerly occupied the basin floors. These shrubs are not regarded as true halophytes, but some are tolerant of moderate salinity and sodicity; for example, *Carmichaelia compacta* grows on the margins of salt knolls (Rogers et al., 2000).

The salt-tolerant plants endemic to the inland South Island survive on patches of saline substrate that are dated as younger than 100,000 ka (Craw et al., 2013). Salination of rain shadow substrates in the central South Island is the result of marine aerosols being blown inland, the same process that has caused the salinity in Australian deserts (Heads, 2014). In New Zealand, the salination has continued since the rain shadow was formed by the Kaikoura orogeny mountains (Craw et al., 2013). Saline habitat of some sort was probably always available in New Zealand, for example, in drier areas (as in the current rainshadow) or along the shores of the inland seas. The present salination processes allow the continued survival of the earlier communities but probably did not cause their origin.

Ranunculus

Two of the New Zealand genera of Ranunculaceae discussed so far—*Caltha* and *Anemone*—have been attributed to mid-Cretaceous evolution (Schuettpelz et al., 2002; Schuettpelz and Hoot, 2004). In contrast, the worldwide genus *Ranunculus* has a high diversity in alpine New Zealand, and it has been suggested that this "radiation" evolved only in the last 3–5 Myr (Lockhart et al., 2001). The very different conclusions—Cretaceous evolution versus Pliocene evolution—are the direct result of the different calibrations and assumptions that were adopted by the respective authors. The study by Lockhart et al. (2001) was based on ITS rates inferred for *Dendroseris*, based on the assumption that *Dendroseris*, endemic to Juan Fernández Islands, is no older than the islands. This is questionable though, as other Juan Fernández Islands endemics (such as *Lactoris*) are known to be much older than the current islands. Thus, the young age proposed for New Zealand *Ranunculus* (Lockhart et al., 2001) is dubious, along with the inference of long-distance dispersal and the idea that differentiation in the New Zealand alpine ranunculi accompanied the Neogene orogeny.

For the tribe Ranunculeae as a whole, Emadzade and Hörandl (2011) and Emadzade et al. (2011) calibrated a phylogeny with fossils and island age (Juan Fernández Islands), treated the clade ages as maximum ages, and used these to rule out Mesozoic vicariance. Based on center-of-origin analyses (DIVA and DEC), they deduced a center of origin for Ranunculeae in the northern hemisphere, and a center of origin for *Ranunculus* in Eurasia. From these northern centers, the authors inferred multiple colonizations of different continents and the southern hemisphere by long-distance dispersal. They suggested that "the presence of endemic species of *Ranunculus* in some oceanic islands, far away from the continents (e.g., Hawaii Islands, Juan Fernández Islands, and Canarian Islands), also confirms that long-distance dispersal is possible in this tribe." This would be true if there had never been any other islands near the cited islands, but it is probable that there were (Heads, 2012a,b).

Within *Ranunculus*, Lehnebach (2008) described many interesting affinities. One group has a "Pacific triangle" distribution (as in a clade of *Caltha*), with representatives in New Zealand (*R. cheesemanii*,

R. brevis, R. ternatifolius, R. maculatus, and *R. membranifolius*), Argentina to Mexico, and North America. Trans-Tasman breaks are also evident (cf. *Anemone*); *R. recens* of New Zealand is nested among Australian species, as is *R. kirkii* of New Zealand. The Tasmanian *R. decurvus* is nested among *R. carsei* and *R. foliosus* of New Zealand. In a typical Tethyan connection, *R. sessiliflorus* of Australia is sister to *R. trilobus* of Europe and the Mediterranean. Another Tethyan clade in *Ranunculus* (clade 1 in Lehnebach, 2008) is structured as follows (Hörandl et al., 2005):

> Europe, lowlands (clade 8).
>> Malesia, mountains (clade 15).
>> Australia and NZ, mountains (including *R. lappaceus*—not the main NZ alpine clade) (clade 14).

In addition to the trans-Pacific, trans-Tasman, and trans-Tethyan patterns, *Ranunculus* also displays trans-Atlantic affinities between Africa and South America (Emadzade et al., 2011). The authors attributed these intercontinental connections to chance dispersal, but this depended on treating the calculated clade ages as maximum clade ages.

To summarize, the New Zealand Ranunculaceae show trans-Tasman, trans-Tethyan, and trans-Pacific connections, confirming the idea of the Tasman region as a biogeographic edge or margin of intercontinental significance. New Zealand Ranunculaceae also have interesting patterns *within* the country, and some of the best documented are in the alpine members of *Ranunculus*, discussed next.

The New Zealand Alpine Ranunculi: Ranunculus Sect. Pseudadonis

Sect. *Pseudadonis* (formerly known as the "New Zealand alpine" group) is the most conspicuous in New Zealand *Ranunculus*. It occurs in New Zealand (17 species) and southeastern Australia/Tasmania (two species) and is sister to a worldwide complex (Hörandl and Emadzade, 2012). The two clades make up *Ranunculus* "clade III." This means that a vicariance origin of sect. *Pseudadonis* is straightforward, while deriving the group from elsewhere requires additional, ad hoc hypotheses.

Within New Zealand, *Ranunculus* section *Pseudadonis* has a typical southern distribution, extending from the subantarctic islands north to the Egmont–East Cape line (Chapter 1). The section comprises four groups (Lockhart et al., 2001; Hörandl et al., 2005; Heenan et al., 2006; Lehnebach, 2008: Fig. 3C), numbered 1–4 on the maps shown here. One of the groups is a narrow endemic in the Eyre Mountains, while the other three have distributions that overlap in many parts of the New Zealand alpine zone. The groups themselves have good statistical support, but their mutual relations are not yet well resolved.

For group 1 (Figure 11.4), Lockhart et al. (2001) proposed a center of origin on the New Zealand mainland, followed by dispersal to Australia and the subantarctic islands. This is incongruent with the phylogeny though, as the subantarctic islands species (*Ranunculus pinguis*) is sister to the rest of the group, not nested in it, and so a vicariance history is more likely. The distributions suggest that the primary break in group 1 was caused by the rifting that formed the Great South Basin. The second node (the Australian species versus the three species in South and Stewart Islands) developed later, and so it can be related to the opening of the Tasman Basin.

A vicariance history for group 1 suggests that Fiordland populations have been separated from those in southeastern Australia and Tasmania. This trans-Tasman affinity is less common than the usual one, between Nelson and Australia, but there are precedents. These include the moss *Dicranoloma eucamptodontoides* of Tasmania and Fiordland (Percy Saddle) (Klazenga, 2003), and the angiosperm *Sprengelia incarnata* (Ericaceae) of mainland Australia, Tasmania, and Fiordland (West Cape to Five Fingers Peninsula).

Finally, in group 1 of *Ranunculus* sect. *Pseudadonis*, there are two allopatric clades with a typical "Southern Alps/Stewart Island + Central Otago" pattern (Figure 11.4). There is a narrow zone of overlap in the Humboldt Mountains.

Group 2 of sect. *Pseudadonis* is mapped in Figure 11.5 (phylogeny from Lockhart et al., 2001: Fig. 6). Lockhart et al. (2001) suggested that the group dispersed from New Zealand to Australia, but this does not account for the high level of allopatry among the four clades. In particular, the whole

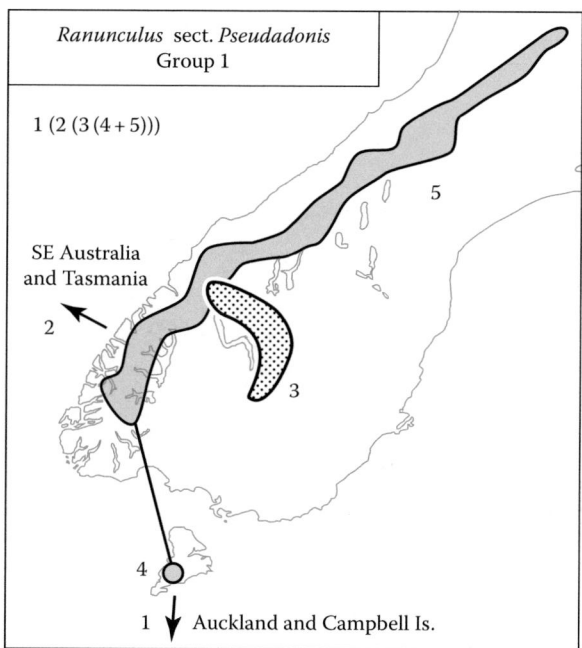

FIGURE 11.4 Distribution of *Ranunculus* sect. *Pseudadonis* Group 1 (Ranunculaceae). 1, *R. pinguis*; 2, *R. gunnianus*; 3, *R. pachyrhizus*; 4, *R. viridis*; 5, *R. sericophyllus*. (Data from Lockhart, P.J., *Annls. Missouri Botanical Garden* 88, 458, 2001; Heenan, P.B., *N. Z. J. Bot.*, 44, 425, 2006; Lehnebach, C.A., Phylogenetic affinities, species delimitation and adaptive radiation of New Zealand *Ranunculus*. PhD thesis, Massey University, Palmerston North, New Zealand.)

trans-Tasman group has a primary division between clade 1, with a South Island "central strip" distribution, and the rest. The breaks are located at the Waihemo fault zone, the Mount Cook node, and the Wairau fault. This break developed at some time before the Late Cretaceous, trans-Tasman break in the sister of clade 1 (between *R. lyallii/buchananii* and *R. anemoneus* in Australia). This is explicable if the cause of the first break was mid-Cretaceous movement on the Waihemo fault zone and on a precursor of the Alpine–Wairau fault.

Lockhart et al. (2001) proposed that the high-alpine species *R. haastii* and *R. grahamii* (Figure 11.5) were derived from *R. lyallii*, a low-alpine species, but there is no genetic evidence for this. *R. haastii* and *R. grahamii* (along with *R. acraeus*, another high-altitude species) form the "central strip" sister group of the rest of group 2 (including *R. lyallii* and five other species); they are not sister to, or nested in, *R. lyallii*.

The next break in group 2 (Figure 11.5) is between the northern clade 2 (*R. verticillatus* and *R. nivicola*) and its southern South Island–Australian sister, clades 3 + 4. This break occurs at the Paparoa Range and its core complex.

Next in group 2 (Figure 11.5), clade 3 of Central Otago and the Takitimu Mountains is separated from its sister, the western clade 4 (of Kosciuszko and southwestern South Island), at the Moonlight tectonic zone. This was *before* seafloor spreading in the Tasman rifted apart clade 4. Clade 4 displays a typical Paparoa Range–southeastern Australia connection (Chapter 7).

The next major clade in section *Pseudadonis* is group 3 (Figure 11.6). This includes *R. crithmifolius*, with a southern limit at the Waihemo fault zone, and *R. godleyanus*, endemic to the mountains between Mount Cook and Arthur's Pass. The Mount Cook–Arthur's Pass sector seen here as a center of endemism acts, conversely, as a zone of absence in the many groups that are disjunct between Mount Cook and Nelson, such as *Celmisia* series *Lignosae* (Chapter 9).

Finally, group 4 of sect. *Pseudadonis* comprises a single, local endemic (*R. scrithalis*) in the Eyre Mountains (Figure 11.6). These mountains are well known as a center of distinctive local endemics, such as *Celmisia philocremna* and *C. thomsonii* (Asteraceae), and are located next to the Moonlight tectonic zone and the Taieri–Wakatipu synform (Figure 3.14).

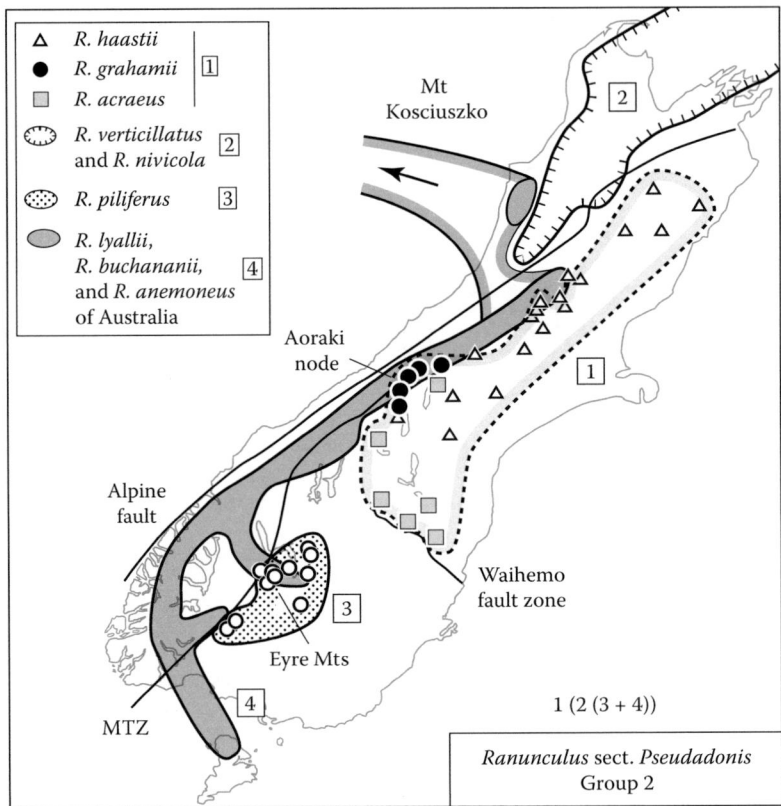

FIGURE 11.5   Distribution of *Ranunculus* sect. *Pseudadonis* group 2 (Ranunculaceae). MTZ, Moonlight tectonic zone. (Data from Lockhart, P.J., *Annls. Missouri Botanical Garden* 88, 458, 2001; Lehnebach, C.A., Phylogenetic affinities, species delimitation and adaptive radiation of New Zealand *Ranunculus*. PhD thesis, Massey University, Palmerston North, New Zealand.)

The New Zealand Brooms: *Carmichaelia* (Fabaceae) and Allies

Legumes (Fabaceae) are represented in New Zealand by *Canavalia* (one species), *Swainsona* (one species), *Sophora*, and the New Zealand brooms—*Carmichaelia* and allied genera. In *Sophora*, the New Zealand species belong to sect. *Edwardsia*, a group found in parts of the southern hemisphere and in Hawaii. Section *Edwardsia* shows neat allopatry with its sister group, *S. tomentosa*, consistent with an origin of both by simple vicariance (Heads, 2014: 105).

The members of the *Carmichaelia* group include lianes, shrubs, and small trees. They are widespread through mainland New Zealand and on Norfolk and Lord Howe Islands. The group belongs to an Old World clade with one main component in Africa–Eurasia, and the other, its sister group, in Australasia. The phylogeny and distributions are as follows (Wagstaff et al., 1999; Wojciechowski, 2005):

> **Africa and Eurasia** (*Sutherlandia*, southern Africa; *Lessertia*, southern Africa; *Colutea arbore-scens*, Europe and North Africa; *Astragalus cysticalyx*, Kazakhstan; *Phyllolobium chinense* and *Astragalus sinicus*, China).
>
>> "*Swainsona*" clade: Pilbara region of NW Australia (four species), extending to central New South Wales (one species).

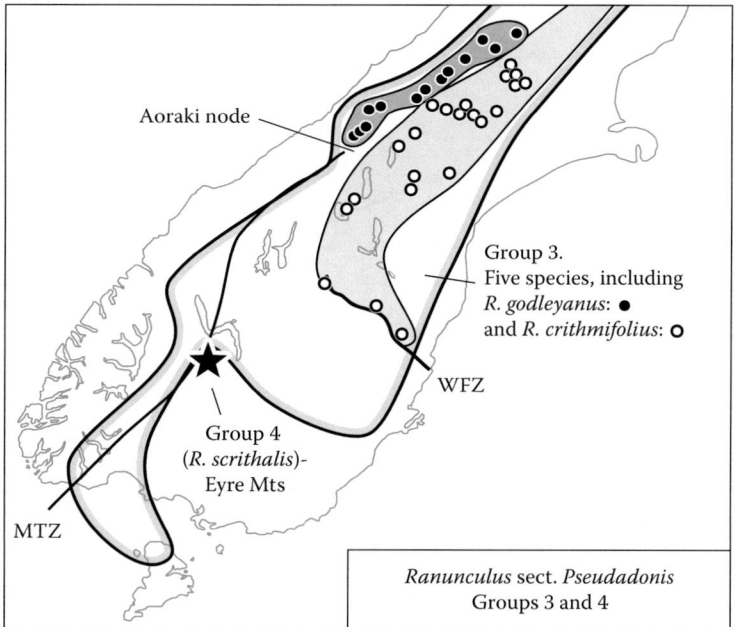

FIGURE 11.6 Distribution of *Ranunculus* sect. *Pseudadonis* groups 3 and 4 (Ranunculaceae). (The members of group 3 not mapped separately are *R. enysii*, *R. gracilipes*, and *R. insignis*.) MTZ, Moonlight tectonic zone; WFZ, Waihemo fault zone. (Data from Lockhart, P.J., *Annls. Missouri Botanical Garden* 88, 458, 2001; Lehnebach, C.A., Phylogenetic affinities, species delimitation and adaptive radiation of New Zealand *Ranunculus*. PhD thesis, Massey University, Palmerston North, New Zealand.)

> *Swainsona* s.str. (including *Montigena*): Pilbara to central Australia, southeast and east coast Australia, Tasmania, and New Zealand (South Island).
>
> *Carmichaelia* group of genera: New Zealand mainland, Lord Howe I., and Norfolk I.

The reciprocal monophyly of the clade in Africa–Eurasia and the one in Australasia (the last three groups listed) means they can be derived from a widespread Old World ancestor by simple vicariance.

The Australasian clade comprises one group that is most diverse in northwestern Australia, and its sister group, widespread through Australia and New Zealand. The Diplodactylidae, a family of geckos, has a similar arrangement (their second clade is also in New Caledonia) (Heads, 2014; Figure 4.8). The diplodactylids and the Australasian clade of legumes can each be derived from a widespread Australasian ancestor that had an original center of differentiation around northwestern Australia, with overlap in Australia developing later.

The significance of the Pilbara node in the Australasian legume clade would be consistent with Tethyan connections in the group between Australasia and the Mediterranean. Trans-Indian Ocean connections with Africa are possible but less likely, given the group's absence from Madagascar and rarity in southwestern Australia. In either case, the complete allopatry of the Australasian clade and the Africa-Eurasian clade is consistent with Mesozoic vicariance.

The clade comprising *Swainsona* s.str. and the *Carmichaelia* group is found through Australia, Lord Howe and Norfolk Islands, and mainland New Zealand (Figure 11.7). The clade is disjunct across the Tasman Basin, but the basin itself only corresponds to an infrageneric split (within *Swainsona*), and the primary phylogenetic break, between *Swainsona* and the rest, is located instead in or around the South Island (cf. *Ranunculus*, Figure 11.5).

The six main clades in the *Swainsona–Carmichaelia* complex (Figure 11.7) are all allopatric, with the one exception of *Carmichaelia*, which overlaps all the other genera on mainland New Zealand. This can be attributed to range expansion in a single group, *Carmichaelia*, following the origin of all the groups by vicariance. In one possible reconstruction (Figure 11.8), the eastern distribution of *Carmichaelia* resembles the distribution (living and fossil) of groups such as the tuatara (*Sphenodon*). In this model,

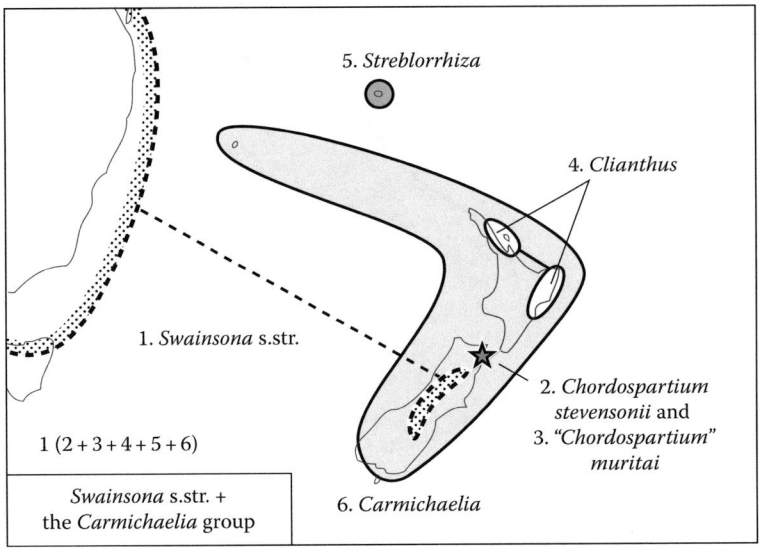

FIGURE 11.7 Distribution of the *Carmichaelia* group and part of the distribution of its sister group *Swainsona* (throughout Australia) (Fabaceae). (Data from Purdie, A.W., *N. Z. J. Bot.*, 23, 157, 1985; Heenan, P.B., *N. Z. J. Bot.*, 34, 299, 1996b; Heenan, P.B., *N. Z. J. Bot.*, 36, 41, 1998b; Heenan, P.B., *N. Z. J. Bot.*, 38, 361, 2000.)

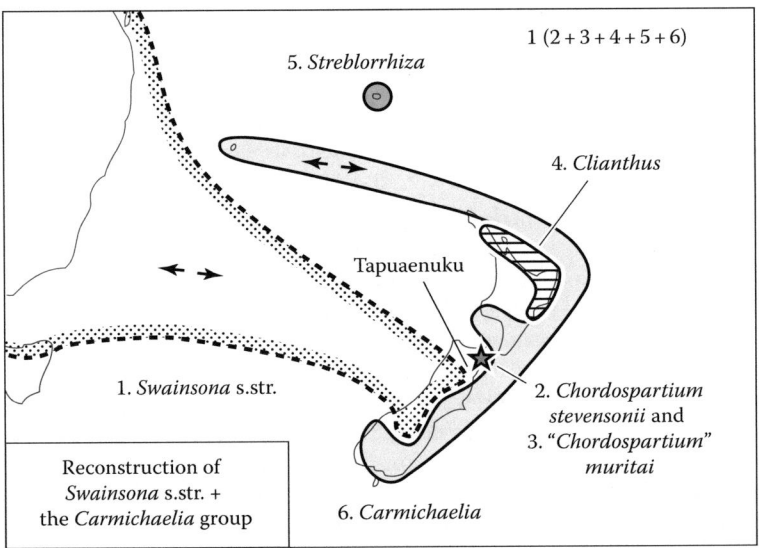

FIGURE 11.8 Reconstructed distribution of *Swainsona* and the *Carmichaelia* group (Fabaceae), assuming that all genera were originally allopatric.

the initial break, between *Swainsona* and the rest, lies along or near the central strip in the South Island that runs through Canterbury to Marlborough.

In mainland New Zealand, the *Swainsona-Carmichaelia* complex as a whole has a strong center of diversity on the Torlesse terrane (Figure 11.9; *Carmichaelia* itself is widespread throughout and is not shown). This is also the case in other Tethyan groups in New Zealand, such as *Ceratocephala* (Ranunculaceae) and *Myosotidium* (Boraginaceae).

Swainsona s.str. is absent in the far north of Australia, but it forms a continuous band from the west coast of Australia to the east coast. Despite this, the genus is absent from Lord Howe and Norfolk Islands and

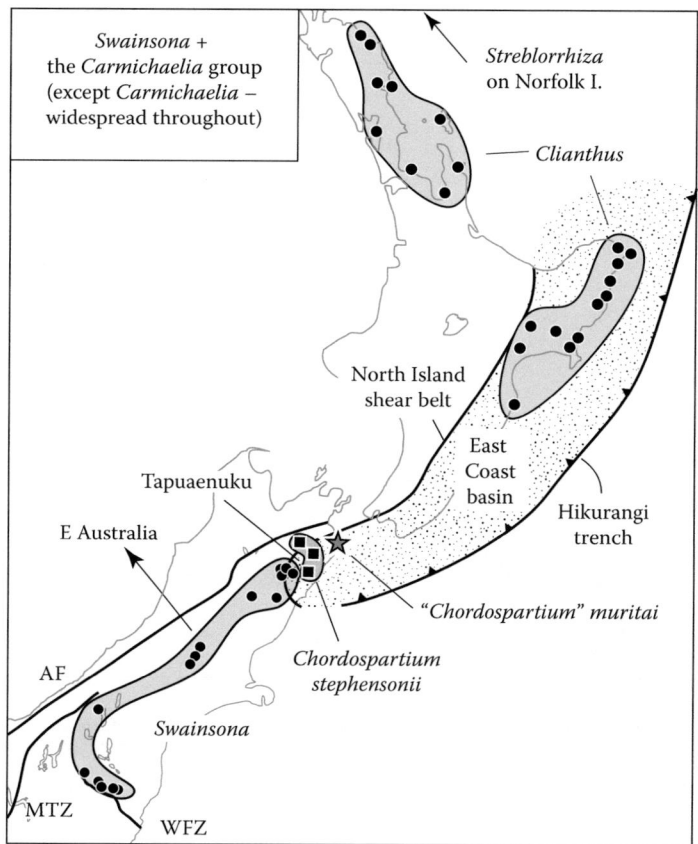

FIGURE 11.9 Distribution of the genera in the *Carmichaelia* group (Fabaceae) except *Carmichaelia*, which is widespread throughout New Zealand. AF, Alpine fault; MTZ, Moonlight tectonic zone; WFZ, Waihemo fault zone. (Data from Purdie, A.W., *N. Z. J. Bot.,* 23, 157, 1985; Heenan, P.B., *N. Z. J. Bot.,* 34, 299, 1996b; Heenan, P.B., *N. Z. J. Bot.,* 36, 41, 1998b; Heenan, P.B., *N. Z. J. Bot.,* 38, 361, 2000.)

has only a single species in New Zealand (Figure 11.9). This is in contrast with its sister, the *Carmichaelia* group, which is most diverse in New Zealand and also occurs on Lord Howe and Norfolk Island.

In the South Island, *Swainsona* has a distribution lying along the western margin of the central strip. It forms a narrow arc that skirts the great Canterbury–Kaikoura diversity in its sister group, *Carmichaelia* and allies (Figure 11.9). At the northern end of its range, *Swainsona* has a distinctive boundary near Tapuaenuku, and northeast of here it is replaced by *Chordospartium stevensonii* and *"Chordospartium" muritai*.

As mentioned, the distributions suggest that the earliest breaks in the phylogeny took place before seafloor spreading in the Tasman and New Caledonia basins rifted both *Swainsona* and the *Carmichaelia* group. The distributions of the clades *within* New Zealand coincide with pre-breakup rifting around the eastern South Island–Lord Howe Rise area.

In New Zealand, *Swainsona* has a range delimited by the Waihemo fault zone (a mid-Cretaceous extensional fault) and Tapuaenuku (a mid-Cretaceous, multiphase igneous complex also associated with extension) (Figure 11.10). The single New Zealand species, *S. novae-zelandiae*, is sister to *S. galegifolia* of eastern Australia (Canberra to Cairns). *S. novae-zelandiae* is distinct from the Australian members of the genus as it has a rhizome and acrotonic (distal) branching; these enable it to survive in its scree habitat. Its ecology distinguishes the species from most of its New Zealand relatives (the whole *Carmichaelia* group), as it only occurs in subalpine and alpine sites. It is true that *Carmichaelia monroi* and *C. crassicaule* extend from the lowlands and montane zone respectively into the alpine zone (Allan, 1961; Mark and Adams, 1973), but all other members of the *Carmichaelia* group are restricted to lower altitudes. This means there is altitudinal vicariance as well as a large component of geographic allopatry between *Swainsona* and the *Carmichaelia* group.

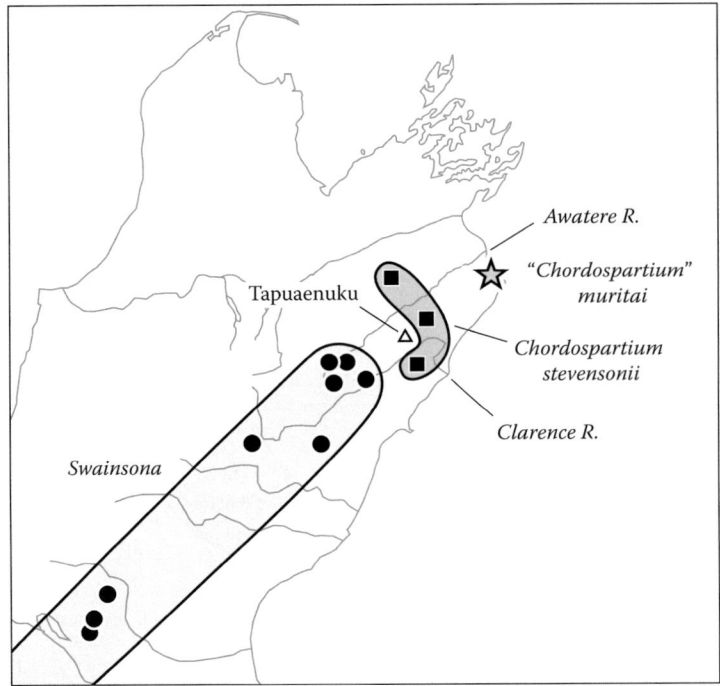

FIGURE 11.10 Distribution of *Swainsona*, *Chordospartium stevensonii*, and "*Chordospartium*" *muritai* (Fabaceae) in Marlborough. (Data from Purdie, A.W., *N. Z. J. Bot.*, 23, 157, 1985; Heenan, P.B., *N. Z. J. Bot.*, 36, 41, 1998b.)

The *Carmichaelia* Group of Genera (Carmichaeliinae)

This group is widespread through New Zealand, and molecular studies indicated four main clades (Figure 11.9; Wagstaff et al., 1999, Figure 2):

>   *Carmichaelia* (including *Notospartium* and *Corallospartium*): North Island, South Island, and Stewart Island, also Lord Howe Island.
>   *Clianthus* (one species: *C. puniceus* including *C. maximus*; P. Heenan, pers. comm. 2013): north-eastern North Island.
>   *Chordospartium stevensonii*: Marlborough, mainly montane (Wairau, Awatere, and Clarence Valleys, also near the Clarence River mouth); upper branches and inflorescences pendulous ("weeping").
>   "*Chordospartium*" *muritai*: Marlborough, on a coastal cliff near Cape Campbell, 40 m above the sea (12 plants only; IUCN, 2016), and "two other sites between the Wairau and Clarence rivers" (Dawson and Lucas, 2011); upper branches and inflorescences erect (orthotropic).

An extinct liane from an islet off Norfolk Island, *Streblorrhiza*, is probably a fifth member of the group, but it has not been sequenced.

The four sequenced clades appear in the phylogeny as an unresolved polytomy. Morphological analyses suggested a clade: *Clianthus* (*Carmichaelia* including *Chordospartium* + *Steblorrhiza*) (Heenan, 2001).

Apart from *Streblorrhiza*, Wagstaff et al. (1999) accepted only two genera for the group, *Clianthus* and *Carmichaelia* (including *Chordospartium*). Yet the molecular phylogeny implies either one genus or four; either *Clianthus* should be sunk in *Carmichaelia* or both should be retained, along with *Chordospartium* (*C. stevensonii*) and a separate, endangered genus for "*Chordospartium*" *muritai*. (*Chordospartium stevensonii* and *C. muritai* did not form a monophyletic clade in the molecular study or in morphological analyses; Heenan, 1998a.)

Purdie (1985) noted that "in some respects *Chordospartium muritai* appears to combine the characteristics of the other broom genera." It differs from *Chordospartium stevensonii* in its flowers, fruits, and seeds, which are less than half the size, and in its final branchlets and racemes, which are orthotropic (not pendulous or "weeping"). *Chordospartium stevensonii* combines the weeping habit of *Carmichaelia* group "*Notospartium*" with the fibrous cortical ridges of *Carmichaelia* group "*Corallospartium*." But the molecular phylogeny shows the last two groups forming a clade within *Carmichaelia*, and so neither of the distinct *Chordospartium* lineages can simply be a hybrid of these.

The recombination of characters in this group is "contradictory and anomalous" (Heenan, 1998c), as it cannot be expressed in a nested hierarchy. Nevertheless, similar character recombination is common in all plant and animal groups. In itself, it does not necessarily imply hybridism, and it does not mean that the groups should be synonymized in the taxonomy. It *may* be the result of chance parallelism or simple hybridism, but in many cases these are unlikely, and the pattern is now often attributed to the inheritance of ancestral polymorphism. This model ("incomplete lineage sorting") accepts that there was prior diversity in the ancestor before divergence of the modern clades began, and that this variation was inherited (Heads, 2009a). For example, the character distributions suggest that an ancestral complex of the *Carmichaelia* group already included plants with orthotropic, plagiotropic, and pendulous forms before the modern genera began to diverge. The modern genera have maintained the ancient variation in different clades. The fibrous, cortical ridges, still present in two clades that are not sister groups—*Chordospartium stevensonii* and *Carmichaelia* group *Corallospartium*—are also likely to represent an old polymorphism, a last relic of earlier lines of foliage in plants that are otherwise more or less leafless.

Within the *Swainsona-Carmichaelia* group, four of the genera—*Swainsona, Chordospartium stevensonii*, "*C.*" *muritai*, and *Clianthus*—are allopatric along a central belt (Figure 11.9). This coincides with a zone of Late Cretaceous–Paleogene basins located along the central South Island strip, the East Coast Basin of the Hikurangi subduction zone, and the extinct New Caledonia subduction zone (East Cape to Northland and New Caledonia). Four of the genera have distributional breaks in Marlborough, where the geography is dominated by the Kaikoura Ranges, but these are young, Neogene features, and the generic differentiation probably developed on earlier, rifting landscapes (Chapter 8).

Carmichaelia is absent from the subantarctic islands, most of Stewart Island and also southwestern Fiordland. (The latter is an important center of endemism for distinctive groups such as the "heteroblastic" *Brachyglottis bifistulosa*—Asteraceae.) Apart from these western absences, the whole *Carmichaelia* group includes only four species in the Western Province, and none are endemic there. Thus, the diversity has a strong concentration on Eastern Province terranes. This contrasts with other diverse groups in New Zealand, such as *Abrotanella* (Chapter 10) and *Ranunculus* (earlier this chapter), which have significant endemism in both Eastern and Western provinces.

Carmichaelia

Carmichaelia is of special interest for New Zealand biogeography as it is diverse and occurs almost throughout the three main islands, as well as on Lord Howe Island (Figure 11.7).

It is also intriguing from a morphological point of view as the members show such a wide range of architecture, and include lianes, shrubs, and trees. These are all different forms of "broom"; the plants only have leaves in the seedling stage. Adults are more or less leafless (leaves are sometimes present at the beginning of new spring growth or in shady sites), and the stems form photosynthetic cladodes. Most species occur in drier habitats, but some are in wet areas, and "so there is not a clear correlation between the plant's [xeromorphic] form and its habitat" (Dawson and Lucas, 2011: 174).

Carmichaelia *Subgenus* Kirkiella

An initial molecular study found little phylogenetic structure within *Carmichaelia* (Wagstaff et al., 1999). However, morphological features suggest that *C. kirkii*, making up subgenus *Kirkiella*, is sister to all the others (Heenan, 1998a). *C. kirkii* is the sole liane in the genus—the other species are shrubs and

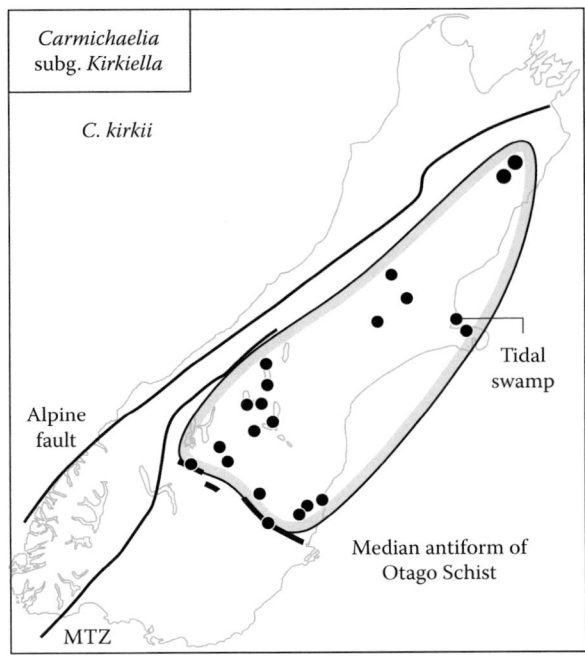

FIGURE 11.11 Distribution of *Carmichaelia* subg. *Kirkiella* (Fabaceae). MTZ, Moonlight tectonic zone. (Data from Heenan, P.B., *N. Z. J. Bot.*, 34, 157, 1996a; Sow Burn record from Petrie, D., *Trans. N. Z. Inst.*, 28, 540, 1896.)

small trees—and it is the only member of the genus in which the stem is circular, not flattened, in cross section. The species is found through inland and coastal parts of the South Island, from Otago north to Tapuaenuku (Figure 11.11; Heenan, 1996a; cf. *Swainsona*, Figure 10.10).

   C. kirkii often grows with divaricate shrubs on alluvial terraces in the lowlands. Simpson (1945) described its ecology: "Terrace scrub, growing through *Coprosma*, sometimes in similar association in swamp." He also cited a significant record from "tidal swamp near New Brighton, Christchurch" (Simpson, 1945: 267). Most of the localities of *C. kirkii* are at inland sites in Otago, Canterbury, and Marlborough. Nevertheless, its distribution intersects the modern coastline at Christchurch and also near Shag Point, Otago, where the Waihemo fault zone meets the coast. Other inland and even montane taxa also have unusual records near sea level at Shag Point (Heads and Patrick, 2003). *C. kirkii* has its southern limit at the median antiform of the Otago Schist and its western limit at the Nevis-Cardrona fault system.

   The ecology and distribution of *C. kirkii* can be derived from an ancestral complex that inhabited tidal swamps and river banks. Following marine transgression and regression, populations have been stranded inland, where they survive around alluvial terraces and still favor wetter sites.

   Apart from *C. kirkii*, with its own subgenus, several groups in *Carmichaelia* have been accepted in morphological revisions (Simpson, 1945; Allan, 1961; Heenan, 1995, 1996a). Their mutual relations are unresolved, and Heenan stressed that the groups display challenging recombinations of characters. Subgenus *Kirkiella* and the five other groups all overlap in the central strip that extends from Central Otago to Marlborough on Eastern Province terranes.

Carmichaelia *Subgenus* Suterella

This group (Figure 11.12) includes two species of alluvial river terraces and outwash fans. Both have rhizomes that become thick and woody with age. *Carmichaelia uniflora* grows in the west, in what is currently a wetter, montane environment. It is the smallest plant in the genus; it has slender, grass-like cladodes 6 cm long and disproportionately large flowers that are borne singly (Wardle, 1991). The other species in the group, *C. corrugata*, occurs further east, in what are now drier lowlands and coastal areas.

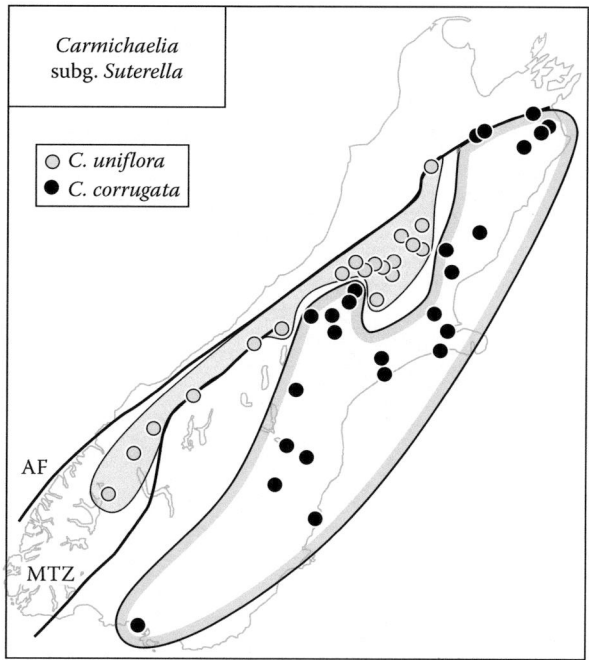

FIGURE 11.12 Distribution of *Carmichaelia* subg. *Suterella* (Fabaceae). AF, Alpine fault; MTZ, Moonlight tectonic zone. (Data from Heenan, P.B., *N. Z. J. Bot.*, 33, 455, 1995.)

The overall distribution of subgenus *Suterella* extends north and east to a precise boundary at the Alpine–Wairau fault, a Cretaceous break but not a current climatic barrier. This suggests that the phylogeny and distribution are related to early activity on the Alpine fault and its precursors, rather than the mountains that later arose along it. This would be consistent with a break between the eastern and western species of *Suterella* at the Moonlight tectonic zone (active in the late Cretaceous–Paleogene). All these events took place before the Neogene uplift of the modern mountains and the development of an orographic climate. *Carmichaelia corrugata* survives at maritime sites in the east, including stabilized sand dunes at Kaitorete Spit where it grows with the unrelated, local endemic *C. appressa* (subgenus *Carmichaelia*). These records are consistent with the idea that the ancestors of *Suterella* survived around the margins of inland seas.

Carmichaelia *Subgenus* Monroella

This group (Figure 11.13) extends from Central Otago (at the median antiform of the Otago Schist) north to the Wairau fault. The main distribution follows the usual central strip, but there are also coastal records near Lake Ellesmere and in Marlborough at the Flaxbourne River mouth. As discussed earlier (Chapter 8), similar patterns are seen in many groups.

The western species in the group, *C. vexillata*, is disjunct between southern Canterbury and Marlborough, and the simplest explanation for this is that populations in the west have been wiped out by Neogene uplift.

In Marlborough, subgenus *Monroella* intersects the current coastline, where it is represented by, a local, calcicole endemic, *C. astonii*. This occurs near the mouth of the Flaxbourne River, at 150 m elevation, "on limestone outcrop near sea, among assemblage of 'alpine' plants" (*F.J.F. Fisher* Dec. 1952, Otago University herbarium). It is also recorded inland along the Clarence Valley.

In a parallel case, *Gentianella astonii* (Gentianaceae) is a calcicole that grows at sea level at the Flaxbourne River mouth and otherwise occurs further south and inland (Marlborough and northern Canterbury), at up to 1050 m elevation (Glenny, 2004). Its sister species is *G. calcis*, distributed from northern Canterbury (Weka Pass) south to the St Mary Range at the Waihemo fault zone.

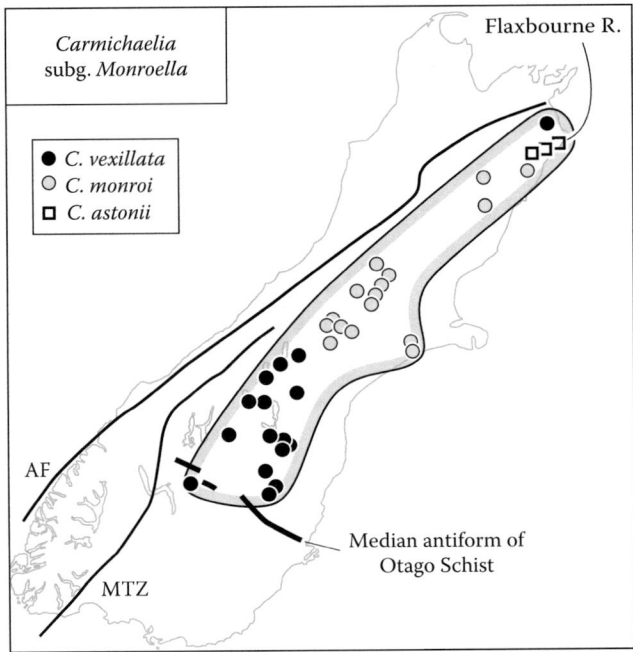

FIGURE 11.13 Distribution of *Carmichaelia* subg. *Monroella* (Fabaceae). AF, Alpine fault; MTZ, Moonlight tectonic zone. (Data from Heenan, P.B., *N. Z. J. Bot.*, 33, 455, 1995.)

The anomalous Flaxbourne River populations near sea level could maintain the ancestral ecology, while populations further south have been stranded inland, with some surviving uplift into the mountains. As an alternative, *Carmichaelia astonii*, *Gentianella astonii*, and the others could have been lowered with tectonic subsidence. The complex coastline of the Marlborough Sounds to the north indicates that a system of mountains and valleys has been partly drowned. This has been caused by Pleistocene subsidence with northward tilting (Trewick and Bland, 2012). In any case, the Cape Campbell–Flaxbourne River mouth area has low rainfall, and so the fact that the alpine groups "descend" there is even more anomalous in ecological terms—they occur there despite the local climate, not because of it.

Whether populations of *C. astonii*, *G. astonii*, and the others at the Flaxbourne River mouth have been lowered, or whether their related populations inland have been uplifted, groups with lower elevations in the north than in the south are incompatible with a simple ecological cause and suggest that historical factors are responsible for the pattern. Many widespread New Zealand groups survive in the cooler, southern parts of their range only at *lower*, warmer elevations. This means they occupy habitats with similar climate, as predicted in a simple niche model of distribution. Yet the *C. astonii* group grows in Marlborough at lowland and coastal sites, while further south it occurs inland and at *higher* elevation. None of the revisions of *Carmichaelia* (Simpson, 1945; Allan, 1961; Heenan, 1995, 1996a) cite elevational ranges for the species, and so a more detailed account of the pattern is not possible here. Nevertheless, it is repeated in many other groups that are subalpine in the southern or central South Island but also include northern populations on the Marlborough coast, and so it probably reflects a general cause, such as marine transgression and regression.

These inland and montane groups whose distributions intersect the coast in the north display an interesting, "antiecological" geographic trend. A parallel case is that of the odd-nosed monkeys, a clade that ranges from equatorial mangrove forest in Southeast Asia to 4700 m at temperate latitudes in China. Most groups increase their altitude in the other direction, *toward* the tropics, and yet certain butterflies repeat the trend exhibited by the odd-nosed monkeys (Heads, 2012a).

None of these patterns are explicable by ecology, but all occur in areas where there has been dramatic tectonic activity, with the raising and lowering of entire mountain ranges. In ecophyletic series such as these, with different members in different habitats, the overall trend needs to be explained with reference to normal processes. Explaining each group by a separate migration and proposing how each one has been "wedged" into the local ecology by ad hoc adaptation leaves the overall pattern unaccounted for.

Carmichaelia *Subgenus* Huttonella

Huttonella (Figure 11.14) and *Monroella* show a high level of allopatry, and overlap is restricted to three very localized sites. *Huttonella* includes three species. *C. compacta* and *C. curta* are endemic to central and north Otago, respectively, and are separated by the Waihemo fault zone. The group is also represented on the coast just north of the Waihemo fault zone as there are early records of *C. curta* at the "Otepopo" (i.e., Waianakarua River) (Petrie, 1896). This is also seen in subg. *Suterella* (Figure 11.12), but not in subg. *Monroella*, which is on the coast only in Canterbury and the Kaikoura region.

The third species in *Huttonella*, *C. juncea* s.lat., is distributed in both eastern and western provinces. Heenan (1995) treated five species accepted by Allan (1961) as straight synonyms under *C. juncea*, but whether these entities form a smooth or a stepped cline, the geography of the variation is of special biogeographic interest. *C. "lacustris"* is restricted to stony shores at Lake Te Anau and Lake Manapouri. *C. "nigrans"* is recorded from both sides of the Alpine fault and has long, black pods (Allan, 1961). Plants from the Torlesse node (*C. "prona"* at Lakes Marymere and Lyndon) and northwest Nelson (*C. "fieldii"*) have smaller pods, are lighter in color, and have 2–4 (–6) seeds (Heenan, 1995). Plants from Marlborough (*C. "floribunda"*) and the North Island (*C. juncea* s.str.) have short, light-tan pods with 1–2 (–4) seeds. If the distributions are mapped on a reconstruction for 27 Ma, the result is a simple system of parallel arcs (Figure 11.15).

Carmichaelia *Subgenus* Carmichaelia

This subgenus differs from all the others as it is present outside New Zealand, on Lord Howe Island (Figure 11.16). The Lord Howe species, *Carmichaelia exsul*, is closest in its morphology to *C. williamsii* (Heenan, 1998a), a characteristic member of the northeastern biogeographic arcs. Subgenus *Carmichaelia*

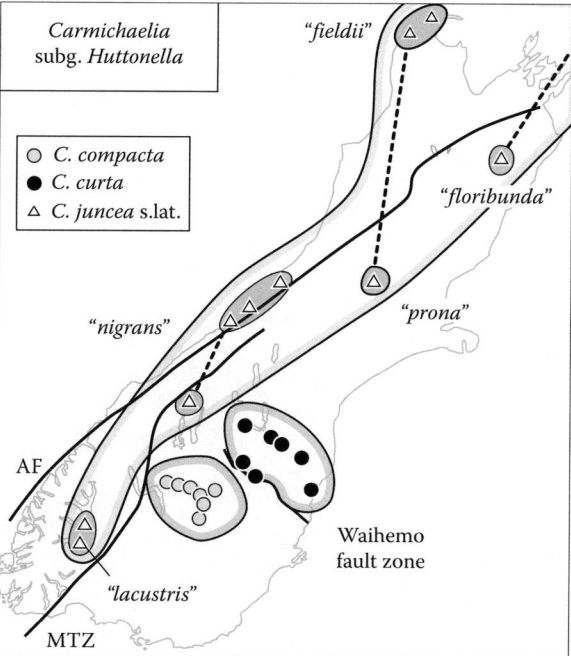

FIGURE 11.14 Distribution of *Carmichaelia* subg. *Huttonella* (Fabaceae). (Species now treated under *C. juncea* are also indicated.) AF, Alpine fault; MTZ, Moonlight tectonic zone. (Data from Heenan, P.B., *N. Z. J. Bot.*, 33, 455, 1995.)

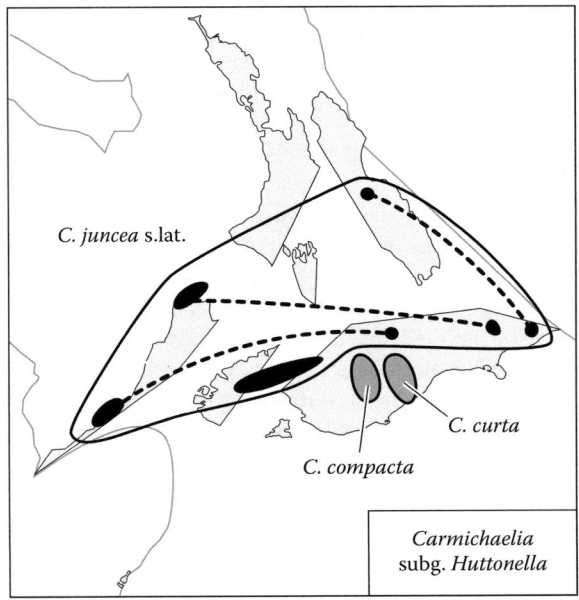

FIGURE 11.15   Reconstructed distribution of *Carmichaelia* subg. *Huttonella* (Fabaceae) at 27 Ma (cf. Figure 11.14).

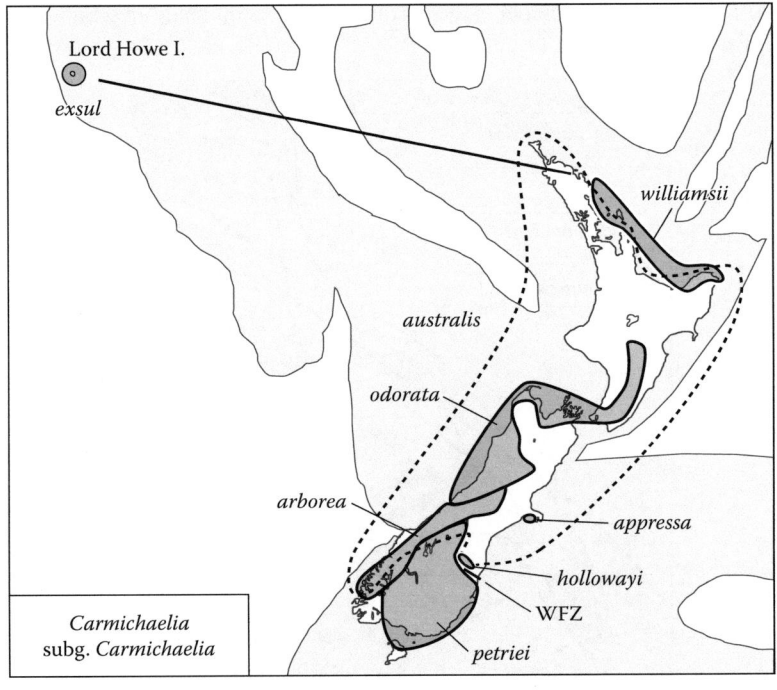

FIGURE 11.16   Distribution of *Carmichaelia* subg. *Carmichaelia* (Fabaceae). (Data from Heenan, P.B., *N. Z. J. Bot.*, 34, 157, 1996a.)

is also widespread through the New Zealand mainland, although it is absent in most of Stewart Island and southern Fiordland. With one exception, the species are allopatric or at least parapatric. (There is minor, local overlap among *C. petriei*, *C. arborea*, and *C. odorata*, but at too small a scale to be shown on the map.) The exception is *C. australis*, which overlaps with all the other species. The simplest explanation is that all the species evolved in allopatry, and *C. australis* was originally restricted to the parts of the North and South Islands where the other species do not occur. At some stage, *C. australis* populations have expanded into the range of the other species, although there is still significant allopatry with *C. petriei* and *C. williamsii*.

The most distinctive member of subgenus *Carmichaelia* is *C. williamsii*, which has a typical "horstian" distribution off Northland and the East Coast (Figure 11.17). It is a coastal species and has large, yellow, bird-pollinated flowers unlike those of any other *Carmichaelia*. This is the only *Carmichaelia* on the Poor Knights, Hen and Chickens, and Aldermen Islands.

Other coastal species in subgenus *Carmichaelia* include *C. appressa*, mentioned earlier. This prostrate subshrub occurs by the sea on stabilized sand dunes at Kaitorete Spit, Banks Peninsula. Closely related populations are known above high coastal cliffs between the Rakaia and Rangitata rivers, and in disturbed alluvial gravel in the upper Rangitata Valley (Heenan, 1996).

Other members of subgenus *Carmichaelia* occur in similar habitats. For example, the populations of *C. australis* that were formerly treated as *C. arenaria* grow on sandy soil over coastal limestone below the Paparoa Range. Further south, *C. hollowayi* (Figure 11.16) is a procumbent, rhizomatous, suckering shrub found only on Oligocene limestone in the St Mary Range, along the northern margin of the Waihemo fault zone. Without seeing the pods, *C. hollowayi* could be confused with prostrate, rhizomatous, and suckering forms of the variable *C. petriei*. If the latter are grown on in a garden, they develop an erect growth form, and the species is regarded as an upright or spreading shrub that can become rhizomatous when browsed (Heenan, 1996a, 1998b). *C. arborea* is another variable species; plants are usually spreading or upright shrubs with several main stems, but low forms with a rhizomatous or suckering habit are sometimes seen (Heenan, 1996a). In all these cases, rhizome development is barely suppressed, if at all, suggesting that there was considerable architectural plasticity in the direct ancestor of the group.

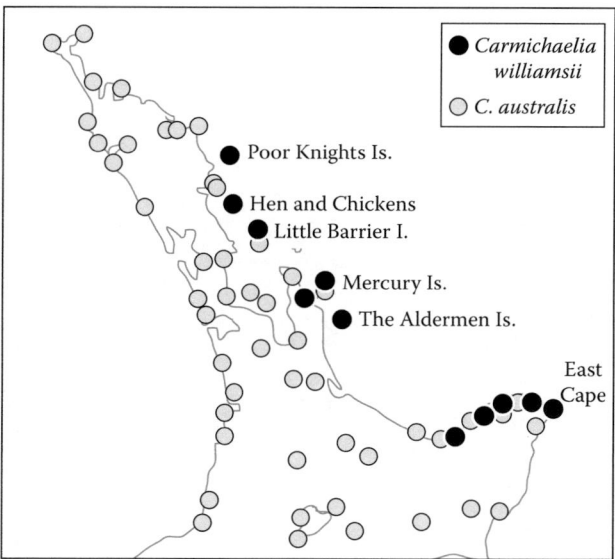

FIGURE 11.17 Distribution of *Carmichaelia williamsii* and distribution (northern part only) of *C. australis* (Fabaceae). (Data from Heenan, P.B., *N. Z. J. Bot.*, 34, 157, 1996a.)

Carmichaelia *Subgenus* Enysiella

This group, mapped in Figure 11.18, has another "central strip" distribution. It is closest to subg. *Carmichaelia* (Wagstaff et al., 1999), but unlike that group, it is not found west of the Alpine fault, and it has a southern limit at the Waihemo fault zone. In the South Island, it extends north to Marlborough, but as with *Swainsona* and *Carmichaelia* subgenus *Kirkiella*, it is absent from the Seaward Kaikoura Range and the northern end of the Inland Kaikoura Range. It occurs in the central North Island, on active volcanoes (Ruapehu and Tongariro) as well as on the nearby Kaimanawa Mountains (Chapter 5).

Carmichaelia *Clade:* "Notospartium *Group"* and "Corallospartium *Group"*

These two groups (Figure 11.19) were formerly treated as genera, but now appear to be sister groups within *Carmichaelia* (Wagstaff et al., 1999). They form an unnamed clade that extends from the Eyre Mountains at the Taieri–Wakatipu synform to the Wairau fault.

The "*Notospartium* group" comprises three species that are separated by the Awatere and Hope faults, and display a disjunction between the Hope fault and south Canterbury (Figure 11.20; Heenan, 1996b). The "*Corallospartium* group" includes a single, distinctive species, *C. crassicaulis* or coral broom, which extends from the Eyre Mountains north to a disjunct record at the Hope fault.

The Kaikoura Ranges have been raised at the Hope, Awatere, and Clarence faults, and these three, along with the Wairau fault, form a splay system at the northern end of the Alpine fault (Figure 8.8). The central species of the "*Notospartium* group," *Carmichaelia glabrescens*, is endemic in the Kaikoura region on calcareous parent material and includes coastal populations around the mouths of the Clarence River and Waima River. The group as a whole surrounds Tapuaenuku, while the southern species, *Carmichaelia torulosa*, extends from Hope fault south to southern Canterbury.

Within the "*Notospartium* group," there is a species break at the Pleistocene Hope fault, suggesting that the break was a result of the Kaikoura orogeny. The group as a whole extends south of the fault, while

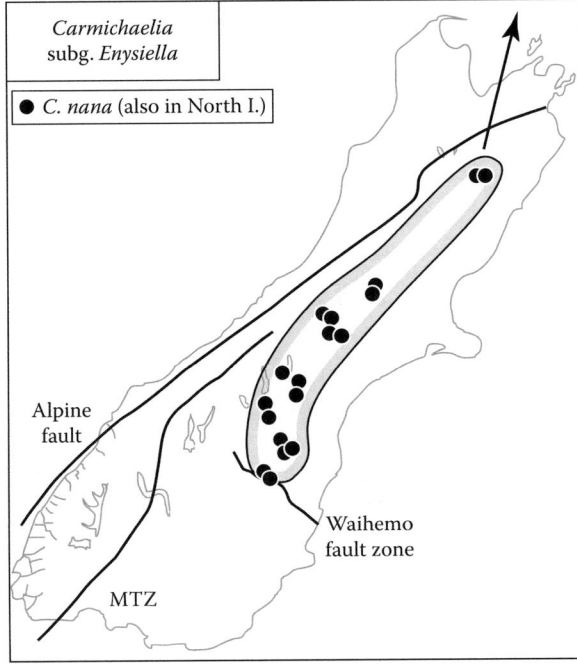

FIGURE 11.18 Distribution of *Carmichaelia* subg. *Enysiella* (Fabaceae). MTZ, Moonlight tectonic zone. (Data from Heenan, P.B., *N. Z. J. Bot.*, 33, 455, 1995.)

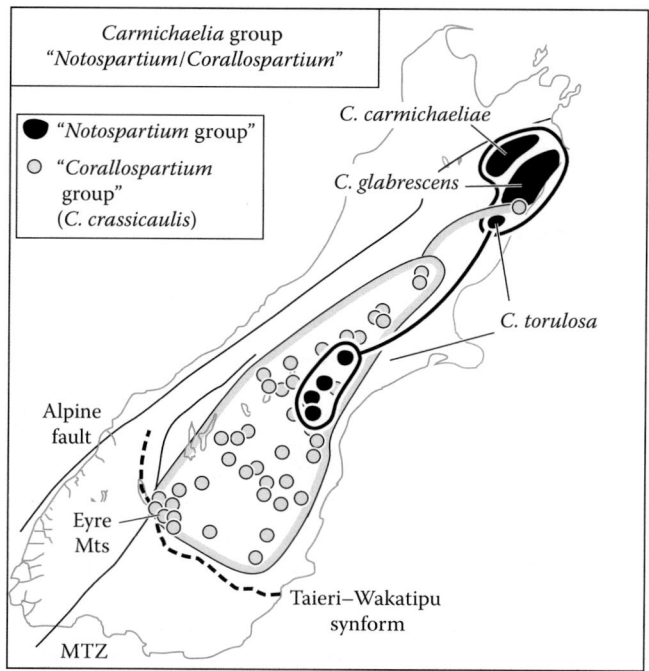

FIGURE 11.19 Distribution of *Carmichaelia* clade ("*Notospartium* group" and "*Corallospartium* group") (Fabaceae). MTZ, Moonlight tectonic zone. (From Heenan, P.B., *N. Z. J. Bot.*, 34, 299, 1996b; Heenan, P.B. and Barkla, J.W., *N. Z. J. Bot.*, 45, 265, 2007; Lammerlaw record from Petrie, D., *Trans. N. Z. Inst.*, 17, 272, 1884.)

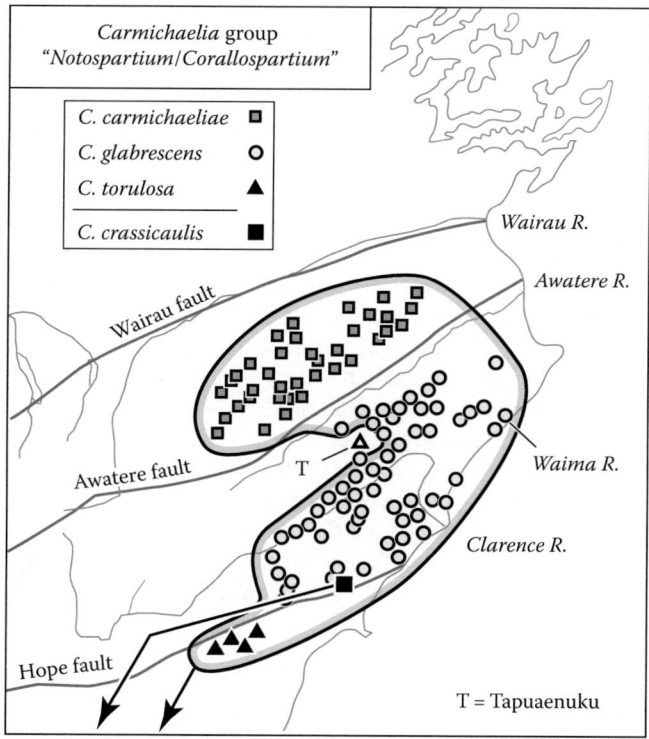

FIGURE 11.20 Distribution of *Carmichaelia* clade ("*Notospartium* group" and "*Corallospartium* group") (Fabaceae) in Marlborough. (Data from Heenan, P.B., *N. Z. J. Bot.*, 34, 299, 1996b.)

its sister, the "*Corallospartium* group," extends north to Mount Fyffe, 2 km north of the fault. Thus, the fault, striking east–northeast, does not coincide with the break between the two, which strikes northeast along the eastern South Island.

Ecology of Carmichaelia

Carmichaelia grows in open sites, shrubland and forest margins, but is rare in forest, except in clearings. The genus ranges from the coast to subalpine shrubland, with only two species (*C. crassicaulis* and *C. monroi*) in alpine vegetation. Outside the alpine zone and forest, *Carmichaelia* is found in a wide variety of open habitat types.

   C. kirkii is recorded in tidal swamps. In Westland, with very high rainfall, *Carmichaelia* can be abundant in "shrub swamps" (Simpson, 1945; Wardle, 1975). Many species are associated with rivers and streams. In subgenus *Huttonella*, *C.* "*lacustris*" in Fiordland is "no doubt" submerged during floods (Simpson, 1945). *C.* "*prona*," in the different environment of central Canterbury, grows at Lake Lyndon "where the tussock grassland merges into the mud of the lake." At nearby Lake Marymere, it "fills openings amongst stones submerged in times of flood." *C.* "*nigrans*" occurs in river shingle on "old flood beds." These three populations (now treated under *C. juncea*; Heenan, 1995) tolerate periodic inundation and high levels of disturbance. Many other species also grow along streamsides (*C. carmichaeliae* and *C. glabrescens* are strictly riparian), on river beds, river terraces, lake shores, and stable but unconsolidated alluvial gravels.

   Carmichaelia also occupies very well-drained sites, such as cliffs, rock outcrops, shingle slopes, glacial moraines, and outwash. It is an early colonizer of disturbed surfaces created by landslips and erosion, and other disturbed sites that it favors include montane slopes that are swept by avalanches.

   C. compacta grows in light, sandy soil (Petrie, 1891), on the edges of salt knolls (Rogers et al., 2000) and on rocky sites (Simpson, 1945). *C. astonii*, *C. hollowayi*, and *Notospartium glabrescens* grow on basic substrates such as limestone (the first species is restricted to Amuri Limestone) and apatite-rich volcanics. On West Dome (south of the Eyre Mountains), *C. petriei* grows on the Dun Mountain ophiolite belt, with ultramafics, basic volcanics and greywacke (McIntosh and Lee, 1986). In the central North Island, *C. nana* grows in the Kaimanawa Mountains and on active volcanoes, in bare, arid sites on young tephra.

   The range of ecology in the modern genus can be derived from a plastic, weedy ancestor that occupied tidal swamps, shingle beaches, limestone beaches, ophiolites, active volcanoes, and other disturbed sites. These have all been available in Zealandia through the Late Cretaceous and Cenozoic.

Summary of the Carmichaelia Group + Swainsona

Within this complex, there is one trans-Tasman group in Australia and New Zealand (*Swainsona*), one in New Zealand and Lord Howe Island (*Carmichaelia*), and one genus on Norfolk Island (*Streblorhiza*, possibly sister to *Carmichaelia*). These disjunctions can all be explained by Late Cretaceous–Paleogene seafloor spreading and the subsidence of Zealandia.

   Within New Zealand, seven widespread groups in *Carmichaelia*, as well as *Swainsona*, all overlap in the standard eastern South Island sector between the Waihemo fault zone and the Wairau fault. The groups have different southern boundaries at different Cretaceous features. The distributions of the *Carmichaelia* clades, along with traces of an earlier, maritime ecology, suggest that the history of inland seas in eastern New Zealand, including the Late Cretaceous–Paleogene basins, led to the early differentiation of the group. Following the origin of the main clades, allopatric speciation has been important in subgenera *Huttonella*, *Monroella*, and *Suterella* (Heenan, 1995) and also in subgenus *Carmichaelia*.

Apiaceae in New Zealand

The New Zealand genera of Apiaceae, the carrot family, belong to four subfamilies. (Earlier treatments included *Hydrocotyle* in Apiaceae, but this genus has now been transferred to Araliaceae.)

Subfamily Mackinlayoideae

Apart from *Centella*, most members of this subfamily are located around the margins of the South Pacific (Stevens, 2016). New Zealand has one species of *Actinotus* and one of *Centella*, but the affinities with their congeners have not been resolved. *Actinotus* is in western Australia, eastern Australia, and New Zealand. *Centella* includes the pantropical *C. asiatica* but has its main center of diversity in South Africa (Stevens, 2016).

Subfamily Saniculoideae

The single New Zealand species of this subfamily, *Eryngium vesiculosum*, is endemic to the North and South Islands (in coastal sand, gravel, and rock clefts), Tasmania, and the southeastern quarter of mainland Australia, inland to the MacDonnell Ranges. The boundary with other *Eryngium* species at breaks in central Australia (the site of inland Cretaceous seas) and the trans-Tasman affinity are both consistent with a Cretaceous origin.

Eryngium is the largest genus in the family Apiaceae, and has 230–250 species. Its sister is *Sanicula* (Calviño et al., 2008a), and while the two genera overlap in parts of the Americas, Europe, and western Asia, they are allopatric elsewhere (Figure 11.21; cf. shags, Figure 4.2). The simplest explanation for the distributions is vicariance of a global ancestor, followed by limited overlap. The break between *Sanicula* in Timor and the Moluccas, and *Eryngium* in Australia, for example, coincides with a well-known biogeographic boundary (Heads, 2001).

In studies on *Eryngium*, Kadereit et al. (2008) sampled 52 of the ~250 species, and Calviño et al. (2008b) sampled 117. The latter authors presented a phylogeny:

> Subgenus *Eryngium*: All examined species from Africa, Europe, and Asia, except four species of the western Mediterranean.

> Subgenus *Monocotyloidea*: The four western Mediterranean species—a paraphyletic, basal grade—and a trans-Pacific clade from the New World (North, Central, and South America) plus Australia and New Zealand. The subgroups in this Pacific clade are in the following:

> 1. **Central-eastern South America** (southern Brazil, Uruguay, and northeastern Argentina).

> 2. **Mexico and central-eastern United States**.

> 3. Australasia, South America, and North America. There are two main clades:

>> A. Two species (*E. nudicaule*, *E. echinatum*) of **central-eastern South America** (Argentina, Brazil, Paraguay, Uruguay), with *E. nudicaule* also extending northwestward into Bolivia and Peru. The species do not reach the Pacific coast.

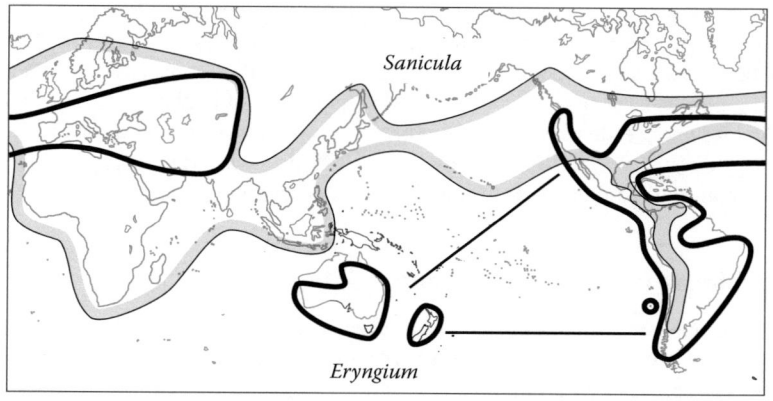

FIGURE 11.21 Distribution of a clade in Apiaceae: *Eryngium* and *Sanicula*. (Data from Kadereit, J.W. et al., *Taxon*, 57, 365, 2008.)

B. All species sampled from the Pacific coasts of **Australia, New Zealand, Chile, Juan Fernández Islands, and the United States** (California and, with *E. petiolatum*, Oregon and Washington; Kadereit et al., 2008). This comprises a typical "Pacific triangle" clade (cf. clades of *Caltha* and *Ranunculus*; see "Ranunculaceae," earlier in this chapter).

Within the "Pacific triangle" clade B, the species from the Juan Fernández Islands (*Eryngium inaccessum*, *E. fernandezianum*, and *E. bupleuroides*) form a well-supported group, sect. *Fruticosa*, and were described as "bizarre" by Calviño et al. (2008b). Their peculiar characters include a woody habit, in contrast with most *Eryngium* species, which are herbaceous perennials or annuals. Calvino et al. (2008a) argued that the Juan Fernández Islands *Eryngium* species "clearly exemplify" a shift from herbaceous continental ancestors to woody insular species, although this assumes that the ancestor was monomorphic. Woody species in the central Pacific that have herbaceous relatives elsewhere are seen in many groups (see the section "Chatham Islands" in Chapter 6).

Calviño et al. (2008b) and Kadereit et al. (2008) suggested that the Australasian *Eryngium* species originated by trans-Pacific dispersal from Chile. Calviño et al. (2008b) regarded the origin of the Californian species as ambiguous as they could have originated by dispersal from either Australasia or Chile. Yet it is significant that *Sanicula*, the sister of *Eryngium*, is absent from Australasia and Juan Fernández Islands. A vicariance origin would account for this, and it would also remove the necessity for any trans-Pacific dispersal events in *Eryngium*.

Subfamily Azorelloideae

The New Zealand members of Azorelloideae belong to a clade that is found around the Southern Ocean on the subantarctic islands, and also around the Pacific margin in China, Australasia, Patagonia, and the Andes (Nicolas and Plunkett, 2012). There are three main clades and these show considerable allopatry (areas occupied by just one of the three main clades are in **bold**):

Southwestern China (*Dickinsia*); Tasmania and Mount Kosciuszko region (*Diplaspis*).

New Zealand + Australia (*Schizeilema* p.p.); NZ, mainly on subantarctic islands (*Stilbocarpa*; see Chapter 6); subantarctic Chile and Argentina to central Chile (*Huanaca, Azorella* p.p., *Schizeilema ranunculus*).

Andes from central Chile to Costa Rica (*Azorella* p.p., *Mulinum, Laretia*); subantarctic Chile and Argentina (*Azorella lycopodioides*); subantarctic islands (*Azorella selago* and *A. macquariensis*).

There is local overlap between the first clade and the second in the Mount Kosciuszko region, while the last two clades overlap in the subantarctic islands and in subantarctic Chile and Argentina. Outside these areas the clades are allopatric (in the areas listed in bold type). Nicolas and Plunkett (2012) analyzed the results with a center-of-origin program and suggested a center of origin (in subantarctic South America and Australia), but this does not account for the extensive allopatry; China, New Zealand, and the Andes north of central Chile each have only a single clade. As usual, an explanation for this based on home advantage following chance dispersal is contradicted by the areas of overlap. In a simple vicariance model, an original circum-Pacific/Southern Ocean distribution was fragmented by main breaks in the south (in the Tasman–Patagonia region) followed by limited overlap there.

Subfamily Apioideae (Carrots, Celery, and Relatives)

Members of this group are often conspicuous in open vegetation in New Zealand, especially in the mountains. The group is diverse in species numbers and in morphological features. The crushed foliage of many members has an aniseed aroma; the spines of *Aciphylla* leaves can easily draw blood.

A phylogeny for the core group of the subfamily indicated many intercontinental relationships (Spalik et al., 2010). The authors calibrated the phylogeny with fossils and also constrained the age of Apiaceae to a maximum of 84 Ma. This date is from Bremer et al. (2004), a study that was also calibrated with fossils, and so the date gives a useful minimum age. Only by treating it as a maximum were Spalik et al. (2010) able to rule out vicariance as a mode of speciation in any of the five apioid clades represented in New Zealand. These clades are as follows.

Lilaeopsis

These small herbs are known in New Zealand from the maritime zone to the montane. The following phylogeny has been proposed for the genus (Bone et al., 2011):

Western United States, western Canada and western South America (not Mexico or Central America) (*L. occidentalis* s.lat.).

Eastern United States, Mexico, South America (not Central America), Mauritius, and Madagascar (five species).

SE Australia and New Zealand (five species).

The first clade is disjunct in western parts of North and South America and is sister to a trans-Atlantic + trans-Indian Ocean group. (Unsequenced populations from the Kerguelen Islands probably belong in the latter.) The same pattern is seen in other worldwide groups such as *Epilobium* (Onagraceae) (Chapter 10).

Spalik et al. (2010) concluded that because *Lilaeopsis* is nested within the "North American Endemics" group of tribe Oenantheae (*Limnosciadium*, etc.), its distribution developed from a North American center of origin. From there, the plant migrated by trans-oceanic dispersal to South America. And, Bone et al. (2011) proposed one dispersal to Australia or New Zealand, two dispersals between Australia and New Zealand, and three dispersals from South America to North America. None of these events are necessary if a widespread *Lilaeopsis* ancestor differentiated first at a boundary between western and eastern America. The only dispersal that is needed in the modern groups is local range expansion in North America, accounting for overlap between *Lilaeopsis* and its relatives, the "North American Endemics." Spalik et al. (2010: 13) wrote that their study permitted the rejection of a Gondwanan (vicariance) hypothesis for *Lilaeopsis* "without any doubts," but their conclusion depended on the treatment of minimum dates as maximum dates.

Tribe Aciphylleae: Anisotome *and* Aciphylla

The relationships of the tribe are as follows (Spalik et al., 2010):

Tribe Smyrnieae: Iran to Europe and the Mediterranean.

Unnamed tribe ("*Acronema* clade"): Himalayas and China.

Tribe Aciphylleae: New Zealand (including the subantarctic, Chathams, and Three Kings Islands), Tasmania, and southeastern Australia. Most of the ~65 species are South Island endemics.

The three groups have allopatric ranges along the Tasman–Tethys belt and can be explained by simple vicariance in a Tethyan group. Instead, Spalik et al. (2010) proposed a dispersal model for the Aciphylleae, with a center of origin in eastern Asia. The authors commented that many members of Aciphylleae are montane plants and are thought to be derived from recent immigrants that underwent rapid radiation (Winkworth et al., 2005). Nevertheless, the clock dates of Spalik et al. (2010) suggested that the ancestor of the tribe colonized New Zealand in the Eocene, and this is only a minimum age.

Calibrating the timeline of evolution in Aciphylleae using tectonic events (rather than fossil ages) is straightforward. For example, the *Gingidia montana* complex has a trans-Tasman, disjunct distribution in New Zealand and at the McPherson–Macleay Overlap (Heenan et al., 2013b), as in *Teucridium–Oncinocalyx* of the Lamiaceae (Chapter 4). Alpine fault-style disjunction occurs in *Aciphylla stannensis* (Stewart Island) + *A. trifoliolata* (northwest Nelson) (Figure 7.5) and also in *Gingidia baxterae* (Heads, 1998a: Fig. 3).

Preliminary results suggest that the tribe Aciphylleae, found in New Zealand, Tasmania, and south-eastern Australia, has two main groups (Radford et al., 2001; Spalik et al., 2010):

Anisotome (15 spp.). New Zealand: Subantarctic islands and the three main islands, north to the Egmont–East Cape line; not on the Chatham Islands.

Aciphylla s.lat. (including *Gingidia*, *Coxella*, *Scandia*, and *Lignocarpa*) (50 spp.). Southeastern Australia, Tasmania, and New Zealand: The three main islands, plus the Three Kings and Chatham Islands; not on the subantarctic islands.

Aciphylla s.lat. is present on the Chatham Islands but is absent from the subantarctic islands, although the genus includes many alpines. In contrast, *Anisotome* is absent on the Chathams, but is well represented in the subantarctic islands by three endemics (*A. latifolia* on Auckland and Campbell Islands, *A. antipoda* on Auckland, Campbell and the Antipodes, and *A. acutifolia* on the Snares and Solander Island). In addition to these differences, *Anisotome* ranges north only to the Egmont–East Cape line (Dawson, 1961; NZPCN, 2015), in contrast with *Aciphylla*, which extends to the Three Kings. Again, this is consistent with early vicariance between the southern *Anisotome* and the northern *Aciphylla*, with subsequent overlap between the two restricted to mainland New Zealand.

For *Aciphylla*, the most detailed molecular study so far was based on incomplete sampling, and the clades had poor statistical support (Radford et al., 2001). Yet while the results are only preliminary, in conjunction with morphological studies they are of special interest for biogeography.

The basal group in *Aciphylla* was suggested to be a distinctive assemblage of alpine species from Tasmania ("*Anisotome*" *procumbens*), Central Otago (*Aciphylla simplex*), and Central Otago, North Otago (Mount St Mary), and southern Canterbury (Two Thumb Range) (*Aciphylla dobsonii*). Four other species have been placed with *A. dobsonii* in the past, but none of these have been sequenced. They are all alpine plants from Fiordland and western Otago localities: Darran Mountains (*A. leighii*), Matukituki River and Humboldt Mountains (*A. congesta*), northern Fiordland and Eyre Mountains (*A. spedenii*), and eastern Fiordland, from Lake Hauroko north to at least the Murchison Mountains (*A. crosby-smithii*) (Oliver, 1956; cf. Allan, 1961; Dawson and Le Comte, 1978).

To summarize, the putative basal group of *Aciphylla* is endemic to Tasmania and the southwestern South Island (west of a line: Lake Hauroko–Mount St Mary–Two Thumb Range). This trans-Tasman clade is sister to the rest of *Aciphylla* in Australia and New Zealand. The possible connection between Tasmania and Fiordland/Otago is of special interest (cf. *Ranunculus* and *Sprengelia*).

In the molecular phylogeny, the next basal group to branch off after the Tasmania–southwestern South Island group comprises two species of southeastern Australia, *Aciphylla simplicifolia* and *A. glacialis*. *A. simplicifolia* (mountains between Melbourne and Canberra) has been described as most closely related to the unsequenced *A. townsoni*, of northwest Nelson to Arthur's Pass (Oliver, 1956).

The last main clade of *Aciphylla*, and the most species rich, is in New Zealand and at the MMO, where it is represented by the *G. montana* complex.

This indicates a possible overall phylogeny for *Aciphylla*:

Trans-Tasman (south): Tasmania–Fiordland–south Canterbury (7 spp.).

Trans-Tasman (central): SE mainland Australia–Nelson to Arthur's Pass (3 spp.).

Trans-Tasman (north): MMO and New Zealand (Three Kings, North, South, Stewart and Chatham Islands) (~40 spp.).

The three clades show considerable allopatry, consistent with original vicariance and subsequent overlap in the South Island.

Oreomyrrhis (Chaerophyllum *Clade*)

Apart from the well-known, trans-South Pacific distribution pattern that is exemplified by *Nothofagus*, many taxa distributed around the South Pacific basin also extend to the North Pacific. One example is *Oreomyrrhis* (Heads, 2014; Figure 4.9). It is present in Australasia and South America and is also recorded north to Taiwan and central Mexico. *Oreomyrrhis* was regarded by earlier workers as a genus but is now treated as an unranked group in *Chaerophyllum* (Chung et al., 2005; Chung, 2007; Spalik et al., 2010).

Chung et al. (2005) derived a very young age for *Oreomyrrhis* (2.23 Ma), but this was based on the dubious assumption that *Dendroseris* on Juan Fernández Islands was no older than the islands (Chapter 10). The authors also assumed a northern center of origin followed by dispersal to the southern hemisphere. However, the group is probably much older than the Juan Fernández Islands calibration suggests, and so it could have originated by vicariance with its sister, a clade of *Chaerophyllum* found in North America. Within *Oreomyrrhis*, Spalik et al. (2010) recovered the phylogeny:

Borneo (Mount Kinabalu) + Taiwan.
New Guinea.
New Zealand.
Australia + Mexico + South America.
Australia + Patagonia.

The phylogenetic sequence is consistent with a series of vicariance events that moved eastward across the Pacific plate, causing repeated fracturing of a widespread ancestor. The Australasia–South America disjunctions in each of the last two clades imply a minimum Cretaceous age for these two groups and also for the earlier phylogenetic breaks in the west Pacific.

In contrast with this vicariance model, dispersal theory suggests that *Oreomyrrhis* originated in the southern hemisphere following long-distance dispersal from North America to Malesia (>10,000 km) (Spalik et al., 2010). From Malesia, the genus then dispersed to Australia and New Zealand, and then to Patagonia (~9000 km). In a final step, the genus completed a circum-Pacific migration by dispersing from South America to Mexico. *Oreomyrrhis* is absent north of the trans-Mexican volcanic belt (a standard boundary), even though it favors cooler conditions, but its sister group is endemic there (and so the break can be explained by vicariance between the two).

The dispersal scenario for *Oreomyrrhis* requires a series of very long-distance dispersal events, although the group is characterized by local endemics that show no evidence for any special mobility. For example, *O. ramosa* in New Zealand has not crossed from the South Island to Stewart Island, across Foveaux Strait (26 km wide); *O. rigida* has also not crossed from the South Island to Stewart Island, or across Cook Strait (also 26 km) to the North Island. Members of *Oreomyrrhis* have no apparent means of dispersing more than a few meters, and a dispersal history is unlikely and unnecessary. Instead of a counterclockwise migration around the entire Pacific, the phylogeny probably represents a counterclockwise sequence of differentiation events at the Pacific margin.

Chung et al. suggested a date of 2.23 Ma for *Oreomyrrhis*, but this was based on the flawed assumption that island endemics can be no older than their current island. Spalik et al. (2010) suggested that *Oreomyrrhis* evolved at 3–4 Ma, but this is a fossil-based, minimum date. The biogeographic patterns and disjunctions are instead consistent with a Cretaceous age for the clade.

Groups with pan-Pacific distributions, such as *Oreomyrrhis*, are represented in all four quadrants of the Pacific margin. Another example, the freshwater fish group Percichthyoidea is known from Australia and Patagonia (Percichthyidae), China (Sinipercidae), and North America (Centrarchidae) (Chen et al., 2014b).

Likewise, the dragonfly family Petaluridae occurs in Australia, New Zealand, Chile, Japan + western United States, and the eastern United States. Each area has one endemic genus, with the southern groups forming a clade that is sister to the north Pacific clade (Ware et al., 2014). The authors found a center of origin for Petaluridae (in New Zealand), but only because they used a center-of-origin program (DEC implemented in LAGRANGE; Ree and Smith, 2008).

Petaluridae have a good fossil record, with the oldest crown group fossil dated as Early Cretaceous. Ware et al. (2014) concluded that the family originated at 157 Ma, and that the New Zealand *Uropetala* split from the Australian and Chilean genera at 127 Ma (both dates are minimum ages). All of the speciation events were calculated to have taken place before 70 Ma. The authors concluded that the family

Petaluridae has an "exceptionally deep evolutionary history," and the members are "living fossils." Nevertheless, the pan-Pacific distribution is a standard one; an alternative interpretation would be that the Petaluridae have a similar age and distribution to that of many groups (including *Oreomyrrhis* and Percichthyoidea), but have a much better fossil record.

Apium

In *Apium*, one New Zealand clade is sister to a group in South America, while the other belongs to a clade also found in Australia (Bass Strait Islands) and the Juan Fernández Islands (Spalik et al., 2010). *Euphrasia* (Orobanchaceae) is another plant group with independent New Zealand–Juan Fernández Islands, and New Zealand–South America affinities (Heads, 2014; Figure 6.10).

The affinity of Juan Fernández Islands with Australasia rather than with mainland South America is a standard pattern, seen in plants such as *Coprosma* and *Haloragis*, and animals such as the beetle family Trogossitidae (Chapter 1). Other beetles from Juan Fernández Islands include *Pycnomerodes* (Zopheridae), with its sister group in New Zealand, and *Lancetes* (Dytiscidae), with its sister group in Australasia (Leschen and Lackner, 2013). The great difference between Juan Fernández Islands and mainland South America is sometimes accounted for by dispersal. Instead, it can be explained by a biogeographic break that coincides with a major tectonic feature (a plate boundary) located between the two areas.

Daucus

Daucus is represented in Australia and New Zealand by a single, widespread species, *D. glochidiatus* (North, South, Chatham, and Three Kings Islands, widespread in the southern half of Australia). Its sister group is *D. montanus* of Mexico to central Chile, and so the pair repeats the Australasia–South America–North America "Pacific triangle" seen in *Eryngium* (Figure 11.21). For *Daucus*, Spalik et al. (2010) inferred dispersal from America to Australasia, but the Pacific triangle clade *D. glochidiatus* + *D. montanus* is sister to *D. durieua* of Africa and Europe, so a sequence of simple vicariance in a circumglobal ancestor would also explain the pattern.

To summarize, the New Zealand Apiaceae show the same intercontinental affinities as other groups: trans-Indian (*Lilaeopsis*), trans-Tethyan (Aciphylleae), and trans-Pacific (*Eryngium*, Azorelloideae, *Lilaeopsis*, *Oreomyrrhis*, *Apium*, *Daucus*). There is no need to invoke idiosyncratic histories and unique colonizing events for each separate taxon.

Brassicaceae in New Zealand

This is a diverse, global family of herbaceous plants that includes cabbage and mustard. The group does not have high levels of diversity in New Zealand, but the genera that are present have all been studied in recent molecular work and display interesting distributions.

The New Zealand Brassicaceae grow in open sites from the lowlands to the alpine zone. The family as a whole is well known for its ability to tolerate high levels of disturbance, and these occur through much of New Zealand as the country straddles a plate margin. On some mobile scree slopes in the northern South Island mountains, the Brassicaceae genus *Notothlaspi* (endemic to New Zealand) is one of the few plants present.

The Age of Brassicaceae

Clades in eudicot families such as Apiaceae and Brassicaceae are often interpreted as young, but this idea is not based on solid evidence. Using fossils of Brassicaceae to calibrate nodes on the phylogeny is problematic (Franzke et al., 2009) because the group's fossil record is so scanty, and also because identifying the fossils that do exist is so difficult.

In one example, *Rorippa* is a close relative of *Cardamine* and *Barbarea*, and *Rorippa* fossils are known from the Pliocene (2.5–5.0 Ma). "Therefore," Koch et al. (2000) wrote, "we assume that *Cardamine* and

Barbarea diverged about 6.0 MYA." But it is not clear why the authors accepted only 1 Myr difference between the oldest fossil and the age of the group—why not 2, 10, or 100 Myr? Koch et al. (2000) were frank about the problems that their methods faced, and they referred to the "great uncertainty associated with dating speciation events in the fossil record." Yet they did not refer to the main problem with fossil calibrations, the fact that they provide only minimum ages for clades and can rule out later, but not earlier events as relevant.

Based on the evolutionary rate derived by Koch et al. (2000), Franzke et al. (2009) calculated a mean age for the Brassicaceae of 64 Ma, at the Mesozoic/Cenozoic boundary. The authors regarded this as "unreasonably high," although they did not indicate why. As a fossil-calibrated date, it provides a useful minimum age.

In their own preferred analysis, Franzke et al. (2009) chose another calibration point, namely the split at ~70 Ma between Brassicaceae and the Moringaceae (another member of Brassicales, but not an immediate sister group). The date was obtained from a fossil-calibrated study of angiosperm families (Wikström et al., 2001), and so, again, it is a minimum age. Franzke et al. (2009) wrote that "there is at least 'indirect' fossil evidence, that this node estimate has a correct order of magnitude. *Dressiantha*, which is the oldest known fossil flower (Turonian, Upper Cretaceous: ca. 90 mya) that has been clearly assigned to the Brassicales, shows strong affinities to *Moringa*...." Nevertheless, Ronse de Craene and Haston (2006) concluded that *Dressiantha* is not related to Brassicales, but to Anacardiaceae (Sapindales).

Using the 70 Ma calibration for the split between Brassicaceae and Moringaceae (not sister groups), Franzke et al. (2009) calculated that Brassicaceae originated by diverging from its sister, Cleomaceae, at 19 Ma. Nevertheless, other published estimates for the family are much older: 30–60 Ma (the age of the basal split in the family, calibrated with fossil pollen; Koch et al., 2000), 40 Ma (the basal split in the family, Koch et al., 2001, calibrated using the *Chs* rate from Koch et al., 2000, and adding *matK* rates derived from this), and 37 Ma (the age of the family, calibrated using *Dressiantha*; Couvreur et al., 2010). These are all fossil-calibrated minimum estimates and so are compatible with an origin of the family in the Mesozoic, as suggested by the biogeography (see later text).

Beilstein et al. (2010) pointed out that the fossil material assigned to *Rorippa* and used for calibration by Koch et al. (2000, 2001) is not linked to a physical specimen, published image, or description. Instead, they used Oligocene fossil seeds (*Thlaspi primaevum*, 30.8–29.2 Ma) attributed by some authors to Brassicaceae. These were assigned a phylogenetic position at the coalescence of the extant species *Thlaspi arvense* and *Alliaria petiolata*, "as a minimum age constraint for this split." Beilstein et al. (2010) calculated that Brassicaceae originated at 65 Ma, also a minimum age. Nevertheless, the attribution of *T. primaevum* fossil seeds to the family, let alone the genus, was regarded as problematic by Franzke et al. (2011), who cited the high occurrence of homoplasy in fruit characters throughout the Brassicaceae.

To summarize, all of the fossil material used for calibration—*Rorippa*, *Dressiantha*, and *T. primaevum*—is controversial and in any case provides only minimum ages. As Franzke et al. (2011) concluded: "the age of the family is still not clear."

Biogeography of Brassicaceae

The distribution of Brassicaceae and its sister group is as follows (Stevens, 2016):

> Brassicaceae. More or less worldwide, but absent in much of the tropics (large parts of Brazil, central Africa, Madagascar, Borneo, and northern Australia).
> Cleomaceae. Throughout the tropics and in some warmer temperate areas.

There is a high level of allopatry between the two groups, and the simplest explanation is that they originated by vicariance of a worldwide ancestor. Early workers proposed instead that Brassicaceae were derived from Cleomaceae (for review, see Franzke et al., 2011), but molecular studies have shown that the former is not nested in the latter, as this theory would predict; instead the two families are sister groups.

Following its origin, the first differentiation in Brassicaceae split the group into two clades, the Tethyan tribe Aethionemeae and its widespread sister (Franzke et al., 2009). The tribe Aethionemeae comprises two genera. *Aethionema* has a center of diversity in Turkey and extends eastward into Turkmenistan, and westward into Spain and Morocco. *Moriera* is in Afghanistan, Iran, and Turkmenistan. Franzke et al. (2009) treated their own fossil-calibrated, minimum age for the family as an actual age: "The diversification of Aethionemeae versus the rest must have occurred during the very early evolutionary history of the family at the end of Miocene." The authors also assumed that the site of the basal phylogenetic break represents a center of origin (cf. Franzke et al., 2011). They suggested that:

> The vast majority of Aethionemeae occurs in Turkey and perhaps not too far from the ancestral area of the family. The centres of greatest diversity of Brassicaceae in the Old World are the Irano-Turanian [central Asian] and adjacent Mediterranean regions... and almost all of Turkey falls within these two regions. In fact, Turkey is one of the richest crucifer countries in the world and has about 560 species... On the basis of early phylogenetic branching discussed above and current distributional data, we agree [with earlier authors] in considering the Irano-Turanian as the cradle of the family. (Franzke et al., 2009: 433)

Instead of being the center of origin for the Brassicaceae, central Asia can be interpreted as the initial center of differentiation in a group that was already widespread throughout temperate regions of the world, following its vicariance from Cleomaceae. Franzke et al. (2009) inferred that Brassicaceae have effective dispersal capacities, "as evidenced by their spread from the Old-World center of origin into the current worldwide distribution." Yet there is no compelling evidence for an Old World center of origin for Brassicaceae, and the allopatry with Cleomaceae suggests an alternative—an origin in both Old and New Worlds, mainly outside the tropical regions.

Franzke et al. (2009) continued: "Long distance dispersal events, intercontinental dispersal included, have been well documented for several Brassicaceae genera...." Yet long-distance dispersal has been inferred for these genera only because (1) low levels of divergence have been attributed to short times of divergence, (2) fossil-calibrated dates have been treated as maximum dates, and (3) the sites of basal breaks have been treated as centers of origin. Alternative explanations of global differentiation in Brassicaceae clades are discussed next, with reference to the five genera indigenous in New Zealand.

Notothlaspi

This New Zealand endemic is restricted to the South Island, south to the Waihemo fault zone (Mount Ida: Petrie, 1896). It is a characteristic plant of alpine scree slopes and fell-field. The morphology of *Notothlaspi* is unusual, and its precise affinities are still unresolved. In two studies, it appeared as sister to the South African-Namibian tribe Heliophileae, although as yet there is little statistical support (Couvreur et al., 2010; Warwick et al., 2010). If it is confirmed, the relationship would be compatible with a standard disjunction between the Tasman region and South Africa, seen in other plants such as *Dietes* (Iridaceae), *Bulbinella* (Asphodelaceae), and *Cunonia* (Cunoniaceae).

Lepidium

Lepidium is a global genus with ~230 species, including 20 in New Zealand (all but three endemic) (de Lange et al., 2013). Mummenhoff et al. (2001) used fossil calibrations (*Rorippa* fossils from 2.5 to 5 Ma) to infer an origin of *Lepidium* in Pliocene/Pleistocene time. By treating this as a maximum date, the authors were able to rule out earlier vicariance. Mummenhoff et al. (2004) suggested that the center of origin of the Brassicaceae and "presumably" of *Lepidium* was located in the Mediterranean–Irano-Turanian region, but the rationale for this is confused (see earlier discussion), and a different origin for the family—vicariance with Cleomaceae—is accepted here.

Likewise, the patterns of variation within *Lepidium* (discussed next) suggest that the genus *inherited* its global distribution from a widespread ancestor, rather than attaining it by dispersal from a center of origin. In this vicariance model, a global ancestor diverged into *Lepidium* and its southern Andean sister group, *Lithodraba* (Couvreur et al., 2010), around a node in what is now southern South America. *Lithodraba* is endemic in Mendoza, Argentina (near the border with Chile and at the latitudes of Santiago and Concepción), and this region is well documented as a basal node in Pacific groups (Heads, 2014).

Lepidium itself comprises three main clades in an unresolved tritomy: one in Australia, one in Asia and Europe, and one with a global range (Mummenhoff et al., 2001). In the global clade, chloroplast DNA sequences (Mummenhoff et al., 2004) indicate geographic subclades endemic to the following regions (Figure 11.22a):

South and east Africa, Eurasia, Hawaii, and North and South America (broken line).

Eurasia (five separate clades) (broken line).

South America (broken line).

Australia/New Zealand (coastal and inland) + California (black line).

Nuclear DNA sequences from *Lepidium* instead divide the widespread clade into separate Old and New World groups. Likewise, the Australia/New Zealand + California clade is separated into two

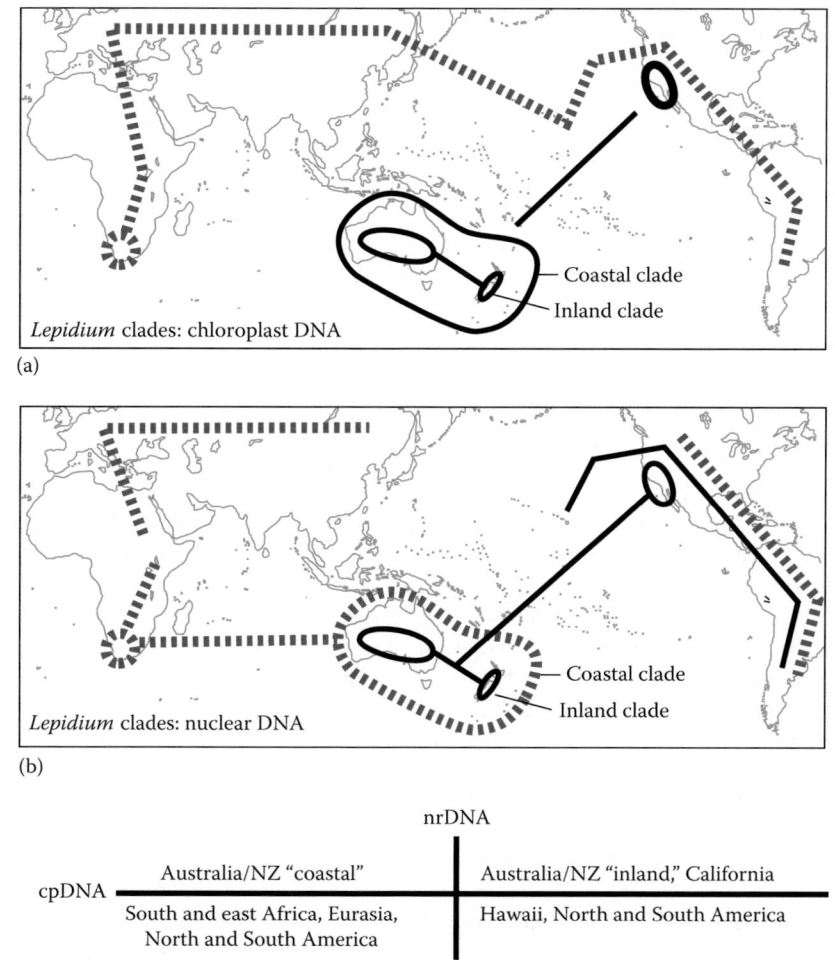

(a)

(b)

cpDNA	nrDNA	
	Australia/NZ "coastal"	Australia/NZ "inland," California
	South and east Africa, Eurasia, North and South America	Hawaii, North and South America

FIGURE 11.22 Distribution of the clades in *Lepidium* (Brassicaceae), as indicated by (a) chloroplast DNA and (b) nuclear DNA. Distributions outside Australasia are diagrammatic. (Data from Mummenhoff, K. et al., *Amer. J. Botany*, 91, 254, 2004.)

components: an inland group related to American plants, and a coastal group related to other Old World species (Mummenhoff et al., 2004). The nuclear data indicate the following clades:

North and South America (one clade) and Eurasia (three clades) (broken line).

Australia/New Zealand "inland clade" + California (black line); immediate relatives in Hawaii, North America, and South America (second black line); many in arid, inland sites.

Australia/New Zealand "coastal clade" + South Africa and east Africa; immediate relatives in North Africa and Eurasia; many in coastal areas (broken line).

The nuclear data show that the Australia–New Zealand plants have both trans-Indian Ocean and trans-Pacific Ocean affinities. Mummenhoff et al. (2004) and Dierschke et al. (2009) explained the pattern by invoking one long-distance dispersal event from South Africa to Australia/New Zealand, and one from California to Australia/New Zealand (both involving distances of >10,000 km), followed by hybridization in Australasia and extinction of the parents there.

This model leaves several questions unanswered: Why was there only one long-distance dispersal event from each of the two centers of origin? Why did this happen so early in the history of the widespread group? Why did the ancestral forms establish in New Zealand but then go extinct? The observed pattern is very simple; it is what would be expected in a case of vicariance and ancestral polymorphism (cf. *Pseudopanax ferox*, Figure 8.13), or in a case of vicariance followed by hybridization. The long-distance dispersal model is only necessary if the meager fossil record is regarded as more reliable than the phylogenetic and biogeographic data.

Instead of a "Californian" group and an "African" group dispersing to Australasia, the pattern is also consistent with the existence of a California–Australasia group (or set of characters), and an Africa–Australasia group (or set of characters). The two groups could have originated by differentiation in a widespread Africa–Australasia–California ancestor, for example, at a break in Australasia during Cretaceous marine transgressions. Subsequent hybridization in Australasia, perhaps following retreat of the inland seas, and isolation in Australia following Gondwana breakup would give the current arrangement. In this model, the Australasian "hybrid" group has developed as an early vicariant in Australasia and is not a secondary derivative of "Californian" and "African" groups. The pattern could also be explained by inheritance of ancestral polymorphism—the two sets of characters—by incomplete lineage sorting.

The putative hybrids (in Australia and New Zealand) display simple allopatry with the supposed parents (in Africa and California). This is not explained in a dispersal model without additional, ad hoc hypotheses, as the hybrids would be expected to occur where the parents overlap. Nevertheless, the same pattern, with precise allopatry between proposed hybrids and proposed parents, has been suggested for other groups. These include *Microseris* (Asteraceae) and *Santalum* (Santalaceae) around the Pacific (Heads, 2014), and *Codia* (Cunoniaceae) in New Caledonia (Pillon et al., 2008). In *Codia*, the "hybrids" have distinctive morphological characters not present in either of the "pure" parents, for example, leaves arranged in whorls of three rather than opposite pairs.

It is not a question of denying that hybridism exists; hybrids are a feature of many nodes, and most plant lineages have probably undergone phases of hybridism in their history. But it cannot be assumed that recombination of characters is the result of hybridism between *extant* groups—it is also explicable by ancient hybridism or incomplete lineage sorting (inheritance of ancestral polymorphism). New Zealand botanists have often invoked hybridism among extant species to account for character recombination, but this is sometimes questionable. For example, Burrows (2011b) proposed extensive hybridism among New Zealand *Pimelea* (Thymelaeaceae) species, although "a notable aspect of the Australian *Pimelea* flora is the lack of hybridization. According to Barbara Rye (pers. comm., 2009, cited in Burrows, 2011b: 408), there are no known instances of the phenomenon."

In the nuclear DNA phylogeny of *Lepidium*, the Australasia–Africa clade occurs mainly in coastal areas, whereas the Australasia–California clade occurs mainly in arid, inland sites (Mummenhoff et al., 2004). Many taxa of dry, inland Australia are derived from taxa of coastal, saline habitats that were stranded inland following the recession of epicontinental seas (Heads, 2014), and this probably applies to the inland clade of *Lepidium*.

In New Zealand, the "coastal clade" includes species maritime localities, including the offshore islands (Auckland Islands to Three Kings, Chatham and Kermadec Islands). The clade includes *Lepidium oleraceum*, *L. banksii*, *L. obtusatum*, *L. flexicaule* (Mummenhoff et al., 2004), and 10 new species (de Lange et al., 2013: Fig. 2).

In New Zealand, the "inland clade" has two coastal endemics (*Lepidium naufragorum* and *L. tenuicaule*) and three inland endemics, landlocked in Otago and Canterbury. The inland species are *L. solandri* and *L. sisymbrioides*, both on limestone and in dry, rocky areas, and *L. kirkii*, on salt flats in tussock grassland (Maniototo to Alexandra) (Mummenhoff et al., 2004; Heenan et al., 2007; de Lange et al., 2013: Fig. 2). DNA sequences in *L. sisymbrioides* and *L. solandri* are most variable near Alexandra, where Central Otago endemism is often focused. As in Australia, the common ancestor of the inland and coastal clades probably expanded its range inland during marine incursions. Later, when the seas receded, some forms were left behind and stranded inland, where they have survived on limestone, salt flats, rocky sites, and in drier grassland.

Cardamine

Cardamine is another worldwide genus, with 160–200 species. The New Zealand species are diverse and range from the lowlands to the high-alpine zone (Mark and Adams, 1973). They also occur in the subantarctic islands, where *C. subcarnosa* is endemic to Auckland and Campbell Islands, as is *C. depressa* var. *stellata*.

A study of the southern hemisphere, montane species, found that *Cardamine glacialis* of South America "remarkably" formed a clade with the montane species of Australia and New Zealand (Bleeker, et al., 2002a). In contrast, the montane species of New Guinea formed a clade with northern hemisphere species. This indicates an intercontinental break between the two groups, located somewhere around the Tasman or Coral Sea. A subsequent study of *Cardamine* sampled more species and confirmed the break between New Guinea on one hand and Australia/New Zealand on the other (Carlsen et al., 2009).

With respect to the timing of evolution in *Cardamine*, Carlsen et al. (2009) accepted the young, fossil-based date of origin for the genus (6.0 Ma) that Koch et al. (2000) calculated (see the earlier section "The Age of Brassicaceae"). With respect to the spatial component of evolution, Carlsen et al. (2009) found that *Cardamine* as a whole comprises three main clades (their mutual relationship is unresolved):

Group A: Eurasia and Beringia.
Group B: Africa and East Asia.
Groups C–J: A global clade comprising all the other groups, including those in Australasia.

Carlsen et al. (2009) thought that *Cardamine* had a center of origin in Eurasia, because of the diversity of *Cardamine* species there, especially diploid ones (these are assumed to be primitive, and primitive species are assumed to indicate a center of origin). On this basis, the authors concluded that "dispersal between Eurasia and North America may have occurred stepwise via the Tertiary Beringian land bridge that existed until 5.4–5.5 Ma…, but dispersal over longer distances must have occurred between these *Cardamine*-rich continents and Oceania, South America, and Africa" (Carlsen et al., 2009: 216).

Instead, if ancestral *Cardamine* was already widespread by the time the modern groups differentiated, the basal breakup of the genus into the three main clades would reflect allopatric differentiation into groups north of Tethys (Group A), south of Tethys (Group B), and along and east of Tethys (widespread clade). This would then need to be followed by subsequent range expansion of the last clade through Eurasia and Africa.

In the study by Carlsen et al. (2009), group H comprised three European species nested among four species from New Guinea, and this is a standard *Tethyan* affinity. Group J (with little statistical support) included all the species in Australia/New Zealand, along with forms of South America, North America, Beringia, and east Asia. This would give a typical *circum-Pacific* range, as seen in the plant

Gaultheria (Ericaceae) (Heads, 2014; Figure 3.16) and the harvestman *Acropsopilio* (Caddidae), known from New Zealand, Chile, Venezuela, North America, and Japan (Giribet and Sharma, 2015).

Carlsen et al. (2009) argued that the patterns in the two groups H and J mean that Australasia, including New Guinea, was colonized twice from the northern hemisphere. (For group J, the authors proposed "very long-distance colonization" from Beringia.) A vicariance explanation is much simpler: Australasian *Cardamine* groups originated by differentiation along two sectors, one Tethyan (group H), and the other Pacific (group J). This intersection of Pacific and Tethyan distributions in Australasia is seen in many other groups, both fossil (e.g., the Mesozoic bivalve *Monotis*) and living (e.g., the angiosperm *Coriaria*: Coriariaceae) (Heads, 2014; Figure 3.11).

Rorippa

This is another worldwide genus, although it is less diverse than *Lepidium* or *Cardamine*. There are three species in New Zealand (de Lange et al., 2009). Bleeker et al. (2002b) sampled 25 of the 50–80 species and found four clades:

1. North and South America.
2. Widespread globally (Africa, Eurasia, Australia and New Guinea, North and South America). The group includes *Rorippa palustris*, present in Europe, Asia, Australia, New Zealand, North America, and South America. It is uncertain whether or not this is indigenous in New Zealand, but it was collected there in 1769 by Banks and Solander.
3. Europe, Asia, and North America.
4. Australia, New Guinea, and New Zealand (in the last locality with the endemic *Rorippa divaricata*). The group also includes *R. laciniata*, present in Australia (widespread in the southeast) and Northland.

Clades 2 and 4 overlap in Australia, New Guinea, and, probably, New Zealand. Yet the distinctive, restricted range of clade 4 and its presence in New Zealand with an endemic species (unlike group 2) suggest it originated around the Tasman and Coral basins by vicariance with one or more of the other three clades. If this was the case, the overlap with group 2 (in New Zealand with *R. palustris* only) was a subsequent event.

Bleeker et al. (2002b) inferred a center of origin of *Rorippa* in the north (with clade 2 colonizing Australasia via Malaysia, clade 4 colonizing Australasia via Malaysia or possibly South America), but this does not account for the standard distribution pattern of clade 4.

Pachycladon *(Including* Cheesemania *and* Ischnocarpus*)*

Pachycladon s.lat. comprises 10 species in the South Island and one in Tasmania (Heenan and Mitchell, 2003; Joly et al., 2014). All species grow in open, rocky habitat. *P. exile* is restricted to lowland sites, and there are some lowland populations of *P. cheesemanii*, but the genus is mainly alpine. *Pachycladon* belongs to the tribe Microlepidieae, a group with the following relationships (phylogeny from Heenan et al., 2012; distributions from Al-Shehbaz, 2012). The statistical support for the clades is still weak, but the geography is intriguing.

Microlepidieae: Southern Australia, Tasmania, and South Island.

Crucihimalayeae (*Crucihimalaya*, *Transberingia*): Central Asia (including Himalayas), one species extending via arctic eastern Russia and Beringia into North America and Greenland.

Physarieae: Mainly the United States, with fewer species in N Mexico, in SW Canada, and, disjunct in Argentina and Bolivia.

The three tribes are more or less allopatric in Australasia, Asia, and the Americas, respectively, consistent with vicariance in an ancestor that was widespread along the Tethys (in central Asia) and around the Pacific margin. The Australasian tribe Microlepidieae in turn comprises three, more or less allopatric clades in an unresolved tritomy (the support for these clades is weak) (Heenan et al., 2012):

Southern half of Australia and Tasmania (a clade of 15 genera).

Tasmania (*Pachycladon radicatum*).

South Island (all the other *Pachycladon* species).

In a detailed phylogeny of *Pachycladon* itself, discussed later, the Tasmanian species groups with the New Zealand ones, and so the distribution of the tribe is western Australia and eastern Australia/Tasmania versus Tasmania and New Zealand. This is a standard pattern and was illustrated earlier in Loganiaceae and Ericaceae (Figure 7.6). The simplest explanation accounts for all these patterns with a single event. One possibility is a vicariance event that divided Australia–New Zealand groups of Loganiaceae, Ericaceae, and Brassicaceae in two, at a boundary in or around Tasmania. This was followed by rifting of the trans-Tasman clade and local overlap in Tasmania.

Mandáková et al. (2010a) suggested that *Pachycladon* is the sole New Zealand genus in its tribe "and therefore an *in situ* origin seems unlikely." Yet the high level of allopatry between *Pachycladon* and the other members of its Australasian tribe, along with the tribe's allopatry with its own relatives in Asia and America, is what would be expected following an *in-situ* origin.

Mandáková et al. (2010a) and Heenan and Mitchell (2003) suggested that *Pachycladon* originated in Australia and later dispersed to New Zealand. Yet this does not explain why dispersal to New Zealand would happen just once, why *Pachycladon* would then die out in Australia, and why it would then disperse back again, just once, from New Zealand to Tasmania (giving rise to *P. radicatum*). The proposal implies an effective trans-oceanic colonizer capable of traveling in both directions across the Tasman (1500 km), but *Pachycladon* shows little sign of range expansion; for example, it has not crossed Cook Strait (26 km wide) to colonize the North Island or crossed Foveaux Strait (also 26 km wide) to reach Stewart Island. These problems are removed if the genus evolved, not by dispersal, but by the *in-situ* development of its particular recombination of characters. These evolved over the region that later became Tasmania and New Zealand, out of an ancestral complex that was already distributed through central Asia and on both sides of the Pacific basin.

Whole-Genome Duplication in Microlepidieae

The species of the Australasian tribe Microlepidieae, including those of *Pachycladon*, have two copies (paralogues) of each gene that occur singly in two highly divergent evolutionary lineages, both restricted to the northern hemisphere (Joly et al., 2009; Mandáková et al., 2010b; all tribe distributions from Al-Shehbaz, 2012):

Genome "a1" of Microlepidieae also occurs in a clade comprising tribes Boechereae (far eastern Russia and North America north of Mexico), Crucihimalayeae (central Asia to Greenland), and Camelineae s.str. (mainly Asia and Europe, also northern North America) (Mandáková et al., 2010b).

Genome "a2" of Microlepidieae also occurs in the tribe Smelowskieae of central Asia, eastern Asia, and northern North America (Zhao et al., 2010).

Mandáková et al. (2010b) suggested that the genome duplication (polyploidy) in the Australasian Microlepidieae is the result of hybridization between the two northern hemisphere groups, producing an allopolyploid. Probably all groups (including humans) have had phases of hybridism in their history, but this does not mean that groups recombining characters of others are simply hybrids of the latter. The case of Microlepidieae is a further example (matching *Lepidium, Microseris, Santalum,* and *Codia*; see the earlier section "*Lepidium*"), in which a supposed hybrid of extant groups is allopatric with its putative parents. The simple allopatry is instead consistent with a vicariance event developing in an already polymorphic ancestor.

Mandáková et al. (2010b) concluded that the hybridization producing Microlepidieae took place in Australia after long-distance dispersal of both parental genomes from Eurasia or North America to Australia. This requires the extinction of both parents in Australia following the hybridism. As another, less likely alternative, Mandáková et al. (2010b) suggested that the hybridism occurred in Eurasia/North America, with the polyploid alone migrating to Australia (and going extinct in Eurasia/North America). As precedents for the very long-distance dispersal, Mandáková et al. (2010b) cited similar dispersal events that have been inferred for other Brassicaceae genera, *Lepidium* and *Cardamine* (discussed earlier). Yet the probability of such events happening in one group is vanishingly small, and the probability of their occurring in three is even lower. Instead, vicariance provides a general mechanism that explains the repetition of similar geographic patterns in these genera and in others such as *Microseris* (Asteraceae) (Heads, 2014).

Dating Evolution in Pachycladon

Heenan et al. (2002) accepted a young age for *Pachycladon*, but this was based on the controversial rate from Koch et al. (2000) (see the earlier section "The Age of Brassicaceae" and the critique by Beilstein et al., 2010). Heenan et al. (2002) inferred that diversification within the genus began at 1.0–3.5 Ma. They wrote (p. 554) that "rock bluff montane habitats suitable for the *Pachycladon* complex" would have been created in the Pliocene (5.2–1.6 Ma) with the Kaikoura orogeny. Yet while rocky sites are prime *Pachycladon* habitat, there is no evidence that montane *elevations* are necessary. The genus occurs at several lowland localities, and this is consistent with the ancestral populations of the genus having remained more or less in place as the mountains rose under them.

Joly et al. (2009) dated the origin of *Pachycladon* at 0.8–1.6 Ma, using two independent rate estimates (Yang et al., 1999; Koch et al., 2001). Both of these were based on fossil calibrations, and so the dates are minimum clade ages (the Koch et al., 2001, rate was taken from Koch et al., 2000—see earlier). Joly et al. (2014) fixed the age of the crown group of *Pachycladon* at 0.82 Ma (Joly et al., 2009) and, as a consequence, calculated high diversification rates for *Pachycladon*. They concluded that "these rates compare to that of the highest rates observed among flowering plants," but the rates are inferred, not observed, and are based on treating a minimum clade age as a maximum age. The biogeographic patterns in the genus discussed later indicate an alternative timeline for evolution.

The standard, clearcut distributions in *Pachycladon* mean that tectonic calibrations can be used instead of relying on a patchy, controversial fossil record. The boundary of the genus at Bass Strait, where there was major rifting and volcanism in the Cretaceous (Heads, 2014), along with breaks within the genus at the Tasman Sea, the Alpine fault, the Waihemo fault zone, and the Moonlight tectonic zone (see later) all suggest that the actual age of the genus is one or two orders of magnitude greater than the fossil-based estimates. Joly et al. (2014) proposed a model of adaptive radiation and ecological speciation in *Pachycladon*, but this was based on the inferred rates of evolution and other calculations, and the distributions of the main clades were not mapped or compared.

Seven Geographic Patterns in Pachycladon

Several species are confined to standard areas of local endemism, for example, *Pachycladon latisiliquum* in northwest Nelson (Mount Arthur, Mount Owen, etc.), *P. stellatum* in the Inland Kaikoura Range (with a northeastern limit near Tapuaenuku; Heenan and Garnock-Jones, 1999), *P. wallii* in the Eyre Mountains and nearby ranges, and *P. exile* in the region just north of the Waihemo fault zone.

Detailed studies of *Pachycladon* (Heenan and Mitchell, 2003; Joly et al., 2014) have also documented many other interesting features of its distribution, and the following seven patterns are of special significance. (For critique of niche model analysis of *Pachycladon* distributions, see Chapter 1.)

1. *Absence of the Genus from the North Island and the Seaward Kaikoura Range*   *Pachycladon* (Figure 11.23) has a center of species diversity in Marlborough south of the Wairau fault, but despite this the genus is absent in the nearby North Island. Many other South Island groups are absent in the North

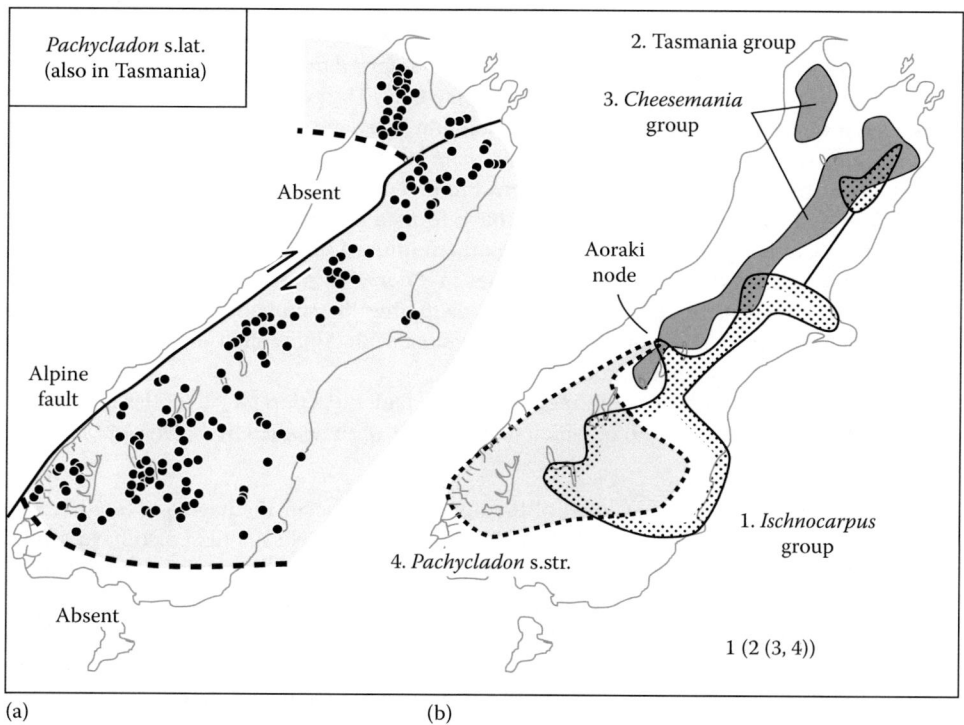

FIGURE 11.23 Distribution of *Pachycladon* (Brassicaceae). The Tasmanian species is not shown. (a) All New Zealand records. (b) Distribution of the three main clades. (Data from Heenan, P.B. and Mitchell, A.D., *J. Biogeogr.*, 30, 1737, 2003; Joly, S., *Syst. Biol.* 63, 192, 2014.)

Island (including another Brassicaceae genus, *Notothlaspi*), and it is difficult to explain this general pattern as the result of current ecological factors. Instead, boundaries at Cook Strait, phylogenetic differentiation between North and South Islands, and local endemism around the strait can all be attributed to evolution at an old break or series of breaks in the Cook Strait area. This is more likely to reflect the complex tectonic history of the region rather than the modern strait itself, which is a recent feature (Chapter 8). In the same way, the absence of *Pachycladon* in mainland Australia can be related to Cretaceous activity in the Bass Strait basins rather than the modern strait itself.

Heenan and Mitchell (2003) commented on the interesting absence of *Pachycladon* from the Seaward Kaikoura Range, eastern Marlborough. They wrote that this absence is "most likely to be influenced by ecological or environmental factors" (p. 1745), but they did not suggest what these factors might be. The absence is probably related not to ecological factors, but to the absence of the genus from the North Island.

2. Connections between Inland Canterbury (Torlesse and Aoraki Areas) and Coastal Marlborough   In Nelson and Marlborough, *Pachycladon* approaches the coast at a single region, the Ben More and Chalk Ranges south of Cape Campbell (Figure 11.23; Druce and Williams, 1989). *P. fasciarium* (in the *Cheesemania* group) is endemic to these two ranges, 5 km from the sea, and is a strong calcicole on Late Cretaceous and Paleogene Amuri Limestone (Figure 11.23b; Heenan, 2009). Overall, the *Cheesemania* group connects this coastal locality with the Torlesse and Aoraki areas of central Canterbury. The pattern repeats that of *Carmichaelia* subg. *Monroella*, which intersects the coast near Cape Campbell with the calcicole *C. astonii* (Flaxbourne River mouth, also present in the Chalk Range area, Figure 11.13). As discussed in the treatment of *Carmichaelia*, these distributions can be interpreted as relictual, and tectonic in origin.

3. *Absence from Westland (including Paparoa Range) and Stewart Island: Alpine Fault Disruption of the Southern Boundary*   The overall distribution of *Pachycladon* in New Zealand (Figure 11.23) includes another pattern of presence and absence that has not been referred to or accounted for in earlier studies, but is compatible with strike-slip movement on the Alpine fault. The genus is present in northwest Nelson, but is absent from the mountains of southern Nelson and Westland (the Paparoa and Victoria Ranges, Mount Tuhua, etc.). In the same way, further south, there is an endemic species in central Fiordland, but the genus is absent from the mountains of southern Fiordland and Stewart Island. The two main centers of absence, Westland and Stewart Island, would have formed a single center before movement on the Alpine fault, and there is a marked "step" in the southern limit of the genus along the fault. The pattern is repeated in many groups and was mapped earlier in *Coprosma fowerakeri* (Figure 9.19a); the step can be explained by movement on the Alpine fault disrupting the southern boundary of the genus. In *Pachycladon,* the trans-Tasman disjunction occurs together with Alpine fault disjunction or disruption, and this combination is seen in many groups (Chapter 9).

Heenan and Mitchell (2003: 1746) examined the Alpine fault biogeographic hypothesis and wrote that it makes the following predictions, all of which they thought were falsified by *Pachycladon*:

- "Sister taxa of Nelson endemics west of the divide should occur in Otago and Southland east of the divide." This is not a predicted by the model though. Nelson taxa also have affinities in several other places, such as Tasmania, Marlborough, Westland, and the North Island, and these connections may have been established before fault movement.

- "There should be sharp phylogeographical breaks across the Alpine Fault." Again, the model does not suggest that this is true for all groups. For example, many taxa that were widespread and undifferentiated before fault movement would have remained widespread and undifferentiated after fault movement. In any case, *P. cheesemanii* and *P. novae-zelandiae* have their western limits at the fault, and *P. latisiliquum* has its eastern limit at the fault.

- "Molecular divergence between disjunct taxa should reflect 25 My of divergence." This is not necessary, as most of the strike-slip movement on the fault (420 km of the total 470 km) is thought to have occurred since 11–16 Ma (Chapter 9).

These objections to the idea of communities being displaced along the fault are not substantial, and no other explanations for the intriguing "stepped" distribution of *Pachycladon* in the South Island have been given.

4. *Overlap of the Three Main New Zealand Clades of* Pachycladon *at the Aoraki Node*   In *Pachycladon*, there are four clades, and these are, in large part, allopatric; the three clades in New Zealand meet at the Aoraki/Mount Cook node (Figure 11.23; Heenan et al., 2002; Heenan and Mitchell, 2003; Joly et al., 2014). The phylogeny is 1 (2 (3 + 4)):

  1. *Ischnocarpus* group (formerly a separate genus). This occurs in the southeast. It comprises two generalist species that occupy a range of rock types (*P. cheesemanii* on greywacke, schist, and plutonics; *P. exile* on limestone, tuff, breccias, and alluvium).
  2. This group includes a single species, *P. radicatum*, endemic to Tasmania.
      3. *Pachycladon* s.str. (plus *P. wallii*). This occurs in the southwest. It comprises species mainly found on schist (*P. novae-zelandiae*, *P. wallii*), as well as *P. crenatum* that grows on gneiss and volcanics.
      4. *Cheesemania* group (formerly a separate genus). This occurs mainly north and west of group 1. The species occur on greywacke and limestone (*P. latisiliquum*, *P. enysii*, *P. fastigiatum*, *P. stellatum*, *P. fasciarium*). *P. latisiliquum*, the only species in northwest Nelson, is sister to the other four.

Heenan and Mitchell (2003) suggested that at least some of the factors driving the radiation of *Pachycladon* "are almost certainly related to the geological parent material," but this overemphasizes the substrate specificity of the plants. It is true that *P. enysii* is only found on greywacke, but this is the most widespread rock type in the South Island (and so is more likely to include endemics), and 7 of the 10 New Zealand species occur on more than one rock type. The two others are more or less local endemics, meaning they are more likely to be restricted to a particular rock type anyway.

Group 1 is "basal," but this just means it is the smaller of two sister groups. There is no reason to assume it is more or less primitive than its sister group in morphology or ecology. Nevertheless, McBreen and Heenan (2006) suggested that group 1 "may represent the ancestral type of the genus," with groups 3 and 4 being derived. Groups 3 and 4 (along with group 2) are sister to group 1, not nested in it, and so there is no phylogenetic support for assuming that group 1 is ancestral.

With respect to ecology, Heenan and Mitchell (2003) proposed a model of "adaptive radiation" in *Pachycladon*, based on an ecological center of origin. In this model, the basal group (group 1) occupied the ancestral habitat, while groups 3 and 4 underwent "radiation into mountain habitats" and "specialization onto different rock substrates" (schist and greywacke). Nevertheless, the high level of allopatry among the four groups suggests a simple vicariance history rather than adaptive radiation from a geographic and ecological center of origin.

Although Heenan and Mitchell (2003) compared the *Pachycladon* clades in terms of the main rock types they occupy, they did not refer to the groups' distinctive geography (Figure 11.23). There is considerable overlap, but the basic allopatry is obvious; the Aoraki/Mount Cook area is the single locality where all three clades meet. This area is the site of the highest mountains in New Zealand, but the breaks and disjunctions in *Pachycladon* are probably older than this Neogene development. Breaks at Bass Strait, the Tasman Sea, the Waihemo fault zone, and the Moonlight tectonic zone (Figure 11.23 and see next section) suggest that the Aoraki node was already active in the evolution of *Pachycladon* long before strike-slip movement at the Alpine fault and uplift of the Southern Alps began.

To summarize, New Zealand *Pachycladon* comprises three main groups, with one each in the north, southwest, and southeast, and this is consistent with their differentiation by vicariance around the Aoraki node. A similar pattern occurs in the *Lobelia macrodon* group (Figure 7.15), and the differences between the two are minor. For example, in the *Lobelia* clade the southeastern component does not extend north through eastern Canterbury.

5. Distribution Limits at the Moonlight Tectonic Zone Panbiogeographic analyses of New Zealand taxa have related many distribution patterns to terrane boundaries, which are old subduction zones and major faults in their own right (Heads, 1990a). Heenan and Mitchell (2003) adopted this approach in their study of *Pachycladon*, but attributed the coincidence between clade boundaries and terrane boundaries to differences in lithology—the different rock types and soil chemistry—rather than to tectonic history. Substrate type is an important ecological factor, but structures such as faults can delimit distributions, even if they do not mark changes in parent material. Distribution boundaries at such features often coincide with tectonic, rather than stratigraphic or lithological breaks, and the Moonlight tectonic zone is an example of this (Figures 7.8 through 7.12).

The three species of *Pachycladon* s.str. are mapped in Figure 11.24. *Pachycladon crenatum* of Fiordland has its eastern limit at the Moonlight tectonic zone, while *P. wallii* has its western limit there. The *Ischnocarpus* group of *Pachycladon* (fine broken line in Figure 11.24) also has a western limit there.

Heenan and Mitchell (2003) demonstrated that *P. wallii* is restricted to the Caples terrane, but it is present only on a small portion of the terrane and reaches its western limit at the Moonlight tectonic zone (Figure 11.25). This tectonic break cuts across the Caples terrane and does not mark any change in lithology. The northern boundary of *P. wallii* occurs at the Caples terrane/Torlesse terrane boundary, which is both a tectonic and a lithological break. Thus, overall, the range can be explained by tectonic events.

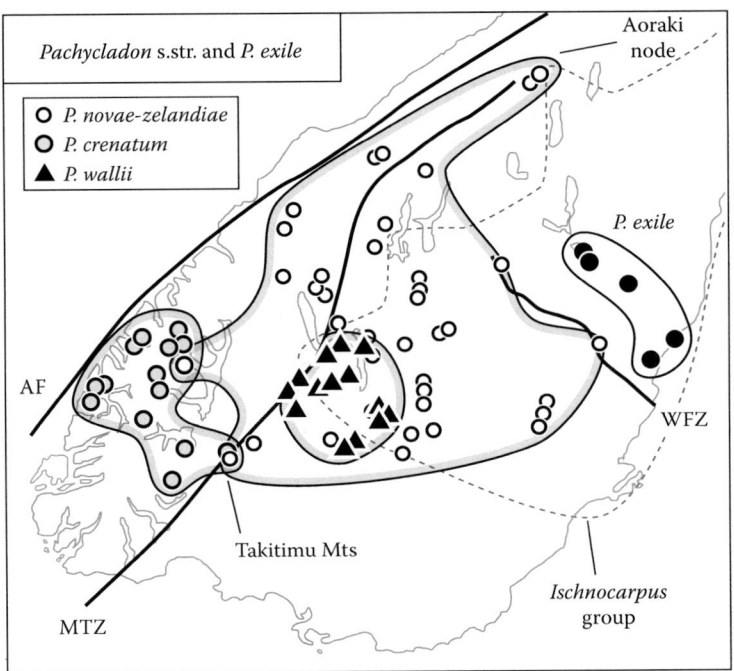

FIGURE 11.24 Distribution of one clade in *Pachycladon* (*Pachycladon* s.str.) plus *P. exile* (Brassicaceae). AF, Alpine fault; MTZ, Moonlight tectonic zone; WFZ, Waihemo fault zone. (Data from Heenan, P.B. and Mitchell, A.D., *J. Biogeogr.*, 30, 1737, 2003; Joly, S., *Syst. Biol.* 63, 192, 2014; Otepopo record from Petrie, D., *Trans. N. Z. Inst.* 28, 540, 1896.)

6. *Distribution Limits at the Waihemo Fault Zone*   The third species in the southwestern group (*Pachycladon* s.str.), *P. novaezealandiae*, extends the range of the group north to Mount Cook (Figure 11.24). The species does not appear to have any particular relationship with the Moonlight tectonic zone, but in east Otago its northern boundary corresponds with the Waihemo fault zone. Figure 11.24 also shows one of the species in the southeastern *Ischnocarpus* group, *P. exile*. This has its southern limit at Otepopo (Waianakarua), near the Waihemo fault zone (record from Petrie, 1896, who cited it as "*Sisymbrium novæ-zealandiæ*, var.—a very small, slender form").

7. *Montane Taxa at Sea Level*   *P. cheesemanii* is recorded on the Otago Peninsula, near Dunedin, from only 10 m elevation, and this is unusual as most populations of *Pachycladon* are montane or alpine (Figure 11.23). Other taxa with anomalous, low-elevation records at Dunedin include the plant *Epilobium pictum* and the moth *Eurythecta leucothrinca* (Tortricidae). The latter is known from a number of sites from Canterbury to Southland, all in montane to subalpine grassland, but it also has a population in coastal saltmarsh at Blueskin Bay, Dunedin (Patrick, 1990). Patrick attributed the population at sea level to the tectonic lowering of the Central Otago peneplain to below sea level around Dunedin. He wrote that the phenomenon of alpine or upland species occurring at coastal sites is well known, "although from a biogeographic standpoint the subject has not been adequately addressed" (p. 305).

Other sites of anomalous, lowland populations in otherwise montane or inland taxa, occur at Shag Point, where the Waihemo fault zone meets the coastline (Heads and Patrick, 2003; Patrick et al., 2010), and there are records of *Pachycladon exile* near the coast, 20 km north of the Waihemo fault zone at the Waianakarua River (Figure 11.24). Anomalous coastal records of inland groups also occur at Christchurch, as in *P. cheesemanii* and *C. kirkii* (Figure 11.7). At its northeastern limit, *Pachycladon* occurs near the coast in the Chalk Range, above the Waima River, but does not occur below 900 m.

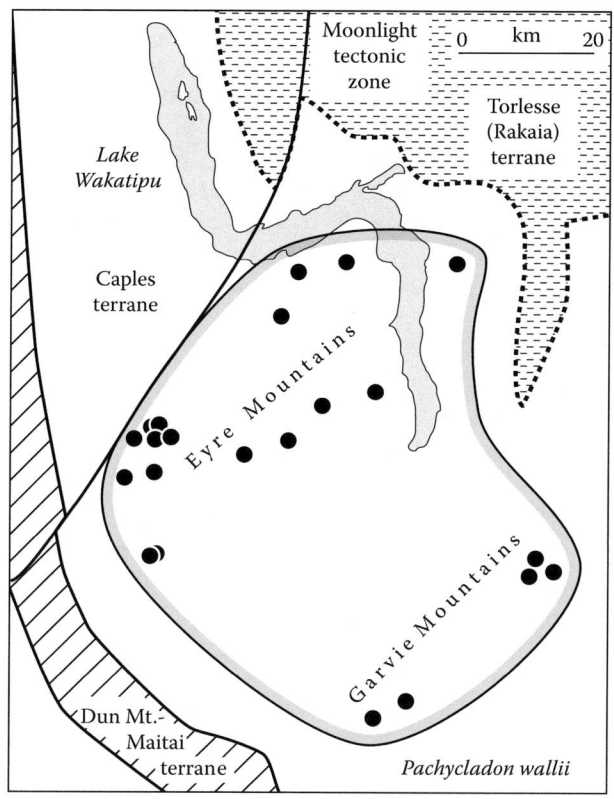

FIGURE 11.25 Distribution of *Pachycladon wallii* (Brassicaceae). (Data from Heenan, P.B. and Mitchell, A.D., *J. Biogeogr.*, 30, 1737, 2003; Joly, S. et al., *Syst. Biol.*, 63, 192, 2014.)

Summary: *Pachycladon* and Brassicaceae

Heenan and Mitchell (2003) assumed that *Pachycladon* evolved at a localized center of origin with a restricted ecological range, and that there was subsequent radiation from there. Nevertheless, they concluded that "the factors driving the radiation of *Pachycladon* are difficult to determine...." They suggested that "present distributions of *Pachycladon* species may relate to Pleistocene climate change," but none of the distinctive patterns in the genus that are cited earlier can be accounted for in this way.

An alternative, vicariance model rejects the ideas of center of origin and radiation as unnecessary, and instead, it focuses attention on phylogenetic and geographic breaks in the group, at localities such as the Waihemo fault zone, the Moonlight tectonic zone, the Mount Cook/Aoraki node, and the Kaikoura region.

To summarize: more molecular work is needed. For example, the well-sampled phylogenies that included *P. radicatum* (Heenan and Mitchell, 2003; Heenan et al., 2012) were not well supported, while the well-supported phylogeny (Joly et al., 2014) did not sample *P. radicatum*. Nevertheless, the present state of knowledge suggests that the history of the genus involves phylogenesis associated with documented tectonic activity at the following zones:

1. An Early Cretaceous boundary of *Pachycladon* at Bass Strait.
2. Mid-Cretaceous breaks in the genus at the Caples/Torlesse terrane boundary, the Moonlight tectonic zone, and the Waihemo fault zone.
3. Late Cretaceous breaks in the genus at the Tasman Sea.

This history would mean that the genus had already diversified by the time the Alpine fault disrupted its southern boundary in the Miocene and the Kaikoura orogeny lifted many populations into the alpine zone.

To summarize this section on the Brassicaceae, the idea of a recent origin was based on particular interpretations of the family's scanty, controversial fossil record. Instead, the detailed biogeographic and phylogenetic data now available suggest that members of Brassicaceae evolved and had their early differentiation much earlier than has been thought—in the Mesozoic, rather than the Cenozoic. This allows simple explanations for the intercontinental affinities of all the New Zealand genera. Within New Zealand, correspondence of clade boundaries with tectonic features that were active in the Cretaceous is still evident in the genus *Pachycladon*.

12

Case Studies of New Zealand Animals

This chapter examines distributions in some of the best-documented groups of New Zealand animals. The treatments demonstrate that distributions are structured at both intercontinental and local scales around the same phylogenetic/biogeographic nodes that have already been illustrated.

Onychophora, Freshwater Mollusks, and Predrift Differentiation

In global groups of terrestrial animals and plants, traditional models attribute any differentiation either to vicariance resulting from continental drift or to more recent, trans-oceanic dispersal. Both models are similar insofar as both assume that biogeographic differentiation in Gondwana and Pangaea either did not exist or is irrelevant to modern patterns. Nevertheless, some studies have found that many modern distributions are best explained by *predrift* rifting, with continental drift rupturing the distribution of lineages that had already evolved (Ribeiro and Eterovic, 2011; Heads, 2012a, 2014). Because of this, many modern distributions and phylogenies do not show a simple correlation with the breakup sequence.

Predrift Vicariance in Onychophorans

The importance of predrift events in the evolution of onychophorans was confirmed in the global study by Murienne et al. (2014). The onychophoran faunas of *geographic* areas such as "southern Africa" and "New Zealand" are not monophyletic, and so the overall pattern is incompatible with an origin by continental drift alone. Yet Murienne et al. (2014) concluded that "trans-oceanic dispersal does not need be invoked to explain contemporary distributions [in onychophorans]." Instead, they suggested that the diversification of the group "*pre-dates* the break-up of Pangaea," and their results "corroborate a growing body of evidence… depicting *ancient biogeographic regionalization over the continuous landmass of Pangaea*" [italics added]. Onychophora have several old fossils, but other groups in which such old fossils do not exist (or are not recognized) present similar patterns and can be attributed to similar processes.

Onychophora include two examples of trans-Tasman endemics:

> *Ooperipatellus*: In New Zealand and, with a single species, in Tasmania and mainland southeastern Australia.
>
> A clade comprising the Tasmanian *Tasmanipatus* and the New Zealand *Peripatoides*.

These three genera together form a Tasman basin clade in *New Zealand, Tasmania,* and *southeastern Australia,* that is itself more or less allopatric with its sister group, comprising six genera of *eastern and western Australia*. This implies a predrift break in a pan-Australia–New Zealand ancestor, somewhere around southeastern Australia, as in clades of Loganiaceae, Ericaceae, and Brassicaceae (Figure 7.6). Prebreakup evolution in the region has also been invoked for differentiation between Australia–South America groups and New Zealand–South America groups (Ribeiro and Eterovic, 2011).

Predrift Vicariance in Freshwater Bivalves

The freshwater mussel family Hyriidae is sister to all other freshwater mussels and comprises three main clades (Marshall et al., 2014; Graf et al., 2015):

New Zealand (*Echyridella*).

Western Australia, New Guinea, and eastern Australia (Velesunioninae: *Westralunio, Microdontia, Lortiella, Alathyria, Velesunio*).

Southeastern Australia and South America (Hyriinae: *Diplodon, Castalia, Triplodon, Hyridella, Cucumerunio*).

(There are also fossils in western North America.) The phylogeny implies that following the separation of the New Zealand representative from the rest, a break in a trans-Pacific ancestor took place in southeastern Australia, separating a pan-Australia group from its sister in southeastern Australia and South America. Based on minimum clade ages from a clock analysis, Graf et al. (2015) supported predrift vicariance: "We hypothesize that early diversification of the Hyriidae was driven by terrestrial barriers on Gondwana rather than marine barriers following disintegration of the super-continent." This acknowledges the importance of predrift rifting, and it contrasts with the more familiar idea that the presence of the "basal" *Echyridella* in New Zealand is the result of continental drift or dispersal.

New Zealand Oribatid Mites: Intercontinental Affinities and Local Distributions

Some of the main phylogenetic events in groups such as *Abrotanella* and Onychophora are attributed here to the predrift rifting phase of Gondwana history, and a similar history would also account for breaks in intercontinental clades of other animals. For example, in Oribatida, an order of mites, *Trichonothrus* of Australasia and South Africa is related to *Novonothrus* of Australasia and Chile (Figure 12.1; morphological revision by Colloff, 2011b). (Plant groups with similar Indian Ocean/Pacific Ocean differentiation include *Astelia* and *Leptinella*.) Colloff (2011b) mapped *Trichonothrus* and *Novonothrus* on a prebreakup reconstruction, as shown here, and this indicates that the affinities are arranged in parallel belts. The phylogeny implies that the belts formed before the breaks at the Indian, Tasman, and Pacific

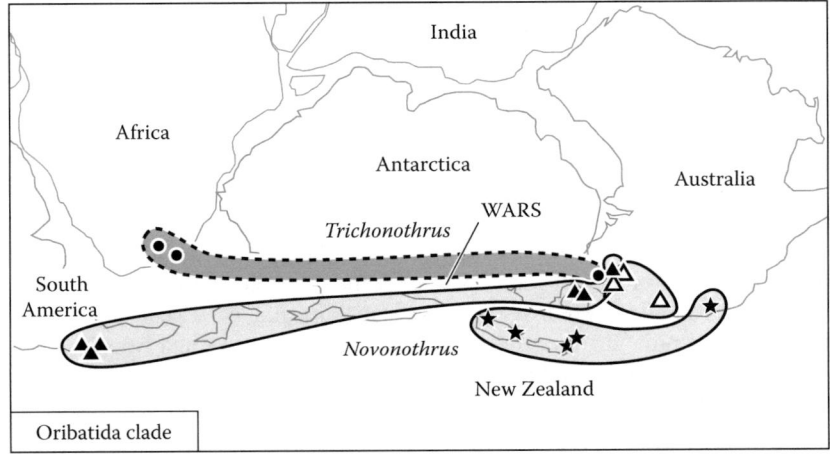

FIGURE 12.1 Distribution of two of the three Australian genera of Nothridae (Oribatida), shown on a prebreakup reconstruction. *Trichonothrus* species, dots; *Novonothrus flagellatus* species group, stars; *N. puyehue* species group, black triangles; *N. barringtonensis* species group, white triangles. WARS, West Antarctic Rift System. (Data from Colloff, M.J., *Zootaxa*, 2828, 19, 2011b.)

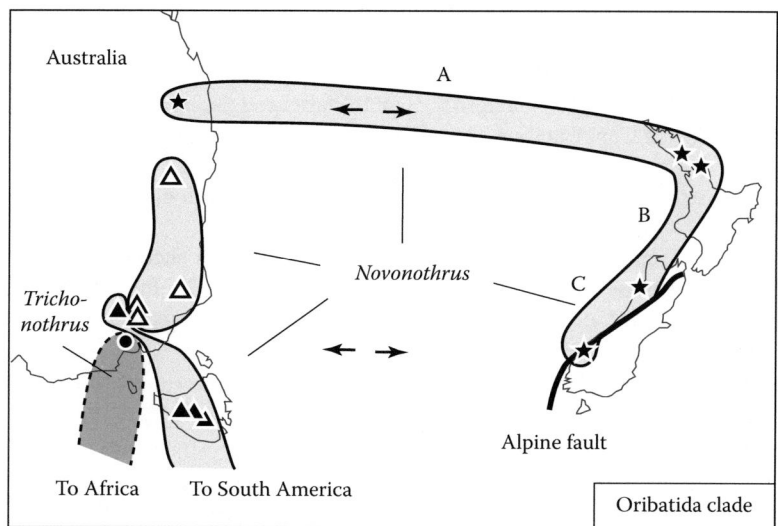

FIGURE 12.2 Distribution of *Trichonothrus* and *Novonothrus* (Oribatida: Nothridae) in the Tasman region. Apparent disjunctions in *Novonothrus* indicated at A, B, and C. (Data from Colloff, M.J., *Zootaxa*, 2828, 19, 2011b.)

basins, and so they could have resulted from predrift extension and magmatism. One of the breaks coincides in part with the West Antarctic Rift System, the other occurs in southeastern Australia (between the McPherson–Macleay Overlap and Tasmania).

In New Zealand, records of *Novonothrus* include a disjunction along the Alpine fault (Figure 12.2; the southern record is from Milford Sound) (Colloff, 2011b). The genus also appears to be disjunct between northern Westland and Auckland/Northland. This may be a collecting artifact, but a similar gap occurs in many groups. Examples include the lichen *Megalospora bartletti* (Heads, 2014: Fig. 6.13), the angiosperm *Metrosideros* (*Mearnsia*) *parkinsonii* (Myrtaceae) (NZPCN, 2016), the therevid fly *Anabarhynchus gibbsi* (Lyneborg, 1992), the micropterigid moth *Sabatinca calliarcha* (Gibbs, 2014), the hydraenid beetle *Podaena maclellani* (Delgado and Palma, 2010), and the hemipteran *Trypetocoris* (Hemiptera: Lygaeidae). *Trypetochoris* is also disjunct between Southland and Nelson along the Alpine fault (Malipatil, 1977). The northern disjunction can be explained by subsidence of the continental crust west of the North Island.

In another oribatid mite, *Crotonia*, southwest Pacific distributions have been explained as the result of metapopulation vicariance and juxtaposition during the rifting and accretion of island arcs (Colloff and Cameron, 2014). The New Zealand members have all been mapped together with the Alpine fault (Colloff, 2015), and in 5 of the 10 species, major aspects of the distribution are associated with the fault. These are either distributional boundaries that coincide with the fault (*Crotonia cupulata*, *C.unguifera*, *C. longibulba*) or disjunctions that are located along the fault (*C. brachyrostrum*, *C. cervicorna*).

Some More New Zealand Cicadas: Overlap in the North Island, Traces of Allopatry, and Niche Models

The New Zealand cicadas (Hemiptera: Cicadidae) comprise two clades. One, including *Maoricicada* and *Kikihia*, has affinities with New Caledonia. The other is endemic to North and South Islands (Figure 12.3) and has its sister in Australia, Asia, and North America (Chapter 6). This second group of New Zealand cicadas has the following phylogeny (Marshall et al., 2012):

Notopsalta sericea: Widespread in North Island (but not in the North Cape area or on offshore islands).

Amphipsalta strepitans: Central Otago mountains to Banks Peninsula, Kaikoura coast, and Wellington coast (south of Banks Peninsula the distribution forms a central strip that is more or less allopatric with the range of *A. zelandica*).

A. *cingulata*: Widespread in North Island (including North Cape) and on Great Barrier and Mayor Islands.

A. *zelandica*: Northern Stewart Island, South Island (but *absent* from large parts of Fiordland, inland Otago, Canterbury, and Marlborough), widespread in North Island (including North Cape), and on Three Kings, Poor Knights, and Little Barrier Islands.

The distributions illustrate several standard patterns that are repeated in many groups.

- In the south of its range, the *Notopsalta–Amphipsalta* clade as such extends only to eastern Fiordland and northern Stewart Island (cf. *Carmichaelia petriei*, Figure 7.5), although it is represented in western Fiordland and southern Stewart Island by members of its sister group, *Kikihia* and *Maoricicada* (Marshall et al., 2008).
- Three of the four species have a geographic limit around the northern shores of Cook Strait (cf. Chapter 8).
- Three of the four species show significant overlap in the North Island, while the two species present in the South Island show significant allopatry. This is consistent with an allopatric origin of the four species, followed by a phase or phases of mobilism in the North Island. The mobilism could have been caused by marine transgressions, volcanism, or both.
- The three groups that overlap in the North Island also present significant differences in their distributions there. *Notopsalta* is absent on the Aupouri Peninsula and from all the offshore

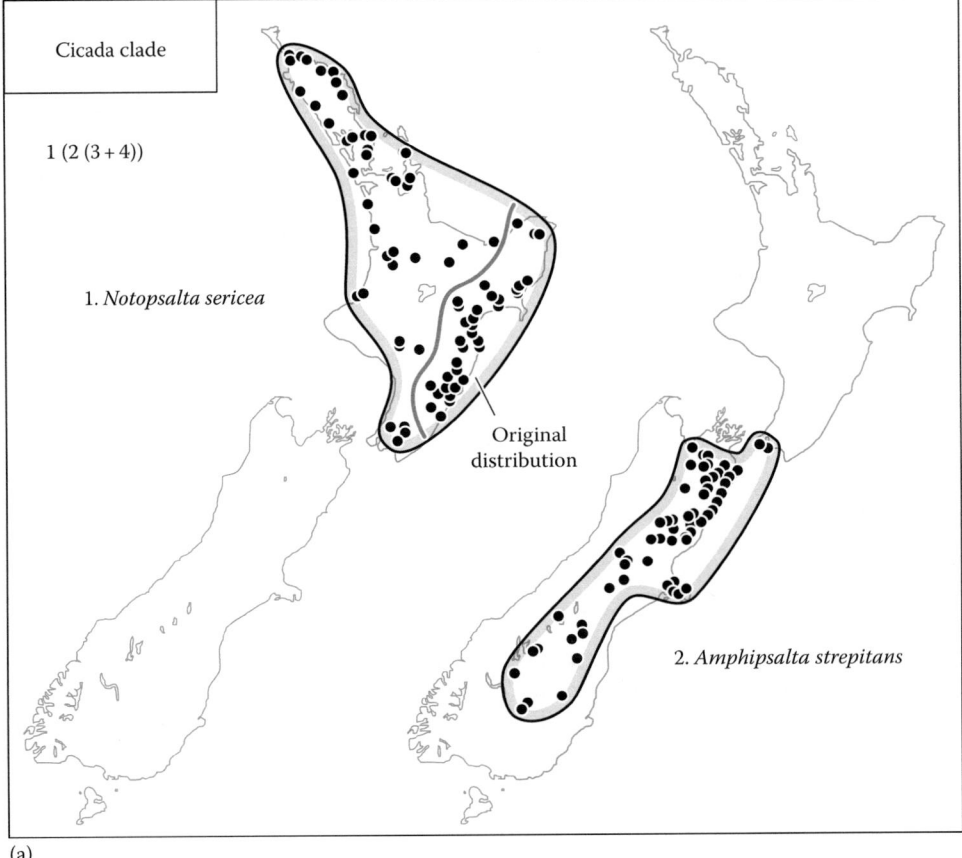

(a)

FIGURE 12.3   Distribution of a clade of cicadas (Hemiptera: Cicadidae). (a) Distribution of *Notopsalta* and *Amphipsalta* (part). (Data from Marshall, D.C., *BMC Evol. Biol.*, 12(177), 1, 2012.)   (*Continued*)

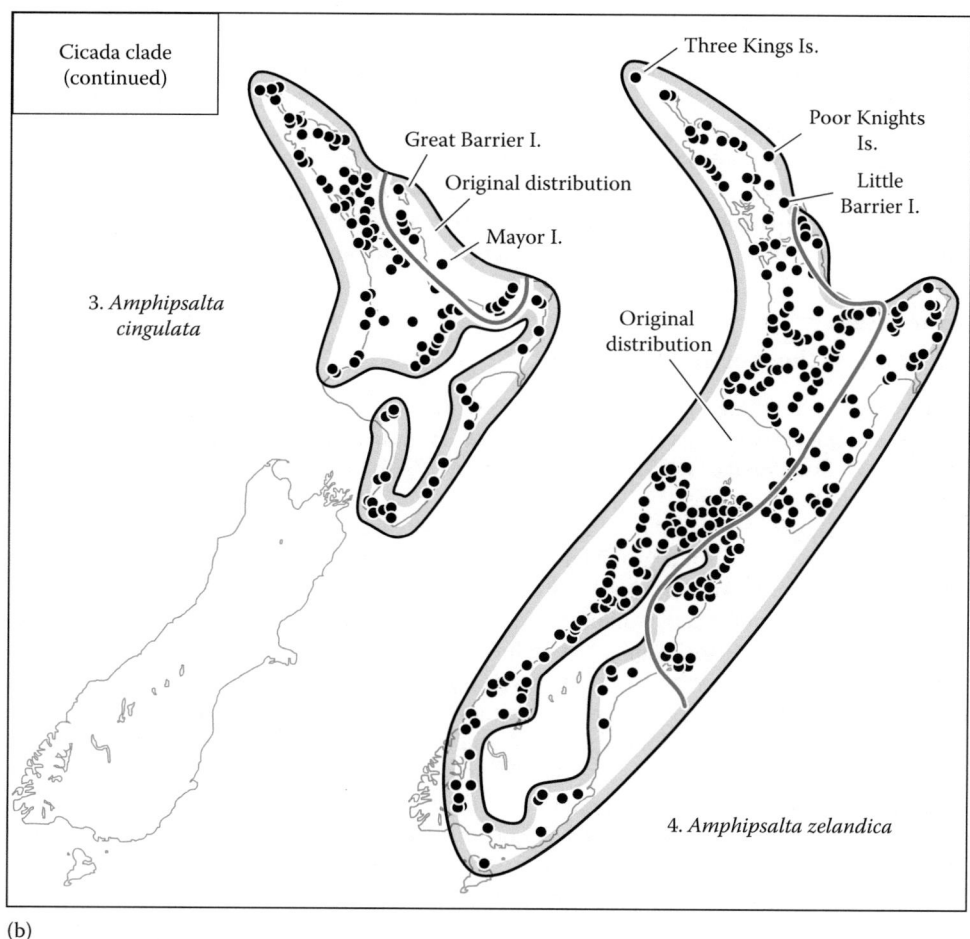

FIGURE 12.3 (*Continued*) Distribution of a clade of cicadas (Hemiptera: Cicadidae). (b) Distribution of *Amphipsalta* (part). (Data from Marshall, D.C., *BMC Evol. Biol.*, 12(177), 1, 2012.)

islands, and *A. zelandica* is on Three Kings, Poor Knights, and Little Barrier Islands, while *A. cingulata* is on Great Barrier and Mayor Islands. This complete allopatry on the islands can be explained as traces of the original allopatry that have persisted through a phase of mobilism. This last phase has led to extensive clade overlap through the North Island mainland and in Marlborough. One possible reconstruction of the original species distributions, assuming original allopatry, is indicated in Figure 12.3 (gray lines).

- The central South Island distribution of *A. strepitans* (Central Otago to coastal Marlborough) and the corresponding gap in the distribution of *A. zelandica* conform to the "central strip" pattern discussed earlier (Chapter 8).

Niche Models for the Cicadas

Marshall et al. (2012) prepared niche models for each of the four species in *Notopsalta* and *Amphipsalta*. The models were hindcast to the Pleistocene, and it was found that none of New Zealand would have contained suitable conditions for any of the species during the last glacial maximum. Nevertheless, the species are all old endemics: *Notopsalta sericea* was dated at 12.5 Ma, *A. strepitans* at 9 Ma, and *A. cingulata* and *A. zelandica* at 7 Ma. Even as minimum ages, these indicate that all four species survived through the Pleistocene glaciations, contradicting the niche models. This is another interesting case of the failure of the niche model approach (cf. Chapter 1; Worth et al., 2014; Heads, 2015c). The results imply that the modern conditions at the sites the species inhabit are not essential for their survival. This in turn suggests that ecological conditions do not explain the distribution patterns.

Marshall et al. (2012) concluded that either (1) the species have only recently occupied their current climatic niche (following niche evolution), (2) the species survived the ice ages in undetected microrefugia, or (3) their current realized niches inadequately represent the fundamental niche of each species. The authors suggested that the first hypothesis is less likely as there is little evidence for insects undergoing shifts in climate exploitation following the last glacial maximum. The second and third hypotheses are both plausible, but they are difficult to disentangle and both may be correct.

Forest Beetles, the Aoraki Node, and the Alpine Fault

The fungus beetle *Agyrtodes* (Leiodidae) has a standard distribution: eastern Australia (Tasmania to northeastern Queensland), mainland New Zealand (North and South Islands), and central Chile (Valparaíso to Chiloé) (Seago, 2009). This suggests an origin no later than the Late Cretaceous rifting in the Tasman and Pacific basins.

One species of *Agyrtodes*, *A. labralis*, has a wide distribution in the South Island and inhabits both *Nothofagus* forest and podocarp forest. Marske et al. (2009) used molecular phylogenetic analyses and niche modeling to identify areas (refugia) that the species occupied during the last glacial maximum. Both methods suggested refugia in Kaikoura, Nelson, and along much of the West Coast. This last area is the Westland biotic gap ("beech gap"), where the absence of many forms has been attributed to extinction caused by glaciation. Nevertheless, local genetic differentiation in *A. labralis* and other forest groups suggested to Marske et al. (2009) that forest was not eliminated from the biotic gap in the Pleistocene. This is consistent with the idea that the groups' absences there were not caused by the ice ages, and that many groups were absent there long before the Pleistocene.

The refugia that Marske et al. (2009) proposed for *Agyrtodes labralis* are characterized by high genetic diversity and many unique haplotypes. Indeed, the authors found that the South Island is "littered with unique haplotypes," and they supported the popular model of multiple refugia in the South Island (Chapter 9).

Three of the main clades in *A. labralis*, namely, the "Otago-Southland" group, its subclade at Canavans Knob, and the "MacKenzie-Canterbury" group, are mapped in Figure 12.4 (from Marske et al., 2009: Fig. 3). The status of the Canavans Knob clade (near Franz Josef Glacier and west of the Alpine fault) is uncertain—in different analyses, it appears as the sister of the Otago-Southland clade, of the MacKenzie-South Canterbury clade, or of both. In any case, it is separated from its sister group by the Alpine fault, and differentiation at the fault or a precursor structure would explain the break.

The population of the Otago-Southland clade that is *geographically* closest to the Canavans Knob clade is at Governor's Bush, Mount Cook, east of the fault and the main divide. This juxtaposition is probably secondary though, and before Alpine fault movement the Canavans Knob clade would have been adjacent to central Fiordland. (Central Fiordland is difficult to access and Marske et al., 2009, did not sample *Agyrtodes* there.)

Canavans Knob is a small hill formed of quartz monzonite intruded at 101 Ma (mid-Cretaceous: Albian; Tulloch et al., 2009). It is a roche moutonnée that was planed to its present form during the Pleistocene advance of Franz Josef Glacier. The endemic clade of *A. labralis* found there indicates that the species has survived through the glacial advances, somewhere in the region. The clade could be more widespread in the vicinity, although at nearby Lake Matheson it is replaced by another clade in *A. labralis*. Again, this suggests *in-situ* survival through the glaciations. The Australasia–Chile distribution of the genus is consistent with a Late Cretaceous break in the genus and a pre-Late Cretaceous origin. This in turn is consistent with the break between the Canavans Knob group and its sister being caused by the same mid-Cretaceous, extension-related intrusion that produced Canavans Knob itself. Other endemics in this area include *Zemacrosaldula pangare* (Hemiptera: Saldidae), with records on and west of the Alpine fault (junction Waikukupa River and Highway 6; Franz Josef; junction Clearwater River and Gillespies Beach Road) (Larivière and Larochelle, 2015).

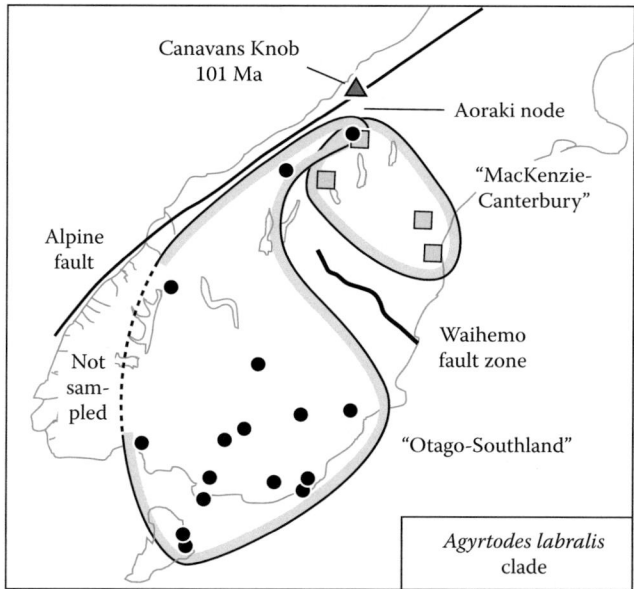

FIGURE 12.4 Distribution of a clade in *Agyrtodes labralis* (Coleoptera: Leiodidae). The relationship among the three subclades is not resolved. (Data from Marske, K.A. et al., *Mol. Ecol.*, 18, 5126, 2009.)

The New Zealand Satyrines and the Waihemo Fault Zone

There are four genera of butterflies endemic to New Zealand, and these form a single, well-supported clade in the tribe Satyrini (Nymphalidae) (Peña et al., 2011). Peña et al. showed that the clade has an allopatric sister, either in Australia or in Australia + New Caledonia + New Guinea.

The distributions of the satyrines within New Zealand are also of special interest. One genus, sister to the rest, is in the North Island and (with a disjunct distribution) in the South Island. The other three genera are all South Island endemics. The phylogeny (Peña et al., 2011) and distributions (Gibbs, 1980; Patrick and Patrick, 2012) can be summarized as follows:

Dodonidia (forest ringlet): North Island (Northland–Coromandel Peninsula) and South Island (Nelson, Marlborough Sounds, south to the Paparoa Range and Lewis Pass; also one old, disjunct record in the Humboldt Mts.); forest, from **sea level to treeline** (up to 1200 m in the North Island); larvae feed on tall sedges (*Gahnia*: Cyperaceae) and sometimes on *Chionochloa* (Poaceae).

    Erebiola (Butler's ringlet): South Island (800)–**1200–1800 m**; flies over, and lands on, subalpine and alpine vegetation; larvae feed on *Chionochloa*.

        Percnodaimon (black mountain ringlet): South Island **800–3100 m**; flies over, and lands on, rocks and scree; larvae feed on *Poa* (Poaceae).

        Argyrophenga (tussock butterfly): South Island; grasslands; **sea level** (including salt marshes)–**1950 m**; larvae feed on *Chionochloa*, *Poa*, and other grasses.

The genera form an ecophyletic series that ranges from sea-level forest and salt marsh (*Dodonidia*, *Argyrophenga*) to high-alpine vegetation and bare rock (3100 m; *Percnodaimon*). The fundamental ecological difference in the group is the basal split between a forest clade (*Dodonidia*) in which larvae feed mainly on forest sedges, and an open-habitat clade (the other three genera) in which larvae feed on grasses.

Dodonidia has a major gap in its South Island range, at least if the old Humboldt Mountains record is accepted. This is an unconfirmed, early sighting, but it was made by a very experienced entomologist,

G.V. Hudson. The gap corresponds to a break at the Alpine fault. In any case, whether or not *Dodonidia* occurs in the south, the area of its absence in the South Island is occupied by the three remaining genera.

Work on *Percnodaimon* is ongoing (B. Patrick, pers. comm., 2016), but already it is evident that two major clades have a break at the Waihemo fault zone: *Percnodaimon pluto* and *Percnodaimon* "sp. 4" (Patrick and Patrick, 2012).

New Zealand Freshwater Fishes

The New Zealand freshwater fish fauna has interesting intercontinental affinities as well as local endemism. Fish can swim, and McDowall and Whitaker (1975: 297) wrote that the derivation of the New Zealand freshwater fish fauna "seems clearly to be by trans-oceanic dispersal." Yet some aspects of the distribution and ecology suggest other possibilities.

Investigating the origin of the New Zealand freshwater fish fauna raises one of the most fundamental questions in evolution and ecology: what is the relationship between life on land and at sea? What are the causes of trans-realm biogeography and ecology? Some fish clades are restricted to freshwater and others are marine, but estuaries and mangrove swamps represent intermediate habitats. Species that live in both freshwater and the sea also blur the land–sea boundary.

About 250 fish species worldwide are "diadromous"; they spend part of their life cycle in freshwater and part in the sea and undergo significant migrations between the two (Bloom and Lovejoy, 2014). Diadromy itself can be variable within species. For example, in the New Zealand fish *Galaxias brevipinnis*, there are diadromous populations in rivers near the coast, as well as landlocked, nondiadromous populations further inland.

In the standard, neodarwinian model of evolution, migrations in general, and diadromy in particular, are thought to have evolved because they are advantageous for a group: "everything is for the best." Thus, "migration might be expected to occur wherever individuals benefit more… if they move seasonally between different areas than if they remain in the same area year-round" (Newton, 2008: 16). In an alternative model, migration has developed as the result of tectonic and evolutionary change, not because of any advantage it may provide. It can develop, for example, if seafloor spreading takes place between feeding and breeding sites.

Bloom and Lovejoy (2014) asked the question: "Why do some diadromous species migrate down from freshwater to oceans to reproduce or feed (anadromy), while other species migrate up from the ocean to freshwater habitats (catadromy)?" Temperate regions have more anadromous species and tropical regions have more catadromous species. This has led to the popular hypothesis that differences in productivity between marine and freshwater biomes have determined the different modes of diadromy. In temperate regions, the oceans have higher productivity than freshwaters, while in the tropics freshwaters have higher productivity than oceans. This suggests that temperate anadromous fishes are derived from freshwater ancestors that began migrating to the ocean to exploit the higher productivity, while tropical catadromous fishes are derived from marine ancestors that began migrating to rivers for the same reason.

The "greater productivity" hypothesis is elegant, but it was contradicted by the results from Bloom and Lovejoy's (2014) study. The authors found that the different modes of diadromy "do not have predictable ancestry based on latitude." Instead, the authors concluded that the paleogeography and geological history of certain areas may have played an important role in determining the evolutionary patterns of diadromy. For example, uplift occurring under a part of the range of a migrating marine species could lead to catadromy. To summarize, diadromy has not evolved because it is advantageous; instead, it is an inevitable by-product of normal geological and biological processes.

Diadromy in the New Zealand Fish Fauna and the Problem of Ancestral Habitat

In the New Zealand freshwater fish fauna, as many as 18 of the 42 species (43%) are diadromous (McDowall, 2010b). The significance of this is controversial. Some authors have considered that the diadromy is a key to understanding the origins and distribution of the fauna (Leathwick et al., 2008; McDowall, 2010a,b), and the topic is worth examining here in some more detail.

The question of diadromy and its history raises the general question of habitat evolution and evolutionary ecology in fish. Many authors studying this have mapped the habitat of clades—marine or freshwater—onto phylogenies and used the patterns to infer *either* marine *or* freshwater ancestry for a group as a whole. A group of five species with either marine (M) or freshwater (F) ecology and a phylogeny M (M (M (M, F))) will be assumed to have an ancestral marine habitat.

Parenti (2008) cautioned that "this [ecological origin] hypothesis results from optimizing habitat or, at the least, inferring that the habitat of the basal taxon is the ancestral habitat. It rests on the assumptions that the habitat has changed (e.g., from marine to freshwater), even when there may be no evidence for such a transformation." Instead, Parenti (2008) argued that for many groups, "ancestral habitat may be reconstructed as epicontinental seas, spanning marine and freshwater habitats... A lineage with marine and freshwater representatives... may be interpreted to have a marine and freshwater ancestral distribution...." In Parenti's alternative model, the ancestor was already present in both habitats before the descendants began to differentiate, and this means that the descendants have not migrated from one habitat into the other.

As Parenti (2008) pointed out, these *ecologically variable* ancestors are analogous to geographically widespread ancestors that inhabit different regions. She wrote:

> If the majority of species in a taxon are marine and just a handful are freshwater, it is often assumed that the group originated in marine habitats and several taxa dispersed to freshwater... The reverse is also assumed. Such an assumption is not an analysis of biogeographic data, but an implicit and untestable center of origin hypothesis... Optimizing habitat on the nodes of an areagram to invoke an ancestral habitat, here marine or freshwater, is equivalent to invoking a center of origin for areas.

Parenti (2008) also discussed migration in fish, and concluded that "the answer to 'Why migrate?' is straightforward when earth history is considered: migration evolves as the earth evolves." For example, migrations associated with feeding or breeding can be extended with the opening of ocean basins or the incursion of inland seas. Parenti (2008) concluded that the migration of fish is a result, not the cause, of their evolution.

Diadromous species undertake regular migrations between specific areas, one at sea and one on land, in freshwater. Dispersal biogeographers have assumed that any plant or animal group in the sea can, and will, disperse far and wide, and so they have proposed chance dispersal, rather than vicariance, to explain distribution in both marine and diadromous clades. Yet, while the *Galaxias maculatus* populations of Australia, New Zealand, and South America are each diadromous *within their regions*, each of the three regions maintains its own, genetic clade (Burridge et al., 2012). Even within New Zealand, diadromous species show interesting distribution patterns. For example, three New Zealand–endemic diadromous fishes (*Galaxias postvectis* [Figure 9.9], *G. fasciatus*, and *Cheimarrichthys*) are absent in Fiordland, while the diadromous *Stokellia* reaches its western limit at the Moonlight tectonic zone and the Alpine fault.

In New Zealand fishes, marine and diadromous species tend to have wider distributions than inland, nondiadromous species. Yet the difference in distribution area is not necessarily the result of the diadromous or nondiadromous habit. Instead, both the habit and the distribution may have been caused by phylogenetic breaks separating localized, inland forms from widespread coastal and marine forms. New Zealand diadromous species tend to occur in coastal rivers. Two-thirds of the species are caught most often in rivers within 30 km of the coast, and only three are caught most often at distances of 50 km or more (Leathwick et al., 2008). The diadromous species have lower rates of occurrence in rivers above obstacles to upstream migration. In contrast, nondiadromous species occur inland, and just one, *Galaxias cobitinis*, occurs most often in more coastal rivers.

Many fishes are known to abandon diadromy when the lakes in which they occur become landlocked, preventing migration to and from the sea. Moreover, one long-term consequence of isolation by landlocking is the evolution of distinct new species (McDowall, 1997, 2010b). In this model, the widespread distribution of the diadromous form and the narrow distribution of the nondiadromous form are not caused by their respective ecology, but by an original vicariant break in a widespread ancestor.

Diadromy has behavioral, rather than physiological or morphological, causes. It can vary among the individuals of a species (McDowall, 1997), and it is not caused by greater dispersal ability. Hicks (2012) examined six diadromous species in New Zealand and found that in all six the diadromy was facultative. For example, in *G. brevipinnis*, all fish upstream of lakes were nondiadromous, whereas fish downstream of lakes were diadromous.

Dispersal theory bases biogeographic interpretation on the ecology and behavior of individual groups, rather than on general patterns. One discussion of fishes concluded that "for species that are diadromous an explanation of their broad-scale distributions lies explicitly in their behaviour, and relates substantially to historical and contemporary dispersal through the sea... For those that are not diadromous, explanations for their local-scale distributions may lie largely in vicariance, and... responses to geology" (McDowall, 2010b: 127). Instead, it is suggested here that geology and phylogeny account for distribution patterns in both nondiadromous and diadromous clades, as well as their behavior. Geology explains the landlocking that has led to nondiadromy, as well as aspects of ecology such as the elevational range of groups.

These and other effects of tectonic history can be seen in the main biogeographic features of the New Zealand freshwater fish fauna, which includes representatives of seven families.

Lampreys (Geotriidae: Petromyzontiformes)

The lampreys are not true "fish," but are one of the two extant clades of jawless vertebrates (the hagfishes are the other). A morphological analysis found that there are three groups of lampreys, arranged in an unresolved tritomy (Gill et al., 2003):

> Northern hemisphere. **Marine and freshwater** (*Petromyzon* and other genera).
>
> South Pacific disjunct (SW and SE Australia, Tasmania, New Zealand, Patagonia). **Diadromous** (*Geotria*).
>
> South Pacific disjunct (SE Australia, Chile). **Diadromous** (*Mordacia*).

Geotriidae are one of the four freshwater fish families in New Zealand that show the standard trans-South Pacific disjunction. In all four families, the disjunction can be explained by the same event—rifting of populations during Gondwana breakup.

Apart from the lampreys, the freshwater fishes in New Zealand are all teleosts. Fossils of living teleost lineages date back to the Late Jurassic (~150 Ma; Parenti, 2008). Even groups nested within the apomorphic percomorph clade (such as Tetraodontiformes) have fossils dating back to 95 Ma, in the mid-Cretaceous. These dates are useful to keep in mind as minimum ages for the clades.

Freshwater Eels (Anguillidae: Anguilliformes)

Most eels (Anguilliformes) are marine, but the Anguillidae are diadromous. Anguillids make long-distance migrations between their usual freshwater habitat and spawning grounds in deep, offshore basins. The migrations have probably increased by a few centimeters per year, as seafloor spreading has continued. The European eel (*Anguilla anguilla*) and American eel (*A. rostrata*) are sister species and both migrate, converging on adjacent breeding grounds in the Sargasso Sea. Biogeographers have attributed the eels' migrations to the gradual geological opening of the Atlantic (review in Heads, 2005a).

In New Zealand, there are three freshwater eels. *Anguilla australis* (spawning in the Coral Sea) and *A. reinhardtii* (possibly with episodic immigration to New Zealand) are shared with eastern Australia and the western Pacific islands. *A. dieffenbachii* is endemic to New Zealand. Its spawning site is unknown but is thought to lie somewhere in the Fiji basins, with the fish migrating from freshwater feeding grounds in New Zealand (Righton et al., 2012). This migration could have been extended by the Cenozoic opening of the Fiji basins.

A main theme in McDowall's (2010b) work on New Zealand fishes is that freshwater groups had a marine center of origin before making an "inland penetration." *Anguilla dieffenbachii*, for example, shows greater "inland penetration" than *A. australis*. Yet the pattern is also consistent with simple vicariance in a widespread ancestor, with *A. dieffenbachii* evolving in inland New Zealand and the Fiji basin, and *A. australis* in outer New Zealand, the Coral Sea, and Australia. Overlap of the two in modern New Zealand is a subsequent, local development within the broad, regional pattern of allopatry.

The clade formed by *Anguilla dieffenbachii* + *A. australis* occurs in Australasia and Pacific islands, and is sister to a clade comprising *A. anguilla* of Europe and *A. rostrata* in the eastern United States and the Caribbean (Teng et al., 2009). The whole clade represents either a fractured Tethyan distribution, or a connection between Australasia and the Caribbean across the Pacific. In any case, the clade's relatives are *A. mossambica* of southern Africa, Madagascar, and the Mascarenes, and *A. malgumora* of Borneo. All these distributions illustrate simple allopatry in a widespread global group.

Galaxiidae (Galaxiiformes)

Galaxiids are well known to biogeographers because of their disjunct, austral distribution in Australasia, Patagonia, and South Africa. The family includes freshwater species and also diadromous forms that occur in coastal seas. Juveniles have been found up to 700 km off the New Zealand mainland (a single individual was recorded at the Bounty Islands; McDowall et al., 1975), but are unknown in the deep water off the New Zealand continental plateau.

The austral family Galaxiidae is the sole member of the order Galaxiiformes. It is "basal" in a diverse, global complex that also includes five other orders (Li et al., 2010):

> Stomiiformes (lightfishes and other groups): Deep sea, antarctic to subtropical.
>
> Argentiniformes: Deep sea, oceans throughout the world.
>
> Osmeriformes: Freshwater and marine, widespread in the northern hemisphere and the Tasman region (Retropinnidae).
>
> Salmoniformes: Freshwater, some diadromous; northern hemisphere (fossil record back to Late Cretaceous).
>
> Esociformes: Freshwater, northern hemisphere (fossil record back to Late Cretaceous).

Within this diverse complex, the galaxiids have an undistinguished ecology (freshwater and diadromous), but their panaustral distribution is unique. There is no overlap with the first two, deep sea groups, and galaxiids overlap the distributions of the last three groups only in southeastern Australia, Tasmania, and New Zealand (with Retropinnidae: Osmeriformes). The "basal" position of galaxiids in the phylogeny, sister to all the rest, means that they cannot be derived from any one of their northern hemisphere relatives. (The pattern is similar to that of the South Pacific Nothofagaceae, basal in the global Fagales.) Instead, the pattern indicates simple vicariance between the panaustral galaxiids and their sister group, widespread in the northern hemisphere, the Tasman region, and the deep oceans of the world. Secondary overlap is restricted to comparatively small areas bordering the South Tasman Sea. As in the icefish, a marine group with a center of diversity around Antarctica (Chapter 6), basal evolution in the ancient group Galaxiidae is likely to have been mediated by events during and just prior to seafloor spreading.

Galaxiids are the southern representative of a clade that has an ecological range extending from the deepest seas (e.g., lightfishes) to alpine streams (e.g., "pencil galaxiids"). The simplest explanation for this impressive ecological range, over thousands of meters of elevation, is that it was inherited from the clade's common ancestor, which probably also had a global geographic range. Likewise, the simplest explanation for the split between the galaxiids and their sister group is vicariance in a global complex, and this means there is no need for galaxiids to have invaded either their southern distributional area or their freshwater habitat by physical movement. In this model, members of the global complex ancestral to the six orders *already* occupied what became the central part of the New Zealand Plateau, as well as many other regions globally, before the Galaxiiformes and its sister group differentiated from each other.

In their classic account of *Modes of Reproduction in Fishes*, Breder and Rosen (1966) discussed the diadromy of galaxiids (see also Rosen, 1974; McDowall, 2010a). McDowall (2010b) wrote that Rosen (1974) rejected marine stages in the life cycle of galaxiids, but this is not correct. Rosen accepted diadromy in galaxiids; what he rejected was the idea that this explained the disjunct distribution of the family.

McDowall (2010b: 197) concluded that "clearly dispersal must have been very important for a substantial part of the very wide geographical range of the family Galaxiidae," but this idea is controversial. For example, in the 1980s, one widely used textbook accepted that galaxiid distribution is the result of dispersal (Brown and Gibson, 1983), but in the 1990s, the second edition of the same book attributed the pattern to plate tectonics (Brown and Lomolino, 1998).

McDowall (2010b) suggested that the coastal marine stage in the life cycle of some galaxiids explains the intercontinental disjunctions in the family. Nevertheless, the idea that marine groups with pelagic larvae all have a wide distribution has now been contradicted by sequencing studies of many marine groups (Chapter 1). Many clades that were assumed, because of their pelagic larvae, to be widespread and homogeneous, show high levels of molecular geographic structure (Heads, 2005a). The distinct, diadromous forms of *G. maculatus* in Australia, New Zealand, and South America provide a typical example of this (Burridge et al., 2012).

Galaxiid Phylogeny

The phylogeny for Galaxiidae (Burridge et al., 2012) is given next. Australia-Tasmania and southern South America are the centers of diversity for the "basal" clades in the family; all other localities are in bold. (Numbers = numbers of species per clade. Aus = Australia and Tasmania. NZ = New Zealand.)

Aus (*Lovettia*, 1) and South America (*Aplochiton*, 2); PACIFIC OCEAN.

Aus (*Galaxiella*, 3).

South America (*Brachygalaxias*, 2).

South America (*Galaxias platei*).

NZ (Aus, **South Africa**) (*Neochanna*, 6, and *G. zebratus* s.lat.); INDIAN OCEAN.

Aus (*Paragalaxias*, 4, *G. fuscus,* and two other species).

Aus, **NZ**, South America (*G. maculatus* and two other species); PACIFIC OCEAN.

Aus, **NZ**, **NZ subantarctic islands**, **New Caledonia** (29 species); TASMAN SEA.

Burridge et al. (2012) carried out an ancestral area analysis of Galaxiidae using a center-of-origin model (DEC) and located a center of origin in Australia + South America. Instead, the family's allopatry with its diverse, worldwide sister group suggests a widespread common ancestor, with the phylogeny involving breaks around Pacific Ocean and Indian Ocean groups. The last, large clade in the galaxiid phylogeny is restricted to the margins of the Tasman Basin, but includes just over half of the species. It has a different distribution from all the other clades as it is the only one present in the New Zealand subantarctic islands and New Caledonia.

The phylogeny of Galaxiidae, as given earlier, can be read as a series of dispersal events between Australia, New Zealand, and the other areas. As an alternative, it can be interpreted as a series of vicariance events, with differentiation, not individuals, "jumping" between localities in the southern hemisphere. There is no need for any range expansion between the regions, just within them. In fact, many of the clades do show subsequent range expansion within regions, but less than the phylogeny suggests. This is because the areas cited here (from Burridge et al., 2012) are geographic, not biogeographic, entities and their use obscures any allopatry within regions. For example, *Lovettia* and *Galaxiella* are both listed with a distribution, "Australia," but in fact they are more or less allopatric there: *Lovettia* is in Tasmania (but not in the northeast) and has local records on the adjacent Victoria coast. *Galaxiella* is in southwestern Australia and northeastern Tasmania, and is also widespread in Victoria.

The Neochanna Clade of Galaxias

As indicated in the phylogeny, the New Zealand galaxiids belong to three main clades. One of these is the *Neochanna* clade. It comprises a New Zealand group of five, strictly freshwater species, and its sister in southern Australia + South Africa. The Australian species is diadromous. The first break corresponds

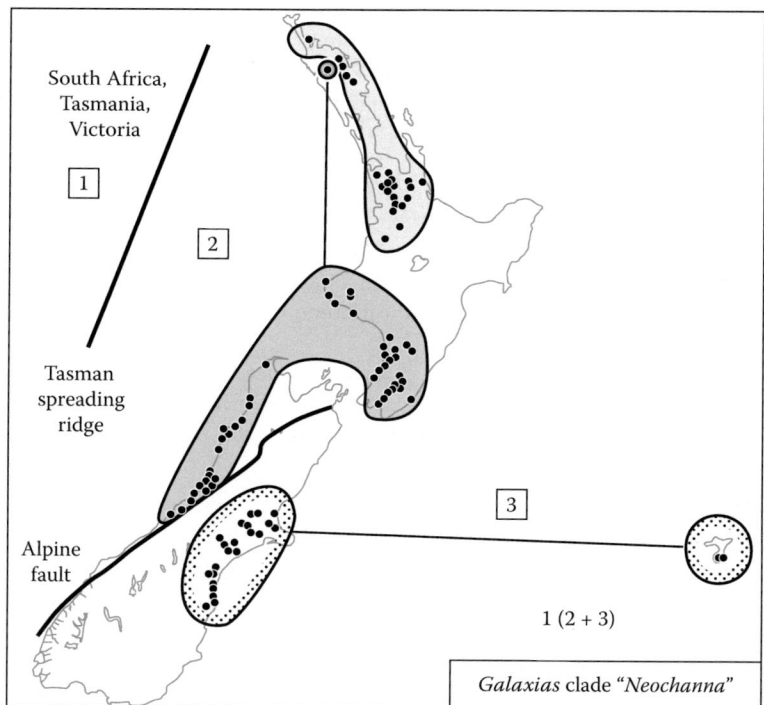

FIGURE 12.5 Distribution of *Galaxias* "*Neochanna* group" (Galaxiidae). (Data from Waters, J.M. and McDowall, R.M., *Mol. Phylogenet. Evol.*, 37, 417, 2005; Burridge, C.P. et al., *J. Biogeogr.*, 39, 306, 2012.)

to the Tasman Sea and its spreading ridge (or a precursor), while the main break *within* New Zealand is at another plate boundary, the Alpine fault. This separates a northern clade from one in Canterbury and the Chatham Islands (Figure 12.5; Waters and McDowall, 2005). (The connection along the Chatham Rise is repeated, for example, in a sister pair of cockroaches: *Celatoblatta peninsularis* of Banks Peninsula and *C. brunni* of the Chatham Islands; Chinn and Gemmell, 2004.)

The northern clade in *Neochanna* illustrates a direct connection between western Northland and Taranaki; this disjunction can be attributed to the subsidence of land west of the North Island (see *Metrosideros parkinsonii*, etc., cited in the section "New Zealand Oribatid Mites," near the start of this chapter). The presence of two subclades in Northland is consistent with the composite tectonic nature of the peninsula.

Galaxias maculatus *and Allies*

The *G. maculatus* group occurs in Australia, New Zealand, and *South America* and is sister to a diverse clade found in Australia, New Zealand, *the subantarctic islands, and New Caledonia*. In each of three regions—South America, the subantarctic islands, and New Caledonia—only one of the two groups is present. This phenomenon can be explained as a remnant of the original allopatry. Following an origin of both groups by vicariance in a widespread, South Pacific ancestor, there has been some overlap, but this is restricted to Australia and New Zealand.

Since the 1970s, panbiogeographic work has supported a vicariance interpretation for galaxiids, despite the widespread distribution of *G. maculatus* and the diadromy seen in this and other species (Rosen, 1974). Dispersal theory has always supported recent migration to explain the distribution of the species. Nevertheless, molecular studies of *G. maculatus* found "extremely high divergences" (up to 14.6%) among the populations of Tasmania, New Zealand, and South America (Waters et al., 2000a), and these are "more typical of interspecific and even intergeneric comparisons" (Waters et al., 2000b). Waters et al. (2000a) also reported "extremely strong inter-continental geographical structure" in the species.

Similar, strong differentiation among these same regions also occurs in strictly marine fishes, for example, between the disjunct New Zealand and Patagonian populations of *Micromesistius australis* (Gadidae) (Ryan et al., 2002). In *G. maculatus*, Waters et al. (2000a: 1819–1820) concluded that "divergences may be consistent with a vicariance model," and that the "dispersal powers of *G. maculatus* may be more limited than previously suggested... A vicariant role in the divergence of eastern and western Pacific *G. maculatus* cannot be rejected."

Within the trans-Pacific *G. maculatus* (including *G. rostratus*; Burridge et al., 2012) the phylogeny is: (South America (Australia + New Zealand)). Although Waters et al. (2000a) agreed that vicariance was possible in principle, they described the sequence of the phylogeny as "a relationship that conflicts with a vicariance model." Nevertheless, it really only conflicts with the breakup sequence of Gondwana. The phylogeny and the allopatry suggest that the main differentiation occurred with tectonic extension and vicariance soon before breakup (see Chapter 3).

Dispersal theory suggests that if a freshwater group is disjunct across an ocean basin and shows even traces of an ecological association with the sea, its distribution has probably been caused by transoceanic dispersal (McDowall, 2010b). Yet even strictly marine groups can be the result of vicariance.

The Species-Rich Clade of Galaxias

The last clade listed in the phylogeny of Galaxiidae given earlier is a purely Australasian group with more species (29) than all the other clades combined (26). It is structured as follows (Burridge et al., 2012), (Aus, Australia; NZ, New Zealand). Areas outside Australia and mainland New Zealand are in bold.

    Aus (*G. auratus*, *G. truttaceus*, and *G. tanycephalus*).

      NZ (*G. argenteus* and *G. fasciatus*).

        NZ (*G. postvectis*; Figure 9.9).

          NZ (*G. prognathus*, *G. cobitinis*, and *G. macronasus;* "pencil galaxiids" in part [Figure 12.6]).

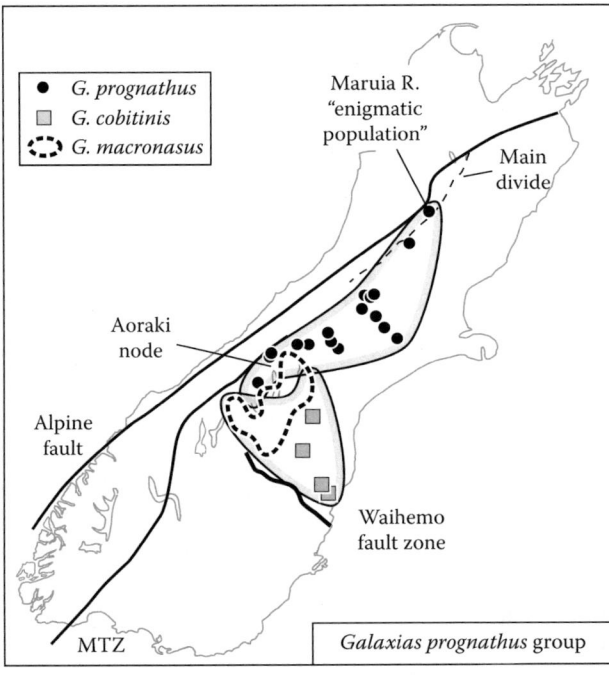

FIGURE 12.6   Distribution of the *Galaxias prognathus* group (Galaxiidae). Symbols for *G. prognathus* and *G. cobitinis* are not shown in the region of overlap with *G. macronasus*. MTZ, Moonlight tectonic zone. (Data from McDowall, R.M. and Waters, J.M., *N. Z. J. Zool.* 29, 41, 2002; Waters, J.M. and Craw, D., *Freshwater Biol.* 53, 521, 2008; Burridge, C.P. et al., *J. Biogeogr.* 39, 306, 2012.)

FIGURE 12.7 Distribution of *Galaxias divergens* and *G. paucispondylus* (Galaxiidae). (Data from McDowall, R.M., *New Zealand Freshwater Fishes: An Historical and Ecological Biogeography*, Springer, Berlin, Germany, 2010b.)

NZ (*G. paucispondylus* and *G. divergens;* "pencil galaxiids" in part [Figure 12.7]).

Aus (*G. fontanus*).

Aus (*G. johnstoni* and *G. pedderensis*).

Aus (*G. niger* and *G. brevipinnis* in part).

NZ including **Auckland and Campbell Islands** (*G. brevipinnis* s.str.) and **New Caledonia** ("*Nesogalaxias*").

NZ (*G. vulgaris* and allies: 10 species).

The phylogeny repeats a pattern seen in the first part of the galaxiid phylogeny: differentiation appears to "jump" backward and forward between Australia and New Zealand, consistent with a widespread southern Australasian ancestor that has undergone repeated phases of differentiation and overlap. Through this proposed phase of predrift differentiation, breaks developed alternately in the area that became New Zealand and in the area that became Australia (cf. Heads, 2014; Figure 1.11 for eastern and western Australia). This continued until Australia and the New Zealand Plateau broke apart, causing the break between the Australian clade *G. niger* + *G. brevipinnis* p.p. and its New Zealand–New Caledonia sister group.

As in the first part of the galaxiid phylogeny, different clades assigned to "Australia" in fact show a high level of allopatry. For example, the *Galaxias fontanus* clade is endemic to eastern Tasmania, and the *G. johnstoni* + *G. pedderensis* clade is endemic in southwestern Tasmania.

The first New Zealand clade, *Galaxias argenteus* plus *G. fasciatus*, is widespread in streams near the mainland coasts (*G. fasciatus* is also on the Chatham Islands), but is much rarer inland, and this is where most of the diversity in New Zealand *Galaxias* is located.

The next clade, *G. postvectis*, has been mentioned already as an example of Alpine fault disjunction (Figure 9.9), with the gap in Fiordland filled by part of its sister group. The Alpine fault movement can date the disjunction in the clade, but it only gives a minimum date for the origin of the clade. The trans-Tasman distributions in the sister group of *G. postvectis* indicates that this occurred before rifting in the Tasman Basin.

Members of the next two clades, the *G. prognathus* group, and *G. paucispondylus* + *G. divergens*, have been termed "pencil galaxiids" because of their slender shape. They now appear to be a paraphyletic grade, basal to the remaining Australian and New Zealand species (Burridge et al., 2012). Although McDowall long argued that the distribution of the galaxiids reflects a dispersal history, later he accepted that the "pencil galaxiids" "may reflect ancient Gondwanan origins" (McDowall, 2010a: 186).

The *G. prognathus* group has a typical eastern South Island distribution, ranging west to the Alpine fault and south to the Waihemo fault zone (Figure 12.6; Burridge et al., 2012). It occurs east of the main divide, but it also has a record in the upper tributaries of the Maruia River (Figure 12.6). This is "surprising" (McDowall and Waters, 2002) and "enigmatic" (McDowall, 2010b: 291) because the locality lies west of the main geographic divide. Nevertheless, it is close to the Alpine fault, suggesting that this western limit has been caused by movement at the fault itself rather than the uplift of the main divide.

The next clade on the phylogeny comprises the two other "pencil galaxiids" (Figure 12.7). *G. paucispondylus*, or "alpine galaxias," inhabits South Island rivers east of the Alpine fault (McDowall and Waters, 2003). Its sister is *G. divergens* of the northern South Island and south-central North Island. The two species are mainly allopatric, and overlap is restricted to inland southern Marlborough. The overlap can be attributed to the range expansion of *G. divergens* into Marlborough from an original boundary between the two species at the Alpine–Wairau fault. McDowall and Whitaker (1975) accepted a vicariance history for "pencil galaxiids," with ancestors being fragmented by the uplift of the Southern Alps. Vicariance probably is the best explanation, but with breaks at the Alpine fault (e.g., around the Maruia River) rather than the main divide.

G. paucispondylus and *G. divergens* together define a "central strip" that extends from the margin of Fiordland (Mararoa River) to the central North Island. *G. paucispondylus* is restricted to mid- and high-elevation rivers. Most populations of *G. divergens* are also submontane, and, as McDowall (2010b) wrote, it is tempting to attribute its distribution pattern to a preference for cooler climate. Nevertheless, as McDowall concluded, the idea is not tenable. This is because in the north of the South Island, the distribution of *G. divergens* intersects the current coast line, with populations recorded at sea-level in Abel Tasman National Park and in the Marlborough Sounds (McDowall, 2010b: Fig. 17.6). If distribution reflected climate, the species would occur at higher elevations further north, but in fact these northern populations are lower.

The same antiecological pattern is seen in groups such as *Carmichaelia* subg. *Monroella* (Figure 11.13). Instead of being caused by climate, the overall distributions of the species can be explained if the present elevation of the populations, and thus many aspects of their ecology, has been determined by (1) their prior distribution and (2) tectonic changes. The anomalous, low elevation of populations around the Marlborough Sounds can then be explained by the tectonic subsidence known to have occurred there (Trewick and Bland, 2012).

In the remaining New Zealand clades of *Galaxias*, *G. brevipinnis* s.str. is widespread in New Zealand, including Auckland and Campbell Islands where it is the only freshwater fish present. Its sister is "*Nesogalaxias*" of New Caledonia, treated earlier as a separate genus.

G. brevipinnis s.str. and "*Nesogalaxias*" form the widespread sister group of the *G. vulgaris* complex. This last group is distinct in its high diversity (10 species, including several local and regional endemics) and its restricted total distribution: eastern South Island and Stewart Island. The group as a whole is not mapped here, but the largest clade, with six species, is shown in Figure 12.7. *G. vulgaris* s.str. extends from the Waihemo fault zone north across the Canterbury plains to the Clarence River mouth. Its sister group, *Galaxias* "species D," ranges south of the Waihemo fault zone from Central Otago to the Catlins, where the distribution intersects the current coastline (McDowall, 2010b; Waters et al., 2010).

The *G. vulgaris* complex as a whole comprises 10 species: the six mapped in Figure 12.8, plus two restricted to Dunedin–Waipori; the polyphyletic *G. gollumoides* of the Catlins and the Nevis valley, west to the Von and Waiau Rivers, and Stewart Island; and possibly *G. anomalus* of northeastern Central Otago (Manuherikia–Maniototo) (Waters et al., 2010; Burridge et al., 2012).

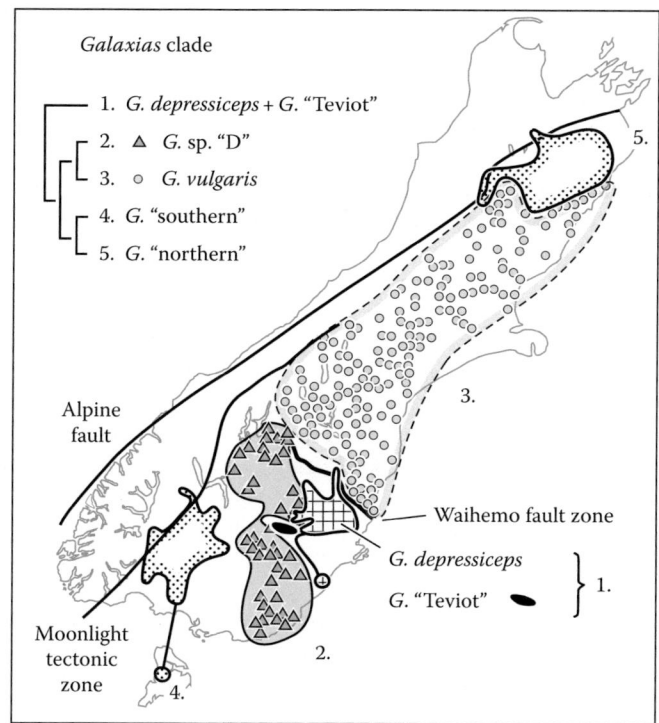

FIGURE 12.8 Distribution of the clade comprising *Galaxias vulgaris* s.str. and its immediate allies (Galaxiidae). (Phylogeny from Waters, J.M. et al., *Evolution*, 55, 1844, 2011; Waters, J.M. et al., *Syst. Biol.*, 59, 504, 2010; distributions from McDowall, R.M., *New Zealand Freshwater Fishes: An Historical and Ecological Biogeography.* Springer, Berlin, Germany, 2010b.)

The *G. vulgaris* complex is almost restricted to areas east of the main divide and, to be more precise, east of the Alpine fault. Nevertheless, it has an anomalous western population (of *G.* "northern") in the same tributary of the Maruia River as the "enigmatic" population of *G. prognathus* (Figure 12.6), west of the main divide but close to the Alpine fault (McDowall, 2010b: 291).

McDowall (2010b: 273) noted that the western limit of the *G. vulgaris* group (Figure 12.8) coincides with a major tectonic feature, the Fiordland boundary fault (the Hauroko-Hollyford fault). (McDowall, 2010b, cited Craw et al., 1999, for this interesting observation, but in fact he seems to have been the first to point it out.) A possible alternative (Figure 12.8) is a boundary at the Moonlight tectonic zone, which the *G. vulgaris* group only crosses by about 10 km at two points, in the upper Mararoa River (above Mavora Lakes) and in the lower Mararoa River. This could be the result of secondary range expansion. McDowall's (2010b) maps are detailed, but further fieldwork should clarify the best alternative.

The *G. vulgaris* complex includes a great disjunction, the gap between *G.* "southern" of Southland and Stewart Island and its sister group, *G.* "northern," of Marlborough (Waters et al., 2010). These are lowland-submontane species (unlike the alpine "pencil galaxiids"), and the gap can be attributed to uplift and extirpation of western populations that extended around the western margin of *G. vulgaris* (cf. *Carmichaelia vexillata*, Figure 11.13).

The gap in the range of the *G. vulgaris* group between the Teviot River and Dunedin (Figure 12.8) is filled by a locally endemic clade comprising *G. eldoni* and *G. pullus*. The precise relationships among this pair, the *vulgaris* group, and two other Otago species, *G. gollumoides* and *G. anomalus*, is not yet resolved, but the high level of local endemism in Otago matches that of other, less mobile groups.

G. vulgaris and its allies (Figure 12.8), form a series of nested distributions: *G.* "southern" and *G.* "northern" in the west and then *G.* species "D" and *G. vulgaris* along the east, with the last two surrounding *G. depressiceps* and *G.* "Teviot" in the southeast. The last two species in turn enclose *G. eldoni* and *G. pullus* of Dunedin and Waipori (not mapped).

Wallis and Trewick (2009) suggested that members of the *G. vulgaris* group existed on the West Coast in the past and have been extirpated by glaciations, but, as McDowall (2010b: 263) concluded, there is no real evidence that the group was ever there. A vicariance origin of the group is indicated, as the sister of the eastern *G. vulgaris* complex is the New Caledonian "*Nesogalaxias*" plus *G. brevipinnis*, the latter widespread in New Zealand and perhaps most abundant in Westland. Conversely, *G. brevipinnis* is not recorded in many parts of the eastern South Island, for example, in the Maniototo region and in the Oreti and Mataura rivers, where members of the *vulgaris* complex are present (McDowall, 2010b: 209). In other parts of the eastern South Island the two groups have overlapped, following their initial allopatry.

Summary on Galaxiids

In conclusion, dispersal theorists have admitted a possible Gondwanan history for the Australian *Galaxiella*, the South American *Brachygalaxias* and *G. platei*, the South African *G. zebratus* (McDowall, 2010b), and the New Zealand "pencil galaxiids" (McDowall, 2010a). Likewise, Waters et al. (2000a) suggested that the high levels of divergence within *G. maculatus* s.lat. and the correlation of the clades with geography are consistent with a vicariance history. Within galaxiids as a whole, the distributions indicate that main clades differentiated around features such as the Waihemo fault zone during predrift extension. Continental breakup itself caused further rupturing of clade distributions.

The loss of a migratory phase is likely to be an important cause of speciation in galaxiids (Waters et al., 2000b). The question is then: "Why has the migratory phase been lost in the first place?" This could have resulted from a geological event, such as the retreat of inland seas with worldwide sea-level change, or local uplift and the damming of lakes. As McDowall (2010b: 229) wrote, "widespread diadromous species seem almost preadapted to providing the foundations for the proliferation of derived species with restricted geographic ranges."

Retropinnidae (Osmeriformes)

Osmeriformes are one component of the diverse, cosmopolitan sister group of Galaxiidae and include marine, freshwater, and diadromous populations. The phylogeny (Li et al., 2010) is:

Retropinnidae: SE Australia, Tasmania, and New Zealand.
  Plecoglossidae: China, Taiwan, Korea, and Japan.
  Salangidae: China, Taiwan, Korea, and Japan.
  Osmeridae: North Pacific, North American Great Lakes, North Atlantic, and Europe.

The basal clade, Retropinnidae, is a Tasman endemic and is allopatric with the others, which form a widespread northern hemisphere clade. As with the lampreys and Galaxiidae, this is compatible with an origin of the southern group Retropinnidae *in situ*, somewhere in the Tasman region. The Retropinnidae and Galaxiidae probably originated in allopatry at a break in the region, but they now show complete overlap there, which is perhaps not surprising for separate orders. The family Retropinnidae has three genera:

Retropinna: SE Australia, Tasmania and New Zealand (Hammer et al., 2007).
Prototroctes: SE Australia, Tasmania and New Zealand (extinct).
Stokellia: New Zealand (Southland to Marlborough).

The single species of *Stokellia* is diadromous, and so in dispersal theory it should occur throughout New Zealand. In fact though, it extends west only to the Moonlight tectonic zone and north only to the Wairau fault, the northern extension of the Alpine fault (Figure 12.9; Patrick and Woods, 1995; McDowall, 2010b). Although the eastern, Southland–Marlborough distribution has been regarded as an "anomaly" (McDowall, 2002), and "highly idiosyncratic" with "no obvious explanation" (McDowall, 2010b: 249), it is repeated in many other groups, such as *G. vulgaris* and allies (Figure 12.8), and the eleotrid fish *Gobiomorphus breviceps* "A" (discussed later).

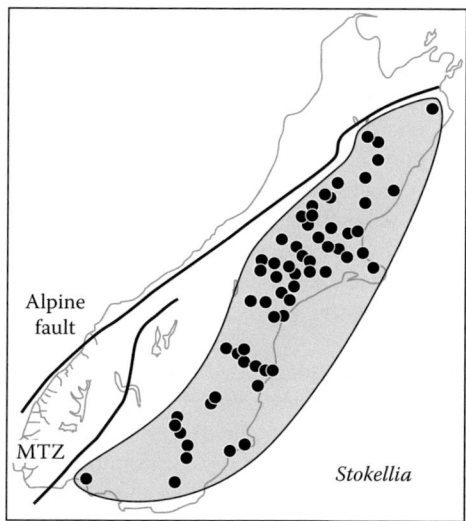

FIGURE 12.9 Distribution of *Stokellia* (Retropinnidae). MTZ, Moonlight tectonic zone. (Data from Patrick, B.H. and Woods, C.S., *J. Roy. Soc. N. Z.* 25, 93, 1995; McDowall, R.M., *New Zealand Freshwater Fishes: An Historical and Ecological Biogeography. Springer,* Berlin, Germany, 2010b.)

Pleuronectidae (Pleuronectiformes)

This flatfish family inhabits ocean floors worldwide, down to a depth of 2000 m. Many forms migrate, as larvae, from offshore sites into estuaries, and a few inhabit freshwater only. The one flatfish in New Zealand freshwaters, *Rhombosolea retiaria*, is the sole diadromous member of an otherwise marine genus. *Rhombosolea* has a standard Tasman Sea range, in southeastern Australia, Tasmania, and New Zealand. Its subfamily, Rhombosoleinae (nine genera, mainly marine), extends in addition to northwestern Australia, Indonesia, and Argentina/Uruguay (*Oncopterus*). This is a typical South Pacific pattern and is repeated in other freshwater fish (clades in Petromyzontidae, Galaxiidae, and Retropinnidae).

Cheimarrichthyidae (Perciformes)

Cheimarrichthys forsteri, the torrentfish, is a New Zealand endemic found in rivers with shingle beds, in the broken water of the swiftest rapids. It makes up a monotypic family. It is diadromous but is restricted to the North and South Islands; despite its ecology, it is not found in Fiordland or Stewart Island. McDowall (2010b: 368) wrote that this is "perhaps surprising," as he assumed that the diadromy would lead to widespread distribution.

For the torrentfish, McDowall (2000) wrote that "the fact that it has a marine-living juvenile has a strong impact on the species' distribution. Upstream/inland penetration in river systems is substantial, and the torrentfish reaches 700 m elevation and 289 km upstream from the sea." But this assumes a marine center of origin. Instead, if the most recent common ancestor of *Cheimarrichthys* and its sister group already occupied the central New Zealand region before *Cheimarrichthys* evolved, inland penetration is unnecessary.

McDowall (1964: 61) wrote that while the immediate affinities of the monotypic *Cheimarrichthys* are unclear, it "seems to be a recent local marine derivative." Later, he compared its osteology with that of blue cod, *Parapercis colias* (Pinguipedidae), of New Zealand coastal seas (McDowall, 1973). He found some differences but many similarities between the two and concluded that they are "very closely related." McDowall (2000) confirmed that *Cheimarrichthys* "has its closest common ancestry with the blue cod…," and McDowall (2010b: 113) wrote: "It seems clear that two New Zealand species, the torrentfish and the black flounder, are species that have a local derivation from fishes that are found in NZ coastal seas (the torrentfish is related to the marine blue cod…)" (cf. McDowall, 2010b: 158, 161, 341).

In contrast with these results, a cladistic morphological study found that *Cheimarrichthys* was not sister to *Parapercis colias* or to the genus *Parapercis*, and did not even belong in the family Pinguipedidae (*Parapercis* and three to six other genera) (Imamura and Matsuura, 2003). Smith and Craig (2007) quoted Rosen's view that studies of percomorph groups "tended to be too narrowly focused in their view of a problem and not ready enough to cast the net of investigation more widely." They provided a wide-ranging molecular study of percomorphs that supported the conclusions of Imamura and Matsuura (2003)—*Cheimarrichthys* is sister, not to a single species of the New Zealand coast, *Parapercis colias* (blue cod), but to a diverse, global assemblage of inshore marine forms that *includes P. colias*. This assemblage was represented in Smith and Craig's (2007) sample by *Lesueurina*, that is, Leptoscopidae, plus *Parapercis*, that is, Pinguipedidae. Leptoscopidae occur in Australia and New Zealand, while Pinguipedidae are widespread around the Pacific, Atlantic, and Indian Oceans (e.g., Randall and Yamakawa, 2006).

Thus, despite McDowall's (1964, 2010b) long-term advocacy of the idea, *Cheimarrichthys* cannot be derived from the local marine form *Parapercis colias*. Instead, the break giving rise to the torrentfish occurs at the base of a diverse, global group. The break has isolated a central New Zealand clade, now diadromous, from a worldwide, marine sister. It is typical example of a Tasman endemic with a global sister (see Chapter 4) and reflects a primary phylogenetic/biogeographic break in a global group, not a local invasion.

Eleotridae (Perciformes: Gobioidei)

Eleotrids are gobies (members of suborder Gobioidei) that are widespread in tropical–subtropical seas and freshwaters. The family is represented in New Zealand freshwaters by eight endemic species of *Gobiomorphus*. Four are restricted to freshwater, three are diadromous, and *G. cotidianus* populations can be either.

The phylogeny of the New Zealand members and their immediate relatives are as follows (phylogenies from Stevens and Hicks, 2009; Thacker, 2009; distributions from GBIF, 2016):

> Pantropical, including Queensland (*Eleotris* including *Erotelis* + *Calumia*).
>
> SW Australia, Tasmania, eastern Australia north to central Queensland (*Philypnodon*).
>
> Victoria to New South Wales/Queensland border (*Gobiomorphus australis*).
>
> Victoria to New South Wales/Queensland border (*G. coxii*).
>
> New Zealand (clade of 8 *Gobiomorphus* species).

The pattern is similar to one seen in certain angiosperms (Figure 7.6), with a group in western and eastern Australia (*Philypnodon*) breaking from a Tasman group around a node in southeastern Australia. Further differentiation has occurred there before the last break, a trans-Tasman split, has separated the southeastern *G. coxii* from the New Zealand species.

Stevens and Hicks (2009) noted that "basal taxa of gobioid fishes are usually found in fresh or brackish water and the genus *Rhyacichthys* (basal within the Gobioidei) is exclusively freshwater-dwelling. Consequently, it has been proposed that gobioids speciated in freshwater from marine ancestors, and returned to marine habitats one or more times." Instead, as Parenti (2008) suggested, it is just as likely that the ancestor occurred in both marine and freshwater habitats.

Four New Zealand species of *Gobiomorphus* are restricted to freshwater and these form a clade, together with the widespread, diadromous *G. cotidianus* (Figure 12.10; Smith et al., 2005; Stevens and Hicks, 2009). *G. cotidianus* groups with *G. basalis* and *G. alpinus*. *G. breviceps* includes two undescribed species, A and B, which are sister groups in Bayesian analyses (and sister to the first three species), but not in maximum likelihood analyses. The four, nondiadromous species have phylogenetic/geographic breaks around the Alpine fault, the associated Awatere fault (with the local endemic *G. alpinus*), and at Cook Strait. *G. basalis* defines the North Island, but the South Island does not appear as a monophyletic area. *G. basalis* and *G. breviceps* show widespread overlap in the North Island. One possible site for the original split between them is indicated at the line a–b in Figure 12.10, with arrows

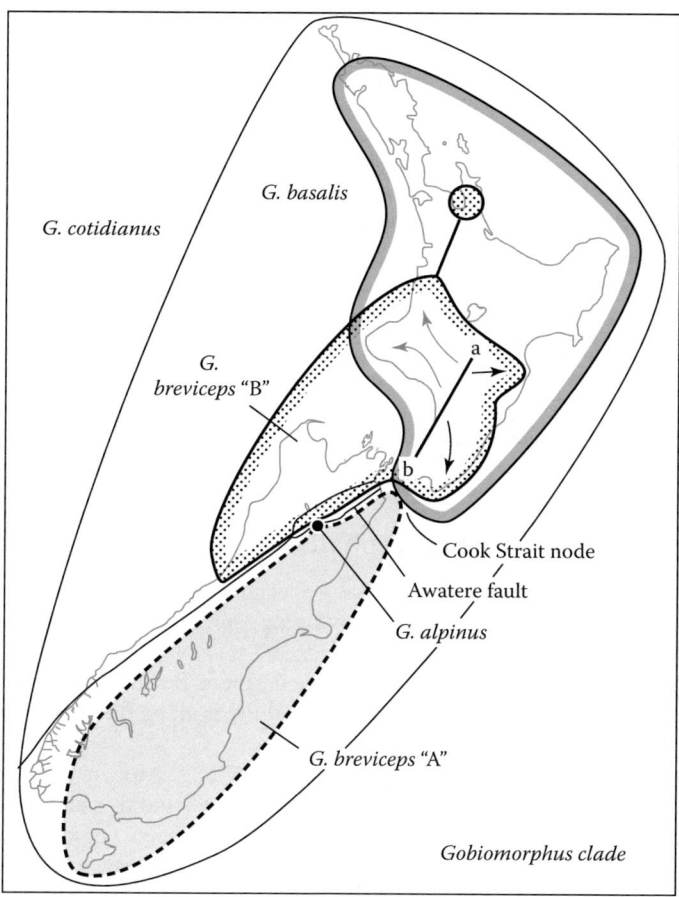

FIGURE 12.10 Distribution of a clade in *Gobiomorphus* (Eleotridae). (Data from Stevens, M.I. and Hicks, B.J., *Evol. Ecol. Res.*, 11, 109, 2009; McDowall, R.M., *New Zealand Freshwater Fishes: An Historical and Ecological Biogeography*, Springer, Berlin, Germany, 2010b.)

indicating range expansion from here in both species. In this group, as in many others (cf. cicadas, Figure 12.3) there is more overlap among clades in the North Island than in the South Island. This suggests that the overlap developed during the extensive Neogene disturbance in the North Island caused by crustal shortening and volcanism.

A Notable Absence in the New Zealand Fish Fauna: Otophysi

The large clade Otophysi is made up of Cypriniformes (carp, minnows), Characiformes (piranhas, tetras), Siluriformes (catfishes), and Gymnotiformes (electric eels). The group includes 67 families and 30% of all fish species (9,740 of the 32,000 total) (Chen et al., 2013). With two exceptions, members of the group occur in freshwater habitat in the lands around the Atlantic and Indian Ocean basins, but do not occur on Madagascar or east of Borneo and the Philippines. (The two exceptions are both small families of marine and freshwater catfishes: Ariidae of pantropical coasts and also inland Australia, and Plotosidae of coasts from East Africa to the Pacific islands, also inland Australia.)

The absence of most Otophysi from Madagascar and east of the Philippines is a standard pattern, seen in groups such as haplorhine primates and hystricomorph rodents, and it can be explained in terms of Earth history (Heads, 2014). Likewise, Betancur-R (2009) studied the main group in the catfish family Ariidae, and he concluded that the area cladogram is congruent with the Gondwana breakup sequence, as in other fish groups such as cichlids, aplocheiloid killifishes, and melanotaeniids (Heads, 2012).

Most groups of Otophysi are absent from Australasia, and there are none at all in New Zealand, either in the sea or in freshwater. Other groups that occur more or less worldwide but are absent from New Zealand include the plant family Ceratophyllaceae. The pattern complements that of New Zealand endemics with worldwide sisters, such as the torrentfish *Cheimarrichthys*. The absence of Otophysi from New Zealand is also matched by the diversity of Galaxiiformes there. This indicates that the absence, as with the absence of most Otophysi east of the Philippines, is the result of historical, geological factors rather than ecological ones (such as the freshwater ecology of most Otophysi). Thus, the absence can be explained as the result of phylogenetic/biogeographic breaks in global groups around Australasia.

Lundberg et al. (2000) discussed freshwater fishes in Australia and New Guinea and criticized two unwarranted assumption that have prevailed in literature on the topic. The assumptions are that (1) the fauna is impoverished and (2) it is dominated by species that have evolved in recent times from marine ancestors. Lundberg et al. (2000) concluded that both assumptions "serve to devalue the significance of the fauna, and both are unquestionably premature." The same assumptions underlie much of the work carried out on New Zealand fishes, but again they cannot be justified and are unnecessary. Rejecting them would clear the way for a reevaluation of the fauna.

The Tuatara, *Sphenodon* (Rhynchocephalia)

The tuatara of New Zealand, genus *Sphenodon*, has a special significance for studies of ecology, biogeography, and phylogeny in the country. It is the only living representative of Rhynchocephalia, and the modern-day sister group (not the ancestor) of all lizards and snakes (Cree, 2014).

Populations of tuatara are known from islands off the northern North Island and from islands around the Cook Strait region. They were also known in historical times along the East Coast region, and fossil material extends the range to South and Stewart Islands.

It has been suggested that the tuatara is "highly threatened" (e.g., Mooers et al., 2005). It is true that there was a population decline caused by human-associated pests, and this also brought about the final stages of extinction on the mainland. But conservation work has been carried out, and there are now about 45,000 individuals on the Cook Strait islands and 10,000 on the northeastern islands. The common species, *Sphenodon punctatus*, is regarded as "lower risk/least concern" by the IUCN (2016).

A form of tuatara endemic to the Brothers Islands in Cook Strait was recognized as a second, rare species (*S. guntheri*; Daugherty et al., 1990), but subsequent studies by Hay et al. (2010) found that it is nested in a clade containing all Cook Strait populations. Hay et al. concluded that there are two main clades of tuatara, one endemic to the northern islands and one to the Cook Strait region, as in the early, two species classification (Buller, 1876). Buller (1877) later recognized a third form from an intermediate site, East Cape Island, located along the standard biogeographic arc: North Cape–East Cape–Cook Strait, but the population is now extinct.

Several of the endemic groups on the northern islands have fossil records on the mainland, and it is sometimes assumed that the groups were widespread throughout the country. But it cannot be assumed that all island endemics once existed on the mainland. In addition, some groups now restricted to offshore islands do have mainland fossil records, but the fossils are often restricted to certain areas. The tuatara is a typical example of this. It is now restricted to offshore islands, but once had a more extensive range around the east coast. According to Colenso (1885), "the old Maoris always said that the tuatara formerly inhabited the headlands of the New Zealand coast (as well as the islets lying off it)." Dieffenbach, quoted in Gray (1842), wrote that it "lives in holes, especially on the slopes of the sand hills of the shore."

Tuatara fossils are known from sandy, coastal sites (including areas that were once covered in forest) as well as sand or gravel banks of large, braided rivers. Fossil records are widespread around the east coast of the country and as far south as Stewart Island. They extend inland only in Southland, Central Otago, and South Canterbury (specimens from Lake Taupo may be the result of human transport) (Cree, 2014: Fig. 2.8). In addition, confirmed records from the West Coast occur only along the sector between northwest Nelson and Charleston, below the Paparoa Range (Jones et al., 2009; Miller et al., 2012; Cree, 2014). The evidence suggests that the mainland distribution of the group was focused along the eastern seaboard, and the abundance (and diversity) along the northeastern islands and headlands reflects this.

The Restriction of Extant Rhynchocephalia to New Zealand and the Paucity of Its Sister Group There

Members of the order Rhynchocephalia first appear in the fossil record in the Early Mesozoic, and by the Late Triassic they were diverse and widespread around the globe. Nevertheless, they disappear from the Asian record in the Early Jurassic, from Europe and North America in the Early Cretaceous, and from South America in the Late Cretaceous (Cree, 2014: 53). The reasons for this demise, except for *Sphenodon* in New Zealand, "are unclear," and most members of the order seem to have disappeared *before* the mass extinction that took place at the end of the Cretaceous. A similar pattern is seen in dinosaurs, as these had their greatest global diversity in the Late Jurassic (Mortimer and Campbell, 2014: 198); they were already declining long before the end of the Cretaceous.

The question here is, "Why did the Rhynchocephalia survive in the New Zealand region?" The idea that groups tend to go extinct in areas where they were less diverse to start with, and survive where they are more diverse, was illustrated earlier using the example of the *Nothofagus* subgenera (Chapter 2). These plants have an excellent fossil record, but the same idea can be applied to a group with a much sparser fossil record, the Rhynchocephalia. The modern restriction of Rhynchocephalia to New Zealand can thus be explained if the region was the original center of diversity for the group. Later, Rhynchocephalia went extinct in areas of secondary diversity outside New Zealand region, in the same way that *Nothofagus* subgenus *Brassospora* went extinct in New Zealand following its secondary range expansion into the region.

The idea that *Sphenodon* is a relic of the original main massing of Rhynchocephalia does not imply anything about its phylogenetic position, or status as primitive or derived, within the order. The particular members in a center of diversity that do survive probably owe this to aspects of their ecology, for example, a tolerance of high levels of disturbance. The known fossil record of Rhynchocephalia in New Zealand is very poor; it consists of three partial dentaries of a derived form from the Miocene (Jones et al., 2009). The model suggested here would be consistent with further fossil finds indicating greater prior diversity.

Rhynchocephalia are sister to the order Squamata (lizards and snakes), and fossil evidence suggests that the two groups diverged in the Triassic (Jones et al., 2009). The two orders together make up the lepidosaurs. As discussed, the reason for the restriction of extant Rhynchocephalia to the country is one of the classic problems of New Zealand biogeography. Another concerns the great poverty of higher-level squamate clades in the region compared with Australia—there is one skink genus, one clade of geckos, and no known snakes. (Putative fossil material of a snake has been reidentified as a galaxiid fish; Worthy et al., 2013a.) Cree (2014: 68) described the absence of land snakes as "puzzling." The phylogenetic poverty of the lizard fauna compared with that of Australia also requires explanation—if the tuatara was able to survive, why are there not large numbers of relictual lizard lineages?

The paucity of squamates in New Zealand and the extant presence of their sister group, rhynchocephalians, in the same locality can both be explained as aspects of a single phenomenon if they reflect the original vicariance between the two orders. If both living and fossil records are considered, the two orders overlap throughout the world. Thus, assuming that they both originated in allopatry, both must have later undergone range expansion into the territory of their sister group; secondary dispersal of Rhynchocephalia out of New Zealand and of Squamata into New Zealand would explain the pattern.

Of course, the distribution of the extant lepidosaurs could instead be the result of chance: chance dispersal of *Sphenodon* to New Zealand, chance extinction of Rhynchocephalia everywhere except New Zealand, chance dispersal of two squamate clades to New Zealand (just once each), and chance failure of all other squamates to disperse there.

New Zealand Skinks

Skinks are one of the most species-rich groups of vertebrates in New Zealand. There is a single, endemic genus, *Oligosoma*, and its sister genus, *Cyclodina*, is on Lord Howe and Norfolk Islands (Chapple et al., 2009). (The affinities of the New Zealand clade with *Cyclodina* and with relatives in New Caledonia were discussed on in Chapter 5.)

A detailed phylogeny is now available for *Oligosoma* (Chapple et al., 2009). There are also distribution maps of extant and subfossil records (New Zealand Department of Conservation, 2016), and detailed studies of individual species (cited later). Because the group is now known so much better than most, it is considered here in more detail.

There are eight main clades in *Oligosoma*, with a phylogeny: 1 (2 (3 ((4, 5, 6) (7,8)))). (The numbering system used here to name the clades reverses that used by Chapple et al., 2009, so as to number the clades from the base of a phylogeny, the convention adapted in this book). (NI, North Island; SI, South Island; 3K, Three Kings Islands).

  1. Northern and central NI and 3K (including offshore islands) (3 species).
  2. Northern NI (mainly offshore islands) (1 species, *O. moco*).
  3. Northern NI and 3K (mainly offshore islands) (1 species, *O. suteri*).
  4. Northern SI to northern NI (Great Barrier Island) (3 species).
  5. NI (widespread, including offshore islands) (3 species).
  6. NI and 3K (widespread, including offshore islands) (7 species).
  7. SI (widespread; 10 species), NI north to Waikato (2 extant species) and Northland (subfossil).
  8. Stewart Island (4 species), SI (11 species), southern NI (north to the Egmont–East Cape line; two species), and Chatham Islands (1 species).

The phylogeny shows a north-to-south progression, with a paraphyletic basal complex in the north that has most diversity on the northeastern islands (clades 1–3), another paraphyletic basal group in the North Island and offshore islands (clades 1–6), and a diverse, distal clade (7–8) mainly in the South Island.

As usual, the overall pattern can be explained in two ways, either by southward dispersal or by a southward sequence of differentiation in a widespread ancestor. Many of the clades overlap in the North Island (cf. cicadas, Figure 12.3), and subtle differences in their distribution there, for example between clades 2 and 3, suggest that this occurred after the clades' origin in allopatry. Further details on the clades' distributions and phylogenies are given next.

Clade 1. *Oligosoma smithi* and *O. microlepis* (Figure 12.11). Within this clade, there are two main sub-clades, but these do not correspond with the two species of current taxonomy. The first subclade occurs in the northern and northeastern North Island, including the eastern side of Northland. The second is distributed along the western side of Northland, from west Auckland to the Three Kings Islands (Hare et al., 2008). Similar west/east breaks are seen in many Northland groups, and these can be explained by differentiation during the south-westward rollback of the subduction zone through the region. This would also account for the subsequent break between *O. microlepis* s.lat. and the eastern clade of *O. smithi*.

Clade 2. *O. moco* (Figure 12.12). This group is a typical example of "horstian" distribution. The basal split is between the population on the Hen and Chickens, and the rest (Hare et al., 2008), consistent with simple vicariance.

Clade 3. *O. suteri* (Figure 12.13). As with *O. moco*, this is an example of "horstian" distribution, but it is centered further north, being present on the Three Kings Islands and absent on the Bay of Plenty islands (Mayor Island, Motiti Island, etc.). The Three Kings population is sister to the rest of the species (Hare et al., 2008).

The "horstian" distribution of *O. moco* and *O. suteri*, clades 2 and 3, has often been discussed. Hare et al. (2008) wrote that the two species: "occur only at a limited number of localities on the mainland and have a largely relictual distribution on offshore islands...." For *O. moco*, Hare et al. (2008) wrote: "...subfossil deposits indicate that it was previously more widespread across the mainland throughout most of the Pleistocene" (Towns and Daugherty, 1994; Towns et al., 2002). For *O. suteri*, Hare et al. (2008) wrote that in earlier times it was "more widespread across coastal mainland regions of the northeastern North Island" (Towns et al., 2002). Nevertheless, Towns et al. (2002) did not provide any evidence that the two species were more widespread on the mainland in the past, and a current treatment of all *Oligosoma* species, living and fossil, does not refer to any extralimital fossils of *O. moco* or *O. suteri* (New Zealand Department of Conservation, 2016). Extralimital fossils of other *Oligosoma* species are known, and some of these are discussed later.

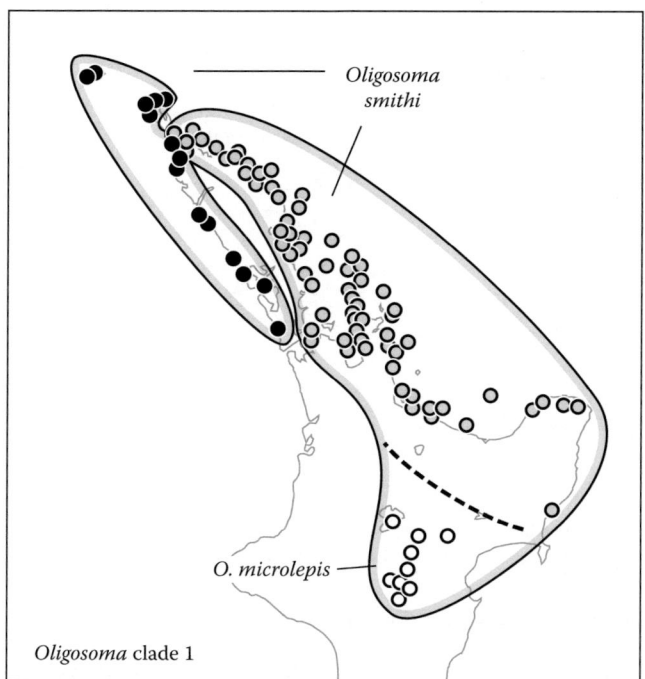

FIGURE 12.11 Distribution of *Oligosoma* clade 1 (Scincidae) and its three main clades. (Data from New Zealand Department of Conservation, *Atlas of the amphibians and reptiles of New Zealand*, www.doc.govt.nz/our-work/reptiles-and-frogs-distribution/atlas/, 2016; Hare, K.M. et al., *Mol. Phylogenet. Evol.*, 46, 303, 2008.)

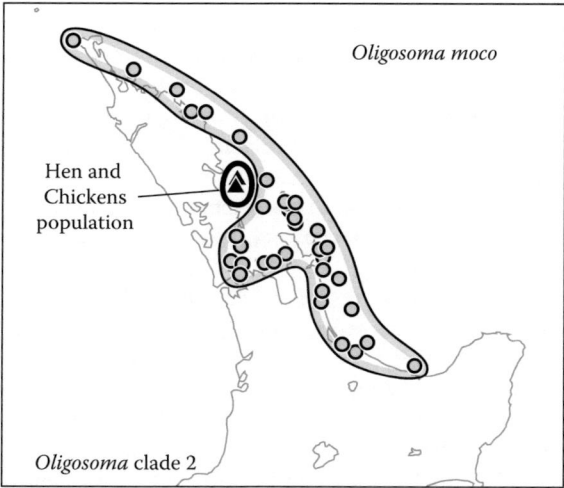

FIGURE 12.12 Distribution of *Oligosoma* clade 2 (Scincidae) and its two main clades, one restricted to the Hen and Chicken Islands. (Data from New Zealand Department of Conservation, *Atlas of the amphibians and reptiles of New Zealand*, www. doc.govt.nz/our-work/reptiles-and-frogs-distribution/atlas/, 2016; Hare, K.M. et al., *Mol. Phylogenet. Evol.*, 46, 303, 2008.)

There is no question that extinctions on the mainland have occurred in many groups, but this cannot be *assumed* to be the cause of island endemism, and in cases of allopatric pairs of mainland versus island vicariants it is probably not the explanation. The default idea that all endemism on the northeastern New Zealand islands is the result of extinction on the mainland is unjustified and downplays the significance of many of the endemics.

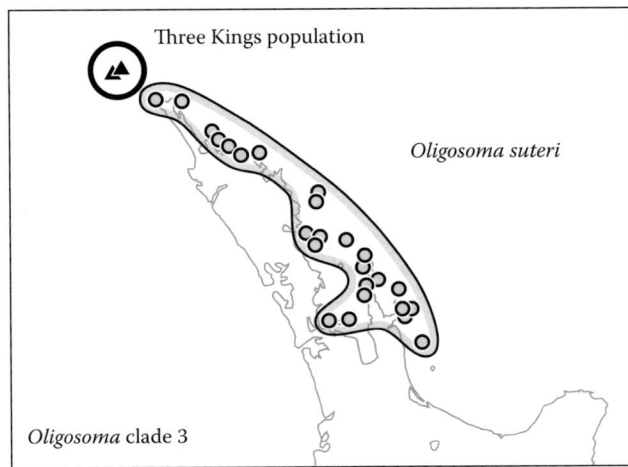

FIGURE 12.13 Distribution of *Oligosoma* clade 3 (Scincidae) and its two main clades, one restricted to the Three Kings Islands. (Data from New Zealand Department of Conservation, *Atlas of the amphibians and reptiles of New Zealand*, www. doc.govt.nz/our-work/reptiles-and-frogs-distribution/atlas/, 2016; Hare, K.M. et al., *Mol. Phylogenet. Evol.*, 46, 303, 2008.)

Clade 4. *O. zelandicum*, *O. striatum*, and *O. homalonotum*. Northern South Island to northern North Island (not mapped here).

Clade 5. *O. aeneum* group (Figure 12.14, from Chapple et al., 2008b). In this group, a clade of the North Cape region + the Poor Knights Islands is sister to *O. aeneum*, which is widespread through the North Island (but absent from the Poor Knights Islands). This break coincides with the faults of the Vening Meinesz transform margin, discussed earlier (Figure 5.5). The activity on this belt can also account for the basal breaks between the Hen and Chickens and the mainland in clade 2, and between the Three Kings and the mainland in clade 3.

Clade 6. In this clade (not mapped here), *O. fallai* is endemic to the Three Kings Islands. It is sister to the remaining species in the clade (*ornatum, oliveri, whitakeri, townsi, macgregori, alani*), which occur throughout the North Island and its offshore islands, including the Three Kings, Poor Knights, and Hen and Chickens Islands (Chapple et al., 2009). The primary break in the clade, located between the Three Kings and the North Islands, is the same one seen in clade 3. If *O. fallai* on the Three Kings was derived from the mainland, it would be sister to a group at a particular point on the mainland (the center of origin), not to the mainland as a whole. It is true that a dispersal scenario can be maintained if particular ad hoc hypotheses are invoked, but this is unnecessary, and the fact that similar patterns are repeated in different groups also argues against this.

One of the clades in group 6 is *O. ornatum* (Figure 12.15). The basal break in this widespread North Island species is between the Poor Knights Islands population and all the others (Chapple et al., 2008a).

Among the remaining species in clade 6, *O. alani* is extant on islands off Northland (Matapia Island and the Moturoa Islands) and off the Coromandel Peninsula (Mercury Islands and Castle Island). It is also recorded on the mainland as subfossils at 14 sites scattered through the North Island from Northland to Wellington. Another species in clade 6, *O. macgregori*, has extant records on the Cavalli Islands and Bream Islands off Northland, Sail Rock in the Hauraki Gulf, and Mana Island north of Wellington, while subfossil bones are recorded on the mainland from Northland and Waikato. As with *O. alani*, the extant populations have been restricted to offshore islands by secondary extinction. It is significant, though, that fossils of the other island endemics do *not* occur at the mainland sites where material of *O. macgregori* and *O. alani* is found. The island groups without mainland fossils (especially *O. moco* and *O. suteri*) branch from more basal nodes in the phylogeny than the island species with mainland subfossils in clade 6, and so this pattern conforms to the overall southward shift in the phylogeny.

Clade 7. *O. infrapunctatum* and allies. This group was mapped earlier (Figures 9.15 and 9.26).

Clade 8. *O. grande* and allies. The level of speciation in *Oligosoma* reaches maximum levels in this mainly southern group (not mapped here), which includes many interesting distributions. For example,

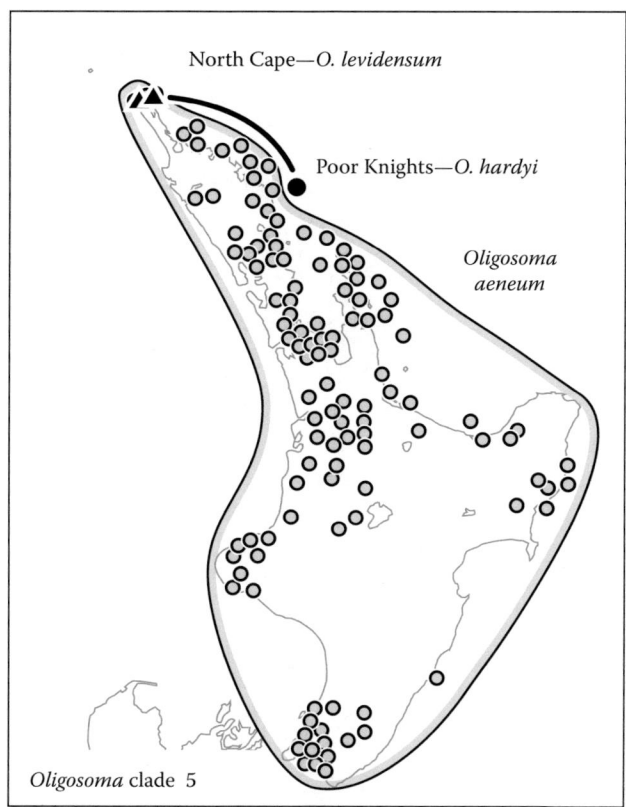

North Cape—*O. levidensum*

Poor Knights—*O. hardyi*

Oligosoma aeneum

Oligosoma clade 5

FIGURE 12.14 Distribution of *Oligosoma* clade 5 (Scincidae) and its two main clades, one restricted to the Poor Knights Islands and North Cape. (Data from New Zealand Department of Conservation, *Atlas of the amphibians and reptiles of New Zealand*, www.doc.govt.nz/our-work/reptiles-and-frogs-distribution/atlas/, 2016; Chapple, D.G. et al., *J. Herpetol.*, 42, 437, 2008b.)

O. chloronoton ranges north from Stewart Island to the Waihemo fault zone, while its sister *O. lineoocellatum* is found from the Waihemo fault zone north to Hawkes Bay (Greaves et al., 2007). *O. maccannii* extends inland from Otago Peninsula and Banks Peninsula to Broken River, Mount Cook and the Otago lakes (O'Neill et al., 2008), a distribution similar to that of the tree *Olearia fimbriata* (Chapter 8).

Another clade in group 8 comprises *O. grande* and three, locally endemic allies and is found in Central Otago (Figure 12.16; from Chapple et al., 2011). *O. grande* itself has a distribution duplicating that of *O. otagense* (Figure 9.15). The *O. grande* complex as a whole ranges from the Moonlight tectonic zone to the Waihemo fault zone, and its main axis of distribution lies along the inner, most highly metamorphosed belt of the Otago Schist, the garnet–biotite zone (Figure 3.14). Three of the species lie south of the median antiform in the Otago Schist. All these features have been produced by mid-Cretaceous extension. The localized endemism in the group also suggests subsequent evolution around the changing shores of the inland seas present in Central Otago from the Late Cretaceous to the Miocene. Other Central Otago groups found, like *O. burganae* + *O. repens* + *O. toka*, south of the schist antiform and west to the Moonlight tectonic zone include the alpine plants *Kelleria childii* (Thymelaeaceae) (Heads, 1990b) and *Celmisia semicordata* subsp. *aurigans* (Asteraceae) (Given, 1980).

Hickson et al. (2000) asked the question: "Why does New Zealand have so many skinks?" They concluded that in the skinks, "phylogeny recapitulates geography"; the diversification began no later than 23 Ma, and rapid, allopatric speciation took place during the Oligocene, when New Zealand was fragmented into many low-lying islands. This fits a model in which the southward sequence of the phylogeny as a whole reflects a sequence of tectonic events. Most of the early differentiation was in northern Zealandia (leading to the more basal phylogenetic nodes), but from the Cretaceous through to the Miocene, differentiation became progressively restricted to Central Otago.

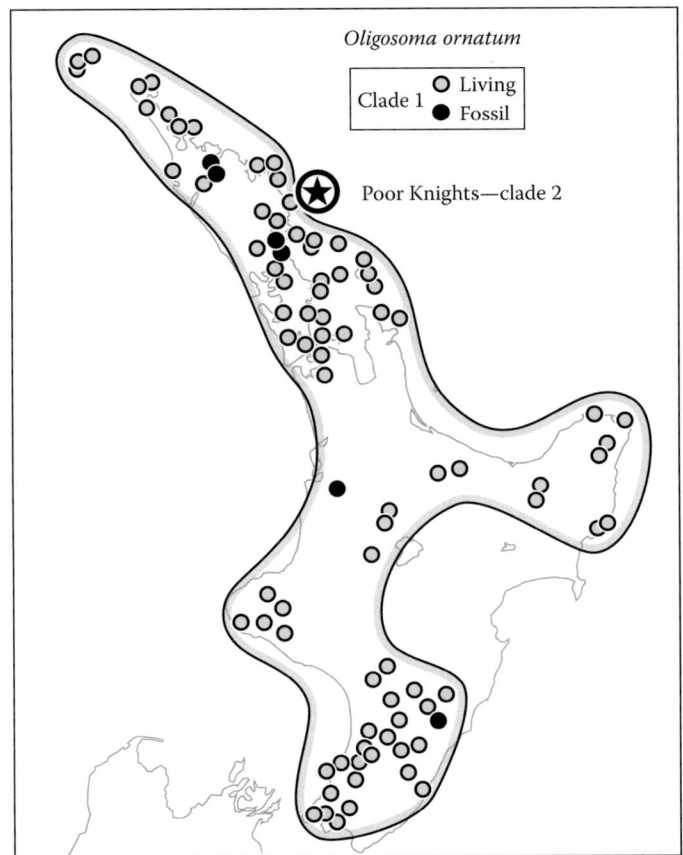

FIGURE 12.15 Distribution of *Oligosoma ornatum* (Scincidae) and its two main clades, one restricted to the Poor Knights Islands (star). (Data from New Zealand Department of Conservation, *Atlas of the amphibians and reptiles of New Zealand*, www. doc.govt.nz/our-work/reptiles-and-frogs-distribution/atlas/, 2016; Chapple, D.G. et al., *Biol. J. Linnean Soc.*, 95, 388, 2008a.)

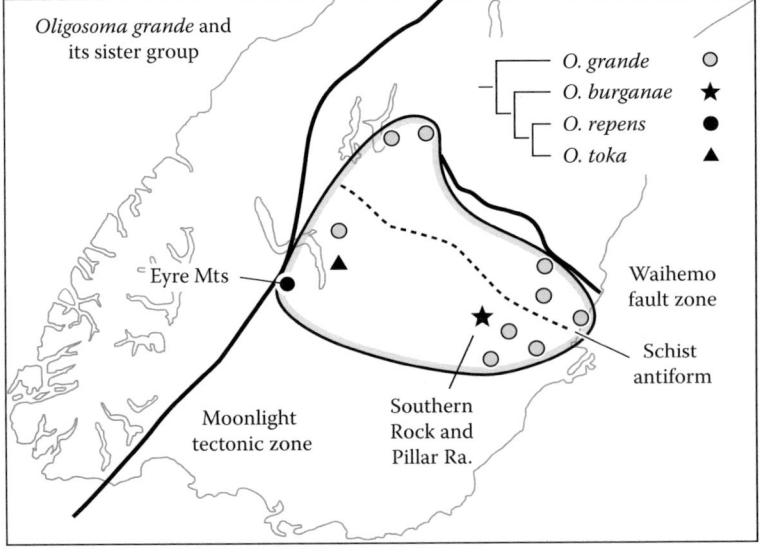

FIGURE 12.16 Distribution of *Oligosoma grande* (Scincidae) and its sister group, comprising three local endemics. (Data from Berry, O. and Gleeson, D.M., *Biol. Conserv.*, 123, 197, 2005; Chapple, D.G. et al., *Zootaxa*, 2782, 1, 2011.)

New Zealand Geckos

The diplodactyloid geckos are a diverse group restricted to Australasia. There are three families, and two show significant allopatry, at least outside Australia (Oliver and Sanders, 2009; Heads, 2014: 159):

- Carphodactylidae: Australia.
- Pygopodidae: Australia and New Guinea.
- Diplodactylidae: Australia, New Zealand, and New Caledonia.

The allopatry outside Australia in the last two clades is consistent with an origin of the three families by vicariance. This has developed at a node somewhere in central Australia, no later than the rifting of the Tasman Basin. Even using fossil calibrations, Oliver and Sanders (2009) found that "basal divergences within the diplodactyloids significantly pre-date the final break-up of East Gondwana."

The New Zealand geckos are restricted to the three main islands and make up a monophyletic clade of Diplodactylidae. Nielsen et al. (2011) presented a detailed phylogeny of the New Zealand geckos, and this is the basis of the following discussion.

The New Zealand clade's sister group (*Diplodactylus*, *Oedura*, *Strophurus*, and *Rhynchoedura*) is diverse and widespread throughout the Australian mainland, including the central deserts (it is absent from Tasmania) (GBIF, 2016). This indicates that the New Zealand clade is not a recent derivative from an Australian genus.

The New Zealand geckos are divided into three main clades. "Clade 1" comprises two genera, *Hoplodactylus* and *Woodworthia* (Figure 12.17). *Hoplodactylus* occurs along the northeastern ("horstian") biogeographic arcs, in the central North Island (as subfossils), and around Cook Strait. A northern clade in the genus, the *H. pacificus* group, has the following phylogeny (see Figure 5.1 for the localities) (Chong, 1999; Nielsen et al., 2011):

Mokohinau Islands (unnamed form).

Poor Knights Islands (unnamed form) + northern Aupouri Peninsula (*H.* "Matapia").

Eastern North Island to Kaitaia (*H. pacificus* s.str.).

Three Kings Islands + North Cape (unnamed forms).

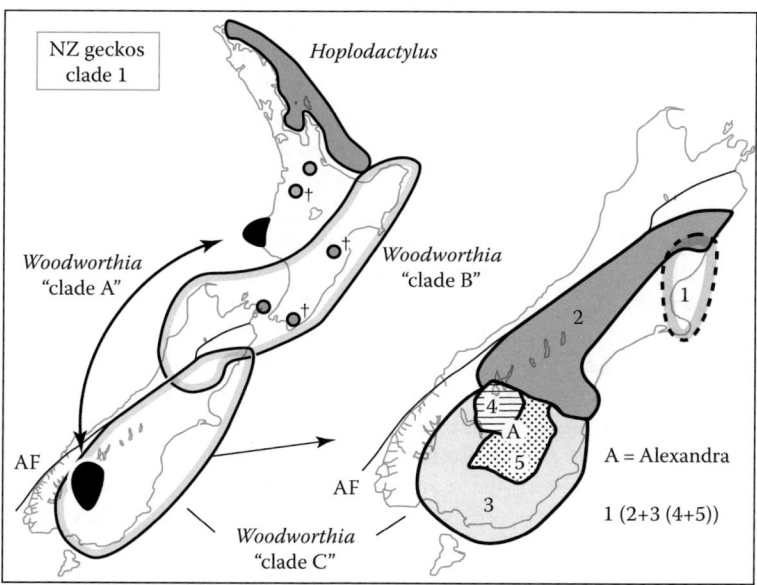

FIGURE 12.17 Distribution of "New Zealand geckos" clade 1 (Diplodactylidae). Dagger symbols indicate fossils. AF, Alpine fault. (Data from Nielsen, S.V. et al., *Mol. Phylogen. Evol.*, 59, 1, 2011.)

Here, the two main groups are allopatric, with the first extending from near North Cape south through the eastern North Island, and the other restricted to the far north, on the Three Kings Islands and North Cape. The Mokohinau Islands clade is sister to a clade that is on the Poor Knights Islands and also widespread through the eastern and northern North Island. This is consistent with differentiation caused by tectonic activity along the biogeographic "horstian" arc, but not with derivation of the Mokohinau group from a point center of origin on the mainland. The Poor Knights Islands clade connects with one of northern Aupouri Peninsula (Te Paki, Matapia Island, and Houhora), following the trend of the Vening Meinesz transform.

In *Woodworthia*, the basal clade (*Woodworthia* "clade A" in Figure 12.17) has a spectacular disjunction along the Alpine fault. The other two clades, "B" and "C," occur throughout the eastern parts of the country, except in Northland (where *Hoplodactylus* has its main distribution), and they fill the gap in the disjunct "clade A."

Woodworthia "clade C" has a focus of distribution and repeated differentiation around Central Otago. The first break is between the clade 2 + 3, which surrounds Central Otago, and 4 + 5, which is endemic there. The next breaks separate 2 from 3 at north Otago boundaries, and 4 (Cromwell, etc.) from 5 (Alexandra, etc.). This concentric sequence of phylogeny can be explained by evolution around former inland seas as these were gradually shrinking. This would have taken place at about the same time (Early Miocene) that *Woodworthia* "clade A" was being rifted apart by the Alpine fault.

In its overall distribution and phylogeny, *Woodworthia* "clade C" can be compared with the cicadas of *Kikihia* "clade 5" (Marshall et al., 2008). The distribution of *Hopolodactylus* and *Woodworthia* can also be compared with that of the skinks. For example, both show basal groups around the northeast horsts and Cook Strait, disjunction along the Alpine fault, and local endemism in Central Otago (Figures 9.15 and 12.16).

Hoplodactylus and *Woodworthia* make up "clade 1" of the New Zealand geckos (Nielsen et al., 2011). "Clade 2" and "clade 3" overlap through the North Island, Nelson, and northern Westland (Figure 12.18). However, in Stewart Island and much of the South Island, they are allopatric, with clade 2 lying south and west of clade 3. The greater overlap of clades in the North Island is a standard pattern (Figure 12.3) and can be attributed to marine transgressions and volcanism. Within "clade 2" of the geckos, two of the three genera overlap in the North Island, and in "clade 3" both genera overlap there.

In "clade 2" of New Zealand geckos (Figure 12.18), genus *Tukutuku* of *southern* Stewart Island (on and south of a line: Doughboy Hill–Kopeka River; Thomas, 1981) is restricted to the Western Province. It is sister to the rest of the clade, which extends from Stewart Island to the Three Kings Islands. A similar pattern, with a Stewart Island clade sister to a group that is widespread through mainland New Zealand, occurs in the sap beetle *Hisparonia hystrix* (Nitidulidae) (Marske et al., 2012).

In traditional theory, Stewart Island endemics colonized the island from some particular area of the southern mainland, but in the geckos this is contradicted by the phylogeny. The distributions of *Tukutuku* in southern Stewart Island and its sister group, throughout Stewart Island and elsewhere, suggest that the two diverged at an original break between southern and northern Stewart Island, and this coincides with the Western Province/Eastern Province boundary (Figure 7.5).

The other two genera in "clade 2," *Dactylocnemis* and *Mokopirirakau*, overlap along the western North Island, but there are significant differences in their distributions. On the Three Kings Islands and in Northland north of Hokianga, *Dactylocnemis* occurs alone and with significant endemism—one of its two basal clades is endemic to the Three Kings Islands and North Cape. In contrast, *Mokopirirakau* occurs alone in the eastern North Island and the South Island.

"'Clade 2" of New Zealand geckos is diverse in Westland and Stewart Island. In contrast, its sister group, "clade 3," is absent from both these areas, but is present through Canterbury and east Otago (with *Naultinus gemmeus*), where "clade 2" is absent. This allopatry probably reflects part of the original break between the two groups.

Clade 3 of New Zealand geckos comprises the widespread *Naultinus* and the rare *Toropuku*, disjunct across the central North Island. The connection suggests former records near Lake Taupo (Figure 12.18),

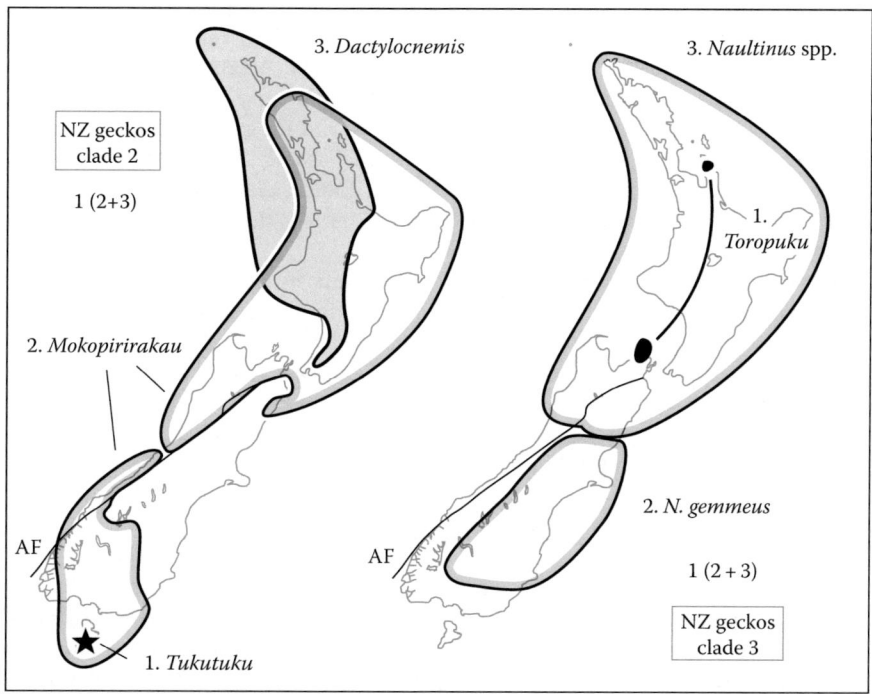

FIGURE 12.18 Distribution of "New Zealand geckos" clades 2 and 3 (Diplodactylidae). AF, Alpine fault. (Data from Nielsen, S.V. et al., *Mol. Phylogen. Evol.*, 59, 1, 2011.)

at the center of the Pleistocene volcanism. This track also acts as an eastern boundary in *Dactylocnemis* and as a western boundary in *Woodworthia* "clade B" (Figure 12.17). East/west boundaries also occur in endemic North Island plants, for example, the two genetic forms of *Dactylanthus* (Balanophoraceae) (Holzapfel et al., 2006). The break can be compared with the rotational deformation of the eastern North Island around the west (Figure 9.3).

Nielsen et al. (2011) suggested that the New Zealand geckos originated as a clade in the Eocene. This date was calculated using fossil calibrations and the assumption that New Caledonia emerged from complete submersion at 37 Ma. Although the latter is doubtful (Heads, 2011, 2014: 284), an Eocene age is useful as a minimum age because it indicates that the New Zealand gecko clade is older than the Oligocene peak of marine incursions (~25 Ma). It also means that the disjunction at the Alpine fault in *Woodworthia* "clade A" was probably caused by the strike-slip displacement that has taken place since 25 Ma.

Central Otago as a Center of Endemism and a Phylogenetic Boundary

Central Otago is well known as an important center of endemism for alpine groups as well as lowland ones. Geckos (Figure 12.17) and skinks (Figures 9.15 and 12.16) both include concentrations of local endemics in the region. The complex geological history there has involved reactivation of Cretaceous structures, and this means that dating evolutionary events is difficult. The main tectonic events include the Early Cretaceous collision of the Caples and Torlesse terranes, and this led to orogeny and regional metamorphism until convergence was replaced by extension. Inland seas developed through the Cenozoic and reached their maximum extent in the Oligocene. By the Miocene, they had become large, freshwater lakes, and with the uplift of the Kaikoura orogeny these dried up completely.

Earlier phases of differentiation would have been caused by mid-Cretaceous extension and local uplift along the line of the schist antiform: Pisa Range–Rock and Pillar–north Dunedin (whether this formed as a metamorphic core complex or a forearc high) (Figure 3.14). A broader, central strip comprising the most highly metamorphosed Otago Schist, the garnet–biotite zone, includes many endemics, such as the fish *Galaxias* "Teviot" (Teviot River, by Roxburgh), and the clade formed by *Galaxias pullus* and *G. eldoni* (Teviot River to Opoho Creek, Dunedin) (McDowall, 2010). Others include *Carex alla-nii* (Cyperaceae) at the Old Man Range, Rock and Pillar Range, and nearby Macraes Flat (NZPCN, 2016), and *Gingidia enysii* var. *spathulata* (Apiaceae) at Roaring Meg (Pisa Range), Maniototo Plain, and Naseby (Dawson, 1967; NZPCN, 2016).

Endemism in Central Otago includes many concentric patterns, as in the geckos, and these can be explained by evolution around shrinking inland seas. A similar, concentric pattern occurs in the grass *Elymus apricus*, endemic to the Central Otago and mapped by Connor (1954) (as *Agropyron scabrum* "Group Otago"). It is allopatric with, and surrounded by, *E. rectisetus*, which ranges from southern Central Otago north to the Wairau River (Edgar and Connor, 2000). These examples, along with other cases of endemism and phylogenetic breaks in Central Otago, can be explained by differentiation around the shifting coastlines of inland seas (Oligocene) and lakes (Miocene).

Differentiation around Nuclear Central Otago

The towns of Cromwell and Alexandra are located in the garnet–biotite zone of the Otago Schist and mark the geographic center of Central Otago (Figure 3.14). Apart from the endemism there (often attributed to the dry climate), nuclear Central Otago is also an important break zone. This is seen in the gecko *Woodworthia* (Figure 12.17), with a boundary at Alexandra, and the same break is documented in the *Sigaus australis* complex of grasshoppers (Orthoptera: Acrididae). This group has a standard range extending from Otago to central Canterbury (Craigieburn Range). Sequencing studies of a broad sample of populations show differentiation around a node near Alexandra (Figure 12.19; Trewick, 2008), with western, northern, and eastern groups overlapping there.

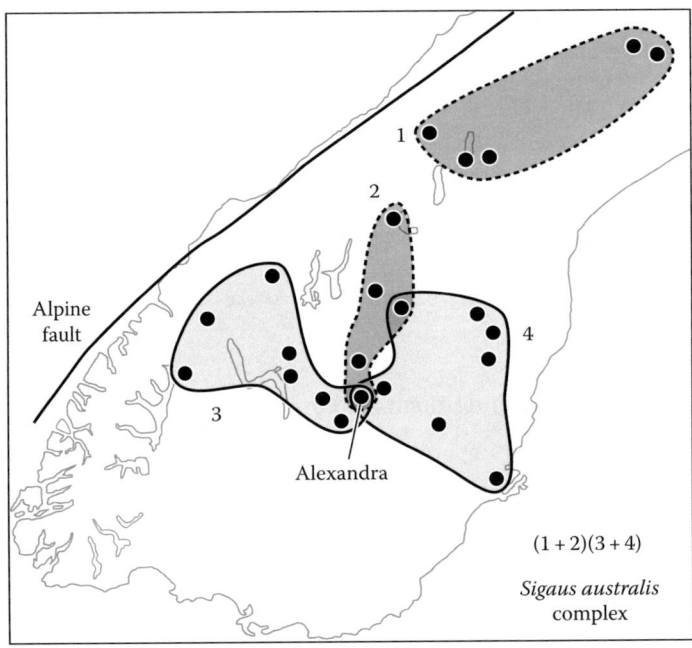

FIGURE 12.19   The *Sigaus australis* complex of grasshoppers (Acrididae), with the four groups and a phylogeny. (Data from Trewick, S.A., *Cladistics*, 24, 240, 2008.)

Many nodes act as both margins and centers of distribution. In the *S. australis* complex, Alexandra represents a breakpoint in a regional group, as well as being a center of local endemism (Trewick, 2008). The northern clade ("2" in Figure 12.19) includes a local *lowland* endemic, *S. childi*, at 200–500 m altitude on the outskirts of Alexandra (Graveyard Gully, Little Valley Road, Earnscleugh). In addition, the western clade ("3" in Figure 12.19) includes a local *alpine* endemic, *S. obelisci*, at 1600–1700 m in the Old Man Range, 15 km southwest of Alexandra. This resembles the lowland/alpine center of endemism at Cromwell (Chapter 1).

Trewick (2008) described the meeting of the *S. australis* groups near Alexandra as "intriguing." He compared it with similar patterns in other grasshoppers (*Phaulacridium*: Acrididae) and also in the diverse beetle *Prodontria* (Scarabaeidae). *Prodontria* ranges from the Snares Islands north to the Ben Ohau Range (near Mount Cook), with as many as seven species locally endemic in the region between the Pisa Range and the Rock and Pillar Range (Heads and Patrick, 2003). These include endemics around Cromwell (*P. lewisi*), southern Pisa Range–Kawarau Gorge (*P. jenniferae*), and Alexandra (*P. modesta* incl. *bicolorata*) (Heads and Patrick, 2003). Emerson and Wallis (1995) found *P. truncata*, from Mount Burns in southeastern Fiordland, to be sister to the remaining species. They also found a Central Otago clade, present at Rock and Pillar Range (*P. capito* p.p.), Hector Mountains (*P. pinguis*), and the Crown Range + Pisa Range (sp. 4—i.e., *P. regalis*). This can be compared with the *Oligosoma burganae* clade of skinks (Figure 12.16).

The clades of the *Sigaus australis* complex illustrate three of the biogeographic affinities of Alexandra. A fourth, the north–south connection Alexandra–Tautuku, is evident in groups such as the fish *Galaxias* sp. D (Figure 12.8) and the stick insect *Niveaphasma* (O'Neill et al., 2009). The southern limit of the plant genus *Kunzea* (Myrtaceae) follows a similar boundary (de Lange, 2014: Fig. 59). These patterns perhaps correspond to an arm of the sea or a lake that was a precursor of the modern Clutha River.

Ecology in Central Otago

Community ecologists as well as systematists have confronted the problem of differentiation in Central Otago. In the montane grasslands around Alexandra and the Otago lakes, floristic composition at the ecological-plot scale shows significant variation, but there is little correlation with current environmental factors (Wilson and Meurk, 2011). The authors were reluctant to explain the results by chance processes, arguing that

> The invocation of chance is long established in ecology… and has been emphasized in recent years…, but it is often invoked when other explanations have failed, and is impossible to falsify. In the final analysis, chance does not exist…, and we must admit that the control of species composition is often by historical and other factors beyond our knowledge…. (p. 2394)

These authors also wrote: "It is clear that the distribution patterns of many species in New Zealand are related to its very complicated geological history over millions of years, rather than to the geography of the present day…." Establishing which historical factors have been relevant for biogeography is complex, but at least ecologists are now acknowledging development over evolutionary time— tens of millions of years rather than thousands of years or less.

Summary

To summarize this chapter: sequencing studies are consistent with some of the new ideas on evolutionary processes in New Zealand introduced in earlier panbiogeographic work (Craw et al., 1999). The processes include the following:

Incorporation of new clades in the biota with terrane accretion, obduction and the "scraping off" of metapopulations at subduction zones.

Uplift of communities in orogens.

Modernization of the extant biota during the Rangitata orogeny.

Lowering of communities with subsidence and erosion.

Rifting of populations by normal faults, seafloor spreading, and transform faults.

Range expansion, extinction, and speciation during the Oligocene marine transgressions.

Stranding of coastal taxa inland with marine transgression and regression.

Widespread survival of populations through the Pleistocene ice ages in multiple microrefugia.

13

---

*Structural Evolution and Ecology**

Evolution takes place in space, time, and form. Aspects of evolution in space and over time have been discussed in the previous chapters, with reference to centers of origin, dispersal, vicariance, and extinction. This chapter and the next three concentrate on evolution in form—morphological evolution and adaptation—and on synthesizing these with biogeography. This chapter gives a general discussion of trends in structural evolution. Chapter 14 deals with a particular trend in plants and animals—reduction and fusion. Chapters 15 and 16 consider aspects of morphology in New Zealand's birds and mammals in conjunction with their biogeography.

Traditional theory has developed specific ideas on the interpretation of space, time, and form in evolution based on centers of the origin, dispersal, and adaptation, and so it has been termed the CODA model (Lomolino and Brown, 2009). There are many problems with the approach, and an alternative is described in this chapter. One of the main questions in biology as a whole concerns the fundamental cause of morphological evolution, and recent work shows that this is more problematic than was thought.

The correlation between the environmental factors of an area and the morphology of the species that live there is well known, and it is often assumed that the former causes the latter. In the plant family Proteaceae, for example, Jordan et al. (2013) concluded that "climate drives vein anatomy." Instead, the morphology of a group may determine the environment that the group occupies. In this model, morphology is not caused by extrinsic factors, but by intrinsic factors such as nonrandom mutation. This idea of mutation-driven evolution accounts for many phenomena that are otherwise difficult to explain, and, although it has not yet been discussed by ecologists, it is supported by a growing number of geneticists (Lynch, 2007a,b,c; Stoltzfus, 2012, 2014a,b; Nei, 2013). The traditional Modern Synthesis approach and also the new, alternative ideas are discussed next in more detail.

The Modern Synthesis View: Structural Evolution Is Determined by Extrinsic *Needs*

The neodarwinian Modern Synthesis accepted that the essence of a morphological feature is its function or its purpose—its "role." The purpose of a feature comes before the structure itself, and the structure exists because it addresses a *need* of an organism. In the CODA model, the need determines the function of an organ, and the function determines its structure. Evolution optimizes structure, making it the best possible in terms of answering the need. If there is a change in a group's habits or its habitat, this will generate new needs and thus new, appropriate structure. The CODA model proposes that evolution is driven by extrinsic needs, rather than intrinsic, genetic factors. Needs and competition generate forces that "wedge" the species and its morphology into a preexistent niche (Chapter 1). In this model, the evolutionary sequence is:

1. An organism develops a new **habit**, or disperses into a new **habitat** or locality.
2. The new habit or habitat means the organism has new **needs**.
3. These needs cause the evolution of new, adaptive **structure**.

---

* This chapter incorporates material first published in Heads (2012a) and reproduced here with permission of the University of California Press, Berkeley, CA.

This raises the questions: Why did the group develop the new habit or move into the new habitat to begin with? And how did it survive there before it developed the required morphology and physiology? In contrast with the Modern Synthesis view, mutation-driven evolution (Nei, 2013) proposes that the last of the three steps listed earlier—the evolution of new, adaptive structure—comes *first*, and that this, in itself, allows the invasion of new niche space.

For example, the Modern Synthesis suggests that: "Sponges live by passing a ceaseless current of water right through their body, from which they filter food particles. Consequently, they are full of holes" (Dawkins, 2004/2005: 497). Instead, the holes probably came first: sponges are full of holes, and consequently a ceaseless current of water passes through their body. In this model, structure determines function, and the structure itself is caused not by its function, or the organism's needs, but by prior, long-term trends in structural genetic and morphological evolution.

In an example from mammals, Gunnell and Simmons (2005) wrote that: "ancestral archontan groups [primates and relatives] as well as proto-bats apparently were exploiting similar arboreal habitats, which may have led to concurrent development of homoplasic morphological attributes." In this scenario, as in the case of the sponges, the animals were exploiting the new habitat, and *then* developed the attributes that would enable this. Again, this does not explain how or why they were exploiting the new habitat to begin with.

In another example, the Modern Synthesis suggests that the twisting of the head evolved in flatfishes "because their ancestors lay down on one side" (Dawkins, 1996/2006: 121). Why they lay down on one side to begin with is left unanswered. Yet if the head were twisted first, the new habit would be explained; in this model, the structural change is a by-product of prior trends in morphology; it has not evolved in order to fulfill a need.

The standard view suggests that behavioral changes in animals can lead to a species exploring a new niche, while the slower process of natural selection then adapts the phenotype (Hardy, 1965). According to this interpretation, many key morphological features would never have evolved had the animal not first *explored new habitat* and/or *developed new habits* (Hardy, 1965; Maynard Smith, 1987). Yet this "behavioral change" theory does not explain why the behavior changed to begin with; it also does not explain evolution in plants, which do not have behavior or habits that allow exploration. As Darwin (1859: 183) himself stated, "It is difficult to tell... whether habits generally change first and structure afterwards; or whether slight modifications of structure lead to changed habits...."

Although Lister (2014) supported the traditional model, he also admitted that:

> *Mutation affecting morphology* might lead the animal to adopt novel behaviours to survive, which, if successful, could result in selective spread of the new phenotype: the inverse to the "behavioural lead" model. A celebrated example is the goat born without front legs that learnt to walk bipedally (West-Eberhard, 2003: 51). Dover's (2000) concept of "adoptation" postulated genetic processes leading to anatomical change, organisms then seeking out an appropriate habitat or niche. (Lister, 2014: 14; italics added)

While animals and plants do show flexibility and can change their habit or habitat—for example, plants can occupy new, anthropogenic habitat, and animals can feed on new plant introductions—it is not obvious that this can lead to major structural changes. On the other hand, it is almost certain that major structural changes will lead to changes in habit, habitat, or both.

Lamarck (1809) advocated what became the standard, Modern Synthesis view; changes in habit or habitat come first, and these induce structural changes. This was supported in Baldwin's (1896) "organic selection." Organic Selection was supported by Simpson (1953b), who renamed it the "Baldwin effect" (playing down the Lamarckian connection). Mayr (1963) rejected the Baldwin effect, but (1960) admitted that: "The new habit often serves as the pacemaker that sets up the selection pressure that shifts the mean of the curve of the structural variation." For example, Mayr (1960) suggested that the New Zealand glow worm (the larva of the dipteran *Arachnocampa*) evolved from being a fungus eater to being a carnivore, and *then* developed suitable adaptations for its new diet. Dawkins (2004/2005: 406) agreed: new structure evolves because of "A change of habit by an adventurous individual...." The process is very similar to the invasion of new habitat by chance dispersal. Dawkins suggested that the "adventurous individual" hypothesis accounts for bird flight, for the first fish to come out onto the land, and for the first whale ancestor to return to the sea. Modern advocates of this Lamarckian–neodarwinian model include Lister (2014) and Corning (2014).

The idea that changes in behavior lead to structural evolution would not seem to apply to plants, although Gilroy and Trewavas (2001) argued that it does, as even plants make decisions. For example, "The ubiquity of light over the surface of the globe is thought to have been responsible for a major evolutionary decision by the primordial plant eukaryotic cell; to remain sessile...." Instead, in the model suggested here, the early "plants" did not have a choice. They became sessile, not because of an adventurous individual, but because of a structural change, for example, the reduction of their cilia and flagellae.

Teleological Models of Evolution: The Origin and Evolutionary Development of a Structure Are Explained by an End Result

In the CODA model, an evolutionary development begins at a biogeographic, ecological, and morphological center of origin, with radiation from there. Radiation takes place by dispersal and by adaptation. The model accepts teleological explanation as a viable mode of accounting for morphological evolution: the *origin* of a biological structure is explained by its *end* result (*telos*), its current function. In this model, an eye sees, therefore, it evolved *in order to* see (sight was needed); a heart pumps blood, and so it has evolved *for this purpose*; and so on. This style of reasoning reverses the normal chronological sequence of causation, in which the cause comes first and is followed by the effect.

In the CODA model, morphological structure is described as designed or planned for a particular purpose. In New Zealand examples, the plant *Mearnsia* (Myrtaceae) has juvenile climbing shoots that "are designed to grow quickly in length" (Simpson, 2005: 44). "Most [New Zealand] flowers are designed to attract and be pollinated by a wide range of generalist insects and birds..." (Dawson and Lucas, 2011: 20). In the gastropod family Turritellidae, "The mantle entry has fringing tentacles, and the outer edge of the operculum bears minute bristles, both designed to sift out excessive sediment..." (Powell, 1979: 125). In birds, the adzebill (*Aptornis*) "had short, strong feet designed for scratching in the ground" (Tennyson and Martinson, 2006). But are these structures of shoots, flowers, mantles, and feet really "designed"?

The end result of a process can explain the reason for its origin in the case of conscious aims; for example, the idea of passing an end-of-year exam can motivate a student to study. But the application of teleology to a natural process such as evolution is more problematic. Despite this, teleological explanation even for natural structure is a typical mode of thinking in the general population. It is most common in infants ("rocks are pointy so that animals won't sit on them") (Kelemen, 1999a,b, 2012; Kampourakis et al., 2012), in people with Alzheimer's disease (Lombrozo et al., 2007; Kelemen and Rosset, 2009), in education programs at primary, secondary, and tertiary levels (Gregory, 2009), and in the mass media (Aldridge and Dingwall, 2003; Thomas, 2004; Dingwall and Aldridge, 2006). In popular literature, such as Wikipedia or *The Economist* magazine, organic structure is always explained using teleology.

Many teleological explanations for *inorganic* structure would be regarded as unjustified by science or adult common sense—we understand that rocks are not, in fact, pointed in order to prevent animals sitting on them. In contrast, teleological explanations for organic structure—"eyes are for seeing," and so on—have been dominant ever since they were championed by Socrates (as quoted in Xenophon) and Aristotle (against Empedocles). They were favored by the Neoplatonists, the medieval Scholastics (e.g., St Thomas Aquinas, *Summa Theologica* 1, 91, 3), and the philosophers of the Enlightenment (e.g., Kant, 1790/1978). In the modern era, creationists use teleology to explain organic structure, but this does not come from the Bible (which does not use this type of argument); it is derived from Aristotle, via the Neoplatonists and Aquinas. These are the same sources of the teleology used in the Modern Synthesis.

Despite this history, the rejection of teleological reasoning in biology was a fundamental component of the scientific revolution, as advanced by Bacon (1605/1966: II, 7) and William Brouncker, the first president of the Royal Society (see Samuel Pepys' diary entry for July 28, 1666). This critique of teleology was developed in the work of subsequent authors, such as Goethe (Eckermann, 1836/1970: 388), Nietzsche (1881/2006: 37, 125; 1887/1996: 57, etc.; 1901/1968: sect. 647, etc.), Croizat (1964), Gould (2002), and Chomsky (2002). (The history of these ideas is discussed in Heads, 2009a.)

There are several different schools of thought within the neodarwinian, teleological approach to biology. In the extreme, ultradarwinian view, life, and adaptations really do show purposeful design (Pinker, 1994; Dennett, 1996; Williams, 1996). A second, more nuanced view holds that evolutionary products might not be designed for a purpose, but accepts that *teleological thinking* has heuristic value, as a powerful mode—in fact, the only possible mode—of analyzing organic structure (Kant, 1790/1978; Ruse, 2003). Another, even milder variant holds that teleological thinking is not justified, but accepts that *teleological language* is useful as a convenient, short-hand mode of expression (Dawkins, 1996/2006, 2004/2005). Yet there is little doubt that almost all biologists who write or speak teleologically also think teleologically; for example, they will agree that the eye evolved for sight.

Whether or not teleology is useful, no alternative has been tried or tested, so there is no way of knowing if it is the *most* useful. It is just a habit, and the principle of multiple working hypotheses suggests that having only a single explanatory model for any phenomenon can be dangerous (Chamberlin, 1965). In any case, despite centuries of investigation, only a minute fraction of the countless morphological features described in any group have been explained by "purpose" and "need," and so the practical value of the approach is limited.

In the teleological view, the purpose of an organ determines its function, and its function determines its structure. For example, "Snakes are either gigantic or dwarf depending on the size of their prey" (Raia, 2009). This model suggests that the snakes were eating a certain-sized prey and then developed an appropriate structure for this, that is, suitable size (gigantic or dwarf, depending on the size of the prey). The snake is a certain size *in order to* eat its prey.

Instead of function determining structure in this way, the alternative model suggests that evolution is mutation-driven. In this case, the structure comes first—for a reason but without any *purpose*—and determines the function; the size of the snake determines the size of the prey. Likewise, giraffes did not develop long necks in order to feed on tall trees; they are more or less obliged to feed on tall trees because they have long necks. In old age, people's teeth may fall out, and unless they can find false teeth they will need to eat soft food. This soft food diet did not cause the teeth to fall out; instead, structural changes determined the functional and behavioral shifts. The structural changes themselves did not occur for a purpose in order to achieve an end result, but because of prior trends in mutation.

The Panadaptationist Position: Selection Is the Only Force Directing Evolution

Neodarwinism proposes that all adaptive structure has evolved by natural selection; a neodarwinian is "someone who takes natural selection as the absolutely fundamental mechanism of evolutionary change" (Ruse, 2003: 197). In this view, selection is the only force determining the *direction* of evolution. The variation present in the gene pool is, in effect, unlimited, and individual mutation events have no effect on the evolutionary trajectory. This model was accepted by Fisher (1930) in a core text of the panadaptationist approach: "Natural Selection is the only means known to biology by which *complex adaptations* of structure to function can be brought about" (p. 156, italics in the original). Yet much modern work suggests that organic structure, along with its function (whether this is adaptive, neutral, or maladaptive), results from prior genetic constraints and trends, not selection.

Following Fisher's (1930) lead, Mayr (1980: 3) redefined Darwinism as "the theory that selection is the only direction-giving factor in evolution." (This is very different from Darwin's own view; Heads, 2009a.) Other statements of this neodarwinian paradigm also stress the gene pool and selection, and downplay the significance of mutation:

> Evolution is not primarily a genetic event. Mutation merely supplies the gene pool with genetic variation; it is selection that induces evolutionary change. (Mayr, 1963: 613)
>
> Mutations are rarely if ever the direct source of variation upon which evolutionary change is based. Instead, they replenish the supply of variability in the gene pool which is constantly being reduced by selective elimination of unfavorable variants... Consequently, we should not expect to find any relationship between rate of mutation and rate of evolution. (Stebbins, 1971: 30)

> Mutation rates are so low that mutation pressure by itself will not bring about significant changes in gene frequencies in populations… mutation pressure, whatever its strength, cannot be an orienting force in evolution… Orientation or directedness in evolution is brought about by other evolutionary forces. (Grant, 1985: 52–53)

Similar views were expressed by J. Huxley (1942: 56), Mayr (1960: 355), and Simpson (1967: 159).

Despite these assertions, the question as to whether or not the course of evolution is determined solely by selection has remained controversial. For example, Wilson (2009: vii) proposed that "All elements and processes defining living organisms have been generated by evolution through natural selection," but in the same book, Ruse and Travis (2009: x) wrote that: "natural selection is not the only evolutionary force…."

Modern neodarwinians sometimes admit that selection has been overemphasized in the past, yet they still insist that selection is the only way *adaptations* evolve. Dawkins (2004/2005: 464) wrote: "Natural selection is the only explanation we know for the functional beauty and apparently 'designed' complexity of living things." Rosenberg and McShea (2008) agreed: "… natural selection—random variation and environmental filtration—is the only mechanism known in nature that can produce adaptation, that can produce the 'in order to' that characterizes so many of the features of organisms. In fact… it is hard to even think of an alternative." (p. 21)—it is "the only game in town" (p. 21).

Nevertheless, despite these opinions, alternatives to panadaptationism are now being taken seriously. Sarkar (2015) wrote: "While adaptationist 'just so' stories have been offered (as typically occurs in every area of biology), recent theoretical analyses based on mathematical population genetics strongly suggest that non-adaptive processes dominate genome architecture evolution." Koonin (2012: viii) concluded that the new views on genomic evolution signify "The demise of (pan)adaptationism as the paradigm of evolutionary biology." One of the main alternatives to panadaptationism is the idea that evolution is driven not by selection, but by mutation.

Mutation-Driven Evolution: Intrinsic Genetic Factors Cause Evolutionary Direction and Evolutionary Trends

The Modern Synthesis was based on Darwin's (1859) early views, and it interpreted organic structure as the result of either chance (random drift) or design (natural selection). In contrast, Darwin's (1872, 1874) later, more thoughtful analyses stressed "laws of growth" and moved away from a panselectionist viewpoint:

> I now admit that… in the earlier editions of my "Origin of species" I perhaps attributed too much to the action of natural selection… I was not… able to annul the influence of my former belief, then almost universal, that each species had been purposely created; and this led to my tacit assumption that every detail of structure, excepting rudiments, was of some special, though unrecognised service. Any one with this assumption on his mind would naturally extend too far the action of natural selection… (Darwin, 1874: 61).
>
> We may easily err in attributing importance to characters, and in believing that they have been developed through natural selection. We must by no means overlook the effects… of the complex laws of growth… (Darwin, 1872: 157–158). In many… cases, modifications are probably the direct result of laws of variation or of growth, independently of any good having been thus obtained. But even such structures have often, as we may feel assured, been subsequently taken advantage of… (Darwin, 1872: 166). It is so necessary to appreciate the important effects of the laws of growth (Darwin, 1872: 173). We thus see that with plants many morphological changes may be attributed to the laws of growth and the inter-action of parts, independently of natural selection… (Darwin, 1872: 175).
>
> I sh[d] be inclined to attribute the characters in both your cases [ammonites] to the laws of growth & quite secondarily to natural selection. It has been an error on my part & a misfortune to me, that I did not largely discuss what I mean by laws of growth at an early period in some of my books. I have said something on this head in a new Chapt. in the last Edit. of the Origin [6th edition, 1872]. (Darwin, letter to A. Hyatt, December 4, 1872, in Burckhardt et al., 2013: 550).

The work of the first geneticists corroborated Darwin's idea that selection is not the only cause of nonrandom evolution and concluded that intrinsic, genetic factors (Darwin's "laws of growth") are more important (Heads, 2009a). A similar critique of selection theory has been developed in recent work in genetics (see discussion in Heads, 2012a).

Selection, Mutation, and the Modern Synthesis

In the Modern Synthesis, natural selection is the only source of direction in evolution. In particular, "The whole group of theories which ascribe to hypothetical physiological mechanisms, controlling the occurrence of mutations, a power of directing the course of evolution, must be set aside" (Fisher, 1930: 20; cf. Haldane, 1932). In the Fisher–Haldane model, adaptation takes place at many loci simultaneously by means of infinitesimal *variation so abundant that the process does not depend on the rate of new mutations* (Stoltzfus, 2006a). In this model, mutation in itself is not an effective evolutionary force, and evolution can be reduced to shifting the frequencies of alleles that are already in the gene pool. When evolution is redefined in this way, the introduction of mutation is irrelevant. Evolution is so far removed from the process of mutation, with so many complex dynamic processes interceding, that the outcome of evolution does not depend on specific mutation events.

Stoltzfus (2006a) summarized the history of these ideas as follows. In Darwin's early writings, there was *variation on demand*: "altered conditions of life" automatically turn on the flow of variation, producing abundant, infinitesimal, hereditary fluctuations leading to adaptation by selection. In neodarwinism, the model is similar; evolution is an engine with a tank of fuel, the gene pool, that automatically keeps itself full. The abundance of raw materials "ensures that selection may spring into action to build anything, anywhere, anytime" (Stoltzfus, 2014a). (There are obvious parallels with chance dispersal, which can also operate in any direction, anywhere, and anytime.) In the Modern Synthesis, evolutionary direction is determined not by internal constraints—genetics—but by the needs imposed by the external environment.

Stoltzfus (2014b) wrote:

> A molecular evolutionist cannot believe, following Darwin, that infinitesimal variations suffice for the incipient stages of useful molecular structures – instead, the incipient stages often involve duplications, fusions, lateral transfers, etc. A duplication mutation (for instance) does not supply a bag of formless infinitesimal raw materials (A, T, C, G), but a nonarbitrary sequence whose specific properties dictate its likely fates… (p. 58)

> Those who imagined that the role of variation is merely to supply random infinitesimal noise around the current value needed no biological theory of form and its variation. Instead, a theory of phenotypic change based on infinitesimal deviations (classical quantitative genetics) was sufficient. Because we now know that the role of generative processes in evolution [mutation in the broad sense] is not limited to supplying raw materials, we now know that evolutionary theory is incomplete without a theory of form and variation. Incorporating such a theory will require us… to reject the false metaphor of selection as a creative agent… (p. 59)

The Geneticists: Mendelian and Mutationist

In contrast with the neodarwinians, the early geneticists—the Mendelians—recognized mutation as a source of discontinuity, initiative, direction, and creativity in evolution. Before the Modern Synthesis appeared, the Mendelians had already synthesized genetics and selection; they proposed that genetics produces the variation, while selection can only modify this. In the debate between the two schools of thought, advocates of the Modern Synthesis saw themselves as rescuing evolutionary biology from this Mendelian heresy. They did this by arguing that genetics is consistent with panselectionism (and thus teleology), and that mutation must be denied any direct importance.

In Modern Synthesis accounts of the history of biology, the first "geneticists" are portrayed as naive and foolish, rejecting selection because they did not understand it (Berry, 1982; Ridley, 1985; Dawkins, 1986: 305). In fact, the geneticists, led by T.H. Morgan, had an excellent understanding of selection, and they did not reject it. What they did do was integrate mutation with selection and deny that selection was the only directional force in evolution. For Morgan (1916, 1932) and the other geneticists, mutation produced the variation while selection pruned off subviable forms, and evolution was a synthesis of the two forces. This is similar to the ideas on mutation-driven evolution that have been proposed by biogeographers (Croizat, 1964) and are now becoming accepted by geneticists (Nei, 2013).

Internal Genetic Causes of Evolutionary Direction: The Significance of Available Variation and Mutation

Panadaptationism and teleological explanations for structure would both be justified if variation was inexhaustible. If any organism was able to evolve any feature at any time, evolution would produce perfect adaptation, and an organism's structure would be a perfect reflection of its "needs." Yet views on genetics have changed since the heyday of the Modern Synthesis in the 1950s–1980s, and constraints on variation are now recognized as fundamental. It is no longer thought that any structure can arise in any organism.

Mutation-Driven Evolution: Evolution as "Climbing Mount Probable"

In the Modern Synthesis, mutation is the ultimate source of variation, but it is neither a proximate cause of evolution (evolution does not begin with a new mutation), nor an effective agent of change (Stoltzfus, 2012). Instead, the sole cause of evolutionary direction is natural selection. The Modern Synthesis is distinguished from the earlier work on mutation "by its utter denial of any internal causes of directionality…" (Yampolsky and Stoltzfus, 2001: 73); any directionality is assumed to be imposed by selection pressure from the environment.

In contrast with this standard view, Yampolsky and Stoltzfus (2001: 73) showed that mutational bias in the introduction of variation "can strongly influence the course of evolution [and] may be expected to contribute to homoplasy, parallelism and directionality." The authors discussed evidence indicating that "Non-randomness in mutation has a *predictable* effect on the outcome of an evolutionary process. This predictable effect may be said to represent an *orienting*, *directional* or *shaping* influence… Bias in the introduction of novelty by mutation is a prior bias on the course of evolution, and may be said to be an 'internal' cause of orientation or directionality" (p. 81). In the mutationist view, the course of evolution is not determined by selection alone, but by a two-step, origin-fixation process. In this model, mutation is not just a source of raw materials, but an agent that introduces novelty, while selection is not an agent that shapes features, but a sieve (Stoltzfus, 2012).

The idea that mutation is random is a common misconception in popular accounts. For example, the Wikipedia article "Introduction to evolution" states (June 2016): "Among offspring there are variations of genes due to the introduction of new genes via *random* changes called mutations" [italics added]. But in the more advanced article "Evolution," random mutation is not mentioned. Instead, there is a whole section on "Biased [non-random] mutation," citing authors such as Lynch (2007a) and Stoltzfus and Yampolsky (2009). The difference between the two Wikipedia articles reflects some confusion in the general understanding of the subject, but at least the topic is now being debated.

Neodarwinian authors who accepted random mutation and selection as the basis for evolution have interpreted evolution as "climbing Mount Improbable" (Dawkins, 1996/2006). (This is repeated in the importance attributed to very unusual events in dispersal theory.) In contrast with the idea that the course of evolution is improbable, Stoltzfus and Yampolsky (2009: 643) concluded that evolution has "tendencies or propensities," and that "these tendencies are predictable…." These authors saw intrinsic bias in mutation as a cause of *non*randomness, and so they interpreted evolution as "climbing Mount *Probable*."

Many other evolutionists also accept that there are valid alternatives to standard neodarwinian theory. Olson (2012: 278) admitted that adaptation is hard to study because "multiple, even non-adaptive, explanations can often account for the same set of observations." Cutter et al. (2009: 1199) agreed: "natural selection is only one of several core evolutionary forces that can lead to nonrandom patterns in genomes… it is essential to generate both neutral, nonadaptive hypotheses as well as adaptive hypotheses as testable alternatives."

In mutation-driven evolution, "Internal mutation pressure plays a central role in genomic evolution, often completely overwhelming the eternal forces of natural selection and driving change in a directional manner" (Lynch, 2007c: 375). Filler (2007) reviewed homeotic evolution in mammals and supported "an enlarged role for a mutational view (Stoltzfus, 2006a) of evolutionary drive to update classic Darwinian and New Synthesis models." Chouard (2010) wrote: "Single-gene changes that confer a large adaptive value do happen: they are not rare, they are not doomed and, when competing with small-effect mutations, they tend to win... As the molecular details unfold, *theory badly needs to catch up*" [italics added].

Aspects of Genetics Consistent with Mutation-Driven Evolution

Different observations in genetics and evolution support, or are consistent with, mutation-driven evolution. Some of these are discussed next as an introduction to the following three chapters, which propose mutation-driven interpretations of morphology in New Zealand groups.

Nonrandom Mutation

Nucleic acids comprise chains of nucleotides, with each of the latter characterized by one of four nitrogenous bases: adenine (A), guanine (G), thymine (T), and cytosine (C). The genomes of animals and land plants are almost invariably A+T rich (Lynch, 2007c). In other words, there is a high incidence of G:C → A:T mutations (the colon denotes a base pair bond between DNA strands) that is seen in all well-characterized eukaryotes. The finding is contrary to the expectation for a genome in mutation equilibrium, which would exhibit equal numbers of mutations in both directions. This is an example of a simple, broad trend in genomic evolution that contradicts core assumptions of the Modern Synthesis.

Biased Gene Conversion

Recombination is a type of mutation (Nei, 2013) that occurs at meiosis, with large-scale, reciprocal genetic exchanges taking place between homologous chromosomes by crossover. It is often thought of as a symmetrical process, but in fact recombination events are also accompanied by unidirectional exchanges. This process is known as gene conversion. A large amount of evidence suggests that in many eukaryotes, including humans and other mammals, gene conversion is GC-biased (Duret and Galtier, 2009), the reverse of the baseline mutational pattern. For example, in birds, Backström et al. (2013) found there is higher proportion of GC in the genome of passerines than in galliforms. Many fast-evolving genes and noncoding sequences in the human genome have GC-biased substitution patterns (Kostka et al., 2012). GC-biased gene conversion (gBGC) affects the probability of fixation of GC alleles and so resembles selection for increasing GC. This can mislead several tests designed to detect positive selection: "By definition, gBGC results in the non-random transmission of alleles to the next generation..." (Duret and Galtier, 2009; see also Ratnakumar et al., 2010).

Lartillot (2013) wrote that gBGC:

> ...is a major evolutionary force shaping genomic nucleotide landscapes, distorting the estimation of the strength of selection, and having potentially deleterious effects on genome-wide fitness.... Across placental mammals, variation in gBGC strength spans two orders of magnitude, at its lowest in apes, strongest in lagomorphs, microbats or tenrecs, and near or above the nearly neutral threshold in most other lineages. Combined with among-gene variation, such high levels of biased gene conversion are likely to significantly impact midly selected positions, and to represent a substantial mutation load.

Evolution of Structural Complexity

Mutation-driven evolution provides an alternative explanation for changes in individual genes, and over the last decade, geneticists have also attributed many other aspects of evolution to internal genetic processes rather than selection. For example, Lynch (2007a) has described the frailty of adaptive hypotheses for the origins of organismal complexity. He wrote:

> The vast majority of biologists engaged in evolutionary studies interpret virtually every aspect of biodiversity in adaptive terms. This narrow view of evolution has become untenable in light of recent observations from genomic sequencing and population genetic theory. Numerous aspects of genomic architecture, gene structure, and developmental pathways are difficult to explain without invoking the nonadaptive forces of genetic drift and mutation. In addition, emergent biological features such as complexity, modularity, and evolvability... may be nothing more than indirect by-products of processes operating at lower levels of organization... The origins of many aspects of biological diversity, from gene-structural embellishments to novelties at the phenotypic level, have roots in nonadaptive processes, with the population genetic environment imposing strong directionality on the paths that are open to evolutionary exploitation. (Lynch, 2007a: 8597)

In this way, nonadaptive forces—not just selection—can result in complex organisms that are viable and adapted in their environment. If carried far enough, though, many nonadaptive trends will, over time, lead to extinction. Lynch (2007a) wrote: "The hypothesis that expansions in the complexity of genomic architecture are largely driven by nonadaptive evolutionary forces is capable of explaining a wide range of previously disconnected observations...." (p. 8600).

With respect to multicellularity, King (2004) wrote: "the historical predisposition of eukaryotes to the unicellular lifestyle begs the question of what selective advantages might have been conferred by the transition to multicellularity." But Lynch (2007a: 8600) rejected the idea that the structure was caused by any advantage, and wrote: "where is the direct supportive evidence for the assumption that complexity is rooted in adaptive processes? No existing observations support such a claim."

Lynch (2007a: 8603) concluded:

> ...many aspects of biology that superficially appear to have adaptive roots almost certainly owe their existence in part to nonadaptive processes. Moreover, if the conclusion that nonadaptive processes have played a central role in driving evolutionary patterns is correct, the origins of biological complexity *should no longer be viewed as extraordinarily low-probability outcomes of unobservable adaptive challenges, but expected derivatives of the special population-genetic features of DNA-based genomes.* [italics added]

Again, there are obvious parallels with dispersal events of "extraordinarily low-probability," and vicariance events that are "expected," given the tectonic history.

Evolution of Diversity and Evolvability

The prior structure of genomes, rather than natural selection of advantageous traits, can predispose groups to evolve in different ways and to diverge to different degrees (Burns et al., 2002). This intrinsic, clade-specific "propensity" (Lovette et al., 2002; Davies et al., 2004) is discussed in developmental genetics as "evolvability" (Arthur, 2002) or "tendency to evolve" (Frohlich, 2006). The degree of evolvability in a group is sometimes thought to evolve by natural selection, but "comparative genomics provides no support for the idea that genome architectural changes have been promoted in multicellular lineages so as to enhance their ability to evolve" (Lynch, 2007a: 8603). In the model adopted here, it is prior genome architecture and tectonics, not natural selection, that govern whether a group produces 10 species or 1000.

If laws of growth and biased mutation, rather than selection, determine morphological trends, does selection have a role in speciation? Many authors argue that it has, but Venditti et al. (2010) concluded that speciation is not the result of natural selection—it is not a never-ending race in which species are always coping with a changing environment. Instead, the authors suggested, speciation is the outcome of events that cause reproductive isolation, such as the rise of a new mountain range.

Evolution of Gene Regulatory Networks

During the development of an organism, gene expression is controlled by complex pathways and networks of interacting regulator genes. The evolution of these genes is of critical importance for phenotypic evolution. Lynch (2007b: 804) described evidence suggesting that a considerable amount of evolution in the regulatory pathways is driven by nonadaptive processes. For example, the regulatory machinery underlying complex adaptations is capable of undergoing frequent and dramatic changes without altering the expressed phenotype. In addition, one of the most striking and puzzling aspects of many genetic pathways is their baroque, unnecessarily complex structure (Lynch, 2007c: 383, citing Wilkins, 2002, 2005). Wilkins (2007) concluded that "...although organisms often seem designed efficiently for one trait, much is clearly suboptimal and many morphological/anatomical traits are baroque in their construction, defying the simplest notions of what constitutes good design."

Long-Term Genetic Conservation

If a sequence or property appears to be conserved in a lineage over evolutionary time, it is often interpreted as having functional importance. Yet Koonin (2009a) concluded that even "this 'sacred', central tenet of evolutionary biology... is not an absolute and the nonadaptive alternative is to be taken seriously."

Evolution of the Genetic Code

Massey (2010) cited increasing evidence for the emergence of beneficial traits in biological systems in the absence of direct selection, and he explored the case of the standard genetic code. The code is structured in such a way that single nucleotide substitutions are more likely to result in changes between similar amino acids. This "error minimization" is often assumed to be an adaptation. Yet Massey (2010: 81) wrote:

> ...direct selection of the error minimization property is mechanistically difficult. In addition, it is apparent that error minimization may arise simply as a result of code expansion, this is termed the "emergence" hypothesis. The emergence of error minimization in the genetic code is likened to other biological examples, where mutational robustness arises from the innate dynamics of complex systems; these include neutral networks and a variety of subcellular networks... the term "pseudaptation" is used for such traits that are beneficial to fitness, *but are not directly selected for.* [italics added]

Again, although the structure—in this case, the code—appears to have been adapted for its eventual function, it is more likely to be an inevitable by-product of prior evolutionary trends.

Genome Reduction

Rho et al. (2009) wrote that: "The evolutionary patterning of genome architecture by nonadaptive forces is supported by population genetic theory, estimates of the relative power of the major forces of evolution, and comparative analyses of whole-genome sequences." The authors described "...a broad syndrome of genomic changes being driven by apparently nonadaptive events...," and one of the most obvious is a simple change in size of the genome. Janes et al. (2010) described "the events leading from an ancestral amniote genome – predicted to be large... to the *small and highly streamlined* genomes of birds" [italics added]. The nonadaptive streamlining in the bird genome parallels the extensive reduction and fusion seen in bird morphology, for example, in the skeleton.

Masatoshi Nei (2013) and Mutation-Driven Evolution

Although most biologists still see evolution as a process dominated by ecological forces external to the organism, the idea that internal mutation pressures can be a directional force in evolution is not a new one (Lynch, 2007c, citing Dover, 1982; Nei, 1987, 2005; Cavalier-Smith, 1997; Yampolsky and Stoltzfus, 2001; Stoltzfus, 2006b). It has also been supported by Masatoshi Nei, whose work on molecular biology and evolution helped found the field (his work has been cited more than 240,000 times).

With respect to evolution in general, Nei (2007: 12235) concluded that: "the driving force of phenotypic evolution is mutation, and natural selection is of secondary importance." Nei's views are summarized in a recent book, titled *Mutation-Driven Evolution* (Nei, 2013). In it, he developed a detailed critique of Fisher's (1930) panselectionism. He also criticized neodarwinian comparisons of natural selection to an "artist" working with the "raw material" generated by mutation. Nei suggested that these models are too simplistic, as most assume a constant selective pressure in space and time, and large amounts of preexisting genetic variation. He argued instead that what really matters in evolution is where and when a specific mutation occurs.

Darwin (1859: 466) thought, incorrectly, that mutations were caused by the direct action of the physical conditions of life. But the topic was of secondary interest for him, as he assumed that sufficient amounts of variation always exist within populations for any required adaptation. He was not too concerned about the ultimate origin of the variation. Yet this neglect of a fundamental process was unsatisfactory to many biologists, including Thomas Huxley and Francis Galton (Nei, 2013: 2). Nei wrote that "Around 1910, it was often said that 'natural selection may explain the survival of the fittest, but it cannot explain the arrival of the fittest' (de Vries, 1912: 827)." The same criticism—that selection theory overlooks the direct importance of mutation—was stressed by Morgan (1903, 1932), and it has become more frequent in recent work (e.g., Ohno, 1970; Nei, 1987, 2013; Kirschner and Gerhart, 2005; Stoltzfus, 2014a,b).

Goodwin (2009) gave an excellent summary that stressed the distinction between the *origin* of a form and the process whereby natural selection can later change the *frequency* of the form. He wrote: "natural selection cannot explain how any form originates... There is tendency among evolutionary biologists to assume that any form can be generated as a result of random variation in the genes... [this] is mistaken" (p. 299).

Nei (2013: 197) concluded that "[Darwin] overemphasized the importance of selection apparently because he did not understand how new variations are generated to begin with. This was only clarified by Morgan et al. working on mutations (1910s–1930s)." Morgan (1916, 1932) "presented a clear form of mutation-selection theory" (Nei, 2013: 7), termed "mutationism," in which mutation provides variation. This is then "pruned" by selection. "In mutationism, the driving force of evolution is mutation, and natural selection is merely a sieve..."; natural selection is a passive process that saves beneficial mutations and eliminates deleterious ones (Nei, 2013: 192). This stands in contrast with the Modern Synthesis, where "it is natural selection that has the power of creating innovative characters (e.g., Mayr, 1963: 201; Dobzhansky, 1970)" (Nei, 2013: 38). Nevertheless, as Nei noted (p. 38): "In the later years of his life Mayr (1997: 2093) changed his view and stated that natural selection is an elimination process."

Mutation-driven evolution "can explain organismal evolution in a more logical way than selection-driven evolution" (Nei, 2013: 189), and, while evolution cannot occur without mutation, it can occur without natural selection (Nei, 2013: 181). Darwin's (1859) explanation for the evolution of the eye: "would have been much simpler if he had assumed that new variations are generated by mutations" (Nei, 2013: 178).

Nei (2013: 14) agreed that Morgan's (1916, 1932) mutation-selection theory, or mutationism, has been either overlooked or denigrated in the Modern Synthesis. This research program is based instead on the work of three population geneticists (Fisher, 1930; Wright, 1931, 1932; Haldane, 1932), all of whom concluded that natural selection is much more important than the original cause of the standing variation (mutation). In their model, mutation provides the raw material, but the actual course of evolution is determined by environmental changes and natural selection.

Nei (2013) argued that "Neo-darwinism has generated an idea of panselectionism in its extreme form," in which *all* aspects of morphology, physiology, and so on are seen as the results of natural selection. Fisher (1930) was "essentially a panselectionist," while "At the present time, behavioural biologists [for example], are largely panselectionists and pay little attention to mutation, genetic drift and gene co-option" (Nei, 2013: 178). Nei concluded (p. 175): "Evolution by omnipotent natural selection is similar to creationism, in which natural selection is replaced by God."

Toward a Postmodern Synthesis in Evolution

As Lynch (2007a: 8597) observed, "because the concept of selection is easy to grasp, a reasonable understanding of comparative biology is often taken to be a license for evolutionary speculation...." Lynch referred to "the myth that all of evolution can be explained by adaptation...," and he wrote that "Dawkins' [1976, 1986, 1996/2006] agenda to spread the word on the awesome power of natural selection has been quite successful, but it has come at the expense of reference to any other mechanisms, a view that is in some ways profoundly misleading." Lynch (2007a: 8599) rejected this "religious adherence to the adaptationist paradigm," and instead proposed "the passive emergence of genome complexity by nonadaptive processes."

Fisher (1930) and Haldane (1932) argued that mutation operating in a direction contrary to selection could never overcome the force of selection. But this assumed that mutation is always a weak force relative to selection, and it ignores the complications that arise in finite populations. Geneticists now know that "Nucleotide composition is influenced considerably by biases in mutation and gene conversion, and [that] many other aspects of genomic architecture, including mobile element proliferation, arise via internal drive-like mechanisms" (Lynch, 2007c: 373).

Despite the inertia in views on evolution, some ecologists are beginning to make similar observations to those of the geneticists. Odling-Smee (2010) noted that the Modern Synthesis is an "externalist" theory, seeking to explain the internal properties of organisms in terms of their external environment alone. Odling-Smee argued that because of this underlying assumption, and others, the Modern Synthesis is "the source of conceptual barriers that are currently making further progress in some areas stubbornly difficult."

The work in genetics discussed here has serious implications for theory and practice in the evolutionary sciences. Koonin (2009b) concluded that: "The edifice of the Modern Synthesis has crumbled, apparently, beyond repair." Koonin (2009a) wrote that "genomes show very little if any sign of optimal design... In the postgenomic era, all major tenets of the Modern Synthesis have been, if not outright overturned, replaced by a new and incomparably more complex vision of the key aspects of evolution...." In the new models, structural evolution can be the result of long-term genetic trends rather than local adaptation, and so the nature of the trends becomes a major focus of attention.

Summary on Mutation-Driven Evolution: Modern Critique of Teleological Explanation in Biology

Applications of the neodarwinian model have tended to neglect trends and series in morphology, and instead have focused on particular, individual morphologies and their ecology. Because of this, individual morphologies are often interpreted as "wonders of nature," adaptations that have come into being in order to answer extrinsic "needs."

In this view, a particular adaptation has evolved, not because it represents a point on a long-term evolutionary trajectory, but because it is *necessary* for the creature, given its habit or habitat. How and why the creature already had the habit or habitat before the feature evolved is not explained. Among the many examples cited by Dawkins (2004/2005) are the following [emphasis in bold is added]:

> In *Homo*, "The **needs** of speech call forth tiny changes in the skeleton" (p. 70). (Instead, in mutation-driven evolution, tiny changes in the skeleton contributed to making speech inevitable.)
>
> In tarsiers, "The reason for their huge eyes is the same as in owls and night monkeys—tarsiers are nocturnal. They rely on moonlight, starlight and twilight, and **need** to sweep up every last photon they can" (p. 162). (This does not explain how or why tarsiers were nocturnal to begin with.)
>
> A fish can change in evolutionary time to whatever unfishy shape is **required** for its way of life (p. 340). (Thus, it had a way of life before it had the shape that was required for this.)
>
> Why do plants stay still? Maybe it has something to do with the **need** to be rooted in order to suck nutrients out of the soil... I'm not sure. (p. 519)

In another book, Dawkins (1996/2006: 14) wrote that: "Animals with similar *needs* often resemble each other...." [italics added]. He cited the case of the spider monkey's prehensile tail, which can, on its own, support the weight of the suspended body. This feature is shared with several other New World monkeys.

Dawkins wrote: "When a tree-climbing animal *needs* a fifth limb it doesn't grow it afresh but presses into service what is already there" (Dawkins, 2009: Caption to Fig. A on color page 26; italics added). In the alternative model of mutation-driven evolution, underlying trends in the evolution of prehensile tails have taken place over large, geographic scales and through geological time. For example, for mammals in general, prehensile tails are much more common in the New World and Australasia than in Africa and Southeast Asia (Organ, 2007). This sort of pattern cannot be explained by ecology and selection.

Instead of focusing on individual morphologies, with each one seen as a perfect adaptation for whatever it does, a model of mutation-driven evolution considers *series* of related morphologies, evolutionary *trajectories*. Individual morphologies of the most bizarre kind have related structures that are less bizarre, and individual structures can be interpreted as components of morphological series. In this model, inherent, long-term structural trends (Darwin's 'laws of growth'), such as tail reduction in mammals, are the primary factors determining morphological and molecular evolution, with selection pruning off forms that are not viable.

As for teleology in general, Bock (2009: 7) wrote that: "The concept of design is inappropriate in biology and should be eliminated from all biological explanations." Reiss (2009: xiii) was just as blunt: "Many evolutionary biologists to-day are in the rather peculiar position of denying design in their battle with 'intelligent design' proponents over the teaching of evolution in the schools, while at the same time they embrace a design metaphor for understanding the features of organisms… This position seems uncomfortable, if not absurd."

The best book on the history of teleology in biology (Reiss, 2009) argued that evolutionary biology can be purged of teleology, and that it would benefit from this. Reiss is a biologist, and a philosopher argued in a similar way: "natural selection, I believe, is inessential to the existence of functions" (Davies, 2001). "Functions are contributions to systemic capacities and while selection can preserve or eliminate those functions, selection is not their source" (p. xiii). "I… argue directly against the attempt to understand functions in terms of ancestral selective success… One rather central reason for rejecting [this] is its retention of the notion of design." Davies' approach "aims to dispense with the notion that natural traits are 'properly' functional or 'for the sake of' some end" (p. 8), and he suggested that the habit of conceptualizing the living world in terms of design is a habit "we ought to break" (p. 215). In conclusion, "To explain away the appearance of design without appeal to a designer [as Darwin did] is not to show that we can think of design without a designer [a difficult task]; it is rather to show that we ought to cease thinking of the natural world as designed" (p. 59). There are other, productive ways of looking at it.

Trends in Morphological Evolution: Some General Aspects

Studies in the comparative morphology of both living and fossil groups often stress "trends" (Sidor, 2001, 2003), "tendencies" (Frohlich, 2006; Douglas and Manos, 2007), and "iterative themes" (Rudall, 2003). Characters and clades demonstrate general, parallel tendencies to evolve in certain ways rather than others. These tendencies can lead to a simple increase or decrease in size, or to more complex changes. For example, "Ever since the appearance of the first land vertebrates, the skull has undergone a simplification by loss and fusion of bones in all major groups. This well-documented evolutionary trend is known as 'Williston's Law'" (Esteve-Altava et al., 2013).

Trends in morphological evolution, such as Williston's law, are often played out in lineages over long time scales—tens or even hundreds of millions of years—and involve developments that are repeated many times in many lineages. A single evolutionary trajectory, such as reduction and fusion of bones in the skull, may also, over tens of millions of years, result in hundreds or thousands of particular by-products, including new species, genera, and so on. The emphasis here, as in comparative biogeography, is on the underlying trends, not on the particular points along these.

The details of evolution in biological form are, if anything, more complex those of evolution in space and time, but while the details of morphological variation are endless, most are based on a small number of simple trends that are repeated in many groups. As Dawkins (1986: 73) noted: "The actual animals that have lived on Earth are a tiny subset of the theoretical animals that could exist. These real animals are the products of a very small number of evolutionary trajectories through genetic space."

Darwin's insightful later work (e.g., Darwin, 1872, 1874) stressed the primary importance of these trends (his "laws of growth"), rather than morphological "centers of origin" (homogeneous ancestors) and natural selection. Likewise, after the 50-year detour of the Modern Synthesis, current authors are returning to the concept of "long-term genetic evolutionary trends" that are not explicable by random mutation and natural selection (Livnat, 2013).

Trends and Ecology

In nonteleological, mutation-driven evolution, structure determines function. Morphological trends produce series of particular morphologies and these determine the ecological space that is viable for the species. If suitable ecology is available in the biogeographic region, the organism will survive. For example, the trend toward reduction and fusion in the vertebrate skull means that reduction in the jaw, including loss of teeth, is common. Members of the genus *Dasypeltis* (Colubridae) are among the very few snakes that are specialist egg-eaters. During the course of evolution, they have lost most of their teeth. "Consequently," as Adriaens and Herrel (2009) reasoned, "they cannot capture and transport other prey types and are stuck in an ecological and evolutionary dead end in being obligate egg eaters." They did not lose their teeth in order to become egg-eaters—they lost their teeth, and so they have no choice in their diet.

Trends and Parallelism

The significance of parallelism has been controversial. The early geneticists—the mutationists—interpreted parallel evolution as the result of nonrandom tendencies in variation. In contrast, the Modern Synthesis assumed that minor details of structure were determined by chance, and that complex structures were unlikely to evolve more than once without natural selection. Nevertheless, morphologists have always known that parallelisms are widespread in plant and animal groups, and molecular studies indicate that they are even more common than was thought; even the most complex morphologies can evolve many times. (Note that if a structure develops by parallel evolution in *all* the members of a group, and only in that group, it will appear to have evolved just once.)

A typical example of parallelism is seen in sponges, where: "despite their complexity, chelae morphology can evolve independently in different poecilosclerid lineages…" (Vargas et al., 2013). In insects, all the defining characteristics of the group have evolved more than once in different parts of the animal kingdom (Conway Morris, 2003). In the plant family, Brassicaceae, molecular analyses indicate that almost every character that was used as the basis for classical taxonomy exhibits substantial homoplasy (Franzke et al., 2011). For example, with respect to their architecture, most Brassicaceae are herbaceous, but shrubs occur in all major lineages and at least 12 tribes.

Parallelisms are often localized in particular areas, and in Brassicaceae, woody lianes (these are 1–9 m tall) have the following distribution (Franzke et al., 2011):

South America (three species of *Polypsecadium*: tribe Thelypodieae, and three species of *Cremolobus*: tribe Cremolobeae).
South Africa (*Heliophila scandens*: tribe Heliophileae).
Southwestern Australia (*Lepidium scandens*: tribe Lepidieae).

Here, the parallelism occurs in different tribes but is restricted to a specific, southern hemisphere range. Distinctive structure that is shared among groups that are located in a single region but are not immediate relatives is often assumed to indicate adaptation to the regional environment. Instead, it may reflect ancestral polymorphism or ancestral morphogenetic trends. For example, many groups that are herbs, shrubs, or lianes elsewhere, produce trees in the Pacific (and the Chatham Islands, Chapter 6). This can be explained either by secondary convergence, or by ancestral polymorphism and a prior concentration of "tree genes" in Pacific angiosperms, before the modern clades existed. In another case of localized parallelism, the floral mutation termed peloria is recorded in the New Zealand flora in *Mazus arenarius* and *Hebe perbella*. Both plants inhabit the Murihiku terrane or rocks deposited on it (Chapter 7).

Trends Seen in the Fossil Record, and "Orthoselection"

Paleontologists have recognized many morphological trends in the evolution of fossil groups over long periods of time. Until Matthew's (1915) work and the Modern Synthesis, these trends were attributed to internal, genetic causes, or "laws of growth." Instead, the Modern Synthesis attributed them to directional natural selection operating over long time scales, a process termed "orthoselection" (Simpson, 1944). Most modern paleontologists have followed Simpson in using orthoselection, not laws of growth, to explain trends (e.g., Conway Morris, 2003, although he did not refer to these terms explicitly).

Long-term morphological trends, such as reduction and fusion in the vertebrate skull, can take tens or even hundreds of millions of years to unfold. It is difficult to imagine how trends that last so long and are displayed in thousands of species, from all conceivable ecologies, could be the result of natural selection. It is more probable that they are the result of internal factors and reflect aspects of genome architecture.

Trends and Extinction: Evolutionary Constraint and Maladaptation

Perhaps 99% of all species that have ever existed have gone extinct (Dawkins, 2004/2005: 255), and so extinction is a fundamental aspect of biology that must be explained. Yet most accounts of evolution include very little on the topic. For example, one large book reviewing evolution (Ruse and Travis, 2009) is 979 pages long, but has just 11 pages on extinction: five on mass extinctions (pp. 715–720), four on extinction by bolide impact (pp. 411–412, 716–718), and two on Cuvier's views of extinction (pp. 501–502).

Trends in evolution can lead to new morphology that has enhanced fitness (adaptation), or they can lead to maladapted forms and even extinction. Many trends will be neutral in terms of their adaptive value, at least to begin with. Some trends are maladaptive from the outset, and all trends, if carried far enough, can end up being pathological. Dercole and Rinaldi (2008: 23) wrote that "The debate on the possible causes of extinction remains wide open." They considered the possibility of "evolution toward extinction in a constant abiotic environment," with this leading to "evolutionary runaway" and eventual "evolutionary suicide." An obvious mechanism for this would be a mutation-driven process.

Futuyma (2010) examined the concept of evolutionary constraint—a type of trend—and its consequences for ecology and extinction. Evolutionary stasis is observed in many fossil lineages, and although some groups can persist for long periods with little if any change, many groups that exhibit reduced levels of change go extinct. Futuyma stressed the observed frequency of the *failure* of adaptation, along with *constraints* on variation, and he regarded this as a paradox for the Modern Synthesis.

Futuyma (2010) gave the same quote from Stebbins (1971: 30) that was cited earlier as an example of the orthodox view: "Mutations are rarely if ever the direct source of variation upon which evolutionary change is based. Instead, they replenish the supply of variation in the gene pool." Many evolutionary geneticists have agreed with this and have assumed that adaptation is *not limited* by genetic variation. "But in this light," Futuyma observed, "myriad observations, of many kinds, appear utterly paradoxical, for they suggest that there are *substantial limits* on the adaptability – the evolvability… of populations and species" [italics added]. Futuyma (2010) wrote that the most extreme evidence of failure of adaptation is extinction—if the variation in the gene pool is so abundant, why does any group ever die out? If a group is threatened with extinction, why does it not simply evolve into a more adapted form? This paradox explains why neodarwinism, which rejects genetic constraints and endogenous morphological trends, so often neglects the general causes of extinction.

With respect to distribution, Futuyma (2010) argued that limits on the geographic and habitat ranges of species "remain an unresolved problem"—why do groups not just adapt and invade new territory? Futuyma cited A.D. Bradshaw's studies on rapid adaptation in some plants to metal-contaminated soils on old mine tailings. This work describes a classic example of evolution by natural selection and is often cited. Yet in a discussion of "genostasis and the limits to evolution," Bradshaw (1991) emphasized that only a minority of species in the original communities evolved high levels of tolerance. Bradshaw also discussed another example; he asked how a species of tree could be abundant near the edge of a marginal habitat such as a salt marsh, raining millions of seeds into it over the course of centuries, yet *fail* to adapt. Bradshaw (1991) argued that such "evolutionary failure is commonplace," and that "there are limits to evolution."

Futuyma (2010) noted that as a consequence of adaptation failure and limits in range, there are many lacunae in the economy of nature, "empty niches" implied by geographic and temporal imbalances in the distribution of clades. He wrote: "The 'unbalanced' biota of islands – even of large, ancient islands such as New Zealand (Gibbs, 2006) – is only an extreme case of a more general condition. Is there really no ecological space in the Atlantic for sea snakes, which are limited to the Indo-Pacific region?" He also cited a terrestrial example: the canopy of wet forests in the Old World is thick with epiphytic ferns and orchids, but only in the New World is it festooned with water-holding Bromeliaceae; these provide a habitat for many animals.

The frequent failure of adaptation is a paradox for the Modern Synthesis and for its proposal that the course of evolution is determined by selection. Futuyma (2010) wrote that "Although the traditional theory of population and quantitative genetics mostly concerned evolution based on standing variation and largely ignored evolution based on de novo mutation (Stoltzfus, 2006a), recent theory has increasingly explored the role of mutation" (e.g., Orr and Unckless, 2008).

Futuyma (2010) concluded that biogeographic and ecological patterning is not due to environment and selection alone, but also to evolutionary constraints—phylogenetic and historical factors operating at regional and global scales. Communities located in similar environments but having different evolutionary histories "have far less convergent structure than ecologists once optimistically expected" (Futuyma, 2010). In this approach, "Historical biogeography, by tracing the origin and spread (and vicariance) of lineages and thus of their membership in regional biotas, regains its position as a key contributor to community ecology."

Methods of Inferring Ancestral Morphology

Examples of morphological evolution are discussed next, but some analytical methods are discussed here first, as the topic can be confusing. For example, one study suggested that "the characteristics of the common ancestor of Ecdysozoa [arthropods and related taxa] are of particular interest and it can be safely assumed to have possessed the synapomorphies of the group" (Telford et al., 2008). In fact, an entity that has a group's synapomorphies already belongs to the group itself, not its ancestor.

In addition, "basal" clades are just small sister groups and are not necessarily any more primitive than their diverse sisters (Chapter 1). Adams (2013) suggested that *Amborella*, the "basal" angiosperm "represents the equivalent of the duck-billed platypus in mammals (in the earliest branch of the mammalian family tree)." But at the basal node in a group there are *two* "earliest branches"; in angiosperms, these are *Amborella* and all the others; in mammals, they are monotremes and all the others. In a phylogeny, there are no basal groups, only basal nodes. In many of their features, "basal" forms such as amphioxus, hagfish, fishes (basal in the vertebrates), and *Amborella* (basal in the angiosperms) are just as reduced and derived as members of their sister groups, if not more so.

If one group in area A and its sister in area B differentiated by vicariance, the ancestral area will have been something like A + B. Likewise, in ecology, a group with marine and freshwater species can have occupied both habitats since its origin. In the same way, in morphology, if a feature in the two groups has character states *a* and *b*, the ancestral state will often be neither *a* nor *b*, but a third character, *c*, in which both *a* and *b* are synthesized. Many "ancestral state reconstructions" map extant characters, say *a* and *b*, on a phylogeny and infer that one or more of these *extant* characters is primitive—that either *a* evolved into *b*, or *b* evolved into *a*. Neither of these may be correct.

As discussed already, paraphyletic basal grades in an *area a* are thought to occupy a group's center of origin, and paraphyletic basal grades with an *ecology a* are thought to occupy the original habitat type (Chapter 1). Likewise, if groups A and B with morphological state *a* and *b* have a phylogeny: A (A (A, B)), optimization models infer that *a* evolved into *b*. Instead, both *a* and *b* are likely to have evolved from a common ancestral character *c*. For example, a group that comprises some species with white tails and some with black could be derived from an ancestor with black and white striped tails, tails that were white in summer and black in winter, or grey tails. Another possibility is that the ancestor was polymorphic, with both black tailed and white tailed forms already present *before* the ancestor started to diverge into the modern species. Black tails and white tails would then represent old characters preserved in a young clade. This concept of ancestral polymorphism is well-known to geneticists (as incomplete lineage sorting), but it has been neglected by morphologists.

Finally, if three groups with morphologies *a*, *b*, and *c*, have a phylogeny: *a* (*b* (*c*)), this does not indicate that morphology *a* evolved first, then *b* and *c*. Interpreting the sequence of morphological evolution from a phylogeny is not always straightforward. Ancestral character state reconstruction requires morphological analysis of the characters themselves (including those of fossil forms), as well as phylogenetic analysis of the clades that bear them. In the same way, ancestral area reconstructions (Figures 4.12, 10.4, 11.8, 12.3) require consideration of regional tectonic processes, not just phylogeny. For example, in a group present on two volcanic islands at a subduction zone, the ancestral area probably included prior islands in the region, rather than the current islands. This principle also applies to ancestral habitat analysis; two sister groups, each present in a different habitat, need not have originated in either one.

One thing molecular phylogenies do imply is that many morphological characters that were thought to be synapomorphies are, in fact, parallelisms. Molecular evidence indicates that morphological parallelism is even more widespread than was accepted; many characters have evolved many times as the result of widespread trends and tendencies. Just because a character is found throughout, or almost throughout, a group does not mean it has evolved just once in the group; the same pattern would result if the character evolved many times, in all the main branches of the group.

To summarize: modern procedures for inferring morphological evolution often rely on simplistic algorithms that assume that the ancestor was monomorphic, and that the morphology of a basal paraphyletic grade is primitive (e.g., Avise, 2006). This approach is derived from the old ideas that "common is primitive" and "rare is derived," and it resembles the idea that the location of a basal, paraphyletic grade is a center of origin.

In an example from plants, "epiphyllous" flowers growing from leaves are very rare; they are found in just a handful of unrelated species among the ~300,000 angiosperms (Dickinson, 1978). A consideration of phylogeny alone would suggest that epiphyllous flowers have been derived by parallelism. Yet, a broader consideration of the origin of the angiosperm leaf itself suggests that epiphylly is an unreduced condition, a relic of prior structure. This would date back to before the origin of the modern leaf by the fusion of a branching, fertile component with a simple scaliform component, along with reduction and sterilization.

An example from animal life concerns reconstructions of ancestral eagles (Accipitridae), and the question as to how large they were. Haast's eagle, named as *Harpagornis moorei* but nested in *Hieraaetus*, is an extinct eagle of New Zealand that weighed 10–15 kg and preyed on moas. It is one of the largest flying birds known. In contrast, its closest extant relatives, *Hieraaetus morphnoides* of Australia and New Guinea, and *H. pennatus* of Africa and Eurasia, are among the smallest eagles (Bunce et al., 2005). Other relatives of *H. moorei* are also smaller species, and therefore, Avise (2006: 48) reasoned, "it now seems probable that *Harpagornis moorei* evolved from much smaller eagles." This may or may not be correct; the large size could instead represent an old, relict character preserved in a much younger species (Chapter 15). Eagles and their ancestors have probably always included large and small forms. In the same way, just because the New Zealand eagle has relatives forming a paraphyletic basal grade in Africa, Eurasia, and Australia does not necessarily mean it dispersed from any of these. Instead, the allopatry of the three species is consistent with an ancestor that was already widespread through the Old World before the origin of *H. moorei*.

Some Examples of Trends in Morphology

A small number of morphological trends can generate large numbers of particular morphologies. For example, reduction and telescoping in features such as the vertebrate skull and the angiosperm flower, with differential suppression and fusion of different sectors and layers, has resulted in hundreds of thousands of particular morphologies. Of course, reduction and fusion are not the only trends, but they are among the most important and are discussed here as illustrative examples.

Example of a Trend: Reduction

Morphological "reduction series" can lead not only to reduction in size (a very fluid aspect of morphology) but also to a reduction in complexity. The process often involves the fusion and partial suppression of parts. In a well-known example, the nervous systems of animals were at first diffuse and then evolved to be centralized, with a concentration at the anterior end of the animal (Miller, 2009).

Consider a similar, hypothetical series of stages in the concentration and reduction of a plant organ:

1. A complex, branched, fertile (gamete- or spore-bearing) structure.
2. A complex, branched structure that is sterile.
3. A simpler, less branched, spiniferous structure.
4. A glandular emergence producing a secretion (such as a physiologically active hormone, a scent, a nutritive substance, or a poison).
5. A simple, nonglandular emergence.
6. A colored patch formed from a single layer of pigmented cells.

A teleological approach to this series would concentrate on the adaptations of each individual stage and what each individual structure does (what it is "good for"). The functions of the six stages might be, for example, (1) reproduction; (2–3) defense; (4) regulation of internal physiology, nutrition of offspring, or defense; (5) nothing; and (6) attracting mates or pollinators. The structure of each stage is explained by these functions.

This static, adaptationist approach focuses on the particular morphologies and is important for description of current ecology. Yet, it fails as a method for interpreting evolutionary morphogenesis as it misses the most important fact, namely, that the series follows a single trajectory of reduction. This simple trend could be under the control of a simple genetic variation that has developed many times over tens of millions of years.

Example of a Trend: The Origin of Curved Structure by Fusion

On a plant shoot or ancestral, preshoot structure that is being reduced and telescoped in phylogenetic time, the series of lateral organs will be condensed. At a certain point in the trend, primordia of the lateral organs may fail to separate and the organs are referred to, somewhat misleadingly, as "fused." This sort of fusion can have many consequences in both plants and animals. For example, if two rod-shaped organs that grow at different rates become fused, the result is a curved structure, as in a bimetallic strip that is heated (Figure 13.1; Fermond, 1858). If the growth continues, hooked or even helical structure

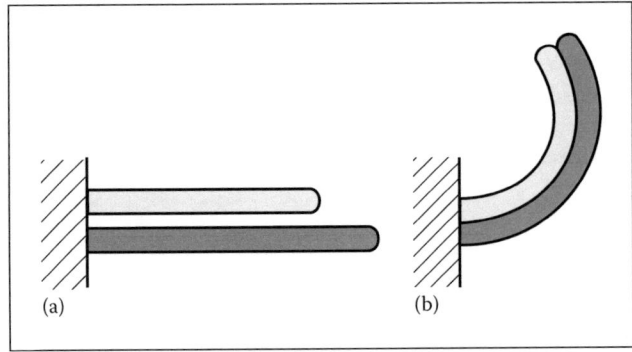

FIGURE 13.1   (a) Two rod-shaped organs of differing lengths, borne on a main axis. (b) Following condensation in the main axis, the two organs, of the same length as in (a), have fused at the primordial stage and this has resulted in a compound, curved structure.

will be produced. In the same way, a complex, branched structure can fuse with a simple scale that subtends it and give rise to a laminar organ with complex venation. Each of these new structures will have implications for the ecology of the group and will appear to be a perfect adaptation and design for whatever it does. For example, in a plant, hooks are bound to catch on other plants in the surrounding vegetation and so bring about a climbing habit. A flat, photosynthetic structure supplied by a complex system of venation will enable or enhance growth in different habitats.

In the eighteenth-century system of morphology that is still taught in most universities, the plant is composed of three, irreducible elements: leaf, stem, and root. Nevertheless, morphologists over the years have realized that this does not explain many critical aspects of plant form (review by Cusset, 1982). An alternative model has been proposed in which "root," "stem," and "leaf" are not basic entities, but are each compounds. A shoot is composed not of a stem plus its leaves, but of radial sectors or modules, each comprising a leaf, its axillary bud, and an associated strip of cortex and vasculature. These modules form the stem, which is a compound, fused structure. If a particular sector or strip in the stem grows at a different rate from the others, a curve or spiral will result. The curvature in the stem has not developed in the modern structure—a composite—*after* it formed, but *as* it formed by the fusion and reduction of its components. Again, this simple phylogenetic-morphogenetic process can generate large numbers of particular morphologies. Fusion of organs with differential growth can produce curvature, and in addition, if the curving structure is itself fused to another structure curving in the opposite direction, tension will be set up. This will be released suddenly when the cohesive forces are overcome, as in the "explosive" fruits of many legumes (Fabaceae).

Another legume, *Canavalia*, is a tropical group represented in New Zealand on beaches in the Kermadec Islands. In this genus, the pedicel twists and the flowers become resupinate. This character is unusual in the family. The usual assumption is that uncommon characters are derived from common characters in a group, and so the *twisting* pedicel would be derived from a *nontwisting* pedicel. Instead, both conditions can be derived from a "pre-pedicel" structure or structures. The main components in these became fused and reduced in different ways during the origin of the modern pedicel. These alternative pathways have resulted in a twisting structure in *Canavalia*, and an "even," nontwisting structure in other legumes. The twisting in the pedicel goes back to before the origin of the pedicel as such, to its components, and to their differential reduction and fusion in the evolution of the modern flower. If they do not fuse evenly, or if the different radial sectors of a shoot grow at different rates, the pedicel twists.

Daniel Dennett is a philosopher of science and an ultra-darwinian panadaptationist. Nevertheless, he wrote that: "A good adaptationist should not just rest content with a plausible story... At the very least, an effort should be made to consider, and rule out, alternative hypotheses" (Dennett, 1996: 309). Dennett (1996) began and ended his book with a biological question contained in the lyrics of a favorite song: "Tell me why the ivy twines..." (pp. 17, 520). Yet Dennett overlooked Darwin's (1872, 1874) concept of laws of growth, and did not cite any work on the particular trends that determine the morphology of twining plants (e.g., Darwin, 1875). On the other hand, he did cite Thompson's (1917) "laws of form" (p. 220), and he admitted that "there may be more hidden constraints than [neodarwinian] theorists often assume" (p. 228). If there are *any* constraints though, selection is not a "universal acid" (Dennett, 1996: 61), able to sculpt any kind of organism at all, at any time.

Example of a Trend: Paedomorphosis/Neoteny

Changes in structure can be attained by changes in the sequence of morphological stages in the life cycle (heterochrony), with some stages being suppressed or combined. For example, the human skull is more similar to the skull of a juvenile chimpanzee than to that of an adult. The adult chimp stages are suppressed in human skulls, which are interpreted as having juvenile morphology carried into the adult (paedomorphosis or neoteny). In the same way, the flower is often regarded as a neotenous organ (Takhtajan, 1976, 2009). In a New Zealand example, Laurent et al. (1998) suggested that the flower of *Oreostylidium* (now placed in *Stylidium*) is paedomorphic and "became fertile at a morphologically immature or reduced stage."

Noble (1931: 104) pointed out that one of the trends of evolution in amphibia is the reduction in the number of skeletal elements and the increase in cartilage, and he described the tendency as "progressive fetalization." The New Zealand frog *Leiopelma archeyi* has the skeletal characteristics of a juvenile *L. hamiltoni*, suggesting that *L. archeyi* is a neotenic form of *L. hamiltoni* (Stephenson, 1960; Bull and Whitaker, 1975: 232). Neoteny has probably also played a part in the evolution of *L. hochstetteri*.

In birds, at the basal phylogenetic node in the group there is a major morphological distinction between the paleognathous palate of the ratites and the neognathous palate of all other extant birds. The paleognathous condition has been interpreted as a result of neoteny (Kaley, 2007), and many other examples of neoteny in ratites, parrots, pigeons, and Rallidae have also been proposed (Chapter 15).

Trends in the Evolution of Symmetry: Plant Shoots and Vertebrate Limbs

Reduction series and other morphological trends that develop in organisms over phylogenetic time often lead to changes in basic symmetry. As complex, multipartite systems with high orders of symmetry are reduced, and the parts fuse and coagulate, the symmetry changes. This takes place in an orderly way, with the process following certain trends. The evolution of symmetry is best seen in plants, as most animals display the same sort of simple, bilateral symmetry. In contrast, plant shoots show a great range of variation in their symmetry. Shoots may be complex, spiral systems with many, crowded parts, and high orders of symmetry; this is a characteristic of many succulent plants and cushion plants. At the other extreme are shoots with simple, bilateral symmetry and distichous leaf arrangement, as in the flat branch complexes of New Zealand *Nothofagus* species.

The angiosperm shoot has been reduced from a complex system with many parts and high orders of symmetry, to a structure with much simpler symmetry. For example, parallel reductions in the ancestral shoot system have led to many unrelated species having leaves arranged in cycles of five, with the sixth leaf standing above the first. The pattern was first described by Leonardo da Vinci (MacCurdy, 1954: 286), and it can be observed in any terrestrial vegetation, whether tropical, temperate or arctic/alpine; "The overall evidence is that phyllotactic pattern is selectively neutral in plants" (Goodwin, 2009: 306). As with leaves that are borne on a stem in cycles of five, many flowers (most eudicots) have five petals, and this can be seen in all types of terrestrial vegetation. The fivefold or pentaradial symmetry seen in so many shoots and flowers is one of the last, simplest members of a reduction series.

As in the plant shoot, some of the main trends in the evolution of the vertebrate limb are reductions and fusions. In pinnipeds, cetaceans, and sirenians, these trends have continued to the point where the animals can no longer function on land, and they are forced to live in water. The reductions in vertebrate limbs have led to the distal parts—the hands and feet—having five parts or fewer (but never more), and more or less flat, bilateral structure.

In the usual pentaradial pattern seen in many plant shoots and flowers, the ontogenetic series of leaves or petals takes *two* turns around the stem to produce a cycle of *five* leaves or petals, with the sixth standing above the first. This phyllotactic mode is referred to as 2/5. The most common spiral symmetries are 1/2 (a distichous or zig-zag arrangement, with two leaves in one turn), 1/3, 2/5, 3/8, 5/13. This series forms the "primary chain" of phyllotactic modes (Hirmer, 1922). The last three modes are all seen in at least one New Zealand group, the ericoid subshrubs and cushion plants of *Kelleria* (Thymelaeaceae) (Heads, 1990b).

A 1/2 phyllotaxis or distichy, with leaves in two rows, occurs in New Zealand plants such as *Phormium*, *Oreobolus pectinatus*, *Nothofagus*, *Carmichaelia*, *Clianthus*, *Carpodetus* (lateral branches only), *Paratrophis banksii*, and *Griselinia*. This type of phyllotaxis is associated with flat, plagiotropic branch complexes.

A 1/3 phyllotaxis (tristichy) occurs in most Cyperaceae, also *Freycinetia*, and *Astelia*.

A 2/5 phyllotaxis is seen in countless plants of the temperate zone (such as the oak *Quercus robur*) and the tropics (such as the coconut palm). In New Zealand, the 2/5 arrangement is best seen in the naturalized broom *Cytisus scoparius*, as the decurrent leaf bases connect a leaf with another, five leaves below.

Modes of spiral phyllotaxis in New Zealand plants are not well known; information is only available for the following 42 taxa (those listed without citations are based on my personal observations). The sample is probably more or less random and indicates the frequency of the 2/5 mode.

2/5: *Prumnopitys taxifolia, Alectryon excelsus, Alseuosmia* (all species; Gardner, 1976), *Brachyglottis compacta, B. repanda, Dracophyllum* sp. aff. *longifolium, Epilobium* sp., *Gaultheria crassa, Haastia pulvinaris* (Low, 1900), *Hymenanthera novae-zelandiae, H. obovata, H. chathamica, Kelleria dieffenbachii, K. laxa* (Heads, 1990b), *Knightia excelsa, Melicytus ramiflorus, M. micranthus, Mida salicifolia* (Gardner, 1997b), *Myoporum laetum, Olearia ilicifolia* (Haase, 1986), *O. albida, O. capillaris, O. coriacea, Pennantia* (all species; Gardner and de Lange, 2002), *Pittosporum eugenioides, P. dallii, P. tenuifolium, P. ralphii, Planchonella costata, Raoulia lutescens* (Beauverd, 1910), *Sophora microphylla, S. tetraptera.*

2/5 and 3/8: *Fuchsia excorticata, Phebalium nudum, Pseudopanax crassifolius.*

3/8: *Entelea arborescens, Euphorbia glauca, Donatia novae-zelandiae, Phyllachne rubra, P. colensoi, Forstera sedifolia* (the last four from Rapson, 1953).

5/13: *Forstera* sp., *Hectorella caespitosa* (Skipworth, 1962), *Kelleria croizatii, K. childii* (Heads, 1990b), *Meryta sinclairii.*

Although fivefold symmetry is a basic aspect of plant structure, it also occurs in animals and animal organs (Breder, 1955). For example, it is seen in many diatoms and radiolarians, and in most echinoderms (Dawkins, 1996/2006: 213–214). In vertebrates, all extant tetrapods and all but a few fossil tetrapods have limbs with five digits or fewer. Among the earliest known tetrapods, fossil forms from the Devonian have limbs with eight (*Acanthostega*), seven (*Ichthyostega*), and six (*Tulerpeton*) digits (Coates and Clark, 1990). *Casineria*, from the Carboniferous, already had five digits and was the first, fully terrestrial tetrapod (Paton et al., 1999).

The reduction series from high modes down to pentaradial arrangements continues in many organisms until it reaches the minimum possible symmetries, trilateral (triangular in section), and bilateral (often more or less flat). These minimal symmetries occur in highly reduced, fused structures, such as leaves and hands. Nevertheless, a trace of the ancestral, higher orders of symmetry often remains, even in bilateral forms.

In the teleological model of evolution, it is assumed that the different patterns of leaf arrangement (phyllotaxis) have developed because of the advantages they bestow on the plants. Yet, experimental work has demonstrated that different phyllotactic arrangements, with leaf divergence angles of 100°–154°, give light interception efficiencies that are indistinguishable (Valladares and Brites, 2004; Goodwin, 2009). The list just given shows that the crowded, 5/13 mode occurs in alpine cushion plants such as *Hectorella* (Montiaceae), as well as megaphyllous, subtropical trees such as *Meryta* (Araliaceae) that live in a very different environment. Rose (1997: 243) discussed the regular spirals in pine cones, sunflowers and others, and wrote that: "Even if one could find an ingenious just-so story to account for the pattern, the sensible conclusion is that the adaptation is built around the structural constraint, and not vice versa." The structural constraints, or laws of symmetry, apply to both plants and animals, as in the 5-partite structure that is dominant in angiosperm flowers and phyllotaxis, as well as in the hands and feet of vertebrates.

14

Case Studies of a Trend: Morphological Reduction and Fusion Series

Some of the most important trends in morphological evolution are reduction, suppression, and fusion of parts. This chapter describes some examples in which these processes account for particular aspects of form in plants and animals.

Case Studies of Reduction Series in Plants

Suppression of Runners, Rhizomes, and Roots

Horizontal (plagiotropic) systems of shoot and root axes occur in many plants. The presence of plagiotropic rhizome systems in some clades, and their suppression in related groups, was discussed earlier with reference to *Carmichaelia* (Chapter 11) and *Epilobium* (Chapter 10). This variation in plagiotropic shoot axes is also an important morphological cline dominating the whole-plant architecture in many other New Zealand groups. The genus *Kelleria* (Thymelaeaceae), for example, displays the sequence from semiaquatic plants with prostrate, rooting runners (horizontal, leafy shoots growing along the soil surface) (*K. paludosa*), through to mats and cushion plants with runners that are less prolific and less prolifically rooting (*K. dieffenbachii, K. childii*), through to erect, ericoid shrubs in which the runners are suppressed and the plant has a short trunk (*K. ericoides*) (Heads, 1990b: Fig. 5).

The conifer family Podocarpaceae (sister to Araucariaceae) illustrates part of the same trend seen in *Kelleria*. Most podocarps are trees with large, single trunks. However, *Lepidothamnus laxifolius* of New Zealand is quite different. It is a prostrate subshrub and only 10 cm tall, with sub-lianoid stems up to 1 m long but not rooting. (This distinctive plant occurs in lowland to alpine vegetation through the main islands north to the Egmont–East Cape line.)

Another variation in podocarp architecture is seen in the New Zealand tree *Lagarostrobos colensoi* (silver pine), which is distinguished from its sister, the taller Tasmanian tree *L. franklinii*, by spreading, horizontal, underground stems. These are best developed in swampy sites, and they give rise to prolific, leafy, sucker shoots (Molloy, 1995; Dawson and Lucas, 2011: 79). *L. colensoi* (sometimes treated as a monotypic genus, *Manoao*) is a characteristic of swamp margins, where it acts as a freshwater mangrove and forms a transition to dryland forest with other podocarps. In waterlogged soils, spongy aerenchyma tissue is always present in both the roots and the underground stems, and this allows the passage of oxygen.

Molloy (1995) wrote that the underground stems of *Lagarostrobus colensoi* are unusual in conifers and they are present in just one other New Zealand species, *Phyllocladus alpinus* (Podocarpaceae). Rhizomes also occur in *Podocarpus micropedunculatus* of Borneo, found from sea level to 500 m on raised beaches, along the margins of clearings, in peat-swamp forest, and in *Agathis* (Araucariaceae) forest (de Laubenfels, 1988). The series is similar to that occupied by *L. colensoi*.

Most species of subalpine shrubland do not produce rhizomes in the strict sense, but they can produce adventitious roots on branches lying along the ground. This is a natural form of layering. Examples include *Kelleria dieffenbachii, Gaultheria rupestris, Parahebe macrantha, Coprosma serrulata*, and *Podocarpus nivalis* (Burrows, 1963). In this process, the plant develops in the form of a layer, spreading out horizontally rather than growing vertically, and the growth of a single trunk is suppressed. In contrast, a tree develops a single trunk, and all plagiotropic, basal shoots are suppressed.

A variant is seen in the prostrate, alpine subshrub *Hymenanthera alpina* (Violaceae), in which the shoot complex is formed from two systems (Arnold, 1959). One is plagiotropic and comprises horizontal, underground stems that are leafless and bear adventitious roots. The other is formed from orthotropic, abortive shoots that form blunt, suberized thorns and are similar to the pneumatophores ("roots") of mangroves.

Plagiotropic systems of root axes are well known, and lack of suppression of adventitious roots is documented even in the large forest tree, *Podocarpus totara*. Large totara trees will be submerged by extreme volumes of alluvial deposition (Chapter 9), but they can sometimes survive burial by quite deep deposits of river silt. This is because the trees send out adventitious roots near the new ground level (Foweraker, 1929, as discussed in Burrows, 1963). Foweraker's photographs are said to illustrate trees with as many as three series of roots girdling the stem, each separated by several feet. Christensen (1923) showed that a number of other New Zealand trees and shrubs survived burial by river shingle in the same way. This lack of suppression of root growth is important in enabling the plants to inhabit the flood plains of rivers.

In a similar pattern to that of the adventitious roots, some New Zealand trees and larger shrubs produce adventitious shoots from root systems which lie near the ground surface. Again, this results in a plagiotropic system. Burrows (1963) recorded it in *Halocarpus bidwillii*, *Lagarostrobos colensoi*, *Hoheria glabrata*, *Griselinia littoralis*, and *Olearia ilicifolia*. In the New Zealand mangrove, *Avicennia marina*, the root apex is suppressed in the embryo, and the entire root system is adventitious (Baylis, 1950). It develops from the hypocotyl (a seedling structure lying between the root and the shoot) and later from branches lying in the mud.

"Inflorescence Plants" with Divaricate, Ericoid, and Broom Architecture: Reduction by Suppression and Sterilization

Plants with architecture described as divaricate, ericoid, and broom-like are abundant in New Zealand vegetation. They not only have much smaller foliage than their relatives but they also have distinctive architecture and are not just "normal" plants with reduced leaves. The explanation for their evolution and diversity in New Zealand has been controversial, and Dawson (1988) devoted Chapter 6 of his book to the "small-leaved shrub problem" in New Zealand vegetation.

Two simple morphogenetic changes will produce the small-leaved growth forms: more or less complete suppression of the vegetative shoot (leaving just an inflorescence) and suppression of most of the flowers in the inflorescence (leaving just bracts) (Heads, 1994e). This double reduction will result in a plant that is equivalent to a more or less sterilized inflorescence, with the foliage made up of bracts. The process is a simple one, but it accounts for several different aspects of morphology. These "inflorescence plants" or "bract plants" dominate the understory of many New Zealand forests, as well as open sites in the lowlands and many subalpine and alpine communities.

Architecture of Divaricate Shrubs

Divaricate shrubs and the different theories proposed for the architecture are discussed elsewhere (Heads, 2014: 79). The architecture of divaricate shrubs is diverse, as seen in the different modes of orientation in the shoot axes. The shoot axes can be all plagiotropic (more or less horizontal, e.g., *Coprosma acerosa*), all orthotropic (more or less vertical, e.g., *Pittosporum* species), orthotropic and then plagiotropic by primary growth (e.g., many *Coprosma* species), plagiotropic then orthotropic by primary growth (e.g., *Paratrophis microphylla*), or plagiotropic by primary growth with basal sections becoming orthotropic by secondary growth (e.g., *Sophora microphylla*) (personal observations).

Despite this variation, the plants described as "divaricate" by New Zealand botanists all differ from related tree species in their shoot systems. Divaricates all have two types of shoots: short shoots (brachyblasts or spur shoots) with much reduced internodes, and long shoots in which the apices abort (Heads, 1998b). These features are often overlooked, but they were described by Philipson (1978). (McGlone et al., 2004, commented that the long shoots in divaricates have indefinite growth, but this is not correct.)

The divaricate shrub can be interpreted as a plant in which the main shoot system has been suppressed, leaving a relictual, basal inflorescence that has been largely sterilized. This model is illustrated here with a diagram of tree and divaricate-shrub members of *Coprosma* (Figure 14.1). The model accounts for many morphological features in the divaricate, "inflorescence plant." These include the determinate shoot apices (as in an inflorescence), the cymose shoot architecture (the same as that of cymose inflorescences), and the fact that the flowers appear to be borne singly, not in branched inflorescences as in the tree species.

In addition to the long-shoot–short-shoot dimorphism, other features are common in divaricates, but they do not occur in all species. For example, in most divaricate plants, but not all, the branches are borne at high angles to the main axis (as in many inflorescences), and they are described as spreading or divaricate. Many divaricates have thin stems and are leptocaul ("filiramulate" is a new term for this), but some are pachycaul (e.g., *Hymenanthera alpina*).

In "normal" plants, the overall shoot architecture is non-divaricate, but the branching in the *inflorescence* is often divaricate. This is seen in *Ackama rosifolia* (Cunoniaceae) and some forms of nikau palm (*Rhopalostylis sapida*: Arecaceae), for example. If the shoot system of these plants underwent complete suppression, leaving only the inflorescence and some roots, and if many of the inflorescence bracts were sterilized, leaving flowers only near the ends of the branches, the result would be a divaricate plant.

One example of this "reduction" is *Chamaesyce*, one of the four main clades in the giant genus *Euphorbia* (Euphorbiaceae), and a diverse, more or less cosmopolitan group in its own right. The branching architecture of the whole plant shows close resemblances to that of a cymose inflorescence (Yang and Berry, 2007), and this can be explained if the whole plant *is* an inflorescence, largely sterilized.

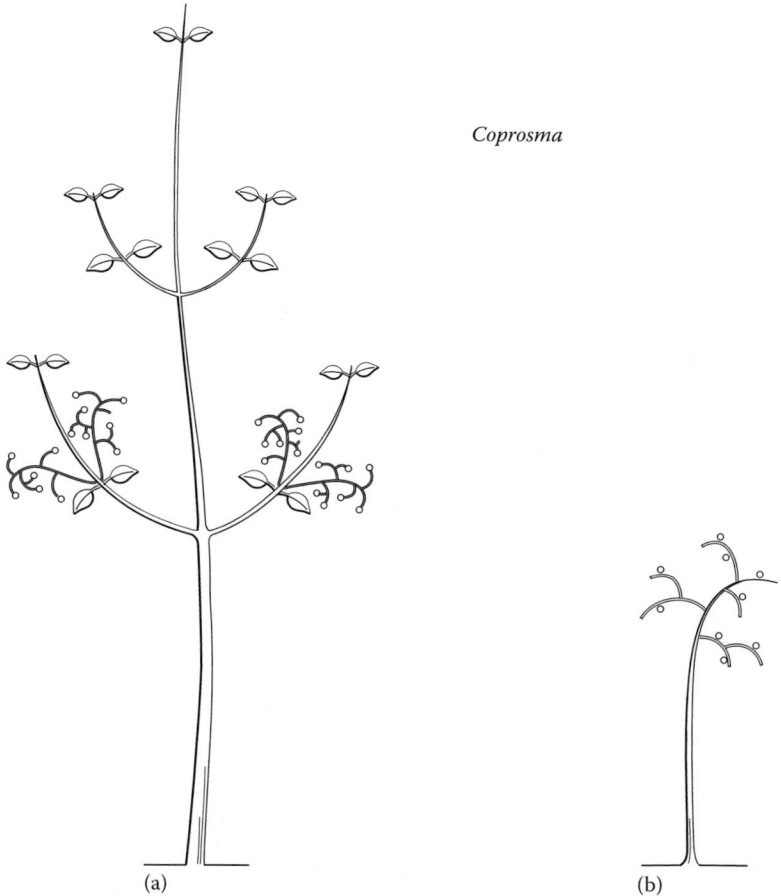

Coprosma

(a)   (b)

FIGURE 14.1 Architecture in *Coprosma* (Rubiaceae). (a) A large-leaved, tree with cymose (determinate) inflorescences. (b) A divaricating shrub, equivalent to an inflorescence of a large-leaved species. Inflorescence bracts are not shown.

Among the variations of divaricate architecture, in heterophyllous New Zealand plants such as *Elaeocarpus*, *Pennantia* (Pennantiaceae), and *Sophora* the divaricate, sterilized inflorescence phase is in a basal ("juvenile") position on the plant. Flowering at the base of the trunk, or basiflory, is rare in temperate trees, but it is conspicuous in many tropical species (cf. Figure 14.1).

Architecture of Ericoid Shrubs and Trees, Including Cushion Plants and "Whipcords"

In addition to divaricates, other small-leaved plants are abundant in New Zealand, and these include "ericoid" species with foliage of needle- or scale leaves. At least some hard cushion plants of the alpine zone can be included here, as they grow out into ericoid plants if cultivated in the lowlands (Heads, 1990b). (Many alpine cushion plants are woody and hard enough to bear a person's weight.) More normal ericoid plants are represented in the New Zealand flora by many Podocarpaceae, Thymelaeaceae, Ericaceae, Asteraceae, and Plantaginaceae. The latter include the members of the *Hebe* complex known as "whipcord hebes" (species of *Hebe* and *Leonohebe*) (Figure 14.2). As with the divaricates, the ericoid or whipcord shrub can be interpreted as a normal plant that has been reduced to its inflorescence, and

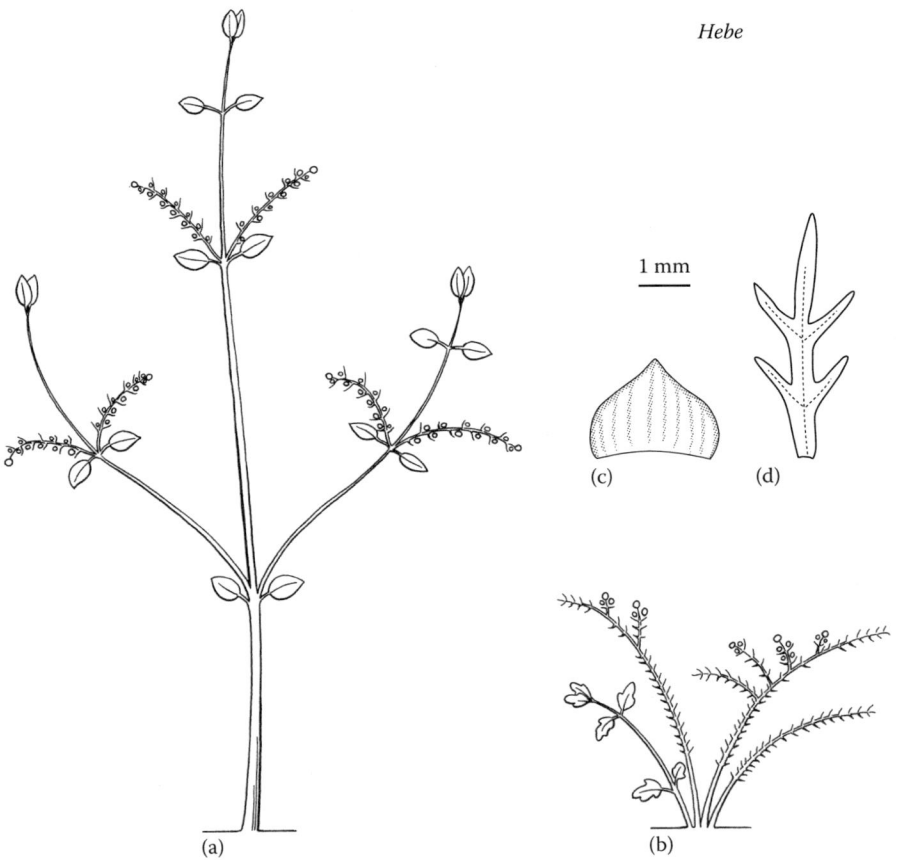

FIGURE 14.2 Architecture in the *Hebe* complex (Plantaginaceae). (a) A large-leaved tree species with racemose (indeterminate) inflorescences. Vegetative axes all orthotropic, with two kinds of foliage: "vegetative" leaves and inflorescence bracts. (b) Whipcord shrubs. Axes often plagiotropic, with two kinds of foliage: "reversion" or "juvenile" foliage, and scale leaves (inflorescence bracts), the latter is mostly sterile. (Reprinted from Heads, M., *Bulletin du Muséum national d'Histoire naturelle* Paris 4e sér., 16, sect. B., *Adansonia*, 163, 1994e. With permission from the Muséum national d'Histoire naturelle, Paris.) (c) Scale leaf of a whipcord species, *Hebe lycopodioides*. (d) "Reversion leaf" of *H. lycopodioides*. (From Wagstaff, S.J. and Wardle, P., *N. Z. J. Bot.*, 37, 17, 1999.)

that in addition, large parts of the inflorescence have been sterilized, leaving bracts without flowers (Heads, 2003). Flowering is more or less restricted to the ends of the branches.

The suppression of a plant's vegetative shoot system, with the above-ground part reduced to an inflorescence, also accounts for the great diversity of form in *Dracophyllum* (Ericaceae). In the model proposed here, the whole shoot system of the cushion species is homologous with the inflorescence of the tree forms (Heads, 2003: 326). The inflorescence components of large-leaved, *tree* species, such as *Dracophyllum involucratum* and *D. verticillatum* (Figure 14.3a, from Virot, 1975), are similar to, and can be equated with, the whole shoot systems of ericoid, *cushion* species, such as *D. muscoides* and *D. minimum* (Figure 14.3b, from Buchanan, 1881). In *D. involucratum*, the bracts are 2.5–4 mm long and the leaves are about 100 times larger (250–300 mm long) (Venter, 2009). In contrast, the bracts of *D. muscoides* are 3.5 mm long and the leaves are about the same (1.5–3 mm long). The proposed model suggests that leaves and bracts are homologous in the ericoid cushion species but not in the tree species. Flowering on the peduncle of the tree species and on the flowering branch of the cushion species is restricted to the distal end of the axis, and the rest of the bracts (or scale foliage) on the axis are sterile.

Many Ericaceae and other plants with ericoid habit are heteroblastic or heterophyllous—they have two different forms of shoots that bear quite different forms of foliage. The "juvenile" (or "reversion") shoots bear normal leaves, while the "adult" shoots bear scale leaves and flowers. The two types of shoot system can be seen in more neutral terms as basal or distal systems on a shoot complex.

In fact, most seed plants can be seen as heterophyllous if the transition to the inflorescence is considered—the shoot systems of the vegetative plant and the inflorescence have very different foliage (normal leaves versus inflorescence bracts), symmetry and architecture. If enough of the basal inflorescence bracts is sterilized, with the flowers suppressed, the result is a plant with heterophyllous foliage.

Most of the leaves on the whipcord hebes are scaliform, but often the plants also bear much larger leaves with very different morphology and venation (similar to "normal" angiosperm foliage) on basal shoots (Figure 14.2). This is referred to as "juvenile" or "reversion" foliage, but in the interpretation given here, these leaves represent a remnant of the normal, adult foliage of the tree forms (Heads, 1994d), most of which is suppressed in ericoid forms. (Bayly and Kellow, 2006, gave only a brief mention of this interesting topic in their monograph on hebes. For illustrations of "juvenile" or "reversion" leaves of whipcord *Hebe*, see Wagstaff and Wardle, 1999.) In the tree forms of the hebe complex, in *Hebe* s.str., there are also two distinct types of foliage: the normal leaves and the inflorescence bracts.

Garnock-Jones (2006) objected that "There are no anatomical differences by which the claimed bract-derived leaves of *Leonohebe* (sensu Heads) [whipcords, etc.] can be distinguished from true leaves in *Hebe*" (p. 37). Nevertheless, in the scale leaves of the whipcords, the palisade mesophyll is *abaxial* to the spongy mesophyll (below it), while in the normal leaves of tree *Hebe*, as in most angiosperms, it is adaxial (Adamson, 1912; Garnock-Jones, 2006: 31). This is a profound, qualitative difference, but it is not unique to the *Hebe* complex. A similar inversion of the mesophyll layers occurs in whipcord species of *Helichrysum* and *Raoulia* (Asteraceae) (Breitwieser, 1993; Breitwieser and Ward, 1998).

In the whipcord hebes, the seedling leaves and reversion leaves "sometimes closely resemble the adult leaves" of *Hebe* s.str., *Parahebe*, and *Hebe* and so on (Garnock-Jones, 2006: 37). The structures are regarded here as equivalent. The question is: why are the seedling and reversion leaves of whipcords so different from the adult, scale leaf in morphology (size, shape, basal attachment, texture, parallel venation), and anatomy (with the tissue layers inverted)? What *are* the adult scale leaves of the whipcords and why are they so similar to the inflorescence bracts? Garnock-Jones (2006) did not give any answer, other than citing Cockayne's (1898) conclusion: the scale leaves of the whipcords are an "adaptation." But of what and by what processes?

Whipcord Architecture in Other New Zealand Groups

In conifers, most trees in the clade Araucariaceae + Podocarpaceae have foliage composed either of large, more or less spreading leaves, as in the New Zealand *Agathis australis*, or small-scale leaves. But the group also includes heterophyllous forms, such as the endemic New Zealand genus *Halocarpus* (Podocarpaceae).

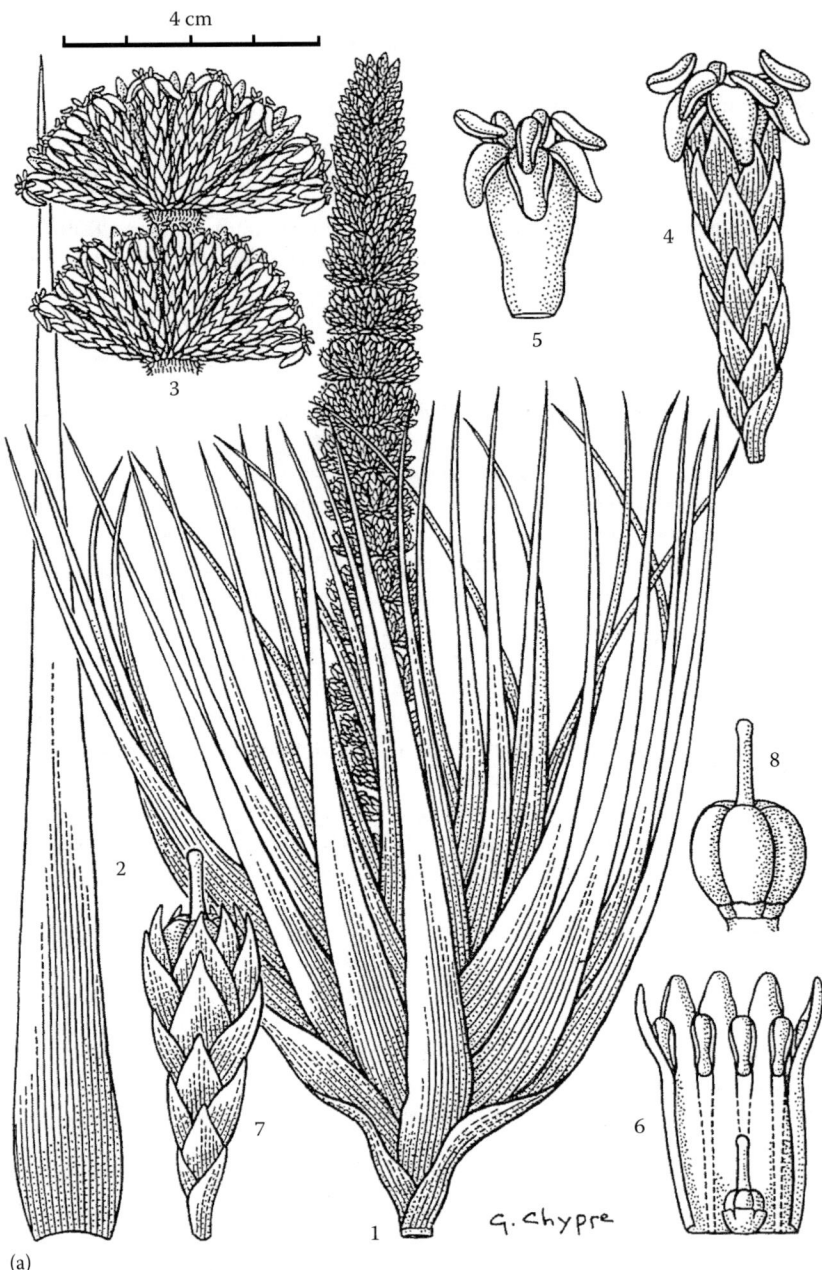

4 cm

FIGURE 14.3 Architecture in *Dracophyllum* (Ericaceae). (a) The tree *D. involucratum*. (Reprinted from Virot, R., Épacridacées, in *Flore de la Nouvelle-Calédonie et Dépendances*, Vol. 6, eds. A. Aubréville, and J.F. Leroy, Muséum national d'Histoire naturelle, Paris, France, 1975, pp. 5–160. With permission from the Muséum national d'Histoire naturelle, Paris.) 1, flowering branch; 2, leaf; 3, floral verticils; 4, branch of inflorescence with bracteoles and single flower; 5, flower; 6, opened flower; 7, branch of infructescence with fruit; 8, fruit with calyx removed. The scale bar (calculated from Venter, 2009) applies only to 3. *(Continued)*

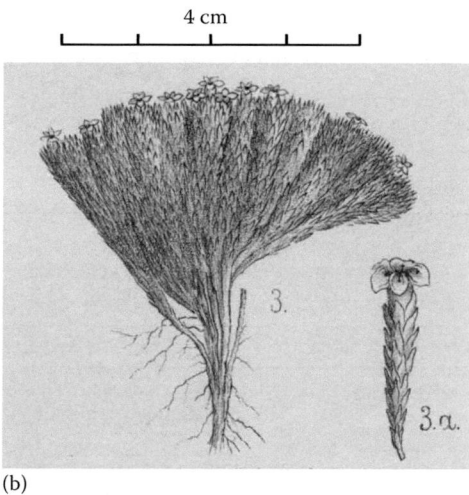

4 cm

(b)

FIGURE 14.3 (*Continued*) Architecture in *Dracophyllum* (Ericaceae). (b) The alpine cushion plant *D. muscoides*, showing a rooting piece of a cushion and a flowering branch. (Reprinted from Buchanan, J., *Trans. N. Z. Inst.*, 14, 342, 1881. With permission from the Royal Society of New Zealand.) The scale bar applies only to the larger piece.

Here the trees have ericoid scale leaves and also quite different "juvenile" foliage that is much larger, up to 4 cm long in *H. kirkii*. The "juvenile" foliage can persist in *H. kirkii* until the tree is 10 m tall (Dawson and Lucas, 2011).

In *Halocarpus*, branches with juvenile foliage give rise to adult branches with scale foliage, but these in turn can produce branches with "juvenile" (now "reversion") foliage. These then give rise to "adult" branches, and "This uncertain state of switching back and forth may continue for a few years" (Dawson and Lucas, 2011: 66). Eventually the crown bears only adult foliage. The authors described this as "an intriguing and puzzling see-saw pattern."

Nevertheless, while this see-sawing is not seen in the usual, simple transition from juvenile plant to adult—and so is puzzling—it is exactly what is seen in the normal cycles of vegetative growth and inflorescence growth. Again, this can be explained if the large, "juvenile" foliage is normal, adult foliage, while the branches with "adult," scale foliage represent a more or less sterilized inflorescence with its bracts. Even the large trees in Araucariaceae + Podocarpaceae with scale leaves (some with traces of "juvenile" foliage) can be interpreted as sterilized inflorescences.

A similar pattern to that of *Halocarpus* is observed in many Cupressaceae (junipers and others); young plants have linear (long, narrow) leaves, while mature plants have small, scalelike leaves. In some species, the "juvenile" leaves only occur in seedlings, while in others they are retained throughout the life of the tree. In some species, "juvenile" foliage occurs on mature plants on shaded shoots, while shoots in full sunlight bear "adult" foliage.

Brachyglottis bifistulosa (Asteraceae), endemic to south-western Fiordland (Buller terrane), is another New Zealand plant with two different leaf forms, each with a distinctive shape and anatomy. The leaves that are borne at the base of a shoot are spreading, while those borne distally are linear and ericoid, and resemble bracts (Drury, 1973; Simpson, 1974). In the model presented here, the distal leaves *are* bracts. Drury placed the species in his Group V, in which all the species are more or less heterophyllous, with toothed "juvenile" leaves.

A similar heterophylly is seen in *Helichrysum dimorphum* (Asteraceae), a liane or sprawling shrub restricted to the northern margin of the Torlesse node (Poulter, Esk, and Broken Rivers). The lower leaves are spreading and "normal," but the distal leaves are scalelike and more or less imbricate. Likewise, in the *Coprosma polymorpha* form of *C. rhamnoides*, spreading, laminate leaves are borne at the base of a shoot, with linear, bract-like leaves beyond this. These distal shoot zones are interpreted here as sterilized inflorescences with foliage of bracts.

In its origin, the structural differentiation between a "leafy," "juvenile," vegetative phase of shoot and an inflorescence (sporogenous) phase also involved the origin of the leaf. Traditional morphology maintains that the leaf is one of three irreducible elements of plant structure, along with stem and root (Goethe, 1790), but this is controversial. It is more likely that the modern leaf is a compound structure and has resulted from the fusion of a bract with an inflorescence equivalent. The inflorescence has been transfused into the bract, where it is now sterilized (except in plants with epiphyllous flowers) but is represented by the leaf venation and the leaf meristems.

In the interpretation of the small-leaved plants proposed here, *most* plants are interpreted as heterophyllous; in many, scale leaves are only present as inflorescence bracts, while in others, they constitute the main foliage. The homologies are as follows:

Ericoid/Divaricate Species	Large-Leaved Species
Whole shoot system	= inflorescence (more or less sterilized)
"Adult" foliage	= inflorescence bracts (sterilized)
"Juvenile" and "reversion" leaves	= normal vegetative foliage

Following the differentiation of ericoid and large-leaved clades, subsequent hybridism has often produced intermediate forms that obscure the original morphological distinction. Nevertheless, in many cases, the phylogenetic break remains evident in the biogeography. For example, *Erica* species (shrubs with ericoid foliage, interpreted here as inflorescence plants) have a famous concentration of diversity in South Africa comprising hundreds of species, while *Rhododendron* (larger-leaved plants including many trees) has its strongest concentration of diversity—again, with hundreds of species—in the region extending from Burma to New Guinea.

Architecture in Monocots That Can Be Interpreted as "Inflorescence Plants"

Reduction of the vegetative shoot and the partial sterilization of the inflorescence can also explain the range of forms seen in monocots such as the diverse genus *Astelia* (Asteliaceae) (Figure 4.7). Here, the extremes of the morphological range are represented by *A. subulata*, a small, turf-forming alpine plant, and *A. trinervia* ("kauri grass"), a large, forest species. In *A. subulata*, the leaves are 1–2 cm long and flowers are often borne *singly* (Moore and Edgar, 1970; Mark and Adams, 1973). Plants of *A. trinervia* have leaves up to 3 m long, and *large, branched* inflorescences that each bears hundreds of flowers (Moore and Edgar, 1970). The inflorescence can be compared with a plant of *A. subulata*, and the bracts (spathes), 3–6 cm long, can be compared with the leaves of *A. subulata*.

The small, simple plants of *Astelia subulata* belong to the Indian Ocean clade of *Astelia*, while the large, complex plants of *A. trinervia* belong to the Pacific Ocean clade (Figure 4.7). This overall pattern repeats that of the other Indian and Pacific Ocean groups in which the western forms are "inflorescence plants," with foliage equivalent to bracts, while their relatives further east are orthotropic, larger-leaved plants. In *Astelia*, as in the others cited earlier, both kinds of architecture overlap in New Zealand.

Brooms in Carmichaelia (Fabaceae): Another Group of Reduced "Inflorescence Plants"

The biogeography of the *Carmichaelia* group was discussed earlier (Chapter 11). The different members show a wide range of architecture, including lianes, rhizomatous subshrubs, taller shrubs with suppressed rhizomes, and trees. Many have a broom habit, in which the only foliage present in the adults comprises minute bract scales, each equivalent to the two stipules of a fully developed leaf (Slade, 1952: Fig. 1). The 22 species include the following forms (Allan, 1961):

Lianes or scrambling semi-lianes, up to 3 m tall, branching at right angles. Leaves are present in the inner part of the mass of shoots, especially in summer (*Carmichaelia* subg. *Kirkiella*).

Depressed shrubs <15 cm tall, with rhizomes. Leaves are present only in seedlings (*C.* subg. *Suterella*).

Similar depressed shrubs, but without rhizomes. Leaves are mostly restricted to young plants (*C.* subg. *Monroella* and *C.* subg. *Enysiella*).

Taller shrubs that are prone, decumbent, or straggling, without rhizomes. Leaves are present only on young plants (*C.* subg. *Huttonella*).

Erect shrubs or small trees up to 10 m tall, usually without rhizomes. Leaves are often present, at least in summer (*C.* subg. *Carmichaelia*, *Carmichaelia* group *Notospartium*, *Chordospartium* s.str., *"Chordospartium" muritai*).

Most other diverse New Zealand groups of woody plants also include subshrubs just a few centimeters tall, as well as trees. This entire range is seen in the following clades: *Lepidothamnus–Phyllocladus–Halocarpus–Lagarostrobos–Prumnopitys* (Podocarpaceae), *Mearnsia–Metrosideros* (Myrtaceae), *Hymenanthera–Melicytus* (Violaceae), *Myrsine* (Myrsinaceae), *Dracophyllum* (Ericaceae), *Coprosma* (Rubiaceae), the *Hebe* complex (Plantaginaceae), and *Celmisia–Olearia* (Asteraceae). In many, if not all of these, the reduced morphology of the smaller plants can be explained if they represent inflorescences.

Tree architecture in the New Zealand brooms probably conforms to the pattern of growth that is usual in its family, Fabaceae (legumes). The same architectural model (Troll's model; Hallé et al., 1978) is also evident in most species of Myrtaceae and Nothofagaceae. The shoot axes are more or less horizontal (plagiotropic) during the primary phase of growth, but their basal parts are then reoriented and raised into a vertical, orthotropic position by secondary growth. The distal part remains plagiotropic. Superposition of these orthotropic–plagiotropic shoots forms an orthotropic trunk that is sympodial—formed by the growth of several buds. This contrasts with architecture in groups such as Winteraceae, in which the trunk is orthotropic and monopodial.

Troll's model is evident in perhaps 20%–30% of all trees and is the most widespread of any architectural model (Hallé et al., 1978). Nevertheless, it has been overlooked in many New Zealand plants in which it occurs (personal observations), for example, *Metrosideros* (Simpson, 2005) and *Sophora* (Carswell and Gould, 1998; Heenan et al., 2001). It is likely to be present in the tree species of the *Carmichaelia* group, but observations over successive years are needed.

Many *Carmichaelia* species are described as heterophyllous. In these, the "juvenile," nonflowering stems are plagiotropic, bear distichous, laminate leaves, and resemble a normal, leafy plant (Slade, 1952). In contrast, the "adult," flowering stems are orthotropic, flattened cladodes that are leafless for all or much of the year. Figure 14.4 shows the two shoot phases in *C. corrugata* (from Heenan, 1997: Fig. 2; cf. *C. petriei* illustrated by Atkinson and Greenwood, 1989: Fig. 8). In the leafless phase, the only foliage comprises bract scales. The two shoot systems, "juvenile" and "adult," are quite different in their morphology and if separated would be difficult to recognize as the same species. "Reversion" shoots with "juvenile," laminate foliage appear at times in most of the *Carmichaelia* groups, even the leafless ones (especially if plants are cultivated under high humidity), but not in *Carmichaelia* group *Notospartium*.

In the *Carmichaelia* group, as in the divaricates, ericoids, and others, the "adult" plant can be interpreted as an inflorescence in which many of the bracts have been sterilized. The "juvenile" and "reversion" stems are normal vegetative shoots. The flattening in the cladodes of *Carmichaelia* resembles the conspicuous flattening in the main inflorescence axes (peduncles) of many Orchidaceae, Arecaceae, Bromeliaceae, Olacaceae, Myrtaceae, and *Coprosma*.

Other features common to both divaricate shrubs and many inflorescences include wide-angle branching and determinate shoot apices. Both are widespread in adult plants of the *Carmichaelia* group, and members are sometimes described as divaricate (e.g., *C. arborea*, *C. odorata*, and *C. kirkii*; Heenan, 1996a). Secondary thickening of the aborted shoot apices can result in blunt spines in *Carmichaelia petriei* and sometimes *C. australis* (Wardle, 1991; Heenan, 1996a). Similar structures occur in divaricate species of New Zealand *Hymenanthera* and *Pittosporum*, and in many divaricate plants of Madagascar and East Africa.

Morphological Differentiation of *"Inflorescence Plants"*

Many plant structures have been interpreted as sterilized, reduced inflorescences. Examples include the thorns present in *Bougainvillea* (Nyctaginaceae) (Hallé et al., 1978: 239), *Gleditsia* (Fabaceae) (Brunaud, 1970), and *Carissa* (Apocynaceae) (Brunaud, 1970; Cohen and Arzee, 1980). In a New Zealand example,

FIGURE 14.4   Architecture in *Carmichaelia corrugata* (Fabaceae) showing long, decumbent, leafy, juvenile stems, and short, erect, leafless, adult cladodes (2 years old). (Reprinted from Heenan, P.B., *N. Z. J. Bot.*, 35, 243, 1997. With permission from Taylor & Francis.)

the spines in *Discaria toumatou* (Rhamnaceae) are almost always sterile, but on rare occasions, they can produce two, fully developed flowers (J. Keogh, pers. comm., 1984). This suggests their origin as reduced inflorescences.

In other groups, tendrils have originated in the same way. The tendrils in Vitaceae, which are extra-axillary, are replaced at the time of flowering by inflorescences in the same position. Parts of the inflorescence are often differentiated as tendrils, and so a tendril can be interpreted as a sterilized inflorescence (Shah and Dave, 1970). The tendril of the New Zealand *Tetrapathaea* (Passifloraceae) is usually sterile but is recorded "rarely bearing a flower" (Allan, 1961); in addition, inflorescences often bear short tendrils (personal observations in Ngaio Gorge, Wellington).

At first glance, complex structural changes would seem to be required in order to convert a large-leaved form, such as a typical *Rhododendron*, into an ericoid shrub, such as a typical *Erica*; an erica is nothing like a smaller version of a rhododendron. Yet the two would be quite simple, morphogenetic variants if the whole shoot system of a divaricate or ericoid plant represents the inflorescence, more or less sterilized, of the tree forms. In structural terms it would be straightforward to derive the divaricate and the ericoid/whipcord/cushion forms from a tree by suppressing most of the vegetative shoot, leaving just the inflorescence, and sterilizing most of the latter, leaving only branches, empty bracts, and a few flowers. The flowers would then appear to be borne in single-flowered inflorescences.

Shoot axes in divaricate and ericoid forms are often at least partly plagiotropic, and most have solitary flowers. This architecture differs in many ways from the vegetative architecture of their relatives that are "normal" orthotropic trees, with distinct, multiflowered inflorescences. Apart from the characteristic branching pattern, the foliage in the divaricate and ericoid plants is often scale- or bract-like and differs from the foliage of their large-leaved relatives in size, shape, margin, venation, and anatomy. Nevertheless, in the divaricate and ericoid forms (but not in the tree forms), the "leaves" are similar to the inflorescence bracts. This can be explained if the ericoid and divaricate plants are homologous to inflorescences and the leaves are equivalent to sterilized inflorescence bracts.

The same reduction pattern seen in New Zealand heterophyllous conifers, ericoid shrubs, whipcord shrubs, divaricate brooms, and *Astelia* also occurs in overseas groups such as Combretaceae in Africa and Madagascar, Elaeocarpaceae (tremands) in Australia (Heads, 2014: 16), and many others. The process means that the bulk of the plant develops much reduced foliage (sterilized inflorescence bracts). This creates an increased tolerance of harsh growing conditions, whether these are in dry woodlands, swamp forests, or in extremes of cold or hot.

The reduction of the vegetative shoot and sterilization of most of the inflorescence account for several key features of the New Zealand small-leaved shrubs, including the following:

The great difference between the small-leaved "inflorescence plants" and their large-leaved relatives: the former are not just miniature versions of the latter.

The similarity of the foliage leaves and the inflorescence bracts in the ericoid and divaricate plants, and the great difference between the leaves and bracts of related plants with tree architecture.

The presence of heterophylly in some ericoid forms, but not in their tree relatives; if the inflorescence bracts are taken into account, both forms have heterophyllous foliage.

The fact that the flowers are borne singly in the ericoid and divaricate plants (inflorescences), but in distinct, multi-flowered inflorescences in their tree relatives.

The presence of empty, sterile bracts extending some distance down the shoot below the inflorescence in small-leaved groups such as *Hebe* sect. *Connatae*. If this sterilization extended further down the shoot, and the normal leaves were suppressed, an ericoid plant would result.

To summarize, the idea that divaricate, ericoid, and broom plants represent more or less sterilized inflorescences with roots explains many aspects of their morphology. This is another example of the evolution of new morphology by reduction. The small-leaved, resilient forms that have resulted have flourished in the disturbed, high-stress environments with open vegetation that are widespread in New Zealand.

Biogeography of Divaricate Plants

The divaricate syndrome occurs in a few species in each of many different families, and the group is nothing like monophyletic. Despite this, it has a distinctive geographic range (Heads, 2014: 79–81). Most divaricates occur around the south-west Indian Ocean (East Africa, South Africa, and Madagascar) and the south-west Pacific (Australia and New Zealand), with some in America, but the architecture is very rare in New Guinea, the central Pacific islands, and West Africa (personal observations). The distribution of this parallelism is a standard Indian (–Atlantic) Ocean pattern, as seen in many monophyletic clades. Within New Zealand, divaricates in the Chatham Islands are notable for their rarity (Chapter 6), and the islands lie east of the main centers of diversity in the "group." Divaricate architecture has been explained as a defense against browsing by ratite birds, including the extinct elephant birds (Aepyornithidae) on Madagascar, and moas in New Zealand. Yet this does not explain the absence of divaricate plants in New Guinea, as ratite birds (cassowaries) are widespread there.

The very low numbers of divaricates on and east of a line—New Guinea–Chatham Islands—can best be explained by the prior distribution of "divaricate genes" before the modern groups existed, rather than by physical or biological aspects of the present environment. With their small leaves, divaricates are preadapted to survival in harsh environments, such as desert, mountains, swampy, or rocky sites, and their scarcity on the Chatham Islands is an ecological anomaly.

The traditional explanation for divaricates was accepted by Cowen (2013: 241): "Moas coevolved with New Zealand plants so that 10% of the native woody plants have a peculiar branching pattern called *divarication*—they branch at a high angle to form a densely growing plant with interlaced branches...." But while divarication might be peculiar in whole shoot systems, it is common in inflorescences. (Many species are named *divaricata*, and in most of these, the name refers to the inflorescence.) If the divaricate plant is homologous with an inflorescence, the branching is quite normal. Cowen (2013: 241) concluded: "The only reasonable explanation of divarication is that it evolved as a defense against browsing moas."

But the idea that divarication is a defense against ratite, rather than ungulate, herbivores is contradicted by the absence of divaricate plants in New Guinea, where ratites are common and there are no ungulates (Heads, 2014: 79–81).

Biogeography of "Inflorescence Plants"

In many Indo-Pacific plant groups, the western forms are "inflorescence plants" with divaricate, ericoid, or lianoid architecture; the shoots are often plagiotropic, at least in part, and the leaves are smaller than in related plants. These relatives occur further east, often in and around the central Pacific, and are trees with larger leaves and orthotropic shoots (Heads, 2014). Both kinds of architecture often overlap in New Zealand and this results in high levels of diversity there. Examples of this west–east differentiation occur in the following groups:

> Asteliaceae: *Astelia* "Indian Ocean clade" versus *Astelia* "Pacific Ocean clade" (Figure 4.7).
>
> Ericaceae: *Erica* versus *Rhododendron* (Heads, 2003).
>
> Ericaceae: Ericoid *Dracophyllum* of New Zealand versus large-leaved tree *Dracophyllum* of New Zealand and New Caledonia (Heads, 2014: Fig. 6.22).
>
> Myrtaceae: *Mearnsia* versus *Metrosideros* (Figure 4.6).
>
> Violaceae: *Hymenanthera* versus *Melicytus* (Heads, 2014: Fig. 6.17A).
>
> Plantaginaceae: *Chionohebe*, *Hebejeebie,* and *Leonohebe* versus *Hebe* s.str. (Figure 10.11).
>
> Rubiaceae: In *Coprosma*, a paraphyletic complex of small-leaved, more or less divaricate forms in the south-west Pacific, versus a group of orthotropic trees in the central Pacific, with both occurring in New Zealand (Cantley et al., 2014).

On the Chatham Islands, ericoid plants show a similar paucity to that of the divaricates. The diverse *Hebe* complex is represented there by just three, mesophyllous species, and there are no ericoid ("whipcord") forms. It might be assumed that this absence is because of ecological factors, as the whipcord species are now restricted to mainland sites over 620 m elevation (Wagstaff and Wardle, 1999), and the highest point of the Chatham Islands is only 294 m. Yet whipcords are also absent from the subantarctic islands, where there is extensive suitable habitat. This suggests that their restriction to the New Zealand mainland is not determined by ecological factors.

New Zealand Alpine Plants and Their Morphology

Malcolm and Malcolm's (1988) book includes superb close-up photography of many of New Zealand's alpine plants. The authors wrote: "… the structures that look beautiful to us are strictly functional to the plants, in fact, utterly vital to their survival, because those structures are the solutions to the many survival problems that the plants face in their harsh mountainous home" (p. 2). Some features that bestow survival value on alpine plants, such as smaller size and hairiness of parts, may be the result of selection, but most aspects of their structure have never been explained in adaptive terms. Examples include phyllotaxis, floral symmetry (for example, four stamens in *Kelleria* versus two in *Pimelea*), the monocot style of leaf venation in *Dracophyllum*, the different kinds of leaf teeth, subtle differences in leaf shape that are often used in identification, and many others.

The divaricate growth form occurs in alpine species of *Coprosma* (Rubiaceae), *Pittosporum* (Pittosporaceae), and *Hymenanthera* (Violaceae). It was interpreted by Malcolm and Malcolm (1988) as a solution to the problems of wind, drought, frost, and grazing by moas. Tough, resistant, woody growth is one of the end results of divaricate evolution, but the structural processes of suppression and sterilization that have led to divaricate architecture have also occurred in places such as south-eastern Africa and Madagascar, as well as New Zealand, and so these underlying processes are not the result of environmental factors specific to New Zealand or to alpine conditions.

In the standard center of origin–dispersal–adaptation model for the evolution of alpine floras, plants invade newly risen mountains by dispersal and then develop suitable adaptations for life there. However, Malcolm and Malcolm (1988) described an alternative history for one alpine plant, *Phyllachne* (now included in *Forstera*: Stylidiaceae, see Figure 4.36). They interpreted this as:

> … a truly ancient relict from 60 million years ago when the country was largely flat and covered in evergreen forest. Probably *Phyllachne* at that time lived on infertile soils in forest openings. When the mountains rose up some 6 million years ago, *Phyllachne* rose with them, still living on infertile soils but facing the added hazards of cold winds and frequent freeze-thaw cycles. By good chance, its cushion growth form allowed it to cope with those hazards. (p. 61)

This same process of uplift, rather than invasion, also explains the history of most other alpine plants, including grasses, sedges, rosette herbs, cushions, mats, and other subshrubs. The fact that particular plants had these growth forms before orogeny was "good chance" in terms of their subsequent survival, but the existence of the structure in the first place was the result of prior laws of growth.

Angiosperm Flowers and Anemophily

The angiosperm reproductive structure has evolved from a cone-like organ, or strobilus, in which the individual units have developed into either flowers (the cone becomes an inflorescence) or ovaries (the cone becomes a flower). In the former case, the flowers are small, unisexual, and often wind pollinated (anemophilous). In the latter, the flowers are large, complex, and often insect pollinated. One is not derived from the other; instead, both are derived from a common ancestral structure, a strobilus, by reduction.

The two trends in the reduction of the ancestral strobilus and the evolution of the flower represent parallel developments, and both types of structure occur in many groups. An example indicated by molecular studies is the rose order Rosales, which in the new, wide sense includes the Urticales (Stevens, 2016). Anemophilous, dioecious flowers like those of the Urticales (e.g., nettles) are often thought to be derived from animal-pollinated, hermaphrodite flowers like those of Rosales, but it is difficult to derive the simple, anemophilous flower of a nettle or a mulberry (Urticales type) from the complex, multipartite flower of a rose (Rosales type). In contrast, the anemophilous *inflorescence* of Urticales and the insect-pollinated *flower* of Rosales s.str. share similar strobilar structure.

Anemophilous flowers are common in the New Zealand, for example, in the grasses and in *Coprosma* (Rubiaceae: Anthospermeae). Species of *Coprosma* occur in most New Zealand vegetation, and it has been suggested that their anemophilous flowers are "a response to the lack of specialized insect pollinators in New Zealand" (Malcolm and Malcolm, 1988). Yet similar, anemophilous flowers characterize all members of the tribe Anthospermeae, and outside New Zealand, these are widespread in Africa, Australia, and South America where there are many pollinating insects.

Other authors have also accepted the traditional view that anemophilous flowers are secondary. One textbook suggested that "Some island plants appear to have shifted from ancestral animal pollination to wind pollination (anemophily). Examples include *Rhetinodendron* [Asteraceae] on the Juan Fernández Islands, *Plocama* [Rubiaceae] and *Phyllis* [Rubiaceae: Anthospermeae] in the Canaries, and *Coprosma* in New Zealand" (Whittaker and Fernández-Palacios, 2007: 178). But, again, this overlooks the widespread distribution of Anthospermeae in continental areas that have many insects.

Whittaker and Fernández-Palacios (2007) cited "possible benefits" for anemophily in island plants, including independence from impoverished pollinator faunas, the ability to take advantage of windy conditions, and outcrossing. The authors were cautious though, and they suggested that the anemophily may instead be linked to woodiness or dioecy (separate male and female plants), rather than being selected for directly. A nonselective explanation for anemophily (and dioecy) would help explain why, for example, the anemophilous Rubiaceae tribe Anthospermeae has a disjunct southern hemisphere/tropical distribution and why it is so widespread in New Zealand.

Reduction in the Flowers of Viscaceae

Some of the most reduced of any flowers are those of a mistletoe group, Viscaceae (= Santalaceae tribe Visceae). In the New Zealand species, all in *Korthalsella*, the plants themselves are "greatly reduced" (Stevenson, 1934: 186), with no leaves, and with flowers that are minute and "very simple" (p. 188). There are "none of the typical angiosperm floral organs," such as ovule or ovary. Stevenson considered that the reduction has led to the modular construction of the individual plant by many individuals and also to localized distribution:

> The great reduction has led to restricted and localized occurrence of the plants, because all the seeds are shed very close to the parent plant, and many on to the parent plant itself. [... the plants are compound, consisting of two or more, parasitic the one upon the other" p. 184]. On *K. lindsayi* with its flattened internodes, this "cannibalism" occurs to a much greater extent than in *K. salicornioides*. This complete inability in both species on the part of the plant to distribute itself over any distance has resulted in its becoming confined to isolated districts. Where it does occur, however, it may be exceedingly plentiful. An indication of the slowness with which distribution takes place is seen in the frequent fact that some individual shrubs can be densely covered with one or other of the parasites, whilst others of the same kind within a few feet are quite uninfected. (p. 186)

In other New Zealand plants, Heenan et al. (2013) described the "remarkable reduction" in the flowers of *Cardamine cubita* (Brassicaceae). The flower is "extraordinarily reduced," both in its overall size (just 0.8–0.9 mm diameter) and in its parts (petals are absent, the gynoecium has only two ovules, and the androecium has only two stamens).

Flowers and Peloria

Reduction, suppression, and fusion of parts are evident in the telescoping and condensation that have led to the angiosperm flower and many other structures, such as the vertebrate head. All aspects of the angiosperm flower have been involved in reduction. For example, in the pollen tetrads of Cyperaceae, three of the four microspores abort and all four are enclosed in a single pollen grain, a "pseudomonad." This also occurs in Ericaceae tribe Styphelieae.

A modern analogue of the telescoping–condensation in the origin of the flower is seen in floral peloria, mutant forms that involve the fusion of more than one flower. Fusion of five zygomorphic (bilateral) flowers can result in a single, compound structure with pentaradial symmetry. For example, the normal male flower of *Nothofagus* subgenus *Lophozonia* has been interpreted as "a complex structure, a 'pseudanthium' formed by the failure to ramify of a three-flowered dichasium" (Rozefelds and Drinnan, 1998). (A pseudanthium "is an inflorescence that manifests or functions as a single flower.") Likewise, coalescence of female flowers is sometimes observed in the South American *Nothofagus pumilio* (Hjelmqvist, 1948: 82–83, as cited in van Steenis, 1953: 326). The normal, male flower of *Coprosma foetidissima* (Rubiaceae) is multipartite and suggests a similar origin by peloric fusion. The same process and its many variations have been important in the origin of the angiosperm flower and also its subsequent differentiation.

Cleistogamy

Cleistogamy is a condition in which flowers remain bud-like and closed and are self-pollinated. It has been recorded in as many as 50 angiosperm families and is assumed to have evolved many times. Cleistogamy is a simple morphological tendency that, if fixed at different stages, would in itself generate many new species.

Cleistogamous flowers are recorded in six genera of New Zealand plants, including *Viola* (Violaceae) and *Melicope* (Rutaceae) (Godley, 1979; Barker, 1992). In New Zealand *Viola*, normal, open (chasmogamous) flowers are produced in short days, whereas cleistogamous flowers in longer days. In the grass *Deyeuxia lacustris*, endemic to Northwest Nelson and Lake Tennyson, *all* flowers are cleistogamous, while plants of related species can produce both chasmogamous (fully open) and cleistogamous flowers (Edgar and Connor, 1999). Cleistogamous forms of *Deyeuxia avenoides* are distinguished, in addition to their closed flowers, by differences in their culms, panicles, and anthers (Edgar, 1995). In genera such as *Microlaena*, the distinctive structures of grass flowers called lodicules may be present in chasmogamous flowers, but they are always absent in cleistogamous flowers (Connor, 1979).

Of the several forms of cleistogamy found in grasses, subterranean cleistogamy is the most unusual. In two species of *Amphicarpum*, in *Paspalum amphicarpum* and in *Chloris chloride*, single spikelets are borne underground on short stalks, and normal, aerial inflorescences are also produced (Connor, 1979). In a similar arrangement, several grass genera have clandestine, axillary spikelets ("cleistogenes") borne at ground level (Connor, 1979). In most cases, these are very different in their morphology from the normal, aerial spikelets, and clandestine, cleistogamous spikelets are well developed in *Stipa*. All these variations represent parallel suppressions in development that could generate new species, each with a different morphology and ecology.

Flowers: Glands as Reduced Organs

In many cases, organs whose growth has been inhibited or "hemmed in" over phylogenetic time are not completely eliminated, but leave a trace in the form of glandular, secretory tissue (Schwendt, 1907). The body of most animals, including vertebrates, has determinate growth. The animal body has undergone extensive reduction and suppression of parts, and is much more glandular than the plant body, which has indeterminate growth and is less reduced. The comparative lack of reduction in plants is also seen in their higher orders of symmetry, as most plants are not bilateral. Nevertheless, at least some glandular structure is evident in most angiosperms, in the flower and in the root cap. As in animals, reductions and glandular tissues are concentrated at the posterior and anterior ends of the organism's main axis.

The reductions and fusions that have taken place in plants with the telescoping of the floral axis have resulted in glandular tissue, best known in the nectary disc and the stigma. Many New Zealand plants also exhibit glandular tissue in other parts of the flower. For example, in *Kelleria* (Thymelaeaceae), the reduced petals are represented by nectar glands. The bird-catching tree *Pisonia* (Nyctaginaceae) can trap small birds on its fruit, which are sticky because of secretions from the fruit wall. Likewise, in *Phormium* (Hemerocallidacae), the septal nectaries form a swollen, fleshy structure, a chamber in the lower half of the ovary lined with glandular tissue (Rudall, 2002; Gardner, 2007). Copious nectar is secreted through three small pores and this attracts the honeyeater birds that are the plant's main pollinators. The septal glands are not the result of incomplete fusion between adjacent carpels; they represent reduced structures and their position suggests that earlier in phylogeny they were sterile carpels.

Glandular structures also occur on the ventral side of many petals and sepals, as in *Plagianthus* (Malvaceae). In Fagales, the compound scales of the cupule (the basal cup of an acorn) represent a set of floral axes that have been sterilized, reduced, and fused (Rozefelds and Drinnan, 2002). The apices of the cupule scales are glandular in some species of *Nothofagus*, but they are reduced to a nonglandular condition in Fagales such as oaks (*Quercus*) and *Lithocarpus*.

Glands are also a feature of many leaves, as in the New Zealand *Passiflora tetrandra*. Domatia and other structures in the axils of the leaf veins have a complex diverse morphology (Jacobs, 1966), and as with glands, they can also be interpreted as relictual structures (Heads, 1993).

In the teleological, neodarwinian model, nectar "doesn't have any other *raison d'être* than to feed bees. Nectar is manufactured, in large quantities, purely for bribing bees and other pollinators" (Dawkins, 2006: 238). Instead, in the mutation-driven model of evolution, nectar is produced because of the presence of glands, and glands are the result of reduction trends. The nectar has no *raison d'être* in the sense of purpose; it is a by-product.

Hygrochastic Capsules: The Result of Purposeful Strategy or Prior Trends?

Some New Zealand alpine plants have a hygrochastic capsule that opens when moistened by rain, with the seeds being splashed out by subsequent rain drops. As discussed already (Chapter 1), a study of hygrochastic capsules in the *Hebe* complex (Plantaginaceae) found that the average distance of seed dispersal was 13 cm (Pufal and Garnock-Jones, 2010). The study concluded that: "hygrochasy is a very effective mechanism of restricting seed dispersal to rainfall events and ensuring short-distance dispersal within a small habitat patch." The authors suggested that this system has evolved because the plants have "an antitelechoric strategy" that reduces far-flung dispersal. (The antitelechoric strategy has not been a complete success, as two separate lineages in the group have sisters in south-eastern Australia and Tasmania. The authors attributed this to long-distance dispersal.)

Pufal and Garnock-Jones (2010) concluded: "It appears that [hygrochasy] is an adaptation for directed dispersal to safe sites." This explains the evolutionary development of the feature by its end result; instead, the end result can be explained as a function of the evolutionary trajectory. The immediate reason for the hygrochastic dehiscence is the structure of the capsule, in particular, the structure of the different layers in the fruit wall that set up differential tensions when moistened and the different break zones in the fruit wall. The structure itself reflects a general trend: the overall reduction of the flower from a complex, multipartite system into a more or less fused mass, by telescoping, reduction, and fusion. The abscission (break) layers represent meristems or growth zones that have undergone almost complete suppression; they produce no structure except a thin layer of small, weak cells.

Hygrochastic capsules are well known in many desert plants, such as Aizoaceae, and are regarded as adaptations to the desert climate. In New Zealand, fruits that function in a similar way are present in the plants of very wet mountains, but they are still regarded as adaptive. Is there anywhere plants live where they would not be interpreted as an adaptation? Hygrochasy is a function of differential expansion of different tissue layers when wet, and the long-term evolution of this feature has little to do with the environmental conditions at any particular time. The fruit wall or "carpel" is a complex of many different layers in different stages of parallel reduction and fusion in different groups. This process has developed over hundreds of millions of years and is a trend that is much older than any individual landscape it has occurred in. The trajectory determines the structure of the fruit wall at any one point in time and so affects the means of dispersal; if the morphology lies above a minimum threshold of viability for its environment, the lineage survives, otherwise it goes extinct. To summarize, hygrochasy has been seen as a "strategy," driven by an end or purpose, but it can also be seen as the result of prior tendencies in the evolution of the fruit wall.

An antitelechoric strategy or a morphological tendency that results in dispersal over shorter and shorter distances can lead to the group's extinction. Ronce (2007) described the imminent "evolutionary suicide" of the rare, localized plant *Centaurea corymbosa* (Asteraceae) in southern France (cf. Chapter 13). Ronce wrote: "Because of the high risk of dispersal in this cliff species, seed traits enhancing long-distance dispersal have been counterselected, resulting in the absence of colonization and exchange between populations." If evolution is driven by long-term trends in mutation rather than strategy, seed traits that reduce dispersal can develop as by-products, not responses to the risks of dispersal, and these same trends, if continued, will lead to the plant's extinction.

Ecological Interaction as the Result of Intersecting Trends in Plant and Animal Evolution

Over the flowering season of 2012/2013, a plant of *Parsonsia heterophylla* (Apocynaceae) growing in a Wellington garden in Ngaio trapped at least one individual of *Dilophus* sp. (Diptera) per day (personal observations). While visiting the flowers, the flies caught their legs in the V-shaped gap formed by the anther tails and could not escape. The flowers also included snapped-off limbs of some flies that had escaped. The *Dilophus* is one of the three insects that visited the flowers. The others—a second nematoceran fly and a third hymenopteran which only made rare visits to the flowers—were not caught. In 2014 dead specimens of another small (~2 mm) fly and a butterfly, *Lycaena* sp. (caught by its proboscis), were found trapped in the same way (F.M. Climo, pers. comm., 2014), and a small moth was observed caught in a *Parsonsia* flower on nearby Mt Kaukau. In Taiwan, small lycaenid butterflies visiting the flowers of *Parsonsia alboflavescens* are trapped and dismembered in the same way (T. Livshultz, pers. comm., January 30, 2015).

Why does *Parsonsia* trap the *Dilophus* and the others? Is it adaptive? The phenomenon may have consequences or implications, such as reduced pollination, but these are not the cause of it. The relationship is an epiphenomenon, the result of prior trends in stamen structure, on one hand, and insect leg structure, on the other. It is likely that the relationship is an old one: the Eocene fossil *Dilophus campbelli* from North Otago is "close to a common extant species *D. nigrostigma*" (Harris, 1983), while the plant *Parsonsia* has a typical, trans-tropical Pacific distribution that can be attributed to mid-Cretaceous tectonics (Heads, 2014: Fig. 3.14).

A similar case to that of *Parsonsia* and its insects is seen in some other New Zealand Diptera that visit orchid flowers. Dead fungus gnats—adults of the glow worm *Arachnocampa* (Keroplatidae)—have been observed trapped in a narrow tunnel in the flowers of the orchid *Nematoceras trilobum* (C. Lehnebach, pers. comm. in Hansford, 2013: 51). Likewise on Swampy Summit, Dunedin, Thomson (1890) described his observations on an unnamed *Corysanthes* orchid [probably the plant now treated as *Nematoceras orbiculare*]: "often when a flower is opened an unfortunate little fly is found inside, glued by its head to the sticky gland of the column…." As Lehnebach commented: "Killing your pollinator isn't a very good strategy… we may be looking at a maladaptation." Some of the by-products of an evolutionary trend will be adaptive, some neutral, and some maladaptive—all of these are secondary to the evolutionary trajectory itself.

Case Studies of Reduction Series in Animals

Reduction in Crustaceans

The reduction of the abdomen (the "tail") in decapods forms a familiar series (discussed and illustrated by Glaessner, 1960: 44). The sequence includes (1) shrimps (Caridea), which have a tail and use abdominal appendages (pleopods) to swim, (2) lobsters (Astacidea), which have a tail but reduced pleopods, and (3) crabs (Brachyura), which have a very short tail folded under the thorax, and which are benthic, walking on the seafloor rather than swimming. This reduction trend has determined major aspects of locomotion and ecology in decapods, such as the benthic habitat of many groups.

Reduction of the Gills in New Zealand Gastropods and the Transition from Sea to Land

The morphology and physiology of species correlate with their habitat, but does the habitat cause the morphology, or does the morphology determine the habitat? For example, consider two sister species of gastropods on the seashore. One occurs at lower levels and has less ability to tolerate air, while the other lives at the top of the intertidal zone and its morphology allows a subaerial existence. In a standard model, the upper shore species invaded its habitat and *then* developed suitable morphology because of selection pressure. In an alternative model, the structure developed because of prior genetic tendencies in the ancestor and simple genetic and structural trends. *Given their structure*, the two species have selected different environments.

Morton (1952) discussed a group of land snails in New Zealand (*Murdochia*: Cyclophoridae, now in *Cytora*: Pupinidae) and wrote:

> The Cyclophoridae have accomplished the transition from the sea to a land habitat with *relatively few modifications* of their primitive structure… The retention of the operculum closing the shell aperture prevents water loss. The air-filled pallial cavity remains widely open anteriorly and there is, properly speaking, no development of a lung, respiration taking place merely through the smooth vascularised epithelium of the pallial roof; *gill filaments are lost and there are no folded respiratory lamellae.* [italics added]

Similar reduction, fusion, and loss of "gills" and other, elaborate, branched structures are observed in many groups—it is a widespread trend—and the process often has important ecological consequences. For example, in areas of coastal uplift, intertidal organisms are always being raised into terrestrial habitat; any that have viable (pre)adaptations will survive, others will die out.

Reductions in Moths and Weevils of the Three Kings and Poor Knights Islands

The moth *Izatha quinquejacula* of the Three Kings Islands and *I. dulcior* (Oecophoridae) of the Poor Knights Islands are not regarded as sisters, but they share distinctive morphological features and were keyed together by Hoare (2010). One feature they share is a tendency toward the loss or reduction of the deciduous cornuti on the genitalia. This is most obvious in *Izatha dulcior*, which differs from its sister species (*I. epiphanies*) in its complete absence of cornuti. Likewise, *I. quinquejacula*, endemic to the Three Kings Islands, has only 3–5 cornuti compared to 12 or more in its mainland relatives (*I. mesoschista* and *I. haumu*). Hoare (2010: 23) commented: "It is not clear why island life should relax or even reverse the selection for sexual weaponry." Nevertheless, the Aupouri Peninsula endemics *I. haumu* and *I. taingo* show no reduction in cornuti number, although their habitat is a former island. This suggests it is not the island *habitat* that is the key factor in the cornuti reduction in *I. dulcior* and *I. quinquejacula*, but the *location* of these "outer arc" species. The loss of cornuti is a by-product of the (regional) trend, not a cause.

In the same Poor Knights–Three Kings region of reduction in *Izatha*, other "loss features" are displayed in weevils, Curculionidae: "A very clear example of the dramatic effect that loss of flight has on the morphology of these insects is provided by flightless species of *Psepholax* on the Three Kings Islands and Poor Knights Islands..." (Lyal, 1993b). Here again the morphological trend has led to a morphological condition (flightlessness) as a by-product, not because it has been "adapted for."

Reduction and Bilateral Symmetry in Vertebrates

Rosenberg and McShea (2008: 27) wrote: "... the question remains whether every feature of every plant and animal that has emerged in the natural history of this planet is an adaptation, or even whether most are... It might explain appendages in animals, but it is not so obvious that it explains bilateral symmetry...." Vertebrates are, on first appearance, bilateral; this is the fundamental fact of their symmetry. Yet vertebrates also display many traces of underlying spiral symmetry, and left/right asymmetry is deep seated throughout the vertebrate body.

Remnants of Spiral Symmetry in Vertebrates

A detailed study of the spiral pattern of scales in the African reedfish *Calamoichthys calabaricus* (Polypteridae) compared this with phyllotaxis in the palm *Sabal palmetto* (Breder, 1947: Plate 16). Breder concluded that the arrangement of the parts is "basically the same" in the animal and the plant, and that both can be treated from the standpoint of the Fibonacci sequence: 1, 1, 2, 3, 5, 8, 13, ..., with each number the sum of the previous two. (This sequence is evident in the "numerators" and "denominators" of the phyllotactic series, Chapter 13.) The dermal denticles of sharks fall into similar geodesic (spiral) lines, as do the jaw teeth of sharks. In mammals, spiral, rather than bilateral, structures occur from the most deep-seated regions (the mammalian heart is a helix; Buckberg, 2002) to the most superficial (as in the spiral of the hair tracts on the crown of a human head or the nape of the neck in the marsupial *Pseudocheirus*). Breder (1947) noted that the "scales" of the pangolin (*Manis*) "are clearly following geodesics in a rather simple fish-scale fashion...."

Breder (1947: Fig. 32) also found that the scale spirals in fish have a close relationship with the underlying muscle blocks (myomeres) and the axial skeleton. Both scales and myomeres follow related patterns that show intimate connections and form a single structure. In the gut, the spiral valve of sharks is well known. The same deep-seated spiral symmetry is also seen in the plant shoot, which is constructed from spiral sectors comprising a leaf, its axillary bud, and an associated cortical strip, as well as vascular components (see Chapter 13).

Asymmetry in Bilateria and in Vertebrates

In addition to the *spiral* scales of fishes and hair whorls of mammals, left-right *sidedness* (laterality, chirality, handedness, or directional asymmetry) occurs to a greater or lesser extent in all the orders of chordates (including vertebrates) and echinoderms. It is present from the embryo stage onward and is evident even in the most fundamental organs, such as the brain and the heart (Raya and Izpisúa Belmonte, 2006). A vertebrate with pure, bilateral symmetry would not be viable. Despite this, the usual assumption in explaining asymmetric morphology in any group of vertebrates is that the group began with "perfect" bilateral symmetry and has altered this in response to particular needs. Boorman and Shimeld (2002) suggested that internal asymmetry "arose and became fixed early in vertebrate evolution," but asymmetry is present in the other chordates, in echinoderms, and in Bilateria (arthropods, annelids, etc.), suggesting that it was already present in vertebrates before their origin as a distinct group.

In their review, Boorman and Shimeld (2002) stressed that left-right asymmetries have a wide distribution throughout the Bilateria. In many cases they are obvious; most gastropod mollusks have shell coiling with a fixed direction, and many tube-dwelling polychaete worms have a tube with fixed spiral direction. In other groups the asymmetry is more subtle.

Left-right asymmetry is frequent in the copulation organs of insects (Eberle et al., 2015). It is also evident in the wing size and gut morphology of *Drosophila* and the mouthparts of thrips. *Drosophila* is known to possess genes that govern the helical torsion of the body and the rotation of the embryonic gut proventriculus (Levin, 2005). Both of these asymmetries are instances of handedness, and this appears to dominate in the ciliates, mollusks, and other invertebrates. Extreme asymmetry is seen in the claws of the fiddler crabs, *Uca*, with one much larger than the other.

Despite their apparent, superficial, bilaterality, most vertebrates have internal organs, such as the heart and gut that show strong asymmetry. The left side of a vertebrate contains most of the heart, the stomach, the pancreas and the spleen, whereas the right side contains most of the liver and the gall bladder. In addition, the gut coils anticlockwise, and in tetrapods the left lung has fewer lobes than the right lung (Raya and Izpisúa Belmonte, 2006). Asymmetric lateral reduction occurs in the partial or complete suppression of the right lung in amphisbaenians, and the left lung in snakes, various lizards, and amphibians (Butler, 1895; Gans, 1975). These asymmetries are uniform among the individuals of a species, and rare departures from the norm in humans result in severe medical conditions.

In the embryology of all vertebrates, the first major asymmetry is a directional kink in the median, symmetrical heart tube (Boorman and Shimeld, 2002). In all chordates, the right brain controls the left body and the left brain controls the right body. The right brain specializes in visuospatial abilities, while language is controlled by the left hemisphere. Left-right handedness is, of course, well known in humans, but population-level asymmetries in limb preferences are common in vertebrates (Ströckens et al., 2013).

Echinoderms are close relatives of chordates, but the basic symmetry of the adult is pentaradial, not bilateral. The animals have what appears to be a bilateral larva, but this symmetry is not retained throughout larval development. In sea urchin larvae, "the pentaradial adult body plan is laid down in a structure called the adult rudiment, which lies within the larva. This adult rudiment develops on the left of the larva…" (Boorman and Shimeld, 2002).

The phylum Chordata consists of vertebrates and two other subphyla, Cephalochordata (amphioxus) and the Tunicata (ascidians or sea squirts, and others), and left-right asymmetry is evident in all three (Boorman and Shimeld, 2002). The internal organs or viscera are a common site of asymmetry: this is seen in the gut of tunicates, the gut and pharynx of amphioxus, and the heart, gut, and associated organs of vertebrates. Amphioxus larvae display a "radical head asymmetry" until metamorphosis, when:

> the mouth moves to a symmetric midline position… The existing right-sided gill slits now migrate to the left side, and new gill slits open on the right. Combined, these processes re-establish gross pharyngeal symmetry. However, clear morphological asymmetries remain in the adult, including the positioning of Hatschek's pit and of a blind gut diverticulum… (Boorman and Shimeld, 2002: 1006)

In the traditional idea, that symmetrical vertebrates have given rise to asymmetric forms (e.g., Palmer, 1996). Nevertheless, left-right asymmetry in all vertebrates is regulated by a conserved developmental pathway. Three genes, *nodal*, *lefty*, and *Pitx2*, are "left" genes, in that they are preferentially expressed on the left side of embryos of all studied vertebrates. Boorman and Shimeld (2002) concluded that the three genes form a conserved signaling cassette that has regulated left-right asymmetry since the origin of the group. Mounting evidence also supports the existence of a *nodal/Pitx* pathway regulating asymmetry in amphioxus and ascidians (Boorman and Shimeld, 2002). Through the Paleozoic phylogeny of the vertebrates, a trend toward reduction and fusion is evident, producing a "simpler" structure. During this process, fundamental changes toward a simpler symmetry have also occurred; nevertheless, the reduction in itself, if uneven, will produce asymmetry. Slight asymmetries and traces of earlier morphology persist, even in the most bilateral structures.

Asymmetry in Agnatha

In the jawless vertebrates, the two sets of gill slits in the hagfish are sometimes asymmetric, with different numbers of gill slits on the left and right sides. In the hagfish *Polistotrema stouti* (now *Eptatretus stoutii*), Hubbs and Hubbs (1945) recorded the count as asymmetric in about one-sixth of the specimens, and in these the openings were more numerous on the left side almost twice as often as on the right.

Asymmetry in Fishes, with Special Reference to Flatfishes

The spiral symmetry of fishes, as seen in the scales and muscle blocks, was cited earlier. In addition, many fishes show bilateral asymmetry. The asymmetry of flatfishes is well known, but bilateral differences between left and right halves characterize all fish groups and appear in many structures, both superficial and internal. Hubbs and Hubbs (1945) wrote that "Why the two sides of fishes differ from each other is a question which has never been adequately answered." They cited many examples, including the following.

The postlarvae of the eel *Leptocephalus diptychus* (now *Moringua edwardsi*: Moringuidae) are marked on each side by a row of conspicuous spots; the specimens examined had three on the left side and four on the right. Some fish species have more rays in the right pectoral fin than in the left, while others have more rays in the left. In *Xenodexia* of Guatemala (sister to the rest of the family Poeciliidae), the right pectoral fin is "spectacularly modified into a very complex structure," with hooks, pads, and other processes. The left pectoral fin is more or less normal, with "scarcely a trace" of the bizarre specialization of the opposite member (Hubbs, 1950).

In the Asiatic ricefish, *Horaichthys setnai* (Horaichthyidae), most females lack the right pelvic fin; in most cases, there is no trace of either the right pelvic fin or its basal bone at any stage of development. In the same species, the oviduct in the females opens to the left of the median ventral line in about 60% of the individuals and to the right side in only 20%.

In fishes that have the branchiostegal membranes not united with each other or with the isthmus, the left branchiostegal membrane folds over the right one (there are some reversals in the pike, *Esox*: Esocidae). In *Esox*, *Albula* (Albulidae), and *Oncorhynchus* (Salmonidae), the branchiostegal rays in the left membrane outnumber those of the right.

In the sawfishes (Pristidae), bilateral variation often occurs in the number of teeth along the two edges of the rostral blade. In cyprinid fishes with asymmetric dentition, the larger number of pharyngeal teeth occurs most often on the left arch. Asymmetry is also evident in the gonopodium of Anablepidae and the priapium of Phallostethidae.

Fishes in general have the liver on the left side and the intestinal coils lie against the right body wall. In addition, the brain of fishes and also amphibians is asymmetric, as the left and right habenulae differ in size and anatomy, and these differences arise early in ontogeny (Palmer, 2004). Whether the right or the left optic nerve is uppermost in the chiasma is a prime distinction between two families of flatfishes.

Left-right twisting in the jaw occurs in many vertebrates, and it has determined aspects of feeding ecology in some cichlid fishes. In *Perissodus*, the mouth opens either to the right or to the left as a result of an asymmetric joint of the jaw to the suspensorium (Hori, 1993; Stewart and Albertson, 2010). Perfect bilaterality is often assumed to be the original condition for animals; in cichlids, for example, it has been argued that: "While bilaterality is a defining characteristic of triploblastic animals [~Bilateria], several assemblages have managed *to break this symmetry* in order to exploit the adaptive peaks garnered through the lateralization of behaviour or morphology..." (Stewart and Albertson, 2010; italics added). Instead, it is suggested here that all bilaterians have always had a fundamental asymmetry that underlies the bilaterality.

In the Perissodini clade of cichlids, "selection has accentuated a *latent, genetically determined handedness* of the craniofacial skeleton, enabling the evolution of jaw asymmetries in order to increase predation success" (Stewart and Albertson, 2010; italics added). The latent, genetic handedness is the primary, prior trend or tendency, and selection has been a secondary force accentuating this. Despite this, the authors argued that the primary course of evolution has been by "discrete shifts in skeletal anatomy that reflect differences in habitat preference and predation strategies..." (Stewart and Albertson, 2010: 1). Instead, in a mutation-driven model, the shifts in jaw anatomy *caused* the habitat preferences and feeding strategies. Apart from the asymmetry in their jaws, cichlids also exhibit asymmetry in their small intestine, which exits from the left side of the stomach, rather than the right side as in most fishes (Barlow, 2000).

The fundamental asymmetry of the vertebrates, with spiral patterns underlying the apparent bilateral plan, sheds light on the well-known case of the flatfishes. The 400 species in this order, Pleuronectiformes, are regarded as a "wonder of nature" because of the pronounced asymmetry; both eyes are on one side of the head. On the blind side, the muscular development is more or less weakened and the color disappears, while on the eyed side, the teeth in many species become reduced in number or lost (Hubbs and Hubbs, 1945). Some flatfish families are dextral, others sinistral; one family has both dextral and sinistral genera.

During metamorphosis of a flatfish larva into a juvenile, most of the hard tissue in the cranium twists; one eye appears to migrate to the other side of the head, and the jaws develop a strong asymmetry (Stewart and Albertson, 2010). At the same time, the color and scales of the skin on both sides differentiate (Okada et al., 2001). This appears to be the most dramatic right–left asymmetry in the body plan of any vertebrate. Did it develop because of extrinsic needs or because of internal, mutation-driven processes?

In an orthodox neodarwinian model, the habit and habitat of a group come first, before the adaptations for these. The twisting of the head developed in flatfishes "because their ancestors lay down on one side" (Dawkins, 2006: 121). Pinker (1994) wrote that "The Picassoesque face of the flounder was the product of warping the head of a kind of fish that had [already] opted to cling sideways to the ocean floor, bringing around the eye that had been staring uselessly into the sand." Again, this adaptationist approach assumes that the habit—the fish lying on its side—came first and that the structure followed, resolving the tension. It would be more credible if the structure evolved first, and if the new structure determined the new habit.

The neodarwinian model for the flatfish focuses on the idea that the new seafloor habitat and posture meant that the animal had new needs. In particular, the head needed to twist, and Dawkins (2005) emphasized that the "absurd distortion" in the flatfishes is carried "to the point of grotesqueness." He described the flatfish skull as "imperfect" (Dawkins, 2005), and another neodarwinian called it "weird" (Dennett, 1996). But is the same trend in asymmetry in the heart and the brain of mammals also weird and absurd?

Dawkins (2006: 121) suggested that the twist in the flatfish developed "because their ancestors lay down on one side," but this does not explain why they would lie down to begin with. The flatfish appear grotesque or absurd, a freak "wonder of nature" only if it is studied as an isolated case, not as a member of a series or trend. By taking it out of its morphological context, it also appears to be an exquisite adaptation that has evolved "for the best," in order to supply an extrinsic need. This *need* is the key concept in the lamarckian/neodarwinian model (Lamarck, 1809).

As Nei (2013: 161) noted, the flatfish morphology has been explained using Lamarckian ideas (Darwin, 1872) or mutation (Morgan, 1903). Nei favored mutation-driven evolution, and he compared flatfish morphology with left-right asymmetry in the vertebrate gut and in snails. Thus an alternative explanation does not rely on teleology, but instead suggests that the flatfish lay down on one side because the head was twisted to start with. In this situation, lying on the seafloor was the best option available, given the morphology. This second model is supported by a comparative approach to the problem, as left-right asymmetry is a trend evident in many animal phyla. Instead of being a one-off "wonder of nature," the flatfish morphology is a particular point on a trajectory that can be analyzed with respect to phylogeny and morphogenesis. Left-right asymmetry occurs not just in Pleuronectiformes but, to a greater or lesser extent, in *all* orders of chordates and echinoderms. Flatfishes are just a moderate case of a phenomenon that is widespread throughout animal as well as plant life. Vertebrates are more or less bilateral, but this is far from "perfect" or "original."

Asymmetry in Squamates

Snakes, of course, display great reduction in their pectoral and pelvic girdles. But there is also extensive reduction in the viscera, and this has often been asymmetric; for example, in snakes the left lung is vestigial. In other examples of asymmetry, snail-eating snakes in Colubridae subfamily Pareatinae have more teeth on the right mandible than on the left, and, as in the cichlids mentioned already, the asymmetry can determine the aspects of feeding ecology (Hoso et al., 2007). One of the snakes, the snail-eating *Pareas iwasakii*, was found to extract the soft body of a coiled snail faster if the snail had a dextral rather than a sinistral body. The snakes dropped sinistral snails more often, owing to behavioral asymmetry when striking.

In other squamates, *Anolis* lizards display great diversity in the shape, color, pattern and size of their dewlaps, flaps of skin beneath the lower jaw. Asymmetry, with one side of the dewlap differing in pattern or color from the other, has been reported in one species, *Anolis lineatus*, of Aruba and Curaçao (Gartner et al., 2013). The morphology is "remarkable and unique" and the authors asked: "Of what adaptive use might be the uniquely asymmetric dewlap of *A. lineatus*? One hypothesis is that each side of the dewlap in *A. lineatus* serves as a separate signal." As with the flatfishes, a signal effect is possible, but instead of being a "wonder of nature" this function would be a result, not a cause, of the asymmetry.

Asymmetry in Birds

In birds, only the left ovary and oviduct are functional; the only exceptions to this are kiwis (Apterygiformes) and some raptors (Kinsky, 1971). In the owls, Strigiformes, ear asymmetries are present in at least five main clades (for a detailed account of the morphology, see Norberg, 1977).

In Charadriiformes, at least four species of oystercatchers (*Haematopus*: Haematopodidae) include a minority of individuals with bent bills, most of which are bent to the left (Hatch, 1985). The bill in all individuals of the New Zealand wrybill plover, *Charadrius frontalis* (Charadriidae), bends to the right, as in the American genus *Aramus* (Aramidae: Gruiformes). *C. frontalis* feeds by searching under stones and by scything its bill down and to the right in mud.

In other birds, the limbs of the pigeon *Columba livia* and the parrot *Amazona amazonica* are asymmetric, with both right limbs longer in the pigeon and both left limbs longer in the parrot (McNeil and Martinez, 1967). In another parrot, *Aratinga pertinax*, observations of 56 individuals bringing food to the beak indicated that 28 were right handed and 28 were left handed. There was also a slight departure from bilateral symmetry in hindlimb bones and this bore a close relationship with handedness (McNeil et al., 1971). Passeriformes also display asymmetry; for example, in crossbills (*Loxia*: Fringillidae) and one of the Hawaiian honeycreepers (*Loxops*: Fringillidae), the two mandibles cross each other laterally.

Asymmetry in Mammals

Of the two main groups of cetaceans, the baleen whales (Mysticeti) have a double blowhole and a symmetrical skull, while the toothed whales and dolphins (Odontoceti) have a single blowhole (the result of reduction) and an asymmetric skull (Nowak, 1999). Skulls in Eocene fossil cetaceans (protocetids and basilosaurids) are also asymmetric and show curvature and axial torsion of the cranium (Fahlke et al., 2011). The asymmetry in odontocetes is evident in the sound-producing apparatus and the facial part of the skull, and it has been attributed to the need for sound production (Yurick and Gaskin, 1988). Instead, it can be interpreted as another manifestation of the underlying asymmetry seen in all vertebrates. Given the starting point in ancestral pre- and proto-whale morphology and the subsequent evolutionary process of reduction and fusion, sound production was probably inevitable.

Ness (1967) measured the skull asymmetry in 30 of the 35 odontocete genera. One species, *Stenodelphis blainvillei*, proved to be symmetrical, but all other species had a leftward deviation of the nasal prominence. The sequence in degree of asymmetry is as follows: *Kogia* > monodontids > delphinoids (except *Orcinus*) > ziphiids > *Orcinus*. In the narwhal (*Monodon*: Monodontidae), the upper left incisor forms a single, helical tusk that grows up to 3 m long, although the asymmetry of the skull "is independent of the possession of a tusk" (Ness, 1967). Ness suggested "that the leftward deviation of the dorsal aspect of the [odontocete] skull acts to conserve the symmetry of the body surface by countering an opposite asymmetry of the upper breathing passages, an asymmetry which is primary but whose function is unknown." Apart from morphological asymmetry, several examples of lateralized behavior patterns have been detected in odontocetes, and these can be compared with handedness in humans (Galatius and Jespersen, 2005).

In baleen whales (mysticetes), asymmetry is seen in the coloring of the head in the fin whale *Balaenoptera physalus*:

> ...whereas the left side, both dorsally and ventrally, is dark slate, the right dorsal cephalic side is light gray and the right ventral side is white. This asymmetry also affects the baleen plates: those on the whole left side and the rear two-thirds of the right side are gray, whereas those on the front third of the right maxilla are yellowish. (Aguilar, 2009: 433)

Grzimek (1988/1990, vol. 4: 433) suggested: "It may be that the one-sided 'light patch' plays some role (as a lure?) in its feeding...."

In Chiroptera, the São Tomé Island (West Africa) frugivorous bat *Myonycteris brachycephala* has an asymmetric dental formula (Juste and Ibáñez, 1993). The asymmetry results from the absence of a lower internal incisor in either the right or left mandible. It has been argued that "This morphological innovation is possibly related to a developmental instability associated with the colonization by the island species" (Juste and Ibáñez, 1993). Instead, the "developmental instability" in the bat skull can be related to prior trends in bat morphology, associated with phylogeny and mediated by tectonic instability along the Cameroon Volcanic Line (Heads, 2012a, 2015b).

Ontogeny and Phylogeny of Symmetry in Vertebrates

For vertebrates, Raya and Izpisúa Belmonte (2006) wrote that:

> The establishment of LR [left-right] asymmetries is controlled by robust genetic and epigenetic mechanisms... Genetic, pharmacological and microsurgical approaches have identified progressively earlier requirements for the correct establishment of LR asymmetries in frog, chick and zebrafish embryos. Nevertheless, despite years of intensive research, no satisfactory explanation has been provided for the initial symmetry-breaking event in any of these species.

There is some information about the molecular cues that specify the left-right axis in visceral organ development, but how the brain becomes lateralized is not well understood (Halpern et al., 2003). In zebrafish, the *Nodal* signaling pathway functions in both visceral asymmetry and in the embryonic brain, where it biases the laterality of the epithalamus. The asymmetry of the pineal complex means it has differential influences on the adjacent diencephalic nuclei, which acquire distinctive molecular and cellular features.

Levin (2005) wrote that non-conjoined monozygotic twins, while not exhibiting the kinds of visceral laterality defects that occur in conjoined twins, show many subtler kinds of mirror-image asymmetries, as in hand preference, hair whorl direction, and tooth patterns. Nervous system lateralization is widespread in vertebrates, and many animals, such as mice, show paw preference. Nevertheless, consistent preference among all individuals (handedness) only approaches high levels in humans (~90% are right handed).

In a wide-ranging review, Levin (2005) discussed asymmetry in vertebrates and posed a provocative question: "When, during evolution, did handed asymmetry appear, and were there true bilaterally symmetrical organisms prior to the invention of oriented asymmetry... ? Is it connected to chirality in lower forms (such as snail shell coiling and chirality in some plants)?" As suggested earlier, left-right (LR) asymmetry can be derived from spiral symmetry during phylogenetic/morphogenetic reduction.

To conclude, LR asymmetry is a fundamental characteristic of the vertebrates and the whole Bilateria—it is a true Darwinian "law of growth," underlying the secondary bilateral symmetry. This means there is no need to develop adaptive explanations for particular cases of asymmetry. LR symmetry is a derived form of LRLRLR-type symmetry, equivalent to distichous phyllotaxis, with units in two rows. This gives a bilateral structure, with one of the minimal symmetries, and the system can be derived from a more complex, spiral structure during a general process of reduction. Some of the reduction series seen in particular regions of the vertebrate body are cited in the next sections. These are followed in the next two chapters by discussions of examples from New Zealand birds and mammals.

Reduction in Vertebrates

Limbs

In vertebrates, the limbs provide one of the most obvious and best-known cases of a reduction trend. Large-scale reduction of the limbs is common throughout the vertebrates, and it occurs in many groups of fishes, amphibians, squamates, birds, and mammals such as whales and pinnipeds. In the classic Lamarckian and neodarwinian accounts, changes in habit and conditions produced changes in function and structure of the limbs. Instead, in a mutation-driven model, limb reduction itself, as a general trend or law of growth, would have caused great changes in the functions, habits, and habitats of organisms. The reduction would also explain relictual, glandular features, such as the tarsal, metatarsal, and interdigital "scent" glands of many artiodactyls.

In squamate reptiles, a limbless, snakelike body form has evolved at least 25 times (Wiens et al., 2006). In groups of mammals, the Cetacea, Sirenia, and pinniped Carnivora represent some of the most extreme variations in body form because of the dramatic reduction in their limbs. The reductions and fusions have produced hemmed in, more or less flat, bilateral structures, and this meant the animals could no longer function on land. They were obliged to live in the water, either full time or, in the case of Pinnipedia, part time.

Another reduction process in the limb is evident in the stabilization of digit number at five, with many decreases from this, but increases beyond five being very rare (Chapter 13). The reduction leading to one digit is well known in the horse family, Equidae. In the modern horse, *Equus*, the *lateral* hoofs are gone. The second and fourth digits and metacarpals/metatarsals are represented only by rudimentary splint bones, and the limb is supported by the third metacarpal/metatarsal and its hoof. A parallel, extreme reduction is seen in "false horses" such as *Thoatherium* of the fossil order Litopterna. In fossil equids such as *Protohippus* (Miocene–Pliocene) and in Litopterna such as *Diadiaphorus*, the digits and metacarpals are reduced, but less so. In Eocene horses such as *Eohippus*, reduction of digits 2, 4, and 5 is minor, as in human limbs.

In contrast to the horses, in which the *lateral* digits are reduced, in the hind flippers of pinnipeds, it is the *central* digits that are reduced. They are most reduced in Phocidae (earless seals) and show less reduction in Odobenidae (walrus) and Otariidae (eared seals). The common theme in all these mammals—horses, litopterns, and pinnipeds—is symmetrical reduction in the distal part of limb at, or around, a central axis.

Dawkins (2005) discussed reduction in limbs. He cited the Devonian tetrapods with 8, 7, and 6 digits (see Chapter 13) and wrote: "My tentative guess is that... the different species really did *benefit* from their respective numbers of toes...." [italics added], a standard neodarwinian position. But he also wrote: "Later the tetrapod limb design hardened at five digits, probably because some *internal embryological process* came to rely on that number. In the adult, the number is frequently reduced from the embryonic number [e.g., in horses]" [italics added].

A long-term, internal genetic/epigenetic process, rather than speculative "benefit," best explains why the post-Devonian digit number is often reduced from five but never increased, and why the earlier tetrapods had higher and more variable digit numbers. The reduction series has been played out in vertebrate life since the early Paleozoic, over hundreds of millions of years.

Reduction at the Posterior End of Bilaterian Animals

Reduction is evident throughout the vertebrate body but is perhaps most obvious at the anterior and posterior poles. As mentioned earlier, glands occur in zones of reduction, and the posterior end of most birds includes the uropygial gland, with secretions that are used in preening. Reduction at this pole also includes larger-scale changes.

Archaeopteryx, the earliest known fossil bird, had an elongate tail skeleton that supported pairs of tail feathers in a frond-like (pinnate) arrangement. In contrast, all modern birds have a short caudal axis that bears tail feathers in a fanlike (palmate) arrangement (Gatesy and Dial, 1996). The modern form has developed through shortening of the caudal centra, sequestering of proximal caudals into the synsacrum, and fusion of the distal caudals to form the pygostyle. (Any reduction in the total number of caudal segments has been minor.) Evolution has proceeded from long tailed to short tailed and from frond to fan, and Gatesy and Dial (1996) discussed how the fan arrangement enables maneuverable flying. No modern birds have reelongated the caudal vertebrae to resemble *Archaeopteryx*.

The reduction or loss of the tail also occurs in many mammals. Dawkins (2005: 127) wrote: "I don't know why we apes have lost our tail... tail loss is found in moles, hedgehogs, the tailless *Tenrec ecaudatus*, guinea pigs, hamsters, bears, bats, koalas, sloths, agoutis and several others." There are also tailless monkeys (*Cacajao*, *Macaca* spp.) and tailless lemurs.

Dawkins (2005: 127) argued that: "Any organ which is not used will, other things being equal, shrink for reasons of economy if nothing else," but posterior organs including the tail are often reduced anyway, possibly as part of a mutation-driven trend. The teleological model of evolution accepts that tails exist because they supply a need, caused by the organism's adopting an otherwise unhealthy habit or habitat. Tails develop "When a tree-climbing animal *needs* a fifth limb" (Dawkins, 2009).

Nevertheless, Dawkins (2005) also suggested that wolves and many other mammals use their tail for signaling, although "this is likely to be secondary 'opportunism' on natural selection's part." This idea—that function is a secondary, opportunist development *dependent on structure*—also applies to all the other uses the tail is put to. Arboreal animals use it to control their balance, and in Borneo and Sumatra, the long-tailed macaque *Macaca fascicularis* undertakes arboreal travel, while the related pig-tailed macaque *M. nemestrina*, with a short tail, travels on the ground. (This may indicate that tail loss has helped determine the terrestrial habitat, although gibbons and chimps are also tailless and are active in trees.) Posterior reduction is also evident in the tuatara, *Sphenodon*, as the male lacks an intromittent organ. This contrasts with turtles and crocodilians, which have a penis, and squamates, which have paired hemipenes (Cree, 2014: 330).

Reduction in the Vertebrate Head

In the usual, subconscious idea, the evolution of the vertebrate head begins with a smooth, simple, round object, and then different organs—eyes, ears, nose, elaborations of these—are added on or developed as they are needed. Instead, the head is better interpreted as the result of suppression, reduction, and fusion in an already complex, multipartite structure. Traditional theory argues that morphological reduction takes place because a structure is *not* needed, but reduction in the vertebrate head has been a broad trend continuing for hundreds of millions of years. Reduction in the head and its precursors, for example in the pharynx and muzzle/rostrum region, has been one of the most important trends in vertebrate evolution.

Danionella dracula, a miniature cyprinid fish from northern Burma, provides a dramatic example of overall reduction in the vertebrate skull. (It occurs at a key node in Old World biogeography that delimits intercontinental regions; Heads, 2012a.) Compared with its closest relative, *Danio rerio*, *Danionella* lacks 44 bones or parts of bones. The reduction has been attributed to neoteny, and Britz et al. (2009) described *Danionella* as:

> ...one of the most developmentally truncated vertebrates... Absence of the majority of bones appears to be due to developmental truncation via terminal deletion... Larval-like features [provide]... a remarkable example [of] progenetic paedomorphosis via heterochronic change in developmental timing.

In contrast with the suppression of so many bones, *Danionella* has *not* lost what appear to be teeth, which are otherwise absent from all 3700 species of cypriniformes.

The fish family Syngnathidae (seahorses and pipefishes) is another group that displays extreme reduction. As the name indicates, both jaws are fused together, and the fish is forced to feed by suction. This resembles the situation in the anteaters of the mammalian order Pilosa.

Reduction in the vertebrate skull has led to the complete fusion of parts, or at least to the reduction in movement between parts. Cranial kinesis refers to the movement of the skull bones relative to each other, and this is most obvious between the upper jaw and the neurocranium (braincase). Cranial kinesis is rare in mammals, but it has been retained in most fishes. In squamates the quadrate and upper jaw can rotate freely with the neurocranium (streptostyly) (Cree, 2014: 36–37). Scleroglossan squamates such as geckos, skinks, and snakes have further points of movement in the bones of the skull (mesokinesis or prokinesis), whereas in rhynchocephalians, the quadrate bone has become firmly attached to the upper jaw and neurocranium.

Comparative Lack of Reduction in Some Features of the Vertebrate Head

Most parts of the head have been recast during the overall reduction of its evolution, but some organs remain less reduced in some groups than in others. In the platypus skull, the bill shares obvious similarities with that of ducks. The platypus bill has 40,000 electrical sensors arranged in a dozen or so narrow strips along the length of the bill. A large proportion of the brain has structural connections with the bill and processes data from the sensors (Dawkins, 2005: 245). There are also 60,000 mechanical sensors. All the sensors are set into pores, and the electrical pores at least are modified mucus glands (p. 247). The bill is a complex structure within which different parts have already been reduced to glands and "sensors," but in most mammals the bill is reduced even further.

Another case of lack of reduction occurs in the North American star-nosed mole, *Condylura*. The nose is adorned with conspicuous tentacles in which the skin is more sensitive than any other area of skin in mammals, and the tentacles function as a tactile sensory organ. A large proportion of the *Condylura* brain is associated with the nose and processes information from it, as with the platypus bill and its sensors. The question is whether the habit and the function give rise to the tentacles and the associated sector of the brain or vice versa. The fossil record provides little help in understanding this sort of evolution, as glands and other soft tissue structures are seldom preserved.

Changes in Symmetry Accompanying Reduction and Fusion in the Skull

Several examples of reduction in the skull were discussed earlier in this chapter, in the section "Asymmetry in Bilateria and in Vertebrates". Dawkins (2005) noted that the flatfish skull is "distorted" (p. 359), and that in "the wonky-eyed jewel squid of Australia, the left eye is much larger than its right. [This is true for all *Histioteuthis* species]. It swims at a 45° angle, with the larger, telescopic left eye looking upward for food, while the smaller right eye looks below for predators" (p. 397). In other examples of skull asymmetry, "The wrybill [*Charadrius frontalis*] is a New Zealand sandpiper whose bill curves markedly to the right...."

Dawkins (2005) played down these examples of asymmetry and wrote: "... those are all exceptions, mentioned... to make a revealing contrast with the symmetrical world of our primitive worm ['common ancestor 26'] and its descendants." Nevertheless, asymmetry in Bilateria is not an exception; no Bilateria show pure, simple bilateral symmetry, and a bilateral form would not survive. This contradicts the idea that the "ancestral species" of the Bilateria had perfect bilateral symmetry.

Ears

The morphological series in which the gill arches of jawless fishes are represented by ear ossicles and jaws in higher vertebrates is a classic example of a reduction/fusion process. "The driver for this transformation [to ear ossicles] would have been natural selection's favoring enhanced ability to detect vibration" (Travis and Reznick, 2009: 120). In mutation-driven evolution, the ability to detect vibration is just one of the many by-products of the reduction that formed the modern vertebrate ear. Another is the asymmetry of the ears seen in groups such as owls.

The ear in tuatara, *Sphenodon*, displays signs of substantial reduction, even though this has probably had a negative impact on the animals' hearing:

> Tuatara lack an external ear opening, a functional tympanic membrane and an air-filled tympanic cavity. These absences suggest that tuatara are not as good at detecting high-frequency airborne sounds as lizards that have these structures. However, in early rhynchocephalians the skull includes a feature known as the tympanic-quadrato-jugal conch, which indicates that a tympanic membrane was present… The condition of the middle ear in tuatara is therefore now considered degenerate rather than primitive. (Cree, 2014: 48)

Eyes

In animals, eye formation is a standard trend or "law of growth"; life is "almost indecently eager to evolve eyes" (Dawkins, 2005: 603 and color plate 50). The structural aspects of the evolution of the eye have often been overlooked in favor of discussions about the eye's function and purpose. Dennett (1996) wrote that: "The solar system… isn't *for* anything. An eye, in contrast, is *for* seeing" (p. 64); "Does anybody seriously doubt that eyelids evolved to protect the eye?" (p. 246).

In fact, the revolutionary critique of teleology that marked the beginning of modern science cited the eyelashes as an example. Bacon (1605/1966) accepted that the protection of the eye—"safeguard of the sight"—was the final result of having eyelashes, but he argued that there must also be an "efficient cause" for their existence. He made the intriguing, non-teleological suggestion that the eyelashes originated in the first place because "pilosity is incident with orifices of moisture" (Bacon, 1605/1966: 113–114). In this way, he interpreted a particular morphology in terms of a general morphological trend, not a particular functional need.

Despite Bacon's early critique of teleology, the Modern Synthesis retained the teleology of ancient and medieval thinking. One modern author wrote: "natural selection explains why the parts of organisms serve purposes… teleology is an all-pervasive feature of the biological world. There are many obvious examples of biological traits that are for something: eyes are for seeing, hearts are for pumping blood around the body…" (Garvey, 2007: 47).

It is obvious that eyes see and hearts pump blood, but are these functions the cause of the structures? Are their functions what the organs are for, or are the functions just by-products? The heart pumps blood because (1) the heart is part of a vascular network into which blood has been secreted and (2) the heart undergoes repeated contractions. If both (1 and 2) were the result of prior trends in evolution, for example, reduction in vascular and nervous systems, then the heart would inevitably pump blood, without this being the organ's purpose or the cause of its origin.

The teleology of neodarwinism has long historical roots. The neodarwinian Williams (1996: 41) quoted Aristotle with approval: "…each of the parts of the body, like every other instrument, is for some purpose…." In this view, the purpose of an organ can be deduced by examining its current function, and so the purpose of the eye is sight. Williams (1996) argued that "we have eyes because we need to see" (p. 5) and that the eye is "the classic case of plan and purpose in nature" (p. 6).

Nevertheless, even though Williams (1996) was an adaptationist, he admitted that the paucity of human eyes—just two—and the excess of the muscles that control their movement—six in each eye, when three would suffice—"*seem to have no functional explanation*" (Williams, 1996: 6, italics added). The "excess" of muscles indicates relictual structure; the muscles supplied organs that are no longer in existence or have been "transfused" into other organs during reduction and fusion. The orbit–eye–optic nerve system is another zone of reduction, as in the ear and the muzzle. For example, in humans, the sphenoid is a somewhat butterfly-shaped bone that forms the back of the orbits. It is homologous with several bones that remain unfused and separate in other mammals (in dogs the sphenoid is represented by eight bones) and in reptiles.

Thus the eye of Bilateria can be explained, not as an evolutionary elaboration from a morphological "point center of origin," a single cell (Gehring, 2011), but as a typical by-product of reduction and "condensation" in the facial zone; light receptors and also many other glands have developed here. Light receptors are not restricted to well-defined eyes. Arikawa et al. (1980) described the range of sites where extraocular light reception has been recorded in insects. These include the central nervous system (photoreceptors have been found in the brain and the last abdominal ganglion), on the antennae, in dermal regions outside the head (often in the region of terminal abdominal spiracles), and on the genitalia of the Chinese swallowtail butterfly, *Papilio xuthus*. Dawkins (1996/2006: 129) described the last case: "Weirdly, there is good evidence of cells that respond to light in the genitalia of both male and female butterflies… nobody seems to know how the butterflies use them." The structure is unusual or "weird" in an adaptationist model, but it is not unexpected in the context of a zone of reduction at the posterior, as well as the anterior end of the animal.

Fundamentalist ultradarwinians (Pinker, 1994; Dennett, 1996; Williams, 1996) and creationists believe that organic structure was designed for a purpose, to fulfill a need. This view accepts that we have eyes in order to see. Instead, eye formation and continued evolution can be interpreted as an inevitable by-product of correlated mutations. Another by-product of these mutations is the condensation, reduction, and telescoping in the face and pharynx that is evident throughout the vertebrates and other Bilateria. A dramatic example is the reduction in the face of the odd-nosed colobines of south-east Asia (Heads, 2012a: 251). In different vertebrates, this reduction can also leave unreduced, "overadapted" organs of different kinds. Some eyes are too sensitive for daylight and must be used at night, others are so reduced they do not function and so the animal is obliged to live in complete darkness—in caves or underground, for example—where its blindness is shared by all. Many authors have argued that the eye *looks* designed, but this is the medieval argument—*everything* looks designed.

Eyespots occur even in unicellular organisms, and the production of "eyes" is another old trend, along with spiral symmetry and left-right asymmetry. As Dawkins (2006: 176) emphasized, "eyes evolve easily and at the drop of a hat," and the evolution of a lens, condensing out of the vitreous mass "wouldn't have been very difficult structurally" (Dawkins, 1996/2006: 142). These observations are consistent with the idea that "eye-making" occurs as a by-product at certain types of reduction zones, not as an end in itself. Around the eye, the lacrimal gland is a conspicuous structure in which morphological reduction has led to glandular activity (secretion of tears). Between the two normal eyes of vertebrates is the more or less suppressed pineal or third eye and the associated pineal gland, and, again, these indicate great reduction in this part of the face.

The Third Eye of the Tuatara: More Evidence for Reduction

In the tuatara (*Sphenodon*), juveniles have a well-developed pineal or third eye that has a small lens and retina, and is functional. The eye is also known as the parietal eye, as it lies between the two parietal bones. The pineal eye is associated with the pineal gland, which by producing melatonin affects sleep patterns and circadian rhythms; juvenile tuataras are more strictly diurnal than adults, and in adults the pineal eye is reduced (Cree, 2014: 430).

A pineal eye is also present in some fishes, frogs, and lizards, and is inferred in the many fossil reptiles that have a parietal foramen. In a number of lizards the pineal eye shows moderate development, but the nerve connections have degenerated, and the parts that are remaining are secretory (Dendy, 1910). The development of the eye is more or less suppressed in turtles, crocodilians, snakes, birds, and mammals, but these groups retain a trace of the structure in their pineal gland. To summarize, the pineal eye and gland represent an example of more or less suppressed structure being represented by glandular activity that has in turn led to behavioral changes.

The suppression of the third eye is a local by-product of the overall reduction in the face. Dendy (1910: 330) and others have demonstrated that the pineal eye in the tuatara is derived from the *left* member of two pineal outgrowths, while in lizards and in the New Zealand lamprey (*Geotria australis*), it is derived from the *right* member. This suggests a more or less equal development of both members in the common ancestor of the different groups. Thus the ancestral form would have possessed two pairs of photosensory structures—the normal, lateral eyes, and the structures that became the pineal eye.

The Vomeronasal Organ

The vomeronasal organ is located in the roof of the mouth and provides a chemical sense related to smell. It is found in many mammals, but in humans its development is suppressed in the fetus. In adults of *Sphenodon* (tuatara) and squamates (snakes and lizards), it persists as distinct, paired organs (Cree, 2014: 37–38). In squamates it is complex and contains a cartilaginous structure, the mushroom (fungiform) body, that includes an enlarged layer of chemoreceptor cells. Here the vomeronasal organ has a direct opening into the mouth via the vomeronasal duct. In contrast, the vomeronasal organ of the tuatara has a low density of chemoreceptor cells and no fungiform body. Also, its connection with the oral cavity is indirect as the vomeronasal duct has been obliterated. Because of this reduction, the structure is less effective as a chemosensory organ in tuataras than in squamates.

Reduction in the Head of Invertebrates

Similar reductions to those seen at the anterior end of the vertebrate animal are also well known in other Bilateria. Good examples of reduction in insects are observed in Trichoptera, for example, a group described with "Mouthparts reduced: mandibles not functioning" (Imms, 1973: 164). It is not surprising that the reduction has affected the function; caddis-flies take only liquid food (p. 165).

In Diptera larvae it is not just the mouthparts but the entire head that is reduced. In suborder Nematocera the head is fully developed (*eucephalous*). In suborder Brachycera the head is "incomplete posteriorly and partly embedded in the prothorax (*hemicephalous*): the mouthparts are highly modified" (Imms, 1973: 172). Among Cyclorrhapha (now treated in Brachycera) "the larvae are *acephalous* and this condition results from the whole head being invaginated into the thorax… True mouthparts are atrophied and their place is taken by adaptive structures…."

The same reduction–fusion process is also evident in Lepidoptera. In this group Imms (1973: 160) proposed that: "The structural similarity of these insects has led to great uniformity of behaviour," in other words, structure determines function. Yet a page later, Imms reverted to the standard view that function determines structure: "Numerous moths take no food and their mouthparts consequently display varying degrees of atrophy" (p. 161). Instead, the atrophy can be interpreted as one result of a broad, prior trend, and it is this trend that has eventually prevented feeding.

15

Biogeography and Evolution in New Zealand Birds

This chapter examines the biogeography of New Zealand birds, along with aspects of their morphology, and investigates whether the groups can be explained by centers of origin overseas, dispersal to New Zealand, and adaptation to local conditions (the CODA model).

Oliver (1955: 33) concluded that when New Zealand was separated from other continents by Cretaceous seas, the bird fauna already included the direct ancestors of many modern groups. He listed kiwis, moas, pigeon, wekas, Chatham Island rails, petrels, penguins, shags, gannets, terns, gulls, shorebirds, laughing owl, kea, kaka, kakapo, New Zealand wrens, bellbird, tui, stitchbird, huia, kokako, saddleback, piopios, fernbirds, robins, and tomtits. (He omitted albatrosses and Mohouidae.) Oliver accepted that these birds have been in New Zealand ever since the country was isolated by the sea. In contrast, modern dispersal theory, based on the fossil record, argues that all of these groups except the wrens and ratites (Fleming, 1962a,b; Wallis and Trewick, 2009), or all except the wrens (Mitchell et al., 2014a), dispersed into New Zealand, by rare trans-oceanic dispersal. The biogeographic details of some of the individual groups have already been discussed in earlier chapters and in Heads (2014), and others are referred to in the following discussion.

Structural aspects of birds and mammals have been studied in great detail, and New Zealand examples are reviewed here in light of the trends discussed in the last chapter. Oliver (1955: 30) cited "the more obvious features of New Zealand birds which seem to have a more or less definite *relation* to external conditions." Nevertheless, he added: "To explain the type of structure as an *effect* of the conditions to which it is supposed to correspond is not in the present state of biological knowledge a justifiable proceeding" [italics added].

In recent years, valuable studies of functional morphology in New Zealand birds such as ratites, parrots, and ducks have been published, and these allow a reassessment of the structural evolution. Notable aspects of morphology in New Zealand birds, including possible examples of adaptive features, are discussed next, after an introduction to birds in general.

Bird Morphology: Extreme Reduction and Fusion

One of the most obvious features in the structure of birds is the high level of reduction in the skeleton. The bones are thin, delicate, and very light, and show extensive fusion. The reduction is most obvious in the skull, where, for example, the face has a light, horny beak instead of a heavy, toothed muzzle (Brooke and Birkhead, 1991).

de Juana (1992: 37) wrote that the bird skeleton:

> ... forms an admirable combination of strength [for its weight] and lightness, by fusing some parts, eliminating others and making all as light as possible. In accordance with this... many skeletal parts are suppressed, while other are fused... the skull is particularly simple and light, and in adults practically all parts are fused... the mandibles have been shortened and the teeth suppressed...

This profound, overall reduction in the skull has obvious implications for feeding and other aspects of ecology. The overall reduction and fusion in the body as a whole has resulted in a lighter animal, and this, together with other features of the limb (for example, the hands are much *less* reduced than in many mammals), has made the ability to fly inevitable.

Considering the vertebral column, in birds overall there is "a very noticeable shortening" of the regions corresponding to the trunk, and the body is far more compact than in reptiles (de Juana, 1992: 42). Along with this shortening there has been massive fusion: the last thoracic vertebrae and all the lumbar vertebrae are fused to form a single bone, the *synsacrum*; in grebes, cranes, and pigeons other thoracic vertebrae are fused to form the *os dorsale* or notarium. Also, in birds in general the tail is much reduced, although the pectoral girdle and sternum remain large.

Reduction in the Limbs of Birds

Most explanations for the different structures in birds suggest that these have evolved because of the group's special needs, rather than being the result of trends found throughout vertebrates. For example, de Juana (1992: 36) wrote that: "In many... aspects the evolution of birds has been influenced by flight and homoiothermia. Flight imposes singularly important modifications on the general model...."

Instead of being responses to *needs* determined by *habits* such as flight, many features of birds can be interpreted as by-products of prior mutational trends, such as reduction. As indicated, given the structural trends that were operating in birds and their ancestors, flight itself was inevitable along with other consequences. The evolution of flight had advantages and disadvantages for the birds; the changes that permitted flight have also "involved the sacrifice of the hind limbs as well as the forelimbs as efficient organs for terrestrial or arboreal locomotion" (Heather, 1966).

In an evolutionary trajectory, such as reduction in and around the pectoral and pelvic girdles, the particular morphologies produced can be diverse. Reduction in the hindlimbs, as in many birds, has consequences, while reduction in forelimbs, as in kangaroos, will determine a very different style of movement, feeding, and so on. The positioning of the girdles will also determine aspects of ecology. In grebes, for example, the legs are set so far back on the body as to cause instability and clumsiness when walking on land (Fjeldså, 2004: 3).

In most birds the clavicles are fused, and together they form the *furcula* (wishbone). The implications of this fusion for flight, if any, are controversial: the furcula is sometimes thought to be important for flight, but it is absent in some parrots and owls (Brooke and Birkhead, 1991: 14). This suggests the fusion has developed as a by-product of large-scale, overall reduction and fusion in the whole pectoral girdle, not because a furcula answers any particular need.

In the bird forelimb proper, the hand is large, but its individual parts display a high level of fusion and reduction. It has been suggested that the "most striking feature [of wings] is the loss and fusion of the bones in the hand" (Brooke and Birkhead, 1991: 14). Three digits are represented by mere vestiges, and the metacarpals are fused with all but two carpals to form the *carpometacarpus*. In the hindlimbs, some of the tarsals are fused with the tibia to form the *tibiotarsus*. The rest of the tarsals fused with one another and the metatarsals to form the *tarsometatarsus*. Aspects of the reduction-fusion morphocline are also seen in fossil birds. While the tarsometatarsus in modern birds has undergone complete fusion, in early birds such as Enantiornithes (dominant in the Cretaceous period), it is only fused at the proximal end (Zhou et al., 2005).

Extreme Reduction in Forelimbs Leading to Loss of Flight

Apart from the great reduction and fusion observed in normal wings, in many birds the process has continued even further, rendering the birds flightless. Traditional theory argues that this is not the result of a general, underlying trend in vertebrates, but owes its origin to particular events (behavioral changes) in particular lineages: behavioral change led to the structural change. In a classic text of neodarwinism entitled *Plan and Purpose in Nature*, Williams (1996: 130) wrote that some birds "adopted lifestyles that led to the loss of flight." In this model the lifestyle or behavior came first, and this led to changes in the structure—in other words, the birds preferred to walk, even though they had functioning wings. In the alternative, mutation-driven model, the cause of flightlessness is, instead, a prior trend to limb reduction in vertebrates. This is one of the most obvious trends seen in all the main vertebrate groups, and it occurs in many birds.

In one of the two main clades of birds, the ratites, most forms have reduced wings and are flight-less. In penguins, orthodox theory suggests that: "their wings have evolved into flippers for swimming" (Fordyce and Ksepka, 2012), but limb reduction and flightlessness are so widespread that invoking a special "purpose" for this particular example is unnecessary. Ratites and penguins are the best-known flightless birds with reduced wings, but there are many others in at least 15 predominantly volant fami-lies (Kirchman, 2012). In New Zealand, the ducks *Anas aucklandica* (Auckland Islands) and *A. nesiotis* (Campbell Island) have stumpy, vestigial wings (illustrated by Scofield and Stephenson, 2013) and both are flightless. The New Zealand parrot *Strigops* is more or less flightless. Other flightless birds of New Zealand include the extinct adzebill, *Aptornis*, related to Rallidae and Heliornithidae.

Flightlessness is very rare in passerines, but some forms of New Zealand wrens, Acanthisittidae, are flightless (*Traversia*, extinct in historical times, and the fossil genera *Dendroscansor* and *Pachyplichas* were all flightless) (Millener and Worthy, 1991). Dawkins (1996/2006: 114) suggested that ratites became too big to fly and their wings degenerated, but this process would not account for the acanthisittids.

Dawkins (2009: 344) wrote that New Zealand "has more than its fair share of flightless birds, probably because the absence of mammals left wide open niches to be filled by any creature that could get there by flying. But those pioneers, having arrived on wings, later lost them as they filled the vacant mam-mal roles on the ground." Molecular evidence gave no support for this model, at least in the case of the New Zealand flightless rails; the flightless "weka" lineage of New Zealand (and, in some phylogenies, New Caledonia) is basal to a diverse, widespread east Asian-Pacific group, not nested in the widespread, volant *Gallirallus philippensis* as was thought (Kirchman, 2012). In addition, mammals were not absent from New Zealand; Miocene fossils include an archaic form (Worthy et al., 2006) as well as ground-dwelling bats (Hand et al., 2015).

Ventral Reduction

In vertebrates, the greatest reductions have occurred in the ventral, "visceral animal," while less reduc-tion has occurred in the dorsal, "somatic animal" (Romer, 1972). An example of a ventral structure in birds that displays a strong reduction cline is a pair of appendices, the caeca, found in most birds at the lower end of the small intestine (Heather, 1966). In ratites, Galliformes, Anseriformes, owls, rails, and some waders, the caeca are well developed and have a series of lobe-like folds. In pigeons and passerines the caeca are small and have no digestive function, although they bear abundant lymphoid tissue in their walls. In petrels, penguins, hawks, and kingfishers, the caeca are vestigial and functionless.

Feeding

Feeding in birds has been studied in great detail, and, in the traditional model, feeding habits determine the structure of the gut and the mouth. One influential text suggested that: "a change in feeding habits of birds may result eventually in structural modifications of the bill, the tongue, the palate, the jaw muscles, the stomach, and perhaps other features" (Mayr et al., 1953: 123).

Other interpretations are possible though. Birds have a wide range of food sources, and "There are many adaptations that make this possible, but the well known plasticity of their bills is an important factor" (de Juana, 1992: 46). This reasoning suggests that prior aspects of structure (the plasticity) affect diet, rather than the other way around, and this is supported here. de Juana (1992: 48) gave an example of morphology determining behavior and diet: "unusual morphological features can explain extremely specialized diets, as is the case of the Snail Kite (*Rostrhamus sociabilis*—Accipitridae), which taking advantage of its abnormally long, hooked bill, feeds uniquely on water snails of the genus *Pomacea*."

Kiwis and Moas (Ratites)

Merrem (1813) divided extant birds into two groups, Ratitae (ratites, including ostriches, kiwis, moas, etc.) and Carinatae (all other birds, including tinamous). The names mean "raft-like or flat" and "keeled," respectively, and refer to the form of the sternum. Huxley's (1867) influential account followed Merrem's arrangement. Huxley also noted that the tinamous are the most ratite-like of the carinates. Pycraft (1900) took this idea a step further, including tinamous with ratites in one group, that he termed Palaeognathae, and all other birds in Neognathae (although he regarded Palaeognathae as polyphyletic). This agrees with the molecular phylogeny. Nevertheless, Merrem's (1813) names are much older and much better known, and so the name Ratitae Merrem can still be used for the group that now includes tinamous, despite the fact that in tinamous the sternum is carinate (keeled), not "ratite" or "raft-like." Many taxonomic names are inaccurate in terms of morphology or geography, and just because the delimitation of a group is changed does not mean that its name needs to.

Ratites (including tinamous) are the sister group of all other extant birds, and are the largest extant birds. Most ratites lack a keel on the sternum, and there is also a strong reduction in the forelimbs. (In moas even the wing bones have disappeared.) As a result, they are flightless. All birds are flightless when they are small chicks, and in many respects, adult ratites can be shown to resemble embryonic or juvenile carinates (flying birds with a keel on the sternum). Thus it is often suggested that ratites have evolved by paedomorphosis or neoteny from flying ancestors (de Beer, 1956; Heather, 1966), and this paedomorphosis has been the underlying cause, or mode, of the wing and sternum reduction. Chatterjee (1997: 226–227) wrote: "Paedomorphosis has been a major component of the evolution of flightless birds... Flight... has been discarded in favour of flightlessness time and again in both aquatic and terrestrial groups exploiting new adaptive niches." Similar reduction of the keel on the sternum occurs in flightless pigeons (*Raphus*, the dodo) and parrots (*Strigops*, the kakapo).

As with the reduction in the sternum, the large, robust structure of many ratites is not a true defining feature, as it occurs in fossil members of other bird groups. The morphological feature linking all the ratites (including tinamous) is not the size or flightlessness, but the paleognathous structure of the palate, with a lack of fusion between the maxillar process of the nasal bone and the maxillary bone (Mayr, 2009: 15). As with flightlessness, the paleognathous palate represents a stage that all birds pass through in their early development, and its presence in adult ratites is the result of neoteny (de Beer, 1956; Härlid and Arnason, 1999; Kaley, 2007). Again, this suggests that the primary differentiation in living birds—the break between ratites and the rest—reflects old trends in the evolution of the vertebrate "ventral animal" (including palate and sternum), rather than an adaptive response to needs imposed by the environment. In addition to features of the plumage and the palate, the persistence of skull sutures in the adult ratites (in carinates these are obliterated by the fusion) has also been attributed to neoteny (de Beer, 1956).

Ratites are represented in New Zealand by the endemic kiwis (*Apteryx* species, the smallest ratites), and the extinct, endemic moas (order Dinornithiformes). Kiwis, cassowaries, emus, and elephant birds form an Indian Ocean clade, sister to the Pacific Ocean group: moas (New Zealand) and tinamous (South America north to southern Mexico). Following this vicariance, there has been secondary geographic overlap within New Zealand (Heads, 2014, 2015a). Aspects of the distribution of kiwis and moas in New Zealand were discussed earlier (Figures 9.16 and 9.17). This section considers some of their structural aspects.

The morphological features of the kiwi that are distinctive in birds include the location of the nostrils at the end of the bill, the rudimentary, vestigial wings (Oliver, 1955), the absence of tail feathers (rectrices), and the absence of clavicles (Parker, 1889). The egg is larger in relation to the body than in any other bird except storm petrels (Scofield and Stephenson, 2013), and both ovaries function in ovulation, not just the left, as in most birds.

Kiwis are flightless and nocturnal, as are many mammals but few birds. They have small eyes and poor eyesight, but excellent smell and a large brain; they feed on invertebrates by probing soil, litter, and rotting logs with their long bill, like a shorebird. They are often regarded as forest birds, but on Stewart Island *Apteryx australis* also feeds on sand beaches at night in beach wrack (photo in Scofield and Stephenson, 2013). These birds "can often be seen actively feeding during the day and, in the evening, foraging along the tide line for intertidal invertebrates [amphipods, etc.]" (Scofield and Stephenson, 2013). Other ratites recorded from maritime localities include *Rhea americana*, a species distributed from Argentina to Brazil, whose habitat includes flat, sandy, and open country of the littoral zone (Davies, 2002).

Overall, the combination of characters in the kiwi is the most unusual in any New Zealand bird. The traditional explanation is that each of the different characters has evolved because they are advantageous. Parker (1891: 68) wrote: "the kiwi, finding its prey by scent alone, has developed an extraordinarily perfect olfactory sense, while at the same time, having no need to keep watch against beasts of prey, its eyes have diminished in size and efficiency to a degree elsewhere unknown in the bird class."

The character combination in the kiwi—poor eyesight, good sense of smell, and so on—turned out to be a viable combination of characters for the bird, given its environment, but why was it there to start with? As usual, "Natural selection may explain the survival of the fittest, but it cannot explain the arrival of the fittest" (de Vries, 1912: 827, as quoted by Nei, 2013: 2). Developing alternative explanations for the origin of the characters involves a comparative approach, in order to see a character in its structural context.

Flightlessness

The theme of flightlessness in species on islands is often discussed, and the kiwi is a classic example. Yet, with the exception of tinamous, all the ratites are flightless (ostriches in Africa, rheas in South America, emus in Australia, etc.) and tinamous are more or less sedentary—they run away or hide from danger rather than flying.

The ratites occur all around the southern hemisphere and, as fossils, in much of the northern hemisphere, and so kiwis' flightlessness is part of a worldwide trend, expressed throughout one of the two main clades of birds. It does not suggest any adaptation to the New Zealand habitat in particular, and the kiwis, cassowaries, moas, and others were either already flightless when they differentiated from each other, or the flightlessness developed in parallel as part of a general trend. Williams (1996: 130) cited kiwis among birds whose "adopted lifestyles... led to the loss of flight"; instead, the loss of flight in kiwis and the other ratites probably determined many aspects of their lifestyle.

Feathers

The most distinctive feature of birds is the feather. The feathers in some ratites are unusual as they are less reduced than is the norm in birds—each feather has, in addition to the main feather, a large "aftershaft" or hypoptilum. In cassowaries and emus, the aftershaft is similar to the main feather and equals it in length (Davies, 2002). Tinamou feathers also have an aftershaft. However, in moas, only *some* feathers had an aftershaft (Worthy and Holdaway, 2002), and an aftershaft is absent in kiwi, ostrich, and rhea feathers (Folch, 1992). In most bird feathers the aftershaft is a minor relic or has undergone complete suppression, as in pigeons, and cuckoos. Miller (1924) wrote that there is never more than a trace of it in any of the 15 families of suboscine passerines, although a "curious partial exception is found in *Acanthisitta*," as some of the dorsal feathers bear a short, slender aftershaft.

Bill Structure and the Tactile Sense

New Zealand endemic birds such as the kiwi, the kakapo, and the blueduck all show "paradoxical" morphology in their visual system and their face in general, and also in related patterns of ecology and behavior. The New Zealand center of paradox is the result of the birds' holding unusual positions along evolutionary trends, although the trends themselves are global.

An adaptationist model proposes that the kiwi has "an amazing number of mammalian features—basically it's a bird that pretends it's a mammal" (Calder, 1978; cf. Dennett, 1996: 306). In this model, the nocturnal, terrestrial habit of the kiwi is explained by its strategy—its long-term evolutionary aim—rather than tendencies in prior, ancestral structure. Yet many aspects of kiwis' morphology are determined by their bird ancestry; for example, they have a toothless bill, unlike most mammals, and this in turn determines many aspects of their feeding "strategy."

The kiwi's bill is unusual in being so long and thin. The nostrils are unique among birds as they open close to the tip of the maxilla, rather than at the base of the bill or in the roof of the mouth. There is a high concentration of sensory pits clustered around the tips of both the maxilla and mandible; these pits house clusters of mechanoreceptors (Martin et al., 2007b). In this way, kiwi bill tips are the focus of both olfactory and tactile reception, and they are used to probe sand, soil, litter, and old logs for invertebrates.

The bird bill is a composite structure made up of fused parts; these are still visible as more or less distinct structures in Procellariiformes and in penguins such as *Eudyptes*. The kiwi bill represents one possible mode of reduction in the complex ancestral bill. In other birds, reduction led to the nostrils being stranded at the other end of the bill. Different growth zones have been "turned off" in the evolutionary history of different lineages, while others have been reduced to single layers of anomalous cells, small glandular patches, and so on.

Long "bristles" occur at the base of the kiwi bill and on its forehead (Folch, 1992: 104; Cunningham et al., 2011). Oliver (1955) described these as "long hairs," and while they are not true (mammalian) hair, they are hair-like. In fact they are modified, reduced feathers formed from a long rachis without barbs. In the kiwi, these rictal bristles have a tactile function. Rictal bristles are also a characteristic in the same position, at the angle of the jaw, in groups such as fantails (Rhipiduridae), although in these birds the mode of feeding is very different. The general reduction in the facial region would have affected feathers—the epidermal system— as well as other systems, and so the bristles could have originated as a by-product of this.

The Bill-Tip Organ

Cunningham et al. (2007) and Corfield et al. (2008) described how the kiwi detects prey by probing for invertebrates in the soil and litter, burying the tip of the bill in the same way that shorebirds do. Successful probing in kiwi and shorebirds probably depends on tactile sensitivity at the tip of the bill.

The avian bill is a complex sensory structure, containing at least two types of mechanoreceptors: Herbst corpuscles, which detect pressure, and the less common Grandry corpuscles, which detect velocity (Cunningham et al., 2013). Herbst corpuscles are small but complex structures found at nerve endings near the surface of the skin in many parts of a bird. Nerve tissue is a form of glandular tissue—it secretes neurotransmitter—and, as with glandular tissue in general, it is probably the result of morphogenetic–phylogenetic reduction.

At least five clades of birds have bills containing high concentrations of mechanoreceptors: kiwis (Apterygidae), waterfowl (Anseriformes), shorebirds (Charadriiformes: Scolopacidae), ibises and spoonbills (Pelacaniformes: Threskiornithidae), and parrots (Psittaciformes). Passerines are absent from the list, probably because of more complete reduction. The dense clusters of mechanoreceptors contained in pits at the bill tips are termed bill-tip organs. The kiwis, scolopacids, and ibises all share a common foraging strategy: bill probing for invertebrates and small vertebrates buried beneath the substrate, or sweeping for prey in the water column in the case of Threskiornithidae. All three groups have elongated bills and distinct bill-tip organs.

A detailed anatomical study found that "bill-tip organ structure was broadly similar between the Apterygidae and Scolopacidae" (Cunningham et al., 2013). Because the two groups are not close relatives, the authors assumed that the similarities were the result of parallel or convergent evolution. Thus they concluded that the similarities, along with similarities in the associated somatosensory brain regions, "are likely a result of similar ecological selective pressures." Another possibility is that the similarities are symplesiomorphies—unreduced structures that have not undergone the trend toward reduction seen in most other bird groups. The unreduced bill-tip organs and corresponding parts of the brain help determine the feeding behavior and ecology.

With respect to the brain, Cunningham et al. (2013) found hypertrophy of the PrV (principal sensory trigeminal nucleus) in kiwis, scolopacids, Anatidae, and parrots—all tactile-specialists—and inferred that this appears to have co-evolved alongside bill-tip specializations. Yet the bill-tip organ and the PrV are parts of the same system, and so no particular co-evolution is required. The brain is not so much a distinct "organ" as a polyphyletic structure made up of parts of the olfactory system, the auditory system, and others that have become fused. This has occurred during the overall reduction of the anterior end of the bilaterian body that formed the modern head.

To summarize, scolopacid shorebirds and kiwis both have sensory pits concentrated in a bill-tip organ, and they also share related similarities in their brain. This is consistent with the idea that kiwis originated as shorebirds, either on sandy shores in thick beach forest or in mangrove. Maritime populations would have been stranded inland by recession of Cretaceous and Paleogene seas and later uplifted to subalpine altitudes during Neogene orogeny.

Olfaction

The kiwi is well known for its olfactory acuity. The olfactory chamber is large compared with most other birds, and the functional epithelial surface area is further increased by folds on the concha (Portmann, 1961). In some birds the turbinate (turbinal or nasal concha) bone is elaborated to form a scroll, and its most extensive state, with five large folds, occurs in the kiwi (Heather, 1966).

Other parts of the olfactory system also display a similar lack of reduction compared with other birds. For example, in the kiwi brain, the olfactory bulbs in the cerebral hemisphere are the largest of any bird, and are represented in adults by an extensive, olfactory cortical sheet that surrounds the frontal pole of the brain (Martin et al., 2007b). (There is no obvious or pedunculated olfactory bulb, but this is only because the entire frontal telencephalon is covered at its rostral pole by the olfactory sheet; Corfield et al., 2008.) The olfactory bulb is also large in ducks, waders, and gulls, but it is reduced or almost absent in hawks, parrots, and passerines. The following figures indicate the proportion of the cerebral hemisphere that is contributed by the olfactory bulb: kiwi 33%, Procellariiformes 30%, Caprimulgiformes 25%, Apodiformes 19%, Columbiformes 17%, Piciformes 9%, Psittaciformes 6%, and Passeriformes 5% (Brooke and Birkhead, 1991). The reduction series shows a broad correlation with birds' phylogeny, from ratites through to passerines. Corfield et al. (2015) proposed that "Some of the diversity in OB sizes was… undoubtedly due to differences in migratory behavior, foraging strategies and social structure." Instead, the changes in olfactory bulb size are likely to be a cause, rather than an effect, of these aspects of behavior.

Kiwis are probably not the only ratites with a good sense of smell. Moas were unique among birds in the huge development of the olfactory chambers or capsules (Worthy and Holdaway, 2002: 117). They have "the most enlarged olfactory capsule of any bird (and markedly larger than emu, ostrich and other large ratites)." They do not have projecting olfactory bulbs or lobes in the brain, but this is caused by a great anterior and lateral expansion of the cerebrum that does not develop in other ratites, and envelops the entire anterior prominence of the olfactory bulb (Worthy and Scofield, 2012). The idea that moas simply lack olfactory bulbs or other olfactory specializations (Ashwell and Scofield, 2008) is misleading. The olfactory lobes of the brain are largest in *Pachyornis*, and large olfactory bulbs are also found in cassowaries (Tennyson and Martinson, 2006).

To summarize, both kiwis and moas have unique structures in the anterior brain associated with olfaction; in kiwis and possibly in moas the lack of typical olfactory lobes is not indicative of olfactory ability. This is consistent with the idea that the birds inherited their sense of smell from a common ratite ancestor, rather than developing it as an adaptation for New Zealand conditions and a particular lifestyle, probing the litter and so on.

The Kiwi Visual System

Reduction can lead to the evolution of eyes (as in early Bilateria), but continued reduction can cause their suppression, as in the pineal eye. This is a consequence of a main trend in vertebrate evolution—reduction and fusion in the face and pharynx. The facial skeleton is formed from the splanchnocranium or viscerocranium, the part of the skull derived from the ventral, "visceral animal" (Romer, 1972). (As stressed already, this is more reduced than the dorsal, "somatic animal.")

The optic lobes in the brains of most birds are very large (Gill, 1990), and most ratites have large eyes. In contrast, kiwi eyes are very reduced. This paradox is even more unusual because kiwis are nocturnal, and most nocturnal birds, such as owls, have *enhanced* eyesight. "It would be predicted that in flightless birds nocturnality should favour the evolution of large eyes… in Kiwi (Apterygidae), flight-lessness and nocturnality have, in fact, *resulted in the opposite outcome*" (Martin et al., 2007b; italics added). For the kiwi, visual information is of little importance, and this is "probably a *unique situation among birds* (Martin et al., 2007b; italics added). Among birds, kiwis' eyes are unusually small for the bird's weight. As for the visual system in the brain, in birds generally, the major retinal projection is to the optic tectum in the mid-brain. In the kiwi, the optic tectum has reduced size, depressed form, and reduced thickness." These morphological reductions mean that the ability of kiwis to see at low light levels has also been reduced.

Corfield et al. (2008) wrote: "In contrast to other Palaeognaths, kiwi brains are characterized by a reduced midbrain, [and] an enlarged telencephalon [cerebrum, with the olfactory bulb]... A reduction in midbrain size in kiwi is not unexpected, as a loss of visual acuity is accompanied by a decrease in the thickness of the optic tectum, ..." Thus the optic system or sector as a whole, from brain to eye, has been reduced, while the olfactory system has not.

Animals in which there is a structural reduction of the optic system (leading to poor eyesight) or a structural enhancement (leading to eyes that are over-sensitive) will tend to adopt a nocturnal or subterranean lifestyle. The kiwi situation contrasts with that in other nocturnal birds, such as owls (Strigiformes) and oilbirds (*Steatornis*, aff. Caprimulgiformes), in which the eyes are among the largest of flying birds. This enlargement has been taken to be the normal structural "response" to selection pressure, and Martin et al. (2007b) wrote: "the nocturnal habit has a strong effect on eye size relative to body mass." Yet kiwis do not follow this trend; Martin et al. (2007b) wrote that "Paradoxically," in the nocturnal and flightless kiwi the eyes are "exceptionally *small* (relative to body mass)...." Nevertheless, the reduction in the eye is just another aspect of the trend in facial reduction, and the fact that owls and kiwis show the opposite extremes of this trend is only paradoxical in the "adaptation to habit" model.

Nocturnality in Ratites

Kiwis are nocturnal. The nocturnal habit is widespread in reptiles, but rare in birds and so it is assumed to be derived from ancestors that were active in the day-time. Yet nocturnality in ratites is not restricted to kiwis. Feduccia (1996: 286) noted that cassowaries are "somewhat nocturnal." Cassowaries are active at night in montane New Guinea and can be heard as they walk along tracks through the forest (personal observations). In the morning, they are active by 4.30 a.m., long before the dawn at 6 a.m. (Majnep and Bulmer, 1977). In tinamous, the calls of *Tinamus major* are sometimes heard late at night, and *Crypturellus soui* calls throughout the day and night (Davies, 2002). Ostriches are diurnal but are sometimes active on moonlit nights (Davies, 2002). Tennyson and Martinson (2006: 34) raised the possibility that the moa *Pachyornis elephantopus* was partly nocturnal. These observations suggest that the ancestral ratite complex was already polymorphic for this character before the origin of the kiwis.

The nocturnality in kiwis is extreme and is associated with suppressed development in the visual system. With their reduced vision, the birds would not be able to compete in the daylight, and so they have become nocturnal. With respect to nocturnality, as Worthy and Holdaway (2002: 247) observed, "It is difficult to determine the cause and effects of behaviors." In the kiwi, the behavioral patterns are probably caused by structural trends, not because the behavior has any inherent advantage in New Zealand conditions.

In New Zealand, nocturnal vertebrates include the bats, the kiwis, the kakapo *Strigops*, and the owls, while the kaka and kea (*Nestor* species) are often active at night (Worthy and Holdaway, 2002). The gecko *Hoplodactylus* s.lat. is nocturnal, while the skink *Oligosoma* s.lat. is diurnal (Worthy and Holdaway, 2002), and this reflects the general trend in their families (most geckos are nocturnal, while most skinks are diurnal) rather than adaptation to New Zealand conditions.

Regressive Evolution in the Kiwi Visual System

Martin et al. (2007b) discussed the situation in the kiwi, and wrote: "Regressive evolution of visual systems have been described in both vertebrate and invertebrate animals... However, all of these examples have involved a complete loss of vision following colonisation of subterranean habitats devoid of light." Again, the question is whether the loss of vision was a cause or consequence of the colonization.

Martin et al. (2007b) wrote:

> In Kiwi, complete regression of the eye and parts of the brain associated with visual information processing has not occurred... their foraging habitats are not completely devoid of light... We propose that [adaptive] regressive evolution of Kiwi vision is the result of the *trade-off* between the requirement for a large eye to gain information at low light levels, and the metabolic costs of extracting and processing that information... It seems possible that there is an ambient light level

below which the costs of maintaining a large eye and associated visual centres are not balanced by the rate at which information can be gained, and that this occurs in forest floor habitats at night. [italics added]

Yet this does not explain why the kiwi ancestors invaded the nocturnal niche to start with. One possibility is that it was caused by degeneration in the visual system, and, as indicated, this is a common trend in many vertebrates. Reductions in the visual system have helped determine the nocturnal habit of the kiwi, while reductions in the wings and sternum have led to flightlessness. The complex, unreduced structure of the bill and the olfactory system have determined the probing habit.

Another Feature of the Kiwi Visual System: The Pecten

The pecten is a vascular, membranous structure in the bird eye, and differs from the simple peg-like pecten (conus papillaris) of reptiles. The bird pecten shows great interspecific diversity in morphology, and many functions have been proposed for it (Sillman, 1973: 360). For example, it is sometimes suggested to be a nutritive organ. However, these functions have not been accepted as convincing explanations for the structure, and Sillman (1973: 383) concluded that "the riddle of the pecten remains unresolved." In any case, there are three main types of pecten, distributed in birds as follows (Meyer, 1977; Martin, 1985):

1. Conical (found only in the kiwi). The pecten extends forward through the chamber of the eye almost to the lens. It resembles the conus papillaris of reptiles, suggesting that it represents an unreduced condition.
2. Vaned (ostrich, rhea, tinamous).
3. Pleated (all remaining birds).

The distinctive structure of the kiwi pecten indicates an early break between this bird and all other birds. In the kiwi, the eye region as a whole has been reduced, but the pecten has not.

The Auditory System in Ratites: Further Evidence of Reduction

In birds, after the fibers of the auditory nerve enter the brain stem, the axons from the avian auditory ganglion bifurcate to innervate two distinct cochlear nuclei: the nucleus magnocellularis and the nucleus angularis. In contrast with this normal arrangement in the brain, an additional structure occurs in some species (Corfield et al., 2014). It occurs in the kiwi, where it forms a "well developed mass of gray matter corresponding exactly in appearance with that of the angular nucleus, but entirely separate from it" (Craigie, 1930: 342). Its relationship with respect to the two cochlear nuclei remains speculative, and it has no input and output connections within the auditory pathway. The same structure has also been observed in reptiles, and in birds it occurs in emus and Galliformes as well as kiwis. It is absent in Neoaves (all extant birds except ratites, Galliformes, and Anseriformes). This suggests that it is an ancestral character that has been lost in the Neoaves (Corfield et al., 2014).

The Kiwi Brain

Apart from the unusual combination of nocturnal habit and reduced vision, kiwis show another obvious anomaly in their brain structure—its large size compared with the brain of birds in general. This is unexpected, as the brains of ratites other than kiwis, including moas, are small. Corfield et al. (2008) wrote that the large size of the kiwi brain suggests:

> ...that *unique selective pressures towards increasing brain size* accompanied the evolution of kiwi. Indeed, the size of the cerebral hemispheres with respect to total brain size of kiwi is rivaled only by a handful of parrots and songbirds... These findings form an exception to, and hence challenge, the current rules that govern changes in relative brain size in birds. [italics added]

The "rules" cited here propose that ratites do not have large brains, and that increased brain size should lead to increased intelligence. The large brain/small brain split occurs at an important biogeographic boundary, the New Zealand region, that separates the kiwi and moa lineages. The break is anomalous as it cannot be explained by any ecological sorting or selection pressure. Instead, it can best be explained by documented trends in the brain size of birds and mammals. These changes are related to changes in overall body size, and to encephalization, with the latter thought to "mediate the evolution of complex behaviors and cognition" (Jerison, 1973). The enlargement of the neocortex observed in human and non-human primates has given rise to the idea "that larger brains, and telencephala in particular, are associated with enhanced cognitive prowess...." Yet despite their enlarged telencephalon, kiwis do not appear to have any special intelligence (Corfield et al., 2008). They are monitored closely for conservation purposes, and if they exhibited any sophisticated behavior it would have probably been reported, as it has for species such as the New Zealand kea (*Nestor notabilis*).

Corfield et al. (2008) concluded:

> Kiwi have adopted a niche very different to other ratites (including moa) in that they are nocturnal, are probe feeders, are highly territorial and appear to have a complex communication system... The increases in brain size seen in kiwi might be associated with sensory specializations associated with this particular niche.... *there must be some adaptive advantage* to having an enlarged brain that counteracts the costs associated with the increase in size. We can only speculate regarding what selective pressures might have driven the changes in morphology and size of kiwi brains, but it seems likely that these changes must have conferred kiwi with some adaptive advantage. [italics added]

This approach assumes that the structure must be adaptive, but in mutation-driven evolution there is no need for any new, evolved structure to be advantageous or adapted to the current environment. Corfield et al. (2008) themselves wrote that the great difference between moa and kiwi: "suggests that the evolutionary changes in relative brain size of kiwi *were not driven by particular characteristics of the New Zealand ecosystem...*" [italics added]. This is a logical and important conclusion— the current environment does not explain morphological structure.

Encephalization and Brain Evolution in Moas and Kiwis

Ashwell and Scofield (2008) found that the relative brain sizes or encephalization quotients (EQ) of moas (0.2–0.5) were similar to those of other ratites (except the large-brained kiwis), but less than those of most birds. The authors suggested that in New Zealand, the absence of predation by mammals might have contributed toward a reduction of EQ in the moa, but this is contradicted by the high EQ of kiwi. Values of EQ for the three kiwi species examined ranged from 0.721 to 1.260. EQ for the Jurassic bird *Archaeopteryx* is much lower, with a range of 0.198–0.260. Ashwell and Scofield (2008) concluded that the lineage of birds leading to modern kiwi has undergone a *four- or fivefold increase in relative brain size* since the late Jurassic.

Ashwell and Scofield (2008) wrote that it remains uncertain what factors have driven the increase in the kiwi's EQ. They cited two possible explanations. The first is the standard, teleological view: "the remarkable neurological *adaptations of the kiwi to a nocturnal niche* may have driven expansion of central centres associated with the relevant sensory specializations [e.g., olfaction]...." Ashwell and Scofield's second suggestion was not teleological, and it stresses a process rather than an end: "the observed EQ of kiwi might be the result of a *profound reduction in somatic size without a corresponding reduction in brain weight*. A similar explanation has been proposed for the large size of kiwi eggs relative to the body of the female..." [italics added]. This reduction in somatic size, leaving some organs unreduced, is a simple process, but it would lead to fundamental changes in morphology and ecology.

Evolutionary lag in some organs during somatic size change has also been suggested for other birds. Scofield and Ashwell (2009) suggested that the large New Zealand fossil Haast's eagle (*Harpagornis*) evolved to its large size from smaller ancestors. These "appear to have undergone rapid expansion of body size and elements of the hindlimb somatic nervous system at the expense of enlargement of the brain and visual, olfactory, and vestibular apparatuses." An alternative idea is that the body in most eagles *other* than *Harpagornis* has been reduced, while the brain and sense organs have remained the same size, as in Ashwell and Scofield's (2008) account of kiwi evolution.

Kiwis and Their "Remarkable Path of Evolution" as Mammal Equivalents

In kiwis, reduction in eye size, visual fields, and visual processing areas in the brain is balanced by a compensatory increase in olfactory and beak mechanosensory processing. Ashwell and Scofield (2008) cited the "remarkable nasal and olfactory bulb specialisations in the skull and brain" and concluded that: "the kiwi embarked on a *remarkable path* of neurological specialization which allowed the birds to exploit a *niche usually occupied elsewhere by mammals*" [italics added].

There is no doubt that the kiwi olfactory system is unusual in birds, as is the nocturnal, flightless lifestyle. The trend, or the "remarkable path," of morphological change in the sensory systems has "allowed"—perhaps forced—the birds to exploit their ecological niche; but the niche that was, in the end, occupied has not itself determined the path of genetic and morphological change. The niche is the result of the evolutionary path, not its cause.

Kiwis are sometimes thought to have evolved the way they have in order to fill the niche of the absent mammals (Calder, 1978). The idea is based on the nocturnal, flightless habit, the olfactory specialization, and the assumption that kiwis "are descended from a fauna that evolved in the *absence of terrestrial mammals* over a period of 80 million years" (Martin et al., 2007b; italics added). Yet the kiwi would have shared the forest floor with mammals, as the ground-feeding bat *Mystacina* is known from fossils back to the Miocene (Hand et al., 2015), and there is a Miocene fossil of a small, archaic, non-volant mammal (Worthy et al., 2006). In any case, other flightless ratites—the cassowaries—can also feed at night, and they are endemic to an area (New Guinea and north-eastern Queensland) with a diverse mammal fauna that is, for the most part, nocturnal. Ostriches in Africa and rheas in America have managed to survive while flightless, despite the presence of predators such as large cats. Within vertebrates, it would be surprising if the "nocturnal and flightless" niche has ever been the exclusive preserve of mammals, as it will have also been occupied by amphibians, reptiles, and birds at many times through the history of these groups.

Martin et al. (2007b) described the reduced reliance upon visual information in kiwis as an example of "adaptive regressive evolution" (cf. cavefish; Jeffery, 2005). "At some point in the evolution of kiwi, natural selection favoured foregoing visual information in favour of other sensory information. The ecological circumstances favouring this are unclear." The idea of regressive evolution is well supported, but the suggested cause—natural selection and adaptation—is problematic, as the authors admit.

Martin et al. (2007b) noted that:

> Reliance upon tactile and olfactory information over visual information is found in both Kiwi and in nocturnal mammals such as rodents... This suggests the independent evolution in Kiwi and in these mammals of similar sensory performance that is tuned to a common set of *perceptual challenges* presented by the forest floor environment at night that cannot be met by vision. [italics added]

Instead, the similar structures in unrelated groups of mammals and birds can be interpreted as byproducts of similar reduction trends active throughout vertebrates.

Folch (1992: 104) concluded that: "Being basically nocturnal, kiwis are specially adapted to moving around at night...." Instead, their reduced vision *determines* a nocturnal habit. Folch also wrote that "As a result of this [adaptation], their sense of smell and hearing are very sharp, whereas their vision is poor..."; this combination is "more typical of a mammal." As well as these interesting parallelisms between the kiwi and many mammals, there are also similarities with several other groups. In features such as the pecten, the kiwi eye resembles that of the reptiles, while in its bill-tip organ the kiwi resembles some wading birds.

Summary of Kiwi Evolution

The standard model proposes that "circumstances" produced the kiwi "strategy," but the precise causes are often obscure. For example, why are the birds nocturnal in the first place? And why are the eyes so small, when there are so many other forest-floor birds without reduced eyes? Instead of being the result of selection, the different "strategies" in the evolution of the avian face and brain are attributed here to long-term, structural trends in skull and brain evolution. In the same way, changes in the forelimb, pectoral girdle, and sternum have caused flightlessness in ratites worldwide. In the case of the kiwi and the moa, the structural variants in New Zealand represent parts of broad intercontinental complexes, not responses to local conditions. For example, the kiwi is the sole bird in which both ovaries, not just the left, are functional; it would be difficult to construct an adaptive explanation for this, but the differentiation is probably an old one that has occurred early in the history of the ratites and birds in general.

In the CODA model (center of origin + adaptation + dispersal), the arrival of the kiwi to New Zealand took place either by flying or by walking (rather than by evolution), and it occurred just once, as an anomalous dispersal event. In the CODA model, this remarkable dispersal was soon supplemented by an ecological "invasion" into the nocturnal environment, by means of a "remarkable path" in brain evolution (Ashwell and Scofield, 2008). Yet the remarkable dispersal event is unnecessary unless a young age is accepted, and the remarkable course in brain evolution could instead reflect *no* change in brain size while body size has decreased.

In kiwis, Corfield et al. (2008) cited reduction in eye size, visual fields and visual processing areas. They wrote: These features were taken [by authors] to suggest that a regression of the visual system *accompanied* the invasion of a nocturnal niche in this species, ..." [italics added]. It is easy to extend this to the idea, suggested here, that regression and suppression in the visual system accompanied *and caused* the invasion of the nocturnal niche.

The paucity of nocturnal predators would have assisted the survival of a nocturnal kiwi, but this environmental factor is not the reason that kiwis were nocturnal to start with. Nocturnality reflects a trend already present in ratites, and in kiwis it has a simple structural basis; a bird with poor sight cannot compete with sighted birds in the daytime. The morphology of kiwis represents the intersection of several trends: relative increase in the brain, the olfactory system, and the part of the bill proximal to the nostrils, and relative decrease in the somatic body (not the egg), the visual system, the wings, and the tail. The combination of characters is unique, but the trends themselves are common throughout vertebrates.

Quails (Galliformes)

After the ratites separated from all other birds (neognaths), the first split within neognaths separates Galloanserae, that is, Galliformes (chickens, turkeys, and pheasants) plus Anseriformes (ducks and geese), from all the others (Burleigh et al., 2015). The only galliform in New Zealand is the New Zealand quail (*Coturnix novaezelandiae*: Phasianidae). This was common in the three main islands, and inhabited lowland grassland, fernland, and shrubland. It survived into historical times, but then declined and was extinct by around 1875.

McGowan (1994: 441) made an important observation on the main clades of phasianids, writing that they "show quite different patterns in their distribution *and, therefore, habitat use*" [italics added]. Although the main clades in phasianids have been rearranged in the molecular phylogeny (Wang et al., 2013b), McGowan's approach stressed an important point—distribution can itself determine habitat use and ecology, and some habitats are not occupied simply because they are not accessible. In the same way, McGowan (1994: 450) proposed that habitat can determine diet (rather than diet determining habitat, as in the CODA model): "... whether phasianids are herbivorous, insectivorous or omnivorous depends to a large extent on their habitat."

In the model suggested by McGowan (1994) and supported here, evolutionary biogeography determines the broader region that the group occupies, and so it is a primary factor determining the habitat (it provides the "habitat pool" that is available for the species). Likewise, the habitat, along with morphology, determines the birds' diet. Whether or not the bird continues to survive, expand its range, or go extinct depends on inherent trends in its evolution and on any changes in the environment.

The quail *Coturnix novaezelandiae* was distributed through New Zealand, while its sister, *C. pectoralis*, is found more or less throughout Australia (not in the far north) in a wide range of habitats. Until the advent of European agriculture, the two species occupied most of the open, lowland habitats found throughout southern Australasia. Overall, species of *Coturnix* (incl. *Excalfactoria* and *Synoicus*) range throughout the warmer regions of the Old World, except the central Pacific islands, and are present in a wide range of habitats (IUCN, 2016). The last two species in the following list show significant overlap with each other and the remaining species, indicating range expansion in the two at some stage, but the other six species are more or less allopatric. The eight species of *Coturnix* occur in:

New Zealand (*C. novaezelandiae*).

Australia, but not the far north (*C. pectoralis*).

New Guinea and Australia, but not in large parts of south-west Australia (*C. ypsilophora*).

Sub-Saharan Africa and Madagascar (*C. delegorguei*).

India to South-east Asia (*C. coromandelica*).

China and Japan (*C. japonica*).

Widespread in Africa (not Madagascar), India, eastern Asia (not Sumatra) to eastern Australia (*C. chinensis* s.lat.).

Widespread in Africa, Europe, mainly western Asia (*C. coturnix*).

The high level of allopatry suggests that the birds inherited different regions when they originated, and at the same time they would have inherited different habitats. In addition to possible range expansion in *C. coturnix* and *C. chinensis*, the overlap of *C. ypsilophora* and *C. pectoralis* in central Australia appears to be secondary, as the former species is mainly in the north, and the latter (related to the New Zealand species) is in the south.

Ducks (Anseriformes)

In the ducks, as in the duck-billed dinosaurs (Hadrosauridae) and the platypus, the bill results from a distinctive flattening and broadening of the front part of the skull. This probably developed during a general reduction of the region, but the bill in ducks still includes relictual traces of structures that have been more or less lost in most birds and mammals. Several examples are cited next, from Carboneras (1992c: 541–543).

In ducks, lamellae occur on the inside of bill and are "particularly well developed in plankton filtering species"; in *Mergus* they are serrated "for catching fish." Ducks have a shield-shaped horny appendage at the tip of the upper mandible known as the "nail." It is harder than the rest of the sheath and is "specially designed for purposes of grazing or mollusk-eating." Other "special adornments" in ducks include the fleshy protuberance (a "spectacular knob") over the bill of the male comb duck (*Sarkidiornis*). There are some "particularly strange bills" in the family, "none more so" than those of the blue duck, of New Zealand, and *Malacorhynchus*, widespread through Australia, Tasmania, and, as fossils, New Zealand. The strange bill of *Hymenolaimus* and the fact that it is sister to a pantropical group make the bird of equal interest for biogeography, morphology, and ecology.

The Blue Duck (*Hymenolaimus malacorhynchos*)

> We shot several curious Birds, among which was a Duck of Blue grey Plumage with the end of its Bill as soft as the lips of any other animal, as it is altogether unknown I shall endeavor to preserve the Whole in spirits. James Cook, Fiordland, 1773 (Cook, 2003: 263).

The New Zealand blue duck, *Hymenolaimus*, is one of the "high-value" New Zealand endemics that have circumglobal sister groups. The blue duck's sister group, *Sarkidiornis*, occurs in tropical South America,

Africa, Madagascar, and from India to southern China (Robertson and Goldstien, 2012). The simplest explanation for the pattern is that a common ancestor distributed throughout Gondwana broke down into the two genera by simple vicariance, at breaks around New Zealand.

Blue ducks inhabit headwaters of New Zealand rivers and feed on aquatic invertebrates. They detect different food items by tactile and visual cues, and both sensory systems are associated with distinctive morphological features. The use of tactile cues when foraging is suggested by the presence of "specialized flaps of thickened, keratinized epidermis along the ventral margins of the upper mandibles near the bill tip" (Martin et al., 2007a). The species epithet *malacorhynchos* means "soft-billed," and Scofield and Stephenson (2013) presented a good photograph of the pink bill showing the "bizarre, black fleshy flaps of skin hanging from the sides of the tip." These flaps contain concentrations of Herbst corpuscles, sensory nerve endings that detect pressure. Similar flaps are found in just one other duck, the pink-eared duck of Australia (*Malacorhynchus*) that surface filter-feeds on plankton.

The rhamphotheca or horny covering of the beak in procellariiforms is not fused, but consists of several separate plates with distinct sutures between them. This is believed to be a primitive feature (Heather, 1966). The bill in ducks also retains an overall structure that is complex and unreduced compared with the secondarily simple, fused bill of the passeriforms. The bill flaps with Herbst corpuscles found in *Hymenolaimus* and *Malacorhynchus* represent other signs of incomplete reduction.

Specializations in the Blue Duck's Binocular Field

In addition to the "unreduced" aspects of the bill and their ecological consequences, blue ducks show reductions in the skull that have changed the binocular field. Martin et al. (2007a) argued instead that habit (feeding mode) and habitat have determined structure (structural aspects of vision):

> In birds, the position and extent of the region of binocular vision appear to be *determined primarily by feeding ecology...* Of prime importance [in birds] is the degree to which vision is used for the precise control of bill position when pecking or lunging at prey, or when feeding chicks. In species that feed in this way the bill falls either centrally or just below the centre of the frontal binocular field... In birds that do not *require* such precise control of rapid bill movements (probe and filter feeders with precocial self-feeding chicks), the bill falls outside or at the periphery of the binocular field. [italics added]

Instead of the geometry of the bill and head being determined by needs imposed by the habit (feeding ecology), the geometry could in itself determine the habit, with the geometry itself being the result of prior reduction trends in the facial region.

In blue ducks the eyes are frontally placed, and this results in a wide field of overlap where vision is binocular. The narrow, tapering bill intrudes into the binocular field, implying that a blue duck can see its own bill tip. These features are found in a wide range of species that differ in their ecology and phylogeny, but have in common "precision pecking or lunging at prey" (Martin et al., 2007a).

The morphology indicates that a blue duck can see its own bill tip, whereas a pink-eared duck, *Malacorhynchus* of Australia, cannot. The visual field of pink-eared ducks indicates that it is a tactile feeder, whereas blue ducks have visual fields that suggest they feed using visual cues in the water column (requiring clear water); they also take prey from rock surfaces.

Martin et al. (2007a) wrote: "We hypothesize that temporal and spatial heterogeneities of *prey availability* have resulted in the evolution in blue ducks of behavioural, anatomical and physiological adaptations..." [italics added]. Yet the conditions of prey availability in New Zealand are unlikely to be unique. In any case, the structures that underlie the two modes of foraging in blue ducks are particular results of two trends in facial structure: (1) The birds use *tactile cues* from the "specialised flap with Herbst's corpuscles" in the bill-tip—another "bill-tip organ," matching that of the kiwi. These cues are the primary guide in scraping small chironomid and caddis fly larvae from rock surfaces. (2) The birds use *binocular vision*, determined by skull geometry, as the primary foraging guide in capturing larger and more nutritious prey (nymph stages of mayflies and stoneflies).

The modifications and retentions of the sensory systems reflect the broader, global, long-term trends of reduction in the vertebrate head. These trends can account on their own for the variation in the binocular field and relictual tissue in the bill of the blue duck, and there is no real need to invoke unknown aspects of local prey availability.

Pigeons (Columbiformes)

Pigeons (Columbidae) are represented in New Zealand by the endemic *Hemiphaga*. As discussed earlier (Chapter 6), the genus is the New Zealand representative of a Tasman–Coral Sea complex (with *Lopholaimus* and *Gymnophaps*). Within New Zealand, *Hemiphaga* has two allopatric species, one on the mainland, Norfolk and Kermadec Islands, and one on the Chatham Islands. There is allopatry from genus down to species level, consistent with a history of simple vicariance.

One distinctive feature of *Hemiphaga* is its tail, which has just 12 tail feathers or rectrices (Oliver, 1955). This is at the low end of the range for pigeons overall, which is 12–22 (Baptista et al., 1997). The number of tail feathers in birds varies from 32 (Bulwer's pheasant) to 6 (several small songbirds) (Brooke and Birkhead, 1991). As usual, the passerines represent the end of the reduction series. The anterior and posterior ends of the "ventral animal" are the main sites of reduction in the vertebrate body, though, and there is no need to assume that the reduced tail of *Hemiphaga* was adapted to the special environmental conditions of New Zealand.

Other trends seen in pigeons outside New Zealand include flightlessness. Flightless pigeons, including the dodo *Raphus* and the solitaire *Pezophaps*, are often cited as examples of paedomorphosis or neoteny (Baptista et al., 1997: 60), and neoteny can also explain the flightlessness in many New Zealand birds. Characteristics seen in the adults of the flightless pigeons and also in pigeon chicks in general include loose and decomposed plumage, reduced wings, swollen abdomens, naked facial skin, and lateral expansions of the bill (cf. the Samoan tooth-billed pigeon, *Didunculus*).

Owlet-Nightjars (Aegotheliformes)

Aegothelidae consist of a single genus, *Aegotheles*, which is endemic to the Tasman-Coral Sea region. The New Zealand form is known from Pleistocene to sub-Recent fossils, but is now extinct. The family's sister group is a diverse, global complex, the swifts and hummingbirds (Apodiformes), suggesting an early break around the broader Tasman region.

The two main clades in *Aegotheles* show simple allopatry, with one in the west (Tasmania, Australia, New Guinea) and one in the east (New Zealand–New Caledonia) (Heads, 2010). The east-west break in the group is consistent with the late Cretaceous opening of the Tasman Basin, and so the origin of the group could date to the mid-Cretaceous extension and predrift rifting.

Rails (Ralliformes)

In historical times, 17 species of Rallidae inhabited New Zealand, and these occurred in most habitats until widespread, anthropogenic extirpations took place (Holdaway et al., 2001). Some of the most interesting New Zealand rails were endemic and flightless, and much discussion has focused on how these managed to reach the country. In dispersal theory, the birds' volant ancestors flew to New Zealand, where their descendants lost the powers of flight. An alternative idea is that the flightlessness developed in an autochthonous group. As stressed already, limb reduction is one of the most common trends throughout vertebrates and is widespread in birds.

The rail family, Rallidae, has a global distribution, and 30 of its 150 species are flightless. The taxonomy of the group has been very unstable, but flightless forms are conspicuous in New Zealand. Marked variation in the skull is also a feature of the New Zealand contingent, and a diverse clade there

(*Gallirallus*, *Diaphorapteryx*, and others) presents as much diversity in skull morphology as there is throughout the rest of the family (Olson, 1975).

Porphyrio Species: Takahe and Pukeko

Porphyrio includes a clade with the following phylogeny (Garcia-R and Trewick, 2015):

> **Tethys 1**: W Mediterranean (*P. p. porphyrio*), Thailand to Java (*P. p. indicus*).
>
> **Africa and Madagascar** (*P. p. madagascariensis*).
>
> **Tethys 2**: NZ (*P. p. hochstetteri*, the New Zealand takahe) + NZ to Turkey (all other *P. porphyrio* subspecies, including *P. p. melanotus* [known in New Zealand as pukeko]).

The notable "Tethys versus Africa" allopatry seen in the last pair of clades also occurs in bats (*Chalinolobus* and relatives; see next chapter). Garcia-R and Trewick (2015) did not mention the allopatry in the rails, but did suggest that the forms of *Porphyrio* "…demonstrate extraordinary dispersal capabilities, with evidence of multiple invasions…" (p. 141). For example, they inferred that New Zealand was invaded twice. In a vicariance model, both New Zealand forms are instead interpreted as belonging to the Tethys 2 group that has originated *in situ*, by vicariance with an African clade. The ancestors of both New Zealand forms were already in the New Zealand region before the break between Tethys and Africa formed the Tethys 2 clade.

Galllirallus: Wekas and Their Relatives

One of the flightless rails endemic to New Zealand is the weka, *Gallirallus australis* s.lat. of North, South, and Stewart Islands. Wekas were widespread through the mainland and from the coast to the alpine zone, but were extirpated from most areas in the nineteenth century. Many regional forms were discussed and named. Nineteenth century authors thought that the weka forms a clade with the New Caledonian *Gallirallus lafresnayanus*, and they referred to this as the genus *Ocydromus* (Hutton, 1872).

The leading ornithologist Stresemann (1934: 656) was impressed with the weka and saw it as a peculiar, primitive form without close relatives. He treated it as an ancient element, along with the New Zealand ratites, wrens, and basal parrots. A panbiogeographic study of the weka concluded that it evolved by vicariance (Beauchamp, 1989), and molecular work has at least supported an "ancient derivation of the weka lineage" (Trewick, 1997a).

In contrast with vicariance, dispersal theory suggests that the weka, along with other *Gallirallus* species endemic to Pacific islands, was derived from *G. philippensis* (Trewick, 1997a,b). This is because *G. philippensis* is (1) the only more or less volant species in the genus and (2) widespread through the south-west Pacific, from the Philippines to Australia, New Zealand, and Samoa.

Compared with *G. philippensis*, the flightless New Zealand weka *G. australis* appears "strange and distinctive…," but Taylor (1996: 110) suggested that "Early classifiers were deceived by the neotenic characters associated with flightlessness." Taylor argued that wekas are similar to *G. philippensis* "and the differences in the wings, the pectoral girdle, and some plumage features of the adult are simply recently derived neotenic characters." Based on this idea of a recent, derived origin, Taylor (1996) interpreted the weka as the product of an "invasion of New Zealand by *G. philippensis*-like stock." Worthy and Holdaway (2002: 386) agreed, writing that it is "now known" that *Gallirallus australis* and also the extinct Chathams groups *Diaphorapteryx* and *Cabalus*, are "probably all derivatives of *Gallirallus philippensis*." Thus Trewick (1997a,b), Taylor (1996), and Worthy and Holdaway (2002) all supported a dispersal origin for *G. australis* by budding-off from the "widespread weed," *G. philippensis*.

The morphology of *Gallirallus australis* may well be derived by neoteny, but this does not mean it arrived in New Zealand by dispersal, or that it is derived from *G. philippensis*. In fact, the dispersal model for wekas has been contradicted by the molecular phylogeny (Kirchman, 2012; Garcia-R et al., 2014, 2016). The phylogeny shows that *G. australis* is *not* nested in the widespread, volant "weed," *G. philippensis*, or in any Australian species. Instead, *G. australis* plus *G. lafresnayus* of New Caledonia is sister to a diverse, widespread clade of eastern Asia–Australasia–Pacific that includes *G. philippensis* in a deeply nested position. (*Diaphorapteryx* of the Chatham Islands is also placed outside this widespread group.) *Gallirallus australis* of New Zealand and *G. lafresnayanus* of New Caledonia form a distinct weka clade, as indicated by Hutton (1872).

Gallirallus s.lat. as a whole including several species extinct in modern times, has the following phylogeny (Garcia R et al., 2016). The weka lineage and its sister group are highlighted in bold.

Philippines (Calayan I. in the Babuyan group) (*G. calayanensis*).

Indonesia (Maluku Islands/Moluccas) (*G.* ("*Habroptila*") *wallacii*) and the Chatham Islands (*G.* ("*Diaphorapteryx*") *hawkinsi*).

New Zealand (North, South, Stewart Is.)—New Caledonia. (*G. australis* and *G.* ("*Tricholimnas*") *lafresnayanus*): "*Ocydromus*," wekas.

Widespread Pacific. 14 species. Okinawa (Ryukyu Is.), Philippines, Sulawesi, Australia, Lord Howe I. (*G.* (`*Tricholimnas*') *sylvestris*), and New Zealand (Chatham Is.: *G.* "*Cabalus*" *modestus* and *G.* "*Nesolimnas*" *dieffenbachii*), east to Wake I. (south-west of Hawaii), the Cook Is. and Marquesas Is. The clade includes the widespread Pacific species *G. philippensis* (present in New Zealand on North, South, and Stewart Is.).

G. macquariensis of Macquarie Island, extinct by the end of the nineteenth century, was recognized as a full species by Holdaway et al. (2001), but it has not been sequenced. (*G. striatus* is transferred to the *Crex-Dryolimnas* group.)

Garcia-R et al. (2014) estimated that the clade including *G. calayanensis* and clades below it originated at 17–27 Ma, and suggested a history of chance, oversea dispersal. But the age is a fossil-calibrated, minimum clade age, and an older, vicariance history is possible. Vicariance is suggested by the fact that in the main *Gallirallus* clade, beginning with *G. calayanensis*, the species show extensive allopatry, with one exception. This is *G. philippensis*, which overlaps with other species, at least outside the Philippines and the northern New Guinea islands. The pattern indicates that it has expanded its range, although when this occurred is not clear (there are Pleistocene fossils in New Zealand; Holdaway et al., 2001). In any case, the phylogeny demonstrates that the range expansion of *G. philippensis* is not related to the origin of wekas, but happened some time after this.

The widespread sister of the weka group includes the volant *G. philippensis* and 13 other species (including 5 named genera) that have a wide distribution in the Asia-Pacific region. The group ranges from a western limit: Ryukyu Islands (Okinawa), Philippines, Sulawesi, and Australia (widespread), east to Wake Island (north of the Marshall Islands), Solomon Islands, Samoa, and the Cook Islands. Unsequenced fossil forms from the Marquesas Islands also appear to belong here. The widespread *G. philippensis* includes records from all parts of Australia (overlapping with other species in the north), and it extends (overlapping with other species) north to the Philippines, Palau and east to New Zealand and Samoa. It is now evident that *G. philippensis* is not an ancestor of wekas; instead it is a recent or periodic invasive in New Zealand and elsewhere.

The sister group of *Gallirallus* has a very different distribution from *Gallirallus* and mainly occurs further west. It includes a widespread group that ranges from India to south-eastern Australia and the Auckland Islands. This has the following phylogeny, with three clades (Garcia-R et al., 2016). The clades are almost entirely allopatric; overlap is restricted to the northern and eastern Philippines.

Aramidopsis plateni: Sulawesi.

"*Gallirallus*" *striatus*: India to the Philippines, throughout (not Sulawesi).

Lewinia: Philippines (Luzon and Samar only) and New Guinea (not Sulawesi) to New Zealand (Auckland Islands).

The first node is an allopatric break between the Sulawesi clade and its sister. The second break occurs at a node in or around Luzon and Samar in the northern and eastern Philippines.

The whole India–Auckland Islands group forms a clade with *Crex* of Africa, Europe, and Asia (east to China), and *Dryolimnas cuvieri* of the Seychelles. This circum-Indian Ocean group (Africa, Asia, to Auckland Islands) is sister to the mainly Pacific Ocean *Gallirallus* (Philippines to Wake Island, New Zealand, and the Marquesas Islands). The two clades show a high degree of allopatry, and are completely allopatric in New Zealand. No fossils of the Auckland Islands *Lewinia* are known from mainland New Zealand despite the wealth of Holocene bird fossils there (including several other rail species), and the *Gallirallus* clade is known in New Zealand only from the mainland and the Chatham Islands.

Neoteny and Other Trends in the Weka

Although the molecular phylogeny indicates problems with the traditional CODA model for the weka, the inference of neoteny is plausible. Neoteny has already been cited earlier as an important evolutionary trend in ducks and ratites. Taylor (1996: 113) wrote: "Flightlessness has evolved many times within the Rallidae…, selection reduces the flight muscles and pectoral girdle, probably through neoteny… These modifications may involve *only a few genetic changes*… The frequency with which flightlessness is developed by rails suggests that they are *predisposed* to it…" [italics added]. In this model, the trait is a frequent by-product of a simple, prior trend.

Structural trends are also important in the bills of Rallidae, and there is a morphocline from stout- to long-billed forms. At the extreme ends of the spectrum, stout-billed forms of Rallidae dig out vegetation, while long-billed forms probe, and the structure determines the feeding mode. Taylor (1996: 114) wrote that "many species tend to be omnivorous and therefore have unspecialised bills," but the unspecialized bill would in itself determine an unspecialized diet.

An Extinct Chatham Islands Rail, "Cabalus"

Within the diverse Pacific clade of *Gallirallus*, *G. ("Cabalus") modestus* of the Chatham Islands (subfossil), perhaps together with *Gallirallus ["Nesolimnas"] dieffenbachii* of the Chathams (subfossil), is sister to a widespread Pacific group ranging from Okinawa, the Philippines, and Australia east to the Cook and Marquesas Islands (Trewick, 1997b; Kirchman, 2012). There are many Rallidae fossils on mainland New Zealand, and it might be significant that "*Cabalus*" of the Chatham Islands is not represented there (Holdaway et al., 2001).

Another Extinct Rail, Diaphorapteryx

Diaphorapteryx is another extinct, flightless rail from the Chathams (see phylogeny, above). It was "an exceedingly large and robust rail with very reduced wings and a large robust skull and bill" (Worthy, 2004). Earlier authors compared it with "*Ocydromus*" and with *Aphanapteryx* of the Mascarene Islands. Olson (1975) compared it with *Gallirallus ("Tricholimnas") sylvestris* of Lord Howe Island.

Penguins (Sphenisciformes)

Penguins (Spheniscidae) are the second great group of flightless birds, along with ratites. Penguins have a southern hemisphere distribution (Chapter 6), and they are sister to the Procellariiformes, including albatrosses and others (Yuri et al., 2013).

Martínez (1992: 140) suggested that in penguins, "their evolutionary energy has been concentrated on adaptation to an amphibious lifestyle." They have lost their flying wings (the wing bones are robust but flattened) and have a waddling gait but can swim well. Given this reduction in both forelimbs and hindlimbs, penguins have had little choice but to adopt an amphibious lifestyle and feed in the water. The usual view is that "Their structure and physiology have been moulded both by their marine habitat and the climatological peculiarities of their environment" (Martínez, 1992: 141). Instead, the structure has probably determined the lifestyle, and explains why the organism was in the habitat to begin with.

Ksepka and Ando (2011: 171) wrote: "Many of the features that make the extant penguin flipper so efficient in underwater propulsion are incompatible with aerial flight, and thus must have evolved after the loss of aerial flight." Instead, if evolution is mutation-driven, the changes would have *caused* the loss of aerial flight and the subsequent development of underwater swimming. In another example, the penguin bill tends to be long and thin in species that are fish eaters, but shorter and stouter in those that take krill. Martínez (1992: 141) suggested that "The shape is adapted to suit the typical prey"; instead, the bill structure may determine the typical prey.

In penguins, the feathers are peculiar as they are so reduced and are borne in such a dense arrangement that they have been compared with scales (Hutton and Drummond, 1904; Ksepka and Ando, 2011). The feathers are often assumed to have developed as insulation for warmth, yet in the Galapagos Islands, where *Spheniscus mendiculus* is endemic, the seawater temperature reaches 82°F/28°C (Lynch, 1997). In addition, there are penguin fossils dated at 42 Ma from Peru, near the Equator, and "These species lived in one of the hottest places on Earth during one of the hottest times in Earth history…" (Ksepka, 2013: 159). These observations suggest that the origin of the "scale" plumage in penguins was not caused by adaptation to environmental conditions, but was one result of a general trend of reduction in the feather/scale/hair complex that has affected all vertebrates (cf. the rictal bristles of the kiwi).

Penguins are often thought of as "seabirds," but the Fiordland penguin (*Eudyptes pachyrhynchus*) has nesting habitat in true forest dominated by the typical rain forest families Cunoniaceae (*Weinmannia*) and Myrtaceae (*Metrosideros*) (Martínez, 1992: 150, 156). The birds "are good walkers and may nest several hundred metres inland" (Scofield and Stephenson, 2013). (In a similar way, pinnipeds are regarded as "marine mammals," but in New Zealand can be found in forest a kilometer or more inland.)

The sister group of *Eudyptes* is *Megadyptes* of the south-eastern South Island, discussed earlier (Chapter 7). *Eudyptes* has a crest of feathers that passes above the eye. *Megadyptes*, the yellow-eyed penguin, has no crest but does have a yellow band passing through the eyes and across the nape. In this case, a structural crest in one clade is represented in its allopatric sister group by a band of color.

Albatrosses and Their Allies (Procellariiformes)

The great diversity of seabirds around New Zealand is dominated by procellariiforms: Diomedeidae (albatrosses and mollymawks: Chapter 6) and Procellariidae (petrels and shearwaters: Chapter 9). Although procellariiforms and their sister group, penguins, spend most of their time at sea, they also have a close link with one particular area of land where they return each year and breed. This natal philopatry makes it easier to understand the allopatry in these seabirds that have such great powers of flight: rifting will separate breeding localities. Many procellariiform groups of different ranks have allopatric distributions, as seen in the Pacific members of the petrel *Pterodroma* (Heads, 2012a: Fig. 8.10).

Procellariiformes (tube-noses) are unusual in birds as they have a well-developed sense of smell and a large, complex bill made of several horny plates. This resembles the unreduced complexity in the bills of ducks and some penguins. Procellariidae show a range of feeding types from scavenging on seal and penguin carcasses, to catching fish and squid, and filter feeding on zooplankton (euphausiid crustaceans). The last mode of feeding is found in the larger prions (*Pachyptila*), which have comb-like lamellae in the upper mandible. Carboneras (1992a: 223) summarized feeding in the family: "The methods used and the type of prey taken are both related to the morphology of the different species," and that "the particular feeding technique of each species is partly determined by its morphology." This is the view supported here.

The large "seabird clade" includes penguins and procellariiforms, as well as groups such as Sulidae, Phalacrocoracidae, and Threskiornithidae. These are discussed next.

Gannets and Shags (Suliformes)

In the Sulidae (gannets and boobies), "The long tapering bill is stout and conical and specially designed for seizing fish, with its cutting edges serrated; unlike most other Pelecaniformes, however, there is no terminal hook, except in Abbott's booby" (Carboneras, 1992b: 312). In the adaptation model, the absence of a hook appears as a "design flaw," but in structural terms it represents the end-point of a trend toward the suppression of the hook and other parts of the bill. Molecular phylogenies indicate that the shags and cormorants (Phalacrocoracidae), whose biogeography was discussed earlier (Figure 4.2), belong to the Sulidae and Anhingidae. Their bill has a sharp hook.

Spoonbills (Threskiornithidae: Pelecaniformes s.lat.)

In the family Threskiornithidae, the ibises (a group of genera) and spoonbills (*Platalea*) all feed in shallow water. The ibis' bill is long, slender, and decurved, and so it is "perfectly adapted for probing" (Matheu and del Hoyo, 1992: 473). Spoonbills, *Platalea*, were first recorded in New Zealand in 1861, but could have occurred there on previous occasions (cf. Oliver, 1955). The broad, distal end of the bill is "ideal" for its typical feeding system, swinging the head from side to side, in contrast with the ibis' probing. Eyes are positioned in such a way that the birds have binocular vision, although the bird does not exploit this: "…in spite of this development, it is more usual for the birds to feed in a tactile fashion" (Matheu and del Hoyo, 1992). This suggests "a development in binocular vision that is proceeding, but without affecting behavior" (p. 478).

Waders (Charadriiformes)

The long bills of waders such as oystercatchers (Haematopodidae), plovers (Charadriidae), and snipe (Scolopacidae) are well known. Bill tips in these families have high numbers of tactile receptor organs, Herbst's corpuscles, and in the even less reduced bills of Scolopacidae there are also Grandry's corpuscles, as in kiwis.

Haematopodidae (Oystercatchers)

There are three species in New Zealand, all endemic (cf. Chapters 6 and 9). Hockey (1996: 313) wrote that "Diet, to some extent, conditions bill shape: those oystercatchers with more pointed bills will be adept at probing for worms, those with blade-like tips more expert in opening cockles and mussels." As usual, this correlation would also occur if bill shape determines the type of foraging.

Charadriidae (Plovers and Others)

This family is represented in New Zealand by two regular visitors, two self-introduced species, and four New Zealand endemics. The first of the endemics is the endangered shore plover, up until now treated as *Thinornis novaeseelandiae*, in a monotypic genus. Molecular studies instead place the bird in *Charadrius*. "*Thinornis*" has a distinctive biogeography, as it is sister to a pair of species that together extend across the Old World, from New Guinea to Morocco (*C. placidus*: China, Japan, Bangladesh, migrant to Nepal and Vietnam; and *C. dubius*: New Guinea and Japan to Morocco, migrant to tropical Africa and Indonesia) (dos Remedios et al., 2015). Neither "*Thinornis*" nor its sister group is in Australia, although there are other *Charadrius* species there, and this is consistent with the standard New Zealand–New Guinea connection.

The three remaining Charadriidae endemic to New Zealand make up a distinctive clade:

Charadrius bicinctus (banded dotterel).
C. ("Pluvialis") obscurus (New Zealand dotterel).
C. ("Anarhynchus") frontalis (wrybill plover).

Charadrius bicinctus (North, South, Chatham, and Auckland Islands) and *C. obscurus* (North, South, and Stewart Islands) nest on sea coast beaches, shingle riverbeds, and also at high altitudes: *C. bicinctus* at 4500′/1370 m on Mount Tongariro, and *C. obscurus* formerly at 6000′/1830 m on Mount Egmont/ Taranaki and at 7–8000′/2130–2430 m in the Southern Alps (Oliver, 1955). *Charadrius frontalis* breeds only on the shingle riverbeds of Canterbury (Waiau to Waitaki Rivers) and north-west Otago (Glenorchy).

The clade comprising these last three species is of special interest because of the strange morphology in the wrybill, because of the wrybill's restricted range, and because the group of three species as a whole is sister to a worldwide clade comprising at least 14 species (dos Remedios et al., 2015).

The wrybill, *Charadrius frontalis*, breeds on sand and shingle beds around braided rivers, "a biotope in which the odd, right-curved bill can be fully exploited" (Piersma, 1996a: 387). It uses its curved bill "to probe and sweep under stones." It winters around northern North Island estuaries and mudflats "where the curved bill is probably of no adaptive significance, since the species usually feeds on small prey in muddy substrates" (p. 387). Oliver (1955) quoted Potts (1871): the bird "is enabled to follow up retreating insects by making the circuit of a waterworn stone with far greater ease than if it had [a straight bill]." But he also cited Stead (1932), who was skeptical about the advantage of the wrybill and thought this would be "very slight." Stead (1932: 91) suggested: "A bill with an upward curve, one would have thought, would have been of greater use." (The related family Recurvirostridae includes the avocet genus, *Recurvirostra*, which has an upcurved bill.) Piersma (1996a: 388) described *C. frontalis* and wrote:

> With a shorter, straight bill, a plover would not be able to catch insects under stones... On inter-tidal mudflats, Wrybills also feed by tilting the head to the left. They then make sideways sweeps with the bill... the Magellanic plover of Patagonia [*Pluvianellus*] preferentially pecks slightly sideways, with its head tilted in one direction. This is correlated with a slight lateral deflection of the tip of the horny covering of the bill, but not the bone; in 80% of examined specimens the twist was to the right.

Most Charadriidae feed by waiting for prey to reveal itself, and both the wrybill and Magellanic plover are unusual as they undertake active foraging (Piersma, 1996a: 396). This may be related to the bill deflection, but even if the deflection does bestow an advantage, this does not explain its origin or its geographic restriction. Here it is suggested that it is the result of uneven reduction during the formation of the modern bill and is an old feature, as with the twisting in the skull of the flatfish.

Scolopacidae (Snipe)

The New Zealand members of this family include regular visitors that breed in the far northern hemisphere (knots, sandpipers godwits, and others), and also the non-migratory snipe *Coenocorypha*. This genus is a regional endemic, known from the New Zealand subantarctic islands, the New Zealand mainland, and north to New Caledonia and Fiji (Heads, 2014: Fig. 8.10). *Coenocorypha* of the south-west Pacific is sister to *Gallinago imperialis* of the northern Andes (Peru, Ecuador, and Colombia). This trans-Pacific pair is sister to the remaining species of *Gallinago*, widespread on all continents including Australia, but absent from Zealandia, except as vagrants (Gibson and Baker, 2012). As with the New Zealand endemic clade of *Charadrius*, the trans-Pacific clade of Scolopacidae indicates early evolution in a global complex around the New Zealand-Pacific region, and there is no reason to think that the modern representative in the region, *Coenocorypha*, is present because it dispersed here.

Scolopacidae have longer, thinner bills than the related Charadriidae, and the "bill tips contain particularly high densities of tactile receptors" (Piersma, 1996b: 464). The bill tips include both Herbst corpuscles and Grandry corpuscles. Piersma (1996b: 464) wrote that "The great morphological radiation within the Scolopacidae reflects the fact that different species use different habitats and food types." The correlation between morphology and habitat/food type is clear. But the distributions suggest that the morphological diversity arose by vicariance rather than radiation, and the similarities in bill morphology of kiwis and scolopacids (discussed earlier in this chapter) could have been inherited from a polymorphic ancestor.

Eagles and Hawks (Accipitriformes)

Haast's eagle is a South Island endemic that went extinct sometime after 1300. It was described as *Harpagornis moorei*, in a new, monotypic genus, but according to molecular studies, it is nested in *Hieraaetus* (Bunce et al., 2005). It is the largest eagle known, living or fossil, and moas were one of its main food sources.

The large size of Haast's eagle, discussed earlier in this chapter, is often attributed to "relaxation from mammalian competition" (Worthy and Holdaway, 2002: 254), although there is nothing similar in New Guinea where large mammals are also absent. It had the claws of a predator, and the head of a vulture, which Worthy and Holdaway (2002: 324) argued is "not surprising," as it needed to eat such a large carcass. The lack of competition explains why the bird survived, but not why it had such a large body to begin with.

Owls (Strigiformes)

There are two New Zealand owls: one species of *Ninox* (morepork) and one of *Sceloglaux* (laughing owl; extinct by 1914). Both are members of Strigidae. *Sceloglaux* has been mentioned already, as its sister group may be *Uroglaux* of New Guinea. This New Zealand–New Guinea disjunction is also seen in the honeyeaters and in *Gerygone*, where the groups concerned appear to have evolved by vicariance with Australian groups. The south-west Pacific biogeography of *Ninox* and its allies *Sceloglaux* and *Uroglaux* is of great interest. Wink et al. (2009) have made a start at resolving the molecular phylogeny, but further sampling is desirable.

For the owls in general, Marks et al. (1999: 87) wrote that "Most of the morphological features distinctive of strigids [including *Ninox* and *Sceloglaux*] arise from the fact that they are predatory and nocturnal. These two forces have produced specialised plumage, ears, eyes, bill and feet." The trend to night vision is not complete; strigid owls cannot see well in very dark conditions and have "no difficulty whatsoever in seeing during daylight hours." Yet in barn owls, Tytonidae, the retina is dominated by rod cells (more sensitive than cone cells), "*restricting* [the birds] *to a nocturnal lifestyle*" (Schwab, 2003; italics added). Other significant reductions have occurred in the eye. In some birds the eyeballs are quite mobile, but in general the movement of eyeballs in birds is restricted (Brooke and Birkhead, 1991: 48). Owls cannot move their eyes at all; the extraocular muscles have atrophied during evolution, and are rudimentary (Schwab, 2003).

As discussed earlier, vertebrates and bilateral plant parts often retain a trace of earlier ancestral structures with higher orders of symmetry, in spiral modes (Chapter 13). This is often expressed in deviations from perfect bilateral symmetry, as in the reproductive system of birds in general, and in the sideways deflected bill of birds such as *Charadrius frontalis*. Five genera of owls (out of a total of 27) have asymmetric external ears—the right and left have different size, shape, or both (Marks et al., 1999: 90). In *Aegolius* and *Strix* species, the asymmetry extends to the bony structure. *Aegolius funereus* has the right ear as much as 50% larger than the left, and directed upward, while the left is directed downward. Owls are characterized by "the development, almost unique in vertebrates, of spatial binaural [two ears] asymmetry" (Kühne and Lewis, 1985: 262). In the standard model, "Binaural asymmetry [in *Aegolius*, etc.] is… correlated with *the need* for localisation in the vertical plane" (Kühne and Lewis, 1985: 266; italics added). Nevertheless, only 5 out of 27 owl genera have binaural asymmetry; if the need is so strong, how do the others cope? It is not necessary to invoke needs and strategies though. It is possible to explain both the usual morphology (bilateral symmetry) and the unusual condition (asymmetry) as by-products of a single process of reduction in the skull and its precursors. If this has been even, the result is bilateral symmetry; if it has been uneven, the result is asymmetry as in flatfishes, owls, and others.

Kingfishers (Coraciiformes)

The biogeography of *Todiramphus* in the south-west Pacific is discussed elsewhere, with stress on the widespread endemic forms in Micronesia and central Polynesia (Heads, 2012a: Fig. 6-10). *T. sanctus* is distributed through Indonesia and Australasia, to New Zealand and American Samoa. It is replaced in western Samoa, the central Pacific, and much of Micronesia by *Todiramphus* s.str.

Falcons (Falconiformes)

Falcons are represented in New Zealand by a single species, the endemic *Falco novaeseelandiae*. The sister group of *Falco novaezeelandiae* is *F. femoralis*, of South America, Central America, and Mexico (Fuchs et al., 2015). This gives a standard, trans-South Pacific disjunction.

The New Zealand–South and Central America pair, *Falco novaezeelandiae* and *F. femoralis*, is sister to a widespread clade of 11 species, found worldwide (breeding records) *except in New Zealand, Central America, and most of South America*. (Within South America, *Falco femoralis* breeds throughout, except in the Amazon basin and southern Chile. In contrast, in South America the widespread clade only has breeding records in the Peruvian Andes and Patagonia, with a single, globally widespread species, *F. peregrinus* (IUCN, 2016). This indicates a large degree of allopatry between the two groups, and a possible origin of both by vicariance.

Falco peregrinus is distributed more or less worldwide, but has conspicuous areas of absence: New Zealand, Central America, and large parts of South America. White (1994) described these as "unexpected," while Trewick and Gibb (2010: 235) wrote that the species "perversely" does not occur in New Zealand. Nevertheless, the absences reflect a normal pattern of fundamental allopatry between the *novaeseelandiae* + *femoralis* pair and its sister group. The allopatry suggests that a global group has divided into the two clades by vicariance, with subsequent overlap restricted to local parts of South America and northern Mexico.

Allopatry also occurs *within Falco novaeseelandiae*. This is evident among the three recognized forms, now sometimes treated as subspecies (Marchant and Higgins, 1993; White et al., 1994; Scofield and Stephenson, 2013):

Falco n. ferox. ("Bush form"; small and dark): North Island and north-western South Island, south to Greymouth.

F. n. australis. ("Southern form"; a morphological intermediate between the other two forms, *but closer to F. n. ferox*): Fiordland, Stewart Island (now extinct there), and Auckland Islands.

F. n. novaeseelandiae. ("Eastern form"; large and pale): Remaining parts of the South Island, east of Nelson and Fiordland.

(The species is also represented on the Chatham Islands by an egg collected before 1888 and by subfossil bones [Heather and Robertson, 2005], but the particular form there is unknown.) The three forms differ in size, plumage, ecology, habitat, and range, and although there is some intergrade, the overall pattern is well defined. In particular, "Bush and southern forms are similar to each other" in plumage and in size (Marchant and Higgins, 1993: 290), giving a standard Alpine fault disjunction with the gap filled by the "Eastern form."

To summarize, an ancestral complex of falcons with a global distribution has divided into the New Zealand–South America clade and the widespread clade. Later, the New Zealand and South American groups have been separated by continental drift. Next, the New Zealand representative has been divided into three forms, and in a last step, two of these have been separated by movement on the Alpine fault.

Parrots (Psittaciformes)

Parrots are the sister group of passerines (Yuri et al., 2013), and it is significant that in both orders the "basal" members are New Zealand endemics. Parrots comprise two main clades. One is a New Zealand-endemic clade made up of the kakapo (*Strigops*), the kea and kaka (*Nestor*), and a Miocene fossil genus *Nelepsittacus* from Central Otago, regarded as closest to *Nestor* (Worthy et al., 2011b). These three genera form the sister group of all other parrots. The last parrot genus in New Zealand (with eight species there) is *Cyanoramphus*, a typical member of "all other parrots." (For the distribution of *Cyanoramphus* species in New Zealand, see Chapter 6.)

New Zealand has a low diversity of parrots at the species and genus level, but it is the only area that has both of the main parrot clades, and so in phylogenetic terms the diversity there is higher than anywhere else. Deriving the *Nestor* + *Strigops* clade from any particular area outside New Zealand is not straightforward, as its sister group is pantropical, and this is consistent with a vicariance origin for both the main clades of parrots.

Nestor and *Strigops* were always recognized as distinctive, and a review of cranial osteology in parrots concluded that the peculiarities of *Strigops* "are greater than those of any other form (except perhaps *Nestor*)" (Thompson, 1899: 9). However, before the molecular era *Nestor* and *Strigops* were not regarded as allied, and they were each placed in separate clades with other genera. Their immediate ancestors were thought to have dispersed to New Zealand—"The ancestor of the three species of parrot in the genus *Nestor*... probably came from Australia" (Diamond and Bond, 1999: 21).

One review of parrot systematics summarized the standard dispersal + adaptation explanation for *Nestor* and *Strigops* (Smith, 1975: 60): "It seems reasonable to suppose that their unique development could only have taken place in the exceptional circumstances which apply to *oceanic islands*. New Zealand's extreme *geographical isolation* kept its original fauna to those few species able to cross *considerable stretches of sea*. The complete lack of competitive mammals and the few land birds reduced selection pressure... on the parrots, allowing them to *radiate* into *extreme forms*..." [italics added].

Now, following the molecular discoveries, authors have instead located the origin of *Nestor* + *Strigops* on a continent, Gondwana, not on an oceanic island. At this time, New Zealand was not geographically isolated, and its biota was the result of normal, *in-situ* inheritance, not chance, trans-oceanic dispersal. Mammals were present and land birds were diverse, and there is no evidence that *Nestor* and *Strigops* radiated from any particular point (or monomorphic ancestor). As the sister group of the other parrots, not derived from within them, *Nestor* and *Strigops* have features that are no more or less extreme than those of their sister—they are simply alternatives. In any one character, the ancestor may have resembled *Nestor*, *Strigops*, the "other parrots," or any combination of these.

The molecular discovery of the basal break in parrots and its location in New Zealand (not between New Zealand and Australia) is a completely new result, despite Thompson's (1899) hints. It repeats the finding that many Australian and New Zealand passerines (discussed in following sections) are "basal" clades, not derived from within northern hemisphere groups as had been thought.

Morphology of Parrots

Psittaciformes have a characteristic bill; it is often strong and broad, and the two mandibles display a distinctive counter-curvature. In addition, "the tongue and jaw muscle structure is complex, diverse, and highly developed, closely related to the great power and control *needed* in the handling and ingestion of food items" (Collar, 1997: 286; italics added). "Access to the high nutritional content of hard seeds and nuts is the prime explanation for the evolution of the powerful counter-curved bills of all parrots" (Collar, 1997: 306).

In the alternative model, the structure did not come into existence because it answers an extrinsic need. Instead, the typical, robust morphology represents one of the ancestral conditions, and the counter-curvature is a simple structural alternative to the "normal," uncurved bill. The reduction that produced the modern bill followed two pathways: in one, the reduction of the components was even, and a straight bill resulted, in the other, the reduction was uneven, leading to a counter-curvature (cf. Figure 13.1). Both variations have turned out to be suitable for taking different types of food. In several parrots the bill is much reduced, and the diet does not include large hard seeds and nuts. The upper mandible of *Nestor notabilis* is less curved than most parrots, and the lower mandible is almost straight (cf. Oliver, 1955), as in the vulturine parrot *Psittrichas* of New Guinea.

Although most parrots are seed eaters, there are many exceptions. *Strigops* feeds on podocarp "fruit" when this is available, but most often feeds by crushing leaves and stems and extracting the juice (Collar, 1997: 309). *Nestor notabilis* is a generalist, opportunist feeder; it is recorded eating fruit (including *Nothofagus* nuts), other plant material, insects, and chicks of Hutton's shearwater (*Puffinus huttoni*), a colonial, burrow-nester (Greene, 1999). Keas also feed on sick or dying sheep (Worthy and Holdaway, 2002: 485). In the subantarctic islands, *Cyanoramphus* species feed on leaves of large tussocks and on much smaller quantities of flowers, berries, other vegetation, bird corpses, and eggs. On Antipodes Island they also hunt, kill, and feed on grey-backed storm petrels (*Oceanites nereis*) at their nests (Greene, 1999).

Other parrots that are not seed eaters include pygmy parrots (*Micropsitta*, of New Guinea), which "glean principally for lichen along the trunk and branches of trees" (Collar, 1997). In *Lathamus*, *Loriculus philippensis*, and members of the Loriinae, nectar and pollen are the chief source of nutrition. "*As a consequence*," Collar (1997: 309) wrote, "their gizzards are much less muscular and their intestines shorter than other parrots of equivalent size...." [italics added]. Collar (1997: 286) also argued that "The gizzard or ventriculus is weak in the Loriinae, which ingest mainly nectar and pollen, but highly developed in other parrots, which *need* to break down often extremely hard vegetable material" [italics added]. Instead, the different structure of the intestine in different groups helps determine the different diets in these parrots.

The Loriinae have distinctive, brush-tipped tongues that facilitate feeding on nectar, and *Nestor* has a similar hair-like fringe. The "frayed" tongue present in these and many other birds means that nectar is taken up at a rapid rate by capillary action. In all these cases—seed eaters and the others—the feeding method depends in large part on the structure of the bill, the tongue, and the gut. The bill varies from large and broad at one extreme, to much smaller in groups such as the lichen-eating pygmy parrots, *Micropsitta*.

The Kakapo (*Strigops habroptilus*: Strigopidae)

The kakapo represents a monotypic genus, *Strigops*, known from South and Stewart Islands. It is the largest parrot (males weigh up to 3.6 kg) and the only flightless parrot. Compared with *Nestor* it has a smaller pectoral skeleton and a larger pelvic skeleton. In addition, the carina (the keel on the sternum where the main flight muscles are attached) is vestigial. Livezey (1992) suggested that these and other morphological features in the kakapo represented "corollaries" of flightlessness, and Collar (1997: 290) agreed: "All of its distinctive anatomical features stem from its loss of flight...." Yet the morphological changes in themselves would have *caused* flightlessness, and reduction in the pectoral skeleton is a common trend in vertebrates. Thus, although Livezey (1992) argued that "Flightlessness of *Strigops habroptilus* clearly represents a complex of selective tradeoffs...," it is not necessary to invoke selection. With respect to the mode of reduction, Livezey (1992) wrote that the sternum is one of the last skeletal elements to ossify in birds, and so it is reasonable to interpret the "underdevelopment" of the sternum and carina as the result of neoteny.

Strigops individuals spend most of their time on the ground, but the birds are skilled tree climbers and often climb up to 10 m above the ground, sometimes up to 30 m (Higgins, 1999: 633); the bird is flightless, but arboreal. Functioning wings and the ability to fly would, of course, be useful at 30 m, but the pectoral limb and girdle, and the sternum, have all been caught up in the great reduction trend. These arboreal birds do not realize they have lost the power of flight, and so they persist in their behavior.

The Latin name of kakapo, *Strigops* (owl-face), refers to the conspicuous facial disc of hair-like feathers which resembles that of owls (Strigidae) (Higgins, 1999; Powlesland et al., 2006). The disc in kakapo includes elongate rictal bristles (Collar, 1997), and these are also present in feathers of the lores, cheeks, and ear-coverts (Higgins, 1999).

In owls the facial disc is often interpreted as a means of directing sound toward the ears, while rictal bristles are interpreted in birds in general as an adaptation for insect-feeding. Nevertheless, *Strigops* is a strict vegetarian (even chicks are not fed animal food; Collar, 1997), and so these explanations for the facial disc and the rictal bristles are unconvincing. A general reduction in the facial region is a more probable cause, and the same structural trend would also explain the structure in owls. In juveniles of *Strigops*, the "irises are encircled by a ring of short feathers resembling 'eye-lashes' not evident on adults" (Powlesland et al., 2006: 5), and this ring is also explicable as a relic structure.

Strigops is vegetarian, and Livezey (1992) suggested a simple structural reason for the lack of carnivory: "Flightlessness limits the importance of 'pursuit tactics' for foraging by terrestrial birds, and may have precluded reliance on insects and vertebrates as prey in S. *habroptilus*." In addition, the crop is enlarged ("to cope with its bulky vegetarian diet"; Powlesland et al., 2006), providing another structural basis for the diet.

The nocturnality of *Strigops* is regarded as an adaptation for avoiding avian predators. Nevertheless, Smith (1975) wrote that "The nocturnal or crepuscular activity of the Kakapo is not exceptional, for all the New Zealand parrots feed and move in dim light... [In addition] *Neophema* [now *Neopsephotus*] and *Geopsittacus* [now *Pezoporus occidentalis*] [both from central Australia] are almost nocturnal..., while captive cockatoos, in the author's experience, fly and feed by moonlight." In addition, the idea that nocturnality in *Strigops* is an adaptation to avoid predators conflicts with its flightlessness, and with the fact that the sister group of *Strigops*, the widespread *Nestor*, is an excellent flier and more or less diurnal. These aspects suggest that the features of the kakapo are the result of mutational trends rather than natural selection.

The structural basis of kakapo vision has been addressed by Corfield et al. (2011), in a study of the bird's specializations "for nocturnality." The authors described how a shift in vertebrates from a diurnal to a nocturnal lifestyle is, in most cases, associated with either *enhanced* visual sensitivity or a *decreased* reliance on vision. Yet the kakapo shows neither a simple increase nor a decrease in light sensitivity, and instead "has a unique combination of traits" (Corfield et al., 2011: 6). Corfield et al. (2011: 1) summarized the confusing situation:

> The Kakapo's orbits are significantly more convergent than any other parrot, suggesting an *increased binocular overlap* in the visual field [and enhanced light sensitivity].... [But] With respect to the brain, the Kakapo has a significantly *smaller* optic nerve and tectofugal visual pathway. Specifically, the optic tectum, nucleus rotundus and entopallium were significantly reduced in relative size compared to other parrots... the Kakapo possesses *a visual system unlike that of either strictly nocturnal or diurnal birds and therefore does not adhere to the traditional view of the evolution of nocturnality in birds....* (p. 1)

There is no evidence that the results of the different morphological trends in kakapo morphology are adaptive, beyond the fact that the parrot has continued to survive. The unusual combination of features—reduced forelimbs and reduced vision, but increased binocular vision—suggests that the kakapo has survived despite the changes, not because of them.

Animals that undergo a significant increase or decrease in their sensitivity to light often prefer to, or are forced to, live a nocturnal lifestyle. Instead, the CODA model suggests the nocturnal habit developed first, before the development of adaptations for it (Corfield et al., 2011: 1):

> Living in a scotopic, or low light, environment poses significant *challenges* for the visual system [the new habit created new needs]... the visual systems of animals that live in scotopic environments have evolved in one of two ways. Firstly, they can evolve mechanisms to *increase* the sensitivity of the eye to light.... Alternatively, animals can *decrease* their emphasis on the visual system and enhance the sensitivity of other sensory systems to provide equivalent information about their environment.... shifting from a diurnal to a nocturnal lifestyle can either be associated with the enlargement of the visual system to enhance light sensitivity or the reduction of the visual system... [italics added]

This model accepts that the birds occupied a scotopic niche before developing adaptations for it, and changing its visual system. Yet the individual changes in the bird, including reductions in the visual system and reduction in the skull (giving increased binocular vision), all conform to standard trends that are repeated countless times through the vertebrates.

The standard view is that:

> The evolution of flightlessness, folivory and nocturnality on a largely predator-free island may have reduced the Kakapo's reliance on vision in favor of enhancing other sensory modalities. (Corfield et al., 2011: 6)

The bird might have survived because of the lack of predators, but this does not explain why it was nocturnal to start with, and this question concerns the morphological trends ("laws of growth") that were the immediate cause.

Reduction in the visual system has probably led to extinction in some lineages of birds, while others already had structures that enabled survival at night or in caves (e.g., the kiwi bill, the robust bill of the kakapo, the structures enabling bats to echolocate), and these groups have survived. This does not mean that these structures developed in the first place in order to fulfill an extrinsic need. They developed because of prior trends in the evolution of the ear and face, and because these produced a viable morphology.

The Kea (*Nestor notabilis*: Nestoridae)

The second genus of the "basal parrot" clade is *Nestor*, with four species: the kaka, *N. meridionalis*, in North, South, and Stewart Islands (lowland to montane); the kea, *N. notabilis*, in the South Island and (fossil) in the North Island (montane to alpine); one endemic, extinct species on Norfolk Island; and another on the Chatham Islands.

Dussex et al. (2014) summarized recent molecular studies indicating that the moa genus *Megalapteryx didinus* and the kea *Nestor notabilis*, both *upland* species, survived the last glacial maximum in restricted, glacial refugia. The authors suggested that the birds recolonized the South Island from these refugia, after the glaciers retreated. In contrast, during the last glacial maximum, "the brown kiwis *A. rowi* and *A. australis* [both *forest* species] likely occupied the areas of scrub and grassland present *over much* of the South Island at the time…." Dussex et al. (2014) suggested that species responded differently to the glaciations, "mainly due to their habit requirements." Nevertheless, the birds' habitats suggest that the opposite process to that suggested took place: during glaciations, birds such as *Megalapteryx* and kea that are tolerant of cool temperatures and alpine vegetation would have been *more widespread*, not less. Forest birds such as brown kiwis would have been *less widespread* during glaciations, not more. There seems to be a problem with the ecological interpretations of the molecular data, in particular the interpretation of spatial distributions and calculation of clade ages.

In the kea, genetic studies found distinct, geographic clusters (Dussex et al., 2014). Microsatellite data showed three genetic clusters: Fiordland/Mt Aspiring, Mt Cook/Westland, and Nelson/Kaikoura/Arthur's Pass. Mitochondrial data "identified a clear separation between a north and south cluster (Fig. 4) in the vicinity between the Mt Cook and Aspiring populations." This break coincides with the Aoraki node (Chapter 7). Mitochondrial haplotype C was restricted to the south (Fiordland and Aspiring regions) and north (Nelson and Marlborough), but was absent in the central South Island (Westland, Arthur's Pass, and Mount Cook). The geographic structure in the kea is clear-cut and follows typical lines. Further, more detailed sampling would give interesting results, and focusing on the boundary zones would be useful.

As for dating, the calibration used by Dussex et al. (2014) was obtained by assuming that the kea and the kakapo, *Strigops*, diverged following the breakup of Zealandia and Gondwana at 80–65 Ma (Wright et al., 2008). Calibrating the timeline in this way, tying the basal node to a tectonic event in Earth history, avoids the use of fossils to give actual (not just minimum) clade ages, and, as a general method, this is supported here. Nevertheless, the particular tectonic explanation that the authors have offered is debatable. For one thing, *Nestor* + *Strigops* is not allopatric with its sister group at a break in the Tasman Basin, as this model would suggest, as the clade "all parrots except *Nestor* and *Strigops*" also inhabits New Zealand. There is one genus there, *Cyanoramphus*, and this is also in Polynesia, but not in Australia (Heads, 2012a: 297). The differentiation between *Cyanoramphus* and its Australian relatives *can* be explained by the late Cretaceous breakup of Gondwanan Australia and Zealandia, and this implies that the basal break in all parrots occurred during a prior tectonic phase, such as the 15 m.y. of predrift rifting. The overlap, in New Zealand, between *Nestor* + *Strigops* and its global sister group, may have occurred early on, long before the origin of *Cyanoramphus* itself. As always, calibration is just one part of the calculation of clade age, and serious errors can also be introduced by other evolutionary assumptions.

For the kea, Dussex et al. (2014) supported a scenario of "postglacial divergence from a single ancestral glacial refugium… from this… the species recolonizes its range at the end of the LGM and then diverges into three subpopulations via founder effect." This is based on the dating. Nevertheless, the authors admitted: "Our data could not resolve the location of the refugium from where recolonization of the range originated." Perhaps this is because there was no single refugium, but multiple microrefugia, and the clades are older than the study assumed.

Dussex et al. (2014: 2205) concluded that:

> Ancient divergence... consistent with the "beech-gap" hypothesis was clearly rejected (PP < 0.1%), which means that kea certainly did not survive in northern and southern refugia isolated by glaciers during the LGM. Rejection of this scenario is *somewhat surprising as...* *a consistent north-south break* in both mitochondrial and microsatellite data suggest that the recent evolutionary history of kea could be in agreement with the "beech-gap" hypothesis.

This last sentence is important. In particular, the disjunct distribution of mitochondrial haplotype C is a related, standard pattern that can be explained by strike-slip displacement along the Alpine fault. Problems with dating can also explain the apparent anomaly described earlier, with upland birds thought to have declined (rather than expanded) in the last glaciation, and lowland birds thought to have expanded (rather than declined).

Dussex et al. (2014) concluded: "The recent evolutionary origin of this genetic structure suggests that each genetic cluster does not need to be considered as independent conservation units." Yet of all their results, the clade dates are the ones that rely most on assumptions and extrapolations. The raw data—the clades and their geography—are more interesting and provide key information for the conservation of keas.

Perching Birds: Passeriformes

The passerines include about half of all bird species (~5,000 out of 10,000). The phylogeny of this global group follows a simple geographic sequence, as follows (Jønsson et al., 2011; Aggerbeck et al., 2014) (NZ = New Zealand, NG = New Guinea):

NZ (Acanthisittidae).
   Pantropical, main diversity in the Americas (Suboscines, or Tyranni).
      Australia (Menuridae and Atrichornithidae).
         Australia, NG (Ptilonorhynchidae and Climacteridae).
            Australia, NG, **Micronesia, Melanesia and Polynesia** (east to New Zealand and Samoa), **west to Thailand** (Meliphagidae, Acanthizidae, Maluridae, Dasyornithidae).
               Australia, NG (Orthonychidae and Pomatostomidae).
                  NZ (Callaeidae and *Notiomystis*) + **cosmopolitan** (all others, incl. NZ: Petroicidae, Zosteropidae, Motacillidae, Locustellidae). PASSERIDA, ~3500 spp.
                  NZ (Mohouidae) + **cosmopolitan** (all others, incl. NZ: Rhipiduridae, Oriolidae). CORVIDA, ~750 spp.

Of these 10 clades, Australia, New Guinea, and New Zealand each have 6, more than any other area. Southeast Asia and parts of Oceania have three or four, other areas have just two, the two cosmopolitan clades Passerida and Corvida.

The Passerida + Corvida form a hyperdiverse, cosmopolitan complex that includes 85% of passerine species. Of the four main clades in this cosmopolitan complex, all four occur together only in New Zealand and two are endemic there. So although New Zealand passerines are depauperate at the species level, they are diverse in the basal groups, and in the distal Passerida + Corvida clade they are twice as rich as the rest of the world put together. This suggests that the low species diversity is not caused by the vagaries of chance dispersal, but by extinction. This would have been caused by Cenozoic flooding, volcanism, uplift, climatic change, and human activity, with the first perhaps being the most significant.

Center-of-origin analyses estimate that oscines (i.e., all passerines except the Acanthisittidae and Suboscines) had a center of origin in Australia/New Guinea because of the basal groups there (see the passerine phylogeny). But basal groups are just small sister groups, and a comparative study suggests that the Tasman-Coral Sea region represents a single, shared break-zone in many widespread global ancestors, not a center of origin. A similar pattern occurs in the parrots (the sister group of passerines), in the passerines, in Passerida and in Corvida; each group has its "basal" member endemic to New Zealand.

Based on a center-of-origin analysis using LAGRANGE (Ree and Smith, 2008), Aggerbeck et al. (2014) concluded that Corvida (their "core Corvoidea") had a center of origin in proto-Papua and from there colonized all continents except Antarctica. Nevertheless, the pattern is easier to explain if the global range expansion of Corvida occurred *before* the basal break in the group took place between Mohouidae of New Zealand and its global sister group. If this were the case, the basal break in Corvida and the similar basal break in Passerida could have occurred at the same time and have been caused by the same factor.

The basal breaks in passerines as a whole (the New Zealand Acanthisittidae versus its global sister-group) and parrots (*Nestor* and *Strigops* versus the rest) imply another, earlier phase of differentiation around Zealandia, perhaps only soon before the one in Passserida and Corvida. Both phases probably developed during the period between the inception of extension and rifting across Zealandia, and the eventual seafloor spreading.

The Timeline of Evolution in New Zealand Birds

Molecular work carried out so far has proposed a basal position of New Zealand forms in parrots, passerines, Corvida, and Passerida, and this contradicts some traditional ideas on the timing of bird evolution in the region. An influential, earlier study by Fleming (1962b) concluded: "The systematic differences which distinguish New Zealand birds from their relatives overseas give us a rough-and-ready yardstick to the time that has lapsed since their colonisation, unreliable in particular cases, but useful in default of other evidence." This approach, relying on degree of differentiation, is still adopted in many molecular clock studies but is problematic, and there is plenty of other evidence that can be used instead. The evidence used here comes from dated geological events and biological distributions.

Fleming (1962b: 271) inferred the following ideas on the age of New Zealand bird groups depending on their degree of differentiation:

> *"strong" subspecies* have developed since Cook Strait last became a barrier, some 15,000 years ago, and since forest came to the formerly glaciated Auckland Islands, perhaps less than 10,000 years ago, so there seems no reason to assume that any of the seventeen New Zealand subspecies of Australian species is much older than late Pleistocene (say 20,000 years).

Yet the last flooding of Cook Strait was, indeed, Pleistocene (Lewis et al., 1994), and most North Island/South Island differentiation now appears to be much older than this (see section "The Cook Strait node" in Chapter 8).

Fleming (1962b: 271–272) continued:

> Somewhat older [than the subspecies], we may infer, are thirteen [endemic] *full species* classed in overseas genera... Using the yardstick of "strong subspeciation in about 15,000 years," I see no need to put their origin further back than Early Pleistocene, a million years or so.
>
> The next systematic category is the *endemic genus* in an overseas family... The date of probable colonisation ranges over a long period, here equated with the 25 million years of the Neogene (Late Tertiary) but perhaps reaching back into the Eogene for some genera.
>
> Eight species, in six genera, are grouped in three *endemic families*... Their distinctness implies an ancient origin, certainly Tertiary and probably in part Eogene (early Tertiary). I personally think it most unlikely that any of the colonisations so far mentioned was as old as Upper Cretaceous.
>
> Finally, two *endemic orders* of Ratites, the Kiwis and the Moas... we are probably on safe ground in attributing these two colonisations to the Upper Cretaceous....

The significance of the "degree of differentiation" among clades is problematic, and in most cases will not be a simple reflection of time since the groups' divergence (Chapter 2). Fleming (1962b) himself presented his ideas as tentative: "There are, of course, many pit-falls in this kind of 'educated guessing'."

In such a situation, alternative approaches can be attempted. The calibrations used in this book (and Heads, 2014) indicate older ages for the clades than those suggested in the fossil-based studies of Fleming (1962a,b) and many molecular workers. For example, the New Zealand–New Guinea affinities among genera of Meliphagidae and within *Gerygone* (Acanthizidae) correlate with tectonic connections that are Paleogene—consistent with the ideas of Fleming (1962b) on the meliphagid genera, but much older than Pleistocene date he would predict for the *Gerygone* species.

The following sections give brief reviews of the New Zealand passerine families, with notes on their biogeographic and morphological evolution. Most distinguishing features of New Zealand groups remain unexplained in the usual model of adaptation, but some of the explanations that have been offered are cited here.

Acanthisittidae

These are small birds, 7–10 cm long. Any significant morphological variation in this "basal" group is of special interest for passerines in general, which make up half of all bird species. Apart from the suboscine-type syrinx structure in Acanthisittidae, one of its most distinctive features is in the ear. The opening is unusual—"it is a narrow horizontal slit that opens to a pocket, which extends downward to the inner ear" (Gill, 2004: 464, based on Pycraft, 1905: 608). (Gill also wrote that the stapes has a unique form and cited Feduccia, 1975, but in fact Feduccia wrote that it has the primitive form shared by most non-passerines and oscines, and that it is the suboscines that are specialized.) This unusual variation in the ear evident here at the phylogenetic base of passerines occurs in the same region as the critical reduction series in the ear-jaw of the higher vertebrates.

The family Acanthisittidae is also characterized by a trend to flightlessness, which is unusual but not unique in passerines. Of the acanthisittids, *Traversia* (Stephens Island wren, recently extinct) and the fossil genera *Dendroscansor* and *Pachyplichas* were all flightless. Only one other flightless passerine is known—a fossil *Emberiza* from the Canary Islands. In *Traversia*, the keel on the sternum (the carina) has undergone an "extraordinary degree" of reduction, and the sternum is "almost acarinate" (Millener, 1989), as in ratites and *Strigops*. The carina also shows marked reduction in *Pachyplichas* (Millener, 1988) and *Dendroscansor* (Millener and Worthy, 1991). Along with these reductions in the sternum, the tails in the Acanthisittidae are also reduced. Acanthisittidae have 10 rectrices, while most other passerines have 12 (Gill, 2004).

Krull et al. (2009) described ultrasonic sound production in these small birds and concluded that it may not function as a means of intraspecific communication. These authors wrote: "It is more likely that the presence of ultrasonic harmonics in the rifleman [*Acanthisitta*] is either targeted at prey species or *represents merely an epiphenomenon* and that rifleman cannot hear the ultrasonic harmonics of their calls" [italics added]. In other words, as with the reduction in the tail, the feature is a by-product of a general trend, such as a reduction in size, not because it fulfills a particular need.

Meliphagidae and Acanthizidae

Both these families belong to a basal grade of passerines with its main diversity in Australasia. This paraphyletic, regional complex of families is basal to the main, global group of Passerines (Corvida plus Passerida). The New Zealand genera of Meliphagidae—*Anthornis* (bellbird) and *Prosthemadera* (tui)—are sister groups and share similar egg patterning (Higgins et al., 2008: 501). Their closest relative is in New Guinea, with this New Zealand–New Guinea clade related to groups in Australia. In Acanthizidae, the New Zealand and Norfolk Island grey warblers (*Gerygone modesta* and *G. igata*) also have their closest relative in New Guinea (maps in Heads, 2014: Figs. 6.5 and 6.6), with the sister group of this clade in Australia. The phylogenetic break between New Zealand–New Guinea in the east and Australia in the west coincides with the spreading center in the Tasman-Coral Seas.

Anthornis melanura, the bellbird, has subspecies on the mainland and Auckland Islands (*A. m. melanura*), the Three Kings Islands (*A. m. obscura*), the Poor Knights (*A. m. oneho*), and the Chatham Islands (*A. m. melanocephala*) (Bartle and Sagar, 1987). The last three (especially the Chathams form) are larger than the mainland form. The Chathams subspecies has a yellow eye, in all the others it is red. A distinct, immature (post-juvenile) plumage phase occurs in the Poor Knights and Chathams subspecies, and these also share blue head iridescence, rather than violet as in the other two. The characters are consistent with a basal grade on the islands, and a molecular study of the group would be valuable.

The most obvious differentiation in the tui, *Prosthemadera*, is that between the subspecies known from the Kermadec to Auckland Islands, including the mainland, and the large Chatham Islands subspecies. It is interesting that the tui is only recorded as a rare, casual visitor on the Poor Knights Islands, while *Anthornis* is present there with a local endemic, possibly related to Chatham Islands birds.

Honeyeaters (Meliphagidae) have a specialized, brush-tipped tongue and internal structural modifications to the bill that are "adapted to a nectarivorous diet... while many are adapted to feeding on nectar...others have a diet consisting largely or mainly of fruit, insects, lerp, honeydew or manna... *Consequently*, they show a diversity of sizes and shapes, and vary particularly in the shape and length of the bill" (Higgins et al., 2008: 512; italic added). This suggests that the diet has determined the bill's morphology. Nevertheless, Higgins et al. (2008) also suggested that the New Zealand tui, *Prosthemadera*, has a "stocky, slightly decurved bill, and *therefore prefers to* extract nectar from fairly short flowers with little curvature" (Higgins et al., 2008: 537; italics added). In this approach, bill morphology instead determines diet, with structure determining function and behavior.

Callaeidae and Notiomystidae

The three genera of Callaeaidae are all New Zealand endemics and the family is sister to *Notiomystis*, another New Zealand endemic. The clade as a whole, with four genera, has special phylogenetic significance as it forms the sister group of the cosmopolitan group Passerida (Aggerbeck et al., 2014).

Notiomystis is a monotypic genus that was widespread through the North Island and its offshore islands into historical times, but declined to a single population on Little Barrier Island before it was reintroduced elsewhere. The tongue is brush-tipped and, in part, tubular. (The main food is nectar, although fruit and small invertebrates are also taken; Higgins and Christidis, 2009.) Because of this, *Notiomystis* was thought to be a meliphagid (honeyeater) until molecular work indicated that the similar tongue structure in the two groups is a parallelism.

Callaeidae comprise three monotypic genera. *Callaeas* (kokako) and *Philesturnus* (saddleback) were widespread in the three main islands up until historical times, and both survive, while *Heterolocha* (huia) was restricted to the North Island (from Wellington north to the Raukumara Range, East Cape) and is now extinct. All three species have a distinctive, fleshy wattle at the angle of the mouth (Scofield, 2009). In *Callaeas*, the wattle is bright blue in the North Island form, orange in the South Island form. Wattles in general occur in zones of reduction, especially around the gape, but also, for example, on the upper eyelid of the white-cheeked turaco (Musophagidae: Cuculiformes; Turner, 1997: 483) and in supraorbital position in the passerine family Platysteiridae.

In Callaeidae, "the main components of the diet vary among the species, and this is reflected in the morphology of their bills" (Turner, 1997: 232). In *Callaeas* (kokako), the bill is rather short and robust, with a decurved upper mandible, and the bird feeds on fruits and leaves. *Philesturnus* (saddleback) has a longer bill that is almost straight, and this is used in probing, tearing, and gleaning insects from leaf litter.

The third callaeid genus, *Heterolocha* (the extinct huia), had a "bizarre specialization of differing male and female bills" (Scofield, 2009). In fact, while the sexual difference in the bills is well marked in the huia, it also occurs in other birds. In the huia, males had well-developed cranial musculature and a short, powerful bill (average length: 60 mm). Females had a much longer bill (average length: 96 mm) that was slender and decurved (Jamieson and Spencer, 1996).

Observations were made on a pair of birds held in captivity, and "the manner in which the birds assisted each other in their search for food, … appeared to explain the use, in the economy of nature, of the differently formed bills in the two sexes" (Buller, 1888: 10). The males broke up rotten wood and the females probed in holes and crevices. The sexes hunted together (Oliver, 1955) and specialized on larvae of the huhu beetle (*Prionoplus reticularis*: Cerambycidae, New Zealand's largest beetle), and also on wetas. Jamieson and Spencer (1996) were skeptical about cooperative foraging, and concluded: "There is no firm evidence that indicates that male and female Huia assisted each other in extracting grubs from wood, although Buller's original comment (and one by J. M. Wright, quoted in Oliver, 1955: 518) are suggestive." Nevertheless, it is clear that the two sexes of huia showed extensive divergence in the skeleton and musculature of their head and neck. Each sex would have had its own mode of bill use, with the male chiseling and gaping (or prying), and the female probing (Jamieson and Spencer, 1996, citing Burton, 1974).

Many male/female differences are known in birds, and the male and female bill differences in huias "are only unique in the magnitude of the bill difference—the largest known" (Selander, 1966). In other birds, the male green woodhoopoe (*Phoeniculus purpureus*: Bucerotiformes) is 5%–8% larger than the female, but its bill is 36% longer. In this species, Radford and du Plessis (2004) concluded: "There is little evidence that sexual selection is currently acting on bill dimorphism… the extreme sexual dimorphism in Green Woodhoopoe bill length is maintained by ecological separation to reduce foraging competition." Ecological differences between the forms may maintain the dimorphism, but they do not explain its origin, or why it does not occur in all birds.

The subspecies of *Callaeas* appear to have had different ecologies, as well as differences in plumage and wattle color (Holdaway and Worthy, 1997). South Island kokako inhabit high altitude beech forest with low plant diversity. According to early records, the bird was "rarely found below a altitude of two or three thousand feet, and indeed, is found in greatest numbers at and above the higher of these altitudes, in the glens of the *Fagus* [*Nothofagus*] forest" (Travers, 1871). In contrast with the southern form, the North Island kokako inhabits high diversity podocarp-angiosperm forest at low altitude (Clout and Hay, 1981). This conforms to a standard, biogeographic-elevational pattern, in which forms of northern New Zealand occur at *lower* altitudes than forms of southern areas (Chapter 11). The pattern conflicts with ecological determinism, as southern localities have ecological equivalents (with cooler temperatures) at *higher* altitudes further north.

Petroicidae

The overall biogeography of this Australasian family is discussed elsewhere (Heads, 2014: 262). *Petroica* is the only genus in New Zealand. It is widespread in the south-west Pacific, but has a conspicuous absence from New Caledonia, although it surrounds this archipelago to the north, south, east, and west. (It is in Australia, Tasmania, New Guinea, the Bismarck Archipelago, the Solomon Islands, Vanuatu, Fiji, Samoa, New Zealand [including the Auckland, Snares and Chatham Islands], and Norfolk Island.) The traditional dispersal hypothesis proposed that *Petroica* migrated to New Zealand and the Pacific islands by chance dispersal from a center of origin in Australia (Fleming, 1950). This does not account for the most striking feature of the distribution—the absence from New Caledonia. Distribution surrounding New Caledonia but not on it is also displayed in the plant genus *Coprosma*.

Petroica has two main New Zealand clades, the robin (*P. australis*) and the tomtits (*P. macrocephala* and *P. traversi*, formerly the "black robin"). Fleming (1950) suggested that these are the results of two separate dispersal events from Australia.

Miller and Lambert's (2006) molecular analysis included the different forms of robins and tomtits, along with Fiji and Norfolk subspecies of *P. multicolor*. In a maximum parsimony analysis, the New Zealand species form a clade (*P. australis* sister to *P. macrocephala* + *P. traversi*), while a maximum likelihood analysis showed the tomtits *P. australis* + *P. traversi* sister to a clade including the robins and *P. multicolor* of southern Australia to Fiji and Samoa.

Miller and Lambert (2006) concluded: "There is some evidence to support the hypothesis that two invasions of *Petroica* from Australia have occurred." Nevertheless, the study sampled just one *Petroica* species from outside New Zealand, no Australian samples were sequenced, and the idea of a double invasion was contradicted by the maximum parsimony analysis. A subsequent review summarized Miller and Lambert's (2006) results, but concluded in stronger terms: "Molecular analysis is consistent with the hypothesis that New Zealand was colonized twice by *Petroica*" (Trewick and Gibb, 2010: 238). Thus Fleming's (1950) conclusion was maintained, although there was little evidence to support it.

In contrast with the earlier molecular work just cited, a more detailed study with a main focus on the Australian species found that the New Zealand tomtit, *Petroica macrocephala*, was sister to all six other species of *Petroica* that were sampled (including *P. multicolor*) (Christidis et al., 2011). This interesting and unexpected result contradicts the idea that *P. macrocephala* was derived from any Australian group. Further work on the phylogeny of *Petroica* is bound to give interesting results.

The New Zealand robin, *Petroica australis* of North, South, and Stewart Islands, is "By far the largest-bodied members of this genus" (Boles, 2007b). It is about three times the mass of the tomtit *P. macrocephala* (North, South, Stewart, Snares, Auckland, and Chatham Islands) and its melanistic sister, *P. traversi* of the Chatham Islands. In addition, while most petroicids have slender legs, in the New Zealand robin they are thickened and have been interpreted as "an adaptation to its predominantly terrestrial lifestyle" (Boles, 2007b: 445). Instead, the structure may have determined the terrestrial habit.

Zosteropidae

Zosterops lateralis lateralis occurs in Tasmania, on islands in Bass Strait and throughout the New Zealand region: Norfolk Island, the New Zealand mainland, Macquarie, Auckland, Campbell, Chatham, Three Kings, and Kermadec Islands (but absent on the Poor Knights) (Worthy and Holdaway, 2002). It is also a non-breeding, annual migrant from Tasmania to continental south-eastern Australia.

van Balen (2008: 437) wrote that "White eyes (*Zosterops* species) are generally highly sedentary, yet there is the contradictory tendency for flocks to disperse over considerable distances." For example, *Z. erythropleurus* migrates over several thousand kilometers, between the Vladivostok region and Vietnam. In his influential paper, Mayr (1954) wrote that a small flock of *Z. lateralis* "found its way in 1856 from Australia to New Zealand," and this has been accepted by many authors. For example, van Balen (2008) wrote that *Z. lateralis*, the commonest bird in New Zealand, provides one of the most extraordinary examples of dispersal, as it "invaded in the 1850s from Australia…." What is the evidence for this extraordinary invasion?

Clegg et al. (2002) wrote that a "detailed record" exists for *Zosterops* in New Zealand species back to the early nineteenth century, but in fact its ecology and distribution at this time are obscure. van Balen (2008: 468) wrote "In June 1856, large flocks that appeared on the Wellington coast in New Zealand may have represented the beginning of the spread of this species throughout New Zealand (scattered records exist before that time, the earliest in 1832 [at Milford Sound])."

Nevertheless, the earliest records, from 1832, do not mean that the species must have invaded the country at that time, as ornithology in New Zealand was then still in its infancy. The first scientific collections of such common birds such as the New Zealand quail (now extinct), the wrybill (*Charadrius frontalis*), the fernbird, and the grey warbler were only made in 1827. Thus the absence of any *Zosterops* records before 1832 need not mean that the bird really was absent.

In fact Buller, in 1888, already realized that that the early history of *Zosterops* in New Zealand was problematic:

> Whether it came over to us originally from Australia, or whether it is only a species from the extreme south of New Zealand, which has of late years perceptibly increased and has migrated northwards, is still a matter of conjecture. The evidence… is somewhat conflicting; but I have myself arrived at the conclusion that the Silvereye, although identical with the Australian bird, is in reality an indigenous species. (Buller, 1888/1967: 46)

Despite this uncertainty, also emphasized by Mees (1969), the invasion of New Zealand by *Zosterops lateralis* has been adopted as a classic example of long-distance, trans-oceanic dispersal.

In New Zealand *Zosterops lateralis* favors areas of secondary, disturbed vegetation with at least some trees, including domestic gardens and rough farm country. This preference suggests that the bird would have expanded its range into the large areas of new habitat created by European clearing through the 1850s. The range expansion was not a "chance dispersal" event, but was instead determined by ecological factors.

Zosterops l. lateralis is the southern Tasman Basin representative in a diverse, widespread species that ranges from Western Australia to Fiji, with 16 allopatric subspecies. *Zosterops*, the genus of whiteeyes, extends from Africa to Samoa. van Balen (2008: 436) described the birds' propensity "… to colonize small islands and subsequently speciate there," but this does not account for many aspects of the distribution. For example, why does the genus not extend east of Savaii in Samoa (*Z. samoensis*)? Neither the genus nor its family occur in American Samoa, the Cook Islands, or French Polynesia. In the last region, *Z. lateralis* has been introduced in the Society Islands, and it is now common there (van Balen, 2008: 468). This indicates that the family was not absent east of Samoa because of ecological reasons.

Other boundaries in the group are also difficult to explain by chance dispersal. Moyle et al. (2009) observed that in the Solomon Islands, *Zosterops rennellianus* of Rennell Island has not made the 20-km jump to Bellona, and *Z. rendovae* s.str. and *Z. teteparius* have not been recorded from their respective neighboring islands, just 2 km away. Moyle et al. (2009) suggested that there has been rapid loss of dispersal ability in *Zosterops* species following hypothesized invasions, but this is ad hoc and does not explain how the groups dispersed as far as they did. In fact, the problem does not concern *Zosterops* alone; the eastern limit of the genus and family at Samoa is a standard break, and marks the boundary between two widespread clades of *Todiramphus* kingfishers, for example (Heads, 2012a: Fig. 6-10).

The trans-Tasman migration proposed for *Zosterops* was regarded as a paradox by Trewick and Gibb (2010), as the birds do not appear to have adequate means for long-distance, trans-oceanic migration. But trans-oceanic flight is not necessary if the Chatham Islands—Samoa boundary is the old eastern limit of the family, and was established near the Pacific plate boundary before the opening of the Tasman.

To summarize, there is little evidence that *Zosterops* colonized New Zealand in the nineteenth century (cf. Buller, 1888; Mees, 1969), although it may have been restricted to remote parts of the country before then. There are other birds that provide much better evidence for trans-Tasman dispersal, such as the cattle egret *Bubulcus ibis* (Chapter 1), or the welcome swallow, *Hirundo tahitica*, which started breeding in New Zealand in 1958.

With respect to morphology in *Zosterops*, van Balen (2008: 413) wrote that: "The plumage of whiteeyes [green, for the most part] seems primarily an adaptation to their favourite environment of green foliage." Green is the predominant color in other groups, such as parrots, but many birds found in green foliage are not green, and so the adaptive explanation seems ad hoc and unconvincing.

One of the most characteristic features of the Zosteropidae is the distinctive eyering. It is composed of small, scale-like feathers, and in most species it is white, but it can also be gray, black, or yellow. In some species it is well developed, while in others it is absent. As with other structures around the eye, including the wattle of some turacos and wattle-eyes (Turner, 1997), and feathers around the eye in *Strigops*, the eyering probably represents a highly reduced trace of a former structure.

Motacillidae

The family is represented in New Zealand by two species of *Anthus*, one on the mainland and one on the subantarctic and Chatham Islands (Chapter 6). The New Zealand birds belong to a clade, *A. novaezealandiae* s.lat., that is also present in Australia and New Guinea. In one molecular study, this appeared as sister to *A. nyassae* of Africa (Mozambique to Gabon) (Voelker, 1999). The affinity indicates a standard trans-Indian Ocean connection, but further sampling is needed. No particular adaptations seem to have been cited for the New Zealand forms.

Locustellidae

Bowdleria, the fern bird, is endemic to New Zealand, but is now often included in Locustellidae as a member of *Megalurus* (Alström et al., 2011). In molecular studies, the mainland New Zealand species, *M. punctatus*, is sister to *M. gramineus*, widespread in Australia (Alström et al., 2011). Nevertheless, one key member of the family, *Amphilais* ("*Dromaeocercus*") of Madagascar, has not yet been sequenced. In its morphology, the skull of *M. punctatus* is most similar to that of *Amphilais*, and the two also share spiny, decomposed rectrices not seen in other *Megalurus* species (Olson, 1990b).

Megalurus punctatus is distinguished by its "extremely robust hindlimb" and a "bizarre modification of the pelvis" in which "the anterior iliac shields are grotesquely expanded..." (Olson, 1990b). Olson compared these structures with similar features in *Orthonyx* (Orthonychidae, Australia, and New Guinea) and *Mohoua* (see next) and wrote: "The functional complex appears to be correlated with the use of the feet in foraging, particularly in moving vegetation and detritus to expose prey." The correlation is clear enough, but what is the explanation for it? Does moving vegetation really require such grotesque and bizarre modifications of the pelvis? Or is such baroque structure a byproduct of prior trends?

Mohouidae

Mohouidae comprise a single genus, *Mohoua* (incl. *Finschia*), with three species, all endemic to New Zealand. When the species were first described, in the nineteenth century, they were not recognized as related and instead were each placed in different families: brown creeper in Paridae, yellowhead in Muscicapidae, and whitehead in Fringillidae (Fleming, 1982: 342). This illustrates a repeated theme in early studies, in which Australasian groups were initially regarded as unrelated offshoots of northern ancestors, in line with Matthewian dispersal theory (Heads, 2014).

Contradicting the earlier ideas, molecular studies now show that the three species of Mohouidae form a distinct clade, and that the group is not nested within any other family. Instead, its best position at the moment is sister to a cosmopolitan group, the Corvida ("core Corvoidea"; Aggerbeck et al., 2014; Jønsson et al., 2016). This indicates that the original break in the Corvida occurred in or around New Zealand.

It is interesting that families of Corvida such as Oreoicidae, Campephagidae, and Artamidae are absent as such from mainland New Zealand, where *Mohoua* is endemic. A simple vicariance model for Corvida would account for the absence of these families from New Zealand. At the same time, it would also account for the absence of *Mohoua* from Lord Howe and Norfolk Islands, as the group is replaced by artamids on Lord Howe and campephagids on Norfolk.

The species of Mohouidae have the following phylogeny and ecology (Oliver, 1955; Soper, 1976; McLean and Gill, 1988; Elliott, 1990; Elliott et al., 1996; Marchant and Higgins, 2002; Boles, 2007a; Aidala et al., 2013):

> **M. novaeseelandiae (brown creeper)**: South and Stewart Islands. Mature forest, second growth forest and shrubland, also exotic pine plantations. Occurs from sea level to subalpine shrubland (~1000 m). Feeds in the canopy and upper understorey. Nests at a mean height of 4.7 m, in upright forks, vine tangles, and shrub canopies.
>
>> **M. ochrocephala (yellowhead)**: South and Stewart Islands (extinct in the latter since the 1960s). Tall forest on fertile, low altitude valley floors, up to 900 m. Feeds within the canopy, but also roots through litter accumulated in the forks of branches. Nests in the largest trees, at a mean height of 14.5 m. Nests are built in cavities in trunks and large branches.
>>
>> **M. albicilla (whitehead)**: North Island. Forest and dense shrubland, also exotic conifer plantations. Recorded from sea level to 1300 m. Feeds at ~3.1–4.5 m. Most nests are built at ~2.5 m above the ground (but one has been recorded at 30 m on Little Barrier Island).

The yellowhead of South and Stewart Islands tends to occur higher up in the forest canopy, while the whitehead of the North Island is lower down. This ecological vicariance complements the geographic vicariance between the species. The northern whitehead extends to higher elevations than the two southern species. This is in line with the expected trend in the elevational belts, with higher elevations in the warmer north, and these occur in most groups (but not all; Chapter 11).

The three species of Mohouidae are the sole hosts of the long-tailed cuckoo *Eudynamys taitensis*, which lays its eggs in their nests. One study found that 6.3% of yellowhead nests were parasitized by cuckoos; in whiteheads, 5% of low-elevation nests and 36% of high-elevation nests were parasitized (Elliott, 1990).

Unlike brown creepers and whiteheads, yellowheads nest in holes in tree trunks and large branches. Hole-nesting in birds is often regarded as an adaptation to reduce nest predation (Lack, 1968). Soper (1976) and Elliot (1990) suggested that hole-nesting in the yellowhead was an adaptation that reduces nest parasitism by *Eudynamys*, as the hole is too small for the cuckoo to enter.

All three species of mohouids are often recorded hanging upside-down and climbing along the underside of branches while feeding (Moon, 1992; Marchant and Higgins, 2002; Boles, 2007a; Chamber, 2007). Because of this, Worthy and Holdaway (2002: 429) suggested that "All the species need strong legs and powerful toes," but this morphology is just as likely to be a cause of the habit. *Mohoua ochrocephala* is larger than *M. albicilla* and also differs in skeletal specializations "for foraging" (Olson, 1990a,b; Aidala et al., 2013). Olson (1990a) concluded that the three species of *Mohoua*:

> ...show increasing specialization for use of the hindlimb in foraging and in order of most primitive to most derived should be listed as *M. novaeseelandiae, M. albicilla,* and *M. ochrocephala.* [p. 157]...the pelvis and hindlimb in *Mohoua* have become specialized for use of the feet in moving vegetation and litter while foraging... [In *Mohoua*] *novaeseelandiae* the pelvis is not markedly different from that of most passerines... In [*Mohoua ochrocephala* and *M. albicilla*] the pelvis is much more specialized, with the anterior portions of the ilia enlarged and compressed laterally so that they meet on the midline and present a humped appearance in lateral view... The femur and tibiotarsus in *M. ochrocephala* are very broad and stout... (p. 159)

Apart from gleaning along branches, yellowheads also feed in debris accumulated in branch angles, gripping the bark with one foot and scratching with the other. While the legs are indeed robust and powerful, there is no particular evidence that they evolved because of an imposed need. The pelvis is unusual, and this recalls the grotesque, bizarre pelvis of the fernbird *Megalurus* (see last family; Worthy and Holdaway, 2002: 429). (The two groups are not direct relatives.)

Oriolidae

Until a few decades ago, this family was represented in New Zealand by two species of New Zealand thrush, *Turnagra*, with one in the South Island and one in the North Island. Both species became extinct through the twentieth century. In different molecular studies, *Turnagra* has been paired with either *Sphecotheres* of Australia and New Guinea, or with *Oriolus* of Africa, Europe and Asia, to Australia and New Guinea (Heads, 2014: 323). Behavior and ecology in *Turnagra* is not well documented, but it is known to have frequented the undergrowth and forest floor, and to have been omnivorous, hawking insects, and eating spiders, fruit, and oats (Worthy and Holdaway, 2002).

Rhipiduridae

The single New Zealand species, *Rhipidura fuliginosa*, occurs on the mainland, Chatham Islands and Lord Howe Island. It belongs to subgenus *Rhipidura*, which consists of a basal, paraphyletic grade of six species in the Melanesian islands (New Guinea to Fiji), and a clade of three species in Australia, southern New Guinea, and New Zealand (Heads, 2014: 344). In the usual model, this would be explained by dispersal from a center of origin in Melanesia, to Australia and New Zealand. Instead, the pattern can be explained by a widespread, common ancestor that underwent initial differentiation around the Melanesian arcs.

Rhipidura fuliginosa is a typical member of its family, feeding as an aerial insectivore or by gleaning along branches (Boles, 2006: 208). Rictal bristles, reduced, hair-like feathers that occur in kiwis, *Strigops* and *Notiomystis*, are also a prominent characteristic of rhipidurids, where they are often as long as the bill (Boles, 2006). Brooke and Birkhead (1991: 23) wrote that "The mouths of birds that catch flying insects are often surrounded by bristles...," and these unusual feathers are often interpreted as adaptations for catching insects (Cunningham et al., 2011). Nevertheless, they also occur in non-insectivorous birds (Lederer, 1972), and in an alternative model, the bristles developed around the mouths of some birds as a by-product during reduction in the face and the facial feathers.

In rhipidurids, as in many passerines, the bill is small, and compared with sea-birds or ratites it has almost disappeared. Boles (2006: 212) wrote that "The major component of the diet of fantails is small insects and other invertebrates. This is because the rhipidurid bill is not strong enough to handle larger, more robust prey." In the same way, Brooke and Birkhead (1991: 124) wrote that: "A bird's bill greatly influences its feeding habits." This line of argument used here for bird bills can be adopted as a general principle of morphological evolution; instead of prior function determining structure, as in the adaptation model, the structure of an organ determines its function.

Corvidae

The two New Zealand members of this family are both extinct species of *Corvus*. One is a widespread, unnamed species on North, South, Stewart, and Auckland Islands; the second is *C. moriorum* of the Chatham Islands (Worthy and Holdaway, 2002). No special adaptations have been proposed.

The New Zealand Passerines as a Relict Fauna

The New Zealand passerine fauna shows a distinctive parallel with the angiosperms. As many as 10, diverse angiosperm families are represented in New Zealand by just one or two species (see Chapter 8). A similar pattern is seen in the passerines. The passerine families thought to have been breeding in the New Zealand region at the time of human discovery are listed next, with their total numbers of New Zealand mainland species, both extant and subfossil (Worthy and Holdaway, 2002). Families endemic to New Zealand are indicated with an asterisk.

> *Acanthisittidae: **Seven** (**four in historical times**; two extant). (There is also a monotypic fossil genus from the Miocene; Worthy et al., 2010.)
>
> Meliphagidae: **Two** (plus one on the Chatham Is.).
>
> Acanthizidae: **One** (and one on the Chatham Is.).
>
> *Callaeidae: **Five**.
>
> *Notiomystidae: **One**.
>
> Petroicidae: **Three**. (This is controversial; Worthy and Holdaway, 2002, accepted five), plus one each on the Snares and Chatham Is.)
>
> *Mohouidae: **Three**.
>
> Oriolidae (*Turnagra*): **Two**.
>
> Locustellidae (*Megalurus*, formerly *Bowdleria*): **One** (plus one on the Chatham Is.).
>
> Rhipiduridae: **One**.
>
> Zosteropidae: **One** (plus one on Norfolk Is.).
>
> Motacillidae: **One**.
>
> Corvidae: **One** (plus one on the Chatham Is.).

In other bird groups, New Zealand has just one pigeon species on the mainland (plus one on the Chathams and two on Norfolk), and just two cuckoos. A similar pattern is seen in bats (see next chapter).

In the passerines, it is significant that the Acanthisittidae, sister to all other passerines, are the richest in species. This suggests that the group was always diverse in New Zealand, ever since the initial differentiation between it and the other passerines.

As with the flowering plant families, the low overall species diversity and the high number of endemic families indicate that the passerine families are relictual. They have undergone much extinction (likely factors are repeated marine transgressions, orogeny, and Pleistocene cooling), and current research on the New Zealand Miocene fossil fauna should provide further evidence for this. In contrast, marine groups such as shags and procellariiforms are much more diverse.

Heather (1966) wrote that New Zealand's long isolation has meant that its fauna is "characterised by a paucity of avian types... which have colonised the area by chance dispersal or drift from without." Heather proposed that most reached New Zealand during the Recent, with a few endemic genera being the result of adaptive radiation in late Pliocene or early Pleistocene (*Hymenolaimus*, *Notornis*, *Nestor* and *Strigops*, Acanthisittidae, *Turnagra*, Callaeidae). The modern terrestrial avifauna "cannot be traced to the remote past...."

Heather (1966) stressed the observed colonizations of New Zealand from the west, as in the white-faced heron (*Egretta novaehollandiae*: Ardeidae) of Australia, New Caledonia and Fiji, first observed in New Zealand in 1865 (Oliver, 1955); the spur-winged plover (*Vanellus miles*: Charadriidae) of Australia, first recorded in New Zealand in 1886; and the welcome swallow (*Hirundo neoxena*: Hirundinidae) of southern Australia, first recorded in New Zealand in 1943, and breeding there from 1958. Yet the habitat of these species, and the fact that the invasions occurred at the same time as European settlement, indicate that the species invade whenever suitable (open, disturbed) environments are present (see "Trans-Tasman Sea affinities," Chapter 1). Michaux (2014) discussed the case of the white-faced heron, and wrote: "This is not an example of chance dispersal but of periodic range expansion brought about by ecological disruption that has opened new areas for the bird to exploit. The birds would have dispersed to New Zealand from time immemorial..." (p. 56).

In the chance dispersal model, the poverty of land birds in New Zealand "may be attributed to its isolation and the strong element of chance in the number and type of immigrant forms..." (Heather, 1966: 16). In contrast, a vicariance model would be consistent with a biota that was much *more* diverse in the past but was decimated during the climatic and geological changes of the Cenozoic.

To summarize, there is little evidence that any of the morphological distinctions of New Zealand birds represent adaptations to any particular environmental conditions in the country. There is also little evidence that the endemic New Zealand birds owe their distribution to dispersal from a center of origin elsewhere. Thus, overall, the CODA model of evolution (center of origin + adaptation + dispersal) for these groups is questionable.

16

Biogeography and Evolution in New Zealand Bats

This section introduces the biogeography and morphology of the New Zealand terrestrial mammals—two genera of bats. As in other groups, structural evolution in mammals can be interpreted in terms of trends or tendencies, although in standard theory, the structure and behavior of mammals are always interpreted in terms of adaptation. For example, a classic text on primates argued that evolution consists of "a molding of... structure... for various biological roles...," and that "Most evolutionary changes in teeth are the result of selective forces derived from a specific dietary regime" (Szalay and Delson, 1979). Thus the diet causes the structure of teeth, rather than vice versa, as argued here. In the traditional approach, the focus is on understanding the morphology of a particular group in terms of its ecology, rather than interpreting it with respect to broader structural trends outside the group itself.

In another well-known example from primates, the average size of social groups in different primate species shows a direct relationship with the degree of neocortex development (Fleagle, 1999). One school of thought suggests that the evolution of brain size in a lineage is driven by social interactions. Instead, increasing social interactions are probably driven by increasing brain size and functionality. Complex social interactions are impossible without the ability to remember individuals, but if this ability is gained, such interactions become almost inevitable. The functions of the neocortex can be interpreted, not as purposes or essences, but as epiphenomena, by-products of trends in encephalization found throughout birds and mammals.

Many of the main trends in primate evolution, such as reduction of the tail, and reduction, suppression, and fusion of components in the skull, were already taking place in the early vertebrates, and the modern groups represent the results and continuations of these ancient tendencies. One important trend in the evolution of the vertebrates—reduction and fusion in the pharyngeal region—led, among other things, to the closure of the gills. The overall reduction series, of which gill closure was just one part, was not an adaptation "for" conquering the land, but this was one of its results. Forms with any sort of viable "lung" would have survived there. Forms with suppressed gills were eliminated from the sea, but some were viable on land. There was no need for any active invasion of the land, as animal populations are often transferred from the sea to land by passive processes. This takes place by tectonic uplift or by stranding, following marine incursion and retreat.

In mammals, the reduction and fusion in the components of the face are evident in the major phylogenetic break in the primates. This separates the moist-nosed (less reduced) strepsirrhines from the dry-nosed (more reduced) haplorhines. Again, the end results of the evolution, at least as seen today, did not develop as adaptations for current conditions, but represent the playing out of simple trends over very long-time scales. The best-known trends in mammals include reduction in the face and muzzle and enlargement of the neocortex, and both of these are well documented in primates. Instead of answering particular "needs" imposed by the animal's ecology, the clines in primates, for example, from the long-muzzled strepsirrhines to the short-muzzled haplorhines, have developed over geological time, as tendencies caused by repeated mutations.

The New Zealand Mammal Fauna

New Zealand land-breeding mammals include three pinnipeds that breed in the region: the elephant seal *Mirounga leonina* (Phocidae), the fur seal *Arctocephalus forsteri* (Otariidae), and Hooker's sea lion, the monotypic *Phocarctos* (Otariidae) (King, 2005). *Phocarctos* is endemic to New Zealand and is sister to *Neophoca* of Australia (Chapter 6).

Phocarctos breeds on Auckland, Campbell, the Snares Islands, and the South Island. Its terrestrial habitat includes grassy sward, herb field with megaherbs, and *Metrosideros* forest. On the Snares, Stewart Island, Otago Peninsula (Harcourt, 2005), and Auckland Islands (personal observations), sea lions can move inland in forest up to 1 km from the sea, challenging the usual idea that the only indigenous land mammals in New Zealand are bats.

Bats are represented in the extant New Zealand fauna by the endemic *Chalinolobus tuberculatus* (Vespertilionidae) and two species of the endemic *Mystacina* (Mystacinidae) (one not seen since the 1960s). In addition, Early Miocene fossils from St Bathans include material of another mystacinid, another vespertilionoid (not *Chalinolobus*), and "an archaic bat most closely resembling bats that died out elsewhere >45 Ma" (Hand et al., 2006; cf. Hand et al., 2013, 2015). This indicates that the extant bat fauna of New Zealand is relictual, and so it resembles the angiosperms and passerines with their many monotypic and oligotypic genera. The fossil record of New Zealand bats is growing, and this is consistent with the idea that bats are ancient elements in New Zealand, rather than Pleistocene or even Miocene immigrants.

The New Zealand land mammal fauna—the bats, the pinnipeds, and the archaic "mouse-sized" mammal known from Miocene fossils—is one of the most unbalanced mammal faunas known, and its origin is of special interest. How terrestrial mammals colonize islands and why insular mammal faunas are unbalanced are questions that are often debated.

The usual explanation for mammals on islands is "sweepstakes dispersal." Simpson (1940: Fig. 6) did not accept continental drift, and instead he introduced sweepstakes dispersal to explain faunas such as the mammals of Madagascar. This has remained the standard model, and Teeling et al. (2003) used it to explain the origin of the New Zealand bat *Mystacina*. Nevertheless, some mammalogists are questioning the idea. Mazza et al. (2013) wrote: "…sweepstake dispersal theory is currently enjoying vast popularity and has been readily embraced—perhaps too enthusiastically—by both neontologists and paleontologists. In fact, doubts are starting to emerge… Natural rafting raises many more problems than it solves…" (see also Mazza, 2014a,b).

Mazza et al. (2013) contended that the sweepstakes model is an oversimplification: "Insular terrestrial mammal communities, either living or fossil, can be found to be unbalanced for many more reasons than a sweepstake invasion by new species." For example, extinction has been important in islands. Mazza et al. (2013) cited the rapid disappearance of over 75% of the original vertebrate species from the islands that were created by flooding when the Panama Canal was opened. Similar extinction would have taken place in New Zealand during the Oligocene flooding.

Instead of being populated by long-distance, sweepstakes dispersal from continents thousands of kilometers away, new islands are more likely to be populated from other islands nearby. Mazza et al. (2013) cited modern work in geology suggesting that the Galápagos biota, which includes endemic mammals, "would have been inherited from a whole series of ancestral Galápagos Islands," rather than by sweepstakes dispersal from the South American mainland. The same principle applies to the New Zealand region and its many islands, past, present, and future.

Evolution and Biogeography in Bats

Most New Zealand bats live in extensive, old-growth forest, where the colonies occupy holes in tree trunks. Their biogeography is discussed in more detail in the following sections, but given that bats can fly, it may seem strange to discuss their biogeography at all. Bat distributions are often neglected in modern interpretive work, but they display precise, structured patterns that are also seen in other groups that do not fly. For example, *Rhinolophus* (= Rhinolophidae s.str.) is widespread through the Old World (Africa, Eurasia, and Australia) but is not on Madagascar. *Pteropus* (Pteropodidae) is widespread through Madagascar, tropical Asia, and the Pacific but is not on mainland Africa. Pteropodidae (fruitbats or flying foxes) are the largest bats and the strongest fliers, but they show many other striking patterns, for example, nested, disjunct arcs between New Caledonia and New Guinea (Heads, 2010: Fig. 7). These patterns suggest that, despite appearances, the flying ability of bats is not a major factor accounting for their distribution.

Andersen's (1912) landmark review of the fruitbats concluded as follows:

> The evidence afforded by the geographical distribution of Bats has generally been considered of doubtful value; hence they have either been entirely excluded from the material worked out by zoogeographers or at least treated with pronounced suspicion, as likely to be more or less unreliable documents of evidence. This unwillingness or hesitation to place Bats on an equal zoogeographic footing with nonflying Mammalia would seem to be due partly to the preconceived idea that owing to their power of flight Bats must evidently have been able easily to spread across barriers... [This] may in theory appear plausible enough, but when tested on the actual distribution of the species and subspecies it proves to be of much less importance than commonly supposed; it rests, in reality, on a confusion of two different things: the power of flight would no doubt *enable* a Bat to spread over a much larger area than non-flying Mammalia, but, as a matter of fact, only in very few cases is there any reason to believe that it has *caused* it to do so [...]. Rather, local differentiation tends to show that the present distribution of Megachiroptera has not been influenced to any great, and as a rule not even to any appreciable, extent by their power of flight; if it had, the Fruit-bat fauna of one group of islands could not, so commonly as is actually the case, differ from that of a neighboring group or continent, and the tendency to differentiation of insular species or forms would have been neutralised by the free intercourse between neighbouring faunas. (Andersen, 1912: lxvii–lxviii)

The oldest known fossil bat is dated at 55 Ma, and a fossil-calibrated phylogeny indicated earliest molecular divergences among bats at ~64 Ma (Teeling, 2009). Yet there are problems with this fossil-based approach. Bats have one of the poorest fossil records of any mammalian order (Teeling et al., 2012), and "the majority of the bat fossil record is fragmentary" (Teeling, 2009). In addition, "all fossil bats, even the oldest, are clearly fully developed bats" (Hill and Smith, 1984: 33). The absence of transitional forms (from bat ancestors to bats) suggests that bats could be much older than their earliest known fossils (Smith et al., 2012). The oldest known bat fossils are, in some ways, as advanced as many living species, again suggesting that the group itself originated much earlier, perhaps in the Mesozoic (O'Donnell, 2005).

With respect to bats, Hill and Smith (1984: 182) wrote that "the naive, yet common, practice of ascribing a date of origin to the oldest fossil representative of a group is truly an illogical exercise...." Likewise, modern molecular dating methods using Bayesian analyses stipulate, without justification, that a group can be no more than, say, 5 or 10 million year older than the oldest fossil, and so these methods are also flawed (Chapter 2). Paleontologists have suggested that of all the bat genera that have ever existed, just 12% have left a fossil record (Eiting and Gunnell, 2009), but the arguments just cited suggest that the true percentage could be much less than this.

Morphological evolution in bats is discussed in more detail in the following sections. The wings of bats are among their most distinctive features and are formed by a patagium. This is a membranous structure that stretches between the limbs and the trunk in different animals and enables gliding or flight. It is found in frogs (*Rhacophorus*, etc.), pterosaurs (pterodactyls and relatives), lizards (*Draco*: Agamidae; *Ptychozoon* and *Hemidactylus*: Gekkonidae), birds, bats, and gliding mammals. The latter include marsupials (*Acrobates* in Acrobatidae; *Petaurus* in Petauridae, and *Petauroides* in Pseudocheiridae), rodents (members of Anomaluridae, and Petauristini in Sciuridae), and colugos or flying lemurs, members of the Order Dermaptera. The patagium thus represents a "trend" that is widespread through tetrapods, but in groups that do not form a monophyletic unit. The feature can be interpreted as a relictual structure that persists in several groups but has been eliminated in most tetrapods.

The Long-Tailed Bat, *Chalinolobus* (Vespertilionidae)

Chalinolobus is widespread in Australasia, through Australia, Tasmania, New Guinea, New Caledonia, Norfolk Island, and the New Zealand mainland. Australia has most species (five), and these have a marked concentration in the east of the country. All five species occur together near the McPherson–Macleay Overlap at the New South Wales/Queensland border (Parnaby, 1992). There are two endemics in eastern Australia but none in the west. Outside Australia, *Chalinolobus* occurs in south-eastern

New Guinea (Port Moresby, Milne Bay, and the D'Entrecasteaux Islands; *C. nigrogriseus*, shared with northern Australia), New Caledonia (*C. neocaledonicus*; related "particularly with the small northern *C. gouldi venatoris*" of Queensland), Norfolk Island (perhaps a distinct species; Simmons, 2005), and New Zealand (*C. tuberculatus*) (Flannery, 1995a).

Chalinolobus belongs to a clade of hypsugine bats with the following phylogeny and distribution (Koubínová et al., 2013):

Chalinolobus: Australasia, + *Vespadelus*: mainland Australia (widespread, most species in the south-east) and Tasmania.

Hypsugo cadornae: Vietnam to north-eastern India (Sikkim), + *H. savii*: north-western India (Kashmir) and western China to the Middle East, the Mediterranean region, and Madeira.

The clade as a whole is endemic to the standard Tethyan sector: New Zealand–central Asia–Mediterranean. The sister group of this Tethyan clade comprises the following four groups:

Hypsugo eisentrauti: Africa.

Nycticeinops: Africa.

Laephotis: Africa.

Neoromicia: Africa and Madagascar.

More sampling is required (above all in the polyphyletic *Hypsugo*), but the pattern is consistent with an origin of the Tethyan clade and its African sister group by simple vicariance in a widespread Old World ancestor. The same Tethys–Africa vicariance occurs in the rallid bird *Porphyrio* (Chapter 15).

Differentiation within *Chalinolobus* and the precise affinities of the New Zealand species are not well understood, and there is no modern study. Dobson (1875) proposed that the New Zealand *C. tuberculatus* included *C. morio* of the southern half of Australia. Koopman (1971) instead wrote that *C. tuberculatus* "appears to be a well-marked species probably most closely related to the *C. picatus* group." As another possibility, Hand et al. (2005) wrote that *C. tuberculatus* "…is thought to be most closely related to *C. gouldii* of Australia and *C. neocaledonicus* of New Caledonia." In this way, different authors have allied the New Zealand species with each of the three main groups that Koopman (1971) proposed in the genus, and a molecular analysis is desirable.

As for the origins of the New Zealand species, O'Donnell (2005: 102) wrote that it was "probably derived from an ancestral form of *Chalinolobus* that was windblown across the Tasman Sea during the Pleistocene," although he did not cite any evidence for this. Hand et al. (2009) suggested that both New Zealand bat genera were derived by dispersal from Australia. The Australian and New Zealand representatives of *Chalinolobus* are distinct at *species* level, while the Australian (fossil) and New Zealand representatives of Mystacinidae are separate *genera*, and so Hand et al. (2009) inferred a more recent dispersal for *Chalinolobus*. This assumes that the degree of differentiation is related to time elapsed, but in many cases this will not be correct.

Chalinolobus tuberculatus is endemic to the New Zealand mainland and inshore islands. There are two main clades separated by the usual Cook Strait boundary: one clade is in the North Island and the other is in South and Stewart Islands (Winnington, 1999; O'Donnell, 2001).

The Snout and Mouth in *Chalinolobus*

The morphological distinctions between *Chalinolobus tuberculatus* of New Zealand and the other members of the genus are not well documented. Koopman (1971) described *C. tuberculatus* as "clearly distinct" from any of the Australian species, and most authors have followed this, but Koopman did not specify distinguishing characters. Dobson (1875) observed that *C. gouldii* and *C. nigrogriseus* of Australia have an "internal basal lobe of the ear forming a distinct lobule at the base projecting backward…," while *C. tuberculatus* (Dobson included *C. morio* of Australia here) is without a distinct lobule. This ear lobule (and its absence in some species) seems to be of phylogenetic significance, but what is its morphological meaning? In order to answer this question, instead of looking in detail at any ecological and "biological" implications of the lobules, we can stand back from the species and see the "lobules" in a broader morphological and phylogenetic context. Approaching the question in this way involves a consideration of the genus as a whole.

The muzzle in *Chalinolobus* is very short. Dobson (1875) emphasized the "remarkable obtuseness of the muzzle and shortness of the head," and in New Zealand the muzzle of *C. tuberculatus* is much shorter than in *Mystacina* (see the profile illustration in O'Donnell, 2001). Reduction of the snout is an important trend in bats, as in primates and many other vertebrates, but it has not affected all the component structures in the same way. While the muzzle of *Chalinolobus* may be shortened and the eyes are small, many features have persisted as relictual lobes, glands, and other cryptic "outgrowths." The most obvious examples of this incomplete suppression include parts of the mouth, nose, and ear that have retained a more complex structure than the very reduced equivalents of most other mammals, including humans.

Chalinolobus is distinguished from other genera of bats by distinct fleshy lobes or wattles near the corner of the mouth that project outward or downward from the lower lip (Dwyer, 1960: Fig. 1; Chruszcz and Barclay, 2002; these are labeled "horizontal lobe" in Parnaby, 1992: Fig. 15). (The name *Chalinolobus* means a bridle, or bit, with lobes.) In addition to these lip lobes are the ear lobes mentioned earlier. The margin of the ear descends down the side of the face, beneath the eye, and terminates in another lobe at the corner of the mouth, next to the lip lobule. The lobes (labeled "downward projecting ear margin" in Parnaby, 1992) are of different sizes in different species—largest in *C. gouldii* and *C. dwyeri*, reduced in *C. picatus*, and most reduced in *C. nigrogriseus*, *C. morio*, and *C. tuberculatus* (Daniel, 1990: Fig. 25; Parnaby, 1992: Fig. 15). In other bats, wart-like outgrowths around the mouth and lips occur in *Amorphochilus* (Ecuador to northern Chile; Furipteridae) (McDade, 2003).

The lip and ear lobes in *Chalinolobus* can be compared with the wattles found in the same position in many birds, such as Callaeidae. In both groups, the structures reflect similar trends in morphological change that have occurred in phylogeny and represent "leftovers" from reductions in the muzzle or rostrum.

In addition to the lip lobes, the nostrils in *Chalinolobus* are separated by distinct grooves from well-developed glandular elevations on the side of the muzzle (Dobson, 1875). At least one species, *C. gouldii*, also has two glandular elevations under the chin, on either side of midline (Chruszcz and Barclay, 2002). Glandular "scent organs" occur on the face, ear, and throat region of many bats, and, as with glands in general, appear to represent growth that has been "hemmed in."

Of all the mammals, bats show the greatest variation in the face and the ear, and differential reduction has left baroque structures with elaborate ornamentation. A page from Haeckel's (1904) classic *Kunstformen der Natur* (reproduced in the Wikipedia article, "Bats") illustrates the variation in the faces of bats, including those of *Anthops* and *Mormoops*. *Anthops* (Hipposideridae) is the flower-faced bat of the Solomon Islands and is named for the spectacular "floral" elaborations in the central part of the face, including the entire nasal region. Mormoopidae (part of the sister group of Mystacinidae) have bizarre, funnel-shaped ears, and lips and chin that "possess leaf-like dermal outgrowths so intricately arranged as to baffle description" (Dalquest and Werner, 1954). This morphology is difficult to explain if the head begins as a smooth, ovoid object to which different organs are added on as they are needed. Instead, the head can be interpreted as the result of suppression, reduction, and fusion in an already complex structure.

The Short-Tailed Bat, *Mystacina* (Mystacinidae)

The family that *Chalinolobus* belongs to, Vespertilionidae, is diverse and widespread around the world. In contrast, Mystacinidae has a narrow distribution, and its two extant species, *Mystacina tuberculata* and *M. robusta*, are restricted to New Zealand. *M. robusta* has not been seen since the 1960s. There are also fossil mystacinids in both Australia (*Icarops*) and New Zealand (*Mystacina*) (Hand et al., 2013, 2015).

Mystacina is known from the North and South Islands, and from islets off Stewart Island. Off the west coast of Stewart Island, *M. tuberculata* is found on Codfish Island, and both *M. tuberculata* and *M. robusta* are recorded on the South-west Muttonbird (Titi) Islands. Off north-eastern Stewart Island, *M. tuberculata* is found on Jacky Lee Island (Daniel, 1990; Lloyd, 2005). Yet there are no records of *Mystacina*, either historical or current, from Stewart Island itself, or from the south-eastern South Island, including the Catlins region (Daniel and Williams, 1984; Lloyd, 2005). Subfossil and fossil material is restricted to Central Otago and north of there (Daniel, 1990; Hand et al., 2015).

Differentiation in *Mystacina tuberculata* is more complex than the simple North Island versus South and Stewart Island differentiation seen in *Chalinolobus turberculatus*, as there are several distinct clades. These show extensive geographic overlap in the North Island, and Lloyd (2003) gave the phylogeny as: North I. (North I. (North I. (North I. + South I.))). This is consistent with initial differentiation of a widespread New Zealand ancestor at breaks in or around the North Island.

The only fossil species of *Mystacina* that has been described is *M. miocenalis* from the Bannockburn Formation at Manuherikia River, Central Otago (19–16 Ma) (Hand et al., 2015). This bat was much larger than the extant *Mystacina* species and weighed an estimated 39 g compared with 14 g for *M. tuberculata*. (This contrasts with the very *small* fossil kiwi from the same formation; Chapter 9.) Other Mystacinidae fossils from the same formation are from smaller bats than *M. miocenalis*, but it is not clear which genus they belong to.

Mystacina miocenalis appears to have been semi-terrestrial, as are its extant congeners, and based on the similar molar morphology, it also had a similar, broad diet (Hand et al., 2015). Associated plant fossils indicate that *M. miocenalis* inhabited tall forests. These were dominated by Podocarpaceae in lowland areas around Lake Manuherikia and by Nothofagaceae in higher elevation hinterland, as seen in many parts of modern New Zealand.

By now, many fossil plants and terrestrial arthropods have been documented from the Bannockburn Formation. Many of these are "remarkably similar to those in the modern endemic New Zealand biota, and suggest remarkably long-term ecological associations with *Mystacina*" (Hand et al., 2015: 14). In particular, "The majority of the plants inhabited, pollinated, dispersed or eaten by modern *Mystacina* were well-established in southern New Zealand in the early Miocene, based on the fossil record… Similarly, many of the arthropod prey of living *Mystacina* are recorded as fossils in the same area" (Hand et al., 2015: 1).

Biogeography of Mystacinidae

Most accounts portray *Mystacina* as a classic example of the CODA (center of origin–dispersal–adaptation) model. The genus is thought to have had a center of origin in Australia, dispersed to New Zealand, and then developed appropriate adaptations for the unique conditions there. Assessing this idea requires a consideration of the fossil record.

The family Mystacinidae consists of two genera—*Mystacina* has its extant and fossil records restricted to New Zealand (back to the Early Miocene) (Hand et al., 2015) and *Icarops* is known from Oligocene–Miocene fossils of Australia (South Australia, Queensland and the Northern Territory) (Hand et al., 2009). Other mystacinid fossil material is also known from the Early Miocene of Central Otago, but the genus is undetermined (Hand et al., 2009, 2013).

The Mystacinidae are an Australia–New Zealand group, while their sister group is a diverse complex in tropical and subtropical America. Based on fossil calibrations, the trans-Pacific pair diverged at 35–68 Ma (Lloyd, 2003) or 41–54 Ma (Teeling et al., 2003), and as minimum dates, these are compatible with Cretaceous vicariance.

Mystacinidae and its sister group belong to the Noctilionoidea, one of the five superfamilies of bats. The group has the following phylogeny (Teeling et al., 2012; Gunnell et al., 2014):

Myzopodidae (sucker-footed bats): Madagascar and, as fossils, in Tanzania and Egypt.

Mystacinidae: New Zealand and, as fossils, in eastern Australia.

Phyllostomidae and the much smaller families Mormoopidae, Noctilionidae, Furipteridae, and Thyropteridae: Northern Argentina to southernmost United States.

This whole group, the Noctilionoidea, also contains the Speonycteridae, an Oligocene fossil family from Florida (Czaplewski and Morgan, 2012). Thus the phylogeny indicates that Mystacinidae are the Tasman Basin representative of a series of three allopatric clades distributed around the southern hemisphere and further north in America.

Spatial analyses of the Noctilionoidea have taken a center-of-origin approach. Gunnell et al. (2014) wrote: "…the presence of a 37 million year old basal noctilionoid [Myzopodidae] in North Africa suggests that the origins of Noctilionoidea may well be found in eastern Gondwana with a subsequent dispersal south into Australia (mystacinids) and then westward onto South America…." The authors did not state why they thought the fossil suggests this. They were, perhaps, swayed by the fact that the fossil from Egypt is the oldest in the group, but this does not mean the locality is a center of origin, unless a literal reading of the fossil record is adopted. Or the authors may have felt that because Myzopodidae are "basal" in the noctilionoids, they occupy the center of origin. Yet the basal position is also irrelevant, as the sister group of Myzopodidae, in Australasia and America, is just as basal and has the same age.

The ages of the families in Noctilionoidea are also unresolved. Gunnell et al. (2014) wrote that: "dating analyses unambiguously place the origin of Myzopodidae in the Eocene," but the analyses were fossil-calibrated, and so the date is a minimum age. The Mormoopidae-Phyllostomidae node has been dated as Oligocene, although the confidence limits overlap the boundaries of the Eocene (Teeling et al., 2005).

While the Vespertilionidae are the most diverse bat family in terms of species numbers (and the second most diverse family of mammals), Phyllostomidae have the largest number of genera (56) and are more diverse in terms of morphology and ecology. Phyllostomids are cited several times in the discussion here, and aspects of their morphology and ecology are compared with those of their relative, *Mystacina*.

Mystacinobia, a Bat Fly Associated with *Mystacina*

The dipteran family Mystacinobiidae consists of a single, wingless genus, *Mystacinobia*, which lives in association with *Mystacina* and feeds on the guano. The sister group of *Mystacinobia* is the undescribed "McAlpine's fly" from south-eastern Australia, and the two form a Tasman Basin clade that is sister to a worldwide complex, the family Sarcophagidae (Chapter 4).

The two trans-Tasman groups—the bat family Mystacinidae and the pair of flies—can both be derived by simple vicariance in widespread ancestors. In the case of the bats, the ancestor (the common ancestor of Noctilionoidea) was panaustral, while in the case of the flies, the ancestor was global. If the ancestors attained their wide global range during mid-Cretaceous phases of high sea level, they could then have been fractured during the pre-breakup rifting of Gondwana. This would also explain the case of the Australasian *Chalinolobus*, if the ancestor was widespread along the Tethys seaways and through Africa.

Zealandofannia, Another Bat Fly Associated with *Mystacina*

The dipteran family Faniidae occurs more or less worldwide, except for an interesting, trans-Indian Ocean zone: Madagascar, India, New Guinea, and New Caledonia, where it appears to be absent (Domínguez and Roig-Juñent, 2011). Other groups are endemic along a similar Madagascar–New Caledonia arc (e.g., the plant family Nepenthaceae), and this suggests that the absence in Faniidae is real and important.

One member of Faniidae, the monotypic *Zealandofannia*, is endemic to New Zealand and is restricted to the same habitat as *Mystacinobia*—the guano of *Mystacina*. It is sister to a diverse, worldwide clade made up of all the remaining Faniidae except *Australofannia*, *Piezura*, and *Euryomma* (Domínguez and Pont, 2014).

Domínguez and Roig-Juñent (2011) considered the family as a whole and wrote:

> The southern hemisphere species of Faniidae indicate a clear pattern of vicariance and dispersal consistent with the rupture of Gondwana… The ancestor of the Faniidae was widely distributed across different regions of the world, which along with the subsequent separation of two clades that correspond to the Laurasic and Gondwanan landmasses allow the proposal of an

older age than in previous hypothesis (Late Jurassic or early Cretaceous times instead of upper Cretaceous) and a Pangeic origin for the Fanniidae... (p. 85)

[The distribution of] the South American, as well as the Australian and New Zealand species of Fanniidae... could be explained on the basis of the Gondwanan fragmentation scheme instead of the north to south migration waves proposed by Chillcott (1961) and Hennig (1965). (p. 75)

Chirophagoides, a Mite Associated with *Mystacina*

Along with the bat flies *Mystacinobia* and *Zealandofannia*, the mite *Chirophagoides* (Sarcoptidae) is an obligate associate of *Mystacina*. Its sister group comprises *Chirnyssoides* of tropical America, plus *Notoedres*, a diverse group found worldwide (but in New Zealand restricted to introduced mammals) (Klompen, 1992). As with *Mystacinobia* + "McAlpine's fly," and also *Zealandofannia*, *Chirophagoides* cannot be derived from anywhere outside the Tasman region without invoking extra, ad hoc hypotheses. Instead, its affinities are consistent with simple vicariance between the Tasman and the rest of the world.

Did *Mystacina* Disperse to New Zealand or Did It Evolve There from a Widespread Proto-Noctilionoid Ancestor? The Significance of the Fossil *Icarops*

Carter and Riskin (2006) wrote that "The fossil record provides strong evidence for an Australian origin for Mystacinidae," but a closer examination shows that the evidence is equivocal. The Australian genus of fossil mystacinids was named *Icarops*, after Icarus of Greek myth, "in reference to the ancient mystacinid that flew eastwards from Australia to New Zealand" (Hand et al., 1998: 540). Hand et al. did not present an explicit phylogenetic analysis, and their idea that *Icarops* is more primitive than *Mystacina* was supported by a single, ambiguous character: "Based on the [high] degree of crowding of the anterior dentition [caused by reduction in the snout], *Icarops aeanae*... is interpreted here to be the most plesiomorphic of *Icarops* species... There is nothing known about *Icarops aenae* that would preclude it from ancestry of Quaternary species of *Mystacina*" (Hand et al., 1998: 543). Nevertheless, even if *I. aenae* is accepted as more primitive than *Mystacina*, this does not mean it was ancestral. In any case, *Icarops* has its own synapomorphies not shared by *Mystacina* (Hand et al., 1998, 2001), and these confirm its status as the sister group, not the ancestor, of *Mystacina*.

Hand et al. (1998) wrote that during the Oligocene much of New Zealand was submerged, "and any Gondwanan mammals it might have retained were evidently lost... It is likely that the first intrepid mystacinids arrived on the islands of New Zealand, from Australia, sometime after the late Oligocene." This was written before the discovery of the archaic, nonvolant fossil mammal from the New Zealand Miocene (Worthy et al., 2006), and also before the discovery of Early Miocene mystacinids in New Zealand (Hand et al., 2009, 2013, 2015). These findings, along with the global affinities of the extant bats, the batflies, and the mite *Chirophagoides*, indicate that New Zealand did not undergo complete submergence and that the groups can all be explained by vicariance.

Hand et al. (2001) described further material of *Icarops* and they supported the earlier hypothesis that *Icarops* species were plesiomorphic with respect to *Mystacina*. They listed aspects in which *Icarops paradox* differs from *Mystacina* species and wrote "All of these dental features appear to be less specialized than in New Zealand's *Mystacina* species," but no more detailed analysis was provided.

Hand et al. (2005) again suggested that mystacinids dispersed from Australia to New Zealand. They accepted that New Zealand and Australia separated ca. 82 million years ago, "long before the oldest known earliest Eocene [fossil] bats," but as mentioned earlier, the fossil record of bats is very fragmentary. Hand et al. (2005) wrote: "The close similarity in dentition (and postcranials) between Australia's *I. aenae* and New Zealand's *Mystacina* species suggests that the most likely time of separation of the lineages, via a trans-Tasman dispersal event, was the early Miocene, ca. 20 million years ago," but this was based on the age of the *I. aenae* fossils, not the age of the clade itself.

Hand et al. (2009) wrote: "Although they are generally more plesiomorphic than Quaternary mystacinids, *Icarops* taxa also exhibit dental apomorphies of their own, suggesting that the Australian *Icarops* and NZ *Mystacina* lineages diverged *at least* 26 Ma [the age of the oldest *Icarops* material]" [italics added]. They noted that mystacinids are also known from the Early Miocene (19–16 Ma) fauna of

Central Otago "but as yet cannot be referred to either lineage." Either of the two fossil mystacinid species from New Zealand (Miocene) "exhibits a mixture of what we interpret… to be relatively plesiomorphic features, such as typically found in species of *Icarops*, and more-derived features, such as in species of *Mystacina*" (Hand et al., 2013).

Hand et al. (2009) concluded: "The fossil record sheds little light on whether or not NZ might have been colonized before Australia by noctilionoids, but Australia's Oligo-Miocene mystacinids are closely related to those of New Zealand and are less derived, which may be an indication that Australia was the source of NZ's mystacinids…." For any two sister groups, establishing which one is "plesiomorphic" by counting characters is problematic. Two sister groups have the same age, but unless they each have the exact same number of apomorphies, one will always appear to be "more plesiomorphic" than the other. Nevertheless, this does not mean that it is older than the other, or ancestral to it. Thus the data presented so far (there is no formal phylogeny) provide no reason to assume that Australia is a center of origin for Mystacinidae. If *Mystacina* were derived from *Icarops*, as Hand et al. (2009) suggested, it would be nested in *Icarops* (which has its own synapomorphies), and not stand, as it does, as a separate group.

Hand et al. (2009) concluded: "Exactly when and from where mystacinids first colonized NZ is not yet clear…." Mystacinidae are allopatric with the other components of Noctilionoidea, and so a simple vicariance history with *in-situ* development explains the distribution. Within Mystacinidae, *Icarops* and *Mystacina* are allopatric sister groups, again suggesting that they "colonized" Australia and New Zealand by evolving there, not by dispersing there.

Did *Mystacina* Develop Its Distinctive Features as Adaptations to the New Zealand Environment?

In the traditional model, a group's structure develops in response to needs imposed by its habit or its habitat. This is exemplified by the idea that "*Mystacina* has a mosaic of morphological adaptations… which arose as a result of its diverse lifestyle" (Lloyd, 2005: 110). Instead, a model of mutation-driven evolution proposes that the lifestyle arose because of the "adaptations." The adaptations arose in the first place, not because they answered "needs," but because of prior trends in mutation bias.

Terrestrial Behavior in Mystacina

One of the most distinctive features of *Mystacina* is its terrestrial behavior, and this has been proposed as a possible adaptation to New Zealand conditions. Most bats move awkwardly on the ground, but *Mystacina* species are agile crawlers. Daniel (1979) cited their "astonishing rodent-like agility on the ground and on tree trunks and branches." Individuals run with a "curious stiff action and quite fast, using their folded wings as forelegs, the wrist-joint coming in contact with the ground" (Stead, 1937). *Mystacina* takes prey on the ground, on tree branches, and also in flight, and is at home in all three habitats. While most bats capture prey in the air, or by making brief gleaning forays on surfaces, *M. tuberculata* searches for both plant material and insects while on the ground (Jones et al., 2003). It spends about 30% of its time crawling on the ground and tree trunks as a gleaner, 30% as an aerial insectivore, 20% as a frugivore, and 20% as a nectarivore (Daniel, 1979). In addition, the bats burrow into deep litter and use their teeth to excavate, or at least enlarge, tunnels in rotten wood for use as roosts. When foraging under leaf litter, humus, or snow, individuals often disappear for some time (Lloyd, 2005). Colonies roost in hollow trees or caves, and less often in abandoned seabird burrows, holes in pumice cliffs, and bat-excavated tunnels in fallen, hollow trees.

Explanations for Terrestrial Behavior in Mystacina

Daniel (1979) summarized the conventional wisdom on the crawling behavior of *Mystacina*: "It is likely that this special ability, unique among the Chiroptera, has evolved in New Zealand in response to the same factors that have enabled 28% of the endemic, terrestrial bird fauna to become flightless or have reduced powers of flight…." He concluded: "The absence of mammalian predators and the lack of competition from other mammals throughout the Tertiary are thought to have significantly influenced the evolution of its terrestrial and arboreal adaptations."

As with many aspects of the New Zealand biota, the absence of mammalian predators is unusual. Nevertheless, this environmental feature cannot be the cause of flightlessness in New Zealand birds such as ratites, as these are also flightless in other regions, such as Africa and tropical America, where there are many mammalian predators. A similar argument applies to terrestrial behavior in bats. Like *Mystacina*, the common vampire of tropical America (*Desmodus rotundus*: Phyllostomidae) is an agile crawler with a true walking gait; it approaches its prey by crawling up to it quietly. Yet unlike *Mystacina*, it lives in an area with many terrestrial predators (Riskin et al., 2006). In addition, a few vespertilionid bats are good crawlers, and many molossids are "especially good at walking on the ground" (Neuweiler, 2000: 24; Riskin et al., 2006). Morphological studies also suggest that the extinct sister group of *Mystacina*, the Australian *Icarops*, was "facultatively terrestrial," despite the presence of many mammalian and other terrestrial predators. Hand et al. (2009) concluded that "mystacinids were already terrestrially-adapted prior to their isolation in NZ..." and that the specialized terrestrial locomotion did *not* develop in New Zealand due to the absence of ground-dwelling mammalian predators and competitors. Hand et al. (2009) suggested that the terrestrial locomotion of *Mystacina* represented an "exaptation"—in other words, the feature did not evolve as an adaptation but turned out to be adaptive in its predator-free New Zealand environment.

In *Mystacina*, the skeletal features discussed by Hand et al. (2009) are probably the key structures enabling terrestrial locomotion, but others have also been suggested. The proximal parts of the wings show a conspicuous thickening, so that when the bats are moving on the ground, the wings are furled in a protective, leathery sheath (Dwyer, 1960). In addition, the propatagium (the part of the wing between the neck and the first digit of the hand) and the basal part of the uropatagium (between the two hindlimbs) are reduced, and this enables the free movement of both the forelimbs and hindlimbs.

The claws on the thumbs and the toes of *Mystacina* have minute denticles or talons on the inside curve at the base that are unique in Chiroptera (Daniel, 1979; illustrated by Nowak, 1999). Carter and Riskin (2006) wrote that these denticles "presumably aid in crawling and climbing," although this is speculative. With respect to their origin, they appear to represent relictual structures left over during the reduction of the limb precursor, as with the reduced lateral digits in the horse.

Reduction in the Vertebrate Face

In vertebrates as a whole, extensive reduction and telescoping have occurred in the face and pharynx regions. This has resulted in great condensation and fusion in structures such as the muzzle (cf. *Chalinolobus*), the lips, the musculature around the mouth, the jaws and teeth, the tongue, and internal features, such as the vomeronasal organ. The different conditions of these features in the modern forms have been derived from much more complex predecessors. The new structures have naturally resulted in changes in diet, behavior, and other aspects of ecology.

Reduction in the vertebrate mandible (lower jaw) is often cited as a classic example of a morphological trend, as the different stages are preserved in fossil forms. The reduction has taken place over many millions of years and in groups found in all environments. Reduction in primate jaws, for example, has resulted in the wisdom teeth of humans often becoming "impacted." In early vertebrate forms, the mandible is formed from several bones. In teleosts these are reduced to three, in amphibians to two, and there is just one in birds (by fusion) and in mammals (by loss of all the others). A study of cranial evolution in synapsids (mammals, therapsids, pelycosaurs) concluded that: "Although commonly ascribed to the effects of long-term selection [orthoselection], evolutionary trends can alternatively reflect an underlying intrinsic bias in morphological change" (Sidor, 2001). This agrees with ideas on biased mutation in genetics (Nei, 2013).

The face and pharynx of modern vertebrates show great variability among the different groups. This can be interpreted as a trace of the more or less suppressed complexity in the ancestral structures and also to variation in the degree of reduction. The reduction has seldom been complete, and there are many leftover, relictual structures that are often glandular. For example, in tetrapods, the parathyroid glands in the pharynx are derived from pharyngeal pouches in the embryo. In fish, this region develops into gills, and there are no parathyroid glands. Genetic studies provide further evidence that the parathyroid glands represent reduced, glandular relics of the larger, elaborated structures, the gills (Okabe and Graham, 2004). The glands produce secretions that have important consequences for the biology and ecology of the organism, but these are the end results of the reduction, not its cause.

The Snout in *Mystacina*

Cheeseman (1893) cited the "curious projecting muzzle" of *Mystacina*; in contrast with the short snout of *Chalinolobus* it is elongated and "trunk-like" (Grzimek, 1990 [1988]: 586). Likewise, in *Mystacina* the premaxillary bones (bearing the upper incisors) are not reduced in the way they are in many other bats (Vespertilionidae, Megadermatidae, Emballonuridae; Giannini and Simmons, 2007).

The snout in *Mystacina* is obliquely truncated and terminates in a small nose disc, or narial pad, covered with stiff hairs (Dwyer, 1962; Grzimek, 1990 [1988]: 586). The name *Mystacina* (*mustax* = moustache) refers to the bristles found on the upper lip (see McDade, 2003: 456) and encircling the nostrils (Worthy and Holdaway, 2002: Fig. 13.3). The narial disc is perhaps equivalent to the moist rhinarium, seen in the nose of dogs, cows, strepsirhine primates, and bats such as Pteropodidae. In other bats there are fleshy masses around the nostrils (Plecotini and *Antrozous*: Vespertilionidae; *Craseonycteris*: Craseonycteridae; and *Rhinopoma*: Rhinopomatidae), and the location suggests that these also represent cryptic traces of previous structures. In the Craseonycteridae of Thailand, the common name of the single member—the hog-nosed bat—refers to its fleshy, pig-like snout. Furipteridae of tropical South America also have a pig-like snout (McDade, 2003).

Complex, elongated snouts occur in monotremes (platypus, echidnas), and elongated, more or less mobile snouts similar to those of *Mystacina* also occur in the following mammals: many marsupials (most obvious in Didelphimorphia, Paucituberculata, Microbiotheria, and Peramelemorphia; less so in Diprotodonta except Tarsipedidae), Afrotheria (elephants, tenrecs), Xenarthra (the giant anteater, *Myrmecophaga*, but very short in sloths), Eulipotyphla (= Insectivora s.str.) (shrews, hedgehogs, moles, solenodons), Scandentia (tree shrews), some rodents (such as mice) and some ungulates (such as tapirs). In Carnivora the snout is trunk-like in elephant seals, and a longer snout is retained in groups such as dogs and viverrids, but it is much shorter in cats.

The trunk or proboscis, seen in mammals such as elephants, is an extreme development of the fused nose and upper lip, with the nostrils at the end. Although it is not preserved in fossils, associated cranial features provide strong indications that many fossil mammals had a proboscis, including members of Proboscidea (elephants and relatives), Perissodactyla (palaeotheres, amynodontid rhinoceroses), Artiodactyla (some oreodonts), and "South American ungulates" (astrapotheres, pyrotheres, Litopterna such as *Macrauchenia*) (Wall, 1980).

Reduction in the snout is widespread in mammals, and in primates it is well illustrated by the two main groups. Strepsirrhines (e.g., lemurs) have a well-developed snout and also a rhinarium, a patch of moist, hairless skin around the nostrils that is present in most mammals. Haplorhines (including monkeys and humans) have a more or less reduced snout and there is no evident rhinarium; it is reduced and infolded within the nostrils. In many mammals the rhinarium is connected, via the philtrum (a cleft in the upper lip), to the vomeronasal organ (Chapter 14). This is used in many animals to detect pheromones, but in humans it is suppressed, or at least non-functional, beyond the embryo stage, and the philtrum is closed. In pigs, the disc-shaped rhinarium retains extensive musculature and a nerve supply that appears to have been condensed and concentrated during reduction in the snout.

In traditional accounts, bats were classified into two main groups—Megachiroptera ("megabats"), comprising the family Pteropodidae (fruitbats or flying foxes), and Microchiroptera ("microbats"), including all the other bat families. Molecular studies have suggested new delimitations (along with new names) for the two main groups—Yinpterochiroptera comprise Pteropodidae and four other families, and Yangochiroptera comprise the remaining 13 bat families (Figure 16.1).

Yinpterochiroptera include *Pteropus* fruitbats in which the face resembles that of a fox or a dog, with a long muzzle and large eyes. In contrast, many groups of both Yinpterochiroptera and Yangochiroptera have a shortened muzzle and a face that is "abnormal" in some way. For example, *Emballonura beccarii* (Emballonuridae) of New Guinea and neighboring islands is "a very distinctive

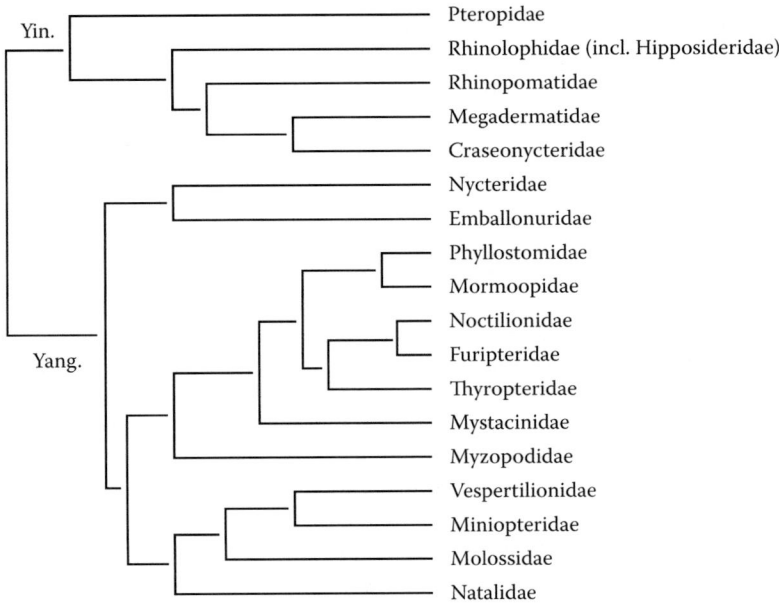

FIGURE 16.1   Phylogeny of extant bat families. Yin., Yinpterochiroptera; Yang., Yangochiroptera. (From Teeling, E.C. et al., Phylogenies, fossils and functional genes: The evolution of echolocation in bats, in *Evolutionary History of Bats: Fossils, Molecular and Morphology*, eds. G.F. Gunnell and N.B. Simmons, Cambridge University Press, Cambridge, UK, pp. 1–22, 2012.)

species… with a 'pushed-in' face resembling that of a pug" (Flannery, 1995b). Facial shortening in bats occurs to the greatest extent among Phyllostomidae of the subfamily Stenodermatinae. In this group, the wrinkle-faced bat *Centurio senex* is the most extreme, and it has "the most bizarre and grotesque face of all bats." The nose is almost flat and its face is covered with "a complex array of wart-like outgrowths, folds and flaps" (Hill and Smith, 1984: 15, 25, and Fig. 2.7). The folds are associated with secretory glands. Dumont et al. (2009) suggested that the short, wide skull is "specialized for generating high bite forces" and is "built to bite," but the high bite force can be interpreted as an effect of the shortening in the skull, rather than a cause, and the folds and glands can also be explained as by-products of the reduction.

The reduction of the snout in vertebrates has led to fusions among many prior structures, such as the premaxillary bones that bear the upper incisors. In birds, the fusion of premaxillary bones forms the bill. The form and structure of the premaxillary bones vary more in bats than in any other group of vertebrates except, perhaps, some bony fishes and snakes, and "The reasons for this variation are not fully understood" (Hill and Smith, 1984: 26). In humans, the premaxilla is fused with the maxilla, giving a derived condition. In Yinpterochiroptera, the premaxilla and maxilla are not fused, but they are attached in different ways by ligaments, while in Yangochiroptera they have become fused (Giannini and Simmons, 2007). As a member of the latter clade, *Mystacina* displays the fusion, but the premaxilla is not reduced (as it is in vespertilionids, etc.), and *Mystacina* has retained an elongate muzzle.

Reduction in the Bat Snout and Its Implications

Bats such as *Chalinolobus* and other vespertilionids have a snout that is more reduced than in groups such as *Mystacina*. In other bats, similar reduction is known to have direct, ecological implications. Examples include members of Phyllostomidae, in the South American sister group of *Mystacina*. Here "Rostral length in glossophagines closely parallels the length of the tubular flowers on which a particular species feeds" (Hill and Smith, 1984: 67). This is probably an oversimplification, but it illustrates the general principle of ecological/morphological correlation.

Consider a hypothetical series of bat species with snouts 5, 10, 15, and 20 mm long, with each feeding on flowers with tubes of a corresponding length. Each of the five species could be interpreted in a teleological way, as having evolved the optimum length for its flower. In this model, the length of the rostrum in each species was modified by adaptation answering the needs imposed by the bat's food flower and its feeding habit. In an alternative model, the range of feeding types is interpreted as the result of a single trend or trajectory in rostrum length, with this giving rise to the particular morphologies. A bat that developed a rostrum 20 mm long was able to find flowers of the same length to feed on, and likewise for the other species. If no flowers of the right length were available, the bat species might become extinct.

Within Phyllostomidae, the snout with the longest relative length occurs in the monotypic *Musonycteris*, which feeds on nectar. The long snout might represent a secondary development, but it cannot be assumed to be, just because it is unusual. The bat is endemic to the Guerrero terrane of central western Mexico (Colima, Michoacán, and Guerrero), a center of old endemics (Heads, 2009c), and there is no reason to think that the morphology of *Musonycteris* represents an adaptation to local conditions. Tschapka et al. (2008) wrote that its long jaws exclude *Musonycteris* from the use of nonliquid food, such as fruit and insects, because of the effects of leverage. They predicted that the food spectrum of *Musonycteris* would be dominated at least for a part of the year by plants with long-tube flowers that are not accessible to other nectarivorous bats and that "The geographical distribution of such plants should also provide an explanation for the unusually small distribution area of the bat...." In fact, though, the authors' study showed that most of the main food plants used by *Musonycteris* (Capparaceae and Bombacaceae) did *not* have long-tube flowers, but brush-type flowers that presented nectar openly to approaching visitors. This means there is little evidence that the long snout evolved in order to fulfill a need.

The Noseleaf and Echolocation

It was suggested earlier that the glandular elevations on the snout of *Chalinolobus* along with the narial disc and bristles in *Mystacina* represent incomplete reduction around the nose. In other bats, more elaborate, unreduced structure is evident in the nose region, and this is most obvious in the "noseleaf." This organ is conspicuous and often complex in most members of Rhinolophidae, Hipposideridae, and Megadermatidae (all in Yinpterochiroptera) and many Phyllostomidae (Yangochiroptera). (The names Rhinolophidae, Hipposideridae, and Phyllostomidae all refer to the noseleaf; the name Megadermatidae refers to both the noseleaf and the tragus, a similar flap of tissue in the ear.) A noseleaf is absent in many bats, including *Chalinolobus* and *Mystacina*; either it never evolved in these lineages or it has been suppressed.

In many ways the noseleaf is an enigmatic structure, and while its morphological variation is often used in bat systematics, its structural and functional relationships are "not well understood" (Hand, 1998). The homology of the noseleaf is still under debate, but its presence or absence reflects deep differences in the structure of the head. In many bat embryos, the noseleaf primordia appear even before the eyes and external ears are visible (Pedersen and Müller, 2013).

In rhinolophids and hipposiderids, the structure of the noseleaf ranges between simple and "truly bizarre and ornate" (Hill and Smith, 1984). It is often associated with glandular structures of different kinds. In the trident-nosed bat *Cloeotis* (Hipposideridae), it consists of three pointed protrusions (McDade, 2003: 402). In some groups it is very large; in *Lonchorhina aurita* (Phyllostomidae) it is about as long as the ears (Nowak, 1999), and it is also very long in *Macrophyllum* (Phyllostomidae). Some phyllostomids and megadermatids can move their noseleaf.

In Nycteridae, the slit-faced bats, the articulated facial cleft is unique, but its lateral components are thought to be homologous with the noseleaves of hipposiderids (Pedersen and Müller, 2013). In Mormoopidae, the noseleaf is "little more than a bump on the nose" (Hill and Smith, 1984). In Thyropteridae, there is no nose leaf, although there is "a small wart-like projection between and above the nostrils."

In Natalidae, a small tropical American family, all forms lack a true nose leaf. Nevertheless, at the tip of the snout, there is a "hairy protuberance that resembles a nose leaf" (McDade, 2003). Natalidae also show signs of reduction within the large ear, as this bears "distinctive papillae" which, at least in *N. stramineus*, are glandular. In addition, adult male natalids also have a gland-like structure in the center of the forehead. While its function was reported as unclear (McDade, 2003), it indicates the same trend (elaborated–reduced–glandular–absent) along the central line of the face that has resulted in many other features. These include the "noseleaf" of phyllostomids and others, the "hairy protuberance" of natalids, the rhinarium of many mammals, the "flower face" of *Anthops*, the "slit face" of nycterids, the "tentacles" of the star-nosed mole (*Condylura*), the horns of Rhinocerotidae and diverse extinct mammals (such as the rodent *Ceratogaulus* and the armadillo *Peltephilus*), the spectacular fleshy protuberance over the bill of the male comb duck (*Sarkidiornis*), and the horns of dinosaurs such as *Triceratops*.

To complete this brief survey of the bat families, snout elaborations other than small, glandular elevations are absent in the "normal" faces of Pteropodidae and in nine families of echolocating Yangochiroptera: Emballonuridae, Noctilionidae, Furipteridae, Thyropteridae, Molossidae, Craseonycteridae, Natalidae, Myzopodidae, Miniopteridae. They are also absent in most members of the largest bat family, Vespertilionidae, although Nyctophilinae retain a "small, rudimentary and fleshy nasal flap" (Hill and Smith, 1984: 213), and a similar structure occurs in Rhinopomatidae (McDade, 2003: 352).

The great range in noseleaf variation recalls the dramatic cline in nasal morphology in the "odd-nosed" clade of Southeast Asian monkeys (Heads, 2012a: 251). The cline extends from the proboscis monkey (*Nasalis*) in the Borneo mangrove, which has a long, pendulous nose, to the snub-nose monkeys (*Rhinopithecus*) of alpine China, in which the nose has disappeared. This involves the extreme reduction or even absence of the nasal bones.

Evolution of the Noseleaf

In the standard model, the irregular occurrence of noseleaves across the bat phylogeny—present in some Yinpterochiroptera and in some Yangochiroptera—means that the noseleaf has evolved several times, and this is attributed to selection pressure. What could cause the selection pressure? In some cases, the noseleaf has indeed turned out to be useful for the bats, as it is employed in focusing or directing the sounds emitted through the nose by many echolocators (Pedersen and Müller, 2013). But the standard model does not account for its early origin in ontogeny; its great variation among families, genera, and species; its trend in variation from a baroque, elaborate ornamentation to a reduced glandular structure; or its vestigial morphology or absence in many echolocators and non-echolocators. All these aspects are consistent with a complex structure that has been reduced, telescoped, fused, and more or less suppressed, in a long-term, mutational trend.

As in most other vespertilionids, *Chalinolobus* uses echolocation but does not have a noseleaf. *Mystacina* is an echolocator but lacks a noseleaf, while many other members of its superfamily also use echolocation and do have a noseleaf. The sister group of Mystacinidae is an American complex dominated by members of the Phyllostomidae (56 genera, 192 species). This is the most diverse bat family in tropical America, and its members are among the most common mammals there (McDade, 2003: 413). Phyllostomids include many long-snouted groups, and many have distinctive noseleaves, but these tend to be reduced in non-echolocators. In carnivorous, vertebrate-feeding forms (e.g., *Chrotopterus*: Phyllostominae) and insectivorous forms (e.g., *Lonchorhina*: Glossophaginae), the noseleaf and the ear are both large, while in the blood-feeding vampires (Desmodontinae), the noseleaf is much reduced. The size and complexity of the noseleaf is also reduced in the frugivorous *Centurio* (Stenodermatinae) and Brachyphyllinae, and in many nectarivorous Glossophaginae. Nevertheless, the frugivorous *Platyrrhinus* (Stenodermatinae) has a conspicuous noseleaf.

In agreement with the idea of mutation-driven evolution, Sterelny and Griffiths (1999: 43) wrote that: "Selection can operate only on the variation that is available, and history, development and genetics determine the range of variation." In this model, mutation produces the variation, which may or may not be pruned by the *secondary* process of natural selection. Yet Sterelny and Griffiths (1999) also accepted that echolocating bats are a "wonderful example" of adaptation, with "elaborately structured facial architectures to maximise chances of receiving returning echoes..." (p. 29). They proposed that horseshoe bats (Rhinolophidae), with a well-developed noseleaf, have a "splendidly bizarre" face and that "The architecture's role in helping the ancestors... explains its existence in bats today" (p. 44). Nevertheless, members of the largest family of bats, Vespertilionidae, use echolocation but do not have noseleaves, and many have small ears. Even if the "bizarre" facial architecture of rhinolophids is beneficial in some way, echolocation does not explain the particular structural changes underlying its historical development.

In the adaptation model, many questions are left unanswered. For example, why is the rhinolophid noseleaf so complex compared with that of the very successful group, the phyllostomids? Sterelny and Griffiths (1999) described the neodarwinian theory as uncontroversial: "few will deny that bat echolocation is the product of natural selection" (p. 47). Yet this overlooks the alternative: mutation-driven evolution. The adaptive value of echolocation explains why it has not been eliminated, but it does not explain why it was there in the first place. About a quarter of all mammal species use it; bats, shrews, tenrecs, toothed whales, and dolphins are the most notable groups (Müller and Hallam, 2004). Blind people can develop great skills in echolocation. Given the prior trend of evolutionary telescoping in the vertebrate face, which has involved the ear, the nose, and the larynx, in many groups the ability to echolocate would have been inevitable.

As emphasized, noseleaves are not present in all echolocating bats—they are absent in both the New Zealand genera, for example—and the wide *range* in bat faces, along with the overall trends in their morphological evolution, needs to be explained. Some bats have faces that are just as bizarre as those of the horseshoe bats cited by Sterelny and Griffiths (1999), but have a very different morphology. There is no evidence that the great range of "bizarre" faces in *Mormoops, Anthops, Centurio,* and others indicates adaptation to local conditions. Instead, these structures, along with the "trunk-like" snout of *Mystacina* and many other mammals, and the horns of different groups, suggest stages of an overall reduction trend active in the vertebrates ever since their origin. In the bats that have an elaborated noseleaf or tragus, these structures will have consequences for echolocation, but echolocation is not the reason they exist in the first place. They are variable, or absent, in the different groups, and this may help determine different types of echolocation, although the subject is still obscure. The reduction of the noseleaf and its homologues in *Mystacina, Chalinolobus,* many other bats, and most mammals is part of the general trend toward reduction in the median, ventral part of the vertebrate body (the "visceral animal" of Romer, 1972).

Ears in Bats

The ears of bats display great variation, and some are "quite bizarre" (Hill and Smith, 1984: 10). In many species, the outer, visible part of the ear (the pinna) is small, while others have "enormous ears extending almost the length of their bodies" (McDade, 2003). There is substantial variation even within echolocating genera. For example, in New Guinea Emballonuridae, *Emballonura dianae* has large ears, while those of *E. beccarii* are much smaller (Flannery, 1995b). Vespertilionidae include groups such as *Idionycteris, Plecotus, Corynorhinus,* and *Euderma,* all with very large ears (a third of the body length). In the Australian species of *Chalinolobus,* the ears are large in *C. dwyeri* and of decreasing size in *C. gouldii, C. nigrogriseus, C. picatus,* and *C. morio* (Parnaby, 1992: Fig. 15). Those of *C. tuberculatus* are similar to those of *C. morio,* and about a quarter the size of those of *C. dwyeri.* Similar variation is seen in other New Zealand mammals, as the "eared seals," Otariidae (including *Arctocephalus* and *Phocarctos*), have external ears, while the true seals, Phocidae (including *Mirounga*), do not.

A teleological model explains the origin of the large ears in bats in terms of their end result: "For such structures to have evolved there must be a significant benefit. For example, it has been shown that bats with long ears are superior at avoiding thin wires stretched across their flight paths when compared to other bats with smaller ears" (Gardiner et al., 2008). If, instead, the particular morphologies have resulted from a prior, general trend, this would explain why echolocating vespertilionids, such as *Chalinolobus* species, have such a wide range in ear size. It seems significant that the *Chalinolobus* species with the smallest ears (*C. morio* and *C. tuberculatus*) also have the smallest lip and ear lobes. This indicates correlated reduction over a broad region of the face.

In many bats, the pinnae have conspicuous transverse ribbing, giving a "washboard" pattern. In most mammals, this has been eliminated, but similar ribbing still occurs in the ears of many strepsirrhine primates, in tarsiers (Wood Jones, 1951), in members of the order Dermoptera (colugos or flying lemurs), and in rodents such as *Pogonomys.* Dermoptera are the sister group of primates but are best known for their patagium, the flap of skin that extends between their four limbs and allows them to glide. The distribution of all these features among bats, primates, and Dermoptera suggests that the groups have evolved by recombination of ancestral characters. Patagia also occur in pterosaurs, birds, lizards, marsupials (Petauridae), and rodents (Anomaluridae and Sciuridae).

As in bats, the ears of primates include many signs of extensive reduction. For example, in great apes, the complex musculature moving the outer ear is reduced to a more or less vestigial state. Another reduced component of the human ear is Darwin's tubercle. This is a small thickening on the inner side of the outer rim of the ear (the helix), about two-thirds of the way up. It is present in many humans and is the infolded homologue of the pointed ear tip of other mammals, which is often large and conspicuous (Darwin, 1874: 15–17).

Another structure in the ear that has undergone great reduction is the tragus. In humans, this is a small, rounded, cartilaginous knob at the front of the outer ear, above the notch at the base, but in many bats it is much longer. As with the noseleaf, it is variable and is important in identification. *Mystacina* has a pointed tragus that is 10 mm long (Lloyd, 2005). In *Chalinolobus tuberculatus*, the tragus is smaller—narrow at the base but widened and rounded toward the tip and curved inward (O'Donnell, 2005). In contrast with bats, in most mammals the tragus is reduced. Nevertheless, it is well developed in some rodents (Pocock, 1922). Wood Jones (1951) compared the ears of tarsiers with those of bats and noted that tarsiers have a conspicuous tragus (cf., for example, *Tarsius dianae*; Niemitz et al., 1991).

The tragus affects hearing and has implications for echolocation, as the sounds bouncing off it generate spatial information. Yet, as with the noseleaf, this effect of the bat tragus does not account for its great variation, and in some echolocating forms (Rhinolophidae, Hipposideridae) it is absent.

Phyllostomidae show a great range of diversity in the tragus: it can be very large (*Lonchorhina*), medium-size (*Macrotus*), or very small (*Chrotopterus*, *Choeronycteris*). In the insectivorous *Lonchorhina* (the name means "lance-nose"), the noseleaf and the tragus are both long and slender (over half the length of the outer ear). Other groups with a well-developed tragus include Craseonycteridae.

In many groups, including both New Zealand genera, the tragus is a simple, strap-shaped to lanceolate structure (McDade, 2003). However, it can take a wide range of other forms, including the following:

Bifurcate (Megadermatidae).

Square, with an "accessory appendage" (Molossidae: *Tadarida aegyptiaca*).

With notched, ridged, or serrate margins (Phyllostomidae such as *Chrotopterus*, *Phyllostomus*, *Erophylla*, and *Artibeus*; also Noctilionidae).

Kidney-shaped and prominent (Miniopteridae: *Miniopterus majori*).

Mushroom-shaped (Myzopodidae; Vespertilionidae: *Nyctalus*).

Hatchet-shaped (Vespertilionidae: *Neoromicia*).

Spathulate with a shelf-like fold (Mormoopidae: *Pteronotus*).

Small, truncate, and inconspicuous (most Molossidae, Mormoopidae, Rhinopomatidae, Emballonuridae).

Absent (in the non-echolocating Pteropidae and also in the echolocating Rhinolophidae and Hipposideridae). Some Rhinolophidae instead have a much-enlarged antitragus, a lobe located opposite the tragus (see photograph of *Rhinolophus yunanensis* in McDade, 2003: 393). This is broader than long, unlike the tragus of most bats. Hipposideridae always lack a tragus; the antitragus may be inconspicuous, or large as in *Hipposideros fulvus* (illustration in McDade, 2003: 406).

It is possible that the different morphologies are all adapted to subtle environmental differences, but none have been proposed, and an alternative is that the morphological variations represent different points on mutation-driven evolutionary trajectories.

Feeding in New Zealand Bats

The diversity of diet in bats is unparalleled among the other orders of mammals (Kunz and Pierson, 1994: 1). In New Zealand, *Chalinolobus tuberculatus* is an aerial insectivore, while *Mystacina* feeds the year round on flying and resting arthropods, and in some seasons on nectar, pollen, and fruit (Daniel,

1979; Lloyd, 2005). In addition, *Mystacina robusta* often fed on the dead seabirds (*Puffinus griseus*) that were hung up to dry by nineteenth-century hunters (Daniel, 1990). The dentition of *Icarops*, the Australian fossil mystacinid, suggests that it was also omnivorous (Hand et al., 2009).

It has been stated that *Mystacina* has a "surprisingly broad" diet (Daniel, 1979), that it is "remarkable among Chiroptera for the diversity of its diet" (Lloyd, 2005: 116), that its diet is "broader than that of any bat recorded" (Hand et al., 2009), and that mystacinids "are the only bats demonstrated to be omnivorous" (McDade, 2003: 453). Holdaway (1989) wrote that *Mystacina* "has exploited an unusually diverse range of foods in the absence of mammalian competitors," and, as discussed earlier, adaptation in the absence of mammals is the standard explanation for many aspects of structure and ecology in the New Zealand biota.

Nevertheless, the uniqueness of the omnivorous diet in *Mystacina*, as with its walking ability, has been exaggerated. The sister group of *Mystacina* in America includes many phyllostomids that are just as omnivorous, despite living in an area with a diverse mammal fauna. For example, *Phyllostomus hastatus* of Central America and northern South America eats vertebrate prey, as well as insects, fruit, pollen, nectar, and flower parts (Freeman, 1988). Twenty-four other phyllostomid genera all eat insects, nectar, and fruit, and five genera all eat vertebrates, insects, and fruit (Monteiro and Nogueira, 2011). Thus the omnivorous diet of *Mystacina* is quite normal in the context of its superfamily, and it need not have originated by natural selection caused by environmental conditions in New Zealand.

Of the 19 families of bats, 13 are limited to one food source, and 5 are limited to two food sources (Baker et al., 2010, perhaps overlooking Mystacinidae). Phyllostomidae are exceptional, as the members show six distinct feeding styles: insectivory, frugivory, nectivory, carnivory (feeding on vertebrates), sanguivory, and omnivory. The usual model for the evolution of bat feeding overlooks the possibility of a polymorphic ancestor. Instead, it proposes a restricted initial ecology followed by adaptive radiation from that point, and because strict insectivory is so widespread in bats, this is often reconstructed as the primitive mode. Thus, for Phyllostomidae, "Starting from an insectivore ancestor around 30 Mya, different lineages specialised into carnivores, sanguivores, frugivores and nectarivores…. Some trade-offs are clear, such as the mandibular elongation observed in specialised nectarivores, which supports a longer tongue but decreases bite force changing the scope of usable resources (increased variety of flowers versus harder fruit and insects)" (Monteiro and Nogueira, 2011). In contrast, a model of mutation-driven evolution suggests that a longer snout is not a "trade-off," as nothing is sacrificed for the sake of something else, and snout length evolves instead according to particular laws of growth (see the earlier section "Reduction in the Bat Snout and Its Implications"). If the morphology turns out to be viable, the lineage survives; if not, it goes extinct.

The teeth in mammals present a great range in morphology, and the dentition of a species is often related to its diet. The standard model proposes that the diet determines the type of teeth. Instead, if evolution is driven by mutation, the teeth determine the diet, and so "a bat eats fruit because its teeth are constructed in a certain way…" (Findley and Wilson, 1982: 243). The basic number of teeth in placental mammals is 44, but bats have 20–38 as the result of reduction (shorter jaws tend to have fewer teeth). The amount of dental variation in bats exceeds that of all other mammals (Hill and Smith, 1984: 22).

Dentition in *Mystacina* is typical of insectivorous bats, but the upper premolar and a lower premolar and incisor have all been lost, and the remaining lower incisor is small. The resulting gap in the teeth in the lower jaw facilitates use of the extensile tongue to feed on nectar (Lloyd, 2005). The tongue is ~12 mm in length, almost as long as the head (~15 mm). It has a small patch of fine, brush-like papillae on the tip, "suggesting modification and specialisation for this diet [including nectarivory]" (Daniel, 1979). Papillae are often glandular and represent the last trace of earlier structure before complete suppression (cf. the tongue of birds, Chapter 15). It is doubtful that the incompletely reduced tongue and reduced dentition of *Mystacina* are adaptations for New Zealand conditions, as similar structures occur in Phyllostomidae. In the subfamily Glossophaginae, which includes nectar-feeding forms, the tongue is long and can exceed two-thirds of the body length, while the dentition is reduced. In some glossophagine species, the lower incisors are missing, resembling the reduction in *Mystacina* (Tschapka et al., 2008). Differential reduction in tongue and teeth has created the distinctive new morphology.

17

Conclusions

Nay, now thou goest from Fortune's office to Nature's; Fortune reigns in gifts of the world, not in the lineaments of Nature.

Shakespeare, *As You Like It.*

Some of the main distributional boundaries and connections in New Zealand have been described in the earlier chapters, and the reader can test these out in groups that he or she has a special knowledge of. The reader can also test the methods used in this book, which are based on a close study of the distributions and the regional geology rather than on fossil ages, clock ages, and centers of origin. Distributions in any group can be explained either by chance dispersal, or by vicariance in a widespread common ancestor (possibly followed by secondary range expansion). Despite this, most authors currently rely on center-of-origin programs that will find a center of origin wherever there is a basal, paraphyletic grade.

The two models of biogeography, dispersal theory and panbiogeography, differ in their explanations for the main aspects of distribution—allopatry and overlap—as indicated in Table 17.1.

Assuming that two models are conceivable, which model is the best? Many groups show similar break zones, and this supports a general mechanism rather than idiosyncratic processes for each taxon. Nevertheless, most current studies instead use traditional dispersal methodology. They assign fossil-calibrated clade ages (using programs for chronological analysis such as BEAST) and calculate centers of origin and dispersal events (using methods for spatial analysis such as DIVA). The results are not based on evidence but on theory; they are determined by the priors in the first case and by the center-of-origin assumptions of the programs in the second case.

Tectonics and Biogeography

In the typical, trans-South Pacific genus *Abrotanella*, morphology-based phylogeny gave a set of clade distributions that was fairly straightforward, with interlocking clades, but the molecular phylogeny gave an arrangement that was even simpler, with a few main pieces that all dovetail together neatly (Chapter 10). Similar results are found in many groups. The implication here, and from the molecular work in general, is that biogeography reflects evolution even more closely than was thought. This provides one obvious reason for the close association between the main features of tectonic geology and of organic distribution.

The underlying tectonic structure and history of a region determine the nature of the environment and provide the theater for the evolutionary play. In New Zealand, biogeographic distributions have developed on geographies of the past that were very different from that of today, and the same phenomenon is observed in many parts of the world. For example, the endemic fauna of Lake Ohrid, in Macedonia (the oldest lake in Europe), has close affiliations with the local fauna, and its phylogenetic and ecological roots are more ancient than the lake itself (Wysocka et al., 2014).

Many biologists have examined the biological implications of continental drift, but there is much more to geology than this process. In particular, geographic entities such as New Zealand and all the continents are geological and biological hybrids, composed of series of accreted terranes. The relationship between terrane tectonics and biogeography was first considered in New Zealand (Craw, 1988; Heads, 1990a), and studies in this field are now being carried out around the Mediterranean. One key process, slab rollback, has led to the opening of backarc basins in both regions, and it has been linked to biological diversification in the Mediterranean region (Bidegaray-Batista and Arnedo, 2011).

TABLE 17.1

Differing Explanations for Allopatry and Overlap in Dispersal Theory and Panbiogeography

	Explanation for Allopatry	**Explanation for Overlap**
Dispersal theory	Chance dispersal	Normal dispersal
Panbiogeography and vicariance theory	Vicariance	Normal dispersal

Source: Heads, M., *Aust. Syst. Bot.*, 27, 282, 2015.

Many different geological processes have been discussed in earlier chapters as possible mechanisms for associated biogeographic patterns. In general, long-term, episodic phases of tectonism and volcanism (these often persist for a million years or more) are more important for biology than particular volcanic eruptions, fault movements, and so on that do not affect populations in a permanent way.

The brief analyses given in earlier chapters suggest that aspects of phylogeny and distribution in and around New Zealand have developed in association with a wide range of geological processes. These include the following:

Uplift of communities in orogens and on growing volcanic edifices.

Lowering of communities with subsidence and erosion. In this way, tectonics (uplift, subsidence) and prior biogeography determine many aspects of ecology.

Incorporation of clades into the Zealandia region with terrane accretion, obduction, and the "scraping off" of populations at subduction zones.

Modernization of the extant biota in the Early Cretaceous phases of the Rangitata orogeny. Phylogenetic divergence developed during tectonic convergence (up until 105 Ma) at belts of uplift (and sometimes extension), metamorphism, and magmatism.

Phylogenetic differentiation with predrift rifting (cf. Ribeiro and Eterovic, 2011; Murienne et al., 2014; Graf et al., 2015). Differentiation occurred with rifting, half-graben formation, and uplift at metamorphic core complexes. There is much evidence suggesting that the Gondwanan biota was already heterogeneous at the time of final breakup (Buckley et al., 2015).

Differentiation with seafloor spreading and continental drift. This is well documented, but it appears that rifting between two large regions, such as Australia and Zealandia, does not always need to be complete for organic differentiation to occur.

Range expansion, *extinction*, and *speciation* during marine transgressions from the late Cretaceous to the Oligocene.

Paleogene rifting of groups by normal faulting and *Neogene differentiation and disjunction* caused by strike-slip and reverse faulting.

Stranding of coastal taxa inland with marine transgression and regression.

Widespread survival of metapopulations through the Pleistocene ice ages in multiple microrefugia.

Centers of Origin, Dispersal, and Adaptation: The CODA Model of Evolution

All biologists are familiar with two fundamental evolutionary concepts. The first proposes that a taxon evolves at a biogeographic center of origin and attains its distribution by dispersing from there. The second concept is that morphological structures have developed by adaptation in order to supply the needs of the organism. These two concepts make up the center of origin–dispersal–adaptation (CODA) model (Lomolino and Brown, 2009), the core of the Modern Synthesis. In this approach, differentiation is interpreted in terms of *radiation* from biogeographic, morphological, and ecological centers of origin.

Comparative Study in Biology and the Phylogenetic Context

Studies using the CODA paradigm explain the properties of a group by focusing on the group itself. With respect to geography, CODA analysis aims to find a group's center of origin, *within* the group's range, by studying the phylogeny, diversity, and fossils of the group. In morphological studies, CODA analyses aim to find the "need" of the group that has caused a particular structure to arise.

Instead, the group can be seen as just one member of a series, an evolutionary trajectory. A group's geography can be examined not just with reference to the group itself, but in comparison with its relatives. For example, if a group is distributed in an area A, the CODA model examines the phylogeny of the group and will find a center of origin within A. In contrast, a broader, comparative analysis may show that the group's three closest relatives have distributions in areas B, C, and D, respectively, and this is consistent with a simple vicariance history. The explanation is fuller and more complex than a simple proposal of "chance dispersal" at any time, and it allows the possibility of a synthesis with geology.

In the same way, if a group has morphology A and its relatives have morphologies B, C, and D, the series needs to be explained before any of its particular products. Again, this sort of approach is more complex than the simple "adaptive" explanation, as it requires the study of relatives and of general trends in morphological evolution. In most cases, similar trends also occur in other groups that are not immediate relatives, suggesting even broader underlying trends.

For example, in its basic structure, the human heart is a helix. But why? The helix forms part of a general pattern with the other spirals seen in the vertebrate body, from the gut to the hair whorls, and the bilaterality of vertebrates is secondary. The distribution and morphology of an organism or a structure—the heart as a helix, for example—can be *described*, but to be understood, their meaning needs to be *interpreted*, and this requires a comparative study beyond the group itself. The comparative approach stresses the phylogenetic and morphological context of a morphology or a geographic distribution and the recurrence of repeated trends in both.

The Wedge Model of Evolution

In the CODA model, evolution proceeds by adaptation and the "penetration" or "wedging" of new species into an empty niche by selection pressure. The new needs of the organism are supplied by new features of structure and physiology that evolve by natural selection. Since the 1950s, studies explaining most aspects of evolution have adopted the wedge idea and the CODA model. For example, levels of biological diversity are thought to be limited by "species packing and the filling of available niche space" (Ricklefs and Jenkins, 2011).

In the first edition of the *Origin of Species*, Darwin (1859: 67) proposed that species are packed or "wedged" into what we would call their ecological niche: "The face of Nature may be compared to a yielding surface, with ten thousand sharp wedges [species] packed close together and driven inwards by incessant blows." In his notebooks, Darwin often described species driving out others by competition as "wedging" (Gould, 1989). (For a more detailed version of the wedge, as cited in Darwin's earlier manuscript notes, see Stauffer, 1975: 208.)

The "wedge" metaphor of evolution was an early version of the CODA model: species evolve by being wedged into their new niche, following dispersal. This remains the standard view of competition, selection pressure, the niche, dispersal, and so on, as accepted in the Modern Synthesis. Dennett (1996: 75), for example, referred to "the wedge of natural selection." "Wedge species" are recognized in groups such as South American primates (Boubli et al., 2008), and a recent book is titled *Evolution's Wedge: Competition and the Origins of Diversity* (Pfennig and Pfennig, 2012).

In contrast with the wedge model, the vicariance model proposes that a new species *inherits* its initial geographic region and habitat, and so there is often no need for a species, even a widespread one, to have ever expanded its range. For example, a population uplifted during an orogeny does not "invade" the new mountain range or the group's new montane niche. Instead, the species, the mountains, and the niche all evolve together.

The wedge theory has been the usual model adopted in explaining New Zealand groups. It is seen, for example, in the idea that freshwater fishes have evolved by "penetrating" inland (McDowall, 2010b).

In twentieth-century biology, the metaphor of the wedge has been supplemented with a new suite of mechanical references, with authors citing evolutionary "drivers," "engines," "motors," "pumps," and "explosions." Nevertheless, the original idea of the wedge remains at the core of the CODA model.

Centers of Origin in Biogeography

Darwin (1859) proposed that a new species evolves at a localized center and attains its distribution by dispersal from there. This idea is the precursor and main assumption of modern center-of-origin/ancestral area programs. For a phylogeny (A (A (A (A, B)))), these programs will always conclude that the group had a center of origin (usually in A, but see Cook and Crisp, 2005), rather than a widespread, common ancestor in A and B (Figure 1.1). In a vicariance interpretation, clades expand their ranges (disperse) because of general environmental change, in times of general mobilism, not because of their particular attributes or through chance dispersal.

Centers of Origin in Ecology

The idea of a center of origin has also been applied to ecological habitats and the evolutionary ecology of groups. In this approach, clades that are widespread in "ecological space," occurring in more than one habitat type, are assumed to have originated in one habitat and colonized other habitat from there. Yet this is problematic.

Consider a genus of five species, with four found in gullies (G) and one on ridges (R). If the phylogeny is G (G (G (G, R))), then a center-of-origin model will calculate that the genus had its origin in gullies and dispersed onto ridges. However, the phylogeny is also consistent with an ancestor that already inhabited both gullies and ridges before the origin of the modern clades.

A particular new habitat, such as a landslide, will be colonized by pioneer plants and animals. However, in many cases, different habitat types and the taxa in them have evolved together, by differentiation, as in geographic vicariance. For example, the Southern Alps have risen together with the populations that were already in the area before uplift. Here there is *in-situ* evolution and changing ecology (from seashore to alpine in a million years or so), but no "empty niche" or speciation by chance dispersal.

Centers of Origin in Morphology

A group's evolutionary trajectory can also be considered in "morphological space" (with "morphology" taken to include all aspects of form, from molecular to macroscopic). In the Modern Synthesis, the ancestor of any group was a single individual, parent pair, or species, and was, above all, homogeneous in its morphology. From this morphological "point," the descendants evolved new forms to result in, say, a diverse genus occupying a broad sector of morphological space.

Instead, there is growing evidence that the ancestor of a diverse group was already polymorphic (for example, with northern and southern forms) before the modern groups (for example, eastern and western forms) began to differentiate. Inheritance of ancestral polymorphism gives rise to character incongruence that can resemble the results of introgression.

Evolution by Radiation (Center of Origin–Dispersal–Adaptation)

In the traditional view, a new species arises at a geographic–ecological–morphological point. It is then "wedged" into a new geographic space through dispersal and into a new morphological and ecological space through adaptation and natural selection. The processes occur together, and evolution comprises radiations away from centers. In the alternative view of *spatial* evolution supported here, differentiation is caused by vicariance. Here there is no center of origin and no radiation, as new taxa evolve by the splitting of widespread ancestors. The origin of new taxa does not result from movement, but from the cessation of movement (or at least a decrease), and differentiation takes place *in situ*, over a wide area. An alternative view of evolution in *form*, supported here, is emerging in molecular studies. It suggests that biased mutation is the primary cause of evolution, and that selection and adaptation are secondary processes (Lynch, 2007c; Stoltzfus, 2012; Nei, 2013). The primary causes of particular morphologies are the long-term evolutionary trends played out over geological time, and these trends are the result of prior genome architecture and biased mutation rather than adaptation.

All the patterns of differentiation discussed in this book are explicable in a straightforward way, if the idea of radiation from a geographic, ecological, and morphological center of origin (by chance dispersal and random mutation) is rejected. The regional geology and biology can be synthesized if the organic groups have each developed from an ancestral complex that was *already* widespread in its geography, was polymorphic, and occupied different habitats before the formation of the modern clades began.

A few groups with distinctive features have been proposed as examples of dispersal to New Zealand followed by adaptation to local conditions, but these models do not stand up under closer examination. Divaricate plants (with long shoots that abort at the apex and short shoots in which the internodes are suppressed) are common in New Zealand. Yet plants with the same architecture have a widespread geographic distribution and are also found in a wide range of habitat types, from lowland swamp forest to desert and the alpine zone. The morphological features in this group, and in others such as ratites and bats, can be explained with reference to long-term trends alone, without invoking adaptation or design. The trends are simple processes, such as reduction, fusion, and sterilization, and these have probably resulted from simple genetic changes.

New Zealand Groups as Unique Wonders of Nature

Center-of-origin theory proposes that every monophyletic group has a unique history, beginning with a single origin from a single parent pair, at a single locality. But groups and characters are not unique; instead, everything in biology is repeated, replicated, and reproduced, and this gives rise to the standard patterns—nothing happens just once. For example, if one group originated by a phylogenetic break at such and such a locality, another group will be found to have done the same.

It is often claimed that New Zealand groups are exceptional and extraordinary (Baker et al., 1995), or weird and wonderful (Trewick and Morgan-Richards, 2014). This book suggests instead that New Zealand groups, while they vary in their details, are not in the end unusual freaks or wonders of nature, but conform to the same main patterns of biogeographic and morphological evolution that are repeated in many other groups around the world.

Summary of the CODA Model and an Alternative

Simple trends in morphological evolution, such as reduction–fusion series, can be played out over tens or even hundreds of millions of years, over intercontinental or global scales and in hundreds of thousands of taxa. In the Modern Synthesis, these trends are attributed to natural selection (orthoselection), but the environments they develop in are so varied that this is implausible. They are more likely to be the result of biased mutation.

The structural changes caused by genetic and genomic trends will themselves lead to shifts in habits and habitats. For example, if an ungulate's neck grows longer, it will tend to feed on taller trees. Animals whose sight is compromised by a reduction of the optic system will be forced to adopt a nocturnal or subterranean lifestyle. Animals whose teeth are reduced or lost will be forced to live on softer food.

Most studies interpret the distribution, morphology, habitats, and diversity patterns of a group in terms of the CODA model. Consider the plant genus *Coprosma*, for example. It is distributed across New Zealand in a wide range of lowland and montane habitats: mangrove edge in Northland, arid semidesert in Central Otago, active volcanoes in the central North Island, high mountains in the South Island, and rainforest and swampland in many regions. The CODA model proposes that *Coprosma, following its origin*, has invaded the extreme habitats of the mangrove, the volcanoes, the Southern Alps, and the semidesert by dispersing into them, later developing suitable adaptive features to survive there.

Instead, it is suggested here that *Coprosma* was already widespread at the time of its origin. The ecology of the different clades is determined by regional biogeography and tectonics, along with secondary adaptations in populations. For example, the elevational range of a metapopulation determines many aspects of its ecology and is itself the result, in the first place, of uplift, subsidence, and erosion.

When *Coprosma* originated—when it first became recognizable as *Coprosma*—the populations were already stranded in certain habitats, and they inherited these, they did not invade them. Of course, secondary range change may or may not occur later.

Biogeographic patterns in *Coprosma* are shared by many groups and can be used to date the main phases of evolution. The patterns in *Coprosma* and its immediate relatives, found in Africa, Australasia, the Pacific, and South America, suggest that polymorphic ancestral complexes already occupied ancestral landscapes long before extant groups, such as *Coprosma*, or their landscapes existed. Many modern accounts of groups stress their "biology"—their ecology and behavior—and use these to explain their structure and their distribution. Instead, it is easier to explain the groups' ecology using the distribution of groups and the evolution of their structure, interpreted in terms of broad, mutational trends.

Some Practical Implications of Biogeographic Studies

Dispersal theory depends on chance dispersal and does not solve concrete, basic problems in distribution. These take the form: "Why is such and such a distributional feature located where it is, rather than somewhere else?," and dispersal theory can only answer: "chance." Likewise, the Modern Synthesis view of structural development is based on random mutation, and so it does not solve concrete problems in morphology. The only reason given for a structure is: "because it is for the best"; it is "for the good of the organism." Overall, the CODA model is based on chance dispersal and random mutation, and so it cannot provide useful answers to general problems in evolution.

Apart from the theoretical issues, the ideas and methods discussed here have practical implications. One obvious example concerns the significance of distribution maps. In geology, mapping is a fundamental part of research and training. Biologists have not been so interested in mapping, and distribution maps did not become a normal part of revisions until the 1960s. Many monographs of biological groups (including most monographs of fossil groups) still have no maps. Given this history, it is not too surprising that so many phylogenetic studies fail to map, or even mention, the striking distributions of their clades. These are often the most impressive results, and their neglect has held up progress.

Advances in biogeography have practical implications for work in systematics at many levels. For example, a grid with cells of 10′ longitude × 10′ latitude is often used for mapping New Zealand groups in revisions (e.g., in Heads, 1998b, and in many treatments in the *Fauna of New Zealand*), but the grid cells are too large to show critical details. The cells are ~14 × 18 km, while many tectonic and biogeographic patterns are evident at finer scales than this. Maps with dots of actual localities are better. Outline maps are too vague for primary documentation. (They are useful in comparative studies, as they allow several clades to be mapped together.) Hardly any detailed maps of clade boundaries exist, and so in future work, producing these would be more valuable than making further, poorly sampled molecular surveys. As with geological mapping, detailed biological mapping of clades will often require fieldwork in areas that are rugged and remote.

The message here is that making maps is worthwhile. In contrast, dispersal theory takes a nihilist position, arguing that there is no point in making maps, or in documenting or discussing distributions in any detail, as they are the result of chance and there are no common patterns. If there appear to be any, they are in reality pseudopatterns, with many different causes and origins. In any case, distribution data and phylogenetic data are imperfect (because of sampling issues and so on), and we have to wait until they are perfect before examining them further.

Of course, every distribution map is out-of-date as soon as it is made, because of births, deaths, and movements. In the same way, no phylogeny is a perfect depiction of reality. A dot map is a more accurate portrayal of distribution than an outline, but even a dot map of all individuals at one point in time would not remain accurate for long. A group is structured as a dynamic complex of populations—a metapopulation—and its distribution undergoes constant change *in detail*. A distribution map does not attempt to represent details at this scale, and it is just a summary, an integration. But despite its flaws, it is one of the most valuable documents about a group for both theoretical and practical purposes, including conservation management. Conservation relies on biogeography in its everyday work, and the next section addresses some implications of the new biogeographic ideas.

Biogeography and Conservation in New Zealand

Description and Interpretation in Conservation

Every year brings a flood of new data in comparative biology, as seen in publications and online databases. Yet new data in itself does not always represent a true advance in knowledge. Cotterill and Foissner (2010) wrote that "A biotic inventory is the inaugural step in knowledge generation … It follows that biotic inventories constitute core discovery processes…." This is true, but an inventory in itself—a simple list—is raw data. It is of little scientific value except for specialists who are able to interpret it for themselves. The development of deeper knowledge about a region and its biota requires analysis and interpretation, not just data. Cotterill and Foissner (2010) were optimistic, and they predicted that "Unprecedented advances in knowledge are set to follow on consummate inventories of biodiversity." But advances in knowledge depend on scientific interpretation—they are not an automatic consequence of gathering information.

Biogeographic Theory and Conservation in New Zealand

The different methods of biogeographic interpretation—dispersal theory and vicariance—have different implications for conservation practice. For example, Cook and Crisp (2005: 747) wrote: "It has been argued that centres of origin, such as source populations, should be given higher conservation priority because such areas have spawned species in the past and may therefore be likely to do so in the future… Clearly this has implications for conservation management." Yet there would be no justification for preserving "centers of origin" deduced from phylogenies if, as suggested here, they are artifacts of analysis.

If the primary aim of conservation were to protect centers of origin, in a dispersal model preserving New Zealand's biota would be of secondary importance, as it is all derived from a center of origin elsewhere. Even groups such as kiwis and tuataras are thought to have come from somewhere outside the country. Dispersal theory suggests that many groups endemic to New Zealand have arrived in recent times from Australia, South America, and other areas. If the groups are nothing more than minor variants and can disperse here again, their conservation is a low priority.

In dispersal theory, the New Zealand flora (McGlone et al., 2001) and the New Zealand avifauna (Trewick and Gibb, 2010) are interpreted as subsets of Australia's biota. If, for example, New Zealand's biota were to be destroyed and Australia's left intact, New Zealand would be repopulated from Australia (by the prevailing winds and currents) and New Zealand–adapted groups similar to the original ones would develop. This would not be predicted to happen in vicariance theory, as New Zealand's biota is not interpreted as a subset of Australia's.

Dispersalists in New Zealand find themselves in the position of having to play down the significance of their groups. For example, Perrie and Brownsey (2007) argued for the importance of recent, long-distance dispersal in the New Zealand fern flora and, in support, they emphasized that "less than half" the New Zealand species are endemic. In contrast with this approach, the biogeographic methods used here interpret *any* endemism in an area as a useful marker indicating deeper structure. In fact, New Zealand fern species overall show the high figure of 47% endemism. In the more diverse genera, this rises to 67% in both *Hymenophyllum* and *Blechnum*, and 71% in the tree fern *Cyathea* (Brownsey and Smith-Dodsworth, 1989).

In another example, dispersal theory proposed that the New Zealand torrentfish, *Cheimmarichthys*, is no more than a blue cod—an inshore, marine species—that "penetrated inland." Instead, molecular studies indicate that the torrentfish is sister to a diverse, worldwide complex (Chapter 12). In vicariance theory, this indicates that the torrentfish is an ancient group of global significance for evolution and that its freshwater habitats are well worth conserving.

The Penguin *Megadyptes* (Spheniscidae): A Case Study

Another example of the different implications that dispersal theory and vicariance theory have for conservation concerns two sister genera of penguins, *Megadyptes* and *Eudyptes* (Chapter 7). Both have breeding localities that overlap in the New Zealand subantarctic islands and western Stewart Island, but

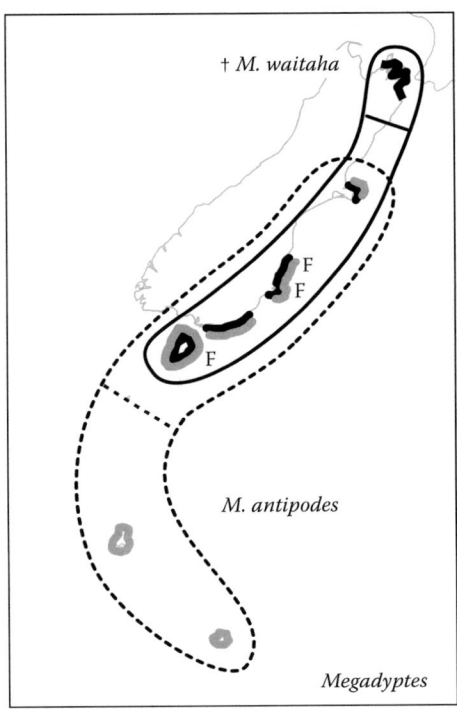

FIGURE 17.1 Distribution of both mainland species in the genus *Megadyptes* (Spheniscidae). *M. antipodes*: extant records, gray; mainland subfossil records, F. *M. waitaha*: subfossil records, black (species extinct by the year 1500). The main genetic breaks within each of the two species are also indicated. (From Boessenkool, A. et al., *Proc. R. Soc. Lond. B*, 276, 815, 2009; Boessenkool, S. et al., *Mol. Ecol.*, 18, 2390, 2009.)

north of here, in the South Island, *Eudyptes* breeds in the west (Solander Island, Fiordland, Westland), while *Megadyptes* breeds in the east (shores of Foveaux Strait north to Banks Peninsula) (Figure 17.1; Robertson et al., 2007). The *Megadyptes/Eudyptes* split indicates a possible break at the Great South Basin and the Titri–Waitati fault, features formed by mid-Cretaceous extension (Chapter 6).

Megadyptes has just one extant species, *M. antipodes*, but a second species, *M. waitaha*, went extinct around 1500 (There is also a fossil species on the Chathams). The map in Boessenkool et al. (2009a, Figure 1bi) shows *M. waitaha* restricted to the South Island and prehistoric specimens of *M. antipodes* restricted to the subantarctic islands. Nevertheless, the text cites three prehistoric specimens of *M. antipodes* from the mainland. The authors wrote that "These specimens probably represent non-breeding vagrants from the Subantarctic [no evidence was given for this]... Therefore, it seems almost certain that the entire extant yellow-eyed penguin population on the South Island is derived from a Subantarctic stock" (p. 819). This inference is not justified, and it is not accepted here.

Boessenkool et al. (2009a: 815) wrote that their studies "reveal that [*M. antipodes*] expanded its range to the New Zealand mainland only in the last few hundred years," and that this occurred after *M. waitaha* was extirpated on the mainland by humans. (Yet the new, mainland population of *M. antipodes* was not extirpated when it invaded, and the bird is 10% larger than *M. waitaha*.) Rawlence et al. (2015) proposed that *M. antipodes* colonized the mainland "probably within a few decades of *M. waitaha*'s extinction...."

Although the prehistoric, mainland specimens of *M. antipodes* mean there is no actual evidence for colonization from the subantarctic islands, Boessenkool et al. (2009b) suggested that "This dramatic expansion... clearly indicates strong potential for long distance dispersal" (pp. 2390–2391). The dispersal is "remarkable" (p. 2398) and "provides clear evidence of the species' ability to cross vast stretches of open ocean" (p. 2397). In fact there is no evidence for any range expansion in *Megadyptes*, apart from local overlap with its western sister group, *Eudyptes* in Auckland, Campbell and Stewart Islands, and local overlap between *M. antipodes* and *M. waitaha*. Breeding localities of *M. waitaha* are not known (bones are recorded from Stewart Island to Marlborough Sounds), but any overlap with *M. antipodes* could be explained by local range expansion *within* the South Island; there is no need to invoke dispersal from the subantarctic islands. Of course, the great ability of penguins to swim long distances is not in question, but the precise endemism of the groups means that this ability in itself is not relevant to the evolution of the distributions.

Boessenkool et al. (2009a: 820) concluded that "the conservation status of South Island *M. antipodes* might be questioned on the basis of these results." Time, effort and large sums are spent on the conservation of this species in the South Island, but there would indeed be little point if the species migrated there just a few hundred years ago. The "recent dispersal" model suggests that if the mainland *M. antipodes* went extinct it would be replaced in a few decades.

On the other hand, a vicariance model for *Megadyptes* instead stresses that subantarctic groups often have outlying populations on the south-east Otago coast. This was exemplified earlier by the subantarctic *Poa* clade mapped in Figure 6.4, and albatrosses, breeding on the New Zealand mainland at a single locality, Otago Peninsula. The biogeographic evidence suggests that the mainland populations of *Megadyptes* are much older than the ages suggested by Boessenkool et al. (2009) and Boessenkool et al. (2009a,b), and there is every reason to conserve the mainland population of this distinctive genus.

Because of the (inferred) recent dispersal of *Megadyptes antipodes* from the subantarctic islands to the South Island, Boessenkool et al. (2009b) hypothesized that the species would constitute a single panmictic population. In fact, they found that "Contrary to our hypothesis... [there is] surprisingly strong evidence for the presence of two distinct genetic clusters" (p. 2396): South Island populations and subantarctic populations. The two clades "differ substantially," and this is consistent with a vicariance history. Again, the evidence conflicts with a "recent dispersal" model and indicates that conserving the mainland *Megadyptes* is important. Groups such as *Megadyptes*, the albatrosses, and the grass *Poa* provide the key to understanding early differentiation between the mainland and the Campbell Plateau.

Conservation and the Globally Basal Endemics

The Modern Synthesis proposed a theory of historical biogeography that was based on fossil-calibrated chronology and a clock-like model of evolution. The theory interpreted the plants and animals of the south-west Pacific as recent, secondary derivatives from the northern hemisphere. Molecular phylogenies now indicate that south-west Pacific groups, such as the New Caledonian angiosperm *Amborella*, the *Celmisia* group, the New Zealand torrentfish, the New Zealand wrens (Acanthisittidae), and many others have diverse, global sister groups. The "globally basal" endemics cannot be derived from elsewhere without invoking additional, ad hoc hypotheses, and are best explained as the result of vicariance in global groups (Chapter 4). The globally basal groups indicate that the Tasman region represents a parallel universe inhabited by sister angiosperms, sister reptiles, sister frogs, sister parrots, and sister passerines, that together represent at least one important phase of early differentiation in worldwide groups.

The local or regional Tasman endemics that have global sister groups, along with their habitat, should be a top priority for conservation. Some of the groups, such as snowgrass (*Chionochloa*), tuatara (*Sphenodon*), and wattlebirds (Callaeidae), already have a high profile in conservation work, but others, such as athoracophorid slugs, *Craterostigmus* centipedes, the torrentfish, and New Zealand wrens, do not. In the two extant New Zealand wrens, for example, the rifleman *Acanthisitta* shows suggestive patterns of geographic variation through the South Island in its morphology (Oliver, 1955). Given the bird's phylogenetic position, detailed molecular studies of the variation would be of special interest, and the geographic patterns of variation should be conserved, whether they are clinal or include distinct clades.

Many of the globally basal endemics in the Tasman region have not even been mapped. Detailed mapping of taxa such as *Teucridium* (Lamiaceae) and molecular surveys of groups such as the rifleman *Acanthisitta* would be of more value for science and conservation than further clock studies.

Conservation and Translocation

New Zealand includes many centers of endemism, and most of the endemics are plants and invertebrates. Because of this, choosing sites where rare and threatened vertebrates can be translocated outside their original range can be problematic. Translocation of vertebrates in New Zealand has a long history. For example, tuataras were translocated to the Kermadec Islands at some time between 1870 and 1951, although they did not establish there (Cree, 2014: 454).

In the modern era, Towns et al. (1990) and Gibbs (1990b) have warned about translocating vertebrates onto islands, but the issues have been neglected. Translocations to Codfish Island, off western Stewart Island, provide an illustrative example. Four birds have been introduced to the island as a conservation measure: Campbell Island teal (*Anas nesiotis*), Snares Island snipe (*Coenocorypha huegeli*), kakapo (*Strigops habroptilus*), and yellowhead (*Mohoua ochrocephala*) (Scofield and Stephenson, 2013; Birdlife International, 2016). None of these birds were recorded on Codfish Island before their introduction. (Whitehead, 2007, wrote that kakapo were first transferred to Codfish Island in 1987 from Stewart Island and "There is no evidence that kakapo previously inhabited Codfish Island" [D. Eason, pers. comm.].)

The kakapo is vegetarian, while yellowheads are vigorous predators on litter invertebrates; a shower of dead leaves, bark and lichen falls to the forest floor beneath a feeding flock (Worthy and Holdaway, 2002: 429). This level of activity must have noticeable effects on the ecology, and the biota of Codfish Island is of special biogeographic significance. Endemics include the harvestman *Pantopsalis phocator* (Taylor, 2004), the staphylinid beetle *Agnosthaetus stenomastax* (Clarke, 2011), and the fernbird *Bowdleria punctata wilsoni* ("probably the most distinct" subspecies in the species; Scofield and Stephenson, 2013).

Other groups, such as the weevil *Microcryptorhynchus suillus* (Chapter 6) and the land snail *Alsolemia* (Chapter 7), are present on Codfish and other islands, but not on Stewart Island itself. A similar pattern, skirting Stewart Island's west coast, is displayed in the beetle genus *Stenosagola* (Staphylinidae) (Park and Carlton, 2013). This New Zealand endemic is widespread through North and South Islands with 20 species, but the only members in the Stewart Island area are *S. stewartensis* of Big South Cape Island (off South Cape) and *S. nunni* of Big South Cape Island, Codfish Island, Fiordland, and Takitimu Mountains. The genus is not on mainland Stewart Island. This pattern suggests horstian connections, and it highlights the significance of the Codfish Island invertebrates for understanding evolution in the Stewart Island–Fiordland region.

Translocations of large invertebrates can also be problematic. In northern New Zealand, all known records of the endangered Mercury Islands weta (*Motuweta isolata*) are from Middle Island (Figure 5.6), but conservation programs have now introduced it to two of the other Mercury Islands (Watts et al., 2008). In the same way, the giant weta *Deinacrida rugosa* from the Cook Strait islands has been introduced to localities where it was unknown before, such as Maud Island in the Marlborough Sounds and Matiu-Somes Island in Wellington Harbour. The giant wetas are the largest insects in New Zealand and are omnivorous, so they must have a significant effect on the smaller invertebrates. These have often been neglected, but many have high levels of local endemism and are much more informative about the history of a region than larger and better-known, but less diverse, organisms.

Ensuring That the Main Aims of Conservation Are Met

In general, dispersal theory concludes that the taxa in a region have evolved long after the major Earth history events in the region took place. Different groups have differentiated at different times and have different histories, and so there is no underlying pattern to conserve. In contrast, panbiogeographic work suggests that Earth and life have evolved together. This means that the focus in conservation should not be on individual taxa or even habitats, but instead on the fundamental phylogenetic and biogeographic structure of a region, the overall network of nodes. This can then guide the selection of taxa and habitats that should be protected. A group with one species ranging from Fiordland to Mount Arthur and another ranging from Mount Arthur to Auckland can be conserved with a single reserve on Mount Arthur. The best way of ensuring conservation of the main aspects of diversity—both known and unknown—is to protect the main biogeographic nodes.

Glossary of Geological Terms

Allochthonous: Not formed in place, formed elsewhere.

Angular unconformity: An *unconformity* in which horizontal strata are deposited on tilted and eroded layers, producing an angular discordance between the two.

Antiform: An upward fold.

Arc: A chain of volcanoes at a convergent plate margin, forming either a continental arc or an island arc.

A-type granite: A type of granite with high iron and zirconium content, emplaced in an intraplate setting (contrast with *I-type granite*).

Autochthonous: Formed in place.

Backarc: Behind an arc, on the side away from the trench.

Backarc basin: An extensional basin formed by spreading behind an *arc*. The formation of the basin may separate an arc from a mainland (e.g., Japan from mainland Asia) or may split an island arc lengthwise along its axis (e.g., the Lau arc separating from the Tonga arc).

Bivergent: Divergent in two places.

Dextral (right-lateral) displacement: Relative movement in which the block on the other side of the fault from the observer moves to the right (contrast with *sinistral [left-lateral] displacement*).

Dextral (right-lateral) fault: A *strike-slip fault* with *dextral (right-lateral) displacement* (contrast with *sinistral fault*).

Dip: The dip of a tilted bed, fault surface, or other planar feature is the steepest angle of descent of the feature relative to a horizontal plane. The dip direction is at 90° to the strike.

Dip-slip fault: A fault where the relative movement (slip) on the fault plane parallels the *dip* of the fault (contrast with *strike-slip fault*).

Footwall: If a *dip-slip fault* plane is not exactly vertical, the lower block is the *footwall* (contrast with *hanging wall*).

Forearc: In front of an *arc* (between the arc and its trench).

Graben: An elongate depression between two adjacent *normal faults* that dip toward each other (contrast with *horst*) (cf. *half-graben*).

Half-graben: A basin bounded by a normal fault on one side, unlike a *graben*, which is bordered on both sides by two, parallel faults.

Hanging wall: If a dip-slip fault plane is not exactly vertical, the upper block is the hanging wall ("hanging over" the fault) (contrast with *footwall*).

Horst: A high, elongate footwall block formed between two normal faults that dip away from each other—the land on either side of the horst has dropped (contrast with *graben*).

Island arc (or arc): A chain of volcanic islands formed in oceanic crust along the convergent plate margin.

I-type granite: A type of granite with high sodium and strontium content, emplaced in a subduction zone setting (contrast with *A-type granite*).

Metamorphic core complexes: Domes of metamorphic rock formed in an orogen as the result of major continental extension. The middle and lower crusts are dragged out from beneath the fracturing, extending the upper crust. The effect on the topography is a rapid, localized uplift in the region of extension.

Metamorphic facies: Groups of mineral assemblages found in rocks metamorphosed under particular pressure–temperature regimes.

Metamorphic grades: Another scale used to measure the degree of metamorphism (cf. *metamorphic facies*). With increasing metamorphism, shale, a sedimentary rock formed from mud, is converted into slate, phyllite, schist, and gneiss. This sequence has been described using metamorphic grades based on key minerals. Metamorphic grades in the New Zealand Torlesse and Caples terranes range from chlorite, the lowest, to biotite and garnet.

Mylonite: Deformed rocks formed in ductile fault zones.

Nappe: A sheet-like body of rock that is *allochthonous* and has been moved by more than 5 km from its original position.

Normal fault: A *dip-slip fault* in which the *hanging wall* moves downward, relative to the *footwall*. Normal faults develop when the crust is extended (contrast with *reverse fault*).

Obduction: The overthrust of continental crust by rocks of the oceanic crust or mantle.

Ophiolite: A suite of rocks from oceanic crust and the mantle that has been uplifted and often emplaced onto continental crust.

Orogen (orogenic belt): A mountain belt formed by crustal contraction, with folding and faulting.

Orogeny (orogenesis): The formation of an orogen.

Peridotite: An intrusive *ultramafic* rock formed in the mantle and found in *ophiolite* suites.

Reverse fault: A dip-slip fault in which the hanging wall moves up, relative to the *footwall*. Reverse faults develop where the crust is shortened or contracted (contrast with *normal fault*).

Serpentinite: A rock formed by the weathering of *ultramafic* rock.

Shear: Deformation in which parallel internal surfaces slide past one another. Shear can occur within brittle, brittle-ductile, and ductile rocks and can give rise to faults and folds.

Sinistral (left-lateral) displacement: Relative movement in which the block on the other side of the fault from the observer moves to the left (contrast with *dextral [right-lateral] displacement*).

Sinistral (left-lateral) fault: A strike-slip fault with *sinistral (left-lateral) displacement* (contrast with *dextral fault*).

Slab roll-back (trench roll-back): The retrograde movement of a *subduction zone* toward the subducting plate. The descending slab and its trench fall or roll back away from the arc instead of subducting forward and downward. This can cause extension in the crust above.

Strike: The strike line of a bed, fault, or other planar feature is a line representing the intersection of the feature with a horizontal plane.

Strike-slip fault (= wrench fault): A fault in which the slip is parallel to the strike of the fault (in contrast with a *dip-slip fault*). The fault surface is usually more or less vertical (cf. *transcurrent fault*).

Subduction: The sinking of one tectonic plate under another at a convergent plate margin.

Syncline: A fold in which the younger layers lie closer to the central core of the structure.

Synform: A downward fold.

Terrane (= tectonostratigraphic terrane): A fault-bounded crustal block with its own, independent history.

Transcurrent fault: A strike-slip fault that is confined to the crust (cf. *transform fault*).

Transform fault: A strike-slip fault that cuts the lithosphere, forming a plate boundary (cf. *transcurrent fault*).

Ultramafic rocks: Rocks containing <45% SiO_2, e.g., *peridotite*.

Unconformity: A surface dividing superposed strata of different ages. (See also *angular unconformity*.)

References

Adams, C.J. 2010. Lost Terranes of Zealandia: Possible development of late Paleozoic and early Mesozoic sedimentary basins at the southwest Pacific margin of Gondwanaland, and their destination as terranes in southern South America. *Andean Geology* 37: 442–454.

Adams, C.J., H.J. Campbell, I.J. Graham, and N. Mortimer. 1998. Torlesse, Waipapa and Caples suspect terranes of New Zealand: Integrated studies of their geological history in relation to neighbouring terranes. *Episodes* 21: 235–239.

Adams, C.J., H.J. Campbell, and W.L. Griffin. 2007. Provenance comparisons of Permian to Jurassic tectonostratigraphic terranes in New Zealand: Perspectives from detrital zircon age patterns. *Geological Magazine* 144: 701–729.

Adams, C.J., D. Cluzel, and W.L. Griffin. 2009a. Detrital zircon ages and geochemistry of sedimentary rocks in basement Mesozoic terranes and their cover rocks in New Caledonia, and provenances at the Eastern Gondwanaland margin. *Australian Journal of Earth Sciences* 56: 1023–1047.

Adams, C.J., N. Mortimer, H.J. Campbell, and W.L. Griffin. 2009b. Age and isotopic characterisation of metasedimentary rocks from the Torlesse Supergroup and Waipapa Group in the central North Island, New Zealand. *New Zealand Journal of Geology and Geophysics* 52: 149–170.

Adams, K. 2013. Genomic clues to the ancestral flowering plant. *Science* 342: 1456.

Adamson, R.S. 1912. On the comparative anatomy of the leaves of certain species of *Veronica*. *Journal of the Linnean Society, Botany* 40: 247–274.

Adriaens, D. and A. Herrel. 2009. Functional consequences of extreme morphologies in the craniate trophic system. *Physiological and Biochemical Zoology* 82: 1–6.

Aggerbeck, M., J. Fjeldså, L. Christidis, P.-H. Fabre, and K.A. Jønsson. 2014. Resolving deep lineage divergences in core corvoid passerine birds supports a proto-Papuan island origin. *Molecular Phylogenetics and Evolution* 70: 272–285.

Aguilar, A. 2009. Fin whale (*Balaenoptera physalus*). In *Encyclopedia of Marine Mammals*, 2nd edn., eds. W.F. Perrin, B. Würsig, and J.G.M. Thewissen, pp. 433–436. Amsterdam, the Netherlands: Elsevier.

Ai, H.-A., J.M. Stock, R. Clayton, and B. Luyendyk. 2008. Vertical tectonics of the High Plateau region, Manihiki Plateau, Western Pacific, from seismic stratigraphy. *Marine Geophysical Research* 29: 13–26.

Aidala, Z., N. Chong, M.G. Anderson et al. 2013. Phylogenetic relationships of the genus *Mohoua*, endemic hosts of New Zealand's obligate brood parasitic Long-tailed Cuckoo (*Eudynamys taitensis*). *Journal of Ornithology* 154: 1127–1133.

Aitchison, J.C., G.L. Clarke, S. Meffre, and D. Cluzel. 1995. Eocene arc-continent collision in New Caledonia and implications for regional southwest Pacific tectonic evolution. *Geology* 23: 161–164.

Albach, D.C. and M.W. Chase. 2004. Incongruence in Veroniceae (Plantaginaceae): Evidence from two plastid and a nuclear ribosomal DNA region. *Molecular Phylogenetics and Evolution* 32: 183–197.

Albach, D.C., M.M. Martínez-Ortega, M.A. Fischer, and M.W. Chase. 2004. A new classification of the tribe Veroniceae: Problems and a possible solution. *Taxon* 53: 429–452.

Albach, D.C. and H.A. Meudt. 2010. Phylogeny of *Veronica* in the Southern and Northern Hemispheres based on plastid, nuclear ribosomal and nuclear low-copy DNA. *Molecular Phylogenetics and Evolution* 54: 457–471.

Aldridge, M. and R. Dingwall. 2003. Teleology on television? Implicit models of evolution in broadcast wildlife and nature programmes. *European Journal of Communication* 18: 435–453.

Ali, J.R. and M. Huber. 2010. Mammalian biodiversity on Madagascar controlled by ocean currents. *Nature* 463: 653–656.

Allan, H.H. 1961. *Flora of New Zealand*, Vol. 1. Wellington, New Zealand: Government Printer.

Allibone, A.H., R. Jongens, J.M. Scott et al. 2009a. Plutonic rocks of the Median Batholith in eastern and central Fiordland, New Zealand: Field relations, geochemistry, correlation, and nomenclature. *New Zealand Journal of Geology and Geophysics* 52: 101–148.

Allibone, A.H., L.A. Milan, N.R. Daczko, and I.M. Turnbull. 2009b. Granulite facies thermal aureoles and metastable amphibolite facies assemblages adjacent to the Western Fiordland Orthogneiss in southwest Fiordland, New Zealand. *Journal of Metamorphic Geology* 27: 349–369.

Allibone, A.H. and A.J. Tulloch. 2004. Geology of the plutonic basement rocks of Stewart Island, New Zealand. *New Zealand Journal of Geology and Geophysics* 47: 233–256.

Allibone, A.H. and A.J. Tulloch. 2008. Early Cretaceous dextral transpressional deformation within the Median Batholith, Stewart Island, New Zealand. *New Zealand Journal of Geology and Geophysics* 51: 115–134.

Allibone, A.H., I.M. Turnbull, A.J. Tulloch, and A.F. Cooper. 2007. Plutonic rocks of the Median Batholith in southwest Fiordland, New Zealand: Field relations, geochemistry, and correlation. *New Zealand Journal of Geology and Geophysics* 50: 283–314.

Allwood, J., D. Gleeson, G. Mayer, S. Daniels, J.R. Beggs, and T.R. Buckley. 2010. Support for vicariant origins of the New Zealand Onychophora. *Journal of Biogeography* 37: 669–681.

Almeida, E.A.B., M.R. Pie, S.G. Brady, and B.N. Danforth. 2012. Biogeography and diversification of colletid bees (Hymenoptera: Colletidae): Emerging patterns from the southern end of the world. *Journal of Biogeography* 39: 526–544.

Al-Shehbaz, I.A. 2012. A generic and tribal synopsis of the Brassicaceae (Cruciferae). *Taxon* 61: 931–954.

Alström, P., S. Fregin, J.A. Norman, P.G. Ericson, L. Christidis, and U. Olsson. 2011. Multilocus analysis of a taxonomically densely sampled dataset reveal extensive non-monophyly in the avian family Locustellidae. *Molecular Phylogenetics and Evolution* 58: 513–526.

Anacker, B.L. and S.Y. Strauss. 2014. The geography and ecology of plant speciation: Range overlap and niche divergence in sister species. *Proceedings of the Royal Society B*, 281: 20132980.

Andersen, K. 1912. *A Catalogue of the Megachiroptera in the British Museum.* London, UK: British Museum (Natural History).

Anderson, C.L., J.H.E. Rova, and L. Andersson. 2001. Molecular phylogeny of the tribe Anthospermeae (Rubiaceae): Systematic and biogeographic implications. *Australian Systematic Botany* 14: 231–244.

Ansmann, I.C., G.J. Parra, J.M. Lanyon, and J.M. Seddon. 2012. Fine-scale genetic population structure in a mobile marine mammal: Inshore bottlenose dolphins in Moreton Bay, Australia. *Molecular Ecology* 21: 4472–4485.

Appelhans, M.S., W.L. Wagner, and K.R. Wood. 2014a. *Melicope balgooyi* Appelhans, W.L. Wagner & K.R. Wood, a new species and new record in *Melicope* section *Melicope* (Rutaceae) for the Austral Islands. *PhytoKeys* 39: 77–86.

Appelhans, M.S., J. Wen, and W.L. Wagner. 2014b. A molecular phylogeny of *Acronychia, Euodia, Melicope* and relatives (Rutaceae) reveals polyphyletic genera and key innovations for species richness. *Molecular Phylogenetics and Evolution* 79: 54–68.

Apte, S., P.J. Smith, and G.P. Wallis. 2007. Mitochondrial phylogeography of New Zealand freshwater crayfishes, *Paranephrops* spp. *Molecular Ecology* 16: 1897–1908.

Araújo, M.B. and A. Guisan. 2006. Five (or so) challenges for species distribution modelling. *Journal of Biogeography* 33: 1677–1688.

Araújo, M.B. and A.T. Peterson. 2012. Uses and misuses of bioclimatic envelope modelling. *Ecology* 93: 1527–1539.

Arensburger, P., T.R. Buckley, C. Simon, M. Moulds, and K. Holsinger. 2004a. Biogeography and phylogeny of the New Zealand cicada genera (Hemiptera: Cicadidae) based on nuclear and mitochondrial DNA data. *Journal of Biogeography* 31: 557–569.

Arensburger, P., C. Simon, and K. Holsinger. 2004b. Evolution and phylogeny of the New Zealand cicada genus *Kikihia* Dugdale (Homoptera: Auchenorrhyncha: Cicadidae) with special reference to the origin of the Kermadec and Norfolk Islands species. *Journal of Biogeography* 31: 1769–1783.

Arikawa, K., E. Eguchi, A. Yoshida, and K. Aoki. 1980. Multiple extraocular photoreceptive areas on genitalia of butterfly *Papilio xuthus*. *Nature* 288: 700–702.

Arnold, B.C. 1959. The structure of spines of *Hymenanthera alpina*. *Phytomorphology* 9: 367–371.

Arnold, E.N. and G. Poinar. 2008. A 100 million year old gecko with sophisticated adhesive toe pads, preserved in amber from Myanmar. *Zootaxa* 1847: 62–68.

Arthur, W. 2002. The emerging conceptual framework of evolutionary developmental biology. *Nature* 415: 757–764.

Ashwell, K.W. and R.P. Scofield. 2008. Big birds and their brains: Paleoneurology of the New Zealand moa. *Brain, Behavior, and Evolution* 71: 151–166.

Atkinson, I.A. and R.M. Greenwood. 1989. Relationships between moas and plants. *New Zealand Journal of Ecology* 12: 67–96.

Aubréville, A., J.-F. Leroy, Ph. Morat, and H.S. MacKee (eds.). 1967–present. *Flore de la Nouvelle-Calédonie et dépendances.* Paris, France: Muséum National d'Histoire Naturelle.

Austin, J.J. and E.N. Arnold. 2006. Using ancient and recent DNA to explore relationships of extinct and endangered *Leiolopisma* skinks (Reptilia: Scincidae) in the Mascarene islands. *Molecular Phylogenetics and Evolution* 39: 503–511.

Austin, J.J., V. Bretagnolle, and E. Pasquet. 2004. A global molecular phylogeny of the small *Puffinus* shearwaters and implications for systematics of the Little-Audubon's Shearwater complex. *Auk* 121: 847–864.

Avise, J.C. 1992. Molecular population structure and the biogeographic history of a regional fauna: A case history with lessons for conservation biology. *Oikos* 63: 62–76.

Avise, J.C. 2000. *Phylogeography: The History and Formation of Species.* Cambridge, MA: Harvard University Press.

Avise, J.C. 2006. *Evolutionary Pathways in Nature: A Phylogenetic Approach.* Cambridge, UK: Cambridge University Press.

Axelrod, D.I. 1986. Cenozoic history of some western American pines. *Annals of the Missouri Botanical Garden* 73: 565–641.

Azuma, Y., Y. Kumazawa, M. Miya, K. Mabuchi, and M. Nishida. 2008. Mitogenomic evaluation of the historical biogeography of cichlids: Toward reliable dating of teleostean divergences. *BMC Evolutionary Biology* 8(215): 1–13.

Backström, N., Q. Zhang, and S.V. Edwards. 2013. Evidence from a House Finch (*Haemorhous mexicanus*) spleen transcriptome for adaptive evolution and biased gene conversion in passerine birds. *Molecular Biology and Evolution* 30: 1046–1050.

Bacon, F. 1605/1966. *Of the Advancement of Learning* (Second Book). Oxford, UK: Oxford University Press.

Bahl, J., M.C.Y. Lau, G.J.D. Smith et al. 2011. Ancient origins determine global biogeography of hot and cold desert cyanobacteria. *Nature Communications* 2(163): 1–6.

Baker, A.J., C.H. Daugherty, R. Colbourne, and J.L. McLennan. 1995. Flightless brown kiwis of New Zealand possess extremely subdivided population structure and cryptic species like small mammals. *Proceedings of the National Academy of Sciences of USA* 92: 8254–8258.

Baker, A.J., S.L. Pereira, O.P. Haddrath, and K.-A. Edge. 2006. Multiple gene evidence for expansion of extant penguins out of Antarctica due to global cooling. *Proceedings of the Royal Society, London B* 273: 11–17.

Baker, C.H., G.C. Graham, K.D. Scott, S.L. Cameron, D.K. Yeates, and D.J. Merritt. 2008. Distribution and phylogenetic relationships of Australian glow-worms *Arachnocampa* (Diptera, Keroplatidae). *Molecular Phylogenetics and Evolution* 48: 506–514.

Baker, I.A., J.A. Gamble, and I.J. Graham. 1994. The age, geology, and geochemistry of the Tapuaenuku Igneous Complex, Marlborough, New Zealand. *New Zealand Journal of Geology and Geophysics* 373: 249–268.

Baker, J. and D. Seward. 1996. Timing of Cretaceous extension and Miocene compression in northeast South Island, New Zealand: Constraints from Rb-Sr and fission-track dating of an igneous pluton. *Tectonics* 15: 976–983.

Baker, R.J., O.R. Bininda-Emonds, H. Mantilla-Meluk, C.A. Porter, and R.A. Van Den Bussche. 2010. Molecular timescale of diversification of feeding strategy and morphology in New World leaf-nosed bats (Phyllostomidae): A phylogenetic perspective. In *Evolutionary History of Bats: Fossils, Molecules and Morphology*, eds. G. Gunnell and N. Simmons, pp. 385–409. Cambridge, UK: Cambridge University Press.

Baldwin, J.M. 1896. A new factor in evolution. *American Naturalist* 30: 441–451.

Balen, S. van. 2008. Zosteropidae. In *Handbook of the Birds of the World*, Vol. 13, eds. J. del Hoyo, A. Elliott, and D.A. Christie, pp. 402–485. Barcelona, Spain: Lynx Edicions.

Ball, O.J.-P., P.T. Whaley, A.M. Booth, and S. Hartley. 2013. Habitat associations and detectability of the endemic Te Paki ground beetle *Mecodema tenaki* (Coleoptera: Carabidae). *New Zealand Journal of Ecology* 37: 84–94.

Bannister, P., J.M. Bannister, and D.J. Blanchon. 2004. Distribution, habitat, and relation to climatic factors of the lichen genus *Ramalina* in New Zealand. *New Zealand Journal of Botany* 42: 121–138.

Baptista, L.F., P.W. Trail, and H.M. Horblitt. 1997. Columbidae. In *Handbook of the Birds of the World*, Vol. 4, eds. J. del Hoyo, A. Elliott, and J. Sargatal, pp. 60–245. Barcelona, Spain: Lynx Edicions.

Barboza, C.A.M., R.B. Moura, A.M. Lanna, T. Oackes, and L.S. Campos. 2011. Echinoderms as clues to Antarctic–South American connectivity. *Oecologia Australis* 15: 86–110.

Barker, F.K., K.J. Burns, J. Klicka, S.M. Lanyon, and I.J. Lovette. 2013. Going to extremes: Contrasting rates of diversification in a recent radiation of New World passerine birds. *Systematic Biology* 62: 298–320.

Barker, G.M. 2005. The character of the New Zealand land snail fauna and communities: Some evolutionary and ecological perspectives. *Records of the Western Australian Museum Supplement* 68: 53–102.

Barker, N.P., P.H. Weston, F. Rutschmann, and H. Sauquet. 2007. Molecular dating of the "Gondwanan" plant family Proteaceae is only partially congruent with the timing of the break-up of Gondwana. *Journal of Biogeography* 34: 2012–2027.

Barker, W.R. 1992. New Australasian species of *Peplidium* and *Glossostigma* (Scrophulariaceae). *Journal of the Adelaide Botanic Garden* 15: 71–74.

Barlow, G.W. 2000. *The Cichlid Fishes: Nature's Grand Experiment in Evolution*. Cambridge, MA: Perseus Books.

Barnes, D.K.A. and H.J. Griffiths. 2007. Biodiversity and biogeography of southern temperate and polar bryozoans. *Global Ecology and Biogeography* 17: 84–99.

Barraclough, T.G. and A.P. Vogler. 2000. Detecting the geographical pattern of speciation from species-level phylogenies. *American Naturalist* 155: 419–434.

Barratt, B.I.P. 2007. Conservation status of *Prodontria* (Coleoptera: Scarabaeidae) species in New Zealand. *Journal of Insect Conservation* 11: 19–27.

Barratt, B.I.P. and G. Kuschel. 1996. Broad-nosed weevils (Curculionidae: Brachycerinae: Entimini) of the Lammermoor and Rock and Pillar Ranges in Otago, with descriptions of four new species of *Irenimus*. *New Zealand Journal of Zoology* 23: 359–374.

Barreda, V.D., L. Palazzesi1, L. Katinas et al. 2012. An extinct Eocene taxon of the daisy family (Asteraceae): Evolutionary, ecological and biogeographical implications. *Annals of Botany* 109: 127–134.

Barreda, V.D., L. Palazzesi, M.C. Tellería, E.B. Olivero, J.I. Rainee, and F. Forest. 2015. Early evolution of the angiosperm clade Asteraceae in the Cretaceous of Antarctica. *Proceedings of the National Academy of Sciences of USA* 112: 10989–10994.

Barth, N.C. 2014. The Cascade rock avalanche: Implications of a very large Alpine Fault-triggered failure, New Zealand. *Landslides* 11: 327–341.

Bartish, I.V., A. Antonelli, J.E. Richardson, and U. Swenson. 2011. Vicariance or long-distance dispersal: Historical biogeography of the pantropical subfamily Chrysophylloideae (Sapotaceae). *Journal of Biogeography* 38: 177–190.

Bartle, J.A. and P.M. Sagar. 1987. Intraspecific variation in the New Zealand Bellbird. *Notornis* 34: 253–306.

Baselga, A., J.M. Lobo, J.-C. Svenning, and M.B. Araújo. 2012. Global patterns in the shape of species geographical ranges reveal range determinants. *Journal of Biogeography* 39: 760–771.

Bassett, K.N., D.E. Lee, D.C. Mildenhall, C.S. Nelson, and C.M. Reid (eds.). 2014. Oligocene paleogeography of New Zealand: Maximum marine transgression. *New Zealand Journal of Geology and Geophysics* 57: 107–269.

Baum, D.A., K.J. Sytsma, and P.C. Hoch. 1994. A phylogenetic analysis of *Epilobium* (Onagraceae) based on nuclear ribosomal DNA sequences. *Systematic Botany* 19: 363–388.

Baylis, G.T.S. 1950. Root system of the New Zealand mangrove. *Transactions of the Royal Society of New Zealand* 78: 509–514.

Bayly, M. and A. Kellow. 2006. *An Illustrated Guide to New Zealand Hebes*. Wellington, New Zealand: Te Papa Press.

Beanland, S. and S.A. Barrow-Hurlbert. 1988. The Nevis-Cardrona Fault System, Central Otago, New Zealand: Late Quaternary tectonics and structural development. *New Zealand Journal of Geology and Geophysics* 31: 337–352.

Beauchamp, A.J. 1989. Panbiogeography and rails of the genus *Gallirallus*. *New Zealand Journal of Zoology* 16: 763–772.

Beaulieu, J.M., D.C Tank, and M.J. Donoghue. 2013. A Southern Hemisphere origin for campanulid angiosperms, with traces of the break-up of Gondwana. *BMC Evolutionary Biology* 13(80): 1–17.

Beauverd, G. 1910. Contribution à l'étude des Composées. Suite IV. Recherches sur le tribu des Gnaphaliées. *Bulletin de la Société botanique de Genève Séries 2* 2: 207–252.

Bednarek-Ochyra, H. and R. Ochyra. 2011. *Bucklandiella angustissima* sp. nov. (Grimmiaceae), a new austral amphipacific species with the smallest capsules and the shortest setae in the genus. *Cryptogamie, Bryologie* 32: 13–27.

Beever, J.E. 2014. Fissidentaceae. In *Flora of New Zealand—Mosses*, Fascicle 8, eds. P.B. Heenan, I. Breitwieser, and A.D. Wilton. Lincoln, New Zealand: Manaaki Whenua Press.

Beggs, J.M., F.C. Ghisetti, and A.J. Tulloch. 2008. Basin and petroleum systems analysis of the West Coast region, South Island, New Zealand. *PESA Eastern Australasian Basins Symposium III Sydney*, September 14–17, 2008. Perth, Western Australia, Australia: Petroleum Exploration Society of Australia, pp. 1–7.

Beier, M. 1976. The pseudoscorpions of New Zealand, Norfolk, and Lord Howe. *New Zealand Journal of Zoology* 3: 199–246.

Beilstein, M.A., N.S. Nagalingum, M.D. Clements, S.R. Manchester, and S. Mathews. 2010. Dated molecular phylogenies indicate a Miocene origin for *Arabidopsis thaliana*. *Proceedings of the National Academy of Sciences of USA* 107: 18724–18728.

Bell, T.P. and G.B. Patterson. 2008. A rare alpine skink *Oligosoma pikitanga* n. sp. (Reptilia: Scincidae) from Llawrenny Peaks, Fiordland, New Zealand. *Zootaxa* 1882: 57–68.

Bentham, G. 1873. Notes on the classification, history, and geographical distribution of Compositae. *Journal of the Linnean Society of London, Botany* 13: 335–577.

Berbee, M.L. and J.W. Taylor. 2010. Dating the molecular clock in fungi—How close are we? *Fungal Biology Reviews* 24: 1–16.

Berry, J.A. 1999. Revision of the New Zealand endemic eulophid genus *Zealachertus* Bouček (Hymenoptera: Chalcidoidea). *Invertebrate Taxonomy* 13: 883–915.

Berry, J.A. 2007. Alysiinae (Insecta: Hymenoptera: Braconidae). *Fauna of New Zealand* 58: 1–89.

Berry, O. and D.M. Gleeson. 2005. Distinguishing historical fragmentation from a recent population decline—Shrinking or pre-shrunk skink from New Zealand? *Biological Conservation* 123: 197–210.

Berry, P.E., W.J. Hahn, K.J. Sytsma, J.C. Hall, and A. Mast. 2004. Phylogenetic relationships and biogeography of *Fuchsia* (Onagraceae) based on noncoding nuclear and chloroplast DNA data. *American Journal of Botany* 91: 601–614.

Berry, R.J. 1982. *Neo-Darwinism*. London, UK: Edward Arnold.

Betancur-R, R. 2009. Molecular phylogenetics and evolutionary history of ariid catfishes revisited: A comprehensive sampling. *BMC Evolutionary Biology* 9(175): 1–18.

Beu, A.G., B.A. Marshall, and M.B. Reay. 2014. Mid-Cretaceous (Albian-Cenomanian) freshwater Mollusca from the Clarence Valley, Marlborough, New Zealand, and their biogeographical significance. *Cretaceous Research* 49: 134–151.

Bewick, A.J., F.J.J. Chain, J. Heled, and B.J. Evans. 2012. The pipid root. *Systematic Biology* 61: 913–926.

Bialas, R.W., W.R. Buck, and M. Studinger. 2007. Plateau collapse model for the Transantarctic Mountains–West Antarctic Rift System: Insights from numerical experiments. *Geology* 35: 687–690.

Bice, K.L., B.T. Huber, and R.D. Norris. 2003. Extreme polar warmth during the Cretaceous greenhouse? Paradox of the late Turonian $\delta^{18}O$ record at Deep Sea Drilling Project Site 511. *Paleoceanography* 18(2): 1031.

Bickel, D.J. 1991. Sciapodinae, Medeterinae (Insecta: Diptera) with a generic review of the Dolichopodidae. *Fauna of New Zealand* 23: 1–78.

Bidegaray-Batista, L. and M.A. Arnedo. 2011. Gone with the plate: The opening of the Western Mediterranean basin drove the diversification of ground-dweller spiders. *BMC Evolutionary Biology* 11(317): 1–15.

Biffin, E., R.S. Hill, and A.J. Lowe. 2010. Did Kauri (*Agathis*: Araucariaceae) really survive the Oligocene drowning of New Zealand? *Systematic Biology* 59: 594–602.

Birch, J.L., D.J. Cantrill, N.G. Walsh, and D.J. Murphy. 2014. Phylogenetic investigation and divergence dating of *Poa* (Poaceae, tribe Poeae) in the Australasian region. *Botanical Journal of the Linnean Society* 175: 523–552.

Birch, J.L. and S.C. Keeley. 2013. Dispersal pathways across the Pacific: The historical biogeography of *Astelia* s.l. (Asteliaceae, Asparagales). *Journal of Biogeography* 40: 1914–1927.

Birch, J.L., S.C. Keeley, and C.W. Morden. 2012. Molecular phylogeny and dating of Asteliaceae (Asparagales): *Astelia* s.l. evolution provides insight into the Oligocene history of New Zealand. *Molecular Phylogenetics and Evolution* 65: 102–115.

Bird, C.E., B.S. Holland, B.W. Bowen, and R.J. Toonen. 2011. Diversification of sympatric broadcast-spawning limpets (*Cellana* spp.) within the Hawaiian archipelago. *Molecular Ecology* 20: 2128–2141.

Birdlife International. 2016. Data zone: species. http://www.birdlife.org/datazone/species.

Bishop, D.G. 1994. Extent and regional deformation of the Otago peneplain. Report No. 94/1, 10pp. Lower Hutt, New Zealand: Institute of Geological and Nuclear Sciences.

Bishop, D.G., M.G. Laird, and D.C. Mildenhall. 1976. Stratigraphy and depositional environment of the Kyeburn Formation (Cretaceous), a wedge of coarse terrestrial sediments in Central Otago. *Journal of the Royal Society of New Zealand* 6: 55–71.

Bishop, D.G. and I.M. Turnbull. 1996. Geology of the Dunedin Area. Institute of Geological and Nuclear Sciences 1: 250 000 Geological Map 21. Lower Hutt, New Zealand: New Zealand Institute of Geological & Nuclear Sciences.

Blanchon, D.J., B.G. Murray, and J.E. Braggins. 2002. A taxonomic revision of *Libertia* (Iridaceae) in New Zealand. *New Zealand Journal of Botany* 40: 437–456.

Bleeker, W., A. Franzke, K. Pollmann, A.H.D. Brown, and H. Hurka. 2002a. Phylogeny and biogeography of the Southern Hemisphere high-mountain *Cardamine* species (Brassicaceae). *Australian Systematic Botany* 15: 575–581.

Bleeker, W., C. Weber-Sparenberg, and H. Hurka. 2002b. Chloroplast DNA variation and biogeography in the genus *Rorippa* Scop. (Brassicaceae). *Plant Biology* [Stuttgart] 4: 104–111.

Bloom, D.D. and N.R. Lovejoy. 2014. The evolutionary origins of diadromy inferred from a time-calibrated phylogeny for Clupeiformes (herring and allies). *Proceedings of the Royal Society B* 281: 20132081.

Bock, W. 2009. Design—An inappropriate concept in evolutionary theory. *Journal of Zoological Systematics and Evolutionary Research* 47: 7–9.

Boessenkool, A., J.J. Austin, T.H. Worthy et al. 2009a. Relict or colonizer? Extinction and range expansion of penguins in southern New Zealand. *Proceedings of the Royal Society, London B* 276: 815–821.

Boessenkool, S., B. Star, J.M. Waters, and P.J. Seddon. 2009b. Multilocus assignment analyses reveal multiple units and rare migration events in the recently expanded yellow-eyed penguin (*Megadyptes antipodes*). *Molecular Ecology* 18: 2390–2400.

Boles, W.E. 2006. Rhipiduridae. In *Handbook of the Birds of the World*, Vol. 11, eds. J. del Hoyo, A. Elliott, and D.A. Christie, pp. 200–243. Barcelona, Spain: Lynx Edicions.

Boles, W.E. 2007a. Pachycephalidae. In *Handbook of the Birds of the World*, Vol. 12, eds. J. del Hoyo, A. Elliott, and D.A. Christie, pp. 374–437. Barcelona, Spain: Lynx Edicions.

Boles, W.E. 2007b. Petroicidae. In *Handbook of the Birds of the World*, Vol. 12, eds. J. del Hoyo, A. Elliott, and D.A. Christie, pp. 438–489. Barcelona, Spain: Lynx Edicions.

Bone, T.S., S.R. Downie, J.M. Affolter, and K. Spalik. 2011. A phylogenetic and biogeographic study of the genus *Lilaeopsis* (Apiaceae tribe Oenantheae). *Systematic Botany* 36: 789–805.

Boon, W.M., J.C. Kearvell, C.H. Daugherty, and G.K. Chambers. 2001. Molecular systematics and conservation of kakariki (*Cyanoramphus* spp.). *Science for Conservation* 176: 1–46.

Boorman, C.J. and S.M. Shimeld. 2002. The evolution of left–right asymmetry in chordates. *BioEssays* 24: 1004–1011.

Boothroyd, I. and P.S. Cranston. 1999. The "ice worm"—The immature stages, phylogeny and biology of the glacier midge *Zelandochlus* (Diptera: Chironomidae). *Aquatic Insects* 21: 303–316.

Boubli, J.P., M.N.F. da Silva, M.V. Amado, T. Hrbek, F. Boavista Pontual, and I.P. Farias. 2008. A taxonomic reassessment of *Cacajao melanocephalus* Humboldt (1811), with the description of two new species. *International Journal of Primatology* 29: 723–741.

Boucher, F.C., W. Thuiller, T.J. Davies, and S. Lavergne. 2014. Neutral biogeography and the evolution of climatic niches. *American Naturalist* 183: 573–584.

Boucher, F.C., N.E. Zimmermann, and E. Conti. 2016. Allopatric speciation with little niche divergence is common among Alpine Primulaceae. *Journal of Biogeography* 43: 591–602.

Boyer, S.L., J.M. Baker, and G. Giribet. 2007. Deep genetic divergences in *Aoraki denticulata* (Arachnida, Opiliones, Cyphophthalmi): A widespread 'mite harvestman' defies DNA taxonomy. *Molecular Ecology* 16: 4999–5016.

Boyer, S.L. and G. Giribet. 2009. Welcome back New Zealand: Regional biogeography and Gondwanan origin of three endemic genera of mite harvestmen (Arachnida, Opiliones, Cyphophthalmi). *Journal of Biogeography* 36: 1084–1099.

Boyer, S.L. and G. Giribet. 2007. A new model Gondwanan taxon: Systematics and biogeography of the harvestman family Pettalidae (Arachnida, Opiliones, Cyphophthalmi), with a taxonomic revision of genera from Australia and New Zealand. *Cladistics* 23: 337–361.

Bradshaw, A.D. 1991. Genostasis and the limits to evolution. *Philosophical Transactions of the Royal Society, London B* 333: 289–305.

Bradshaw, J.D. 1989. Cretaceous geotectonic patterns in the New Zealand region. *Tectonics* 8: 803–820.

Brathwaite, R.L., R.J. Sewell, and A.B. Christie. 2008. Nature and tectonic setting of massive sulphide mineralisation and associated sediments and volcanics in the Matakaoa Volcanics, Raukumara Peninsula, New Zealand. *New Zealand Journal of Geology and Geophysics* 51: 349–366.

Breder, C.M. Jr. 1947. An analysis of the geometry of symmetry with especial reference to the geometry of fishes. *Bulletin of the American Musuem of Natural History* 88: 321–412.

Breder, C.M. Jr. 1955. Observations on the attributes of pentagonal symmetry. *Bulletin of the American Museum of Natural History* 106: 177–219.

Breder, C.M. and D.E. Rosen. 1966. *Modes of Reproduction in Fishes*. Garden City, NY: Natural History Press.

Breitwieser, I. 1993. Comparative leaf anatomy of New Zealand and Tasmanian Inuleae (Compositae). *Botanical Journal of the Linnean Society* 111: 183–209.

Breitwieser, I., and J.M. Ward. 2003. Phylogenetic relationships and character evolution in New Zealand and selected Australian Gnaphalieae (Compositae) inferred from morphological and anatomical data. *Botanical Journal of the Linnean Society* 141: 183–203.

Bremer, B. and T. Eriksson. 2009. Time tree of Rubiaceae: Phylogeny and dating the family, subfamilies, and tribes. *International Journal of Plant Science* 170: 766–793.

Bremer, K. 1992. Ancestral areas: A cladistic reinterpretation of the center of origin concept. *Systematic Biology* 41: 436–445.

Bremer, K., E.M. Friis, and B. Bremer. 2004. Molecular phylogenetic dating of asterid flowering plants shows Early Cretaceous diversification. *Systematic Biology* 53: 496–505.

Briggs, J.C. 1974. Operation of marine zoogeographic barriers. *Systematic Zoology* 23: 248–256.

Briggs, R.M., B.F. Houghton, M. McWilliams, and C.J.N. Wilson. 2005. $^{40}Ar/^{39}Ar$ ages of silicic volcanic rocks in the Tauranga-Kaimai area, New Zealand: Dating the transition between volcanism in the Coromandel Arc and the Taupo Volcanic Zone. *New Zealand Journal of Geology and Geophysics* 48: 459–469.

Britz, R., K.W. Conway, and L. Rüber. 2009. Spectacular morphological novelty in a miniature cyprinid fish, *Danionella dracula* n. sp. *Proceedings of the Royal Society, London B* 276: 2179–2186.

Brook, F.J. 1998. The coastal molluscan fauna of the northern Kermadec Islands, southwest Pacific Ocean. *Journal of the Royal Society of New Zealand* 28: 185–233.

Brook, F.J. and B.H. McArdle. 1999. Morphological variation, biogeography and local extinction of the northern New Zealand landsnail *Placostylus hongii* (Gastropoda: Bulimulidae). *Journal of the Royal Society of New Zealand* 29: 407–434.

Brooke, M. and T. Birkhead. 1991. *Cambridge Encyclopedia of Ornithology*. Cambridge, UK: Cambridge University Press.

Brouillet, L., T.K. Lowrey, L. Urbatsche et al. 2009. Astereae. In *Systematics, Evolution and Biogeography of Compositae*, eds. V.A Funk, A. Susanna, T.F. Stuessy, and R.J. Bayer, pp. 589–629. Vienna, Austria: International Association for Plant Taxonomy.

Brower, A.V.Z. 1994. Rapid morphological radiation and convergence among races of the butterfly *Heliconius erato* inferred from patterns of mitochondrial DNA evolution. *Proceedings of the National Academy of Sciences of USA* 91: 6491–6495.

Brown, J.H. and A.C. Gibson. 1983. *Biogeography*. St. Louis, MO: Mosby.

Brown, J.H. and M.V. Lomolino. 1998. *Biogeography*, 2nd edn. Sunderland, MA: Sinauer Associates.

Brownell Jr., R.L., R.L. Delong, and R.W. Schreibe. 1974. Pinniped populations at Islas de Guadalupe, San Benito, Cedros, and Natividad, Baja California, in 1968. *Journal of Mammalogy* 55: 469–472.

Brownsey, P.J., R. Ewans, B. Rance, S. Walls, and L.R. Perrie. 2013. A review of the fern genus *Sticherus* (Gleicheniaceae) in New Zealand with confirmation of two new species records. *New Zealand Journal of Botany* 51: 104–115.

Brownsey, P.J. and J.D. Lovis. 1987. Chromosome numbers for the New Zealand species of *Psilotum* and *Tmesipteris*, and the phylogenetic relationships of the Psilotales. *New Zealand Journal of Botany* 25: 439–454.

Brownsey, P.J. and L.R. Perrie. 2014. Schizaeaceae. In *Flora of New Zealand—Ferns and Lycophytes*, Fascicle 5, eds. I. Breitwieser, P.B. Heenan, and A.D. Wilton, pp. 1–24. Lincoln, New Zealand: Manaaki Whenua Press.

Brownsey, P.J. and J.C. Smith-Dodsworth. 1989. *New Zealand Ferns and Allied Plants*. Auckland, New Zealand: David Bateman.

Brunaud, A. 1970. Une mise à fleur incomplète? La transformation en épine du méristème terminal chez une apocynacée: *Carissa macrocarpa* (Eckl.) A. DC. *Revue générale de Botanique* 77: 97–109.

Bryan, S. and R. Ernst. 2008. Revised definition of Large Igneous Province (LIP). *Earth-Science Reviews* 86: 175–202.

Buchanan, J. 1881. On the alpine flora of New Zealand. *Transactions of the New Zealand Institute* 14: 342–356.

Buchanan, J. and G.C. Zuccarello. 2012. Decoupling of short- and long-distance dispersal pathways in the endemic New Zealand seaweed *Carpophyllum maschalocarpum* (Phaeophyceae, Fucales). *Journal of Phycology* 48: 518–529.

Buckberg, G.D. 2002. The helix and the heart. *Journal of Thoracic and Cardiovascular Surgery* 124: 863–883.

Buckeridge, J.S., D.E. Lee, and J.H. Robinson. 2014. A diverse, shallow-water barnacle assemblage (Cirripedia: Sessilia) from the Oligocene of Southland, New Zealand. *New Zealand Journal of Geology and Geophysics* 57: 253–263.

Buckley, T.R. and S. Bradler. 2010. *Tepakiphasma ngatikuri*, a new genus and species of stick insect (Phasmatodea) from the Far North of New Zealand. *New Zealand Entomologist* 33: 118–126.

Buckley, T.R., S. James, J. Allwood, S. Bartlam, R. Howitt, and D. Prada. 2011. Phylogenetic analysis of New Zealand earthworms (Oligochaeta: Megascolecidae) reveals ancient clades and cryptic taxonomic diversity. *Molecular Phylogenetics and Evolution* 58: 85–96.

Buckley, T.R., M. Krosch, and R.A.B. Leschen. 2015. Evolution of New Zealand insects: Summary and prospectus for future research. *Austral Entomology* 54: 1–27.

Buckley, T.R. and R.A.B. Leschen. 2013. Comparative phylogenetic analysis reveals long-term isolation of lineages on the Three Kings Islands, New Zealand. *Biological Journal of the Linnean Society* 108: 361–377.

Buckley, T.R., K. Marske, and D. Attanayake. 2010. Phylogeography and ecological niche modelling of the New Zealand stick insect *Clitarchus hookeri* (White) support survival in multiple coastal refugia. *Journal of Biogeography* 37: 682–695.

Buckley, T.R. and C. Simon. 2007. Evolutionary radiation of the cicada genus *Maoricicada* Dugdale (Hemiptera: Cicadoidea) and the origins of the New Zealand alpine biota. *Biological Journal of the Linnean Society* 91: 419–435.

Buckley, T.R., C. Simon, and G.K. Chambers. 2001. Phylogeography of the New Zealand cicada *Maoricicada campbelli* based on mitochondrial DNA sequences: Ancient clades associated with Cenozoic environmental change. *Evolution* 55: 1395–1407.

Bull, P.C. and A.H. Whitaker. 1975. The amphibians, reptiles, birds and mammals. In *Biogeography and Ecology in New Zealand*, ed. G. Kuschel, pp. 231–276. The Hague, the Netherlands: Junk.

Buller, W.L. 1876. Notes on the tuatara lizard (*Sphenodon punctatum*), with a description of a supposed new species. *Transactions of the New Zealand Institute* 9: 317–325.

Buller, W.L. 1877. Notice of a new variety of tuatara lizard (*Sphenodon*) from East Cape Island. *Transactions of the New Zealand Institute* 10: 220–221.

Buller, W.L. 1888. *A History of the Birds of New Zealand*, 2nd edn., Vol. 2. London, UK: Published by the Author.

Bunce, M., M. Szulkin, H.R. Lerner et al. 2005. Ancient DNA provides new insights into the evolutionary history of New Zealand's giant eagle. *PLoS Biology* 3(1): e9 (pp. 44–46).

Bunce, M., T.H. Worthy, M.J. Phillips et al. 2009. The evolutionary history of the extinct ratite moa and New Zealand Neogene paleogeography. *Proceedings of the National Academy of Sciences of USA* 106: 20646–20651.

Burckhardt, F., D.M. Porter, S.A. Deane, J.R. Topham, and S. Wilmot (eds.). 1999. *The Correspondence of Charles Darwin*, Vol. 11. Cambridge, UK: Cambridge University Press.

Burckhardt, F., D.M. Porter, J. Harvey, and M. Richmond (eds.). 1994. *The Correspondence of Charles Darwin*, Vol. 9. Cambridge, UK: Cambridge University Press.

Burckhardt, F., J.A. Secord, J. Browne et al. (eds.). 2013. *The Correspondence of Charles Darwin*, Vol. 20. Cambridge, UK: Cambridge University Press.

Burckhardt, F. and S. Smith (eds.). 1987. *The Correspondence of Charles Darwin*, Vol. 3. Cambridge, UK: Cambridge University Press.

Burger, J. and M. Gochfeld. 1996. Laridae. In *Handbook of the Birds of the World*, Vol. 3, eds. J. del Hoyo, A. Elliott, and J. Sargatal, pp. 572–623. Barcelona, Spain: Lynx Edicions.

Burleigh, J.G., R.T. Kimball, and E.L. Braun. 2015. Building the avian tree of life using a large-scale, sparse supermatrix. *Molecular Phylogenetics and Evolution* 84: 53–63.

Burns, K.J., S.J. Hackett, and N.K. Klein. 2002. Phylogenetic relationships and morphological diversity in Darwin's finches and their relatives. *Evolution* 56: 1240–1252.

Burridge, C.P., D. Craw, D. Fletcher, and J.M. Waters. 2008. Geological dates and molecular rates: Fish DNA sheds light on time dependency. *Molecular Biology and Evolution* 25: 624–633.

Burridge, C.P., R.M. McDowall, D. Craw, M.V.H. Wilson, and J.M. Waters. 2012. Marine dispersal as a prerequisite for Gondwanan vicariance among elements of the galaxiid fish fauna. *Journal of Biogeography* 39: 306–321.

Burrows, C.J. 1963. The root habit of some New Zealand plants. *Tuatara* 11: 79–80.

Burrows, C.J. 2008. Genus *Pimelea* (Thymelaeaceae) in New Zealand 1. The taxonomic treatment of seven endemic, glabrous-leaved species. *New Zealand Journal of Botany* 46: 127–176.

Burrows, C.J. 2009a. Genus *Pimelea* (Thymelaeaceae) in New Zealand 2. The endemic *Pimelea prostrata* and *Pimelea urvilliana* species complexes. *New Zealand Journal of Botany* 47: 163–229.

Burrows, C.J. 2009b. Genus *Pimelea* (Thymelaeaceae) in New Zealand 3. The taxonomic treatment of six endemic hairy-leaved species. *New Zealand Journal of Botany* 47: 325–354.

Burrows, C.J. 2011a. Genus *Pimelea* (Thymelaeaceae) in New Zealand 4. The taxonomic treatment of ten endemic abaxially hairy-leaved species. *New Zealand Journal of Botany* 49: 41–106.

Burrows, C.J. 2011b. Genus *Pimelea* (Thymelaeaceae) in New Zealand 5. The taxonomic treatment of five endemic species with both adaxial and abaxial leaf hair. *New Zealand Journal of Botany* 49: 367–412.

Burton, D.W. 1963. A revision of the New Zealand and subantarctic Athoracophoridae. *Transactions of the Royal Society of New Zealand* 3: 47–75.

Burton, D.W. 1980. Anatomical studies on Australian, New Zealand, and subantarctic Athoracophoridae (Gastropoda: Pulmonata). *New Zealand Journal of Zoology* 7: 173–198.

Burton, P.J.K. 1974. Anatomy of head and neck in the Huia (*Heteralocha acutirostris*) with comparative notes on other Callaeidae. *Bulletin of the British Museum (Natural History)* 27: 1–48.

Bush, C.M., S.J. Wagstaff, P.W. Fritsch, and K.A. Kron. 2009. The phylogeny, biogeography and morphological evolution of *Gaultheria* (Ericaceae) from Australia and New Zealand. *Australian Systematic Botany* 22: 229–242.

Butler, G.W. 1895. On the complete or partial suppression of the right lung in the Amphisbaenidae and of the left lung in snakes and snake-like lizards and amphibians. *Proceedings of the Zoological Society, London* 1895: 691–712.

Calder, W.A. III. 1978. The kiwi. *Scientific American* 239(1): 102–110.

Caldwell, M.W., R.L. Nydam, A. Palci, and S. Apesteguía. 2015. The oldest known snakes from the Middle Jurassic-Lower Cretaceous provide insights on snake evolution. *Nature Communications* 6: 5996–6007.

Callmander, M.W., P.B. Philipson, G.M. Plunkett, M.B. Edwards, and S. Buerki. 2016. Generic delimitations, biogeography and evolution in the tribe Coleeae (Bignoniaceae), endemic to Madagascar and the smaller islands of the western Indian Ocean. *Molecular Phylogenetics and Evolution* 96: 178–186.

Calviño, C.I., S.G. Martínez, and S.R. Downie. 2008a. Morphology and biogeography of Apiaceae subfamily Saniculoideae as inferred by phylogenetic analysis of molecular data. *American Journal of Botany* 95: 196–214.

Calviño, C.I., S.G. Martínez, and S.R. Downie. 2008b. The evolutionary history of *Eryngium* (Apiaceae, Saniculoideae): Rapid radiations, long distance dispersals, and hybridizations. *Molecular Phylogenetics and Evolution* 46: 1129–1150.

Cameron, E.K., P.J. de Lange, L.R. Perrie et al. 2006. A new location for the Poor Knights spleenwort (*Asplenium pauperequitum*, Aspleniaceae) on the Forty Fours, Chatham Islands, New Zealand. *New Zealand Journal of Botany* 44: 199–209.

Campbell, H.J. 2008. Geology. In *Chatham Islands: Heritage and Conservation*, ed. C. Miskelly, pp. 35–53. Christchurch, New Zealand: Canterbury University Press.

Campbell, H.J. 2013. *The Zealandia Drowning Debate: Did New Zealand Sink Beneath the Waves?* Wellington, New Zealand: Bridget Williams Books.

Campbell, H.J., C.J. Adams, and N. Mortimer. 2008. Exploring the Australian geological heritage of Zealandia and New Zealand. *Proceedings of the Royal Society of Victoria* 120: 38–47.

Campbell, H.J., J. Begg, A. Beu et al. 2009. Geological considerations relating to the Chatham Islands, mainland New Zealand and the history of New Zealand terrestrial life. *Geological Society of New Zealand Miscellaneous Publication* 126: 5–7.

Campbell, H.J. and G. Hutching. 2007. *In Search of Ancient New Zealand*. Auckland, New Zealand: Penguin, and Lower Hutt, New Zealand: Institute of Geological and Nuclear Sciences.

Campbell, H.J. and C.A. Landis. 2001. New Zealand awash. *New Zealand Geographic* 51: 6–7.

Campos-Enriquez, J.O., E. Hernández-Quintero, J. Rodríguez, M. Martínez, and V. Ramón-Márquez. 2011. Mojave-Sonora Megashear at Trincheras Area, Sonora (northwestern Mexico). High resolution aero-magnetic and radioactive, and gravity data. *American Geophysical Union, Fall Meeting 2011*, San Francisco, CA, abstract #T21A-2314.

Cande, S.C., J.M. Stock, R.D. Müller, and T. Ishihara. 2000. Cenozoic motion between East and West Antarctica. *Nature* 404: 145–150.

Cantino, P.D., S.J. Wagstaff, and R.G. Olmstead. 1999. *Caryopteris* (Lamiaceae) and the conflict between phylogenetic and pragmatic considerations in botanical nomenclature. *Systematic Botany* 23: 369–386.

Cantley, J.T., N.G. Swenson, A. Markey, and S.C. Keeley. 2014. Biogeographic insights on Pacific *Coprosma* (Rubiaceae) indicate two colonizations to the Hawaiian Islands. *Botanical Journal of the Linnean Society* 174: 412–424.

Carboneras, C. 1992a. Procellariidae. In *Handbook of the Birds of the World*, Vol. 1, eds. J. del Hoyo, A. Elliott, and J. Sargatal, pp. 216–257. Barcelona, Spain: Lynx Edicions.

Carboneras, C. 1992b. Sulidae. In *Handbook of the Birds of the World*, Vol. 1, eds. J. del Hoyo, A. Elliott, and J. Sargatal, pp. 312–325. Barcelona, Spain: Lynx Edicions.

Carboneras, C. 1992c. Anatidae. In *Handbook of the Birds of the World*, Vol. 1, eds. J. del Hoyo, A. Elliott, and J. Sargatal, pp. 536–630. Barcelona, Spain: Lynx Edicions.

Carlsen, T., W. Bleeker, H. Hurka, R. Elven, and C. Brochmann, C. 2009. Biogeography and phylogeny of *Cardamine* (Brassicaceae). *Annals of the Missouri Botanical Garden* 96: 215–236.

Carpenter, R.J., P. Wilf, J.G. Conran, and N.R. Cúneo. 2014. A Paleogene trans-Antarctic distribution for *Ripogonum* (Ripogonaceae: Liliales)? *Palaeontologia Electronica* 17(Issue 3): 1–9 (article 17.3.39A).

Carr, L.M., P.A. McLenachan, P.J. Waddell, N.J. Gemmell, and D. Penny. 2015. Analyses of the mitochondrial genome of *Leiopelma hochstetteri* argues against the full drowning of New Zealand. *Journal of Biogeography* 42: 1066–1076.

Carrodus, S.K. 2009. Identification and the role of hybridisation in *Pittosporum*. PhD thesis, University of Waikato, Hamilton, New Zealand.

Carroll, E.L., S.J. Childerhouse, M. Christie et al. 2012. Paternity assignment and demographic closure in the New Zealand southern right whale. *Molecular Ecology* 21: 3960–3973.

Carroll, S.B. 2010. The making of the fittest: The DNA record of evolution. In *Darwin*, eds. W. Brown and A.C. Fabian, pp. 121–134. Cambridge, UK: Cambridge University Press.

Carswell, F.E. and K.S. Gould. 1998. Comparative vegetative development of divaricating and arborescent *Sophora* species (Fabaceae). *New Zealand Journal of Botany* 36: 295–301.

Carter, G.G. and D.K. Riskin, 2006. *Mystacina tuberculata. Mammalian Species* 790: 1–8.

Cas, R.A.F., C.A. Landis, and R.E. Fordyce. 1989. A monogenetic, Surtla-type, Surtseyan volcano from the Eocene-Oligocene Waiareka-Deborah volcanics, Otago, New Zealand: A model. *Bulletin of Volcanology* 51: 281–298.

Case, A.L., S.W. Graham, T.D. Macfarlane, and S.C.H. Barrett. 2008. A phylogenetic study of evolutionary transitions in sexual systems in Australasian *Wurmbea* (Colchicaceae). *International Journal of Plant Science* 169: 141–156.

Cassidy, J. 1993. Tectonic implications of paleomagnetic data from the Northland ophiolite, New Zealand. *Tectonophysics* 223: 199–211.

Castillo, P.R., P.F. Lonsdale, C.L. Moran, and J.W. Hawkins. 2009. Geochemistry of mid-Cretaceous Pacific crust being subducted along the Tonga–Kermadec Trench: Implications for the generation of arc lavas. *Lithos* 112: 87–102.

Castro, L.R. and M. Dowton. 2006. Molecular analyses of the Apocrita (Insecta: Hymenoptera) suggest that the Chalcidoidea are sister to the diaprioid complex. *Invertebrate Systematics* 20: 603–614.

Cavalcanti, M.J. and V. Gallo. 2007. Panbiogeographic analysis of distribution patterns in hagfishes (Craniata: Myxinidae). *Journal of Biogeography* 35: 1258–1268.

Cavalier-Smith, T. 1997. Cell and genome coevolution: Facultative anaerobiosis, glycosomes and kinetoplas-tan RNA editing. *Trends in Genetics* 13: 6–9.

Cawood, P.A., A. Kroner, W.J. Collins, T.M. Kusky, W.D. Mooney, and B.F. Windley. 2009. Accretionary orogens through Earth history. In *Earth Accretionary Systems in Space and Time*, eds. P.A. Cawood and A. Kroner, pp. 1–36. London, UK: Geological Society, Special Publications 318.

Cawood, P.A., C.A. Landis, A.A. Nemchin, and S. Hada. 2002. Permian fragmentation, accretion and subsequent translation of a low-latitude Tethyan seamount to the high-latitude east Gondwana margin: Evidence from detrital zircon age data. *Geological Magazine* 139: 131–144.

Centeno-García, E., M. Guerrero-Suastegui, and O. Talavera-Mendoza. 2008. The Guerrero Composite Terrane of western Mexico: Collision and subsequent rifting in a supra-subduction zone. In *Formation and Applications of the Sedimentary Record in Arc Collision Zones*, eds. A. Draut, P.D. Clift, and D.W. Scholl, pp. 279–308. Boulder, CO: Geological Society of America Special Paper 436.

Chacón, J., M. Camargo de Assis, A.W. Meerow, and S.S. Renner. 2012. From East Gondwana to Central America: Historical biogeography of the Alstroemeriaceae. *Journal of Biogeography* 39: 1806–1818.

Challis, G.A. 1961. Post-intrusion deformation of a dike swarm, Awatere Valley, New Zealand. *Geological Magazine* 98: 441–448.

Chamber, S. 2007. *New Zealand Birds*. Auckland, New Zealand: Reed.

Chamberlin, T.C. 1965. The method of multiple working hypotheses. *Science* 148: 754–759.

Chambers, G.K. and W.M. Boon. 2005. Molecular systematics of Macquarie Island and Reischek's parakeets. *Notornis* 52: 249–250.

Chan, T.Y., K.C. Ho, C.P. Li, and K.H. Chu. 2009. Origin and diversification of the clawed lobster genus *Metanephrops* (Crustacea: Decapoda: Nephropidae). *Molecular Phylogenetics and Evolution* 50: 411–422.

Chapple, D.G., T.P. Bell, S.N.J. Chapple, K.A. Miller, C.H. Daugherty, and G.B. Patterson. 2011. Phylogeography and taxonomic revision of the New Zealand cryptic skink (*Oligosoma inconspicuum*; Reptilia: Scincidae) species complex. *Zootaxa* 2782: 1–33.

Chapple, D.G., C.H. Daugherty, and P.A. Ritchie. 2008a. Comparative phylogeography reveals pre-decline population structure of New Zealand *Cyclodina* (Reptilia: Scincidae) species. *Biological Journal of the Linnean Society* 95: 388–408.

Chapple, D.G. and G.B. Patterson. 2007. A new skink species (*Oligosoma taumakae* sp. nov.; Reptilia: Scincidae) from the Open Bay Islands, New Zealand. *New Zealand Journal of Zoology* 34: 347–357.

Chapple, D.G., G.B. Patterson, T. Bell, and C.H. Daugherty. 2008b. Taxonomic revision of the New Zealand Copper Skink (*Cyclodina aenea*; Squamata: Scincidae) species complex, with description of two new species. *Journal of Herpetology* 42: 437–452.

Chapple, D.G., P.A. Ritchie, and C.H. Daugherty. 2009. Origin, diversification, and systematics of the New Zealand skink fauna. *Molecular Phylogenetics and Evolution* 52: 470–487.

Chatterjee, H., S. Ho, I. Barnes, and C. Groves. 2009. Estimating the phylogeny and divergence times of primates using a supermatrix approach. *BMC Evolutionary Biology* 9(259): 1–19.

Chatterjee, S. 1997. *The Rise of Birds: 225 Million Years of Evolution*. Baltimore, MD: Johns Hopkins University Press.

Cheang, C.C., K.H. Chu, and P.O. Ang. 2010. Phylogeography of the marine macroalga *Sargassum hemiphyllum* (Phaeophyceae, Heterokontophyta) in northwestern Pacific. *Molecular Ecology* 19: 2933–2948.

Cheeseman, T.F. 1893. Notes on the New Zealand bats. *Transactions of the New Zealand Institute* 26: 218–222.

Cheeseman, T.F. 1906. *Manual of the New Zealand Flora*. Wellington, New Zealand: J. Mackay, Government Printer.

Cheeseman, T.F. 1914. *Illustrations of the New Zealand Flora*. Wellington, New Zealand: John Mackay, Government Printer.

Chen, L.-Y., S.-Y. Zhao, K.-S. Mao, D.H. Les, Q.-F. Wang, and M.L. Moody. 2014. Historical biogeography of Haloragaceae: An out-of-Australia hypothesis with multiple intercontinental dispersals. *Molecular Phylogeny and Evolution* 78: 87–95.

Chen, W.-J., S. Lavoué, L.B. Beheregaray, and R.L. Mayden. 2014. Historical biogeography of a new antitropical clade of temperate freshwater fishes. *Journal of Biogeography* 41: 1806–1818.

Chen, W.-J., S. Lavoué, and R.L. Mayden. 2013. Evolutionary origin and early biogeography of otophysan fishes (Ostariophysi: Teleostei). *Evolution* 67: 2218–2239.

Cheng, J. and L. Xie. 2014. Molecular phylogeny and historical biogeography of *Caltha* (Ranunculaceae) based on analyses of multiple nuclear and plastid sequences. *Journal of Systematics and Evolution* 52: 51–67.

Chenoweth, L.B. and M.P. Schwarz. 2011. Biogeographical origins and diversification of the exoneurine allodapine bees of Australia (Hymenoptera, Apidae). *Journal of Biogeography* 38: 1471–1483.

Chillcott, J.G. 1961. A revision of the Nearctic species of Fanniinae (Diptera: Muscidae). *Canadian Entomologist Supplement* 14: 1–295.

Chinn, W.G. and N.J. Gemmell. 2004. Adaptive radiation within New Zealand endemic species of the cockroach genus *Celatoblatta* Johns (Blattidae): A response to Plio-Pleistocene mountain building and climate change. *Molecular Ecology* 13: 1507–1518.

Choi, H.-G., G.T. Kraft, I.K. Lee, and G.W. Saunders. 2002. Phylogenetic analyses of anatomical and nuclear SSU rDNA sequence data indicate that the Dasyaceae and Delesseriaceae (Ceramiales, Rhodophyta) are polyphyletic. *European Journal of Phycology* 37: 551–569.

Chomsky, N. 2002. *On Nature and Language.* Cambridge, UK: Cambridge University Press.

Chong, N.L. 1999. Phylogenetic analysis of the endemic New Zealand gecko species complex *Hoplodactylus pacificus* using DNA sequences of the 16S rRNA gene. B.Sc. Honours thesis, Victoria University of Wellington, Wellington, New Zealand.

Chouard, T. 2010. Evolution: Revenge of the hopeful monster. *Nature* 463: 864–867.

Christenhusz, M.J.M. and M.W. Chase. 2013. Biogeographical patterns of plants in the Neotropics—Dispersal rather than plate tectonics is most explanatory. *Botanical Journal of the Linnean Society* 171: 277–286.

Christensen, C.E. 1923. On the behaviour of certain New Zealand arboreal plants when gradually buried by river shingle. *Transactions of the New Zealand Institute* 54: 546.

Christidis, L. and W.E. Boles. 1994. *The Taxonomy and Species of Birds of Australia and Its Territories.* Victoria, Australia: Royal Australasian Ornithologists Union.

Christidis, L., M. Irestedt, D. Rowe, W.E. Boles, and J.A. Norman. 2011. Mitochondrial and nuclear DNA phylogenies reveal a complex evolutionary history in the Australasian robins (Passeriformes: Petroicidae). *Molecular Phylogenetics and Evolution* 61: 726–738.

Chruszcz, B. and R.M. Barclay. 2002. *Chalinolobus gouldii. Mammalian Species* 690: 1–4.

Chung, K.-F. 2007. Inclusion of the South Pacific alpine genus *Oreomyrrhis* (Apiaceae) in *Chaerophyllum* based on nuclear and chloroplast DNA sequences. *Systematic Botany* 32: 671–681.

Chung, K.-F., C.-I. Peng, S. Downie, K. Spalik, and B. Schaal. 2005. Molecular systematics of the trans-Pacific alpine genus *Oreomyrrhis* (Apiaceae): Phylogenetic affinities and biogeographic implications. *American Journal of Botany* 92: 2054–2071.

Clark, A.G. and 384 others. 2007. Evolution of genes and genomes on the *Drosophila* phylogeny. *Nature* 450: 203–218.

Clark, J.R., R.H. Ree, M.G. King, W.L. Wagner, and E.H. Roalson. 2008. A comparative study in ancestral range reconstruction methods: Retracing the uncertain histories of insular lineages. *Systematic Biology* 57: 693–707.

Clarke, D.J. 2011. A revision of the New Zealand endemic rove beetle genus *Agnosthaetus* Bernhauer (Coleoptera: Staphylinidae). *The Coleopterists' Bulletin* 65 (monograph 4): 1–118.

Clarkson, F.M., B.D. Clarkson, and C.E.C. Gemmill. 2012. Biological flora of New Zealand 13. *Pittosporum cornifolium,* tāwhiri karo, cornel-leaved pittosporum. *New Zealand Journal of Botany* 50: 185–201.

Clegg, S.M., S.M. Degnan, J. Kikkawa, C. Moritz, A. Estoup, and I.P.F. Owens. 2002. Genetic consequences of sequential founder events by an island-colonizing bird. *Proceedings of the National Academy of Sciences of USA* 99: 8127–8132.

Climo, F.M. 1969. Classification of New Zealand Arionacea (Mollusca: Pulmonata). III. *Records of the Dominion Museum* 6: 175–258.

Climo, F.M. 1971. Classification of New Zealand Arionacea (Mollusca: Pulmonata). V. Descriptions of some new phenacohelicid taxa (Punctidae: Phenacohelicinae). *Records of the Dominion Museum* 7: 95–105.

Climo, F.M. 1985. Classification of New Zealand Arionacea (Mollusca: Pulmonata). XI. The new genus *Chaureopa* and description of a new species of *Paracharopa* Climo (Charopidae). *New Zealand Journal of Zoology* 12: 283–296.

Climo, F.M. 1989. The panbiogeography of New Zealand as illuminated by the genus *Fectola* Iredale, 1915 and subfamily Rotadiscinae Pilsbry, 1927 (Mollusca: Pulmonata: Punctoidea: Charopidae). *New Zealand Journal of Zoology* 16: 587–649.

Climo, F.M. and K. Mahlfeld. 2011. *Kokopapa,* a new genus of land snails (Pulmonata: Punctidae). With description of six new species and discussion of South Island, New Zealand biogeography. *Bulletin of the Buffalo Society of Natural Sciences* 40: 29–48.

Clout, M.N. and I. Hay. 1981. South Island kokako (*Callaeas cinerea cinerea*) in *Nothofagus* forest. *Notornis* 28: 256–259.

Cluzel, D., C.J. Adams, S. Meffre, H. Campbell, and P. Maurizot. 2010a. Discovery of Early Cretaceous rocks in New Caledonia; new geochemical and U-Pb zircon age constraints on the transition from subduction to marginal breakup in the Southwest Pacific. *Journal of Geology* 118: 381–397.

Cluzel, D., J.C. Aitchison, and C. Picard. 2001. Tectonic accretion and underplating of mafic terranes in the Late Eocene intraoceanic fore-arc of New Caledonia (southwest Pacific): Geodynamic implications. *Tectonophysics* 340: 23–59.

Cluzel, D., P.M. Black, C. Picard, and K.N. Nicolson. 2010b. Geochemistry and tectonic setting of Matakaoa Volcanics (East Cast Allochthon, New Zealand): Supra-subduction zone affinity, regional correlations and origin. *Tectonics* 29: 1–21 (TC2013).

Cluzel, D., S. Meffre, P. Maurizot, and A.J. Crawford. 2006. Earliest Eocene (53 Ma) convergence in the Southwest Pacific: evidence from pre-obduction dikes in the ophiolite of New Caledonia. *Terra Nova* 18: 395–402.

Coates, M.I. and J.A. Clark. 1990. Polydactyly in the earliest known tetrapod limbs. *Nature* 347: 66–69.

Cockayne, L. 1898. An enquiry into the seedling forms of New Zealand phanerogams and their development. *Transactions of the New Zealand Institute* 31: 354–398.

Cockayne, L. 1899. A sketch of the plant geography of the Waimakariri River Basin. *Transactions of the New Zealand Institute* 32: 95–136.

Cockayne, L. 1901. A short account of the plant covering of the Chatham Islands. *Transactions of the New Zealand Institute* 34: 243–325.

Cockayne, L. 1906. Some observations on the coastal vegetation of the South Island of New Zealand. *Transactions of the New Zealand Institute* 39: 312–359.

Cockayne, L. 1921. *The Vegetation of New Zealand*. Leipzig, Germany: Engelmann.

Cockayne, L. 1928. *The Vegetation of New Zealand*, 2nd edn. Leipzig, Germany: Engelmann.

Cockburn-Hood, T.H. 1877. New Zealand a post-glacial center of creation. *Transactions of the New Zealand Institute* 10: 3–24.

Cohen, B.L., A. Kaulfuss, and C. Lüter. 2014. Craniid brachiopods: Aspects of clade structure and distribution reflect continental drift (Brachiopoda: Craniiformea). *Zoological Journal of the Linnean Society* 171: 133–150.

Cohen, L. and T. Arzee. 1980. Twofold pathways of apical determination in the thorn system of *Carissa grandiflora*. *Botanical Gazette* 141: 258–263.

Colenso, W. 1868. Essay on the botany, geographic and economic, of the North Island of the New Zealand group. *Transactions of the New Zealand Institute* 1. *Essays*: 1–53.

Colenso, W. 1885. Notes on the bones of a species of *Sphenodon* (*S. diversum*, Col.) apparently distinct from the species already known. *Transactions of the New Zealand Institute* 18: 118–128.

Collar, N.J. 1997. Psittacidae. In *Handbook of the Birds of the World*, Vol. 4, eds. J. del Hoyo, A. Elliott, and J. Sargatal, pp. 280–479. Barcelona, Spain: Lynx Edicions.

Collier, K.J. 1992. Freshwater macroinvertebrates of potential conservation Interest. *Science and Research Series* 50: 1–49.

Collins, C.J., N.J. Rawlence, S. Prost et al. 2014. Extinction and recolonization of coastal megafauna following human arrival in New Zealand. *Proceedings of the Royal Society B* 281: 20140097.

Collins, J.A., P. Molnar, and A.F. Sheehan. 2011. Multibeam bathymetric surveys of submarine volcanoes and mega-pockmarks on the Chatham Rise, New Zealand. *New Zealand Journal of Geology and Geophysics* 54: 329–339.

Colloff, M.J. 2011a. A review of the oribatid mite family Nothridae in Australia, with new species of *Novonothrus* and *Trichonothrus* from rain forest and their Gondwanan biogeographical affinities (Acari: Oribatida). *Zootaxa* 3005: 1–44.

Colloff, M.J. 2011b. A new genus of oribatid mite, *Spineremaeus* gen. nov. and three new species of *Scapheremaeus* (Acari: Oribatida: Cymbaeremaeidae) from Norfolk Island, South-west Pacific, and their biogeographical affinities. *Zootaxa* 2828: 19–37.

Colloff, M.J. 2015. The *Crotonia* fauna of New Zealand revisited (Acari: Oribatida): Taxonomy, phylogeny, ecological distribution and biogeography. *Zootaxa* 3947: 1–29.

Colloff, M.J. and S.L. Cameron. 2014. Beyond Moa's Ark and Wallace's Line: Extralimital distribution of new species of *Austronothrus* (Acari, Oribatida, Crotoniidae) and the endemicity of the New Zealand oribatid mite fauna. *Zootaxa* 3780: 263–281.

Connor, H.E. 1954. Studies in New Zealand *Agropyron*. Part I and II. *New Zealand Journal of Science and Technology* 35B: 315–343.

Connor, H.E. 1979. Breeding systems in the grasses: A survey. *New Zealand Journal of Botany* 17: 547–574.

Connor, H.E. 1991. *Chionochloa* Zotov (Gramineae) in New Zealand. *New Zealand Journal of Botany* 29: 219–282.

Connor, H.E. and E. Edgar. 1987. Name changes in the indigenous New Zealand flora, 1960–1986 and Nomina Nova IV, 1983–1986. *New Zealand Journal of Botany* 25: 115–171.

Connor, H.E. and K.M. Lloyd. 2004. Species novae Graminum Novae-Zelandiae II. *Chionochloa nivifera* (Danthonieae: Danthonioideae). *New Zealand Journal of Botany* 42: 531–536.

Connor, H.E., B.P.J. Molloy, and M.I. Dawson. 1993. *Australopyrum* (Triticeae: Gramineae) in New Zealand. *New Zealand Journal of Botany* 31: 1–10.

Conran, J.G., J.M. Bannister, and D.E. Lee. 2009. Earliest orchid macrofossils: Early Miocene *Dendrobium* and *Earina* (Orchidaceae: Epidendroideae) from New Zealand. *American Journal of Botany* 96: 466–474.

Conran, J.G., J.M. Bannister, D.E. Lee et al. 2015. An update of monocot macrofossil data from New Zealand and Australia. *Botanical Journal of the Linnean Society* 178: 394–420.

Conran, J.G., J.M. Bannister, D.C. Mildenhall, D.E. Lee, J. Chacón, and S.S. Renner. 2014a. Leaf fossils of *Luzuriaga* and a monocot flower with in situ pollen of *Liliacidites contortus* Mildenh. & Bannister sp. nov. (Alstroemeriaceae) from the Early Miocene. *American Journal of Botany* 101: 141–155.

Conran, J.G., W.G. Lee, D.E. Lee, J.M. Bannister, and U. Kaulfuss. 2014b. Reproductive niche conservatism in the isolated New Zealand flora over 23 million years. *Biology Letters* 10: 20140647.

Conran, J.G., D.C. Mildenhall, D.E. Lee et al. 2014c. Subtropical rainforest vegetation from Cosy Dell, Southland: Plant fossil evidence for Late Oligocene terrestrial ecosystems. *New Zealand Journal of Geology and Geophysics* 57: 236–252.

Conway Morris, S. 2003. *Life's Solution: Inevitable Humans in a Lonely Universe*. Cambridge, UK: Cambridge University Press.

Cook, J. 2003. *The Journals*. London, UK: Penguin.

Cook, L.D., S.A. Trewick, M. Morgan-Richards, and P.M. Johns. 2010. Status of the New Zealand cave weta (Rhaphidophoridae) genera *Pachyrhamma*, *Gymnoplectron* and *Turbottoplectron*. *Invertebrate Systematics* 24: 131–138.

Cook, L.G. and M.D. Crisp. 2005. Directional asymmetry of long-distance dispersal and colonization could mislead reconstructions of biogeography. *Journal of Biogeography* 32: 741–754.

Cooper, A.F. and T.R. Ireland. 2014. The Pounamu terrane, a new Cretaceous exotic terrane within the Alpine Schist, New Zealand; tectonically emplaced, deformed and metamorphosed during collision of the LIP Hikurangi Plateau with Zealandia. *Gondwana Research* 27: 1255–1269.

Corfield, J.R., A.C. Gsell, D. Brunton et al. 2011. Anatomical specializations for nocturnality in a critically endangered parrot, the Kakapo (*Strigops habroptilus*). *PLoS One* 6(8): e22945 (pp. 1–10).

Corfield, J.R., M.F. Kubke, and C. Köppl. 2014. Emu and kiwi: The ear and hearing in paleognathous birds. In *Insights from Comparative Hearing Research*, eds. C. Köppl et al., pp. 263–287. New York: Springer.

Corfield, J.R., K. Price, A.N. Iwaniuk, C. Gutiérrez-Ibáñez, T. Birkhead, and D.R. Wylie. 2015. Diversity in olfactory bulb size in birds reflects allometry, ecology, and phylogeny. *Frontiers in Neuroanatomy* 9(Article 102): 1–16.

Corfield, J.R., J.M. Wild, M.E. Hauber, S. Parsons, and M.F. Kubke. 2008. Evolution of brain size in the palaeognath lineage, with an emphasis on New Zealand ratites. *Brain, Behavior, and Evolution* 71: 87–99.

Corner, E.J.H. 1962. The classification of Moraceae. *Gardens Bulletin, Singapore* 19: 187–252.

Corning, P.A. 2014. Evolution 'on purpose': How behaviour has shaped the evolutionary process. *Biological Journal of the Linnean Society* 112: 242–260.

Cotterill, F.P.D. and W. Foissner. 2010. A pervasive denigration of natural history misconstrues how biodiversity inventories and taxonomy underpin scientific knowledge. *Biodiversity and Conservation* 19: 291–303.

Court, D.J. 1982. Spiders from Tawhiti Rahi, Poor Knights Islands, New Zealand. *Journal of the Royal Society of New Zealand* 12: 359–371.

Couvreur, T.L.P., A. Franzke, I.A. Al-Shehbaz, F.T. Bakker, M.A. Koch, and K. Mummenhoff. 2010. Molecular phylogenetics, temporal diversification and principles of evolution in the mustard family (Brassicaceae). *Molecular Biology and Evolution* 27: 55–71.

Cowen, R. 2014. *History of Life*, 5th edn. Chichester, UK: Wiley-Blackwell.

Cox, C.B. and P.D. Moore. 2010. *Biogeography: An Ecological and Evolutionary Approach*, 8th edn. Oxford, UK: Blackwell.

Cox, S. and R. Sutherland. 2007. Regional geological framework of South Island, New Zealand, and its significance for understanding the active plate boundary. *Geophysical Monograph Series* 175: 19–46.

Coyne, J.A. and H.A. Orr. 2004. *Speciation*. Sunderland, MA: Sinauer Associates.

Cracraft, J. and F.K. Barker. 2009. Passerine birds (Passeriformes). In *Timetree of Life*, eds. S.B. Hedges and S. Kumar, pp. 423–431. New York: Oxford University Press.

Craig, D.A., R.E.G. Craig, and T.K. Crosby. 2012. Simuliidae (Insecta: Diptera). *Fauna of New Zealand* 68: 1–336.

Craigie, E.H. 1930. Studies on the brain of the kiwi (*Apteryx australis*). *Journal of Comparative Neurology* 49: 223–357.

Crampton, J.S. 1988. Stratigraphy and structure of the Monkey Face area, Marlborough, New Zealand, with special reference to shallow marine Cretaceous strata. *New Zealand Journal of Geology and Geophysics* 31: 447–470.

Crampton, J.S. and M.G. Laird. 1997. Burnt Creek Formation and Late Cretaceous basin development in Marlborough, New Zealand. *New Zealand Journal of Geology and Geophysics* 40: 199–222.

Crampton, J.S., M. Laird, A. Nicol, D. Townsend, and R. Van Dissen. 2003. Palinspastic reconstructions of southeastern Marlborough, New Zealand, for mid-Cretaceous-Eocene times. *New Zealand Journal of Geology and Geophysics* 46: 153–175.

Cranston, P.S., N.B. Hardy, G.E. Morse, L. Puslednik, and S.R. McCluen. 2010. When molecules and morphology concur: The 'Gondwanan' midges (Diptera: Chironomidae). *Systematic Entomology* 35: 636–648.

Craw, D., C. Burridge, R.J. Norris, and J.M. Waters. 2008. Genetic ages for Quaternary topographic evolution: A new dating tool. *Geology* 36: 19–22.

Craw, D., J. Druzbicka, C. Rufaut, and J. Waters. 2013. Geological controls on paleo-environmental change in a tectonic rain shadow, southern New Zealand. *Palaeogeography, Palaeoclimatology, Palaeoecology* 370: 103–116.

Craw, R.C. 1978. Revision of genus *Argyrophenga* (Lepidoptera: Satyridae). *New Zealand Journal of Zoology* 5: 751–768.

Craw, R.C. 1988. Continuing the synthesis between panbiogeography, phylogenetic systematics and geology as illustrated by empirical studies on the biogeography of New Zealand and the Chatham Islands. *Systematic Biology* 37: 291–310.

Craw, R.C. 1999. Molytini (Insecta: Coleoptera: Curculionidae: Molytinae). *Fauna of New Zealand* 39: 1–68.

Craw, R.C., J.R. Grehan, and M.J. Heads. 1999. *Panbiogeography: Tracking the History of Life*. New York: Oxford University Press.

Crawford, A.J., S. Meffre, and P.A. Symonds. 2003. 120 to 0 Ma tectonic evolution of the southwest Pacific and analogous geological evolution of the 60 to 220 Ma Tasman fold belt system. *Geological Society of America Special Paper* 372: 383–404.

Cree, A. 2014. *Tuatara: Biology and Conservation of a Venerable Survivor*. Christchurch, New Zealand: Canterbury University Press.

Creese, R.G. and M.H.B. O'Neill. 1987. *Chiton aorangi* n. sp., a brooding chiton (Mollusca: Polyplacophora) from northern New Zealand. *New Zealand Journal of Zoology* 14: 89–93.

Crisci, J.V. 2006. One-dimensional systematists: Perils in a time of steady progress. *Systematic Botany* 31: 215–219.

Crisp, M.D., S.A. Trewick, and L.G. Cook. 2011. Hypothesis testing in biogeography. *Trends in Ecology and Evolution* 26: 66–72.

Croizat, L. 1964. *Space, Time, Form: The Biological Synthesis*. Caracas, Venezuela: Published by the Author.

Croizat, L. 1976. Biogeografía analítica y sintética ('panbiogeografia') de las Americas. *Biblioteca de la Academia de Ciencias Físicas, Matemáticas y Naturales* [Caracas] 15: 1–890.

Cross, E.W., C.J. Quinn, and S.J. Wagstaff. 2002. Molecular evidence for the polyphyly of *Olearia* (Astereae: Asteraceae). *Plant Systematics and Evolution* 235: 99–120.

Crow, J.F. 2008. Mid-century controversies in population genetics. *Annual Reviews in Genetics* 42: 1–16.

Crow, K.D., H. Munehara, and G. Bernardi. 2010. Sympatric speciation in a genus of marine reef fishes. *Molecular Ecology* 19: 2089–2105.

Cunha, R.L., F. Blanc, F. Bonhomme, and S.Arnaud-Haond. 2011. Evolutionary patterns in pearl oysters of the genus *Pinctada* (Bivalvia: Pteriidae). *Marine Biotechnology* 13: 181–192.

Cunningham, S.J., M.R. Alley, and I. Castro. 2011. Facial bristle feather histology and morphology in New Zealand birds: Implications for function. *Journal of Morphology* 272: 118–128.

Cunningham, S.J., I. Castro, and M.R. Alley. 2007. A new prey-detection mechanism for kiwi (*Apteryx* spp.) suggests convergent evolution between paleognathous and neognathous birds. *Journal of Anatomy* 211: 493–502.

Cunningham, S.J., J.R. Corfield, A.N. Iwaniuk et al. 2013. The anatomy of the bill tip of kiwi and associated somatosensory regions of the brain: Comparisons with shorebirds. *PloS One* 8(11): e80036 (pp. 1–17).

Curran, C., A.R. Gorman, and R.J. Norris. 2010. Using geophysical investigations to understand fault reactivation: Evolution of the Waihemo Fault System, North Otago, New Zealand. In *EGU General Assembly Conference Abstracts*, Vienna, Austria, Vol. 12, p. 662.

Currey, R.J.C., S.M. Dawson, and E. Slooten. 2007. New abundance estimates suggest Doubtful Sound bottlenose dolphins are declining. *Pacific Conservation Biology* 13: 265–273.

Curtis, N.R. 2011. Biogeography of the dune insect fauna of New Zealand and Chatham Island. PhD thesis, Lincoln University, Lincoln, New Zealand.

Cusset, G. 1982. The conceptual bases of plant morphology. In *Axioms and Principles of Plant Construction*, ed. R. Sattler, pp. 8–86. The Hague, the Netherlands: Junk/Nijhoff.

Cutten, H.N.C. 1979. Rappahannock Group: Late Cenozoic sedimentation and tectonics contemporaneous with Alpine Fault movement, *New Zealand Journal of Geology and Geophysics* 22: 535–553.

Cutter, A.D., A. Dey, and R.L. Murray. 2009. Evolution of the *Caenorhabditis elegans* genome. *Molecular Biology and Evolution* 26: 1199–1234.

Czaplewski, N.J. and G.S. Morgan. 2012. New basal noctilionoid bats (Mammalia: Chiroptera) from the Oligocene of subtropical North America. In *Evolutionary History of Bats*, eds. G.F. Gunnell and N.B. Simmons, pp. 162–209, Cambridge, UK: Cambridge University Press.

Dalquest, W.W. and H.J. Werner. 1954. Histological aspects of the faces of North American bats. *Journal of Mammalogy* 35: 147–160.

Daniel, M.J. 1979. The New Zealand short-tailed bat, *Mystacina tuberculata*; a review of present knowledge. *New Zealand Journal of Zoology* 6: 357–370.

Daniel, M.J. 1990. Order Chiroptera. In *The Handbook of New Zealand Mammals*, ed. C.M. King, pp. 114–137. Auckland, New Zealand: Oxford University Press.

Daniel, M.J. and G.R. Williams. 1984. A survey of the distribution, seasonal activity and roost sites of New Zealand bats. *New Zealand Journal of Ecology* 7: 9–25.

D'Archino, R., W.A. Nelson, and G.C. Zuccarello. 2010. *Psaromenia* (Kallymeniaceae, Rhodophyta): A new genus for *Kallymenia berggrenii*. *Phycologia*, 49: 73–85.

Darlington, P.J., Jr. 1964. Drifting continents and Late Paleozoic geography. *Proceedings of the National Academy of Sciences of USA* 52: 1084–1091.

Darlington, P.J., Jr. 1957/1966. *Zoogeography: The Geographical Distribution of Animals*. New York: Wiley.

Darwin, C. 1859. *On the Origin of Species*, 1st edn. London, UK: John Murray.

Darwin, C. 1869. *On the Origin of Species*, 5th edn. London, UK: John Murray.

Darwin, C. 1872. *On the Origin of Species*, 6th edn. London, UK: John Murray.

Darwin, C. 1874. *The Descent of Man*, 2nd edn. London, UK: John Murray.

Darwin, C. 1875. *The Movements and Habits of Climbing Plants*, 2nd edn. London, UK: John Murray.

Daugherty, C.H., A. Cree, J. Hay, and M.B. Thompson. 1990. Neglected taxonomy and continuing extinctions of tuatara (*Sphenodon*). *Nature* 347: 177–179.

Davey, F.J. 2005. A Mesozoic crustal suture on the Gondwana margin in the New Zealand region. *Tectonics* 24. doi:10.1029/2004TC001719.

Davey, F.J. 2010. The West Antarctic Rift System—Some outstanding issues. *American Geophysical Union, Fall Meeting 2010*, San Francisco, CA, abstract #T21D-2195.

Davey, F.J., D. Eberhart-Phillips, M. Kohler et al. 2007. Geophysical structure of the Southern Alps Orogen, South Island, New Zealand. In *A Continental Plate Boundary: Tectonics at South Island, New Zealand*, eds. D. Okaya, T. Stern, and F. Davey, pp. 47–73. Washington, DC: American Geophysical Union, Geophysical Monograph Series 175.

Davies, P.S. 2001. *Norms of Nature: Naturalism and the Nature of Functions*. Cambridge, MA: MIT Press.

Davies, S.J.J.F. 2002. *Ratites and Tinamous*. New York: Oxford University Press.

Davies, T.J., T.G. Barraclough, M.W. Chase, P.S. Soltis, and D.E. Soltis. 2004. Darwin's abominable mystery: Insights from a supertree of the angiosperms. *Proceedings of the National Academy of Sciences of USA* 101: 1904–1909.

Davy, B., K. Hoernle, and R. Werner. 2008. Hikurangi Plateau: Crustal structure, rifted formation, and Gondwana subduction history. *Geochemistry, Geophysics, Geosystems* 9: Q07004.

Dawkins, R. 1976. *The Selfish Gene*. New York: Oxford University Press.

Dawkins, R. 1986. *The Blind Watchmaker*. New York: Norton.

Dawkins, R. 1996/2006. *Climbing Mount Improbable*. London, UK: Penguin.

Dawkins, R. 2004/2005. *The Ancestor's Tale: A Pilgrimage to the Dawn of Life*. London, UK: Phoenix.

Dawkins, R. 2009. *The Greatest Show on Earth*. New York: Free Press.

Dawson, J.W. 1961. A revision of the genus *Anisotome* (Umbelliferae). *University of California Publications in Botany* 33: 1–98.

Dawson, J.W. 1967. The New Zealand species of *Gingidium* (Umbelliferae). *New Zealand Journal of Botany* 5: 84–116.

Dawson, J.W. 1980. *Aciphylla trifoliolata* Petrie and *A. stannensis* sp. nov. *New Zealand Journal of Botany* 18: 115–120.

Dawson, J.W. 1988. *Forest Vines to Snow Tussocks: The Story of New Zealand Plants*. Wellington, New Zealand: Victoria University Press.

Dawson, J.W. and J.R. Le Comte. 1978. Research on *Aciphylla*—A progress report. *Tuatara* 23: 49–67.

Dawson, J.W. and R. Lucas. 2011. *New Zealand's Native Trees*. Nelson, New Zealand: Craig Potton.

Dawson, M.I. and P.B. Heenan. 2004. Morphological variation of the *Leucopogon fraseri* complex (Ericaceae: Styphelieae) in New Zealand, and recognition of a new species, *L. nanum*. *New Zealand Journal of Botany* 42: 537–564.

Dawson, M.N. 2009. Trans-realm biogeography: An immergent interface. *Frontiers of Biogeography* 1: 62–70.

Dawson, M.N. 2012. Parallel phylogeographic structure in ecologically similar sympatric sister taxa. *Molecular Ecology* 21: 987–1004.

Dawson, M.N. and W.M. Hamner. 2008. A biophysical perspective on dispersal and the geography of evolution in marine and terrestrial systems. *Journal of the Royal Society—Interface* 5: 135–150.

Dawson, M.N., K.D. Louie, M. Barlow, D.K. Jacobs, and C.C. Swift. 2002. Comparative phylogeography of sympatric sister species, *Clevelandia ios* and *Eucyclogobius newberryi* (Teleostei, Gobiidae), across the California Transition Zone. *Molecular Ecology* 11: 1065–1075.

de Beer, G. 1956. The evolution of ratites. *Bulletin of the British Museum (Natural History)* 4: 59–70.

Deckert, H., U. Ring, and N. Mortimer. 2002. Tectonic significance of Cretaceous bivergent extensional shear zones in the Torlesse accretionary wedge, central Otago Schist, New Zealand. *New Zealand Journal of Geology and Geophysics* 45: 537–547.

Degnan, S.M., Imron, D.L. Geiger, and B.M. Degnan. 2006. Evolution in temperate and tropical seas: Disparate patterns in southern hemisphere abalone (Mollusca: Vetigastropoda: Haliotidae). *Molecular Phylogenetics and Evolution* 41: 249–256.

de Juana, E. 1992. Class Aves (birds). In *Handbook of the Birds of the World*, Vol. 1, eds. J. del Hoyo, A. Elliott, and J. Sargatal, pp. 35–75. Barcelona, Spain: Lynx Edicions.

de Lange, P.J. 1998. *Hebe perbella* (Scrophulariaceae)—A new and threatened species from western Northland, North Island, New Zealand. *New Zealand Journal of Botany* 36: 399–406.

de Lange, P.J. 2014. A revision of the New Zealand *Kunzea ericoides* (Myrtaceae) complex. *PhytoKeys* 40: 1–185.

de Lange, P.J. and E.K. Cameron. 1999. The vascular flora of Aorangi Island, Poor Knights Islands, northern New Zealand. *New Zealand Journal of Botany* 37: 433–468.

de Lange, P.J. and R.O. Gardner. 2002. A taxonomic reappraisal of *Coprosma obconica* (Rubiaceae: Anthospermeae). *New Zealand Journal of Botany* 40: 25–38.

de Lange, P.J., R.O. Gardner, and K.A. Riddell. 2002. *Ackama nubicola* (Cunoniaceae) a new species from Western Northland, North Island, New Zealand. *New Zealand Journal of Botany* 40: 525–534.

de Lange, P.J., R.O. Gardner, W.R. Sykes et al. 2005. Vascular flora of Norfolk Island: Some additions and taxonomic notes. *New Zealand Journal of Botany* 43: 563–596.

de Lange, P.J., P.B. Heenan, B.D. Clarkson, and B.R. Clarkson. 1999. Taxonomy, ecology, and conservation of *Sporadanthus* (Restionaceae) in New Zealand. *New Zealand Journal of Botany* 37: 413–431.

de Lange, P.J., P.B. Heenan, G.J. Houliston, J.R. Rolfe, and A.D. Mitchell. 2013. New *Lepidium* (Brassicaceae) from New Zealand. *PhytoKeys* 24: 1–147.

de Lange, P.J., P.B. Heenan, and A.J. Townsend. 2009. *Rorippa laciniata* (Brassicaceae), a new addition to the flora of New Zealand. *New Zealand Journal of Botany* 47: 133–137.

de Lange, P.J. and J.R. Rolfe. 2008. *Hebe saxicola* (Plantaginaceae)—A new threatened species from western Northland, North Island, New Zealand. *New Zealand Journal of Botany* 46: 531–545.

de Lange, P.J., J.R. Rolfe, C.S. Liew, and P.B. Pelser. 2014. *Senecio australis* Willd. (Asteraceae: Senecioneae)—A new and uncommon addition to the indigenous vascular flora of New Zealand. *New Zealand Journal of Botany* 52: 417–428.

De Laubenfels, D.J. 1988. Coniferales. *Flora Malesiana I* 10: 367–442.

Delgado, J.A. and R.L. Palma. 2010. A revision of the genus *Podaena* Ordish (Insecta: Coleoptera: Hydraenidae). *Zootaxa* 2678: 1–47.

Dendy, A. 1910. On the structure, development and morphological interpretation of the pineal organs and adjacent parts of the brain in the Tuatara (*Sphenodon punctatus*). *Philosophical Transactions of the Royal Society of London B* 201: 227–331 (Pls. 19–31).

Dennett, D. 1996. *Darwin's Dangerous Idea: Evolution and the Meanings of Life.* London, UK: Penguin.

de Queiroz, A. 2014. *The Monkey's Voyage: How Improbable Journeys Shaped the History of Life.* New York: Basic Books.

Dercole, F. and S. Rinaldi. 2008. *Analysis of Evolutionary Processes: The Adaptive Dynamics Approach and Its Applications.* Princeton, NJ: Princeton University Press.

de Vries, H. 1912. Species and varieties: Their origin by mutation. In *Genes, Cells and Organisms: Great Books in Experimental Biology*, ed. D.T. Macdougal. Chicago, IL: Open Court.

Diamond, J. and Bond, A.B., 1999. *Kea, Bird of Paradox: the Evolution and Behavior of a New Zealand Parrot.* Berkeley, CA: University of California Press.

Diamond, J.M. 1997. *Guns, Germs and Steel: The Fates of Human Societies.* London, UK: Random House.

DiCaprio, L., R.D. Müller, M. Gurnis, and A. Goncharov. 2009. Linking active margin dynamics to overriding plate deformation: Synthesizing geophysical images with geological data from the Norfolk Basin. *Geochemistry, Geophysics, Geosystems* 10, Q01004. doi:10.1029/2008GC002222.

Dickinson, K.J.M., A.F. Mark, B.I.P. Barratt, and B.H. Patrick. 1998. Rapid ecological survey, inventory and implementation: A case study from Waikaia Ecological Region, New Zealand. *Journal of the Royal Society of New Zealand* 28: 83–156.

Dickinson, T.A. 1978. Epiphylly in angiosperms. *Botanical Review* 44: 181–232.

Dieffenbach, E. 1843. *Travels in New Zealand.* London, UK: John Murray.

Dierschke, T., T. Mandáková, M.A. Lysak, and K. Mummenhoff. 2009. A bicontinental origin of polyploid Australian/New Zealand *Lepidium* species (Brassicaceae)? Evidence from genomic in situ hybridization. *Annals of Botany* 104: 681–688.

Dietrich, C.H. 2003. Auchenorrhyncha (cicadas, spittlebugs, leafhoppers, treehoppers and plant hoppers). In *Encyclopedia of Entomology*, eds. V.H. Resh and R.C. Cardé, pp. 66–74. San Diego, CA: Academic Press.

Dingwall, R. and M. Aldridge. 2006. Television wildlife programming as a source of popular scientific information: A case study of evolution. *Public Understanding of Science* 15: 131–152.

Dobson, G.E. 1875. On the genus *Chalinolobus*, with descriptions of new or little-known species. *Proceedings of the Zoological Society of London* 43: 381–388.

Dobzhansky, T. 1937. *Genetics and the Origin of Species.* New York: Columbia University Press.

Dobzhansky, T. 1970. *Genetics of the Evolutionary Process.* New York: Columbia University Press.

Doll, R. 1982. Grundriß der Evolution der Gattung *Taraxacum* Zinn. *Feddes Repertorium* 93: 481–624.

Domínguez, M.C. and A.C. Pont. 2014. Fanniidae (Insecta: Diptera). *Fauna of New Zealand* 71: 1–91.

Domínguez, M.C. and S.A. Roig-Juñent. 2011. Historical biogeographic analysis of the family Fanniidae (Diptera: Calyptratae), with special reference to the austral species of the genus *Fannia* (Diptera: Fanniidae) using dispersal-vicariance analysis. *Revista Chilena de Historia Natural* 84: 65–82.

Donald, K.M., D.J. Winter, A.L. Ashcroft, and H.G. Spencer. 2015. Phylogeography of the whelk genus *Cominella* (Gastropoda: Buccinidae) suggests long-distance counter-current dispersal of a direct developer. *Biological Journal of the Linnean Society* 115: 315–332.

Donoghue P.C.J. and M.J. Benton. 2007. Rocks and clocks: Calibrating the tree of life using fossils and molecules. *Trends in Ecology and Evolution* 22: 424–431.

Dornburg A., J.M. Beaulieu, J.C. Oliver, and T.J. Near. 2011. Integrating fossil preservation biases in the selection of calibrations for molecular divergence time estimation conclusions. *Systematic Biology* 60: 519–527.

dos Remedios, N., P.L. Lee, T. Burke, T. Székely, and C. Küpper. 2015. North or south? Phylogenetic and biogeographic origins of a globally distributed avian clade. *Molecular Phylogenetics and Evolution* 89: 151–159.

Douglas, N.A. and P.S. Manos. 2007. Molecular phylogeny of Nyctaginaceae: Taxonomy, biogeography, and characters associated with a radiation of xerophytic genera in North America. *American Journal of Botany* 94: 856–872.

Dover, G. 1982. Molecular drive: A cohesive mode of species evolution. *Nature* 299: 111–117.

Dover, G. 2000. *Dear Mr. Darwin: Letters on the Evolution of Life and Human Nature*. Berkeley, CA: University of California Press.

Drake, J.M. 2013. A niche for theory and another for practice. *Trends in Ecology and Evolution* 28: 76–77.

Drew, B.T., B.R. Ruhfel, S.A. Smith et al. 2014. Another look at the root of the angiosperms reveals a familiar tale. *Systematic Biology* 63: 368–382.

Drew, J.A., G.R. Allen, and M.V. Erdmann. 2010. Congruence between mitochondrial genes and color morphs in a coral reef fish: Population variability in the Indo-Pacific damselfish *Chrysiptera rex* (Snyder, 1909). *Coral Reefs* 29: 439–444.

Driskell, A., L. Christidis, B.J. Gill, W.E. Boles, F.K. Barker, and N.W. Longmore. 2007. A new endemic family of New Zealand passerine birds: Adding heat to a biodiversity hotspot. *Australian Journal of Zoology* 55: 73–78.

Druce, A.P. and P.A. Williams. 1989. Vegetation and flora of the Ben More—Chalk Range area of southern Marlborough South Island. *New Zealand Journal of Botany* 27: 167–199.

Drummond, A.J. and A. Rambaut. 2007. BEAST: Bayesian evolutionary analysis sampling trees. *BMC Evolutionary Biology* 7(214): 1–8.

Drury, D.G. 1973. Nodes and leaf structure in the classification of some Australasian shrubby Senecioneae-Compositae. *New Zealand Journal of Botany* 11: 525–554.

Dueck, L.A., D. Aygoren, and K.M. Cameron. 2014. A molecular framework for understanding the phylogeny of *Spiranthes* (Orchidaceae), a cosmopolitan genus with a North American center of diversity. *American Journal of Botany* 101: 1551–1571.

Dugdale, J.S. 1974. Alpine moths. *New Zealand's Nature Heritage* 4: 1526–1531.

Dugdale, J.S. 1990. Reassessment of *Ctenopseustis* Meyrick and *Planotortrix* Dugdale with descriptions of two new genera (Lepidoptera: Tortricidae). *New Zealand Journal of Zoology* 17: 437–465.

Dugdale, J.S. 1994. Hepialidae. *Fauna of New Zealand* 30: 1–164.

Dugdale, J.S. and R. Emberson. 1996. Insects. In *The Chatham Islands: Heritage and Conservation*, ed. Anonymous, pp. 93–98. Christchurch, New Zealand: Canterbury University Press.

Dugdale, J.S. and R. Emberson. 2008. Terrestrial invertebrates. In *Chatham Islands: Heritage and Conservation*, ed. C. Miskelly, pp. 116–124. Christchurch, New Zealand: Canterbury University Press.

Dugdale, J.S. and C.A. Fleming. 1978. New Zealand cicadas of the genus *Maoricicada* (Homoptera: Tibicinidae). *New Zealand Journal of Zoology* 5: 295–340.

Duke, N.C. 1991. A systematic revision of the mangrove genus *Avicennia* (Avicenniaceae) in Australasia. *Australian Systematic Botany* 4: 299–324.

Dumont, E.R., A. Herrel, R.A. Medellin, J.A. Vargas-Contreras, and S.E. Santana. 2009. Built to bite: Cranial design and function in the wrinkle-faced bat. *Journal of Zoology* 279: 329–337.

Duret, L. and N. Galtier. 2009. Biased gene conversion and the evolution of mammalian genomic landscapes. *Annual Review of Genomics and Human Genetics* 10: 285–311.

Durham, J.W. 1985. Movement of the Caribbean plate and its importance for biogeography in the Caribbean. *Geology* 13: 123–125.

Dussex, N., D. Wegmann, and B.C. Robertson. 2014. Postglacial expansion and not human influence best explains the population structure in the endangered kea (*Nestor notabilis*). *Molecular Ecology* 23: 2193–2209.

Dwyer, P.D. 1960. New Zealand bats. *Tuatara* 8: 61–71.

Dwyer, P.D. 1962. Studies on the two New Zealand bats. *Zoological Publications of Victoria University of Wellington* 28: 1–28.

Eagle, A. 2006. *Eagle's Complete Trees and Shrubs of New Zealand*. Wellington, New Zealand: Te Papa Press.

Eagles, G., K. Gohl, and R.D. Larter. 2004. High-resolution animated tectonic reconstruction of the South Pacific and West Antarctic Margin. *Geochemistry, Geophysics, Geosystems* 5: Q07002.

Early, J.W. 1995. Insects of the Aldermen Islands. *Tane* 35: 1–14.

Early, J.W., L. Masner, I.D. Naumann, and A.D. Austin. 2001. Maamingidae, a new family of proctotrupoid wasp (Insecta: Hymenoptera) from New Zealand. *Invertebrate Taxonomy* 15: 341–352.

Eberle, J., W. Walbaum, R.C.M. Warnock, S. Fabrizi, and D. Ahrens. 2015. Asymmetry in genitalia does not increase the rate of their evolution. *Molecular Phylogenetics and Evolution* 93: 180–187.

Eckermann, J.P. 1836/1970. *Conversations with Goethe*. London, UK: Dent.

Edgar, A. 1972. Classified summarized notes. *Notornis* 19(Suppl.): 1–91.

Edgar, E. 1986. *Poa* L. in New Zealand. *New Zealand Journal of Botany* 24: 425–503.

Edgar, E. 1995. New Zealand species of *Deyeuxia* P. Beauv. and *Lachnagrostis* Trin. (Gramineae: Aveneae). *New Zealand Journal of Botany* 33: 1–33.

Edgar, E. 1996. *Puccinellia* Parl. (Gramineae: Poeae) in New Zealand. *New Zealand Journal of Botany* 34: 17–32.

Edgar, E. and H.E. Connor. 1998. *Zotovia* and *Microlaena*: New Zealand ehrhartoid Gramineae. *New Zealand Journal of Botany* 36: 565–586.

Edgar, E. and H.E. Connor. 1999. Species novae Graminum. *New Zealand Journal of Botany* 37: 63–70.

Edgar, E. and H.E. Connor. 2000. *Flora of New Zealand*, Vol. 5: Gramineae. Wellington, New Zealand: Government Printer.

Edgecombe, G.D. and G. Giribet. 2008. A New Zealand species of the trans-Tasman centipede order Craterostigmomorpha (Arthropoda: Chilopoda) corroborated by molecular evidence. *Invertebrate Systematics* 22: 1–15.

Edwards, K.J. and P.A. Gadek. 2001. Evolution and biogeography of *Alectryon* (Sapindaceae). *Molecular Phylogenetics and Evolution* 20: 14–26.

Edwards, S.V. 2009. Is a new and general theory of molecular systematics emerging? *Evolution* 63: 1–19.

Eiting, T.P. and G.F. Gunnell. 2009. Global completeness of the bat fossil record. *Journal of Mammalian Evolution* 16: 151–173.

Ekman, S. 1953. *Zoogeography of the Sea*. London, UK: Sidgwick and Jackson.

Eldredge, N., J.N. Thompson, P.M. Brakefield et al. 2005. The dynamics of evolutionary stasis. *Paleobiology* 31: 133–145.

Elith, J. and J.R. Leathwick. 2009. Species distribution models: Ecological explanation and prediction across space and time. *Annual Review of Ecology, Evolution and Systematics* 40: 677–697.

Elliott, G.P. 1990. The breeding biology and habitat relationships of the Yellowhead. PhD thesis, Victoria University of Wellington, Wellington, New Zealand.

Elliott, G.P., P.J. Dilks, and C.F.J. O'Donnell. 1996. Nest site selection by mohua and yellow-crowned parakeets in beech forest in Fiordland, New Zealand. *New Zealand Journal of Zoology* 23: 267–278.

Emadzade, K., B. Gehrke, H.P. Linder, and E. Hörandl. 2011. The biogeographical history of the cosmopolitan genus *Ranunculus* L. (Ranunculaceae) in the temperate to meridional zones. *Molecular Phylogenetics and Evolution* 58: 4–21.

Emadzade, K. and E. Hörandl. 2011. Northern Hemisphere origin, transoceanic dispersal, and diversification of Ranunculeae (Ranunculaceae) in the Tertiary. *Journal of Biogeography* 38: 517–530.

Emberson, R. 1998. The beetle (Coleoptera) fauna of the Chatham Islands. *New Zealand Entomologist* 21: 25–64.

Emerson, B.C. and G.P. Wallis. 1995. Phylogenetic relationships of the *Prodontria* (Coleoptera: Scarabaeidae; subfamily Melolonthinae), derived from sequence variation in the mitochondrial cytochrome oxidase II gene. *Molecular Phylogenetics and Evolution* 4: 433–447.

Endrödy-Younga, S. 1995. Descriptions of two new species of *Sphaerothorax* Endrödy-Younga (Coleoptera: Clambidae) from Chile: Relics in *Nothofagus* forest reflecting ecological zonation in Gondwanaland. *Annals of the Transvaal Museum* 36: 177–182.

Engel, J.J. and D. Glenny. 2008. *A Flora of the Liverworts and Hornworts of New Zealand*, Vol. 1. St. Louis, MO: Missouri Botanical Garden Press.

Engel, J.J. and J. Heinrichs. 2008. Studies of New Zealand Hepaticae. 39. *Dinckleria* Trevis, an older name for *Proskauera* Heinrichs and J.J. Engel. *Cryptogamie, Bryologie* 29: 193–194.

Engler, A. 1882. *Versuch einer Entwicklungsgeschichte der Pflanzenwelt, insbesondere der Florengebiete seit der Tertiärperiode*. Vol. 2: *Die extratropischen Gebiete der Südlichen Hemisphäre und die tropischen Gebiete*. Leipzig, Germany: W. Engelmann.

Ericson, P.G., S. Klopfstein, M. Irestedt, J.M. Nguyen, and J.A Nylander. 2014. Dating the diversification of the major lineages of Passeriformes (Aves). *BMC Evolutionary Biology* 14(8): 1–15.

Esteve-Altava, B., J. Marugán-Lobón, H. Botella, and D. Rasskin-Gutman. 2013. Structural constraints in the evolution of the tetrapod skull complexity: Williston's law revisited using network models. *Evolutionary Biology* 40: 209–219.

Ewen, J.G., I. Flux, and P.G.P. Ericson. 2006. Systematic affinities of two enigmatic New Zealand passerines of high conservation priority, the hihi or stitchbird *Notiomystis cincta* and the kokako *Callaeas cinerea*. *Molecular Phylogenetics and Evolution* 40: 281–284.

Excoffier, L., M. Foll, and R.J. Petit. 2009. Genetic consequences of range expansions. *Annual Review of Ecology, Evolution, and Systematics* 40: 481–501.

Fabre, P.-H., A. Rodrigues, and E.J.P. Douzery. 2009. Patterns of macroevolution among Primates inferred from a supermatrix of mitochondrial and nuclear DNA. *Molecular Phylogenetics and Evolution* 53: 808–825.

Fahlke, J.M., P.D. Gingerich, R.C. Welsh, and A.R. Wood. 2011. Cranial asymmetry in Eocene archaeocete whales and the evolution of directional hearing in water. *Proceedings of the National Academy of Sciences USA* 108: 14545–14548.

Fain, M.G. and P. Houde. 2007. Multilocus perspectives on the monophyly and phylogeny of the order Charadriiformes (Aves). *BMC Evolutionary Biology* 7(35): 1–15.

Fan Q.-H. and Z.-Q. Zhang. 2005. Raphignathoidea (Acari: Prostigmata). *Fauna of New Zealand* 52: 1–400.

Farquhar, H. 1906. The New Zealand Plateau. *Transactions of the New Zealand Institute* 39: 135–137. http://rsnz.natlib.govt.nz/images/rsnz_39/rsnz_39_00_0613_0000f_ac_01.jpg

Farrell, B.D. 2001. Evolutionary assembly of the milkweed fauna: Cytochrome oxidase I and the age of *Tetraopes* beetles. *Molecular Phylogenetics and Evolution* 18: 467–478.

Fay, M.F., P.J. Rudall, S. Sullivan et al. 2000. Phylogenetic studies of Asparagales based on four plastid DNA regions. In *Monocots: Systematics and Evolution*, eds. K.L. Wilson and D.A. Morrison, pp. 360–371. Collingwood, Victoria, Australia: CSIRO.

Feduccia, A. 1975. Morphology of the bony stapes in Menuridae and Acanthisittidae. *Wilson Bulletin* 87: 418–420.

Feduccia, A. 1996. *The Origin and Evolution of Birds*. New Haven, CT: Yale University Press.

Feldheim, K.A., S.H. Gruber, J.D. DiBattista et al. 2014. Two decades of genetic profiling yields first evidence of natal philopatry and long-term fidelity to parturition sites in sharks. *Molecular Ecology* 23: 110–117.

Fermond, C. 1858. Faits pour servir á l'histoire générale de la fécondation chez les végétaux. Troisième partie. Théorie mécanique de la préfloraison et de la floraison. *Comptes Rendus de l'Académie des Sciences Paris* 47: 1059–1061.

Fernández, R. and G. Giribet. 2014. Phylogeography and species delimitation in the New Zealand endemic, genetically hypervariable harvestman species, *Aoraki denticulata* (Arachnida, Opiliones, Cyphophthalmi). *Invertebrate Systematics* 28: 401–414.

Fernández-Triana, J.L., D.F. Ward, S. Cardinal, and C. van Achterberg. 2013. A review of *Paroplitis* (Braconidae, Microgastrinae), and description of a new genus from New Zealand, *Shireplitis*, with convergent morphological traits. *Zootaxa* 3722: 549–568.

Fernandez-Triana, J.L., D.F. Ward, and J.B. Whitfield. 2011. *Kiwigaster* gen. nov. (Hymenoptera: Braconidae) from New Zealand: The first Microgastrinae with sexual dimorphism in number of antennal segments. *Zootaxa* 2932: 24–32.

Fife, A.J. 2014a. Fabroniaceae. In *Flora of New Zealand—Mosses*, Fascicle 7, eds. P.B. Heenan, I. Breitwieser, and A.D. Wilton. Lincoln, New Zealand: Manaaki Whenua Press.

Fife, A.J. 2014b. Archidiaceae. In *Flora of New Zealand—Mosses*, Fascicle 10, eds. P.B. Heenan, I. Breitwieser, and A.D. Wilton. Lincoln, New Zealand: Manaaki Whenua Press.

Fife, A.J. 2014c. Calymperaceae. In *Flora of New Zealand—Mosses*, Fascicle 12, eds. P.B. Heenan, I. Breitwieser, and A.D. Wilton. Lincoln, New Zealand: Manaaki Whenua Press.

Fife, A.J. 2014d. Hylocomiaceae. In *Flora of New Zealand—Mosses*, Fascicle 15, eds. P.B. Heenan, I. Breitwieser, and A.D. Wilton. Lincoln, New Zealand: Manaaki Whenua Press.

Fife, A.J. 2015. Cyrtopodaceae. In *Flora of New Zealand—Mosses*, Fascicle 17, eds. P.B. Heenan, I. Breitwieser, and A.D. Wilton. Lincoln, New Zealand: Manaaki Whenua Press.

Fikáček, M., Y. Minoshima, D. Vondráček, N. Gunter, and R.A.B. Leschen. 2013. Morphology of adults and larvae and integrative taxonomy of southern hemisphere genera *Tormus* and *Afrotormus* (Coleoptera: Hydrophilidae). *Acta Entomologica Musei Nationalis Pragae* 53: 75–126.

Filipowicz, N. and S.S. Renner. 2012. *Brunfelsia* (Solanaceae): A genus evenly divided between South America and radiations on Cuba and other Antillean islands. *Molecular Phylogenetics and Evolution* 64: 1–11.

Filler, A.G. 2007. Homeotic evolution in the mammalia: Diversification of therian axial seriation and the morphogenetic basis of human origins. *PLoS One* 2(10): e1019 (pp. 1–23).

Findley, J.S. and D.E. Wilson. 1982. Ecological significance of chiropteran morphology. In *Ecology of Bats*, ed. T.H. Kunz, pp. 243–260. New York: Plenum.

Finn, C.A., R.D. Müller, and K.S. Panter. 2005. A Cenozoic diffuse alkaline magmatic province (DAMP) in the SW Pacific without rift or plume origin. *Geochemistry, Geophysics, Geosystems*, 6: Q02005.

Fisher, R.A. 1930. *The Genetical Theory of Natural Selection*. Oxford, UK: Clarendon Press.

Fitton, J.G., J.J. Mahoney, P.J. Wallace, and A.D. Saunders. 2004. Origin and evolution of the Ontong Java Plateau: Introduction. In J.G. Fitton, J.J. Mahoney, P.J. Wallace, and A.D. Saunders (eds.), *Origin and Evolution of the Ontong Java Plateau.* Geological Society of London Special Publication 229, pp. 1–8.

Fitzpatrick, B.M., J.A. Fordyce, and S. Gavrilets. 2009. Patterns, processes and geographic modes of speciation. *Journal of Evolutionary Biology* 22: 2342–2347.

Fjeldså, J. 2004. *The Grebes.* New York: Oxford University Press.

Flannery, T.F. 1995a. *Mammals of the South-West Pacific and Moluccan Islands.* Ithaca, NY: Cornell University Press.

Flannery, T.F. 1995b. *Mammals of New Guinea*, 2nd edn. Sydney, New South Wales, Australia: Reed.

Fleagle, J.G. 1999. *Primate Adaptation and Evolution.* San Diego, CA: Academic Press.

Fleming, C.A. 1950. New Zealand flycatchers of the genus *Petroica* Swainson. *Transactions of the Royal Society of New Zealand* 78: 14–47.

Fleming, C.A. 1962a. New Zealand biogeography—A paleontologist's approach. *Tuatara* 10: 53–108.

Fleming, C.A. 1962b. History of the New Zealand land bird fauna. *Notornis* 9: 270–274.

Fleming, C.A. 1971. A new species of cicada from rock fans in Southern Wellington, with a review of three species with similar songs and habitat. *New Zealand Journal of Science* 14: 443–79.

Fleming, C.A. 1976. New Zealand as a minor source of terrestrial plants and animals in the Pacific. *Tuatara* 22: 30–37.

Fleming, C.A. 1979. *The Geological History of New Zealand and Its Life.* Auckland, New Zealand: Auckland University Press.

Fleming, C.A. 1982. *George Edward Lodge: The Unpublished New Zealand Bird Paintings.* Wellington, New Zealand: Nova Pacifica and National Museum of New Zealand.

Foggo, M.N., R.A. Hitchmough, and C.H. Daugherty. 1997. Systematic and conservation implications of geographic variation in pipits (*Anthus*: Motacillidae) in New Zealand and some offshore islands. *Ibis* 139: 366–373.

Foissner, W., S. Hess, and K. Al-Rasheid. 2010. Two vicariant *Semispathidium* species from tropical Africa and central Europe: *S. fraterculum* nov. spec. and *S. pulchrum* nov. spec. (Ciliophora, Haptorida). *European Journal of Protistology* 46: 61–73.

Folch, A. 1992. Apterygidae. In *Handbook of the Birds of the World*, Vol. 1, eds. J. del Hoyo, A. Elliott, and J. Sargatal, pp. 104–111. Barcelona, Spain: Lynx Edicions.

Fontaneto, D., ed. 2011. *Biogeography of Microscopic Organisms: Is Everything Small Everywhere?* Cambridge, UK: Systematics Association and Cambridge University Press.

Ford, V.S. and L.D. Gottlieb. 2007. Tribal relationships within Onagraceae inferred from *PgiC* sequences. *Systematic Botany* 32: 348–356.

Fordyce, E. 2010. Darwin's legacy and New Zealand's fossils. In *Aspects of Darwin: A New Zealand Celebration*, eds. D. Galloway and J. Timmins, pp. 65–83. Dunedin, New Zealand: Hewitson.

Fordyce, R.E. and D.T. Ksepka. 2012. The strangest bird. *Scientific American* 307(5): 56–61.

Forster, M.A. and G.S. Lister. 2003. Cretaceous metamorphic core complexes in the Otago Schist, New Zealand. *Australian Journal of Earth Sciences* 50: 181–198.

Forster, R.R. 1954. The New Zealand harvestmen (sub-order Laniatores). *Canterbury Museum Bulletin* 2: 1–329.

Forster, R.R. 1955. Spiders of the family Archaeidae from Australia and New Zealand. *Transactions of the Royal Society of New Zealand* 83: 391–403.

Forster, R.R. 1975. The spiders and harvestmen. In *Biogeography and Ecology in New Zealand*, ed. G. Kuschel, pp. 493–505. The Hague, the Netherlands: Junk.

Forster, R.R. and L.M. Forster. 1974. Fiordland National Park. *New Zealand's Nature Heritage* 7: 2800–2802.

Forster, R.R., N.I. Platnick, and M.R. Gray. 1987. A review of the spider superfamilies Hypochiloidea and Austrochiloidea (Araneae, Araneomorphae). *Bulletin of the American Museum of Natural History* 185: 1–120.

Forster, R.R. and C.L. Wilton. 1968. *The Spiders of New Zealand, Part 2. Ctenizidae, Dipluridae.* Otago Museum Bulletin No. 2.

Foweraker, C. 1929. The rain forest of Westland. No. 2—Kahikatea and Totara forests. *Te Kura Ngahere* [University of Canterbury] 2: 2, 6.

Fox, K.J. 1982. Entomology of the Egmont National Park. *New Zealand Entomologist* 7: 286–289.

Francis, M. 2012. *Coastal Fishes of New Zealand*, 4th edn. Nelson, New Zealand: Craig Potton.

Francis, M.P. 1995. Geographic distribution of marine reef fishes in the New Zealand region. *New Zealand Journal of Marine and Freshwater Research* 30: 35–55.

Franklin, D.A. 1962. The Ericaceae in New Zealand (*Gaultheria and Pernettya*). *Transactions of the Royal Society of New Zealand* 1: 155–173.

Franzke, A., D. German, I.A. Al-Shehbaz, and K. Mummenhoff. 2009. *Arabidopsis* family ties: Molecular phylogeny and age estimates in Brassicaceae. *Taxon* 58: 425–437.

Franzke, A., M.A. Lysak, I.A. Al-Shehbaz, M.A. Koch, and K. Mummenhoff. 2011. Cabbage family affairs: The evolutionary history of Brassicaceae. *Trends in Plant Science* 16: 108–116.

Fraser, C.I., R. Nikula, D.E. Ruzzante, and J.M. Waters. 2012. Poleward bound: Biological impacts of Southern Hemisphere glaciation. *Trends in Ecology and Evolution* 27: 462–471.

Fraser, C.I., D.J. Winter, H.G. Spencer, and J.M. Waters. 2010. Multigene phylogeny of the southern bull-kelp genus *Durvillaea* (Phaeophyceae: Fucales). *Molecular Phylogenetics and Evolution* 57: 1301–1311.

Freeman, P.W. 1988. Frugivorous and animalivorous bats (Microchiroptera): Dental and cranial adaptations. *Biological Journal of the Linnean Society* 33: 249–272.

Frey, M.A. 2010. The relative importance of geography and ecology in species diversification: Evidence from a tropical marine intertidal snail (*Nerita*). *Journal of Biogeography* 37: 1515–1528.

Frick, H. and N. Scharff. 2013. Phantoms of Gondwana?—Phylogeny of the spider subfamily Mynogleninae (Araneae: Linyphiidae). *Cladistics* 30: 67–106.

Friedman, M., B.P. Keck, A. Dornburg et al. 2013. Molecular and fossil evidence place the origin of cichlid fishes long after Gondwanan rifting. *Proceedings of the Royal Society B: Biological Sciences* 280(1770): article 20131733.

Frodin, D.G., Lowry, P.P, III, and Plunkett, G.M. 2010. *Schefflera* (Araliaceae): taxonomic history, overview and progress. *Plant Diversity and Evolution* 128: 561–595.

Frohlich, M.W. 2006. Recommendations and goals for evodevo research: Scenarios, genetic constraint, and developmental homeostasis. *Aliso* 22: 172–187.

Frost, D. 2010. A marine geophysical study of the Tonga Trench—Louisville ridge collisional system in the South-West Pacific Ocean. Master of Earth Sciences thesis, University of Oxford, Oxford, UK.

Fuchs, J., J.A. Johnson, and D.P. Mindell. 2015. Rapid diversification of falcons (Aves: Falconidae) due to expansion of open habitats in the Late Miocene. *Molecular Phylogenetics and Evolution* 82: 166–182.

Fulthorpe, C.S., R.M. Carter, K.G. Miller,. and J. Wilson. 1996. Marshall Paraconformity: A Mid-Oligocene record of inception of the Antarctic Circumpolar Current and coeval glacio-eustatic lowstand? *Marine and Petroleum Geology* 13: 61–77.

Futuyma, D.J. 2010. Evolutionary constraint and ecological consequences. *Evolution* 64: 1865–1884.

Gage, M. 1970. Late Cretaceous and Tertiary rocks of Broken River, Canterbury. *New Zealand Journal of Geology and Geophysics* 13: 507–536.

Galatius, A. and Å. Jespersen. 2005. Bilateral directional asymmetry of the appendicular skeleton of the harbor porpoise (*Phocoena phocoena*). *Marine Mammal Science* 21: 401–410.

Galloway, D.J. 2000. The lichen genus *Peltigera* (Peltigerales: Ascomycota) in New Zealand. *Tuhinga* 11: 1–45.

Galloway, D.J. 2007. *Flora of New Zealand: Lichens*, 2nd edn. Lincoln, New Zealand: Manaaki Whenua.

Gamerro, J.C. and V. Barreda. 2008. New fossil record of Lactoridaceae in southern South America: A palaeobiogeographical approach. *Biological Journal of the Linnean Society* 158: 41–50.

Gans, C. 1975. Tetrapod limblessness: Evolution and functional corollaries. *American Zoologist* 15: 455–467.

Gao, K.-Q. and Y. Wang. 2001. Mesozoic anurans from Liaoning Province, China, and phylogenetic relationships of archaeobatrachian anuran clades. *Journal of Vertebrate Paleontology* 21: 460–476.

Garcia-R, J.C., G. Elliott, K. Walker, I. Castro, and S.A. Trewick. 2016. Trans-equatorial range of a land bird lineage (Aves: Rallidae) from tropical forests to subantarctic grasslands. *Journal of Avian Biology* 47: 219–226.

Garcia-R, J.C., G.C. Gibb, and S.A. Trewick. 2014. Deep global evolutionary radiation in birds: Diversification and trait evolution in the cosmopolitan bird family Rallidae. *Molecular Phylogenetics and Evolution* 81: 96–108.

Garcia-R, J.C. and S.A. Trewick. 2015. Dispersal and speciation in purple swamphens (Rallidae: *Porphyrio*). *The Auk* 132: 140–155.

Gardiner, J.D., G. Dimitriadis, W.I. Sellers, and J.R. Codd. 2008. The aerodynamics of big ears in the brown long-eared bat *Plecotus auritus*. *Acta Chiropterologica* 10: 313–321.

Gardner, R.C., P.J. de Lange, D.J. Keeling, T. Bowala, H.A. Brown, and S.D. Wright. 2004. A late Quaternary phylogeography for *Metrosideros* (Myrtaceae) in New Zealand inferred from chloroplast DNA haplotypes. *Biological Journal of the Linnean Society* 83: 399–412.

Gardner, R.O. 1976. Studies in the Alseuosmiaceae. PhD thesis, University of Auckland, Auckland, New Zealand.

Gardner, R.O. 1997a. *Macropiper* (Piperaceae) in the south-west Pacific. *New Zealand Journal of Botany* 35: 293–308.

Gardner, R.O. 1997b. *Mida salicifolia*, our native sandalwood. *Auckland Botanical Society Journal* 52: 42–43.

Gardner, R.O. 2007. The nectaries of *Phormium* J.R. & G. Forst. (Hemerocallidaceae). *New Zealand Natural Sciences* 32: 29.

Gardner, R.O. and P.J. de Lange. 2002. Revision of *Pennantia* (Icacinaceae), a small isolated genus of Southern Hemisphere trees. *Journal of the Royal Society of New Zealand* 32: 669–695.

Gardner, R.O. and M. Heads. 2004. *Coprosma macrocarpa* subsp. *minor* (Rubiaceae), a new subspecies from northern New Zealand. *New Zealand Natural Sciences* 28: 67–80.

Garnock-Jones, P.J. 1984. *Ceratocephalus pungens* (Ranunculaceae): A new species from New Zealand. *New Zealand Journal of Botany* 22: 135–137.

Garnock-Jones, P.J. 1986. A new status for the New Zealand mousetail (*Myosurus*, Ranunculaceae). *New Zealand Journal of Botany* 24: 351–354.

Garnock-Jones, P.J. 1993. *Heliohebe* (Scrophulariaceae–Veroniceae), a new genus segregated from *Hebe*. *New Zealand Journal of Botany* 31: 323–339.

Garnock-Jones, P.J. 2006. Morphology. In *An Illustrated Guide to New Zealand Hebes*, eds. M. Bayly and A. Kellow, pp. 27–37. Wellington, New Zealand: Te Papa Press.

Garnock-Jones, P.J. and D.G. Lloyd. 2004. A taxonomic revision of *Parahebe* (Plantaginaceae) in New Zealand. *New Zealand Journal of Botany* 42: 181–232.

Gartner, G.E.A., T. Gamble, A.L. Jaffe, A. Harrison, and J.B. Losos. 2013. Left–right dewlap asymmetry and phylogeography of *Anolis lineatus* on Aruba and Curaçao. *Biological Journal of the Linnean Society* 110: 409–426.

Garvey, B. 2007. *Philosophy of Biology*. Stocksfield, UK: Acumen.

Gaston, K.J. 2003. *Structure and Dynamics of Geographic Ranges*. Oxford, UK: Oxford University Press.

Gaston, K.J. 2009a. Geographic range limits of species. *Proceedings of the Royal Society B* 276: 1391–1393.

Gaston, K.J. 2009b. Geographic range limits: Achieving synthesis. *Proceedings of the Royal Society B* 276: 1395–1406.

Gatesy, S.M. and K.P. Dial. 1996. From frond to fan: *Archaeopteryx* and the evolution of short-tailed birds. *Evolution* 50: 2037–2048.

Gauld, I.D. and D.B. Wahl. 2000. The Labeninae (Hymenoptera: Ichneumonidae): A study in phylogenetic reconstruction and evolutionary biology. *Zoological Journal of the Linnean Society* 129: 271–347.

Gauld, I.D. and D.B. Wahl. 2002. The Eucerotinae: A Gondwanan origin for a cosmopolitan group of Ichneumonidae? *Journal of Natural History* 36: 2229–2248.

Gaussen, H. 1954. *Géographie des Plantes*, 2e éd. Paris, France: Armand Colin.

GBIF. 2016. Global biodiversity information facility. http://data.gbif.org/.Accessed May, 2016.

Gehring, W.J. 2011. Chance and necessity in eye evolution. *Genome Biology and Evolution* 3: 1053–1066.

Geiger, D.L. 2012. *Monograph of the Little Slit Shells*. 2 vols. Santa Barbara, CA: Santa Barbara Museum of Natural History.

Genner, M.J., O. Seehausen, D.H. Lunt et al. 2007. Age of cichlids: New dates for ancient fish radiation. *Molecular Biology and Evolution* 24: 1269–1282.

Gerstein, A.C. and J.-S. Moore. 2011. Small is the new big: Assessing the population structure of microorganisms. *Molecular Ecology* 20: 4385–4387.

Giannini, N.P. and N.P. Simmons. 2007. The chiropteran premaxilla: A reanalysis of morphological variation and its phylogenetic interpretation. *American Museum Novitates* 3585: 1–44.

Gibb, G. and D. Penny. 2010. Two aspects along the continuum of pigeon evolution: A South-Pacific radiation and the relationship of pigeons within Neoaves. *Molecular Phylogenetics and Evolution* 56: 698–706.

Gibbons, K.L., B.J. Conn, and M.J. Henwood. 2014. The Australasian genus *Schizacme* (Loganiaceae): New combinations and new species in the New Zealand flora. *Telopea* 17: 363–381.

Gibbs, G.W. 1980. *New Zealand Butterflies: Identification and Natural History*. Auckland, New Zealand: Collins.

Gibbs, G.W. 1990a. Local or global? Biogeography of some primitive Lepidoptera in New Zealand. *New Zealand Journal of Zoology* 16: 689–698.

Gibbs, G.W. 1990b. The silent majority: A plea for the consideration of invertebrates in New Zealand island management. *Conservation Sciences Publication* 2: 123–127.

Gibbs, G.W. 1999. Four new species of giant weta, *Deinacrida* (Orthoptera: Anostostomatidae: Deinacridinae) from New Zealand. *Journal of the Royal Society of New Zealand* 29: 307–324.

Gibbs, G.W. 2001. Habitats and biogeography of New Zealand's deinacridine and tusked weta species. In *The Biology of Wetas, King Crickets and Their Allies*, ed. L.H. Field, pp. 35–56. Oxford, UK: CABI Publishing.

Gibbs, G.W. 2006. *Ghosts of Gondwana: The History of Life in New Zealand*. Nelson, New Zealand: Craig Potton.

Gibbs, G.W. 2010. Micropterigidae (Lepidoptera) of the Southwestern Pacific: A revision with the establishment of five new genera from Australia, New Caledonia and New Zealand. *Zootaxa* 2520: 1–48.

Gibbs, G.W. 2014. Micropterigidae (Insecta: Lepidoptera). *Fauna of New Zealand* 72: 1–127.

Gibbs, G.W. and D.C. Lees. 2014. New Caledonia as an evolutionary cradle: A re-appraisal of the jaw-moth genus *Sabatinca* (Lepidoptera: Micropterigidae) and its significance for assessing the antiquity of the island's fauna. In *Zoologia Neocaledonica 8. Biodiversity Studies in New Caledonia*, eds. É. Guilbert, T. Robillard, H. Jordan, and P. Grandcolas. *Mémoires du Muséum national d'Histoire naturelle* 206: 239–266.

Gibson, G.M. and T.R. Ireland. 1995. Granulite formation during continental extension in Fiordland, New Zealand. *Nature* 375: 479–482.

Gibson, G.M., I. McDougall, and T.R. Ireland. 1988. Age constraints on metamorphism and the development of a metamorphic core complex in Fiordland, southern New Zealand. *Geology* 16: 405–408.

Gibson, R. and A. Baker. 2012. Multiple gene sequences resolve phylogenetic relationships in the shorebird suborder Scolopaci (Aves: Charadriiformes). *Molecular Phylogenetics and Evolution* 66: 64–72.

Gill, B.J. 2004. Acanthisittidae. In *Handbook of the Birds of the World*, Vol. 9, eds. J. del Hoyo, A. Elliott, and D.A. Christie, pp. 464–473. Barcelona, Spain: Lynx Edicions.

Gill, F.B. 1990. *Ornithology*. New York: Freeman.

Gill, H.S., C.B. Renaud, F. Chapleau, R.L. Mayden, and I.C. Potter. 2003. Phylogeny of living parasitic lampreys (Petromyzontiformes) based on morphological data. *Copeia* 2003: 687–703.

Gillespie, L.J., R.J. Soreng, and S.W.L. Jacobs. 2009. Phylogenetic relationships of Australian *Poa* (Poaceae: Poinae), including molecular evidence for two new genera, *Saxipoa* and *Sylvipoa*. *Australian Systematic Botany* 22: 413–436.

Gillespie, M., S.D. Wratten, R. Cruickshank, B.H. Wiseman, and G.W. Gibbs. 2013. Incongruence between morphological and molecular markers in the butterfly genus *Zizina* (Lepidoptera: Lycaenidae) in New Zealand. *Systematic Entomology* 38: 151–163.

Gillespie, R.G., B.G. Baldwin, J.M. Waters, C.I. Fraser, R. Nikula, and G.K. Roderick. 2012. Long-distance dispersal: A framework for hypothesis testing. *Trends in Ecology and Evolution* 27: 47–56.

Gilroy, S. and A. Trewavas. 2001. Signal processing and transduction in plant cells: The end and the beginning. *Nature Reviews (Molecular Cell Biology)* 2: 307–314.

Gimmel, M.L., R.A.B. Leschen, and S.A. Ślipiński. 2010. Review of the New Zealand endemic family Cyclaxyridae, new family (Coleoptera: Polyphaga). *Acta Entomologica Musei Nationalis Pragae* 49: 511–528.

Gingerich, P.D. and M.D. Uhen. 1998. Likelihood estimation of the time of origin of Cetacea and the time of origin of Cetacea and Artiodactyla. *Paleontologica Electronica* 1: 1–47.

Giribet, G. and S.L. Boyer. 2010. 'Moa's Ark' or 'Goodbye Gondwana': Is the origin of New Zealand's terrestrial invertebrate fauna ancient, recent, or both? *Invertebrate Systematics* 24: 1–8.

Giribet, G. and G.D. Edgecombe. 2006. Conflict between datasets and phylogeny of centipedes: An analysis based on seven genes and morphology. *Proceedings of the Royal Society, London B* 269: 235–241.

Giribet, G. and P.P. Sharma. 2015. Evolutionary biology of harvestmen (Arachnida, Opiliones). *Annual Review of Entomology* 60: 157–175.

Giribet, G., P.P. Sharma, L.R. Benavides et al. 2012. Evolutionary and biogeographical history of an ancient and global group of arachnids (Arachnida: Opiliones: Cyphophthalmi) with a new taxonomic arrangement. *Biological Journal of the Linnean Society* 105: 92–130.

Given, D.R. 1968. Taxonomic studies in the genus *Celmisia* Cass. PhD thesis, University of Canterbury, Christchurch, New Zealand.

Given, D.R. 1969. A synopsis of infrageneric categories in *Celmisia* (Astereae-Compositae). *New Zealand Journal of Botany* 7: 400–418.

Given, D.R. 1971. Two new species of *Celmisia* Cass. (Compositae–Astereae). *New Zealand Journal of Botany* 9: 526–532.

Given, D.R. 1975. Celmisias. *New Zealand's Natural Heritage* 7: 2567–2572.

Given, D.R. 1980. A taxonomic revision of *Celmisia coriacea* (Forst.f.) Hook.f. and its immediate allies (Astereae-Compositae). *New Zealand Journal of Botany* 18: 127–140.

Given, D.R. and M. Gray. 1986. *Celmisia* (Compositae: Asteraceae) in Australia and New Zealand. In *Flora and Fauna of Alpine Australasia: Ages and Origins*, ed. B.A. Barlow, pp. 451–470. Melbourne, Australia: CSIRO.

Glaessner, M.F. 1960. The fossil decapod Crustacea of New Zealand and the evolution of the order Decapoda. *Palaeontological Bulletin of the New Zealand Geological Survey* 31: 1–63.

Glenny, D. 2004. A revision of the genus *Gentianella* in New Zealand. *New Zealand Journal of Botany* 42: 361–530.

Glenny, D. 2009. A revision of the genus *Forstera* (Stylidiaceae) in New Zealand. *New Zealand Journal of Botany* 47: 285–315.

GNS Science. 2016. Basement terranes of New Zealand. http://www.gns.cri.nz/Home/Our-Science/Earth-Science/Regional-Geology/Geological-Origins/Basement-terranes-of-New-Zealand. Accessed May, 2016.

Godley, E.J. 1979. Flower biology in New Zealand. *New Zealand Journal of Botany* 17: 441–466.

Godley, E.J. and P.E. Berry. 1995. The biology and systematics of *Fuchsia* in the South Pacific. *Annals of the Missouri Botanical Garden* 82: 473–516.

Goethe, J.W. 1790. *Versuch die Metamorphose der Pflanzen zu erklären*. Gotha, Germany: C.W. Ettinger.

Goldberg, J., M. Knapp, R.M. Emberson, J.I. Townsend, and S.A. Trewick. 2014. Species radiation of carabid beetles (Broscini: *Mecodema*) in New Zealand. *PLoS One* 9(1): e86185 (pp. 1–13).

Goldberg, J. and S.A. Trewick. 2011. Exploring phylogeographic congruence in a continental island system. *Insects* 2: 369–399.

Goldberg, J., S.A. Trewick, and A.M. Paterson. 2008. Evolution of New Zealand's terrestrial fauna: A review of molecular evidence. *Philosophical Transactions of the Royal Society B* 363: 3319–3334.

Goldberg, J., S.A. Trewick, and R.G. Powlesland. 2011. Population structure and biogeography of *Hemiphaga* pigeons (Aves: Columbidae) on islands in the New Zealand region. *Journal of Biogeography* 38: 285–298.

Golding, R.E. 2012. Molecular phylogenetic analysis of mudflat snails (Gastropoda: Euthyneura: Amphiboloidea) supports an Australasian centre of origin. *Molecular Phylogenetics and Evolution* 63: 72–81.

Goldstien, S.J., D.R. Schiel, and N.J. Gemmell. 2006. Comparative phylogeography of coastal limpets across a marine disjunction in New Zealand. *Molecular Ecology* 15: 3259–3268.

Goodwin, B. 2009. Beyond the Darwinian paradigm. In *Evolution: The First Four Billion Years*, eds. M. Ruse and J. Travis, pp. 299–312. Cambridge, MA: Harvard University Press.

Goremykin, V.V., S.V. Nikiforova, P.J. Biggs, B. Zhong, P. de Lange, W. Martin, S. Woetzel, R.A. Atherton, T. McLenachan, and P.J. Lockhart. 2013. The evolutionary root of flowering plants. *Systematic Biology* 62: 50–61.

Gorman, A., C. McLachlan, and G. Wilson. 2013. Investigating if Saunders Ridges is a drowned shoreline on the Otago continental shelf. *Geoscience Society of New Zealand Conference Abstracts*, Christchurch, New Zealand, p. 36. https://securepages.co.nz/~gsnz/siteadmin/uploaded/gs_downloads/Abstracts/2013Christchurch_abstracts.pdf.

Gould, S.J. 1989. Tires to sandals. *Natural History* April: 8–15.

Gould, S.J. 2002. *The Structure of Evolutionary Theory*. Cambridge, MA: Harvard University Press.

Graf, D.L., H. Jones, A.J. Geneva, J.M. Pfeiffer III, and M.W. Kluzinger. 2015. Molecular phylogenetic analysis supports a Gondwanan origin of the Hyriidae (Mollusca: Bivalvia: Unionida) and the paraphyly of Australasian taxa. *Molecular Phylogenetics and Evolution* 85: 1–9.

Grandcolas, P., J. Murienne, T. Robillard et al. 2008. New Caledonia: A very old Darwinian island? *Philosophical Transactions of the Royal Society B* 363: 3309–3317.

Grandcolas, P., R. Nattier, and S. Trewick. 2014. Relict species: A relict concept? *Trends in Ecology and Evolution* 29: 655–663.

Grant, V. 1985. *The Evolutionary Process: A Critical Review of Evolutionary Theory*. New York: Columbia University Press.

Grant-Mackie, J.A., J.S. Buckeridge, and P.M. Johns. 1996. Two new Upper Jurassic arthropods from New Zealand. *Alcheringa* 20: 31–39.

Gray, D.R. and D.A. Foster. 2004. ^{40}Ar/^{39}Ar thermochronologic constraints on deformation, metamorphism and cooling/exhumation of a Mesozoic accretionary wedge, Otago Schist, New Zealand. *Tectonophysics* 385: 181–210.

Gray, J.E. 1842. Descriptions of two hitherto unrecorded species of reptiles from New Zealand; presented to the British Musuem by Dr Dieffenbach. *Zoological Miscellany* 4: 72.

Gray, R.D. and Q.D. Atkinson. 2003. Language-tree divergence times support the Anatolian theory of Indo-European origin. *Nature* 426: 435–439.

Greaves, S.N.J., D.G. Chapple, C.H. Daugherty, D.M. Gleeson, and P.A. Ritchie. 2008. Genetic divergences pre-date Pleistocene glacial cycles in the New Zealand speckled skink, *Oligosoma infrapunctatum*. *Journal of Biogeography* 35: 853–864.

Greaves, S.N.J., D.G. Chapple, D.M. Gleeson, C.H. Daugherty, and P.A. Ritchie. 2007. Phylogeography of the spotted skink (*Oligosoma lineoocellatum*) and green skink (*O. chloronoton*) species complex (Lacertilia: Scincidae) in New Zealand reveals pre-Pleistocene divergence. *Molecular Phylogenetics and Evolution* 45: 729–739.

Greene, T.C. 1999. Aspects of the ecology of Antipodes Island Parakeet (*Cyanoramphus unicolor*) and Reischek's Parakeet (*C. novaezelandiae hochstetten*) on Antipodes Island, October–November 1995. *Notornis* 46: 301–310.

Gregory, P.A. 2007. Acanthizidae. In *Handbook of the Birds of the World*, Vol. 12, eds. J. del Hoyo, A. Elliott, and D.A. Christie, pp. 544–611. Barcelona, Spain: Lynx Edicions.

Gregory, T.R. 2009. Understanding natural selection: Essential concepts and common misconceptions. *Evolution: Education and Outreach* 2: 156–175.

Grey-Wilson, C. 1987. Plant portraits. 82. *Clematis marmoraria. Curtis' Botanical Magazine* 4: 116–119.

Griesemer, J.R. 1990. Modeling in the museum: On the role of remnant models in the work of Joseph Grinnell. *Biology and Philosophy* 5: 3–36.

Grímsson, F., R. Zetter, and C.-G. Hofmann. 2011. *Lythrum* and *Peplis* from the Late Cretaceous and Cenozoic of North America and Eurasia: New evidence suggesting early diversification within the Lythraceae. *American Journal of Botany* 98: 1801–1815.

Grinnell, J. 1914. Barriers to distribution as regards birds and mammals. *American Naturalist* 48: 248–254.

Grinnell, J. 1917. The niche-relationships of the California Thrasher. *Auk* 34: 427–433.

Grinnell, J. 1924. Geography and evolution. *Ecology* 5: 225–229.

Grobys, J.W.G., K. Gohl, B. Davy, G. Uenzelmann-Neben, T. Deen, and D. Barker. 2007. Is the Bounty Trough off eastern New Zealand an aborted rift? *Journal of Geophysical Research* 112: B03103.

Grobys, J.W.G., K. Gohl, G. Uenzelmann-Neben, B. Davy, and D. Barker. 2009. Extensional and magmatic nature of the Campbell Plateau and Great South Basin. *Tectonophysics* 472: 213–225.

Groom, S.V.C. and M.P. Schwarz. 2011. Bees in the Southwest Pacific: Origins, diversity and conservation. *Apidologie* 42: 759–770.

Grossenbacher, D.L., S.D. Veloz, and J.P. Sexton. 2014. Niche and range size patterns suggest that speciation begins in small, ecologically diverged populations in North American monkeyflowers (*Mimulus* spp.). *Evolution* 68: 1270–1280.

Grosser, S., C.P. Burridge, A.J. Peucker, and J.M. Waters. 2015. Coalescent modelling suggests recent secondary-contact of cryptic penguin species. *PLoS One* 10(12): e0144966 (pp. 1–17).

Grzimek, B. 1988/1990. *Grzimek's Encyclopedia of Mammals*, Vol. 1. New York: McGraw-Hill.

Gunnell, G.F. and N.B. Simmons. 2005. Fossil evidence and the origin of bats. *Journal of Mammalian Evolution* 12: 209–246.

Gunnell, G.F., N.B. Simmons, and E.R. Seiffert. 2014. New Myzopodidae (Chiroptera) from the Late Paleogene of Egypt: Emended family diagnosis and biogeographic origins of Noctilionoidea. *PLoS One* 9(2): e86712 (pp. 1–14).

Gustafsson A.L.S., C.F. Verola, and A. Antonelli. 2010. Reassessing the temporal evolution of orchids with new fossils and a Bayesian relaxed clock, with implications for the diversification of the rare South American genus *Hoffmannseggella* (Orchidaceae: Epidendroideae). *BMC Evolutionary Biology* 10(177): 1–13.

Haase, M., B.A. Marshall, and I. Hogg. 2007. Disentangling causes of disjunction: The Alpine fault theory of vicariance (New Zealand, South Island). *Biological Journal of the Linnean Society* 91: 361–374.

Haase, P. 1986. Phenology and productivity of *Olearia ilicifolia* (Compositae) at Arthur's Pass, South Island, New Zealand. *New Zealand Journal of Botany* 24: 369–379.

Haddrath, O. and A.J. Baker. 2001. Complete mitochondrial DNA genome sequences of extinct birds: Ratite phylogenetics and the vicariance biogeography hypothesis. *Proceedings of the Royal Society, London B* 268: 939–945.

Haeckel, E. 1904. *Kunstformen der Natur*. Leipzig, Germany: Bibliographische Institut.

Hagen, K.B. von and J.W. Kadereit, 2001. The phylogeny of *Gentianella* (Gentianaceae) and its colonization of the southern hemisphere as revealed by nuclear and chloroplast DNA sequence variation. *Organisms Diversity and Evolution* 1: 61–79.

Haldane, J.B.S. 1932. *The Causes of Evolution*. New York: Longmans, Green.

Hallé, F., R.A. Oldeman, and P.B. Tomlinson. 1978. *Tropical Trees and Forests: An Architectural Analysis*. Berlin, Germany: Springer.

Halpern, M.E., J.O. Liang, and J.T. Gamse. 2003. Leaning to the left: Laterality in the zebrafish forebrain. *Trends in Neurosciences* 26: 308–313.

Halvorsen, K.A.T., E. Árnason, P.J. Smith, and J. Mork. 2012. Mitochondrial DNA differentiation between the antitropical blue whiting species *Micromesistius poutassou* and *Micromesistius australis*. *Journal of Fish Biology* 81: 253–269.

Hamilton, A. 1878. The district of Okarito. *Transactions of the New Zealand Institute* 11: 435–438.

Hamilton, K.G.A. 1999. The ground-dwelling leafhoppers Myerslopiidae, new family, and Sagmatiini, new tribe (Homoptera: Membracoidea). *Invertebrate Taxonomy* 13: 207–235.

Hamilton, W.B. 1988. Plate tectonics and island arcs. *Bulletin of the Geological Society of America* 100: 1503–1527.

Hamilton, W.B. 2007. Driving mechanism and 3-D circulation of plate tectonics. *Geological Society of America Special Paper* 433: 1–25.

Hamlin, B.G. 1958. Studies in New Zealand *Carices* IV and V. *Transactions of the Royal Society of New Zealand* 85: 387–396.

Hammer, M.P., M. Adams, P.J. Unmack, and K.F. Walker. 2007. A rethink on *Retropinna*: conservation implications of new taxa and significant genetic sub-structure in Australian smelts (Pisces: Retropinnidae). *Marine and Freshwater Research* 58: 327–341.

Hamner, R.M., F.B. Pichler, D. Heimeir, R. Constantine, and C.S. Baker. 2012. Genetic differentiation and limited gene flow among fragmented populations of New Zealand endemic Hector's and Maui's dolphins. *Conservation Genetics* 13: 987–1002.

Hand, S. 1998. *Xenorhinos*, a new genus of Old World leaf-nosed bats (Microchiroptera: Hipposideridae) from the Australian Miocene. *Journal of Vertebrate Paleontology* 18: 430–439.

Hand, S., M. Archer, and H. Godthelp. 2001. New Miocene *Icarops* material (Microchiroptera: Mystacinidae) from Australia, with a revised diagnosis of the genus. *Memoirs of the Association of Australasian Palaeontologists* 25: 139–146.

Hand, S.J., M. Archer, and H. Godthelp. 2005. Australian Oligocene-Miocene mystacinids (Microchiroptera): Upper dentition, new taxa and divergence of New Zealand species. *Geobios* 38: 339–352.

Hand, S., R. Beck, T. Worthy, M. Archer, and B. Sigé. 2007. Australian and New Zealand bats: the origin, evolution, and extinction of bat lineages in Australasia. *Journal of Vertebrate Paleontology* 27(3, supplement) Abstracts: 86A.

Hand, S.J., D.E. Lee, T.H. Worthy et al. 2015. Miocene fossils reveal ancient roots for New Zealand's endemic *Mystacina* (Chiroptera) and its rainforest habitat. *PLoS One* 10(6), e0128871 (pp. 1–19).

Hand, S.J., P. Murray, D. Megirian, M. Archer, and H. Godthelp. 1998. Mystacinid bats (Microchiroptera) from the Australian Tertiary. *Journal of Paleontology* 72: 538–545.

Hand, S.J., V. Weisbecker, R.M.D. Beck, M. Archer, H. Godthelp, A.J.D. Tennyson, and T.H. Worthy. 2009. Bats that walk: A new evolutionary hypothesis for the terrestrial behaviour of New Zealand's endemic mystacinids. *BMC Evolutionary Biology* 9(article 169): 1–13.

Hand, S.J., T. Worthy, M. Archer, and A. Tennyson. 2006. Placing *Mystacina* in a palaeontological context—First Tertiary bats from New Zealand reveal new implications for the origins of the Australasian bat fauna. *Australasian Bat Society Newsletter* 26: 12.

Hand, S.J., T.H. Worthy, M. Archer, J.P. Worthy, A.J.D. Tennyson, and R.P. Scofield. 2013. Miocene mystacinids (Chiroptera, Noctilionoidea) indicate a long history for endemic bats in New Zealand. *Journal of Vertebrate Paleontology* 33: 1442–1448.

Hansford, D. 2013. Orchidelirium. *New Zealand Geographic* 124: 32–51.

Harcourt, R. 2005. Family Otariidae. In *The Handbook of New Zealand Mammals*, 2nd edn., ed. C.M. King, pp. 225–241. Melbourne, Australia: Oxford University Press.

Hardy, A. 1965. *The Living Stream*. London, UK: Collins.

Hare, K.M., C.H. Daugherty, and D.G. Chapple. 2008. Comparative phylogeography of three skink species (*Oligosoma moco, O. smithi, O. suteri*; Reptilia: Scincidae) in northeastern New Zealand. *Molecular Phylogenetics and Evolution* 46: 303–315.

Härlid, A. and U. Arnason. 1999. Analyses of mitochondrial DNA nest ratite birds within the Neognathae: Supporting a neotenous origin of ratite morphological characters. *Proceedings of the Royal Society of London. Series B: Biological Sciences* 266: 305–309.

Harris, A.C. 1983. An Eocene larval insect fossil (Diptera: Bibionidae) from North Otago New Zealand. *Journal of the Royal Society of New Zealand* 13: 93–105.

Harris, A.C. 2006. Notes on two seldom-collected stiletto-flies, *Anabarhynchus fuscofemoratus* and *A. harrisi* (Diptera: Therevidae). *Weta* 32: 19–22.

Harrison, R.A. 1976. Arthropoda of the Southern Islands of New Zealand (9) Diptera. *Journal of the Royal Society of New Zealand* 6: 107–152.

Harvey, M.S. 1996a. The biogeography of Gondwanan pseudoscorpions (Arachnida). *Revue Suisse de Zoologie* 1: 255–264.

Harvey, M.S. 1996b. Micro-arthropods and their value in Gondwanan biogeographic studies. In *Gondwanan Heritage*, eds. S.D. Hopper, J. Chappill, M.S. Harvey, and A. George, pp. 155–162. Sydney, New South Wales, Australia: Surrey Beatty and Sons.

Hatch, J.J. 1985. Lateral asymmetry of the bill of *Loxops coccineus* (Drepanidinae). *Condor* 87: 546–547.

Hay, J.M., S.D. Sarre, D.M. Lambert, F.W. Allendorf, and C.H. Daugherty. 2010. Genetic diversity and taxonomy: A reassessment of species designation in tuatara (*Sphenodon*: Reptilia). *Conservation Genetics* 11: 1063–1081.

Hayward, B.W., C.J. Hollis, and H.R. Grenfell. 1997. Recent Elphidiidae (Foraminiferida) of the south-west Pacific and fossil Elphidiidae of New Zealand. *Institute of Geological and Nuclear Science Monograph* 16 (= *New Zealand Geological Survey Paleontological Bulletin*, 72): 1–170.

Heads, M. 1990a. Integrating earth and life sciences in New Zealand natural history: The parallel arcs model. *New Zealand Journal of Zoology* 16: 549–585.

Heads, M. 1990b. A taxonomic revision of *Kelleria* and *Drapetes* (Thymelaeaceae). *Australian Systematic Botany* 3: 595–652.

Heads, M. 1992. Taxonomic notes on the *Hebe* complex (Scrophulariaceae) in the New Zealand mountains. *Candollea* 47: 583–595.

Heads, M. 1993. Biogeography and biodiversity in *Hebe*, a South Pacific genus of Scrophulariaceae. *Candollea* 48: 19–60.

Heads, M. 1994a. Biogeographic studies in New Zealand Scrophulariaceae: Tribes Rhinantheae, Calceolarieae and Gratioleae. *Candollea* 49: 55–80.

Heads, M. 1994b. A biogeographic review of *Ourisia* (Scrophulariaceae) in New Zealand. *Candollea* 49: 23–36.

Heads, M. 1994c. A biogeographic review of *Parahebe* (Scrophulariaceae). *Botanical Journal of the Linnean Society* 115: 65–89.

Heads, M. 1994d. Biogeography and evolution in the *Hebe* complex (Scrophulariaceae): *Leonohebe* and *Chionohebe*. *Candollea* 49: 81–119.

Heads, M. 1994e. Morphology, architecture and taxonomy in the *Hebe* complex (Scrophulariaceae). *Bulletin du Muséum national d'Histoire naturelle* Paris 4e sér., 16, sect. B., *Adansonia*: 163–191.

Heads, M. 1996. Biogeography, taxonomy and evolution in the Pacific genus *Coprosma* (Rubiaceae). *Candollea* 51: 381–405.

Heads, M. 1997. Regional patterns of biodiversity in New Zealand: One degree analysis of plant and animal distributions. *Journal of the Royal Society of New Zealand* 27: 337–354.

Heads, M. 1998a. Biogeographic disjunction along the Alpine fault, New Zealand. *Biological Journal of the Linnean Society* 63: 161–176.

Heads, M. 1998b. Biodiversity in the New Zealand divaricating tree daisies: *Olearia* sect. nov. (Compositae). *Botanical Journal of the Linnean Society* 127: 239–285. [For the map of *Olearia solandri* in the North Island, see the Erratum, in *Botanical Journal of the Linnean Society* 131: 101, 1999.]

Heads, M. 1999. Vicariance biogeography and terrane tectonics in the South Pacific: An analysis of the genus *Abrotanella* (Compositae). *Biological Journal of the Linnean Society* 67: 391–432.

Heads, M. 2000. A new species of *Hoheria* (Malvaceae) from northern New Zealand. *New Zealand Journal of Botany* 38: 373–377.

Heads, M. 2001. Birds of paradise, biogeography and ecology in New Guinea: A review. *Journal of Biogeography* 28: 1–33.

Heads, M. 2002. Birds of paradise, vicariance biogeography and terrane tectonics in New Guinea. *Journal of Biogeography* 29: 261–284.

Heads, M. 2003. Ericaceae in Malesia: Vicariance biogeography, terrane tectonics and ecology. *Telopea* 10: 311–449.

Heads, M. 2004. What is a node? *Journal of Biogeography* 31: 1883–1891.

Heads, M. 2005a. Towards a panbiogeography of the seas. *Biological Journal of the Linnean Society* 84: 675–723.

Heads, M. 2005b. Dating nodes on molecular phylogenies: A critique of molecular biogeography. *Cladistics* 21: 62–78.

Heads, M. 2006a. Panbiogeography of *Nothofagus* (Nothofagaceae): Analysis of the main species massings. *Journal of Biogeography* 33: 1066–1075.

Heads, M. 2006b. Seed plants of Fiji: An ecological analysis. *Biological Journal of the Linnean Society* 89: 407–431.

Heads, M. 2008a. Biological disjunction along the West Caledonian fault, New Caledonia: A synthesis of molecular phylogenetics and panbiogeography. *Botanical Journal of the Linnean Society* 158: 470–488.

Heads, M. 2008b. Panbiogeography of New Caledonia, southwest Pacific: Basal angiosperms on basement terranes, ultramafic endemics inherited from volcanic island arcs, and old taxa endemic to young islands. *Journal of Biogeography* 35: 2153–2175.

Heads, M. 2009a. Darwin's changing ideas on evolution: From centres of origin and teleology to vicariance and incomplete lineage sorting. *Journal of Biogeography* 36: 1018–1026.

Heads, M. 2009b. Inferring biogeographic history from molecular phylogenies. *Biological Journal of the Linnean Society* 98: 757–774.

Heads, M. 2009c. Globally basal centres of endemism: The Tasman-Coral Sea region (south-west Pacific), Latin America and Madagascar/South Africa. *Biological Journal of the Linnean Society* 96: 222–245.

Heads, M. 2010. The biogeographical affinities of the New Caledonian biota: A puzzle with 24 pieces. *Journal of Biogeography* 37: 1179–1201.

Heads, M. 2011. Using island age to estimate the age of island-endemic clades and calibrate molecular clocks. *Systematic Biology* 60: 204–218.

Heads, M. 2012a. *Molecular Panbiogeography of the Tropics*. Berkeley, CA: University of California Press.

Heads, M. 2012b. Bayesian transmogrification of clade divergence dates: A critique. *Journal of Biogeography* 39: 1749–1756.

Heads, M. 2012c. South Pacific biogeography, tectonic calibration, and pre-drift tectonics: Cladogenesis in *Abrotanella* (Asteraceae). *Biological Journal of the Linnean Society* 107: 938–952.

Heads, M. 2014. *Biogeography of Australasia: A Molecular Analysis*. Cambridge, UK: Cambridge University Press.

Heads, M. 2015a. Panbiogeography, its critics, and the case of the ratite birds. *Australian Systematic Botany* 27: 241–256.

Heads, M. 2015b. Biogeography by revelation: Investigating a world shaped by miracles. *Australian Systematic Botany* 27: 282–304.

Heads, M. 2015c. The relationship between biogeography and ecology: Envelopes, models, predictions. *Biological Journal of the Linnean Society* 115: 456–468.

Heads, M. and R.C. Craw. 2004. The Alpine fault biogeographic hypothesis revisited. *Cladistics* 20: 184–190.

Heads, M. and B. Patrick. 2003. The biogeography of southern New Zealand. In *The Natural History of Southern New Zealand*, eds. J. Darby, R.E. Fordyce, A. Mark, K. Probert, and C. Townsend, pp. 89–100. Dunedin, New Zealand: University of Otago Press.

Heads, S.W. 2008. The first fossil Proscopiidae (Insecta, Orthoptera, Eumastacoidea) with comments on the historical biogeography and evolution of the family. *Palaeontology* 51: 499–507.

Healy, A.J. 1948. Contributions to the naturalised flora of New Zealand, No. 2. *Transactions of the Royal Society of New Zealand* 77: 172–185.

Heather, B.D. 1966. *A Biology of Birds with Particular Reference to New Zealand Birds*. Lower Hutt, New Zealand: Ornithological Society of New Zealand.

Heather, B.D. and H.A. Robertson. 2005. *The Field Guide to the Birds of New Zealand*. Auckland, New Zealand: Viking.

Heenan, P.B. 1995. A taxonomic revision of *Carmichaelia* (Fabaceae–Galegeae) in New Zealand (part I). *New Zealand Journal of Botany* 33: 455–475.

Heenan, P.B. 1996a. A taxonomic revision of *Carmichaelia* (Fabaceae–Galegeae) in New Zealand (part II). *New Zealand Journal of Botany* 34: 157–177.

Heenan, P.B. 1996b. Identification and distribution of the Marlborough pink brooms, *Notospartium carmichaeliae* and *N. glabrescens* (Fabaceae–Galegeae), in New Zealand. *New Zealand Journal of Botany* 34: 299–307.

Heenan, P.B. 1997. Heteroblasty in *Carmichaelia*, *Chordospartium*, *Corallospartium*, and *Notospartium* (Fabaceae–Galegeae) from New Zealand. *New Zealand Journal of Botany* 35: 243–249.

Heenan, P.B. 1998a. Phylogenetic analysis of the *Carmichaelia* complex, *Clianthus*, and *Swainsona* (Fabaceae), from Australia and New Zealand. *New Zealand Journal of Botany* 36: 21–40.

Heenan, P.B. 1998b. *Montigena* (Fabaceae–Galegeae), a new genus endemic to New Zealand. *New Zealand Journal of Botany* 36: 41–51.

Heenan, P.B. 1998c. An emended circumscription of *Carmichaelia* (Fabaceae–Galegeae), with new combinations, a key, and notes on hybrids. *New Zealand Journal of Botany* 36: 53–63.

Heenan, P.B. 2000. *Clianthus* (Fabaceae) in New Zealand: A reappraisal of Colenso's taxonomy. *New Zealand Journal of Botany* 38: 361–371.

Heenan, P.B. 2001. Relationships of *Streblorrhiza* (Fabaceae), an extinct monotypic genus from Phillip Island, South Pacific Ocean. *New Zealand Journal of Botany* 39: 9–15.

Heenan, P.B. 2005. *Olearia quinquevulnera* (Asteraceae: Astereae), a new species name from New Zealand, and observations on its relationships in *Olearia*. *New Zealand Journal of Botany* 43: 753–766.

Heenan, P.B. 2009. A new species of *Pachycladon* (Brassicaceae) from limestone in eastern Marlborough, New Zealand. *New Zealand Journal of Botany* 47: 155–161.

Heenan, P.B. and J.W. Barkla. 2007. A new combination in *Carmichaelia* (Fabaceae). *New Zealand Journal of Botany* 45: 265–268.

Heenan, P.B. and J. Cartman. 2000. Reinstatement of *Clematis petriei* (Ranunculaceae), and typification and variation of *C. forsteri*. *New Zealand Journal of Botany* 38: 575–585.

Heenan, P.B. and P.J. de Lange. 2004. *Myrsine aquilonia* and *M. umbricola* (Myrsinaceae), two new species from New Zealand. *New Zealand Journal of Botany* 42: 753–769.

Heenan, P.B., P.J. de Lange, G.J. Houliston, A. Barnaud, and B.G. Murray. 2008a. *Olearia telmatica* (Asteraceae: Astereae), a new tree species endemic to the Chatham Islands. *New Zealand Journal of Botany* 46: 567–583.

Heenan, P.B., P.J. de Lange, and A.D. Wilton. 2001. *Sophora* (Fabaceae) in New Zealand: Taxonomy, distribution, and biogeography. *New Zealand Journal of Botany* 39: 17–53.

Heenan, P.B. and P.J. Garnock-Jones. 1999. A new species combination in *Cheesemania* (Brassicaceae) from New Zealand. *New Zealand Journal of Botany* 37: 235–241.

Heenan, P.B., D.F. Goeke, G.J. Houliston, and M.A. Lysak. 2012. Phylogenetic analyses of ITS and *rbcL* DNA sequences for sixteen genera of Australian and New Zealand Brassicaceae result in the expansion of the tribe Microlepidieae. *Taxon* 61: 970–979.

Heenan, P.B., E.B. Knox, S.P. Courtney, P.N. Johnson, and M.I. Dawson. 2008b. Generic placement in *Lobelia* and revised taxonomy for New Zealand species previously in *Hypsela* and *Isotoma* (Lobeliaceae). *New Zealand Journal of Botany* 46: 87–100.

Heenan, P.B., P.J. Lockhart, N. Kirkham, K. McBreen, and D. Havell. 2006. Relationships in the alpine *Ranunculus haastii* (Ranunculaceae) complex and recognition of *R. piliferus* and *R. acraeus* from southern New Zealand. *New Zealand Journal of Botany* 44: 425–441.

Heenan, P.B. and M.S. McGlone. 2013. Evolution of New Zealand alpine and open-habitat plant species during the late Cenozoic. *New Zealand Journal of Ecology* 37: 105–113.

Heenan, P.B. and A.D. Mitchell. 2003. Phylogeny, biogeography and adaptive radiation of *Pachycladon* (Brassicaceae) in the mountains of South Island, New Zealand. *Journal of Biogeography* 30: 1737–1749.

Heenan, P.B., A.D. Mitchell, P.J. de Lange, J. Keeling, and A.M. Paterson. 2010. Late Cenozoic origin and diversification of Chatham Islands endemic plant species revealed by analyses of DNA sequence data. *New Zealand Journal of Botany* 48: 83–136.

Heenan, P.B., A.D. Mitchell, and M. Koch. 2002. Molecular systematics of the New Zealand *Pachycladon* (Brassicaceae) complex: Generic circumscription and relationships to *Arabidopsis* sens. lat. and *Arabis* sens. lat. *New Zealand Journal of Botany* 40: 543–562.

Heenan, P.B., A.D. Mitchell, P.A. McLenachan, P.J. Lockhart, and P.J. de Lange. 2007. Natural variation and conservation of *Lepidium sisymbrioides* Hook. f. and *L. solandri* Kirk (Brassicaceae) in South Island, New Zealand, based on morphological and DNA sequence data. *New Zealand Journal of Botany* 45: 237–264.

Heenan, P.B. and B.P.J. Molloy. 2004. Taxonomy, ecology, and conservation of *Olearia adenocarpa* (Asteraceae), a new species from braided riverbeds in Canterbury, New Zealand. *New Zealand Journal of Botany* 42: 21–36.

Heenan, P.B., B.P.J. Molloy, and P.J. de Lange. 2003. Species of *Convolvulus* (Convolvulaceae) endemic to New Zealand. *New Zealand Journal of Botany* 41: 447–457.

Heenan, P.B., B.P.J. Molloy, and R.D. Smissen. 2013a. *Cardamine cubita* (Brassicaceae), a new species from New Zealand with a remarkable reduction in floral parts. *Phytotaxa* 140: 43–50.

Heenan, P.B. and R.D. Smissen. 2013. Revised circumscription of *Nothofagus* and recognition of the segregate genera *Fuscospora*, *Lophozonia*, and *Trisyngyne* (Nothofagaceae). *Phytotaxa* 146: 1–31.

Heenan, P.B., I.R.H. Telford, and J.J. Bruhl. 2013b. Three new species of *Gingidia* (Apiaceae: Apioideae) from Australia and New Zealand segregated from *G. montana*. *Australian Systematic Botany* 26: 196–209.

Heenan, P.B., C.J. Webb, and P.N. Johnson. 1996. *Mazus arenarius* (Scrophulariaceae), a new, small-flowered, and rare species segregated from *M. radicans*. *New Zealand Journal of Botany* 34: 33–40.

Heinrichs, J., A. Bombosch, K. Feldberg et al. 2011. A phylogeny of the northern temperate leafy liverwort genus *Scapania* (Scapaniaceae, Jungermanniales). *Molecular Phylogenetics and Evolution* 62: 979–985.

Heinrichs, J., M. Lindner, H. Groth et al. 2006. Goodbye or welcome Gondwana?—Insights into the phylogenetic biogeography of the leafy liverwort *Plagiochila* with a description of *Proskauera*, gen. nov. (Plagiochilaceae, Jungermanniales). *Plant Systematics and Evolution* 258: 227–250.

Henderson, R.C. 2011. Diaspididae (Insecta: Hemiptera: Coccoidea). *Fauna of New Zealand* 66: 1–275.

Hendry, A.P. 2009. Ecological speciation! Or the lack thereof? *Canadian Journal of Fisheries and Aquatic Sciences* 66: 1383–1398.

Henig, A. and B.P. Luyendyk. 2007. The Manihiki Plateau, Hikurangi Plateau, Wishbone Scarp, and Osbourn Trough: A review and analysis. *American Geophysical Union, Fall Meeting 2007*, San Francisco, CA, abstract #T12A-01. http://adsabs.harvard.edu/abs/2007AGUFM.T12A..01H.

Henne, A., D. Craw, and D. MacKenzie. 2011. Structure of the Blue Lake Fault Zone, Otago Schist, New Zealand. *New Zealand Journal of Geology and Geophysics* 54: 311–328.

Hennig, W. 1965. Vorarbeiten zu einem phylogenetischen system der Muscidae (Diptera: Cyclorrhapha). *Stuttgarter Beiträge zur Naturkunde* 141: 1–100.

Hennig, W. 1966. *Phylogenetic Systematics*. Urbana, IL: University of Illinois Press.

Hennion, F. and D.W. Walton. 1997. Ecology and seed morphology of endemic species from Kerguelen Phytogeographic Zone. *Polar Biology* 8: 229–235.

He-Nygrén, X., A. Juslén, I. Ahonen, D. Glenny, and S. Piippo. 2006. Illuminating the evolutionary history of liverworts (Marchantiophyta)—Towards a natural classification. *Cladistics* 22: 1–31.

Herzer, R., P.G. Quilty, G.C.H. Chaproniere, and H.J. Campbell. 2009b. Assisted passage—Early Miocene transport of insular terrestrial biota towards New Zealand by a moving tectonic plate. *Geological Society of New Zealand Miscellaneous Publication* 126: 22–24.

Herzer, R.H., B.W. Davy, N. Mortimer et al. 2009a. Seismic stratigraphy and structure of the Northland Plateau and the development of the Vening Meinesz transform margin, SW Pacific Ocean. *Marine Geophysical Research* 30: 21–60.

Herzer, R.H. and J. Mascle. 1996. Anatomy of a continental back-arc transform—The Vening Meinesz fracture zone northwest of New Zealand. *Marine Geophysical Research* 18: 401–427.

Heuret, A. and S. Lallemand. 2005. Plate motions, slab dynamics and back-arc deformation. *Physics of the Earth and Planetary Interiors* 149: 31–51.

Hicks, A. 2012. Facultative amphidromy in galaxiids and bullies: The science, ecology and management implications. PhD thesis, University of Otago, Dunedin, New Zealand.

Hicks, S. 2014. Paleoecology and sedimentology of the volcanically active Late Eocene continental shelf, Northeast Otago, New Zealand. PhD thesis, University of Otago, Dunedin, New Zealand.

Hicks, S. 2013. PhD Proposal: Ecological and sedimentological evolution of the volcanically active Oligocene continental shelf, east Otago, New Zealand. http://www.otago.ac.nz/geology/people/students/hicks/. Accessed July 21, 2013.

Hickson, R.E., K.E. Slack, and P. Lockhart. 2000. Phylogeny recapitulates geography, or why New Zealand has so many species of skinks. *Biological Journal of the Linnean Society* 70: 415–433.

Higdon, J.W., O.R.P. Bininda-Emonds, R.M.D. Beck, and S.H. Ferguson. 2007. Phylogeny and divergence of the pinnipeds (Carnivora: Mammalia) assessed using a multigene dataset. *BMC Evolutionary Biology* 7(216): 1–19.

Higgins, P.J. (ed.). 1999. *Handbook of Australian, New Zealand and Antarctic Birds*, Vol. 4: *Parrots to Dollarbird*. Melbourne, Australia: Oxford University Press.

Higgins, P.J. and L. Christidis. 2009. Notiomystidae. In *Handbook of the Birds of the World*, Vol. 14, eds. J. del Hoyo, A. Elliott, and D.A. Christie, pp. 242–257. Barcelona, Spain: Lynx Edicions.

Higgins, P.J., L. Christidis, and H.A. Ford. 2008. Meliphagidae. In *Handbook of the Birds of the World*, Vol. 13, eds. J. del Hoyo, A. Elliott, and D.A. Christie, pp. 498–691. Barcelona, Spain: Lynx Edicions.

Higgins, P.J. and S.J.J.F. Davies (eds.). 1996. *Handbook of Australian, New Zealand and Subantarctic Birds*, Vol. 3: *Snipe to Pigeons*. Melbourne, Australia: Oxford University Press.

Higgins, P.J., J.M. Peter, and W.K. Steele (eds.). 2001. *Handbook of Australian, New Zealand and Antarctic Birds*, Vol. 5: *Tyrant-Flycatchers to Chats*. Melbourne, Australia: Oxford University Press.

Higgins, S.I., R. Nathan, and M.L. Cain. 2003. Are long-distance dispersal events in plants usually caused by nonstandard means of dispersal? *Ecology* 84: 1945–1956.

Higgins, S.I., R.B. O'Hara, and C. Römermann. 2012. A niche for biology in species distribution models. *Journal of Biogeography* 39: 2091–2095.

Hill, J.E. and J.D. Smith. 1984. *Bats: A Natural History*. London, UK: British Museum (Natural History).

Hill, K.B.R., C. Simon, D.C. Marshall, and G.K. Chambers. 2009. Surviving glacial ages within the Biotic Gap: Phylogeography of the New Zealand cicada. *Journal of Biogeography* 36: 675–692.

Himmelreich, S., I. Breitwieser, and C. Oberprieler. 2012. Phylogeny, biogeography, and evolution of sex expression in the southern hemisphere genus *Leptinella* (Compositae, Anthemideae). *Molecular Phylogenetics and Evolution* 65: 464–481.

Himmelreich, S., I. Breitwieser, and C. Oberprieler. 2014. Phylogenetic relationships in the extreme polyploid complex of the New Zealand genus *Leptinella* (Compositae: Anthemideae) based on AFLP data. *Taxon* 63: 883–898.

Himmelreich, S., M. Källersjö, P. Eldenäs, and C. Oberprieler. 2008. Phylogeny of southern hemisphere Compositae-Anthemideae based on nrDNA ITS and cpDNA *ndh*F sequence information. *Plant Systematics and Evolution* 272: 131–153.

Hipsley, C.A. and J. Müller. 2014. Beyond fossil calibrations: Realities of molecular clock practices in evolutionary biology. *Frontiers in Genetics* 5(138): 1–11.

Hirase S., M. Ikeda, M. Kanno, and A. Kijima. 2012a. Phylogeography of the intertidal goby *Chaenogobius annularis* associated with paleoenvironmental changes around the Japanese Archipelago. *Marine Ecology Progress Series* 450: 167–179.

Hirase, S., M. Kanno, M. Ikeda, and A. Kijima. 2012b. Evidence of the restricted gene flow within a small spatial scale in the Japanese common intertidal goby *Chaenogobius annularis*. *Marine Ecology* 33: 481–489.

Hirmer, M. 1922. *Zur Lösung des Problems der Blattstellungen*. Jena, Germany: G. Fischer.

Hjelmqvist, H. 1948. Studies on the floral morphology and phylogeny of the Amentiferae. *Botaniska Notiser Supplement* 1: 1–171.

Ho, S.Y.W. 2014. The changing face of the molecular clock. *Trends in Ecology and Evolution* 29: 496–503.

Ho, S.Y.W. and S. Duchêne. 2014. Molecular-clock methods for estimating evolutionary rates and timescales. *Molecular Ecology* 23: 5947–5965.

Ho, S.Y.W. and M.J. Phillips. 2009. Accounting for calibration uncertainty in phylogenetic estimation of evolutionary divergence times. *Systematic Biology* 58: 367–380.

Ho, S.Y.W., K.J. Tong, C.S. Foster, A.M. Ritchie, N. Lo, and M.D. Crisp. 2015. Biogeographic calibrations for the molecular clock. *Biology Letters* 11(9): 20150194.

Hoare, R.J.B. 2005. *Hierodoris* (Insecta: Lepidoptera: Gelechioidea: Oecophoridae), and overview of Oecophoridae. *Fauna of New Zealand* 54: 1–100.

Hoare, R.J.B. 2010. *Izatha* (Insecta: Lepidoptera: Gelechioidea: Oecophoridae). *Fauna of New Zealand* 65: 1–201.

Hoare, R.J.B. 2012. A new species of *Hierodoris* Meyrick (Lepidoptera: Oecophoridae) with a telescopic ovipositor, from granite sand plains in Fiordland. *New Zealand Entomologist* 35: 51–57.

Hockey, P.A.R. 1996. Haematopodidae. In *Handbook of the Birds of the World*, Vol. 3, eds. J. del Hoyo, A. Elliott, and J. Sargatal, pp. 308–325. Barcelona, Spain: Lynx Edicions.

Hodge, J.R., C.I. Read, D.R. Bellwood, and L. van Herwerden. 2013. Evolution of sympatric species: A case study of the coral reef fish genus *Pomacanthus* (Pomacanthidae). *Journal of Biogeography* 40: 1676–1687.

Hodgson, C.J. and R.C. Henderson. 2000. Coccidae (Insecta: Hemiptera: Coccoidea). *Fauna of New Zealand* 41: 1–264.

Hoelzer, G.A., R. Drewes, J. Meier, and R. Doursat. 2008. Isolation-by-distance and outbreeding depression are sufficient to drive parapatric speciation in the absence of environmental influences. *PLoS Computational Biology* 4: e1000126 (pp. 1–11).

Hoernle, K., F. Hauff, P. van den Bogaard, R. Werner, N. Mortimer, J. Geldmacher, D. Garbe-Schönberg, and B. Davy. 2010. Age and geochemistry of volcanic rocks from the Hikurangi and Manihiki oceanic Plateaus. *Geochimica et Cosmochimica Acta* 74: 7196–7219.

Hoernle, K., J.D.L. White, P.V.D. Bogaard et al. 2006. Cenozoic intraplate volcanism on New Zealand: Upwelling induced by lithospheric removal. *Earth and Planetary Science Letters* 248: 335–352.

Hogg, I.D., M.I. Stevens, K. Schnabel, and M.A. Chapman. 2006. Deeply divergent lineages of the widespread New Zealand amphipod *Paracalliope fluviatilis* revealed using allozyme and mitochondrial DNA analyses. *Freshwater Biology* 51: 236–248.

Holdaway, R.N. 1989. New Zealand's pre-human avifauna and its vulnerability. *New Zealand Journal of Ecology* 12(Suppl.): 11–25.

Holdaway, R.N., J.M. Thorneycroft, P. McClelland, and M. Bunce. 2010. Former presence of a parakeet (*Cyanoramphus* sp.) on Campbell Island, New Zealand subantarctic, with notes on the island's fossil sites and fossil record. *Notornis* 57: 8–18.

Holdaway, R.N. and T.H. Worthy. 1993. First North Island fossil record of kea, and morphological and morphometric comparison of kea and kaka. *Notornis* 40: 95–108.

Holdaway, R.N. and T.H. Worthy. 1994. A new fossil species of shearwater *Puffinus* from the late Quaternary of the South Island, New Zealand, and notes on the biogeography and evolution of the *Puffinus gavia* superspecies. *Emu* 94: 201–215.

Holdaway, R.N. and T.H. Worthy. 1997. A reappraisal of the late Quaternary fossil vertebrates of Pyramid Valley Swamp, north Canterbury, New Zealand. *New Zealand Journal of Zoology* 24: 69–121.

Holdaway, R.N., T.H. Worthy, and A.J.D. Tennyson. 2001. A working list of breeding bird species of the New Zealand region. *New Zealand Journal of Zoology* 28: 119–187.

Holloway, B.A. 1982. Anthribidae (Insecta: Coleoptera). *Fauna of New Zealand* 3: 1–264.

Holloway, B.A. 2007. Lucanidae (Insecta: Coleoptera). *Fauna of New Zealand* 61: 1–254.

Holloway, J.T. 1936. From the middle Dart to the back of beyond. *New Zealand Alpine Journal* 6: 289–299.

Holt, B.G., J.-P. Lessard, M.K. Borregaard et al. 2013. An update of Wallace's zoogeographic regions of the world. *Science* 339: 74–78.

Holt, K.A. 2008. The Quaternary history of Chatham Island, New Zealand. PhD thesis, Massey University, Palmerston North, New Zealand.

Holt, R.D. 2009. Bringing the Hutchinsonian niche into the 21st century: Ecological and evolutionary perspectives. *Proceedings of the National Academy of Sciences of USA* 106: 19659–19665.

Holzapfel, S., M.Z. Faville, and C.E.C. Gemmill. 2006. Genetic variation of the endangered holoparasite *Dactylanthus taylorii* (Balanophoraceae) in New Zealand. *Journal of Biogeography* 29: 663–676.

Holzapfel, S., H.A. Robertson, J.A. McLennan, K. Sporle, K. Hackwell, and M. Impey. 2008. Kiwi (*Apteryx* spp.) recovery plan 2008–2018. Wellington, New Zealand: Department of Conservation.

Homes, A.M. and D.E. Lee. 2011. Did New Zealand drown—What can ferns tell us? *New Zealand Geological Society Conference Abstracts*, Nelson, New Zealand, p. 51.

Hooker, J.D. 1857. On the botany of Raoul Island, one of the Kermadec group in the South Pacific Ocean. *Journal of the Linnean Society, Botany* 1: 125–129.

Hoot, S.B., K.M. Meyer, and J.C. Manning. 2012. Phylogeny and reclassification of *Anemone* (Ranunculaceae), with an emphasis on austral species. *Systematic Botany* 37: 139–152.

Hörandl, E. and K. Emadzade. 2012. Evolutionary classification: A case study on the diverse plant genus *Ranunculus* L. (Ranunculaceae). *Perspectives in Plant Ecology, Evolution and Systematics* 14: 310–324.

Hörandl, E., O. Paun, J.T. Johansson et al. 2005. Phylogenetic relationships and evolutionary traits in *Ranunculus* s.l. (Ranunculaceae) inferred from ITS sequence analysis. *Molecular Phylogenetics and Evolution* 36: 305–327.

Hori, M. 1993. Frequency-dependent natural selection in the handedness of scale-eating cichlid fish. *Science* 260: 216–219.

Horrocks, M. and J. Ogden. 1998. The effects of the Taupo Tephra eruption of c. 1718 BP on the vegetation of Mt Hauhungatahi, central North Island, New Zealand. *Journal of Biogeography* 25: 649–660.

Hoso, M. 2012. Non-adaptive speciation of snails by left-right reversal is facilitated on oceanic islands. *Contributions to Zoology* 81(2): 79–85. http://dpc.uba.uva.nl/ctz/vol81/nr02/art02.

Hoso, M., T. Asami, and M. Hori. 2007. Right-handed snakes: Convergent evolution of asymmetry for functional specialization. *Biology Letters* 3: 169–172.

Houde, P. 2009. Cranes, rails and allies (Gruiformes). In *The Timetree of Life*, eds. S.B. Hedges and S. Kumar, pp. 440–444. Oxford, UK: Oxford University Press.

Howarth, D.G., M.H.G. Gustafsson, D.A. Baum, and T.J. Motley. 2003. Phylogenetics of the genus *Scaevola* (Goodeniaceae): Implications for dispersal patterns across the Pacific Basin and colonization of the Hawaiian Islands. *American Journal of Botany* 90: 915–923.

Hubbell, S.P. 2001. *The Unified Neutral Theory of Biodiversity and Biogeography*. Princeton, NJ: Princeton University Press.

Hubbs, C.L. 1950. Studies of cyprinodont fishes. XX. A new subfamily from Guatemala, with ctenoid scales and a unilateral pectoral clasper. *Miscellaneous Publications, Museum of Zoology, University of Michigan* 78: 1–39.

Hubbs, C.L. and L.C. Hubbs. 1945. Bilateral asymmetry and bilateral variation in fishes. *Papers of the Michigan Academy of Science, Arts, and Letters* 30: 229–310.

Hug, L.A. and A.J. Roger. 2007. The impact of fossils and taxon sampling on ancient molecular dating analyses. *Molecular Biology and Evolution* 24: 1889–1897.

Humboldt, A. von and A. Bonpland. 1805. *Essai sur la Géographie des Plantes; Accompagné d'un Tableau Physique des Régions Équinoxiales*. Paris, France: Levrault, Schoell and Co.

Hurd, C.L., W.A. Nelson, R. Falshaw, and K.F. Neill. 2004. History, current status and future of marine macroalgal research in New Zealand: Taxonomy, ecology, physiology and human uses. *Phycological Research* 52: 80–106.

Hurley, D.E. 1957. Terrestrial and littoral amphipods of the genus *Orchestia*, family Talitridae. *Transactions of the Royal Society of New Zealand* 85: 149–199.

Hutton, F.W. 1872. On the geographic relations of the New Zealand fauna. *Transactions of the New Zealand Institute* 5: 227–256.

Hutton, F.W. 1884–1885. On the origin of the fauna and flora of New Zealand. *Annals and Magazine of Natural History* 13: 425–448 and 15: 77–107.

Hutton, F.W. 1896. Theoretical explanations of the distribution of southern faunas. *Proceedings of the Linnean Society of New South Wales* 21: 36–47.

Hutton, F.W. and J. Drummond. 1904. *The Animals of New Zealand: An Account of the Colony's Air-Breathing Vertebrates*. Christchurch, New Zealand: Whitcombe and Tombs.

Huxley, J. 1942. *Evolution: The Modern Synthesis*. London, UK: Allen and Unwin.

Huxley, T.H. 1867. On the classification of birds; and on the taxonomic value of the modifications of certain of the cranial bones observable in that class. *Proceedings of the Zoological Society of London* 1867: 415–472.

Huxley, T.H. 1868. On the classification and distribution of the Alectoromorphae and Heteromorphae. *Proceedings of the Zoological Society, London* 1868: 294–319.

Huxley, T.H. 1873. Palaeontology and the doctrine of evolution. (Anniversary Address to the Geological Society, 1870.) In *Critiques and Addresses*, ed. T.H. Huxley, pp. 181–217. London, UK: Macmillan.

Imamura, H. and K. Matsuura. 2003. Redefinition and phylogenetic relationships of the family Pinguipedidae (Teleostei: Perciformes). *Ichthyological Research* 50: 259–269.

Imber, M.J. 1994. Seabirds recorded at the Chatham Islands, 1960 to May 1993. *Notornis (Supplement)* 41: 97–108.

Imms, A.D. 1973. *Outlines of Entomology*. London, UK: Chapman & Hall.

Inda, L.A., J.G. Segarra-Moragues, J. Müller, P.M. Peterson, and P. Catalán. 2008. Dated historical biogeography of the temperate Loliinae (Poaceae, Pooideae) grasses in the northern and southern hemispheres. *Molecular Phylogenetics and Evolution* 46: 932–957.

Ingram, T., A.G. Hudson, P. Vonlanthen and O. Seehausen. 2012. Does water depth or diet divergence predict progress towards ecological speciation in whitefish radiations? *Evolutionary Ecology Research* 14: 487–502.

Irestedt, M., K.A. Jønsson, J. Fjeldså, L. Christidis, and P.G.P. Ericson. 2009. An unexpectedly long history of sexual selection in birds-of-paradise. *BMC Evolutionary Biology* 9(235): 1–11.

IUCN. 2015. The IUCN Redlist of threatened species. www.iucnredlist.org. Accessed May, 2016.

Jabaily, R.S., K.A. Shepherd, A.G. Gardner, M.H.G. Gustafsson, D.G. Howarth, and T.J. Motley. 2014. Historical biogeography of the predominantly Australian plant family Goodeniaceae. *Journal of Biogeography* 41: 2057–2067.

Jacobs, M. 1966. On domatia—The viewpoints and some facts. *Proceedings of the Koninklijke Nederlandse Akademie van Wetenschappen C* 69: 275–316.

Jaffré, T. 1995. Distribution and ecology of the conifers of New Caledonia. In *Ecology of the Southern Conifers*, eds. N.J. Enright and R.S. Hill, pp. 171–196. Melbourne, Australia: Melbourne University Press.

Jakob, S.S., E. Martinez-Meyer, and F.R. Blattner. 2009. Phylogeographic analyses and paleodistribution modeling indicates Pleistocene in situ survival of *Hordeum* species (Poaceae) in southern Patagonia without genetic or spatial restriction. *Molecular Biology and Evolution* 26: 907–923.

Jamieson, I.G. and H.G. Spencer. 1996. The bill and foraging behaviour of the Huia (*Heteralocba acutirostris*): Were they unique? *Notornis* 43: 14–18.

Janečka, J.E., W. Miller, T.H. Pringle, F. Wiens, A. Zitzmann, K. Helgen, M.S. Springer, and W.J. Murphy. 2007. Molecular and genomic data identify closest living relative of primates. *Science* 318: 792–794.

Janes, D.E., C.L. Organ, M.K. Fujita, A.M. Shedlock, and S.V. Edwards. 2010. Genome evolution in Reptilia, the sister group of mammals. *Annual Review of Genomics and Human Genetics* 11: 239–264.

Janssen, T. and K. Bremer. 2004. The age of major monocot groups inferred from 800+ *rbc*L sequences. *Botanical Journal of the Linnean Society* 146: 385–398.

Jeffery, W.R. 2005. Adaptive evolution of eye degeneration in the Mexican blind cavefish. *Journal of Heredity* 96: 185–196.

Jeon, M.-J., J.-H. Song, and K.-J. Ahn. 2012. Molecular phylogeny of the marine littoral genus *Cafius* (Coleoptera: Staphylinidae: Staphylininae) and implications for classification. *Zoologica Scripta* 41: 150–159.

Jerison, H.J. 1973. *The Evolution of the Brain and Intelligence*. New York: Academic Press.

Johns, P.M. 1964. The Sphaerotrichopidae (Diplopoda) of New Zealand. 1. Introduction, revision of some known species and description of new species. *Records of the Canterbury Museum* 8: 1–49.

Johns, P.M. 1974. Arthropoda of the subantarctic islands of New Zealand (1) Coleoptera: Carabidae. Southern New Zealand, Patagonian, and Falkland Islands insular Carabidae. *Journal of the Royal Society of New Zealand* 4: 283–302.

Johns, P.M. 1979. Speciation in New Zealand Diplopoda. In *Myriapod Biology*, ed. M. Camatini, pp. 49–57. London, UK: Academic Press.

Johns, P.M. 2010. Migadopini (Coleoptera: Carabidae: Migadopinae) of New Zealand. *Records of the Canterbury Museum* 24: 39–63.

Johns, P.M. 2015. New Zealand biodiversity. http://nzbiodiversity.com. Accessed December, 2015.

Johns, P.M. and L.D. Cook. 2014. *Maotoweta virescens* new genus and new species; hidden in a moss forest (Orthoptera: Rhaphidophoridae). *Records of the Canterbury Museum* 27: 11–17.

Jolivet, P. 2008. La faune entomologique en Nouvelle-Calédonie. *Le Coléoptériste* 11: 35–47.

Jolivet, P. and K.K. Verma. 2008a. Eumolpinae—A widely distributed and much diversified subfamily of leaf beetles (Coleoptera, Chrysomelidae). *Terrestrial Arthropod Reviews* 1: 3–37.

Jolivet, P. and K.K. Verma. 2008b. On the origin of the chrysomelid fauna of New Caledonia. In *Research on Chrysomelidae*, Vol. 1, eds. P. Jolivet, J.A. Santiago-Blay and M. Schmitt, pp. 309–319. Leiden, South Holland: Brill.

Jolivet, P. and K.K. Verma. 2010. Good morning Gondwana. *Annales de la Société Entomologique de France* (n.s.) 46: 53–61.

Jolly, M.T., F. Viard, F. Gentil, E. Thiebaut, and D. Jollivet. 2006. Comparative phylogeography of two coastal polychaete tubeworms in the Northeast Atlantic supports shared history and vicariant events. *Molecular Ecology* 15: 1841–1855.

Joly, S., P.B. Heenan, and P.J. Lockhart. 2009. A Pleistocene inter-tribal allopolyploidization event precedes the species radiation of *Pachycladon* (Brassicaceae) in New Zealand. *Molecular Phylogenetics and Evolution* 51: 365–372.

Joly, S., P.B. Heenan, and P.J. Lockhart. 2009. A Pleistocene inter-tribal allopolyploidization event precedes the species radiation of *Pachycladon* (Brassicaceae) in New Zealand. *Molecular Phylogenetics and Evolution* 51: 365–372.

Joly, S., P.B. Heenan, and P.J. Lockhart. 2014. Species radiation by niche shifts in New Zealand's rockcresses (*Pachycladon*, Brassicaceae). *Systematic Biology* 63: 192–202.

Jones, G., P.I. Webb, J.A. Sedgeley, and C.F.J. O'Donnell. 2003. Mysterious *Mystacina*: How the New Zealand short-tailed bat (*Mystacina tuberculata*) locates insect prey. *Journal of Experimental Biology* 206: 4209–4216.

Jones, M.E.H., A.J.D. Tennyson, J.P. Worthy, S.E. Evans, and T.H. Worthy. 2009. A sphenodontine (Rhynchocephalia) from the Miocene of New Zealand and palaeobiogeography of the tuatara (*Sphenodon*). *Proceedings of the Royal Society B* 276: 1385–1390.

Jønsson, K.A., P.-H. Fabre, and M. Irestedt. 2012. Brains, tools, innovation and biogeography in crows and ravens. *BMC Evolutionary Biology* 12(72): 1–12.

Jønsson, K.A., P.H. Fabre, J.D. Kennedy, B.G. Holt, M.K. Borregaard, C. Rahbek, and J. Fjeldså. 2016. A supermatrix phylogeny of corvoid passerine birds (Aves: Corvides). *Molecular Phylogenetics and Evolution* 94: 87–94.

Jønsson, K.A., P.-H. Fabre, R.E. Ricklefs, and J. Fjeldså. 2011. Major global radiation of 770 corvoid birds originated in the proto-Papuan archipelago. *Proceedings of the National Academy of Sciences USA* 108: 2328–772.

Jordan, G.J., J.M. Bannister, D.C. Mildenhall, R. Zetter, and D.E. Lee. 2010. Fossil Ericaceae from New Zealand: Deconstructing the use of fossil evidence in historical biogeography. *American Journal of Botany* 97: 59–70.

Jordan, G.J., T.J. Brodribb, C.J. Blackman, and P.H. Weston. 2013. Climate drives vein anatomy in Proteaceae. *American Journal of Botany* 100: 1483–1493.

Jörger, K.M., I. Stöger, Y. Kano, H. Fukuda, T. Knebelsberger, and M. Schrödl. 2010. On the origin of Acochlidia and other enigmatic euthyneuran gastropods, with implications for the systematics of Heterobranchia. *BMC Evolutionary Biology* 10(323): 1–20.

Joseph, L., A. Toon, Á.S. Nyári et al. 2014. A new synthesis of the molecular systematics and biogeography of honeyeaters (Passeriformes: Meliphagidae) highlights biogeographical and ecological complexity of a spectacular avian radiation. *Zoologica Scripta* 43: 235–248.

Jugum, D., R. Norris, and J.M. Palin. 2006. New perspectives on the Dun Mountain Ophiolite Belt. Abstract only. In *Geological Society of New Zealand/New Zealand Geophysical Society Conference 2006: Programme and Abstracts*, eds. B. Stewart, C. Wallace, J. Lecointre, and M. Reyners, p. 53. Palmerston North, New Zealand: Geological Society of New Zealand/New Zealand Geophysical Society.

Just, J. and G.D. Wilson. 2004. Revision of the *Paramunna* complex (Isopoda: Asellota: Paramunnidae). *Invertebrate Systematics* 18: 377–466.

Juste, J. and C. Ibáñez. 1993. An asymmetric dental formula in a mammal, the São Tomé Island fruit bat *Myonycteris brachycephala* (Mammalia: Megachiroptera). *Canadian Journal of Zoology* 71: 221–224.

Kadereit, J.W., M. Repplinger, N. Schmalz, C.H. Uhink, and A. Wörz. 2008. The phylogeny and biogeography of Apiaceae subf. Saniculoideae tribe Saniculeae: From south to north and south again. *Taxon* 57: 365–382.

Kaley, K.J. 2007. Fossil birds. In *Handbook of the Birds of the World*, Vol. 12, eds. J. del Hoyo, A. Elliott, and D.A. Christie, pp. 11–42. Barcelona, Spain: Lynx Edicions.

Kamel, S.J., A.R. Hughes, R.K. Grosberg, and J.J. Stachowicz. 2012. Fine-scale genetic structure and relatedness in the eelgrass *Zostera marina*. *Marine Ecology Progress Series* 447: 127–137.

Kamilar, J., S. Martin, and A. Tosi. 2009. Combining biogeographic and phylogenetic data to examine primate speciation: An example using cercopithecin monkeys. *Biotropica* 41: 514–519.

Kamino, L.H.Y., P. de Marco Jr., T.F. Rangel et al. 2012. The application of species distribution models in the megadiverse Neotropics poses a renewed set of research questions. *Frontiers of Biogeography* 4: 7–10.

Kamp, P.J.J. 1986. The mid-Cenozoic Challenger Rift System of western New Zealand and its implications for the age of Alpine fault inception. *Geological Society of America Bulletin* 97: 255–281.

Kamp, P.J.J. 1987. Age and origin of the New Zealand orocline in relation to Alpine fault movement. *Journal of the Geological Society* (*London*) 144: 641–652.

Kamp, P.J.J. and K.P. Furlong. 2010. Tectono-sedimentary framework for exploration in New Zealand basins. *New Zealand Petroleum Conference 2010 Proceedings*, Auckland, New Zealand. New Zealand Ministry of Economic Development. http://www.nzpam.govt.nz/cms/pdf-library/petroleum-conferences-1/2010-nzpc-technical-posters-papers/P12_Kamp-Furlong_paper.pdf.

Kamp, P.J.J., P.F. Green, and S.H. White. 1989. Fission track analysis reveals character of collisional tectonics in New Zealand. *Tectonics* 8: 169–195.

Kamp, P.J.J., A.R.P. Tripathi, and C.S. Nelson. 2014. Paleogeography of Late Eocene to earliest Miocene Te Kuiti Group, central-western North Island, New Zealand. *New Zealand Journal of Geology and Geophysics* 57: 110–127.

Kampourakis, K., E. Palaiokrassa, M. Papadopoulou, V. Pavlidi, and M. Argyropoulou. 2012. Children's intuitive teleology: Shifting the focus of evolution education research. *Evolution: Education and Outreach* 5: 279–291.

Kant, I. 1790/1978. *The Critique of Judgment*. Transl. J.C. Meredith. Oxford, UK: Clarendon Press.

Kao, M.-H. 2001. Thermo-tectonic history of the Marlborough region, South Island, New Zealand. *Terrestrial, Atmospheric and Oceanic Sciences* 12: 485–502.

Karl, S.A., R.J. Toonen, W.S. Grant, and B.W. Bowen. 2012. Common misconceptions in molecular ecology: Echoes of the modern synthesis. *Molecular Ecology* 21: 4171–4189.

Katinas, L., J.V. Crisci, P. Hochc, M.C. Tellería, and M.J. Apodaca. 2013. Trans-oceanic dispersal and evolution of early composites (Asteraceae). *Perspectives in Plant Ecology, Evolution and Systematics* 15: 269–280.

Katz, H. 1968. Potential oil formations in New Zealand and their stratigraphic position as related to basin evolution. *New Zealand Journal of Geology and Geophysics* 11: 1077–1133.

Kaulfuss, U., A.C. Harris, J.G. Conran, and D.E. Lee. 2014. An early Miocene ant (subfam. Amblyoponinae) from Foulden Maar: The first fossil Hymenoptera from New Zealand. *Alcheringa* 38: 568–574.

Kaulfuss, U., A.C. Harris, and D.E. Lee. 2010. A new fossil termite (Isoptera, Stolotermitidae, *Stolotermes*) from the Early Miocene of Otago, New Zealand. *Acta Geologica Sinica* 84: 705–709.

Kaulfuss, U., T. Wappler, E. Heiss, and M.-C. Larivière. 2011. *Aneurus* sp. from Early Miocene Foulden Maar, Otago, New Zealand: The first Southern Hemisphere record of fossil Aneurinae (Hemiptera). *Journal of the Royal Society of New Zealand* 41: 279–285.

Kelemen, D. 1999a. Why are rocks pointy? Children's preference for teleological explanations of the natural world. *Developmental Psychology* 35: 1440–1452.

Kelemen, D. 1999b. The scope of teleological thinking in preschool children. *Cognition* 70: 241–272.

Kelemen, D. 2012. Teleological minds: How natural intuitions about agency and purpose influence learning about evolution. In *Evolution Challenges: Integrating Research and Practice in Teaching and Learning about Evolution*, eds. K.S. Rosengren, S.K. Brem, E.M. Evans, and G.M. Sinatra. Oxford, UK: Oxford University Press.

Kelemen, D. and E. Rosset. 2009. The human function compunction: Teleological explanation in adults. *Cognition* 111: 138–143.

Kelly, D. and J.J. Sullivan. 2010. Life histories, dispersal, invasions, and global change: Progress and prospects in New Zealand ecology, 1989–2029. *New Zealand Journal of Ecology* 34: 207–217.

Kennedy, M., R.D. Gray, and H.G. Spencer. 2000. The phylogenetic relationships of the shags and cormorants: Can sequence data resolve a disagreement between behavior and morphology? *Molecular Phylogenetics and Evolution* 17: 345–359.

Kennedy, M. and R.D.M. Page. 2002. Seabird supertrees: Combining partial estimates of procellariiform phylogeny. *Auk* 119: 88–108.

Kennedy, M. and H.G. Spencer. 2014. Classification of the cormorants of the world. *Molecular Phylogenetics and Evolution* 79: 249–257.

Kennedy, M., C.A. Valle, and H.G. Spencer. 2009. The phylogenetic position of the Galápagos Cormorant. *Molecular Phylogenetics and Evolution* 53: 94–98.

Kerr, L.C., D. Craw, R.J. Norris, J.H. Youngson, and P. Wopereis. 2000. Structure, geomorphology, and gold concentration in the Nokomai Valley, Southland, New Zealand. *New Zealand Journal of Geology and Geophysics* 43: 425–433.

Killops, S.D., R.A. Cook, R. Sykes, and J.P. Boudou. 1997. Petroleum potential and oil-source correlation in the Great South and Canterbury Basins. *New Zealand Journal of Geology and Geophysics* 40: 405–423.

Kim, H., S. Lee, and Y. Jang. 2011. Macroevolutionary patterns in the Aphidini aphids (Hemiptera: Aphididae): Diversification, host association, and biogeographic origins. *PLoS One* 6(9): e24749 (pp. 1–17).

Kim, S.I. and B.D. Farrell. 2015. Phylogeny of world stag beetles (Coleoptera: Lucanidae) reveals a Gondwanan origin of Darwin's stag beetle. *Molecular Phylogenetics and Evolution* 86: 35–48.

King, C.M. (ed.). 2005. *The Handbook of New Zealand Mammals*, 2nd edn. Melbourne, Australia: Oxford University Press.

King, N. 2004. The unicellular ancestry of animal development. *Developmental Cell* 7: 313–325.

King, T.M., M. Kennedy, and G.P. Wallis. 2003. Phylogeographic genetic analysis of the alpine weta, *Hemideina maori*: Evolution of a colour polymorphism and origins of a hybrid zone. *Journal of the Royal Society of New Zealand* 33: 715–729.

Kinsky, F.C. 1971. The consistent presence of paired ovaries in the Kiwi (*Apteryx*) with some discussion of this condition in other birds. *Journal für Ornithologie* 112: 334–357.

Kirchman, J.J. 2012. Speciation of flightless rails on islands: A DNA-based phylogeny of the typical rails of the Pacific. *Auk* 129: 56–69.

Kirk, T. 1871. On the occurrence of littoral plants in the Waikato district. *Transactions of the New Zealand Institute* 3: 147–148.

Kirk, T. 1872. Notes on the flora of the Lake District of the North Island. *Transactions of the New Zealand Institute* 5: 322–345.

Kirschner, M.W. and J.C. Gerhart. 2005. *The Plausibility of Life: Resolving Darwin's Dilemma*. New Haven, CT: Yale University Press.

Klazenga, N. 2003. A revision of the Australasian species of *Dicranoloma* (Bryophyta, Dicranaceae). *Australian Systematic Botany* 16: 427–471.

Klepeis, K.A. and D.S. King. 2009. Evolution of the middle and lower crust during the transition from contraction to extension in Fiordland, New Zealand. *Geological Society of America Special Paper* 456: 243–266.

Klepeis, K.A., D. King, M. De Paoli, G.L. Clarke, and G. Gehrels. 2007. Interaction of strong lower and weak middle crust during lithospheric extension in western New Zealand. *Tectonics* 26: TC4017.

Klicka, J. and R.M. Zink. 1997. The importance of recent ice ages in speciation: A failed paradigm. *Science* 277: 1666–1669.

Klompen, J.S.H. 1992. Phylogenetic relationships in the mite family Sarcoptidae (Acari: Astigmata). *Miscellaneous Publications Museum of Zoology, University of Michigan* 180: 1–164.

Knapp, M., R. Mudaliar, D. Havell, S.J. Wagstaff, and P.J. Lockhart. 2007. The drowning of New Zealand and the problem of *Agathis*. *Systematic Biology* 56: 862–870.

Knapp, M., K. Stöckler, D. Havell, F. Delsuc, F. Sebastian, and P.J. Lockhart. 2005. Relaxed molecular clock provides evidence for long-distance dispersal of *Nothofagus* (Southern beech). *PLoS Biology* 3: e14 (pp. 1–6).

Knapp, S. 2013. What, where, and when? *Science* 341: 1182–1184.

Knox, E.B., P.B. Heenan, A.M. Muasya, and B.G. Murray. 2008. Phylogenetic position and relationships of *Lobelia glaberrima* (Lobeliaceae), a new alpine species from southern New Zealand. *New Zealand Journal of Botany* 46: 77–85.

Knox, G.A. 1963. Antarctic relationships in Pacific biogeography. In *Pacific Basin Biogeography: A Symposium*, ed. J.L. Gressitt, pp. 465–474. Honolulu, HI: Bishop Museum.

Knox, M.A., I.D. Hogg, and C.A. Pilditch. 2011. The role of vicariance and dispersal on New Zealand's estuarine biodiversity: The case of *Paracorophium* (Crustacea: Amphipoda). *Biological Journal of the Linnean Society* 103: 863–874.

Koch, M.A., B. Haubold, and T. Mitchell-Olds. 2000. Comparative evolutionary analysis of the chalcone synthase and alcohol dehydrogenase loci among different lineages of *Arabidopsis, Arabis* and related genera (Brassicaceae). *Molecular Biology and Evolution* 17: 1483–1498.

Koch, M.A., B. Haubold, and T. Mitchell-Olds. 2001. Molecular systematics of the Brassicaceae: Evidence from coding plastidic *matK* and nuclear *Chs* sequences. *American Journal of Botany* 88: 534–544.

Kodandaramaiah, U. 2011. Tectonic calibrations in molecular dating. *Current Zoology* 57: 116–124.

Kohn, B.P., A.J.W. Gleadow, R.W. Brown, K. Gallagher, P.B. O'Sullivan, and D.A. Foster. 2002. Shaping the Australian crust over the last 300 million years: Insights from fission track thermotectonic imaging and denudation studies of key terranes. *Australian Journal of Earth Sciences* 49: 697–717.

Koonin, E.V. 2009a. Darwinian evolution in the light of genomics. *Nucleic Acids Research* 37: 1011–1034.

Koonin, E.V. 2009b. The *Origin* at 150: Is a new evolutionary synthesis in sight? *Trends in Genetics* 25: 473–475.

Koonin, E.V. 2012. *The Logic of Chance: The Nature and Origin of Biological Evolution*. Upper Saddle River, NJ: FT Press.

Koopman, K.F. 1971. Taxonomic notes on *Chalinolobus* and *Glauconycteris* (Chiroptera, Vespertilionidae). *American Museum Novitates* 2451: 1–10.

Kopp, M. 2010. Speciation and the neutral theory of biodiversity. *BioEssays* 32: 564–570.

Korall, P. and K.M. Pryer. 2014. Global biogeography of scaly tree ferns (Cyatheaceae): Evidence for Gondwanan vicariance and limited transoceanic dispersal. *Journal of Biogeography* 41: 402–413.

Kostka, D., M.J. Hubisz, A. Siepel, and K.S. Pollard. 2012. The role of GC-biased gene conversion in shaping the fastest evolving regions of the human genome. *Molecular Biology and Evolution* 29: 1047–1057.

Koubínová, D., N. Irwin, P. Hulva, P. Koubek, and J. Zima. 2013. Hidden diversity in Senegalese bats and associated findings in the systematics of the family Vespertilionidae. *Frontiers in Zoology* 10(1): 1–16 (article 48).

Krosch, M. and P.S. Cranston. 2013. Not drowning, (hand)waving? Molecular phylogenetics, biogeography and evolutionary tempo of the 'gondwanan' midge *Stictocladius* Edwards (Diptera: Chironomidae). *Molecular Phylogenetics and Evolution* 68: 595–603.

Krosch, M.N., A.M. Baker, P.B. Mather, and P.S. Cranston. 2011. Systematics and biogeography of the Gondwanan Orthocladiinae (Diptera: Chironomidae). *Molecular Phylogenetics and Evolution* 59: 458–468.

Krug, P.J. 2011. Patterns of speciation in marine gastropods: A review of the phylogenetic evidence for localized radiations in the sea. *American Malacological Bulletin* 29: 169–186.

Kruijer, H. 2002. Hypopterygiaceae of the world. *Blumea Supplement* 13: 1–388.

Krull, C., S. Parsons, and M. Hauber. 2009. The presence of ultrasonic harmonics in the calls of the rifleman (*Acanthisitta chloris*). *Notornis* 56: 158–161.

Ksepka, D. 2010. March of the fossil penguins. http://fossilpenguins.wordpress.com/2010/01/30/waimanu-the-first-penguin/. Accessed May, 2016.

Ksepka, D. 2013. March of the fossil penguins: An ancient lineage. In *Penguins: Their World, Their Ways*, eds. T. de Roy, M. Jones, and J. Cornthwaite, pp. 158–159. Auckland, New Zealand: Bateman.

Ksepka, D.T. and T. Ando. 2011. Penguins past, present, and future: Trends in the evolution of the Sphenisciformes. In *Living Dinosaurs. The Evolutionary History of Modern Birds*, eds. G. Dyke and G. Kaiser, pp. 155–186. West Sussex, UK: Wiley-Blackwell.

Ksepka, D.T., S. Bertelli, and N.P. Giannini. 2006. The phylogeny of the living and fossil Sphenisciformes (penguins). *Cladistics* 22: 412–441.

Kühne, R. and B. Lewis. 1985. External and middle ears. In *Form and Function in Birds*, Vol. 3, eds. A.S. King and J. McLelland, pp. 227–271. London, UK: Academic Press.

Kula, J., A. Tulloch, T.L. Spell, and M.L. Wells. 2007. Two-stage rifting of Zealandia-Australia-Antarctica: Evidence from ^{40}Ar/^{39}Ar thermochronometry of the Sisters shear zone, Stewart Island, New Zealand. *Geology* 35: 411–414.

Kunz, T.H. and E.D. Pierson. 1994. Bats of the world: An introduction. In *Walker's Bats of the World*, ed. R.M. Nowak, pp. 1–46. Baltimore, MD: Johns Hopkins University Press.

Kuschel, G. 1982. Apionidae and Curculionidae (Coleoptera) from the Poor Knights Islands, New Zealand. *Journal of the Royal Society of New Zealand* 12: 273–282.

Kuschel, G. 2003. Nemonychidae, Belidae, Brentidae (Insecta: Coleoptera: Curculionoidea). *Fauna of New Zealand* 45: 1–100.

Kuschel, G. and R.A.B. Leschen. 2011. Phylogeny and taxonomy of the Rhinorhynchinae (Coleoptera: Nemonychidae). *Invertebrate Systematics* 24: 573–615.

Kuschel, G. and T.H. Worthy. 1996. Past distribution of large weevils (Coleoptera: Curculionidae) in the South Island, New Zealand, based on Holocene fossil remains. *New Zealand Entomologist* 19: 15–23.

Kutty, S.N., T. Pape, B.M. Wiegmann, and R. Meier. 2010. Molecular phylogeny of the Calyptratae (Diptera: Cyclorrhapha) with an emphasis on the superfamily Oestroidea and the position of Mystacinobiidae and McAlpine's fly. *Systematic Entomology* 35: 614–635.

Lack, D. 1968. *Ecological Adaptations for Breeding in Birds*. London, UK: Methuen.

Ladiges, P.Y. and D. Cantrill. 2007. New Caledonia–Australian connections: Biogeographic patterns and geology. *Australian Systematic Botany* 20: 383–389.

Laird, M.G. and J.D. Bradshaw. 2004. The break-up of a long-term relationship: The Cretaceous separation of New Zealand from Gondwana. *Gondwana Research* 7: 273–286.

Lamarck, J.-B. 1809. *Philosophie zoologique*. Paris, France: Dentu.

Lamb, S. 2011. Cenozoic tectonic evolution of the New Zealand plate-boundary zone: A paleomagnetic perspective. *Tectonophysics* 509: 135–164.

Lamm, K.S. and B.D. Redelings. 2009. Reconstructing ancestral ranges in historical biogeography: Properties and prospects. *Journal of Systematics and Evolution* 47: 369–382.

Landis, C.A., H.J. Campbell, T. Aslund et al. 1999. Permian-Jurassic strata at Productus Creek, Southland, New Zealand: Implications for terrane dynamics of the eastern Gondwana margin. *New Zealand Journal of Geology and Geophysics* 42: 255–278.

Landis, C.A., H.J. Campbell, J.G. Begg, D.C. Mildenhall, A.M. Paterson, and S.A. Trewick. 2008. The Waipounamu Erosion Surface: Questioning the antiquity of the New Zealand land surface and terrestrial fauna and flora. *Geological Magazine* 145: 1–25.

Landis, C.A., H.J. Campbell, J.G. Begg, A.M. Paterson, and S.A. Trewick. 2006. The drowning of Zealandia: Evidence and implications. In *Geology and Genes III*, eds. S.A. Trewick and M.A. Phillips. Extended abstracts for papers presented at the *Geogenes III Conference*, Science House, Wellington, New Zealand, July 14, 2006. Geological Society of New Zealand Miscellaneous Publication 121, p. 21.

Lanfear, R. and L. Bromham. 2011. Estimating phylogenies for species assemblages: A complete phylogeny for the past and present native birds of New Zealand. *Molecular Phylogenetics and Evolution* 61: 958–963.

Langhoff, P., A. Authier, T.R. Buckley, J.S. Dugdale, A. Rodrigo, and R.D. Newcomb. 2009. DNA barcoding of the endemic New Zealand leafroller moth genera, *Ctenopseustis* and *Planotortrix*. *Molecular Ecology Resources* 9: 691–698.

Larivière, M.-C., M.J. Fletcher, and A. Larochelle. 2010. Auchenorrhyncha (Insecta: Hemiptera): A catalogue. *Fauna of New Zealand* 63: 1–232.

Larivière, M.-C. and A. Larochelle. 2004. Heteroptera (Insecta: Hemiptera). *Fauna of New Zealand* 50: 1–330.

Larivière, M.C. and A. Larochelle. 2015. *Zemacrosaldula*, a new genus of Saldidae (Hemiptera: Heteroptera) from New Zealand: Taxonomy, geographic distribution, and biology. *Zootaxa* 3955: 245–266.

Larochelle, A. and M.-C. Larivière. 2001. Carabidae (Insecta: Coleoptera): Catalogue. *Fauna of New Zealand* 43: 1–281.

Larochelle, A., and M.-C. Larivière. 2005. Harpalini (Insecta: Coleoptera: Carabidae: Harpalinae). *Fauna of New Zealand* 53: 1–156.

Larochelle, A. and M.-C. Larivière. 2007. Carabidae (Insecta: Coleoptera): Synopsis of supraspecific taxa. *Fauna of New Zealand* 60: 1–188.

Larochelle, A. and M.-C. Larivière. 2013. Carabidae (Insecta: Coleoptera): Synopsis of species, Cicindelinae to Trechinae (in part). *Fauna of New Zealand* 69: 1–193.

Larter, R.D., A.P. Cunningham, P.F. Barker, K. Gohl, and F.O. Nitsche. 2002. Tectonic evolution of the Pacific margin of Antarctica: 1. Late Cretaceous tectonic reconstructions. *Journal of Geophysical Research* 107(B12): 2345.

Lartillot, N. 2013. Phylogenetic patterns of GC-biased gene conversion in placental mammals and the evolutionary dynamics of recombination landscapes. *Molecular Biology and Evolution* 30: 489–502.

Laurent, N., B. Bremer, and K. Bremer. 1998. Phylogeny and generic interrelationships of the Stylidiaceae (Asterales), with a possible extreme case of floral paedomorphosis. *Systematic Botany* 23: 289–304.

Leaman, D.E. 2003. Discussion and reply: 'Shaping the Australian crust over the last 300 million years: Insights from fission track thermotectonic imaging and denudation studies of key terranes'. *Australian Journal of Earth Sciences* 50: 645–650.

Leathwick, J.R. 1995. Climatic relationships of some New Zealand forest tree species. *Journal of Vegetation Science* 6: 237–248.

Leathwick, J.R., J. Elith, L. Chadderton, D. Rowe, and T. Hastie. 2008. Dispersal, disturbance, and the contrasting biogeographies of New Zealand's diadromous and non-diadromous fish species. *Journal of Biogeography* 35: 1481–1497.

Lederer, R.J. 1972. The role of avian rictal bristles. *The Wilson Bulletin* 84: 193–197.

Lee, D.E., J.M. Bannister, and J.K. Lindqvist. 2007a. Late Oligocene–Early Miocene leaf macrofossils confirm a long history of *Agathis* in New Zealand. *New Zealand Journal of Botany* 45: 565–578.

Lee, D.E., J.G. Conran, J.M. Bannister, U. Kaulfuss, and D.C. Mildenhall. 2013a. A fossil *Fuchsia* (Onagraceae) flower and an anther mass with in situ pollen from the early Miocene of New Zealand. *American Journal of Botany* 100: 2052–2065.

Lee, D.E., J.G. Conran, J.K. Lindqvist,. J. M. Bannister, and D.C. Mildenhall. 2012a. New Zealand Eocene, Oligocene and Miocene macrofossil and pollen records and modern plant distributions in the Southern Hemisphere. *Botanical Review* 78: 235–260.

Lee, D.E., W.G. Lee, and N. Mortimer. 2001. Where and why have all the flowers gone? Depletion and turnover in the New Zealand Cenozoic angiosperm flora in relation to palaeogeography and climate. *Australian Journal of Botany* 49: 341–356.

Lee, D.E., J.K. Lindqvist, A.G. Beu et al. 2013b. Oligocene paleogeography of southern New Zealand: Sedimentological/paleontological evidence for forested land, estuaries, rocky and sandy shores. *VII Southern Connection Congress Abstracts*, University of Otago, Dunedin, New Zealand.

Lee, D.E., J.K. Lindquist, A.G. Beu, et al. 2014. Geological setting and diverse fauna of a Late Oligocene rocky shore ecosystem, Cosy Dell, Southland. *New Zealand Journal of Geology and Geophysics* 57: 195–208.

Lee, D.E., R.M. McDowall, and J.K. Lindqvist. 2007b. *Galaxias* fossils from Miocene lake deposits, Otago, New Zealand: The earliest records of the Southern Hemisphere family Galaxiidae (Teleostei). *Journal of the Royal Society of New Zealand* 37: 109–130.

Lee, K. 1959. The earthworm fauna of New Zealand. *Bulletin of the New Zealand*. Department of Scientific and Industrial Research 130: 387.

Lee, M. 1999. An updated survey of the distribution of the stick insects of Britain. *Phasmid Studies* 7: 18–25.

Lee, M.S.Y., M.N. Hutchinson, T.H. Worthy, M. Archer, and A.J.D. Tennyson. 2009. Miocene skinks and geckos reveal long-term conservatism of New Zealand's lizard fauna. *Biology Letters* 5: 833–837.

Lee, M.S.Y. and A. Skinner. 2011. Testing fossil calibrations for vertebrate molecular trees. *Zoologica Scripta* 40: 538–543.

Lee, W.G. and D.R. Given. 1984. *Celmisia spedenii* G. Simpson, an ultramafic endemic, and *Celmisia markii*, sp. nov., from southern New Zealand. *New Zealand Journal of Botany* 22: 585–592.

Lee, W.G., A.J. Tanentzap, and P.B. Heenan. 2012b. Plant radiation history affects community assembly: Evidence from the New Zealand alpine. *Biology Letters* 8: 558–561.

Lees, A.C. and J.J. Gilroy. 2014. Vagrancy fails to predict colonization of oceanic islands. *Global Ecology and Biogeography* 23: 405–413.

Lehnebach, C.A. 2008. Phylogenetic affinities, species delimitation and adaptive radiation of New Zealand *Ranunculus*. PhD thesis, Massey University, Palmerston North, New Zealand.

LeMasurier, W.E. and C.A. Landis. 1996. Mantle-plume activity recorded by low-relief erosion surfaces in West Antarctica and New Zealand. *Geological Society of America Bulletin* 108: 1450–1466.

Leschen, R.A.B. and T.R. Buckley. 2015. Revision and phylogeny of *Syrphetodes* (Coleoptera: Ulodidae): Implications for biogeography, alpinization, and conservation. *Systematic Entomology* 40: 143–168.

Leschen, R.A.B., M.S. Bullians, B. Michaux, and K.-J. Ahn. 2002. Systematics of *Baeostethus chiltoni*, a subantarctic liparocephaline (Coleoptera: Staphylinidae: Aleocharinae): A Pangean relic or a more recent immigrant? *Journal of the Royal Society of New Zealand* 32: 189–201.

Leschen, R.A.B., E. Butler, T.R. Buckley, and P. Ritchie. 2011. Biogeography of the New Zealand subantarctics: Phylogenetics of *Pseudhelops* (Coleoptera: Tenebrionidae). *New Zealand Entomologist* 34: 12–26.

Leschen, R.A.B. and T. Lackner. 2013. Gondwanan Gymnochilini (Coleoptera: Trogossitidae): Generic concepts, review of New Zealand species and long-range Pacific dispersal. *Systematic Entomology* 38: 278–304.

Leschen, R.A.B., J.F. Lawrence, G. Kuschel, S. Thorpe, and Q. Wang. 2003. Coleoptera genera of New Zealand. *New Zealand Entomologist* 26: 15–28.

Leschen, R.A.B. and B. Michaux. 2005. Biogeography and evolution of New Zealand Priasilphidae (Coleoptera: Cucujoidea). *New Zealand Entomologist* 28: 55–64.

Lester, S.E., B.I. Ruttenberg, S.D. Gaines, and B.P. Kinlan. 2007. The relationship between dispersal ability and geographic range size. *Ecology Letters* 10: 745–758.

Leven, E.J. and J.A. Grant-Mackie. 1997. Permian fusulinid Foraminifera from Wherowhero Point, Orua Bay, Northland. *New Zealand Journal of Geology and Geophysics* 40: 473–486.

Lever, H. 2007. Review of unconformities in the late Eocene to early Miocene successions of the South Island, New Zealand: Ages, correlations, and causes. *New Zealand Journal of Geology and Geophysics* 50: 245–261.

Levin, D.A. 2000. *The Origin, Expansion, and Demise of Plant Species*. New York: Oxford University Press.

Levin, M. 2005. Left–right asymmetry in embryonic development: A comprehensive review. *Mechanisms of Development* 122: 3–25.

Levin, R.A., W.L. Wagner, P.C. Hoch et al. 2004. Paraphyly in tribe Onagreae: Insights into phylogenetic relationships of Onagraceae based on nuclear and chloroplast sequence data. *Systematic Botany* 29: 147–164.

Lewis, D.W. 1992. Anatomy of an unconformity on mid-Oligocene Amuri Limestone, Canterbury, New Zealand. *New Zealand Journal of Geology and Geophysics* 35: 463–475.

Lewis, D.W. and S.E. Belliss. 1984. Mid Tertiary unconformities in the Waitaki Subdivision, North Otago. *Journal of the Royal Society of New Zealand* 14: 251–276.

Lewis, K.B., L. Carter, and F.J. Davey. 1994. The opening of Cook Strait: Interglacial tidal scour and aligning basins at a subduction to transform plate edge. *Marine Geology* 116: 293–312.

Li, J., R. Xia, R.M. McDowall, J.A. López, G. Lei, and C. Fu. 2010. Phylogenetic position of the enigmatic *Lepidogalaxias salamandroides* with comment on the orders of lower euteleostean fishes. *Molecular Phylogenetics and Evolution* 57: 932–936.

Liebherr, J.K. and J.W.M. Marris. 2009. Revision of the New Zealand species of *Mecyclothorax* Sharp (Coleoptera: Carabidae: Psydrinae, Mecyclothoracini) and the consequent removal of several species to *Meonochilus* gen. n. (Psydrinae, Meonini). *New Zealand Entomologist* 32: 5–22.

Liebherr, J.K., J.W.M. Marris, R.M. Emberson, P. Syrett, and S. Roig-Juñent. 2011. *Orthoglymma wangapeka* gen.n., sp.n. (Coleoptera: Carabidae: Broscini): A newly discovered relict from the Buller Terrane, north-western South Island, New Zealand, corroborates a general pattern of Gondwanan endemism. *Systematic Entomology* 36: 395–414.

Liggins, L., D.G. Chapple, C.H. Daugherty, and P.A. Ritchie. 2008a. Origin and post-colonization evolution of the Chatham Islands skink (*Oligosoma nigriplantare nigriplantare*). *Molecular Ecology* 17: 3290–3305.

Liggins, L., D.G. Chapple, C.H. Daugherty, and P.A. Ritchie. 2008b. A SINE of restricted gene flow across the Alpine Fault: Phylogeography of the New Zealand common skink (*Oligosoma nigriplantare nigriplantare*). *Molecular Ecology* 17: 3668–3683.

Linder, H.P. 2008. Plant species radiations: Where, when, why? *Philosophical Transactions of the Royal Society B* 363: 3097–3105.

Linder, H.P., A. Antonelli, A.M. Humphreys, M.D. Pirie, and R.O. Wüest. 2013. What determines biogeographical ranges? Historical wanderings and ecological constraints in the danthonioid grasses. *Journal of Biogeography* 40: 821–834.

Linder, H.P., M. Baeza, N.P. Barker et al. 2010. A generic classification of the Danthonioideae (Poaceae). *Annals of the Missouri Botanical Garden* 97: 306–364.

Linkem, C.W., R.M. Brown, C.D. Siler, B.J. Evans, C.C. Austin, D.T. Iskandar, A.C. Diesmos, J. Supriatna, N. Andayani, and J.A. McGuire. 2013. Stochastic faunal exchanges drive diversification in widespread Wallacean and Pacific island lizards (Squamata: Scincidae: *Lamprolepis smaragdina*). *Journal of Biogeography* 40: 507–520.

Lister, A.M. 2014. Behavioural leads in evolution: Evidence from the fossil record. *Biological Journal of the Linnean Society* 112: 315–331.

Lister, G.S. and G.A. Davis. 1989. The origin of metamorphic core complexes and detachment faults formed during Tertiary continental extension in the northern Colorado River region, USA. *Journal of Structural Geology* 11: 65–94.

Litchfield, N.J. 2001. The Titri Fault System: Quaternary-active faults near the leading edge of the Otago reverse fault province. *New Zealand Journal of Geology and Geophysics* 44: 517–534.

Livezey, B.C. 1992. Morphological corollaries and ecological implications of flightlessness in the kakapo (Psittaciformes: *Strigops habroptilus*). *Journal of Morphology* 213: 105–145.

Livezey, B.C. 2011. Progress and obstacles in the phylogenetics of modern birds. In *Living Dinosaurs: The Evolutionary History of Modern Birds*, eds. G. Dyke, and L. Chiappe, pp. 117–145. Berkeley, CA: University of California Press.

Livnat, A. 2013. Interaction-based evolution: How natural selection and nonrandom mutation work together. *Biology Direct* 8(24): 1–53.

Lloyd, B.D. 2003. Intraspecific phylogeny of the New Zealand short-tailed bat *Mystacina tuberculata* inferred from multiple mitochondrial gene sequences. *Systematic Biology* 52: 460–476.

Lloyd, B.D. 2005. Superfamily Noctilionoidea. In *The Handbook of New Zealand Mammals*, ed. C.M. King, pp. 109–129. Melbourne, Australia: Oxford University Press.

Lloyd, D.G. 1972. A revision of the New Zealand, subantarctic, and South American species of *Cotula*, Section *Leptinella. New Zealand Journal of Botany* 10: 277–372.

Lloyd, D.G. 1982. Variation and evolution of plant species on the outlying islands of New Zealand, with particular reference to *Cotula featherstonii. Taxon* 31: 478–487.

Lloyd, K.M., A.M. Hunter, D.A. Orlovich, S.J. Draffin, A.V. Stewart, and W.G. Lee. 2007. Phylogeny and biogeography of endemic *Festuca* (Poaceae) from New Zealand based on nuclear (ITS) and chloroplast (trnL–trnF) Nucleotide Sequences. *Aliso* 23: 406–419.

Löbl, I. and R.A.B. Leschen. 2003. Scaphidiinae (Insecta: Coleoptera: Staphylinidae). *Fauna of New Zealand* 48: 1–94.

Lockhart, P.J., P.A. McLenachan, D. Havell, D. Glenny, D. Huson, and U. Jensen. 2001. Phylogeny, radiation, and transoceanic dispersal of New Zealand alpine buttercups: Molecular evidence under split decomposition. *Annals of the Missouri Botanical Garden* 88: 458–477.

Lombrozo, T., D. Kelemen, and D. Zaitchik. 2007. Inferring design: Evidence of a preference for teleological explanations from patients with Alzheimer's disease. *Psychological Science* 18: 999–1006.

Lomolino, M.V. 2005. Body size evolution in insular vertebrates: Generality of the island rule. *Journal of Biogeography* 32: 1683–1699.

Lomolino, M.V. and J.H. Brown. 2009. The reticulating phylogeny of island biogeography theory. *Quarterly Review of Biology* 84: 357–390.

Lourteig, A. 1979. Oxalidaceae extra-austroamericanae. II. *Oxalis* L. section *Corniculatae* DC. *Phytologia* 42: 57–198.

Lovette, I.J., E. Bermingham, and R.E. Ricklefs. 2002. Cladespecific morphological diversification and adaptive radiation in Hawaiian songbirds. *Proceedings of the Royal Society B* 269: 37–42.

Low, E. 1900. On the vegetative organs of *Haastia pulvinaris*. *Transactions of the New Zealand Institute* 32: 150–157.

Lowry, P.P., II. 1998. Diversity, endemism, and extinction in the flora of New Caledonia: A review. In *Rare, Threatened, and Endangered Floras of Asia and the Pacific Rim* (Academica Sinica Monograph 16), eds. C.I. Peng and P.P. Lowry II, pp. 181–206. Taipei, Taiwan: Institute of Botany.

Lucas, J.S. 1980. Spider crabs of the family Hymenosomatidae (Crustacea: Brachyura) with particular reference to Australian species: Systematics and biology. *Records of the Australian Museum* 33: 148–247.

Ludwig, L.R. and G. Kantvilas. 2015. Four Tasmanian lichens new to New Zealand. *Australian Lichenology* 77: 18–27.

Lundberg, G., M. Kottelat, G.R. Smith, M.L. Stiassny, and A.C. Gill. 2000. So many fishes, so little time: An overview of recent ichthyological discovery in continental waters. *Annals of the Missouri Botanical Garden* 87: 26–62.

Luyendyk, B.P. 1995. Hypothesis for Cretaceous rifting of East Gondwana caused by subducted slab capture. *Geology* 23: 373–376.

Lyal, C.H.C. 1993a. *Andracalles pani* n. sp. from the Three Kings Islands: Evidence for copulation mechanics (Coleoptera: Curculionidae: Cryptorhynchinae). *New Zealand Entomologist* 16: 4–12.

Lyal, C.H.C. 1993b. Cryptorhynchinae (Insecta: Coleoptera: Curculionidae). *Fauna of New Zealand* 29: 1–305.

Lynch, M. 2007a. The frailty of adaptive hypotheses for the origins of organismal complexity. *Proceedings of the National Academy of Sciences of USA* 104: S8597–S8604.

Lynch, M. 2007b. The evolution of genetic networks by non-adaptive processes. *Nature Reviews Genetics* 8: 803–813.

Lynch, M. 2007c. *The Origins of Genome Architecture*. Sunderland, MA: Sinauer Associates.

Lynch, M. 2010. Rate, molecular spectrum, and consequences of human mutation. *Proceedings of the National Academy of Sciences of USA* 107: 961–968.

Lynch, W. 1997. *Penguins of the World*. Kingston, Ontario, Canada: Bookmakers Press.

Lyneborg, L. 1992. Therevidae (Insecta: Diptera). *Fauna of New Zealand* 24: 1–139.

MacArthur, R.H. 1972. *Geographical Ecology: Patterns in the Distribution of Species*. New York: Harper and Row.

MacCurdy, E. 1954. *The Notebooks of Leonardo da Vinci*. London, UK: Reprint Society.

Macmillan, B.H. 1983. *Acaena profundeincisa* (Bitter) B.H. Macmillan comb. nov. (Rosaceae) of New Zealand. *New Zealand Journal of Botany* 21: 347–352.

Macmillan, B.H. 1991. *Acaena pallida* (Kirk) Allan (Rosaceae) in Tasmania and New South Wales, Australia. In *Aspects of Tasmanian Botany*, ed. M.R. Banks, pp. 53–55. Hobart, Tasmania, Australia: Royal Society of Tasmania.

Maddison, W.P., and D.R. Maddison. 1992. *MacClade: Interactive Analysis of Phylogeny and Character Evolution*, Version 3.0. Sunderland, MA: Sinauer Associates.

Magri, D., S. Fineschi, P. Bellarosa et al. 2007. The distribution of *Quercus suber* chloroplast haplotypes matches the palaeogeographical history of the western Mediterranean. *Molecular Ecology* 16: 5259–5266.

Mahlfeld, K. 2005. Biogeography and taxonomy of Cavellia Iredale 1915 (Pulmonata: Charopidae). PhD thesis, Victoria University of Wellington, Wellington, New Zealand.

Mahlfeld, K. 2008. Land snails. In *Chatham Islands: Heritage and Conservation*, ed. C. Miskelly, p. 124. Christchurch, New Zealand: Canterbury University Press.

Majnep, I.S. and R. Bulmer. 1977. *Birds of My Kalam Country*. Auckland, New Zealand: Oxford University Press and Auckland University Press.

Malcolm, B. and N. Malcolm. 1988. *New Zealand's Alpine Plants Inside and Out*. Nelson, New Zealand: Kel Aiken.

Malipatil, M.B. 1977. Distribution, origin and speciation, wing development, and host-plant relationships of New Zealand Targaremini (Hemiptera: Lygaeidae). *New Zealand Journal of Zoology* 4: 369–381.

Mandáková, T., P.B. Heenan, and M.A. Lysak. 2010a. Island species radiation and karyotypic stasis in *Pachycladon* allopolyploids. *BMC Evolutionary Biology* 10(367): 1–14.

Mandáková, T., S. Joly, M. Krzywinski, K. Mummenhoff, and M. Lysak. 2010b. Fast diploidization in close mesopolyploid relatives of *Arabidopsis*. *Plant Cell* 22: 2277–2290.

Manson, D.C.M. 1984. Eriophyinae (Arachnida: Acari: Eriophyoidea). *Fauna of New Zealand* 5: 1–128.

Mao, K., R.I. Milne, L. Zhang, Y. Peng, J. Liu, P. Thomas, R.R. Mill, and S.S. Renner. 2012. The distribution of living Cupressaceae reflects the breakup of Pangea. *Proceedings of the National Academy of Sciences of USA* 109: 7793–7798.

Marchant, S. 1972. Evolution of the genus *Chrysococcyx*. *Ibis* 114: 219–233.

Marchant, S. and P.J. Higgins (eds.). 1990. *Handbook of Australian, New Zealand and Antarctic Birds*, Vol. 1: *Ratites to Ducks*. Melbourne, Australia: Oxford University Press.

Marchant, S. and P.J. Higgins (eds.). 1993. *Handbook of Australian, New Zealand and Antarctic Birds*, Vol. 2: *Raptors to Lapwings*. Melbourne, Australia: Oxford University Press.

Marchant, S. and P.J. Higgins (eds.). 2002. *Handbook of Australian, New Zealand and Antarctic Birds*, Vol. 6: *Pardalotes to Shrike-Thrushes*. Melbourne, Australia: Oxford University Press.

Marjanović, D. and M. Laurin. 2013. The origin(s) of extant amphibians: A review with emphasis on the "lepospondyl hypothesis." *Geodiversitas* 35: 207–272.

Mark, A.F. 1977. *Vegetation of Mount Aspiring National Park*. Wellington, New Zealand: Department of Lands and Survey.

Mark, A.F. and N.M. Adams. 1973. *New Zealand Alpine Plants*. Wellington, New Zealand: A.H. and A.W. Reed.

Mark, A.F. and L.C. Bliss. 1970. The high-alpine vegetation of Central Otago, New Zealand. *New Zealand Journal of Botany* 8: 381–451.

Markey, A., J.M. Lord, and D.A. Orlovich. 2004. *Coprosma talbrockiei*: An oddball sheds light on the *Coprosminae*. Abstract of paper presented at SYSTANZ Meeting, Whakapapa Village, New Zealand, 2004.

Marks, C. and N. Walsh. 2014. Taxonomic reassessment of *Kelleria* (Thymelaeaceae) in Australia and recognition of a new endemic Victorian species. *Muelleria* 33: 3–11.

Marks, J.S., R.J. Cannings, and H. Mikkola. 1999. Strigidae. In *Handbook of the Birds of the World*, Vol. 5, eds. J. del Hoyo, A. Elliott, and J. Sargatal, pp. 76–243. Barcelona, Spain: Lynx Edicions.

Marris, J.W.M. 2000. The beetle (Coleoptera) fauna of the Antipodes Islands, with comments on the impact of mice; and an annotated checklist of the insect and arachnid fauna. *Journal of the Royal Society of New Zealand* 30: 169–195.

Marris, J.W.M. and P.J. Johnson. 2010. A revision of the New Zealand click beetle genus *Amychus* Pascoe 1876 (Coleoptera: Elateridae: Denticollinae): With a description of a new species from the Three Kings Islands. *Zootaxa* 2331: 35–56.

Marsaglia, K.M., N. Mortimer, C. Bender-Whitaker, M. Marden, and C. Marzengarb. 2014. Ophioliticlastic Ihungia igneous conglomerate (Early Miocene), North Island, New Zealand: Evidence for an island source related to subduction initiation and deposition within a slump-generated submarine slope re-entrant. *New Zealand Journal of Geology and Geophysics* 57: 219–235.

Marshall, B.A. 1995a. A revision of the Recent *Calliostoma* species of New Zealand (Mollusca: Gastropoda: Trochoidea). *The Nautilus* 108: 83–127.

Marshall, B.A. 1995b. Recent and Tertiary Trochaclididae from the southwest Pacific (Mollusca: Gastropoda: Trochoidea). *The Veliger* 38: 92–115.

Marshall, B.A. 1998. A review of the Recent Trochini of New Zealand (Mollusca: Gastropoda: Trochidae). *Molluscan Research* 19: 73–106.

Marshall, B.A. 1999. A revision of the Recent Solariellinae (Gastropoda: Trochoidea) of the New Zealand region. *The Nautilus* 113: 4–42.

Marshall, B.A. 2001. Mollusca Gastropoda: Seguenziidae from New Caledonia and the Loyalty Islands. *Mémoires du Muséum national d'Histoire naturelle* 150: 41–109.

Marshall, B.A. 2003. A review of the recent and late Cenozoic Calyptraeidae of New Zealand (Mollusca: Gastropoda). *The Veliger* 46: 117–144.

Marshall, B.A. 2006. Four new species of Monoplacophora (Mollusca) from the New Zealand region. *Molluscan Research* 26: 61–68.

Marshall, B.A. 2011. A new species of *Latia* Gray, 1850 (Gastropoda: Pulmonata: Hygrophila: Chilinoidea: Latiidae) from Miocene palaeo-lake Manuherikia, southern New Zealand, and biogeographic implications. *Molluscan Research* 31: 47–52.

Marshall, B.A. and G.M. Barker. 2007. A revision of the New Zealand landsnails of the genus *Cytora* Kobelt & Möllendorff, 1897 (Mollusca: Gastropoda: Pupinidae). *Tuhinga* 18: 49–113.

Marshall, B.A. and G.M. Barker. 2008. A revision of the New Zealand landsnails referred to *Allodiscus* Pilsbry, 1892 and *Pseudallodiscus* Climo, 1971, with the introduction of three new genera (Mollusca: Gastropoda: Charopidae). *Tuhinga* 19: 57–167.

Marshall, B.A., M.C. Fenwick, and P.A. Ritchie. 2014. New Zealand Recent Hyriidae (Mollusca: Bivalvia: Unionida). *Molluscan Research* 34: 181–200.

Marshall, B.A. and R.O Houart. 2011. The genus *Pagodula* (Mollusca: Gastropoda: Muricidae) in Australia, the New Zealand region and the Tasman Sea. *New Zealand Journal of Geology and Geophysics* 54: 89–114.

Marshall, C.R. 1990. The fossil record and estimating divergence times between lineages: Maximum divergence times and the importance of reliable phylogenies. *Journal of Molecular Evolution* 30: 400–408.

Marshall, D.C., K.B.R. Hill, K.M. Fontaine, T.R. Buckley, and C. Simon. 2009. Glacial refugia in a maritime temperate climate: Cicada (*Kikihia subalpina*) mtDNA phylogeography in New Zealand. *Molecular Ecology* 18: 1995–2009.

Marshall, D.C., K.B.R. Hill, K.A. Marske, C. Chambers, T.R. Buckley, and C. Simon. 2012. Limited, episodic diversification and contrasting phylogeography in a New Zealand cicada radiation. *BMC Evolutionary Biology* 12(177): 1–18.

Marshall, D.C., K. Slon, J.R. Cooley, K.B.R. Hill, and C. Simon. 2008. Steady Plio-Pleistocene diversification and a 2-million-year sympatry threshold in a New Zealand cicada radiation. *Molecular Phylogenetics and Evolution* 48: 1054–1066.

Marske, K.A., R.A.B. Leschen, G.M. Barker, and T.R. Buckley. 2009. Phylogeography and ecological niche modelling implicate coastal refugia and trans-alpine dispersal of a New Zealand fungus beetle. *Molecular Ecology* 18: 5126–5142.

Marske, K.A., R.A.B. Leschen, and T.R. Buckley. 2011. Reconciling phylogeography and ecological niche models for New Zealand beetles: Looking beyond glacial refugia. *Molecular Phylogenetics and Evolution* 59: 89–102.

Marske, K.A., R.A.B. Leschen, and T.R. Buckley. 2012. Concerted versus independent evolution and the search for multiple refugia: Comparative phylogeography of four forest beetles. *Evolution* 66: 1862–1877.

Martin, G.R. 1985. Eye. In *Form and Function in Birds*, Vol. 3, eds. A.S. King and J. McLelland, pp. 311–373. London, UK: Academic Press.

Martin, G.R., N. Jarrett, and M. Williams. 2007a. Visual fields in Blue Ducks *Hymenolaimus malacorhynchos* and Pink-eared Ducks *Malacorhynchus membranaceus*: Visual and tactile foraging. *Ibis* 149: 112–120.

Martin, G.R., K.J. Wilson, M.J. Wild, S. Parsons, M.F. Kubke, and J. Corfield. 2007b. Kiwi forego vision in the guidance of their nocturnal activities. *PLoS One* 2: e198 (pp. 1–6).

Martin, H.A. 2003. The history of the family Onagraceae in Australia and its relevance to biogeography. *Australian Journal of Botany* 51: 585–598.

Martin, P. and G.C. Zuccarello. 2012. Molecular phylogeny and timing of radiation in *Lessonia* (Phaeophyceae, Laminariales). *Phycological Research* 60: 276–287.

Martin, W. 1938. Notes on the indigenous flora of Marlborough (New Zealand) with special reference to plant distribution. *Transactions of the New Zealand Institute* 67: 414–425.

Martínez, I. 1992. Spheniscidae. In *Handbook of the Birds of the World*, Vol. 1, eds. J. del Hoyo, A. Elliott, and J. Sargatal, pp. 140–161. Barcelona, Spain: Lynx Edicions.

Marvaldi, A.E., C.N. Duckett, K.M. Kjer, and J.J. Gillespie. 2009. Structural alignment of 18S and 28S rDNA sequences provides insights into phylogeny of Phytophaga (Coleoptera: Curculionoidea and Chrysomeloidea). *Zoologica Scripta* 38: 63–77.

Masner, L. and J.L. García R. 2002. The genera of Diapriinae (Hymenoptera: Diapriidae) in the New World. *Bulletin of the American Museum of Natural History* 268: 1–138.

Mason-Gamer, R.J. 2005. The β-amylase genes of grasses and a phylogenetic analysis of the Triticeae (Poaceae). *American Journal of Botany* 92: 1045–1058.

Massey, S.E. 2010. Pseudaptations and the emergence of beneficial traits. In *Evolutionary Biology: Concepts, Molecular and Morphological Evolution*, ed. P. Pontarotti, pp. 81–98. Berlin, Germany: Springer.

Matheu, E. and J. del Hoyo. 1992. Threskiornithidae. In *Handbook of the Birds of the World*, Vol. 1, eds. J. del Hoyo, A. Elliott, and J. Sargatal, pp. 472–507. Barcelona, Spain: Lynx Edicions.

Matschiner, M., R. Hanel, and W. Salzburger. 2011. On the origin and trigger of the notothenioid adaptive radiation. *PLoS One* 6(4): e18911 (pp. 1–9).

Matthew, W.D. 1915. Climate and evolution. *Annals of the New York Academy of Science* 24: 171–318.

Matthews, K.J., M. Seton, and R.D. Müller. 2011. Was there a global-scale plate reorganisation event at 100 Ma? *American Geophysical Union, Fall Meeting 2011*, San Francisco, CA, abstract #T23D-2446.

Matzke, N.J. 2013. Probabilistic historical biogeography: New models for founder-event speciation, imperfect detection, and fossils allow improved accuracy and model-testing. *Frontiers of Biogeography* 5: 242–248.

Mayén-Estrada, R. and R. Aguilar-Aguilar. 2012. Track analysis and geographic distribution of some *Lagenophrys* Stein, 1852 (Protozoa: Ciliophora: Peritrichia) species. *Journal of Natural History* 46: 249–263.

Maynard Smith, J., 1987. How to model evolution. In *The Latest on the Best: Essays on Evolution and Optimality*, ed. J. Dupré, pp. 119–131. Cambridge, MA: MIT Press.

Mayr, E. 1931. Birds collected during the Whitney South Sea Expedition. XIV. *American Museum Novitates* 488: 1–11.

Mayr, E. 1942. *Systematics and the Origin of Species*. New York: Columbia University Press.

Mayr, E. 1944. The birds of Timor and Sumba. *Bulletin of the American Museum of Natural History* 83: 127–194.

Mayr, E. 1954. Change of genetic environment and evolution. In *Evolution as a Process*, eds. J. Huxley, A.C. Hardy, and E.B. Ford, pp. 157–180. London, UK: Allen and Unwin.

Mayr, E. 1960. The emergence of evolutionary novelties. In *Evolution after Darwin: The University of Chicago Centennial*, eds. S. Tax and C. Callender, pp. 349–380. Chicago, IL: University of Chicago Press.

Mayr, E. 1963. *Animal Species and Evolution*. Cambridge, MA: Harvard University Press.

Mayr, E. 1965. Summary. In *The Genetics of Colonizing Species*, eds. H.G. Baker and G.L. Stebbins, pp. 553–562. New York: Academic Press.

Mayr, E. 1980. Some thoughts on the history of the evolutionary synthesis. In *The Evolutionary Synthesis*, eds. E. Mayr and W. Provine, pp. 1–48. Cambridge, MA: Harvard University Press.

Mayr, E. 1982. Review of 'Vicariance Biogeography', by G. Nelson and D.E. Rosen (eds). *The Auk* 99: 618–620.

Mayr, E. 1997. The objects of selection. *Proceedings of the National Academy of Sciences of USA* 94: 2091–2094.

Mayr, E. and J. Diamond. 2001. *The Birds of Northern Melanesia: Speciation, Ecology and Biogeography*. New York: Oxford University Press.

Mayr, E., E.G. Linsley, and R.L. Usinger. 1953. *Methods and Principles of Systematic Zoology*. New York: McGraw Hill.

Mayr, G. 2009. *Paleogene Fossil Birds*. Berlin, Germany: Springer.

Mayr, G. 2013. The age of the crown group of passerine birds and its evolutionary significance—Molecular calibrations versus the fossil record. *Systematics and Biodiversity* 11: 7–13.

Mazza, P. 2014a. Pushing your luck (Review of A. de Queiroz, 2014, *The Monkey's Voyage*). *BioScience* 64: 458–459.

Mazza, P.P. 2014b. If hippopotamuses cannot swim, how did they colonize islands? *Lethaia* 47: 494–499.

Mazza, P.P., S. Lovari, F. Masini, M. Masseti, and M. Rustioni. 2013. A multidisciplinary approach to the analysis of multifactorial land mammal colonization of islands. *BioScience* 63: 939–951.

McAlpine, D.K. 2007. The surge flies (Diptera: Canacidae: Zaleinae) of Australasia and notes on tethinid-canacid morphology and relationships. *Records of the Australian Museum* 59: 27–64.

McBreen, K. and P.B. Heenan. 2006. Phylogenetic relationships of *Pachycladon* (Brassicaceae) species based on three nuclear and two chloroplast DNA markers. *New Zealand Journal of Botany* 44: 377–386.

McCormack, J.E., A.J. Zellmer, and L.L. Knowles. 2009. Does niche divergence accompany allopatric divergence in *Aphelocoma* jays as predicted under ecological speciation? Insights from tests with niche models. *Evolution* 64: 1231–1244.

McCoy-West, A.J., J.A. Baker, K. Faure, and R. Wysoczanski. 2010. Petrogenesis and origins of mid-Cretaceous continental intraplate volcanism in Marlborough, New Zealand: Implications for the long-lived HIMU magmatic mega-province of the SW Pacific. *Journal of Petrology* 51: 2003–2045.

McCulloch, G.A. 2010. Evolutionary genetics of southern stoneflies. PhD thesis, University of Otago, Dunedin, New Zealand.

McCulloch, G.A., G.P. Wallis, and J.M. Waters. 2010. Onset of glaciations drove simultaneous vicariant isolation of alpine insects in New Zealand. *Evolution* 64: 2033–2043.

McDade, M.C. (ed.). 2003. *Grzimek's Animal Life Encyclopedia*, 2nd ed., Vol. 13: *Mammals* II. Farmington Hills, MI: Gale Group.

McDowall, R.M. 1964. The affinities and derivation of the New Zealand fresh-water fish fauna. *Tuatara* 12: 59–67.

McDowall, R.M. 1973. Relationships and axonomy of the New Zealand torrentfish *Cheimarrichthys fosteri* Haast (Pisces: Mugiloididae). *Journal of the Royal Society of New Zealand* 3: 199–217.

McDowall, R.M. 1996. Volcanism and freshwater fish biogeography in the northeastern North Island of New Zealand. *Journal of Biogeography* 23: 139–148.

McDowall, R.M. 1997. The evolution of diadromy in fishes (revisited) and its place in phylogenetic analysis. *Reviews in Fish Biology and Fisheries* 7: 443–462.

McDowall, R.M. 2000. Biogeography of the New Zealand Torrentfish, *Cheimarrichthys fosteri* (Teleostei: Pinguipedidae): A distribution driven mostly by ecology and behaviour. *Environmental Biology of Fishes* 58: 119–131.

McDowall, R.M. 2002. Accumulating evidence for a dispersal biogeography of southern cool temperate freshwater fishes. *Journal of Biogeography* 29: 207–219.

McDowall, R.M. 2005. Historical biogeography of the New Zealand freshwater crayfishes (Parastacidae, *Paranephrops* spp.): Restoration of a refugial survivor? *New Zealand Journal of Zoology* 32: 55–77.

McDowall, R.M. 2008. Process and pattern in the biogeography of New Zealand–a global microcosm? *Journal of Biogeography* 35: 197–212.

McDowall, R.M. 2010a. Historical and ecological context, pattern and process, in the derivation of New Zealand's freshwater fish fauna. *New Zealand Journal of Ecology* 34: 185–194.

McDowall, R.M. 2010b. *New Zealand Freshwater Fishes: An Historical and Ecological Biogeography.* Berlin, Germany: Springer.

McDowall, R.M., D.R. Robertson, and R. Saito. 1975. Occurrence of galaxiid larvae and juveniles in the sea. *New Zealand Journal of Marine and Freshwater Research* 9: 1–9.

McDowall, R.M. and J.M. Waters. 2002. A new longjaw galaxias species from the Kauru River, North Otago, New Zealand. *New Zealand Journal of Zoology* 29: 41–52.

McDowall, R.M. and J.M. Waters. 2003. A new species of *Galaxias* (Teleostei: Galaxiidae) from the Mackenzie Basin, New Zealand. *Journal of the Royal Society of New Zealand* 33: 675–691.

McDowall, R.M. and A.M. Whitaker. 1975. The freshwater fishes. In *Biogeography and Ecology in New Zealand*, ed. G. Kuschel, pp. 277–299. The Hague, the Netherlands: Junk.

McGlone, M.S. 1985. Plant biogeography and the late Cenozoic history of New Zealand. *New Zealand Journal of Botany* 23: 723–750.

McGlone, M.S. 2005. Goodbye Gondwana. *Journal of Biogeography* 32: 739–740.

McGlone, M.S., R.P. Duncan, and P.B. Heenan. 2001. Endemism, species selection and the origin and distribution of the vascular plant flora of New Zealand. *Journal of Biogeography* 28: 199–216.

McGlone, M.S., R.J. Dungan, G.M. Hall, and R.B. Allen. 2004. Winter leaf loss in the New Zealand woody flora. *New Zealand Journal of Botany* 42: 1–19.

McGowan, P.J.K. 1994. Phasianidae. In *Handbook of the Birds of the World*, Vol. 2, eds. J. del Hoyo, A. Elliott, and J. Sargatal, pp. 434–553. Barcelona, Spain: Lynx Edicions.

McIntosh, P.D. and W.G. Lee. 1986. Soil-vegetation relationships on the Dun Mountain Ophiolite Belt at West Dome, Southland, New Zealand. *Journal of the Royal Society of New Zealand* 16: 363–379.

McKenzie, R.J. and N.P. Barker. 2008. Radiation of southern African daisies: Biogeographic inferences for subtribe Arctotidinae (Asteraceae, Arctotideae). *Molecular Phylogenetics and Evolution* 49: 1–16.

McLean, I.G. and B.J. Gill. 1988. Breeding of an island-endemic bird: The New Zealand Whitehead *Mohoua albicilla*; Pachycephalinae. *Emu* 88: 177–182.

McLellan, I.D. 1977. New alpine and southern Plecoptera from New Zealand, and a new classification of the Gripopterygidae. *New Zealand Journal of Zoology* 4: 119–147.

McLellan, I.D. 1993. Antarctoperlinae (Insecta: Plecoptera). *Fauna of New Zealand* 27: 1–67.

McLellan, I.D. 2000. Additions to New Zealand notonemourid stoneflies (Insecta: Plecoptera). *New Zealand Journal of Zoology* 27: 21–27.

McLellan, I.D. 2003. Six new species and a new genus of stoneflies (Plecoptera) from New Zealand. *New Zealand Journal of Zoology* 30: 101–113.

McLellan, I.D. 2006. Endemism and biogeography of New Zealand Plecoptera (Insecta). *Illiesia* 2: 15–23.

McNeil, R. and M.A. Martinez. 1967. Asymmétrie bilatérale des os longs des membres du pigeon *Columba livia* et du perroquet *Amazona amazonica*. *Reviews of Canadian Biology* 26: 273–286.

McNeil, R., J.R. Rodriguez, and D.M. Figuera. 1971. Handedness in the brown-throated parakeet *Aratinga pertinax* in relation with skeletal asymmetry. *Ibis* 113: 494–499.

Mee, J.A. and J.S. Moore. 2013. The ecological and evolutionary implications of microrefugia. *Journal of Biogeography* 41: 837–841.

Mees, G.F. 1969. A systematic review of the Indo-Australian Zosteropidae (part III). *Zoologische Verhandelingen* 102: 1–390.

Meffre, S., A.J. Crawford, and P.G. Quilty. 2007. Arc continent collision forming a large island between New Caledonia and New Zealand in the Oligocene. Extended Abstracts, *Australian Earth Sciences Convention 2006*, Melbourne, Australia, 3 pp.

Meiri, S., N. Cooper, and A. Purvis. 2008. The island rule: Made to be broken?. *Proceedings of the Royal Society B* 275: 141–148.

Mejías, J.A. and S.C. Kim. 2012. Taxonomic treatment of Cichorieae (Asteraceae) endemic to the Juan Fernández Islands and Desventuradas Islands (SE Pacific). *Annales Botanici Fennici* 49: 171–178.

Meléndez, R. and B.S. Dyer. 2010. Review of the southern hemisphere fish family Chironemidae (Perciformes: Cirrhitoidei). *Revista de Biología Marina y Oceanografía* 45(S1): 683–693.

Mennes, C.B., V.K.Y. Lam, P.J. Rudall et al. 2015. Ancient Gondwana break-up explains the distribution of the mycoheterotrophic family Corsiaceae (Liliales). *Journal of Biogeography* 42: 1123–1136.

Merrem, B. 1813. Tentamen systematis naturalis Avium. *Abhandlungen der Köninglichen Akademie der Wissenschaften zu Berlin* 23: 237–259.

Metcalf, R.V. and J.W. Shervais. 2008. Suprasubduction zone ophiolites: Is there really an ophiolite conundrum? *Geological Society of America Special Paper* 438: 191–222.

Meudt, H.M. 2006. A monograph of the genus *Ourisia* Comm. ex Juss. (Plantaginaceae). *Systematic Botany Monographs* 77: 1–188.

Meudt, H.M. 2008. Taxonomic revision of Australasian snow hebes (*Veronica*, Plantaginaceae). *Australian Systematic Botany* 21: 387–421.

Meudt, H.M. 2012. A taxonomic revision of native New Zealand *Plantago* (Plantaginaceae). *New Zealand Journal of Botany* 50: 101–178.

Meudt, H.M., P.J. Lockhart, and D. Bryant. 2009. Species delimitation and phylogeny of a New Zealand plant species radiation. *BMC Evolutionary Biology* 9(111): 1–17.

Meudt, H.M., J.M. Prebble, and C.A. Lehnebach. 2015. Native New Zealand forget-me-nots (*Myosotis*, Boraginaceae) comprise a Pleistocene species radiation with very low genetic divergence. *Plant Systematics and Evolution* 301: 1455–1471.

Meudt, H.M. and B.B. Simpson. 2006. The biogeography of the austral, subalpine genus *Ourisia* (Plantaginaceae) based on molecular phylogenetic evidence: South American origin and dispersal to New Zealand and Tasmania. *Biological Journal of the Linnean Society* 87: 479–513.

Meyer, D.B. 1977. The avian eye and its adaptations. In *Handbook of Sensory Physiology*, Vol. II(5): *The Visual System in Vertebrates*, ed. F. Crescittelli, pp. 549–612. Berlin, Germany: Springer.

Miall, A.D. 2008. The Paleozoic Western Craton margin. In *Sedimentary Basins of the World*, Vol. 5: *The Sedimentary Basins of the United States and Canada*, ed. A.D. Miall, pp. 181–209. Amsterdam, the Netherlands: Elsevier.

Michaux, B. 2014. *Tewkesbury Walks: An Exploration of Biogeography and Evolution*. Cham, Switzerland: Springer.

Michaux, B. and R.A.B. Leschen. 2005. East meets west: Biogeology of the Campbell Plateau. *Biological Journal of the Linnean Society* 86: 95–115.

Michener, C.D. 2007. *The Bees of the World*, 2nd edn. Baltimore, MD: Johns Hopkins University Press.

Mill, J.S. 1869. *On Liberty*. London, UK: Longmans, Green, Reader and Dyer.

Millar, A.J.K. and O. de Clerck. 2007. *Skeletonella nelsoniae* gen. et sp. nov., representing a new tribe of marine macroalgae, the Skeletonelleae (Ceramiaceae, Rhodophyta). *Phycologia* 46: 63–73.

Millener, P.R. 1988. Contributions to New Zealand's Late Quaternary avifauna. I. *Pachyplichas*, a new genus of wren (Aves: Acanthisittidae), with two new species. *Journal of the Royal Society of New Zealand* 18: 383–406.

Millener, P.R. 1989. The only flightless passerine, the Stephens Island wren (*Traversia lyallii*: Acanthisittidae). *Notornis* 36: 280–284.

Millener, P.R. and T.H. Worthy. 1991. Contributions to New Zealand's Late Quaternary avifauna. II. *Dendroscansor decurvirostris*, a new genus and species of wren (Aves: Acanthisittidae). *Journal of the Royal Society of New Zealand* 21: 179–200.

Miller, C. 1999. Conservation of the Open Bay Islands' leech, *Hirudobdella antipodum*. *Journal of the Royal Society of New Zealand* 29: 301–306.

Miller, G. 2009. On the origin of the nervous system. *Science* 325: 24–26.

Miller, H.C. and D.M. Lambert. 2006. A molecular phylogeny of New Zealand's *Petroica* (Aves: Petroicidae) species based on mitochondrial DNA sequences. *Molecular Phylogenetics and Evolution* 40: 844–855.

Miller, K.A., H.C. Miller, J.A. Moore et al. 2012. Securing the demographic and genetic future of tuatara through assisted colonization. *Conservation Biology* 26: 790–798.

Miller, W.DeW. 1924. Variations in the structure of the aftershaft and their taxonomic value. *American Museum Novitates* 140: 1–8.

Milot, E., H. Weimerskirch, and L. Bernatchez. 2008. The seabird paradox: Dispersal, genetic structure and population dynamics in a highly mobile, but philopatric albatross species. *Molecular Ecology* 17: 1658–1673.

Mincarone, M. and A. Stewart. 2006. A new species of giant seven-gilled hagfish (Myxinidae: *Eptatretus*) from New Zealand. *Copeia* 2006: 225–229.

Minckley, W.L., D.A. Hendrickson, and C.E. Bond. 1986. Geography of western North American freshwater fishes: Description and relationships to intracontinental tectonism. In *The Zoogeography of North American Freshwater Fishes*, eds. C.H. Hocutt and E.O. Wiley, pp. 519–613. New York: Wiley.

Mitchell, A.D. and P.B. Heenan. 2002. Genetic variation within the *Pachycladon* (Brassicaceae) complex based on fluorescent AFLP data. *Journal of the Royal Society of New Zealand* 32: 427–443.

Mitchell, A.D., P.B. Heenan, B.G. Murray, B.P.J. Molloy, and P.J. de Lange. 2009. Evolution of the southwestern Pacific genus *Melicytus* (Violaceae): Evidence from DNA sequence data, cytology and sex expression. *Australian Systematic Botany* 22: 143–157.

Mitchell, A.D., R. Li, J.W. Brown, I. Schönberger, and J. Wen. 2012. Ancient divergence and biogeography of *Raukaua* (Araliaceae) and close relatives in the southern hemisphere. *Australian Systematic Botany* 25: 432–446.

Mitchell, A.D., C.D. Meurk, and S.J. Wagstaff. 1999. Evolution of *Stilbocarpa*, a megaherb from New Zealand's subantarctic islands. *New Zealand Journal of Botany* 37: 205–211.

Mitchell, K.J., B. Llamas, J. Soubrier et al. 2014a. Ancient DNA reveals elephant birds and kiwi are sister taxa and clarifies ratite bird evolution. *Science* 344: 898–900.

Mitchell, K.J., J. Wood, B. Llamas, R.P. Scofield, and A. Cooper. 2014b. DNA reveals pre-Oligocene diversification in acanthisittid wrens (Acanthisittidae). GeoGenes V. Conference abstracts. *Geoscience Society of New Zealand Miscellaneous Publication* 138: 19.

Mitchell, K.J., J.R. Wood, B. Llamas et al. 2016. Ancient mitochondrial genomes clarify the evolutionary history of New Zealand's enigmatic acanthisittid wrens. *Molecular Phylogenetics and Evolution* 102: 295–304.

Mitchell, K.J., J.R. Wood, R.P. Scofield, B. Llamas, and A. Cooper. 2014c. Ancient mitochondrial genome reveals unsuspected taxonomic affinity of the extinct Chatham duck (*Pachyanas chathamica*) and resolves divergence times for New Zealand and sub-Antarctic brown teals. *Molecular Phylogenetics and Evolution* 70: 420–428.

Moalic, Y., D. Desbruyères, C.M. Duarte, A.F. Rozenfeld, C. Bachraty, and S. Arnaud-Haond. 2012. Biogeography revisited with network theory: Retracing the history of hydrothermal vent communities. *Systematic Biology* 61: 127–137.

Molloy, B.P.J. 1994. Observations on the ecology and conservation of *Australopyrum calcis* (Triticeae: Gramineae) in New Zealand. *New Zealand Journal of Botany* 32: 37–51.

Molloy, B.P.J. 1995. Two new species of *Leucogenes* (Inuleae: Asteraceae) from New Zealand, and typification of *L. grandiceps*. *New Zealand Journal of Botany* 33: 53–63.

Molloy, B.P.J. 2001. *Pachystegia rufa* and allied rock daisies. Conservation Advisory Science Notes 336. Wellington, New Zealand: Department of Conservation.

Molloy, B.P.J. and B.D. Clarkson. 1996. A new, rare species of *Melicytus* (Violaceae) from New Zealand. *New Zealand Journal of Botany* 34: 431–440.

Molloy, B.P.J., P.J. de Lange, and B.D. Clarkson. 1999a. *Coprosma pedicellata* (Rubiaceae), a new species from New Zealand. *New Zealand Journal of Botany* 37: 383–397.

Molloy, B.P.J., E. Edgar, P.B. Heenan, and P.J. de Lange. 1999b. New species of *Poa* (Gramineae) and *Ischnocarpus* (Brassicaceae) from limestone, North Otago, South Island, New Zealand. *New Zealand Journal of Botany* 37: 41–50.

Molloy, B.P.J. and M.J.A. Simpson. 1980. Taxonomy, distribution and ecology of *Pachystegia* (Compositae): A progress report. *New Zealand Journal of Ecology* 3: 1–3.

Molnar, P., H.J. Anderson, E. Audoine et al. 1999. Continuous deformation versus faulting through the continental lithosphere of New Zealand. *Science* 286: 516–519.

Molvray, M. 1997. A synopsis of *Korthalsella* (Viscaceae). *Novon* 7: 268–273.

Monteiro, L.R. and M.R. Nogueira. 2011. Evolutionary patterns and processes in the radiation of phyllostomid bats. *BMC Evolutionary Biology* 11: e137 (pp. 1–23).

Mooers, A.Ø., S. Heard, and E. Chrostowski. 2005. Evolutionary heritage as a metric for conservation. In *Phylogeny and Conservation*, eds. A. Purvis, J.L. Gittleman, and T. Brooks, pp. 120–138. Cambridge, UK: Cambridge University Press.

Moon, G. 1992. *A Field Guide to New Zealand Birds*. Auckland, New Zealand: Reed.

Moore, L.B. 1957. The species of *Xeronema* (Liliaceae). *Pacific Science* 11: 355–362.

Moore, L.B. 1964. The New Zealand species of *Bulbinella* (Liliaceae). *New Zealand Journal of Botany* 2: 286–304.

Moore, L.B. and E. Edgar. 1970. *Flora of New Zealand*, Vol. 2: *Indigenous Tracheophyta, Monocotyledones Except Gramineae*. Wellington, New Zealand: Government Printer.

Mora, C., E.A. Treml, J. Roberts, K. Crosby, D. Roy, and D.P. Tittensor. 2012. High connectivity among habitats precludes the relationship between dispersal and range size in tropical reef fishes. *Ecography* 35: 89–96.

Moreira-Muñoz, A. 2007. Plant geography of Chile: An essay on postmodern biogeography. PhD thesis, Friedrich-Alexander University Erlangen-Nürnberg, Bavaria, Germany.

Morgan, T.H. 1903. *Evolution and Adaptation*. New York: Macmillan.

Morgan, T.H. 1916. *A Critique of the Theory of Evolution*. Princeton, NJ: Princeton University Press.

Morgan, T.H. 1932. *The Scientific Basis of Evolution*. New York: W.W. Norton.

Morris, J.A. 2007. A molecular phylogeny of the Lythraceae and inference of the evolution of heterostyly. PhD thesis, Kent State University, Kent, OH.

Mortensen, J.K., D. Craw, D.J. MacKenzie, J.E. Gabites, and T. Ullrich. 2010. Age and origin of orogenic gold mineralization in the Otago Schist Belt, South Island, New Zealand: Constraints from lead isotope and ^{40}Ar/^{39}Ar dating studies. *Economic Geology* 105: 777–793.

Mortimer, N. 2003. A provisional structural thickness map of the Otago Schist, New Zealand. *American Journal of Science* 303: 603–621.

Mortimer, N. 2004. New Zealand's geological foundations. *Gondwana Research* 7: 261–272.

Mortimer, N. 2006. Cretaceous-Cenozoic Zealandia: Did Pacific subduction ever stop? Abstract only. In *Geological Society of New Zealand/New Zealand Geophysical Society Conference 2006: Programme and Abstracts*, eds. B. Stewart, C. Wallace, J. Lecointre, and M. Reyners, p. 72. Palmerston North, New Zealand: Geological Society of New Zealand/New Zealand Geophysical Society.

Mortimer, N. 2014. The oroclinal bend in the South Island, New Zealand. *Journal of Structural Geology* 64: 32–38.

Mortimer, N. and H. Campbell. 2014. *Zealandia: Our Continent Revealed*. Auckland, New Zealand: Penguin.

Mortimer, N. and A.F. Cooper. 2004. U-Pb and Sm-Nd ages from the Alpine Schist, New Zealand. *New Zealand Journal of Geology and Geophysics* 47: 21–28.

Mortimer, N., F.J. Davey, A. Melhuish, J. Yu, and N.J. Godfrey. 2002. Geological interpretation of a deep seismic reflection profile across the Eastern Province and Median Batholith, New Zealand: Crustal architecture of an extended Phanerozoic convergent orogen. *New Zealand Journal of Geology and Geophysics* 45: 349–363.

Mortimer, N., P.B. Gans, J.M. Palin, and S. Meffre. 2010. Location and migration of Miocene-Quaternary volcanic arcs in the SW Pacific region. *Journal of Volcanology and Geothermal Research* 190: 1–10.

Mortimer, N., I.J. Graham, C.J. Adams, A.J. Tulloch, and H.J. Campbell. 2005. Relationships between New Zealand, Australian and New Caledonian mineralised terranes: a regional geological framework. *New Zealand Minerals Conference*, pp. 151–159. Wellignton, New Zealand: Crown Minerals.

Mortimer, N., R.H. Herzer, P.B. Gans, C. Laporte-Magoni, A.T. Calvert, and D. Bosch. 2007. Oligocene-Miocene tectonic evolution of the South Fiji Basin and Northland Plateau, SW Pacific Ocean: Evidence from petrology and dating of dredged rocks. *Marine Geology* 237: 1–24.

Mortimer, N., K. Hoernle, F. Hauff, J.M. Palin, W.J. Dunlap, R. Werner, and K. Faure. 2006. New constraints on the age and evolution of the Wishbone Ridge, southwest Pacific Cretaceous microplates, and Zealandia-West Antarctica breakup. *Geology* 34: 185–188.

Mortimer, N. and D.T. Strong. 2014. New Zealand limestone purity. *New Zealand Journal of Geology and Geophysics* 57: 209–218.

Morton, J.E. 1952. A preliminary study of the land operculate *Murdochia pallidum* (Cyclophoridae, Mesogastropoda). *Transactions of the Royal Society of New Zealand* 80: 69–79.

Morton, J.E. 2004. *Seashore Ecology of New Zealand and the Pacific.* Auckland, New Zealand: David Bateman.

Motsi, M.C., A.N. Moteetee, A.J. Beaumont et al. 2010. A phylogenetic study of *Pimelea* and *Thecanthes* (Thymelaeaceae): Evidence from plastid and nuclear ribosomal DNA sequence data. *Australian Systematic Botany* 23: 270–284.

Moura, A.E., S.C. Nielsen, J.T. Vilstrup et al. 2013. Recent diversification of a marine genus (*Tursiops* spp.) tracks habitat preference and environmental change. *Systematic Biology* 62: 865–877.

Moussalli, A. and D.G. Herbert. 2016. Deep molecular divergence and exceptional morphological stasis in dwarf cannibal snails *Nata* sensu lato Watson, 1934 (Rhytididae) of southern Africa. *Molecular Phylogenetics and Evolution* 95: 100–115

Moyle, R.G., M.J. Andersen, C.H. Oliveros, F.D. Steinheimer, and S. Reddy. 2012. Phylogeny and biogeography of the core babblers (Aves: Timaliidae). *Systematic Biology* 61: 631–651.

Moyle, R.G., C.E. Filardi, C.E. Smith, and J. Diamond. 2009. Explosive Pleistocene diversification and hemispheric expansion of a "great speciator". *Proceedings of the National Academy of Sciences of USA* 106: 1863–1868.

Müller, R. and J.C.T. Hallam. 2004. From bat pinnae to sonar antennae: Augmented obliquely truncated horns as a novel parametric shape model. In *From Animals to Animats 8* (*Proceedings of the Eighth International Conference on the Simulation of Adaptive Behaviour*), eds. S. Schaal, A. Ijspeert, A. Billard, S. Vijayakumar, J. Hallam, and J.-A. Meyer, pp. 87–95. Cambridge, MA: MIT Press.

Müller, R.D., M. Sdrolias, C. Gaina, B. Steinberger, and C. Heine. 2008. Long-term sea-level fluctuations driven by ocean basin dynamics. *Science* 319: 1357–1362.

Müller, S., K. Salomo, J. Salazar et al. 2015. Intercontinental long-distance dispersal of Canellaceae from the New- to the Old World revealed by a nuclear single copy gene and chloroplast loci. *Molecular Phylogenetics and Evolution* 84: 205–219.

Mummenhoff, K., H. Brüggemann, and J.L. Bowman. 2001. Chloroplast DNA phylogeny and biogeography of *Lepidium* (Brassicacae). *American Journal of Botany* 88: 2051–2063.

Mummenhoff, K., P. Linder, N. Friesen, J.L. Bowman, J.-Y. Lee, and A. Franzke. 2004. Molecular evidence for bicontinental hybridogenous genomic constitution in *Lepidium* sensu stricto (Brassicaceae) species from Australia and New Zealand. *American Journal of Botany* 91: 254–261.

Munro, J.B., J.M. Heraty, R.A. Burks et al. 2011. A molecular phylogeny of the Chalcidoidea (Hymenoptera). *PLoS One* 6(11): e27023 (pp. 1–27).

Murienne, J. 2009a. Testing biodiversity hypotheses in New Caledonia using phylogenetics. *Journal of Biogeography* 36: 1433–1434.

Murienne, J. 2009b. New Caledonia, biology. In *Encyclopedia of Islands*, eds. R. Gillespie and D.A. Clague, pp. 643–645. Berkeley, CA: University of California Press.

Murienne, J., S.R. Daniels, T.R. Buckley, G. Mayer, and G. Giribet. 2014. A living fossil tale of Pangaean biogeography. *Proceedings of the Royal Society B* 281(1775): Article 20132648.

Murphy, D.B. 2010. Metamorphism and the P-T history of Alpine Schist from the Newton Range, Southern Alps, New Zealand. MSc thesis, Victoria University of Wellington, Wellington, New Zealand.

Murray, B.G., J.E. Braggins, and P.D. Newman. 1989. Intraspecific polyploidy in *Hebe diosmifolia* (Cunn.) Cockayne et Allan (Scrophulariaceae). *New Zealand Journal of Botany* 27: 587–589.

Mutch, A.R. 1963. Sheet 23—Oamaru. Geological map of New Zealand 1: 250 000. Wellington, New Zealand: Department of Scientific and Industrial Research.

Myers, N., R.A. Mittermeier, C.G. Mittermeier, G.A. da Fonseca, and J. Kent. 2000. Biodiversity hotspots for conservation priorities. *Nature* 403: 853–858.

Nakano, T., B.A. Marshall, M. Kennedy, and H.G. Spencer. 2009. The phylogeny and taxonomy of New Zealand *Notoacmea* and *Patelloida* species (Mollusca: Patellogastropoda: Lottiidae) inferred from DNA sequences. *Molluscan Research* 29: 33–59.

Nei, M. 1987. *Molecular Evolutionary Genetics*. New York: Columbia University Press.

Nei, M. 2005. Selectionism and neutralism in molecular evolution. *Molecular Biology and Evolution* 22: 2318–2342.

Nei, M. 2007. The new mutation theory of phenotypic evolution. *Proceedings of the National Academy of Sciences of USA* 104: 12235–12242.

Nei, M. 2013. *Mutation-Driven Evolution*. Oxford, UK: Oxford University Press.

Neiman, M. and C. Lively. 2004. Pleistocene glaciations is implicated in the phylogeographical structure of *Potamopyrgus antipodarum*, a New Zealand snail. *Molecular Ecology* 13: 3085–3098.

Nelson, G. 1975. Reviews: Biogeography, the vicariance paradigm, and continental drift. *Systematic Zoology* 24: 489–504.

Nelson, J.B. 2005. *Pelicans, Cormorants, and Their Relatives: The Pelecaniformes*. Oxford, UK: Oxford University Press.

Nelson, W.A. and J.E.S. Broom. 2008. New Zealand Gigartinaceae (Rhodophyta): Resurrecting *Gigartina grandifida* endemic to the Chatham Islands. *New Zealand Journal of Botany* 46: 177–187.

Nelson, W.A., J.E. Sutherland, T.J. Farr, D.R. Hart, K.F. Neill, H.J. Kim, and H.S. Yoon. 2015. Multi-gene phylogenetic analyses of New Zealand coralline algae: *Corallinapetra novaezelandiae* gen. et sp. nov. and recognition of the Hapalidiales ord. nov. *Journal of Phycology* 51: 454–468.

Nesom, G.L. 1994. Subtribal classification of the Astereae (Asteraceae). *Phytologia* 76: 193–274.

Ness, A.R. 1967. A measure of asymmetry of the skulls of odontocete whales. *Journal of Zoology* 153: 209–221.

Neuweiler, G. 2000. *The Biology of Bats*. New York: Oxford University Press.

Newnham, R.M., M.J. Vandergoes, C.H. Hendy, D.J. Lowe, and F. Preusser. 2007. A terrestrial palynological record for the last two glacial cycles from southwestern New Zealand. *Quaternary Science Reviews* 26: 517–535.

Newton, I. 2008. Bird migration. In *Handbook of the Birds of the World*, Vol. 13, eds. J. del Hoyo, A. Elliott, and D.A. Christie, pp. 15–47. Barcelona, Spain: Lynx Edicions.

New Zealand Department of Conservation. 2016. Atlas of the amphibians and reptiles of New Zealand. www.doc.govt.nz/our-work/reptiles-and-frogs-distribution/atlas/. Accessed May, 2016.

Nicholson, K.E., B.I. Crother, C. Guyer, and J.M. Savage. 2012. It is time for a new classification of anoles (Squamata: Dactyloidae). *Zootaxa* 3477: 1–108.

Nicholson, K.N. and P.M. Black. 2004. Cretaceous—Early Tertiary basaltic volcanism in the Far North of New Zealand: Geochemical associations and their tectonic significance. *New Zealand Journal of Geology and Geophysics* 47: 437–446.

Nicholson, K.N., P.M. Black, C. Picard, P. Cooper, C.M. Hall, and T. Itaya. 2007. Alteration, age, and emplacement of the Tangihua Complex Ophiolite, New Zealand. *New Zealand Journal of Geology and Geophysics* 50: 151–164.

Nicholson, K.N., P.M. Black, and K.B. Spörli. 2008. Cretaceous-Oligocene multiphase magmatism on Three Kings Islands, northern New Zealand. *New Zealand Journal of Geology and Geophysics* 51: 219–229.

Nicholson, K.N., P. Maurizot, P.M. Black et al. 2011. Geochemistry and age of the Nouméa Basin lavas, New Caledonia: Evidence for Cretaceous subduction beneath the eastern Gondwana margin. *Lithos* 125: 659–674.

Nicolas, A.N. and G.M. Plunkett. 2012. Untangling generic limits in *Azorella*, *Laretia*, and *Mulinum* (Apiaceae: Azorelloideae): Insights from phylogenetics and biogeography. *Taxon* 61: 826–840.

Nicolas, A.N. and G.M. Plunkett. 2014. Diversification times and biogeographic patterns in Apiales. *Botanical Review* 80: 30–58.

Nielsen, S.V., A.M. Bauer, T.R. Jackman, R.A. Hitchmough, and C.H. Daugherty. 2011. New Zealand geckos (Diplodactylidae): Cryptic diversity in a post-Gondwanan lineage with trans-Tasman affinities. *Molecular Phylogenetics and Evolution* 59: 1–22.

Niemitz, C., A. Nietsch, S. Warter, and Y. Rumpler. 1991. *Tarsius dianae*: A new primate species from Central Sulawesi (Indonesia). *Folia Primatologica* 56: 105–116.

Nietzsche, F. 1881/2006. *Daybreak*. Cambridge, UK: Cambridge University Press.

Nietzsche, F. 1887/1996. *On the Genealogy of Morals*. Oxford, UK: Oxford University Press.

Nietzsche, F. 1901/1968. *The Will to Power*. New York: Vintage.

NIWA. 2016. New Zealand mean annual rainfall (mm), 1971–2000. www.niwa.co.nz/education-and-training/schools/resources/climate/overview. Accessed May, 2016.

Noble, G.K. 1931. *The Biology of the Amphibia*. New York: McGraw-Hill.

Nontachaiyapoom, S., S. Sasirat, and L. Manoch. 2010. Isolation and identification of *Rhizoctonia*-like fungi from roots of three orchid genera, *Paphiopedilum*, *Dendrobium*, and *Cymbidium*, collected in Chiang Rai and Chiang Mai provinces of Thailand. *Mycorrhiza* 20: 459–471.

Norberg, R.A. 1977. Occurrence and independent evolution of bilateral ear asymmetry in owls and implications on owl taxonomy. *Philosophical Transactions of the Royal Society of London B* 280: 375–408.

Norris, R.J. and A.F. Cooper. 2001. Late Quaternary slip rates and slip partitioning on the Alpine Fault, New Zealand. *Journal of Structural Geology* 23: 507–520.

Norton, D.A. and P.J. de Lange. 2003. A new species of *Coprosma* (Rubiaceae) from the South Island, New Zealand. *New Zealand Journal of Botany* 41: 223–231.

Nowak, R.M. (ed.). 1999. *Walker's Mammals of the World*, 6th edn. Baltimore, MD: Johns Hopkins University Press.

Nuryanto, A. and M. Kochzius. 2009. Highly restricted gene flow and deep evolutionary lineages in the giant clam *Tridacna maxima*. *Coral Reefs* 28: 607–619.

Nyári, A.S. and L. Joseph. 2012. Evolution in Australasian mangrove forests: Multilocus phylogenetic analysis of the *Gerygone* warblers (Aves: Acanthizidae). *PLoS One* 7(2): e31840 (pp. 1–9).

Nylinder, S., U. Swenson, C. Persson, S.B. Janssens, and B. Oxelman. 2012. A dated species-tree approach to the trans-Pacific disjunction of the genus *Jovellana* (Calceolariaceae, Lamiales). *Taxon* 61: 381–391.

NZPCN. 2016. New Zealand plant conservation network. www.nzpcn.org.nz. Accessed May, 2016.

Odling-Smee, J. 2010. Niche inheritance. In *Evolution: The Extended Synthesis*, ed. M. Pigliucci and G.B. Müller, pp. 175–207. Cambridge, MA: MIT Press.

O'Donnell, C.F.J. 2001. Advances in New Zealand mammalogy 1990–2000: Long-tailed Bat. *Journal of the Royal Society of New Zealand* 31: 43–57.

O'Donnell, C.F.J. 2005. New Zealand long-tailed bat. In *The Handbook of New Zealand Mammals*, 2nd edn., ed. C.M. King, pp. 98–109. Melbourne, Australia: Oxford University Press.

O'Grady, P.M., G.M. Bennett, V.A. Funk, and T.K. Altheide. 2012. Retrograde biogeography. *Taxon* 61: 699–705.

Ohlsen, D., L. Perrie, L. Shepherd, P. Brownsey, and M. Bayly. 2015. Phylogeny of the fern family Aspleniaceae in Australasia and the south-west Pacific. *Australian Systematic Botany* 27: 355–371.

Ohno, S. 1970. *Evolution by Gene Duplication*. New York: Springer.

Okabe, M. and A. Graham. 2004. The origin of the parathyroid gland. *Proceedings of the National Academy of Sciences of USA* 101: 17716–17719.

Okada, N., Y. Takagi, T. Seikai, M. Tanaka, and M. Tagawa. 2001. Asymmetrical development of bones and soft tissues during eye migration of metamorphosing Japanese flounder, *Paralichthys olivaceus*. *Cell and Tissue Research* 304: 59–66.

Oliver, P.M. and K. Sanders. 2009. Molecular evidence for multiple lineages with ancient Gondwanan origins in a diverse Australian gecko radiation. *Journal of Biogeography* 36: 2044–2055.

Oliver, W.R.B. 1925. Biogeographical relations of the New Zealand Region. *Journal of the Linnean Society of London, Botany* 47: 99–140.

Oliver, W.R.B. 1948. The flora of the Three Kings Islands. *Records of the Auckland Institute and Museum* 3: 211–238.

Oliver, W.R.B. 1955. *New Zealand Birds*. Wellington, New Zealand: Reed.

Oliver, W.R.B. 1956. The genus *Aciphylla*. *Transactions of the Royal Society of New Zealand* 84: 1–18.

Olson, M.E. 2012. The developmental renaissance in adaptationism. *Trends in Ecology and Evolution* 27: 278–287.

Olson, S.L. 1975. A review of the extinct rails of the New Zealand region. *National Museum of New Zealand Records* 1: 63–79.

Olson, S.L. 1990a. Comments on the osteology and systematics of the New Zealand passerines of the genus *Mohoua*. *Notornis* 37: 157–160.

Olson, S.L. 1990b. Osteology and systematics of the fernbirds (*Bowdleria*: Sylviidae). *Notornis* 37: 161–171.

O'Malley, M.A. The nineteenth century roots of 'everything is everywhere'. *Nature Reviews Microbiology* 5: 647–651.

O'Neill, S.B., T.R. Buckley, T.R. Jewell, and P.A. Ritchie. 2009. Phylogeographic history of the New Zealand stick insect *Niveaphasma annulata* (Phasmatodea) estimated from mitochondrial and nuclear loci. *Molecular Phylogenetics and Evolution* 53: 523–536.

O'Neill, S.B., D.G. Chapple, C.H. Daugherty, and P.A. Ritchie. 2008. Phylogeography of two New Zealand lizards: McCann's skink (*Oligosoma maccanni*) and the brown skink (*O. zelandicum*). *Molecular Phylogenetics and Evolution* 48: 1168–1177.

O'Quinn, R. and L. Hufford. 2005. Molecular systematics of Montieae (Portulacaceae): Implications for taxonomy, biogeography and ecology. *Systematic Botany* 30: 314–331.

Orchard, A.E. and J.B. Davies. 1985. *Oreoporanthera,* a New Zealand 'endemic' plant genus discovered in Tasmania. *Papers and Proceedings of the Royal Society of Tasmania* 119: 61–63.

Ordish, R.G. 1989. A new species of *Rhantus* from the Chatham Islands of New Zealand (Coleoptera: Dytiscidae). *New Zealand Journal of Zoology* 16: 147–150.

Organ, J.M. 2007. The functional anatomy of prehensile and nonprehensile tails of the Platyrrhini (Primates) and Procyonidae (Carnivora). PhD dissertation, Johns Hopkins University School of Medicine, Baltimore, MD.

Orr, H.A. 2005. The genetic basis of reproductive isolation: Insights from *Drosophila*. *Proceedings of the National Academy of Sciences of USA* 102: 6522–6526.

Orr, H.A. and R.L. Unckless. 2008. Population extinction and the genetics of adaptation. *American Naturalist* 172: 160–169.

Orta, J. 1992. Phalacrocoracidae. In *Handbook of the Birds of the World*, Vol. 1, eds. J. del Hoyo, A. Elliott, and J. Sargatal, pp. 326–353. Barcelona, Spain: Lynx Edicions.

Pacheco, M.A., F.U. Battistuzzi, M. Lentino, R.F. Aguilar, S. Kumar, and A.A. Escalante. 2011. Evolution of modern birds revealed by mitogenomics: Timing the radiation and origin of major orders. *Molecular Biology and Evolution* 28: 1927–1942.

Palmer, A.R. 1996. From symmetry to asymmetry: Phylogenetic patterns of asymmetry variation in animals and their evolutionary significance. *Proceedings of the National Academy of Sciences of USA* 93: 14279–14286.

Palmer, A.R. 2004. Symmetry breaking and the evolution of development. *Science* 306: 828–833.

Panter, K.S., J. Blusztajn, S.R. Hart, P.R. Kyle, R. Esser, and W.C. McIntosh. 2006. The origin of HIMU in the SW Pacific: Evidence from intraplate volcanism in Southern New Zealand and subantarctic islands. *Journal of Petrology* 47: 1673–1704.

Papadopoulou, A., I. Anastasiou, and A.P. Vogler. 2010. Revisiting the insect mitochondrial molecular clock: The mid-Aegean trench calibration. *Molecular Biology and Evolution* 27: 1659–1672.

Papadopulos, A.S.T., W.J. Baker, D. Crayn et al. 2011. Speciation with gene flow on Lord Howe Island. *Proceedings of the National Academy of Sciences of USA* 108: 13188–13193.

Parenti, L. and M. Ebach. 2009. *Comparative Biogeography: Discovering and Classifying Biogeographical Patterns of a Dynamic Earth*. Berkeley, CA: University of California Press.

Parenti, L.R. 2008. Life history patterns and biogeography: An interpretation of diadromy in fishes. *Annals of the Missouri Botanical Garden* 95: 232–247.

Parham, J.F., P.C.J. Donoghue, C.J. Bell et al. 2012. Best practices for justifying fossil calibrations. *Systematic Biology* 61: 346–359.

Park, J.-S. and C.E. Carlton. 2013. A revision of the New Zealand genus *Stenosagola* Broun, 1921 (Coleoptera: Staphylinidae: Pselaphinae: Faronitae). *The Coleopterists' Bulletin* 67: 335–359.

Park, J.-S. and C.E. Carlton. 2014. A revision of the New Zealand species of the genus *Sagola* Sharp (Coleoptera: Staphylinidae: Pselaphinae: Faronitae). *The Coleopterists' Bulletin* 68, monograph 4: 1–156.

Park, J.-S. and C.E. Carlton. 2015a. *Chandlerea* and *Nunnea* (Coleoptera: Staphylinidae: Pselaphinae), two new genera from New Zealand with descriptions of three new species. *Florida Entomologist* 98: 588–592.

Park, J.-S. and C.E. Carlton. 2015b. *Aucklandea* and *Leschenea*, two new monotypic genera from New Zealand (Coleoptera: Staphylinidae: Pselaphinae), and a key to New Zealand genera of the supertribe Faronitae. *Annals of the Entomological Society of America* 108: 634–640.

Parker, T.J. 1889. Observations on the anatomy and development of *Apteryx*. *Proceedings of the Royal Society of London* 47: 454–459.

Parker, T.J. 1891. On the history of the kiwi. *New Zealand Journal of Science* 1: 2–68.

Parnaby, H. 1992. An interim guide to identification of insectivorous bats of south-eastern Australia. *Technical Reports of the Australian Museum* 8: 1–33.

Parris, B.S. and D.R. Given. 1976. A taxonomic revision of the genus *Grammitis* Sw. (Grammitidaceae: Filicales) in New Zealand. *New Zealand Journal of Botany* 14: 85–111.

Parrish, G.R. and P.J. Anderson. 1999. Lizard transfers from Matapia Island to Motuopao Island, Northland and observations on other fauna. *Tane* [Auckland] 37: 1–14.

Paton, R.L., T.R. Smithson, and J.A. Clack. 1999. An amniote-like skeleton from the Early Carboniferous of Scotland. *Nature* 398: 508–513.

Patrick, B.H. 1990. Record of an upland moth in a coastal salt marsh in Otago, New Zealand. *Journal of the Royal Society of New Zealand* 20: 305–307.

Patrick, B.H. 1996. The status of the bat-winged fly, *Exsul singularis* Hutton (Diptera: Muscidae: Coenosiinae). *New Zealand Entomologist* 19: 31–33.

Patrick, B.H., B.I.P. Barratt, and M. Heads. 1985. Entomological survey of the Blue Mountains. Unpublished report for the New Zealand Forest Service, Invercargill, New Zealand, 28 pp.

Patrick, B.H., R.J.B. Hoare, and B.E. Rhode. 2010. Taxonomy and conservation of allopatric moth populations: A revisionary study of the *Notoreas perornata* Walker complex (Lepidoptera: Geometridae: Larentiinae), with special reference to southern New Zealand. *New Zealand Journal of Zoology* 37: 257–283.

Patrick, B.H. and H. Patrick. 2012. *Butterflies of the South Pacific*. Dunedin, New Zealand: Otago University Press.

Patrick, B.H. and C.S. Woods. 1995. Reaffirmation of the type locality of Stokell's smelt in Southland. *Journal of the Royal Society of New Zealand* 25: 93–97.

Patterson, G.B. and T.P. Bell. 2009. The Barrier skink *Oligosoma judgei* n. sp. (Reptilia: Scincidae) from the Darran and Takitimu Mountains, South Island, New Zealand. *Zootaxa* 2271: 43–56.

Paul, J.R., C. Morton, C.M. Taylor, and S.J. Tonsor. 2009. Evolutionary time for dispersal limits the extent but not the occupancy of species' potential ranges in the tropical plant genus *Psychotria* (Rubiaceae). *American Naturalist* 173: 188–199.

Paulin, C.D. and C. Roberts. 1992. *The Rockpool Fishes of New Zealand*. Wellington, New Zealand: Museum of New Zealand/Te Papa.

Pawson, S.M. and R.M. Emberson. 2001. *Oregus inaequalis* Castelnau, its distribution and abundance at Swampy Summit Otago. DOC Science Internal Series 6. New Zealand Department of Conservation, Wellington, New Zealand.

Pawson, S.M., R.M. Emberson, K.F. Armstrong, and A.M. Paterson. 2003. Phylogenetic revision of the endemic New Zealand carabid genus *Oregus* Putzeys (Coleoptera: Carabidae: Broscini). *Invertebrate Systematics* 17: 625–640.

Peat, N. and B. Patrick. 1999. *Wild Central: Discovering the Natural History of Central Otago*. Dunedin, New Zealand: University of Otago Press.

Peat, N. and B. Patrick. 2014. *Wild Dunedin: The Natural history of New Zealand's Wildlife capital*. Dunedin, New Zealand: Otago University Press.

Pedersen, S.C. and R. Müller. 2013. Nasal-emission and nose leaves. In *Bat Evolution, Ecology, and Conservation*, eds. R.A. Adams and S.C. Pedersen, pp. 71–91. New York: Springer.

Pelser, P.B., A.H. Kennedy, E.J. Tepe et al. 2010. Patterns and causes of incongruence between plastid and nuclear Senecioneae (Asteraceae) phylogenies. *American Journal of Botany* 97: 856–873.

Peña, C., S. Nylin, and N. Wahlberg. 2011. The radiation of Satyrini butterflies (Nymphalidae: Satyrinae): A challenge for phylogenetic methods. *Zoological Journal of the Linnean Society* 161: 64–87.

Pennington, R.T., Q.C.B. Cronk, and J.A. Richardson. 2004. Introduction and synthesis: Plant phylogeny and the origin of major biomes. *Philosophical Transactions of the Royal Society, London B* 359: 1455–1464.

Pepper, M.R., S.Y. Ho, M.K. Fujita, and J.S. Keogh. 2011. The genetic legacy of aridification: Climate cycling fostered lizard diversification in Australian montane refugia and left low-lying deserts genetically depauperate. *Molecular Phylogenetics and Evolution* 61: 750–759.

Perrie, L.R. and P.J. Brownsey. 2005. Insights into the biogeography and polyploid evolution of New Zealand *Asplenium* from chloroplast DNA sequence data. *American Fern Journal* 95: 1–21.

Perrie, L.R. and P.J. Brownsey. 2007. Molecular evidence for long-distance dispersal in the New Zealand pteridophyte flora. *Journal of Biogeography* 34: 2028–2038.

Perrie, L.R. and B.S. Parris. 2012. Chloroplast DNA sequences indicate the grammitid ferns (Polypodiaceae) in New Zealand belong to a single clade, *Notogrammitis* gen. nov. *New Zealand Journal of Botany* 50: 457–472.

Perrie, L.R. and L.D. Shepherd. 2009. Reconstructing the species phylogeny of *Pseudopanax* (Araliaceae), a genus of hybridising trees. *Molecular Phylogenetics and Evolution* 52: 774–783.

Petersen, G., O. Seberg, M. Yde, and K. Berthelsen. 2006. Phylogenetic relationships of *Triticum* and *Aegilops* and evidence for the origin of the A, B, and D genomes of common wheat (*Triticum aestivum*). *Molecular Phylogenetics and Evolution* 39: 70–82.

Peterson, A.T. 2009. Phylogeography is not enough: The need for multiple lines of evidence. *Frontiers of Biogeography* 1: 19–25.

Peterson, A.T. 2011. Ecological niche conservatism: A time-structured review of evidence. *Journal of Biogeography* 38: 817–827.

Peterson, A.T. and B.S. Lieberman. 2012. Species' geographic distributions through time: Playing catch-up with changing climates. *Evolution: Education and Outreach* 5: 569–581.

Peterson, A.T., J. Soberón, R.P. Anderson et al. 2011. *Ecological Niches and Geographic Distributions*. Princeton, NJ: Princeton University Press.

Peterson, A.T., J. Soberón, and V. Sánchez-Cordero. 1999. Conservatism of ecological niches in evolutionary time. *Science* 285: 1265–1267.

Petrie, D. 1884. Description of a new species of *Carmichælia*, with notes on the distribution of the species native to Otago. *Transactions of the New Zealand Institute* 17: 272–274.

Petrie, D. 1891. Botanical Notes. *New Zealand Journal of Science* 1: 68–69.

Petrie, D. 1896. List of the flowering plants indigenous to Otago, with indications of their distribution and range in altitude. *Transactions of the New Zealand Institute* 28: 540–591.

Petterson, J.A. 1997. Revision of the genus *Wahlenbergia* (Campanulaceae) in New Zealand. *New Zealand Journal of Botany* 35: 9–54.

Peucker, A.J., P. Dann, and C.P. Burridge. 2009. Range-wide phylogeography of the Little Penguin *Eudyptula minor*: Evidence of long-distance dispersal. *Auk* 126: 397–408.

Pfennig, D.W. and K.S. Pfennig. 2012. *Evolution's Wedge: Competition and the Origins of Diversity*. Berkeley, CA: University of California Press.

Philipson, W.R. 1978. Araliaceae: Growth forms and shoot morphology. In *Tropical Trees as Living Systems*, eds. P.B. Tomlinson and M.H. Zimmermann, pp. 269–284. Cambridge, UK: Cambridge University Press.

Phillips, L.E. 2001. Morphology and molecular analysis of the Australian monotypic genera *Lembergia* and *Sonderella* (Rhodomelaceae, Rhodophyta), with a description of the tribe Sonderelleae trib. nov. *Phycologia* 40: 487–499.

Piersma, T. 1996a. Charadriidae. In *Handbook of the Birds of the World*, Vol. 3, eds. J. del Hoyo, A. Elliott, and J. Sargatal, pp. 384–443. Barcelona, Spain: Lynx Edicions.

Piersma, T. 1996b. Scolopacidae. In *Handbook of the Birds of the World*, Vol. 3, eds. J. del Hoyo, A. Elliott, and J. Sargatal, pp. 444–533. Barcelona, Spain: Lynx Edicions.

Pillon, Y., J. Munzinger, H. Amir, H.C.F. Hopkins, and M.W. Chase. 2008. Reticulate evolution on a mosaic of soils: Diversification of the New Caledonian endemic genus *Codia* (Cunoniaceae). *Molecular Phylogenetics and Evolution* 18: 2263–2275.

Pinker, S. 1994. *The Language Instinct*. London, UK: Penguin.

Pirie, M.D. and J.A. Doyle. 2012. Dating clades with fossils and molecules: The case of Annonaceae. *Botanical Journal of the Linnean Society* 169: 84–116.

Pirie, M.D., A.M. Humphreys, and H.P. Linder. 2009. Reticulation, data combination, and inferring evolutionary history: An example from Danthonioideae (Poaceae). *Systematic Biology* 58: 612–628.

Pirie, M.D., K.M. Lloyd, W.G. Lee, and H.P. Linder. 2010. Diversification of *Chionochloa* (Poaceae) and biogeography of the New Zealand Southern Alps. *Journal of Biogeography* 37: 379–392.

Plant, A.R. 2010. Hemerodromiinae (Diptera: Empididae): A tentative phylogeny and biogeographical discussion. *Systematic Entomology* 36: 83–103.

Platnick, N.I. and G. Nelson. 1978. A method for analysis of historical biogeography. *Systematic Zoology* 27: 1–16.

Pliny the Elder. 1938–1963. *Natural History.* Transl. H. Rackham. London, UK: Heinemann.

Pocock, R.I. 1922. On the external characters of some hystricomorph rodents. *Proceedings of the Zoological Society of London* 92: 365–427.

Pole, M. 1994. The New Zealand flora—Entirely long-distance dispersal? *Journal of Biogeography* 21: 625–635.

Pole, M. 2001. Can long-distance dispersal be inferred from the New Zealand plant fossil record? *Australian Journal of Botany* 49: 357–366.

Pole, M. 2010. Discussion of 'The Waipounamu Erosion Surface: Questioning the antiquity of the New Zealand land surface and terrestrial fauna and flora'. *Geological Magazine* 147: 151–155.

Poncet, V., F. Munoz, J. Munzinger et al. 2013. Phylogeography and niche modelling of the relict plant *Amborella trichopoda* (Amborellaceae) reveal multiple Pleistocene refugia in New Caledonia. *Molecular Ecology* 22: 6163–6178.

Pons, J.-M., A. Hassanin, and P.-A. Crochet. 2005. Phylogenetic relationships within the Laridae (Charadriiformes: Aves) inferred from mitochondrial markers. *Molecular Phylogenetics and Evolution* 37: 686–699.

Porter, C.C. 1981. *Certonotus* Kriechbaumer (Hymenoptera: Ichneumonidae), an Australian genus newly recorded in South America. *Florida Entomologist* 64: 235–244.

Portmann, A. 1961. Sensory organs: Skin, taste and olfaction. In *Biology and Comparative Physiology of Birds*, Vol. 2, ed. A.J. Marshall, pp. 37–48. New York: Academic Press.

Potts, T.H. 1870. On the birds of New Zealand. Part II. *Transactions of the New Zealand Institute* 3: 59–109.

Powell, A.W.B. 1979. *New Zealand Mollusca.* Auckland, New Zealand: Collins.

Powlesland, R.G., D.V. Merton, and J.F. Cockrem. 2006. A parrot apart: The natural history of the kakapo (*Strigops habroptilus*), and the context of its conservation management. *Notornis* 53: 3–26.

Pratt, R.C., M. Morgan-Richards, and S.A. Trewick. 2008. Diversification of New Zealand weta (Orthoptera: Ensifera: Anostostomatidae) and their relationships in Australasia. *Philosophical Transactions of the Royal Society, London B* 363: 3427–3437.

Pratt, S.J. 2013. Evolution of the genera *Vitex* (Lamiaceae) and *Zygogynum* (Winteraceae) on New Caledonia. MSc thesis, University of Waikato, Hamilton, New Zealand.

Prebble, J.M., C.N. Cupido, H.M. Meudt, and P.J. Garnock-Jones. 2011. First phylogenetic and biogeographical study of the southern bluebells (*Wahlenbergia*, Campanulaceae). *Molecular Phylogenetics and Evolution* 59: 636–648.

Prebble, J.M., H.M. Meudt, and P.J. Garnock-Jones. 2012. An expanded molecular phylogeny of the southern bluebells (*Wahlenbergia*, Campanulaceae) from Australia and New Zealand. *Australian Systematic Botany* 25: 11–30.

Price, R.C. and D.S. Coombs. 1975. Phonolitic lava domes and other features of the Dunedin Volcano, East Otago. *Journal of the Royal Society of New Zealand* 5: 133–152.

Procopius. AD 550/1966. *The Secret History.* Transl. G.A. Williamson. Harmondsworth, UK: Penguin.

Puente-Lelièvre, C., M. Harrington, E. Brown, M. Kuzmina, and D. Crayn. 2012. Cenozoic extinction and recolonization in the New Zealand flora: The case of the fleshy-fruited epacrids (Styphelieae, Styphelioideae, Ericaceae). *Molecular Phylogenetics and Evolution* 66: 203–214.

Pufal, G. and P. Garnock-Jones. 2010. Hygrochastic capsule dehiscence supports safe site strategies in New Zealand alpine *Veronica* (Plantaginaceae). *Annals of Botany* 106: 405–412.

Purdie, A.W. 1985. *Chordospartium muritai* (Papilionaceae)—A rare new species of New Zealand tree broom. *New Zealand Journal of Botany* 23: 157–161.

Pushcharovsky, Y.M. 2011. First-order linear tectonovolcanic ridges in oceans. *Geotectonics* 45: 101–112.

Pycraft, W.P. 1900. On the morphology and phylogeny of the Palaeognathæ (Ratitæ and Crypturi) and Neognathæ (Carinatæ). *Transactions of the Zoological Society of London* 15: 149–290.

Pycraft, W.P. 1905. Some points in the anatomy of *Acanthidositta chloris* with some remarks on the systematic position of the genera *Acanthidositta* and *Xenicus*. *Ibis* 47: 603–621.

Pyron, R.A. and J.J. Wiens. 2011. A large-scale phylogeny of Amphibia including over 2800 species, and a revised classification of extant frogs, salamanders, and caecilians. *Molecular Phylogenetics and Evolution* 61: 543–583.

Quenouille, B., N. Hubert, E. Bermingham, and S. Planes. 2011. Speciation in tropical seas: Allopatry followed by range change. *Molecular Phylogenetics and Evolution* 58: 546–552.

Quiroga, M.P., P. Mathiasen, A. Iglesias, R.R. Mill, and A.C. Premoli. 2016. Molecular and fossil evidence disentangle the biogeographical history of *Podocarpus*, a key genus in plant geography. *Journal of Biogeography* 43: 372–383.

Radcliffe-Smith, A. 1983. *Euphorbia glauca* is not a pachyclad. *Kew Bulletin* 38: 307–308.

Radford, A.N. and M.A. du Plessis. 2003. Bill dimorphism and foraging niche partitioning in the green wood-hoopoe. *Journal of Animal Ecology* 72: 258–269.

Radford, E.A., M.F. Watson, and J. Preston. 2001. Phylogenetic relationships of species of *Aciphylla* (Apiaceae, subfamily Apioideae) and related genera using molecular, morphological, and combined data sets. *New Zealand Journal of Botany* 39: 183–208.

Rahl, J.M., M.T. Brandon, H. Deckert, U. Ring, and N. Mortimer. 2011. Tectonic significance of deformation in low-grade sandstones in the Mesozoic Otago subduction wedge. *American Journal of Science* 311: 27–62.

Raia, P. 2009. Gigantism. In *Encyclopedia of Islands*, eds. R.G. Gillespie and D.A. Clague, pp. 372–376. Berkeley, CA: University of California Press.

Rait, G.J. 2000. Amounts and rates of transport of the Northland and East Coast Allochthons (abstract). *Geological Society of New Zealand Annual Conference*, Wellington, New Zealand. Geological Society of New Zealand Miscellaneous Publication 108A, p. 129.

Randall, J.E. and T. Yamakawa. 2006. *Parapercis phenax* from Japan and *P. banoni* from the southeast Atlantic, new species of pinguipedid fishes previously identified as *P. roseoviridis*. *Zoological Studies* 45: 1–10.

Rapson, L.J. 1953. Vegetative anatomy in *Donatia*, *Phyllachne*, *Forstera* and *Oreostylidium* and its taxonomic significance. *Transactions of the Royal Society of New Zealand* 80: 399–402.

Ratnakumar, A., S. Mousset, S. Glémin, J. Berglund, N. Galtier, L. Duret, and M.T. Webster. 2010. Detecting positive selection within genomes: The problem of biased gene conversion. *Philosophical Transactions of the Royal Society B* 365: 2571–2580.

Rattenbury, M.S., D.B. Townsend, and M.R. Johnston. 2006. *Geology of the Kaikoura area*. 1:250,000 geological map 13. 1 sheet and 70pp. Institute of Geological and Nuclear Sciences, Wellington, New Zealand.

Raven, P.H. and D.I. Axelrod. 1974. Angiosperm biogeography and past continental movements. *Annals of the Missouri Botanical Garden* 61: 539–673.

Raven, P.H. and T.E. Raven. 1976. The genus *Epilobium* (Onagraceae) in Australasia: A systematic and evolutionary study. *New Zealand Department of Scientific and Industrial Research Bulletin* 216: 1–321.

Rawlence, N.J. and A. Cooper. 2013. Youngest reported radiocarbon age of a moa (Aves: Dinornithiformes) dated from a natural site in New Zealand. *Journal of the Royal Society of New Zealand* 43: 100–107.

Rawlence, N.J., G.L.W. Perry, I.W.G. Smith et al. 2015. Radiocarbon-dating and ancient DNA reveal rapid replacement of extinct prehistoric penguins. *Quaternary Science Reviews* 112: 59–65.

Rawling, T.J. and G.S. Lister. 2002. Large-scale structure of the eclogite-blueschist belt of New Caledonia. *Journal of Structural Geology* 24: 1239–1258.

Raya, A. and J.C. Izpisúa Belmonte. 2006. Left–right asymmetry in the vertebrate embryo: From early information to higher-level integration. *Nature Reviews Genetics* 7: 283–293.

Reay, A. 2003. Geology. In *The Natural History of Southern New Zealand*, eds. J. Darby, R.E. Fordyce, A. Mark, K. Probert, and C. Townsend, pp. 1–16. Dunedin, New Zealand: University of Otago Press.

Ree, R.H. and I. Sanmartín. 2009. Prospects and challenges for parametric models in historical biogeographical inference. *Journal of Biogeography* 36: 1211–1220.

Ree, R.H. and S.A. Smith. 2008. Maximum likelihood inference of geographic range evolution by dispersal, local extinction, and cladogenesis. *Systematic Biology* 57: 4–14.

Reichgelt, T., E.M. Kennedy, D.C. Mildenhall, J.G. Conran, D.R. Greenwood, and D.E. Lee. 2013. Quantitative palaeoclimate estimates for early Miocene southern New Zealand: Evidence from Foulden Maar. *Palaeogeography, Palaeoclimatology, Palaeoecology* 378: 36–44.

Reid, D.G., K. Lal, J. Mackenzie-Dodds, F. Kaligis, D.T.J. Littlewood, and S.T. Williams. 2006. Comparative phylogeography and species boundaries in *Echinolittorina* snails in the central Indo-West Pacific. *Journal of Biogeography* 33: 990–1006.

Reiss, J.O. 2009. *Not by Design: Retiring Darwin's Watchmaker*. Berkeley, CA: University of California Press.

Reisser, C.M.O., A.R. Wood, J.J. Bell, and J.P.A. Gardner. 2011. Connectivity, small islands and large distances: The *Cellana strigilis* limpet complex in the Southern Ocean. *Molecular Ecology* 20: 3399–3413.

Renner S.S., J.S. Strijk, D. Strasberg, and C. Thébaud. 2010. Biogeography of the Monimiaceae (Laurales): A role for East Gondwana and long-distance dispersal, but not West Gondwana. *Journal of Biogeography* 37: 1227–1238.

Rey, P.F. and R.D. Müller. 2010. Fragmentation of active continental plate margins owing to the buoyancy of the mantle wedge. *Nature Geoscience* 3: 257–261.

Reyners, M., D. Eberhart-Phillips, and S. Bannister. 2011. Tracking repeated subduction of the Hikurangi Plateau beneath New Zealand. *Earth and Planetary Science Letters* 311: 165–171.

Rheindt, F.E., L. Christidis, S. Kuhn, S. de Kloet, J.A. Norman, and A. Fidler. 2014. The timing of diversification within the most divergent parrot clade. *Journal of Avian Biology* 44: 140–148.

Rho, M., M. Zhou, X. Gao, S. Kim, H. Tang, and M. Lynch. 2009. Independent mammalian genome contractions following the KT boundary. *Genome Biology and Evolution* 2009: 2–12.

Ribas, C.C., R.G. Moyle, C.Y. Miytaki, and J. Cracraft. 2007. The assembly of montane biotas: Linking Andean tectonics and climatic oscillations to independent regimes of diversification in *Pionus* parrots. *Proceedings of the Royal Society London B*, 274: 2399–2408.

Ribeiro, G.C. and A. Eterovic. 2011. Neat and clear: 700 species of crane flies (Diptera: Tipulomorpha) link southern South America and Australasia. *Systematic Entomology* 36: 754–767.

Richards, A.M. 1971. The Rhaphidophoridae (Orthoptera) of Australia Part 10. A new genus from southeastern Tasmania with New Zealand affinities. *Pacific Insects* 13: 589–595.

Richards, A.M. 1972. Revision of the Rhaphidophoridae (Orthoptera) of New Zealand. Part XIV. Three alpine genera from the South Island. *Journal of the Royal Society of New Zealand* 2: 151–174.

Richardson, L.R. 1979. On two land-leeches labelled as from New Zealand (Hirudinea: Haemadipsoidea). *Tuatara* 24: 41–47.

Ricklefs, R.E. and D.G. Jenkins. 2011. Biogeography and ecology: Towards the integration of two disciplines. *Philosophical Transactions of the Royal Society B* 366: 2438–2448.

Ridley, M. 1985. *The Problems of Evolution*. Oxford, UK: Oxford University Press.

Righton, D., K. Aarestrup, D. Jellyman, P. Sébert, G. van den Thillart, and K. Tsukamoto. 2012. The *Anguilla* spp. migration problem: 40 million years of evolution and two millennia of speculation. *Journal of Fish Biology* 81: 365–386.

Riina, R., J.A. Peirson, D.V. Geltman et al. 2013. A worldwide molecular phylogeny and classification of the leafy spurges, *Euphorbia* subgenus *Esula* (Euphorbiaceae). *Taxon* 62: 316–342.

Ring, U., M. Bernet, and A. Tulloch. 2015. Kinematic, finite strain and vorticity analysis of the Sisters Shear Zone, Stewart Island, New Zealand. *Journal of Structural Geology* 73: 114–129.

Ring, U., M. Herd, R. Bohlar, J. Glodny, and M. Palin. 2006. Dating mylonitisation in the Paparoa Core Complex, Westland. Abstract for *the Geological Society of New Zealand Conference*, Palmerston North, New Zealand, December 4–7, 2006.

Riordan, N.K., C.M. Reid, K.N. Bassett, and J.D. Bradshaw. 2014. Reconsidering basin geometries of the West Coast: The influence of the Paparoa Core Complex on Oligocene rift systems. *New Zealand Journal of Geology and Geophysics* 57: 170–184.

Riskin, D.K., S. Parsons, W.A. Schutt, G.G. Carter, and J.W. Hermanson. 2006. Terrestrial locomotion of the New Zealand short-tailed bat *Mystacina tuberculata* and the common vampire bat *Desmodus rotundus*. *Journal of Experimental Biology* 209: 1725–1736.

Rivadavia, F., K. Kondo, M. Kato, and M. Hasebe. 2003. Phylogeny of the sundews, *Drosera* (Droseraceae), based on chloroplast *rbcL* and nuclear 18S ribosomal DNA sequences. *American Journal of Botany* 90: 123–130.

Roberts, C.D. 1991. Fishes of the Chatham Islands, New Zealand: A trawl survey and summary of the ichthyofauna. *New Zealand Journal of Marine and Freshwater Research* 25: 1–19.

Roberts, C.D., A.L. Stewart, C.D. Paulin, and D. Neale. 2005. Regional diversity and biogeography of coastal fishes on the West Coast South Island of New Zealand. *Science for Conservation* [Department of Conservation, Wellington] 250: 1–70.

Robertson, B.C. and S.J. Goldstien. 2012. Phylogenetic affinities of the New Zealand blue duck (*Hymenolaimus malacorhynchos*). *Notornis* 59: 49–59.

Robertson, C. and G. Nunn. 1998. Towards a new taxonomy for albatrosses. In *Albatross Biology and Conservation*, eds. G. Robertson and R. Gales, pp. 13–19. Sydney, New South Wales, Australia: Surrey Beatty and Sons.

Robertson, C.J.R., P. Hyvönen, M.J. Fraser, and C.R. Pickard. 2007. *Atlas of Bird Distribution in New Zealand* 1999–2004. Wellington, New Zealand: Ornithological Society of New Zealand.

Robeson, M.S., A.J. King, K.R. Freeman, C.W. Birky, Jr., A.P. Martin, and S.K. Schmidt. 2011. Soil rotifer communities are extremely diverse globally but spatially autocorrelated locally. *Proceedings of the National Academy of Sciences of USA* 108: 4406–4410.

Roelants, K., D.J. Gower, M. Wilkinson et al. 2007. Global patterns of diversification in the history of modern amphibians. *Proceedings of the National Academy of Sciences of USA* 104: 887–892.

Rogers, G.M., A. Hewitt, and J.B. Wilson. 2000. Ecosystem-based conservation strategy for Central Otago's saline patches. *Science for Conservation* 166: 1–38.

Romer, A.S. 1972. The vertebrate as a dual animal—Somatic and visceral. *Evolutionary Biology* 6: 121–156.

Ronce, O. 2007. How does it feel to be like a rolling stone? Ten questions about dispersal evolution. *Annual Review of Ecology, Evolution and Systematics* 38: 231–253.

Ronquist, F. 1997. Dispersal–vicariance analysis: A new biogeographic approach to the quantification of historical biogeography. *Systematic Biology* 46: 195–203.

Ronse de Craene, L.P. and E. Haston. 2006. The systematic relationships of glucosinolate-producing plants and related families: A cladistic investigation based on morphological and molecular characters. *Botanical Journal of the Linnean Society* 151: 453–494.

Rose, S. 1997. *Biology beyond Determinism*. Oxford, UK: Oxford University Press.

Rosen, D.E. 1974. Phylogeny and zoogeography of salmoniform fish and relationships of *Lepidogalaxias salamandroides*. *Bulletin of the American Museum of Natural History* 153: 265–326.

Rosenberg, A. and D.W. McShea. 2008. *Philosophy of Biology: A Contemporary Introduction*. New York: Routledge.

Ross, P.M., I.D. Hogg, C.A. Pilditch, C.J. Lundquist, and R.J. Wilkins. 2012. Population genetic structure of the New Zealand estuarine clam *Austrovenus stutchburyi* (Bivalvia: Veneridae) reveals population subdivision and partial congruence with biogeographic boundaries. *Estuaries and Coasts* 35: 143–154.

Rowden, A.A., M.R. Clark, and I.C. Wright. 2005. Physical characterisation and a biologically focused classification of "seamounts" in the New Zealand region. *New Zealand Journal of Marine and Freshwater Research* 39: 1039–1059.

Rozefelds, A.C. and A.N. Drinnan. 1998. Ontogeny and diversity in staminate flowers of *Nothofagus* (Nothofagaceae). *International Journal of Plant Science* 159: 906–922.

Rozefelds, A.C. and A.N. Drinnan. 2002. Ontogeny of pistillate flowers and inflorescences in *Nothofagus* subgenus *Lophozonia* (Nothofagaceae). *Plant Systematics and Evolution* 233: 105–126.

Rudall, P. 2002. Homologies of inferior ovaries and septal nectaries in monocotyledons. *International Journal of Plant Sciences* 163: 261–276.

Rudall, P.J. 2003. Unique floral structures and iterative evolutionary themes in Asparagales: Insights from a morphological cladistic analysis. *Botanical Review* 68: 488–509.

Ruiz-González, A., M.J. Madeira, E. Randi, A.V. Abramov, F. Davoli, and B.J. Gómez-Moliner. 2013. Phylogeography of the forest-dwelling European pine marten (*Martes martes*): New insights into cryptic northern glacial refugia. *Biological Journal of the Linnean Society* 109: 1–18.

Rupp, H.M.R. and E.D. Hatch. 1945. Relation of the orchid flora of Australia to that of New Zealand. *Proceedings of the Linnean Society of New South Wales* 70: 53–61.

Ruse, M. 2003. *Darwin and Design: Does Evolution Have a Purpose?* Cambridge, MA: Harvard University Press.

Ruse, M. and J. Travis (eds.). 2009. *Evolution: The First Four Billion Years*. Cambridge, MA: Harvard University Press.

Ryan, A.W., P.J. Smith, and J. Mork. 2002. Genetic differentiation between the New Zealand and Falkland Islands populations of southern blue whiting *Micromesistius australis*. *New Zealand Journal of Marine and Freshwater Research* 36: 637–643.

Ryan, M.T., G.B. Dunbar, M.J. Vandergoes et al. 2012. Vegetation and climate in Southern Hemisphere mid-latitudes since 210 ka: New insights from marine and terrestrial pollen records from New Zealand. *Quaternary Science Reviews* 48: 80–98.

Sagar, M.W. and J.M. Palin. 2011. Emplacement, metamorphism, deformation and affiliation of mid-Cretaceous orthogneiss from the Paparoa Metamorphic Core Complex lower-plate, Charleston, New Zealand. *New Zealand Journal of Geology and Geophysics* 54: 273–289.

Salmon, B. 2001. *Carnivorous Plants of New Zealand*. Manukau, New Zealand: Ecosphere.

Sancho, G., P.J. de Lange, M. Donato, J. Barkla, and S.J. Wagstaff. 2014. Late Cenozoic diversification of the austral genus *Lagenophora* (Astereae, Asteraceae). *Botanical Journal of the Linnean Society* 177: 78–95.

Sanmartín, I. and F. Ronquist. 2004. Southern Hemisphere biogeography inferred by event-based models: Plant versus animal patterns. *Systematic Biology* 53: 216–243.

Sanmartín, I., L. Wanntorp, and R.C. Winkworth. 2007. West Wind Drift revisited: Testing for directional dispersal in the Southern Hemisphere using event-based tree fitting. *Journal of Biogeography* 34: 398–416.

Sarkar, S. 2015. The genomic challenge to adaptationism. *British Journal for the Philosophy of Science* 66: 505–536.

Sastre-De Jesús, I. 1987. Revision of the Cyrtopodaceae and transfer of *Cyrtopodendron* to the Pterobryaceae. *Memoirs of the New York Botanical Garden* 45: 709–721.

Sauquet, H., S.Y. Ho, M.A. Gandolfo et al. 2012. Testing the impact of calibration on molecular divergence times using a fossil-rich group: The case of *Nothofagus* (Fagales). *Systematic Biology* 61: 289–313.

Savolainen, V., M.-C. Anstett, C. Lexer et al. 2006. Sympatric speciation in palms on an oceanic island. *Nature* 441: 210–213.

Schellart, W.P. 2007. North-eastward subduction followed by slab detachment to explain ophiolite obduction and Early Miocene volcanism in Northland, New Zealand. *Terra Nova* 19: 211–218.

Schellart, W.P. 2012. Comment on "Geochemistry of the Early Miocene volcanic succession of Northland, New Zealand, and implications for the evolution of subduction in the Southwest Pacific" by M.A. Booden, I.E.M. Smith, P.M. Black and J.L. Mauk. *Journal of Volcanology and Geothermal Research* 211/212: 112–117.

Schellart, W.P., B.L.N. Kennett, W. Spakman, and M. Amaru. 2009. Plate reconstructions and tomography reveal a fossil lower mantle slab below the Tasman Sea. *Earth and Planetary Science Letters* 278: 143–151.

Schellart, W.P., G.S. Lister, and V.G. Toy. 2006. A Late Cretaceous and Cenozoic reconstruction of the Southwest Pacific region: Tectonics controlled by subduction and slab rollback processes. *Earth-Science Reviews* 76: 191–233.

Schuchert, P. 2014. High genetic diversity in the hydroid *Plumularia setacea*: A multitude of cryptic species or extensive population subdivision? *Molecular Phylogenetics and Evolution* 76: 1–9.

Schuettpelz, E. and S.B. Hoot. 2004. Phylogeny and biogeography of *Caltha* (Ranunculaceae) based on chloroplast and nuclear DNA sequences. *American Journal of Botany* 91: 247–253.

Schuettpelz, E., S.B. Hoot, R. Samuel, and F. Ehrendorfer. 2002. Multiple origins of Southern Hemisphere *Anemone* (Ranunculaceae) based on plastid and nuclear sequence data. *Plant Systematics and Evolution* 231: 143–151.

Schuettpelz, E. and K.M. Pryer. 2009. Evidence for a Cenozoic radiation of ferns in an angiosperm-dominated canopy. *Proceedings of the National Academy of Sciences of USA* 106: 11200–11205.

Schulte, D. 2011. Kinematics of the Paparoa metamorphic core complex, West Coast, South Island, New Zealand. MSc thesis, University of Canterbury, Christchurch, New Zealand.

Schulte, D.O., U. Ring, S.N. Thomson, J. Glodny, and H. Carrad. 2014. Two-stage development of the Paparoa Metamorphic Core Complex, West Coast, South Island, New Zealand: Hot continental extension precedes sea-floor spreading by ~25 m.y. *Lithosphere* 6: 177–194.

Schurr, F.M., J. Pagel, J.S. Cabral et al. 2012. How to understand species' niches and range dynamics: A demographic research agenda for biogeography. *Journal of Biogeography* 39: 2146–2162.

Schuster, T.M., J.L. Reveal, and K. Kron. 2011a. Phylogeny of Polygoneae (Polygonaceae: Polygonoideae). *Taxon* 60: 1653–1666.

Schuster, T.M., K.L. Wilson, and K.A. Kron. 2011b. Phylogenetic relationships of *Muehlenbeckia, Fallopia*, and *Reynoutria* (Polygonaceae) investigated with chloroplast and nuclear sequence data. *International Journal of Plant Sciences* 172: 1053–1066.

Schwab, I.R. 2003. Double crossed. *British Journal of Ophthalmology* 87: 1442.

Schwarzhans, W., R.P. Scofield, A.J.D. Tennyson, J.P Worthy, and T.H. Worthy. 2012. Fish remains, mostly otoliths, from the non-marine early Miocene of Otago, New Zealand. *Acta Palaeontologica Polonica* 57: 319–350.

Schwendt, E. 1907. Zur Kenntnis der extrafloralen Nektarien. *Botanisches Centralblatt* 22: 245–286.

Scofield, P. and B. Stephenson. 2013. *Birds of New Zealand: A Photographic Guide*. Auckland, New Zealand: Auckland University Press.

Scofield, R.P. 2009. Callaeidae. In *Handbook of the Birds of the World*, Vol. 14, eds. J. del Hoyo, A. Elliott, and D.A. Christie, pp. 228–241. Barcelona, Spain: Lynx Edicions.

Scofield, R.P. and K.W. Ashwell. 2009. Rapid somatic expansion causes the brain to lag behind: The case of the brain and behavior of New Zealand's Haast's Eagle (*Harpagornis moorei*). *Journal of Vertebrate Paleontology* 29: 637–649.

Scott, J.M. 2013. A review of the location and significance of the boundary between the Western Province and Eastern Province, New Zealand. *New Zealand Journal of Geology and Geophysics* 56: 276–293.

Scott, J.M. and A.F. Cooper. 2006. Early Cretaceous extensional exhumation of the lower crust of a magmatic arc: Evidence from the Mount Irene Shear Zone, Fiordland, New Zealand. *Tectonics* 25: TC3018.

Scott, J.M., D.E. Lee, R.E. Fordyce, and J.M. Palin. 2014. A possible Late Oligocene—Early Miocene rocky shoreline on Otago Schist. *New Zealand Journal of Geology and Geophysics* 57: 185–194.

Scott, J.M., J.M. Palin, A.F. Cooper, Å. Fagereng, and R.P. King. 2009. Polymetamorphism, zircon growth and retention of early assemblages through the dynamic evolution of a continental arc in Fiordland, New Zealand. *Journal of Metamorphic Geology* 27: 281–294.

Seago, A.E. 2009. Revision of *Agyrtodes* Portevin (Coleoptera: Leiodidae). *Coleopterists' Bulletin* 63(sp7): 1–73.

Seago, A.E., R.A.B. Leschen, and A.F. Newton. 2015. Two new high altitude genera of Camiarini (Coleoptera: Leiodidae: Camiarinae) from Australia and New Zealand. *Zootaxa* 3957: 300–312.

Sebastian, P., H. Schaefer, R. Lira, I.R.H. Telford, and S.S. Renner. 2012. Radiation following long-distance dispersal: The contributions of time, opportunity and diaspore morphology in *Sicyos* (Cucurbitaceae). *Journal of Biogeography* 39: 1427–1438.

Seehausen, O. 2013. Integrating ecology and genetics in speciation research. *Trends in Ecology and Evolution* 28: 12–13.

Selander, R.K. 1966. Sexual dimorphism and differential niche utilization in birds. *Condor* 68: 113–151.

Selden, P.A. and D. Penney. 2010. Fossil spiders. *Biological Reviews* 85: 171–206.

Seldon, D.S. and R.A.B. Leschen. 2011. Revision of the *Mecodema curvidens* species group (Coleoptera: Carabidae: Broscini). *Zootaxa* 2829: 1–45.

Seldon, D.S., R.A.B. Leschen, and J.K. Liebherr. 2012. A new species of *Mecodema* (Carabidae: Broscini) from Northland, New Zealand, with notes on a newly observed structure within the female genitalia. *New Zealand Entomologist* 35: 39–50.

Selvatti, A.P., L. Pedreira Gonzaga, and C.A. de Moraes Russo. 2015. A Paleogene origin for crown passerines and the diversification of the Oscines in the New World. *Molecular Phylogenetics and Evolution* 88: 1–15.

Shafer, A.B.A. and J.B.W. Wolf. 2013. Widespread evidence for incipient ecological speciation: A meta-analysis of isolation-by-ecology. *Ecology Letters* 16: 940–950.

Shah, J.J. and Y.S. Dave. 1970. Morpho-histogenetic studies on tendrils of Vitaceae. *American Journal of Botany* 57: 363–373.

Sharma, P. and G. Giribet. 2009. A relict in New Caledonia: Phylogenetic relationships of the family Troglosironidae (Opiliones: Cyphophthalmi). *Cladistics* 25: 279–294.

Sharma, P.P. and W.C. Wheeler. 2013. Revenant clades in historical biogeography: The geology of New Zealand predisposes endemic clades to root age shifts. *Journal of Biogeography* 40: 1609–1618.

Shepherd, L.D. and L.R. Perrie. 2011. Microsatellite DNA analyses of a highly disjunct New Zealand tree reveal strong differentiation and imply a formerly more continuous distribution. *Molecular Ecology* 20: 1389–1400.

Shepherd, L.D., L.R. Perrie, and P.J. Brownsey. 2007. Fire and ice: Volcanic and glacial impacts on the phylogeography of the New Zealand forest fern *Asplenium hookerianum*. *Molecular Ecology* 16: 4536–4549.

Shepherd, L.D., T.H. Worthy, A.J.D. Tennyson, R.P. Scofield, K.M. Ramstad, and D.M. Lambert. 2012. Ancient DNA analyses reveal contrasting phylogeographic patterns amongst kiwi (*Apteryx* spp.) and a recently extinct lineage of spotted kiwi. *PLoS One* 7(8): e42384 (pp. 1–9).

Shervais, J.W. 2001. Birth, death and resurrection: The life cycle of suprasubduction zone ophiolites. *Geochemistry, Geophysics, Geosystems* 2: Paper 2000GC000080.

Sherwood, A.R. and R.G. Sheath. 1999. Biogeography and systematics of *Hildenbrandia* (Rhodophyta, Hildenbrandiales) in North America: Inferences from morphometrics and *rbc*L and 18S rRNA gene sequence analyses. *European Journal of Phycology* 34: 523–532.

Sherwood, A.R. and R.G. Sheath. 2003. Systematics of the Hildenbrandiales (Rhodophyta): Gene sequence and morphometric analyses of global collections. *Journal of Phycology* 39: 409–422.

Siddoway, C.S. 2008. Tectonics of the West Antarctic Rift System: New light on the history and dynamics of distributed intracontinental extension. In *Antarctica: A Keystone in a Changing World*, eds. A.K. Cooper, P.J. Barrett, H. Stagg, B. Storey, E. Stump, W. Wise, and the 10th ISAES editorial team, pp. 91–114. Washington, DC: National Academies Press.

Sidor, C.A. 2001. Simplification as a trend in synapsid evolution. *Evolution* 55: 1419–1442.

Sidor, C.A. 2003. Evolutionary trends and the origin of the mammalian lower jaw. *Paleobiology* 29: 605–640.

Signor, P.W. and J.H. Lipps. 1982. Sampling bias, gradual extinction patterns, and catastrophes in the fossil record. In *Geological Implications of Impacts of Large Asteroids and Comets on the Earth*, eds. L.T. Silver and P.H. Schultz. pp. 291–296. Boulder, CO: Geological Society of America. Special Paper 190.

Sillman, A.J. 1973. Avian vision. *Avian Biology* 3: 349–387.

Simmons, M.P. and J. Gatesy. 2015. Coalescence vs. concatenation: Sophisticated analyses vs. first principles applied to rooting the angiosperms. *Molecular Phylogenetics and Evolution* 91: 98–122.

Simmons, N.B. 2005. Order Chiroptera. *Mammal Species of the World: A Taxonomic and Geographic Reference*, 3rd edn., eds. D.M. Wilson and D.E. Reeder, pp. 312–529. Baltimore, MD: Johns Hopkins University Press.

Simpson, G. 1945. A revision of the genus *Carmichaelia*. *Transactions of the Royal Society of New Zealand* 75: 231–287.

Simpson, G.G. 1940. Mammals and land bridges. *Journal of the Washington Academy of Sciences* 30: 137–163.

Simpson, G.G. 1944. *Tempo and Mode in Evolution*. New York: Columbia University Press.

Simpson G.G. 1953a. *The Major Features of Evolution*. New York: Columbia University Press.

Simpson, G.G. 1953b. The Baldwin effect. *Evolution* 2: 110–117.

Simpson, G.G. 1967. *The Meaning of Evolution*, revised edn. New Haven, CT: Yale University Press.

Simpson, M.J.A. 1974. *Senecio bifistulosus* Hook. f.—A rare shrub endemic to New Zealand. *New Zealand Journal of Botany* 12: 567–573.

Simpson, P. 2005. *Pōhutukawa and Rātā: New Zealand's Iron-Hearted Trees*. Wellington, New Zealand: Te Papa Press.

Skelley, P.E. and R.A.B. Leschen. 2007. Erotylinae (Insecta: Coleoptera: Cucujoidea: Erotylinae): Taxonomy and biogeography. *Fauna of New Zealand* 59: 1–58.

Skipworth, J.P. 1962. The primary vascular system and phyllotaxis in *Hectorella caespitosa* Hook. f. *New Zealand Journal of Science* 5: 253–258.

Skottsberg, C. 1906. Zur Flora des Feuerlandes. *Wissenschaftliche Ergebnisse der Schwedisch Südpolar-Expedition* IV: 4.

Skottsberg, C. 1922. Recent researches in *Astelia* B. and S. *Transactions of the Royal Society of New Zealand* 67: 218–226.

Slade, B.F. 1952. Cladode anatomy and leaf trace systems in New Zealand brooms. *Transactions and Proceedings of the Royal Society of New Zealand* 80: 81–96.

Smissen, R. and P. Heenan. 2008. Unravelling the fibres of harakeke evolution. *New Zealand Garden Journal* 11(1): 25–27.

Smissen, R.D., P.J. Garnock-Jones, and G.K. Chambers. 2003. Phylogenetic analysis of ITS sequences suggests a Pliocene origin for the bipolar distribution of *Scleranthus* (Caryophyllaceae). *Australian Systematic Botany* 16: 301–313.

Smit, F.G.A. 1979. The fleas of New Zealand (Siphonaptera). *Journal of the Royal Society of New Zealand* 9: 143–232.

Smith, A.C. 1979–1996. *Flora Vitiensis Nova: A New Flora of Fiji*, 6 vols. Lawai, Kauai, HI: Pacific Tropical Botanical Garden.

Smith, G.A. 1975. Systematics of parrots. *Ibis* 117: 18–68.

Smith, P.J., S.M. McVeagh, and R. Allibone. 2005. Extensive genetic differentiation in *Gobiomorphus breviceps* from New Zealand. *Journal of Fish Biology* 67: 627–639.

Smith, S.A., J.M. Beaulieu, and M.J. Donoghue. 2010. An uncorrelated relaxed-clock analysis suggests an earlier origin for flowering plants. *Proceedings of the National Academy of Sciences of USA* 107: 5897–5902.

Smith, T., J., Habersetzer, N.B., Simmons, and G.F. Gunnell. 2012. Systematics and paleobiogeography of early bats. In *Evolutionary History of Bats: Fossils, Molecules and Morphology*, eds. G.F. Gunnell and N.B. Simmons, pp. 23–66. Cambridge, UK: Cambridge University Press.

Smith, W.H.F. and D.T. Sandwell. 1997. Global sea floor topography from satellite altimetry and ship depth soundings. *Science* 277: 1956–1962.

Smith, W.L. and M.T. Craig. 2007. Casting the percomorph net widely: The importance of broad taxonomic sampling in the search for the placement of serranid and percid fishes. *Copeia* 2007: 35–55.

Smithsonian Institution. 2016. Flora of the Hawaiian Islands. http://botany.si.edu/pacificislandbiodiversity/hawaiianflora/index.htm. Accessed May, 2016.

Sobel, J.M., G.F. Chen, L.R. Watt, and D.W. Schemske. 2010. The biology of speciation. *Evolution* 64: 295–315.

Soberón, J. and M. Nakamura. 2009. Niches and distributional areas: Concepts, methods, and assumptions. *Proceedings of the National Academy of Sciences of USA* 106: 19644–19650.

Sokolov, I.M. 2015. Review of the species of *Pelodiaetodes* Moore (Coleoptera: Carabidae: Bembidiini: Anillina) of New Zealand. *Zootaxa* 3963: 561–582.

Soligo, C., O. Will, S. Tavaré, C.R. Marshall, and R.D. Martin. 2007. New light on the dates of primate origins and divergence. In *Primate Origins: Adaptations and Evolution*, eds. M.J. Ravosa and M. Dagosto, pp. 29–49. New York: Springer.

Song, D. and Q. Wang. 2003. Systematics of the longicorn beetle genus *Coptomma* Newman (Coleoptera: Cerambycidae: Cerambycinae). *Invertebrate Systematics* 17: 429–447.

Soper, M.F. 1976. *New Zealand Birds,* 2nd ed. Christchurch, New Zealand: Whitcoulls.

Sorenson, M.D. and R.B. Payne. 2005. Molecular systematics: Cuckoo phylogeny inferred from mitochondrial DNA sequences. In *Bird Families of the World: Cuckoos*, ed. R.B. Payne, pp. 68–94. Oxford, UK: Oxford University Press.

Sota, T., Y. Takami, G.B. Monteith, and B.P. Moore. 2005. Phylogeny and character evolution of endemic carabid beetles of the genus *Pamborus* based on mitochondrial and nuclear gene sequences. *Molecular Phylogenetics and Evolution* 36: 391–404.

Souza, V., L. Espinosa-Asuar, A.E. Escalante et al. 2006. An endangered oasis of aquatic microbial biodiversity in the Chihuahan desert. *Proceedings of the National Academy of Sciences of USA* 103: 6565–6570.

Spalik, K., M. Piwczyński, C.A. Danderson, R. Kurzyna-Młynik, T.S. Bone, and S.R. Downie. 2010. Amphitropic amphiantarctic disjunctions in Apiaceae subfamily Apioideae. *Journal of Biogeography* 37: 1977–1994.

Speight, R. 1917. An ancient buried forest near Riccarton: Its bearing on the mode of formation of the Canterbury Plains. *Transactions of the New Zealand Institute* 49: 361–364.

Spencer, H.G., F.J. Brook, and M. Kennedy. 2006. Phylogeography of Kauri Snails and their allies from Northland, New Zealand (Mollusca: Gastropoda: Rhytididae: Paryphantinae). *Molecular Phylogenetics and Evolution* 38: 835–842.

Spörli, K.B. 2006. The Waipapa terrane of New Zealand: Geology of a treasure trove of Permian—Mesozoic ocean floor faunas. *Geological and Nuclear Sciences Miscellaneous Series* 2: 124.

Sprung, P., S. Schuth, C. Münker, and L. Hoke. 2007. Intraplate volcanism in New Zealand: The role of fossil plume material and variable lithospheric properties. *Contributions Mineralogy Petrology* 153: 669–687.

Stauffer, R.C. (ed.). 1975. *Charles Darwin's Natural Selection: Being the Second Part of His Big Species Book Written from 1856 to 1858*. Cambridge, UK: Cambridge University Press.

Stead, E.F. 1932. *The Life Histories of New Zealand Birds*. London, UK: Search Publishing.

Stead, E.F. 1937. Notes on the short-tailed bat (*M. tuberculatus*). *Transactions of the Royal Society of New Zealand* 66: 188–191.

Stebbins, G.L. 1971. *Processes of Organic Evolution*, 2nd edn. Englewood Cliffs, NJ: Prentice-Hall.

Steele, C.A., J. Baumsteiger, and A. Storfer. 2009. Influence of life-history variation on the genetic structure of two sympatric salamander taxa. *Molecular Ecology* 18: 1629–1639.

Steens, M.I., D.J. Winter, R. Morris, J. McCartney, and P. Greenslade. 2007. New Zealand's giant Collembola: New information on distribution and morphology for *Holacanthella* Börner, 1906 (Neanuridae: Uchidanurinae). *New Zealand Journal of Zoology* 34: 63–78.

Steffen, S., P. Ball, L. Mucina, and G. Kadereit. 2015. Phylogeny, biogeography and ecological diversification of *Sarcocornia* (Salicornioideae, Amaranthaceae). *Annals of Botany* 115: 353–368.

Stegman, D.R., J. Freeman, W.P. Schellart, L. Moresi, and D. May. 2006. Influence of trench width on subduction hinge retreat rates in 3-D models of slab rollback. *Geochemistry, Geophysics, Geosystems* 7: Q03012.

Stelbrink, B., C. Albrecht, R. Hall, and T. von Rintelen. 2012. The biogeography of Sulawesi revisited: Is there evidence for a vicariant origin of taxa on Wallace's "anomalous island"? *Evolution* 66: 2252–2271.

Stelbrink, B., T. von Rintelen, G. Cliff, and J. Kriwet. 2010. Molecular systematics and global phylogeography of angel sharks (genus *Squatina*). *Molecular Phylogenetics and Evolution* 54: 395–404.

Stephens, A.E.A., G.W. Gibbs, and B.H. Patrick. 2007. Three new species in the *Pseudocoremia modica* (Philpott, 1921) complex (Lepidoptera: Geometridae: Ennominae) and their evolutionary relationships. *New Zealand Entomologist* 30: 71–78.

Stephenson, E.M. 1960. The skeletal characters of *Leiopelma hamiltoni* McCulloch, with particular reference to the effects of heterochrony in the genus. *Transacations of the Royal Society of New Zealand* 88: 473–488.

Sterelny, K. and P.E. Griffiths. 1999. *Sex and Death: An Introduction to Philosophy of Biology*. Chicago, IL: University of Chicago Press.

Stern, R.J. 2002. Subduction zones. *Reviews of Geophysics* 40(1012): 1–38.

Stern, R.J. 2004. Subduction initiation: Spontaneous and induced. *Earth and Planetary Science Letters* 226: 275–292.

Stern, R.J. 2010. The anatomy and ontogeny of modern intra-oceanic arc systems. *Geological Society of London Special Publication* 338: 7–34.

Stevens, M.I. and B.J. Hicks. 2009. Mitochondrial DNA reveals monophyly of New Zealand's *Gobiomorphus* (Teleostei: Eleotridae) amongst a morphological complex. *Evolutionary Ecology Research* 11: 109–123.

Stevens, P.F. 2014. Angiosperm phylogeny website. www.mobot.org/MOBOT/Research/APweb/. Accessed February 2014.

Stevens, P.F. 2016. Angiosperm phylogeny website, version 13. www.mobot.org/mobot/research/apweb/

Stevenson, G.B. 1934. The life history of the New Zealand species of the parasitic genus *Korthalsella*. *Transactions of the Royal Society of New Zealand* 64: 175–190.

Stewart, T.A. and R.C. Albertson. 2010. Evolution of a unique predatory feeding apparatus: Functional anatomy, development and a genetic locus for jaw laterality in Lake Tanganyika scale-eating cichlids. *BMC Biology* 8(8): 1–11.

Stilwell, J.D. and C.P. Consoli. 2012. Tectono-stratigraphic history of the Chatham Islands, SW Pacific— The emergence, flooding and reappearance of eastern 'Zealandia'. *Proceedings of the Geologists' Association* 123: 170–181.

Stilwell, J.D., C.P. Consoli, R. Sutherland et al. 2006. Dinosaur sanctuary on the Chatham Islands, Southwest Pacific: First record of theropods from the K–T boundary Takatika Grit. *Paleogeography, Palaeoclimatology, Palaeoecology* 230: 243–250.

Stöckler, K., I.L. Daniel, and P.J. Lockhart. 2002. New Zealand kauri (*Agathis australis* (D.Don) Lindl., Araucariaceae) survives Oligocene drowning. *Systematic Biology* 51: 827–832.

Stoltzfus, A. 2006a. Mutationism and the dual causation of evolutionary change. *Evolution and Development* 8: 304–317.

Stoltzfus, A. 2006b. Mutation-biased adaptation in a protein NK model. *Molecular Biology and Evolution* 23: 1852–1862.

Stoltzfus, A. 2012. Constructive neutral evolution: Exploring evolutionary theory's curious disconnect. *Biology Direct* 7(35): 1–13.

Stoltzfus, A.2014a. The curious disconnect. Introduction. http://www.molevol.org/cdblog

Stoltzfus, A. 2014b. In search of mutation-driven evolution. (Review of Masatoshi Nei, 2013. *Mutation-Driven Evolution*. Oxford University Press, Oxford.) *Evolution and Development* 16: 57–59.

Stoltzfus, A. and L.Y. Yampolsky. 2009. Climbing Mount Probable: Mutation as a cause of nonrandomness in evolution. *Journal of Heredity* 100: 637–647.

Stone, B.C. 1973. Materials for a monograph of *Freycinetia* Gaudich. XIV. On the relation between *F. banksii* A. Cunn. of New Zealand and *F. baueriana* Endl. of Norfolk Island, with notes on the structure of the seeds. *New Zealand Journal of Botany* 11: 241–246.

Strabo. 1917–1933. *Geography*. Transl. H.L. Jones. London, UK: Heinemann.

Stresemann, E. 1934. Aves. In *Handbuch der Zoologie*, Vol. 7: part 2, eds. W. Kükenthal and T. Krumbach. Berlin, Germany: Walter de Gruyter.

Ströckens, F., O. Güntürkün, and S. Ocklenburg. 2013. Limb preferences in non-human vertebrates. *Laterality: Asymmetries of Body, Brain and Cognition* 18: 536–575.

Strogen, D.P., K.J. Bland, A. Nicol, and P.R. King. 2014. Paleogeography of the Taranaki Basin region during the latest Eocene–Early Miocene and implications for the 'total drowning' of Zealandia. *New Zealand Journal of Geology and Geophysics* 57: 110–127.

Strugnell, J.M., A.D. Rogers, P.A. Prodo, M.A. Collins, and A.L. Allcock. 2008. The thermohaline expressway: The Southern Ocean as a centre of origin for deep-sea octopuses. *Cladistics* 24: 853–860.

Stuessy, T. 2006. Evolutionary biology: Sympatric plant speciation in islands? *Nature* 443: E12–E13.

Suggate, R.P. 2004. South Island, New Zealand: Ice advances and marine shorelines. In *Quaternary Glaciations—Extent and Chronology*, Part III, eds. J. Ehlers and P.L. Gibbard, pp. 285–291. Amsterdam, the Netherlands: Elsevier.

Sun, Y., X. He, and D. Glenny. 2014. Transantarctic disjunctions in Schistochilaceae (Marchantiophyta) explained by early extinction events, post-Gondwanan radiations and palaeoclimatic changes. *Molecular Phylogenetics and Evolution* 76: 189–201.

Sundue, M.A., B.S. Parris, T.A. Ranker et al. 2014. Global phylogeny and biogeography of grammitid ferns (Polypodiaceae). *Molecular Phylogenetics and Evolution* 81: 195–206.

Sutherland, J.E., S.C. Lindstrom, W.A. Nelson et al. 2011. A new look at an ancient order: Generic revision of the Bangiales (Rhodophyta). *Journal of Phycology* 47: 1131–1151.

Sutherland, R. 1996. Transpressional development of the Australia-Pacific boundary through southern South Island, New Zealand: Constraints from Miocene-Pliocene sediments, Waiho-1 borehole, South Westland. *New Zealand Journal of Geology and Geophysics* 39: 251–264.

Sutherland, R. 1999. Cenozoic bending of New Zealand basement terranes and Alpine Fault displacement: A brief review. *New Zealand Journal of Geology and Geophysics* 42: 295–301.

Sutherland, R., C.J. Hollis, S. Nathan, C.P. Strong, and G.J. Wilson. 1996. Age of Jackson Formation proves late Cenozoic allochthony in South Westland, New Zealand. *New Zealand Journal of Geology and Geophysics* 39: 559–563.

Swenson, U. 1995. Systematics of *Abrotanella*, an amphi-Pacific genus of *Asteraceae* (Senecioneae). *Plant Systematics and Evolution* 197: 149–193.

Swenson, U. and K. Bremer. 1997. Pacific biogeography of the Asteraceae genus *Abrotanella* (Senecioneae, Blennospermatinae). *Systematic Botany* 22: 493–508.

Swenson, U., S. Nylinder, and J. Munzinger. 2013a. Sapotaceae biogeography supports New Caledonia being an old Darwinian island. *Journal of Biogeography* 41: 797–809.

Swenson, U., S. Nylinder, and J. Munzinger. 2013b. Towards a natural classification of Sapotaceae subfamily Chrysophylloideae in Oceania and Southeast Asia based on nuclear sequence data. *Taxon* 62: 746–770.

Swenson, U., S. Nylinder, and S. Wagstaff. 2012. Are the Asteraceae 1.5 billion years old? A reply to Heads. *Systematic Biology* 61: 522–532.

Sykes, W.R. 1998. *Scaevola gracilis* (Goodeniaceae) in the Kermadec Islands and Tonga. *New Zealand Journal of Botany* 36: 671–674.

Szalay, F.S. and E. Delson. 1979. *Evolutionary History of the Primates*. New York: Academic Press.

Takhtajan, A. 1976. Neoteny and the origin of flowering plants. In *Origin and Early Evolution of Angiosperms*, ed. C.B. Beck, pp. 207–219. New York: Columbia University Press.

Takhtajan, A. 2009. *Flowering Plants*, 2nd edn. Berlin, Germany: Springer.

Tanentzap, A.J., A.J. Brandt, R.D. Smissen, P.B. Heenan, T. Fukami, and W.G. Lee. 2015. When do plant radiations influence community assembly? The importance of historical contingency in the race for niche space. *New Phytologist* 207: 468–479.

Tank, D.C. and M.J. Donoghue. 2010. Phylogeny and phylogenetic nomenclature of the Campanulidae based on an expanded sample of genes and taxa. *Systematic Botany* 35: 425–441.

Tarran, M., P.G. Wilson, and R.S. Hill. 2016. Oldest record of *Metrosideros* (Myrtaceae): Fossil flowers, fruits, and leaves from Australia. *American Journal of Botany* 103: 754–768.

Tay, M.L., H.M. Meudt, P.J. Garnock-Jones, and P.A. Ritchie. 2010. DNA sequences from three genomes reveal multiple long-distance dispersals and non-monophyly of sections in Australasian *Plantago* (Plantaginaceae). *Australian Systematic Botany* 23: 47–68.

Taylor, C.K. 2004. New Zealand harvestmen of the subfamily Megalopsalidinae (Opiliones: Monoscutidae)— The genus *Pantopsalis*. *Tuhinga* 15: 53–76.

Taylor, C.K. 2011. Revision of the genus *Megalopsalis* (Arachnida: Opiliones: Phalangioidea) in Australia and New Zealand and implications for phalangioid classification. *Zootaxa* 2773: 1–65.

Taylor, E.B. and J.D. McPhail. 2000. Historical contingency and ecological determinism interact to prime speciation in sticklebacks, *Gasterosteus*. *Proceedings of the Royal Society, London B* 267: 2375–2384.

Taylor, G.A. 2000. Action plan for seabird conservation in New Zealand. Parts A and B. Threatened Species Occasional Publications 16 and 17. Wellington, New Zealand: Department of Conservation Biodiversity Recovery Unit.

Taylor, P.B. 1996. Rallidae. In *Handbook of the Birds of the World*, Vol. 3, eds. J. del Hoyo, A. Elliott, and J. Sargatal, pp. 108–209. Barcelona, Spain: Lynx Edicions.

Teeling, E.C. 2009. Bats (Chiroptera). In *The Timetree of Life*, eds. S.B. Hedges and S. Kumar, pp. 499–503. New York: Oxford University Press.

Teeling, E.C., S. Dool, and M.S. Springer. 2012. Phylogenies, fossils and functional genes: The evolution of echolocation in bats. In *Evolutionary History of Bats: Fossils, Molecular and Morphology*, eds. G.F. Gunnell and N.B. Simmons, pp. 1–22. Cambridge, UK: Cambridge University Press.

Teeling, E.C., O. Madsen, W.J. Murphy, M.S. Springer, and S.J. O'Brien. 2003. Nuclear gene sequences confirm an ancient link between New Zealand's short-tailed bat and South American noctilionoid bats. *Molecular Phylogenetics and Evolution* 28: 308–319.

Telford, I.R.H., P. Sebastian, P.J. de Lange, J.J. Bruhl, and S.S. Renner. 2012. Morphological and molecular data reveal three rather than one species of *Sicyos* (Cucurbitaceae) in Australia, New Zealand and Islands of the South West Pacific. *Australian Systematic Botany* 25: 188–220.

Telford, M.J., S.J. Bourlat, A. Economou, D. Papillon, and O. Rota-Stabelli. 2008. The evolution of the Ecdysozoa. *Philosophical Transactions of the Royal Society, Biological Sciences* 363: 1529–1537.

Teng, H.-Y., Y.-S. Lin, and C.-S. Tzeng. 2009. A new *Anguilla* species and a reanalysis of the phylogeny of freshwater eels. *Zoological Studies* 48: 808–822.

Tennyson, A.J.D. 2010. The origin and history of New Zealand's terrestrial vertebrates. *New Zealand Journal of Ecology* 34: 6–27.

Tennyson, A.J.D. and P. Martinson. 2006. *Extinct Birds of New Zealand*. Wellington, New Zealand: Te Papa Press.

Tennyson, A.J.D., T.H. Worthy, C.M. Jones, R.P. Scofield, and S.J. Hand. 2010. Moa's Ark: Miocene fossils reveal the great antiquity of moa (Aves: Dinornithiformes) in Zealandia. *Records of the Australian Museum* 62: 105–114.

Tennyson, A.J.D., T.H. Worthy, R.P. Scofield, and S.J. Hand. 2009. The Miocene St Bathans fauna: An update. *Joint Geological and Geophysical Societies Conference, 2009*, Oamaru, New Zealand. Programme & Abstracts, p. 206.

Teske, P.R., I. Papadopoulos, N.P. Barker, C.D. McQuaid, and L.B. Beheregaray. 2014. Mitonuclear discordance in genetic structure across the Atlantic/Indian Ocean biogeographical transition zone. *Journal of Biogeography* 41: 392–401.

Teulon, D.A.J., M.A.W. Stufkens, C.D. von Dohlen, and J. Kean. 2003. *Status of New Zealand Indigenous Aphids, 2002*. DoC Science Internal Series. Wellington: Department of Conservation.

Tezanos-Pinto, G., C.S. Baker, K. Russell et al. 2009. A worldwide perspective on the population structure and genetic diversity of bottlenose dolphins (*Tursiops truncatus*) in New Zealand. *Journal of Heredity* 100: 11–24.

Thacker, C.E. 2009. Phylogeny of Gobioidei and placement within Acanthomorpha, with a new classification and investigation of diversification and character evolution. *Copeia* 2009: 93–104.

Théry, T. and R.A.B. Leschen. 2013. Pselaphinae (Coleoptera: Staphylinidae) of the Three Kings Islands. *New Zealand Entomologist* 36: 37–64.

Thomas, B.W. 1981. *Hoplodactylus rakiurae* n.sp. (Reptilia: Gekkonidae) from Stewart Island, New Zealand, and comments on the taxonomic status of *Heteropholis nebulosus* McCann. *New Zealand Journal of Zoology* 8: 33–47.

Thomas, G.H., C.D.L. Orme, R.G. Davies, V.A. Olson, P.M. Bennett, K.J. Gaston, I.P.F. Owens, and T.M. Blackburn. 2008. Regional variation in the historical components of global avian species richness. *Global and Ecological Biogeography* 17: 340–351.

Thomas, J.N. 2004. Communicating concern via TV; our view of nature. In Scientific knowledge and cultural diversity. Forum of Cultures, Barcelona, Spain, pp. 251–253. http://www.pantaneto.co.uk/issue42/thomas.htm.

Thomas, N., J.J. Bruhl, A. Ford, and P.H. Weston. 2014. Molecular dating of Winteraceae reveals a complex biogeographical history involving both ancient Gondwanan vicariance and long-distance dispersal. *Journal of Biogeography* 41: 894–904.

Thompson, D.W. 1899. On characteristic points in the cranial osteology of the parrots. *Proceedings of the Zoological Society of London* 1899: 9–46.

Thompson, D.W. 1917. *On Growth and Form*. Cambridge, UK: Cambridge University Press.

Thompson, I.R. 2010. A new species of *Leptostigma* (Rubiaceae: Coprosminae) and notes on the Coprosminae in Australia. *Muelleria* 28: 29–39.

Thompson, N.K., K.N. Bassett, and C.M. Reid. 2014. The effect of volcanism on cool-water carbonate facies during maximum inundation of Zealandia in the Waitaki–Oamaru region. *New Zealand Journal of Geology and Geophysics* 57: 149–169.

Thomson, A.D. 1990. A comparison of the approach to taxonomic botany by T.F. Cheeseman and L. Cockayne. In *History of Systematic Botany in Australasia*, ed. P.S. Short, pp. 235–238. Melbourne, Victoria, Australia: Australian Systematic Botany Society.

Thomson, G.M. 1890. Botany of neighbourhood. In *Picturesque Dunedin: Or Dunedin and Its Neighbourhood in 1890*, ed. A. Bathgate, pp. 110–119. Dunedin, New Zealand: Mill, Dick and Co.

Thomson, G.M. 1909. Botanical evidence against the recent glaciation of New Zealand. *Transactions of the New Zealand Institute* 42: 348–353.

Thordarson, T. 2004. Accretionary-lapilli-bearing pyroclastic rocks at ODP Leg 192 Site 1184: A record of subaerial phreatomagmatic eruptions on the Ontong Java Plateau. *Geological Society of London, Special Publication* 229: 275–306.

Thornhill, A.H., S.Y. Ho, C. Külheim, and M.D. Crisp. 2015. Interpreting the modern distribution of Myrtaceae using a dated molecular phylogeny. *Molecular Phylogenetics and Evolution* 93: 29–43.

Timm, C., K. Hoernle, R. Werner et al. 2010. Temporal and geochemical evolution of the Cenozoic intraplate volcanism of Zealandia. *Earth-Science Reviews* 98: 38–64.

Timm, C., K. Hoernle, R. Werner et al. 2011. Age and geochemistry of the oceanic Manihiki Plateau, SW Pacific: New evidence for a plume origin. *Earth and Planetary Science Letters* 304: 135–146.

Timmers, M.A., C.E. Bird, D.J. Skillings, P.E. Smouse, and R.J. Toonen. 2012. There's no place like home: Crown-of-Thorns outbreaks in the central Pacific are regionally derived and independent events. *PLoS One* 7(2): e31159 (pp. 1–12).

Tinto, K.J., G.S. Wilson, and H. Morgans. 2011. Dating mid-Oligocene current activity in the South Pacific— The Marshall Paraconformity from the Canterbury Basin, South Island, New Zealand. *American Geophysical Union, Fall Meeting 2011*, San Francisco, CA, abstract #PP13A-1804.

Tippet, J.M. and P.J.J. Kamp. 1993. Fission track analysis of the late Cenozoic vertical kinematics of continental Pacific crust, South Island, New Zealand. *Journal of Geophysical Research* 98: 16119–16148.

Todd, M.J. 1996. The taxonomy of *Haastia* (Compositae—Asteraceae). MSc thesis, University of Canterbury, Christchurch, New Zealand.

Toft, R. 2012. *Terrestrial Invertebrates in the Mt William North Project Area*. Entecol, Nelson, New Zealand. www.crl.co.nz/downloads/geology/Mt_William_North/environmental/MtWilliamNorth_Enviro_ RToft_TerrestrialInvertebrates_TechnicalReports.pdf.

Tolmachev, A.I. 1970. The roles of migration and autochthonous development in the formation of the high-mountain flora. In *Studies on the Flora and Vegetation of High Mountain Areas*, ed. V.N. Sukachev. Jerusalem, Israel: Academy of Sciences USSR/Israel Program for Scientific Translations.

Tomlinson, A.I. 1976. Climate. In *New Zealand Atlas*, ed. I. Wards, pp. 82–89. Wellington, New Zealand: A.R. Shearer Government Printer.

Toon, A., M. Pérez-Losada, C.E. Schweitzer, R.M. Feldmann, M. Carlson, and K.A. Crandall. 2010. Gondwanan radiation of the Southern Hemisphere crayfishes (Decapoda: Parastacidae): Evidence from fossils and molecules. *Journal of Biogeography* 37: 2275–2290.

Toussaint, E.F.A., R. Hall, M.T. Monaghan et al. 2014. The towering orogeny of New Guinea as a trigger for arthropod megadiversity. *Nature Communications* 5(4001): 1–10.

Towns, D.R. and C.H. Daugherty. 1994. Patterns of range contractions and extinctions in the New Zealand herpetofauna following human colonisation. *New Zealand Journal of Zoology* 21: 325–339.

Towns, D.R., C.H. Daugherty, and P.L. Cromarty. 1990. Protocols for translocation of organisms to islands. *Conservation Sciences Publication* [Wellington] 2: 240–254.

Towns, D.R., K.A. Neilson, and A.H. Whitaker. 2002. North Island *Oligosoma* spp. skink recovery plan 2002–2012. *Threatened Species Recovery Plan* [Department of Conservation, Wellington] 48: 1–61.

Townsend, J.I. 2010. Trechini (Insecta: Coleoptera: Carabidae: Trechinae). *Fauna of New Zealand* 62: 1–101.

Townson, W. 1906. On the vegetation of the Westport District. *Transactions of the New Zealand Institute* 39: 380–433.

Toy, V.G. and K.B. Spörli. 2008. Stratigraphic and structural evidence for an accretionary precursor to the Northland Allochthon: Mt Camel Terrane, northernmost New Zealand. *New Zealand Journal of Geology and Geophysics* 51: 331–347.

Travers, W.T.L. 1871. Notes on the habits of some of the birds of New Zealand. *Transactions of the New Zealand Institute* 4: 206–213.

Travis, J. and D.N. Reznick. 2009. Adaptation. In *Evolution: The First Four Billion Years*, eds. M. Ruse and J. Travis, pp. 105–131. Cambridge, MA: Harvard University Press.

Trewick, S.A. 1997a. Sympatric flightless rails *Gallirallus dieffenbachii* and *G. modestus* on the Chatham Islands, New Zealand: Morphometrics and alternative evolutionary scenarios. *Journal of the Royal Society of New Zealand* 27: 451–464.

Trewick, S.A. 1997b. Flightlessness and phylogeny amongst endemic rails (Aves: Rallidae) of the New Zealand region. *Philosophical Transactions of the Royal Society, London B* 352: 429–446.

Trewick, S.A. 2000. Molecular evidence for dispersal rather than vicariance as the origin of flightless insect species on the Chatham Islands, New Zealand. *Journal of Biogeography* 27: 1189–1200.

Trewick, S.A. 2001. Scree weta phylogeography: Surviving glaciations and implications for Pleistocene biogeography in New Zealand. *New Zealand Journal of Zoology* 28: 291–298.

Trewick, S.A. 2008. DNA barcoding is not enough: Mismatch of taxonomy and genealogy in New Zealand grasshoppers (Orthoptera: Acrididae). *Cladistics* 24: 240–254.

Trewick, S.A. and K.J. Bland. 2012. Fire and slice: Palaeogeography for biogeography at New Zealand's North Island/South Island juncture. *Journal of the Royal Society of New Zealand* 42: 153–183.

Trewick, S.A. and G.C. Gibb. 2010. Vicars, tramps and assembly of the New Zealand avifauna: A review of molecular phylogenetic evidence. *Ibis* 152: 226–253.

Trewick, S.A. and M. Morgan-Richards. 2005. After the deluge: Mitochondrial DNA indicates Miocene radiation and Pliocene adaptation of tree and giant weta (Orthoptera: Anostostomatidae). *Journal of Biogeography* 32: 295–309.

Trewick, S.A. and M. Morgan-Richards. 2009. Evolution in New Zealand: Getting it in perspective. *Geological Society of New Zealand Miscellaneous Publication* 126: 34–36.

Trewick, S.A. and M. Morgan-Richards. 2014. *New Zealand Wildlife: Introducing the Weird and Wonderful Character of Natural New Zealand*. Auckland, New Zealand: Penguin.

Trewick, S.A., A.M. Paterson, and H.J. Campbell. 2007. Hello New Zealand. *Journal of Biogeography* 34: 1–6.

Trewick, S.A. and G.P. Wallis. 2001. Bridging the 'beech-gap': New Zealand invertebrate phylogeography implicates Pleistocene glaciations and Pliocene isolation. *Evolution* 55: 2170–2180.

Trewick, S.A., G.P. Wallis, and M. Morgan-Richards. 2000. Phylogeographical pattern correlates with Pliocene mountain building in the alpine scree weta (Orthoptera, Anostostomatidae). *Molecular Ecology* 9: 657–666.

Troughton, J.H. and K.A. Card. 1974. Leaf anatomy of *Atriplex buchananii*. *New Zealand Journal of Botany* 12: 167–177.

Tschapka, M., E.B. Sperr, L.A. Caballero-Martínez, and R.A. Medellín. 2008. Diet and cranial morphology of *Musonycteris harrisoni*, a highly specialized nectar-feeding bat in western Mexico. *Journal of Mammalogy* 89: 924–931.

Tulloch, A., M. Beggs, J. Kula, T. Spell, and N. Mortimer. 2006. Cordillera Zealandia, the Sisters Shear Zone, and their influence on the early development of the Great South basin. Paper presented at *the New Zealand Petroleum Conference 2006*, Crown Minerals, Auckland, New Zealand. http://www.nzpam.govt.nz/cms/pdf-library/petroleum-conferences-1/2006/papers/Papers_59.pdf.

Tulloch, A.J., J. Ramezani, N. Mortimer, J. Mortensen, P. van den Bogaard, and R. Maas. 2009. Cretaceous felsic volcanism in New Zealand and Lord Howe Rise (Zealandia) as a precursor to final Gondwana break-up. *Geological Society, London, Special Publications* 321: 89–118.

Tulloch, A.J., T.L. Spell, N. Mortimer, and J. Ramezani. 2014. A possible mid-Cretaceous metamorphic core complex under Taranaki basin. In K. Holt (ed.) Abstract volume, GeoSciences 2014. *Geoscience Society of New Zealand Miscellaneous Publication* 139A: 107.

Tunnicliffe, V., B.F. Koop, J. Tyler, and S. So. 2010. Flatfish at seamount hydrothermal vents show strong genetic divergence between volcanic arcs. *Marine Ecology* 31(Suppl. 1): 1–9.

Turnbull, I.M. 1980. Geological map of New Zealand. 1: 50 000 series. Sheet E42A,C Walter Peak (West). Lower Hutt, New Zealand: New Zealand Geological Survey.

Turnbull, I.M. 1981. Contortions in the schists of the Cromwell district, Central Otago, New Zealand. *New Zealand Journal of Geology and Geophysics* 24: 65–86.

Turnbull, I.M. and A.H. Allibone. 2003. Geology of the Murihiku area. 1:250 000 map 20. 1 sheet and 74pp. Lower Hutt, New Zealand: Institute of Geological and Nuclear Sciences.

Turnbull, I.M., J.M. Barry, R.M. Carter, and R.J. Norris. 1975. The Bobs Cove beds and their relationship to the Moonlight Fault Zone. *Journal of the Royal Society of New Zealand* 5: 355–394.

Turner, B.L. 1977. Fossil history and geography. In *The Biology and Chemistry of the Compositae*, Vol. 1, eds. V.H. Heywood, J.B. Harborne, and B.L. Turner, pp. 21–39. London, UK: Academic Press.

Turner, D. 1997. Musophagidae (Turacos). In *Handbook of the Birds of the World*, Vol. 4, eds. J. del Hoyo, A. Elliott, and J. Sargatal, pp. 480–508. Barcelona, Spain: Lynx Edicions.

Turner, G.M., D.M. Michalk, and T.A. Little. 2012. Paleomagnetic constraints on Cenozoic deformation along the northwest margin of the Pacific-Australian plate boundary zone through New Zealand. *Tectonics* 31(TC1005): 1–16.

Umhoefer, P.J. 2011. Why did the Southern Gulf of California rupture so rapidly?—Oblique divergence across hot, weak lithosphere along a tectonically active margin. *GSA Today* 21: 4–10.

Valladares, F. and D. Brites. 2004. Leaf phyllotaxis: Does it really affect light capture? *Plant Ecology* 174: 11–17.

van der Pluijm, B.A. and M. Marshak. 2004. *Earth Structure: An Introduction to Structural Geology and Tectonics*, 2nd edn. New York: Norton.

van der Putten, N., C. Verbruggen, R. Ochyra, E. Verleyen, and Y. Frenot. 2009. Subantarctic flowering plants: Pre-glacial survivors or post-glacial immigrants? *Journal of Biogeography* 37: 582–592.

van Steenis, C.G.G.J. 1953. Results of the Archbold Expeditions: Papuan *Nothofagus*. *Journal of the Arnold Arboretum* 34: 301–374.

van Steenis, C.G.G.J. 1977. Bignoniaceae. *Flora Malesiana I* 8: 114–186.

van Wijk, J.W., J.F. Lawrence, and N.W. Driscoll. 2008. Formation of the Transantarctic Mountains related to extension of the West Antarctic Rift. *Tectonophysics* 458: 117–126.

Vargas, S., D. Erpenbeck, C. Göcke et al. 2013. Molecular phylogeny of *Abyssocladia* (Cladorhizidae: Poecilosclerida) and *Phelloderma* (Phellodermidae: Poecilosclerida) suggests a diversification of chelae microscleres in cladorhizid sponges. *Zoologica Scripta* 42: 106–116.

Vaughan A.P.M., P.T. Leat, and R.J. Pankhurst. 2005. Terrane processes at the margin of Gondwana: Introduction. *Geological Society of London, Special Publication* 245: 1–21.

Vaughan, A.P.M. and R.A. Livermore. 2005. Episodicity of Mesozoic terrane accretion along the Pacific margin of Gondwana: Implications for superplume-plate interactions. *Geological Society of London, Special Publications* 246: 143–178.

Vazačová, K. and Z. Münzbergová. 2014. Dispersal ability of island endemic plants: What can we learn using multiple dispersal traits? *Flora* 209: 530–539.

Veevers, J.J. 2000. Change of tectono-stratigraphic regime in the Australian plate during the 99 Ma (mid-Cretaceous) and 43 Ma (mid-Eocene) swerves of the Pacific. *Geology* 28: 47–50.

Veevers, J.J. 2012. Reconstructions before rifting and drifting reveal the geological connections between Antarctica and its conjugates in Gondwanaland. *Earth-Science Reviews* 111: 249–318.

Venditti, C., A. Meade, and M. Pagel. 2010. Phylogenies reveal new interpretation of speciation and the Red Queen. *Nature* 463: 349–352.

Venter, S. 2002. *Dracophyllum marmoricola* and *Dracophyllum ophioliticum* (Ericaceae), two new species from north-west Nelson, New Zealand. *New Zealand Journal of Botany* 40: 39–47.

Venter, S. 2004. *Dracophyllum elegantissimum* (Ericaceae), a new species from north-west Nelson, New Zealand. *New Zealand Journal of Botany* 42: 37–43.

Venter, S. 2009. A taxonomic revision of the genus *Dracophyllum* Labill. (Ericaceae). PhD thesis, Victoria University of Wellington Wellington, New Zealand.

Vilhelmsen, L. 2004. The old wasp and the tree: Fossils, phylogeny and biogeography in the Orussidae (Insecta, Hymenoptera). *Biological Journal of the Linnean Society* 82: 139–160.

Vilhelmsen, L. 2007. The phylogeny of Orussidae (Insecta: Hymenoptera) revisited. *Arthropod Systematics and Phylogeny* 65: 111–118.

Villarreal, J.C. and S.S. Renner. 2014. A review of molecular-clock calibrations and substitution rates in liverworts, mosses, and hornworts, and a timeframe for a taxonomically cleaned-up genus *Nothoceros*. *Molecular Phylogenetics and Evolution* 78: 25–35.

Vink, C.J. and N. Dupérré. 2010. Pisauridae (Arachnida: Araneae). *Fauna of New Zealand* 64: 1–60.

Virot, R. 1975. Épacridacées. In *Flore de la Nouvelle-Calédonie et Dépendances*, Vol. 6, eds. A. Aubréville and J.F. Leroy, pp. 5–160. Paris, France: Muséum national d'Histoire naturelle.

Voelker, G. 1999. Dispersal, vicariance, and clocks: historical biogeography and speciation in a cosmopolitan passerine genus (*Anthus*: Motacillidae). *Evolution* 53: 1536–1552.

Voelker, G., J.V. Peñalba, J.W. Huntley, and R.C.K. Bowie. 2014. Diversification in an Afro-Asian songbird clade (*Erythropygia-Copsychus*) reveals founder-event speciation via trans-oceanic dispersals and a southern to northern colonization pattern in Africa. *Molecular Phylogenetics and Evolution* 73: 97–105.

von Konrat, M.J., J.E. Braggins, and P.J. de Lange. 1999. *Davallia* (Pteridophyta) in New Zealand, including description of a new subspecies of *D. tasmanii*. *New Zealand Journal of Botany* 37: 579–593.

Vry, J.K., J. Baker, R. Maas, T.A. Little, R. Grapes, and M. Dixon. 2004. Zoned (Cretaceous and Cenozoic) garnet and the timing of high grade metamorphism, Southern Alps, New Zealand. *Journal of Metamorphic Geology* 22: 137–157.

Vyverman, W., E. Verleyen, K. Sabbe et al. 2007. Historical processes constrain patterns in global diatom diversity. *Ecology* 88: 1924–1931.

Wada, S. and S. Chiba. 2011. Seashore in the mountain: Limestone-associated land snail fauna on the oceanic Hahajima Island (Ogasawara Islands, Western Pacific). *Biological Journal of the Linnean Society* 102: 686–693.

Wade, C.M., P.B. Mordan, and F. Naggs. 2006. Evolutionary relationships among the pulmonate land snails and slugs (Pulmonata, Stylommatophora). *Biological Journal of the Linnean Society* 87: 593–610.

Wagner, W.L., P.C. Hoch, and P.H. Raven. 2007. Revised classification of the Onagraceae. *Systematic Botany Monographs* 83: 1–240.

Wagstaff, S.J., M.J. Bayly, P.J. Garnock-Jones, and D.C. Albach. 2002. Classification, origin, and diversification of the New Zealand hebes (Scrophulariaceae). *Annals of the Missouri Botanical Garden* 89: 38–63.

Wagstaff, S.J. and I. Breitwieser. 2002. Phylogenetic relationships of New Zealand Asteraceae inferred from ITS sequences. *Plant Systematics and Evolution* 231: 203–224.

Wagstaff, S.J., I. Breitwieser, and M. Ito. 2011. Evolution and biogeography of *Pleurophyllum* (Astereae, Asteraceae), a small genus of megaherbs endemic to the subantarctic islands. *American Journal of Botany* 98: 62–75.

Wagstaff, S.J., I. Breitwieser, C. Quinn, and M. Ito. 2007. Age and origin of enigmatic megaherbs from the subantarctic islands. *Nature Precedings*. http://precedings.nature.com/documents/1272/version/1/files/npre20071272-1.pdf.

Wagstaff, S.J., I. Breitwieser, and U. Swenson. 2006. Origin and relationships of the austral genus *Abrotanella* (Asteraceae) inferred from DNA sequences. *Taxon* 55: 95–106.

Wagstaff, S.J. and P.J. Garnock-Jones. 2000. Patterns of diversification in *Chionohebe* and *Parahebe* (Scrophulariaceae) inferred from ITS sequences. *New Zealand Journal of Botany* 38: 389–407.

Wagstaff, S.J., P.B. Heenan, and M.J. Sanderson. 1999. Classification, origins, and patterns of diversification in New Zealand Carmichaelinae (Fabaceae). *American Journal of Botany* 86: 1346–1356.

Wagstaff, S.J. and F. Hennion. 2007. Evolution and biogeography of *Lyallia* and *Hectorella* (Portulacaceae), geographically isolated sisters from the Southern Hemisphere. *Antarctic Science* 19: 417–426.

Wagstaff, S.J. and J.A. Tate. 2011. Phylogeny and character evolution in the New Zealand endemic genus *Plagianthus* (Malveae, Malvaceae). *Systematic Botany* 36: 405–418.

Wagstaff, S.J. and P. Wardle. 1999. Whipcord hebes—Systematics, distribution, ecology and evolution. *New Zealand Journal of Botany* 37: 17–39.

Wagstaff, S.J. and J. Wege. 2002. Patterns of diversification in New Zealand Stylidiaceae. *American Journal of Botany* 89: 865–874.

Walcott, R.I. 1998. Modes of oblique compression: Late Cenozoic tectonics of the South Island of New Zealand. *Reviews of Geophysics* 36: 1–26.

Wall, A. 1926. Some problems of distribution of indigenous plants in New Zealand. *Transactions of the New Zealand Institute* 57: 94–105.

Wall, W.P. 1980. Cranial evidence for a proboscis in *Cadurcodon* and a review of snout structure in the family Amynodontidae (Perissodactyla, Rhinocerotoidea). *Journal of Paleontology* 54: 968–977.

Wallace, A.R. 1881. *Island Life*. New York: Harper and Brothers.

Wallace, L.M., P. Barnes, J. Beavan et al. 2012. The kinematics of a transition from subduction to strike-slip: An example from the central New Zealand plate boundary. *Journal of Geophysical Research* 117: B02405.

Wallis, G.P. and S.A. Trewick. 2001. Finding fault with vicariance: A critique of Heads (1998). *Systematic Biology* 50: 602–609.

Wallis, G.P. and S.A. Trewick. 2009. New Zealand phylogeography: Evolution on a small continent. *Molecular Ecology* 18: 3548–3580.

Wandres, A.M. and J.D. Bradshaw. 2005. New Zealand tectonostratigraphy and implications from conglomeratic rocks for the configuration of the SW Pacific margin of Gondwana. *Geological Society of London, Special Publication* 246: 179–216.

Wang, I.J., R.E. Glor, and J.B. Losos. 2013a. Quantifying the roles of ecology and geography in spatial genetic divergence. *Ecology Letters* 16: 175–182.

Wang, N., R.T. Kimball, E.L. Braun, B. Liang, and Z. Zhang. 2013b. Assessing phylogenetic relationships among Galliformes: A multigene phylogeny with expanded taxon sampling in Phasianidae. *PLoS One* 8(5): e64312 (pp. 1–12).

Warcup, J.H. 1971. Specificity of mycorrhizal association in some Australian terrestrial orchids. *New Phytologist* 70: 41–46.

Ward, D. 2011. *Poecilocryptus zealandicus* sp. n. (Hymenoptera: Ichneumonidae: Labeninae) from New Zealand. *New Zealand Entomologist* 34: 37–39.

Ward, J.B. and I.M. Henderson. 2004. Eleven new species of micro-caddisflies (Trichoptera: Hydroptilidae) from New Zealand. *Records of the Canterbury Museum* 18: 9–22.

Ward, J.B., R.A.B. Leschen, B.J. Smith, and J.C Dean. 2004. Phylogeny of the caddisfly (Trichoptera) family Hydrobiosidae using larval and adult morphology, with the description of a new genus and species from Fiordland, New Zealand. *Records of the Canterbury Museum* 18: 23–43.

Ward, J.M. 1993. Systematics of New Zealand Inuleae (Compositae-Asteraceae)—1 A numerical phenetic study of the species of *Raoulia*. *New Zealand Journal of Botany* 31: 21–28.

Wardle, J. 1971. The forests and shrublands of the Seaward Kaikoura Range. *New Zealand Journal of Botany* 9: 269–292.

Wardle, P. 1968. Evidence for a pre-Quaternary element in the mountain flora of New Zealand. *New Zealand Journal of Botany* 6: 120–125.

Wardle, P. 1975. Vascular plants of Westland National Park (New Zealand) and neighbouring lowland and coastal areas. *New Zealand Journal of Botany* 13: 497–545.

Wardle, P. 1978. Origin of the New Zealand mountain flora, with special reference to trans-Tasman relationships. *New Zealand Journal of Botany* 16: 535–550.

Wardle, P. 1980. Ecology and distribution of silver beech (*Nothofagus menziesii*) in the Paringa District, South Westland, New Zealand. *New Zealand Journal of Ecology* 3: 23–36.

Wardle, P. 1991. *Vegetation of New Zealand*. Cambridge, UK: Cambridge University Press.

Ware, J.L., C.D. Beatty, M. Sánchez Herrera et al. 2014. The petaltail dragonflies (Odonata: Petaluridae): Mesozoic habitat specialists that survive to the modern day. *Journal of Biogeography* 41: 1291–1300.

Warnock, R.C.M., J.F. Parham, W.G. Joyce, T.R. Lyson, and P.C.J. Donoghue. 2015. Calibration uncertainty in molecular dating analyses: There is no substitute for the prior evaluation of time priors. *Proceedings of the Royal Society* 282(1798): article 20141013.

Warwick, S.I., K. Mummenhoff, C.A. Sauder, M.A. Koch, and I.A. Al-Shehbaz. 2010. Closing the gaps: Phylogenetic relationships in the Brassicaceae based on DNA sequence data of nuclear ribosomal ITS region. *Plant Systematics and Evolution* 285: 209–232.

Waters, J.M., R.M. Allibone, and G.P. Wallis. 2006. Geological subsidence, river capture, and cladogenesis of galaxiid fish lineages in central New Zealand. *Biological Journal of the Linnean Society* 88: 367–376.

Waters, J.M. and D. Craw. 2006. Goodbye Gondwana? New Zealand biogeography, geology, and the problem of circularity. *Systematic Biology* 55: 351–356.

Waters, J.M. and D. Craw. 2008. Evolution and biogeography of New Zealand's longjaw galaxiids (Osmeriformes: Galaxiidae): the genetic effects of glaciation and mountain building. *Freshwater Biology* 53: 521–534.

Waters, J.M., D. Craw, J.H. Youngson, and G.P. Wallis. 2001. Genes meet geology: Fish phylogeographic pattern reflects ancient, rather than modern, drainage connections. *Evolution* 55: 1844–1851.

Waters, J.M., L.H. Dijkstra, and G.P. Wallis. 2000a. Biogeography of a southern hemisphere freshwater fish: How important is marine dispersal? *Molecular Ecology* 9: 1815–1821.

Waters, J.M., C.I. Fraser, and G.M. Hewitt. 2013. Founder takes all: Density-dependent processes structure biodiversity. *Trends in Ecology and Evolution* 28: 78–85.

Waters, J.M., A. López, and G.P. Wallis. 2000b. Molecular phylogenetics and biogeography of galaxiid fishes (Osteichthyes: Galaxiidae): Dispersal, vicariance, and the position of *Lepidogalaxias salamandroides*. *Systematic Biology* 49: 777–795.

Waters, J.M. and R.M. McDowall. 2005. Phylogenetics of the Australasian mudfishes: Evolution of an eel-like body plan. *Molecular Phylogenetics and Evolution* 37: 417–425.

Waters, J.M., D.L. Rowe, C.P. Burridge, and G.P. Wallis. 2010. Gene trees versus species trees: Reassessing life-history evolution in a freshwater fish radiation. *Systematic Biology* 59: 504–517.

Watkins, R.L.S. 2012. The biogeography, ecology and endophyte mycorrhiza of the New Zealand *Corybas* alliance (Orchidaceae): Specifically, *Nematoceras iridescens* (Irwin et Molloy) Molloy, D.L. Jones & M.A. Clem. PhD thesis, Massey University, Palmerston North, New Zealand.

Watt, J.C. 1971. Entomology of the Aucklands and other islands to the south of New Zealand: Coleoptera: Scarabaeidae, Byrrhidae, Ptinidae, Tenebrionidae. *Pacific Insects Monograph* 27: 193–224.

Watt, J.C. 1974. Chalcodryidae: A new family of heteromerous beetles (Coleoptera: Tenebrionoidea). *Journal of the Royal Society of New Zealand* 4: 19–38.

Watt, J.C. 1982. Terrestrial arthropods from the Poor Knights Islands. *Journal of the Royal Society of New Zealand* 12: 283–320.

Watt, J.C., J.W.M. Marris, and J. Klimaszewski. 2001. A new species of *Platisus* (Coleoptera: Cucujidae) from New Zealand, described from the adult and larva. *Journal of the Royal Society of New Zealand* 31: 327–339.

Watts, C., I. Stringer, G. Gibbs, and C. Green. 2008. History of weta (Orthoptera: Anostostomatidae) translocation in New Zealand: Lessons learned, islands as sanctuaries and the future. *Journal of Insect Conservation* 12: 359–370.

Weaver, S.D. and R.J. Pankhurst. 1991. A precise Rb-Sr age for the Mandamus Igneous Complex, North Canterbury, and regional tectonic implications. *New Zealand Journal of Geology and Geophysics* 34: 341–345.

Webb, C.J. 1988. Notes on the *Senecio lautus* complex in New Zealand. *New Zealand Journal of Botany* 26: 481–484.

Weersing, K. and R.J. Toonen. 2009. Population genetics, larval dispersal, and connectivity in marine systems. *Marine Ecology Progress Series* 393: 1–12.

Weigend, M., F. Luebert, F. Selvi, G. Brokamp, and H.H. Hilger. 2013. Multiple origins for Hound's tongues (*Cynoglossum* L.) and Navel seeds (*Omphalodes* Mill.)—The phylogeny of the borage family (Boraginaceae s.str.). *Molecular Phylogenetics and Evolution* 68: 604–618.

Weintraub, J.D. and M.J. Scoble. 2004. Lithinini (Insecta: Lepidoptera: Geometridae: Ennominae). *Fauna of New Zealand* 49: 1–48.

Wen, J., R.H. Ree, S.M. Ickert-Bond, Z. Nie, and V. Funk. 2013. Biogeography: Where do we go from here? *Taxon* 62: 912–927.

West, K.R. and P.H. Raven. 1977. Novelties in Australian *Epilobium*. *New Zealand Journal of Botany* 15: 503–509.

West-Eberhard, M.-J. 2003. *Development, Plasticity and Evolution*. New York: Oxford University Press.

Weston, K.A. and B.C. Robertson. 2015. Population structure within an alpine archipelago: Strong signature of past climate change in the New Zealand rock wren (*Xenicus gilviventris*). *Molecular Ecology* 24: 4778–4794.

Whattam, S.A. 2009. Arc-continent collisional orogenesis in the SW Pacific and the nature, source and correlation of emplaced ophiolitic nappe components. *Lithos* 113: 88–114.

Whattam, S.A., J. Malpas, J.R. Ali, C.-H. Lo, and I.E.M. Smith. 2005. Formation and emplacement of the Northland ophiolite, northern New Zealand: SW Pacific tectonic implications. *Journal of the Geological Society, London* 162: 225–241.

Whattam, S.A., J. Malpas, J.R. Ali, I.E.M. Smith, and C.-H. Lo. 2004. Origin of the Northland ophiolite, northern New Zealand: Discussion of new data and reassessment of the model. *New Zealand Journal of Geology and Geophysics* 47: 383–389.

Whattam, S.A., J. Malpas, J.R. Ali, and I.E.M. Smith. 2008. New SW Pacific tectonic model: Cyclical intra-oceanic magmatic arc construction and near-coeval emplacement along the Australia-Pacific margin in the Cenozoic. *Geochemistry, Geophysics, Geosystems* 9: Q03021.

Whattam, S.A., J. Malpas, I.E.M. Smith, and J.R. Ali. 2006. Link between SSZ ophiolite formation, emplacement and arc inception, Northland, New Zealand: U–Pb SHRIMP constraints; Cenozoic SW Pacific tectonic implications. *Earth and Planetary Science Letters* 250: 606–632.

Wheat, C.W. and N. Wahlberg. 2013. Critiquing blind dating: The dangers of over-confident date estimates in comparative genomics. *Trends in Ecology and Evolution* 28: 636–642.

Whewell, W. 1847/1967. *History of the Inductive Sciences, from the Earliest to the Present Time*, Part 3. London, UK: Frank Cass.

White, C.M. 1994. Peregrine falcon—*Falco peregrinus*. In *Handbook of the Birds of the World*, Vol. 2, eds. J. del Hoyo, A. Elliott, and J. Sargatal, pp. 274–275. Barcelona, Spain: Lynx Edicions.

White, C.M., P.D. Olsen, and L.F. Kiff. 1994. Falconidae. In *Handbook of the Birds of the World*, Vol. 2, eds. J. del Hoyo, A. Elliott, and J. Sargatal, pp. 216–277. Barcelona, Spain: Lynx Edicions.

White, S.R. 2002. The Siberia Fault Zone, northwest Otago, and kinematics of mid-Cenozoic plate boundary deformation in southern New Zealand. *New Zealand Journal of Geology and Geophysics* 45: 271–287.

Whitehead, J.K. 2007. Breeding success of adult female kakapo (*Strigops habroptilus*) on Codfish Island (Whenua Hou): Correlations with foraging home ranges and habitat selection. MSc thesis, Lincoln University, Lincoln, New Zealand.

Whitlock, M.C. 2009. Founder effects. In *Encyclopedia of Islands*, eds. R. Gillespie and D. Clague, pp. 326–330. Berkeley, CA: University of California Press.

Whittaker, R.J. and J.M. Fernández-Palacios. 2007. *Island Biogeography: Ecology, Evolution, and Conservation*, 2nd edn. Oxford, UK: Oxford University Press.

Wiens, J.J. 2004. What is speciation and how should we study it? *American Naturalist* 163: 914–923.

Wiens, J.J. 2011a. The niche, biogeography and species interactions. *Philosophical Transactions of the Royal Society B* 366: 2336–2350.

Wiens, J.J. 2011b. The causes of species richness patterns across space, time, and clades and the role of 'ecological limits'. *Quarterly Review of Biology* 86: 75–96.

Wiens, J.J., D.D. Ackerly, A.P. Allen et al. 2010. Niche conservatism as an emerging principle in ecology and conservation biology. *Ecology Letters* 13: 1310–1324.

Wiens, J.J., M.C. Brandley, and T.W. Reeder. 2006. Why does a trait evolve multiple times within a clade? Repeated evolution of snakelike body form in squamate reptiles. *Evolution* 60: 123–141.

Wiens, J.J. and M.J. Donoghue. 2004. Historical biogeography, ecology, and species richness. *Trends in Ecology and Evolution* 19: 639–44.

Wikström, N., V. Savolainen, and M.W. Chase. 2001. Evolution of the angiosperms: Calibrating the family tree. *Proceedings of the Royal Society, London B* 268: 2211–2220.

Wiley, A.E., A.J. Welch, P.H. Ostrom et al. 2012. Foraging segregation and genetic divergence between geographically proximate colonies of a highly mobile seabird. *Oecologia* 168: 119–130.

Wilf, P. and I.H. Escapa. 2015. Green Web or megabiased clock? Plant fossils from Gondwanan Patagonia speak on evolutionary radiations. *New Phytologist* 207: 283–290.

Wilf, P., I.H. Escapa, N.R. Cúneo, R.M. Kooyman, K.R. Johnson, and A. Iglesias. 2014. First South American *Agathis* (Araucariaceae), Eocene of Patagonia. *American Journal of Botany* 101: 156–179.

Wilkins, A.S. 2002. *The Evolution of Developmental Pathways*. Sunderland, MA: Sinauer Associates.

Wilkins, A.S. 2005. Recasting developmental evolution in terms of genetic pathway and network evolution… and the implications for comparative biology. *Brain Research Bulletin* 66: 495–509.

Wilkins, A.S. 2007. Between "design" and "bricolage": Genetic networks, levels of selection, and adaptive evolution. *Proceedings of the National Academy of Sciences of USA* 104(Suppl. 1): 8590–8596.

Wilkinson, R.D., M.E. Steiper, C. Soligo, R.D. Martin, Z. Yang, and S. Tavaré. 2011. Dating primate divergences through an integrated analysis of palaeontological and molecular data. *Systematic Biology* 60: 16–31.

Willett, R.W. 1951. The New Zealand Pleistocene snowline, climatic conditions, and suggested biological effects. *New Zealand Journal of Science and Technology* B32: 18–48.

Williams, D.M. 2011. Historical biogeography, microbial endemism and the role of classification: Everything is endemic. In *Biogeography of Microscopic Organisms: Is Everything Small Everywhere?*, ed. D. Fontaneto, pp. 11–31. Cambridge, UK: Cambridge University Press.

Williams, G.C. 1996. *Plan and Purpose in Nature*. London, UK: Weidenfeld and Nicolson.

Williams, M., A.J.D. Tennyson, and D. Sim. 2014. Island differentiation of New Zealand extinct mergansers (Anatidae: Mergini), with description of a new species from Chatham Island. *Wildfowl* 64: 3–24.

Williams, P.A. 1989. Vegetation of the Inland Kaikoura Range Marlborough. *New Zealand Journal of Botany* 27: 201–220.

Wilson, C.J.N., B.F. Houghton, M.O. McWilliams, M.A. Lanphere, S.D. Weaver, and R.M. Briggs. 1995. Volcanic and structural evolution of Taupo Volcanic Zone, New Zealand: A review. *Journal of Volcanology and Geothermal Research* 68: 1–28.

Wilson, E.O. 2001. *The Diversity of Life*. London, UK: Penguin.

Wilson, E.O. 2009. Foreword. In *Evolution: The First Four Billion Years*, eds. M. Ruse and J. Travis, pp. vii–viii. Cambridge, MA: Harvard University Press.

Wilson, G.D.F. 2008. Gondwanan groundwater: Subterranean connections of Australian phreatoicidean isopods (Crustacea) to India and New Zealand. *Invertebrate Systematics* 22: 301–310.

Wilson, H.D. 1987. Vascular plants of Stewart Island (New Zealand). *New Zealand Journal of Botany, Supplement* 1: 81–131.

Wilson, J.B. and C.D. Meurk. 2011. The control of community composition by distance, environment and history: A regional-scale study of the mountain grasslands of southern New Zealand. *Journal of Biogeography* 38: 2384–2396.

Wilson, P.G. 2011. Myrtaceae. In *The Families and Genera of Vascular Plants*, Vol. X: *Flowering Plants Eudicots: Sapindales, Cucurbitales, Myrtaceae*, ed. K. Kubitzki, pp. 212–271. Heidelberg, Germany: Springer.

Wilson, P.G., M.M. O'Brien, M.M. Heslewood, and C.J. Quinn. 2005. Relationships within the Myrtaceae sensu lato based on a *matK* phylogeny. *Plant Systematics and Evolution* 251: 3–19.

Wink, M., A.A. El-Sayed, H. Sauer-Gürth, and J. Gonzalez. 2009. Molecular phylogeny of owls (Strigiformes) inferred from DNA sequences of the mitochondrial cytochrome *b* and the nuclear *RAG-1* gene. *Ardea* 97: 581–591.

Winkworth, R.C., J. Grau, A.W. Robertson, and P.J. Lockhart. 2002a. The origins and evolution of the genus *Myosotis* L. (Boraginaceae). *Molecular Phylogenetics and Evolution* 24: 180–193.

Winkworth, R.C., F. Hennion, A. Prinzing, and S.J. Wagstaff. 2015. Explaining the disjunct distributions of austral plants: The roles of Antarctic and direct dispersal routes. *Journal of Biogeography* 42: 1197–1209.

Winkworth, R.C., A.W. Robertson, F. Ehrendorfer, and P.J. Lockhart. 1999. The importance of dispersal and recent speciation in the flora of New Zealand. *Journal of Biogeography* 26: 1323–1325.

Winkworth, R.C., S.J. Wagstaff, D. Glenny, and P.J. Lockhart. 2002b. Plant dispersal N.E.W.S. from New Zealand. *Trends in Ecology and Evolution* 17: 514–520.

Winkworth, R.C., S.J. Wagstaff, D. Glenny, and P.J. Lockhart. 2005. Evolution of the New Zealand mountain flora: Origins, diversification and dispersal. *Organisms, Diversity and Evolution* 5: 237–247.

Winnington, A. 1999. Ecology, genetics and taxonomy of peka peka (Chiroptera: *Mystacina tuberculata* and *Chalinolobus tuberculatus*). PhD thesis, University of Otago, Dunedin, New Zealand.

Winsor, L. 2011. Some terrestrial flatworm taxa (Platyhelminthes: Tricladida: Continenticola) of the Subantarctic Islands of New Zealand. *Tuhinga* 22: 161–169.

Wiser, S.K., J.M. Hurst, E.F. Wright, and R.B. Allen. 2011. New Zealand's forest and shrubland communities: A quantitative classification based on a nationally representative plot network. *Applied Vegetation Science* 14: 506–523.

Wojciechowski, M.F. 2005. *Astragalus* (Fabaceae): A molecular perspective. *Brittonia* 57: 382–396.

Wood, H.M., N.J. Matzke, R.G. Gillespie, and C.E. Griswold. 2013. Treating fossils as terminal taxa in divergence time estimation reveals ancient vicariance patterns in the palpimanoid spiders. *Systematic Biology* 62: 264–284.

Wood, J., K.J. Mitchell, G.C Gibb et al. 2014a. Extinct Holocene birds of the Chatham Islands: Ancient DNA provides new taxonomic and phylogenetic insights. *GeoGenes V*. Abstracts, pp. 29–30.

Wood, J.R. 2009. Two Late Quaternary avifaunal assemblages from the Dunback district, eastern Otago, South Island, New Zealand. *Notornis* 56: 154–157.

Wood, J.R., K.J. Mitchell, R.P. Scofield et al. 2014b. An extinct nestorine parrot (Aves, Psittaciformes, Nestoridae) from the Chatham Islands, New Zealand. *Zoological Journal of the Linnean Society* 172: 185–199.

Wood Jones, F. 1951. The external characters of a foetal tarsier. *Proceedings of the Zoological Society of London* 120: 723–730.

Wörheide, G., L.S. Epp, and L. Macis. 2008. Deep genetic divergences among Indo-Pacific populations of the coral reef sponge *Leucetta chagosensis* (Leucettidae): Founder effects, vicariance, or both? *BMC Evolutionary Biology* 8(24): 1–18.

World Spider Catalog. 2016. World Spider Catalog, version 17. Natural History Museum Bern. http://wsc. nmbe.ch. Accessed May, 2016.

Worth, J.R.P., G.J. Williamson, S. Sakaguchi, P.G. Nevill, and G.J. Jordan. 2014. Environmental niche modelling fails to predict Last Glacial Maximum refugia: Niche shifts, microrefugia or incorrect palaeoclimate estimates? *Global Ecology and Biogeography* 23: 1186–1197.

Worthington, T.J., R. Hekinian, P. Stoffers, T. Kuhn, and F. Hauff. 2006. Osbourn Trough: Structure, geochemistry, and implications of a mid-Cretaceous paleospreading ridge in the South Pacific. *Earth and Planetary Science Letters* 245: 685–701.

Worthy, T.H. 1997. Quaternary fossil fauna of South Canterbury, South Island, New Zealand. *Journal of the Royal Society of New Zealand* 27: 67–162.

Worthy, T.H. 2004. The fossil rails (Aves: Rallidae) of Fiji with descriptions of a new genus and species. *Journal of the Royal Society of New Zealand* 34: 295–314.

Worthy, T.H., S.J. Hand, J.M.T. Nguyen, A.J.D. Tennyson, J.P. Worthy, R.P. Scofield, W.E. Boles, and M. Archer. 2010. Biogeographical and phylogenetic implications of an Early Miocene Wren (Aves: Passeriformes: Acanthisittidae) from New Zealand. *Journal of Vertebrate Paleontology* 30: 479–498.

Worthy, T.H. and R.N. Holdaway. 2002. *The Lost World of the Moa*. Bloomington, IN: Indiana University Press.

Worthy, T.H. and R.P. Scofield. 2012. Twenty-first century advances in knowledge of the biology of moa (Aves: Dinornithiformes): A new morphological analysis and moa diagnoses revised. *New Zealand Journal of Zoology* 39: 87–153.

Worthy, T.H., A.J.D. Tennyson, M. Archer, A.M. Musser, S.J. Hand, C. Jones, B.J. Douglas, J.A. McNamara, and R.M.D. Beck. 2006. Miocene mammal reveals a Mesozoic ghost lineage on insular New Zealand, southwest Pacific. *Proceedings of the National Academy of Sciences of USA* 103: 19419–19423.

Worthy, T.H., A.J.D. Tennyson, S.J. Hand, H. Godthelp, and R.P. Scofield. 2011a. Terrestrial turtle fossils from New Zealand refloat Moa's Ark. *Copeia* 2011: 72–76.

Worthy, T.H., A.J.D. Tennyson, and R.P. Scofield. 2011b. An early Miocene diversity of parrots (Aves, Strigopidae, Nestorinae) from New Zealand. *Journal of Vertebrate Paleontology* 31: 1102–1116.

Worthy, T.H., A.J.D. Tennyson, R.P. Scofield, and S.J. Hand. 2013a. Early Miocene fossil frogs (Anura: Leiopelmatidae) from New Zealand. *Journal of the Royal Society of New Zealand* 43: 211–230.

Worthy, T.H., J.P. Worthy, M. Archer, S.J. Hand, R.P. Scofield, B.A. Marshall, and A.J.D. Tennyson. 2011c. A decade on, what the St Bathans Fauna reveals about the Early Miocene terrestrial biota of Zealandia. *New Zealand Geological Society Conference Abstracts*, Nelson, New Zealand, p. 120.

Worthy, T.H., J.P. Worthy, A.D. Tennyson, S.W. Salisbury, S.J. Hand, and R.P. Scofield. 2013b. Miocene fossils show that kiwi (*Apteryx*, Apterygidae) are not phyletic dwarves. In *Proceedings of the 8th Meeting of the Society of Avian Paleontology and Evolution*, eds. U.B. Göhlich and A. Kroh, pp. 63–80. Vienna, Austria: Naturhistorisches Museum.

Wouts, W.M. 2006. Criconematina (Nematoda: Tylenchida). *Fauna of New Zealand* 55: 1–232.

Wright, A. 1980. Vegetation and flora of Fanal Island, Mokohinau group. *Tane* 26: 25–43.

Wright, A.E. 1984. *Beilschmiedia* Nees (Lauraceae) in New Zealand. *New Zealand Journal of Botany* 22: 109–125.

Wright, S. 1931. Evolution in Mendelian populations. *Genetics* 16: 97–159.

Wright, S. 1932. The roles of mutation, inbreeding, cross-breeding, and selection in evolution. *Proceedings of the Sixth International Congress in Genetics* 1: 356–366.

Wright, S.D., R.D. Gray, and R.C. Gardner. 2003. Energy and the rate of evolution: Inferences from plant rDNA substitution rates in the western Pacific. *Evolution* 57: 2893–2898.

Wright, S.D., C.G. Yong, J.W. Dawson, D.J. Whittaker, and R.C. Gardner. 2000. Riding the ice age El Niño? Pacific biogeography and evolution of *Metrosideros* subg. *Metrosideros* (Myrtaceae) inferred from nuclear ribosomal DNA. *Proceedings of the National Academy of Sciences of USA* 97: 4118–4123.

Wright, T.F., E.E. Schirtzinger, T. Matsumoto et al. 2008. A multilocus molecular phylogeny of the parrots (Psittaciformes): Support for a Gondwanan origin during the Cretaceous. *Molecular Biology and Evolution* 25: 2141–2156.

Wynen, L.P., S.D. Goldsworthy, S.J. Insley et al. 2001. Phylogenetic relationships within the Eared Seals (Otariidae: Carnivora): Implications for the historical biogeography of the family. *Molecular Phylogenetics and Evolution* 21: 270–284.

Wysocka, A., M. Grabowski, L. Sworobowicz, T. Mamos, A. Burzyński, and J. Sell. 2014. Origin of the Lake Ohrid gammarid species flock: Ancient local phylogenetic lineage diversification. *Journal of Biogeography* 41: 1758–1768.

Xie, L., W.L. Wagner, R.H. Ree, P.E. Berry, and J. Wen. 2009. Molecular phylogeny, divergence time estimates, and historical biogeography of *Circaea* (Onagraceae) in the Northern Hemisphere. *Molecular Phylogenetics and Evolution* 53: 995–1009.

Yago, M., N. Hirai, M. Kondo, T. Tanikawa, M. Ishii, M. Wang, M. Williams, and R. Ueshima. 2008. Molecular systematics and biogeography of the genus *Zizina* (Lepidoptera: Lycaenidae). *Zootaxa* 1746: 15–38.

Yampolsky, L.Y. and A. Stoltzfus. 2001. Bias in the introduction of variation as an orienting factor in evolution. *Evolution and Development* 3: 73–83.

Yang, E.C., G.Y. Cho, K. Kogame, A.L. Carlile, and S.M. Boo. 2008. RuBisCO cistron sequence variation and phylogeography of *Ceramium kondoi* (Ceramiaceae). *Botanica Marina* 51: 370–377.

Yang, Y. and P.E. Berry. 2007. Phylogenetics and evolution of *Euphorbia* subgenus *Chamaesyce* (Euphorbiaceae). *Botany and Plant Biology Joint Congress 2007*, Chicago, IL. http://2007.botanyconference.org/

Yang, Y.-W., K.-N. Lai, P.-Y. Tai, and L. Wen-Hsiung. 1999. Rates of nucleotide substitution in angiosperm mitochondrial DNA sequences and dates of divergence between *Brassica* and other angiosperm lineages. *Journal of Molecular Evolution* 48: 597–604.

Yang, Z.-Y., J.-H. Ran, and X-Q. Wang. 2012. Three genome-based phylogeny of Cupressaceae s.l.: Further evidence for the evolution of gymnosperms and Southern Hemisphere biogeography. *Molecular Phylogenetics and Evolution* 64: 452–470.

Yonezawa, T., N. Kohno, and M. Hasegawa. 2009. The monophyletic origin of sea lions and fur seals (Carnivora; Otariidae) in the Southern Hemisphere. *Gene* 441: 89–99.

Yu, Y., A.J. Harris, and H.J. He. 2010. S-DIVA (statistical dispersal-vicariance analysis): A tool for inferring biogeographic histories. *Molecular Phylogenetics and Evolution* 56: 848–850.

Yu, Y., A.J. Harris, and H.J. He. 2011. RASP (Reconstruct Ancestral State in Phylogenies) 1.103. http://mnh.scu.edu.cn/soft/blog/RASP. Accessed May, 2016

Yuri, T., R.T. Kimball, J. Harshman et al. 2013. Parsimony and model-based analyses of indels in avian nuclear genes reveal congruent and incongruent phylogenetic signals. *Biology* 2: 419–444.

Yurick, D.B. and D.E. Gaskin. 1988. Asymmetry in the skull of the harbour porpoise *Phocoena phocoena* (L.) and its relationship to sound production and echolocation. *Canadian Journal of Zoology* 66: 399–402.

Zaldivar-Riverón, A., S.A. Belokobylskij, V. León-Regagnon, R. Briceño-G., and D.L.J. Quicke. 2008. Molecular phylogeny and historical biogeography of the cosmopolitan parasitic wasp subfamily Doryctinae (Hymenoptera: Braconidae). *Invertebrate Systematics* 22: 345–363.

Zhao, B., L. Liu, D. Tan, and J. Wang. 2010. Analysis of phylogenetic relationships of Brassicaceae species based on *Chs* sequences. *Biochemical Systematics and Ecology* 38: 731–739.

Zhou, Z., L.M. Chiappe, and F. Zhang. 2005. Anatomy of the Early Cretaceous bird *Eoenantiornis buhleri* (Aves: Enantiornithes) from China. *Canadian Journal of Earth Sciences* 42: 1331–1338.

Zielske, S. and M. Haase. 2014. New insights into tateid gastropods and their radiation on Fiji based on anatomical and molecular methods (Caenogastropoda: Truncatelloidea). *Zoological Journal of the Linnean Society* 172: 71–102.

Zink, R.M. 2013. Homage to Hutchinson, and the role of ecology in lineage divergence and speciation. *Journal of Biogeography* 41: 999–1006.

Index